PROCEEDINGS
SEVENTH INTERNATIONAL CONGRESS
INTERNATIONAL ASSOCIATION OF ENGINEERING GEOLOGY
VOLUME 5

COMPTES-RENDUS
SEPTIEME CONGRES INTERNATIONAL
ASSOCIATION INTERNATIONALE DE GEOLOGIE DE L'INGENIEUR
VOLUME 5

Comptes-rendus Septième Congrès International Association Internationale de Géologie de l'Ingénieur

5–9 SEPTEMBRE 1994 / LISBOA / PORTUGAL

Rédacteurs
R. OLIVEIRA, L. F. RODRIGUES, A. G. COELHO & A. P. CUNHA
LNEC, Lisboa, Portugal

VOLUME 5

Thème 4 Matériaux de construction
Thème 5 Études de cas dans les travaux de surface

A.A. BALKEMA / ROTTERDAM / BROOKFIELD / 1994

Proceedings
Seventh International Congress
International Association
of Engineering Geology

5–9 SEPTEMBER 1994 / LISBOA / PORTUGAL

Editors
R.OLIVEIRA, L. F.RODRIGUES, A.G.COELHO & A.P.CUNHA
LNEC, Lisboa, Portugal

VOLUME 5

Theme 4 Construction materials
Theme 5 Case histories in surface workings

A.A.BALKEMA / ROTTERDAM / BROOKFIELD / 1994

INTERNATIONAL ASSOCIATION OF ENGINEERING GEOLOGY
ASSOCIATION INTERNATIONALE DE GEOLOGIE DE L'INGENIEUR

The publication of these proceedings has been partially funded by
La publication des comptes-rendus a été partiellement supportée par

Junta Nacional de Investigação Cientifica e Tecnológica (JNICT), Portugal
Fundação Calouste Gulbenkian, Portugal

*The texts of the various papers in this volume were set individually by typists under the supervision of each of the authors
concerned.*

Les textes des divers articles dans ce volume ont été dactylographiés sous la supervision de chacun des auteurs concernés.

Complete set of six volumes / Collection complète de six volumes: ISBN 90 5410 503 8
Volume 1: ISBN 90 5410 504 6
Volume 2: ISBN 90 5410 505 4
Volume 3: ISBN 90 5410 506 2
Volume 4: ISBN 90 5410 507 0
Volume 5: ISBN 90 5410 508 9
Volume 6: ISBN 90 5410 509 7

Published by:
© 1994 A.A.Balkema, Postbus 1675, 3000 BR Rotterdam, Netherlands (Fax: +31.10.413.5947)
Distributed in the USA & Canada by: A.A.Balkema Publishers, Old Post Road, Brookfield, VT 05036, USA
(Fax: +1.802.276.3837)
Printed in the Netherlands

Publié par:
© 1994 A.A.Balkema, Postbus 1675, 3000 BR Rotterdam, Pays-Bas
Distribué aux USA & Canada par: A.A.Balkema Publishers, Old Post Road, Brookfield, VT 05036, USA
Imprimé aux Pays-Bas

SCHEME OF THE WORK
SCHÉMA DE L'OUVRAGE

Table of contents
Table des matières

4 Construction materials
Matériaux de construction

5 Case histories in surface workings
Études de cas dans les travaux de surface

Keynote lecture: Engineering geology of pedocretes and other residual soils

Conférence spéciale: La géologie de l'ingénieur des croûtes pédogénétique et autres sols résiduels

Frank Netterberg
Pretoria, South Africa

ABSTRACT: The engineering geology of pedocretes (duricrusts) and other residual soils is discussed . As pedocretes are simply residual or transported soils which have been variably cemented and replaced in-situ, their geotechnical properties depend largely upon the nature of the original host soil and the extent and nature of this cementation and replacement. The geotechnical properties of residual soils in general depend largely upon the nature of the original rock and the geomorphic and climatic history of the site. Although they may present unusual problems, especially in-situ, their performance as construction materials is usually better than expected. Modifications and additions to our traditional ways of dealing with soils are necessary, and some of these are discussed.

RÉSUMÉ: La géologie de l'ingénieur des croûtes pédogénétique et autres sols résiduals est discuteé ici. Des croûtes pédogénétique soyant simplement des sols résiduals ou transportés qui ont été cimentés variablement et remplacés sur place, les propriétés geotechiques dépendent largement de la nature du sol original qui joue le rôle du hôte, et l'étendu et la nature de cette cimentation et remplacement. En general, les propriétés geotechniques des sols résiduals dépendent plutôt de la nature du rocher original et l'histoire géomorphique et climatique du site. Bien qu'ils présentent des problémes peu commun, surtout sur place, leur performance en tant que materiaux utilisés pour la construction est généralement mieux qu'attendue. Des modifications et des additions à nos manières tradionelles de traiter des sols sont necessaires, et on en discute quelques-unes.

1 INTRODUCTION

Much of our knowledge of soil and rock behaviour originated in the countries of the Northern Hemisphere where temperate transported soils are the rule. In contrast, in more tropical areas not subjected to the glaciations of the Pleistocene the scene is rather one of deep weathering and advanced pedogenesis. Under such conditions residual soils and variably cemented soils called duricrusts or pedocretes are common founding and construction materials. It is these materials - mostly the latter - which is the subject of this address.

Although this paper is weighted towards southern African experience much of it is applicable to at least the fragments of Gondwanaland.

2 LITERATURE

A computer search of GEOREF in 1994 for the last 15 years using keywords turned up the following number of references: Duricrust : 323, Residual soil: 58, Tropical soil : 88, Laterite : 239, Ferricrete : 108, Calcrete : 342, Caliche : 165, Silcrete : 105. With a total of 1 428 it is therefore impossible to do justice to all this work here. Fortunately, recent comprehensive world-wide reviews on the engineering geology of tropical, lateritic and saprolitic soils are available (Committee on Tropical Soils 1985), laterites in road pavements (Charman 1988) and tropical residual soils (including calcretes, silcretes and gypcretes) (Geological Society 1990). Other comprehensive works include Morin and Todor (1976) on laterite and other problem soils of the tropics, the Laboratório Nacional de Engenharia Civil (LNEC) et al (1959, 1969), Persons (1970), Gidigasu (1976), Townsend et al (1982), Morin (1982) and Bagarre 1990) on laterites, Netterberg (1969a, 1971) on calcretes and Horta (1980,1989) on calcretes and gypcretes, as well as many more local works. In southern Africa we are fortunate in having a comprehensive work in four volumes - "Engineering Geology of Southern Africa" (Brink 1979, 1981, 1983, 1985) and a book on the natural road construction materials (Weinert 1980). All these five volumes contain a wealth of information on residual soils, weathered rocks and pedocretes. Other valuable sources include the three international conferences on Geomechanics in Tropical and Residual Soils (Brasilia, Singapore and Maseru) and the ASCE conference in Honolulu, and the Arid Soils conference in London, as well as speciality sessions at other conferences, both engineering and engineering geological. The more recent comprehensive geological works include Wilson (1983) and Goudie and Pye (1983) on weathering and duricrusts, Goudie (1973a) on duricrusts, Maignien (1966) and McFarlane (1976) on laterites, Reeves (1976) and Milnes and Hutton (1993) on calcretes, and Langford-Smith (1978) on silcretes in Australia and South Africa.

"It may confidently be expected that engineering characteristics comparable with those described" here "will be encountered in contemporaneous stratigraphic units in the other fragments of Gondwanaland, namely South America, Australia, Antarctica and India, in areas of like geomorphic surface and similar climate" (Brink 1979).

3. TERMINOLOGY

The terminology of residual soils and pedocretes is often conflicting and an adequate discussion would require a paper on its own. This thorny issue will be bypassed in this paper and only the simpler and wider meanings will generally be intended. An attempt will be made to focus on the broader principles rather than the specifics.

The term soil in this paper is generally intended in its loosest engineering sense (ie it does not require blasting) unless prefixed by pedological descriptors. It would therefore include both the solum and saprolite as used by the Geological Society (1990) and sometimes some weathered bedrock.

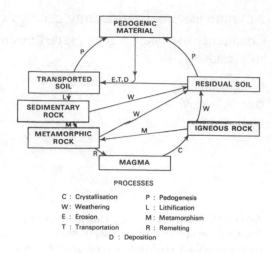

PROCESSES

C : Crystallisation   P : Pedogenesis
W : Weathering   L : Lithification
E : Erosion   M : Metamorphism
T : Transportation   R : Remelting
D : Deposition

Figure 1. Soil-rock cycle (Brink and Williams 1964)

A residual soil is one which has been formed by in-situ weathering and pedogenesis (in the pedological sense). During pedogenesis - and some processes which some pedologists would term geological rather than pedological - segregations of minerals may accumulate to form horizons dominated by nodules, concretions, hardpans, etc, which have variously been termed pedogenic soils, pedogenic materials (Brink and Williams 1964), pedocretes (Jennings and Brink 1978; Netterberg 1980) or duricrusts (Woolnough 1927). These pedocretes may form in transported or residual soils (Figure 1) and even in fractured rock, but are themselves also residual soils.

The basic divisions of tropical residual soils are vertisols, fersiallitic soils, ferruginous soils, ferrallitic soils, and duricrusts (Geological Society 1990).

4 RESIDUAL SOILS IN GENERAL

4.1 *Deficiency of temperate zone engineering soil classifications*

Owing to the deficiencies of the traditional engineering soil classification methods developed largely for temperate zone soils of transported origin, additions to them (eg Horta 1980, 1989; Netterberg and Weinert 1983) or new methods (eg Committee on Tropical Soils 1985; Nogami *et al* 1989; Vaughan *et al* 1988) have been proposed. The basic problem is that most residual soils are weakly bonded in some way and that these bonds may be destroyed in laboratory tests on disturbed samples, resulting in an unduly pessimistic classification.

4.2 *Importance of the soil profile*

The residual soil profile has formed from rock by the processes of in-situ weathering and pedogenesis. The profile normally therefore consists of (from the bottom upwards) nominally fresh rock grading through various degrees of weathering into what might be termed residual soil (solum) in its strictest sense. On ancient landsurfaces such as the African this weathering may extend to depths of 50 m or more. However, due to erosion the full classic lateritic weathering profile of ferrallitic solum, laterite crust, mottled zone and pallid zone is not always preserved. In Africa, Australia and South America (Brink 1979, 1985) the boundary between the soil above and any overlying transported soil is often marked by a thin layer of gravel termed the pebble marker (Jennings et al 1973). The residual soil can usually be recognised by its gradation downwards into weathered material in which the structure of the parent rock is at least partly preserved.

The most common engineering problems associated with residual soils (eg Morin and Todor 1976; Mitchell and Sitar 1982; Blight 1982, 1988; Committee on Tropical Soils 1985; Geological Society 1990) are heave on wetting up, shrinkage on drying out, collapse of grain fabric on wetting under load, erosion, dispersion, pseudokarst formation, excavation, piling and founding problems with core stones, veins and dykes, fissures, relict joints and structures, change in properties on drying or working, aggregate durability, weathering, variability, cracking of compacted layers, and piling and possible foundation (punching) problems with pedocrete hardpans and, not least, that their behaviour is not always predictable by traditional methods developed on temperate transported soils. Many of these problems are associated with residual soils developed from specific rock types. The likely soil profile and engineering problems can therefore usually be predicted from the rock type, a knowledge of weathering, the geomorphic surface and the climate (Brink et al 1982). This has led in southern Africa to the strong emphasis on the accurate recording of the soil profile, including the geological origin of each layer, from which a useful first approximation can be made as to the founding and excavation characteristics (Table 1) and the potential as construction materials. In many cases this may be all that is necessary.

A similar approach to the use of residual soils and weathered rocks as road construction materials may be applied using a simple grouping of rock types with similar engineering behaviour considered together with the climate (Weinert 1980), ie
- acid crystalline rocks (eg granite, gneiss)
- basic crystalline rocks (eg dolerite, basalt)
- argillaceous rocks (eg mudrock, schist)
- high silica rocks (eg quartize, hornfels)
- carbonate rocks (eg dolomite, limestone)
- metalliferous rocks (eg ironstone)
- diamictites (eg tillite)
- pedocretes (eg calcrete, ferricrete)

Table 1. Likely engineering problems of residual soils (mostly after Jennings and Brink 1978).

Residual soil from:	Common example	Soil Texture	Likely problems
Acid crystalline rock	Granite	Clayey sand, boulders	Collapsible grain fabric, erodibility, core stones
Basic crystalline rock	Diabase	Clay, silt	Heave, compressibility
Calcareous rock	Dolomite	Chert, rubble, wad	Sinkhole, doline, compressibility
Argillaceous rock	Mudrock	Silt or silty clay	Heave, slope instability
Arenaceous rock	Sandstone	Sand	Collapsible grain structure from highly felspathic sandstone

Each group has a characteristic range of properties and road problems and specifications must take these into account (Weinert 1980; Netterberg and Paige-Green 1988). Other useful works on construction materials including residual soils, weathered rock and pedocretes are Metcalf (1991), Toll (1991),Queiroz et al (1991), Nogami and Villibor (1991), Gidigasu (1991), Bullen (1992), Smith and Collis (1993) and Millard (1993). A short guide to southern African information is Netterberg (1994). The durability, both of the aggregate and of any cement or lime stabilization used, is a major concern when using residual soils and weathered rocks as pavement materials (Netterberg 1994). Tropical processes can be quite rapid (Blight 1991).

4.3 *Pedological classification*

The pedological classification of tropical soils has been recommended most recently by the Committee on Tropical Soils (1985) and the Geological Society (1990). The latter have recommended the French

(Duchaufor) system which is widely used in Africa, ie fersiallitic soils, ferruginous soils, ferrallitic soils, podzols, andosols and vertisols. Whilst useful, pedological classifications also have their limitations for engineering purposes: they are classifications of the whole *pedological* soil profile, not of materials (horizons), only the top 1-2 m of the engineering soil profile is considered, and the terminology and test data are unfamiliar to most engineers - and many geologists. Although they can be interpreted for engineering purposes (eg Von M. Harmse and Hattingh 1985) the emphasis in southern Africa has been on the engineering soil profile and the geological formation. Some pedological soils may include a pedocrete *horizon*.

5 PEDOCRETES

5.1 *Introduction*

Pedocretes (duricrusts) are soils which have been to a greater or lesser extent cemented and/or replaced by calcite (calcrete), dolomite (dolocrete), iron oxides (ferricrete, plinthite or laterite), silica (silcrete), manganese oxides (manganocrete), phosphate (phoscrete), gypsum (gypcrete) or magnesite (magnesicrete). In accordance with the recommendation of the Speciality Session (1976) terms such as laterite, calcrete, or ferricrete (and therefore also pedocrete), should only apply to materials containing more than about 50 per cent of the cementing or replacing material. Adjectives such as lateritic, calcified, or ferruginised are preferable for materials with less than about 50 per cent cementing or replacing mineral.

In terms of consistency these materials are of three types: indurated (eg hardpans and nodules), non-indurated (soft or powder forms), and mixtures of the two (eg nodular pedocretes).

The term laterite is often used for tropical soils which have been produced by advanced weathering accompanied by a relative enrichment in iron and aluminium sesquioxides due to the decomposition of primary minerals and the removal of bases and silica. Some such soils are indurated and some are capable of induration on exposure. As originally defined, Buchanan's laterite was simply a clayey material rich in sesquioxides which could be cut and shaped with hand tools into blocks and would harden irreversibly on exposure (Buchanan 1807 in Maignien 1966). Since then the term has been used so widely as to be almost meaningless and has been abandoned by pedologists and some geologists and engineers. In an attempt to overcome this confusion the term plinthite has been introduced in the United

States as a synonym for Buchanan's laterite (Soil Survey Staff 1975). In the South African binomial system the terms hard and soft plinthite are used, but the latter is not required to possess self-hardening properties (McVicar *et al* 1977). The term ferricrete has been widely used in South Africa where it is commonly restricted to indurated iron-rich materials. The term laterite is most often used loosely both in southern Africa and on a world-wide basis to include what are here called ferricretes. It will be used in this loose sense here. The differences in origin and composition implied in the use of many of these terms should not be of overriding concern to the engineer as they are not necessarily accompanied by differences in their geotechnical properties.

South African dorbanks are indurated, usually reddish brown, massive or platy horizons up to about 1,2 m thick which are cemented by silica and sometimes some calcium carbonate and/or iron oxides, to produce a consistency up to that of hard rock (Ellis and Schloms 1981, 1984). They are equivalent to the duripans of the United States and the red and brown hardpans of Australia.

Calcareous pedocretes are commonly known as caliche in the United States (Reeves 1976). Synonyms for the different pedocretes used elsewhere in the world have been provided by Goudie (1973a). Pedology has its own nomenclature which will not be discussed here.

5.2 *Age, thickness and rate of formation*

Pedocretes appear to form at rates of mostly between about 20 and 200 mm thickness per thousand years (Netterberg 1985). Calcretes in southern Africa may attain thicknesses of over 30 m (Du Toit 1954) but, except for the Neogene occurrences, they are rarely homogenous over depths exceeding 1-2 m. Up to 10 m of laterite has been recorded beneath the African erosion surface in South Africa (Netterberg 1985) and elsewhere (Geological Society 1990). Silcrete of similar thickness also occurs beneath the African surface in South Africa and its equivalent in Australia. However, the younger Pleistocene pedocretes seldom exceed 2 m. Calcified alluvial gravels (valley calcretes) may exceed a thickness of 10 m in places.

As pedocretes are not sediments they are not subject to the laws of stratigraphy. They are composed of material of at least two ages, namely the host material and the authigenic cement. Several different generations of both host material and cement may be present in a single pedocrete occurrence. Consequently pedocretes are difficult to

Table 2. Typical composition of southern African pedocretes (Netterberg 1985).

Component	Calcrete %	Laterite[1] %	Silcrete %	Phoscrete %	Gypcrete %	Mineralogy
SiO_2	1-60	5-70	85-100	10-60	0-60	Quartz, feldspar, clays, opal, chalcedony
Al_2O_3	0-5	5-35	0-5	5-30	0-60	Feldspar, clays, gibbsite
Fe_2O_3	0-5	5-70	0-10	0-10	0-60	Goethite, haematite
TiO_2	0-1	0-5	0-5	-	0-60	Anatase, rutile
$CaCO_3$	40-100	0	0-5	0-50	0-60	Calcite, dolomite, apatite
$Ca_3(PO_4)_2$	<0,2	0-2	0-2	20-60	-	Apatite, collophane, dahllite
$CASO_4.2H_2O$	0-2	0	0-2	0-2	40-90	Gypsum
H_2O+	0-5	5-20	0-2	0-5	10-20	Clays, gibbsite, goethite, gypsum
Organic	0-1	0,2-2	-	-	-	Organic matter
NaCl	0-1	0	-	-	0-4	Halite, apatite

(1) Includes ferricretes but excludes bauxites. CBD-extractable Fe_2O_3 of South African ferricrete hardpans is 43-77% (Fitzpatrick, 1978; Fitzpatrick and Shwertmann, 1982).

date, to correlate with one another and to place within the stratigraphic column (Netterberg 1978a; Partridge *et al* 1984), but can be used to date faults.

Fossil pedocrete palaeosols at the tops of unconformities dating back as far as the Precambrian (eg Retallack 1990) are being increasingly recognised.

5.3 *Composition*

Pedocretes are mixtures of the original host or parent material and the authigenic cement, which may either have been introduced or relatively concentrated by leaching. As the pedocrete develops the authigenic mineral content increases until it may constitute almost the whole mass. It is not uncommon for one pedocrete to replace another, and mixtures do occur. For example, both calcretes and ferricretes may become silicified.

Typical constituents of different southern African pedocretes are summarised in Table 2. Calcite and dolomite are the usual carbonate minerals in calcretes and palygorskite the most common clay mineral. Goethite and haematite are the most common iron oxides present in laterites and ferricretes, gibbsite the most common aluminium oxide and kaolinite the most common clay mineral. Other clay minerals such as the troublesome

smectite may also occur in both calcretes and ferricretes.

The old term lateritic soil is used in engineering for a material of any particle size distribution with a silica/ sesquioxide *molecular* ratio of the clay fraction of less than 2,0 (LNEC *et al* 1969). This ensures that the clay minerals present are largely of the kaolinite group (Van der Merwe and Heystek 1952) and therefore inactive. In modern pedology the term lateritic soil has been replaced by terms such as ferrallitic and fersiallitic soils.

South African duripans are composed of quartz, feldspars, rock fragments and up to 15 per cent clay, with variable amounts of silica, iron oxide and carbonate cement (Ellis and Schloms 1981, 1984).

The cementing minerals in pedocretes are usually extremely fine grained and may appear amorphous to the naked eye. However, except in the case of some silcretes they are usually crystalline in thin section or to X-rays.

5.4 *Origin*

Most pedocretes appear to go through the stages of development shown in Figure 2 (Weinert 1980, Netterberg and Caiger 1983) - mostly after Netterberg (1969a,b, 1971, 1980). After formation they will weather like any rock if conditions become

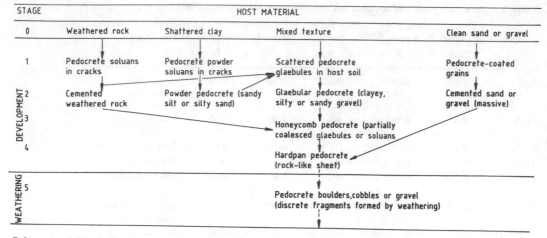

STAGE	HOST MATERIAL			
0	Weathered rock	Shattered clay	Mixed texture	Clean sand or gravel
1	Pedocrete soluans in cracks	Pedocrete powder soluans in cracks	Scattered pedocrete glaebules in host soil	Pedocrete-coated grains
2	Cemented weathered rock	Powder pedocrete (sandy silt or silty sand)	Glaebular pedocrete (clayey, silty or sandy gravel)	Cemented sand or gravel (massive)
3			Honeycomb pedocrete (partially coalesced glaebules or soluans	
4			Hardpan pedocrete (rock-like sheet)	
5			Pedocrete boulders, cobbles or gravel (discrete fragments formed by weathering)	

(Left margin labels: DEVELOPMENT for stages 1–4, WEATHERING for stage 5)

Soluan = soluble cutan (cutan = fissure-filling). Glaebules = nodules, concretions, septaria, etc.
For geotechnical purposes use the terms nodules or nodular unless the other forms are dominant.

Figure 2. A suggested sequence of development and simple morphogenetic classification of pedocretes (Netterberg 1985).

unfavourable for their further development to the hardpan stage and for its preservation. Such boulders, cobbles and gravels may be transported as colluvium or alluvium. They may also be incorporated into a younger pedocrete.

Calcretes are thought to be of two basic origins; groundwater or pedogenic. In the groundwater type carbonate is precipitated above a shallow water table or by lateral seepage of groundwater, in the pedogenic type the carbonate is transported downwards through the soil by rainwater. The origin of the carbonate may be the soil or weathering or it may be transported as dust and in the rainwater itself.

Ferruginous pedocretes form by absolute accumulation, in which sesquioxides are added, or by relative accumulation in which sesquioxides are concentrated by removal of the most soluble constituents under conditions of intense weathering and free drainage (D'Hoore 1954). Although both processes can take place within the same profile one is usually dominant in a particular horizon. Relative accumulations tend to be depleted in weatherable minerals and enriched in aluminium oxides and to contain only clay minerals of the kaolinite group. The ultimate product of this process would be bauxite, which is an ore-grade aluminous laterite. Absolute accumulations tend to be enriched in iron and manganese and poor in aluminium. Depending on the nature of the host materials and the drainage, they may or may not contain significant amounts of weatherable minerals and active clays. Absolute

accumulation appears to be necessary for induration. Iron is thought to be mobilised under reducing conditions in the wet seasons and precipitated under oxidising conditions in the dry seasons and during droughts. Under continuous wet conditions, such as rain forest cover, no induration can take place. Relative accumulation thus leads to the formation of an earthy, scoriaceous, uncemented material with a white to rose colour (Maignien 1966). Although they are primarily relative accumulations, most laterites and bauxites exhibit signs of absolute accumulation. Ferricrete forms by absolute accumulation through upward, downward or lateral migration of iron and may form in a laterite or bauxite. Calcrete, gypcrete and phoscrete all appear to be absolute accumulations. Silcretes may possibly be either.

5.5 Distribution

Maps showing the distribution of pedocretes on a world (Morin and Todor 1976, Petit 1985) and regional (Netterberg 1969a, 1971, 1985; Goudie 1973a; Reeves 1976; Carlisle et al 1978; Milnes and Hutton 1983) basis are available. As most pedocretes are soil horizons in the pedological sense, their distribution is best explained in terms of the five soil-forming factors. Thus, although a pedocrete may form faster over a particular parent material, it may in fact, given suitable drainage conditions and sufficient time, form over any parent material. Drainage conditions favourable for the formation of calcretes appear to be ephemeral or seasonal water courses and pans, or areas with shallow water tables in arid or semi-arid climates. Ferricretes appear to need at least a subhumid climate where Weinert's (1980) N-value is less than 5 for chemical decomposition to release the necessary iron from the parent material. The resultant ferricrete usually occurs at gully heads, hillslope-pediment junctions and on pan and vlei (dambo) side slopes (Partridge 1975) - the groundwater laterites of many other authors (see eg Charman 1988). Under rainfall of more than about 800 mm in a warm climate the extent of chemical decomposition of the parent rock is still greater, leading to relative accumulation of sesquioxides, as bases, silica and some iron are leached out. A special situation exists in the case of ancient and well-placed erosion surfaces where extensive pedocrete cappings usually develop over a deep, kaolinised residual soil - these are laterites in humid areas and silcretes in semi-arid areas (Partridge 1985a; Maud 1985; Partridge and Maud 1987).

The correlation between the distribution of calcretes and climate is perhaps the most marked of all. In southern Africa well-developed calcretes generally only occur where the mean annual rainfall is less than about 550 mm (Netterberg 1969a, 1971, 1985). The N = 5 line also affords a good correlation with this limit of calcrete occurrence. Silcretes occur in two separate geographic areas - the coastal belt and the inland region (Du Toit 1954). The coastal silcretes occur as mesa cappings (ie they are plateau pedocretes)(Schloms and Ellis 1984). They are invariably associated with deep weathering beneath the African erosion surface. The inland silcretes are essentially limited to the area of the Kalahari Group (Du Toit 1954; Summerfield 1982). In southern Africa phoscretes occur only in the area around Saldanha Bay (Weinert 1980). Duripans or dorbanks (Ellis and Schloms 1984) are most common in the drier parts of the Cape and Namibia where the rainfall is less than 300 mm. In southern Africa gypcretes occur mostly in the Namib desert and in the Upington region (Carlisle 1980), but the wetter limit of their occurrence seems to be close to that of calcretes. They extend from Angola (Watson 1983) in the north to near Cape Town in the south (Visser *et al* 1963).

5.6 Description and classification for geotechnical purposes

The problems of applying traditional soil mechanics classifications to pedogenic and weathered materials have received considerable attention. Traditional classifications envisage soils as being composed of discrete, hard, durable, solid particles which can be classified reliably on the basis of grain size and Atterberg limits. Attempts to apply these criteria to weathered, cemented and aggregated materials can be misleading as the gradings and Atterberg limits obtained depend very much on the methods of excavation, processing and testing employed. For this reason it is advisable to add two extra symbols to the Unified Soil Classification of a material to show the geological origin (Horta 1980) and method of excavation used (Netterberg and Weinert 1983).

Because simple laboratory tests do not adequately reflect the special properties of many pedocretes or partially cemented soils, reliance must be placed on careful description of the *in situ* characteristics of these materials when recording the soil profile eg as in Jennings et al (1973).

Some general comments on the occurrence of pedocretes may aid the formulation of the somewhat fuller description which these materials require. The permanent water table often occurs just below a thick calcrete horizon, while ferricretes are commonly associated with fluctuating water tables. Calcretes are therefore usually encountered in a relatively dry condition, while ferricrete horizons are often wetter than the adjacent horizons. The properties of pedocretes vary according to the nature of the host or parent material and the cementing or replacing minerals, and the extent of cementation or replacement. The colours of calcretes and silcretes generally vary between grey and white, but greenish, reddish and brownish materials are also known, and black mottling and staining are sometimes present. Ferricretes are generally red, brown or yellow. Consistencies of the soil *mass* vary from loose or soft soil in the case of weakly cemented, powder or glaebular varieties to very hard rock in the case of well-cemented hardpans. A variety of structures may be present. Jointing is sometimes present, either inherited from the host material or developed subsequent to formation of the pedocrete. A variety of hardpan structures such as vesicular, concretionary, pisolitic, slaggy, brecciated and terrazzo have also been recognised. All textures are presented from clay to rock. Glaebular pedocretes generally classify as gravels or sands and powder calcretes as sands or silts. Inclusions of rock fragments and roots are common and stone artifacts are not rare.

The mature pedocrete profile generally consists of a hardpan horizon overlying a much weaker, less well developed horizon. The sequence in which calcretes form has led to a generalised, simple morphogenetic classification of calcretes (eg Netterberg 1968b) which can be extended to pedocretes in general (Weinert 1980; Netterberg and Caiger 1983)(Figure 2). This classification accommodates those forms which can be easily recognised in the field. The only term which may be unfamiliar is the glaebule (Brewer 1964), which is a collective term for concretions, nodules, pisolites, etc.

The different varieties of pedocretes may be briefly defined as follows (Netterberg 1985):

Calcareous (or ferruginous, etc) soil - a soil (described in terms of texture as clay, silt, sand or gravel) which exhibits little or no glaebular development or massive cementation, but which contains some possibly authigenic mineralisation which is not sufficient to indurate the soil significantly.

Calcified (or ferruginised, etc) soil - a relatively massive to platy soil which has been indurated by cementation to a firm or stiff consistency.

Powder pedocrete - chiefly loose silt and fine sand-sized cemented or aggregated particles or nearly pure authigenic minerals. The grading modulus is less than 1,5. In the case of calcretes the content of particles finer than 2 mm and 0,425 mm usually exceeds about 75 per cent and 50 per cent respectively.

Glaebular (nodular or concretionary) pedocrete - a naturally occurring mixture of silt to gravel-sized glaebules of cemented and aggregated finer particles. The horizon (ie the soil *mass*) is usually of a loose consistency, but the glaebules may be quite hard. The grading modulus is 1,5 or greater. In the case of calcretes the material usually has less than about 25 per cent passing the 2 mm and less than about 50 per cent passing the 0,425 mm sieves.

Honeycomb pedocrete - a partly coalesced glaebular pedocrete representing an intergrade between the glaebular and hardpan stages. It usually still contains loose or soft host soil particles filling the voids between the coalesced glaebules. Ripping is usually required for excavation.

Hardpan pedocrete - an indurated and strongly cemented, usually relatively massive, rock-like horizon cemented to a consistency which varies between stiff soil and very hard rock. Except in the case of the older (Tertiary) materials, individual hardpans seldom exceed about 0,5 or 1 m in thickness and are normally underlain by much softer or looser material.

Pedocrete boulders, cobbles and gravels - discrete or partially connected, usually boulder or cobble-sized fragments formed by the weathering and breakdown of hardpan.

Although some forms, such as powder ferricrete, are rare, this simple classification appears to cater for all the likely types of pedocrete. Charman's (1988) plinthite would be a ferruginous or possibly ferruginised soil in this classification. Although not normally regarded as pedocretes, the related materials beachrock and dunerock (aeolianite) would classify as calcrete hardpan or calcified sand. The geotechnical properties of these materials in Israel have been discussed by Frydman (1982).

In its simplest form, terms such as calcrete sand, ferricrete gravel and silcrete rock can be used in addition to the normal soil profile and classification descriptors. More detailed description is discussed by eg Netterberg (1980), Netterberg and Caiger (1983), Charman (1988) and Geological Society (1990).

5.7 *Differences between traditional and pedocrete materials*

The geotechnical properties of pedogenic materials depend on three factors: the nature of the host or parent material, the stage of development (ie the extent to which it has been cemented or replaced), and the nature of the cementing or replacing mineral. For example, the properties of calcareous soils are closely akin to those of the host soil, whereas hardpan calcretes essentially behave as limestone rock. During pedocrete development, clay and silt become aggregated and cemented into larger silt- to gravel-sized complexes of varying strength and porosity. These particles or aggregations may or may not break down during laboratory testing and under compaction. Moreover, the mineralogy of the cementing material and of the clay fraction is different from those of normal, temperate zone soils from which much of our geotechnical experience and many of our specifications have been derived. Pedocretes can therefore be expected to exhibit certain differences in behaviour from those of traditional materials (Table 3). For example, in traditional soil mechanics it is usually assumed that all the water is outside the particles, whereas porous pedocrete aggregates retain moisture and this affects conventional moisture content and Atterberg limit determinations. Palygorskite (attapulgite), which is the dominant clay in many calcretes (Netterberg 1969a, 1971; Watts 1980), has approximately the same plasticity index as some smectites (Hough 1969; Grim 1962), but has a non-expansive lattice and a hollow needle-like particle shape, typically matted into a haystack structure (Haden and Schwindt 1967; Grim 1968), instead of the usual flaky particle shape of most other clays. It has the lowest shrinkage limit and dry density and the highest optimum moisture content and shear strength

Table 3. Differences between traditional and pedogenic materials (Netterberg 1976).

Property	Traditional	Pedogenic
Composition	Natural or crushed aggregate with fines	Varies from clay to rock
Aggregate	Solid, strong rock	Porous, weakly cemented fines
Fines	Rock particles with or without clay	Cemented, coated & aggregated clay and/or silt particles
Clay minerals	Mostly illite or smectite	Wide variety e.g. halloysite. palygorskite
Cement	None (usually)	Iron oxides, $CaCO_3$ etc.
Hydration	None	Variable
Chemical reactivity	Inert	Reactive
Solubility	Insoluble	May be soluble
Weathering	Weathering or stable	Forming or weathering
Atterberg limits	Stable	Sensitive to drying & mixing
Grading	Stable	Sensitive to drying & working
Salinity	Non-saline	May be saline
Self-stabilisation	Non-self-stabilising	May be self-stabilising
Stabilisation (cement)	Increases strength	Usually increases strength
Stabilisation (lime)	Decreases plasticity	Usually decreases plasticity &/or increases strength
Variability	Homogeneous	Extremely variable
Climate	Temperate to cold	Arid, tropical, temperate
Traffic	High	Low

of all the clays, while its compressibility coefficients are comparable to those of illite (Hough 1969). The sesquioxides in laterites may be hydrated or amorphous, while troublesome clays such as hydrated halloysite and allophane may be present. The possible effects of these minerals have been reviewed by Morin and Todor (1976), Gidigasu (1976), the Geological Society (1990) and others.

In essence, these differences render the geotechnical behaviour of pedocretes less predictable by means of traditional classification tests.

Table 4 provides a generalised summary of some typical properties of southern African calcretes in comparison with calcified and calcareous soils. The properties of the other pedogenic materials at similar stages of development appear to be generally similar to those of calcretes, except that laterites and ferricretes never seem to become as strong or as rich in iron and aluminium oxides as calcretes do in carbonates; in addition, their content of soluble salts and pH - typical values of pH are 5-7,5 (LNEC *et al* 1969) - are always lower than those of calcretes (usual pH range 7,5-9). Laterite hardpans seldom exceed 10 MPa in uniaxial compressive strength (LNEC *et al* 1969), while the mean strength of

Kalahari hardpans is 32 MPa (Goudie 1983), and the hardest (silicified) calcretes probably approach 80 MPa (Netterberg 1980). Silcretes on the other hand, become stronger than calcretes and may behave like quartzites, but duripans (dorbanks) can probably be regarded as silicified soils for geotechnical purposes. Less is known of the properties of phoscretes and gypcretes, but they do not appear to become as hard as calcretes.

5.8. *Variability*

Numerous authors have provided some geotechnical properties of typical pedocretes used for road construction in southern Africa. Perhaps the most striking feature of these results is their variability. A general conclusion is that their CBR swell and Texas triaxial classes are generally good in spite of apparently unfavourable plasticity indexes and grading characteristics. Variability, both lateral and vertical, is an unavoidable feature of most pedocrete occurences. Pedocretes tend to deteriorate in quality with depth (sometimes sharply), due to decreased cementation, whilst weathered rocks generally become less weathered.

Table 4. Summary of some properties of calcretes in comparison with calcareous and calcified soils (mostly after Netterberg 1982a).

Material type	Total Carbonate as CaCO$_3$ %	Grading modulus %	PI[1,2] %	Classification		Nat./crushed coarse agg.		Whole horizon *in situ*	
				ASTM D 3282	ASTM D 2487	10% FACT kN	Mohs hardness[3]	Overall consistency[4]	Workability
Calcareous soil	1-10?[5]	Varies	Varies	Varies	Varies	Varies	Varies	Varies	Varies
Calcified sand[6]	10?-50	1,5?-1,8?	NP-20	A-1-b to A-2-b	SC,SM,SP	18?-70?	2-3	Med dense - dense or firm - stiff	Doze-rip
Calcified gravel[6]	10?-50	>1,8?	<8?	A-1-a to A-1-b	GC to GW?	70?-135?	≥3?	Med dense - very dense or firm - v. stiff	Rip-blast
Powder calcrete	70-99	0,4-1,5	SP-22	A-2-4 to A-7-5	CL,MH,ML, SM,SC	18-90?	2-3	Loose - stiff	Doze-shovel
Nodular calcrete	50-75	1,5-2,3	NP-25	A-1-a to A-6	SC,SM, GC,GW	9-178	1-5	Loose - med. dense	Doze-shovel
Honeycomb calcrete[6]	70-90	>2,0	SP-8?	Rock?	Rock? R[7]	80?-205	3-6	Stiff - very stiff	Rip & grid roll
Hardpan calcrete[6]	50-99	>1,5?	NP-7	Rock	Rock R[7]	27-196	2-6	Stiff - very hard	Rip-blast
Calcrete Boulders and Cobbles[6]	50-99[1]	>2,0	NP-3	Boulders	Boulders & Cobbles B[7]	98-205	3,5-5	Very stiff - very hard	Rip or doze & crush

(1) Without the loose soil between calcrete boulders and cobbles.
(2) On the LAA fines in the case of honeycombs, hardpans and boulders.
(3) Of the carbonate or silicified carbonate cement.
(4) According to method of Jennings, Brink and Williams (1973) and Core Logging Committee (1978).
(5) Up to about 50% when many nodules present.
(6) Generally after crushing or compaction in the case of tests on disturbed material and soil classification.
(7) Suggested (non-standard) symbol.
(8) Figures in italics represent limits used to distinguish one stage of development from the next.

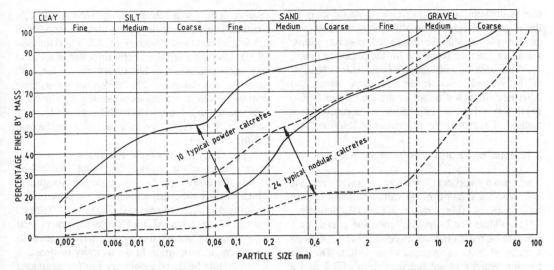

Figure 3. Typical grading envelopes for powder and nodular calcretes (After Netterberg and Caiger 1983).

5.9 *Particle size distribution*

Grading analyses are meaningful only in the case of powder and glaebular pedocretes and uncemented or very weakly cemented soils. All other forms of pedocrete comprise either boulders and cobbles which are too coarse, or indurated horizons which require excavation and processing before they can be said to have a grading. Most powder and glaebular calcretes (Figure 3) (Netterberg 1969a, 1971) and laterites (Gidigasu 1976; Charman 1988, Millard 1993) are relatively poorly graded by mass, displaying an apparent deficiency of coarse sand and an excess of fine sand. However, if the mass gradings are corrected to a volumetric basis (which is what really matters) they may be greatly improved (Netterberg and Caiger 1983). This is because the bulk density of the individual coarse sand particles is lower than that of the fine sand particles (Netterberg 1969a; Gidigasu 1976). In addition, grading analyses tend to be further influenced by the variable and often rather low strength of the cemented and aggregated particles in most pedocretes. Sodium hexametaphosphate is the most effective dispersing agent for lateritic soils (LNEC et al 1969, Gidigasu 1976), and probably for calcretes.

5.10 *Atterberg limits, shrinkage and swell*

Southern African calcretes plot on both sides of the Casagrande A-line. They tend to possess higher liquid limits and linear shrinkages relative to their plasticity indexes than other soils (Netterberg 1971). The linear shrinkage is often less than the commonly assumed value of half the plasticity index while the shrinkage limit is often higher than the plastic limit. This latter feature appears to be unique to calcretes, and may be due to a combination of particle porosity and high silt content, as well as to the presence of diatoms, palygorskite or sepiolite. Laterites and lateritic soils also plot on both sides of the A-line (Mitchell and Sitar 1982). Those falling below the A-line are likely to be troublesome (Gidagasu 1976) as they might contain hydrated halloysite.

Some laterites – and even some calcretes (Netterberg 1982a) – exhibit lower Atterberg limits after oven-drying than they do when air-dried. When plotted on the Casagrande plasticity chart the oven-dried material is found to have moved downwards several percentage points roughly parallel to the A-line in a similar fashion to organic soils. Gypcretes can be affected due to partial dehydration of gypsum on oven-drying. Even air-drying can affect some soils.

It is well known that materials with friable aggregated particles are sensitive to the degree and period of mixing. For this reason the mixing time specified in test methods (eg 10 minutes – NITRR 1979) should be adhered to.

It might be expected that the activity of lateritic soils would be low. Although this is generally the case, values between 0,3 and 3 have been recorded on a world-wide basis (Nixon and Skipp 1957; Gidigasu 1976). This appears to be due to underestimation of the true percentage clay fraction due to difficulties in disaggregating and dispersing these soils. All of these difficulties suggest that greater use should be made of the bar linear shrinkage test as a substitute for the plasticity index.

The swell of compacted lateritic soils (LNEC *et al* 1969) and calcretes is low even when the Atterberg limits are high and the De Castro (1969) swell test offers an alternative or supplementary method of assessing the properties of the fines.

5.11 *Apparent relative density of the soil fines*

The mean apparent relative density of the <0,425 mm fraction of 44 calcretes was found to be 2,66 (S = 0,06; range 2,47-2,80)(Netterberg 1971). Values for 28 laterite sands and gravels on a world-wide basis ranged between 2,67 and 3,46 with a mean of 3,06 (Townsend *et al* 1982). The size fraction tested was not stated. Values of 2,2 to 4,6 for the < 2 mm fraction are quoted by Gidigasu (1976). The apparent relative densities of calcretes do not therefore vary significantly from those of most other soils, but those of laterites are higher, reflecting their content of iron minerals.

An increase in density after oven-drying may indicate the presence of hydrated halloysite (Newill 1961).

5.12 *Aggregate properties*

The strength of the coarse and fine aggregate fraction varies from very low in the case of calcified, silicified and ferruginised soils to generally high in the case of hardpans, boulders and cobbles. The strength of the particles generally tends to increase with particle size (Netterberg 1971) and percentage cementing agent (De Graft-Johnson 1975). The dry 10 per cent FACT value of 9-13 mm calcrete nodules of crushed aggregate varies from about 20 to about 200 kN (Table 4). Soaked values are typically two-thirds of the dry values (Netterberg 1971). Few data are available for southern African ferricretes, the dry values for which range between about 25 and 100 kN (Van der Merwe 1971; Weinert 1980), while Los Angeles abrasion values (LAA) in excess of 50 per cent are common (LNEC *et al* 1969). Pedocrete glaebules and especially honeycombs may include clay which can affect both aggregate strength (Novais-Ferreira and Correira 1965; Van der Merwe 1971) and plasticity indexes and gradings after compaction.

Aggregate particles are often porous and the water absorption of calcrete nodules and crushed aggregate has been found to range between 0,5 and 25 per cent (Netterberg 1982a). The particle bulk relative density was found to range from 1,6 to 2,6. Crushed laterite hardpans have water absorptions of 0,5-9 per cent and densities of 2,6-3,0 (Gidigasu 1980). Useful correlations between water absorption and aggregate strength have been found for both calcretes and laterites. The difference in particle density as well as variations of density with particle size (Netterberg 1969a, 1971; Gidigasu 1976), may need to be taken into account in mechanical stabilisation work.

The shape of pedocrete aggregate varies from round and smooth to irregular and rough. According to Van der Merwe (1971) it is aggregate roughness which has been the major factor in the outstanding performance of laterite bases in Zimbabwe. Triaxial testing of mixtures of rough laterite gravel and smooth quartz gravel showed the triaxial class to deteriorate in proportion to the quantity of quartz.

A simple field and laboratory test for aggregate strength called the aggregate pliers test (APT) measures the percentage of 13-19 mm aggregate particles which cannot be broken with the fingers (AFV) and the percentage which cannot be broken with a standard pair of 180 mm pliers (APV) and has been found useful in evaluating calcretes for wearing courses of unpaved roads (Netterberg 1978a). This test has been correlated with the 10 per cent FACT value, and should also be applicable to other pedocretes. A good correlation also exists between the 10 per cent FACT value and the Mohs hardness of the carbonate cement of calcretes (Netterberg 1971).

The variable and often rather low strength of the particles making up pedocretes creates problems both in laboratory testing and during compaction. Laboratory results may not give a true reflection of the density of the layer or the material itself, while gradings and plasticity indexes may improve or deteriorate after compaction, or even after stockpiling. Degradation during handling and compaction such as that reported by Sandford *et al* (1980) is common. As a result of the porous nature of the particles it is advisable to cure samples for compaction and CBR testing overnight to permit moisture equilibration (LNEC *et al* 1969). Curing of the material on the road is also advisable before compaction for this reason (Cocks and Hamory 1988). Water held in the aggregate may cause blistering of prime coats and loosening of the upper base when it vaporises on a hot day (Rossouw 1982).

5.13 *Hardening and self-stabilisation*

Some pedocretes possess the ability to undergo self-hardening. Indeed, the term laterite was first coined for such a material, although this type is rare (McFarlane 1983). Laterite building blocks, which have been dug out by spade and left to harden, have been used for centuries and are still used today (Maignien 1966). A type of firm to stiff calcrete hardpan (vlei limestone) possesses similar properties and has also been widely used as building blocks in Namibia (Netterberg 1969). Induration of laterites (and presumably calcretes) also takes place if the laterite is exposed by excavation or erosion or if its surface is deforested (Maignien 1966). Reasons advanced for the induration of laterites include impregnation and accumulation of iron and aluminium, oxidation of iron, dehydration of sesquioxides and the development of a continuous structure of crystallised sesquioxides. As pointed out by Grant and Aitchison (1970) only actively forming pedocretes can be expected to possess this self-hardening property.

Test methods for potentially self-stabilising materials include the petrification degree test (Nascimento *et al* 1965) and the soaked CBR test, carried out after allowing the compacted material to dry in the mould or after moist curing, or following wetting and drying cycles (Netterberg 1975). As the CBR sometimes decreases after treatment, this is probably also a useful indication of durability. The development of self-stabilisation in a road layer cannot, however, be predicted at this stage and this should be regarded as an added (and uncertain) safety factor.

Although there is clear evidence that some pedocretes harden on exposure or that large increases in soaked CBR or uniaxial compressive strength may be attainable in the laboratory after several wetting and drying cycles, or even after simple curing (Rossouw 1982), documented evidence of the value of self-stabilisation in road construction is lacking (Netterberg 1975), but does occur (Cocks and Hamory 1988). Morin and Todor (1976) concluded that the property of self-hardening can rarely be used to advantage in road construction.

Although laterite building blocks have lasted for centuries in Asian buildings, in different environments induration is not necessarily permanent and pedocretes are subject to weathering like any rock. Disintegration and solution may result, and the softening of laterite crusts has been observed after reforestation (Maignien 1966; Morin and Todor 1976). In essence, the response of a pedocrete to any change in its environment needs to be assessed for all important works.

5.14 *Slope stability and erosion*

Sesquioxides generally have a positive influence on slope stability and erosion (Gidagasu 1976), and De Castro (1969) swell values in excess of 11 per cent give a very good indication of an erodible material.

5.15 *Foundations on pedocretes*

In contrast to the normal weathering profile in rocks, pedocretes usually deteriorate with depth. Hence it is often dangerous to found heavy structures on strong pedocrete layers unless these have been proved to be of adequate thickness. Such pedocrete horizons can be impenetrable to picks or to light truck-mounted augers, as well as to tube piles but may be underlain by loose or soft and sometimes potentially expansive or collapsible soil. Light structures can often take advantage of the rafting action of such a layer both for spreading the load and for minimising any differential heave caused by underlying materials.

Lateral variability of pedocrete horizons has also been a problem in several cases. In such variable materials trenching is advisable during site investigations in addition to the usual pitting or augering. The use of *in situ* tests (Whittaker 1980; Beckwith and Hansen 1982) has been recommended for evaluation of variable materials such as calcretes. Losses of strength of up to 90 per cent have been measured in a powder calcrete by this means after wetting.

A collapsible fabric has been suspected in some powder and nodular calcretes and cemented soils, while any nodular pedocrete with a clay matrix should be checked for potential heave. In the hinterland of Natal, laterites capping the African erosion surface have been observed by R.R. Maud to produce soils with exceptionally high void ratios of up to 3,0 and dry densities of less than 1 000 kg/m^3. The upper horizons of such profiles are frequently highly organic and they suffer massive consolidation when wetted under load. Owing to the high organic content these soils do not lend themselves to compaction; such treatment can therefore not be applied as a solution to the founding even of light loads such as single-storey houses.

Small-scale karst features are common in weathered calcretes (Netterberg 1980) and evidence of small sinkholes has been presented by King (1964) and Goudie (1973b). Pseudokarst is known in some laterites.

Gypcretes and to a lesser extent dorbanks and calcretes are likely to be saline. The possibility of corrosion of galvanised piping and culverts, damage to concrete, and dissolution must be considered.

Sabkha is not discussed here, but is relevant.

5.16 *Pedocretes in road construction*

The most important geotechnical use of pedocretes everywhere is in road construction, and calcrete is particularly widely used (Netterberg 1982b). In southern Africa as a whole, laterite (LNEC *et al* 1959), calcrete and silcrete have been used for wearing courses for unpaved roads and for all layers of the road prism, including surfacing chippings (Weinert 1980). Similar uses have been recorded elsewhere, such as Texas, Australia and Brazil. Crushed phoscrete with a soil binder has apparently only been used for base (Weinert 1980), while duripans have been used fairly extensively for all layers up to subbase. Gypcrete compacted with water nearly saturated with salt makes excellent unpaved roads along the arid west coast of Namibia.

The only pedocrete used to any extent in the developed countries of the Northern Hemisphere is calcrete (caliche) in the southwestern United States.

Problems which have been experienced with pedocretes can be listed as follows: difficulty in locating suitable supplies; variability, both laterally and vertically in the borrow pit as well as on the road; sensitivity to drying and manipulation; high apparent plasticity and poor apparent grading, often accompanied by high CBR values; salt damage to roads and corrosion of pipes in gypcretes, dorbanks and calcretes; sulphate attack on concrete in gypcretes; aggregate degradation; segregation; compaction problems; determination of cement and lime stabilisation requirements and contents; stabilisation cracking; failure to stabilise; loss of stabilisation; blistering and variable absorption and penetration of binder and prime coat; excessive absorption (calcretes), crushing (laterites) and stripping (silcretes) when used in bituminous surfacings; and perched water tables.

Potential problems include those of alkali-silica, carbonate and alumina reactions, organic matter inhibition of stabilisation (probably only in some laterites), unusual reactions with cement and lime, heave, collapse, dispersion, dissolution, and small to medium scale karst and pseudokarst phenomena. Methods of predicting some of these problems have been given by De Graft-Johnson (1975), Gidigasu (1976) and Netterberg (1976) (Table 5).

In general, pedocretes have performed better as pavement materials than would have been expected on the basis of their gradings and Atterberg limits. This is at least partly due to the development of fabric and suction in unsaturated materials (Toll

1991). Shear failures and rutting are rare and have generally been confined to areas of poor drainage or where the more clayey materials have been used.

Traditional materials specifications have generally been found to be too conservative for pedocretes in terms of their strict grading and plasticity requirements. It has therefore been necessary to devise local empirical specifications for these materials. However, these cannot be discussed here. An important point is to ensure that one is actually dealing with a pedogenic material. It is suggested that the molecular silica/sesquioxide ratio of the fraction passing 2 μm (LNEC et al 1959, 1969), 425 μm, or 2 mm (DNER 1974) should be $\leq$ 2,0, or that the $Al_2O_3 + Fe_2O$ content of the fraction passing 425 μm should be $\geq$ 10% (Cocks and Hamory 1988) for laterites and lateritic soils and that a calcrete should preferably possess at least about 10 per cent equivalent $CaCO_3$ in the fraction passing 425 μm (Netterberg 1971). The correct test methods for these determinations should be followed.

Table 5. Recognition of problem pedogenic materials
(mostly after Netterberg 1976)

Small	Large
Aggregate strength	Water absorption
Maximum dry density	Natural moisture content
Reserve of weatherable minerals?	Linear shrinkage
	Potential swell
SiO_2/R_2O_3?	Aggregation index
Fe_2O_3/Al_2O_3?	Optimum moisture content
$SiO_2/CaCO_3$	Salinity
Calcite/dolomite	Gypsum content
	Mica content
	Organic matter content
	Amorphous mineral content
	Content of hydrated halloysite, palygorskite, sepiolite, glauconite
	Variability
	Shrinkage limit
	Change in engng properties after drying

5.17 *Use as building blocks and concrete aggregate*

Buchanan's laterite has been used for centuries in Asia and self-hardening calcrete hardpan has been used as building blocks in Namibia.

Crushed silcrete hardpan and boulders are used as coarse aggregate in concrete work in parts of Botswana and Namibia, while crushed calcrete boulders and hardpan were used as coarse aggregate for a road bridge in Namibia in 1965 (Netterberg 1969a). While no problems have so far been experienced, future use of calcretes and silcretes should consider the possibility of alkali-aggregate reactions.

Laterite can be used as coarse aggregate in concrete work, but not as fine aggregate, and no-fines concrete made of laterite achieves good strengths (LNEC *et al* 1959). However, slight weakening of the aggregate can take place in certain environments (Gidigasu 1980).

5.18 *Pedocretes in earth dam construction*

The possibility of gypcretes, calcretes, duripans and even lateritic soils (Anagosti 1969) going in solution must always be considered. Lateritic soils have been used for earth dams (Gidigasu 1976) and the problems involved in this use are well known. This is a situation in which self-stabilisation is not desired. Some pedocretes may be dispersive.

5.19 *Materials location*

In prospecting for pedocretes in southern Africa interpretation of aerial photography (King 1964; Brink and Williams 1964; Mountain 1967b; Brink *et al* 1982; Lawrance *et al* 1993) and the association of pedocretes with certain landforms has proved most useful.

The oldest and thickest pedocretes occur as cappings to remnants of Tertiary land surfaces and are easy to locate. The younger pedocretes and buried fossil deposits are more difficult to locate and are also the smallest and most variable. The younger calcretes tend to be associated with the edges of pans and ephemeral water courses. The younger ferricretes tend to occur at hillslope-pediment junctions, at gully heads, and at the edges of vleis or other areas of impeded drainage.

Vegetation indicators have also proved useful for both calcretes (Netterberg 1978c) and laterites (Netterberg 1985) all over southern Africa.

A useful more recent adjunct to airphoto interpretation has been the interpetation of satellite imagery (eg Lawrance and Toole 1984; Lawrance *et al* 1993).

Prospecting for calcretes in southern Africa has been discussed by Mountain (1967b), Netterberg (1967, 1969a, 1971, 1978c), Caiger (1968, 1980),

Netterberg and Overby (1983) and Lawrance and Toole (1984). The use of a simple probing device (Netterberg 1971) and, when time is short, aerial (preferably helicopter) reconnaissance have proved valuable.

Prospecting for laterite has been discussed by Schofield (1957), Clare (1960), Mountain (1967b), Holden (1967), Persons (1970), Morin and Todor (1976), Charman (1988) and others.

Prospecting for silcretes in southern Africa has been discussed by Mountain (1967b) and Caiger (1980).

6 CONCLUSIONS

Residual soils including pedocretes are of common occurrence in those areas of the globe not subjected to the Pleistocene glaciations. Their origin, mineralogy and fabric differ from those of the mostly transported, temperate zone soils of the developed countries in the Northern Hemisphere from which much of our knowledge and practice comes. It is therefore not surprising that their engineering properties and behaviour also differ. Our traditional test methods, soil classifications, specifications and construction practices must therefore be modified and/or new ones developed in order to make the best use of these materials. Residual soils and pedocretes may exhibit certain problems (especially in-situ) such as heave or collapse, but generally perform better as construction materials than expected. Not all residual soils and pedocretes are problem materials. The problem materials are only those whose behaviour is difficult to predict. Much progress has been made in this respect, but much still remains to be learned.

7 REFERENCES

Bagarre, F. 1990. *Utilisation des graveleux latéritiques en technique routiere*. Inst. Sci. Tech. l'Equipement l'Environment Développment. (ISTED). Paris.

Beckwith, G.H. & Hansen, L.A. 1982. Calcareous soils of the southwestern United States. In Demars, K.R. & R.C. Chaney (eds), *Geotechnical properties, behaviour and performance of calcareous soils*. Amer. Soc. Testing Mat. Spec. Tech. Publ. 777: 16-35.

Blight, C.E. 1982. Residual soils in South Africa, engineering and construction in tropical soils. *Proc. ASCE Geotech. Engng Div. Spec. Conf.* Honolulu, 147-171.

Blight, C.E. 1988. Construction in tropical soils. *Proc. 2nd Int. Conf. Geomech. Tropical Soils, Singapore*: 449-467.

Blight, C.E. 1991. Tropical processes causing rapid geological change. *Geol. Soc. Engng Geol. Spec. Publ.* 7:459-471.

Brewer, R. 1964. *Fabric and mineral analysis of soils*. New York: Wiley.

Brink, A.B.A. 1979. *Engineering Geology of Southern Africa*. Vol. 1, Building Publications, Pretoria.

Brink, A.B.A. (ed) 1981. *Engineering Geology of Southern Africa*. Vol. 2, Building Publications, Pretoria.

Brink, A.B.A. (ed) 1983. *Engineering Geology of Southern Africa*. Vol. 3, Building Publications, Pretoria.

Brink, A.B.A. (ed) 1985. *Engineering Geology of Southern Africa*. Vol. 4, Building Publications, Pretoria.

Brink, A.B.A., T.C. Partridge & A.A.B. Williams 1982. *Soil Survey for Engineering*. Clarendon Press, Oxford.

Brink, A.B.A. & A.A.B. Williams 1914. *Soil engineering mapping for roads in South Africa*. CSIR Rep. 227, Nat. Inst Road Res. Bull. 6.

Bullen, F. 1992. The use of local materials in pavement construction in tropical regions. *Proc. 7th Conf. Road Engng Assoc. Asia Australasia, Singapore.* 2:564-572.

Caiger, J.H. 1968. Aerial photography and aerial reconnaissance applied to gravel road construction in the Okavango Territory of South West Africa. *Civ. Engr. S. Afr.* 10:99-106.

Caiger, J.H. 1980. The search for road building material in the sands of Owambo. *Proc. 7th Reg. Conf. Afr. Soil Mech. Fndn Engng.* 2:11-9, Accra.

Carlisle, D. 1980. *Possible variations on the calcrete-gypcrete uranium model*. U.S. Dept Energy, Open File Rep. GJBX. 53(80), Los Angeles: Univ. Calif.

Carlisle, D., P.M. Merifield, A.R. Orme, M.S. Kohl & O. Kolker 1978. *The distribution of calcretes and gypcretes in southewestern United States and their uranium favorability based on a study of deposits in Western Australia and South West Africa (Namibia)*. Open File Rep. GJBX-29(78). Los Angeles: Univ. Calif.

Charman, J.M. (ed.) 1988. *Laterite in road pavements*. CIRIA Spec. Publ. 47. London.

Clare, K.E. 1960. *Roadmaking gravels and soils in Central Africa*. Road Res. Lab Overseas Bull. Number 12, Harmondsworth.

Cocks, G.C. & G. Hamory 1983. Road construction using lateritic gravel in Western Australia. *Proc. 2nd Int. Conf. Geomechanics Tropical Soils*. Singapore. Unpaginated preprint.

Committee on Arid Soils 1993. *Proc. Symp engineering characteristics of arid soils, London*. Rotterdam: Balkema.

Committee on Tropical Soils 1985. *Peculiarities of geotechnical behaviour of tropical, lateritic and saprolitic soils*. Progress Report (1982-85) Sao Paulo: Associacao Brasileria de Mecânica dos Solos.

Core Logging Committee 1978. A guide to core logging for rock engineering. *Bull. Ass. Engng Geol.* XV:295-328.

De Castro, E. 1969. A swelling test for the study of lateritic soils. *Proc. Spec. Ses. Lat. Soils.* 1:97-196. *Mexico City: 7th Int Conf. Soil Mech. Fndn Engng.*

De Graft-Johnson, J.W.S. 1975. Laterite soils in road construction. *Proc. 6th Reg Conf. Afr. Soil Mech. Fndn Engng.* Durban: 1:89-98.

Departamento Nacional de Estradas de Rodagem 1974. *Base e subbase estabilizada granulometricamente com utilizacao de solos lateriticos,* Rio de Janiero, Especificacaoes de Servico DNER-ES-P 47-74 & DNER-ES-P 48-74.

D'Hoore, J. 1954. Proposed Classification of the accumulation zones of free sesquioxides on a genetic basis. *Afr. Soils.* 3:66-81.

Du Toit, A.L. 1954. *The Geology of South Africa.* 3rd Ed, Edinburgh: Oliver and Boyd.

Ellis, F. & B.H.A. Schloms 1981. A note on the dorbanks (duripans) of South Africa. *Palaeocology of Africa and the Surrounding Islands.* 15:149-158.

Ellis, F. & B.H.A. Schoms 1984. Distribution and properties of dorbanks (duripans) in South and South Western South Africa. *Poster 12th Cong. Soil Sci. Soc. S. Afr.* Bloemfontein.

Fitzpatrick R.W. 1978. *Occurrence and properties of iron and titanium oxides in soils along the eastern seaboard of South Africa*. PhD thesis, University of Natal.

Fitzpatrick, R.W. & U. Schwertmann 1982. Al-substituted goethite - an indicator of pedogenic and other weathering environments in South Africa. *Geoderma,* 27:335-47.

Frydman, S. 1982. Calcareous sands of the Israeli coastal plain. In K.R., Demars & R.C. Chaney (eds), *Geotechnical properties, behaviour and performance of calcareous soils. Am. Soc. Testing Mat. Spec. Tech Publ.* 777: 226-251.

Gidigasu, M.D. 1976. *Laterite soil engineering.* Amsterdam: Elsevier.

Gidigasu, M.D. 1980. Notes on the potential use of crushed laterite rock for concrete aggregate. *Proc. 7th Reg. Conf. Afr. Soil Mech. Fndn Engng.* 1:467-78, Accra.

Gidigasu, M.D. 1991. Characterization and use of tropical gravels for pavement construction in West Africa. *Geotech. Geol. Engng.* 9(3-4):219-260.

Goudie, A.S. 1973a. *Duricrust in tropical and subtropical landscapes.* Oxford: Clarendon Press.

Goudie, A.S. 1973b. Geomorphic and resource significance of calcrete. *Prog. Geo.* 5:79-118.

Goudie, A.S. 1983. Calcrete. *In* Goudie, S.A. & K. Pye. *Chemical sediments and geomorphology: precipitates and residua in the near-surface environment: 93-131.* London: Academic Press.

Grace, H. 1991. Investigations in Kenya and Malawi using as-dug laterite as bases for bituminous surfaced roads. *Geotech. Geol. Engng.* 9(3-4):183-195.

Grant, K. & G.D. Aitchison 1970. The engineering significance of silcretes and ferricretes in Australia. *Engng Geol.* 4:93-120.

Grim, R.E. 1962. *Applied clay mineralogy.* New York: McGraw-Hill.

Grim, R.E. 1968. *Clay mineralogy.* New York; McGraw-Hill.

Haden, W.L. & I.A. Schwint 1967. Attapulgite, its properties and uses. *Ind. Eng. Chem.* 59, No 9:59-69.

Holden, A. 1967. Laterites: Identification and interpretation by airphotos. *Proc. 4th Reg. Conf. Afr. Soil Mech. Fdn Engng.* 1:27-9, Cape Town.

Horta, J.C. de O.S. 1980. Calcrete, gypcrete and soil classification in Algeria. *Engng. Geol.* 15 (1-2):15-52.

Horta, J.C. de O.S. 1989. Carbonate and gypsum soil properties and classification. *Proc. 12th Int Conf. Soil Mech. Fndn Engn.* Rio de Janeiro. 1:53-56.

Hough. B.K. 1969. *Basic soil engineering.* 2nd ed. New York: Ronald Press.

Jennings, J.E. & A.B.A. Brink 1978. Application of geotechnics to the solution of engineering problems - essential preliminary steps to relate the structure to the soil which provides its support. *Proc. Instn Civil Engrs.* Part 1. 64:571-589.

Jennings, J.E., A.B.A. Brink & A.A.B. Williams 1973. Revised guide to soil profiling for civil engineering purposes in southern Africa. *Trans S. Afr. Inst. Civ. Engng.* 15, No 1:3-12.

King, R.B. 1964. Airphoto interpretation for road construction. *J. Photogram.* 2, No 2:76-107.

Laboratório Nacional de Engenharia Civil, Laboratório de Engenharia de Angola, and Laboratório de Ensaio de Materiais e Mecânica dos Solos 1959. *As laterites do Ultramar Português.* Lab. Nac. Eng. Civil Mem. 141, Lisbon.

Laboratório Nacional de Engenharia Civil (Lisboa), Laboratório de Engenharia de Angola (Luanda), Laboratório de Ensaio de Materiais e Mecânica dos Solos (Lourenco Marques), and Junta Autonoma de Estradas de Angola (Luanda) 1969. Portuguese studies of engineering properties of lateritic soils. Proc. Spec. Session Eng. Prop. Lat. Soils. 2:85-96, *7th Int. Conf. Soil Mech. Fndn Engng.* Mexico City.

Langford-Smith, T. (ed.) 1978. *Silcrete in Australia.* Armidale: Univ. New England.

Lawrance, C., R. Byard & P. Beaven 1993. Terrain evaluation manual. *Transp. Res. Lab. State. of the Art Review 7.* London: HMSO.

Lawrance, C.J. & T. Toole 1984. *The location, selection and use of calcrete for bituminous road construction in Botswana.* Trans Road Res. Lab, Lab Rep 112, Crowthorne.

MacVicar, C.J.J.N., J.M. de Villiers, R.F. Loxton, E. Verster, J.J.N. Lambrecht, F.R. Merriweather, J. le Roux, R.H. van Rooyen & H.J. von M. Harmse 1977. *Soil classification - a binomial system for South Africa.* Soil Irr. Res. Inst, Pretoria.

Maignien, R. 1966. *Review of research on laterites.* Paris: UNESCO.

Maud, R.R. 1971. The occurrence and properties of ferricretes in Natal, South Africa. *Proc. 5th Conf. Afr. Soil Mech. Fndn Engng.* 1, Theme 2:3-5, Luanda.

Maud, R.R. 1986. Neogene palaeoclimates in the eastern and south-eastern coastal areas of South Africa. *S. Afr. J. Sci.* 82:65-66.

McFarlane, M.J. 1983. Laterites. In A.S. Goudie & K. Pye (eds), *Chemical sediments and geomorphology: precipitates and residua in the near-surface environment: 7-58.* London: Academic Press.

Metcalf, J.B. 1991. Use of naturally-occurring but non-standard materials in low-cost road construction. *Geotech. Geol. Engng.* 9(3-4):155-165.

Millard, R.S. 1993. Road building in the topics. *Transp. Road Res. Lab. State-of-the-Art Review 9.* London: HMSO.

Milnes, A.R. & J.T. Hutton 1983. Calcretes in Australia. In Anon. *Soils, an Australian viewpoint:* 119-162. London: Academic Press.

Mitchell, J.K. & N. Sitar 1982. Engineering properties of tropical residual soils. *Proc. Conf. Engng Constr. Trop. Residual Soils.* 30-57. Honolulu.

Morin, W.J. 1982. Characteristics of tropical red residual soils. *Proc. Conf. Engng Constr. Trop. Residual Soils.* 172-198, Honolulu.

Morin, W.J. & P.D. Todor 1976. *Laterite and lateritic soils and other problem soils of the tropics.* US Agency Int. Dev. Rep. AID/csd 3682. Baltimore: Lyon Associates.

Mountain, M.J. 1967a. Pedogenic materials. *Proc. 4th Reg. Conf. Afr. Soil Mech. Fndn Engng.* 1:65-70. Cape Town.

Mountain, M.J. 1967b. The location of pedogenic materials using aerial photographs, with some examples from South Africa. *Proc. 4th Reg. Conf. Afr. Soil Mech. Fndn Engng.* 1:35-40. Cape Town.

Nascimento, U., F. Branco & E. De Castro 1965. Identification of petrification in soils. *Proc. 6th Int. Conf. Soil Mech. Fndn Engng.* 1:80-1. Montreal.

National Institute for Transport and Road Research 1979. *Standard methods of testing road construction materials.* NITRR, Technical Methods for Highways No 1. Pretoria.

Netterberg, F. 1969a. *The geology and engineering properties of South African calcretes. 4 vols.* PhD Thesis, Johannesburg: Univ. Witwatersrand.

Netterberg, F. 1969b. The interpretation of some basic calcrete types. *Proc 1st S. Afr. Quat. Conf. Cape Town. S. Afr. Archaeol. Bull.* 24:(95 & 96). 88-92.

Netterberg, F. 1971. *Calcrete in road construction.* CSIR, Res. Rep. 286, NIRR Bull. 10. Pretoria.

Netterberg, F. 1975. Self-stabilization of road bases: fact or fiction? *Proc. 6th Reg. Conf. Afr. Soil Mech. Fndn Engng.* 1:115-9, Durban.

Netterberg, F. 1976. Some contrasts between pedogenic and traditional materials. *Proc. 6th Reg. Conf. Afr. Soil Mech. Fndn Engng.* 2:195-8, Durban.

Netterberg, F. 1978a. Dating and correlation of calcretes and other pedocretes. *Trans. Geol. Soc. S. Afr.* 81:379-91.

Netterberg, F. 1978b. Calcrete wearing courses for unpaved roads. *Civ. Eng. S. Afr.* 20, No 6:129-38.

Netterberg, F. 1978c. Prospecting for calcrete road materials in South and South West Africa. *Civ. Eng. S. Afr.* 20, No,. 1:3-10.

Netterberg, F. 1980. Geology of southern African calcretes: 1. Terminology, description and classification. *Trans. Geol. Soc. S. Afr.* 83, No. 2:255-83.

Netterberg, F. 1982a. Geotechnical properties and behaviour of calcretes in South and South West Africa. In K.R. Demars & R.C. Chaney (eds), *Geotechnical Properties, Behaviour and Performance of Calcareous Soils.* Am. Soc. Test. Mat. Spec. Pub. No 777: 296-309. Philadephia.

Netterberg, F. 1982b. Behaviour of calcretes as flexible pavement materials in southern Africa. *Proc. 11th Aus. Road Res. Board Conf.* 11, No. 3:60-9. Melbourne.

Netterberg, F. 1985. Pedocretes. In A.B.A. Brink (ed), *Engineering geology of southern Africa.* 4: 281-307. Silverton: Building Publications.

Netterberg, F. 1994. Low-cost local road materials in southern Africa. *Geotech. Geol. Engng.* 12:35-42.

Netterberg, F. & J.H. Caiger 1983. A geotechnical classification of calcretes and other pedocretes. In J.C.L. Wilson (ed), *Residual deposits: Surface related weathering processes and materials.* Geol. Soc. Spec. Publ. No 11: 235-43. London: Blackwell.

Netterberg, F. & C. Overby 1983. Rapid materials reconnaissance for a calcrete road in Botswana. *Proc. 7th Reg. Conf. Afr. Soil Mech. Fndn Engng.* 2:765-8. Accra.

Netterberg, F. & H.H. Weinert 1983. Towards a Unified soil and rock classification system. *Bull. Ass. Engng Geol.* 20, No. 1:98-100.

Newill, D. (1961). A laboratory investigation of two red clays from Kenya. *Geotechnique.* 11:302-318.

Nixon, I.K. & B.O. Skipp 1957. Airfield construction on overseas soils. Part 5: Tropical red clays. *Proc. Inst. Civil Eng.* 8:254-97.

Nogami, J.S., V.M.N. Cozzolino & D.F. Villibor 1989. Meaning of coefficients and index of MCT soil classification for tropical soils. *Proc. 12th Int. Conf. Soil Mech. Fndn Engng.* Rio de Janeiro. 1:547-550.

Nogami, J.S. & D.F. Villibor 1991. Use of lateritic fine-grained soils in road' pavement base courses. *Geotech. Geol. Eng.* 9(3-4):167-182.

Novais-Ferreira, H. & J.A. Correira 1965. The hardness of lateritic concretions and its influence in the performance of soil mechanic tests. *Proc. 6th Int. Conf. Soil Mech. Fndn Engng*. 1:82-6. Montreal.

Partridge, T.C. 1975. Some geomorphic factors influencing the formation and engineering properties of soil materials in South Africa. *Proc. 6th Reg. Conf. Afr. Soil Mech. Fndn Enging*. 1:37-42. Durban.

Partridge, T.C. 1985a. The palaeoclimatic significance of Cainozoic terrestrial stratigraphic and tectonic evidence from southern Africa: a review. *S. Afr. J. Sci*. 81:245-47.

Partridge, T.C. & R.R. Maud 1987. Geomorphic evolution of southern Africa since the Mesozoic. *S. Afr. J. Geol*. 90:179-208.

Partridge, T.C., F. Netterberg, J.C. Vogel & J.P.F. Sellschop 1984. Absolute dating methods for the southern African Cainozoic. *S. Afr. J. Sci*. 80:394-400.

Persons, B.S. 1970. *Laterite genesis, location and use*. New York: Plenum Press.

Petit, M. 1985. A provisional map of duricrusted areas at the 1:20 000 000 scale. In I. Douglas & T. Spencer. *Environmental change and tropical geomorpholgy: 269-279*. London: Allen and Unwin.

Queiroz, C., S. Carapetis, H. Grace & W. Paterson 1991. Observed behaviour of bituminous-surfaced low-volume laterite pavements. *Trans. Res. Record 1291*. 2:126-136.

Reeves, C.C. jr. 1976. *Caliche origin, classification, morphology and uses*. Lubbock: Estacado Books.

Retallack, G.J. 1990. *Soils of the past*. Boston: Union Hyman.

Rossouw, B. 1982. Use of ferrricrete for road construction in the South-western Cape Province. *Civ. Engr. S. Afr*. 24, No. 9:493-99.

Sandford, T.C., C.K. Adams & D.R. Khaemba 1980. Particle breakdown in pavement layers during construction., *Proc. 7th Reg. Conf. Afr. Soil Mech. Fndn Engng*. 1:179-90. Accra.

Schloms, B.H.A. & F. Ellis 1984. Distribution of silcretes and properties of some soils associated with silcretes in Cape Province, South Africa. *Proc. 12th Cong. Soil Sci. Soc. S. Afr*. Plant and Soil. Bloemfontein.

Schofield, A.N. 1957. *Nyasaland laterites and their indications on aerial photographs*. Road Res. Lab. Overseas Bull. No 5, Harmondsworth.

Smith, M.R. & L. Collis 1993. *Aggregates*. Geol. Soc. Engng Geol. Spec. Publ. 9. London.

Soil Survey Staff 1975. *Soil taxonomy. A basic system of soil classification for making and interpreting soil surveys*. Agric Handbook No 436, Soil Conser. Ser., US Dept. Agr.

Speciality Session D 1976: Pedogenic Materials. *Proc. 6th Reg. Conf. Afr. Soil Mech. Fndn Engng*. 2:193-212. Durban.

Summerfield, M.A. 1982. Distribution, nature and probable genesis of silcrete in arid and semi-arid southern Africa. In D.H. Yaalon (ed), *Aridic Soils and Geomorphic processes*. Catena Suppl. 1: 38-65. Braunschweig.

Summerfield, M.A. 1983. Silcrete. In A.S. Goudie & K. Pye (eds), *Chemical sediments and geomorphology: precipitates and residua in the near-surface environment: 59-91*. London: Academic Press.

Tankard, A.J. 1974. Petrology and origin of the phosphorite and aluminium phosphate rock of the Langebaanweg-Saldanha Bay area, South-western Cape Province. *Ann. S. Afr. Mus*. 68, Part 8:217-49.

Toll, D.G. 1991. Toward understanding the behaviour of naturally-occurring road construction materials. *Geotech. Geol. Eng*. 9(3-4):197-217.

Townsend, F.C., E.L. Krinitsky & D.M. Patrick 1982. Geotechnical properties of laterite gravels. *Proc. Cong. Engng Constr. Trop. Residual Soils*. 236-62. Honolulu. Am. Soc. Civil Engrs. New York.

Van der Merwe, C.P. 1971. The properties and use of laterites in Rhodesia. *Proc. 5th Reg Conf. Afr. Soil Mech. Fndn Engng*. 1, Theme 2:7-15. Luanda.

Van der Merwe, C.R. & H. Heystek 1952. Clay minerals of South African soil groups: I. Laterites and related soils. *Soil Sci*. 74:383-401.

Vaughan, P.R., M. Maccarini, & S.M. Mokhtar (1988). Indexing the engineering properties of residual soil. *Q.J. Engng Geol*. 21:69-84.

Visser, H.N., J.W. Von Backström, U. Keyser, J.M. Van der Westhuizen, J.A.H. Marais, C.B. Coetzee, F.W. Schumann, W.L. Van Wyk, S.B. De Villiers, F.J. Coertze, P.P. Wilke, D.H. De Jager, M.H.P. Rilette, & D.K. Toerien 1963. *Gips in die Republiek van Suid-Afrika*. Dept Mines, Geol. Surv., Handbook 4. Pretoria.

Von M. Harmse, H.J. & J.M. Hattingh 1985. Pedological profile classes. In A.B.A. Brink (ed.) *Engineering geology of southern Africa: 272-285*. 4. Silverton: Building Publications.

Watson, A. 1983. Gypsum crusts. In A.S. Goudie & K. Pye. (eds), *Chemical sediments and geomorphology: precipitates and residua in the near-surface environment:* 133-61. London: Academic Press.

Watts, N.L. 1980. Quaternary pedogenic calcretes from the Kalahari (southern Africa): Mineralogy, genesis, diagenesis. *Sedimentology*. 27:61-86.

Weinert, H.H. 1980. *The natural road construction materials of southern Africa*. Cape Town: Academica.

Whittaker, A.J. 1980. Bilateral plate-jacking tests on calcretes. *Proc. S. Afr. Geotech. Conf.* 13-9. Silverton: A.A. Balkema.

Wilson, R.C.L. (ed.) 1983. *Residual deposits: Surface related weathering processes and materials*. Geol. Soc. Spec. Publ. 11. London.

Woolnough, W.G. 1927. The duricrust of Australia. *J. Proc. Roy. Soc.* New S. Wales. 61:1-53.

Keynote lecture: Engineering geological impacts on the design and construction of dams

Conférence spéciale: L'importance des aspects géologiques sur le projet et construction des barrages

M.A. Kanji
University of São Paulo, Brazil

ABSTRACT: Some general considerations on dam foundations are initially presented, concerning their basic requirements with respect to different types of dam, alongwith the importance of the engineering geological study in the different phases of a project. Common geological problems of dam foundations and respective typical solutions that have been utilized are mentioned. Finally, three case histories of Brazilian dams (Jaguara, Agua Vermelha and Itaipu) are presented, to illustrate the impact of geological adverse features on the design and construction of these dams.

RESUMÉ: Quelques considerations basiques sur les exigences des fondations par les diferents types de barrages son inicialment presentés. Le rôle de l'ingenieur géologue et des principaux aspects géologiques des fondations sont mencionés. Le principal objectif de ce travail est la presentation de trois examples de barrages Brésiliens (Jaguara, Agua Vermelha, et Itaipu) oût l'interference des conditions géologiques encontrées sur son projet et construction a eté très fort.

1. INTRODUCTION

The basic objective is to present the importance of significant engineering geologic features and their impact on the design of dam foundations and in the construction phase. Three case histories are presented, each of them designed and built basically in a different decade, since the 60's.

Although this subject is not new in pertinent congresses or symposia, it is thought opportune to present some considerations on the trends of dam construction, basic requirements for the foundations of different dam types, as well as to emphasize certain aspects of the engineering geology activity on different phases of a dam project.

Additionally, the principal geologic features that may represent a conditioning factor of the design or present undesirable problems during construction are also mentioned.

2. BASIC CONSIDERATIONS CONCERNING DAM FOUNDATIONS

2.1 Relationship between height and types of existing dams.

It is very instructive this relationship as it demonstrates the historical and present trends in dam engineering, as shown in

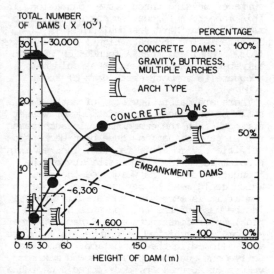

Figure 1. Number of existing dams, according to their heights, and percentage of each type with respect to total number of dams (data from Goguel 1991).

Figure 1, built with data obtained from Goguel (1991). It can be seen that the number of dams decreases rapidly with their heights, but the most interesting aspect is the proportion of the different types of dam with respect to its heights.

For small heights the embankment dams represent the absolute majority (90%) of all dam types, decreasing its frequency to about 40% for heights greater than 150 m.

The reverse trend is observed for concrete dams, reaching 60% of the total number of dams for heights greated than 150m. The arch dams show an increasing frequency for heights greater than 30m, reaching about 50% of the total number of dams for heights greater than 150m. The gravity, buttress and multiple arches types are more frequent among the concrete dams up to heights of about 60 to 80m, but decrease their frequency to some 10% of the total number of dams for heights greater than 150m.

The reasons for these trends can be interpreted as follows: a) the geomorphologies represented by wide valleys frequently only allow the placement of low dams and those that permits the placement of high dams are usually narrow valleys; b) since narrow gorges are geologically constituted of more resistant rocks, the existence of soil deposits for embankment dams is less probable; c) the more economic solutions generally favors the adoption of embankment, or concrete gravity, buttress, and multiple arches types of dams on sites presenting high or moderate values of valley width to height (W/H) ratios and for arch dams on sites of low W/L ratios (commonly less than 5 but most preferrably less than 3). Some additional factors that determine the height of dam could be of non-geologic nature, as space for construction operations, availability of local technology, hauling distances of construction materials, hydrological safety against overtopping during construction, construction experience of the contractor, and vacation or experience of the designer. With respect to this last factor, Souto (1982) points out that sites for which an arch dam would be best fitted were utilized for construction of a rockfill dam and vice-versa.

It is worth mentioning that the choice of dam type has no fixed rules, since each site has its particular characteristics of topography, geology, hydrology, access, etc., leading to a most convenient solution. This is a most interesting aspect about dams, since never two sites present the same problem, demanding of the engineering geologist a good experience and ability to take decisions on always a new problem.

Lastly, as Figure 1 shows, about half of the highest dams are the arch type, which presents the greater demands with respect to the rock foundations.This aspect is of importance in view of the facts that most of the best geological sites have been already utilized, leaving potential sites with geological problems or weak rock, and that there is an increasing tendency to buid even bigger works to attend scale economy and unavoidable human demands, although present world econominy has decelerated momentaneously the rates of construction in most countries.

2.2 Main dam types and foundation requirements.

The importance given to strength, deformation or permeability characteristics of the rock mass is dependant on the engineering structure concerned, as well emphazised by Lombardi (1997). In some cases other factors also play an important role, as the in-situ state of stress and regional seismicity. This is also true for dam foundations. In general terms, the usual foundation requirements for the most common types of dams are mentioned in continuation.

2.2.1 Earth Dams.
Ideally they require a foundation with a strength at least slightly greater than the strength of the lower layers of the embankment, but the stability has to be checked to avoid potential failure surfaces passing through the foundation. In this manner, a foundation corresponding to saprolites or hard residual soils is acceptable for moderate hight. Van Schalkwyk (1982) considers that a foundation with an allwable stress of about 1MPa would be adequate to support earth dams of about 100m high.

There are cases, however, of earth dams built or designed on foundations of weaker strength than the embankment. In such cases the slope inclinations have been decreased sufficiently to avoid failure. The remaining problem is of the compressibility of the foundation under the weight of the dam, which could lead to serious cracking of the dam if it is excessive and greatly unequal. In the case of the La Esperanza Dam (60m high, under construction in Ecuador), great economy was achieved as compressible layers were allowed in the foundation, since the settlement analysis by FEM showed maximum values of about 1.5m, considered acceptable to the dam.

Earth dams are highly susceptible to

piping either through soil or weathered fractures, which has to be avoided by efficient foundation treatment in most cases (drainage, cut-offs, impervious blankets, grouting, etc.)

2.2.2 Rockfill dams

Since they are more rigid than earth dams, they transmit greater stresses to the foundations which, therefore, should be somewhat better than for earth dams. The concrete faced rockfill dams are susceptible to foundation deformation (and of certain parts of the rockfill itself, of course) and demand moderately good rock for its plynth.

For this type of dam also, the foundation should present geotechnical characteristics at least somewhat superior than the base layers of the rockfill. Hard saprolite is sometimes accepted but most often the cut-off or the plynth are extended to moderately weathered rock, provided there is not a weaker plane adversely oriented within the foundation. Van Schalkwyk (1982) considers that a foundation corresponding to an allowable stress of 1MPa is required a rockfill dam of 50m high or a little higher. Although less sensitive to foundation piping, the rockfill dams and their foundations also must be designed against piping.

2.2.3 Concrete gravity dams.

The foundation of a concrete gravity dam must offer enough safety with respect to its sliding and overturning, as well as must present limited unequal deformability to avoid cracking of the rigid structure or by overcoming the acceptable deformations of the joints between adjoinning concrete blocks.

The contact plane between dam and foundation surface is a potential failure plane. However, the great number of concrete -rock direct shear tests already performed allows a safe adoption of its strength; furthermore, several design measures are taken to improve the characteristics of this contact, as concrete teeth and inclined excavation towards upstream.

More often, weak beds or planes of weaknesses adversely oriented that could allow the sliding of the dam through the rock foundation are the most critical geological features, which have to be taken into account in the stability analysis. To demonstrate its degree of criticality, suppose a dam of moderate heigh (50m) on a hard and sound rock foundation, but containing a horizontal weak joint at a depth of 10m, which is intercepted by downstream excavation (powerhouse, water fall, etc.), consisting in a common situation. A simple analysis by conventional equations shows that in order to get a Factor of Safety

of 1.5 that joint must have a necessary friction angle ($\emptyset$ nec) on the order of 50° (for full uplift condition) or 45° (for drained condition according to the USBR criterion), as show in Figure 2. For the calculation of the uplift pressures acting along the joint, vertical open fissures both at heel and toe of the dam were considered, but if the fissure at toe do not exist, occuring farther downstream, the uplift pressures will be still more unfavorable. A friction angle of 45° corresponds to a very rough joint with ondulations of about 10°, and a 50° friction angle corresponds to a plane which could not be considered a continuous joint, a approaching the angle of intact rock of moderate strength. Similar values were found by Borovoy (1982) in an equivalent situation.

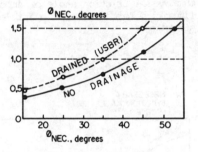

Figure 2. Factors of safety and necessary friction angles for the sliding of a 50m high concrete gravity dam (L/H=0.7), along a 10m deep horizontal joint within the foundation.

The influence of the depth of a horizontal weak plane within the foundation can be appreciated in Figure 3, drawn with values obtained from adimentional graphs presented by Cruz et al. (1978) for variations of dip and depth of the weak plane, position of drainage curtain, downstream water level, and base width to height of dam (W/H). It can be seen that the influence of the plane depth is almost none from 0 to 0,2H, significant reduction from 0,2 to 0,8H, and neglegible beyond 0,8H.

On the other hand, the dam must be stable with respect to overturning, but this aspect is seldom a problem, depending basically on the weight and on the width to height dimensions of the structure itself, and usually the FS related to overturning is much higher than that of sliding stability through a joint in the foundation.

Notwithstanding, an adequate study of stress distribution in the foundation should be performed, to identify eventual unacceptable tension or compression stresses, respectively at the upstream and downstream portions of the dam base. The rock must show limited deformation compared to the structure acceptancy. On this respect, Marchuk et al. (1987) mention several Siberian concrete dams that show open tension planes in greater extensions than expected at the upstream portion of dam bases, due to high compression of the rock at the dam toe, which effect extends to some distance downstream from the toe. This fact lead to the recognition that the rock mass compressibility is very important to avoid openning tension planes which could promote progressive failure, and to the recommendation of keeping some distance between the dam and powerhouse.

The above shows that concrete gravity dams normally require a good foundation, with a minimum overal friction angle of about 45, and limited deformability, compatible with the structure.

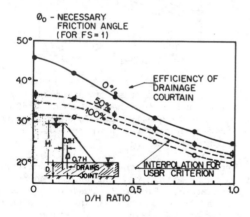

Figure 3. Necessary friction angles of a horizontal joint at various depths within the foundation, for different drainage efficiencies (data after Cruz et al. 1978).

2.2.4 RCC dams.
The "rolled compacted concrete" dam is a relatively new type of dam, with about a dozen dams of limited height built or under construction in the last 15 years, approximately. The construction methods develloped very fastly, but there are no specific studies integrating the dam with foundation characteristic. The values of the mechanical properties of RCC are about the half or even lower than those of conventional concrete. Considering the lower modulus of elasticity of RCC, one could expect greater acceptance of foundation deformations, but since the compressive strength also decreases, so does the tensile strength and its resistance to cracking. In the lack of specific studies or other criteria, foundation requirements for RCC dams have been considered, on the safe side, similar to those of conventional gravity dams. This was the criterion adopted for the foundations of the Nova Olinda dam (St.of Paraiba, Brazil), 60m high (the highest of this type in operation in 1992), built on sericitic schists and quartzites, excavated to practically sound rock, and treated with consolidation grouting, and impermeabilization and drainage curtains.

2.2.5 Buttress dams.
They present basically the same foundation requirements as concrete gravity dams, but with same more restraints. Under the buttresses, which are zones of concentrated loadings, the strength and deformability properties must be very good. The deformation differences for adjoining buttresses should be limited, to avoid anacceptable distortions of the upstream concrete wall. Some remedial measures to bring the foundation rock the desired standards are usually consolidation grouting, or strengthening with tension or passive anchors.

Additionally, it must be considered that the percolation path though the foundation is short, requiring a rock mass of moderate to low permeability, which enforces the construction of grouting and drainage curtains.

2.2.6 Arch dams
Due to the concentrated loading they exert on the foundations, this type of dam demands the best foundation characteristics of all types of dam. The most critical aspects are the stability and the deformability of the abutments to avoid, respectively, failure through the rock mass in the gorge walls or failure of the dam structure by excessive deformation of the foundations.

The normal modern pratice in the study of such dams is to resort to numerical and/or physical modelling of dam and foundation ensamble, including joint sets or individual joint elements recognized in the geological investigations. Through these analysis, excessive deformations and plastification or near plastification zones can be identified, and design modifications or remedial measures can be undertaken. As one of numerous exemples, that has been the approach for the desing of the Ertan dam in the P.R.China (Yoan et al 1991) where moderately weathered rock was accepted in certain zones of the foundation, resulting in considerable economy for the project as furthr excavation was avoided.

2.3 Needs of engineering geological investigation.

The ingineering geologist is responsible for the adequate definition of all geological aspects that intervene in a dam project, mainly those that constitute geological conditioning factors. He has to be able to interact with other involved epecialists, as geotechnical, hydraulic, structural and dam engineers, geophysicists, seismologists, environmentalists, etc., and supply them with their needs. For this he must have sound basic concepts of these fields, but aforemost of soil and rock mechanics. His tasks vary according to the phases of a project, as inventory, feasibility, design, construction, behaviour observation, and reahabilitation. In the present context, however, it is intended to emphazise only the aspects that follow.

2.3.1 Identification of the adverse geological features that condition the design.

It is the responsibility of the engineering geologist to program and follow up the site investigation, to be carried on under adequately specified techniques, to correctly interpret the foundation conditions and to indicate its interference with the project. The investigation intensity should be sufficient to allow him to assertain that the foundation do not present geological features more critical than those already encountered. Some authorities tend to minimize the site investigation programme for economical reazons, but it is also his task to clearly demonstrate its necessity and consequences, including the risks involved, which could be hazardous. In hydroelectrical projects, the cost of the site investigation is usually on the order of 0,5 to 1,5% of the total final cost, being usignificant considering the guaranty of safety and avoidance of unexpected cost during construction that it can bring.

The definition of potential modes of foundation failure of a dam foundation is a basic condition for its design. An illustration of typical potential modes of failurs through the foundation of a concrete gravity dam is shown in Figure 4 (Kanji 1990), corresponding to failure through rock of low strength or through joints under different situations, or still by diverse combinations of these modes.

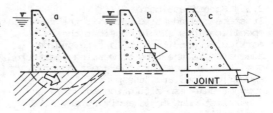

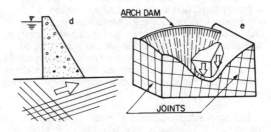

Figure 4. Possible modes of foundation failure of a concrete gravity dam (Kanji 1990).

2.3.2 Definition of design parameters.
Depending on his experience, the engineering geologist should be able to define and recommend design parameters of pertinent soil or rock masses, or at least to give necessary support for such definition. For this role, he has to be experienced or familiar with lab and in-situ testings, geotechnical classifications, and selection of representative sites or samples for the testings. The adopted parameters will entirely determine the results of the analysis, which could conduct to the adoption of costly and time consuming treatments, or to the disconsideration of necessary remedial measures. In order to get reliable results from those analysis, the parameters must be realistic.

2.3.3 Design of foundation treatment.
Very often the design of the dam and of the foundation are performed independently by different teams. However, being necessarily integrated, their designs should be compatible to each other. The standards of the final product to be obtained in the treatment or reinforcement of the foundation or parts of it, must be established by mutual undestanding. The engineering geologist must be able to evaluate to which degree and effort the foundation can be improved from its natural existing condition to suit the design requirements or to recommend design modifications.

2.3.4 Specifications.

It seems redundant to say that the technical specifications should consider the particular conditions of a site and be adequated to the type of working considered, but it is a fact that the great majority of specifications are "unespecific", being a copy of same especification which is so generalist that could apply to any site. The specifications, besides being important to define the technologies that shold be used for a determinate work and the conditions for their utilization, may have a heavy influence on the supervision attitude and economical aspects regarding the contractor. Typical and frequent topics are the support of underground excavations, overbreak of excavation surfaces, acceptable foundation rock, surface and deep foundation treatment, tension anchors acceptance, among others.

2.3.5 Supervision during construction.

Usually it is the responsability of the supervision (being direct or independent) to accept services performed by the contractor, perform the quality control and proceed to design modifications when the conditions forecasted in the design are not met. The supervision geologist is in charge of the acceptance of the final foundation surface, excavation walls, grouting, natural construction materials, etc., to mention a few items. His knowledge of the design considerations and analysis is of outmost importance, to know the acceptable range of parameter variations and allow adequate decisions, without disfiguring the designer responsability nor imposing unjustifiable economical penalties on the contractor.

3. TYPICAL GOLOGICAL PROBLEMS AND USUAL SOLUTIONS.

It is so great the number of possible particular geologic situations that could impair dam foundations characteristics, that their detailled description would be exhautive. For practical reasons they can be classified and grouped as follows. It should be reminded that Deere and Varde (1986) considered weak rocks for foundation purposes in the broader sense of weak rock masses, adding hard rock with weak discontinuities and soluble rocks to those of weak intact strength.

3.1 Weak Rocks.

3.1.1 Unsufficient intact rock strength.

The stresses applied by the dam may be too high with respect to the strength of the intact or weathered rock, when the foundation would undergo general failure, as illustrated in Figure 4(a). In this situation , if the foundation rock mass corresponds to a weathered rock that improves with the depth, some additional excavations is necessary to reach rock strong enough to support the dam. When the case corresponds to a rock that is homogeneously weak regardless of the depth and when only some decrease of the contact pressures may be enough, the dam width may be enlarged, as described by Isambert et al. (1982). However, if no one of these alternatives is applicable, the design must be considered inadequate to that foundation, and another type of dam should be adopted. Anyway, it is not recommendadle in weak rocks to apply stresses in excess of about 30% of their compressive strength, as at this level they may start to show dilatancy.

3.1.2 Loss of strength due to saturation.

Weak rocks with clayey matrix or cementation may show considerable loss of strength with saturation. The strength could drop bellow the adopted allowable pressure, as occured in the well known case of the St. Francis dam (USA, 1929), settled on clayey conglomerates. The drop in stregth varies greatly from rock to rock. For these reasons, the saturated strength sould be tested and considered for design.

3.1.3 Dissecation

Rocks with high clay contend, as shales, mudstones, and alike, undergo cracking by dissecation at and near the foundation surface when exposed, with loss of strength. The preservation of the water contend of the rock is achieved by surface protection or by quick covering with the first layers of the embankment.

3.1.4 Durability

Many weak rocks undergo rapid weathering when exposed to air and to cycles of wetting and drying, mainly those more argillaceous and with expansive minerals. Here also quick covering is the best solution. Temporarily protection can be done by shotcrete, layers of earthfill, or even plastic sheets. Dobereiner (1989) calls the attention to the fact that complete drainage of clayey rock to promote its stabilizations, may also promote the rapid weathering of the rock.

3.1.5 Expansivity.

Clayey rocks containning montmorillonite or other expansive mineral show expansivity whenstress relieved by the excavations and when they gain in water contend by exposition to air or rain. The phenomenon is potentially more intense when initial water contend is low since the expansion is promoted by water

absoption. Similarly, the rock surface
should be protected to avoid changes in water
contend.

3.1.6 Anisotropy.
Most weak rocks of sedimentary origin, but
also some highly atratified metamorphic
rocks, show strong contrasts between the
hardness of the layers as, for example,
sandstone/shale, quartzite/schist, etc. When
the shear stress is applied parallel to the
stratification, the strength of the weaker
layer will condition the strength of the
rock mass. On the other hand, the largest
deformability will occur under loading
applied normally to the layers. This type
of deformability will occur under loading
applied normally to the layers. This type
of anisotropic and heterogenous rock mass
presents serious problems in the adoption
and acceptance of foundation levels when the
stratification is horizontal or nearly, as
clearly one of the materials is satisfactory
and the other not, making it necessary to
reach excavation levels in wich the weaker
material is in accordance to the design
parameters. An unusual treatment of 1 to 15
cm thick clay seams intercalated in compacted
sandstones and siltones at the foundations
of the Feitsui dam in Taiwan is reported by
Cheng (1987), consisting of high pressure
churn grouting to replace the caly seams by
cement.

3.1.7 Rock blasting.
Very often weak rocks are of very difficult
rippability, requiring blasting for rock
excavation. However, the blasting must be
very careful in order to avoid damaging the
rock and opening the natural joints close to
the final excavation surface. This problem
is aggravated in rock masses of hard layers
intercalated to weak ones or containing weak
joints, when the excavation surfaces are at
a low angle to the layers, producing
unavoidable highly irregular surfaces and
overbreaks. Figure 5 shows a good example
of these aspects, as observed at the
excavations in highly jointed sericitic
quartzites for the Corumba Hydroelectrical
Project, presently under construction in
Brazil.

3.1.8 Cleanning of foundation surface.
From the practical point of view, this is
a task of major difficulty in weak rocks of
a friable and crumble nature, as they
release particles or small blocks very easily.
Due to the low strength, they crush under
the loads of heavy equipment and caterpillar
edges, and only light equipment with rubber
wheel should be utilized. Nieto (1982)
mentions that clayey fine sandstones loose
their structure under the load of heavy

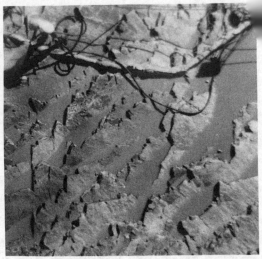

Figure 5. View of the irregularities of the
foundation surface in very anisotropic rock.

equipment and turn into muddy material due
to their high natural water contend. A
confirmation of this fact is given by Idel
and Stoehr (1982), with respect to the
observed behaviour of a test fill compacted
with heavy equipment, for the Aabach dam in
Germany, in which the siltsone desintegrated
completely into a muddy silt under
saturation. Usually the final cleanning of
the foundation surface is done by air
blowing and brooming.

3.2 Hard rocks

3.2.1 Single weak discontinuity
The presence of a discontinuity within a
hard rock mass may allow the sliding of the
whole fondation along that feature,as
illustrated by Figure 4(c). Of course the
kinematic condition is possible when the
discontinuity is exposed by down stream
excavation (e.g. for powerhouse or pre-
existing waterfall) or when it is relatively
shallow allowing the sliding associated to
passive failure downstream. In stratified
sedimentary rocks the presence of weak
layers or bedding planes corresponds to an
equivalent situation. The adopted solutions
vary according to the strength and position
of the weak plane with respect to the
engineering structure.
  For gravity type dams, the most
unfavorable position is when the weak planes
are horizontal or of low dips. When they are
relatively shallow, their entire removal by
excavation has been the more economical
solution, even considering the volume
replacement by compacted fill or concrete.

hen it is at moderate depths, its excavation can be total or partial (under the most critical portions of the structure), according to the economic feasibility in each project. As an example, partial removal of a weak argillaceous layer intercalated to hard sandstones at the feet of both abutments was the solution adopted at the Kapanda dam in Angola (Mgalobelow and Solovjena 1991). In other instances the weak feature is at such depth that its removal by excavation is totally uneconomical, but its presence is still unacceptable for the foundation stability. In this situation, the solution may be the adoption of concrete shear keys by means of galleries excavated along the discontinuity, as made in the Itaipu dam in Brasil-Paraguai. Another type of shear key may be of the "cut-off" type, excavated from the surface and filled with concrete, as adopted for the Dudhanga dam in India (Deshnukh et al. 1982), due to the existence of horizontal interlayers of weak shales within hard sandstones; complementary, the rock mass immediately downstream of the shear key was reinforced by consolidation grouting and rock bolting

For arch dams, the presence of weak planes of high dips at the abutments may represent a critical situation. Some clay-infilled discontinuities (faults) present at one of the abutments of the Longtan dam in the P.R.China conducted to the construction of shear keys consisting of vertical and horizontal concrete-filled galleries (Carrere et al. 1987).

Besides drainage, another method of increasing the strength of discontinuities is the instalation of tensioned anchors, as made at the Cachoeira Dourada dam in Brazil, in the 60's, with the instalation of 120 tons capacity tendon anchors, to allow the highering of the dam. This method, however, seems a last resort, when no other method is possible, as often is the case of increasing the safety of an existing dam.

3.2.2 Faults and shear zones.
Faults and shear zones may occur as zones of intense parallel fracturing, accompanied or not by intense weathering, representing weak and permeable zones. Their treatment varies according to their conditions as well as to its type and degree of interference with respect to the engineering structure. Since they generally present a limited width, the common treatment consists of the removal by excavation of the most dammaged surficial zone and replacement by adequate compacted fill or poor concrete, complemented with grouting from the surface or from galleries. A good example of grouting from galleries to seal the

contact between dam and foundation is given by Innerhofer and Loacker (1982) in relation to the Bolgenach dam in Germany. The possibility of using shear keys has already been mentioned but it is practically restricted to thin planes, in order to limit the cross-sectional dimentions of the galleries.

3.2.3 Intensely jointed zones.
Certain zones of a rock mass may be intensely jointed, showing high permeability and compressibility, and low shear strength along the predominant directions of jointing. The common treatments consist of consolidation grouting and in some cases reinforcement by rock bolting or tensioned anchors. Egger and Spang (1987) report the study made for a dam site in Sardinia, in which the rock conditions encountered were poorer than antecipated; some plastification zones were identified by FEM analysis, which were reinforced by 40mm diameter high strength rock bolts for each 1.25 m^2 of surface area.

3.3 Soluble rock
The solubility of some rocks results in the formation of cavities, weakening the rock mass and making it highly permeable. In the well known case of limestones, the karstic phenomena usually take a geological time to occur. However, for rocks that present a faster solubility rate, as gypsum and other evaporites like salt rock, the formation of cavities or seepage paths may develop during the operational life-time of the dam. Although unpublished, it is known that the foundations of the 100m high S.Houssein dam in Irak include gypsum layers which undergo dissolution with the reservoir level variations, requiring frequent grouting treatment. This aspect, also present at other dam sites, requires the development of dissolution rates prediction, as started by Ferrero et al. (1986) for the foundation of the Casa de Piedra dam in Argentina, including gypsum layers.

Besides the great difficulty of the detection of the cavities, they present the problem of the cost for their filling by impermeable material, from poor concrete or mortar, to cement and clay grouting.

4. CASE HISTORIES (LONG STORIES MADE SHORT)

4.1 Jaguara Hydroelectrical Project.

The site selected for this project by the inventory and preliminary studies in the 50's, corresponds to a wide valley in the Grande River (Brazil), with abrupt shoulders and a 500m wide river bed, relativelly shallow but with a central narrow main channel about 40m deep. The geology consists of Pre-Cambrian folded layers of vitreous and sericitic quartzite, closely jointed. The river bed was inaccessible due to the water velocity, but its geology was deduced from visual inspection of the sound outcrops popping out from the waters, as being beds striking normally to the river direction and dipping from 20 to 40 degrees upstrem.

In the early 60's the basic design was started, conceiving concrete structures (intake, powerhouse and spillway) next to the left bank, taking advantage of a natural river channel, minimizing excavations for the tailrace channel. A 300m long and 60m high earth dam would close the rest of the river. The situation and the geology seemed so favorable, that the late Prof. A. Casagrandeexpressed that it was the best dam site he had ever seen.

At the same time, detailed geologic investigation was started, mainly by rotary drillings, since the weathering of the rock at the abutments and the permanent inundation of the whole river bed made the surface geological mapping very difficult. The drillings, distributed according to geological criteria, showed geotechnical conditions highly favorable. However, two drillings showed short streches with core losses at depths of 30 to 40m, which could not be explained not correlated taking into account the geologic structure.

After an insistent search, some stereographic pairs of airphotos were found, old but taken in exceptionally dry season and low waters. Their interpretation showed surprizingly that the geologic structure corresponded to folded strata plunging gently towards the left abutment and that what was considered previously as strike normal to the river axis was actually a strong and persistent joint system which mislead all involved geologists untill then. As a consequence, the deep channel at the middle of the river could correspond to a weaker and more erosionable layer, a thaught that was supported by compatible correlation of the drillings with missing rock cores. For a confirmation of this hypothesis a provisional bridge 100m long was built on the river bed, with great difficulty and reinforcement against the water velocity, to allow a new drillhole at a strategic point. The weaker material was met at the forecasted depth, when the drilling was proceeded with great care and special bits, allowing the complete recovery of a red plastic clay. The clay was a result from complete weathering of a chlorite-epidote schist, probably a basic concordant

intrusion later metamorphosed. Lab testings on undisturbed samples revealed average values of unconfined strength of 0,1MPa and friction angles of about 20 degrees for the clay. The foundation of the intake structure would be unstable, obliging to the analysis of solution alternatives, which included deep excavation or the underground "mining" of the clay and its replacement by concrete. The most convenient alternative, however, was the displacement of the structures 50m downstream, where the excavations for the elimination of the clay were minimized, although the design was delayed and most drawings had to be remade. The project is in operation since 1971. A detailed account of the investigations was presented by Kanji and Brito (1971).

4.2 Agua Vermelha Hydroelectrical project.

This project, also on the Grande River but on its terminal portion, is settled on basalts in several and successive flows. The dense basalt outcropping in the river bed and waterfall gave a very favorable impression of the site geology, at the begining of the basic design, in the mid 70's. Here too the conception was of a dam about 60m high, generating 1400Mw, with concrete structures at the mid portion of the river (the tailrace channel coincides with existing deep river channel)and a 4 km long earth dam at both sides of the concrete structures.

Several geological features of great engineering significance were however found during the detailed site investigation, obliging to substantial adjustment of the initially concieved desing:
a) a circular structure of the basalt was identified by airphoto interpretation in the middle of the river, which was interpreted as a possible volcanic chimney, where the basalts flows were of only 3 to 5m thick, mainly of the vesicular and amygdaloidal type, highly fractured and moderately weathered. Its modulus of deformability was about 2,000 MPa, showing great contrast with the modulus of the adjoining dense basalt, estimated as 10,000 MPa. The spillway, that was originally designed partly on each type of rock, was displaced to be sited entirelly on the rocks of the circular structure, to avoid differential settlements. A transition wall, able to accomodate these settlements was included in between the spillway and intake structures. The flipbucket of the spillway was also modified to avoid the incidence of the water jet on the weak rocks of the circular structure, but rather on the sound dense basalt, to minimize erosion downstream

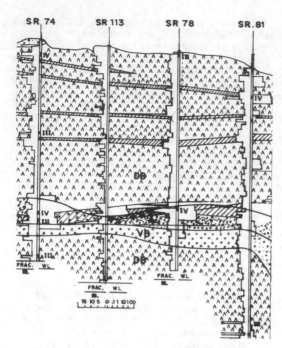

Figure 7. open subhorizontal joint exposed by partial excavation

Figure 6. Geotechnical section showing infered jointed levels or continuous discontinuities DB, Dense Basalt; VB, Vesicular-amygdaloidal Basalt; A, Agglomeratic Basalt; WL, water loss in l/min.m.atm; III,IV, rock mass classification; SR, notary drilling.

of the spillway.

b) The rotary drillings were carefully made, but of the conventional type, with HW and NW diameters. The very high core recoveries and RQD's obtained were an indication of excellent conditions of the dense basalt. Nevertheless, some fractured thin zones of the rock cores, from 5 to 20cm wide, lead to the suspition of the existence of continuous horizontal joints, that could endanger the structures stability. A correlation of these thin zones showed very good agreement, as shown in Figure 6. As a result, an excavation of about 7m of sound rock was recommended to eliminate a very unfavorable joint under part of the structures. Believing only partially on this recommendation, the client decided to excavate initially only a downstream trench to that level, to inspect directly the feature and eventually safe further excavation. The aspect of the joint after excavation is depicted in Figure 7, making the client to decide for its complete excavation.

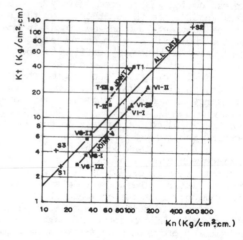

Figure 8. Tangential (Kt) and normal (Kn) stiffness for the tested discontinnity (after Infanti Jr.&Kanji, 1991)

c) A much more continuous joint was identified at about 30m depth, which would be intersected by powerhouse excavations, allowing the sliding of the foundation. Since conventional drillings did not indicate possible weak filling material nor its thickness, it was decided to realize some integral sampling drilling (as developed by the LNEC). Their results confirmed the existence of a 3 to 6cm thin discontinuity infilled with a highly plastic clay (IP approx. 50%), and the consistency of the correlations permitted the assumption it was

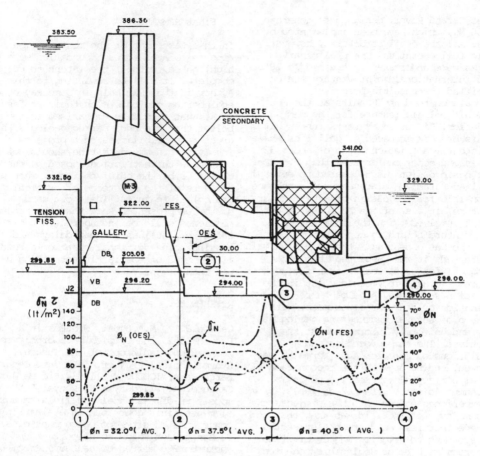

Figure 9. Agua Vermelha intake-powerhonse structures showing original (OES) and final (FES) excavation surfaces, with respectively mobilized friction angles at Joint 1 (J1); DB and VB as in Fig. 6 (after Infanti Jr. et al 1978).

very extense. For a total confirmation, 3 shafts were excavated down to that level, showing great influx of water due to the high permeability of the joint. A 600m long gallery was then drilled from the shafts, following continuously the joint under the entire length of the concrete structures. Several in-situ direct shear tests run on this discontinuit showed average values of 22,5 degrees, incompatible with the dam stability. For these special investigations, a great support was given by Prof. M. Rocha's endorsement.

d) As a consequence, a FEM analysis, including a joint element to simulate the continuous joint was made, as reported by Infanti Jr. et al (1978), to investigate possible solutions. The joint element analysis required the adoption of tangential and normal stiffnesses (Kt, Kn) for the joint, a matter with very little experience at that time, which required a thorough analysis of testing data. The values obtained are shown in Figure 8, which indicated thar for this case Kn is about 4 to 5 times Kt. Taking in consideration that Kt is dependent on normal stress and that the normal stresses varied along the joint, specific values of Kt were adopted for different portions of the joint. The optimized solution consisted of the excavation of about 30.000 m³ of rock with removal of the downstrem portion of the weak joint, as ilustrated by Figure 9, with the concrete filling working as a structural support for the intake structure. The rock at the new foundation levels was then able to resist the applied shear stresses. The project is in operation since the late 70's.

4.3 Itaipu Main Dam

The Itaipu Hydroelectrical Project is a joint undertaking of Brazil and Paraguai, on the Parana River, designed to generate 12,000 Mw, which have been implemented by steps. All the civil structures, however, are in full operation. The preliminary studies were undertaken in the middle 70's by an international joint venture, which identified the basaltic breccias intercalated to dense basalts as the most critical geologic feature for the dam foundation. Direct shear tests run in-situ on this breccia accessed by galleries showed consistently a friction angle of about 32 degrees and cohesions greater than 0.3 MPa.

The basic design was delegated to several Brazilian-Paraguayan joint ventures, each one responsible for the design of a particular structure, in view of the enormous amount of work involved, with the international joint venture acting as coordinator. Soon after the basic design of the 195m high main dam started, it was suspected that other geological features, weaker than the weakest features considered, could exist in the foundation rock mass. A complete re-examination of rock cores was made, with the help of several engineering geologists duely and uniformly informed about the objectives. The same procedure and correlations as made for Agua Vermelha was followed, ending up with the recognition of several horizontal (or nearly) potential continuous joints or fractured zones. From the experience gained on other projects on the same type of rock, it was knew that some of these features could present strengths corresponding to friction angles of 20 to 25 degrees with no or neglegible cohesion.

A shaft 140m deep was recommended for direct inspection and assesment of these features, which was finally agreed by the owner, again after the opportune endorsement by Prof. M.Rocha, a member of the Board of Consultants at the time. The shaft confirmed practically all the zones forecasted and allowed the adequate foundation design decisions for each concrete block of the dam, with excavation and removal of the more important joints. There was one particular joint, however, which excavation would not be practical nor economical, as it was 20m bellow the deepest foundation level of the dam. Its geological genesis is attributed to high horizontal shear stresses below the river bed along with vertical stress relieve by the river erosion of the canyon.

The adopted solution in the detailed design phase was the reinforcement of this joint by shear keys, consisting of 8 longitudinal and 8 transversal galleries excavated along the joint and filled with concrete, as decribed by Souza Lima et al. (1985), among other publications.

5. FINAL REMMARKS.

In this text some of the most common aspects related to dam foundation were highlighted, without the intend to be complete. It will be rewarding to this Author if they may help in some way other engineering geologists on their professional work. What should be emphazised is the belief that the engineering geologist has great responsibilities on a project involving a dam and that he should not be content with restrictions posed by the client or by the chief designer, when he suspects of the existence of critical geological features. The short case histories presented show that the projects mentioned would be endangered in their safety or, at least, great delays and considerable additional cost would result if the need of special investigation had not been demonstrated.

REFERENCES.

Cadman.J.D., M.A.Buosi, N.S.Bertin & A.C. Bastos 1982. The complex matamorphic rock foundation conditions of the Tucurui Hydroelectical Project. 14th Int. Congr. Large Dams, ICOLD. 2:811-834 Rio de Janeiro.
Carrere, A., C. Nury & P. Pouyet 1987. Un modele tridimensionnel non lineaire aux elements finis contribue aa evaluer la convenance du site geologiquement dissicile de Longtan en Chine au support d' une voute de 220 metres de hauteur. Proc. 6th Int. Congr. Rock Mech. 1:305-311 Montreal: Balkema.
Cheng, Y. 1987. New deselopment in seam treatment of Feitsui arch dam foundation. Proc, 6th Int. Congr. Rock Merch., ISRM 2:319-326 Montreal.
Cruz, P.T., W.A.Lacerda & E.P.Soares 1978. Abacos para a analise parametrica de esta-bilidade ao escorregamento de barragens de concreto de tipo gravidade. Solos e Rochas 1/2:23-53. S. Paulo:ABMS.
Deere, D.U. & O.Verde 1986. Engineering geological problems related to foundations and excavations in weak rocks. Gen.Report. Proc. 5th Int. Congr. Eng. Geol.IAEG.: Buenos Aires:Balkema.
Dobereiner, L. 1989. Construction problems related to excavation on solt rocks. Gen. Report. Proc. 12th ICOSOMEF. :2435-2449. Rio de Janeiro:Balkema.
Egger, P. & K. Spang 1987. Stability investigations for ground improvement by rock bolts at a large dam.Proc. 6th Int. Congr. Rock Mech., ISRM. 1:349-354. Montreal.
Ferrero, J.C., A. Carrillo, O. Varde & E. Capdevila 1986. Geomechanical investigations

Casa de Piedra dam Proc. 6th Int. Congr.
Eng. Geol. IAEG/AAGI. 2:613-628. Buenos
Aires:Balkema.

Goguel, B. 1991. Dam foundations on Rock.
Gen. Report. Proc. 7th Int. Congr. Rock
Mech. ISRM. 3:9pp. Aachen:Balkema.

Idel, K.H. & E.Stoehr 1982. Influence of
geomechanical investigation on comparison
of a rockfill dam with upstream membrane
and an earthfill dam with sealing core. Proc.
14th Congr, Large Dams, ICOLD. 2:309-324.
Rio de Janeiro.

Infanti Jr.,N., M.A.Kanji, F.Kneese, G.Re &
E. Tagliatella 1978. Sliding stability
analysis of Agua Vermelha intake-powerhonse
structure. Proc. Int. Symp. Rock Mech. Rel.
Dam Found., ISRM/ABMS. 2:III.143-163. Rio
de Janeiro.

Infanti Jr., N. & M.A.Kanji 1991. Estimating
the shear stiffness of rock joints. Proc.
Int. Symp. Rock Joint. 2:799-804. Loen:
Balkema.

Innerhofer, G. & H.Loacker 1982. The
connection of the core of the Bolgenach dam
to weathered marl. Proc. 14th Congr. Large
Dams, ICOLD. 2:275-287. Rio de Janeiro.

Lombardi, G. 1977. Behaviour of engineering
structures in structurally complex
formations. Gen. Report. Proc. Int. Symp.
Geotech. Struct. Compl. Form. AGL. 2:330-
335. Capri.

Kanji, M.A. 1990. Fundacion de presas en
rocas blandas y rocas fracturadas.
Problemas y Soluciones. Spec. Conf. Proc.
3rd. South-Am Congr. Rock Mech. ISRM/SVMSIF.
1:235-252. Caracas.

Kanji, M.A. & S.N.A. Brito 1971.
Condicionamentos geologicos da barragem de
Jaguara. Braz. Sem. Large Dams, CBGB. 16pp.
Rio de Janeiro.

Kanji, M.A., G. Re, R.A.Abrahao, O.Moura Fo.
& N.F. Midea 1977. Complex structures as dam
foundation: Geomechanical Chracteristics
(A. Vermelha). Proc. Int. Symp. Geotech.
Struct. Complex Form. ISRN/AGI. 1:297-305.
Capri.

Marchuk, A.N., A.A.Khrapkov, Y.N.Zukerman &
M.A.Marchk 1987. Contact effects at the
interface between rock foundations and
concrete dams with power plants at their
toes. Proc. 6th Int. Congr. Rock Mech.
ISRM. 1:433-435. Montreal:Balkema.

Mgalobelov, Y.B.& L.D.Solovjena 1991. Three
dimensional computational studies of limit
state on rock foundation of concrete dams.
Proc. 7th Int. Congr. Mech. ISRM. 1:779-
784. Aachen:Belkema.

Nieto, A. 1982. Caracterizacion geotecnica
de Rocas Blandas-Estado del Conocimiento.
1st South-Am. Congr. Rock. Mech. ISRM/SCG.
1:1-69. Bogota.

Souza Lima. V.M., J.F.A.Silveira & J.C.
Degaspare 1985. Horizontal and vertical
displacements of the Itaipu main dam; a
study on field measurements and theoretical
predictions. Proc. 15th Congr. Large Dams,
ICOLD. Lausanne.

Van Schalkwyk, A. 1982. Geology and selection
of the dam in South Africa. Proc. 14th Int.
Congr. Large Dams, ICOLD. 2:701-718. Rio
de Janeiro.

Yuan, Z.W., Y.R.Quiong & Y. Quiang 1991. The
monolithic stability analysis of Ertan arch
dam in complex rock formations. Proc. 7th
Int. Congr. Rock Mech. ISRM. 1:827-831.
Aachen:Belkema.

4 Construction materials
Matériaux de construction

Limestone characteristics and geotechnical aspects of Sangardão quarry

Caractéristiques des calcaires et aspects géotechniques de la carrière de Sangardão

A. M. Antão
Instituto Politécnico da Guarda, ESTG, Portugal

M. Quinta Ferreira
Departamento de Ciências da Terra, Universidade de Coimbra, Portugal

F. Castelo Branco
Sebrical, Lda, Sangardão, Condeixa a Nova, Portugal

ABSTRACT: Sangardão quarry is located 10 km south of Coimbra on a thick Jurassic limestone formation. The quarry was studied aiming a better knowledge of the geotechnical characteristics of limestone, of the processed rock materials and to optimize extraction and materials processing. Field work was done in order to define the geological structure (main faults and joints systems), stratigraphy and litologic units. Petrographical analysis, physical and mechanical tests were done. Extracted materials are white and yellowish micritic limestone, white to pinkish oolitic limestone. On the basis of the limestone material properties new potential uses are proposed.

RÉSUMÉ : La carrière de Sangardão est localisée 10 km au sud de Coimbra, dans une formation de roches carbonatées datée du Jurassique moyen. La carrière a eté etudié pour mieux connaître les propriétés géotecniques des calcaires, des matériaux transformées, et aussi pour essayer une optimization de l.' extraction de ces matériaux. Des travaux dans le chantier ont été executéspour définir la structure geológique (failles principalles et joints), la stratigraphie et les types litologiques. Des analyses pétrographiques, des essais physico-mécaniques ont eté executée. Les matériaux exploitées sont des roches calcaires micritiques de couleur claire et calcaires oolitiques. Avec les propriétés du materiel calcaire, des nouveaux usages ont eté proposés.

1 INTRODUCTION

With the objective to optimize the quarry exploration, the production of processed materials, and to seek for new processed materials, a detailed study of Sangardão quarry was done. In this study the geology, the lithology, the stratigraphy and the geological structure were analyzed. Physical, mechanical and chemical characteristics of the rock materials and processed products were determined.

At present extracted materials are used for aggregates, "tout-venant", soil amendment and stone pavement. Other potential uses are searched according to the technological characteristics of the materials, aiming to develop more product types not processed at the moment.

2 LOCATION

The quarry is located in central Portugal, 10 km South of the city of Coimbra, close to Condeixa (Figure 1). The site is served by National road EN1 and the North Highway (A1) is 1200 meters west. The extraction area has approximately 22 ha and the base level is at 90 m of elevation.

3 GEOLOGY

3.1 Geomorphology

The quarry is in the mesocenozoic "West Border" within a horst like zone, limited by faults at west and a monoclinal relief cornice at east. It is formed by carbonate rocks, essentially limestone from medium Jurassic, having a karstic morphology clearly visible in the topmost

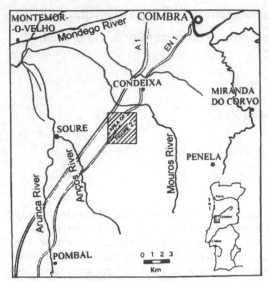

Fig. 1 Sangardão quarry location .

of the Sicó mountain, were the quarry is located.

3.2 Regional geology

At the quarry area the middle Jurassic formations (Figure 2), were mapped as a compact limestone unit of about 200 m thick (Rocha et al., 1981). They contact to the West by an unconformity with the Carrascal sandstone formations (arenitos do Carrascal) of low-medium Cretaceous, and to the East with formations dated from the Lias (Aalenian).

The quarry is in the SE side of the "Alencarce-Rebolia-Relvas" synclinal, with the hard Jurassic limestone making a topographical contrast with the soft sandstone formations.

Structurally the area is located in the north sub-zone of the Lusitanian basin (Ribeiro et al., 1980), with the N65E fault orientation conditioning the fracturing in the zone, according to Parga (1969). The main faults are shown in Figure 2. They have two main directions: SW-NE and NW-SE.

quarry faces. Red residual soil fills the carsification voids mainly along fractured zones. The rock mass is intensely fractured leading to a "high permeability" (Cunha, 1990)

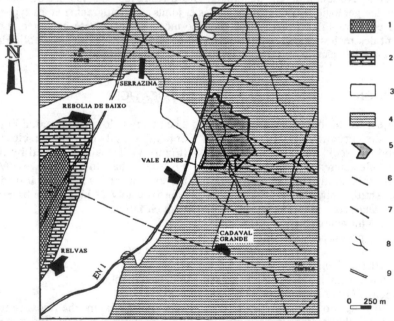

Fig. 2 Geological units in the quarry area. 1 - Fine sandstone from medium Cretaceous; 2 - Limestone from medium Cretaceous; 3 - Sandstone from lower Cretaceous; 4 - Compact limestone from medium Jurassic; 5 - Quarry area; 6 - Faults mapped; 7 - Faults inferred from photo interpretation; 8 - Water lines; 9 - Main roads. (Modified from Rocha et al., 1981, with additament).

3.3 Local geology

The quarry was mapped in detail trying to define the structural features and the lithology of the area. Four main types were considered:
- White compact micritic limestone with conchoidal fracture.
- Yellowish compact micritic limestone.
- White oolitic limestone
- Yellow to pinkish oolitic limestone

Petrologic analysis were done to identify the mineralogy and the texture (fabric) of the rock.

The bedding is only clearly visible in the lower quarry faces due either to the higher carsification of the top zones and to the very intense fracturation of the area. The beds have a thickness varying from 2 to 3.5 m, dipping gently up to 10° west. Sometimes the bedding planes are filled with red residual clay soil.

The structural field mapping was accompanied by geological photo-interpretation of the area. Black and white 1:33 000 aerial photographs were used. A few main faults and an intense jointing were found. All joints were analyzed using the Schmidt net (Figure 3).

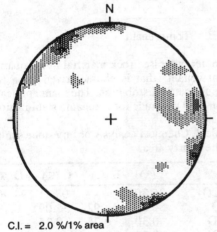

C.I. = 2.0 %/1% area

Fig. 3 Schmidt contour diagram of the joints systems in the quarry area (upper hemisphere).

4 MATERIALS

4.1 Rock material

The rock is a microcristallyne limestone with ooliths. These materials were also studied by other authors (Geometral, 1991; Rocha et al., 1981). In the last reference the rock is considered a compact sublitographyc limestone, an intramicrite, according to Folk's classification (Folk, 1965), to oolitic limestone.

The physical properties of the rock material were studied using selected rock samples collected directly on the quarry faces to avoid contamination. Each sample weighed about 100 to 150 Kg.

In Table 1 and 2 some physical properties of tested samples are presented. The tests were executed according to the ISRM Suggested Methods (ISRM, 1981).

4.2 Processed rock materials

The processed materials were studied in order to know the variation of their main characteristics. According to the technical Portuguese regulations, some new applications were also searched.

Table 1. Description and physical properties of tested samples.

Sample	Material	n (%)	γ_d (kN/m³)	$W_{máx}$ (%)
A1	Yellowish oolitic limestone	1.13	26.34	0.42
A2	Yellowish oolitic limestone	1.49	26.23	0.56
A3	Yellowish micrite limestone	1.42	26.26	0.53
A4	Yellowish compact limestone	1.48	26.24	0.55
A5	Yellow to pinkish compact limestone	1.81	26.16	0.68
A6	Yellowish oolitic limestone	1.05	26.34	0.39
A7	Yellow to pinkish micrite limestone	1.11	26.35	0.41
A9	Yellowish micrite limestone	1.16	26.32	0.43
A10	Yellowish micrite limestone	2.55	25.98	0.96
A11	Yellowish limestone	2.07	26.09	0.78

n - Porosity;
γ_d - Dry Unit Weight;
$W_{máx}$ - Maximum water content.

Table 2. Variation of some physical properties with the direction of testing.

Sample	V_L (m/s)	V_S (m/s)	$\Delta L/L_o$ (x10^{-4})	σ_c (MPa)
11 //	6450	4193	1.06	73.2
11 +	6437	3963	1.60	49.3
12 //	6187	3503	1.35	43.3
12 +	6173	3470	0.60	83.9
13 //	6563	3590	1.06	83.1
13 +	6607	3630	1.20	98.5
14 //	6467	3767	0.27	98.8
14 +	6090	3495	0.31	78.9

V_L - Compressional wave velocity;
V_S - Shear wave velocity;
$\Delta L/L_o$ - Linear expansion at 48 hours;
σ_c - Uniaxial compression strength;
// - Parallel to bedding;
+ - Perpendicular to bedding.

4.2.1 Aggregates

For the characterization of aggregates, tests to assess their mechanical performances were used (Tourenq & Denis, 1982). Some tests are normalized in Portuguese specifications namely the Los Angeles (LNEC, 1971) and the Sand Equivalent (LNEC, 1967) tests, others are index tests recommended by the ISRM (1981).

The tests were done using the collected samples and the results were compared with the average reference values obtain by the company for the different granulometric classes. Table 3 shows the results obtained.

The Los Angeles (LA) and the Sand Equivalent (SE) tests were done on samples crushed on a jaw mill to an aggregate size class G according to the Portuguese specification for the Los Angeles test. In the Slake Durability test the samples showed very low abrasion so, in order to obtain a better comparison of the results, the percentage of lost material is reported (Rodrigues, 1988; Quinta Ferreira, 1990).

4.2.2 Soil amendment

The use of limestone for soil amendment is normalized in Portugal (IPQ, 1973). The main characteristics for this purpose are the

Table 3. Characterization tests on aggregates.

Samples	LA(%)	I_{d2}(%loss)	SE(%)
A1	36	—	86
A2	35	1.10	75
A3	33	1.56	83
A4	32	0.85	80
A5	32	1.20	81
A6	35	0.50	83
A7	33	0.50	72
A8	33	—	79
A9	30	1.45	100
A10	32	0.70	86
A11	33	1.20	67
A13	33	—	80

LA - Los Angeles abrasion test;
I_{d2} - Slake Durability Index;
SE - Sand equivalent test;
— - Not determined.

granulometric size and the chemical composition.

The chemical analysis of a few samples are presented in Table 4. The granulometric distribution of limestone powder for soil amendment is presented in Table 5.

4.2.3 "Tout-venant"

Another processed rock material in the quarry is "tout-venant" that is characterized by a wide granulometric distribution. The main application is on road subgrade for mechanic stabilization.

Table 4. Chemical analysis of limestone samples in the quarry area.

	A (%)	B (%)	C (%)	D (%)
$CaCO_3$	95.30	—	98.31	94.90
CaO	54.88	54.92	55.09	0.35
MgO	0.30	0.25	0.35	—
N_2O	0.54	—	—	—
K_2O	0.09	—	—	0.001
SiO_2	0.72	0.37	—	0.12
R_2O_3	—	0.30	—	—
L.O.I.	—	43.62	—	—

R_2O_3 - Sum of the Al_2O_3 + Fe_2O_3;
L.O.I. - Loss on Ignition;
A - Made by LNEC in 1981;
B - Mannuppela et al. (1982);
C - Made by M.A.P.A., in 1992.
— - Not obtained;

Table 5. Granulometric distribution of limestone amendment.

Sieve n°	Aperture (mm)	I (%)	II (%)	III (%)	IV (%)
25	0.71	56.9	64.4	71.0	95.8
20	0.85	—	—	—	—
18	1.00	—	—	86.7	—
16	1.18	—	—	—	—
14	1.40	—	—	98.5	—
12	1.70	—	—	—	—
10	2.00	100	100	100	100

I - Made in 1991;
II - Made in 1992, 15 January;
III - Made in 1992, 24 January;
IV - Made in 1993.

4.3 Discussion

The properties of the rock materials are influenced by their sedimentary origin. The formation process of the limestone creates a moderate anisotropy due to variations of sedimentation, eventually creating weaker planes along the stratification.

The compressional and shear waves velocities are slightly higher parallel to the bedding planes (Table 2) because in this case the elastic waves can propagate along a path having higher continuity. The samples tested in the uniaxial compression test showed in general higher strength perpendicular to stratification because in this case the stresses are applied perpendicular to the weaker planes, reducing their influence. The exceptions were attributed to the presence of fractures.

The linear expansion is essentially due to the water absorption of the clay minerals in the rock. The results obtained show predominantly higher values perpendicular to the bedding. In this case the clay minerals that also are deposited during the precipitation of the limestone can have an easier expansion perpendicular to bedding. The content of swelling minerals is very small as can be inferred by the low expansion values obtained.

Considering the differences in the values, parallel and perpendicular to the stratification, for the measured properties it can be concluded that the anisotropy of the rock material is small. Despite that it is sufficient to influence the shape of the fragments when crushed in a jaw mill.

In the Slake Durability test all the rock types showed good behavior, presenting almost no slaking. This behaviour would be expected due to the very small porosity of the samples tested. The results obtained (Table 3) showed lower values than those from Machado (1992) for the carbonate rocks from Tomar region, ranging between 2.13 and 11.08 per cent loss.

The petrological analysis made in the quarry formations, showed a quite homogeneous mineralogical composition, and some variations in the rock fabric. This variation is the main responsible for the higher values of Los Angeles abrasion test in the oolithic samples (A1, A2 and A6).

The values of the Los Angeles tests made by the company in a regular basis, on the granulometric class of processed aggregates, have values between 26 and 29%, somewhat lower than those presented on Table 3. The higher values obtained using the samples collected in the quarry, could be related to the influence of anisotropy, the different grading used and to the more elongated shape presented by the material crushed in the jaw mill. The norm used for this test (LNEC, 1971) advise to avoid crushing the material in the laboratory. The samples obtained in the quarry face, and the quantity collected, required that the samples were prepared in the laboratory using a jaw mill, while for the processed aggregates an industrial ball mill was used.

For the aggregates, the comparison of the values obtained with those presented by Machado (1992) for similar rocks of Tomar region, suggest that studied aggregates will have a good behavior when used in bases and subgrade for road pavement. Table 6 shows the standards for the mechanical characteristics of aggregates required in Portugal by JAE (1978), for the different pavement layers.

Table 6. Mechanical characteristics for aggregates required by JAE (1978).

Tests used	Subgrade	Base	Abrasion layer
LA (%)	<35 (F)	<30 (F)	< 20 (B)
SE (%)	> 50	> 50	> 60

(F) - Grading size F;
(B) - Grading size B.

The aggregates used in roads should support the main tension developed in the road surface, and resist to the abrasion due to attrition.

Besides the excessive Los Angeles values, the uniform composition of this limestone, almost 100% of calcite, and the fact that calcite is a mineral of low hardness (3 in the 10 grades of the Mohs scale), makes the rock inappropriate for surface abrasion layers due to the quick wear of this aggregate. The presence of only one mineral also decreases the skid resistance (Mauget, 1984; Pokorny, 1984) because the different mineral hardnesses are responsible for the maintenance of the skid resistance of the pavement.

Table 7 presents a comparison between some test results and the acceptance values for limestone aggregates recommended by some institutions. On the basis of the obtained values it can be concluded that Sangardão limestone satisfies all presented criteria.

Table 7. Comparison between some tests results and acceptance values on limestone aggregates.

Tests	Sangardão limestone	Road Research Lab. (1)	Federal Republic of Germany (2)	Regional Lab. of Cracovy Poland(3)
Bulk density (g/cm^3)	2.67	2.66 CR (2.6-2.9)	2.6 -2.9	
W (%)	0.6	1.0 CR (0.2-2.9)		2.5-4.0
σ_c (MPa)	76	34.5	C 80-180	
ACV (%)		40 30	C R	
AIV (%)		45 30	R	
LA (%)	28			45-50

W - Water absorption;
ACV - Aggregate crushing value;
AIV - Aggregate impact value;
LA - Los Angeles value;
C - For concrete; R - For road;
(1) Dearman (1981); (2) Toussaint (1984);
(3) Paluch (1984).

For limestone amendment (IPQ, 1973) the content of magnesium carbonate should be less than 15 % and the sum of calcium plus magnesium carbonates (expressed as calcium carbonate) should be equal or greater than 85%. All of the material (100%) should pass in the 2 mm sieve and at least 60% on the 0.71 mm sieve. Comparing the results obtained (Table 4) with the required characteristics for soil amendment, it can be seen that the chemical composition of the products is very good due to the high content of carbonates, over 95%, significantly above the minimum values required. Concerning the granulometry of the amendment powder, it can be seen on Table 5 that the percentage passing the sieve number 25 (0.71 mm) has increased in recent years. This is an improvement, when the amendment is used in the soil, because smaller particles allow faster results in the correction of the soil properties.

5 EXTRACTION OPTIMIZATION

In a limestone quarry like Sangardão, the clay contamination is the main factor influencing the quality of the aggregates produced. To reduce the clay contamination to a minimum it is necessary to operate during the stope, selection, crushing and during the storage of processed materials.

The selection of the quarry face to execute the stope has to be done according to the weather. The presence of wet clay soils in the quarry face can turn useless any effort to produce crushed materials with acceptable quality. This is worst when the selection and crushing is done under rainy weather. A reasonable procedure is to explore the more contaminated benches during dry weather, preferably in summer, using the clean or slightly contaminated benches in the more rainy periods. The red color of the contaminated material allows an easy separation from clean materials, that will be conveyed to the crushing mill, but the efficiency of the separation procedure is very dependent on the skills of the operator of the loading equipment.

During the selection and crushing sequence, the use of several crushing stages, allows to obtained cleaner materials, when the smaller fractions are immediately separated as they usually are the most contaminated materials.

The larger materials used in the following crushing stages will be less contaminated. When necessary, to improve the aggregates quality, the number of crushing stages is increased.

The storage of crushed materials is done trying to avoid the contamination from dust carried by the wind or from the muds carried by the superficial rain water. Concerning the wind contamination, the storage areas are whenever possible located away from the main wind direction coming from the crushing mill. The creation of artificial barriers between the stocks and the mill is most of the time difficult to execute. The stoking sites are usually selected in well-drained areas to reduce as much as possible the mud contamination.

6 OTHER POSSIBLE USES

At the moment the limestone rock extracted at Sangardão quarry is totally transformed in granular materials. The main consumer is the civil engineering construction and public works. Aggregates and crushed materials with wide granulometric distribution are the main products, while the fraction smaller than 2 mm is used as agricultural soil amendment to correct acid soils.

Portland cement industry is another important consumer of Sangardão limestone that is used to correct the clinker composition.

The comparison between the ASTM specification for masonry (Dearman, 1981), and the values obtained on the materials tested are presented in Table 8. The results from Sangardão clearly indicate that the limestone has good characteristics to be used in masonry. The better zones to obtain good blocks are the less fractured and thicker beds located in the lower quarry faces. Despite the earlier stated the increment of the limestone uses as masonry has an important drawback that is the high cost of this material due to the important hand labor required.

The limestone powder can also be used as filler for paper industry to substitute caulin. The characteristics of the limestone are suitable for this purpose.

Table 8. ASTM specification for masonry and properties of Sangardão limestone.

Physic properties	ASTM C568-79	Sangardão limestone
Absorption (%) máx.	3	0.6
Density (kg/m^3) min.	2560	2716
Compressive Strength (MPa) min.	55	76

Other possible uses for the limestone are the production of hydrated lime, expanded clay, smelter for ceramics, filler for the industries of paper, glass, paints and rubber or for animal food.

REFERENCES

Cunha, L. 1990. As Serras Calcárias de Condeixa-Sicó-Alvaiázere. Geografia Física. INIC. Coimbra (in Portuguese).
Dearman, W.R. 1981. Engineering properties of carbonate rocks, Bull. Int. Ass. of Eng. Geol. N°24. Essen.
Folk, R.L. 1965. Petrology of sedimentary rocks. Austin, Texas. Hemphill's.
Geometral 1991. Projecto de lavra da Pedreira de Sangardão (in Portuguese).
IPQ 1973. Correctivos agrícolas alcalinizantes calcários. Norma NP-983, Instituto Português da Qualidade (in Portuguese).
ISRM 1981. Rock characterization testing & monitoring - ISRM Suggested methods. Ed. E.T. Brown. Oxford: Pergamon Press.
JAE 1978. Norms relative to the characteristics of granular materials used in pavement layers. Junta Autónoma de Estradas, Portugal (in Portuguese).
Lee, D.Y., Ordemir, O. 1984. Slurry seals texture as affected by aggregate gradation. Bull. Int. Ass. of Eng. Geol. N°30. Paris.
LNEC 1967. Ensaio de Equivalente de Areia. E199. Lisboa (in Portuguese).
LNEC 1971. Ensaio de Desgaste pela máquina de Los Angeles. E237. Lisboa (in Portuguese).
Machado, A.P. 1992. Rochas sedimentares como materiais de construção na região de Tomar. MSc Thesis presented at the New University of Lisbon. Lisbon (in Portuguese).

Manuppella, G., Moreira, J. 1982. Calcários e dolomitos da área da Figueira da Foz-Cantanhede - Coimbra - Montemor - o - Velho - Soure. Estudos, Notas e Trabalhos do Serviço de Fomento Mineiro. Sep.vol.XXV, fasc. 1-2. Porto (in Portuguese).

Mauget, G. 1984. Point de vue d'un producteur sur les specifications "Granulats". Bull. Int. Ass. of Eng. Geol. N°30. Paris.

Paluch, D. 1984. Relation entre les caracteristiques des granulats et les proprietes des materiaux qu'ils permettrent d'elaborer. Recherches en laboratoire et sur chantier, Bull. Int. Ass. of Eng. Geol. N°30.Paris.

Parga, J.R. 1969. Sistemas de fracturas tardi hercinicas del Maciço Hesperico. La Coruna, Crafinsa.

Pokorny, A.G. 1984. Classification des granulats pour les couches de roulement des chaussées. Bull. Int. Ass. of Eng. Geol. N°30. Paris

Quinta Ferreira, M. 1990. Aplicação da geologia de engenharia ao estudo de barragens de enrocamento. PhD Thesis presented at the University of Coimbra, 322 pgs, Coimbra (in Portuguese).

Ribeiro et al. 1980. Introduction a la géologie générale du Portugal. Lisboa. Serviços Geológicos de Portugal.

Rocha, R., Manuppella, G., Mouterde, R. 1981. Carta Geológica da Figueira da Foz (1/50000). Serviços Geológicos de Portugal. Lisboa (in Portuguese).

Rodrigues, J.D. 1988. Proposed geotechnical classification of carbonate rocks based on Portuguese and Algerian examples. Engineering Geology, Vol. 25, N°1, pp. 33-43.

Tourenq, C., Denis, A. 1982. Les essais de granulats. Laboratoire centrale de Ponts et Chaussées. Paris.

Toussaint, A. 1984. Use of aggregates from regional deposits in road construction.

Les ressources en granulats d'Algérie

Aggregate resources in Algeria

K.Chaib & R.Louhab
Organisme National de Contrôle Technique des Travaux Publics, Kouba, Algérie

RÉSUMÉ: Après avoir passé en revue les données actuelles sur le produit granulat en Algérie, les auteurs présentent, par ensemble géographique et géologique, les ressources potentielles en matériaux routiers. La note conclue par des perspectives - plan d'actions susceptibles d'améliorer la situation actuelle.

ABSTRACT: In fact, we give some actual data of the algerian's agregats product. In second hand, we present, by geographical and geological groups, the potential resources in road materials. The future prospects for improving the current situation conclude this paper.

1 INTRODUCTION . DONNEES ACTUELLES.

Le marché algérien du granulat est loin de répondre actuellement , à la demande croissante du réseau routier national (90.000 km) et de l'autoroute est-ouest (1200 km). D'où la nécessité d'un inventaire des ressources en granulats ainsi que l'élaboration d'une carte des gîtes potentiels. Selon des données C.N.A.T/E.N.G (1988-1992) (Ref.3) , il en ressort les éléments suivants :
- Production agrégats 1992 : estimée à 32 millions de m3 sur une capacité installée de 76 millions de m3.(FIG. 1 et 2).

Cette production est repartie comme suit :
- Roches massives: 29 millions de m3 dont 93 % de roches sédimentaires et 7 % de roches ignées.
- Roches alluvionnaires : 3 millions de m3.
- Ecart entre capacité de production et production : de l'ordre de 40 à 50 % , ce qui influe négativement sur le prix de vente.
- Déséquilibre en moyens matériels : se situe au niveau des unités de production entre les capacités d'extraction , de chargement - transport et de concassage. Une répartition rationnelle entre les

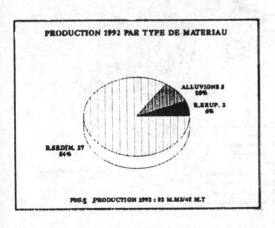

FIG.1 PRODUCTION 1992 : 32 M.M3/45 M.T

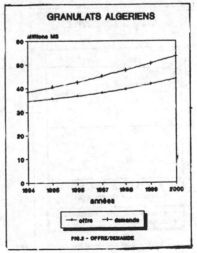

FIG.2 - OFFRE/DEMANDE

unités de production s'avère nécessaire
- Vétusté du matériel : à fin 1992 la
moyenne d'âge des stations était de 10
à 12 ans.
- Prévisions horizon 2000.Offre/Demande :
(Ref.3).
D'autres obstacles entravent le marché
et la qualité du granulat , entre autre :
- Absence d'études géologiques et
géotechniques dans le choix des sites.
- Déficit en personnel qualifié
(carriers).
- Insuffisance des structures de
maintenance.
- Mauvaise exploitation des fronts de
taille.
- Inexistence de laboratoires de contrôle
interne.
- Non respect des spécifications
techniques.
- Absence d'une Banque de Données
Granulats.

LES RESSOURCES EN GRANULATS EN ALGERIE:

L'inventaire des ressources en granulats
reste à faire en Algérie. Cet inventaire,
à différentes échelles , est motivé par
les observations suivantes :
- Les besoins exprimés par les régions ;
- Les ressources en granulats sont non-
renouvelables , leur exploitation devrait
éviter les gaspillages ;
- Les techniques routières exigent , de
plus en plus , une meilleure connaissance
qualitative des caractéristiques des
granulats ;
- La route est constituée de 95 % de
granulats et de 5 % de liant.
En Algérie , la plupart des granulats
sont susceptibles de satisfaire aux
spécifications en vigueur.(Tableau I)
(Ref. 4.5.9.13.14).
Néanmoins , l'adhésivité aux liants
hydrocarbonés faiblit pour certains
type de roches.
(Fig. 3 et 4) (Ref. 5.9.13.14)
2.1 Les ressources en roches massives :
    (Fig. 5   et 6)(Ref. 1.4.10.12.13.14)

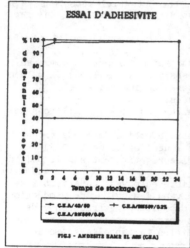

FIG.3 - ANDESITE RAMR EL ANS (CEA)

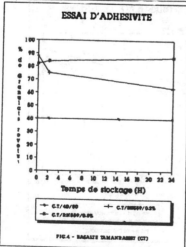

FIG.4 - BASALTE TAMANRASSET (CT)

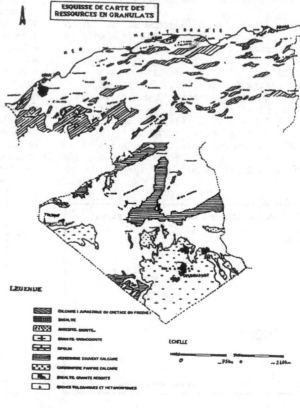

FIG.5

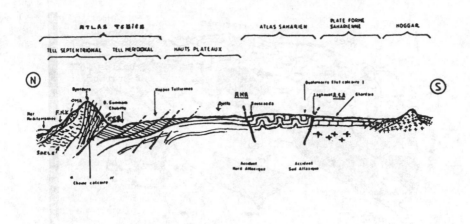

FIG.6 - COUPE GEOLOGIQUE SCHEMATIQUE DE L'ALGERIE (Nord-Sud)

0 ___ 150 km

SUITE :

Tableau I - Quelques caracteristiques des roches en ALGERIE

Roche-Age	Region	L.A	M.D.E	ADH. 24H
Calc.litho. Crétacé	EL-TARF*	21-24	30	-
Calc.litho. Crétacé	GUELMA*	24	22	-
Calc.mas.crist.Jur.	SKIKDA*	10-14	11-14	-
Calc.fin Crétacé	SKIKDA°	11-13	13-14	-
Calc.mas.Jurassique	N.CONST.*	19	11	-
Calc.mas.Eocène	SKIKDA*	20	17	-
Calc.litho.Crétacé	BATNA*	22	32	-
Calc.fin Crétacé	BISKRA*	18-26 (20)	14-21 (17)	-
Calc.fin Crétacé	B.B.A*	13-24 (18)	14-28 (22)	-
Calc.mas.Eocène	BIBANS*	10-25	19-25	-
Calc.litho.Crétacé	SETIF**	24	14-26	-
Calc.mas.Jurassique	BEJAIA**	21-26 (24)	14-18	-
Calc.sublitho.Crét.	BOUIRA*	23	17	-
Calc.mas.Eocène	LAKHDARIA*	20	30	-
Calc.mas.Eocène	KEDDARA*	18-29 (23)	20	-
Calc.sublitho.Crét.	BERROUAG.*	20-25 (22)	32-34 (33)	-
Calc.litho.Crétacé	S.EL GHOZ*	24	16	-
Calc.litho.Crétacé	TABLAT*	24	25	-
Calc.epimet.Crétacé	S. KEBIR*	18-24 (20)	19	-
Calcaire Jurassique	BOGHNI*	23	13	-
Calc.mas.Jurassique	CHENOUA*	23	20	70-80 CB
Calc.mas.Jurassique	TENES**	15	18	-
Calc.mas.Jurassique	ZACCAR*	15-23 (20)	18-24 (20)	-
Calc.mas.Jurassique	AIN DEFLA*	27-30	33-35	-
Calc.dolom.Jurass.	EL ATTAFS*	22-32 (24)	15-27 (22)	-

Tableau I - Quelques caracteristiques des roches en ALGERIE

Roche-Age	Region	L.A	M.D.E	ADH. 24H
Calc.Crétacé	MASCARA*	19	18	-
Calc.Crétacé	TISSEHSI.*	27	21	-
Calc.Jurassique	TIARET*	20	30-35	-
Calc.Jurassique	S.B.A*	21	18-21	-
Calc.mas.Jurassique	ORAN*	24	20-23	-
Calc.Crétacé	ARZEW*	19	19	-
Quartzite Paleozoi.	AIN DEFLA*	16	33	-
Calc.Crétacé	OUARGLA*	22	23	-
Calc.sublitho.Crét.	LAGHOUAT*	27-31	17-20	-
Calc.Carbonifère	IN SALAH**	26	24	-
Cipolin	ALGER*	17-30 (23)	22	85 CB
Basalte	C. DJINET*	13-14	16-21	40B40/50
Granodiorite	S.MUSTAP.*	24	17	-
Andesite	H.EL AIN*	14-23	23	40B40/50
Dacitoïde	H.EL AIN*	> 23	> 32	-
Rhyolite	HADJOUT*	16	20	-
Diabase	CHERCHEL**	11	15	-
Rhyolite	AIN DEFLA*	14	16	-
Basalte recent	HOGGAR**	< 19	21	40B40/50
Granite primaire	DJANET**	14-16	18-21	-
Amphibolite	HOGGAR**	30-32	32	70-80 CB
Cipolin	HOGGAR**	24-32	25	85-100
Gabbro	HOGGAR**	13-18	16-17	90-95
Pyroxenite	HOGGAR**	12-16	15	90-95
Microdiorite	HOGGAR**	13-17	15-17	75-80

* C.T.T.P
**Laboratoires regionaux

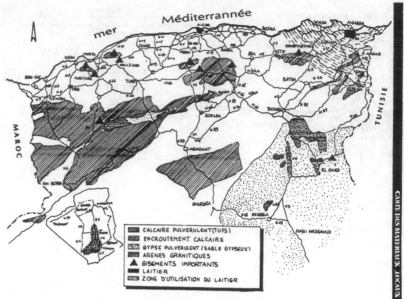

CARTE DES MATERIAUX LOCAUX

Légende :
- CALCAIRE PULVERULENT (TUFS)
- ENCROUTEMENT CALCAIRE
- GYPSE PULVERULENT (SABLE GYPSEUX)
- ARENES GRANITIQUES
- ▲ GISEMENTS IMPORTANTS
- LAITIER
- ZONE D'UTILISATION DU LAITIER

FIG.7. MATERIAUX LOCAUX.

2.1.1 Dans le Tell Septentrional :
D'une structure assez complexe , due pour
l'essentiel à la phase Alpine , le Tell
Septentrional renferme des ressources en
matériaux reparties comme suit :
a) Les Noyaux kabyles :
(les Massifs kabyles, d'Alger, du Chenoua..)
On y reconnait des terrains de nature
cristallophyllienne , volcanique et
sédimentaire datant du Paléozoïque.
- Terrains cristallophylliens
  (métamorphiques)
Cipolin (calcaire métamorphique) :
amphibolites
quartzite métamorphique
- Terrains sédimentaires :
calcaires (notamment)
- Terrains volcaniques :
notamment un volcanisme acide
(porphyroïde).
b) La "Chaîne Calcaire":
(Cap Ténès, le Chenoua, Bouzegza, Palestro,
Djurdjura, Chellala, Nord constantinois...)
La couverture mésozoïque et éocène des
noyaux Kabyles , affectée de plis
longitudinaux plus ou moins serrés , se
rompant en écailles redressées ou plates ,
renferme les ressources en matériaux
suivantes :
- Calcaire massif , et calcaire lité
(parfois à silex) du Lias ;

- Calcaire massif , parfois à Nummulites,
de l'Eocène.
C) Les Flyschs allochtones :
Ils pourraient constituer des ressources
en matériaux , notamment les calcaires
(ressources restreintes).
d) Le volcanisme récent :
Dans les zones internes du Tell , des
phénomènes de relaxation tectonique
favorisent un plutonisme de roches
acides et des venues de roches éruptives
perçantes en s'interstratifiant dans les
formations miocènes. Parmi les
ressources nous pouvons citer :
- Les basaltes de Temouchent.
- Les andésites et dacitoides de Hameur-
el-Ain.
- Les basaltes de Cap Djinet.
- La granodiorite de Thenia.
- Le granite de Skikda.
2.1.2 Dans le Tell Méridional
c'est le domaine d'amples chevauchements
et de charriages opérés vers le Sud au
Miocène inférieur.
a) Les Nappes Telliennes :
Elles sont essentiellement de nature
marneuse (Crétacé) , mais les ressources
en matériaux sont constituées par :
- Un calcaire noirâtre du Crétacé
supérieur.
- Un calcaire blanc , parfois à silex de

3154

l'Eocène.
b) Les Zones Autochtones ou Paraautochtones:
Les terrains mésozoïques (Jurassique et Crétacé) et parfois plus anciens (Paléozoïque) constituent des gîtes appréciables:
- Paraautochtones du littoral Oranais : Massifs d'Oran , d'Arzew et Traras.
* Calcaire massif du Lias
- Paraautochtone du Massif au Cheliff : Djebel Zaccar,Djebel Doui
* Roches paléozoïques : calcaire quartzite,rhyolite et dolérite
* calcaire massif du Jurassique
- Autochtone Nord Tellien Algérois : Atlas de Bou-Maad et de Blida
* Calcaire épimétamorphique du Crétacé
- Autohtone Intra-Tellien de l'Ouarsenis et des Bibans :
* Calcaire du Jurassique
* Calcaire du Crétacé
2.1.3 Dans les Hautes-Plaines ou Hauts-Plateaux.
D'Ouest en Est , les Hautes-Plaines Oranaises , les Monts de Saida , Tiaret les Monts de Hodna , les Hautes-Plaines Constantinoises , correspondent à des zones peu ou pas déformées et constituées de terrains sédimentaires mésozoïques.
Les matériaux exploitables se retrouvent parmi :
* Le Jurassique calcaro-dolomitique.
* Le Crétacé : calcaire dolomitique récifal à rudistes, calcaire noduleux à huîtres.
2.1.4 Dans l'Atlas Saharien :
Coincés entre deux accidents (le Nord Atlassique et le Sud Atlassique) , les plis coffrés renferment des ressources en matériaux appréciables :
- Calcaires dolomitiques et dolomies massives du Lias.
- Calcaires en plaquettes, souvent bitumineux à la base, du Crétacé
- Corniches de calcaires récifaux à rudistes du Crétacé.
2.1.5 Dans la plate-forme Saharienne :
Les roches massives de la plate-forme Saharienne constituent des potentialités en matériaux non négligeables.
- Les calcaires du Crétacé (Ghardaia , El Goléa)
- A moindre degré, les calcaires du Carbonifère (Bechar , In Salah).
- Dans le bouclier Targui (Hoggar) :
La structure globale du Hoggar correspond à la juxtaposition de compartiments longitudinaux disparates mettant en contact des roches dont l'age varie entre l'Archéen (> 2700 MA) et l'Eocambrien (530 MA). Les ressources sont constituées par :
a) Des roches massives anciennes :
- Des roches métamorphiques: Des cipolins des quartzites, des amphibolites, etc...
- Des roches plutoniques: des granites , des diorites, etc...
- Des roches volcaniques: des basaltes, des andésites, des dacites, des rhyolites des gabbros, etc...
b) Des roches massives récentes :
Elles sont représentées essentiellement par un volcanisme récent (tertiaire et récent) du type : basalte et andésite.

3 LES RESSOURCES EN ROCHES MEUBLES :

Ce sont des matériaux meubles, non consolidés.
3.1 Tout-venant d'Oued : Oued Cheliff , Oued Isser, Oued Djemâa...
Les Oueds du Tell algérien remanient très fréquemment une grave de type calcaire. Ce sont des matériaux souvent pollués (présence de fines argileuses) , nécessitant un lavage.
3.2 Eboulis de pente du Tell:(ex: Eboulis du Zaccar, etc.). Ils sont de différentes natures lithologiques, mais il existe une forte proportion de matériaux de type calcaire.
3.3 Regs de la plate-forme Saharienne.

4 LES RESSOURCES EN MATERIAUX LOCAUX (NATURELS) :

Parmi lesquels nous pouvons citer :
4.1 Les encroûtements calcaires : (Ref.2) Hauts-Plateaux, Meseta Oranaise, Sud Constantinois, Sahara...
Ils sont exploitables sur une faible épaisseur (quelques décimètres), mais ils ne peuvent être utilisés qu'en zones climatiques II, III et IV.
4.2 Les Roches Carbonatées tendres:(Ref2) Algérois, Hauts-Plateaux,Meseta Oranaise.
Exploitables en profondeur, elles ne peuvent être utilisées qu'en zones climatiques II, III et IV
4.3 Les sables gypseux et gypso-calcaires (Ref. 2)
Ils ne peuvent être utilisés qu'en zone aride.
4.4 Les arènes granitiques:(Notamment au Hoggar).
L'arène granitique est un sable résiduel produit d'altération des roches de la famille des granites. Les feldspaths

s'altèrent en Kaolins, les micas en argiles. Le quartz demeure dans sa structure originelle. C'est un matériau sans ossature.
Elles sont notamment rencontrées dans le Bouclier Targui (Hoggar) et en de rares pointements au Nord.

5 LES RESSOURCES EN MATERIAUX LOCAUX ARTIFICIELS (REGION DE ANNABA):(Ref.17)

Nous pouvons citer le cas des laitiers des Hauts-Fourneaux qui restent exclusivement concentrés à l'Est du Pays (ANNABA) ; leur utilisation reste également confinée à l'Est (ANNABA EL-TARF).

6 PERSPECTIVES-PLAN D'ACTIONS-CONCLUSION.

6.1 Perspectives-Plan d'actions :
La prise en charge du problème des granulats en Algérie, et de par la même , la qualité de ce produit afin de répondre avec efficacité aux besoins du réseau routier principal et l'Autoroute E-W, nécessitent le lancement d'un ensemble d'actions articulées autour d'un programme à l'échelle nationale.
Ce programme ou " Politique de la qualité du granulat en Algérie " , est constitué de différentes phases :
PHASE 1 : récolte de données sur les carrières existantes en Algérie (Banque de Données Granulats)
Synthèse des inventaires des organismes régionaux.
PHASE 2 : Etude photogéologique des gîtes potentiels à proximité du " Réseau Routier Principal " et de " l'Autoroute E - W ".
PHASE 3 : Investigations in-situ et au laboratoire.
PHASE 4 : Finalisation d'une carte des gisements potentiels avec définition de leurs caractéristiques (par Wilaya, par région...).
PHASE 5 : Elaboration de cahier de charges par gîte et définition de la méthodologie d'exploitation par carrière.
PHASE 6 : Formation du personnel à l'exploitation des carrières.
PHASE 7 : Recherche de formulation optimale des granulats.
PHASE 8 : Elaboration de recommandations d'utilisation des granulats en Algérie.
PHASE 9 : Définition avec les services des mines de la procédure de préservation de ces gîtes pour les besoins routiers.
Enfin , une attention particulière sera

accordée à la valorisation des matériaux locaux (naturels et artificiels).
6.2 Conclusion
Les énormes besoins en granulats induits par les différents plans de développement que connaît le pays , tous secteurs confondus , les déficits croissants entre l'offre et la demande, font qu'il devient primordial de procéder à un diagnostic de la situation actuelle (Programme Relance Production Granulats , en cours), et d'identifier toutes les ressources potentielles , c'est à dire l'élaboration d'une carte des ressources.
Ces actions devraient être complétées par l'amélioration des conditions de gestion technique et économique des moyens matériels et humains. La réussite d'un tel programme, qui revêt un caractère d'intérêt national, impliquera l'adhésion de l'ensemble des partenaires concernés par le produit : administration, entreprises, laboratoires et organismes spécialisés.

REFERENCES

1. Bulletin de service géologique de L'Algérie - N°39(1969).
2. Les tufs en technique routière (PRANDI) - Août 1976.
3. Rapport E.N.G - CNAT 1988 - 1992
4. Maîtrise de la qualité des travaux et équipements routiers - ENPC 1989
5. Recommandations 1990 - Enduits superficiels - C.T.T.P
6. Granulats - Presses ENPC - 1990.
7. Seminaire ENPC(Mai 1991): Maîtrise de la qualité des granulats.
8. Seminaire ENPC(Le Mans - Juin 1991) : "Granulats".
9. Dossier pilote Renforcement - C.T.T.P- Décembre 1991.
10. Cartes géologiques de l'Algérie.
11. Premier Congrès National de la route - ARAL - ALGER 1992.
12. 1éres Journées Nationales sur les Granulats - ENG - CTTP - ALGER 1993. MM.CHAIB/LOUHAB.
13. Les Inventaires de Gîtes à Matériaux - C.T.T.P (Programme de Renforcements/Etudes Routières et Autoroutières) MM.CHAIB/LOUHAB.
14. Les Inventaires de Gîtes à Matériaux - Laboratoires Régionaux.
15. Note d'expérimentation laboratoire sur les problèmes d'adhésivité (Utilisation des dops AKZO) - CTTP MM.DEKALI/BELAKHDAR - 1993.

16. 6th International Congress I.A.E.G
    Amsterdam - 6 , 10 - August 1990.
17. Note technique sur l'utilisation
    routière des laitiers de hauts
    fourneaux - Mr OURAD - C.T.T.P -1990
18. Seminaire sur la qualité des travaux
    routiers - C.T.T.P MM.CHAIB/LOUHAB
    ALGER 1991.

The use of geochemical methods for neutralization of surroundings aggressive to underground structures

L'utilisation des méthodes géochimiques pour la neutralisation des environnements agressifs pour les structures souterraines

N.G. Maximovich & S.M. Blinov
Perm University, Russia

ABSTRACT: Methods are described which are based on the enhancement of chemical resistance of structures and used to combat corrosive media. It is proposed to neutralize such media with various agents. The results of laboratory and experimental works are described which were carried out on the site of an enterprise in the West Urals with the purpose to reduce sulfate-induced corrosion of underground structures when mine waste materials from Kizel Coal Deposit are used for construction. Sulfates were precipitated with barium compounds. The results show that the method proposed by the authors and the new approach based on the geochemical effect on corrosive media are promising.

RÉSUMÉ: Étude des méthodes actuelles de la protection contre les milieux agressifs basées en général sur l'accroissement de la stabilité chimique des constructions. De différents agents sont proposés pour neutraliser de tels milieux. Il y a les résultats d'études de laboratoire et de travaux expérimentaux réalisés dans l'aire d'opération d'une entreprise d'Oural Occidental afin de diminuer l'agressivité de sulfate des eaux souterrains ayant lieu dans le cas de l'utilisation dans les travaux de construction des roches très sulfureux de terrils du bassin houiller de Kizel. Les sulfates sont déposés, avec les composés de baryum. Les travaux effectués ont montré l'avantage de la méthode proposée et de la nouvelle approche basée sur l'action géochimique sur les milieux agressifs.

1 INTRODUCTION

Underground engineering structures are always to some extent subjected to chemical and physico-chemical attacks of the environment. Agressive underground water or soils cause a reduction in the strength properties or even failure of structures. In the practice of constuction the most frequent are the attacks of sulphate and acid corrosive media on concrete structures.

Corrosive media may be both of natural and technogenic origin, the latter being a result of production activities (spillage, leakage, artificial soils) (Maximovich, Gorbunova 1990). Technogenic corrosive media are formed as a result of changes in the hydrodynamic and geochemical parameters of natural media, for example, in the cases of saline soil underflooding, pyrite oxidation, etc. (Rethati 1981; Hawkins, Pinches 1989). Microbiological processes can also be of a certain effect (Boch 1984). The protection of underground structures is mainly restricted to the enhancement of their chemical resistance by some methods: addition of special ingredients to concrete, enhancement of its density, application of water proofing compounds on concrete surfaces. These methods are described in Russian normative documents. Instances are also known when the corrosive soil was replaced or the level of underground water was lowered. All the above methods increase the operating cost of construction considerably. Most of the methods can not be used under the conditions of plants in operation.

The experience of combatting corrosive media shows that it is necessary to find new approaches to this problem with due regard for their economic and technological expediency. One of such methods is the method of geochemical effect on corrosive media.

2 SITE ENGINEERING AND GEOLOGICAL CONDITIONS

Such an approach has been used by the authors on one of the sites of the Industrial Association METANOL in Gubakha (Perm Region, Russia). The site is located on a slope of the river Kosaya (the Kama River basin) in the West zone of folding of the Urals. There the aleurolite and coaly-clay shale of Carbon are overlaied with eluvial clay containing gruss and quartz aleurolite debris 0.5-17.0 m thick. The clay is overlaid with deluvial clay 2.0-16.0 m thick containing inclusions of quartz sand and aleurolite debris and gruss.

During grading and the formatiion of embankment in addition to the soil moved within the site rocks from coal mine dumps of Kizel Coal deposit were used. The soils from the dumps contained high concentrations of sulfur in various kinds of compounds, the concentrations reaching 8.7 wt %. A considerable amount of the sulfur was contained in water-solube compounds. The results of aqueous extract analysis showed that the content of sulfates could be as high as hundreds of grams per kilogram of soil. On the earth surface the rocks were weathered which resulted in a pH decrease in the water contacting with them to 1.0-3.0.

As a result of underflooding in the fill-up soils concrete-corrosive water emerged at elevations above the levels of foundation lower surfaces. The observations conducted starting from 1984 show that there is a tendency of sulfate corrosiveness increase of the water. In some zones the content of sulfates has increased to 4.1 g/l and this is above the admissivible standard value. The following composition is characteristic of the underground water (in g/l): 0.22 HCl, 2.57 SO$_4$, 0.115 Cl, 0.414 Ca, 0.04 Mg, 0.80 Na+K. The value of pH of the water is 6.6 to 7.9. The composition of the water is formed as a result of its interaction with mine dump waste rocks which is confirmed by the results of laboratory analysis. The relatively high values of pH are caused by acidity neutralization in the process of interaction with clay soil and inclusions of carbonate debris. Stripping of the foundations and execution of corrosion proofing work under the conditions when the continuous cycle manufacturing complex is in operation are practically impossible. The lowering of water level and soil replacement under the existing conditions are unacceptable or too expensive.

3 KNOWN METHODS OF MEDIUM CORROSIVENESS REDUCTION

The experience gained in the reduction of sulfate and acid corrosiveness of the environment is relatively small. Studies are known which have been made on the possibility to use the ash resulting from coal burning as an additive to soils to neutralize their acidity and to precipitate deleterious components. Positive results have been obtained from the use of alkaline additives such as lime, limestone and trona (Sandereggen, Donovan 1984). Some investigators propose various additives to be used if there are sulfides in soils which additives are to suppress their microbiological oxidation in the cases of sulfate corrosiveness (Evangelon et al. 1985). However, these processes are difficult to control.

4 CHEMICAL GROUNDS FOR THE METHOD PROPOSED

For the precipitation of sulfates the authors have offered to use soluble barium compounds.

$$SO_4 + Ba = BaSO_4 \text{ (barite)}$$

This reaction is practically instantaneous and does not depend on medium pH. It is expedient to use barium hydroxide and barium chloride as the reagents. Barium chloride is highly soluble in water, so concentrated solutions can be used. The solubility of barium hydroxide is one order of magnitude lower than that of barium chloride but its use neutralizes the acid reaction of the medium and no extra components arc to be added to underground water.

$$Ba(OH)_2 + H_2SO_4 = BaSO_4 + 2H_2O$$

The resulting barite is relatively stable under exogenic conditions and practically does not decompose under weathering. It is not toxic, is used in drilling muds and may be used as a concrete filler and also in medicine fox X-raying of the digestive tract.

5 NATURAL GEOCHEMICAL ANALOGUES FOR THE METHOD

In nature the processes of barite formation are rather widespread. The understanding of these processes is important in the development of methods aimed to combat corrosive sulfate water. There are several processes of barite formation: mixing of waters carrying separately barium ions and SO_4 ions; reaction of solutions containing barium ions with sulfate rocks; reaction of solution containing sulfate ions with barium-containing rocks; oxidation of solutions containing Ba and S ions.

In underground waters occuring at small depths barium ions are rare for they contain one or other quantity of sulfates.

The major barite deposits were formed under hydrothermal conditions. If the zone of hypergenesis barite occurs in the form of nodules in clay and sand deposits in the coastal areas of seas. Barite modules can also occur in the oozes of recent sediments. Barite is formed as a result of chemical erosion of rocks. Sulfuric acid which results from sulfide oxidation reacts with barium ions and forms barite. With the other minerals evacuated, the so-called barite sypuchkas are formed replacing the ore bodies.

6 RESULTS OF LABORATORY AND FULL-SCALE INVESTIGATIONS

Three kinds of soil with different sulfur contents taken on the site were studied in laboratory with the purpose to determine the possibility of use of barium compounds for the neutralization of corrosive media and to define their optimal concentrations. Through the soils placed in a special devices barium chloride and barium hydroxide solutions of various concentrations and also destilled water were filtered. The filtered solutions were analized for the content of sulfate ions and pH. A total of 29 series of experiments were performed. It was found that when distilled water was flittered through a soil, from 19% to 62% of the total sulfur content passed into solution. On the basis of the studies the amount of barium salts required to precipitate sulfate ions was determined. It was found that the processing of soil may precipitate up to 97% of mobile sulfur.

The experiments were carried out on two parts of the site. On the first part where the underground water was of medium sulfate corrosiveness two holes with a diameter of 60 mm were drilled (Figure 1a) in which 30 kg of barium chloride ($BaCl_2 \times 8H_2O$) and 10 l of demineralized water were pourred. In the well located down the stream of underground water the composition of water was analized. A day after the beginning of the experiment the content of sulfates in the well under observation decreased to zero. A sulfate corrosiveness two holes with a short-time increase in the content of chlorine ions was observed and barium ion appeared. 14 days after there were no Ba ions found in the water and the content of chlorine ions and mineralization were close to the their original values and the content of sulfates was 0.55 g/l. At a later time the content of sulfates and miniralization decreased regularly (Figure 1b). By the end of the fourth month of observation the conccentration of sulfate ions was 0.18 g/l, i.e. the water became non-corrosive relative to concrete and its sulfate-calcium composition changed to the chloride-calcium one. During the whole period of observation the content of chlorides was far less than it is required for salt corrosiveness.

On the second experimental part of the site there were two trenches to bury the reagent and four observation holes down the stream of underground water (Figure 2a). The results of analyses of water extracts showed that the content of sulfates in the soils varied within the range of 1.05–7.43 g/kg. The soils in accordance with Russian standards were classified as corrosive. The content of sulfates in the underground water of that site part before the experiments had been 1.09–1.52 g/l and mineralization had ranged from 2.81 to 3.422 g/l.

The experiments on the decrease of sulfates corrosiveness were performed in several stages. At the first stage 45 kg of barium chloride ($BaCl_2 \times 8H_2O$) were buried in a trench. The result of chemical analyses showed that in all the observation holes there was a tendency for a decrease in the content of sulfate ions and by the end of the fourth month of observations their concentration was not higher than 0.36 g/l. The content of chlorides varied from 0.01 to 1.08 g/l. The mineralization decreased to 1.50–2.48 g/l (Figure 2b). The concentration of sulfates varied sharply due to their transport from the zone of aeration by atmospheric precipitation. There was observed a direct relationship between the amount of atmospheric precipitation and the content of SO_4, but during the whole

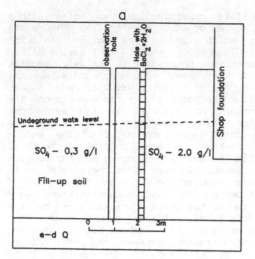

Fig.1 a) Experimental site 1 section view

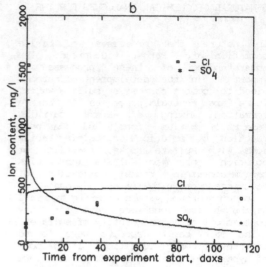

b) Water composition in absorvation hole

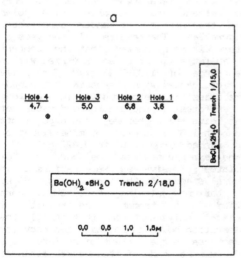

Fig.2 a) Experimental site 2 plan
1/15,0 − N/water level (sm)

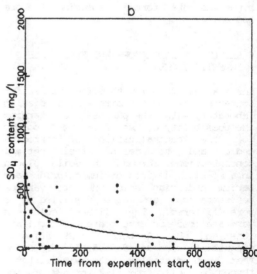

b) Sulfate ion concent in observation holes

period of observations their amount tended to decrease. At the second stage barium hydroxide $(Ba(OH)_2 \times 8H_2O)$ was used to neutralize the corrosiveness of the medium. A year and a half after the beginning of the experiment 60 kg of the reagent were buried in trench 2. The average content of sulfates decreased to 0.04 g/l. A decrease in water miniralisation was observed. At the final stage of observations it was from 0.39 to 1.40 g/l (Figure 2b).

The experiments on the sites part showed that the underground water which had possessed medium and strong corrosiveness became non-corrosive relative to concrete. The content of chlorides and pH were normal during the whole period of observations. As a result of reagent introduction the geochemical activity of the soils changed considerably. Analyses of water extracts showed that the content of soluble salts in the soil of site part 2 decreased by a factor of 2.5 and by the end of observations it was 2.69 g/kg; the contents of sulfides decreased by a factor of 3 and was 1.30 g/kg. The content of soluble salts at a distance of 1 m from the trench with the reagent was not more than 0.07 g/kg and in the water extracts hydrocarbonate and calcium ions were prevailing. To achieve a positive effect 29 kg of barium chloride or 22 kg of barium chloride or 22 kg of barium hydroxide were required per a cubic meter of soil. Those values were close to the results of calculations on the basis of laboratory analyses.

7 SIDE EFFECTS

The treatment of soils with soluble barium compounds caused changes in the mineral composition and properties. A yellow sediment was found at the bottom of the trench filled with barium hydroxide. Roentgenometric analyses showed that the sediment contains 24 % of barite, 15 % of calcite, 30 % of witherite, 30 % of quartz gypsum. The soil in the walls and in the lower part of the trench was cemented and difficult for cracking with a pinch.

To determine the composition of the precipitate separating from the underground water due to the reaction with barium salts, from the hole located near the site part 1 the water was sampled which had the following composition (in g/l): 0.23 HCO_3, 4.08 SO_4, 0.07 Cl, 0.37 Ca, 0.04 Mg, 1.51 Na+K, 0.02 Fe*** with the total mineral content being 6.31 at pH

5.65. An excess of barium hydroxide or barium chloride was added to the water. The resulting precipitate was collected and rhoentgenographed. In the case of barium chloride 99 % of the precipitate was barite and in the case of barium hydrochloride the precipitate consisted of barite (72 %) and witherite.

It is known that if a solution contains sulfate ions, there occurs an exchange reaction for the solubility of barite is much lower than that of witherite (Bisehberg, Plummer 1986):

$$BaCO_3 + SO_4 = BaSO_4 + CO_3.$$

This should be considered as a positive factor for if sulfate ions penetrate into the soil under treatment they will settle out in accordance with the above reaction.

The formation of barite and witherite and the reaction of the alkaline component with the soil results in a considerable increase of its strength. The filling of pores decreases the water permeability of the soil. These side effects should be considered as positive for the decrese in soil permeability to water and in the intensity of water exchange diminishes the effect of corrosive water on concrete structures and the increase of soil strength enhances the reliability of structure foundations. The use of barium chloride and barium hydroxide does not cause negative changes in the composition of underground water. In holes adjacent to the source barium ions are found only during the first moments after the start of the treatment. In the case of barium chloridee the concentration of chlorine ions increases only at the initial stage, but its content is always lower than the admissible level. No increase in pH was observed when barium hydroxide was used. Barium ions are not corrosive for concrete.

8 PROSPECTS FOR THE METHOD

Dependind on the actual geological conditions and the features of structures various ways are possible for the realization of the new method. If underground water occurs at the small depths trenches may be used to introduce the reagents into the soil, the trenches being upstream of the structure to be protected. If underground water occurs at the great depths, the reagents can be injected. If there is a possibility that corrosive media can appear, the reagents may be introduced into the soil in the course of construction. Thus, the experiments performed have shown that to combat media corrosive to structures on the basis of non-traditional approaches now under developments, i.e. by acting on corrosive by geochemical methods, is an effective way to protect the structures.

REFERENCES

Busenberg, E. & L.Plummer 1986. The solubility of $BaCO_3$ (cr.) (witherite) in CO_3 – H_2O solutions between 0 and 90° C, evaluation of the association constants of $BaHCO_3$ + (aq.). Geochim. et cosmochim acta 50, N10: 2225–2233.

Bock, E. 1984. Biologische korrosion. Tiefbaw-Ingenienrbau-Straussenbau, N5: 240–250.

Evangelou, V.P., J.H.Grove & D.Rawlings 1985. Rates of iron sulfide oxidation in coal spoil suspensions. J. of environmental Quality 14 N1:91–94.

Hawkins, A.B. & G.M.Pinches 1987. Cause and significance of heave at Llandough Hospital, Cardiff – a case history of ground floor heave due to gypsum growth. Quarterly Journal of Engineering Geology 20: 41–57.

Maximovich, N.G. & K.A.Gorbunova 1990. Geochemical aspects of geological medium changes in coalfields Proc. 6th IAGE: 1457–1461. Rotterdam: Balkema.

Rethati L. 1981. Geotechnical effects of changes in groudwater level. Proc. 10th ICSMFE: 471–476. Stockholm: Balkema.

Sandereggen J.L. & I.I.Donovan 1984. Laboratory simulation of flu ash as an amenoment to pyritte – rich tailing. Ground water monitoring review 3: 75–80.

Effect of petrography on engineering properties of coarse aggregates, Wadi Millah, Saudi Arabia

Influence de la pétrographie dans les propriétés géotechniques des agrégats grossiers, Wadi Millah, Arabie saoudite

Al-Edeenan, W. Shehata & A. Sabtan
King Abdulaziz University, Jeddah, Saudi Arabia

ABSTRACT: A large percentage of the coarse aggregate in western Saudi Arabia exists in the form of natural gravel in the wadis transacting the Arabian Shield and the Tihama. Rocks of various igneous, metamorphic and sedimentary origin outcrop in western Saudi Arabia and are therefore represented in the wadis natural aggregate.

The coarse aggregate in Wadi Millah, which is located in the central western part of the Tihama, was investigated, as an example, to determine its overall engineering properties and to investigate the effect of petrography on these properties.

Four major rock groups namely; basalt, gabbro, granite and schist, in addition to a fifth group consisting of minor constituents of other rock types, were identified in Wadi Millah natural aggregate. The results of the point load strength, crushing value and Los Angeles abrasion value tests on individual rock groups indicated that the properties of the basalt and schist aggregates are more superior than those for the gabbro and granite. Although, the overall properties of the mixed natural aggregate could meet the standard specifications, except for the elongation index which was unacceptable, these properties vary with the variation of the percentage of each rock group. It is therefore suggested that the study of petrography of these natural aggregates may suggest some range of values for the engineering properties.

RÉSUMÉ: Un grand pourcentage des agrégats grossiers existe en Arabie Saoudite sous la forme des graviers naturel dans les oueds qui traversent la voûte d'Arabie et de Tihama.

Des roches d'origine ignée, métamorphique et sédimentaire se situer en Arabie Saoudite occidental et, alors, sont représentées dans les agrégats naturel des oueds.

Les agrégats grossiers de l'oued de Millah, au centre occidental de Tihama a été étudié, comme par exemple, pour déterminer leurs caractéristiques géotechniques, et pour examiner l'effet de pétrographie sur ces caractères. Quatre groupes ont été identifie du basalte, du gabbro, du granite et du schiste en addition d'un cinquième groupe de constituants mineurs d'autres roches. Les résultats de la résistance à la compression ponctuelle, la valeur de creusement et la valeur d'abrasion de l'essaye de Los Angeles indiquent que les caractéristiques des basaltes et des schistes sont supérieurs à celles du gabbro et du granite. Même que tous les caractères des agrégats naturel mélanges concordent avec les spécifications sauf pour l'indice d'élongation, ces caractéristiques varient selon le pourcentage de chaque groupe de roche. Donc l'étude de la pétrographie de l'agrégat naturel peut indiquer quelques groupements de valeurs en ce que concerne les caractéristiques géotechnique.

1 INTRODUCTION

The scarp mountains of the Arabian Shield in Western Saudi Arabia extend along the full length of the Arabian Peninsula. From Yemen northward to the vicinity of Meccah, the elevations decrease with the development of knife-edge ridges and deep canyons. From Taif and Meccah northward, the height and ruggedness of the mountains decrease (Chapman, 1978). The foot hills of these mountains forms a physiographic province called Tihama (Fig. 1) bound by the scarp mountains from the east and the Red Sea coastal plains from the west (Brown, 1960).

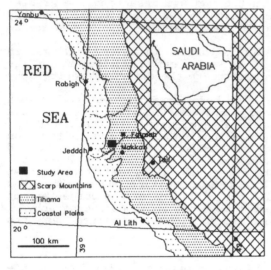

Fig. 1: Physiographic features of east central Saudi Arabia and location map.

The Tihama is hilly and is characterized by steep sided west-flowing valleys separated by upland areas. Wadi Millah is a tributary of Wadi Fatmah which is one of the major wadis transecting central Tihama. During the periods of flood, these valleys bring down appreciable amounts of gravel into the downstream sections. Some of these gravel occurrences are already in use as sources of natural aggregate. The petrography of these aggregates is dependent on the rock types exposed in the watershed area and their engineering properties are dependent on the petrography (Hartley, 1974; Haraldsson, 1984).

Approximately 1.5 million m^3 of natural aggregate are estimated to exist in Wadi Millah. Gravel of igneous, metamorphic and sedimentary origin could be identified in these aggregates. The purpose of this paper is to study the effect of petrography of the gravel on the engineering properties of the material as natural aggregate. Detailed studies have been carried out on the influence of various geological factors on aggregate properties (Sabine et al., 1954; Hartley, 1970, 1974; Ramsay et al., 1974; Lees and Kennedy, 1975; Ghosh, 1980; Kazi and Al-Mansour, 1980; Goswami, 1984; Turk and Dearman, 1988, 1989).

Aggregate samples were collected from five sections along a ten-kilometer stretch in Wadi Millah, spaced approximately two kilometers each. Three samples were collected from each section in order to account for the statistical variations within the section. A full scale laboratory testing program was performed on these samples. In two promising locations, large samples were collected to depths of 1.75-2.00 m. Parts of these samples were separated into their rock constituents and underwent some selected engineering property tests. All the laboratory tests were performed on either ASTM designations or BS standards. A correlation was then made to show the effect of the petrography on the properties of the bulk aggregate.

2 PETROGRAPHY

Cobble and gravel samples were collected, identified and studied in thin sections for their petrographic composition. Rocks such as andesite, basalt, granite, diorite, granodiorite, chlorite schist, quartz, rhyolite, marble, amphibolite and dacite were encountered in Wadi Millah in various proportions. These rocks were accordingly subdivided into several groups having common characteristics on basis of the Rock Group classification presented by the British Standard Institution (1975).

The aggregate samples could be separated into four main groups namely basalt group to include the basalt and andesite; gabbro group to include the diorite; granite group to include the granite and the granodiorite and schist group to include the chlorite schist. A fifth group was also separated to include all the minor constituents of other rock types such as quartz, rhyolite, marble, amphibolite and dacite. Fig. 2 shows the distribution of the different rock groups regarding the existing aggregate size. It is obvious that the basalt group forms the main constituent of the 1½ in. size while the gabbro and the schist groups, more or less, form the main constituents of the other sizes.

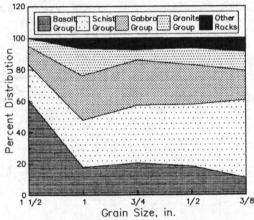

Fig. 2: Distribution of Rock Groups in different aggregate sizes.

3 PROPERTIES OF BULK AGGREGATES

No major variations could be noticed between the average properties of the samples collected from the different sections. The visual texture and particle shape tests were performed according to the BS (1975, BS 812: Part 1) standard. The surface texture is rough. The particle shapes varies between angular and irregular to flaky and angular. The flakiness index ranges between 16.31% to 20.59% while the elongation index varies between 39.12% and 46.79%. The grain size distribution performed according ASTM (1982) designation C136 indicated that the gravel is generally well-graded (GW) but locally poorly graded (GP).

The average unit weight varies from 1.610 to 1.645 g/cm^3 while the average water absorption ranges between 0.92% to 1.01%. These tests were performed according to ASTM (1982) designations C29-78 and C128-79 respectively. The average impact value varies between 19.88% and 22.38% while the average crushing value ranges between 17.24% and 18.20%. These tests were performed according BS (1975, BS 812: Part 3) standards. Los Angeles abrasion test (ASTM, 1982 C535-81) values averaged losses ranging between 31.60% and 33.97%. All these results, except for the elongation index, indicate that the aggregate satisfies both the ASTM and BS requirements. The sodium sulfate soundness test (ASTM, 1982 C88-76), on the other hand, have a relatively wide range of average losses (8.0% to 29.2%) which is partly unacceptable.

4 PROPERTIES OF INDIVIDUAL ROCK GROUPS

Selected mechanical property tests were performed on the individual aggregate rock groups. Point load tests were performed on aggregate particles according to Broch and Franklin (1972) and Turk and Dearman (1985). The point load strength index was converted and reported as uniaxial compressive strength (Fig. 3).

It is clear from Fig. 3 that the basalt group displayed the highest values of the uniaxial compressive strength with an average of 269 MPa while the granite and gabbro groups gave the lowest values of 64 MPa and 108 MPa respectively. The schist group showed an unexpectedly high average value of uniaxial compressive strength of 181 MPa.

The aggregate crushing value test was performed on the individual rock groups according to BS (1975, BS 812: Part 3) procedure. The results of the test, given in Fig. 4, indicate that both the basalt and schist groups have very high to high strength with average crushing values of 13.91% and 15.55% respectively. The gabbro and granite have moderate to low strength with average values of 21.03% and 23.49%

respectively. The other rock types gave an average strength of 18.69%.

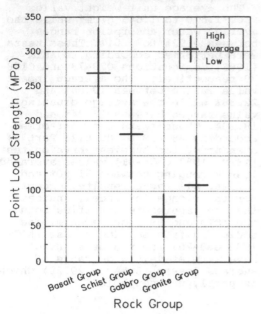

Fig. 3: Point load uniaxial compressive strength of individual rock groups.

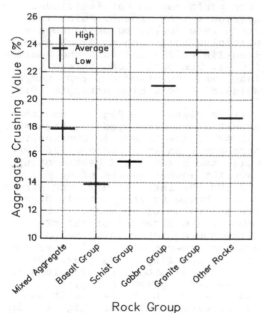

Fig. 4: Aggregate crushing strength of individual rock groups.

The Los Angeles abrasion test was also conducted on samples representing the individual rock groups according to ASTM (1982) designation C535-81. Fig. 5 shows the results for the five rock groups. The basalt and schist groups, with average Los Angeles abrasion values of 31.45% and 33.71% respectively, have very high to high strength. The gabbro and granite groups have moderate to low strength with average abrasion values of 42.17% and 43.02% respectively. The other rock types gave exceptionally high Los Angeles abrasion value (57.85%) possibly because of the concentration of marble grains in the tested samples. The difference in the engineering properties between the different rock groups could be either due to difference in the grain size of the constituent minerals as explained by Goswami (1984) or due to the difference in mineralogical composition.

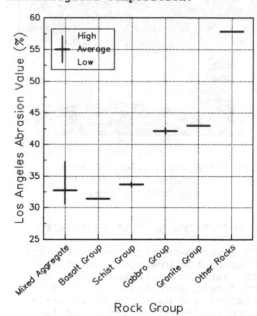

Fig. 5: Los Angeles abrasion value of individual rock groups

A correlation has been made between the aggregate crushing values and the Los Angeles abrasion values for the different individual rock groups (Fig. 6). The values of the other rock types group were ignored since the petrographical composition may vary from one sample to another. A good correlation was obtained so that the value of one parameter can be predicted with the knowledge of the other.

5. DISCUSSION ON THE EFFECT OF PETROGRAPHY ON THE ENGINEERING PROPERTIES

Figs. 4 & 5 show the values of the aggregate crushing values and Los Angeles abrasion values for the mixed natural aggregate in addition to those of the individual rock

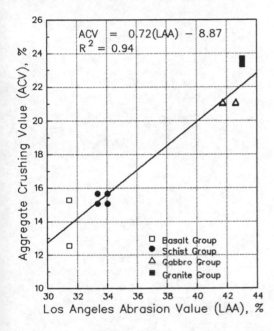

Fig. 6: Correlation between aggregate crushing values and Los Angeles abrasion values.

groups. It is clear that the presence of the basalt improves the overall engineering properties of the aggregate as also found by Haraldsson (1984). The presence of the gabbro and granite, on the other hand, reduces the quality of the aggregate. The effect of the schist, however, is controversial. Its high elongation ratio is expected to decrease the strength of the

concrete. However, Al-Mezgagi et al. (1994) in a parallel study indicated that the elongated particles decrease the strength of concrete until a certain percentage where the strength starts to increase again. They indicated that at higher percentage of the schist particle, the effect of the shape starts to diminish and is substituted by the effect of the particle strength. It can then be concluded that by the relative increase in the content of basalt, all the engineering properties of the natural aggregate from Tihama will improve. The increase in the schist may also improve the strength characteristics but might still have an effect on the workability of the concrete.

6 REFERENCES

Al-Mezgagi, H. A., Khiyami, H. A., Sabtan, A. A. and Ghazal, M. M. 1994. Effect of particle shape on the strength of concrete. This Congress.

ASTM 1982. Concrete and mineral aggregate. American Society for Testing and Materials. Easton, MD.

Broch, E. and Franklin, J. A. 1972. The point-load strength test. Int. J. Rock Mech. Min. Sci. 9:669-697.

Brown, G. 1960. Geomorphology of western and central Saudi Arabia. 21st Intern. Geol. Cong. Copenhagen. 21:150-159.

BS 1975. Methods of sampling and testing of mineral aggregates, sands and fillers. British Standard Institution, London. BS 812:Part 1, 2 & 3.

Chapman, R. W. 1978. General information on the Arabian Peninsula. in Quaternary Period in Saudi Arabia. S. S. Al-Sayari and J. G. Zötl (eds.). p. 4-44. Springer- Verlag, Wien.

Ghosh, D. K. 1980. Relationship between petrological, chemical and geomechanical properties of Deccan basalts, India. IAEG Bull. 22:287-292.

Goswami, S. C. 1984. Influence of geological factors on soundness and abrasion resistance of road surface aggregates - A case study. IAEG Bull. 30:59-61.

Haraldsson, H. 1984. Relations between petrography and the aggregate properties if Icelandic rocks. IAEG Bull. 30:73-76.

Hartley, A. 1970. The influence of geological factors upon the mechanical properties of road surfacing aggregates (with particular reference to British conditions and practice). Proc. 21st Highway Geol. Univ. of Kansas.

Hartley, A. 1974. A review of the geological factors influencing the mechanical properties of road surface aggregate. Q. Jour. Engng. Geol. 7:69-100.

Kazi, A. and Al-Mansour, Z. R. 1980. Influence of geological factors on abrasion and soundness characteristics of aggregate. Engng. Geol. 15:195-203.

Lees, G. and Kennedy, C. K. 1975. Quality, shape and degradation of aggregate. Quart. Jour. Engng. Geol. 8:193-209.

Ramsay, D. M., Dhir, R. K. and Spence, J. M. 1974. The role of rock and clast fabric in the physical performance of crushed rock aggregate. Engng. Geol. 8:267-285.

Sabine, P. A., Morey, J. E. and Shergold, F. A. 1954. The correlation of the mechanical properties and petrography of a series of quartz dolerite roadstones. Jour. Appl. Chem. 4:131-137.

Turk, N. and Dearman, W. R. 1985. Improvements of the determination of point load strength: IAEG Bull. 31:137-142.

Turk, N. and Dearman, W. R. 1988. An investigation into the influence of size on the mechanical properties of aggregate. IAEG Bull. 38:143-149.

Turk, N. and Dearman, W. R. 1989. An investigation of the relation between ten percent fines load and crushing value of aggregates (U.K.). IAEG Bull. 39:145-154.

Effect of particle shape of Wadi Ranyah aggregate on the strength of concrete

Influence de la forme des particules des agrégats de Wadi Ranyah sur la résistance du béton

H.A.Al-Mezgagi, H.A.Khiyami, A.A.Sabtan & M.M.Ghazal
King Abdulaziz University, Jeddah, Saudi Arabia

ABSTRACT: Wadi Ranyah is one of the longest wadis in Saudi Arabia. It is characterized by the presence of large quantities of natural coarse aggregate. The aggregate consists mainly of schist, pyroclastics, andesite, basalt, and other different rock types. The point load strength on the individual rock fragments indicated that the schist has the highest strength values and in the same time has the highest shape factors.
  A parametric study was designed to investigate the effect of particle shape on the strength property of concrete. Concrete cubes were prepared with different mixes having various percentage of schist particles. The strength of concrete behaved normally and decreased as the percentage of schist particles increased. But unexpectedly, as the schist particles exceeded 30% of the coarse aggregate fraction, the strength started to increase. It was obvious that a small percentage of the schist particle could decrease the concrete strength due to the influence of the shape factor. However, as the percentage increases the strength parameter of the schist material starts to dominate.

RÉSUMÉ: L'oued de Ranyah est parmis lel plus longs oueds en Arabie Saoudite. Il est caracterisée par la présence de grandes quantités d'agrégats grossiers naturels. Les agrégats se composent principalement de schiste, de pyroclastiques, d'andesite, de basalte et d'autres types de roches differentes. La resistance en compression ponctuelle sur les fragments de roche individu indique que le schiste a la valeur la plus forte et l'indice de forme le plus élevée.
Une étude paramétrique a été désignée au but d'éxaminer l'éffect de la forme des grains sur la resistance du béton. Des cubes de béton ont été préparés avec des mélanges differents de grains de schiste. La resistance du béton baisse quant le poucentage de schist augment. Mais, au contraire d'habitude, quant les grains de schiste fait plus de 30% de la fraction d'agrégats grossiers, la resistance commence à augmenter. Un petit pourcentage de grains de schiste peut faire baisser la résistance du béton due à l'influence de la morphoscopie. Avec l'augmentation de ce pourcentage le paramètre de resistance du schiste commence à diminuer.

1 INTRODUCTION

The particle shape of the aggregate is determined by the percentage of flaky and elongated particle that they contain. The aggregate shape has a direct effect on the strength of aggregate particle, the bond with cementing materials, and the resistance to sliding of one particle over the other (Collis and Fox, 1985)

  The shape characteristics of aggregate particles are classified in quantitative terms in BS 812. It classifies the flaky or elongated particles in coarse aggregate within a size from 6.3 mm to 63 mm. Many specifications restrict the percentage of long or thin particles allowed in aggregate. It is believed that as percentage of elongated aggregate becomes higher the aggregate strength gets lower. Kaplan (1959) pointed out that the strength of concrete tends to be reduced by increasing aggregate flakiness. The shape characteristics of aggregate can have marked effects on the property of both fresh and hardened concrete (Lees,

1964 and Neville, 1981). Particle shape limits for concrete aggregate can be a major consideration when developing and designing aggregate production for new source (Collis and Fox, 1981).

2 GEOLOGY

The Arabian Shield is located in the western part of the Arabian Peninsula (Figure 1). The rocks consist mainly of

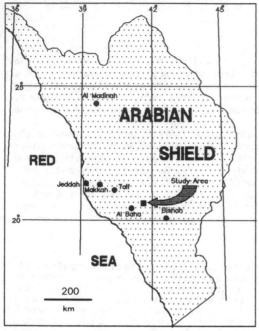

Figure 1.   The location of the Arabian Shield

layered and plutonic groups of rocks. The layered group is composed of volcanic and clastic sediments derived from volcanic sequences or plutonic intrusions. The plutonic groups occupy large areas in the Arabian Shield, and consist of plutonic granitic rocks. Wadi Ranyah is one of the major wadis in south west Saudi Arabia. It cuts different rock types and, during flood, brings huge amounts of gravel. The local geology of the studied area belongs to the Baish, Bahah and Jeddah groups (Greenwood, 1975). It consists of sericite–chlorite schist, gabbro, diorite, and serpentenite. Tufaceous clastic and carbonate units as well as basaltic flows also exist.

A 50 Km section of Wadi Ranyah was selected for this study. The alluvial deposits are composed of coarse aggregate (70%) and fine aggregate (30%) bounded by terraces of clayey silt and fine sand on both sides of the wadi. The aggregate samples were collected and visually classified petrographically as follows:
a – Mafic and felsic schist aggregate (30%): composed of light (felsic) and dark (mafic) minerals and alteration products. The grains are flaky and elongated with a rough surface texture.
b – Pyroclastic aggregate (20%): composed of dark-colored angular to subangular particles of porphyritic tuff.
c – Mafic volcanic aggregate (10%): composed of dark color basalt.
d – Intermediate volcanic aggregate (10%): composed of light and dark elongated gains of andesite.
f – Other different rock types (30%): composed of aggregate particles of granite, rhyolite and gabbro in addition to quartz.

3 GEOTECHNICAL PROPERTIES OF BULK AGGREGATE

The quality of the bulk aggregate in Wadi Ranyah was tested for its several physical and mechanical properties. The tests were conducted according to the ASTM or BS specifications. The water absorption ranges from 0.46% to 1.09% (ASTM (1986) C 127–80). The unit weight is 1.674 gm/cm^3 (ASTM (1986) C 29– 78), and the bulk specific gravity varies from 2.66 to 2.79 (ASTM (1986) C 127–80). The average crushing value (C.V.T) BS 812 (1975) and the Los Angeles value (L.A.V) (ASTM (1986) C 131) are 13.74% and 18.85% respectively. It can be seen that the coarse aggregate in wadi Ranyah is within the acceptable limits (Table 1).

Table 1. The evaluation of the geotechnical properties of coarse aggregates (size 3/4" – 1/2")

Property	Test result
Flakiness Index	17.15
Elongation Index	40.85
Unit Weight	1.68
Specific Gravity	2.78
Absorption	0.85
L.A.V.	8.85
C.V.T.	13.74

4 THE EFFECT OF SHAPE FACTOR

The classification of aggregate into rock groups was performed in order to show the effect of shape alone on the aggregate strength, keeping other factors constant. Different sizes of aggregate from each rock group were prepared and tested for their elongation index. It was obvious that the schist group has the highest elongation index in all aggregate sizes (Figure 2).

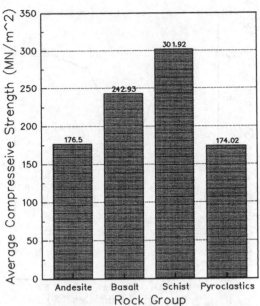

Figure 3. The compressive strength of the tersted rocks

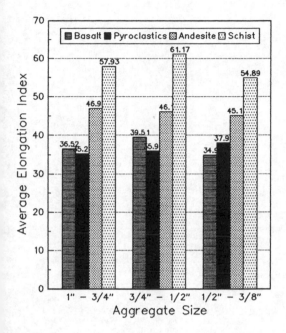

Figure 2. The elongation index for the tested rocks

The point load test was also conducted on aggregate samples of each rock group. Opposite to what was expected, the schist group gave the highest strength values (Figure 3).

5 CONCRETE MIX DESIGN

The efficiency of concrete depends upon its components. The shape parameter of coarse aggregate is one of the factors that influence the durability of concrete. To test that parameter, five samples with the following percentages of elongated particles, 0%, 15%, 30%, 45%, and 60% were prepared. The coarse aggregate used in this study has a size of 3/4"- 1/2" (5-20 mm).

Five cement concrete batches were prepared, each with a different percentage of elongated particles. Three cubes (15x15x15 cm) were prepared from each batch. The mix design was calculated by the weight and absolute volume methods (A.C.I, 1979).

The results indicated that the strength of concrete is inversely proportional to the percentage of elongated particles. However, as the percentage of the elongated particles exceeds 30%, the relation becomes directly propotional. This result shows that the shape factor is significant only when the aggregate contains up to 30% of elongated particles (Figure 4). As the percentage exceeds 30%, the strength of the rock material controls the strength of the concrete.

6 CONCLUSIONS

Wadi Ranyah might be considered as a suitable site for coarse aggregate resource. The physical and mechanical properties are within the specification limits. The shape factor plays a role in the concrete strength within the first 30% elongated particles. As the percentage increases, that influence is balanced by the effect of the rock material itself. More tests are recommended to clarify this point and preferably with finer increment of the percentage of elongated particles.

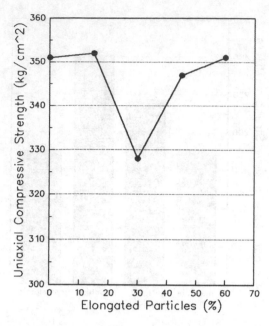

Figure 4. The influence of the aggregate particles on the rock strength

7 REFERENCES

American Concrete Institute (A.C.I.) 1979. Manual of Concrete Practice, Part 1.

American Society for testing and material (ASTM). 1986.

Collis, L. and Fox, R.A. 1985. Aggregate. Geological Society, Engineering Geology Special Publication. 1: 220.

Greenwood, R.W. 1975. Geology of Al-Aqiq quadrangle, Kingdom of Saudi Arabia, Ministry of petroleum and Mineral Resources, Map GM-23.

Kaplan, M.F. 1959. Flexural and ompressive strength concrete as affected by the properties of coarse aggregate, Jour. American Concrete, Inst. 55: 1193- 1208.

Lees, G. 1964. The measurement of particle shape and its influence in engineering material. Jour. of the British Granite and Whinstone Federation. 4.

Neville, A.M. 1981. Properties of concrete. 3rd Edition, Pitman.

Selection of basalt for aggregate and rockfill

Sélection de basalte pour granulat et remblai d'enrochement

John G.Cabrera
Las Cruces, N.Mex., USA

ABSTRACT: When investigating possible sources of rock for concrete aggregate and embankment fill at some of the largest hydroelectric projects in Brazil, located in areas of basalt lava flows, it has been found that some of the basalts undergo rapid disintegration due to a high content of expansive clay minerals. Moreover, some basalts contain minerals with metastable forms of silica, such as opal, chalcedony, volcanic glass or strained quartz, which react adversely with the excess alkalis in cement.

Emphasis is given to field observations of megascopic structural and textural characteristics which permit preliminary elimination of rock that is most likely to have deletereous minerals. One can often observe a strong correlation between vegetation and intraflow zones rich in nontronite or other expansive clay minerals. Tests made to determine the presence of expansive clays and of the minerals that can react with cement alkalis are described.

RÉSUMÉ: Quand on fait des recherches concernant des possibles gisements de granulats pour le béton et de remblais pour les enrochements de certains sites des plus grands projets hydroeléctriques du Brésil, situés en régions des coulées basaltiques, on a découvert que quelques basaltes subissent une désagregation rapide due à un haut contenu d'argiles expansives. D'ailleurs, certains basaltes contiennent des minéraux avec silice métastable, tels que l'opale, la calcedoine, le verre volcanique ou le quartz déformé qui réagissent défavorablement a l'excès des alcalis du ciment.

On insiste sur les observations mégascopiques du terrain et les caractéristiques de texture qui permettent l'élimination initiale des roches plus susceptibles de contenir des minéraux nuisibles. Fréquemment on peut observer une remarquable corrélation entre la végetation et les zones de coulées riches en nontronite et autres argiles expansives. Des essais faits pour déterminer la présence d'argiles expansives et des minéraux qui peuvent réagir avec les alcalis du ciment sont décrits ci-dessous.

INTRODUCTION

During the period from 1960 through 1980 a number of hydroelectric power installations were constructed within the Upper Paraná River Basin in Brazil, Paraguay and Argentina. Since most of these projects are within the area covered by Cretaceous basalt flows, considerable data on the properties of basalt were obtained, including the deletereous minerals that could make it unsuitable for concrete aggregate or rockfill.

The main projects that contributed much of the information in this paper are: Itumbiara and São Simão on the Paranaíba River; Volta Grande, Porto Colombia and Marimbondo on the Rio Grande and Ilha Solteira, Jupiá and Itaipu on the main stem of the Parana River.

In the course of the investigations at those sites during the feasibility phase, it became evident that certain intra-flow zones of basalt frequently underwent a rapid breakdown or slaking when exposed by excavation. Outcrops of these zones were already characterized by extreme weathering, a greyish-yellow color, microfissuring and disintegration to an easily erodable soil that would sustain vegetation, contrasting significantly with the sound basalt.

Fig.1 Itaipu Project. Blocks of dense
basalt just before exposure to natural
wetting and drying cycles.

Fig.2 Itaipu Project. Dense basalt containing swelling clay minerals after
nearly 4 months of exposure.

Petrographic analyses of the basalt that
crumbled rapidly showed the presence of
expansive clay minerals of the smectite
group. In time, the specific member of
this group that occurred in most of the
slaking basalts was identified as nontronite, an iron-rich, shiny, greenish-yellow clay. Throughout the investigations
at several project sites this mineral was
found to be derived from volcanic glass,
probably altered to palagonite by residual hydrothermal solutions.

Another mineral frequently present in
basalt amygdules or as a coating of vesicles, which at first was considered responsible for a rapid disintegration, is
blue-green celadonite, a soft member of
the mica group. However, it is not expansive or deleterious.

Some of the initial projects on the
smaller tributaries of the Paraná River in
Brazil were made with basalt aggregate
that later on was found to contain amorphous silica which reacted with the excess
alkalis in the cement, producing a gel
that expanded and caused cracking of the
concrete.

METHODS OF STUDY

A. Field recognition of basalt containing
expansive clay and/or free silica

Since a considerable amount of time and

expense is dedicated to testing rock for
soundness it is important to carry out a
thorough initial reconnaissance of proposed borrow areas, keeping the following
indications in mind:

a. Generally the slopes formed by the
zones with expansive clay minerals are
less steep than those of basalt not containing them.

b. Often the bases of those zones are
nearly horizontal steps or berms covered
with a gently-sloping talus of disaggre-
gated basalt that constitutes a fertile
soil for certain types of grasses. On
some basalt escarpments it is possible to
trace subhorizontal bands of vegetation
that coincide with the zones of slaking
basalt (Cabrera, 1985).

c. The color of the outcrops of basalt
containing nontronite and other expansive
clay minerals generally is not black or
dark grey as the sound basalt, but varies
from greenish-brown to greyish-yellow.

d. As the basalt slakes, it initially
develops a nearly orthogonal micro-frac-
ture pattern. The corners of each small
prism then begin to round and the disintegration progresses until the outcrop
surface on a bluff acquires a corrugated
profile, with a basal talus of crumbled
fragments.

e. Basalt in which chalcedony has been
altered to chalky or "bony" opal or that
contains much agate filling vesicles or as

siliceous bands can be suspected of reacting adversely with cement due to the free
silica. Often basalt containing chalcedony or opal develops a reticulated vein
pattern of lighter-colored material which
contrasts with the groundmass.

B. Preliminary Diagnostic Tests

Methylene blue adsorption (MBA)

This is a preliminary indicative test to
determine the presence of expansive clays
of the smectite group. At some of the
projects in the Paraná Basin this qualitative diagnostic test was used after initial petrographic examinations. Only the
swelling clay minerals adsorb large a-
mounts of methylene blue, in contrast with
other clay minerals, micas or chlorite,
resulting in a good correlation between
drying shrinkage and MBA (Cole, 1979).

For this test the rock sample is ground
to a fineness that allows passage of a
N° 200 mesh screen, then the powder is
dispersed in an aqueous solution and a
spot of the suspended solid placed on a
filter paper. The MBA value is the number of milliliters of dye solution used
up by 1 gram of that solid. The "end
point" of dye adsorption is reached when
a persistent blue halo surrounds the spot
of solid precipitated from the suspension
(Higgs, 1988).

This dye adsorption test is a fast way
of testing the clay fraction of a rock
suspected of containing expansive clay
minerals, but should follow a petrographic examination.

Quick chemical test for alkali-aggregate reaction (AAR)

If dense basalt contains a high percentage of metastable forms of silica, such
as opal, chalcedony or volcanic glass,
it is likely that enough free silica is
available to react with excess alkalis
in the cement paste. The usual AAR reaction occurs in cements having more than
0.6 percent Na_2O equivalent, as determined by chemical analysis.

The alkali-silica reaction occurs when
the silica becomes an alkali-silica gel
with the capacity to absorb water, swell
and exert pressure on its surroundings to
the extent of cracking the concrete. After swelling and inducing cracks the gel
absorbs so much water that it can flow
and migrate through the cracks it has
formed, filling them and any other void
until it finally exudes from the surface
of the samples that have been kept moist.

It is possible to determine the potential
for reactivity of a natural aggregate
with the alkalis of a cement by means of
a quick chemical test (Amer. Soc. for
Testing and Materials [ASTM] Test C 289).

Briefly, the test consists of immersing
a crushed and sieved sample of the rock
in a solution of sodium hydroxide (NaOH)
at 80° C for 24 hours. Afterwards the
solution is filtered off and an analysis
made for dissolved silica. The results
are then plotted on a semi-logarithmic
graph showing dissolved silica as the
abscissa and reduction in alkalinity as
the ordinate, both expressed in milli-
moles per liter (mMl). Proposed aggregates that plot in the lower right hand
portion of the graph, approximately
having dissolved silica greater than
32 mMl and between reductions in alkalinity values of 0 to 150 mMl, are suspected of being reactive.

This test is not definitive in revealing
the inherent degree of reactivity of an
aggregate primarily because of its limited
duration. The essential distinction
between deletereous and harmless aggregates is not the difference in reactivity
alone but rather the difference in the
rate of release of soluble silica, which
with time becomes available for a reaction.

C. Field exposure tests

During the foundation investigations of
the Porto Colombia Dam site drill cores
of basalt were left in the boxes exposed
to the elements after being carefully
logged. Within a period of two to three
months many of the cores containing nontronite slaked to crumbs and acquired a
greyish-yellow hue.

It became general practice during site
investigations to carry out exposure
tests, allowing representative samples
of basalt intended for aggregate or rockfill to undergo natural wetting and drying cycles for periods of up to six
months. Generally these exposure tests
were made with large, blocky fragments
of basalt (20 cm to 30 cm on the side)
which were piled up in masonry bins
(Figure 1) and exposed to natural precipitation, solar radiation and temperature
variations. Basalt containing expansive
clay can be either dense, with a fine-
grained, ophitic crystal matrix, amygdaloidal or vesicular. Since vesicular or
amygdaloidal basalt often contains other
minerals (e.g. chalcedony or opal) that
may react adversely with cement it is
generally excluded at first from use as

Table 1. Exposure to natural elements of 100 samples of dense basalt.

Duration of Exposure (days)	Cumulative weight loss (%) in 19 mm sieve	Number of samples affected by: Disintegration	Fragmentation	Not affected
0	0	-----	----	100
30	39.4	36	34	30
60	57.9	58	24	18
90	75.2	72	22	6
120	79.5	76	20	4
150	81.1	78	22	0
180	87.2	84	16	0

Table 2. Cyclic saturation and oven drying tests made on 50 samples of dense basalt.

Number of cycles	Cumulative weight loss (%) in 19 mm sieve	Number of samples affected by: Disintegration	Fragmentation	Not affected
0	0	-----	-----	50
12	39.6	19	17	14
24	71.1	27	16	7
36	83.2	38	12	0
48	93.3	44	6	0
60	96.8	46	4	0

aggregate.

Dense basalt containing expansive clay minerals loses much of its moisture content under the sun's rays and develops microfissures upon drying. Petrographic examination of representative samples show development of a reticulate pattern of fissures surrounded by expansive clay, principally nontronite. When it rains, these fissures fill with water, promoting the expansion of the smectitic clay within the rock matrix, ultimately resulting in the breakdown of an apparently sound rock to a soil (Figure 2). The results of a field exposure test made on 100 samples of basalt with a dense matrix are shown in Table 1.

D. Laboratory tests

Cyclic wetting and drying tests

Samples of dense basalt were subjected to alternate cycles of saturation in water by immersion followed by oven drying. Each cycle lasted for 12 hours. The result of a typical test of this nature is shown in Table 2.

Adsorption test

In southern Brazil some field laboratories test for the presence of expansive clays in basalt by subjecting the clay minerals to a water adsorption test, using an Enslin-Schmidt apparatus (Schneider, 1969). This is a device that permits the determination of the adsorbed water in relation to the dry weight of the rock material ground to 200 mesh screen size and dried in an oven at 110°C. Adsorption is expressed as the percentage of adsorbed water in relation to the dry weight of the sample. A comparison of the adsorption curve obtained in the test with a standard curve of a non-expanding mineral (e.g. quartz) permits a qualitative appraisal of the expansive characteristics of the clay mineral present.

It has been found that rocks without expansive clay minerals or those that have very low percentages of such minerals, yield adsorption curves below 50 per cent of the dry weight of the sample for the required time of saturation (1,000 minutes). In contrast, rocks containing expansive clays in excess of 10 per cent by volume experience a much higher adsorption than 50 per cent of the dry weight for the same time duration and can reach values close to 100% when the content of expansive clay minerals is very high.

Determination of expansive breakdown by saturation in ethylene glycol

Another simple qualitative method of determining the presence of expansive clay in rock is carried out by soaking the sample in ethylene glycol. This method was originally developed by the U.S. Army Corps of Engineers. It appears in the Corps' "Handbook for Concrete and Cement" under the designation CRD-C 148.

This test is equivalent to an accelerated wet-dry slaking test. As ethylene glycol consists of large molecules which form an organo-clay complex with swelling clays of the smectite group the clay crystal lattice expands more than with water molecules. The test is based on the premise that if the rock disintegrates due to the expansion of clay in the presence of glycol, breakdown may also occur if similar rock samples are exposed for longer times to natural alternate wetting and drying.

The test is made on a representative sample weighing between 4.5 and 5 kg., which is crushed and screened so that only the fraction passing a 3 inch (76.2 mm) screen and retained on a 3/4 inch (19 mm) screen is used. After washing, the sample is weighed and dried to a constant weight. The sample is then immersed in a glass container full of ethylene glycol, which is tightly capped. At intervals of not over three days the sample is removed and examined. A photographic record is kept. After a period of 15 days the sample is thoroughly washed over a 3/4 inch (19 mm) screen to remove all fragments that pass that sieve. The remaining material is dried to a constant weight and the weight compared to the original dry weight. If the clay-bearing sample of basalt loses less than 10% of its original dry weight as a result of this test, it indicates that the rock may be suitable for concrete aggregate. However, some X-ray diffraction tests should be made as a final confirmation.

Mortar bar expansion

The mortar bar expansion test (ASTM C 227) is a longer and more definitive procedure to determine the effect of alkali-aggregate reaction. The bars are made with the aggregate rock reduced to a well-graded sand and blended into a mortar using a high alkali cement (alkali content exceeding 0.60 per cent) with a water-cement ratio of 0.40 by weight. After curing the specimens and storing them above water in sealed containers their lengths are measured at regular intervals and at the same time the bars are studied for exudations, cracking, warping and spalling. The aggregate is considered deleterious and reactive if linear expansion of the mortar bars exceeds 0.10 per cent within six months or in excess of 0.05 per cent at 3 months.

At some of the power sites in the Upper Paraná River Basin (e.g. Jupiá) mortar bars made with a local coarse aggregate that contained from 40% to 70% deleterious material remained well below the 0.10% expansion limit. In comparison, bars made with a local aggregate that contained only 5% to 30% deleterious material exceeded the 0.10% limit. Expansion due to alkali-aggregate reaction is not directly proportional to either the rate or the amount of reaction as measured by chemical methods (Mielenz, 1962). Instead, the most positive manner of identifying reactive combinations is to study the features of any existing concrete structure made with the suspected aggregate which has had a service record of several years. Feasibility studies for important projects should include early fabrication of some faceted concrete test structures (e.g. tetrahedra) made with the same basalt aggregate intended for use in the construction concrete.

X-Ray diffraction analysis

This is the most definitive test to determine the presence of expansive clay minerals of the smectite group. As the equipment is expensive and requires specially trained technicians, the samples from the Upper Paraná River sites were sent to university laboratories or special testing laboratories. For each test a concentrated suspension of clay minerals less than 2 microns in diameter is smeared on a regular glass petrographic slide and the clay platelets become oriented parallel to the slide surface. The slide is then exposed to radiation so that the angle θ between the X-ray beam and the silica layer (e.g. the slide surface) is known. This angle θ is changed at a constant rate during exposure of the sample to X-ray and a radiation detector is placed so that it detects diffracted radiation when the proper angle θ occurs between the sample and the X-ray beam.

For any one species of layer silicates the number of values of the angle θ at which diffraction occurs is small. Each value of the spacing between planes of high electron density in the crystal structure is characteristic of the clay mineral. For example, when nontronite is treated with ethylene glycol to substitute the water molecule in the interlayers, the distance between planes of highest electron density in consecutive octahedral sheets is approximately 16.3Å (Angstrom units). All other spacings for the nontronite are fractions of the 16.3Å spacing such as 8.1Å. 5.4Å, 4.1Å, 3.26Å (Figure 3). These spacings and relative intensity of the diffraction maximum for each spacing are typical of expansive clays of the smectite group.

CONCLUSIONS

Although it should be excluded from use as concrete aggregate, basalt containing expansive clay mineral can be used as fill material when there is a minimum time lapse between excavation and placement. Test rockfills at Itaipu and other sites in the Paraná River Basin showed that if that basalt is buried rapidly, to prevent it from drying out and fissuring, it will retain particle shape. Pits excavated in the test fills showed very little breakdown of the material when protected by an

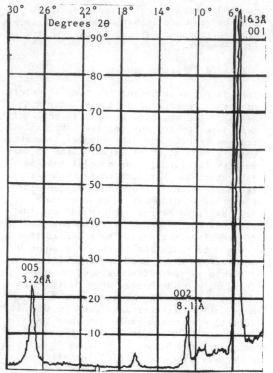

Fig. 3. Representative X-ray diffracto-
gram of glycol-treated clay sample from
basalt core obtained in boring F-4, São
Simão Dam site.

overlying two to three meters thick layer
of sound non-degrading basalt. This was
confirmed in the rockfill embankment of
the Itaipu Dam, where, based on previous
experience, the board of Special Consultants, including Prof. Arthur Casagrande,
allowed the incorporation of a controlled
amount of basalt known to contain expansive clay.

The principal factors influencing the
expansion of concrete resulting from the
reaction between aggregate and the cement
alkalis are: (1) the alkali content of
the cement; (2) the curing temperature of
the environment; (3) the availability of
moisture; (4) the type of cement used in
the mix; and (5) the amount, size and reactivity of the aggregate.

The most practical and effective way of
reducing the deterioration of concrete by
alkali-silica reaction is the use of a low
alkali cement (Na_2O equivalent to less
than 0.5%) or the replacement of a portion of the cement by certain pozzolanic
additives, which generally are made from
siliceous products and can reduce or entirely prevent the alkali-aggregate reaction by reacting with the excess alkalis

in the pore fluid of concrete before the
aggregate reacts with the fluid. Silica
fume, an amorphous silicon oxide which is
a byproduct from the elaboration of silicon and ferrosilicon, has been shown to be
a very efficient pozzolan which will
combine with the cement alkalis, precluding
an alkali-aggregate reaction.

ACKNOWLEDGMENT

The author wishes to thank Engineering
Geologist Francisco A. d'Almeida for his
collaboration during the studies of the
foundation conditions and construction
materials at the Itaipu Power Project,
between Brazil and Paraguay.

REFERENCES

Cabrera, J.G. 1985. Development of some
major discontinuities in basalt flows.
Bulletin of the Association of Engineer-
ing Geologists XXII, N° 1: 114-117.
Cole, W.F. 1979. Dimensionally unstable
grey basalt. Cement and Concrete Re-
search 9: 425-430.
Higgs, N.B. 1988. Methylene blue adsorp-
tion as a rapid and economical method of
detecting smectite. Geotechnical testing
Journal II, N° 1: 68-71.
Mielenz, R.C. 1962. Petrography applied to
Portland cement concrete. In Fluhr and
Legget, Editors, Reviews in Engineering
Geology. Geological Society of America,
Vol. 1: 1-38.
Schneider, A.W. and Pires da Rocha, F.X.
1969. Emprego de basaltos em pavimen-
tacao rodoviaria. Conselho Nacional de
Pesquisas, Instituto de Pesquisas
Rodoviarias. Publ. 433: 1-42.

Frost resistance of cohesive materials with open porosity
Application to compacted sand concrete

Durabilité au gel-dégel des matériaux cohérents à porosité ouverte
Application au béton de sable compacté

Eric Gervreau, Wilhelm Luhowiak & Jean Sicard
Laboratoire d'Énergétique, Matériaux et Sciences des Constructions, IUT Cergy-Pontoise, Université de Cergy-Pontoise, France

ABSTRACT : Frost resistance of clay bricks have been tested. The bricks made of kaolinite and sand have been compacted and stabilized with cement. A mixture with 8% of cement leads to compressive strength higher than 20MPa after 20 freeze-thaw cycles with slight dimensionnal variations. These bricks can be used for building.

RÉSUMÉ : Nous présentons les résultats expérimentaux concernant la résistance mécanique et les variations dimensionnelles des blocs de terre stabilisée au ciment et compactée soumis à des cycles sévères de gel-dégel. Avec un mélange de kaolin et de sable dosé à 8% de ciment nous obtenons des résistances supérieures à 20MPa avec de faibles variations dimensionnelles. Ces blocs sont utilisables pour la construction.

1. INTRODUCTION

La terre stabilisée est utilisée fréquemment pour des ouvrages de génie civil (remblais routiers, barrages en béton compacté aux rouleaux) sous tous les climats, des plus arides jusqu'aux plus humides, des plus chauds aux plus froids. D'autre part l'architecture en terre crue est également assez largement répandue dans les régions sèches de notre planète. On estime qu'actuellement le tiers de la population mondiale habite dans des maisons à base de terre.

Dans les régions humides et froides les constructions en terre crue sont nettement moins courantes. En France par exemple l'habitat en terre, essentiellement composé de constructions anciennes ne représenterait environ que 10% du patrimoine rural.

Pourtant la plupart des pays ont besoin de construire des logements sociaux économiques. La terre est un matériau naturel disponible en abondance ; il n'implique souvent ni achat, ni transport, ni transformation importante, de plus il est facilement recyclable. Ce matériau nécessite donc peu d'énergie pour être produit, mais il ne permet pas l'édification de constructions très hautes car il est relativement peu résistant mécaniquement et réputé peu durable dans les régions pluvieuses. Il est souvent réservé à des constructions basses, faiblement exposées à l'eau et dans des zones où le prix des terrains permet un étalement des logements sur de grandes surfaces.

Le Laboratoire d'Energétique, Matériaux et Sciences des Constructions de l'Université de Cergy Pontoise s'intéresse tout particulièrement aux matériaux de construction faits à partir de terre stabilisée et de béton de sable compacté.

Dans un premier temps nous avons effectué une étude sur le compactage et la stabilisation de briques d'argile. Le compactage permet un accroissement de la compacité. La stabilisation est obtenue en ajoutant un liant hydraulique (ciment ou chaux). Les résultats concernant les résistances mécaniques et la durabilité de ces blocs d'argile ont fait l'objet d'une récente publication [1]. La connaissance de la tenue au gel de ces matériaux à base d'argile est fondamentale comme le prouve des études récentes [2 à 5]. C'est pour cela que dans un second temps, nous nous sommes intéressés à la résistance au gel de ces matériaux poreux et ceci nous a amenés à réaliser des essais de gélivité sur des briques de terre stabilisées compactées dynamiquement.

Nous présentons dans cet article, les résultats expérimentaux concernant la résistance mécanique et les variations dimensionnelles des blocs de terre stabilisée au ciment et compactée soumis à des cycles sévères de gel et de dégel.

2. LES MÉCANISMES DE DÉTÉRIORATION PAR GEL-DÉGEL

Lorsque l'on abaisse la température d'un milieu poreux saturé ou non, l'eau située à l'intérieur des pores passe de l'état liquide à l'état solide et il s'ensuit une augmentation de volume d'environ 9%. Cette expansion de l'eau développe au pourtour extérieur du corps de glace des pressions. Ce phénomène très localisé se remarque tout particulièrement au voisinage d'un pore rempli d'eau (cas d'un milieu saturé).

Ce processus entraîne une microfracturation du milieu poreux appelée gélifraction qui se traduit par une augmentation du volume apparent du milieu lors du gel et un effritement du matériau au dégel caractérisé par une perte de masse.

On constate un abaissement du point de fusion de la glace avec la taille des pores ou capillaires. Lorsque l'on abaisse la température, l'eau gèle donc d'abord dans les gros interstices. L'augmentation de volume résultant du gel de l'eau dans les gros pores entraîne une poussée de l'eau sous forme liquide vers les pores les plus fins et donc une augmentation de la pression interstitielle dans le voisinage du front de gel.

bonne tenue mécanique au gel et au dégel elles doivent présenter les caractéristiques structurales suivantes :
1   faible porosité
2   pores répartis uniformément dans le milieu
3   pores de petites tailles
4   absence de cavités

Ces quatre caractéristiques recherchées conditionnent le choix du mélange granulaire qui va servir à confectionner le matériau et les conditions de fabrication (en l'occurrence la teneur en eau et l'énergie de compactage dépensée pour réaliser des briques).

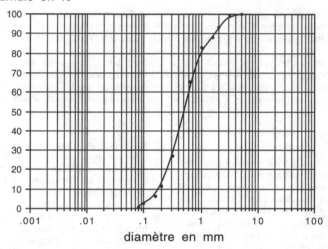

Figure n°1 : Courbe granulométrique du sable de rivière utilisé.

Pour les milieux non saturés, l'abaissement de la température engendre la croissance de cristaux de glace dans les gros pores qui adsorbent des quantités d'eau supplémentaires (la glace attire l'eau); c'est ce que l'on appelle la succion cryogénique. Les teneurs en eau augmentent alors au voisinage du front de gel ce qui diminue fortement les résistances mécaniques lors du dégel. Si le taux de gel est élevé (supérieur à 0,2°C/mn) ce phénomène sera atténué car l'eau à l'état liquide dans les petits interstices, n'aura pas le temps de migrer avant de geler à son tour.

Face au phénomène de gel, un sol fin traité au ciment ou à la chaux peut subir la succion cryogénique, on observera alors la formation de lentilles de glace perpendiculairement au flux de chaleur. Elles vont rompre les liaisons qui avaient été créées par le traitement au liant hydraulique. Cette destruction sera irréversible et le matériau retrouvera un comportement de sol fin non traité donc peu résistant mécaniquement.

Si l'on veut que les terres stabilisées possèdent une

Ce sont ces mêmes caractéristiques, complétées par un dosage optimum en liant hydraulique et une quantité d'eau suffisante pour son hydratation complète, qui vont permettre d'obtenir de bonnes résistances mécaniques.

3. ÉTUDE EXPÉRIMENTALE

Nous avons étudié la résistance au gel-dégel (à fort taux de gel et de dégel) de briques compactées composées d'un mélange de sable et de kaolin stabilisé au ciment portland artificiel (CPA55)

3.1. Les composants
3.1.1. Le sable
Il s'agit d'un sable siliceux de rivière de classe granulaire 0-5mm (figure n°1)

Coefficient d'uniformité $Cu = \dfrac{D_{30}}{D_{10}} = 3$

Coefficient de courbure $Cc = \dfrac{D_{30}^2}{D_{10} \cdot D_{60}} = 0,9$

Selon la classification des Laboratoires des Ponts et Chaussées (France) il s'agit d'un sable propre mal gradué.

Le poids volumique des grains est de 26,2 kN/m^3
La porosité est élevée : 30%
La perméabilité est de 2,8 10^{-3} m/s

3.1.2. Le kaolin : $Al_2O_3\ 2SiO_2\ 2H_2O$
Il s'agit d'une argile plastique peu gonflante assez courante
Limite de liquidité : 58%

3.1.3. Mélange sable+kaolin
Des études antérieures ont montré que le mélange pondéral suivant permet d'obtenir un plus faible indice des vides, donc une meilleure compacité (voir tableaux n°1 et n°2) :

   sable 90%
   kaolin 10%

La perméabilité du mélange est de 0,77x10^{-4} m/s

plus lorsque l'on ajoute du sable et de l'argile au mélange. L'eau excédentaire est indispensable pour permettre une mise en place correcte (optimisation du compactage).

3.2. Confection des éprouvettes
Les dimensions du moule permettent de confectionner des blocs de 10 x 5 x 5cm à partir d'environ 500g d'un mélange de sable, de kaolin, de ciment et d'eau.
Les éprouvettes sont réalisées par compactage dynamique dans un moule dans les conditions de teneur en eau de l'optimum Proctor.
L'énergie de compactage pour fabriquer un bloc est de 200Joules soit 800kJoules par mètre cube de matériau.

3.3. Détermination des teneurs en eau
Trois contraintes conditionnent le choix des teneurs en eau.

 1. une quantité d'eau suffisante pour permettre l'hydratation du ciment
 2. une quantité d'eau (assez faible) pour permettre un démoulage dès la fin du compactage

	sable	sable	sable	sable	sable
ciment	0%	3%	6%	9%	12%
teneur en eau optimale	11%	12%	10%	9,5%	8%
poids volumique sec (kN/m^3)	17,4	18,1	18,4	18,6	19,2
indice des vides	0,5	0,45	0,42	0,41	0,36

Tableau n°1 Essais Proctor sur le mélange sable-ciment

	sable+kaolin	sable+kaolin	sable+kaolin	sable+kaolin	sable+kaolin
ciment	0%	3%	6%	9%	12%
teneur en eau optimale	9%	9%	9,5%	9,5%	10%
poids volumique sec (kN/m^3)	19,6	19	20,1	19,6	20
indice des vides	0,34	0,38	0,31	0,34	0,31

Tableau n°2 Essais Proctor sur le mélange sable-kaolin-ciment

3.1.4. Ciment Portland artificiel CPA55
Le ciment Portland artificiel classe 55 présente des valeurs minimales garanties de résistance à la compression (essai normalisé sur mortier normal) à 2 jours et à 28 jours respectivement de 10MPa et de 45MPa
La présence d'eau au voisinage des grains de ciment est nécessaire à son hydratation qui dure plusieurs semaines. Pour hydrater le ciment, il suffit de 25% de son poids en eau environ, mais il en faudra beaucoup

 3. une quantité d'eau optimale pour assurer une bonne efficacité du compactage.
Nous avons retenu le troisième paramètre comme facteur prépondérant, les deux autres constitueront les bornes supérieure et inférieure permettant de choisir les teneurs en eau du mélange avant fabrication des blocs. A posteriori nous avons pu vérifier que l'hydratation était complète et que le démoulage est possible.
Les résultats des essais Proctor sur les mélanges de

sable-ciment et de sable-kaolin-ciment sont portés dans les tableaux n°1 et n°2.

Nous remarquons la faible teneur en eau optimale pour les forts dosages en ciment du sable sans kaolin ; nous constatons également une augmentation du poids volumique sec (compacité) avec la teneur en ciment du sable. Ceci peut s'expliquer par le fait que l'on a un sable propre ne comportant pas de fines, le ciment intervient avant la prise comme le feraient des fines, équilibrant la courbe granulométrique, ce qui permet d'obtenir un plus faible indice des vides.

En revanche, pour le mélange sable+kaolin la teneur en ciment ne semble pas beaucoup influer sur les teneurs en eau optimales et les poids volumiques secs, donc sur la compacité du mélange après compactage. La teneur en eau optimale reste voisine de 10%.

3.4. Mélange réalisé

Nous avons réalisé quatre mélanges pondéraux (voir tableau n°3).

Les valeurs du rapport E/(C+K) correspondent aux valeurs courantes utilisées pour les BCR (rapport poids d'eau sur poids d'éléments fins).

3.5. Cycles de gel-dégel

Après 10 jours de conservation des éprouvettes en salle humide à 20°C et à 100% d'humidité) les éprouvettes subissent des cycles gel-dégel rapides.

La moitié des éprouvettes est immergée dans l'eau lors du dégel, l'autre moitié est dégelée en salle humide.

Après 10, 15 et 20 cycles on mesure les variations dimensionnelles des éprouvettes avant de les soumettre à des essais de compression simple.

Les taux de gel et de dégel imposés sont très sévères (de l'ordre de 1°C/minute entre +20°C et -18°C) : l'apparition de succion cryogénique et de ses conséquences sont écartées, seule la gélifraction du matériau peut se produire.

4. RÉSULTATS EXPÉRIMENTAUX

4.1 Pertes de masse

On peut voir sur la figure n°2 que seules les éprouvettes dosée à 3% de ciment sont altérées de

ciment (C)	3%	6%	9%	12%
sable	87%	84%	81%	78%
kaolin (K)	10%	10%	10%	10%
eau (E)	8,5%	9%	9,5%	10%
E/C	2,8	1,5	1,06	0,83
E/(C+K)	0,65	0,56	0,50	0,43

Tableau n°3 Dosages des mélanges sable-kaolin-ciment réalisés

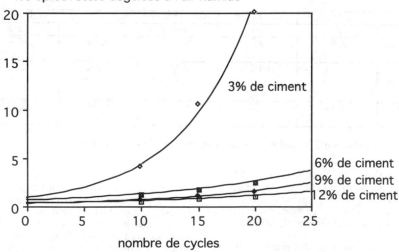

Figure n°2 : perte de masse des éprouvettes en fonction du nombre de cycles et du dosage en ciment.

façon sensible après 20 cycles de gel-dégel.
Les éprouvettes dégelées dans l'eau sont encore plus affectées que celles dégelées à l'air humide.

4.2 Résistance à la compression

Il est courant de relier l'age d'un béton à sa résistance par une formule du type :

$$Fcj = A \cdot Fc_{28} \cdot \log(j+1)$$

Fcj : résistance à la compression à l'age j jours

Les éprouvettes témoins nous permettent de tracer les courbes donnant la résistance à la compression en fonction du temps (figure n°3). On peut en tirer une valeur moyenne de A égale à 0.88
Pour le béton de granulat courant avec des dosages en ciment de l'ordre de 20% et des E/C de 0,45 à 0,50, les règles de calcul utilisées en France donnent une valeur théorique de A=0,685

A la suite d'une vingtaine de cycles gel-dégel on enregistre une diminution sensible des résistances en compression. Cette perte de résistance est plus marquée pour les éprouvettes dégelées immergées dans l'eau et faiblement dosées en ciment.
Nous remarquons sur les figures n°4 et n°5 que la perte de résistance à la compression est d'autant plus importante que le dosage en ciment est faible. Mais il

semble que pour les éprouvettes dégelées à l'air humide les pertes de résistance sont sensiblement égales pour des dosages en ciment supérieurs ou égaux à 6%. En outre les pertes de résistance des éprouvettes dégelées dans l'eau sont toujours supérieures aux pertes de résistance des éprouvettes dégelées dans l'air humide quel que soit le dosage en ciment.
Pour des conditions sévères de gel-dégel les éprouvettes dosées à 9 et 12% de ciment conservent des résistances en compression élevées, supérieures aux valeurs imposées par les normes françaises pour les briques de terres cuites ordinaires (12.5MPa).

4.3. Variations dimensionnelles

Les résistances mécaniques s'avèrant être suffisantes, nous avons alors vérifier que les variations dimensionnelles permettent l'utilisation dans de bonnes conditions de ce matériau en maçonnerie.
Les éprouvettes témoins conservées durant 28 jours à l'air humide présentent un retrait de 1% environ, ce qui peut être expliqué par le retrait classique dû à la prise du ciment. Ce retrait n'est pas mis en évidence pour les éprouvettes soumises à des cycles de gel-dégel. Un gonflement consécutif à la gélifraction du matériau existe donc et il est compensé par le retrait lié à la prise du ciment lors de nos essais de gélivité qui ont eu lieu alors que la prise n'était pas terminée.
Au delà du temps de prise il est raisonnable de s'attendre à un gonflement de gélifraction de l'ordre de 1%.

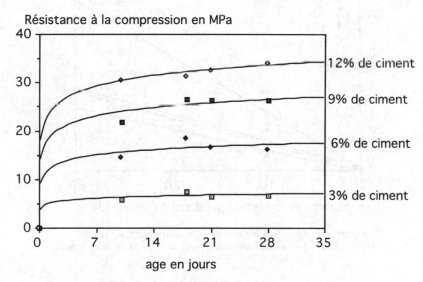

Figure n°3: évolution de la résistance à la compression en fonction de l'age des éprouvettes et du dosage en ciment.

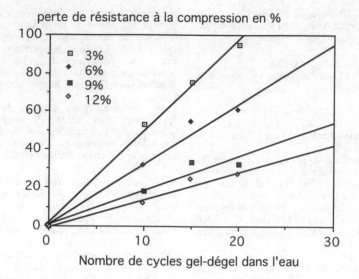

*Figure n°4: perte de résistance des éprouvettes en fonction
du nombre de cycles et du dosage en ciment pour des dégels dans l'eau*

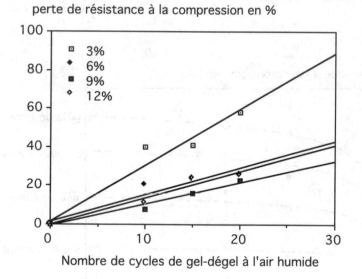

*Figure n°5: perte de résistance des éprouvettes en fonction
du nombre de cycles et du dosage en ciment pour des dégels en salle humide*

5. CONCLUSION

Le travail présenté avait pour but de tester la résistance au gel-dégel de blocs de construction en terre stabilisée fabriqués par compactage dynamique dans un moule rigide. Nous avons fabriqué des blocs à partir d'un mélange sable-kaolin-ciment. Ces blocs ont été soumis à des conditions de gel-dégel particulièrement sévères, à forts taux de gel et de dégel et en présence d'eau.

Pour des dosages en ciment de 9 et 12% nous avons enregistré une diminution des résistances à la compression de 30% seulement, ces résistances restant toujours supérieures à 20MPa.

Nous pensons que les briques fabriquées peuvent trouver leur utilisation dans des ouvrages de maçonnerie courante en extérieur comme en intérieur. Le développement de tels blocs réalisés à base de matériaux locaux et à faible coût énergétique peut permettre la réalisation d'un habitat économique en régions humide et froide moyennant des dispositions constructives adaptées.

REFERENCES

[1] DUVAL Roger, GERVREAU Eric, LUHOWIAK Wilhelm (1993) étude des paramètres de compactage pour la mise au point de blocs d'argile annales de l'ITBTP n°511 février 1993 pp 41-56

[2] GAGNE Richard, PIGEON Michel, AITCIN Pierre Claude (1990) durabilité au gel des bétons de hautes performances mécanique, Matériaux et Construction n°23 pp103-109

[3] BOUTONNET Michel, LIVET Jean (1984) Influence du traitement des limons sur leur comportement au gel, Bulletin de liaison des LPC n°133 sept-oct 1984 pp 83-86

[4] HANSEN W., KUNG J. H. (1988) pore structure and frost durability of clay bricks materiaux et constructions n°21 pp 443-447

[5] MONTARGES R., DUVAL P. (1990) comportement au dégel des sols silteux, Revue de l'IFP VOL 45/N°4 juillet-août 1990 pp 459-473

Triaxial testing of two limestone quarries

Essais de compression triaxial dans deux carrières de calcaire

S. Yıldırım

Yıldız Technical University, Istanbul, Turkey

ABSTRACT: Rocks have been used as a construction material in highway embankments and rockfill dams for a long time. In both cases the standard method of limit equilibrium analysis requires shear strength parameters to be incorporated in a stability program. The choice of these parameters no doubt has a great influence on the economy-stability balance and, in special cases such as those explained in this paper, may be the determining factor. Drained triaxial compression tests have been carried out on two limestone quarries, both considered to be suitable for concrete aggregate. It is concluded that the different behaviour observed in the tests might be related to the mechanical properties such as Los Angeles abrasion and water absorption values.

RÉSUMÉ: Les roches ont été depuis trés longtemps utilisées comme matériaux de remblai dans les travaux routiers. Il est nécéssaire, dans chaque cas, de choisir les paramètres concernant la résistance au glissement afin de les utiliser dans un programme d'ordinateur destiné à établir les calculs selon la méthode de l'équilibre à la rupture. Le choix de ces paramètres a une grande influence sur le bilan économie-stabilité et comme on explique plus loin, il devient un facteur déterminant. Des essais de compression triaxial ont été effectués sur deux roches calcaires provenant de deux mines différentes et étant considerées tous les deux convenables pour être utilisées comme agrégat pour le béton. On peut conclure que la différence des résultats obtenus dans ces essais peut s'expliquer par la différence de leurs caractéristiques de résistance à l'usure, selon la méthode Los Angeles, ainsi que de leurs valeurs d'absorption d'eau.

1 INTRODUCTION

Darlik Dam is a rockfill dam which supplies water to the Istanbul Metropolitan Area. The central core founded on sound rock with slopes 5 vertical to 1 horizontal. Two filter zones on both up and downstream were placed between the core and the shell consisting of limestone rockfill with external slopes 1 vertical to 1.8 horizontal founded again on sound rock over a length of 45 m on both sides. At the design stage a friction angle of 42° was assumed in the stability analysis; therefore, possible sources of rockfill had to be tested to verify the parameters.

It can be argued that plain-strain tests are more suitable techniques than triaxial testing unless the dam construction takes place in a narrow valley. However triaxial testing is simpler and results in 3-5° less friction angle; therefore, it is generally accepted that the results are on the safe side.

Two series of drined triaxial tests were undertaken on the samples from two possible quarries at the beginning of the construction. Rockfill used in the field contains fragments too large for the laboratory specimen; therefore, a method of scaling down the material in view

of the particle size must be adopted. In general two methods are in use:

1. Using laboratory grading which plot a line parallel to the field grading on a semi-logarithmic particle size distribution (Lowe 1964).

2. Removing the oversized particles determined according to the specimen diameter that could be tested on a particular testing rig (Zeller and Wullimann 1957).

The first method was adopted for the investigation explained in this paper, Figure 1.

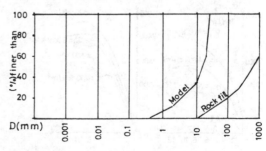

Figure 1. Model gradation

2 LABORATORY WORKS

At the time of this investigation the maximum size of triaxial specimen was of 150 mm diameter; therefore, the maximum size of the modelled aggregate was fixed to 25 mm, which was 1:6 of the specimen diameter. A lorry loaded with the possible quarry production was dumped at the construction site and the gradation of the rockfill was determined by weighing and approximately measuring the blocks; then a parallel was drawn to this curve as shown in Figure 1. The preliminary test showed that maximum unit weight would be obtained if the speciments were vibrated for a certain period at a moisture content of 5 %.

Having transferred to the triaxial machine the speciments were saturated by circulating water through the sample and finally the confining pressure was raised to its final value ranging from 200-800 kN/m² under drained conditions. The end of the consolidation period was determined as the time no air or water was expelled from the sample. The specimens were sheared under increasing deviator stresses at an axial strain rate of 1 % per hour. When the maximum deviator stress was reached, the test was stopped and the gradation was determined in order to find the amount of particle breakage.

In general, no difficulty was experienced in the test under 200-600 kN/m² confining pressures, but the problem of membrane-puncture was overcome by using double membrane in one of the series.

3. TEST RESULTS

Drained triaxial compression tests have been carried out on samples 150 mm in diameter by 300 mm in height. The samples were from two quarries with identical gradation and densities. The test results are shown in Figure 2 in the form of deviator stress versus axial strain. In both series a similar trend was observed: the higher the confining pressure the larger the maximum deviator stresses with increasing axial strains.

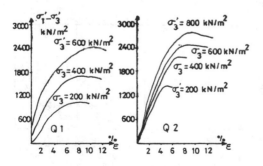

Figure 2. Triaxial test results

The effect of maximum diameter of modelled gradiation on the results has been investigated by several researchers (Marachi et al. 1969). They found that decreasing the maximum size of material from 150 mm to 12 mm resulted in increased angles of shearing resistance of about 4°. The same trend was observed for the cell pressures between 200 and 4500 kN/m². However, others (Tombs 1969 and Charles 1973) measured a decrease of friction angle by 2° when they decreased the maximum particle size from 75 mm to 10 mm on samples of 300 mm diameter. It appears that it is rather difficult to assess the relative influence of maximum particle size on the angle of friction.

It is a common point of view that the measured angle of friction decreases with increased confining pressures. Crushing particles must have an influence beside the depressed tendency for the dilatancy during the shearing stage. Maximum principal stress ratios at a failure are plotted in Figure 3 on a logarithmic scale, which also indicates the range of Marsal's results carried out on sand, gravel and rockfill (Marsal 1973).

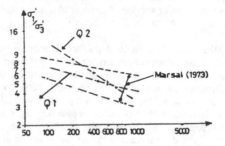

Figure 3. Principal stress ratios

The angle of internal friction expressed as:

$$\phi' = \sin^{-1}\frac{R_f - 1}{R_f + 1} \tag{1}$$

is shown in Table 1 below. R_f denotes principal stress ratio at failure.

Table 1. Measured angles of friction

Quarry	Confining pressure kN/m²	Friction angle degree
1	200	45.6
	400	42.6
	600	41.8
2	200	51.5
	400	46.9
	600	42.4
	800	39.5

It is clear that the measured angle of friction is markedly different at comparable confining pressures although both quarries were nominated as suitable for obtaining concrete aggregate. The other mechanical properties are shown in Table 2.

Table 2. Mechanical properties of quarries

Quarry	Los Angeles Abrasion %	Chemical Soundness %	Water Absorption %
1	38	2.5	1.9
2	29	3.0	1.0

The above values are averaged from 42 tests. It appears that Quarry 2 had distinctly better mechanical properties than Quarry 1 and abrasion value-water absorption had a great influence on the friction angle. The difference between the specimen gradations before and after the tests was similar in both series.

In Figure 4 the test results are replotted in form of Mohr-Coulomb failure envelope showing marked curvature in both series. Over cell pressure range of 200-600 kN/m², the shear strength can be expressed by the equation:

$$\tau = A(\sigma')^b \tag{2}$$

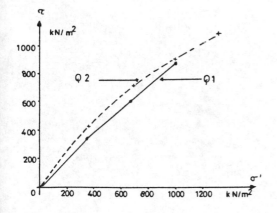

Figure 4. Curved failure envelopes

Table 3 lists the parameters together with Charles and Watts' results.

Table 3. Parameters defining the curvature of envelope

Quarry	A	b
Limestone 1	2.19	0.87
Limestone 2	8.30	0.68
Sandstone	6.80	0.67
Slate	3.00-5.30	0.75-0.77
Bazalt	4.40	0.81

The effect of the curved failure envelope can be demonstrated for a potential shallow slip plane under seismic conditions. It can be easily shown that the pseudo-static factor of safety against failure is:

$$F = \frac{\tan\phi'}{\tan\alpha} \frac{(1 - k\tan\alpha)}{(1 + k\tan\alpha)} \tag{3}$$

where
ϕ': Angle of friction
α: Angle of slope to horizontal
k : Seismic coefficient

In Darlik Dam the use of the minimum friction angle of 39.5°, measured for Quarry 2, would indicate a factor of safety as low as 1.08 with proposed seismic coefficient. However, a stability analysis considering the curved failure envelope would indicate:

$$F = \frac{A}{\gamma^{1-b} z^{1-b}} \frac{(\cos^2\alpha - k\sin\alpha\cos\alpha)^b}{k\cos^2\alpha + \sin\alpha\cos\alpha} \tag{4}$$

where
γ : Unit weight
z : Depth

Equation (4) reduces to that derived by Charles and Watts for the static cases if k is taken as zero. For a unit weight of 22 kN/m² and A,b values of Quarry 2, it can be shown that the factor of safety against sliding is well above 2 for depths within the first 10 meters.

4. CONCLUSION

The influence of confining pressure on the principal stress ratio has been demonstrated for small cell pressures which are expected for Darlik Dam with a height about 50 m. The results showed pronounced curvature at low stresses, which was in accordance with that of previous works. It appears that Los Angeles abrasion value and water absorption have a great influence on the measured values of friction angle whereas chemical soundness seems to be less influential. It is also concluded that

seems to be less influential. It is also concluded that rockfills may have greater safety factors under seismic conditions than those predicted by stability analysis omitting the curved shape of the failure envelope.

REFERENCES

Charles, J.A. 1973. Correlation between laboratory behaviour of rockfill and field performance with particular reference to Scammonden Dam. Ph.D thesis. University of London

Charles, J.A. & Watts, K.S. 1980. The influence of confining pressure on the shear strength of compacted rockfill. *Geotecnique* 30:353-367

Lowe, J. 1964. Shear strength of coarse embankment dam materials. *Proc. 8th Int.Conf.Large Dams* 3:745-761

Marachi, N.D.,Chan, C.K.,Seed, H.B. & Duncan, J.M. 1969. Strength and deformation characteristics of rockfill materials. *Report TE-69-5*. University of California-Berkeley.

Marsal, R.J. 1973. Mechanical properties of rockfill. *Embankment dam engineering*. Casagrande vol. Wiley. New York:109-200.

Tombs, S.G. 1969. Strength and deformation characteristics of rockfill. Ph.D. thesis. University of London.

Zeller, J.,Wullimann, R. 1957. The shear strength of the shell materials for the Goschenealp Dam. *Proc.4th ICSMFE*:399-404

Geology and strength properties of high-quality Precambrian crushed rock aggregate: An example from the Koskenkylä quarry, Finland

Propriétés géologiques et de résistance des granulats de roche concassés précambriens de haute qualité: Un exemple pris dans la carrière de Koskenkylä, Finlande

Hannu Mäkitie & Paavo Härmä
Geological Survey of Finland, Espoo, Finland

Kari Lappalainen
Finnish National Road Administration, Helsinki, Finland

ABSTRACT: Most of the high-quality crushed pavement aggregate at Koskenkylä quarry, southern Finland, is produced from a relatively homogeneous Palaeoproterozoic quartz-feldspathic schist, with a smaller amount from hypabyssal granite and various metavolcanites. Typical features of the felsic rocks are fine grain size ($\emptyset$ 0.05 - 1.0 mm), a variable degree of deformation and, in places, granophyric intergrowth. The strength properties (Los Angeles value, studded tyre value, point load strength index) of aggregate are presented as are the methods used for measuring them. The strength properties of the aggregate are among the highest ever reported in Finland. The very good durability of the quarried rocks is mainly due to the microtexture and mineral composition.

RÉSUMÉ: La plupart des granulats de pavage concassés de haute qualité de la carrière de Koskenkylä, dans le sud de la Finlande, sont produits à partir d'un schiste relativement homogène feldspathique quartzeux paléoprotérozoique avec une petite quantité d'un granit hypabyssal et de métavolcanites variables. Les caractéristiques typiques des roches felsiques sont la dimension des particules fines ($\emptyset$ 0.05 - 1.0 mm), la déformation variable et par endroits une formation à structure enchevêtrée granophyrique. Les propriétés de résistance (valeur Los Angeles, valeur cramponne, indice de résistance à la charge ponctuelle) des granulats sont présentées, ainsi que les méthodes utilisées pour les mesurer. Les granulats possèdent quelques-unes des propriétés de résistance les plus élevées qui aient jamais été signalées en Finlande. La très bonne résistance des roches extraites de la carrière est due principalement à la microtexture et à la conposition minérale qui sont décrites.

1 INTRODUCTION

The aggregate used for road pavements in Finland has to satisfy special requirements for quality owing to the severity of the Nordic climate and the fact that vehicles are fitted with studded tyres in winter. Wearing of road pavements caused by the increasing number of vehicles and the decrease in the availabity of glaciofluvial gravels have added to the importance of high-quality crushed hard rock aggregate. The contribution of rock aggregate to the durability of the asphalt pavement is about 60% (Saarela 1993). The durability of rock aggregate is mainly due to the microtexture and mineral composition of the rock. However, the geology and petrography of the bedrock of hard rock aggregate quarries have seldom been reported.

The hard rock aggregate quarry of Koskenkylä is in the parish of Pernaja, southern Finland, about 70 kilometres east of the city of Helsinki (Fig. 1). The quarry produces high-quality crushed Precambrian aggregate for road pavement. This study describes the petrography of the quarried rocks, the strength properties of the rock aggregate and the methods used for measuring them.

The strength properties (Los Angeles value etc) of the aggregate are among the highest ever reported in southern Finland. Since 1986 the quarry has produced 2 million metric tons of pavement aggregate with a market value of USD 3-4 / ton at the quarry.

2 GEOLOGY

2.1 *General*

The bedrock at Koskenkylä is part of the Palaeo-proterozoic, deeply eroded Svecofennian schist belt

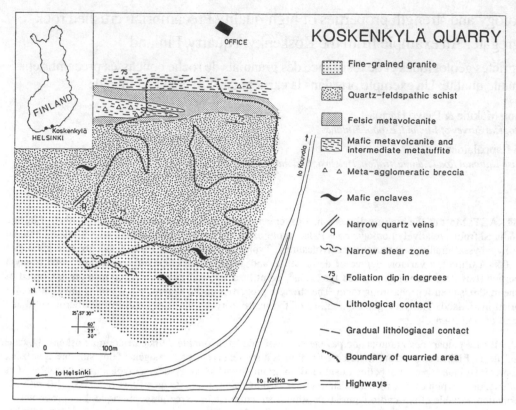

Fig. 1. Geological map of the hard rock aggregate quarry of Koskenkylä.

of southern Finland, with some associated orogenic granitoids (Laitala 1964, 1984). The anorogenic Rapakivi batholith of Wiborg lies 2 km to the east of the quarry.

The quarry area can be divided into fine-grained granitoids in the south, fine-grained, relatively homogeneous quartz-feldspathic schists in the middle, and felsic and mafic metavolcanites in the north (Fig. 1). The felsic rocks grade into plutonic rocks in the south, but often show sharp contacts or breccias with the mafic metavolcanites in the north. The metamorphic grade is that of amphibolite facies. The felsic rocks are fresh, containing a few 0.1 - 2 cm wide quartz veins and relics of mafic dykes or mafic enclaves in the south. Occasionally they are intersected by narrow (20 cm) pegmatite dykes.

2.2 Petrography

The granitoids in the south of the quarry are fine-grained (∅ 0.1 - 1.0 mm), homogeneous granites,

often with a reddish grey tint (Fig. 2). A characteristic feature of the rock is intense granophyric intergrowth (Fig. 3). By granophyric, we mean an intergrowth of quartz and alkali feldspar, which is roughly radiate or less regular than that of micrographic texture (e.g. Shelley 1993).

Quartz and plagioclase occur here and there as single larger grains (∅ 1 mm) surrounded by granophyric intergrowth or very fine-grained feldspar and quartz. The rocks also include granoblastic, microscopic portions composed of even-grained (∅ 0.5 mm) quartz and feldspar. The mafic minerals occur as single ragged hornblende (∅ 1 - 2 mm) crystals and clusters (∅ 1 mm) of very fine-grained biotites. In places hornblende has altered to actinolite and chlorite.

In a few shear zones, 1-3 m wide, granites are intensely foliated and resemble supracrustal banded schist. In the south-western most part of the quarry there are some fine-grained greyish granitoids, more or less granodioritic or tonalitic in composition. Two modal compositions are presented in Table 1 (ana-

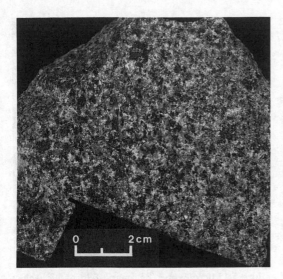

Fig. 2. Fine-grained granite in the south of Koskenkylä quarry, Pernaja.

Fig. 3. Granophyric intergrowth composed of quartz (Qtz) and K-feldspar (Kfs) in fine-grained granite. Koskenkylä quarry, Pernaja.

lyses 1 and 2). According to the IUGS classification (Streckeisen 1976), mots of the rocks are granites. K-feldspar has been distinguished from quartz and untwinned plagioclase by staining techniques.

The middle part of the quarry is occupied by relatively homogeneous quartz-feldspathic schists (Fig. 4), characterized by a lack of sedimentary or volcanic structures. The grain size of the schists is 0.1 - 1.0 mm and the degree of foliation varies. The foliation is often not very conspicuous, probably due to the competence of the rocks. A stronger degree of deformation produces microscopic bands composed of felsic and mafic minerals.

The schists contain some lobate mineral boundaries and undulatory quartz (Fig. 5). The grain size varies but a typical microscopic feature is that small granoblastic areas (∅ 5 mm) composed mostly even-grained plagioclase are surrounded by very fine-grained quartz and K-feldspar (Fig. 6). Hornblende forms single, often ragged (∅ 2 mm), grains and is in places partly altered to actinolite, epidote and chlorite. Biotite is very fine-grained, and occurs in places as clusters. The schists also contain a few very thin epidote-filled cracks. Three modal compositions are presented in Table 1. (analyses 3, 4 and 5).

The felsic metavolcanites in the north are very fine-grained (∅ 0.1 - 0.01 mm). Volcanic origin is shown by quart-filled amygdules. The rocks contain large grains of hornblende, but clusters composed of fine-grained biotite, hornblende and epidote are also

Table 1. Modal composition of quarried rocks. K-feldspar is probable partly a mixture of potassium feldspar and sodium feldspar.

	1.	2.	3.	4.	5.	6.
Quartz	35.6	27.1	29.3	43.4	25.3	33.1
Plagioclase	31.6	36.5	31.3	45.6	37.1	49.8*
K-feldspar	25.6	29.2	33.1	1.2	30.7	
Biotite	5.5	0.8	5.4	7.2	5.7	8.5
Amphibole	0.2	3.3	-	0.4	-	4.8
Chlorite	1.4	1.1	-	1.1	0.1	1.4
Epidote	-	0.3	-	0.1	-	1.0
Opaques	-	1.5	0.4	0.4	0.3	_
Others	0.1	0.2	0.5	0.6	0.8	1.4
Total	100	100	100	100	100	100

1. Fine-grained granite. Southern part of Koskenkylä quarry, Pernaja.
2. Fine-grained granite. Southern part of Koskenkylä quarry, Pernaja.
3. Quartz-feldspathic schist. Middle part of Koskenkylä quarry, Pernaja.
4. Quartz-feldspathic schist. Middle part of Koskenkylä quarry. Pernaja.
5. Quartz-feldspathic schist. Middle part of Koskenkylä quarry, Pernaja.
6. Fine-grained granitic rock (* = includes also K-feldspar). Middle part of Koskenkylä quarry, Pernaja.

Fig. 4. Quartz-feldspathic schist from the middle part of Koskenkylä quarry, Pernaja.

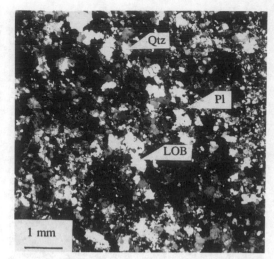

Fig. 5. Microtexture of quartz-feldspathic schist. Variations in grain size of quartz (Qtz) and plagioclase (Pl) and lobate grain boundaries (LOB). Koskenkylä quarry, Pernaja.

common. Near the mafic metavolcanites, the rocks have large epidote-filled inclusions (Ø 5 - 15 cm) indicating primary porosity.

The mafic metavolcanites in the north are composed of banded amphibolites with epidote inclusions (Ø 5 - 15 cm), layered intermediate tuffites and meta-agglomeratic breccias. There are also breccias of biotite- and hornblende-bearing schists surrounded by very fine-grained felsic rocks. Seen from a distance, these breccias on the rock wall resemble migmatites. In places it is difficult to classify and distinguish them from meta-agglomeratic breccias.

2.3 Chemical composition

One chemical whole rock analysis (XRF) (Table 2) was made on a sample taken from the boundary area between the southern granophyric fine-grained granite and the quartz-feldspathic schist in the middle of the quarry. The composition is granitic. The very low sulphur content is striking.

2.4 Genesis

The granophyric microtexture in the southern granites indicates hypabyssal crystallization. Granites emplaced near the surface crystallize more quickly and at low water pressures resulting in a complete

Fig. 6. Microtexture of quartz-feldpathic schist. Plagioclase (Pl) and quartz form small granoblastic areas surrounded by very fine-grained quartz and K-feldspar. Koskenkylä quarry, Pernaja.

solid solution of alkali feldspars. If volatiles are lost, independent crystals do not develop; instead, the simultaneous growth of quartz and alkali feldspar produces granophyric intergrowths (e.g Shelley 1993).

3196

Table 2. Chemical composition of a quarried rock at Koskenkylä quarry. Modal composition is presented in Table 1 (analysis 6).

	1.
SiO_2	71.74
TiO_2	0.36
Al_2O_3	13.63
Fe_2O_3tot	3.60
Mn	0.04
MgO	0.78
CaO	1.24
Na_2O	3.85
K_2O	4.04
P_2O_5	0.05
S	<0.01
Ba	0.08
Cl	0.01
Total	99.42

1. Fine-grained granitic rock from the boundary between quartz-feldspathic schist and fine-grained granite. Koskenkylä quarry, Pernaja.

Petrography and the gradual contacts of the felsic rocks indicate following genesis. The felsic rocks derive from the same hypabyssal magmatic body, which was emplaced at a relatively high level in the crust. The northern felsic metavolcanites with amygdules represent crystallization at the uppermost levels and low pressures. The granophyric inter-growths in the southern granitoids represent a deeper level and slower crystallization. The quartz-feldspathic schists with no sedimentary or volcanic textures in the middle of the quarry were formed under intermediate conditions.

The quarry is located about 2 km from the contact of Wiborg Rapakivi granite batholith, which has a thermometamorphic aureole, 5 km wide, in which the initial microcline was transformed into orthoclase owing to heat generated by the rapakivi granite (Vorma 1972). Here, the granophyric microtexture is interpreted as a result of primary magmatic crystallization. Some microtextures in the quartz-feldspathic schist resemble mylonites healed by thermal metamorphism (cf. Fig. 6).

3 QUALITY CLASSIFICATION AND STRENGTH PROPERTIES

3.1 The Finnish quality classification of road pavement aggregate

The quality of the mineral aggregate used in road pavements is given in terms of strength and shape classes determined from a sample (PANK ry 1993). The strength class is established from the results of the Los Angeles test, the studded tyre test (Nordic ball mill test) and the point load strength test (Table 3). Grain shapes are measured to determine the aggregate shape class (Table 4). The strength and shape classes depend on the weakest property. For locations subject to the most severe wear, the aggregate must be of the best quality, i.e. classes I A in strength and I in shape (PANK ry 1993). For comparison: aggregate crushed from coarse-grained granites in southern Finland usually have a strength class of about I D.

Table 3. Finnish strength classes of road pavement aggregate (PANK ry 1993).

Strength classes	Point load strength index (Is_{50})	Studded tyre value	Los Angeles value
I A	≥ 13	≤ 7	≤ 20
I B	≥ 10	≤ 11	≤ 25
I C	≥ 8	≤ 14	≤ 30
I D	≥ 6	≤ 17	≤ 35
II	≥ 4	≤ 30	-

Table 4. Finnish shape classes of aggregate grains (PANK ry 1993).

Shape classes	Elongation (c/a) fractions, mm 8-12	12-16	Flakiness (b/a) fractions, mm 8-12	12-16
I	≤ 2.5	≤ 2.3	≤ 1.5	≤ 1.4
II	≤ 2.6	≤ 2.4	≤ 1.7	≤ 1.6
III	≤ 2.7	≤ 2.5	≤ 1.8	≤ 1.7
IV	≤ 2.9	≤ 2.7	≤ 1.9	≤ 1.8

3.2 Point load strength test

In the point load strength test, a cylindrical core sample (diameter = 32 to 62 mm, length = diameter x (1.3.....1.9)) is compressed to failure between two conical metal points (TIE 241, 1992). The central angle of the conical points is 60°, and the radius of curvature of each point is 5 mm. The points must be of a hard alloy with a Vickers hardness of over 1200. Constant pressure is applied perpendicularly to the longitudinal axis of the sample to produce failure within 10-60 seconds. The force at failure (P) is recorded. Before measurement, the core samples are kept for at least 24 hours at room temperature (15-30°C) at a humidity of 60-80%. The result is rejected if failure occurs only on one side of the two points, or is caused by micro-fracturing due to a force significantly lower than that required by reference samples. At least 16 measurements are needed to calculate the point load strength index Is_{50}. For oriented aggregates there must be at least eight accepted recordings in the direction of foliation, and at least eight accepted recordings perpendicular to the foliation. There must be an equal number of accepted recordings in both directions. The point load strength index (Is_{50}) is calculated from:

$$Is_{50} = P / D^2 \ \text{x} \ (D)^{0.45} / (50)^{0.45}$$

where Is_{50} = the point load strength index (MPa), P = the maximum of the point load (N) and D = the distance between the conical metal points (mm)

The point load strength index (Is_{50}) is given as the average of the recordings. The value is obtained by deleting the two highest and the two lowest recordings and, then calculating the average of the remainder (not less than 12 recordings).

Finland is the only country that uses the point load strength test for evaluating the strength of road pavement aggregate. It cannot be used on gravel or crushed material; only on samples cored from solid rock.

3.3 Studded tyre test

The studded tyre test (= Nordic ball mill test) is employed to find the ball mill value K, which is used to assess the durability of the pavement aggregate (draft prEN 1097-9, 1993). The test subjects on aggregate sample weighing 1 kg to rotation with metal balls for one hour in a cylindrical mill. The sample contains grain fractions 11.2 - 13.2 mm and 13.2 - 16.0 mm in size. The quantities of

each fraction are calculated from the formula:

mass of the test sample = 500 g x density of the sample (g/cm³) / 2.66 (g/cm³)

The aggregate is washed and dried for the test. To reduce the effect of the grain shape value, both fractions are screened through a 5.6 mm harp sieve.

The cylindrical ball mill (219.1 x 6.3 mm), which can be opened at one end, has an internal diameter of 206.5 ± 2 mm and an internal length of 335 ± 1 mm. There are three steel skew-profiled raising beams, 8 mm high and 15 mm wide, on the internal circumference of the mill drum at equal distances from each others in a longitudinal direction.

The test is started by filling the mill with two aggregate fractions, 2 litres ± 10 ml of water, and 7 kg ± 10 g of ball bearings 15.0 ± 0.5 mm in diameter. The mill rotates at 90 ± 3 rpm. After 5 400 rotations the aggregate sample and the metal balls are washed on a 2 mm sieve, then dried. The test result is given as the percentage by weight of the aggregate finer than 2 mm as against the original sample. The ball mill value K is calculated from the formula:

$$K = 100 \ \text{x} \ (m_1 - m_2) / m_1$$

where m_1 = the mass of the original sample (g) and m_2 = the mass retained on a 2 mm sieve (g)

3.4 Los Angeles test

The Los Angeles test specifies a procedure for measuring the resistance to fragmentation of a sample of aggregate (ASTM C 131, 1989). In the test a sample of aggregate is rolled with steel balls in a rotating drum.

The sample contains grain fractions 9.5 - 12.7 mm (2 500 g) and 12.7 - 19 mm (2 500 g) in size. The total weight of the sample is 5 000 g ± 5 g. The cylindrical drum has an internal diameter of 710 ± 3 mm and an internal length of 508 ± 3 mm. An opening 150 ± 3 mm wide permits insertion and removal of the sample. The cylindrical inner surface is interrupted by a 100 mm wide projecting plate. The plate, which is rectangular in cross-section, lies in a diametrical plane, along a generating line. The total ball load, consisting of 11 steel balls, each with a diameter of 46 - 48 mm, weighs between 4 650 and 4 820 g. The drum rotates 5 000 revolutions at a speed of 30 - 33 rpm. When the rolling is complete, the aggregate sample is screened with a

1.6 mm sieve. The test result is expressed as the Los Angeles coefficient (LA), which is the weight (m) of aggregate finer than 1.6 mm. The Los Angeles coefficient (LA) is calculated from the formula:

$$LA = (5\ 000 - m) / 50$$

where LA = the Los Angeles coefficient and m = the mass retained on a 1.6 mm sieve (g).

3.5 Shape determination of aggregate grains

The shape of the grains, as measured from the aggregate sample, is given as the relative values c/a and b/a, described with thickness (a), width (b) and length (c) (Kannisto et al. 1979). The shape value is measured from grains either 8 - 12 mm or 12 - 16 mm in sieve. Each fraction has a different quality requirement. The shape value is determined from 100 random grains in the aggregate sample, by measuring thickness, length and width and calculating the averages. The shape value is obtained by dividing the average length (c) and the average width (b) by the average grain thickness (a). The ratio c/a is called the elongation, and the ratio b/a the flakiness. For isometric grains the axial ratio values are low, but for elongated grains they are high.

3.6 Strength properties of aggregate from Koskenkylä quarry

The average strength properties of aggregate from Koskenkylä quarry are listed in Table 5. The standard deviation is not high except for the point load strength index.

Table 5. Average strength and shape data on aggregate from Koskenkylä quarry. Standard deviation is in parentheses. Quality class after PANK ry (1993).

Test	Value		Quality class
Los Angeles	14.2	(1.3)	I A
Studded tyre	4.9	(0.4)	I A
Point load strength index (Is_{50})	15.8	(4.4)	I A
Elongation (c/a)	2.6	(0.3)	II
Flakiness (b/a)	1.7	(0.2)	II

4 ENVIRONMENTAL ASPECTS IN PRODUCTION

Environmental protection is an integral part of both quarring and production. Quarring is a dusty operation, and airborne dust is controlled and when possible eliminated by wetting the dust and the crushed aggregate at the primary crushing. There is no blasting or crushing at Koskenkylä quarry in winter, because the water used in the system would freeze.

Finland's Institute of Occupational Health has measured the amount and concentration of airborne dust at Koskenkylä (Pyy et al. 1993). At times the occupational exposure limits for total dust (10 mg/m^3) and quartz in fine-dust fraction (0.2 mg/m^3) were exceeded at the secondary crushing. The fibrous mineral concentration varied from 0.1 fibres/cm^3 to 0.8 fibres/cm^3, the average concentration being 0.2 fibres/cm^3. Most of the fibres were hornblende and actinolitic hornblende.

Even though the quarry is in an area of no important for groundwater supply, attention will still be paid to groundwater protection. The nearest house is at a distance of about 500 metres from the quarry.

5 CONCLUSIONS

The Finnish National Road Administration spends about USD 10 million every year on repairing road pavements (FinnRA 1990). Considerable savings could be achieved if high-quality rock aggregates were used in the construction of road pavements. Finnish research indicates that the durability of a road's asphalt pavement depends on the quality of the rock aggregate used (Saarela 1993). The quality of aggregate can often be slightly improved by sieving away the structurally damaged and badly-shaped final product size fraction produced by blasting and the first crushing stages (Heikkilä 1990). An asphalt pavement made of good-quality rock aggregate will last twice as long as one made of poor-quality rock aggregate.

The high strength properties of the felsic rock aggregate at Koskenkylä quarry are due to the characteristic features of the rocks: a fine grain size associated with a large amount of quartz, K-feldspar and plagioclase. The rocks, however, show slight variations in grain size, and the grain boundaries are, at least partly, lobate due to deformation. The mafic mineral content is low. These geological factors are typical of the durable rock aggregate, crushed from Finnish Precambrian bedrock (Uusinoka et al. 1990).

Quartz and feldspars are hard minerals that resist the abrasion of studded tyres. According to laboratory tests and findings from test roads, the content of mica, chlorite and other minerals with Mohs' hardness less than 3 should be under 30% in the rock (PANK ry 1993). These minerals weather easily and cannot resist abrasion. The proportion of these minerals in the felsic rocks at Koskenkylä quarry is less than 10%, and they are usually in fine-grained and ragged form. As a result the rock is not so brittle, as if it were composed entirely of quartz and feldspars. Moreover, there are no sulphides or other minerals with poor resistance to chemical and physical weathering. There should not be any pyrrhotite in aggregate and the content of other sulphides should be under 8% (Saarela 1993).

How can we find new rocks suitable for high-quality aggregate production? In the case of Koskenkylä quarry, the local bedrock map indicates the presence of granodiorite and quartz diorite (Laitala 1964), which occur abundantly and mostly as medium- or coarse-grained rocks in southern Finland. We should investigate the geological data for references to a fine grain size or intense deformation among rocks of suitable mineral composition. Felsic and mafic metavolcanites are often fine-grained and thus have potential for aggregate production. The strength properties do not always depend on the rock type, and the strongest rocks tested in Finland have been intermediate and felsic metavolcanic rocks. Because hard rock aggregate quarries occupy relatively small areas in a geological sense and the local geology can vary within short distances – as at Koskenkylä – accurate field observations should be made by a qualified geologist.

REFERENCES

ASTM C 131, 1989. Standard Test Method for Resistance to Degradation of Small-Size Coarse Aggregate by Abrasion and Impact in the Los Angeles Machine. Designation: C 131 -81 (reapproved 1987). *1989 Annual Books of ASTM Standards*. Sec 4. Vol. 04.02. p. 72-75.

Draft pr EN 1097-9, 1993. *Method for determination of the resistance to wear by abrasion from studded tyres: Nordic test*. (not published).

FinnRA, 1990. Finnish Road Statistics 1990. Transport. Finnish National Road Administration 1990. *Statistical Reports 5/1991*. 108 p.

Heikkilä, P., 1991. Improving the Quality of Crushed Rock Aggregate. *Acta Polytechnica Scandinavica*, Ci 96. 169 p.

Kannisto, P., Saarinen , L., Niemi, A, Partanen, E., Eerola, M. & Sistonen, M., 1979. Testing methods for aspfalt pavements (In Finnish). Asfaltti-päällysteiden testatusmenetelmiä. Tie-menetelmät. *Valtion teknillinen tutkimuskeskus / Tiedonanto* 50. Espoo. 226 p.

Laitala, M., 1964. Suomen geologinen kartta - Geological map of Finland 1:100 000. *Kallioperä-kartta - Pre-Quaternary rocks*, sheet 3021 Porvoo. Geological Survey of Finland.

Laitala, M., 1984. Pellingin ja Porvoon kartta-alueiden kallioperä. Summary: Precambrian rocks of the Pellinki and Porvoo map-sheet areas. Suomen geologinen kartta, 1:100 000. *Kallioperä-karttojen selitykset - Explanation to the maps of Pre-Quaternary rocks*, 3012 Pellinki, 3021 Porvoo. Geological Survey of Finland. 53 p.

PANK ry, 1993. *Quality classification of road pavement aggregates* (In Finnish). Asfalttinormit 94. Päällystealan neuvottelukunta. Helsinki. 58 p.

Pyy, L., Hartikainen, T., Härmä, P., Junttila, S., Korhonen, K., Suominen, V. & Tossavainen, A. 1993. Occurrence of fibrous minerals in limestone mines and rock aggregate quarries in Finland (In Finnish). Final raport. *Fund of Occupational Safety and Health*, Helsinki, 202 p.

Saarela, A. 1993. *Asphalt pavements. Part 1: Design*. ASTO, The Finnish Asphalt Pavement Research Program 1987-1992. Technical Research Centre of Finland. Espoo, 56 p.

Shelley, D., 1993. *Igneous and metamorphic rocks under the microscope*. Classification, textures, microstructures and mineral preferred orientations. Chapman & Hall. London. 445 p.

Streckeisen, A., 1976. To each plutonic rock its proper name. *Earth Sci. Rev.* 12: 1-33.

TIE 241, 1992. Testing methods for asphalt pavements (In Finnish). Asfalttipäällysteiden testausmenetelmiä. *VTT/ Tie- ja geo- ja liikennelaboratorio*, Espoo. 9 p.

Uusinoka, R., Ihalainen, P. & Peltonen, P., 1990. Petrographic and textural analyses of crystalline silicate aggregate for road pavement subjected to wear by studded tires. *Proc. 6th Int. Congr. IAEG*. vol. 4: 3231-3135.

Vorma, A., 1972. On the contact aureole of the Wiborg rapakivi granite massif in southeastern Finland. *Geol. Surv. Finland, Bull.* 255, 28 p.

Effect of sulfates and material finer than sieve 200 on the hardened cement concrete

Influence des sulfates et du matériel inférieur au tamis 200 sur le béton durci

Y. Abou-Seedah, A. Al-Harthi, M. Abousafiah & M. Al-Tayeb
FES, KAAU, Jeddah, Saudi Arabia

ABSTRACT : The engineering properties of the natural coarse and fine aggregates of Wadi Alaf, western Saudi Arabia, were found to be within the acceptable specification limits. However, the high percent of dust and the relatively high sulfate content in the groundwater, under arid conditions, suggested this investigation. The natural aggregate at Wadi Alaf and regular cement were used to prepare several concrete mix batches each with different percentage of magnesium sulfate. Other batches were prepared using different percentages of material finer than sieve #200. The uniaxial compressive strength of the hardened concrete decreases with the increase of the sulfate content. It also increases with the increase in the percentage of the material finer than sieve #200. The uniaxial compressive strength of the hardened concrete decreases with the increase of the sulfate content. It also increases with the increase in the percentage of the material finer than sieve #200 up to 7.5% beyond which the strength starts to decrease. The expansion and drying shrinkage increase with the increase of sulfate content, while in case of the material finer than sieve #200, the expansion and shrinkage are minimum when the percentage is around 2.5%. The study suggests that the percentage of the material finer than sieve #200 should be around 2.5% in order to improve the strength and to minimize the expansion and shrinkage of the hardened concrete. It also suggests the use of sulfate resistant cement in areas where the groundwater show possible sulfate attack.

RÉSUMÉ: Les propriétés géotechniques des agrégats grossiers et des agrégats fins du Wadi Alaf qui est situé à l'ouest de l'Arabie Saoudite sont trouvés dans les limites acceptables des spécifications. Le pourcentage élevé de limon et relativement celui de sulfate dans les eaux souterraines sous des conditions arides a suggéré ces examens. Les agrégats naturels du Wadi Alaf ont été usés pour préparer des mélanges différentes de béton chacun avec du pourcentage différent de sulfate et de magnesium. D'autres mélanges ont été préparés utilisant des matériaux plus fins que le tamis 200. La résistance en compression uniaxiale du béton duré baisse avec l'augmentation du sulfate. Elle augmente aussi avec l'augmentation du pourcentage des matériaux plus fins que le tamis 200 jusqu'à 7,5 %, et à partir de là elle commence à baisser. L'expansion et la rétraction augmentent selon le contenu de sulfate. Dans le cas des matériaux plus fines que le tamis 200 l'expansion et la rétraction sont minimales pour le pourcentage environ 2,5%. Cette étude indique que le pourcentage de matériaux plus fins que le tamis 200 doit être environ 2,5% pour développer la résistance et baisser l'expansion et la rétraction du béton durci. Elle suggéré aussi l'utilisation du cément résistant aux sulfates dans les régions où les eaux souterraines font prévoir d'attaque possible par les sulfates.

1 INTRODUCTION

1.1 General

The sulfate present in soils, aggregates and water are know to have caused deterioration of many concrete structures. For example the reaction (chemically) of sulfates with different materials in the cement paste to form calcium sulfates and/or calcium sulfoaluminate, these reactions being accompanied by considerable expansion and disruption of the concrete (ACI, 1979).

The deposition of sulfate crystals in the pores of concrete also tends to disintegrate concrete. Alkali waters entering concrete may evaporate and deposit their salts. The growing crystals resulting from alternate wetting and drying may eventually fill the pores and develop pressures sufficient to disrupt the concrete.

The presence of materials finer than sieve #200, which are usually form of clay and plastic fines are not acceptable when they exceed a certain limit ranging from 1-5% depending upon the size of aggregates. (ASTM C-33). These materials may produce spots of weak zones within the hardened cement concrete and they require a large amount of water. Also, the presence of these materials may weakening the strength of the cementations between larger particles.

In the present study, the effect of sulfate contents (with no material finer than sieve #200) and the material finer than sieve #200 (with zero sulfate contents) on the hardened cement concrete were examined. The aggregates of Wadi Alaf at the western province of Saudi Arabia were selected because huge quantities of aggregates are utilized daily for either road and concrete construction or as filling materials in Jeddah and Makkah Cities.

1.2 Location and Geology

Wadi Alaf is located in the central part of the Arabian Shield which consists of Precambrain igneous and metamorphic rocks. It is located about 115 km North - east of Jeddah city as shown in figure 1.

Field investigation and petrographic studies pointed out that, the studied area consists of diorite, basalt and rhyolite with some minor granite.

2 PREPARATION OF TEST SPECIMENS

Ten concrete mix batches were prepared to determine the effect of sulfates and the materials finer than #200 on the hardened cement concrete. The sulfate contents for five of the mixes were 0,2.5,5,7,.5 and 10% of the mixing water. The material finer than sieve No.200 for the

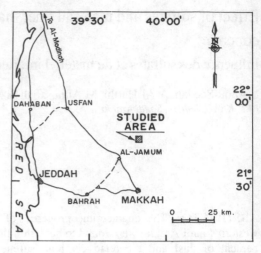

Figure 1   Location map showing the studied area

other five mixes were chosen to be 0,2.5,5,7.5 and 10% of the weight of sand.

A total of one hundred eighty (180) samples from the ten mix batches were made, eighteen mold samples were made from each mix batch, twelve of them were cubic molds (10 x 10 x 10 cm^3). The other six were cylindrical molds (196.35 cm^3).

It should be noted that all of the mix batches were designed to have a maximum uniaxial compressive strength of about 300 kg/cm². The strength under uniaxial compression and the change of length of the hardened cement concrete are main tests conducted on this research.

3 EFFECT OF SULFATES

3.1 Strength under uniaxial compression.

The relationship between compressive strength, curing time (days) and sulfate contents is given in figure 2. It can be seen that, the strength increases with the curing time due to the hydration of cement paste and then the strength starts to decrease with time due to the presence of sulfate. The reduction in strength will increase with the increase of percentage of sulfate contents (fig. 2).

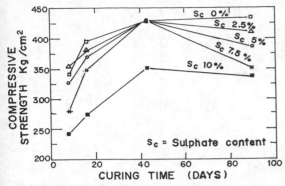

Figure.2 Relationship between strength under uniaxial comparession, curing time and sulfate contents.

3.2 Change in length of the hardened cement concrete.

3.2.1 Expansion

There cylindrical specimen of each sulfate concentration were placed in a humid chamber and the expansion was measured at non fixed interval period. Test results are given in figure 3. The expansion increases with time and also the expansion increases with the increase in the percentage of sulfates. After a limited period of time the expansion stops due to the complete formation of crystalline materials and calcium silicate gel.

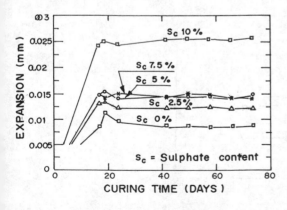

Figure 3 Relationship between sulfate contents, curing time and expansion.

3.2.2 Shrinkage

Three cylindrical specimen of each sulfate concentration were placed to dry in a room temperature (20°C). The drying shrinkage was recorded at non fixed interval period. The results of this test were shown in figure 4.

It was found that, the drying shrinkage increases with time and also with the increase in sulfate content except for 2.5% and 7.5% sulfate contents were drying shrinkage decreases. The drying shrinkage is steady after a limited period of time due to the compete settlement of the solids and the bleeding of the free water.

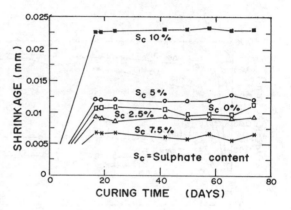

Figure 4 Relationship between sulfate content, curing time and drying shrinkage.

4 EFFECT OF MATERIALS FINER THAN SIEVE NO.200

4.1 Strength under uniaxial compression

The strength of twelve specimen from each mix batch was determined at intervals of times. Results are given in figure 5.

It was found that, the strength increase with the increase of percentage of materials finer sieve No.200 and with the increase of curing period up to 7.5%, beyond which the strength starts to decrease. The gain of strength with the presence of filler up to 7.5% might be due to the filling of voids, beyond this the clay particles starts to form a zone of weakness.

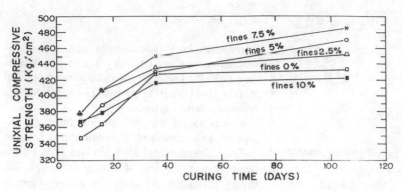

Figure 5   Relationship between strength under uniaxial compression, curing time and percentage of materials finer than sieve No. 200

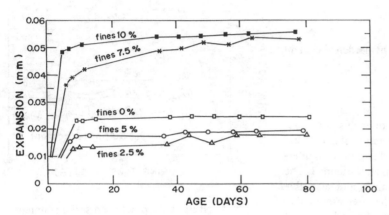

Figure 6   Relationship between percentage of materials finer than sieve No.200, curing time and expansion.

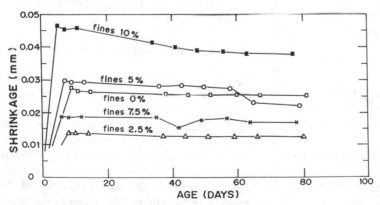

Figure 7   Relationship between drying shrinkage, curing time and percentage of materials finer than sieve No.200.

4.2 Change of length of the hardened cement concrete

4.2.1 Expansion

Three cylindrical expeceimens of each mix batch having different percentage of materials finer than sieve No.200 were placed in a humid chamber, the expansion of the concrete was measured at non fixed interval of times. Results are plotted in figure 6.

It can be seen from this figure that, the expansion increase with time and this expansion is minimum when the percentage of materials finer than sieve No.200 is 2.5% up to 5% beyond which the expansion increases.

4.2.2 Shrinkage

Three cylindrical specimen of each mix having different percentage of materials finer than sieve No.200 were placed in a room temperature (20°C). The drying shrinkage was recorded at non fixed interval period. Test results are shown in figure 7.

It was observed that, the drying shrinkage increases with time and their will be minimum drying shrinkage from 2.5% up to 7.5% beyond which the shrinkage increases.

It should be noted that the mix batch having 5% of materials finer than sieve No.200 shows abnormally up to 60 days of curing.

5 CONCLUSIONS

This study revealed that, the engineering properties of the natural coarse and fine aggregates of Wadi Alaf are within acceptable specification limits. Except that, the percent of dust is relatively high. The sulfate (So_4) contents of the groundwater of Wadi Alaf were equal 0.5% in the year 1992 at the time of study this percentage had risen to 0.75% within one year.

This study shows that, the uniaxial compressive strength of the hardened cement concrete decreases with the increase of the sulfate content. It increases with the increasing percentage of material finer than sieve #200 up to 7.5% beyond which the strength starts to decrease.

The expansion and drying shrinkage increase with increasing of sulfate content, while in case of material finer than sieve No.200, the expansion and drying shrinkage are minimum when the percentage is around 2.5%.

This study suggests that, the percentage of the material finer than sieve No.200 should be around 2.5% in order to improve the strength and to minimize the expansion and drying shrinkage of the hardened cement concrete. It also suggests the use of sulfate resistant cement in areas where the groundwater show possible sulfate attack.

REFERENCES

American Concrete Institute (ACI) 1979.
    Manual of concrete practice Part 1.

American Society for testing and materials
    (ASTM) 1986. Copyright in Eosten, MD,
    U.S.A, Section 4, Volume 0.4.02.

Laboratory methods applied in predicting the durability of railway ballast in natural conditions of Finnish Precambrian bedrock

Méthodes de laboratoire pour prédire la durabilité du ballast de chemin de fer provenant des conditions géologiques de l'assise rocheuse précambrienne en Finlande

R. P. J. Uusinoka, P. E. Ihalainen & P. J. Uusi-Luomalahti
Tampere University of Technology, Finland

ABSTRACT: The shape properties of the tested ballast materials proved to have a great influence on the strength parameters. The grain shape of the aggregate seemed to depend both on the textural features of the rock and the number of the crushings. The observed influence of the shape properties can be substantial. This phenomenon is more important to the test results of the materials with a firm texture than to those with a weak texture. The influence of grain shape might even prevent from revealing the real strength of the rock, which sometimes may lead to an unnecessary rejection of some usable ballast materials.

RESUME: La forme des grains dans les ballasts étudiés exerce une grande influence sur les paramètres de durabilité. La forme des grains dans l'agrégat semble dépendre aussi bien des propriétés de la texture de la roche qu'au nombre des concassages. L'influence observée de la forme des grains peut être essentielle. Ce phénomène joue un rôle plus important dans les résultats obtenus en testant les matériaux durs qu'en testant les matériaux d'une texture tendre. L'influence de la forme des grains peut même empêcher de découvrir la durabilité réelle de la roche, qui peut quelquefois résulter à un rejet d'un ballast utilisable.

1. INTRODUCTION

The Precambrian bedrock of the Fennoscandian Shield consists of hard and unweathered crystalline rocks. Numerous local variations in composition and strength, however, often weakens the quality of rock aggregate, such as railway ballast. In Finland the rock aggregate quarries are most generally small and often owned by private persons or small private companies. Thus, due to the difficulties in arranging a great-scale preliminary crushing, reliable results from the preliminary investigations concerning the quality of the material are claimed. Finnish State Railways, as the only purchaser of railway ballast in Finland, can thus practically independently determine the quality standards of this material.

2. METHODS OF TESTING THE DURABILITY OF RAILWAY BALLAST

The durability of the railway ballast is determined according to the properties like abrasion and impact resistance as well as brittleness. Combination of the test methods usually consists of the determination of the Los Angeles value and certain variations of wet abrasion tests (see Selig & Boucher 1990).

In Finland the criteria of the durability of railway ballast are based on the brittleness and the abrasion resistance of the rock material (Turunen 1986). The brittleness of rock is determined by Swedish Impact Test and the abrasion resistance either by the dry scouring test or the Nordic Ball Mill.

The Swedish Impact test measures the resistance of rock aggregate to crushing due to repeated weight-drop impacts. The mass of the drop-weight in the equipment is 14 kg and the drop-height is 0.25 m. The particle size of the sample tested is 8 - 12 mm. The percentage of the tested material passing through 4 mm sieve represents the Swedish impact value (see Heikkilä 1991).

Abrasion tests are used for estimating the resistance of rock aggregate to scraping and mechanical abrasion. The test has been carried out in Finland by scouring a special preparate on a rotating cast iron disc. The preparate consists of aggregate mounted on a plate of sulphur plaster. Aluminiun-titaniun oxide is used as an abrasive. The volume of rock material worn out of the grains during the test represents the abrasion resistance of the rock.

In the near future, however, the abrasion resistance will be determined by a testing method called Nordic Ball Mill. In this method about 1 kg of aggregate material is milled in a steely cylindrical mill with a diameter of 206 mm and length 335 mm. The rotation axis is horizontal and the number of revolutions is 5400 at 90 rpm in one determination. Small (ø 15 mm) steel balls (7kg) and water (2 kg) are used as the abrasive agents.

The samples for laboratory tests are most commonly detached from bedrock by light blasting. In some cases core drilling has also been used. The materials needed for tests are prepared by crushing the boulders or core samples in laboratory using a two-phase crushing procedure.

3. RELIABILITY OF THE LABORATORY TESTS

Initiative procedures in estimating the durability of the railway ballast include the observations on the comminution of the grains in some of the tracks whose cumulative traffic load as well as origin of the ballast material are known. By comparing the degree of comminution between different ballast materials and expressing the results in relation to the cumulative traffic loads of the corresponding tracks one could discover the durable and less durable ballast materials. By determining the strength parameters of corresponding virgin ballast materials the testing methods described above could be ranked according to each ones reliability in estimating the durability of ballast.

The reliability of the strength determinations themselves was examined by applying methods to produce and process the test aggregate. The influence of these applied producing methods on the test results as well as on the deviation of the results were also examined.

Two aggregate materials from Central Finland were herewith selected for examples, the one belonging to the most durable rocks in Finland, i.e. a fine-grained metamorphic volcanite (uralite-porphyrite) from Riitiala and the other in turn belonging to the worst material still accepted for ballast, i.e. a medium-grained granodiorite from Siitama. The latter belongs to the most common rocks in the Fennoscandian Shield.

The shape of the grains obtained from laboratory crushing depends on the textural features of the rock, the grain size of the crusher feed, and the number of crushings used in producing the test aggregate. The shape properties of the aggregate material used in the laboratory tests proved to be an important factor in evaluating the significancy of the numerical values of the strength parameters. To find out the role of the grain shape the rock materials were crushed by one-, two- or four-phase crushing method. The feed consisted either on blocks blasted from bedrock or virgin ballast material. The crushing method used in producing material for the strength tests proved to have a significant influence on elongation and flakiness of the individual grains whose average shape approaches to a cubical or isomorphic form with the increase of the number of crushing cycles. According to numerous analyses, it is obvious that the shape of aggregate used in the tests has a great influence on the strength parameters (Uusi-Luomalahti & Ihalainen 1994).

The average grain shape of the test aggregate seemed to depend on the textural features of the rock and the number of the crushings. The size of the feed had an influence on the grain shape practically in the one-phase crushing only. The shape properties were changed towards isomorphic form by riddling of the test aggregate. The influence of riddling on the strength parameters is shown in Figs. 1 and 2. As to the Riitiala volcanite, by using the four-phase crushing of the ballast feed the result of the Swedish Impact Value, determined from riddled material, improved even 70 % as compared with material of boulder feed crushed with one-phase process and determined without riddling. The corresponding improvement of the Swedish Impact Test Value was 50 % in the case of Siitama granodiorite. By replacing the Swedish Impact Value with the Nordic Ball Mill the percentages of the same improvements were 50 % (Riitiala) and 40 % (Siitama), respectively.

According to numerous analyses, it is obvious that like the impact value also the Nordic Ball Mill value will have a significant correlation with the values of elongation and flakiness of the aggregate tested (Figs. 1 and 2). The influence of the shape properties on the results of the Ball Mill test is, however, less important, because the aggregate will be riddled before testing, which procedure gives a slightly more isomorphic shape to the test material as compared with that used in the Swedish Impact test.

SWEDISH IMPACT VALUE

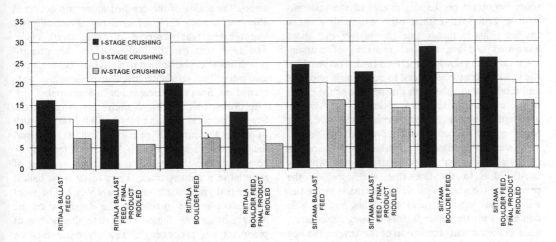

Fig. 1. The influence of feed size, riddling, and the number of crushings on the Swedish Impact Value

NORDIC BALL MILL VALUE

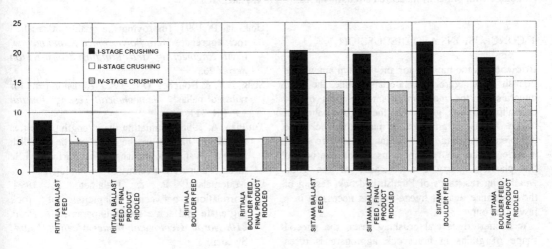

Fig. 2. The influence of feed size, riddling, and the number of crushings on the Nordic Ball Mill Value

The influence of the shape properties is relatively more important on the test results of the materials with a firm texture than on those with a weak texture. This is mostly due to the fact that more elongated and angular final products of crushing were rather obtained from a firm feed. Furthermore, the weakest components tend to comminute into dust and disappear together with the increase of the cycles of the crushing of the weak rocks. This is why the apparent firmness of the weak rocks improves together with the number of the cycles of crushing.

The shape properties had no influence on the results of dry scouring (=abrasion resistance) test, which is mostly due to the fact that while making the preparates the grains with elongated shape properties will be automatically ruled out. This method also contains a host of other factors causing uncertainty, such as the manual preparation of the sample, feeding rate of the abrasive, and relative humidity of the air in the testing room.

The reliability of Swedish Impact Test as well as Nordic Ball Mill Test proved to show a remarkable increase, if the shape indices of test aggregates are either standardized before testing or mathematically combined with primarily measured strength values.

4. CONCLUSIONS AND DISCUSSION

Brittleness is the most important problem among the Finnish rock aggregate materials, while the hard minerals, as the most important members of the crystalline rocks, guarantee their good abrasion values. In case of gneissose or migmatitic rocks, the complete sampling and testing procedure must include all the main components of the bedrock of the planned quarry area. Because of the good weathering resistance of Finnish bedrock, testing of the resistance against freeze-thaw is necessary in a few cases only.

According to several crushing tests, the general shape of grains in firm rock aggregate is often elongated and angular, which in turn tends to deteriorate the test results. Quite the reverse, crushing the weaker rocks will produce cubical aggregate material with strength parameters better than deserved. Thus, according to the results of the preliminary tests, a firm aggregate might possibly be ruled out, although it might be usable for ballast as a coarser-grained product crushed with another method.

The greater shift toward an isomorphic shape among the grains of the crushed aggregate occurred after the second cycle of crushing. Practically no improvement was observed after the fourth cycle. The feed size had no influence on the strength properties after the second cycle. These observations guarantee the reliability of the preliminary tests carried out from the processed ballast material.

According to the present study, the values of the Swedish Impact Test and Nordic Ball Mill suit well to predict the durability of railway ballast. However, the physical backgrounds of laboratory tests never completely coincide with the stress and wearing of material in a railway track. Tests made from ballast with a real grain diameter will probably be the most convenient way in predicting the durability of the material in the near future. Otherwise, the method of producing and processing of test aggregate must be determined so strictly, that all the non-essential factors in testing procedure do not hide the determination of the real strength parameters of rock material.

REFERENCES

Heikkilä, P. 1991. Improving the quality of crushed rock aggregate. *Acta Polytechnica Scandinavica, Civil Engineering and Building Construction Series* 96.

Selig, E.T. & Boucher, D.L. 1990. Abrasion tests for railroad ballast. *Geotechnical Testing Journal* 13: 301-311.

Turunen, A. 1986. Evaluating the strength properties of railway ballast on the basis of the Swedish Impact Test (in Finnish). Unpublished M.Sc thesis, Tampere University of Technology.

Uusi-Luomalahti, P. & Ihalainen, P. 1994. Correlation between elongation of rock aggregate and some strength parameters. *Proc. 21th Nordic Geological Winter Meeting*, Luleå, Sweden.

New highly efficient construction materials from industrial wastes
Matériaux de construction nouveaux et très efficients provenus des déchets industriels

V. A. Mymrin
Scientific Centre for Engineering Geology and Environment, Russian Academy of Sciences, Moscow, Russia

ABSTRACT: A series of slag-soil materials for construction of road and airfield bases, dams and foundations has been developed and patented. The materials consist of local soils or burnt coal enrichment process wastes, of dumped ferrous metallurgical slags , wastes of engineering works, dry or washed wastes of thermal power stations, wastes of chemical, textile, petrochemical, paint and varnish, and other industial wastes. The construction of more than ten roads during almost 20 years in different climatic regions of the former USSR show that all the materials are technologically effective, ecologically clean, reduce the cost of road base construction by 4-5 times, and increase construction rate in 2-3 times, the strengt of them continously improves during long period of exploitation.

INTRODUCTION

It is a well known fact that national economy of former USSR has urgently to build about 2.5 million kilometers of roads. Even if expenditure of inert materials per one kilometer of road is 5 thousand tons total requirement comes to at least 12 million tons. Extraction of such quantity of sand, crushed stone, sand and gravel mixtures,etc. from the guarries willbring the country to an ecological disaster and delivery ofeven one half of these materials by railway will require about 125 million railway cars, which will overload and ultimately destroy the main transport network of the country.

From another side the continuous growth in the output of different industrial wastes create grave ecological problems and should be gradually eliminated.

1 THE AIMS, SUBJECTS ANDMETHODS OF THE RESEARCH

1.1 The aims of the research:
1.working out new economically effective building materials of the founations of roads, airfields, dams, foudations ofseveral storeys building and sites for various purposes.For making thesematerials industrial wastes are used, mainly as nontraditional binding substances;

2.working out genuine wasteless technologies and ccmprehensive approach to ecological improvement of industrial regions by utilization of a wide range of liquid and solid wastes which previously were not used at all or used on small scale for construction purposes;

3.to research the processes of chemical interaction of initial components with the help of modern phisical and chemical methods.

In the process of achieving the latter goal it has been established that there existed a contradiction between experimental actual datas and generally accepted theory of strengthening binding materials by synthesis and mutual ingrowing of newly formed crystal substances, reinforcing the volume of the samples. Therefore the necessity arose

4.of elaborating a theory of reinforcement of materials under study by synthesizing new gel substances which turn into stone-like matter as a result of syneresis.

1.2 Subject of research

In the course of almost 30-years-long research the most typical dump slags of ferrous metallurgy (blast furnace, open-hearth furnace, converter, electric steel smelting and foundry), mixtures of machine engineering wastes(foundry slags and burnt foundry sand), ashes and slags of thermal power industry,beginning with active ashes of oil shales and up to non-active ashes of hydro-removal after burning Donetsc antracite were studied.

Objects of reinforcement were burned rocks in coal production dumps, soils from different geographical, geobotanical and climatical belts of the USSR. They greatly differed in genesis, granulometric, microaggregate, chemical and mineral compositions, structure, etc.

As activators of chemical interaction of ground and industrial wastes were used both liquid and powdered forms of wastes of various plants (chemical,petrochemical,machine engineering, textile, varnish and dyes, etc.)

Non-ferrous slags almost were not such subjects because of our firmness in soon technological improvements of non-ferrous processes. In such a case modern dumps became deposits of non-ferrous metals.

This publication does not include positive results of utilization of phosphogypsyms since they have not yet been patented.

1.3 Methods of researchs

In studying composition of initial components and temporalchanges of composition of newly formed substances during hydration,activation and interaction a wide rang of traditional and latest methods of research was used,wich complemented each other. Namely:difining the limit of strenght in uniaxial compression and rupture by the Brasilian method, temporal chandes of moisture and linear deformation, water and frost resistance, roengeno-phasial analyses by method of powder, the quantity of bound water and content ofcarbonates by TG and DTA, scanning electron microscopy and chemical analyses (free CaO, SiO_2, Al_2O_3, Fe_2O_3 non-bound SO_3, pH and others), infra-red spectroscopy, roentgeno-spectral analysis on "Cameca", "Edax" and "Link system" and lazer micromass analysis on "LAMMA-1000".

1.4 Requirements of Russian standards

The table below shows the requirements of russian standards to building materials from strengthening soils.

Features	Grade of strenght 1 2 3 after 90 days		
Strenght of water saturated, MPa	6-4	4-2	2-1
Frost resistance coeff., at least	0,75	0.70	0.65

2 DISCUSSION AND NEW OBSERVATION

As a result of these studies, for the firsttime in the world practic the possibility was proved of usin mixtures of foundry wastes(burnt sand andslag) as high-strenght materials of road bed. For this purpos material was selected from the dum of machine engineering plant in Nizhni Novgorod. Local waste of petrochemical production was used as a hardening activator. After 90days strenght of the samples reached 6 MPa, that is the upper limit of strenght of reinforced so by Russian standards (see table), and after a year it almost doubled The materials are highly water and frost resistant, their coefficient often exceed 1.

Even better indices of strenght, water and frost resistance were obtained on the basis of mixtures, similar in composition, of machine engineering plant in Bratsk, Irkut region of Siberia. Agolution of chmical waste of this plant was used as an activator. After 90 days the material exceeded the upper limit of strenght of the 1st degree(4-6 MPa) and after a year it more than doubled. Water and frost resistanc of the materials makes it possible torecommend for using in upper lay ers of road covers.

Slag-soil materials, consisting of local soils and dispersed dumpe slags of different processes, hav different rates of reinforcing depending on chemical and mineral compositions. To study processes of their hyd-ratration and chemical interaction of all components we had to use the whole set of the above mentioned methods.It was established that slag-soil mixtures become reinforced due to synthesis of sol porous solutions, their transformation into gels as a result of numerous stages of gel syneresis and transition to a stone-like state. Components of dissolved surfaces of both grounds (especially the thinnest clay-colloid fractions) and dump ferrous metallurgy slags took part in forming of gel's roengeno-amorphic new formations. A small quantity of carbonates crystals appears on at later stages as a result of adsorption of CO_2 of the air by alkaline-earth ions of slag-soils.

Ash-soils, based on local soils and active ashes of oil shals, on the contrary, become reinforced due to synthesis of non-gel roentgeno-amorphous new formation, but thanks to formations of crystal bodies. The analysis with the scanning electron-microscope of new formations which are only seven days old seem to resemble strongly hypostatic gel forms while they are covert crystal masses, which is proved by roengeno-phase research. The most dispersible clay components of the soils are dissolved in the porous solution of the hydrated mixtures, forming chemosorption compounds on the surface of particles already at the earliest stages of structure formation. Multistage phase transition of new crystal formations and their mutual ingrowing and reinforcing of volyme of the samples determine high physico-mechanical properties of ash-soil meterials.

In reinforcement of loam of Ukraine by hydraulically removed ashes and slags of heat rower station a a considerable number was dicovered of both crystal and gel newly formed substances reiforcing materil. However, with relatively high strenght (4,5MPa) of 90-day-old samples the material lacks in high frost resistance. Nevertheless for southern regions of Russia and Ukraine they can be successfully used as foundation of roads.

At some metallurgical works, in view of limited areas for slag dumping, liquid slags are poured in one place. Thus, mixtures are formed of blast furnace and steel smelting slags, the ratio of which is quite accidental. Such mixture from Lugansk (Ukraine) was used for stabilizing of dumped burnt coal mining substandard rocks. The solution of liquid wastes of nearest chemical plant was used as activator. After 90 days the strenght, water and frost resistance of the materils met the requirements of standards for the second (2-4MPa) and third (1-2MPa) grades of reinforced grounds. Like in the case of all slowly solidifying materials strenght continued increasing during a period of some years after. The strenght of activated slag-soils reaches 12 MPa by the 90th day, 15 MPa by the 3^d year and up to 19MPa bythe 9^{th}year.

3.PRACTICAL USE OF RESULTS

3.1 Road ctructure

The figure below shows for comparison the simplest conventional and the proposed road base structures:

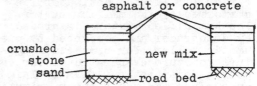

asphalt or concrete

crushed stone sand new mix road bed

Two layers (crushed stone and sand) are replaced by a layer of the new mix (slag-soil or industrial wastes mix) which is 10-20% thinner.

3.2 Road construction technology from new building materials

The technology of constuction of road bases with the use of the developed slag-soil material does not call for employment of costly special-purpose machines ormechanisms.

The construction of the road bases layers are made by mixing the soil (natural or technogeneous) with ferrous slags, activator and water directly on the road, or the mixture is prepared in stationary mixers.

3.2.1 The preparation of the mixture

in stationary mixers comprises: charging the crushed soil, slag and activator (powder or water solution) and water into the batching devices, mixing the components, transportation of mixture on preliminarily compacted road bed alongthe axe of the roadway. A roll of constant volume is made by means of a motor grader and uniformly spreads it. Compaction of mixture is performed by a self-propelled roller to a density of at least 98% of maximum, which is determined by the method of standart compaction of optimum humidity.

3,2.2 Technology of mixing the components directly on the road

comprises: soil to be reinforced is delivered to the planned and copacted earth road bed. The soil is unloaded along the axis, a grader makes a roll of constant volume and uniformly spreads it over the entire width. The slag-activator (if powder) binder is introduced into the crushed soil by two runs of milling cutter.

Simultaneously the humidification of slag-soil mixture to optimum humidity by water or activator solution (if liquied activator) is performed by the milling cutter. The humidified mixture is graded by autograder and compacted by pneumatic roller. For protection against moisture evaporation after compaction avaporatight film is laid on surface (molten bitumen, bituminous emulsion, etc.)

One or two layers of asphalt or concrete pavement are laid over new slag-soil road base materials.

3.3 Practical using of results

Construction and use of the airfield in Belorussia (1975), roads in Krasnodar(1976-78), Vologda(1980), Lipetsk(1983), Samotlor(Siberia,1987), Tynghiz(the Caspian Sea, 1989), Belgorod(1988),Oryol(1985), Bryansk (1988),Pervouralsk(1993) and other places have confirmed the high physico-mechanical properties of the materials produced. The period between repair work has been greatly prolonged due to many-years-long process of hardening of the materials.

The four-five fold reduction of costs roads construction is achieved thanks to the following factors:

1. use of cheap binding and activating materials(industrial wastes) reinforcing local soils.
2. three-five fold reduction of transported materials due to use of local soil and elimination of the layer of sand serving as means of drainage.

CONCLUSION

Replacement of natural materials by offered compositions reduces of materials transported at least three times, lowers the cost of transportation more than two times, improves quality and accelerates the rate of construction and prolongs the period of using roads without repair.

Considerable reduction in the level of noise during the use of road present thanks to higher elasticity of gel neoformations compared to the rigid usual materials.

However, the most important effect of this method are utilizing of long list of industrial wastes that contaminates the environment and reducing open-quarry extraction of natural building materials.

The high economic efficiency of material produced, which often exceed requirements of construction standards,along with acute need of national economy in road building, considering sky-rocketing prices of natural inert materials and the ever increasing urgency to solve ecological problems, allthis gives hope for wide-scale utilization ofindustrial wastes in the practice of construction in Russia. Parallel to this we are looking for foreign partners.

Le béton retaillable ou Pierre de La Possonnière – Cas particulier d'un matériau sculptable à base de liants hydrauliques

The sculptural concrete or Possonnière stone – The case of a sculptural material made of hydraulic binders

G. Martinet & H. Hornain
Laboratoire d'Études et de Recherches sur les Matériaux, Bagnolet, France

ABSTRACT : "sculptural" concrete, which can be described as a material made of hydraulic binders and natural minerals, offers the particular advantage of being able to be worked on, after hardening, whith the usual stone cutting tools. With its simple mixture composition and a consistency that allows for it to be placed easily, this material presents a homogeneous and cohesive microstructure. Despite high water and low binder contents, the characteristics of the fresh concrete are similar to those of a stable hydraulic grout without segregation. The properties of the hardened concrete, namely its color, texture and mechanical behavior are similar to those of soft limestone.

RESUME : issu d'un mélange à base de liants hydrauliques et de matériaux minéraux naturels, le béton retaillable possède la particularité de pouvoir être retravaillé à l'aide des outils classiques de taille de la pierre après durcissement. De composition simple et de mise en œuvre aisée, le matériau possède une microstructure homogène et cohérente. Malgré une forte teneur en eau et un faible dosage en liant, les caractéristiques du béton fluide sont proches de celles d'un coulis stable sans ségrégation. Le béton durci obtenu possède une texture, une couleur et des propriétés mécaniques proches de celles d'un calcaire tendre.

PRESENTATION GENERALE

Le matériau, objet du présent article, se caractérise par une composition simple à base de liants hydrauliques et de sables naturels, par une certaine souplesse de formulation et par une mise en œuvre aisée.

Toutefois, malgré sa rusticité, les créateurs du produit, les frères Cruaud, artisans tailleurs de pierre et sculpteurs, ont démontré que ce matériau était unique en son genre : une fois décoffré, il peut être retravaillé à l'aide des outils classiques de taille de la pierre, d'où son assimilation à une pierre.

L'utilisation de la Pierre de la Possonnière ainsi baptisée par ses inventeurs en référence au village du Maine et Loire où leur entreprise est située, devrait trouver de nombreuses applications : matériau porteur armé, revêtement de façade en béton, décoration intérieure, ect. De par son grain, sa couleur et son homogénéité le produit possède un aspect presque identique à celui des pierres calcaires abondamment utilisées en France. Ces propriétés devraient intéresser les architectes.

Plusieurs chantiers ont déjà été réalisés dans les villes de Nantes, d'Angers et de Paris : réhabilitation de bâtiments anciens, réalisation de parements sur des immeubles récents en béton, rénovation d'un ensemble de façades à Paris à proximité de l'Opéra Garnier.

La Pierre de la Possonnière a été récompensée par un premier prix au salon des Inventions, Techniques et Produits Nouveaux de Genève en 1992 et a fait l'objet d'un dépôt de brevet.

Messieurs Cruaud ont fait appel à plusieurs reprises au LERM pour préciser certaines des caractéristiques de leur produit dans le but de l'améliorer. Bien qu'encore incomplets, les résultats des différents essais effectués ont semblé mériter d'être rassembler dans un même article qui décrit la nature et la texture du produit, en précise l'intérêt et propose un programme d'étude complémentaire.

1. PRINCIPE DE FABRICATION

La fabrication du mélange est assez traditionnelle. Après préparation des ingrédients,

- conditionnement du liant en sacs. Ce liant peut être différent d'une formulation à l'autre : ciment blanc ou mélange chaux/ciment dans des proportions variant selon le temps de durcissement désiré et la dureté de la "pierre" recherchée,
- criblage et tamisage de la partie sableuse calcaire,
- choix des proportions liant, sable calcaire et sable siliceux,

le malaxage des différents produits se fait à l'aide d'un agitateur de forme hélicoïdale à grande vitesse. On notera que selon les proportions relatives des composants, le malaxage peut aussi être réalisé à l'aide d'une bétonnière. Il se déroule selon les étapes suivantes :

- malaxage eau + liant. Le rapport eau/liant (E/L), le plus souvent très élevé, peut être supérieur à 3,
- ajout de la charge siliceuse,
- ajout de la charge calcaire, constituant principal du mélange.

Après cette dernière adjonction, la pâte prend un aspect proche de celle d'un coulis. La viscosité du mélange est faible. Le malaxage est poursuivi quelques minutes afin de bien homogénéiser le mélange. A ce moment de la préparation, il semble que l'entraînement d'air soit important : formation de bulles.

Le produit ainsi préparé peut être coulé dans un coffrage soit directement, soit par pompage. Dans un délai de six heures à deux jours, le matériau peut être décoffré et travaillé à l'aide des outils classiques de taille de la pierre.

2. CARACTERISATION DU MELANGE PENDANT LA PRISE

2.1. *Rhéologie du mélange*

Après la mise en place et dès les premières minutes, un surnageant limpide mais néanmoins peu abondant apparaît : phénomène de ressuage. Le mélange se comporte comme un coulis hydraulique stable (Miltiadou 1991).

Le surnageant, réabsorbé par le matériau durci, a le plus souvent disparu au moment du démoulage.

Après malaxage le produit n'est affecté par aucune sédimentation. L'ensemble des particules du matériau y compris les fragments millimétriques de calcaire restent en suspension. Ce comportement est confirmé par l'observation du béton durci où aucune ségrégation n'est observée. Le mélange se comporte comme un corps thixotropique de type vase ou coulis bentonitique. La nature des composants et le système de malaxage semblent permettre la formation d'une solution colloïdale dont les particules fines englobent les débris plus grossiers. Les éléments responsables de la stabilisation du coulis pourraient être les fines créées lors du criblage de la charge calcaire.

L'analyse granulométrique laser montre que les calcaires utilisés sont constitués de particules dont le diamètre médian est voisin de 15 µm. Les micelles ou colloïdes sont des particules inférieures à 3 µm. Mais on note dans l'histogramme la présence de trois sous-classes de particules centrées sur 2 µm (dix pour-cent des particules sont inférieures à 1,8 µm), 5 µm et 40 µm (Fig. 1).

Les proportions notables de particules très fines du même ordre de grandeur que celles des micelles et des colloïdes sont probablement à l'origine des propriétés de la suspension. Ces fines sont vraisemblablement de nature minéralogique argileuse ou calcaire (calcite finement divisée).

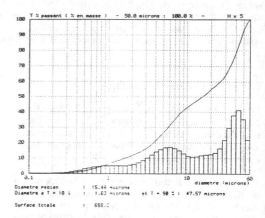

Figure 1 : Analyse granulométrique du sable calcaire

De plus, l'adjonction de carbonates dans des proportions telles que celles qui sont utilisées dans le mélange est susceptible d'avoir une incidence sur le pH de la solution qui peut dépasser une valeur critique (point isoélectrique) et entraîner la formation de colloïdes électropositifs ou électronégatifs (Aubouin & al. 1975). Il est possible que le mélange réalisé soit composé de particules exerçant entre elles des forces de répulsion électrique suffisantes empêchant la sédimentation.

2.2 Caractéristiques aux jeunes âges

La rhéologie des suspensions a été caractérisée par mesure du seuil de cisaillement (Couarraze, Grossiard 1983) pendant la prise à l'aide du rhéographe automatique du LERM (Bombled 1974).

Le principe de la méthode consiste à mesurer à des temps déterminés l'effort nécessaire à l'enfoncement à une profondeur constante d'une sonde cylindrique de diamètre déterminé dans le coulis en cours de prise. L'appareil est piloté par un ordinateur qui permet de calculer la courbe du seuil de cisaillement en fonction du temps entre l'instant initial et la fin de prise du coulis.

Les paramètres mesurés sont :

- la période d'induction, λ, (Fig. 2) pendant laquelle le seuil de cisaillement évolue peu : la pâte garde sa fluidité initiale. Les anhydres du ciment s'hydrolysent et les premiers hydrates apparaissent. Dans le cas présent, il semble que l'ettringite (tri-sulfo-aluminate de calcium hydraté) soit la phase la plus précoce. La période d'induction du produit testé ici est égale à 40 minutes ;

- les valeurs du début et de la fin de prise Vicat sont respectivement 6 heures et 11 heures. Les temps obtenus situeraient le matériau dans la famille des liants traditionnels à prise lente malgré la fluidité de la suspension et le rapport E/L très élevé utilisé (Papadakis, Venuat 1964) ;

- la vitesse de prise, caractérisée par le coefficient α, est liée à la vitesse de formation des hydrates : formation d'ettringite, d'hydroxyde de calcium $Ca(OH)_2$ et de silicates de calcium hydratés C-S-H. La vitesse de prise, exprimée en Pa/mn, est ici égale à 162.

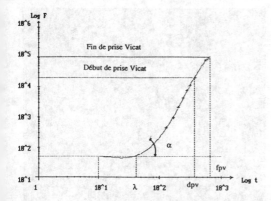

Figure 2 : Rhéologie du mélange aux jeunes âges

Au total, les mesures rhéologiques se traduisent par une courbe de prise tout à fait habituelle bien que le rapport E/L soit supérieur à 2,6 et que la quantité de liant hydraulique soit faible (10 % en volume). Aucune fausse prise n'est observée ; le comportement lors de la prise apparaît tout à fait satisfaisant.

Les résistances mécaniques en compression du produit ont été mesurées aux échéances de 2, 7 et 28 jours. Les résultats obtenus sont les suivants :

- 2 jours   3,0 MPa
- 7 jours   4,5 MPa
- 28 jours   6,0 MPa

L'évolution entre 7 et 28 jours apparaît assez faible. L'acquisiton des résistances finales du produit est plus lente que dans le cas d'un mortier normal. Après plusieurs mois la résistance de produits comparables au produit testé varie de 15 à 25 MPa selon la quantité de liant. Par ailleurs, la porosité totale du matériau mesurée par pesée hydrostatique est de l'ordre de 25-30 %.

3. CARACTERISATION DE LA MICROSTRUCTURE DU MATERIAU

3.1. *Examens microscopiques du matériau proprement dit*

L'examen, après attaques sélectives, de sections polies au microscope optique en lumière réfléchie montre que la matrice du matériau est composée de fins cristaux de calcite $CaCO_3$, d'hydroxyde de calcium $Ca(OH)_2$ à aspect persillé et d'un liant silicaté renfermant des résidus anhydres de ciment.

L'examen de lames minces au microscope optique en lumière transmise indique que le béton retaillable présente une texture homogène et une porosité générale élevée essentiellement représentée par des bulles d'air entraîné de dimensions variables entre 10 μm et 1 mm. La présence d'air entraîné est due au malaxage énergique réalisé lors de la préparation du mélange d'une part, et à la granulométrie étroite de la charge sableuse, d'autre part.

Les particules sableuses sont constituées de grains de quartz anguleux à sub-arrondis dont la taille est comprise entre 50 et 500 μm et de fragments de carbonates issus d'un calcaire oolitique. Il s'agit soit d'oolites libres de diamètre compris entre 0,5 et 1 mm, composés de cristaux micritiques disposés en couches concentriques organisées autour d'un nucléus de calcite spathique, de bioclastes ou de quartz, soit de fragments de calcaires polyphasés de dimension supérieure (≥ 2 mm) constitués d'assemblages d'oolites ou de foraminifères cimentés par une matrice sparitique.

La proportion sable quartzeux / calcaire oolitique dans la fraction granulats du béton retaillable est en relation avec la dureté recherchée du matériau.

Le liant est microscristallisé. Quelques résidus anhydres du ciment (C2S essentiellement) sont visibles en lame mince mais l'essentiel de la matrice apparaît formée de microcristaux de calcite provenant soit des fines calcaires soit de la carbonatation des silicates de calcium hydratés C-S-H et de la chaux $Ca(OH)_2$, composants habituels des pâtes de ciment. La diminution de pH provoquée par la carbonatation du liant hydraté serait éventuellement à prendre en considération lorsque des problèmes de corrosion d'armatures se posent.

L'examen au microscope électronique à balayage confirme l'homogénéité remarquable du matériau, sa bonne cohésion et sa micro et macroporosité notable (Fig. 3). Il montre par ailleurs que la liaison liant-sable est satisfaisante : la plupart des grains de sable sont totalement ou partiellement recouverts de cristaux de silicates de calcium hydratés précipités à la surface des grains et dont le faciès varie de fibreux à alvéolaire (Fig. 4).

De nombreux cristaux de calcite néogéniques associés aux C-S-H et à des cristaux massifs ou en tablettes de $Ca(OH)_2$, sont observés (Fig. 5). De plus, la présence de cristaux de carbonates dont la surface montre des figures de dissolution nettes (Fig. 6) semble indicative d'un milieu pouvant générer la dissolution partielle des carbonates. Une succession de phénomènes de dissolution-cristallisation des fines calcaires provenant des oolites est possible.

Figure 4 : C-S-H à faciès fibreux et cristaux d'ettringite recouvrant les grains de sable - MEB

Figure 5 : fissure partiellement remplie de cristaux de calcite néoformés (1) - MEB

Figure. 3 : Microstructure du matériau - MEB
1 = granulat ; 2 = fines calcaires enrobées par les C-S-H

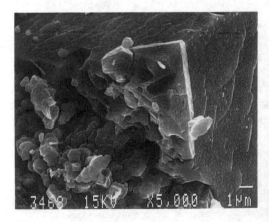

Figure 6 : Cristaux de calcite montrant de nettes figures de dissolution - MEB

3.2. *Examens microscopiques de l'interface béton retaillable-support béton ordinaire*

La qualité de l'interface entre le matériau béton retaillable et son support est déterminante, le but principal du produit étant une utilisation en parement sur support de béton ou de pierre. Plusieurs essais de traction directe, exercée sur un revêtement en béton retaillable d'épaisseur pluricentimétrique coulé sur support en béton traditionnel, ont été réalisés. La plupart des ruptures obtenues sont adhésives (rupture au contact enduit/support) et les résistances mesurées sont variables. Comprises entre 0,2 et 0,9 MPa, elles peuvent être considérées comme peu élevées à satisfaisantes (Martinet & *al*. 1991). Toutefois, dans le cas présent, le matériau était pénalisé par une application sur une surface de béton relativement lisse. Des résistances supérieures seraient obtenues sur support rugueux.

L'examen microscopique de l'interface montre une surface de contact continue et régulière. Aucune solution de continuité n'est mise en évidence ; les particules fines du béton retaillable pénètrent nettement dans la faible microrugosité du support. D'un point de vue microscopique, la qualité du contact est très satisfaisante.

4. CONCLUSIONS

En raison de ses caractéristiques de "retaillabilité", la Pierre de la Possonnière devrait permettre de nombreuses réalisations intéressantes par application sur support très divers : mise en valeur de parements extérieurs ou éléments de décoration intérieure. La fabrication et la mise en œuvre par coulage sans vibration sont simples. L'absence d'addition de produits de type résines ou adjuvants est également un avantage rendant le produit très compatible avec les ensembles pierreux par exemple.

Les analyses réalisées par le LERM ont permis de mieux connaître le comportement rhéologique du matériau à l'état fluide ainsi que ses caractéristiques de microstructure à l'état durci. Elles montrent que le mélange se comporte comme un coulis stable. Malgré un rapport eau/liant élevé et une faible teneur en liant, le produit se comporte comme un mortier traditionnel à prise lente.

La microstructure du béton retaillable est remarquablement homogène et cohérente. Constituée d'assemblages microporeux de silicates de calcium hydratés et de cristaux aciculaires d'ettringite primaire, le liant enrobe parfaitement les oolites et les grains de quartz de la fraction sableuse. Les hydrates semblent subir une carbonatation précoce engendrant la néoformation de microcristaux automorphes de calcite favorables

à la cohésion du matériau. Toutefois, dans le cas du béton retaillable armé, la carbonatation et la chute de pH qu'elle entraîne nécessiterait éventuellement de prendre en compte le risque de dépassivation des aciers. Ce point est à l'étude.

Au total, l'ensemble des examens et des essais présentés ici donne une impression très favorable quant aux caractéristiques minéralogiques et microstructurales du matériau et quant à son potentiel d'utilisation. Néanmoins, des investigations complémentaires sont nécessaires pour le caractériser plus complètement notamment en ce qui concerne sa résistance aux agressions chimiques en environnement urbain (pluies acides, sulfates, ect.) et son comportement aux cycles de gel et de dégel.

De plus, deux autres points importants mériteraient une étude plus approfondie :

- caractérisation physico-chimique de la suspension au moyen d'analyses spécifiques du produit pendant les périodes d'induction et de prise : identification des composés formés en fonction du temps, électrophorèse, conductivité électrique, ect,

- caractérisation fine des carbonates par analyse isotopique du carbone dans le produit durci afin de distinguer et de quantifier les carbonates néoformés et les carbonates anciens non mobilisés provenant de la charge sableuse calcaire.

REMERCIEMENTS

Le LERM tient à remercier vivement Messieurs Cruaud, créateurs et propriétaires de la formule Béton Retaillable, pour leur confiance et pour leur collaboration à la rédaction de cet article.

REFERENCES

Aubouin, J., Brousse, R., Lehmann, J.P. 1975. Précis de pétrologie, Dunod Ed., 718 p.

Bombled, J.P. 1974. Rhéologie des mortiers et bétons frais, étude de la pâte interstitielle de ciment. Revue des matériaux de construction, n°688, mai-juin pp. 137-155.

Couarraze, G. & Grossiard, J.L. 1983. Initiation à la rhéologie, Lavoisier, Tec & Doc, 217 p.

Martinet, G., Le Roux, A., Martineau, F. 1991. Mise au point et optimisation de mortiers pour la restauration des monuments de Haute-Egypte, Bulletin de l'Association Internationale de Géologie de l'Ingénieur, 44, pp. 63-75.

Miltiadou, A.E. 1991. Etude des coulis hydrauliques pour la réparation et le renforcement des structures et des monuments historiques en maçonnerie. Etudes et Recherches LPC. Série Ouvrages d'Art OA8, 277 p.

Papadakis, M. & Venuat, M. 1964. Fabrication et utilisation des liants hydrauliques. Eyrolles Ed., 312 p.

Étude expérimentale du comportement mécanique de granulats calcaires
Experimental study of the mechanical behaviour of crushed limestone

M. Jrad, F. Masrouri & J. P. Tisot
Laboratoire de Géomécanique, École National Supérieure de Géologie de Nancy, INPL, France

RÉSUMÉ : Devant la diminution des ressources en matériaux d'origine alluvionnaire qui constituent l'essentiel des matériaux exploités en génie civil, il convient de mettre en place des solutions de remplacement. Les roches carbonatées pourraient certainement fournir les quantités nécessaires si nous avions une meilleure connaissance du comportement des granulats obtenus à partir de ces roches. Dans ce sens après avoir décrit et caractérisé trois types de granulats calcaires, nous étudions leur comportement mécanique à partir d'essais en compression triaxiale en condition consolidées drainées, d'essais de cisaillement direct ainsi que d'essais type œdométrique.

ABSTRACT : Because of limitations in the extraction of gravels, which constitute an essential material in Civil Engineering, in some alluvial sites, it is important to find substitutes. Crushed limestone would be able to replace traditional materials, when its mechanical behaviour is known. This paper reports on the results of consolidated and drained triaxial, direct shear, and consolidation tests on 3 types of aggregates.

1 PRÉSENTATION DES MATÉRIAUX ÉTUDIÉS

Les trois matériaux retenus pour cette étude proviennent de la région Lorraine en France et sont :

  1. le calcaire de Dugny : de la région de Verdun en Meuse ;
  2. le calcaire de Jaumont : de la commune de Montois la montagne en Meurthe et Moselle ;
  3. le calcaire de Gudmont : de la commune de Gudmont en Haute Marne.

Les courbes granulométriques de ces matériaux sont données sur la figure 1. La porosité interne de la roche d'origine (n) et le poids spécifique des grains (γ_s) sont mesurés. Nous avons effectué des opérations destinées à déterminer les poids volumiques les plus faibles (γ_{dmin}) et les plus forts (γ_{dmax}) qu'on puisse obtenir au laboratoire. Dans le tableau 1 les valeurs relatives à tous ces essais sont récapitulées. Pour quantifier la dureté de ces matériaux des essais Micro-Deval, et des essais Los Angeles (Tableau 2) ont été effectués.

Ces trois matériaux sont différents par la porosité interne de leurs roches d'origines ainsi que par la dureté des grains qui les constituent.

Le fait de prendre en compte un poids volumique des grains corrigé de la valeur de la porosité interne permet d'aboutir pour les trois granulats étudiés à des indices des vides corrigés identiques et donc uniquement fonction de la granulométrie.

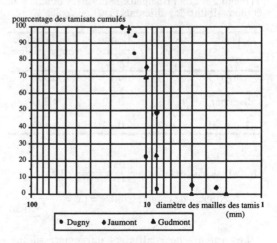

Fig. 1 - Courbes granulométriques des granulats calcaires naturels.

Tableau 1 - Caractéristiques géotechniques et mécaniques des matériaux étudiés.

Matériau	γ_s (kN/m^3)	γ_s corrigé (kN/m^3)	γ_{dmin} (kN/m^3)	γ_{dmax} (kN/m^3)	e_{min}	e_{max}	n
Dugny	26.70	21.36	11.25	13.70	0.56	0.90	25.50
Jaumont	27	22.22	11.40	14.10	0.57	0.95	21.50
Gudmont	27	24.68	12.70	15.50	0.59	0.95	9.40

Avec :

γ_s : poids spécifique des grains ;

γ_s corrigé : poids spécifique des grains avec la prise en compte de la porosité interne aux granulats ;

γ_{dmin} : poids volumique minimal ;

γ_{dmax} : poids volumique maximal ;

e_{min} : indice des vides minimal corrigé ;

e_{max} : indice des vides maximal corrigé;

n : porosité interne de la roche d'origine.

Tableau 2 - Valeurs des Micro Deval et Los Angeles pour les deux matériaux étudiés.

Matériaux	Classe granulaire	MDE	Los Angeles
Dugny	6.3 - 10	70	
	10 - 14	70.3	60
Jaumont	4 - 6.3	63	40
	6.3 - 10	64	
	10 - 14	59	38
Gudmont	4 - 6.3	17	28
	6.3 - 10	18	
	10 - 14	20	24

Tableau 3 - Les principaux paramètres relatifs à la compressibilité des différents matériaux étudiés.

Calcaire	Indice des vides initial eo	Densite relative Dr (%)	Coefficient de compressibilité Cc
Dugny	1.375	0.25	0.300
Jaumont	1.368	0.40	0.308
Gudmont	1.132	0.23	0.110

avec :

Dr (%) : densité relative $= \dfrac{e_{max} - e}{e_{max} - e_{min}} * 100$;

Cc : coefficient de compressibilité $= Cc = - \dfrac{\Delta e}{\log \dfrac{\sigma o + \Delta \sigma}{\sigma o}}$

Les valeurs de coefficient de compressibilité montrent que c'est le matériau le plus tendre qui est le plus compressible.

2 - SOLLICITATIONS APPLIQUÉES

2.1 Essais œdométriques

Pour étudier la compressibilité de ces matériaux, nous avons appliqué des sollicitations de type œdométrique sur des échantillons de 205 mm de diamètre et 110 mm de hauteur. Ces sollicitations se traduisent par des contraintes axiales variables de 100 à 900 kPa par incrément de 100 kPa. Après chaque incrément les déformations verticales ($\Delta h/h$) qui en résultent sont mesurées, les déformations latérales sont nulles. Sur la figure 2 nous avons présenté l'évolution de l'indice des vides en fonction des contraintes appliquées au cours des essais œdométriques. Les principaux paramètres calculés à partir de ces essais sont regroupés dans le tableau 3.

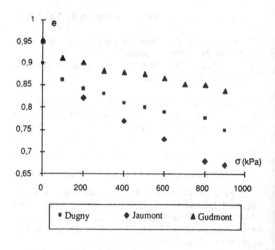

Fig. 2 - Evolution de l'indice des vides au cours des essais œdométriques.

2.2 Les essais triaxiaux

Sur des éprouvettes de 145 mm de diamètre et d'élancement égal à deux nous avons effectué des essais triaxiaux en conditions consolidées drainées avec des pressions de confinement égales à 100, 200, 400, 800 et 1200 kPa. Pour chaque essai et chaque matériau les résultats d'essais triaxiaux sont données sur les figures 4, 5 et 6 sous la forme de courbes représentant l'évolution du rapport de contrainte η en fonction de la déformation volumique ($\Delta v/v$). η est considéré comme le rapport de la contrainte déviatorique $q = \sigma_1 - \sigma_3$ sur la contrainte moyenne $p = \dfrac{\sigma_1 + 2\sigma_3}{3}$.

L'analyse de ces courbes montre que suivant la pression de confinement σ_3 nous avons deux types de comportement :

- un comportement contractant suivi d'une phase de dilatance pour les faibles pressions de confinement ;
- un comportement exclusivement contractant pour les pressions de confinement plus importante.

D'après Luong (1978) le niveau de contrainte correspondant au début de la phase de dilatance ($\eta = \eta_c$) définit l'angle de frottement caractéristique (ϕ_c) du matériau. Cet angle caractéristique défini par la relation : $\phi_c = \text{Arcsin}\ \dfrac{3\eta_c}{6+\eta_c}$ permet de déterminer le seuil de désenchevêtrement des grains. Dans le tableau 4 nous avons regroupé les valeurs des niveaux de contrainte et angle de frottement caractéristique pour les 3 granulats calcaires étudiés.

Nous constatons que l'angle de frottement caractéristique augmente avec la dureté des grains du matériau.

Tableau 4 - Détermination des angles de frottement caractéristiques des différents matériaux étudiés.

Calcaire	Gudmont	Jaumont	Dugny
Poids volumique (kN/m^3)	15	14	13.60
Niveau de contrainte caractéristique(η_c)	1.77	1.57	1.55
Angle de frottement caractéristique(ϕ_c)	43	38.50	38

Ceci peut être expliqué par le fait que plus les grains sont durs plus les contacts intergranulaires sont stables. Par ailleurs, les niveaux de contraintes (η_{max}) correspondant au maximum du rapport (q/p) définissent les rapports de contrainte maximale que peut supporter le matériau et donc son enveloppe de rupture. Dans les tableaux 5, 6 et 7 nous avons regroupé pour les différents matériaux étudiés les valeurs des niveaux de contraintes maximales (η_{max}) ainsi que les angles de frottement correspondant en fonction des contraintes de confinement appliquées. La figure 7 donne dans le plan (p,q) l'allure des courbes correspondant à ces valeurs.

Nous observons une diminution de la résistance du sol au fur et à mesure que la contrainte de confinement augmente et que la dureté des granulats constituant le matériau diminue. L'hypothèse simplificatrice qui consiste à assimiler l'enveloppe de rupture à une droite est donc inapplicable dans le cas de nos matériaux. Fumagalli (1969) à partir des résultats d'essais triaxiaux en compression réalisés sur des calcaires a constaté que l'angle de frottement diminue avec l'augmentation de la pression verticale.

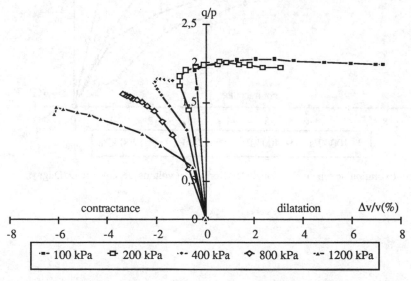

Fig 3 - Evolution de q/p en fonction de la déformation volumique, calcaire de Gudmont.

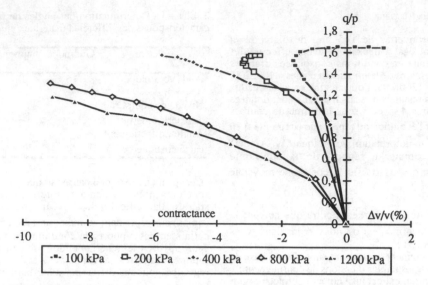

Fig 4 - Evolution de q/p en fonction de la déformation volumique, calcaire de Jaumont.

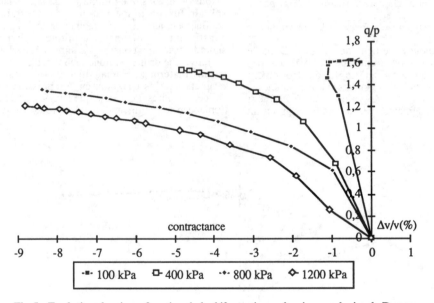

Fig 5 - Evolution de q/p en fonction de la déformation volumique, calcaire de Dugny.

Tableau 5 - Calcaire de Gudmont.

contrainte de confinement (σc) en kPa	100	200	400	800	1200
η_{max}	2.045	2.011	1.795	1.600	1.431
ϕ (degré)	49.70	48.90	43.70	39.20	35.30

Tableau 6 - Calcaire de Jaumont.

contrainte de confinement (σc) en kPa	100	200	400	800	1200
η_{max}	1.647	1.575	1.560	1.310	1.247
ϕ (degré)	40.30	38.60	38.30	32.50	31

Tableau 7 - Calcaire de Dugny.

contrainte de confinement (σc) en kPa	200	400	800	1200
η_{max}	1.631	1.546	1.360	1.215
ϕ (degré)	39.90	37.90	33.70	30.30

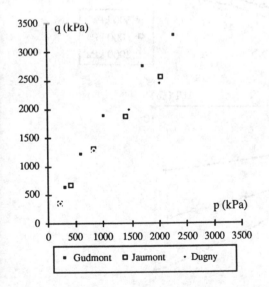

Fig.6 - Enveloppes de rupture des différents matériaux étudiés.

Ceci est dû à l'écrasement des granulats qui sont de plus en plus affectés au fur et à mesure que la pression verticale augmente. Certains auteurs comme Mohkam (1983) explique ce phénomène par le fait que sous l'effet de l'augmentation de la pression de confinement les aspérités des grains se brisent, diminuant l'imbrication de ceux ci, ce qui entraîne une diminution de l'angle de frottement.

2.3 Les essais à la boite de cisaillement

A l'aide d'une boite de cisaillement de 150*150 mm de section nous avons effectué des essais de cisaillement direct à une vitesse de déformation fixée à 1.33% par min et sous des contraintes normales de : 500, 1000 et 2000 kPa. Lors des essais nous avons constaté qu'après un certain temps de cisaillement le chapeau de chargement commence à s'incliner et cette inclinaison devient de plus en plus importante au fur et à mesure que le déplacement relatifs des deux demi-boite augmente. La contrainte normale appliquée n'est plus donc répartie d'une manière uniforme sur toute la surface de cisaillement et au niveau du plan de rupture le milieu est plus serré du côté de l'enfoncement du chapeau. Pour limiter les erreurs dues à l'inclinaison du chapeau de chargement on s'est limité alors à un pourcentage de déformation de l'ordre de 10%.

Pour tous les échantillons étudiés, nous avons constaté que la résistance du matériau continue à évoluer d'une façon non linéaire sans aucune tendance à la stabilisation. Quand à la variation de volume tous les échantillons (indépendamment de la nature du matériau) sont caractérisés par un comportement exclusivement contractant. Sur la figure 8 nous avons représenté pour le calcaire de Dugny l'évolution du rapport (τ/σ_n) : rapport de la contrainte tangentielle sur la contrainte normale, en fonction des déformations tangentielles ($\Delta l/l$) et des déformations axiales($\Delta h/h$).

Pour calculer les angles de frottement interne nous avons considéré les valeurs des résistances au cisaillement (τ/σ) correspondant à 10% de déformation tangentielle. Dans le tableau 8 nous avons calculé le poids volumique initial ainsi que l'angle de frottement interne correspondant à chaque matériau.

Tableau 8 - Angles de frottement des différents matériaux étudiés.

Calcaire	Dugny	Jaumont	Gudmont
Poids volumique initial (kN/m^3)	11.70	12	13
Angle de frottement interne (degré)	29.50	29	31

En comparant les valeurs des résistances des différents matériaux déterminées à partir de la boite de cisaillement avec celles de la cellule triaxiale, nous constatons que la résistance des matériaux est nettement plus élevée dans le cas d'essais réalisés en compression triaxiale. Les raisons de cet écart sont diverses, en effet :

- dans le cas des essais à la boite de cisaillement, la résistance au cisaillement d'un sol n'est bien définie que dans le cas où la courbe contrainte-déformation présente un maximum. En absence du maximum, la résistance du sol est prise à une déformation arbitraire, ainsi la résistance du sol est plus ou moins importante suivant la déformation considérée et il s'ensuit que la résistance calculée peut ne pas être représentative de la vrai résistance à la rupture du matériau. D'autre part dans le cas de la boite de cisaillement conventionnelle, seules sont mesurées les contraintes normales (σ_n) et de cisaillement (τ). Plusieurs études ont montré que l'angle de frottement mesuré à la boite de cisaillement dépend aussi de la valeur de la contrainte horizontale (σ_h) (Atkinson, 1991 ; De Josselin de Jong, 1988). Tandis que pour l'essai triaxial conventionnel la résistance du sol est entièrement déterminée par la mesure de la contrainte radiale et axiale ;

- la différence de comportement de l'échantillon au cours des essais. Ainsi la libre installation de la surface de rupture et la possibilité de dilatation de l'échantillon radialement dans le cas d'essais triaxiaux entraînent un comportement plus réaliste que dans le cas d'essais à la boite de cisaillement où les dilatations sur les côtés sont empêchées, le plan de rupture est imposé, et les sollicitations réelles sont mal connues.

- dans le cas des essais à la boite de cisaillement , Monnet (1977) a constaté (à partir d'essais faits sur des échantillons de sable) que l'effort de cisaillement n'est pas uniforme dans le plan de rupture, il est maximum vers les bords de la boite et minimum au centre. Cette répartition est tout à fait différente de la répartition uniforme utilisée pour calculer la résistance du sol .

2.4 Étude de l'évolution de la granulométrie

Une étude expérimentale du comportement des sols granulaires soumis à de fortes contraintes demande en plus du comportement en contrainte déformation le contrôle d'un autre paramètre rarement pris en compte et qui peut jouer un rôle important dans le comportement mécanique des sols granulaires soumis à des fortes sollicitations, c'est l'attrition des grains.

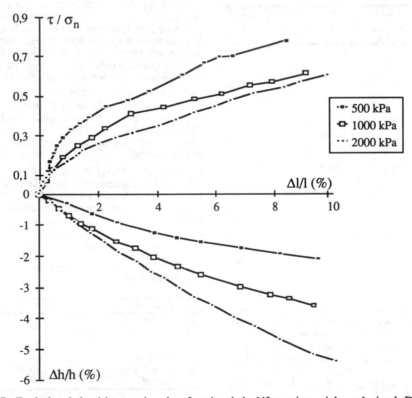

Fig. 7 - Evolution de la résistance du sol en fonction de la déformation axiale , calcaire de Dugny.

Pour quantifier cette évolution de la granulométrie nous avons utilisé le paramètre Bg introduit par Marsal (1977). Bg est définit comme étant la somme des valeurs positives de Δwk avec :

$$\Delta wk = wki - wkf$$

- wki : le pourcentage du poids retenu par le tamis de diamètre k avant essai ;
- wkf : le pourcentage du poids retenu par le même tamis après essai ;
- k prend les valeurs des diamètres de tous les tamis utilisés.

Ainsi défini, Bg représente le pourcentage du poids des grains affectés au cours de l'essai. Pour calculer Bg nous avons effectué une analyse granulométrique des échantillons qui ont été soumis aux essais triaxiaux et œdométriques. Dans les tableaux 9 et 10 nous avons calculé pour chaque matériau et chaque essai le paramètre Bg et les valeurs des coefficients d'uniformité C_u et D_{10} (le diamètre du tamis à travers lequel passent 10% du poids des grains).

Les paramètres Bg, D10 et Cu, montrent qu'il y a eu un grand changement de la structure granulaire. En effet on constate que le pourcentage du poids des grains affectés (Bg) évolue avec l'augmentation de la contrainte de confinement et devient important pour les fortes valeurs de σ_c. les valeurs de wkf montrent que ce sont surtout les grosses particules qui sont affectées au cours des essais. Les coefficients d'uniformité montrent qu'au fur et à mesure de l'augmentation de la contrainte de confinement dans le cas des essais triaxiaux l'étendue de la granulométrie devient de plus en plus importante. En effet, initialement la granulométrie était serrée et avec l'augmentation de σ_c elle est devenue étalée.

Marsal (1977) et Cambou (1979) à la suite de leurs nombreux essais concluent que la rupture des grosses particules est plus importante que la rupture des petites particules. Marsal pense que le facteur déterminant de la rupture des grains est l'intensité des forces intergranulaires, et la grandeur de ces forces est liée au nombre de contacts qui augmente avec la grosseur de la particule. Selon Marsal, le paramètre Bg dépend essentiellement de la dimension des particules, de la distribution granulométrique et de l'indice des vides de l'échantillon. Bg pourrait dépendre aussi de la résistance des grains, car le matériau le plus tendre (calcaire de Dugny) est beaucoup plus affecté que le calcaire de Gudmont de dureté plus élevée.

Pour les essais de cisaillement direct nous n'avons pas pu effectuer une analyse quantitative de l'évolution de la granulométrie, ceci à cause d'une part de la difficulté de préciser l'épaisseur de la zone de cisaillement et d'autre part de la difficulté de prélever du matériau situé au niveau de cette zone.

3 - CONCLUSION

L'étude expérimentale effectuée sur trois types de granulats calcaires a permis de cerner les particularités physiques et mécaniques de ces matériaux sous diverses sollicitations à partir de trois appareils expérimentaux différents : l'appareil œdométrique, la cellule triaxiale, la boite de cisaillement.

Tableau 9 - Évolution de la granulométrie du calcaire de Gudmont au cours des essais triaxiaux et œdométriques.

		Essais triaxiaux				Essais œdométriques	
σ (kPa)	0	200	400	800	1200	200	900
Bg(%)	0	8.41	14.53	21.56	22.91	2.35	11.54
Cu	1.5	1.75	2.22	7	7	1.66	1.81
D10 (mm)	7	6	4.5	2.86	2.86	6	5.5

Tableau 10 - Évolution de la granulométrie du calcaire de Dugny au cours des essais triaxiaux et œdométriques.

		Essais triaxiaux				Essais œdométriques	
σ(kPa)	0	100	400	800	1200	200	900
Bg(%)	0	14.08	21.84	37.66	43.65	5.65	15.76
Cu	1.22	1.60	2.62	5	9.5	1.41	1.83
D10 (mm)	9	7	4	2	1	8.5	6

À partir des essais relatifs à ces appareils nous avons pu mettre en évidence l'influence des propriétés de base des grains d'une part sur la compressibilité et la dilatance de ces matériaux, et d'autre part sur le seuil de désenchevêtrement des grains et leur résistance au cisaillement. D'autre part l'étude de la granulométrie avant et après essai a permit de mettre en évidence le phénomène de l'attrition des grains qui a entraîné des déformations importantes du matériau en activant le mécanisme de réarrangement des grains. Ce phénomène est aussi à l'origine de la décroissance de la résistance des matériaux au fur et à mesure de l'augmentation de la pression de confinement. Le choix d'un critère de rupture de type Mohr-Coulomb est donc inadapté pour ce genre de matériau (Jrad 1993).

4 - RÉFÉRENCES BIBLIOGRAPHIQUES

Cambou B (1979) - Approche du comportement d'un sol considéré comme un milieu non continu. Thèse Université Claude Bernard, Lyon.

Jrad M. (1993) - Étude expérimentale du comportement mécanique de granulats calcaires. Thèse. Institut National Polytechnique de Lorraine de Nancy, France.

Luong M.P. (1980) - Phénomènes cycliques dans les sols pulvérulents. Revue française de géotechnique, n°10, Février, pp. 39-53.

Marsal R.J.(1977) - Research on granular materials. Exp work compiled for the IX Int.Conf on Soils. Mec and Found.Eng.June.Tokyo, pp 1-78.

Mohkam M. (1983). - Contribution à l'étude expérimentale et théorique du comportement des sables sous chargements cycliques. Thèse. Université Scientifique et Médicale et Institut National Polytechnique de Grenoble.

Monnet J. (1977). - Détermination d'une loi d'écrouissage des sols et utilisation de la méthode des éléments finis. Thèse Docteur Ingénieur, I.N.S.A, Lyon.

Atkinson J.H. (1975). - Anisotropic elastic deformations in laboratory tets on undisturberd London clay. Géotechnique 25, pp. 357-374.

De Josselin de Jong G. (1988). - Elasto-plastic version of the double sliding model in undrained simple shear tests. Géotechnique 38, pp. 533-555.

Fumagalli E. (1969). - Tests on cohesionless materials for rockfill dams. Journal of the soil mechanics and foundations division, A.S.C.E. Vol 95, n°SM1. Jan, pp. 313-330.

Study of petreous aggregates for concrete in Zimapan arch dam in Mexico
Étude des agrégats rocheux pour le béton du barrage voûte de Zimapan au Mexique

Roberto Uribe-Afif
Comisión Federal de Electricidad, Mexico

ABSTRACT: Initially, the different parts of the projects are described, including dimensions, volumes and type of concrete. The geological studies of the quarry sites are also described together with the reaction produced by the cement alkalis upon some chemical components of carbonated rock aggregates. The borrow banks description is presented with an analysis of the mineralogical, petrological, textural and chemical charactristics of each lithological formation. The relation between the different rocks and the different types of cement available at the dam site was an important factor for the selection of the borrow banks to use. Field geological sampling and laboratory tests for the analysis of alkali-carbonate reactivity formed an important part of this study. Finally we recommend the procedure for the proper quarry exploitation and the design properties of the concrete using the aggregate of the borow banks studied.

RÉSUMÉ: On describe initialment la plupart des project avec les dimensions, volume et type de béton. Les études geologiques des diverses carriéres sont aussi presentés avec les études de la reaction des alcalis du cement sur quelques components chimiques des agregats carbonatés de la roche on present la description de les zones d'emprunt avec l'analyse des caractéristiques chimiques, mineralogiques, petrologiques et de texture de chacune formation litologique. La relations entre les differnets roches et le différents types du ciment disponible dans la zone du barrage fut un factor important pour la selection de las zones d'emprunt á utilizer. Le pré lé vement d'echuntillons geologiques et de épreuves au laboratoire pour l'analyze de reactivité alcali-carbonate ont été tres importants pour le present étude. Finalement, on recommande le procédure pour la correcte explotation de la carriére et les propertés du béton pour le project utilizant les agregats de les zones d'emprunt étudies.

1. ANTECEDENTS

The hydroelectric project of Zimapán is found in the state of Hidalgo, Mexico. It is constituted by a concrete arch dam of 200 m in height, spillway, conduction tunnel and powerhouse, all structures covered with concrete.

The importance and volume of all works of this project, makes necessary to eliminate any possibility of reaction, between the aggregates and the cement.

2. GEOLOGICAL CHARACTERISTICS OF POSSIBLE BORROW BANKS

All the banks sampled belong to different lithologies from El Doctor Formation. This formation has the following characteristics: calcareous rocks that constitute an important group of major structure basically composed by anticlines and sinclines with a variable stratification and a predominant orientation NW-SE, its wide is between 900 and 1500 m. These calcareous bodies have been assosciated to platform environments

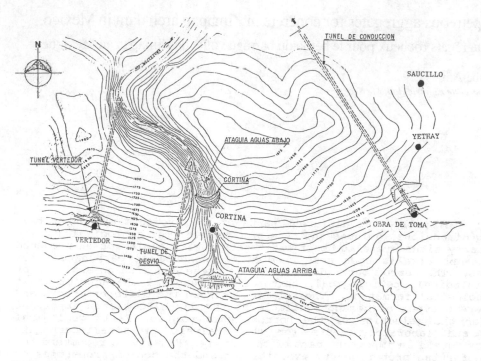

FIGURE 1. BANKS LOCALIZATION

and depth sea, where it has been observed the following lithologies: limestone conglomerate, dolomitic breccias, dolostones and reef limestones to platform conditions and thin beds limestones with chert highly folded to depth sea conditions. The age of this formatiaon is low cretaceus.

2.1 PETROGRAPHIC DESCRIPTION

2.1.1 "OBRA DE TOMA" BANK

Limestone gray to dark gray of fine grain texture with crack filled by calcite, clay minerals in variable proportions, but always less than 5% volume. The rock presents a stratified structure medium to coarse wide, presence of microstylolites.

2.1.2 "SAUCILLO" BANK

Limestone gray to dark gray of fine grain texture, abundant cracks filled with calcite. Chert and clay minerals as filling in stylolites, the major structures are thin beds (10-30 cm).

2.1.3 "YETHAY" BANK

Same to "Saucillo" bank.-

2.1.4 "CORTINA" BANK

Dolomitic limestone and dolostones dark gray, crystalline texture from fine to medium grain, abundant cracks filled with calcite. The main structure has medium to coarse beds (30-60 cm,)

2.1.5 "VERTEDOR" BANK

Limestones, dolomitic limestones and dolomitic breccias light gray color to dark with different textures (fine grain and clastic), cracks filled with calcite and sylolits filled with clay minerals. The major structure is an estratification medium to coarse from 30 to 50 cm.

2.2 CHARACTERIZATION OF EL DOCTOR FORMATION

As it can be observed on description of rocks and the physical characteristics of the El Doctor Formation, the samples obtained are representative to establish some difference on its reactivity grade with cement.

Reactive material from petrological point of view was collected (dolostone, dolomitic limestone and dolomitic breccias), also there were selected inoffensive materials (limestones). This was carried out to have completely characterized the Doctor Formation, the last is very important to mention due to the geological formations around the project, where it has been observed the Doctor Formation is the widely distributed and it has the best physical properties.

3. EL DOCTOR FORMATION AND ALKALI-CARBONATE REACTIVITY (ACR)

Having in consideration that El Doctor Formation has been detected as the more viable possibility to be the source of aggregates for concrete to the project it is necessary to make a detailed analysis of its charactristics.

The formation has some important lithological variations associated to a particular environment, and in this case the presence of dolostone and/or clay materials induce a possibility to occurrence of ACR.

Dolomite is a mineral frequently associated with calcite and some limestones and generally its hard is used to distinguish of calcite. Its origin is not a primary in a common way and most of the times it is the result of a replacement of calcite, for so much is a product almost post-depositional (Pettijohn, 1972). The process of replacement of calcite by dolomit is known under the name of dolomitization. The process has an accidental character. By the effect of replacement it can ocur in beds or a single changed can be occurred a long certain cracks, fractures, superficial covers or stylolites.

The irregular character of the phenomenon can permit that one rock is constituted only by calcite and in other portion can contain cracks and stylolites with high proportion of insoluble residual with inclusion

TABLE 1.- PETROGRAPHIC CHARACTERISTIC

BANK	OBRA DE TOMA	EL SAUCILLO	TRITURADORA	CORTINA	VERTEDOR	REACTIVE MATERIAL
LITHOLOGY	MICRITIC LIMESTONE	MICRITIC LIMESTONE	MICRITIC LIMESTONE	DOLOMITIC LIMESTONE	DOLOMITIC BRECCIAS	DOLOMITIC LIMESTONE ARGILACEUS
PHISIC QUALITY	GOOD	GOOD	GOOD	GOOD	GOOD	GOOD
INSOLBLE RESIDUAL %	1.20	1.20	1.36	1.00	1.38	5 – 50
RELATION CAL/DOL	100/0	100/0	100/0	40/60	30/70	50/50
MICROTEXTURE	MICRITIC MUD	MICRITIC MUD AND STYLOLITES	MICRITIC MUD AND STYLOLITES	EQUAL CRYSTALS OF CALCITE AND DOLOMITE	CLASTIC, DOLOMITIC LIMESTONE AND ESPATIC CRYSTALS IN A MICRITIC MUD MATRIX	DOLOMITIC ROMBS IN A MICRITIC MUD MATRIX WITH CLAYS DISPERSES
THIN SECTION	DARK	DARK	DARK	LIGHT	DARK	DARK
SAMPLE OF HAND	FINE GRAIN, CONCOIDAL CRACK, LIGHT	FINE GRAIN, CONCOIDAL CRACK, LIGHT	FINE GRAIN, CONCOIDAL CRACK, LIGHT	MEDIUM TO COARSE GRAIN, DARK	COARSE GRAIN, DARK	FINE GRAIN, CONCOIDAL CRACK, DARK

TABLE 2.- CHEMICAL CHARACTERISTICS

LITHOLOGY	CaO %	MgO %	ALKALIS Na2O %	K2O %	FORMATION (AGE)
MICRITIC LIMESTONE OBRA DE TOMA	56.7	0.60	0.02	0.08	F. EL DOCTOR (CRETACEUS)
MICRITIC LIMESTONE EL SAUCILLO	56.1	0.50	0.04	0.08	F. EL DOCTOR (CRETACEUS)
MICRITIC LIMESTONE TRITURADORA	58.1	0.50	0.02	0.06	F. EL DOCTOR (CRETACEUS)
DOLOMITIC LIMESTONE CORTINA	36.6	1.70	0.38	0.12	F. EL DOCTOR (CRETACEUS)
DOLOMITIC BRECCIAS VERTEDOR	37.0	1.75	0.12	0.08	F. EL DOCTOR (CRETACEUS)

dolomite rombs in a typic reactive texture. Dolor-Mantuani (1983) observed that a limestone has an expansion, enterely produced by a small quantity of expansive material contained in a very fine stylolite.

The dolomitization process on the rocks of El Doctor Formation, this has been mainly detected in dam and spillway zones, in both sides have an irregular distribution (González, 1991).

In the Table 1 a resume of the different lithologies is presented.

An important support on the evaluation of ACR of these banks, are the chemical characteristics that allow to carry out interesting relations as can be observed in Table 2, where it is clear different the contain of CaO. That shows an important diminishing when there exists the presence of dolomite or clay minerals in the rocks and this brings the raising on the content of MgO. As already mentioned it is an interest point, thus the samples of Obra de Toma, Saucillo and Yetahy, can be considered as limestones, constituted entirely by calcite without dolomitic fraction, which is recognized as the activity part of ACR.

4. DESIGN CONSIDERATIONS

For concretes of the project it has been carried out a design with the following characteristics:

- Maximum size of aggregate = 3"

- Relation gravel/ sand = 70/30

- Relation water/ cement = 0.68

- Cement consumption = 200 kg/m^3

- Additive Pozzolith 322R = 6 cm^3/kg of cement.

The gravel will be distributed in sizes according to the following proportion:

Gravel 3/4"   =   30%
Gravel 1 1/2" =   30%
Gravel 3"   =   40%

About the ACR Walker (1978) observed employing the aggregates in bigger size they increased the expansion grade, so it was observed that low relation water/cement give higher expansions. If it is considered that cement is the material that gives the more alkalis, it must be considered a mixture with a lower cement consumption by m^3.

5. CONCLUSIONS

Having in consideration the accumulated information for this paper, it is possible to establish conditions that would give a positive result to have the adequate materials for the fabrication of concrete in a short time. It is necessary to follow the next indications to produce aggregates of very good quality.

The Doctor Formation is characterized for having excellent physical properties to produce aggregates for concrete of good quality for the project.

The chemical properties of this formation with its ACR potential makes indispensable for its exploitation taking in consideration the recommendations from this paper. They become an answer to the possible problems that might derive if an inadequate exploitation of the formation is carried out.

The geological conditions of the El Doctor Formation give it a heterogeneous body due to its deposit characteristic and diagenetic process which has been origin. This complex formation needs a particular attention to understand these variations.

This lighology diversity permits to make considerations of the materials since the rock varies from micritic limestone, dolomitic limestone, dolostones, limestone with clays and dolomitic breccias. For the productions of aggregates can be divided into inocuous material (micritic limestone) and reactive potentially with the alkalies of cement (dolomitic limestone, domlomitic breccias and dolostones).

This subdivision permits in case of the field conditions a selective exploitation of material.

An important point to consider in selecting explotation of the banks, is the presence of an irregular phenomenon on the El Doctor Formation, which is known as dolomitization. It has an aleatoric distribution as was commented in this paper and the place where it is shown, converse the material in a potential reactivity bank. Another problem to consider is this might happened in a rock mass zone or follow differnt ways as faults, cracks, fissures, etc.

In a particular way the Cortina and Vertedor banks have been observed zones that can be considered as reactive potential by have dolimitization.

The samples without any clear sign of dolomitization are Obra de Toma, Saucillo and Yethay, and are basically constituted by micritic limestone, with an adequate process it is possible its use.

Thre are two particular evidences to indicate the reactive materials with the cement alkalis. these are: microtexture of dolomitic rombs in a micritic limestone and clay material disseminate and a typic mineralogical association represented by 3 members micritic calcite-dolomite-illite.

Only the banks Cortina and Vertedor showed these characteristics reactive potentially.

Another important point to comment the petrological aspect is the percentage of acid insoluble residuals it is between 5 and 10% of reactivity rocks, any of the samples had more than 1.4 of residuals and, the relation calcicte/dolomite is equal to 1 or very close to it. This value relation was shown only in Cortina and Vertedor banks.

6.   RECOMMENDATIONS

The Vertedor and Cortina banks with the actual knowledge from them, are shown with the major conditons of risk.   For so much can not be considered as explotable.   The Obra de Toma banks is conditioned by exploitation due to its proximity to the mentioned banks which are considered as potentially reactive.

The Saucillo and Yethay banks can be actually exploited if they follow these conditions:

a)   The geological residence of the project will carry out a detailed follow of the exploitation bank, to detect any lithological change towards reactive materials (dolomitic limestones, dolostones and dolomitic breccias.

b)   In presence of reactive material in front of the bank, this can continue to be exploited if it does not exceed the 15% of the total.

c)   The aggregate product of exploitation, must be an adequate process to eliminate clay minerals. The percentage of fine material will not exceed the 7% of the total, because in this portion it is highly probably to find reactive material.

Referring the volume, these two banks have the capacity to cover the needs of the project.

REFERENCES

1.   Dolar-Mantuani, L., 1983., "Handbook of Concrete and Aggregates". Noyes Publications, USA.

2.   González, C., 1991, "Personal Communication".

3.   Pettijohn, F.J., 1975, "Sedimentary Rocks". Harper and Row, Publishers, Third edition.

4.   Walker, H.N., 1978, "Significants of Test and Properties of Concrete and Concrete-Making Materials" Special Technical Publications 169 B, ASTM, Chapter 41.

Geological and geotechnical characteristics of tills for road construction

Caractéristiques géologiques et géotechniques des matériaux morainiques pour la construction de routes

Y.Yu & A.Jacobsson
Luleå University of Technology, Sweden

ABSTRACT: In some regions of Sweden, the exploitation of glacio-fluvial gravel and sand, together with the conservancy of these sediments, has resulted in a shortage of this kind of material for road construction. As an alternative, till can be used for this purpose which is very rich in the country. In order to effectively utilize till for construction, two important issues are addressed here. One is to find the geological factors that are essential when searching for such till. Common search techniques are also discussed. The other concerns how and what extent the geotechnical properties of the processed till material that is to be used for the construction have been changed in comparison with the natural till. It appears that the decision concerning how to produce the gravel surface material must be a question of costs together with the quality of the available till.

RESUMÉ: Dans certaines régions de la Suède, l'exploitation de gravier et de sable d'origine glacio-fluvial associée à la politique de protection de ce type de sédiments, a provoqué une pénurie de ces matériaux destinés à la construction de routes. Une alternative serait d'utiliser du materiel d'origine morainique, très abondant dans ce pays. Afin d'exploiter efficassement les moraines comme sources de matériaux pour la construction, deux problèmes importants sont à considérer. Le premier est d'identifier des caractéristiques géologiques essentielles lors de la recherche de telles moraines. Les techniques de recherche courantes sont aussi abordées. Le second est de savoir comment et dans quelle mesure les propriétés du matériel morainique initial sont affectées par l'extraction. Il apparait que la méthode de production de ces matériaux doit être décidée en fonction des coûts et de la qualités du materiel morainique disponible.

1 INTRODUCTION

Due to the glacial geology, different types of till cover more than 75% of the Swedish land area. There are certain till types whose natural texture is such that the material after crushing and - if necessary - separation of fines can be used as a high quality material in road construction and maintenance. Such till material, often called as the coarse-grained till, can be used as a substitute for glaciofluivial material - esker and delta gravel, especially for gravel surface on small or secondary roads.

However, in order to effectively utilize the till material, two important issues should be addressed, i.e. how to find the till in the land and what changes in geotechnical properties take place after the crushing process. Based on these a suitable road construction can surely be performed. This paper briefly reports previous investigations on searching till materials for road construction, and discuss results from a detailed experimental investigation on the natural and processed till taken from northern Sweden. To begin with, a general description on the production is presented.

2 PRODUCTION OF ROAD MATERIAL FROM TILL

2.1 Raw material

The road superstructure consists of a subbase, a road base and a gravel wearing courses or surfacing. Variations in the thickness of these layers depend partially on the natural   pes of soil in the so-called terrace, Fig. 1.

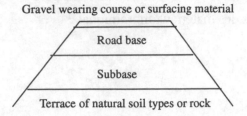

Gravel wearing course or surfacing material

Road base

Subbase

Terrace of natural soil types or rock

Fig. 1 Composition of the road structure

Rock, natural gravel or till can constitute the raw material in the foregoing road structure. Rock is usually an expensive raw material and is required in large quantities in order to be profitable. There is already a shortage of natural gravel in most places in the country, while till, the most common and rich soil type, poses certain technical problems in the production process. This significantly affects profitability. As shown in Fig. 2, the stone content is pushed steadily downwards as the amount of material to be crushed increases. It lies even under the gravel wearing course zone for till of which 60% has to be crushed.

When combining the grain size distribution of material with sufficient bearing capacity and corrugation-resistant material from the surface of gravel roads, the so called perfect gravel zone was obtained, Fig. 3 (Beskow 1934). It is possible to get perfect gravel by mixing till and gravel. The specific proportions to get the optimal grain size distribution for material in a road pavement are shown in Fig. 4.

2.2 Production technique

Jaw crushers, cone crushers and a screen are normally used. Variations in the set-up depend on the composition of the till as well as on the type of material which is to be produced. When producing subbase material, it is often sufficient to merely sort (i.e. screen) the till. However, the profitability can be increased if a jaw crusher is used in order to utilize the cobbler stones and boulders. If the stone content is correct, a jaw crusher and cone crusher can suffice when producing wearing course gravel. A screen is, however, also required on most occasions. A jaw crusher, a cone crusher and a scree are normally required when producing the road base.

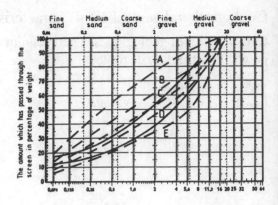

Fig. 2 The importance of the stone content in till crushing. A = basic composition, B = 20% material to be crushed, C = 40%, D = 50%, E = 60%.

3 GEOLOGICAL CHARACTERISTICS

In recent years a few investigations have been carried out in Sweden to effectively search gravelly-sandy till for road construction. Geological characteristics of natural tills are crucial for such a search.

3.1 Geological factors

Geological factors that could simplify the search for coarse-grained till were particularly investigated by Johansson and Enkell (1980) and may briefly be summarized in the following.
(1) The topography: Coarse-grained till is often deposited in valleys and other depressions of the ground.
(2) The bedrock: A hard bedrock e.g. pophyrite, quartzite has been an important source of coarse-grained till.
(3) The position: Most of the coarse-grained till has been deposited at or above the highest shoreline. The coarse-grained till is often situated close to between glaciofluvial accumulations.
(4) The morphology: Coarse-grained till is very often observed in hillocks or ridges showing an irregular pattern in the terrain.

3.2 Searching techniques

There are two methods commonly used for searching for tills, i.e. aerial photographic interpretation method, field and survey method. They are not independent and always combined in use.

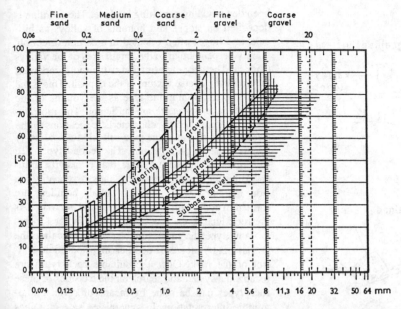

Fig. 3 The "perfect gravel zone" obtained by combining the variable zones of wearing course gravel (corrugation-resistant) and the subbase gravel (moisture-saturated).

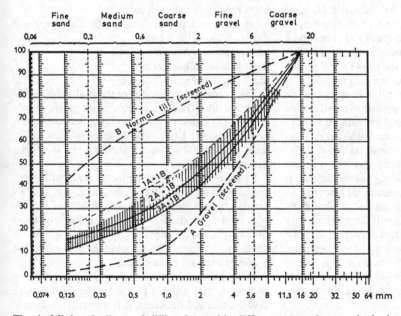

Fig. 4 Mixing the "normal till" and gravel in different proportions to obtain the "perfect gravel".
Modified from Beskow (1934).

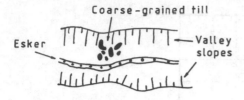

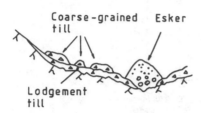

Fig. 5 Occurrence in principle of coarse-grained till (moraine type) in a valley.

When searching for till materials within large areas aerial photographic interpretation is the most rational method both from practical and economical point of view. Especially in areas where the geological maps are obsolete or are lacking the aerial photographic technique is in practice the only method.

The method to map coarse till is in principle the same as for geotechnical and geological aerial photographic interpretation in general including inventory of existing materials, aerial photographic interpretation, field control and presentation. As for all other types of aerial photographic interpretation it is necessary to have a thorough knowledge of the formation to be mapped. For coarse-grained till one has to know how the moraine type looks and in what characteristic sites in the terrain it will occur. These factors are the most important criteria for the interpretation. The methodology for localizing coarse-grained till has been discussed by e.g. Viberg & Inganäs (1978), Viberg (1984) and Johansson & Enkell (1980), which is mainly based on the following parameters.

Genesis
Features - surface forms
Position in the terrain
Surrounding geology

There are general two moraine types existing in the land. The first type - ridges and hillocks - has been deposited by a melting land ice. The moraine is deposited on slopes and bottoms of valleys as hillocks and ridges often oriented in a certain direction, mainly perpendicular to the length axes of the valleys (Fig. 5). The way of formation is the most important factor for the coarse texture of the deposit. This is the most interesting moraine type as the volume of the deposits in many cases are sufficient. The morphological features and the characteristic locations in the terrain make this moraine type suitable to localize by means of aerial photographs. The second type - lee side moraine - present texture that is heavily influenced by the local rock type. Sandy or gravelly lee side moraines with relatively small boulders are formed in areas with e.g. porphyrite and quartzite. As the rock type is the most dominating factor for this moraine type the most suitable way to localize the moraine is by means of bedrock maps and field control of the rock. Once the suitable rock types are localized,a systematic search on the lee sides of these rock types can be done by means of aerial photographic interpretation. In general lee side moraines are difficult to identify in aerial photographs. In cultivated areas, however, they are easily detected.

Further interpretation is based on the identification of the following criteria.
Identification of valley/depression
Identification of surface forms (hillocks/ridges) and their position in the terrain
Identification of borrow pits/cuts
Identification of boulderness
Identification of glaciofluvial deposits

The interpretation consists of the qualitative possibilities to identify these criteria in the aerial photographs. The identification of every criterion is more or less reliable. Black and white air photos in the scale of 1:30,000 was used, although this is not the optimal material.

By the interpretation, a number of areas with varying degree of favourable conditions for coarse-grained till are localized. Clearly, increased knowledge on the genesis and occurrence of coarse-grained till and the use of high quality infrared colour air photos will improve the possibilities to localize the coarse-grained till in the future.

Field survey method including seismic-refraction technique, trial pits and boreholes can yield deposit composition and thickness. It is normally performed after the aerial photographic method, and may be used to verify or correlate the aerial photographic interpretation. The seismic-refraction technique may be used to get an indication of the layer sequence, if any, of a moraine deposit and the depth to the bedrock. In addition, the relative density of till and the

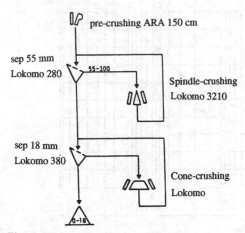

Fig. 6 The crushing process for the till.

location of the ground-water table can sometimes be estimated from the seismic velocity. Due to some uncertainties, a seismic investigation is generally supplemented by sounding, sampling, test pits and test trenches.

4 GEOTECHNICAL CHARACTERISTICS

For requirements to road construction, experimental studies have been performed on coarse-grained till to find out whether the geotechnical properties change after the till has been processed in a crushing mill. In order to make a fair comparison between the processed and the natural material, all grains coarser than 20 mm in diameter were separated from the till and, thus, the two materials have the same maximum particle size. In this study these two materials were taken from Älvsbyn, northern Sweden, and the crushing process via machine is shown in Fig. 6. The unprocessed or separated and processed till are marked by M1 and B1, respectively. Concerned geotechnical properties for the investigated till include particle size distribution, compaction, frost-susceptibility, permeability and capillarity.

4.1 Particle size distribution

The sieve method was used to determine the particles sizes down to the fine sand (0.06 mm), whereas the Granulometer was used for the particle sizes between 0.125 mm and 0.001 mm. The particle sizedistribution of the natural or unprocessed till is given in Fig. 7, with 15% silt and clay (< 0.06 mm), 30% sand (0.06 - 2 mm) and 55% gravel (2 - 60 mm). It may thus be termed as sandy silty gravel moraine.

It is shown from Fig. 7 that the processed or crushed material B1 has increasing contents of sand and gravel, which is expected. Also, the fine particle percentage of silt and clay increases from 15% to 20%. On the other hand, the particle sizedistribution curve for the material M1, the separated till, are relatively similar to the one of the material B1, Fig. 8, with a 5% higher sand content for the material M1. The gravel content is relatively higher for the material B1, whereas the fine content is almost same i.e. 20% for both materials. The processed material B1 is supposed to be used as gravel surface. Since its particle size distribution is similar to one of the separated material, one would question if the crushing process is necessary to produce an economically better material.

Referred to the regulation for the curve of fraction 0 - 20 mm for the gravel surface from Swedish National Road Board (BYA 84), both of these materials are not proper, Fig. 9. However, the processed material B1 has already been used as the gravel surface (Minell 1986), and shown that it worked well, implicating that the current regulation or requirement might not be necessary for secondary roads. In fact, the regulation is largely based on experience, and the gravel surface material has rarely processed from natural till. Also according to Fig. 9, the material M1 in the content range of larger sand particles actually satisfies the regulation, and it is thus possible to utilize an effective sand sieve process for the material more satisfying the regulation.

In short, the till material with a simple separation of pebbles and boulders, eventually combined with a sand sieve procedure, has a particle size distribution curve with only a marginal difference from one of the processed till. However, this conclusion may be limited to the investigated till which has edge-shaped particles from crushing stones. There are a lot of methods which may be used to adjust the particle size distribution of a till material in order to satisfy a specific requirement. One method is to mix the till with a heavily gravel material in a suitable proportion, e.g. 2 shares of till + 1 share of gravel, which can itself be reproduced through a suitable separation from the till.

4.2 Compaction

The compaction of a soil is described by the maximum dry density which is primarily dependent on soil (mineral) type, particle size distribution, water content and compaction method. In the case of till, unsorted soil, the water content is the most important

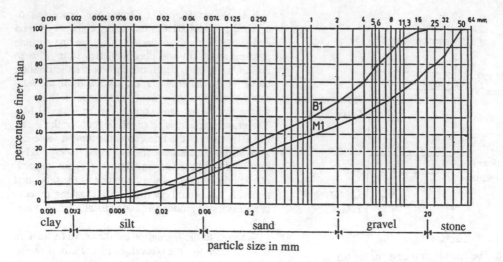

Fig. 7　The particle size distributions for the non-processed M1 amd processed B1.

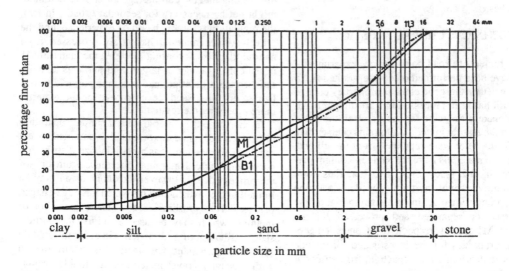

Fig. 8　The particle size distributions for the non-processed M1 and processed B1

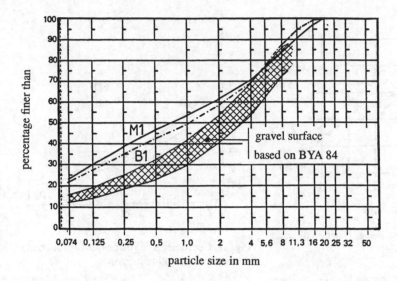

Fig. 9   The gravel surface according to Swedish National Road Board (BYA 84).

factor to govern the compaction behaviour. A till material needs certain water content to reach its maximum dry density. Here, the water content works as a grease film between particle cones, and such water content is called the optimal water content.

The compaction behaviour of the till materials are investigated by the modified Proctor method. The results from the experiments on the materials are given in Fig. 10 and Table 1. It can be seen from Fig. 10 that the material B1 has a curve over the one of the material M1, indicating that the material B1 has a higher maximum dry density and thus has somewhat better compaction behaviour than the material M1. The relatively top curve for the material M1 implies that it is sensible to variation in water content. On the other hand, the curve for the material B1 is flatter and shows a trend to a second maximum, which can be explained by the insensitivity of the material to variation in water content.

Table 1 Results from compaction tests

material	ρ_s g/cm^3	w, % optimal	ρ_d, g/cm^3 max	e
M1	2.68	6.0	2.18	0.23
B1	2.71	5.5	2.20	0.23

'Thus, the crushed or processed till has a better compaction behaviour than the unprocessed till. Both materials are well compacted with no notable difference, however, the processed till has a somewhat higher grain density, ρ_s, than the unprocessed. Note also that both materials have the same void ratio, Table 1.

4.3 Frost-susceptibility

The frost-susceptibility concerns the frozen heave, thawing bearing capacity and thawing-induced slope stability. Traditionally in Sweden, a soil may be classified into one of the three groups for frost-susceptibility, i.e. frost-susceptibility class I, non-frost soil, frost-susceptibility class II, intermediate frost-susceptible soil, and frost-susceptibility class III, very frost-susceptible soil.

According to Fig. 7, the weight percentage of particle sizes less than 0.002 mm is about 2%, the weight percentage of particle sizes less than 0.074 mm is 22% and the weight percentage of particle sizes less than 0.25 mm is 32 - 36%, for both materials M1 and B1 in the range of 0 - 20 mm. Thus, these materials can be classified as intermediate frost-susceptible soil, i.e. frost-susceptibility class II in the Swedish classification system (BYA 1984). It is also noted that both materials have a tendency to a more frost-susceptible property, and may even be classified

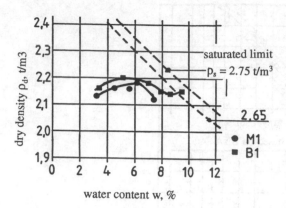

Fig. 10 Compacted dry densities as function of water content for both till materials.

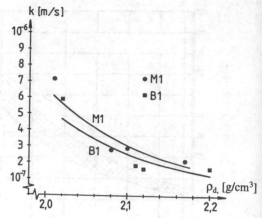

Fig. 11 Coefficients of permeabilit as function of dry density for both till materials.

as medium to high frost-susceptible, i.e. between frost-susceptibility class II and III. Therefore, certain mixing method may be needed to reduce the frost-susceptibility of the till, also referred to Fig. 3.

4.4 Permeability and capillarity

Like sands or silts, the permeability of the concerned till is primarily governed by its particle size distribution and void ratio. Based on a number of permeability tests on the till, it is found that the modified Hazen equation (Broms 1981) may be valid

$$k = 200 \cdot d_{10}^{2} \cdot e^{2} \tag{1}$$

where k is the coefficient of permeability (cm/s) and d_{10} the particle size (cm) corresponding to the particle size distribution curve against 10% passed weight.

Laboratory permeability tests on both materials consist of packing, water-saturation and testing processes. Results from the tests are given in Fig. 11 as a function of dry density. As expected, the coefficients of permeability decrease with increasing the dry density ρ_d, and vary in the range of 10^{-6} - 10^{-7} m/s.

The regression analysis is made to the results from both materials, Fig. 11. The regression curve for the non-processed till M1 stands over the one for the processed till, which indicates that the material M1 has some higher permeability than the material B1. This is reasonable if one checks the particle size distribution curves for both materials in Fig. 7. The material B1 has a somewhat higher content of fine soil particles than the material M1, which normally

leads to a lower permeability. However, from a practical point of view, the difference is negligibly small, and their averaged coefficients may be fitted by Eq. (1) with $d_{10} = 0.002$ cm and e.g. $e = 0.28$.

A few capillarity tests on the materials with the particle sizes less than 2.0 mm. A very simple empirical equation appears to be useful

$$h_c = \frac{C}{e \cdot d_{10}} \tag{2}$$

where h_c the capillary height (cm) and C the empirical constant, 0.1 - 0.5, (cm²), which is mainly dependent on e.g. the particle form. For the studied materials, the capillary height is found to be about 8.0 m, a value that is accordant to the dominating fraction of silt particle size. The relevant parameters for Eq. (2) are $C = 0.2$ cm², $e = 0.25$ and $d_{10} = 0.001$ cm.

5 CONCLUDING REMARKS

Laboratory experimental results show that the geotechnical properties of the studied till materials change only to a minor degree during the processing procedure compared to the procedure of a simple separation of pebbles and boulders. To facilitate the identification of tills in the future, it is logically necessary to create a closer cooperation between the geologist and the geotechnical engineer. It is anticipated that till will be able to compete both economically and technically with natural gravel in the production of road construction material.

REFERENCES

Beskow, G. 1934. Gruvsvägbanors sammansättning. Teknisk-Ekonomiska utredningar rörande vägväsendet. Del I. Vägar. *Stat.Offent.Utredn.* 27: 365-388.

Broms, B. 1981. *Soil and rock mechanics.* Royal Institute of Technology, Stockholm.

BYA 1984. Byggnadstekniska föreskrifter och allmänna råd, Swedish National Road Board TU 154, Borlänge.

Eyles, N. 1983. Glacial geology. Toronto: Pergamon.

Johansson, H.G. & Enkell, L. 1980. Geologiska kriterier av betydelse vid sökande av grovkorniga moräner. Summary: Important geological factors used in the search for gravelly sandy till. *Road and Traffic Institute Report.* 194, Linköping.

Knutz, Å. 1984. The production of Road Material from till. Proc. of ten years of Nordic till research, *Striae* 20: 99-100, Uppsala.

Minell, H. 1986. Bearbetning av några moräntyper i Norrbotten för framställning av vägöverbyggnadsmaterial, *SGAB IRAP* 86010, Luleå.

Viberg, L. & Inganäs, J. 1978. Studies of aerial photographic interpretation for two test areas. Research Report to Swedish National Roard Board.

Viberg, L. 1984. Geobildtolkning av grova moräner. *Swedish Geotechnical Institute Report* 23, Linköping.

7th International IAEG Congress / 7ème Congrès International de AIGI, © 1994 Balkema, Rotterdam, ISBN 90 5410 503 8

Aggregate selection criteria for Drakensberg basalt, Kingdom of Lesotho

Critères de sélection d'agrégats de basalte de Drakensberg, Royaume du Lesotho

J.L.van Rooy
Department of Geology, University of Pretoria, South Africa

ABSTRACT: The Lesotho Highlands Water Project in the Kingdom of Lesotho is being constructed in an area underlain by Triassic and Jurassic basalts of the Lesotho Formation and Triassic sandstones and mudrocks of the Karoo Sequence. The project will consist of a series of dams and tunnels to convey water from the Lesotho Highlands to the industrial centre of the Republic of South Africa.

The geotechnical properties of the Drakensberg basalt is discussed with special reference to their use as aggregate for concrete, road building and rip-rap. The durability of the basalts is primarily determined by the amount of secondary smectite clays present in the rock. The basalts are subdivided into five groups depending on the amygdale and secondary clay content. A selection procedure is proposed to determine the suitability of the different basalt types for use as aggregate.

RESUMÉ: Le projet de captage des eaux dans les hauts plateaux du royaume du Lesotho est en voie de réalisation dans un environnement de basaltes triassiques et jurassiques de la formation du Lesotho, et de grès et argilites de la séquence du Karoo. Une fois réalisé, ce projet consistera en une série de barrages et tunnels qui détourneront les eaux des hauts plateaux vers les centres industriels de la république d'Afrique du Sud.

Les propriétés géotechniques des basaltes des Drakensbergs sont discutées en fonction de leur utilisation comme agregats pour béton, macadam et enrochement. La qualité des basaltes est essentiellement fonction de la teneur en smectite secondaire dans la roche. Les basaltes sont subdivisés en 5 groupes d'après les vésicules et la teneur en argile. Une méthode de sélection est proposée, afin de déterminer la qualité des différents types de basaltes destinés à être utilisés comme agregats.

1 INTRODUCTION

The Kingdom of Lesotho is a land locked mountainous country in Southern Africa. The largest resource is the abundance of water as a result of annual rain- and snowfalls of approximately 1 000 mm. Half the flow of the Senqu-/Orange River (150 m³/s) originates in Lesotho.

The so-called Pretoria-Witwatersrand-Vereeniging (PWV) complex is the industrial centre of the Republic of South Africa with approximately 60 per cent of the industrial production of South Africa sourced from this area. The water demand has been satisfied by the Vaal River supplemented by water from a number of water transfer schemes since 1974. A major untapped source was the Orange River which originates in the Lesotho Highlands, approximately 300 km to the south of the PWV area.

The Lesotho Highlands Water Project (LHWP) was decided upon for the extraction and gravity transfer of water from the upper reaches of the Orange River in Lesotho to the Republic of South Africa (RSA). The project will be constructed in four phases, over a period of 25 years, depending on the water demand of the RSA. A final treaty was signed by the Kingdom of Lesotho and the Republic of South Africa on 24 October 1986 which approved the LHWP and resulted in the commencement of construction work on phase IA of the project.

There will be five big dams with a combined storage capacity of 6.5 km³, a hydro power station with a capacity of 110 MW, tunnels with a combined length of 225 km and new and upgraded roads totalling 650 km at the completion of the project (JPTC 1991).

The project area is confined to the northern part of the Kingdom of Lesotho and the north eastern part of the Orange Free State, as indicated on Figure 1.

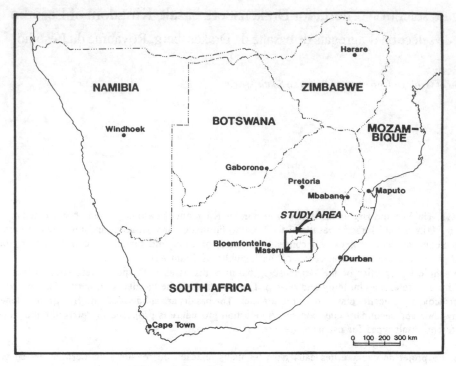

Figure 1. Locality plan of the LHWP area

The Phase IA design started in 1986 and will include the construction of the Katse Dam in the Malibamatso River, a 45 km long Transfer Tunnel 4.95 m in diameter from Katse Dam to the Muela Hydro-electric power station, the Muela Dam which will be part of the Hydro-electric scheme and a Delivery Tunnel, 36 km long from the Muela Dam to the Ash River a tributary of the Vaal River. The Katse Dam will be one of the largest dams in the Southern Hemisphere and, with a height of 180m, it will be the highest in Africa.

The level of interest in the geotechnical properties and behaviour of the Karoo Sequence rocks has increased dramatically since the LHWP became a reality.

Concern about the durability of the sedimentary rocks as well as Lesotho basalts, and Karoo dolerites existed, due to the known presence of secondary clay minerals in these rocks and was further underlined after the observation of hair-line cracks in the basalts, which sometimes run through the amygdaloidal zones of the specimens causing a reduction in strength and previous experience with slaking mudrocks. This paper deals only with the durability of the basalts.

2 GEOLOGY OF THE LHWP

The project area is located in the Great Karoo Basin, a large shallow basin of mainly sedimentary rocks and between 200 to about 160 million years old. The formation of the basin took place while Africa was still part of the supercontinent, Gondwanaland.

The present day remnant of the Karoo Basin underlies the whole of Lesotho as well as about 75 per cent of the surface area of the RSA.

This basin is a near horizontal layered sequence of sedimentary rocks of continental origin capped by thick flood basalts of the Lesotho Formation, also referred to as the Drakensberg Basalts (Figure 2).

Katse Dam, the upgraded and new infrastructure, as well as the transfer tunnel will be situated in the Drakensberg basalts and associated dolerite dykes.

An enormous igneous event, extending from the late Triassic, through the Jurassic and into early Cretaceous, occured on the fragmenting Gondwanaland supercontinent and is referred to as the Karoo flood basalt province. (Irwin et al 1980).

The Drakensberg basalts are intersected by dolerite dykes which formed the channels along which the lava was fed to the surface. The lavas consist almost totally of basalt, petrographically very similar to the Karoo dolerites.

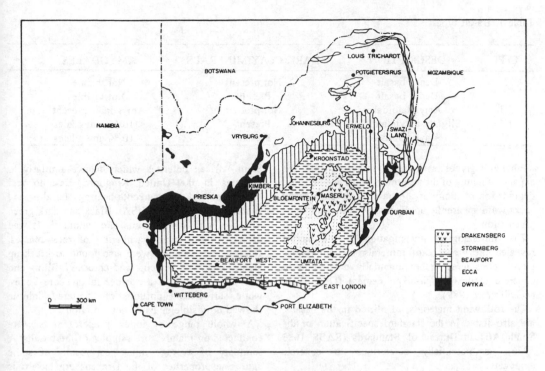

Figure 2. Geology of the Karoo Basin

In the main part of the lava pile, the basalt flows are mostly evenly superposed. The average thickness of the flows is about 6 metres (OSC - Olivier Shand Lahmeyer MacDonald Consortium, 1986) although the median value of 3.2 m and the upper and lower quartile values of 6.2 m and 1.4 m are more indicative of the usual flow thickness (Galliers et al 1991).

The zoning within each lava flow has been described in detail by a number of authors (OSC 1986; Melvill et al 1989; Lesotho Highlands Consultants 1989; Van Rooy 1992; Galliers et al 1991). In all but the thinnest flows, a vertical zonation based on the division of the basalts into amygdaloidal and non-amygdaloidal varieties is evident. Thin flows are typically amygdaloidal throughout but the thick flows consist typically of a massive, fine to relatively coarse grained, central zone bounded at the top and bottom by amygdaloidal zones.

These non-amygdaloidal basalt zones show the most prominent development of dark disseminated clay minerals, consisting, at least partly, of smectite clay minerals. These clay minerals originate from the deuteric alteration of primary interstitial glass and pyroxene (Van Rooy and Nixon 1990) or the filling of small angular voids in the rock as a result of

secondary deposition from circulating ground water.

Five different types of basalt have been identified by Van Rooy (1992) based on the visible amygdale content and the presence of dark disseminated clay minerals. The amygdaloidal types were further subdivided into slightly and moderately to highly amygdaloidal basalts (Table 1).

During the deuteric alteration of the basalts pseudomorphous replacement of olivine, localized but intensive attack on the plagioclase, and the replacement - partial in some places, complete in others - of the original glass occurred. Chlorite is a common product of primary mineral alteration with accompanying carbonates, and montmorillonite is a common product in vesicles of basalt which have been subjected to the actions of solutions and gases originating from a cooling basaltic magma.

The minerals produced by the post-consolidation processes are of key importance to the durability of the basalts.

3 GEOTECHNICAL PROBLEMS ASSOCIATED WITH THE BASALTS

The geotechnical problems in basalt are largely related to its mineralogical composition and texture.

3247

Table 1. Basalt types

TYPE	DESCRIPTION	DARK CLAY MINERALS	AMYGDALES
I	Dense basalt	Not present	Not visible
II	Dense basalt	Present	Not visible
III	Amygdaloidal	Not present	Amygdales present
IV	Slightly amygdaloidal	Present	< 10 % amygdales
V	Amygdaloidal	Present	> 10 % amygdales

Ethylene glycol soak tests indicated that the expansive nature of the montmorillonite would cause degradation of many of the basalts, although many rocks were apparently unaffected even by prolonged glycolation.

The mineralogical investigation of the basalts revealed the presence of minerals and amorphous phases considered to be potentially deleterious to concrete made of high-alkali cement (OSC 1986; Melvill et al 1989).

The following materials, identified in the basalts, are also listed in the standard specification of the South African Bureau of Standards (SABS 1083 1976) as potentially deleterious to concrete aggregate:
Clay minerals (both expansive and non-expansive)
Glass (in its original state or devitrified)
Calcite
Zeolites
Silica minerals (quartz, opal, chalcedony, although fairly rare in the basalts).

The distinction between dolerite and basalt is of importance in the selection of aggregate. It appears that the reliance on a basic petrographic description (i.e. dolerite or basalt) as the principle means of distinguishing between sound and unsound aggregate is unwise and that the assumption that dolerites constitute high quality aggregates is not necessarily valid (Orr 1979).

Orr (1979) described rapid-weathering

South-African dolerites which behave similarly to dolerites in the United Kingdom, Lesotho and Mauritius and to smectite-bearing basalts and gabbros in the Western USA. Black and dark green soft clay infilling of joints and within ill-defined patches caused rapid weathering of rock masses. Dolerite core samples were also found to break up along joints with smectite and chlorite infilling and later disintegrated to a powder in the core boxes, while surrounding portions of unaltered dolerite showed no visible signs of deterioration.

A whole range of index property tests were conducted on mainly core samples of the basalts. A summary of average index test results of some aggregate properties of the Drakensberg basalts is given in Table 2.

The results of the tests follow the expected trends according to the mineralogical composition for the different basalt types. The visual classification of the basalt into the different types according to Table 1 is therefore an essential part of the durability evaluation.

4 SELECTION OF SOUND BASALT AGGREGATE

It is evident from the test results that the basalt types containing visible secondary clay minerals (Types II, IV and V) have higher clay mineral percentages and

Table 2. Average index test results of some aggregate properties of the Drakensberg basalts

TYPE	% CLAY (smectite)	WATER ABSORPTION	S.G.	DENSITY	WET/DRY % loss	MgSO$_4$	UCS (MPa)
I	8-10	0.9	2.88	2844	0.1	5.5	189
II	33-44	2.2	2.75	2684	1.2	9.5	124
III	10-26	2.5	2.48	2545	1.1	5.3	129
IV	21-49	2.3	2.6	2792	0.65	4.7	151
V	15-45	3.0	2.46	2696	12	10.8	122

water absorption values than the Types I and III basalts without visible secondary clay minerals. The only basalt type with an acceptably low water absorption value is the dense non-amygdaloidal basalt Type I. The water absorption of all the other types is influenced by the amygdale content and the presence of clay minerals.

The specific gravity seems to be useful only in distinguishing between non-amygdaloidal types with clay minerals and non-amygdaloidal types without clay minerals (Types I and II), the latter having the higher specific gravity values. The role of the amygdales override the influence of the clay minerals when determining this property. The specific gravity is therefore of virtually no relevance when trying to determine, for instance, the durability or degree of alteration of the amygdaloidal basalts. The same applies to the density values where the amygdale content again greatly influences the test results.

The wetting and drying test results are also influenced by the amygdales present and the clay minerals seem to play a subordinate role. The basalt Type V, with clay minerals and amygdales present showed the highest per cent loss during this test. The magnesium sulphate test results also seem to follow the same trends.

The values of the unconfined compressive strength test depend to a large extend on the intergrain bonds between the individual minerals. This is influenced by the presence of cavities (amygdales) and weak minerals (secondary clay minerals) and the lower values for this test of Types II, III and V basalts is proof thereof.

The basalt were also subjected to the ethylene glycol soaking test. Although this test is regarded as being severe, it shows the critical amount of secondary clay minerals necessary to cause degradation of the rock under variable moisture conditions. The value of this test, as an indicator of the critical amount of clay, is enhanced if the results are correlated with the percentage secondary clay minerals from a XRD analysis.

Figure 3 is a schematic presentation of a proposed basalt aggregate selection procedure for sound road building, concrete and rip-rap materials. The tests listed in Table 3 with their corresponding acceptance limits should be used as guideline values in selection the procedure. Due to the limited knowledge of the long term behaviour of the basalt aggregates a fairly conservative approach was followed and it is suggested that these criteria be followed until further in-service behaviour patterns suggest otherwise.

If all the test results obtained during this research project is taken into account it is clear that only the

Table 3. Compliance limits for sound aggregate

Test Method	Acceptance limits
Per cent smectite	< 20 %
Water absorption (%)	< 2 %
Wetting and drying (% loss)	< 10 %
Magnesium sulphate	< 18 %
UCS (MPa)	> 160

dense basalt without clay minerals (Type I) and the slightly amygdaloidal basalt (less than 10 % amygdales) without visible secondary clay minerals (Type III) will comply with the selection criteria for sound basalt aggregate. The Type IV basalt might also be suitable depending on the secondary clay content. The Katse Dam wall is being constructed with a Type IV basalt with a low percentage secondary clay.

Due to the occurence of the flood basalts as consecutive flows of varying thicknessess it is extremely difficult to obtain a thick enough flow from which rock with fairly homogeneous properties and quality can be quarried. Normally a mixing of different types with property values close to the acceptance limits are utilised.

5 SUMMARY

A proper characterization of materials used or intended to be used for construction purposes should include petrographic description, prior to any mechanical testing. The most important factors determining the durability are the amount of clay minerals present, as well as their distribution within the rock. The petrography, therfore, plays an overriding role in the chemical durability of the basalt.

6 ACKNOWLEDGEMENTS

The following organizations are thanked for permission to use information from various unpublished reports and documents: Lesotho Highlands Developing Authority, Lesotho Highlands Consultants and Highlands Delivery Tunnel Consultants.

7 REFERENCES

Galliers, R.M.; Brackley, I.J.A.; Lephoma, M. and Sympson, A.B., 1991. Expected engineer – ing geology of the transfer tunnel. *Proc. 10th*

Reg. Conf. for Africa on Soil Mech.and Found. Eng., Maseru, Lesotho. September 1991, 351-358.

Irwin, P.; Akhurst, J.; Irwin, D., 1980. *A Field guide to the Natal Drakensberg*. Natal Branch of the Wildlife Society of Southern Africa.

JPTC, 1991. Joint Permanent Technical Committee. Brochure on the Lesotho Highland Water Project.

Lesotho Highlands Consultants, 1989. *Tender document for Katse dam and appurtenant works*. Data for tenderers. **3, 4 and 5**. Lesotho Highlands Water Project. February 1989. (Unpublished)

Melvill, A.L.; Lephoma, M.; Hartford, D.; Eagles, M., 1989. Concrete in the Lesotho Highlands Water Project. *Concrete Society of Southern Africa*. National Convention. Concrete into the 90's. October 1989. Sun City, Bophuthatswana.

OSC (Olivier Shand and LahMeyer MacDonald Consortiums), 1986. *Lesotho Highlands Water Project, Feasibility Study*. April 1986.

Orr, C.M., 1979. Rapid weathering dolerites. *The Civil Eng in South Africa*, 7, 161-167.

SABS, 1976. (South African Bureau for Standards) Aggregate from natural sources. SABS Specification 1083-1976 (as ammended 1979), Pretoria, SABS.

Van Rooy, J.L.; Nixon, N., 1990. The relationship between petrography and durability of Drakensberg basalts. *S. A. Jnl. of Geology*, **93** no. 5/6. 729-737.

Van Rooy, J.L., 1992. *Some rock durability aspects of Drakensberg basalts for civil engineering construction*. Unpublished PHd thesis. University of Pretoria.

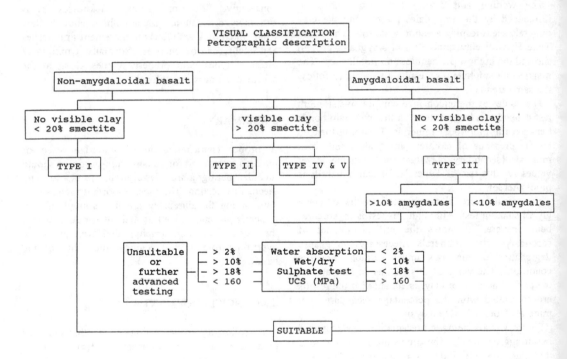

Figure 3. Schematic presentation of selection procedure

3250

Alkali-silica potentially reactive minerals in Lazio and Campania alluvial deposits (Italy)

Minéraux de silice potentiellement réactifs aux alcalis dans les depôts alluvionaires de Lazio et Campania (Italie)

G. Barisone, G. Bottino & G. Restivo
Dipartamento Georisorse e Territorio, Politecnico di Torino, Italy

ABSTRACT: This work studies the presence and the distribution of alkali-silica reaction potentially bearing minerals in recent and actual alluvial deposits of the Lazio and Campania regions (Central-Western Italy). These two regions, with some hydrogeologically connected surrounding territories, cover a total surface of about 30,000km2; the alluvial deposits were investigated by collecting and analyzing almost 30 samples, mainly from actual alluvia and collected directly in river beds or - due to their prevailingly coarse granulometry - in the stock piles of crushed sand produced in active quarries. The quantitative evaluation of the potentially reactive minerals was made by means of psammographic analysis (sometimes integrated by means of semi- quantitative diffrattometric determinations), identifying (on the petrographic microscope) the grains of previously crushed (mainly to 2-0.04mm), HCl etched (to eliminate carbonates) and sieved (8 granulometric classes) samples. The main reactive minerals identified are flint (by far the most important), chalcedony and opal (microfossiles); the undulatory extinction quartz was not taken into account because of its weak and uncertain reactivity degree. The results obtained are synthetized in two schematic maps (originally 1:500,000 scale), one for each region; in these maps, which also show the main outcropping flint bearing formations, alluvial deposits are divided into 5 classes, according to their reactive mineral contents.

RÉSUMÉ: L'étude ici présentée concerne la détermination quantitative - dans les alluvions récentes et actuelles du Lazio et de la Campania - des minéraux susceptibles de déclencher dans les bétons la réaction alkali-silice. Le deux régions étudiées couvrent, avec quelques territoires hydrogéologiquement reliés, une surface globale d'environ 30,000km2; l'étude des alluvions a comporté l'acquisition et l'analyse d'une trentaine d'échantillons, en directement récoltés dans le lit des fleuves. La teneur en minéraux réactifs des granulats examinés a été déterminée par analyse psammo-pétrographique (méthode optique), particulièrement rapide et precise, intégrée par des analyses diffracttométriques semi-quantitatives; les déterminations ont été conduites sur les échantillons préalablement brisés (2-0.04mm), traités avec HCl pour éliminer les carbonates et subdivisés aux tamis en 8 classes granulométriques. Le principaux minéraux réactifs identifiés sont la silex (largement le plus commun), le calcédoine et l'opale (ce dernier, en prévalence sous forme de microfossiles); le quartz a extinction ondulé n'a pas été considéré, à cause de son degré de réactivité, faible et incertain. Les résultats obtenus sont synthétisés en deux cartes (échelle a l'origine 1:500,000), une pour chaque région; dans ces cartes, qui montrent aussi les affleurements principaux de roches contenant silex, les dépôts alluvionnaires sont divisés en 5 classes, suivant leur teneur en minéraux réactifs.

1 INTRODUCTION

The third and final part of a study carried out, over the last 8 years (Barisone et al., 1986; Barisone et al., 1989) on the alluvial deposits in the Italian peninsula (south of the river Po) is here described. The study was concentrated on recent and present alluvia, with the aim of determining their content in alkali-silica reaction potentially bearing minerals; in particular, in this last phase the deposits of two regions on the western coast, Lazio and Campania (and of some idrogeologically connected surrounding territories) were examined (for a total area of about 30.000km2).

2 GEOLOGICAL NOTES

A geological-petrographic study on the outcropping formations has been carried out, in order to better define the origin of the potentially reactive minerals found in the studied alluvia (mainly flint, with varied amounts of calcedony, opal, microfossiles, etc.); due to the large area involved, the data was obtained mainly by the critical analysis of recent existing scientific works, integrated by in situ surveys on the most relevant or contradictory zones. It was therefore possible to group the pre-quaternary rocks into three classes. according to the absence, the lesser presence and the greater presence of flint.

The more extensively outcropping geological formations in the here considered area are mainly composed of volcanic formations, the age of which varies from Pliocene (northern Lazio) to Pleistocene (central Campania); sedimentary formations of the Mesozoic and Cenozoic age are also very common, particularly in northern Lazio and in southern Campania. These sedimentary formations contain some lithologies characterized by a relatively high content of potentially reactive silica; the most extensively outcropping flint bearing formations in the studied area are listed in Tab.1. These formations cover a wide period of time, from Triassic to Miocene; on average, however, flint occcurs more frequently in Mesozoic formations representing "complete series" and in some Miocenic stages.

It should be noted that the area covered by the outcropping flint bearing formations is only one of the parameters which influence the presence and the quantity of flint in recent and present alluvia: the aptitude of "mother" rocks to be eroded and the different ways of the detritical transport are also of very great importance.

The flint bearing formations are represented, in the basins of the Noce, Sele and Calore rivers (southern Campania), by carbonatic rocks of the Triassic and Jurassic age (Lagonegro and M. Facito formations), outcropping mainly in the higher part of the water courses.

In the Volturno basin these formations are mainly formed by marly limestones of slope facies (Upper Cretaceous-Palaeogene); clays of the "Argille Scagliose" complex and marly limestones of the Triassic age are less important, due to their reduced outcropping.

In the Liri and Garigliano basins (southern Lazio) flint mainly occurs in limestones and marly limestones of the Palaeogene- Cretaceous age; clays of the "Argille Scagliose" complex also outcropp along the middle course of the Liri river.

The wide basin formed by the Tevere river and its main tributaries (Clitunno, Paglia, Aniene) contain many flint bearing rocks (mainly limestones and marls), whose age varies from Jurassic (Rosso ammonitico and Carniola formations), Cretaceous (Maiolica and Scaglia formations) to Miocene (marls of pelagic facies, Bisciaro formation).

3 ALLUVIAL DEPOSIT ANALYSIS

A short description of the techniques adopted for the collection and the analysis of representative samples from the alluvial deposits studied is given in the following sections.

3.1 Sampling methodology

A correct sampling of alluvial deposits, often characterized by a prevailingly coarse granulometry, would have required the in situ collection of huge amounts of material (Barisone, 1985). The problem was partially solved by collecting, whenever possible, the samples directly in active quarries (obviously quarrying only the alluvial deposits of a single river); in such cases, due to the homogenization produced by the industrial crushing, it was enough to take a total amount of about 5kg of crushed sand (obtained from different places within the stock piles) for each sample.

In the case of the absence of quarries, the samples were directly collected in river beds; these samples represent, obviously, only the natural sand composition (often forming only a reduced fraction of total alluvia).

3.2 *Psammo-petrographic analysis*

The so collected samples were analysed as follows (Barisone, 1984):
1. If necessary, crushing of the coarse material collected from river beds, reducing it to dimensions >2mm;
2. Successive quarterings, in order to obtain a final representative sample weight of 100-300g (conformably to a prevailing silicatic or carbonatic composition;
3. Warm etching in diluted HCl (8-24h), in order to eliminate the soluble minerals (carbonates, sulphates, etc.);

Lithology	Facies	Age	Formation
Marls with chert	Pelagic	Middle-lower Miocene	Bisciaro, Schlier
Sandstones, marls and shales		Middle-lower Miocene	Marnoso-Arenacea, Daunia , Faeto Flysch
Marls with chert, with detritical supplies	Slope (transition)	Middle-lower Miocene	
Sandstones, shales and marls		Palaeogene	Macigno, Varicoloured Shales
Limestones and marly limestones,often with chert	Pelagic	Palaeogene- Upper Cretaceous	Scaglia rossa, Scaglia bianca
Limestones and marly limestones, sometimes cherty, with detritical supplies	Slope (transition)	Palaeogene- Upper Cretaceous	
Micritic limestones and clayey micrites, often with chert	Pelagic	Cretaceous- Upper Jurassic	Maiolica, Calcare Rupestre, M.ti Lepini
Limestones and marly limestones, locally with chert, with detrital supplies	Slope (transition)	Cretaceous- Upper Jurassic	
Limestones,marly limestones and marls (cherty)	Pelagic	Jurassic	Corniola, Rosso Ammonitico, Lagonegro
Limestones and marly limestones, sometimes cherty, with detrital supplies	Slope (transition)	Jurassic	
Limestones often cherty and marly limestones	Pelagic	Upper Triassic	
Marly limestones, sometimes with chert	Pelagic	Middle Triassic	M. Facito
Shales and limestones of chaotic complexes		Various	Argille Scagliose

Table 1. Main flint-bearing formations outcropping on the East side of the Appenines in the Lazio and Campania regions.

4. Wet sieving, in order to obtain 8 granulometric classes: >2mm, 2-1mm, 1-0.500mm, 0.500-0.250mm, 0.250-0.125mm, 0.125-0.075mm, 0.075-0.040mm, <0.040mm;

5. Microscopic analysis: using a stereoscopic microscope for the classes >1mm, a petrographic microscope for those from 1mm to 0.04mm (with the grains previously immersed in eugenole). For each granulometric class a minimum number of grains was identified, (from 50 in the >2mm class up to 600 in the 0.075- 0.040mm class), reaching a total number of about 2000 countings for each sample.

In order to verify the results of the psammographic analysis, only a few thin sections of sand, previously agglomerated with epoxy resins, (Christensen et al., 1983) were used, the better definition being obtained at the price of a hard dilatation of time and cost of the analysis itself.

Diffractometric analysis were also carried out on about a half of the samples, in order to verify the kinds of soluble minerals which occurred and to compare the mineralogical composition (obviously, only concerning cristalline minerals) of 0.075-0.040mm and <0.040mm classes (this last being very difficult to analyse by optical methods). Finally the <0.040mm class, which always represents only a very small percentage of the total and with diffractometric curves normally very similar to the 0.075- 0.040mm class, was mineralogically assimilated to this last class.

Statistical data on the mineralogical composition of the samples are shown in Tab.2; only four groups are dealt with for the purposes of this work: flint, opal and chalcedony microfossiles, quartz plus feldspars or feldspathoids, carbonates.

4 CONCLUSIONS

On the basis of the analysis results, the alluvial deposits examined were subdivided into five classes, according to their reactive mineral contents and, consequently, to their potential dangerosity; these classes correspond to reactive mineral contents of, respectively:

1. <0.5% (not reactive)

2. 0.5-2% (reactivity uncertain, because of the contained species of reactive minerals; this might require further investigations, i.e by means of the mortar bar test or similar)

3. 2-7% (almost surely reactive, the intensity of the reaction depending on the species of the contained reactive minerals)

4. 7-12% (surely reactive, the intensity of the reaction depending on the species of the contained reactive minerals)

5. >12% (reactivity uncertain, because of the contained reactive minerals; this requires further investigation, i.e. by means of the mortar bar test or similar).

Table 2. Statistical data on natural aggregates composition

District	Samples N^	Zone	Samples N^	Flint					Quartz and feldspars or feldspath oids		Carbonates	
				Average %	Standard deviation	Values %		Micro fossiles	Average %	Standard deviation	Average %	Standard deviation
						Min	Max					
Campania	11	South	4	6,1	5,5	13,4	0,5	-	31,7	18	41,5	20,9
		Center South	4	3	2,1	5,6	1,1	0	13,8	15	63,1	23,7
		North	3	8,8	1,8	9,9	6,7	+	16	17,7	68,1	19,1
Lazio	16	South	4	5,1	1,6	6,1	2,7	++	33,1	13,7	48,3	13,8
		Center South	4	9,8	2,1	12,9	8,1	+	3,5	2,8	81,9	3,2
		North	8	4,7	1,7	7,2	2,4	-	35,7	16,7	42,6	20,5

Microfossiles: 0 Absent; - Uncertain or scarce (<1% of the total flint); + Normal (1-5 %); ++ Abundant (5-20 %); +++ Very abundant (>20 % of the total flint).

3254

Fig. 1. Schematic map of flint distribution in the Campania region

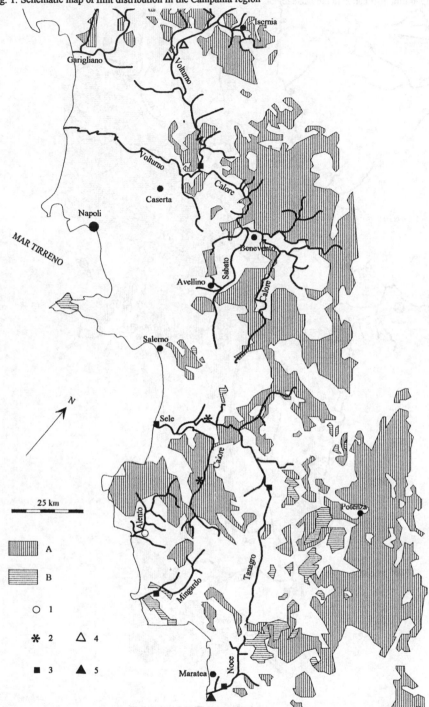

Rocks with flint content: A=low B=high
Flint content of alluvia: 1=<0.5 % 2=0.5-2 % 3=2-7 % 4=7-12 % 5=>12 %

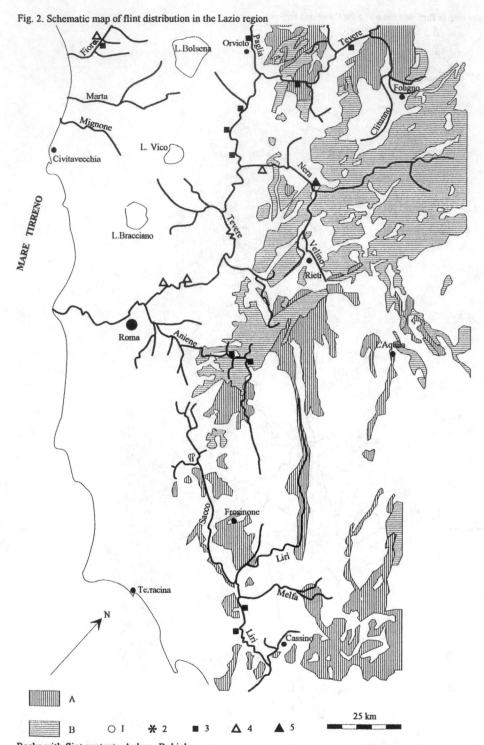

Fig. 2. Schematic map of flint distribution in the Lazio region

Rocks with flint content: A=low B=high
Flint content of alluvia: 1=<0.5 % 2=0.5-2 % 3=2-7 % 4=7-12 % 5=>12 %

This subdivision was made on the basis of both bibliographic data and the real reactivity of Italian flints, as stated according to studies now near to conclusion; it does not yet appear to be well calibrated for Italian sands, but the authors have preferred to pospone the definition of more detailed intervals until the work they are carrying out on the reactivity rating of the different reactive minerals in Italy is completed.

The alluvia content in reactive minerals in Campania shows a very irregular distribution; however, it should be noted that the rivers of the central-southern part currently have low flint contents (0.5-4%), due to the prevalence, in their basins, of rocks (volcanites in the center, limestones in the south) without flint, whilst in the northern part of the region flint in alluvia can reach 6-10%.

The north-west part of the Lazio region is mainly formed by volcanic rocks; here the rivers moreover have reduced floods and lengths, and their deposits are almost only loams and clays: consequently, only one river (having, moreover, the largest part of its basin in the Toscana region) in this zone was sampled, with sand and gravel deposits suitable for aggregates. The largest part of the territory has, however, large outcroppings of sedimentary rocks, often flint bearing; here the alluvia flint contents vary from 2 % to 10 %. It should be noted that here the largest number of samples come from the deposits of the Tevere river, whose basin extends over almost all the region.

The reactive mineral contents of the tested alluvia is summarized in two schematic maps (Figs.1 and 2), here reduced from the previous 1:500,000 scale, where different symbols indicate the location of the samples and their flint content.

REFERENCES

Barisone, G. 1984. Petrographic analysis of aggregates related to alkali-silica reaction. Proc. Symposium International sur les granulats, Nice.

Barisone, G. 1985. Gli inerti per calcestruzzo ed il fenomeno dell'alkali-silica reaction sul versante adriatico dell'Italia centro-meridionale. Proc. III Convegno Nazionale Attivita` estrattiva dei minerali di 2a categoria, Bari.

Barisone, G. & Bottino, G. & Cardu, M. 1986. Italian peninsular aggregates and the alkali-silica reaction. Proc. V International Congress of I.A.E.G., Buenos Aires.

Barisone, G. & Bottino, G. & Pavia, R. 1989. Alkali-silica reaction potentially bearing minerals in alluvial deposits of Calabria, Umbria and Toscana regions (Italy). Proc. VIII International Conference on Alkali-aggregate reaction, Kyoto.

Christensen P., Brandt I., Henriksen K.R., 1983. Quality control of sand by means of point-counting in light-optical microscope. Proc. VI International Conference on Alkalis in concrete, Copenhagen.

La dégradation des granulats dans le matériau béton: Un exemple d'alcali-réaction dans un ouvrage de Génie Civil

The aggregate degradation in concrete: An example of alkali reaction in an engineering work

H. Hornain, G. Martinet & F. Jehenne
Laboratoire d'Études et de Recherches sur les Matériaux, Bagnolet, France

A. Le Roux, J.-S. Guédon-Dubied & F. Martineau
Laboratoire Central des Ponts et Chaussées, Paris, France

ABSTRACT: Alkali-aggregate reaction in concrete is related to an alteration process of siliceous rocks in the wet and very alkaline environment constituted by the hydrated cement paste. The alteration product is in the most cases an alkali-silica gel or an alkali-silica gel containing calcium. The imbibition pressure induced by water penetration in the gel can lead to the cracking of the concrete. The rocks affected by the reaction are those which contain, in given proportions, ill-crystallized and more or less scaterred silica.

Alkali-reaction can lead to significant degradations of concrete buildings and its economic weight is of importance. Therefore it has been necessary to develop method of diagnosis and evaluation of the potentiel risks due to this reaction.

The first stage is the evaluation of the aggregates by reliable tests allowing the prevision of the long term behaviour of a construction. Such a procedure is illustrated by the comparison between the present state of a building in service for ten years and the response of the rocks sampled in the quarry of origin to the rapid tests standardized in France.

RESUME: La dégradation des bétons par alcali-réaction met en jeu des phénomènes d'altération de roches silicatées dans le milieu humide et fortement alcalin que constitue la pâte de ciment hydratée. Le produit d'altération est le plus souvent un gel silico-alcalin ou silico-calco-alcalin dont l'imbibition par l'eau induit des pressions de gonflement suffisantes pour entraîner la fissuration des bétons. Les roches concernées par la réaction sont celles qui renferment des silices amorphes ou mal cristallisées sous une forme plus ou moins dispersée et dans des proportions déterminées.

L'alcali-réaction peut entraîner des dégradations importantes des ouvrages en béton et son importance économique est grande. C'est pourquoi il a été nécessaire de développer des méthodes de diagnostic et d'évaluation des risques potentiels dus à cette réaction.

La qualification des granulats par des essais fiables permettant de prévoir le comportement d'un ouvrage à long terme constitue la première étape. L'illustration d'une telle démarche est faite à l'aide d'une comparaison entre l'état actuel d'un ouvrage en service depuis plusieurs dizaines d'années et la réponse des matériaux échantillonnés dans la carrière d'origine aux essais rapides normalisés en France.

1. INTRODUCTION

Le béton est un système chimique hétérogène évolutif dans lequel les roches introduites comme granulats peuvent subir des altérations très rapides comparées à celles qui se produisent dans leur milieu naturel où elles sont dans un état proche de leur état d'équilibre et où les mécanismes de transformation sont en général extrèmement lents. Hors de ce milieu naturel, réduites en gravillons et en sables puis incluses dans l'environnement fortement basique que constitue le béton, certaines roches sont susceptibles de s'altérer dans des temps de l'ordre de une à plusieurs dizaines d'années en provoquant des désordres plus ou moins importants dans les ouvrages.

En règle générale, la stabilité des roches dans les bétons est tout à fait compatible avec la durée de vie, de l'ordre du siècle, prévue pour les ouvrages les plus importants. Toutefois l'utilisation du béton comme moyen de stockage et de confinement des déchets d'origine nucléaire où les échelles de temps à prendre en compte sont beaucoup plus grandes, peut reposer le problème de la stabilité à très long terme des roches dans le béton. Des phénomènes très lents tels que ceux qui sont observés dans les massifs rocheux infiltrés par les eaux pourraient avoir lieu tels que, par exemple, la dissolution des roches à feldspaths acides au contact des solutions basiques. A moins que les phénomènes de carbonatation ne prennent le pas sur les phénomènes de dissolution et ne ramènent le liant calcique du béton dans un état proche des roches dont il est issu. Il existe là un vaste domaine d'exploration où les connaissances des géologues et celles des physico-chimistes du béton pourraient être combinées avec profit.

Parmi les altérations rapides (de quelques années à quelques dizaines d'années) des roches dans le béton on peut citer à titre d'exemple l'oxydation des pyrites contenues dans certains granulats qui, par réaction avec les aluminates du ciment, peut conduire à la formation de sulfoaluminates de calcium expansifs. On peut citer également les réactions de dédolomitisation ou de dissolution de la silice en milieu fortement alcalin. Ces deux dernières réactions, dont la seconde est de loin la plus fréquente, constituent les "alcalis-réactions", phénomènes particuliers d'altération des roches non encore totalement élucidés et qui peuvent engendrer des désordres graves dans les bétons (Hornain 1993).

2. CARACTERISTIQUES GENERALES DU MILIEU BETON

Schématiquement le béton durci analogue à un conglomérat le plus souvent polygénique, est formé d'un squelette granulaire de roches, d'un liant hydraté assurant la cohésion du squelette granulaire et d'une phase liquide interstitielle.

La phase liquide interstitielle provient en majeure partie de l'eau de gachage toujours en excès par rapport à l'eau nécessaire pour hydrater le ciment. Cette eau dont la quantité peut être de l'ordre de 120 litres / m^3 dans les bétons traditionnels, est distribuée de manière plus ou moins homogène dans la porosité capillaire du matériau. Son pH, de l'ordre de 13,5 dans les bétons à base de ciment Portland, est fortement basique. Cette basicité est due principalement aux oxydes alcalins libérés par le ciment lors de son hydratation. C'est ainsi que la concentration en hydroxydes alcalins dans la phase liquide interstitielle du béton est très couramment de l'ordre de 0,7 N. La concentration en hydroxyde de calcium, insolubilisé par la présence des alcalins, n'est que de 0,001 N.

La teneur en hydroxydes alcalins de la phase liquide est un paramètre décisif des alcalis-réactions.

Le liant durci, avec lequel la phase liquide interstitielle se trouve en équilibre, est principalement constitué de silicates de calcium hydratés (C-S-H selon l'abréviation généralement utilisée en chimie du ciment) de formule chimique moyenne $3CaO.2SiO_2.xH_2O$. C'est le feutrage des fibres micrométriques de C-S-H qui donne au matériau l'essentiel de sa cohésion. Aux C-S-H sont associés l'hydroxyde de calcium $Ca(OH)_2$, constituant sensible des bétons, divers aluminates de calcium et des résidus de clinker non hydraté. L'ensemble constitue un milieu plus ou moins microporeux dont la microstructure conditionne les transferts par diffusion ou par perméation et par conséquent la sensibilité du matériau aux différentes agressions d'origine interne ou externe.

La microstructure au niveau de l'interface entre le liant et le granulat revêt une grande importance et détermine également le comportement mécanique et les propriétés de transferts du béton. Dans les bétons traditionnels l'interface est caractérisée par l'existence d'une zone microporeuse souvent appelée "auréole de transition" constituée d'assemblages d'hydrates de faciès et de compacité différents de ceux du liant dans la masse. Cette auréole de transition constitue une zone de faiblesse mécanique et une zone privilégiée de transferts et d'échanges physico-chimiques. Elle est limitée ou inexistante dans la nouvelle génération de béton dits "à hautes ou très hautes performances" (BHP ou BTHP).

Au total le béton peut être assimilé à un milieu chimique hétérogène caractérisé par une très forte basicité, et une microstructure plus ou moins poreuse permettant des échanges importants au sein de matériau lui-même et avec le milieu extérieur.

3. UN PHENOMENE PARTICULIER D'ALTERATION DES ROCHES: L'ALCALI-REACTION

Les alcali-réactions peuvent être considérées comme des réactions chimiques hétérogènes solide-liquide où le solide est représenté par le granulat et le liquide par la solution interstitielle fortement alcaline du béton.

On peut distinguer deux grandes catégories d'alcalis-réactions : les réactions de dédolomitisation d'une part et les réactions alcalis-silice ou alcalis-silicates d'autre part.

- La réaction de dédolomitisation, rarement observée en France, consiste en une décomposition du minéral dolomite en brucite $Mg(OH)_2$, carbonate de calcium et carbonate alcalin soluble qui permet la régénération de l'hydroxyde alcalin et à la réaction de se poursuivre. La réaction, qui se décompose en deux étapes, peut s'écrire :

1) $CaMg(CO_3)_2 + 2MeOH \Longrightarrow$
$Mg(OH)_2 + CaCO_3 + Me_2CO_3$

2) $Me_2CO_3 + Ca(OH)_2 \Longrightarrow CaCO_3 + 2MeOH$

avec $Me = K^+$, Na^+,...

Elle se traduit par une déstructuration de l'interface liant-granulats et peut provoquer un affaiblissement des ouvrages.

- Les réactions alcali-silice, de loin les plus fréquentes, consistent en une dissolution de certaines formes de silice contenues dans les granulats sous l'action de la solution interstitielle basique du béton. Elles conduisent à la formation de gels silico-alcalins ou silico-calco-alcalins expansifs.

Les trois conditions principales de développement des alcalis-réactions sont les suivantes :

* humidité relative au moins égale à environ 80 % ;
* présence dans les roches utilisées comme granulats de variétés réactives de silice en proportions données telles que les opales, les calcédoines, la tridymite, la cristobalite, les microquartz de recristallisation à réseau déformé ou encore les verres siliceux plus ou moins dévitrifiés. Ces silices réactives peuvent être contenues dans un grand nombre de roches d'origine magmatique (granites, andésites, tufs volcaniques...), métamorphiques (gneiss, quartzites, cornéennes...) et sédimentaires (grés, silex, calcaires à silice diffuse...) (Le Roux 1991). La cinétique de la réaction dépend du degré de désordre du réseau cristallin de la silice ainsi que de l'accessibilité de cette dernière à la solution alcaline agressive. Ceci conduit souvent à distinguer les réactions alcalis-silice à cinétique rapide, d'une part et les réactions alcalis-silicates à cinétique lente, d'autre part ;
* présence d'hydroxydes alcalins en quantité suffisante dans le béton. En deçà d'une limite voisine de 3 kg d'oxydes alcalins par m^3 de béton les risques d'alcali-réaction sont limités .

D'autres paramètres tels que la température, les cycles d'humidification-dessiccation peuvent influencer la cinétique des réactions.

Le mécanisme général de la réaction alcali-silice peut être schématisé par une réaction en deux étapes :

--> neutralisation des groupes silanols $\equiv Si - OH$ présents dans les variétés de silice réactives, suivant une réaction du type acide-base :

$$\equiv Si - OH + OH^- \Longrightarrow \equiv Si - O - Na + H_2O$$

--> attaque des ponts siloxanes $\equiv Si - O - Si \equiv$ par les ions OH^- :

$$\equiv Si - O - Si \equiv + 2OH^- + 2Na^+ \Longrightarrow 2(\equiv Si - O - Na) + H_2O$$

Cette réaction conduit à la destruction de la structure du minéral et à la formation dans un premier temps de gels silico-alcalins polymérisés. Les ions calcium présents dans le béton par l'intermédiaire de l'hydroxyde de calcium $Ca(OH)_2$ en équilibre avec la phase liquide interstitielle, jouent également un rôle déterminant dans le développement de la réaction : maintenus à la périphérie des gels aux premiers stades de la réaction par suite de leur encombrement, les ions Ca^{++} pourraient y pénétrer lorsque le réseau est suffisamment distendu et remplacer les ions K^+ et Na^+ qui redeviendraient ainsi disponibles. C'est ainsi que les gels rencontrés dans les bétons affectés par l'alcali-réaction sont le plus souvent silico-calco-alcalins.

Les conditions de formation des gels et leur caractère expansif dépendent des conditions de concentration en silice réactive et en oxydes alcalins. Il existe un rapport pessimal SiO_2 réactif / Na_2O (ou K_2O) de la solution pour lequel les risques d'expansion et de désordres sont maximaux. Différentes études ont montré que ce rapport était voisin de 4,5. Dans la pratique cela se traduit par la prise en compte d'une teneur pessimale en minéraux réactifs voisine de 3%. Cette teneur correspond à l'opale, c'est-à-dire à un minéral pour lequel la réactivité est maximum. Pour un minéral de réactivité plus faible, la teneur pessimale serait sans doute différente. Celà se traduit également par la distinction qu'il y a lieu de faire entre les notions de réactivité et de désordres dans les ouvrages : certaines combinaison se trouvant en dehors de la zone pessimale peuvent réagir sans provoquer de désordres. Ainsi il est possible de mettre en œuvre sans risque des formulations de béton où au moins 60 % des granulats sont potentiellement réactifs.

L'origine des contraintes de gonflement des gels d'alcali-réaction est encore controversée. Deux mécanismes distincts sont proposés : d'une part pression osmotique ou pression d'imbibition exercée au sein du gel par suite de la différence d'activité chimique entre le liquide contenu dans le gel lui-même et la phase liquide interstitielle environnante; d'autre part pression de "cristallisation" liée aux fortes sursaturations locales en ions silicates en particulier. En tout état de cause les contraintes développées par la formation du gel, de l'ordre de 6 à 12 MPa, sont supérieures à la résistance en traction du béton et peuvent donc engendrer des fissurations dommageables pour les ouvrages. L'étude que nous avons présentée à Londres (9th ICAAR 1992) situe le seuil à 5 MPa.

En France l'alcali-réaction bien que relativement peu fréquente, reste un phénomène préoccupant. Elle touche des ouvrages de génie civil tels que des barrages et des ponts. Suivant un inventaire réalisé par le LCPC (Godart 1993), une demi-douzaine de barrages et environ 180 ponts sont plus ou moins affectés. De nombreuses autres constructions telles que des chateaux d'eau, des appontements, des immeubles ont également été affectées. Les désordres constatés sont en relation avec:
- l'utilisation de granulats classés potentiellement réactifs par les essais normalisés de qualification vis-àvis du risque alcali-réaction,
- la présence d'oxydes alcalins en excès,
- et un environnement généralement humide.

Actuellement les problèmes de diagnostic et de dépistage des alcalis-réaction dans les ouvrages sont assez bien résolus. De même les méthodes normalisées mises au point récemment en France permettent de caractériser assez surement la réactivité potentielle des granulats. Les recommandations publiées par le Laboratoire central des Ponts et Chaussées (Document Recommandations 1991) permettent également de prendre en connaissance de cause, toutes les mesures nécessaires à la limitation des risques de désordres par alcali-réaction.

L'exemple qui va suivre présente un cas concret d'alcali-réaction dans un barrage construit il y a environ cinquante ans et illustre *a posteriori* la démarche qui serait adoptée aujourd'hui pour déterminer les mesures à prendre en fonction des conditions locales tant en ce qui concerne les matériaux mis en œuvre que les conditions environnementales.

4. CAS CONCRET D'ALTERATION DES ROCHES PAR ALCALI-REACTION

Il y a lieu de rappeler d'abord brièvement la démarche fixée par le document LCPC "Recommandations provisoires pour la prévention des désordres dus à l'alcali-réaction" (Document Recommandations 1991, 9th ICAAR 1992), rédigé à la demande du Directeur des Routes du Ministère de l'équipement et pris actuellement comme référence en France. Ce document dont une version réactualisée est en cours de parution, fixe le niveau de prévention à atteindre en fonction de la nature de l'ouvrage et des conditions d'environnement. Il précise les vérifications correspondantes à effectuer.

La démarche préventive se fait en deux temps .

Premier temps: détermination du niveau de prévention à atteindre, parmi trois niveaux possibles A, B, C: le choix du niveau de prévention fait intervenir à la fois la catégorie de l'ouvrage et sa classe d'exposition.

Deuxième temps: orientation vers la (ou les) solutions possibles en fonction du niveau de prévention.

Les barrages font partie de la catégorie III regroupant les ouvrages de Génie Civil, pour lesquels le maître d'ouvrage juge l'apparition du risque d'alcali-réaction inadmissible. En conséquence le niveau de prévention est maximum, et les granulats employés doivent être classés comme non réactifs vis-à-vis de l'alcali-réaction. Cependant dans les cas où l'approvisionnement en granulats non réactifs est particulièrement difficile, il pourra être utilisé des granulats reconnus potentiellement réactifs à condition de procéder à une étude approfondie de la formule envisagée, sur des bases expérimentales définies contractuellement. Le schéma suivant donne un aperçu de la démarche à mener pour la détermination de la réactivité du granulat considéré.

Caractérisation des granulats vis à vis de l'Alcali-Réaction
Annexe du fascicule de documentation P 18 542

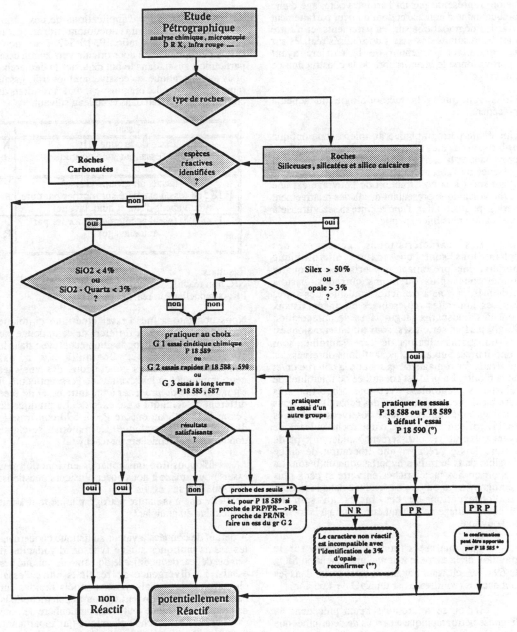

* Intéresse uniquement les sables

** La zone de doute correspond à la zone comprise entre la valeur fixée pour le seuil dans la norme et cette valeur moins 10 %

Le second essai pratiqué, quels que soient cet essai et la valeur obtenue, sera considéré comme décisionnel

(°) dans le cas ou on sera conduit à pratiquer l' essai P 18 590 on aura une information incomplète, seuls P18 588 et P 18 589 distinguent actuellement les PRP

(°°) Reconfirmer soit au niveau de la pétrographie soit en réalisant un autre essai

La démarche adoptée dans ce paragraphe est celle du post diagnostic.
En effet nous considérons: d'une part un ouvrage en béton, représenté par un barrage voûte, agé d'une cinquantaine d'années, et dont on peut parfaitement voir les dégradations sur les parements, et d'autre part les résultats des essais normalisés réalisés sur les granulats provenant de la carrière ayant approvisionné le chantier lors de la construction de l'ouvrage.

Nous appliquons la méthodologie du schéma précédent.

Une étude pétrographique au microscope optique polarisant a été réalisée sur les échantillons provenant de la carrière, dans le but de déterminer la présence ou l'absence d'espèces réactives. La roche ayant servi à la construction de l'ouvrage est une roche volcanique présentant des faciès relativement variés, pouvant aller d'une texture assez vitreuse à une texture plus microgrenue.

Les caractéristiques générales des échantillons sont: une texture microlithique porphyrique présentant des cristaux de quartz automorphes plus ou moins corrodés, parfois disloqués, parfois à extinction roulante. Les autres espèces minérales représentées en phénocristaux sont des feldspaths plagioclases de composition An$_{20}$, parfois séricitisés, souvent aussi disloqués; les feldspaths alcalins de type sanidine, sont minoritaires. Suivant les échantillons observés, les minéraux ferro-magnésiens sont de la chlorite ou/et de la biotite. La pâte est constituée de microlites de feldspaths plagioclase avec une proportion très variable en verre en fonction des échantillons. Dans certaines lames minces ont été observées des figures de dévitrification typiques d'une roche volcanique ayant subie un refroidissement rapide, ce type de texture laisse présager une libération de silice possible dans le milieu hyperbasique du béton. La proportion des faciès riches en verre et ceux plus microgrenus n'a pu être déterminée dans la mesure où l'observation de ces faciès au sein de l'échantillonnage ne pouvait être étendu à la totalité de l'ouvrage.

Une analyse chimique pratiquée sur le gravillon de la carrière a donné une teneur en SiO$_2$ de 69,67 % classant nettement cette roche dans les rhyodacites (confirmée par un K$_2$O de 4,65%).

Si l'on se réfère au schéma précédent: la diagnose pétrographique a permis de déterminer que le granulat est une rhyodacite présentant un certain nombre d'espèces réactives au regard de l'alcali-réaction comme des quartz corrodés, avec quelques extinctions roulantes, une certaine proportion de verre avec ou non des figures de dévitrification. La teneur en SiO$_2$ étant d'environ 70%, nous nous trouvons dans le cas de figure où nous pouvons pratiquer au choix, parmi les essais rapides:

- l'essai cinétique chimique P. 18 589
- ou les essais de variation dimensionnelle P. 18 588 et P. 18 590.

Les domaines d'applications de ces essais sont précisés dans un document normatif, le fascicule de documentation P. 18 542. Dans notre cas, la pétrographie ne nous oriente vers aucun essai particulier; aussi, dans le but de tester la démarche, nous avons pratiqué successivement les trois essais rapides existants. Le diagnostic issu des résultats de chaque essai apparait dans le tableau suivant.

Essai	
Essai du groupe G1: P.18 589 (essai par voie chimique)	N R
Essais du groupe G2: P.18 588 (essai accéléré sur mortier par cure de vapeur d'eau et autoclavage)	P R
P.18 590 (essai accéléré sur mortier par autoclavage)	P R

Résultats:
NR: non réactif
PR: potentiellement réactif

Nous constatons que l'essai cinétique chimique conduit à un diagnostic différent de celui donné par la pétrographie. Cette opposition est prévue dans le document P.18 542. Lorsqu'il n'y a pas concordance entre les conclusions des analyses pétrographiques et les résultats de l'essai effectué, il est nécessaire de procéder à un autre essai de type différent de celui qui a été exécuté. La pratique de l'essai P.18 588 ou encore P. 18 590 confirme le diagnostic pétrographique. Ce matériau se classe donc comme potentiellement réactif.

L'opposition qui apparait en fonction des essais nous amène à nous poser plusieurs questions:
- est-elle due aux essais?
- est-elle due à la nature pétrographique tout-à-fait particulière du granulat?

A défaut de données à valeur statistique concernant les essais pratiqués sur ce type de rhyodacite, il semble délicat de modifier le domaine de validité des essais. La divergence de résultats entre l'essai chimique et les essais dimensionnels rapides sur barres de mortier provient vraisemblablement de la nature pétrographique tout-à-fait particulière de cette rhyodacite. Cette roche volcanique qui appartient à la famille des rhyolites au sens large, est, de par sa génèse, acide et peut posséder une fraction non négligeable de verre, plus ou moins en état de dévitrification, c'est cette originalité qui fait qu'elle réagit de façon non prévisible à l'essai chimique.

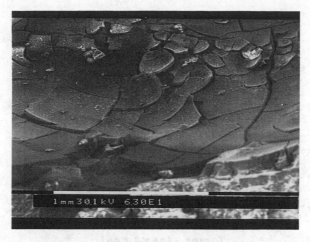

Photographie 1: Aspect du gel d'alcali-réaction, (grossissement X63).

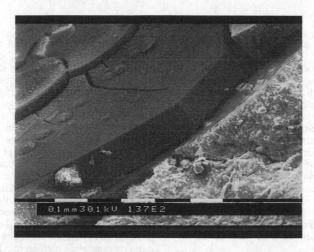

Photographie 2: Aspect du gel d'alcali-réaction, (grossissement X137).

Il se trouve donc que cet ouvrage a vu le jour dans les années 1947-1948, avec tous les aléas de contruction et d'approvisionnement en ciment(s) dus à cette période (construction étalée entre 1939 et 1948). nous en voyons l'état actuellement, et cet état confirme totalement le caractère réactif des granulats, puisque des fissures et du gel ont été décelées sur de nombreux échantillons (voir photographies 1 et 2, prises au Microscope électronique à balayage).

Il semble donc que évident que la nature pétrographique du granulat soit directement responsable des désordres, en particulier à cause d'une proportion de verre en phase de dévitrification permettant une libération de silice dans le système.

Si la décision devait être prise aujourd'hui de construire l'ouvrage, il est rassurant de constater que la maître d'œuvre aurait son attention attirée par le risque de voir à moyen terme l'ouvrage se dégrader. Il serait donc conduit à revoir l'approvisionnement en granulats.

Notre étude est allée au delà du simple diagnostic; après une étude systématique des propriétés mécaniques sur les échantillons, nous avons été confrontés à une dispersion des résultats entre les parements amont et aval. Les valeurs de l'amont étant toujours plus faibles que celles de l'aval. Nous en concluons que les désordres sont plus importants quand le béton est plus hydraté. Une série d'essais de mesure du potentiel résiduel

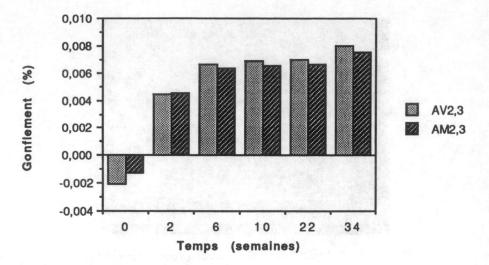

Figure 1: Potentiel de gonflement résiduel d'éprouvettes issues des parements amont (AM) et aval (AV) du barrage étudié.

de gonflement nous a permis de confirmer ces observations.

On constate qu'après une période de réhydratation d'une quinzaine de jours, on enregistre un gonflement résiduel sensiblement plus élevé pour le parement aval. Au moment du prélèvement, les échantillons des parements amont présentent systématiquement des caractéristiques mécaniques inférieures à celles des parements aval. Ces mesures laissent présager la présence de désordres qui affaiblissent le matériau, désordres que nous pouvons relier à la présence de gel d'alcali-réaction plus importante.

Au bout d'un laps de temps de huit mois, l'écart entre les valeurs amont et aval semble s'être stabilisé.

Les mesures de potentiel de gonflement résiduel confirme le fait que les échantillons du parements aval étaient moins affectées par l'alcali-réaction et que celle-ci a pu s'exprimer lorsque les échantillons ont été plongés dans un milieu favorisant leur vieillissement accéléré, (38°c et 100% d'humidité relative).

5. CONCLUSIONS

La démarche de post diagnostic menée ici a permis de tester les essais actuels et de pouvoir les confronter avec les résultats cinquante ans après la mise en œuvre des granulats dans le béton. La pétrographie assortie d'essais rapides, nous a permis de savoir très tôt que le granulat serait un granulat "à risques" et par là même à éviter son utilisation dans le cas d'un ouvrage de nature exceptionnelle. La dispersion des résultats nous a aussi permis de montrer que celle-ci pouvait provenir de la nature pétrographique de la roche, que cette rhyolite au sens large présente outre une acidité importante, la présence de verre siliceux plus ou moins en voie de dévitrification la plaçant à part de toutes les autres roches volcaniques. Son évolution au sein du béton ne peut en aucun cas être comparée à celle qu'aurait un basalte classique.

Cette présentation montre que la durabilité des bétons est en partie conditionnée par la nature du granulat et que vouloir l'ignorer peut conduire à l'apparition de désordres qu'il aurait été facile de prévenir en appliquant les règles d'usage qui veulent que l'on connaisse bien ce que l'on utilise, d'autant plus qu'il existe une méthodologie et une panoplie d'essais qui, s'ils sont appliqués doivent éviter l'apparition de désordres.

6. REFERENCES

Godart, B 1993.Progression dans les connaissances sur les phénomènes d'alcalis-réactions. Evaluation et surveillance des ouvrages. Annales de l'ITBTP, n°517, Série Béton 303.

Hornain, H 1993. Mécanismes physico-chimiques d'alcalis-réactions. Etat des connaissances à travers les derniers colloques congrès et publications. Annales de l'ITBTP, n°517, Série Béton 303.

Le Roux, A 1991. Méthodes pétrographiques d'étude de l'alcali-réaction. Bull. A.I.G.I, n°44, p. 47-54.

Recommandations provisoires pour la prévention des désordres dus à l'alcali-réaction. LCPC, Janvier 1991.

The 9th International Conference on Alkali-Aggregate Reaction in Concrete, 1992. Londres, 1128 p.

Characteristics of tertiary limestones in the Madrid area for use as construction materials

Caractéristiques des calcaires tertiaires dans la région de Madrid pour utilisation comme matériaux de construction

J.E. Dapena – *CEDEX Geotechnical Laboratory, Politechnical University Madrid, Spain*

A. Garcia del Cura – *Institute of Economic Geology, CSIC, Alicante, Spain*

S. Ordoñez – *Faculty of Geology, Alicante University, Spain*

ABSTRACT: This work studies the Tertiary limestones of the Tagus depression in the Madrid area. Basic tests were carried out using cylindrical samples on blocks extracted from sound rock masses in the zone, so the characteristics obtained are for sound limestone, except the coefficient of "Los Angeles", which was obtained from crushing at the quarry face, without making a preliminary selection.

The results obtained indicate that this sound limestone has characteristics with a wide range of variables, which requires them to be classified into two of the three groups distinguished by ASTM Standard C-568-79, when categorizing limestone rock for buildings, and into two of the five general classification groups for rocks based on their strength.

RÉSUMÉ: Dans ce travail, on étudie les calcaires tertiaires de la dépression du Tajo, situés dans les environs de Madrid. Dans ce but, on réalise notamment des essais sur éprouvettes cylindriques taillées sur des blocs de roche extraits du massif dans la zone de roche que l'on peut considérer comme saine. Conséquemment, les caractéristiques que l'on obtient ici correspondent aux calcaires sains, saufle coefficient de "Los Angeles" qui a été obtenu à partir d'échantillons provenant du concassage du front de la carrière, sans sélection préalable.

Les résultats obtenus indiquent que cette roche calcaire saine a des caractéristiques ayant des valeurs variables dans une large fourchette, qui obligent à les classer dans deux des trois groupes utilisés par la ASTM C-568-79 pour le classement des roches calcaires pour bâtiments, et dans deux des cinq groupes du classement général des roches selon leur résistance.

1 INTRODUCTION

The Tertiary limestones in the South of Madrid are used for many construction-related purposes: on the one hand they constitute raw materials for cement and lime factories, and on the other, are used to prepare aggregates for granular layers of road surfaces and aggregates for concretes. There is also a quarry located in Colmenar de Oreja, that provided limestone blocks for the construction of different 18th century monuments in Madrid, such as the Royal Palace, statues of the Spanish monarchs and the Puerta de Alcalá. In recent times this source has been used for the construction of the Almudena Cathedral, whose study was conducted earlier. (Dapena, Ordoñez and Garcia del Cura (1988)). This work studies the different limestone rock characteristics, from the viewpoint of their use as aggregate-type construction material and building blocks, for which samples were taken in the quarries of Hoyón, Morata and Campo Real.

2 QUARRY GEOLOGY

The quarries studied are located in the Neogene Basin (Madrid), in the so-called "Upper Unit" (Megías et al. 1983 and Calvo et al. 1989), but others exist above this unit (Ordoñez et al. 1986 and García del Cura et al. 1991).

The origins of these limestones have been attributed to lacustrine environment, but the latest petrological and sedimentological studies (Ordoñez and García de Cura, 1983), together with comparisons with existing environments in Central Spain, suggest a low-energy river environment with incidence of Tobaceaic, stomatolithic and oncolithic faciae, alternating with stagnant marshy faciae. Extensive diagenetic micro-karstification and cementation processes related to temporary emersion, caused a lithification of the material and the development of some kinds of characteristic porosity.

Incidences of clay of paligorsquitic composition exist on these levels of limestone, and alternate with decimetric levels of limestones.

3 DATA-TAKING IN THE QUARRIES

The three quarry faces have similar alteration profiles. The following can be distinguished, according to the classification given by Deere and Patton (1970), a considerably altered upper zone with a residual soils, where limestone fragments appear mixed with an abundance of reddish clayey soils.

An intermediate zone of altered rock, containing cracks, joints and stratification planes stained red and a deeper zone from which sound limestone blocks can be extracted.

Samples of two types, block and granular, were taken. The blocks were collected from the quarry face and selected from sound limestone rock, so the characteristics are for this type of material. The granular samples were taken from the stocks of materials already crushed to between 20 and 40 mm. At a first glance, this sample appeared to be more heterogeneous, as would be expected from crushed rock materials all taken from quarry faces, especially when considering that no attempt was made to select the largest and soundest blocks. The list of samples taken, together with the name of the quarry of origin, and shown in Table 1.

Table 1. Samples taken from different quarries

Refer.	Quarry	Sample
7708	Hoyón	Block
7709	Hoyón	Block
7710	Hoyón	Granular
7711	Morata	Block
7712	Morata	Block
7713	Morata	Block
7714	Morata	Granular
7715	Campo Real	Block
7716	Campo Real	Block
7717	Campo Real	Block
7718	Campo Real	Granular

4 PREPARATION OF SAMPLES AND TESTS

4.1 Sample preparation

Either cylindrical or granular samples are required for the tests to be performed. The granular samples, ranging from 20 to 40 mm., were taken directly from the quarry extraction, but as coarse fragments, between 25 and 40 mm. were predominant, some had to be crushed to prepare grain-size G for the "Los Angeles" test (5 Kg between 20 and 25 mm., and 5 Kg with sizes ranging from 25 to 40 mm).

Cylindrical test samples, 4.78 cm. in diam., were extracted from each of the blocks, at least one being 10 cm. high, and the rest between 2 and 5 cm.

4.2 Tests

The composition and fissuring of each sample were studied through a microscope, to observe the thin flakes prepared from the fragments separated from the cylindrical samples; chemical analyses were also used.

The rest of the studies were conducted in accordance with the procedure contained in the following ASTM Standard:

- Test to determine the absorption, performed in accordance with ASTM Standard C-97-83.

- Test to determine apparent specific weight, conducted in accordance with ASTM Standard C-97-83.

3270

- Test to determine propagation velocity of ultrasounds, conducted in accordance with ASTM Standard D-2845-69 (1976).

- Test to determine the simple compressive strength according to ASTM Standard D-3148-80, with a strain speed below 1.3 mm./minute.

- Test to determine the modulus of elasticity and Poisson's coefficient, carried out following the procedure described in Standard D-3148-80.

- Tensile strength test. Brazilian Method.

- Test to determine the point load

5 COMPOSITION

On analyzing the thin flakes through the microscope, it was observed that practically all the existing carbonate was calcium carbonate, so in a chemical analysis of the carbonate-content, it could be said that all the CO_2 obtained came from CO_3Ca.

The results obtained from analyzing the carbonate-content can be seen in Table 2, and their distribution in Fig. 1.

Table 2. Calcium carbonate content and specific weight of material

Sample	% CO_3Ca	ρ gr/cm^3
7708 A	98,20	2,73
7708 B	97,92	2,71
7789 A	97,92	2,71
7709 B	97,31	2,72
7711 A	98,80	2,71
7711 B	97,60	2,70
7712 A	98,80	2,72
7712 B	99,40	2,71
7713 A	99,08	2,74
7713 B	98,80	2,71
7715 A	94,04	2,72
7715 B	94,32	2,70
7716 A	97,32	2,70
7716 B	96,40	2,71
7717 A	98,52	2,73
7717 B	98,52	2,73

From the values obtained, it could be stated that the rocks studied are relatively pure limestones, with a carbonate-content generally greater

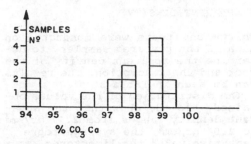

Fig. 1 Distribution of the calcium carbonate content in the samples studied

than 97%, the average representative value being 98% CO_3Ca.

Furthermore, with regard to composition, the limestones from all three quarries can be considered homogeneous

6 SPECIFIC WEIGHT OF THE MATERIAL

Sixteen tests were performed to determine the specific weight of the rock. Two representative fragments were taken from each block and crushed until they passed through a 40 ASTM seive, after which the specific weight of the particles was determined.

The test results are shown in Table 2, and their distribution in Fig. 2.

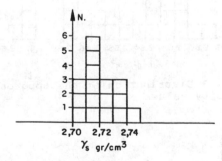

Fig. 2 Distribution of the specific weight of the material

The values obtained range from 2.70 gr/cm^3 to 2.74 gr/cm^3, the average representative value being $\rho = 2.72$ gr/cm^3, which is the specific weight of the calcite.

7 APPARENT DENSITY

Thirty one tests were conducted on each of the prepared samples, to determine the apparent density of the rock and the absorption. The results can be seen in Table 3.

The distribution of the values obtained, Fig. 3, show that the apparent density ranges from 2.47 gr/cm³ to 2.64 gr/cm³, the average representative value for limestones, as a whole, being $\gamma = 2.55$ gr/cm³.

ASTM Standard C-568-79, classifies limestone rocks for buildings into three groups. These rocks can be considered to fall into two of these categories, medium and high density.

8 ABSORPTION

Absorption makes it possible to know where the hollows are in the tested rock samples. The test results are shown in Table 3, and their distribution in Fig. 4.

The absorption values of these limestones are nearly all between 0.6 and 2.5 and, although there are

Tabla 3. The apparent specific weight, γ_{ap}, absorption, a, and accessible porosity, na, of each of the cylindrical samples

Ref.		γ_{ap} g/cm³	γ gr/cm³	a %	n_{pa} %
7708	I	2.537	2.642	1.57	3,86
7708	II	2.539	2.646	1.60	3,93
7708	III	2.491	2.650	2.40	5,85
7708	IV	2.544	2.652	1.60	3,97
7709	I	2.632	2.681	0.70	1,80
7709	II	2.593	2.678	1.24	3,13
7709	III	2.581	2.672	1.32	3,35
7711	I	2.595	2.663	0.98	2,50
7711	II	2.628	2.672	0.63	1,61
7711	III	2.626	2.670	0.62	1,61
7711	IV	2.623	2.671	0.69	1,76
7712	I	2.550	2.700	2.17	5,51
7712	II	2.551	2.677	1.85	4,63
7712	III	2.586	2.693	1.53	3,93
7712	IV	2.594	2.682	1.27	3,23
7713	I	2.537	2.651	1.69	4,19
7713	II	2.571	2.670	1.44	3,64
7713	III	2.503	2.615	1.71	4,12
7713	IV	2.514	2.600	1.32	3,16
7715	I	2.519	2.687	2.48	6,18
7715	II	2.536	2.690	2.25	5,66
7715	III	2.475	2.680	3.08	7,54
7715	IV	2.478	2.673	2.95	7,17
7716	I	2.579	2.684	1.51	3,86
7716	II	2.518	2.650	1.98	4,85
7716	III	2.526	2.665	2.06	5,11
7716	IV	2.593	2.682	1.28	3,27
7717	I	2.566	2.676	1.61	4,07
7717	II	2.531	2.653	1.81	4,49
7717	III	2.527	2.652	1.86	4,60
7717	IV	2.384	2.616	3.72	8,53

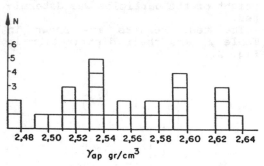

Fig. 3 Distribution of the apparent density values

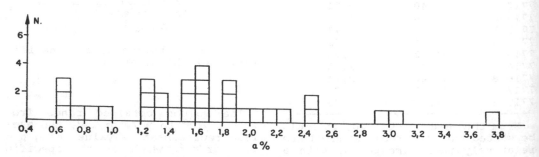

Fig. 4 Distribution of the absortion values

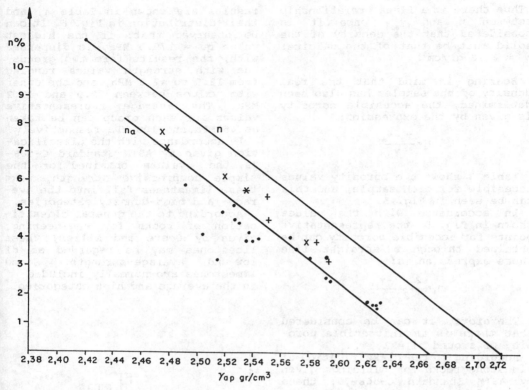

Fig. 5 Relationship between apparent density, total porosity, n, and accessible porosity, na.

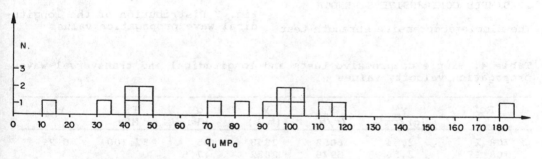

Fig. 6 Distribution of the values obtained from the simple compressive tests

some values of 3.8, the average representative value can be taken as a = 1.60%.

According to ASTM Standard C-568-79, the absorption values obtained show medium and high-density limestone rocks.

9 POROSITY

On the basis of its apparent density and the density of the solid, the porosity of a rock fragment is given by the expression:

$$n = \frac{\gamma_s - \gamma_{ap}}{\gamma_s}$$

Thus there is a linea relationship between n and γ_{ap}, once it is considered that the density of the solid must be that of the calcite: $\gamma_s = 2.72$ gr/cm³.

Bearing in mind that the real density of the samples has also been determined, the accesible porosity is given by the expression:

$$n_a = \frac{\gamma_r - \gamma_{ap}}{\gamma_s}$$

Table 3 shows the porosity values accesible for each sample, and this can be seen in Fig. 5.
In accordance with the values shown in Fig. 5, the representative points for accesible porosity can be obtained through a straight line whose expression is:

$$n_a = \frac{2.67 - \gamma_{ap}}{2.72}$$

Therefore, it can be considered that there is an inaccesible porosity of around ni = 1.8%.
According to the classification of limestone rocks for buildings given by ASTM Standard C-568-79, these rocks fall into the high-density category, Type III.

10 SIMPLE COMPRESSIVE STRENGTH

The simple compressive strength test

results are shown in Table 4, and their distribution in Fig. 6. It can be observed that, if the highest value qu = 179.5 MPa, is dispensed with, the results form two groups, one with strength values ranging from 12.5 to 47.5 MPa, and the other with values between 72.9 and 19.2 MPa. The average representative values for each group can be taken as 45 MPa and 100 MPa respectively.
In accordance with the classification given by ASTM Standard C-568-79, the values obtained for the simple compressive strength, make these limestones fall into the average- and high-density categories.
According to the general classification of rocks for engineering, given by Deere and Miller, these limestones may be regarded as of low-and average-strength. Sound limestones are normally included in the average and high categories.

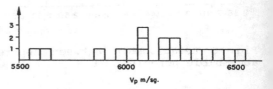

Fig. 7 Distribution of the longitudinal wave propagation values

Table 4. Simple compressive tests and longitudinal and transversal wave propagation velocity values

Ref	γ_{ap} (gr/cm³)	V_p (m/s)	V_s (m/s)	q_u MPa	E_{t50} MPa	ν_{t50}
7708 I	2,54	6053	2858	96	57.100	0,29
7708 II	2,56	5976	2822	47,5		
7709 I	2,60	5889	2865	12,5		
7709 II	2,59	6306	2897	100		
7711 I	2,56	6419	2853	41,2		
7711 II	2,59	6531	2903	98,9		
7712 I	2,57	6212	2933	113,8	61.538	0,23
7712 II	2,58	6282	2967	119,2		
7713 I	2,50	6083	2808	33,0		
7713 II	2,49	6005	2883	45,1		
7715 I	2,51	5589	2655	93,9	44.444	0,39
7715 II	2,49	5647	2683	81,3	44.444	0,19
7716 I	2,56	6206	2773	40,6		
7716 II	2,59	6159	2908	179,5	61.538	0,23
7717 I	2,50	6072	2876	72,9		
7717 II	2,52	6178	2851	102,4	52.174	0,30

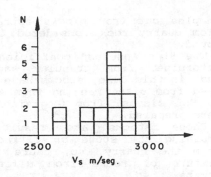

Fig. 8 Distribution of the transversal wave propagation values

11 WAVE PROPAGATION VELOCITY

Test results for determining the longitudinal and transversal wave propagation velocity are shown in Table 4, and their distribution in Figs. 7 and 8.

The longitudinal wave propagation velocity has a value dispersion greater than that of the transversal waves. The former virtually all range from V_p = 5.550 m/sec., to V_p = 6.500 m/sec., and the average representative value can be taken as V_p = 6.200 m/sec.
The values obtained for the shear wave propagation velocity, range from V_s = 2.650 m/sec. and V_s =3.000 m/sec., and the average representative value can be taken as 2.850 m/sec.

12 POINT LOAD INDEX

The point load test results are shown in Table 5. The values obtained range from I_s (50) = 7.01, Fig. 9, and the average representative value can be taken as I_s (50) = 5.7.
According to Beniawski's classification, the values obtained on the point load index of these rocks fall into the medium and high-strength categories.

13 TENSILE STRENGTH

The tensile strength was obtained by conducting the Brazilian test on cylindrical test pieces 47.8 mm in diameter with heights ranging from 22.9 to 48.8 mm.

The results obtained can be seen in Table 6, and their distribution in Fig. 10. The values range from q_t = 4.6 MPa to q_t = 12.8 MPa, and the average representative value can be taken as q_t = 7 MPa.

Table 5. Results of point load test

Sample	Height (cm)	Load (kN)	I_s (MPa)	I_s(50) (MPa)
7708 III	9.27	11.31	4.95	4.85
7708 IV	8.07	11.94	5.23	5.12
7709 III	5.50	13.83	6.05	5.93
7709 IV	7.55	13.20	5.78	5.66
7711 III	6.40	14.46	6.33	6.20
7711 IV	6.84	13.20	5.78	5.66
7712 III	8.47	16.34	7.15	7.01
7712 IV	7.96	13.83	6.05	5.93
7713 III	6.80	8.80	3.85	3.77
7713 IV	7.70	6.91	3.03	2.97
7715 III	8.15	15.08	6.60	6.47
7715 IV	8.97	15.08	6.60	6.47
7716 III	8.62	13.83	6.05	5.93
7716 IV	7.08	12.57	5.50	5.39
7717 III	9.87	13.83	6.05	5.93
7717 IV	8.79	15.08	6.60	6.47

$$. I_s(50) = I_s \cdot \left\{ \frac{47,8}{50} \right\}^{0,45}$$

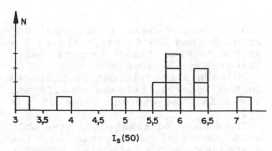

Fig. 9 Distribution of the point load index

14 MODULUS OF ELASTICITY AND POISSON'S COEFFICIENT

Table 4 shows the values of the modulus of elasticity and Poisson's coefficients obtained from the stress-strain tangents at 50% of the failure load.

Table 6. Results of the brazilian test, q_t

Sample	Length (cm)	q_t (KPa)
7708 V	4.23	6926.8
7708 VI	4.62	6832.2
7709 V	3.96	6591.9
7711 V	4.05	8056.8
7711 VI	4.63	5436.7
7712 V	4.35	9093.3
7712 VI	4.87	6290.0
7713 V	4.62	8014.1
7713 VI	3.11	5481.5
7715 V	4.71	4580.8
7715 VI	4.38	7236.9
7716 V	2.29	12794.9
7716 VI	4.88	10643.8
7717 V	4.63	7795.4
7717 VI	2.37	9497.1

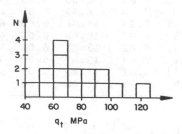

Fig. 10 Distribution of the results of the Brazilian test

Six values have been obtained, ranging from E = 44.444,4 MPa to E = 61.538,5 MPa, their average value being: E = 53.540,0 MPa.

Poisson's coefficient is included in the interval between ν= 0.19 and ν = 0.39. The average value is ν = 0.27.

The moduli of elasticity have only been obtained for rocks in the average-strength category, probably because of failure due to fissuring in the low-strength samples. The relationship between the modulus of elasticity and the strength, makes it possible to classify the medium-strength limestones as being of average relative modulus, see Fig. 11.

15 "LOS ANGELES" COEFFICIENT

The aforementioned properties are, generally, determined using test samples cut from blocks extracted from quarry rock considered to be sound.

The "Los Angeles" coefficient is determined from granular samples and, in this case, stocks have been used from sizes ranging from 40 to 20 UNE sieves, from G grain-sizes were prepared.

These stocks are result of crushing the stone on extraction from the quarry, so, there is a mixture of rocks from different strata.

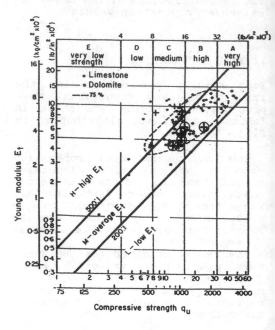

Fig. 11 Situation of the limestone rocks in the intact rock classification given by Deere and Miller, 1966 q_u = Simple compression strength. E_t = modulus tangent for 50% of the failure load.

The results obtained from the "Los Angeles" tests were:

Reference	C.L.A.
7710	31.5%
7714	29%
7718	34%

which means that these limestones fall into the good and average categories.

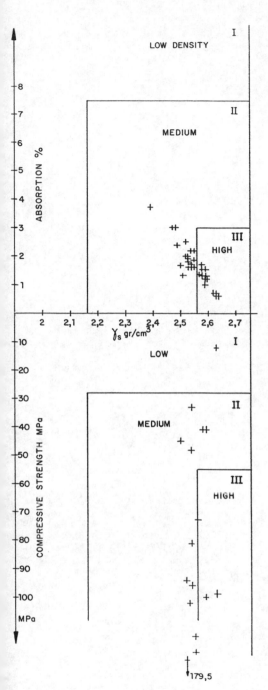

Fig. 12 Limestone buildings stone classification ASTM C-568-79

16 ROCK CLASSIFICATION

The characteristics of the Madrid Tertiary limestones have been determined using the criteria habitually used to classify the petrified material to be used in public works and buildings.

The values obtained lie within an extensive interval into which it is advisable to classify these limestones, i.e., what ASTM Standard C-568-79 defines as medium to high-density, Fig. 12, and what the general rock classifications of Deere and Miller as well as Beniawski, define as low- to medium-strength.

17 CONCLUSIONS

The study of Tertiary limestones in the Tagus depression, close to Madrid, was mainly carried out with test pieces cut from blocks, representative of the soundest material existing in the quarries. Therefore, their properties will show the characteristics of this type of material.

Table 7 shows the characteristics of the sound Tertiary limestone rock from the Madrid area:

These limestone rocks, when sound, can be classified as average to high density (ASTM C-568-79) and as rocks of low to medium strength (Deere and Miller).

Table 7. Limestone Rocks characteristics

	Interval	Average
CO_3Ca %	94,0–99,4	98
ρ gr/cm^3	2,70–2,74	2,72
γ_{ap} gr/cm^3	2,47–2,64	2,55
a %	2,5–0,6	1,60
n %	8,8–3,0	6,20
q_u MPa	10,0–170	85
I_s (50)	3–7	5,70
q_t MPa	4,60–12,8	7
E GPa	44.4–61.5	53,5
v	0,19–0,39	0,27
V_p Km/seg	5,5–6,5	6,20
V_s Km/seg	2,60–3,0	2,85
C.L.A. %	29–34	32

ACKNOWLEDGEMENTS

Our thanks to the Centro de Estudios y Experimentación de Obras Públicas and the Comunidad de Madrid for the support they gave in carrying out this work.

A suggested improvement to the Mortar Bar method of evaluation of the alkali-aggregate reaction

Une amélioration supérée pour la méthode du prisme de mortier pour l'évaluation de la réaction alkali-agrégat

María del Pilar Mejía Vallejo & Fabián Hoyos Patiño
Universidad Nacional, Medellín, Colombia

Julián Vidal Valencia
Universidad EAFIT, Medellín, Colombia

ABSTRACT: The potential for alkali-aggregate reaction in concrete structures has been so far evaluated by petrological, chemical and physical methods. The physical standard test method. ASTM C-227, requires the curing of mortar bars at 37,8°C, during a period of 3 to 6 months, and the measurement of any increase in length of the mortar bars at the end of the curing period; an increase of 0,05% at 3 months, or 0,1% at 6 months, is indicative of potential haarmful alkali reactivity of the aggregate cement combination used in the mortar bars. In this paper we do propose a modification of the test method which allows to get comparable results in a period of 1,5 to 3 months.

The modified version of the ASTM C-227 test method prescribes a curring temperature of 60°C instead of the standard temperature of 37,8°C. The results gotten so far on 16 sets of samples (64 specimens) from combinations of 4 aggregate sources and 4 Portland cement brands, show a high correlation (r⨦0,92) between the expansion of the mortar bars at 3 months, as per the ASTM C-227 test method, and the expansion at 1,5 months, when the mortar bars were cured at 60°C.

Further investigation on a wider spectrum of aggregate cement combinations should be needed in order to validate the results presented in this paper.

RÉSUMÉ: La réactivité potentiel des agrégats avec le ciment dans les structures du béton a été évaluée jusqu'ici par des méthodes pétrologiques, chimiques et physiques. La méthode physique plus étendue, d'accord avec la normalisation de l'ASTM exige la préparation des échantillons prismatiques de mortar et la mesure de quelque élongation plus grande que 0,05% en 3 mois ou 0,1% en 6 mois quand les échantillons sont maintenues à 37,8°C. Les élongations que surpassent ces limites sont considérées indicatives de l'occurrence de la réaction alkali-agrégat. Nous avons changé la température de l'essai à 60°C pour arriver à des résultats semblables à ceux obtenus avec la méthode normalisée à 45 jours. Le coefficient de corrélation des résultats à 37,8°C et 3 mois, et ceux à 60°C et 1,5 mois, sur 64 échantillons (4 agrégats et 4 ciments commerciales) fût r=0,86.

INTRODUCCION: The alkali-aggregate reaction, in concrete structures has been known and studied during the last half of the century, and several standard methods have been proposed to evaluate the potential reactivity of concrete aggregates. The alkali aggregate reaction involves high alkali cement, with more than 0,6% of sodium oxide equivalent, and disordered forms of silice (opal, chert, calcedony and deformed quartz in crystaline rocks) or argillaceons dolomitic limestores. There are petrological and chemical approaches, ruled by the standard test methods under de ASTM designation C-289 and C-295 (ASTM, 1983) and analogous standard methods of other agencies. These methods are considered only as screening rather than confirmatory of the potential reactivity, since their realiability is questionable (ASTM, 1983); in the last term, the reaction occurs only when the "thrigh" mixture of reactive aggregates and cement is made, and the potential for reactivity should be evaluated for such a mixture.

The ASTM C-227 test method gives quantitative information on the suscceptibility of a given cement aggregate combination to the alkali-silice reacction (ASTM, 1983). This method allows the determination of the potential for alkali-aggregate reaction by measuring the expansion of mortar bars during a term of 3-6 months. Expansion is generally considered harmful if it is larger than 0,05% at 3 months ar larger than 0,1% at 6 months.

The test procedure requires a temperatura that shall not vary from 37,8°C by more than 1,7°C, and a relative humidity not less than 50%. The reaction rate is temperature dependent. In general temperature increase should give place to a correlated expansion increase; such a phenomenon can be exploited in order to shorten the period over which it is necessary to perform the test. In this paper we do present experimental results that can be used to look for a modification of the standard test methods, allowing to get comparable information on the alkali-aggregate reactivity in half the period required in the standard test method.

The aggregates and cement we have worked with are common construction materials in Pereira, a mid size city in western Colombia, which have been evaluated for their reactivity as a part of a broader study on construction materials sources. We have evaluated sixteen aggregate-cement combinations, four aggregate and four cement types.

THE MATERIALS

a) Aggregates: The source of stone aggregates for Pereira are alluvial deposits and active sediments from four rivers, Cauca, Risaralda, Mapa, and La Vieja. The petrographic examination of samples from these sourcces as per the test method ASTM C-295 indicate a high potential for alkali-aggregate reaction given the occurrence of a large fraction of minerals known to be reactive. The petrographic composition of the samples is give on Table 1.

TABLE:1. Petrographic composicion of aggregates

Source\Mineral	Cauca	Risaralda	Mapa	La Vieja
Chert	33,2	46,6	12,6	49,8
Quartz	32,6	32,7	68,7	23,2
Basalt	1,6	0,8	2,7	13,0
Andesite	16,1	-	3,0	2,7
Diorite	0,8	11,1	-	5,1
Sandstone	11,7	5,8	9,9	2,3
Mudstone	2,3	-	-	-
Schist	1,7	3,0	3,1	4,9

Samples from the same sources, tested as per the ASTM C-289 test method, in order to evaluate their alkali-aggregate reactivity by the chemical method, gave a positive result for all of them. The results of these tests are shown in a conventional way in Figure 1.

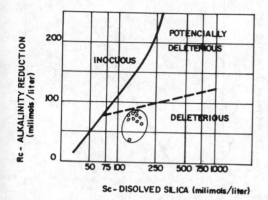

Figure 1.
Results of Alkalinity tests on samples from Pereira área.

b) Cement: Although most of the cements used in Colombia could be classified as low-alkali cement, with less than 0,6% of sodium oxide equivalent; some cement brands have a high alkali content. Four commonly used cement brand names were selected to perform the alkali-aggregate reactivity evaluation; their alkali content determined as per the ASTM test method C-114 are shown in Table 2.

TABLE 2. Alkali content in four colombian cements

-

Cement	Na₂O	K₂O	Na Equivalente
A	0,21	0,69	0,66
B	0,12	0,82	0,66
C	0,25	0,27	0,43
D	0,11	0,27	0,29

THE CONVENTIONAL TEST

The alkali-aggregate reactivity of the above described materials was evaluated as per the ASTM test method C-227. The results are sumarized in Table 2, where the expansions measured at 3 months are recorded as a percentage. Each of the figures of this table corresponds to the average of 4 mortar bars, with their standard deviation in parentesis.

TABLE 3. Expansion of mortar bars at 3 months- as per test method ASTM C-227

-

Aggregate Cement	1	2	3	4
A	0,058	0,061	0,061	0,063
	(0,0022)	(0,0010)	(0,0006)	(0,0003)
B	0,052	0,063	0,059	0,069
	(0,0022)	(0,0028)	(0,0013)	(0,0013)
C	0,039	0,040	0,033	0,041
	(0,0018)	(0,0017)	(0,0028)	(0,0008)
D	0,021	0,019	0,039	0,041
	(0,0010)	(0,0042)	(0,0013)	(0,0013)

While both the petrographic and chemical analyses indicated the potential reactivity of the aggregates, the mortar bar test shows that harmful expansion should occurs only when the "right" combination of high alkali cement and reactive aggregate is given. It is apparent from Table 1, 2 and 3 that for deleterious change in length to occur it is necessary the combination of reactive agregate and high alkali cement.

THE 60-45 APROACH TO THE MORTAR BAS TEST METHOD

Given the temperature dependence of the chemical reactions it is logic to expect that an increase in temperature increase the rate of length changes in

3281

the mortar bas test method and, therefore, a decrease in the time to get reliable results.

In the course of this study a second set of samples were tested as per the ASTM C-227 test method except that the temperature was maintained at 60°C ±2°C instead of 37,8±1,7°C. The rate of expansion of the mortar bars was consistently higher at 60°C than at the conventional temperatur of 37,8°C.

The bar expansion at 37,8°C and 60°C are closely correlated in such a way that reliable results can be obtained at 1,5 or 2 months, if the bars are cured at 60°C, rather than at 3 months, working at the standard temperature, thus shortening the period over which measurements are required. The length changes of the mortar bars, given as a percentage of the initial length at 37,8°C and 3 months, at 60°C and 2 months, and at 60°C and 1,5 months, are presented in Table 2 and Figures 2 and 3.

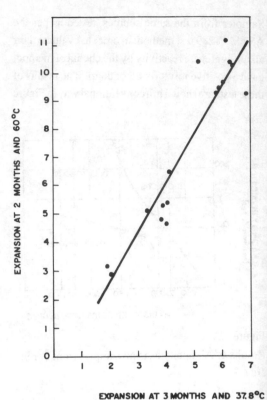

EXPANSION AT 3 MONTHS AND 37.8°C

Figure 3.
Correlation of change length of mortar bars at 37.8°C and 3 months to 60°C and 2 months

The length changes of the mortar bars at 37,8°C and 3 months are correlated with the change lengths at 60°C and 2 months by the equation

$$_{60,2} = 1,77 \quad _{37,3} - 0,01$$

with a regresion coeficient $r^2 = 0,87$.

If the length changes at 60°C and 1,5 months (interpolated) are compared with the length changes at 37,8°C and 3 months the correlation is given by the equation

$$_{60,15} = 1,136 \quad _{37,3} - 0,007$$

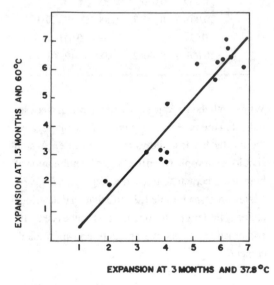

EXPANSION AT 3 MONTHS AND 37.8°C

Figure 2.
Correlation of change length of mortar bars at 37.8°C and 3 months to 60°C and 1.5 months.

with a regression coefficient $r^2 = 0,85$. Both correlations are valid at a significance level of 5%.

Based on these results it can be proposed a research line in order to improve the mortar bar standard test methods to determine the susceptibility of cement-aggregate combinations to expansive reactions. The increase of the temperature, at which the test is performed, to 60°C, would allow to reduce to a half the time needed to get results comparable to the standard method at 3 months.

References

ASTM, 1983, Annual Book of ASTM standards. Vol. 04. 02, 680 p. Philadelphia

TABLE 4. Length Changes of Martar Bars ar different times and temperatures

Combination	$t_{37,3}$	$t_{60,2}$	$t_{60, 1,15}$
A1	0,0605	0,0967	0,0633
A2	0,0626	0,1038	0,0675
A3	0,0580	0,0930	0,0565
A4	0,0613	0,1118	0,0704
B1	0,0588	0,0945	0,0629
B2	0,0685	0,0925	0,0607
B3	0,0520	0,1040	0,0620
B4	0,0633	0,1025	0,0645
C1	0,0333	0,0508	0,0309
C2	0,0410	0,0648	0,0481
C3	0,0390	0,0528	0,0313
C4	0,0403	0,0533	0,0307
D1	0,0385	0,0480	0,0287
D2	0,0405	0,0460	0,0277
D3	0,0208	0,0285	0,0195
D4	0,0193	0,0318	0,0206

Should the correlation proposed in this paper become validated with a larger number of tests, the criterion to determine the reactivity of a given cement-aggregate combination can be changed. A cement aggregate combination should be considered reactive if a mortar bar, cured at 60 ± 2°C and relative humidity more than 50%, get a change length of 0,05% in 45 days.

Influence of montmorillonite group bearing aggregates on concrete

Influence des agrégats avec montmorillonites dans le béton

Y. Wakizaka & K. Ichikawa
Public Works Research Institute, Ministry of Construction, Tsukuba, Japan

Y. Fujiwara & K. Uji
Technology Research Center, Taisei Co., Yokohama, Japan

J. Hayashi & S. Mitani
Tsukuba Institute of Construction Technology, Kumagai Gumi Co., Ltd, Japan

T. Maeda & H. Sasaki
Technical Research Institute, Hazama Co., Tsukuba, Japan

ABSTRACT: Accelerated setting is observed in concrete made of montmorillonite group bearing aggregates. This phenomenon is typical of the montmorillonite group. The mechanism of accelerated setting is as follows. First, ion exchange occurs between calcium ions in the montmorillonite group and potassium ions in the mortar pore solution just after mixing. Dissolution of gypsum in cement clinker minerals is artificially controlled. The concentration of calcium ions in the mortar pore solution is already saturated, affecting hydration reactions of C_3A and C_3S. The exothermic reaction is activated between one and four hours after mixing, causing accelerated setting and reducing the workability of the concrete construction. Evaluation of montmorillonite group bearing aggregates requires information on both content and mineralogical properties.

RÉSUMÉ: Une prise accélérée est observée dans le cas de bétons composés par des agrégats portant du groupe montmorillonite. Ce phénomène est caractéristique du groupe montmorilonite. Le processus de prise accélérée est le suivant: un échange d'ions se produit d'abord entre les ions calcium du groupe montmorillonite et les ions potassium de la solution interstitielle du mortier immédiatement suite au mélange. La dissolution du gypse dans les minéraux du clincker de ciment est controlée artificiellement. La concentration des ions calcium dans la solution interstielle de mortier est saturée au départ, ce qui affecte les réactions d'hydratation de C_3A and C_3S. La réaction exothermique est activée entre une et quatre heures après le mélange, cause de la prise accélérée. Elle réduit la maniabilité du béton. L'évaluation des agrégats portant du groupe montmorillonite demande des informations à la fois sur leur nature et leur propriétés minérales.

1. INTRODUCTION

Recently, low quality rocks such as clay minerals bearing rocks are being utilized as concrete aggregates for dams in Japan. There are many problems with the use of concrete aggregates for dams. For example, poor geological condition around dam sites including quarry sites, and it is difficult to change the position of quarry sites in terms of environment.

Montmorillonite group (smectite) is a clay mineral. In Japan, the montmorillonite group often occurrs in volcanic rocks as hydrothermal alteration minerals and deuteric alteration minerals, and is widely distributed in green tuff regions.

Some reports have described concrete collapses under volume change of montmorillonite group bearing aggregates due to repeated drying and wetting, and freezing and thawing (e.g. Maru and Yanagida, 1981; Shigekura, 1981). Maru and Yanagida (1981) reported that the montmorillonite group bearing aggregates popouts on the concrete surface.

However, the experimental bases of these reports were poor. We conducted a number of experiments in order to investigate the influence of montmorillonite group bearing aggregates on concrete, and discovered that concrete containing montmorillonite group bearing aggregates hardens earlier than concrete made of montmorillonite group free aggregates. We have already reported this (Wakizaka et al.,

Table 1 Properties of M and S-aggregates.

aggregate	rock type	cation exchangeable capacity (CEC) (meq/100g)	exchangeable cation (meq/100g)					montmorillonite content (%)	water absorption (%)[1]	specific gravity[1]	percentage of abrasion (%)[2]	soundness (%)[3]	d(060) (Å)	d(001) (Å)
			Na	K	Ca	Mg	total							
Ma-1	andesite, dolerite	23.3	7.0	0.3	16.0	2.7	26.0	10.8	5.03	2.51			1.49	17.7
-2	andesite, dolerite	16.6	3.5	0.3	25.2	8.4	37.4	--	3.35	2.55	20.6	7.2	--	--
-3	andesite, dolerite	15.2	0.9	0.4	16.1	4.8	22.2	6.8	4.40	2.53			1.49	17.3
-4	andesite, dolerite	20.7	0.9	0.3	16.7	4.0	21.8	8.8	4.33	2.53			1.49	--
Mb-1	dolerite	24.4	0.6	0.3	72.9	3.4	77.2	9.8	--	--			1.55	14.7
-2	dolerite	31.0	0.4	0.2	24.0	4.2	28.8	13.5	4.03	2.68	15.4	20.4	1.54	--
-3	dolerite	22.0	0.4	0.3	20.9	4.0	25.6	11.9	3.26	2.68			?	--
Mc	andesite	24.6	0.6	0.2	28.7	1.1	30.6	13.0	3.57	2.61	18.4	4.4	1.50	14.7
S	basalt	3.9	0.7	0.4	2.9	2.1	6.1	0	2.86	2.79	--	--	--	--

[1] Value of fine aggregate in accordance with JIS A 1109.

[2] Value of coarse aggregate (40 mm) in accordance with JIS A 1121. Value was obtained for mixed samples of Ma (Ma-1, Ma-2, Ma-3 and Ma-4) and Mb (Mb-1, Mb-2 and Mb-3), respectively.

[3] Value of coarse aggregate (40 mm) in accordance with JIS A 1122. Value was obtained for mixed samples as same as abrasion tests.

1990a,b; Wakizaka et al., 1993). In this article we will review the phenomenon and discuss evaluation and utilization methods of the montmorillonite group bearing aggregates for concrete.

2. DESCRIPTION OF AGGREGATES SAMPLES

2.1 Samples

Three samples of main montmorillonite group bearing aggregates (M-aggregates;Ma, Mb, Mc). Ma and Mb-aggregates were subdivided into four and three types by their properties respectively. Fourty samples were used as reference montmorillonite group bearing aggregates (R-aggregates). Standard aggregate (S-aggregate) was taken from montmorilonite group free rocks. Limestone, coral sand and brick (A-aggregates) were used as reference samples for water absorption.

2.2 Mineralogical properties

Table 1 shows mineralogical and physical properties of M, S and A-aggregates.

Cation exchange capacity (CEC) of M-aggregates varied from 15.2 to 31.0 meq/100g. Among all M-aggregates, the most dominant exchangeable cation was calcium ions. Ma-1-aggregate had more sodium ions than other aggregates. CEC of S-aggregate is 3.9 meq/100g. Whereas CEC of R-aggregates ranged from 4.08 to 29.0 meq/100g. The most dominant exchangeable cation in all R-aggregates was also calcium ions.

Montmorillonite content of the aggregates was determined by the X-ray diffraction method according to Sakurai et al. (1986). Montmorillonite group content of M-aggregates was 6.8 to 13.5%. S and A-aggregates did not contain the montmorillonite group. The content of typical R-aggregates varied from 0.3 to 16.5%.

Observations of d(060) by X-ray diffraction indicated that the montmorillonite group in M and R-aggregates was dioctahedral type. Greene-Kelly treatment on M-aggregates showed d(001) of 17.7 to 14.7 Å, so this montmorillonite group was considered to be beidellite or near beidellite. X-ray diffraction analysis under 40% relative humidity on M-aggregates showed the exchangeable cation of Ma-aggregates was a mixture of sodium and calcium ions, and those of other aggregates were calcium ions.

2.3 Physical and engineering properties

Water absorption did not satisfy the standards defined by Japan Society of Civil Engineers (1986) was water absorption. Water absorption of all M-aggregates were over the minimum standard (3.0%). Water absorption value of S-aggregate was 2.86%, and A-aggregates between 1.80 and 9.54%. Water absorption of R-aggregates, determined from fourteen samples, was over 3.0% for two samples.

Specific gravity of all M and S-aggregates satisfied the standard (2.5). The specific gravity of A-aggregates ranged from 2.26 to 2.56, and R-aggregates (fourteen samples) 2.37 to 2.64. Four samples of R-aggregates did not satisfy the standard value.

Abrasion and soundness tests were performed only for M-aggregates. Abrasion of all M-aggregates satisfied the standard (40%). Whereas, soundness of Mb-aggregate was over the minimum value of the standard (12%).

2.4 *Relationship between mineralogical and physical properties*

The montmorillonite group content in aggregates was directly measured by X-ray diffraction. The montmorillonite group had the largest CEC of the aggregates we used. Other minerals had the very small CEC. It was considered that CEC level gives an indication of montmorillonite group content. The montmorillonite group usually has higher water absorption than other minerals. It was also considered the water absorption indicates relative montmorillonite content. It turns out that CEC and the water absorption of the aggregates roughly correlated with the montmorillonite group content of the aggregates (Wakizaka et al. 1993).

3. EXPERIMENTAL METHODS FOR FRESH CONCRETE

3.1 *Consistency and setting properties*

Fine aggregates containing montmorillonite group had more effect on concrete than coarse aggregates (Wakizaka et al. 1990a). We made a mortar examination in spite of the concrete examination, and examined consistency and setting properties of mortar through flow tests (JIS R 5201) and setting tests (ASTM C403). Mixing of the mortar was carried out five times.

The cement was ordinary portland cement. Producing time of the cement was different with each mortar mixing. The chemical composition of all cement was about the same; SiO_2:21.3-21.7%, Al_2O_3:5.0-5.4%, Fe_2O_3:2.7-2.9%, CaO:63.0-64.0%, MgO:1.7-2.1%, SO_3:2.0-2.1%, Na_2O:0.34-0.35%, K_2O:0.48-0.51%.

The mortar mix proportion was water/cement:50% and sand (fine aggregate)/cement:2.0. Grain size distribution of the fine aggregate was 4.75-2.36 mm:9.5% (oven-dry weight ratio), 2.36-1.18 mm:23.0%, 1.18-0.6 mm:23.0%, 0.6-0.3 mm:23.0%, 0.30-0.15 mm:14.0%, under 0.15 mm:7.5%.

3.2 *Interaction of system cement-water-aggregates*

1) *Time dependent change of chemical compositions of mortar pore solution*

Time dependent changes of chemical compositions of mortar pore solution were analyzed for solution extracted from mortar at a given time. Mix proportion of the mortar and grain size distribution were the same as in the flow and setting tests. Analyzed chemical components were calcium, magnesium, potassium, sodium, sulphuric acid and hydroxyl ions.

2) *Setting properties of ion-exchanged aggregates*

To examine the effect of ion exchange properties by the montmorillonite group on mortar setting properties, setting tests were made on the mortar containing ion-exchanged aggregates (runs 6, 7). Ion exchange treatment was carried out as follows. First, fine aggregates were immersed in 1N-NaCl, 1N-KCl and 1N-CaCl$_2$ respectively for 24 hours under a vacuum. Second, the aggregates were washed with distilled water until chloric ions were not detected. Mix proportion of the mortar and grain size distribution of fine aggregates were also the same as for the flow and setting tests.

3) *Properties of hydration process*

To assess the hydration of mortar containing montmorillonite bearing aggregates, hydration heat and hydration products were observed. Hydration heat was measured with a conduction calorimeter on mortar containing 0.08-0.105 mm aggregates. Mix proportion of the mortar was water/cement:65% and sand/cement:1.0 (2g/2g).

Hydration products were analyzed for mortar with water removed by methyl ethyl ketone. X-ray diffraction and differential thermal analyses were made on the dried mortar.

3.3 *Effect of chemical admixtures and an admixture on fresh concrete*

We examined effect of chemical admixtures and an admixture on flow and setting properties of fresh concrete. The chemical admixtures were a water reducing agent (naphthalenesulfonic acid salt), an AE water reducing agent (lignin sulfonic acid salt), a superplasticizer (melanin sulfonic acid salt) and an air entraining agent (abietic acid salt). And fly ash was used as an admixture. Mix proportion of the chemical admixtures and the admixture was as follows; the water reducing agent: 0.6% (an agent/cement), the AE water reducing agent: 0.3%, the superplasticizer: the air entraining agent: 0.03%, the flay ash: 20%. Then, flow and setting tests were performed (run 8).

4. RESULTS AND DISCUSSION

4.1 Consistency and setting properties

Results of flow and setting tests (runs 4 and 5) are summarized in Fig. 1 and 2. Flow values of all M-aggregates are lower than S-aggregates, particularly Ma-aggregates.

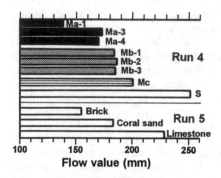

Fig. 1 Flow values of mortar (run 4 and 5)

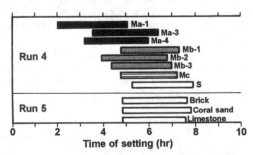

Fig. 2 Setting time of mortar (run 4 and 5). Left and right ends of each bar represent initial and final setting time respectively.

Setting time of all M-aggregates is shorter than S-aggregates. Initial setting time of Ma-aggregates is less than two hours, similar to false setting. But in these cases, the set could not be recovered by remixing. So it is considered that this phenomenon is not actually false setting.

Fig. 3 shows the relationship between montmorillonite group content and flow value, and Fig. 4 the relationship between CEC and flow value. It is clear that flow value is negatively correlated with montmorilonite group content. Unevenness in the correlation can be attributed to differences in mineralogical and chemical properties of the montmorillonite group and also the analytical error.

Fig. 5 and 6 show relationships between montmorillonite group content, CEC and initial setting time. There is a rough negative correlation in Fig. 5 and 6. Unevenness is caused by the same factors as above.

Fig. 7 and 8 show the relationships between water absorption and flow value, initial setting time. Flow value and initial setting time correlate negatively with water absorption. As mentioned previously, water absorption indicates relative content of the montmorillonite group in the aggregates. We wanted to clarify whether changes in flow value and setting time

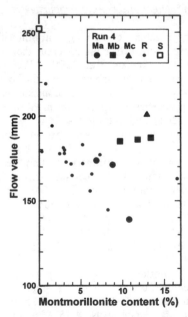

Fig. 3 Relationship beyween montmorillonite group content in aggregates and flow value of mortar (run 4).

3288

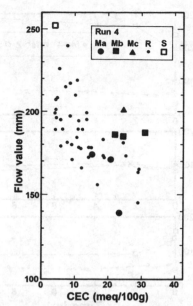

Fig. 4 Relationship between cation exchange capacity (CEC) of aggregates and flow value of mortar (run 4).

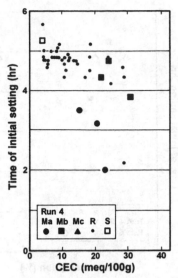

Fig. 6 Relationship between CEC of agregates and initial setting time of mortar (run 4).

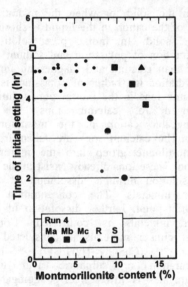

Fig. 5 Relationship between montmorillonite group content in aggregates and initial setting time of mortar (run 4).

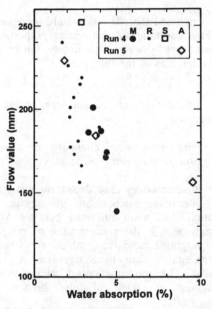

Fig. 7 Relationship between water absorption of aggregates and flow value of mortar (run 4 and 5).

are caused by the mineralogical properties of montmorillonite or by water absorption of the aggregates. We mixed the mortar containing A-aggregates and obtained flow values and setting properties (see Fig. 7 and 8). Flow values of A-aggreates correlate negatively with water absorption. Correlations are slightly different to montmorillonite bearing aggregates, while initial setting time is not affected by water absorbibility. From these relationships,

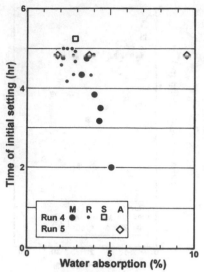

Fig. 8 Relationship between water absorption of aggregates and setting time of mortar (Run 4 and 5).

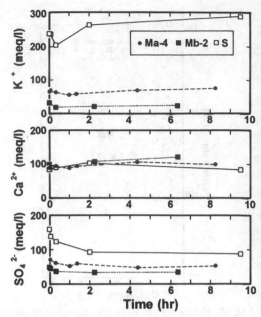

Fig. 9 Time dependent change of chemical compositions of mortar pore solution.

we concluded that flow is mainly governed by water absorption and that setting properties are affected not by water absorbility, but by other characteristics of the aggregates.

4.2 Interaction of system cement-water-aggregates

1) *Time dependent changes of chemical composition of mortar pore solution*

Fig. 9 summarizes time dependent changes of the chemical composition of mortar pore solution. The main difference between M and S-aggregates is the concentration of potassium and sulphuric acid ions, which are lower in M-aggregates than in S-aggregate just after mixing. The concentration of calcium ions in M-aggregates rises slightly higher than S-aggregate.

This suggests that cations produced by hydration of cement are exchanged for other cations in the montmorillonite group. If the solid has ion exchange capacity, and the concentration of cations in the liquid is higher than in the solid, then the cation in the solid can exchange with the cation in the liquid. Ion exchange takes place between the cation in the liquid and another cation in the solid when the charge of the cation in the liquid is larger than

that in the solid, or when the hydration ion radius of the cation in the liquid is shorter than in the solid. In motar pore solution the concentration of potassium and sodium ions is higher than calcium and magnesium ions, and the hydration ion radius of potassium is small. Potassium ions in the mortar pore solution are exchanged for calcium ions, the main exchangeable cation in the montmorillonite group. The calcium ions are taken from the montmorillonite group into the mortar pore solution. These ions already exsist in the mortar pore solution through dissolution of cement clinker minerals. The concentration is at saturation point. Further dissolution of cement clinker minerals is controlled. Controlled cement clinker minerals is considered to be gypsum ($CaSO_4 \cdot 1/2H_2O$) because concentration of sulphuric acid ion of M-aggregates is lower than S-aggregate (Fig. 9).

2) *Setting properties of mortar containing ion-exchanged aggregates*

Fig. 10 shows setting times for mortar made of ion-exchanged aggregates (run 7). In all M-aggregates, the setting time for calcium and sodium-exchanged aggregate is shorter than potassium-exchanged aggregates. For

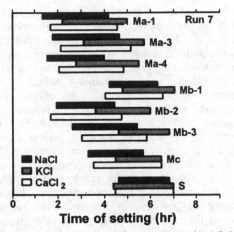

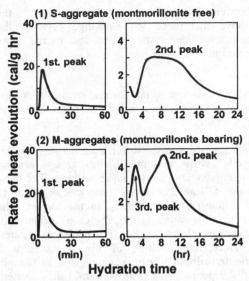

Fig. 10 Setting time of mortar made of ion exchanged aggregates (run 7).

Hydration time

Fig. 11 Profiles of exothermic peaks of mortar containing montmorillonite group free aggregates (1) and montmorillonite group bearing aggregates (2).

S-aggregate the setting time are the same for all elements. The setting time of potassium-exchanged Mb-1, Mb-3 and Mc-aggregates is very close to that for S-aggregate.

These results indicate that setting properties of the mortar are significantly affected by the ion exchange capacity of the montmorillonite group. Potassium ions in the mortar pore solution are probably exchanged with another cation in the montmorillonite group, because the setting time of the potassium is longer than the others. This ion exchange accelerates the setting time of the mortar containing montmorilonite group bearing aggregates.

3) *Properties of hydration process*

We can determine the hydration process of mortar by measuring hydration heat. Fig. 11 shows profiles of rate of hydration heat evolution. Usually two peaks are observed. The first peak is produced by the initial stage hydration of C_3S ($3CaO \cdot SiO_2$) and/or formation of ettringite. The second peak correspond to the point when ettringite is converted to monosulphate. The profile of S-aggregate mortar is the same as normal types, as shown in Fig. 11 (1).

M-aggregate mortar shows a third peak between the first two (see Fig. 11(2)). The calorific value of the third peak correlates with the montmorillonite content in the aggregates (Wakizaka et al., 1993). The second peak has a sharp profile. The presence of

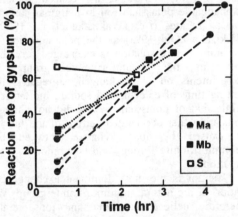

Fig. 12 Time dependent change of gypsum reaction rate of mortar.

montmorillonite group in the aggregates clearly affects the hydration of the mortar.

Next we analyzed hydration products in the mortar. M-aggregates differ from S-aggregate in terms of gypsum content in the mortar. Fig. 12 shows the gypsum reaction rate of the mortar: for S-aggregate this does not change with time, but for M-aggregates it does. The reaction rate is 10 to 40% at 30 minutes after the beginning of hydration and 50 to 100% at two to four hours. The difference in gypsum reaction rate between M and S-aggregate is attributed to the above mentioned ion exchange between the montmorillonite group and the mortar pore solution.

4.3 Mechanism of accelerated setting of mortar which contain montmorillonite bearing aggregates

We investigated the differences between mortar made with montmorillonite group bearing aggregates and mortar made without. The flow value fell and the setting time was short due to the montmorillonite group in the mortar. The small flow value is thought to be because of the water absorption capacity of the aggregates.

Accelerated setting of mortar made with montmorillonite bearing aggregates is thought to proceed as follows in the system cement-water-aggregates. Just after mortar mixing the concentration of potassium and sulphuric acid ions in the mortar pore solution drops and the concentration of calcium ion rises. During this stage, the concentration of exchangable potassium ion in marginal parts of the aggregates rises (Wakizaka et al., 1990b; Wakizaka et al.,1993) as the potassium ion in the mortar pore solution is exchanged for other ions in the montmorillonite group. In experiments on ion-exchanged aggregates, the setting time of calcium and sodium mortar fell, while that of potassium mortar did not change, indicating that accelerated setting of the mortar is caused by ion exchange between the montmorillonite group and the mortar pore solution.

Calcium is the most dominant exchangeable cation in the natural montmorillonite group. We therefore believe that potassium ions in the mortar pore solution are exchanged for calcium ions in the montmorillonite group, which are taken from the montmorillonite group to the mortar pore solution until the concentration of calcium ions in the mortar pore solution reaches saturation point. Under these conditions, gypsum in cement clinker minerals does not dissolve in the mortar pore solution. This is indicated by the reaction rate of gypsum in the mortar. In the montmorillonite group free aggregate mortar, the reaction rate of the gypsum is constant at around 60%. The reaction rate of the gypsum in montmorillonite bearing agregates mortar is low (10 to 40%) at 30 minutes after mixing, but rises thereafter.

From 0.5 to 4 hours after mixing the third peak in hydration process of the mortar is observed. The calorific value of the third peak is normally correlated with the content of the montmorillonite group in the aggregates. It is clear that ion exchange between the montmorillonite group and the mortar pore solution affects the hydration reaction of cement clinker minerals. We believe that the exothermic reaction of the third peak is created by a hydration reaction between gypsum and $C_3 S$, which accelerates mortar setting.

It is proposed that the accelerated setting of mortar containing montmorillonite bearing aggregates proceeds as follows. First, calcium ion in the montmorillonite group is exchanged with potassium ion in the mortar pore solution. Dissolution of the gypsum in cement clinker minerals is artificially controlled. Hydration reactions of the $C_3 A$ and the $C_3 S$ are affected. The exothermic reaction is triggered between one and four hours after mixing, leading to accelerated setting.

In the experiments on ion-exchanged aggregates, sodium ion mortar also showed accelerated setting. The sodium ions in the montmorillonite group are exchanged for the potassium ions in the mortar pore solution. Sodium hydroxide (NaOH) is then created by reaction between the sodium ions and hydroxyl ions in the mortar pore solution. We already know that sodium hydroxide accelerates the hydration reaction.

4.4 Evaluation of montmorillonite bearing aggregates for concrete

Workability of concrete is reduced by accelerated setting. We must select good aggregate for concrete, but not in terms of montmorillonite content alone, since accelerated setting is caused by ion exchange between the montmorilonite group and the mortar pore solution. Mortar setting is governed by the mineralogical properties of the montmorillonite group controlling ion exchange as well as montmorillonite group content.

Good aggregates for concrete are selected as follows. First, the quarry site is divided into zones according to montmorillonite group generation. Mineralogical analyses of X-ray diffraction and exchangable cation are performed on samples from each zone to determine montmorillonite group species and type and quantity of exchangable cations. Zones may be rivised if the analyses differ. Next, we measure montmorillonite group content by X-ray diffraction. Relative content is sufficient. Content can be correlated with CEC and/or water absorption to find relative content in terms of CEC and water absorption. Setting tests are also conducted to find correlation with content and initial setting time. The minimum permissible initial setting time is determined by concrete construction schedule. Finally, we find the maximum permissible content of montmorillonite group from the correlation and select good aggregates for concrete construction.

4.5 *Utilization of montmorillonite group bearing agregates*

In some cases we may not be able to avoid aggregates with more montmorillonite group than maximum permissible level. These aggregates can be used for concrete as follows:

a) Montmorillonite group is reduced to fine grained powder during the grinding process more quickly than other minerals. Montmorillonite group content can be reduced by removing this powder from the aggregate. Coarse aggregates do not affect the setting properties of concrete as much as fine aggregates with the same montmorillonite group content (Wakizaka et al., 1990a).

b) Setting tests on mortar made from ion-exchanged aggregates found that potassium ion- exchanged mortar takes a long time to set. Some mortar made of the potassium ion-exchanged aggregates took as long as montmorillonite group free-aggregates mortar. Montmorillonite group bearing aggregates can be utilized for concrete if treated first with potassium ion exchange. But this treatment is not actually used because the potassium chloride (KCl) or potasium hydroxide (KOH) required for the treatment can cause salt damage or alkali aggregate reactions respectively on the concrete.

c) The third method, the most realistic uses. Chemical admixture studies have already been made on the effect of chemical admixture on concrete containing the montmorillonite group bearing aggregates. Use of an AE water reducing agent (lignin sulfonic acid salt) delays somewhat the setting time (Fig. 13), but creates two problems: the agent increases the air

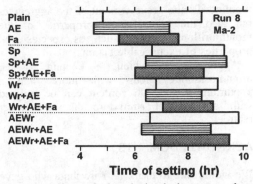

Time of setting (hr)

Fig. 13 Effect of chemical admixtures and an admixture on setting time of mortar (run 8). AE:an air entraining agent, Fa:fly ash, Sp:a superplasticizer, Wr:a water reducing agent, AEWr: an AE water reducing agent.

content of the concrete, and prevents the concrete from consolidating property. Super retarder (carboxylic acid salt, chiefly gluconic acid) is better than the AE water reducing agent. The super retarder delays setting time, and produces satisfactory hardened concrete (Kawano and Takahashi, 1992; Nagayama et al., 1993; Nagayama and Katahira, 1993; Kawano et al.,1994).

5. CONCLUSIONS

The flow value and setting time drop for mortar with montmorillonite group bearing aggregates. The small flow value is due to the high water absorption of the aggregates. Accelerated setting, a marked characteristic of montmorillonite group bearing aggregates occurs as fpllows. First, calcium ions in the montmorillonite group are exchanged with potassium ions in the mortar pore solution. Concentration of calcium ions in the mortar pore solution reaches saturation. Dissolution of the gypsum in the cement clinker minerals is controlled. Hydration reactions of the C_3A and the C_3S are affected. The exothermic reaction (hydration reaction between gypsum and C_3S) is activated from one to four hours after mixing, leading to accelerated setting.

Workability of concrete construction is reduced by accelerated setting. So, we must select good aggregates for concrete construction. Evaluation of montmorillonite group bearing aggregates requires not only the montmorillonite group content of the aggregates but also the mineralogical properties of the montmorillonite group such as species and properties of exchangable cations.

Aggregates which contain the montmorillonite group with more than a maximum permissible content can be utilized with a chemical admixture such as super retarder.

ACKNOWLEDGMENTS

We would like to thank K. Kuwabara who gave the opportunity of this study to us. Thank is also due to Prof. T. Watanabe for instruction on montmorillonite group mineralogy. We also thank S. Takahashi for measurment and evaluation of the hydration processes and the hydration products of the mortar. The senior author wishes to thank H. Kawano for instruction on concrete technology.

REFERENCES

Maru, A. and Yanagida, C. 1981. Selection of aggregates in terms of petrology, mineralogy and geology. *Concrete Journal*, Vol. 19, No. 11, pp. 85-89. (J)

Japan Society of Civil Engineer 1986. *Standard methods of concrete construction.* p.74, Tokyo, Japan Society of Civil Engineers. (J)

Kawano, H. and Takahashi, H. 1992. Properties of seting time of RCD concrete using clay mineral bearing aggregates. *Proceedings of the 47th Annual conference of the Japan Society of Civil Engineers,* 5, pp. 514-515.(J)

Kawano, H. Takahashi, H. and Kato, S. 1994. Effect of clay mineral in aggregate on setting time of fresh concrete. *Civil Engineering Journal,* Vol. 36, No. 2, pp. 48-53. (J)

Nagayama, I., Yamashita, T., Katahira, H. and Morita, S. 1993. Effect of super retarder and setting properties of RCD concrete using montmorillonite bearing aggregates. *Proceedings of the 48th Annual conference of the Japan Society of Civil Engineers.* 5, pp. 944 -945. (J)

Nagayama, I. and Katahira, H. 1993. Pilot construction of the RCD method on Miyagase dam (6)–Effect of super retarder on accelerated setting of concrete–, *Engineering for dams,* No. 86, pp. 23-34. (J)

Sakurai, T. Tatematsu, H. and Mizuno, K. 1986 Simple quantitative methods for sweling clay mineral content. *Railway Technical Research Report,* No. 1312, p. 27. (JwE)

Shigekura, Y. 1981. Influence of harmful material in aggregates on concrete. *Concrete Journal,* Vol. 19, No. 11, pp. 105-108. (J)

Wakizaka, Y., Uji, K., Hayashi, J. and Sasaki, H. 1990. Physical properties of concrete using montmorillonite group bearing aggregates. *Proceedings of the Japan Concrete Institute,* 12-1, pp. 733-738. (J)

Wakizaka, Y., Fujiwara, Y., Maeda, T. and Mitani, S. 1990. Chemical mineralogical reaction between montmorillonite group bearing aggregates and cement. *Procedings of the Japan Concrete Institute.* 12-1, pp. 739-744. (J)

Wakizaka, Y., Ichkawa, K., Fujiwara, Y., Uji, K., Hayashi, J., Mitani, S., Maeda, T. and Sasaki, H. 1994. Setting properties of concrete containing montmorillonite group. *Journal of the Japan Society of Engineering Geology,* Vol. 34, pp. 209-222. (JwE)

(J):written in Japanese
(JwE):written in Japanese with English abstract

Weathered basalts from Southern Brazil: Use in low-volume roadway

Basaltes altérés du Sud du Brésil: Leur utilisation dans les routes secondaires

J.A.P.Ceratti & G.P.Arnold
Federal University of Rio Grande do Sul, Brazil

J.A.de Oliveira
Highway State Department, Rio Grande do Sul, Brazil

ABSTRACT: This paper describes a study of mechanical characterization and degradation of weathered volcanic rocks from ten sites, typical of basalt spills in Rio Grande do Sul State, Brazil. These rocks were used as base and sub-base material in seven sections of secondary roads. The mechanical behaviours and structural response of these pavements were evaluated by means of surface damages surveys and deflectometric investigations.

RÉSUMÉ: Ce travail décrit une étude de caractérisation et dégradation de dix occurrences de roches volcaniques altérées, typiques des apanchements basaltiques de l'état du Rio Grande do Sul, Brésil. Ces roches ont été utilisées comme matériaux de couches de base et sous-base de sept sections de chaussées secondaires. Le comportement mécanique et la réponse structurelle de ces chaussées ont été évalués au moyen de l'observation de imperfections superficielles et d'investigations deflectométriques.

1 INTRODUCTION

The Rio Grande do Sul State Government has prioritized the paving of secondary roads, mainly in agricultural regions, most of them located first over the thick basalt spills which cover 40% of the state's area. A paved road allows traffic all over the year, lowering maintenance and vehicle operation costs. All these factors contribute in reducing the fuel cost of the goods transported.

Local roads, generally of short length and low traffic volume connect rural properties to main roads. They have more tolerance towards the physical properties of the materials which constitute their pavements. These roads require the search for reducing costs through the use of alternative materials. In Rio Grande do Sul State the most recent experience has been the construction of 12 experimental sections with lengths varying from 5 to 20 km, by the end of the '80s.

The structural layers of those pavements were made of weathered basalt, extracted from sites next to the roads, with the use of bulldozers instead of explosives.

Twelve months late some sections showed premature failure but others still bore good traffic conditions.

These results stimulated the interest in characterizing the mechanical behaviour of those materials in order to rationalize its use in low cost pavements.

In 1990 the School of Engineering of the Federal University of Rio Grande do Sul (UFRGS) and the Highway State Department (DAER) signed an agreement with the purpose of widening the knowledge of the behaviour of such materials. Seven sections were chosen for the study an extensive test program was implemented.

Arnold (1993) has characterized the mechanical behaviour of representative materials fron the sites. Surface damages and deflections of the sections were surveyed.

This work describes and presents the results of the characterization study and structural evaluations.

2 MATERIAL DESCRIPTIONS

The volcanic manifestation which originated basalt spills in southern Brazil and parts of Paraguay, Argentina and Uruguay, spreads over an area of 1,200.000 km² and dates from the middle part of the inferior Cretaceous age (120 to 130 million years ago). The spills in Rio Grande do Sul State are shown in Figure 1. Their compositions vary from basic to acid, including several intermediates.

Spill package portions of rocks intensely fractured are frequently found with predominance of degradation process rather than decomposition ones. Such occurrences are treated as altered basalt and its use is normally restricted to nom paved roads. The altered basalt occurs associatedd with a topography varying from rounded to sharp. The surface part of the occurrence is in contact with an incipient layer of weathered soil.

The occurrences are characterized by fractured systems, with fractures spaced from less than 100 mm to more than 800 mm. Therefore poligonal fragments are found varying from angular to subangular with shelly to plane faces and diverse dimensions distributed in a wide grain size distribuition range.

The weathering manifests itself in a generalized way and sometimes in localized pockets, with common presence of spheroidal weathering. The ocurrence of a clayey matrix filling joints and surrounding fragments is characteristic. However, in some occurrences significative portions may exist where the joints are practically free of clayey material. The exploitation is generally possible in working fronts 2 to 10 meters high the extraction in steps being sometimes possible.

2.1 Studied samples

Fourteen sampling points were chosen as being representatives of the ten occurrences used in the seven sections selected for this study. In the sections where materials from occurrences were used, each of them constructed a sampling point and in those where material from an unique occurrence was used two alteration horizons were differentiated, each of them also constituting a sampling point.

Near 400 kg of the material were collected in each sampling point with picks and showels. The sampling points are represented in Figure 1.

3 LABORATORY TEST PROCEDURES AND PRESENTATION OF RESULTS

3.1 Petrographic analysis from examination of thin sections and x-ray diffraction

In the petrographic analysis emphasis was given to the weathered mineralogy. During the description it was possible to estimate the nature of clay minerals. The birefringence colors frequently showed that they were smectitic minerals. A polarized microscope was used for thin section descriptions; a point-counter with interlinear spacing of 1 mm being used for

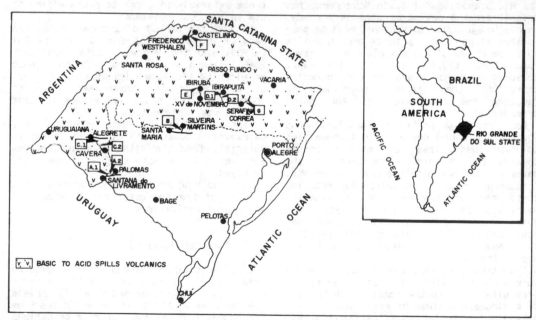

Figure 1. Spills volcanics in the Rio Grande do Sul State and localization of the sampling points.

the modal analysis.

X-ray analysis confirmed the presence of expansive minerals. However, due to diverse crystallization states, the quantification was limited just allowing the estimation that in every sample 90% of the occuring clay minerals were of the expansive type. The results of the petrographic analysis are shown in Table 1.

3.2 Water absorption and specific gravity

The absorption and specific gravity was obtained calculating the mean values of the two determinations for each sample. The DNER-DPT M 81-64 (Brasil-DNER, 1968) method was adopted. Near 1000 grams of aggregated passing sieve size 19mm and retained in sieve size 12.5mm were used. The absorption values ranged from 1.2% to 6.9%. The basalts specific gravity values varied from 2.75 to 3.00 kN/m3 and was of 2.65 kN/m^3 for rhyolites.

3.3 Los Angeles abrasion test

This test was performed according to DNER-DPT M 35-64 (Brasil-DNER, 1968) with crushed aggregate included in range 'A' of that test. The sample was put into a rotating cylinder together with twelve steel spheres. The cylinder rotates 500 times at a frequency of 33 r.p.m.. Weight losses ranged from 15.70% to 51.10%.

3.4 Sodium sulfate soundness

This test involved the recrystalization of sodium sulfate into the aggregate structure. Internal generated stresses tend to fragment

the aggregate particles diminishing grain size. The test was carried out following "Test Method Calif. NQ 214-D - Method of test for the soundness of aggregates by use of sodium sulfate".

Weight losses, evaluated by sieving, were used as index of material soundness. Weight percentages of each tested fraction, were pondered relating to range 'A' grain size distribution (shown in Figure 2), aiming at standardizing the results (Arnold, 1993).

Weight losses varied from 2.6% to 10.7% in most cases. Samples A.2 and C.2 represented the highest weight losses as expected due to their high weathering degree.

3.5 Washington degradation test

This test evaluates the degradation of crushed aggregates resulting from water and air exposure. Fine size particles are analysed through sedimentation test. A high Washington degradation factor (D_y) indicates a good quality aggregate. For the studied samples this factor varied from 13 to 99.

3.6 Point load test

The point load test used to evaluate the mechanical strength of the mineral skeleton, is performed in cylindrical speciments with 1" diameter.

The ratio of saturated strength over dry strength is used as an indicative of the aggregate capability of resisting the clay minerals expansive forces. The Is(50)dry in the studied samples varied from 0.32 to 12.10 MPa.

Test results are summarized in Table 2. The gradation curves of aggregate used in each test are represented in Figure 2.

Table 1. Results of mineralogical analysis.

Mineral type	Samples													
	A.1	A.2	B.1	B.2	C.1	C.2	D.1	D.2	E.1	E.2	F.1	F.2	G.1	G.2
Plagioclase	42	37	33	38	-	46	4	3	22	19	30	25	-	32
Clinopiroxene	30	32	33	34	-	7	-	-	25	15	29	26	-	24
Glass/groundmass	11	8	11	9	-	-	-	-	19	27	-	-	-	13
Clay - minerals	11	13	11	13	-	19	8	9	19	19	10	13	-	18
Alkalic feldspar	-	-	-	-	-	-	46	42	-	-	-	-	-	-
Quartz-feldspar	-	-	-	-	-	-	30	27	-	-	13	16	-	-
Others[a]	6	10	12	6	-	28	12	19	15	20	18	20	-	13
	B	B	B	B	-	B	R	R	B	B	B	B	-	B

Others: Basalt(B): iron oxide and metallic minerals
Rhyolite (R): quartz, metallic minerals and altered mafics

Table 2. Results of technological tests.

Samples	Tests results						
	Specific gravity	Absorption (%)	Los Angeles abrasion (%)	Soundness loss (%)	Washington degradation factor	Is(50) dry condition [MPa]	Is(50)saturated / Is(50) dry
A.1	2.95	1.60	15.70	3.45	94	11.30	0.76
A.2	2.85	5.00	51.10	42.05	13	0.32	0.22
B.1	2.85	1.40	21.20	4.70	79	8.30	0.69
B.2	2.85	1.20	15.70	4.75	98	9.95	0.97
C.1	2.95	1.20	31.00	4.00	52	11.30	0.85
C.2	2.80	2.10	21.00	14.70	80	7.50	0.79
D.1	2.65	6.70	50.00	10.70	39	5.50	0.40
D.2	2.65	6.90	46.20	7.60	59	5.75	0.71
E.1	2.85	6.50	50.20	7.25	39	3.85	0.99
E.2	2.75	6.00	43.90	9.15	17	1.70	0.70
F.1	3.00	3.40	37.30	7.05	47	7.20	0.94
F.2	3.00	3.00	37.80	7.45	35	4.90	0.74
G.1	2.85	2.50	24.50	7.60	62	11.05	0.92
G.2	2.95	1.70	15.90	2.60	99	12.10	0.72

4 DISCUSSION OF RESULTS

In spite of being classified by petrographic testing as basalts, samples E.1, E.2, F.1 and F.2 present levels of quartz-feldspar intergrowth or crystalization residues which remarkably surpass those normally found in basaltic rocks. This fact allied to specific gravity and absorption values seems to indicate that from a chemical point of view, those rocks present an andesite-basaltic composition. However, the confirmation of this assumptiom demands further research. Nevertheless the information is relevant to data interpretation.

4.1 Los Angeles abrasion and absorption

Figure 3 shows the influence of absorption in weight losses occurring in Los Angeles tests. Weight losses tend to increase linearly, the correlation coefficient being 0.81. Samples with andesite-basalt composition noticeably display different behaviour.

4.2 Los Angeles abrasion and soundness test

Figure 4 shows the relationship between Los Angeles abrasion test results and those of loss in sodium sulfate soundness test. Samples with a more acid composition differentiate from the others.

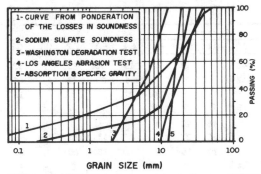

Figure 2. Gradation curves of aggregates used from tests.

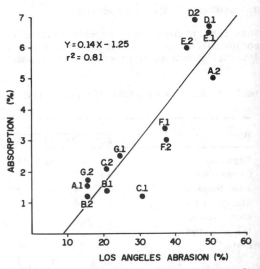

Figure 3. Relationship between Los Angeles abrasion and absorption.

3298

4.3 Los Angeles abrasion test and Washington degradation factor

The relationship between results of Los Angeles abrasion test and Washington degradation factor is shown in Figure 5. Theresults seem to be in inverse relationship and the correlation factor is 0.8.

4.4 Los Angeles abrasion test and point load test

Figure 6 shows the relationship between Los Angeles and Is(50)dry results. Those results are also in inverse proportion and once again two different behaviours can be noted, depending on the rock mechanical composition.

4.5 Point load test and Washington degradation factor

Figure 7 shows a direct relationship between Washington degradation factor and the Is(50)dry, whose correlation coefficient is 0.71.

4.6 Sodium sulfate soundness test and Is(50)satured/Is(50)dry ratio

Figure 8 shows the relationship between

weigth losses in soundness test and the ratio Is(50)satured/Is(50)dry. That figure indicates that a value of 0.7 for the ratio would indicate that the skeleton is strong enough for expansion stresses caused by clay minerals of the expansive type.

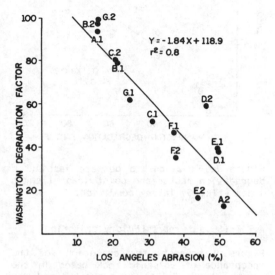

Figure 5. Relationship between Los Angeles abrasion and Washington degradation factor.

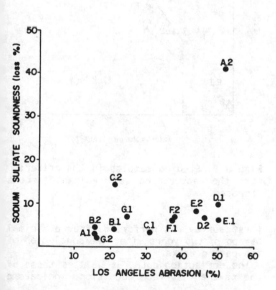

Figure 4. Relationship between Los Angeles abrasion and sodium sulfate soundness.

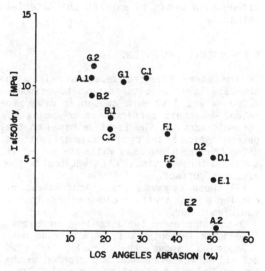

Figure 6. Relationship between Los Angeles abrasion and point load strenght index – Is(50) in dry condition.

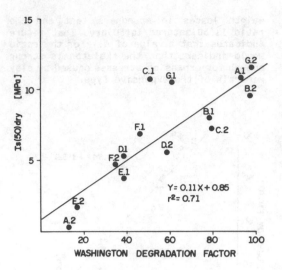

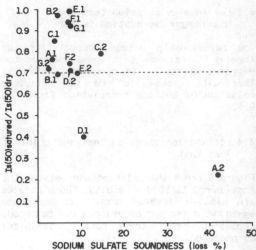

Figure 7. Relationship between Washington degradation factor and point load strength index - Is(50) in dry condition.

Figure 8. Relationship between losses in soundness test and the ratio Is(50)satured/Is(50)dry

4.7 Is(50)satured/Is(50)dry criterion

Figure 9 presents the criterion for the acceptance of volcanic rock based in the ratio between Is(50) in saturated and dry conditions.

A value of 0.7 taken from Figure 8 is adopted as the lowest acceptable ratio. Besides a value of 3.50 MPa for the Is(50)saturated is specified as minimum strength in order to warrant the material integrity.

5 STRUCTURAL EVALUATION

In the seven selected sections, all with sub-base layer of altered basalt, segments 1,000 meters long were chosen in order to attend to their performances by means of periodic surveys. The first is based on the determination of the recoverable deflections using the Benkelman beam while the latter is achieved through visual observations of the pavement surface.

All those segments are characterized by bearing a mean traffic volume of 300 to 600 vehicles a day.

Up to the present two attending campaigns, spaced twelve months out, were carried out.

Table 3 shows the results of two deflection measurement sets performed on the external wheel track of the segments. The

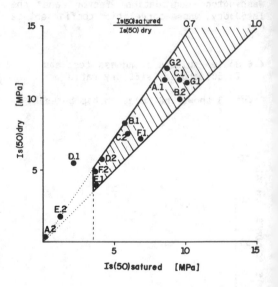

Figure 9. Studied samples in the criterion for the acceptance of volcanic rocks proposed.

first survey was performed after a minimal period of two years after the conclusion of the paving works.

The deflection data, as well as those of the pavement surface condition, indicate a

3300

Table 3. Benkelman beam deflections results.

Road	Borrow pits	Road conclusion date	Mean deflection ($\times 10^{-2}$ mm)					
			1st date	$\bar{D}$	CV	2nd date	$\bar{D}$	CV
1	C.1 and C.2	jul 89	mar 92	78	0.30	apr 93	82	0.23
2	A.1 and A.2	nov 88	apr 92	86	0.36	feb 93	98	0.46
3	B.1 and B.2	oct 89	aug 91	66	0.47	dec 92	64	0.64
4	G.1 and G.2	jan 89	dec 91	53	0.32	jul 93	68	0.19
5	E.1 and E.2	jun 89	sep 91	48	0.22	–	–	–
6	D.1 and D.2	dec 89	sep 91	109	0.18	nov 92	118	0.18
7	F.1 and F.2	jan 89	oct 91	57	0.56	apr 93	70	0.57

$\bar{D}$ – mean deflection
CV – coefficient of variation

process of pavement deterioration according to the expected pattern. This process seems related to drainage deficiencies observed in situ rather than to a possible basalt degradation. However, only a longer attendance period will lead to a more consistent conclusion.

6 CONCLUSIONS

It is possible to infer from a careful interpretation of petrographic analysis the chemical composition of volcanic rocks. Then the interpretation of results can be qualified, taking into account the different compositions in justification of different behaviours.

The characterization of mechanical behaviour of those rocks still requires the evaluation by means of several tests which, analyzed as a whole, will show what can be expected from the material when used in pavements.

ACKNOWLEDGEMENTS

The authors gratefully acknowledges the financial support of FAPERGS and CNPq and the contribution of Washington Perez Núñez for the development of the present paper.

REFERENCES

Arnold, G.P. 1993. Estudo do comportamento mecânico de basaltos alterados do Rio Grande do Sul para emprego em pavimentos rodoviários. M.Sc. Thesis. Universidade Federal do Rio Grande do Sul, 165p.

Brasil. Departamento Nacional de Estradas de Rodagem. 1968. Métodos de ensaios. Rio de Janeiro, 206 p.

Skid resistant aggregates in Greece

Résistance au glissement des agrégats en Grèce

E.A.Charitos & E.S.Sotiropoulos
Geomechaniki Ltd, Athens, Greece

M.C.Prapides
Ministry of Public Works, Ioannina, Greece

A.T.Kazacopoulos
Ministry of Public Works, Thessaloniki, Greece

ABSTRACT: The right quality of aggregate material is the most important factor for the skid resistance of pavements. This paper presents the results of an investigation carried out in order to locate suitable rock formations that can become sources of skid resistant aggregates for the construction of asphalt surfaces on Greek roads and highways.

1 INTRODUCTION

The increasing number of road accidents in Greece, is mainly due to the poor skid resistance of pavements.

The use of proper materials for a skid resistant construction, increases the surface roughness of the pavement. This in turn affects the safety of traffic.

For this reason, good skid resistance must be insured on a long term basis.

The high rate of road accidents in Greece is becoming a major social issue.

The percentages of annual accidents in wet pavement to the total number of accidents in Greece are presented in Fig. 1.

A major improvement of the safety conditions, especially for motion in wet surfaces, will be the construction of a surface dressing which exhibit large skid resistance.

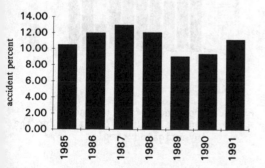

Fig 1. The percentage of annual accidents in wet pavement in Greece.

2 CRITERIA FOR THE AGGREGATE SELECTION

The parent rock which is intended to be used in the aggregate production must have specified mechanical strength, in order to bear the complex distress during the aggregate production as well as during the asphalt mixture dispersion.

The aggregate materials used are 95% in the composition of upper road layers which is the most important factor for the skid resistance of the pavement.

It has been proven, both by in-situ and laboratory testing, that the behaviour of the aggregate materials is entirely different under traffic conditions.

The aggregate material must present adequate resistance in polishing and abrasion (in addition to its common mechanical properties) in order to be suitable for a skid resistant construction.

The performance of the aggregate, can mainly be evaluated by its Aggregate Polished Stone Value, PSV and Aggregate Abrasion Value, AAV (determined according to BS 812/1975 specification) and the Los Angeles value (determined according to ASTM C-131).

Appropriate aggregates for skid resistant constructions are those which combine a high P.S. value with low A.A. and L.A. values.

3   CONDITIONS IN GREECE

Road rolling surfaces constructed in Greece, mainly consist of aggregates of limestone origin.

Limestone and its varieties are the main source of aggregates in Greece, because they can be found in almost every part of the Greek territory in abundance, they can easily be worked and usually their other properties meet the requirements regarding the asphalt concrete so their performance as a skid resistant material is very poor.

The main problem is, that limestone aggregates become easily polished and their surface becomes smooth and slippery.

In Greece presently there are only four sources available for the production of skid resistant aggregates with limited extraction quantities. The use of materials from those sources obviously results in a large increase of the road construction cost.

If European minimum required levels for skid resistance aggregates apply in Greece, new sources of material must be found.

The Greek Ministry of the Environment, Physical Planning and Public Works, has started an experimental skid resistant pavement research program in the Greek Road Network.

The object of this project is the evaluation of skid resistant pavement performance in connection with the characteristics and the particularities of each region.

In Western Greece, for instance, sections of porous asphalt surfacing have been constructed using crushed aggregates from cobbles taken from local river-beds.

The performance of these porous asphalt sections regarding skid resistance is satisfactory.

4   RESULTS

An extensive investigation was carried out, in order to locate new possible sources of skid resistant aggregates for major transportation projects in Attiki region.

The investigation comprised extensive sampling and laboratory testing in 15 quarries and borrow areas around Attiki.

A number of very significant conclusions have been drawn from the results of the above investigation.

Figures 2 and 3 show the frequency of occurrence of Polished Stone Value and Aggregate Abrasion Value, for 31 different samples.

From the date presented in these figures it can be seen that there are aggregates with exceptional specifications and therefore, the respective rock formations can be used as sources for skid resistant aggregates.

Table 1 shows the laboratory results of samples tested within the scope of this investigation.

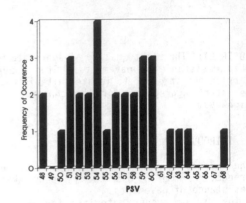

Fig 2.   Frequency of occurrence of Polished Stone Value for the investigated samples.

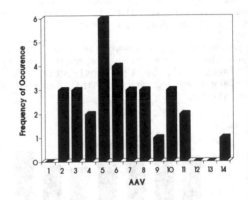

Fig 3.   Frequency of occurrence of Aggregate Abrasion Value for the investigated samples

TABLE 1. Characteristic Mechanical Properties of Aggretates

AGGREGATES	LOCATION	L.A.	P.S.V.	A.A.V
Schists	Votonosi-Epirus	19	68	14.0
Schists	Dervenochoria-Attica	27	63	3.0
Gabro	Metsovo-Epirus	18	54	5.0
Quartzite	Distrato-Epirus	24	60	5.0
Andesite	Peristeri-Epirus	19	60	6.0
Porphyry	Evzoni-Macedonia	12	58	6.0
Quartzite	Nikiti-Macedonia	16	59	5.0
Peridotite	Gerakini-Macedonia	22	56	10.0
Amphibolite	Byronia-Macedonia	21	52	8.0
Granite	Kavala-Macedonia	22	58	4.0
Cobbles	Aoos-Epirus	16	62	8.0
Cobbles	Kalentinis-Epirus	20	64	10.0
Cobbles	Aliakmon-Macedonia	25	54	6.0
Cobbles	Mornos-C. Greece	20	60	7.0
Granite	Xanthi-Thrace	21	54	3.0

5 CONCLUSIONS

The right judgement in selecting construction materials is very important for the structural integrity of a skid resisting construction. It is well known that the desirable result has to be viewed from the economical stand point and is a combination of the above mentioned choice and of proper construction method.

The results of this research proved that in different areas of Greece there are large quantities of rock formation with very good quality that can become sources of skid resistant aggregates.

REFERENCES

Halstead, W.J. 1983. Criteria for use of asphalt friction surfaces: National Cooperative Highway Research Program, T.R.B., 104, Washington.
British Standards Institution, BS 812 : Part 3 : 1975. British Standards Methods for Sampling and testing of mineral aggregates, sands and fillers.
American Society for Testing and Materials, ASTM C 131-89. Standard Test Method for Resistance to Degradation of Small-Size Coarse Aggregate by Abrasion and Impact in the Los Angeles Machine.
Michas N., Sotiropoulos E., Tsitseklis E. 1974. Research of Skid Resistance, Athens, (in Greek).
Prapides M., Siamopoulos G., Stamoulakis G., Tzovas D. 1990. Skid Resistant Aggregates in Epirus. (in Greek).
Athanassopoulou A., Collaros G., Charitos E. 1992. Skid Resistance Properties of Pavements in Thrace. Treatment of the Problem with the Use of Proper Materials Proc. 1rst Hellenic Symposium on Asphalt-Concrete and Flexible Pavement Construction, Thessaloniki, Greece. (in Greek).

Application of unbound materials on a road pavement

Application de matériaux granulaires dans une couche de base de chaussée

Ana Cristina Freire & Luis Quaresma

National Laboratory of Civil Engineering, Lisbon, Portugal

ABSTRACT: Unbound aggregates present a vast applicability namely on base and sub-base layers of flexible pavements or as sub-base of rigid or semi-rigid pavements. These materials are also applicable as capping layers. Materials applied in granular layers of pavement layer may have different sources: natural, rolled or from massive rocks, or marginal, industrial sub-products.

The behavior of unbound materials is complex, because they are not a continuous material, being constituted from individual particles. Therefore physical and mechanical characterization of the particles and of the mixture is important.

In this paper physical and mechanical characteristics obtained in laboratory, namely with repeated load triaxial tests and in situ with Falling Weight Deflectometer are presented and analyzed. Different materials, applied in base layers of a flexible pavement in Portugal were tested.

RÉSUMÉ: Les matériaux granulaires ont une application très vaste dans les chaussées, nommement dans des couches de base ou de fondation des chaussées souples ou dans les couches de fondation des chaussées rigides ou semi-rigides. Ces matériaux peuvent être aussi utilisés dans les couches de forme. Les matériaux appliqués dans les couches granulaires peuvent avoir plusieurs origines: l'origine naturelle, alluvionaire ou de massif rocheux, ou bien marginale, nommement sous-produit industriel. Le comportement des matériaux granulaires est très complexe, car ils ne sont pas un matériau continu, étant composé par des particules individueles. Ainsi, la caracterisation physique et méchanique des particules et du mélange a une importance assez grande.

Dans ce rapport on présente et on analyse les caractéristiques physiques et mécaniques obtenues en laboratoire, nommement par des essais triaxiaux répétés et in situ par des essais de portance effectués avec le FWD. Plusiers matériaux ont été essaiés, appliqués dans les couches de base d'une chaussée souple au Portugal.

1 INTRODUCTION

This paper presents and analyses the results of a study concerning physical and mechanical characterization of unbound granular materials applied on a road base pavement. In what concerns the laboratory tests, repeated load triaxial tests were performed. In situ load tests, with the Falling Weight Deflectometer were also carried out.

2 ROLE OF UNBOUND MATERIALS IN ROAD PAVEMENTS

Unbound aggregates present a vast applicability on highway pavements, mention must be made of three different layers where these materials can be used: i) capping layers; ii) sub-base; iii) base.

The main functions of granular layers in a flexible pavement are as follows (Jones, R. H. et al., 1989):

1. to provide a working platform during road construction;
2. to contribute to the structural performance of the pavement;
3. to provide a drainage layer;
4. to act as a frost blanket.

The first function is important, mainly in sub-base and capping layers, allowing that, during the pavement construction, traffic has a good circulating platform.

As a structural element, granular layers contribute

to the degradation of stress applied to the pavement, by the wheel loads, in order that, for different climates conditions and for various passages of vehicles, stress state verified on foundation or on sub-base layer must be compatible with its support capacity.

Concerning frost-defrost phenomenon, its appearance is limited to some countries, in function of their own clime. In Portugal, this phenomenon occurs only in some restricted areas, without great geographical meaning.

When the drainage characteristics are required for granular layers a high porosity to allow the flow of the water is necessary.

Base layers made of unbound granular materials can be of two differents types: i) dry-bound macadam; ii) well-graded materials.

Construction of base layers with well graded materials allows a major automation of the construction methods, resulting in a high compaction, with the increment of the contact points between particles and the increase of the surface friction angle. For these materials it is recommend a 0/30 or 0/40 mm grading, (JAE, 1978).

3 DEVELOPED STUDY

3.1 Considerations

It was developed a physical and mechanical characterization study of two natural crushed aggregates (*Graywacke*, *Limestone*), applied in the base layers of a flexible pavement in Portugal (Freire, A. C., 1993). Analyzed materials represent geological formations that exist in an extensive area of Portugal. Metamorphic formations of schist and graywacke outcrop in the North and Center of the country, forming part of the setentrional block of Hesperic massif, composing the ante-Ordovician schist-graywacke complex, that is divided into two groups (Douro and Beiras). In the South of Portugal, schist-graywacke rocks compose Baixo Alentejo Carboniferous-Flysh on meridional block of the Hesperic massif, (Ribeiro, A. *et al.*, 1979). Limestone rocks occur in Portugal in two distinct areas belonging to the occidental meso-Cenozoic borders (Estremadura e Beira Litoral) and to the meridional border (Algarve) (Ribeiro, A. *et al*, 1979). These rocks were formed in a maritime or lacustrine environment, since Jurassic to Miocene, occurring intercalated with other type of rocks, such as sandstone, clays and marls.

3.2 Physical characterization

Identification and physical characterization studies of materials presented (Table 1 and 2) were obtained after tests performed on particles and on the mixture of aggregates, in accordance with Portuguese specifications, or when they did not exist, in accordance with international specifications.

The two materials were applied in the base layer of the flexible pavement of Via Infante de Sagres in Algarve with the grading composition presented in Table 3.

Considering the results of characterization tests and by comparing them with the limits presented on national and foreigner specifications, it is possible to verify that they fulfil the limits established. Thus, for the Los Angeles abrasion test, results show that *Graywacke* is harder than *Limestone*, accomplishing however the limit established (<30%) (JAE, 1978).

Results of the two shape tests, flakiness and elongation index, after comparison with the British specifications limits (<35%) (HMSO, 1986), show that the two fractions of *Graywacke* have an elongated form, maybe as the result of the geological nature of the material.

Table 1. Grading curves of the fractions of studied materials

ASTM Sieve	Percentage passing						
	Graywacke				Limestone		
	Crushed aggregate		Coarse sand 0/15	Natural sand 0/5	Crushed aggregate		
	25/50	15/25			25/50	15/25	0/8
2"	100	-	-	-	100	-	-
1 1/2"	65.5	-	-	-	99.5	-	-
1"	10.5	100	-	-	30.4	100	-
3/4"	1.1	70.1	100	-	6.8	99.6	-
1/2"	-	3.9	92.9	-	0.2	35.9	100
3/8"	-	-	65.5	100.0	-	4.3	-
N° 4	-	-	39.9	99.9	-	-	55.7
N° 10	-	-	21.4	99.2	-	-	18.7
N° 20	-	-	12.6	88.1	-	-	7.1
N° 40	-	-	7.9	58.4	-	-	4.1
N° 80	-	-	4.5	5.4	-	-	2.5
N°200	-	-	1.8	0.2	-	-	1.7

Table 2. Physical characterization of studied materials

Characteristics			Material	
			Graywacke	Limestone
bulk density (g/cm³)	> n° 4 ASTM		2.72	2.74
	< n° 4 ASTM		2.64	2.68
water absorption (%)	> n° 4 ASTM		1.7	0.8
	< n° 4 ASTM		1.2	1.4
sand equivalent (%)			49	60
Los Angeles abrasion (%)			20	29
shape	flakiness index (%)	(25/50)	29	20
		(15/25)	25	31
	elongation index (%)	(25/50)	42	35
		(15/25)	43	32
methylene blue test (V_{Bl})			0.65	0.37

Table 3. Ponderal composition of applied materials (%)

Materials	Crushed aggregate			Coarse sand	Natural Sand
	25/50	15/25	0/8	0/15	0/5
Graywacke	20	25	---	40	15
Limestone	30	30	40	---	---

The results of the sand equivalent test of the two aggregate mixtures are also in compliance with the limits established (> 50%).

3.3 Mechanical characterization

3.3.1 Developed studies had as main purpose the quantification of deformability of several layers of the pavement, expressed by deformability moduli, to make it possible to compare them with results obtained by repeated load triaxial tests performed over specimens with the same grading curves.

3.3.2 In order to determinate deformability moduli of all the layers of studied pavement, in situ load tests with the Falling Weight Deflectometer were carried out over the embankment and over the others layers, during construction of the pavement. Tests were carried out over layers of three selected sections defined among kms 1+850 to 2+150, 10+600 to 11+100 and 16+225 to 16+525,

Table 4. Sections characterization

Section	km	Extension (m)	Materials		
			embankment	sub-base	base
T1	1+859 to 2+150	300	calcareous formations		graywacke (crushed aggregate)
T2	10+600 to 11+100	500	schist-graywacke formations	graywacke (natural aggregate)	
T3	16+225 to 16+525	300	calcareous formations		limestone (crushed aggregate)

included on 3th and 4th sections of Via Infante de Sagres (Table 4).

Two alignments were tested, on each outside wheel track. Distance between consecutive test points, on each alignment was 15 m.

Considering that unbound materials behaviour is non linear, it is necessary to perform a mechanical characterization in conditions as close as possible of those imposed by traffic, namely intensity and velocity of load applications, and geometry of the contact area. In this way, load tests of granular layers were performed with the smaller drop height, achieving load forces of 20 kN, applied to a 45 cm diameter plate. Higher forces were applied to bituminous layers; 50 kN load forces, corresponding to intermediate drop height were applied on a 30 cm diameter plate.

Deflections were measured, at the layer surface, in seven different points: in the centre of the plate, and at 30, 45, 60, 90, 150 and 250 cm from the centre.

3.3.3 The interpretation of Falling Weight Deflectometer load tests was made by means of a program (Antunes, M. L., 1993), that associates an iterative process with the use of ELSYM5 program, from Berkeley University (Ahlborn, G., 1972) as a subroutine. This program, through the introduction of seed moduli, calculates theoretical deformations that approach measured values. This process is iterative and ends when the maximum number of adopted iterations is reached or when the difference between calculated and measured values is less than a certain value.

A statistical analysis was made, after the determination of the deformability moduli, through the elaboration of frequency graphics and the determination of moduli averages and standard deviations, for all the sections.

Figure 1 presents, as an example, a relative frequency graphic for the base layer on section T1.

The analysis of relative frequency of deformability moduli allows the definition on a certain interval.

Table 5 presents the variation interval of deformability moduli for base and sub-base layers, obtained from in situ load tests. This interval represents an confidence interval of 85% for the obtained results, after a statistical approach.

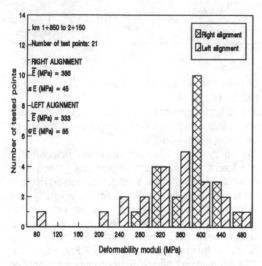

Figure 1. Deformability moduli for the granular base on section T1

Table 5. Interval of deformability moduli of granular layers obtained from FWD load tests

| Date | Tests perform on | Layer | | | | | |
| | | sub-base | | | base | | |
		T1	T2	T3	T1 (*)	T2 (*)	T3 (**)
Mar. 92	sub-base	130 235	290 480	260 370	---	---	---
May 92	base	---	225 355	---	---	245 375	---
Jun. 92	basecourse	---	215 260	265 335	---	230 350	275 400
Jul. 92	base	250 350	---	330 380	350 550	---	475 550
Nov. 92	wearing course	235 280	230 255	260 330	290 350	310 350	390 475
Mar. 93		155 215	170 210	170 200	260 460	350 500	350 500

(*) - Graywacke; (**) - Limestone

From the analysis of results presented on Table 5, it can be concluded that deformability moduli of base layers is greater than deformability moduli of sub-base layer. It also seems that deformability modulus of *Limestone* and *Graywacke* are similar.

Regarding stress levels applied to granular layers, deformability moduli decrease with the application of bituminous layers.

3.3.4 Repeated load triaxial tests were performed over the two granular materials applied on the granular base layers of the studied pavement.

A 600 mm height and 300 mm diameter cylindrical specimen is tested, by using a large triaxial apparatus for the study of granular materials under repeated loading, originally described by Nunes M. and Gomes Correia A. (1990). By means of instrumentation attached directly to the specimen during repeated loading, a more accurate measurement of axial and radial deformations has been achieved.

The deviator stress is applied by means of a hydraulic jack attached to the loading frame which applies a load to the top of the plate. This system can apply a 25.7 kN maximum force, which imposes a 330 kPa maximum deviator stress on a 300 mm diameter specimen.

Specimen loading was of a trapezoidal type, with the regulation of load and unload time (type on/off) (Figure 2).

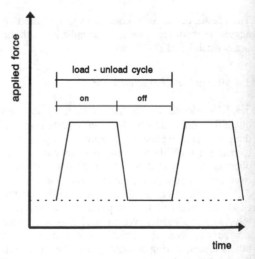

Figure 2. Cyclic triaxial tests - load application form

Due to the large specimen diameter and the low confining pressures there is no triaxial cell around the specimen as is usual with triaxial apparatus. The confining stresses is simulated by applying a sub-atmospheric pressure to the inside of the triaxial specimen.

The axial deformations of the specimen are measured by the relative deformations in four reference points. Three of them are attached to metallic rings. The two rings are placed in the lateral face of the specimen, and the LVDTs (linear variable differential transducers) are used at 120° to one another. The fourth measurement is made on top of the specimen.

Measurement of radial deformations is made by a steel cable threaded through ten sets of wheels and attached to a LVDT holder at each end. The LVDT measures the increase in the circumference of the specimen and thus the radial deformation.

This system was implemented in the middle of the specimen, quantifying deformations in the average section.

3.3.5 The test were carried out on specimens with a 600 mm heigh, 300 mm diameter and 10000 g weight, making it possible to study the full sized aggregates used in road pavements. To achieve the desired density as constant as possible through all the height, the specimen was compacted on 12 layers. Compaction of each layer was made by means of a vibrating hammer, trough a 20 mm thick and 145 mm diameter steel base. Since the base area was smaller than the area of the specimen shear is induced and densities are greater than those obtained by using a full faced base plate vibrator.

Figure 3 shows a general view of the specimen and hydraulic triaxial loading apparatus.

3.3.6 The influence of stress and compaction state in the resilient behaviour of the two studied materials was evaluated trough triaxial tests under repeated loading, on specimens with two different compaction states, by applying increasing confining stresses and several deviator stresses paths.

The stresses applied were mente to simulate those verified in situ.

For each stress path, one hundred cycles are applied. The main purpose of conditioning the specimen is that it allows the stabilization of the large permanent strains occurring during the first one hundred of cycles. After conditioning, the unbound granular material becomes almost entirely elastic.

Table 6. Water content (ω), compaction (GC) and bulk-density of test specimens ($\rho_{d\,ref}$)

Materials	Test identification	$\rho_{d\,ref}$ (g/cm³)	ω (%)	GC (%)
Graywacke	GT1	2.24	6.0	96.9
	GT2		6.6	98.0
Limestone	CT1	2.10	5.2	96.7
	CT2		4.6	100.0

Figure 3. Axial and radial measurement system on a test specimen

The characteristics of specimens are presented on Table 6.

3.3.7 For each one of performed tests resilient moduli were determined (Table 7).

The analysis of results of resilient moduli shows that:

    1. Graywacke resilient modulus is lower than Limestone resilient moduli;

    2. resilient moduli seems to be affected by distortional stress level;

3. resilient moduli obtained in the second stress level does not differ from those obtained in the first cycle; therefore, the stress path does not exert a great influence in the resilient behaviour of tested material;

4. the resilient moduli are high, when compared with the values usually presented for granular materials to be applied as granular base layers (150-300 MPa);

5. specimens with a highest bulk density show better characteristics of deformability by a higher modulus.

Table 7. Resilient moduli M_r of granular materials from triaxial tests

Materials	Test	Cycle	σ_3 (MPa)	M_r (MPa)		
				level 1	level 2	level 3
Graywacke	GT1	1	20	253		
			25	282		
			50	328		
			60	345		
	GT2	1	15	266		
		1	20	280	295	
		2	25	310	300	375
				322	330	380
Limestone	CT1	1	25	461		
			50	582		510
			75	613		
	CT2	1	20	480		
		1	25	550	540	600
		1	50	638	680	590
		2		647	650	660

For the study of the influence of confining stress state in the resilient moduli, an attempt was made to represent the behaviour of granular material by means of mathematic expressions with the possibility of adjusting the stress-deformation ratios.

Figures 4 and 5 represente the adjustment made by two different constitutive laws.

The analysis of figures 4 and 5 show that the confining stress and compaction affect the resilient behaviour of granular materials.

It can also be concluded that the behaviour law that relates the resilient modulus with the confining stress is the one that is best adapted to the results, presenting better correlation coefficients.

Futhermore, it can be observed that specimen with

a greater compaction present higher deformability moduli.

3.6.8 To analyse the development of permanent strain with the number of cycles of loading, two specimens were tested.

The study comprises the application of 10000 cycles without prior conditioning. A 35 kPa confining stress and a 200 kPa deviatoric stress were applied.

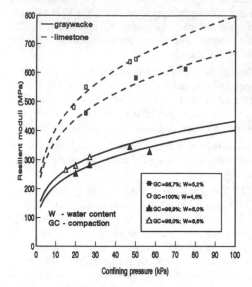

Figure 4. Resilient behaviour of studied materials. Adjustment of M_r-σ_3 model.

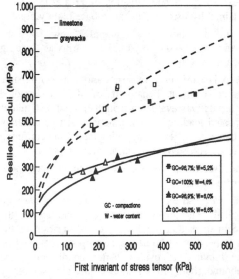

Figure 5. Resilient behaviour of studied materials. Adjustment of M_r-θ model.

Table 8 presents permanent strain specimen characteristics of triaxial tests.

Results of the evolution of the axial permanent strain are represented in figures 4 and 5, versus the logarithm of number of cycles.

Table 8. Water content (ω), compaction (GC) and bulk-density of test specimens ($\rho_{d\,ref}$)

Materials	$\rho_{d\,ref}$ (g/cm³)	ω (%)	GC (%)
Graywacke	2.24	8.5	97.3
Limestone	2.10	6.2	100.0

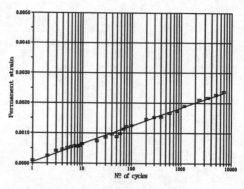

Figure 7. Plastic behaviour of *Limestone*

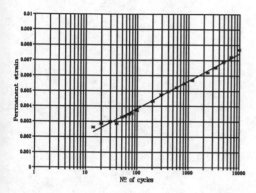

Figure 6. Plastic behaviour of *Graywacke*

The laws obtained allow a good adjustment, for the two studied materials:

Graywacke: $\epsilon_p(\%)=2{,}69\times10^{-4}+1{,}7\times10^{-3}\log N$

Limestone : $\epsilon_p(\%)=4{,}24\times10^{-5}+5{,}9\times10^{-4}\log N$

4. CONCLUSIONS

The results show that the characteristics of deformability of granular studied materials can be influenced by several factors, such as the geological nature and the physical characteristics of aggregates.

It was also observed that the triaxial test of repeat loads presents a high potential as regards the laboratory study of the characteristics of the deformability of granular materials.

ACKNOWLEDGEMENT

The work described in this paper forms part of the research carried out within the framework of a collaboration agreement between the Road Administration (JAE) and LNEC.

REFERENCES

Ahlborn, G. 1972. ELSYM5, Computer program for determining stresses and deformations in five layer elastic system. Phd Dissertation. University of California. Berkeley.

Antunes, M. L. 1993. Avaliação da capacidade de carga de pavimentos utilizando ensaios dinâmicos (Pavement bearing capacity evaluating using dynamic non destructive tests). Doctor Thesis, Universidade Técnica de Lisboa, Lisboa.

Freire, A. C. 1993. Estudos relativos a camadas de pavimentos constituídas por materiais granulares. (Research applied to pavement granular layers). Master Thesis, Universidade Nova de Lisboa, Lisboa.

HMSO, 1986. Specification for highway works, Part 3. Department of Transport, United Kingdom, London.

Jones, R. H.; Dawson, A. R. 1989. Unbound aggregates in roads. Proceedings. International Symposium on Unbound Aggregates in Roads (UNBAR 3). London: Butterworths.

JAE, 1978. Elaboração de projectos. Pavimentação. norma P7-78. Junta Autónoma de Estradas, Lisboa.

Nunes, M.; Gomes Correia, A. 1987. Modelação em laboratório do estado de compacidade *in situ* de balastos (Laboratory modelling of the *in situ* compaction state of ballasts). 4º Congresso Nacional de geotecnia. Vol. 1, Lisboa, pp. 387-396.

Ribeiro, A. *et al.* 1979. Introduction à la Géologie Générale du Portugal. Serviços Geológicos de Portugal, Lisboa.

The durability of Ashfield shale

La durabilité de l'argilite d'Ashfield

M.Ghafoori, D.W.Airey & J.P.Carter
School of Civil and Mining Engineering, University of Sydney, N.S.W., Australia

ABSTRACT: Slake durability tests have been performed on samples of Ashfield shale with varying degrees of weathering to investigate the deterioration of the rock from cycles of wetting and drying. The correlations between the slake durability index and the natural water content, Atterberg limits, uniaxial compressive strength, and the weathering grade of the shale have been investigated. The natural moisture content has been found to be a good predictor of the durability, strength, index properties and clay content of the shale.

RÉSUMÉ: Des essais sur des échantillons des argiles d'Ashfield ayant plusieurs degrés de altération ont été effectués pour déterminer la détérioration de la roche pendant les cycles de mouillage et séchage. Les corrélations entre l'index de durabilité en immersion et une série de propriétés comprenant la proportion d'eau naturelle, la limite d'Atterberg, la résistance à la compression simple, et le degré d'altération des argiles ont été étudiées. L'humidité naturelle est un bon prédicteur de la durabilité, de la résistance, des indices et de la fraction argileuse des argilites.

1 INTRODUCTION

Argillaceous rocks such as shale are characterised by a wide variation in their engineering properties, and because of their complexity and variability shales are one of the most problematic and least understood geological materials. Both engineers and geologists have been interested in shales; geologists primarily in the formation, mineralogy and physical properties, and engineers primarily in the behaviour of shale as a construction material or as a foundation material for structures. In practice, the mineralogy and engineering behaviour are closely related, and this is particularly true for the resistance to short term weathering by wetting and drying.

This paper summarises the findings of a study that has investigated the slaking behaviour and engineering properties of Ashfield shale of the Wianamatta sequence, from the geological region known as the Sydney basin in New South Wales, Australia. This rock outcrops in some parts of the Sydney metropolitan area and has posed some problems for designers of foundations and deep urban excavations. The shale also outcrops over a large area in the Sydney basin and it is often used as a construction material.

Slaking is generally a short-term physical disintegration process that may occur in some geological materials that are exposed at the surface. Material characteristics and local environmental conditions are the two important factors that control this disintegration. In slake durability tests the deterioration of a rock from cycles of wetting and drying is investigated. All rocks will be affected to some extent by cycles of wetting and drying, but the effects on shales can be very pronounced. The test enables the weathering resistance of shales and other rocks to be evaluated. It has been suggested that the slake durability test can also be used to predict the engineering performance of shales (Gamble, 1971). Tests have been performed to determine the slake durability of Ashfield shale. The results of these tests are presented in this paper and the relation of the durability to other properties is investigated.

2 TESTING APPARATUS AND MATERIAL

The slake durability tests have been performed on samples of Ashfield shale using a device similar to that developed by Franklin and Chandra (1972), following the suggested ISRM (1979a) method. The apparatus was constructed in the geotechnical workshop according to the dimensions given by Franklin and Chandra (1972).

Samples for this study came from three widely spaced locations in the Sydney metropolitan area. The majority of the samples were obtained from borehole corings at the Ryde-interchange site in Northern Sydney. At the Ryde site the shale varied from fresh unweathered to highly weathered. The degree of weathering was estimated based on visual assessment following the ISRM guidelines. Other fresh samples were obtained from blocks of shale obtained from excavations at Moorebank, in western Sydney, and from Surry Hills, near the centre of the city. The blocks of intact shale were cored and

the Wianamatta group shales have indicated that clay minerals comprise from 45 to 60 percent of the material and that kaolin and illite are the predominant clay minerals (Loughnan 1960, Dolanski, 1971 and Slansky, 1973). A further 14 samples were analysed as part of this study to determine the clay minerals present in the Ashfield shale. X-ray diffraction and scanning electron microscopy were used. The procedures for X-ray diffraction analysis and interpretation followed Carroll (1970), Grim (1953, 1962) and Brown (1961). Both the bulk material and material finer than 2 microns have been analysed. Ten samples were analysed after treatment with ethylene glycol, and were further analysed after heating to 550°C.

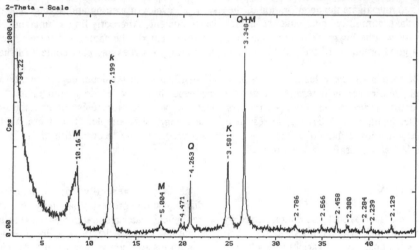

Figure 1 Typical X-ray diffraction trace

cut up in the laboratory. The samples were kept at their natural moisture content before testing by coating the samples with a moisture sealant called "Valvoline Tectyle" and then protected by a thin plastic film "Gladwrap" as soon as possible after coring.

Slake durability tests have been performed on samples varying from fresh intact to highly weathered shale. Tap water and distilled water at room temperature (about 22°C) were used as the slaking fluid. All samples were oven dried usually for 6 hours before the first wetting cycle and for 12 hours after the slaking test.

3 CLAY MINERAL ANALYSES

Clay minerals have a significant influence on the engineering behaviour of shales. Previous studies of

The X-ray diffraction traces of oriented clay samples with particles less than 2 microns are shown in Figure 1. The analyses of the Ashfield shale show no chlorite, no mixed-layer clays, and no smectite. The X-ray trace of the less than 2 micron fraction shows that kaolinite and micaceous clay minerals are the dominant clay minerals in Ashfield shale, and that on average kaolinite is more abundant. Similar results were obtained from the bulk samples. The micaceous clay mineral in the Ashfield shale is illite. The other main mineral in Ashfield shale is quartz. Previous studies have indicated that quartz comprises from 25 to 45 percent of the Ashfield shale (eg. Loughnan, 1960), and the current results, indicating kaolin more abundant than illite and illite more abundant than quartz, are broadly consistent with those presented previously.

Previous studies have indicated that two minor but significant minerals within the Ashfield shale are siderite and feldspar. Electron microscopy revealed the siderite to be finely dispersed in the Ashfield shale, and to comprise about 10 per cent. Previous studies have indicated siderite contents of 10 to 12 per cent (Ferguson and Hosking, 1955). No significant feldspar was detected in the samples used in this study.

4 TEST RESULTS

Tests were performed on four sub-samples of four fresh Ashfield shale specimens from the Ryde site to investigate the effect of slaking time on the durability results. The relationship between slaking time and the slake durability index is shown in Figure 2. The rate of slake loss was found to be nearly linear for these fresh samples, but different samples showed variable resistance to slaking. The longer cycle times show better distinction among the shale samples.

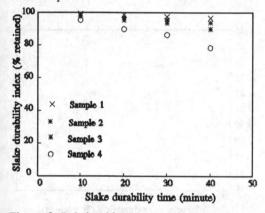

Figure 2 Relationship between slaking time and durability

The effect of increasing numbers of cycles of wetting and drying is shown in Figure 3 for eight samples with varying degrees of weathering from the Ryde site. In these tests a constant slaking time of 10 minutes was used in each cycle. The percentage of the samples retained decreases steadily with number of cycles, and decreases with increasing degree of weathering. Different procedures have been recommended for classifying the durability of shales from these tests. For the Ashfield shale two ten minute cycles, as suggested by Gamble (1971), give reasonable distinction between the samples, but the single cycle results as suggested by Franklin and Chandra (1972) are less satisfactory. However, if the time for the wetting phase of the cycle is increased good distinction can be obtained from a single cycle. Table 1 shows the effects of slaking on samples of fresh shale from the different sites. It can be seen that the durability of fresh shale is not constant but varies from location to location.

Table 1 Comparison of slake durability indeces of fresh Ashfield shale

No. of cycles	Ryde-interchange		Moorebank		Surry Hills	
	1	2	1	2	1	2
1	98.10	96.52	99.05	99.12	99.27	99.23
2	94.22	93.86	98.34	98.41	98.68	98.62
3	90.96	90.93	97.82	97.84	98.21	98.21
4	88.48	88.33	97.13	97.17	97.66	97.64

Following the classification scheme established by Gamble (1971) and Franklin and Chandra (1972), these results indicate that the durability of Ashfield shale varies from high for fresh intact material to low for highly weathered material. The high durability of fresh Ashfield shale is consistent with the mineralogical assessment and in particular in the absence of smectitic clays. Mineralogical investigations revealed that the Ashfield shale from the Ryde site contained a lower quartz and higher clay content than from the other sites which may explain the difference in the durabilities of the fresh shales.

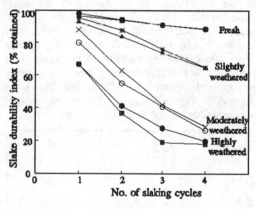

Figure 3 Effect of numbers of cycles

5 CORRELATIONS WITH ENGINEERING INDICES AND PROPERTIES

5.1 Atterberg limits

To determine the liquid and plastic limits of Ashfield shale, the material passing from the drum in the slaking test was used. For the more durable samples the retained portion of material in the slaking tests was crushed by mortar and pestle. The BS 1377:1975 procedure for the determination of the liquid and plastic limits was used.

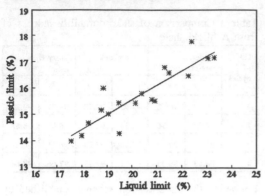

Figure 4 Relationship between liquid limit and plastic limit

Figure 4 shows the relationship between the liquid and plastic limits. There is a clear trend of increasing plastic limit, and plasticity index, with increasing liquid limit. On the basis of the Atterberg limits the Ashfield shale may be classified as silt and clay with low plasticity (CL-ML). Figure 5 shows a plot of liquid limit against natural water content. It can be seen that there is a rough trend for the liquid limit to increase with increasing natural water content. All samples of the Ashfield shale have been found to be essentially fully saturated in-situ so the natural water content gives a direct measure of the porosity of the shale. It is generally found that weathering increases the clay content and porosity of shales and the trend indicated in Figure 5 suggests that weathering may be responsible for the observed variation.

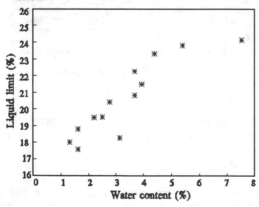

Figure 5 Liquid limit versus natural water content

A quantitative slaking test in combination with the Atterberg limits has been suggested (Gamble, 1971) as a useful way of classifying argillaceous rocks for engineering purposes. Figure 6 shows the plot of liquid limit versus two cycle slake durability index for the samples of Ashfield shale tested. The figure shows a clear correlation between the slake durability index and the liquid limit for the Ashfield shale, with increasing liquid limit, and plasticity index, associated with a reduction in slake durability.

Following the classification scheme proposed by Gamble (1971) the Ashfield shale would be classified as follows: the fresh Ashfield shale from the Ryde-interchange site would be classified as a medium high durability - low plasticity shale; the slightly to moderately weathered Ashfield shale from this site would be classified as a medium durability - low plasticity shale; the Ashfield shale from the other two sites (Surry Hills and Moorebank) would be classified as a very high durability - low plasticity shale.

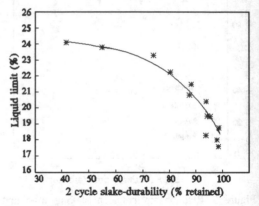

Figure 6 Liquid limit versus two cycle slake durability index

5.2 *Natural water content*

It has been well established that the moisture content in argillaceous rock masses such as mudstone and shale can have significant effects on their engineering behaviours, and as noted above the plasticity of the Ashfield shale appears to be related to the natural moisture content. A review of the engineering properties of Ashfield shale (Ghafoori et al, 1992, 1993) has found that a good correlation exists between uniaxial compressive strength and natural water content for a large number of samples from different locations with different degrees of weathering. The data are reproduced in Figure 7 and

3318

show that uniaxial compressive strength decreases with an increase in the moisture content, as might be expected. It was also found that engineering geologists' estimates of the degree of weathering did not give good correlations with the engineering properties. This suggests that natural moisture content or porosity may provide better predictions of the behaviour than the more subjective geological definitions of weathering.

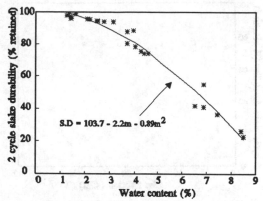

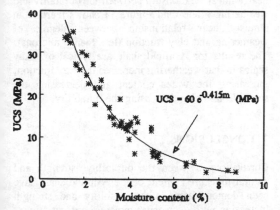

Figure 7 Relationship between uniaxial compressive strength and moisture content

The fresh shale samples from Surry Hills and Moorebank had lower moisture contents and higher strengths than the samples from the Ryde site, and as noted above the former samples were more durable. Figure 8 shows the relation between natural water content and 2 cycle slake durability index. This figure shows a clear trend of increasing water content being associated with reducing durability of the Ashfield shale. The durability decreases rapidly for natural water contents greater than 3%. These results demonstrate that a simple classification based on the natural moisture content, an easily measured quantity, could be adopted for Ashfield shale.

5.3 Clay content

Four samples of highly weathered to fresh Ashfield shale from the Ryde site were used in this study. The samples were ground and passed through a 425 μm mesh sieve. After preparing the samples a sedimentation test was used to determine the particle size distribution. The results of these tests indicated clay fractions, that is material finer than 2 microns, of between 10% and 30% whereas the mineralogical studies indicated clay fractions of greater than 50%

Figure 8 Relationship between two cycle slake durability and moisture content

in all samples. This presumably reflects the presence of significant quantities of large kaolin particles.

Figures 9 and 10 present the variations of the liquid limit and 2 cycle slake durability index versus clay fraction less than 2 microns, respectively, for highly weathered to fresh Ashfield shale. Figure 9 shows that increasing clay fraction (less than 2 microns) is associated with an increase in liquid limit which as noted above is associated with an increase in plasticity index. Figure 10 shows that the increasing clay fraction, which is normally related to increasing porosity, is associated with decreasing durability as expected.

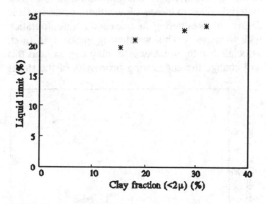

Figure 9 Liquid limit versus clay fraction

Activity is the ratio of the plasticity index to the percent of clay (less than 2 microns) in a sample (Skempton and Northey, 1953). The activity of the four samples tested varied between 0.2 to 0.27. Based on Skempton's classification all four samples

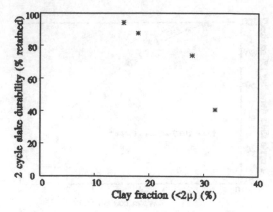

Figure 10 Slake durability versus clay fraction

would be classified as inactive. This low activity was expected because of the predominance of kaolinite and the absence of smectite in the Ashfield shale and the swelling potential can be expected to be very low.

6 WEATHERING

As argillaceous rocks are exposed close to the surface, weathering processes begin to take effect. Both physical and chemical processes, near the surface, become increasingly important and control the weathering of argillaceous rocks. The physical processes are generally a prerequisite for subsequent chemical weathering. Discolouration of the rock material gives evidence of weathering, and as the degree of weathering increases the discolouration will increase. This weathering process is often associated with an increase in clay content and this will change the engineering properties of the shale,

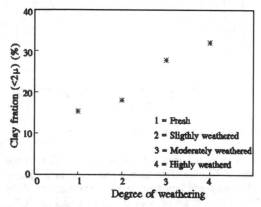

Figure 11 Relationship between weathering grade and clay fraction

in particular it will increase the porosity and decrease the density and strength.

Some tests were run on Ashfield shale with different degrees of weathering to check the effect of weathering on the durability results. The relationship between weathering grade and slake durability index was shown before in Figure 3 and that between weathering grade and clay fraction is shown in Figure 11. Figure 3 shows there is a nearly linear relationship between the durability and weathering grade, and Figure 11 shows there is an almost linear relationship between degree of weathering and clay fraction (less than 2 microns). The results for Ashfield shale are typical of many shales in that weathering increases the clay fraction, increasing the water content and decreasing the strength, and also decreasing the durability.

7 CONCLUSIONS

Correlations between the durability, strength and natural moisture content of the Ashfield shale have been demonstrated, with durability and strength increasing with decreasing moisture content.

The effect of weathering is to increase the clay and moisture contents and thereby reduce the strength and durability. However, different fresh unweathered shales had different moisture contents and different strengths and durabilities.

The variability of the natural moisture content can be correlated with mineralogical factors. Increasing moisture content was associated with increasing clay content (less than 2 microns) and in turn this was associated with an increase in plasticity index and plastic and liquid limits. The higher strengths and durabilities of some fresh shales were associated with higher quartz contents.

ACKNOWLEDGMENTS

The authors are grateful to the New South Wales Roads and Traffic Authority for preparing and providing the core samples.

REFERENCES

B.S.I. 1377 (1975), Methods of test for soils for Civil Engineering purposes.
Brown, G., ed (1961), The X-ray identification and crystal structures of clay minerals, Mineralogical society (clay minerals group), London, 544 p.

Carroll, D.(1970), Clay minerals, a guide to their X-ray identification, Special Paper 126, Geological Society of America, 80 p.

Dolanski, J. (1971), X-ray identification of clay mounts from Wianamatta shale core, Rep. Geol. Surv. New South Wales, GS 1971/757 (unpubl.).

Ferguson, J.A. and Hosking, J.S. (1955), Industrial clays of the Sydney Region, New South Wales: Geology, mineralogy, and appraisal for ceramic industries. Aust. J. Appl. Sci., 6, 380-405.

Franklin, J.A. and Chandra, A., (1972), The slake durability test, International Journal of Rock Mechanics and Mining Science, 9, pp. 325-341.

Gamble, J.C., (1971), Durability-plasticity classification of shales and other argillaceous rocks, Ph.D. thesis, University of Illinois at Urbana-champaign.

Ghafoori, M., Mastropasqua, M., Carter, J.P., and Airey, D.W. (1992), Engineering properties of Ashfield shale, Research Report No. R663, School of Civil and Mining Engineering, University of Sydney.

Ghafoori, M., Airey, D.W. and Carter, J.P. (1993), Correlation of moisture content with the uniaxial compressive strength of Ashfield shale, Australian Geomechanics, pp. 112-114.

Grim, R.E. (1968), Clay mineralogy, McGraw-Hill, New York.

Loughnan, F.C. (1960), The origin, mineralogy and some physical properties of the commercial clays of New South Wales, University of New South Wales, Geological Series, 2, 348 p.

Slansky, E. (1973), Clay mineral identification in Ashfield shale from the Maroota area, Rep. Geological Survey of New South Wales, GS 1973 396 (unpubl.).

Skempton, A.W. and Northey, R.D. (1953), The sensitivity of clays, Geotechnique, Vol. 3, No. 1.

Characteristics of granitic aggregates and its influence on the abrasive wear of metallic parts of drilling and crushing equipment

Caractéristiques des granulats granitiques et leur influence sur les parties métalliques abrasives de l'équipement d'excavation et d'écrasage

L. Soares & F. Fujimura
Polytechnic School, University of São Paulo, Brazil

ABSTRACT: Laboratories tests and site investigations were performed to characterize crushed granitic rocks of two quarries localized near São Paulo city, Brazil. Both quarries are branches of the same Company which have been in activity for more than twenty years, having similar crushing plant and equipment. The performance of drilling and crushing equipment of each quarry is quite different although the similarity of rock mineral components. This paper presents the test results and analyses the geological and geotechnical parameters that control and determine the different behavior between them. In spite of conclusion based only on rocks of two quarries the results are very encouraging and justify the continuity of research for other lithologic units, due its importance for equipment's productivity and production costs of aggregates.

RÉSUMÉ: Des essais de laboratoire et des études de champ ont été réalisés pour la caractérisation à la concassage sur deux carrières de granite places à côté de São Paulo. Tous les deux carrières appartiennent à la même Société en activité depuis 20 ans, avec des installations de concassage similaires. L'accomplissement des équipements de forage et concassage pour chaque carrière il est peu différent entre eux. Cette publication présente les résultats des essais et les analyses géologiques et des paramètres géotechniques de contrôle pour déterminer les différents comportements entre les deux carrières. Les conclusions bases sur les roches des deux carrières sont très encourageants et justifient la continuité de cette recherche pour l'autres unités lithologiques, en fonction de leur influence sur la productivité et sur le coût de production des agrégats.

INTRODUCTION

The wear is normally stated and accepted as a natural degradation process and material consumption by attrition or by other material injury process subjected in the course of use.

This vision of phenomenon has been responsible for principal cause of lack of better scientific knowledge on mechanism of the losses by wear. It's represents an enormous economic importance in the industrial activities.

ZUM - GAHR astonished the industrial world in 1987 when he estimated the annually worn losses in 1 to 5% of GNP (Gross National Products).

The preview study of RABINOWICZ (1984) had already indicated an amount of 58 billion of dollars only in the USA car industries, and it would be possible to reach an expressive value of 180 billion of dollars, or 6% of GNP, as losses produced by wear in the industrial activities.

Despite of this economic importance the efforts developed in Brazil concerning this subject are very sporadic and done by small groups of private company or individual researcher of universities and research institutes.

In the mining industries this matter is less known and substantially aggravated by steady presence of non homogeneous material like as rock and ore with complex behavior and strength.

From several forms of wear the abrasive wear is the most destructive and causes important economic losses as stated by ROTONDARO

(1988). It is intensively verified in size reduction and fragmentation process of rocks and minerals. Metal consumption by wear is usually the second largest single item of expenses in conventional grinding installations and in some cases it may approach or even exceed the power cost (BOND, 1961). In rock drilling wear research by CLARK (1982), the bit wear is a major factor in determining the cost of drilling and may determine the drilling method for a given rock.

The absence almost total of confidence data leads to an insecurity situation on estimating cost of plant operation and in dimensioning reposition pieces.

ROCK FRAGMENTATION AND ENERGY

Size reduction is one of the most important step in the processing minerals and includes blocks of several meters till mineral particles of few microns. In all cases the breakage of rock is often stated is an energy inefficient operation, since the energy utilization during the fragmentation is only about 1% with respect to the new surface created, as experienced by HUKKI (1975).

The energy form applied by drilling or crushing tools determines the type of fragmentation and the efficiency of equipment utilization.

So, the prediction of metallic abrasive bit wear or jaw crusher metallic lining consumption should be based on intact rock characteristic, because the intact rock properties are usually the most useful indexes for performances predictions.

The influences of unconfined compressive strength on penetration rate of percussive drilling tools was well demonstrated by CLARK (op. cit.). But, despite of general decrease in the penetration rate for higher compressive strength, correlation obtained is not so consistent enough for the prediction of drillabillity of all rock types. This fact means the standard physical property of rock (unconfined compressive strength, tensile strength, shear strength, Young' s Modulus, etc.) alone , would possibly not representative of the factors that influence the breakage mechanism of crusher or drilling equipment.

Rock characteristics, such as grain size, hardness, cement bonding, mineral composition, fabric, are additional factors that influence the equipment efficiency and varies from one rock type to

another and for different samples of the same rock. This fact is specially true in respect of abrasive bit wear and metallic consumption of jaw crusher plates.

PERFORMED RESEARCH AND STUDIES

In São Paulo State nearby São Paulo city , as illustrated in Fig.1, two stone-quarries (Embu, Itapeti) 70 km distance from one to other, produce each one about 80,000 m³ of crushed stones for civil construction monthly. These quarries are branches of same Company and so the crushing installations and drilling equipment are very similar and operating under same conditions.

The rock masses in the both cases is classified as granitic rock but the performances of the equipment is quite different specially in regarding of bit wear and abrasive worn of jaw crusher plates.

This fact motivated a comparative study between two quarries involving field operating conditions, geological features, strength properties of intact and crushed rocks and microscopically examination of mineralogical features to analyze factors affecting unequal behaviors.

The existing data from third quarry, named Riuma, was included for a better data comparison with those surveyed from the both quarries

ABRASIVE METAL WEAR

Metal wear in crushing system is usually expressed in g/ton crushed. The pounds of metal worn always varies with the abrasiveness of the mineral and abrasion resistance of the metal.

Based on laboratory test and survey of 83 quarries SIRIANI (1972) concluded that the power consumption (kWh/ton) of crusher and wear of fixed jaw plates (g/ton) correlates well at high confidence level as showed in Fig. 2, for several types of rock.

However it is known the wear in the granitic rock can change from 40g/ton to 70g/ton at same energy level. This fact confirm the needs of a better understanding on rock fragmentation process and variability of mineralogical features in controlling worn mechanism.

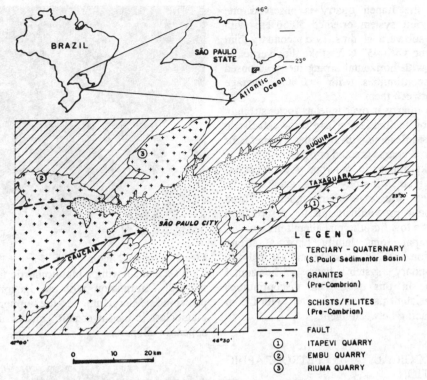

Fig. 1 - Site localization of stone quarries

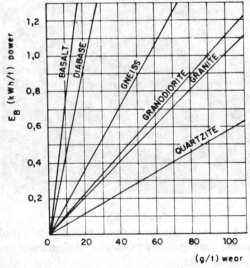

Fig. 2 Wear of fixed jaw plates x power consumption (Siriani, 1972)

Based on same laboratory and field data surveyed by SIRIANI (1972), FUJIMURA (1984) analyzed the influence of mineralogical components of rock and showed a general relationship between quartz content and metal worn. The sepentinite and chlorite minerals indicate the rock weathering and weakening process. The presence of these minerals can reduce rock abrasiveness strongly.

REGIONAL AND LOCAL GEOLOGIC FEATURES

The investigated quarries are located in Pre-Cambrian areas over granitic rocks. The Itapeti quarry is over a big massive granitic rock elongated beyond 30km and oriented E-W direction, concordant with a shear zones of regional structures. Although its homogenous lithology, the massive body shows moderate textural diversity by tectonic deformation at edges. In the Itapeti quarry is observed one principal joint system oriented 50/60°NE with vertical or sub vertical dips. Two secondary joint systems, one with 45° NW strike dips vertically and other with horizontal arrangements imposed heavy discontinuities with 0.80m average spacing between them.

The Embu quarry is over igneous rock strongly deformed by tectonic shear and show cataclastic material meddlesome with irregular fragments or non fractured coarse grained crystals.

These rocks appear in outcrops with structural direction given by porphyriblasts, and could be denominated muscovite-biotite gneiss. These rocks present predominant plastic deformation and exhibit a low hdrothermal alteration sign.

The principal joint family have a strong concentration in a 25°NW strike and vertical dip. The secondary system have a horizontal disposition. In this quarry the discontinuities have a consistent parallel strike and with 1,15 m average spacing between them.

MINERALOGICAL AND PETROGRAPHIC DESCRIPTION

The Itapeti quarry has a homogeneous rock with a light gray color. In a macroscopical analyze, is possible to see a massive structure and its igneous origin. By the microscopical analyses show a small hydrothermal alteration, with a porphyroid texture, fine granulation and slight indication of deformation of the constituent mineral (Fig. 3). The optical evaluation shows feldspars predominance, being 35-40% oligoclase and 20-25% microcline. The quartz content (25%) in this thin section shows a metamorphic action on mineralogical components. The biotite complete about 10% and the other materials (opaque, titanite, apatite, tourmaline) are 5%.

The Embu quarry is composed by dark gray and rose gray rocks, that shows gneissic texture with strong mineralogical and textural similarity. Those rocks shows hydrothermal alteration with a fine hipidiomorfic granular texture.

By the microscopical evaluation is easily detected the original igneous rock texture, heavily changed by tectonic deformation.(Fig. 4)

Fig. 3   Crossed Nicols - 38X
Thin section showing a hipidiomorfic texture and a small deformation with an igneous texture preserved. Microcline in the upper portion; plagioclase in the lower part; quartz and biotite appear in all parts.

Fig. 4   Crossed Nicols - 38X
Thin section showing the original texture of granite changed by tectonic deformation to blastomilonite
In this rock the mineralogical composition is quite similar to the Itapeti composition, showing 35% of microcline, 25-35% of plagioclase, 20% of quartz and biotite between 5 and 10% of the analyzed sample. The grain size is very different

between Itapeti and Embu quarries, where the first one presents a coarse grained mineral.

DRILLING AND CRUSHING PERFORMANCES

There is a creditable amount of scientific studies in literature concerning drilling and crushing practices, costs, uses and maintenance. But almost of them analyses like as an equipment mechanical operating problems point of view.

A detailed knowledge of mechanics of drilling allied to information on the basic mechanism of rock fragmentation, energy transfer process and other basic factors related to confidence mineralogical features of rock would be one principal means by which drilling and crushing technologies can be improved.

Laboratory tests, results of experimentation and survey of actual data are essential activities for increasing practical basis to theoretical developments.

The Figures 5 and 6 show the histogram of total drilled distances of Itapeti and Embu quarries using crawler drill ROC-601, with button bit of 92 mm nominal diameter and 90 psi operating air pressure.

An approximate normal distribution is observed for both quarries with average values of 1200 m for Itapeti and 450 m for Embu.

The total drilled lengths for each bits was determined by summing all partial distances drilled in each perfuration.

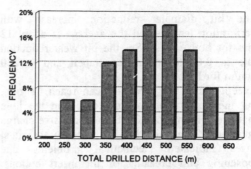

Fig. 6 Histogram of drilled distance - Embu quarry

The great difference is by consequence of high abrasiveness of Embu rock as presented in Figures 7 and 8. The bits wearing data were obtained from each quarry.

The bit life time is represented by bit diameter reduction (mm) measured by difference of diameter before ad after each perforation steps. The sum of all partial reductions results in total bit wear.

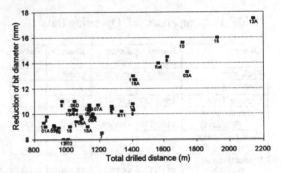

Fig. 7 Reduction of bit diameter by wear - Itapeti quarry

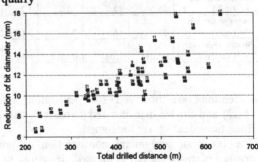

Fig.8 Reduction of bit diameter by wear - Embu quarry

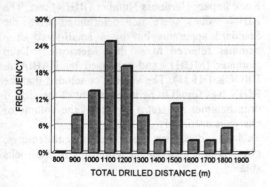

Fig. 5 Histogram of drilled distance - Itapeti quarry

The bit diameter reduction increase with perforation lengths and the average is about 12 mm for both cases. So, the bit wear reduction rate is 0,0088 mm/m for Itapeti and 0,0270 mm/m for Embu.

Exceptionally the reduction can reach 18 mm, but the perforation rate decrease and the hole diameter diminution affects the explosive charge operation. The short bit life time in the most cases indicated mechanical problems due loosening and breaking of bit insert buttons. Operating data for crushing and drilling (Table 1) shows the performance differences between two quarries. The existing data of Riuma quarry was also included to better operating comparisons.

Although the similar crushing production rate of primary jaw crushers - FAÇO 150x120, around 380 m³/h for both, the jaw plates duration (hours or days) for the Itapeti is almost triple (2.86) than the Embu. Approximately the same ratio is verified for total production of each crusher and for average bit life.

Table 1. Comparison of Operating Data

Quarry	Itapeti (A)	Embu (B)	Riuma (C)	Ratio (A/B)
Rock type	granite	granite	granite	
Crushing				
Crusher Type	150X120	150X120	120x90	
Plates duration (hrs/days)	1400/140	490/49	500/50	2.86
Total production (m³ x 1000)	520	190	145	2.74
Production rate (m³ /h)	371.43	387.76	290.00	≈ 1
Drilling				
Penetration rate (m/h)	16.5	12.7	16.0	1.29
Average bit life (m)	1200	450	720	2.67
Bit dia. reduction (mm/m)	0.0088	0,0270	0,0165	0.33

Exceptions are observed for penetration rate (1.29) and for bit diameter reduction rate that is three times less severe for Itapeti.

The bit wear and metal consumption of jaw crusher plates were not expressed by g/ton, but in millimeters of bit diameter reduction and in hours of utilization, respectively. This procedure is usually adopted by quarries due its facility in controlling the time life by abrasive wear.

The severe abrasiveness character of Embu rock was well demonstrated in laboratory crushing tests (Table 2) performed in small jaw crusher specially assembled for this purpose. The abrasive wear was calculated by weighting jaw plates before and after each operations of 100 kg crushing tests, using 15 mm to 20 mm size fragments. From Table 2 it is evident that the value of ratio (0.30) is very close to ratio of bit diameter reduction (0.33) and to inverse value of jaw crusher plates ratio (0.35)

Table 2. Comparison of laboratory tests
Jaw crusher plates wear

Quarry	Itapeti (A)	Embu (B)	Riuma (B)	Ratio (A/B)
Rock type	granite	granite	granite	
Plates wear (g/ ton)	15.7	52.5	52.2	0.30

MINERALOGICAL, PHYSICAL AND MECHANICAL PROPERTIES

In the drilling and crushing processes the dynamic fracture mechanism plays a important role in the rock fragmentation. In order to examine this process the impact hardness test was carried out, using the modified version of original Protodyakonov test described by WOOTTON (1974) and BROOK (1977) as the Rock Impact Hardness Number (RIHN) test. The RIHN values were not determined from the Standard apparatus but by a small-sized drop hammer referred to as the "Metricated" Drop Hammer (MDH) and proposed by RABIA & BROOK (1981). The authors concluded that the RIHN was found to be independent of the size of drop hammer apparatus, provided the volume of charge is accurately chosen.

Table 3 describes the mineralogical features, physical and mechanical properties of the rocks studied.

Table 3. Geotechnical characteristics of rocks

Quarry	Itapeti		Embu		Riuma	
Rock	granite		granite		granite	
Mineral content	Plagioclase Microcline Quartz Biotite Others	: 35-40 % : 20-25 % : 25 % : 10 % : 5%	Plagioclase Microcline Quartz Biotite Others	: 25 -35 % : 35 - 40 % : 20 % : 5 - 10 % : 5 %	Feldspar Quartz Biotite Others	: 20 - 50 % : 20 - 50 % : 25 - 35 % : 5 %
Crushed fragments form	Cubic (70 % of fragments)		Cubic (63 % of fragments)		Cubic (64 % of fragments)	
P - wave velocity (m/s)	5,200		5,600		5,100	
Unconfined compressive strength (Mpa)	140		220		150	
Tangent Young' s Modulus (Mpa)	58,000		68,000		56,000	
Specific gravity (g/cm³)	2.68		2.71		2.65	
RIHN	24.0		45.0		24.5	

The diagram shown in Fig. 9 was obtained using crushed fragments of size 15 - 20 mm, and weights equivalent to 12 cm³ of volume charge. The linear straight lines through origin of axes indicates that confidence values of RIHN was obtained.

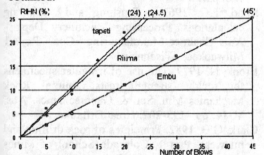

Fig. 9 RIHN graphs of rocks in the "Metricated" Drop Hammer

The RIHN number for Itapeti and Riuma rocks is very similar, not for Embu that is almost double value.

Fig. 10 shows the relationship between the RIHN and the unconfined compressive strength. It is observed that RIHN correlates very well and so the unconfined compressive strength can be estimated by this rock hardness index. Same conclusion was obtained by Matsui & Shimada (1993), using standard impact test equipment.

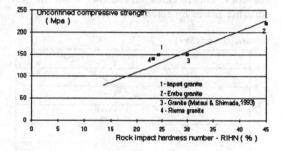

Fig.10 Relation between the RIHN and unconfined compressive strength

Fig. 11 shows the relationship between the bit diameter reduction surveyed in the three quarries and the RIHN.

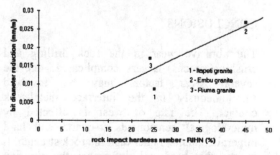

Fig. 11 Relationship between RIHN and bit diameter reduction rate

The scatter in the plotted values means RIHN values do not exactly represent the real rock abrasiveness , although its general relation trend between them (same relationship is verified with unconfined compressive strength). It is one of reasons why the mineralogical features besides the rock strength properties would be considered . The description of the rock's texture and fabric affords a basis for understanding its mechanical properties.

The quartz content and other hard minerals in the sound competent rock represent the most destructive combination for the abrasive wearing process.

The relationship between bit diameter reduction rate and jaw crusher plates wear (laboratory tests) is shown in Fig. 12. Despite of small amount of data a fairly linear relationship is viewed. This means that the bit diameter reduction rate can be used as an initial prediction of rock abrasiveness and metal worn of jaw crusher plates by abrasive wear

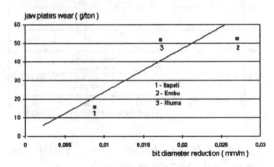

Fig. 12 Relationship between bit diameter reduction rate and metal consumption of jaw crusher plates.

CONCLUSIONS

The abrasive wear in the rock drilling and crushing process is very complicated due the several wear factors may be involved simultaneously in the interface metal-rock contact. The rate of wear is affected by mineralogical components (quartz and hard mineral contents) rather than the rock strength. It is true that hard rock influences the abrasion phenomenon but the high strength affect much more the power consumption than the rock

abrasiveness. In the same type of rock the abrasiveness increases with rock strength in which the grain size plays an important role as shown in this study.

The obtained results suggest a great number of factors affecting the rock abrasiveness and does not exist a simple method to determine it yet. Others lithologic units with different mineral composition must be investigate, including weakening and weathering aspects, combination of RHIN index with mineralogical properties, etc. in order to achieve more realistic conclusion.

ACKNOWLEDGMENTS

The authors would like to thank to Embu S.A Engenharia e Comercio and to Comercial Pavimentadora Riuma for their permissions to realize this investigation.

REFERENCES

Bistrichi, C.A. et al. 1980. Mapa geológico do Estado de São Paulo. Publicação IPT 1184, 126 p. il.

Bond, F.C. 1961. Crushing and grinding calculations. Processing Machinery Dept. - Allis Chalmer Manufacturing Co. - Rev. 1 Milwaukee, Wisconsin.

Brook, N. 1977. The use of irregular specimens for rock strength tests. Journal Rock Mechanics Min. Sci. & Geomechanics. Abs. V. 14, pp: 193-202. Great Britain.

Clark, G.B. 1982. Principles of rock drilling and bit wear. Part 1, Colorado School of Mines quarterly, V. 77 - n.1. Colorado.

Fujimura, F. 1992. O lado nocivo do elemento quartzo no desgaste abrasivo de mandíbula de britadores. Boletim Técnico da Escola Politécnica da Universidade de São Paulo. Depto. de Engenharia de Minas. BT/PMI/023.

Fujimura, F. 1985. Influência de componentes mineralógicos da rocha no desgaste abrasivo de mandíbulas em britadores. I Congresso Brasileiro de Mineração - IBRAM. V.2, p:10-26. Brasilia.

Hukki, R.T. 1975. The principles of comminuition: analytical summary. Eng. Min.J. 176 p.106-110.

Matsui, K; Shimada, H. 1993. Rock impact hardness index for predicting cuttability of road

header. Dept. of Mining Engineering Faculty of Engineering - Kyushu University. Japan

Rabia, H.; Brook, N. 1981. The effects of apparatus size and surface area of charge on the impact strength of rock. Int. Journal Rock Mechanics Min. Sci. & Geomechanics. Abst.,V.18. p: 211-219. Great Britain.

Rabinowicz, E. 1984. The least wear - Wear, 100 -pp 533-541.

Rotondaro, R.G. 1988. Mecanismos de desgaste abrasivo. Tese para obtenção do grau de Doutor em Engenharia. EPUSP. São Paulo.

Siriani, F.A. 1972. Caracteristicas gerais de desgaste de mandíbula. Tese para obtenção do grau de Doutor em Engenharia. EPUSP. São Paulo.

Wooton, D. 1974. Aspects of energy requiriments for rock drilling. Ph.D. Thesis, Leeds

Zun-Gahr, K.M. 1987 Microstruture and wear of metals. Elsevier Science Publishers B.V. pp- 96 Amsterdam.

The use of AL-Malih valley crushed limestone as coarse aggregate in concrete production

L'usage des pierres à chaux dans la région Wadi-el-maleh
comme agrégats à gros grains pour la production du béton

Mohammed A.S. Elizzi
Department of Civil Engineering, University of Al-Mustansiriya & NCCL, Baghdad, Irak

Ali M. Abdul-Husain & Perikhan Fadhil
NCCL, Baghdad, Irak

ABSTRACT: The southern part of Iraq sufferes from shortages of suitable natural coarse aggregate for concrete production. Most of the available natural gravel are contaminated with coating sulphates. Therefore coarse aggregates, for the time being are brought from middle & northern parts of Iraq, to be used in the southern areas.
   This research was carried out to study the suitablity of crushed limestone from AL-Malih valley as coarse aggregate in concrete production.
Geological investigation, evelution of crushed rock aggregates properties, and evalution of concrete properties produced by usin crush limestone as coarse aggregate, have been carried out in this research on limestone of AL-Malih valley area.
   The results showed that, it is possible to use this type of crushed rock aggregates in producing good quality concrete. Therefore it is recommended to use the crushed limestone from the investigated area as coarse aggregated in producing concrete at the southern provience of Iraq.

RÉSUMÉ: La région Sud l'Iraq souffre d'un manque d'agregats dur (à gros grains naturels) pour la production du béton, par-ce-que le gravier disponible dans cette région est impur et comprend des composés sulfurés salins, par conséquent le gravier dur (à gros grains) est apporté maintenant des régions lointaines de La Partie Sud de L'Iraq, ce qui augmente le coût de la production du béton. Le but de cette étude est de se renseigner sur l'applicabilité de l'usage de la pierre à chaux de la région Wadi-el-maleh, située dans la zone administrée par le Najef, pour produire le béton..
Cette étude clarifie l'applicabilité de cette pierre comme graviers à gros grains pour produire le béton,, ce qui, par conséquent, reduira le coût du produit dans la partie Sud de l'Iraq.

1 INTRODUCTION

Natural coarse aggregate at present time, which are gravel, have been intensively used for concrete manufacture in the middle and southern parts of Iraq. Unfortunately most of the available natural gravel are contaminated with sulphates, which influence the quality of concrete. Many studies were carried out to give an alternative coarse aggregate to be used for concrete manufacture. (Elizzi 1990) and (AL-Baldawi 1986).
   This research was carried out to investigate the use of a crushed limestone from AL-Malih valley as coarse aggregate in concrete production. AL-Malih valley quarry, is situated in southern of Iraq, about 15 km to the south of Najaf city, Fig 1. Dolomitic

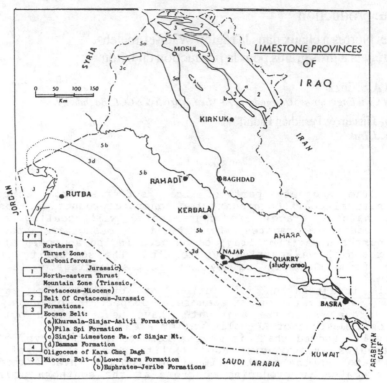

Fig 1. Limestone provincess of Iraq and location of AL-Malih valley area (AL-Baldawi & AL-Naib 1986, modified)

limestone from AL-Malih valley area was used in this study due to the following:

1. Most of dolomitic limestones are exposed at surface (thickness of covering soil is 0-5 m)

2. The ease of transportation and proximity of AL-Malih area to the main roads, will reduce the cost of raw waterials.

3. The distribution of a large reserve in the area with the possibility of future extension in known directions.

4. The reserve of limestone rocks at AL-Malih valley area was estimated equrasls to 75 million cubic meter (AL-Komi 1983).

Chemical & physical analysis on crushed limestone rocks of AL-Malih area were carried out to show their suitability as coarse aggregate for concrete production. ALkali reactivity studies were also concluded on these aggregates.

Finally compressive strength tests were carried out on different concrete mixes. All tests showed acceptable results for the use of limestone rocks as coarse aggregate in concrete production.

2. EXPERIMENTAL WORK

2.1 Chemical analysis of limestone rocks

Chemical analysis was carried out on several rock specimens from different depths. Table(1) gives the results of this analysis.

2.2 Physical properties of limestone rocks

Physical properties of limestone were determined by carrying out several tests, includes: Abrasion

3334

Table 1. Chemical composition of Limestone from Al-Malih Valley area

Oxide	CaO	MgO	SiO_2	Al_2O_3	SO_3	Fe_2O_3	Ins.M	L.O.I	Total
Percent	37.5	13.64	0.62	0.76	0.15	0.12	2.18	44.37	99.34

tests, apparent specific gravity, water absorption etc. All the physical properties results are listed in table(2)

Table 2 Physical properties of limestone aggregate

Test	Value
Abrasion	24%
Water absorption	1.5%
Apparent specific gravity	2.68
Total specific gravity	2.55
Impact	11.5%

2.3 Potential alkali reactivity

Potential alkali reactivity takes place when the alkalies present in cement paste $(Na2O, K2O)$ are attacked by the silica or calcium carbonates present in aggregate. Insoluable complex compounds are formed due to this reaction, and as a result of that, cement paste is subjected to osmotic pressure and then fractures, (Hanson 1944)

To determine the potential alkali reactivity of the aggregate used, the bar mortar method described in ASTM C227-81 was followed to measure the linear expansion of mortar bars over a

Table 3. Compressive strength and workability of different concrete mixes at different ages.

Mix No.	Mix ratio	W/C	Workability Slump (mm)	Compressive strength N/mm2 for given age (days)		
				7	28	90
1	1:1 1/2:3	0.70	80	20.0	22.8	27.3
2	1:1 1/2:3	0.64	75	17.5	23.5	28.8
3	1: 2 :4	1.10	95	10.8	13.5	15.0
4	1: 2 :4	0.90	75	13.0	21.2	23.0
5	1: 2 :4	0.86	70	14.2	20.6	—
6	1: 2 :4	1.65	110	4.4	5.9	7.2
7	1: 3 :6	1.50	100	5.7	7.4	9.4
8	1: 3 :6	1.10	50	7.7	9.9	12.4
9*	1:1 1/2:3	0.42	60	41.2	51.0	57.3
10*	1: 2 :4	0.50	55	33.4	43.1	51.4
11*	1: 3 :6	0.75	65	8.8	11.3	14.2

* Natural gravels were used as coarse aggregate

period of six months, the average value of expansion was 0.052%

2.4 Concrete mixes

Three different mixes namely 1:3:6, 1:2:4 and 1:1 1/2:3, which are commonly used in Iraq, were used. Crushed limestone was used as coarse aggregate, while natural silica sand was used as fine aggregate, different W/C ratios were tried to keep meduim workability. Concrete specimens (150 x 150 x 150 mm), for the mentioned mixes, were made and cured in water until the time of testing. Compressive measurments were made at ages of 7, 28 and 90 days. All results are shown in Fig 2 and table 3.

2.4.1 Workability

Workability of all concrete mixes was determined by slump tests, according to BS 1881-1970. Table No.3 gives all the results.

2.4.2 Compressive strength

Table No.3 and Fig 2 show compressive strength results of the different concrete mixes. It can be seen the effect of W/C ratio and the type of coarse aggregate on compressive strength of similar mixes at a certain age.

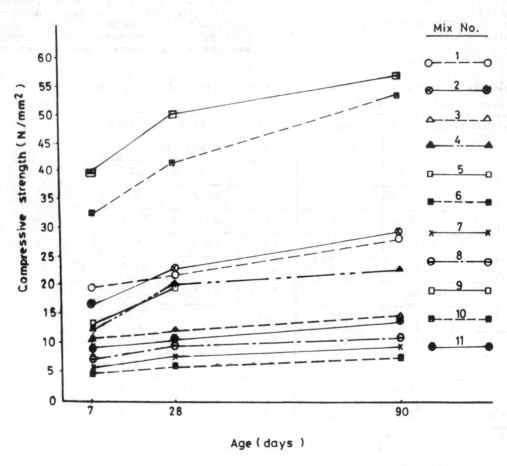

Fig 2.   Compressive strengths of concrete mixes

3. RESULTS AND DISCUSSION

3.1 Properties of crushed rock aggregates

All tests results of crushed rock aggregates properties are presented in table(2). The results show that the properties of crushed rock aggregates used were within the required limits of the standarded specifications for aggregates used in concrete: The water absorption of tested rock is 1.5%, while according to specification (Akroyd 1962), the absorption of aggregate suitable for concrete should not exceed 5% The abrasion of the tested limestone is 24% which is less than 35% required by Iraqi Standard 47-1970. The same standard requires impact resistance for the aggregates not more than 30%, while the impact resistance for the tested limestone aggregates is 11.5%.

3.2 Concrete workability

Table No. 3 shows all results of workability tests of different mix proportions as well as the type of aggregate. It can be seen from this table that for the same Agg/C ratio mixes of crushed rock coarse aggregate required higher W/C ratio than mixes of natural gravel coarse aggregate to maintain the same workability. It, also, shows that the differences in W/C ratios increase as the Agg/C ratio increases. The reson for that is the higher water absorption, angularity and rougher surface texture of crushed rock aggregate. These differences can be reduced by using super plasticizer (Elizzi 1990).

3.3 Concrete strength

Fig 2 and table 3 show compressive strength results of the defferent concrete mixes. It can be seen that the strength of concrete mixes (No. 9,10 & 11), where natural gravels were used as coarse aggregate is higher than other mixes where crushed limestone was used. This was due to the higher differences in W/C ratio required to maintaine, the same workability.

In spite of this reduction in compressive strength, the concrete produced by using crushed limestone as coarse aggregat can be used for many concrete structures, depending on the compressive strength (at age 28 days) required by B.S. standard code of practice (CP110-Part1.1972)and Iraqi Building code for reinforced concrete (Code 1,1987) as follows:

a.It is possible to use concrete mixes No(1,2,4 & 5) for general reinforced concrete (minimum compressive strength required, 20 N/mm).

b.It is possible to use all concrete mixes except mix No.6 for plain concrete work (minimum compressive strength required, 7 N/mm2).

3.4 Alkali reactivity

According to ASTM C227-81 the linear expansion of specimens age 6 months should not exceed 0.10%, while the linear expansion of specimens using crushed limestone was found equals to 0.052% as mentioned in (2.3). So this type of limestone aggregate is suitable for concrete production.

4. CONCLUSIONS

The overall results stemming from the experimental work, has led to the following conclusings:

4.1 The properties of crushed limestone aggregate from AL-Malih valley area are within the required limits of the standard specifications for normal aggregates used in concrete.

4.2 For the same richness, concrete mixes of crushed rocks coarse aggregate required higher W/C ratio than natural gravel aggregate concrete mixes to maintain the same workability.

4.3 It is recommended to use this type of crushed rock aggregate in producing good quality concrete which will solve most of the problem due to the shortage of coarse aggregat at the south of Iraq, which will reduce the cost of concrete production.

REFERENCES

Akoroyd, T.N.W. 1962. concrete properties and manufacture. Pergamon press, London

AL-Baldawi, T.A. and AL-Naib, S.B. 1986 The use of AL -Rahba crushed dolomitic limestone as aggregate in concrete.Jr.Bldg.Res.,V.5, No.2,pp 1 - 28

AL-Komi, M.A.1983. Reserve of limestone and dolomitic limestone in Iraq. Special report, state organization of Minerals, Baghdad.

B.S. Standard code of practice1962. CP110-Part1.

Elizzi, M.A. & Ikzer, B.G. 1990 The use of crushed sedimentary rocks for concrete production in Iraq.Proc. 6th Intr. IAEG Congress pp 2956-2970 Rotterdam Balkema.

Hanson, F.R.S., 1944. Cretaceous and tertiary reef formation and associated sediments in the Middle East.Am.Assoc., Petroleum Geologist, V34, pp 215-238.

Comportement au gel des granulats calcaires: Implication sur les essais normalisés

Frost behaviour of limestone aggregates: Implication on standard tests

J.M. Remy, F. Homand & M. Bellanger
Laboratoire de Géomécanique de l'ENSG, Vandœuvre-lès-Nancy, France

RESUME : La sensibilité au gel des matériaux calcaires est un facteur limitant pour leur utilisation en granulats routiers ou bétons. Une analyse expérimentale du comportement au gel des roches calcaires débouche sur un modèle physique qui tient compte des mécanismes microscopiques du gel et des géométries des réseaux poreux. Les résultats des essais normalisés de sensibilité au gel sur granulats sont confrontés à ce modèle. L'essai normalisé s'avère mal adapté aux matériaux calcaires et ne tient pas compte de la complexité de leur milieu poreux.

ABSTRACT : The frost sensitivity of limestones affects their usefulness as pavement or concrete aggregates. An experimental study of the behaviour of carbonate rocks under frost conditions induces a physical model which takes into account microscopic frost mechanisms and geometries of porous networks. The results of frost sensitivity standard tests on aggregates are compared with this model. Standard test is proved non suitable for limestone materials and doesn't take into account the complexity of their porous media.

1 INTRODUCTION

L'épuisement des gisements alluvionnaires, les exigences d'environnement, l'éloignement des chantiers par rapport à ces gisements et l'incidence sur les coûts sont autant de facteurs qui incitent à une utilisation plus large des matériaux locaux en vue d'une mise en oeuvre en granulats, et notamment des matériaux calcaires issus de roches massives. Les problèmes de durabilité relatifs à ces matériaux, et plus particulièrement leur sensibilité au gel, paramètre limitant leur utilisation dans un grand nombre de régions à climat continental, doivent donc être analysés et évalués d'une manière très précise.

La caractérisation des matériaux calcaires en termes de porosité, saturation et sensibilité au gel est rendue délicate du fait de l'importance et de la complexité de leurs milieux poreux (Bousquié, 1979 ; Remy, 1993). L'essai normalisé français de sensibilité au gel de granulats d/D (NF P 18 593, 1990), inspiré des travaux de Tourenq (1970) et des recommandations RILEM (1980), rend bien compte de cette difficulté. Cet essai, non spécifique aux matériaux calcaires, est fondé sur des caractéristiques externes aux matériaux (évolution du coefficient Los Angeles) et ne tient pas compte des particularités des matériaux calcaires relatives à leurs espaces poreux. De ce fait, un grand nombre de ces matériaux se retrouvent hors-spécifications (Tourenq, 1984).

Les granulats les plus communément considérés comme susceptibles au gel comprennent les cherts, les grès, les matériaux argileux et les calcaires (Gillott, 1980). Cependant, toutes ces roches sédimentaires ne conduisent pas systématiquement à des problèmes de durabilité au gel, les propriétés physiques de ces matériaux, et notamment les propriétés relatives à leurs milieux poreux, étant beaucoup plus significatives que leurs compositions pétrographiques (Bellanger *et al.*, 1993). La durabilité est également influencée par des paramètres externes tels que les conditions d'humidité avant et pendant le gel et le taux de refroidissement lors du gel.

Après un rappel des mécanismes théoriques du gel de l'eau dans un milieu poreux, nous dégagerons l'influence de la géométrie des réseaux poreux sur le comportement au gel des roches calcaires. A la lumière de ces résultats, nous tirerons ensuite quelques conclusions sur la signification réelle des essais normalisés de sensibilité au gel sur granulats calcaires.

2 MECANISMES DU GEL DE L'EAU DANS UN MILIEU POREUX

L'analyse du mécanisme de gel de l'eau dans un milieu poreux à l'échelle microscopique est une étape obligatoire pour comprendre la gélifraction macroscopique des matériaux poreux.

L'augmentation de volume de 9% induite par la transformation eau-glace n'est pas directement responsable des contraintes apparaissant lors du gel d'un matériau poreux. Deux théories proposent des modèles pour l'origine des désordres dus au gel dans un matériau, l'une tient compte des pressions développées dans l'eau non gelée (pressions hydrauliques, pressions osmotiques) par la glace en formation (Powers, 1949 ; Litvan, 1978), l'autre favorise les pressions capillaires développées par l'interface eau-glace lors de la progression du front de gel (Everett, 1961, Aguirre-Puente et al., 1977, Blachère, 1979).

Ces deux théories ne sont pas appliquées dans les mêmes domaines. Les travaux concernant le gel des pâtes de ciments et des bétons reposent sur l'hypothèse des pressions hydrauliques et osmotiques (Pigeon, 1989) ; les notions de facteur d'espacement des bulles d'air (Powers, 1949) et de distance critique (Fagerlund, 1979) découlent de cette théorie : les pressions résultant du déplacement de l'eau non gelée dans le réseau poreux augmentent avec la distance maximale que l'eau doit parcourir pour atteindre l'extérieur du réseau ou des cavités remplies d'air fonctionnant en quelque sorte comme des "soupapes" ; une fissuration apparaît lorsque les pressions engendrées sont supérieures à la résistance de la matrice solide, c'est-à-dire pour une distance critique de déplacement de l'eau.

Les travaux spécifiques au gel des sols, roches ou céramiques favorisent la théorie liée aux développement de pressions capillaires aux interfaces eau-glace (Aguirre-Puente et Bernard, 1978 ; Jones, 1980). Cette théorie s'appuie sur les mécanismes d'initialisation de la formation de la glace et sur sa propagation dans les réseaux poreux et repose sur des bases thermodynamique (notion de potentiel chimique) et capillaire (courbure d'une interface - loi de Laplace) solides. Nous nous attacherons plus particulièrement à cette dernière théorie car elle permet d'expliquer les différents comportements observés sur des roches calcaires lors du gel (Aguirre-Puente et Bernard, 1978).

2.1 Modèle de propagation d'un front de glace

Dans un matériau poreux saturé soumis au gel, la formation de la glace débute dans les macropores de plus larges rayons du fait de l'abaissement du point de congélation dans les pores fins (équation de Thompson - Everett, 1961). La glace ne peut pénétrer dans les capillaires plus fins et envahir le réseau poreux que sous les conditions suivantes :
- extraction continue d'énergie (refroidissement constant),
- interface eau-glace à la jonction macropore-capillaire satisfaisant la loi de Laplace, e.g. pression dans la glace supérieure à la pression dans l'eau :

$$P_g - P_e = \frac{2\sigma}{r} \qquad (1)$$

avec p_g et p_e, pressions dans la glace et dans l'eau, σ, tension interfaciale eau-glace et r, rayon de courbure principal de l'interface.

Tant que le rayon de courbure du ménisque eau-glace est supérieur au rayon du capillaire considéré, la croissance de la glace se poursuit dans le macropore initial par succion cryogénique de l'eau du réseau au travers du capillaire (Figure 1a). Ce mécanisme implique que les contraintes mécaniques imposées au système macropore-capillaire (état de consolidation et résistance mécanique de la matrice solide) permettent une augmentation du volume de macropore (croissance de glace supplémentaire).

La propagation du front de glace dans le capillaire n'est effective que pour un rayon de courbure du ménisque égal au rayon du capillaire ($r = r_c$) (Figure 1b).

L'influence d'une couche d'eau adsorbée, répartie sur la surface des pores, sur les mécanismes de gel est prépondérante (Aguirre-Puente et Bernard, 1979). Dans le cas du mécanisme de succion cryogénique, l'eau excédentaire au niveau du front de glace se répartit dans la couche d'eau liée entre la glace et la paroi du macropore site de formation de la glace. L'eau liée intervient en tant que zone de transfert de masse et induit un phénomène d'aspiration/répulsion à l'interface glace-eau. L'eau liée joue également un rôle de canal d'écoulement. Aguirre-Puente et Bernard (1979) ont mis en évidence une migration d'eau des zones de fort potentiel chimique vers les zones de faible potentiel chimique, e.g. des zones de forte épaisseur d'eau liée non atteintes par l'interface eau-glace vers les zones de faible épaisseur d'eau liée renfermant le front de glace. Le gradient de potentiel chimique dans l'eau adsorbée est compensé par une augmentation de pression dans la glace.

Figure 1 : Mécanismes de progression des fronts de glace et de développement des contraintesdans les milieux poreux. (a) : succion cryogénique dans un macropore. (b) : propagation du front de gel dans un capillaire, surpressions contre les parois solides.

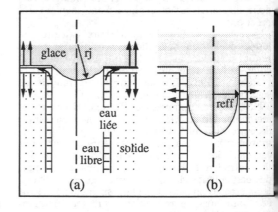

2.2 *Origines des contraintes développées lors du gel*

Le modèle de propagation du gel à l'échelle microscopique met en évidence deux causes principales de désordres dans les réseaux poreux soumis au gel :

(i) un endommagement de type gonflement lié à une succion cryogénique et aux phénomènes d'aspiration/répulsion associés (Figure 1a), favorisé par un refroidissement limité (non propagation du front de glace) et impliquant une faible résistance de la matrice solide (gonflement).

(ii) un endommagement résultant des surpressions occasionnées par le passage du front de glace au travers de vides de dimensions différentes (Figure 1b). Les surpressions sont localisées aux jonctions macropore/capillaire et sur les parois des capillaires. L'augmentation de pression maximale, donnée par la loi de Laplace ($p_g - p_e = 2\sigma/r_c$, r_c rayon du capillaire), doit être supérieure à la résistance locale de la matrice pour qu'un endommagement apparaisse. La surpression est d'autant plus importante que le rapport dimension des macropores/dimension des capillaires associés est élevé.

La couche d'eau adsorbée intervient dans tous les cas, soit comme moteur de l'aspiration (zone de transfert de masse/gonflement), soit comme zone de transfert de contraintes (surpression dans la glace et pression de confinement sur les parois, l'eau liée étant nettement moins compressible que l'eau libre). La couche d'eau liée est répartie sur la surface des vides. La surface de pores spécifique du milieu aura donc une grande influence sur ces mécanismes.

Dans le cas d'une saturation partielle du milieu poreux, la présence d'air permet une dissipation des contraintes induites par le gel (Remy, 1993). La pression maximale dans la glace lors de la propagation du front diminue du fait d'interfaces eau-air situées en aval du front de gel et provoquant une dépression dans l'eau.

3 INFLUENCE DE LA GEOMETRIE DES RESEAUX POREUX SUR LE COMPORTEMENT AU GEL DES ROCHES CALCAIRES

3.1 *Matériaux et méthodes*

L'étude expérimentale du comportement au gel des matériaux calcaires (Remy, 1993) porte sur des roches calcaires de Lorraine (Nord-Est du Bassin de Paris, France) constituant un large éventail de faciès carbonatés.

La porosimétrie au mercure est le principal outil utilisé pour la caractérisation des milieux poreux. Une première injection de mercure dans le matériau fournit la porosité totale N_{Hg} et la distribution de cette porosité en terme de rayon d'accès aux pores.

L'allure de la distribution est analysée qualitativement (unimodale à multimodale) et quantitativement par un coefficient de dispersion C_d. Au cours du retrait du mercure, une partie reste piégée dans le réseau poreux ; ce volume constitue la porosité piégée N_p. Le piégeage, localisé dans les pores de plus grands rayons, est initialisé à l'échelle locale par des ruptures du film de mercure liées à des rapports rayon de pore/rayon des capillaires associés élevés. Un degré d'arrangement du réseau important (répartition en domaines de pores larges et de capillaires fins) favorise le piégeage du mercure (Li et Wardlaw, 1986). La porosité piégée quantifie une porosité de pore *s.s.* avec pores nodaux et capillaires fins associés. Une seconde injection de mercure permet l'envahissement du réseau libre de mercure. Elle quantifie la porosité libre N_l qui représente la part interconnectée du réseau poreux. L'essai porosimétrique fournit également une mesure de la surface spécifique de la porosité totale (S_{po}) et du réseau connecté (libre, S_{pol}).

La caractérisation des matériaux est complétée par:
- une étude expérimentale de la saturation en eau des réseaux poreux comprenant des saturations par immersion à pression atmosphérique (immersion pendant 48 heures, porosité Hirshwald, N_{48h} et degré de saturation associé, S_{48h}) et des imbibitions par capillarité (degré de saturation, S_{cap}). Le coefficient Hirschwald correspond au degré maximum de saturation que peut prendre un matériau en conditions naturelles (Mamillan, 1984).
- des mesures de perméabilités K au perméamètre à l'azote à charge constante.

L'analyse du comportement au gel des matériaux repose sur une simulation des conditions de gel en enceinte climatique à partir de cycles gel/dégel doux (-5°C - taux de refroidissement de 5°C/h) et brutaux (-20°C - 8°C/h) (Figure 2).

Figure 2. Détails des cycles gel/dégel utilisés. Evolution des températures dans l'enceinte et à la surface des échantillons.

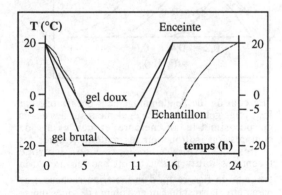

Les échantillons soumis aux cycles gel/dégel ont des degrés de saturation compris entre 5% et 100% (obtenus par séchage) incluant le degré de saturation

Hirschwald (S_{48h}). Une fois le degré de saturation voulu atteint, les échantillons sont enveloppés dans un manchon en caoutchouc étanche et ne sont soumis aux cycles qu'après une période d'homogénéisation de quelques heures.

La mesure de l'endommagement des matériaux consiste en un suivi des propriétés ultrasoniques (vitesses et atténuations des ondes ultrasoniques longitudinales - ondes P) au cours des cycles. L'atténuation des signaux ultrasoniques est déterminée par la méthode du rapport des spectres (Toksöz *et al.*, 1979). Cette méthode repose sur une comparaison de deux spectres résultant d'une transformation de Fourier des signaux bruts, l'un provenant d'un échantillon de référence (aluminium), l'autre de l'échantillon à étudier, et débouche sur le calcul d'un facteur de qualité en atténuation Q. L'utilisation de l'atténuation des ondes ultrasoniques pour suivre l'évolution des matériaux au cours du gel est fondée sur le principe suivant (Figure 3) :
- l'atténuation du signal est la plus faible dans le matériau gelé (facteur de qualité plus élevé) du fait d'une meilleure transmission des signaux dans la glace que dans l'eau,
- le spectre du matériau dégelé est par contre plus atténué (Q plus faible) du fait de l'endommagement.

Cette méthode demande une grande rigueur expérimentale mais fournit plus de renseignements que l'évolution des vitesses seules (Remy *et al.*, 1994).

Figure 3. Spectres des signaux ultrasoniques d'un matériau calcaire à l'état initial, gelé et dégelé.

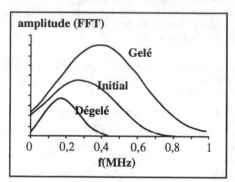

Le contrôle de l'endommagement occasionné par les cycles gel/dégel repose également sur des essais de porosimétrie au mercure et des essais de compression isotrope sur les matériaux après les cycles. L'essai de compression isotrope ("compressibility test") fournit une mesure de la porosité de fissure η_o d'un matériau rocheux ; ce paramètre, correspondant au volume de vides qui se ferme sous pression hydrostatique isotrope, est issu de l'analyse des courbes pression-déformation (Walsh, 1965).

3.2 *Caractérisation du milieu poreux des matériaux calcaires*

Les caractéristiques relatives aux essais porosimétriques et aux comportements hydriques (saturations, perméabilité) des matériaux calcaires étudiés sont résumées dans le tableau 1.

L'analyse porosimétrique fournit une classification des milieux poreux des roches calcaires en trois grands types (Figure 4), chacun de ces types conduisant à des propriétés hydriques particulières (Remy, 1993) :

(i) Calcaires à réseaux poreux unimodaux stricts (faciès D2 et G2) (Figure 4a) :
- un seul point d'inflexion sur les courbes d'injection de mercure : une seule famille d'accès aux pores,
- distribution des accès très resserrée,
- coefficient de dispersion $C_d < 1$,
- N_l (porosité libre) > N_p (porosité piégée).

Ces réseaux sont caractéristiques d'une porosité de type capillaire à degré d'interconnexion élevé (rayons d'accès proches des rayons des capillaires) et conduisent à de fortes saturations en eau (S > 95%) : au cours d'une immersion à pression atmosphérique (S_{48h}) ou lors d'une imbibition par capillarité (S_{cap}), le réseau connecté principal constituant la majeure partie du réseau poreux, se sature totalement. La perméabilité est relativement élevée et augmente avec le rayon de la famille principale de capillaires.

(ii) Calcaires à réseaux poreux unimodaux étalés (faciès D1 et G1) (Figure 4b) :
- un seul point d'inflexion net mais une distribution étalée des accès aux pores,
- $C_d > 1$,
- $N_p > N_l$.

Ces réseaux reflètent une porosité de pores *s.s.* constituée principalement de pores nodaux et à très faible degré d'interconnexion. Cette géométrie conduit à de faibles degrés de saturation et à de faibles perméabilités.

(iii) Calcaires à réseaux poreux multimodaux (faciès J et E) (Figure 4c) :
- plusieurs points d'inflexion : plusieurs familles d'accès aux pores,
- $C_d > 1$,
- $N_p > N_l$.

Ces réseaux reflètent une porosité constituée de pores nodaux et de capillaires. Les pores nodaux représentent les sites de piégeage de mercure et fonctionnent également comme sites de piégeage d'air lors d'un processus de saturation. Par contre, le degré d'interconnexion du réseau de capillaires est important, c'est principalement cette part de la porosité qui se sature lors d'une immersion ou d'une imbibition par capillarité. La perméabilité est commandée par ce réseau connecté et augmente avec la dimension des capillaires le constituant.

Tableau 1. Caractéristiques pétrographiques et pétrophysiques des roches calcaires étudiées.

Faciès	Abr.	N_{Hg} (%)	N_p (%)	N_l (%)	C_d (s.u.)	S_{po} (m²/g)	S_{pol} (m²/g)	S_{48h} (%)	S_{cap} (%)	K (mD)
Calcaire bioclastique - Ciment sparitique Grainstone à packstone	J	22-24	14	8	2,5-3,5	1,2	0,3	60	62	0,3
Calcaire à entroques - Ciment sparitique Grainstone	E	14-15	10	5	15	1	0,2	55	56	2
Calcaire micritique - peu d'éléments figurés Aspect sublithographique / Micrite dense Mudstone	D1	8	5,5	2,5	1	1,3	0,3	82	67	4.10^{-3}
Calcaire micritique sans élément figuré Texture crayeuse - Micrite lâche Mudstone	D2	25	18	7	0,5	1,2	0,7	96	98	1
Calcaire micritique sans élément figuré Aspect sublithographique - Lamination algaire Mudstone	G1	5	4,5	0,5	1,5	1,4	0,08	80	58	$1,5.10^{-3}$
Calcaire oolithique et bioclastique à matrice micritique - Wackestone à packstone	G2	11-12	7	5	0,8	1,7	0,5	97	98	0,08

Figure 4. Courbes porosimétriques des roches calcaires. (a) : réseau poreux unimodal strict. (b) : réseau poreux unimodal étalé. (c) : réseau poreux multimodal.

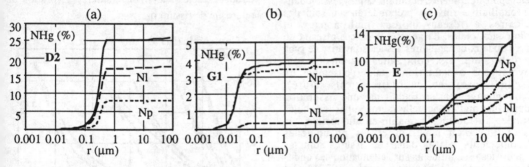

3.3 *Comportement au gel des roches calcaires*

Les matériaux calcaires, soumis aux cycles gel/dégel présentent des évolutions des propriétés ultrasoniques (vitesses et atténuations des ondes P) au cours des cycles qui dépendent de la géométrie de leurs milieux poreux et de leurs degrés de saturation (Figure 5a, cycles brutaux 20°C à -20°C).

Le facteur de qualité en atténuation Q évolue par chutes brutales et paliers pour les calcaires à réseaux poreux multimodaux et unimodaux étalés totalement saturés marquant des augmentations instantannées mais non continues de l'atténuation des signaux ultrasoniques. Le facteur de qualité ne montre pas d'évolution notable si ces matériaux sont partiellement saturés (saturation Hirschwald, S_{48h}).

La diminution du facteur de qualité est plus progressive mais continue pour les calcaires à réseaux poreux unimodaux stricts qu'ils soient totalement saturés ou à saturation Hirschwald. Cette diminution s'atténue nettement à partir d'un degré de

saturation beaucoup plus faible (70% dans le cas de l'exemple présenté, faciès D2).

L'évolution des propriétés ultrasoniques au cours des cycles gel/dégel est également fonction du cycle utilisé (Figure 5b). Pour des cycles doux (T_{min} : -5°C, taux de refroidissement de 5°C/h) ou plus rigoureux (T_{min} : -20°C, 8°C/h), les calcaires à réseaux poreux unimodaux stricts montrent des vitesses des ondes qui évoluent d'une manière similaire : diminution progressive mais continue. Par contre, l'évolution de la vitesse dans les calcaires à réseaux poreux unimodaux étalés et multimodaux qui présente des diminutions rapides et des paliers pour des cycles brutaux devient plus progressive et continue pour des cycles doux.

Ces résultats expérimentaux permettent d'établir un modèle physique du comportement au gel des matériaux calcaires fondé sur les mécanismes théoriques du gel précédemment décrits et qui tient compte de la géométrie des réseaux poreux.

Figure 5. Evolution des propriétés ultrasoniques au cours des cycles gel/dégel mesurées durant la phase de dégel sur les calcaires à réseaux poreux unimodaux stricts (faciès D2) et multimodaux (faciès J). (a) : influence de la saturation sur l'évolution de l'atténuation des ondes P. (b) : influence du type de cycle sur l'évolution des vitesses des ondes P.

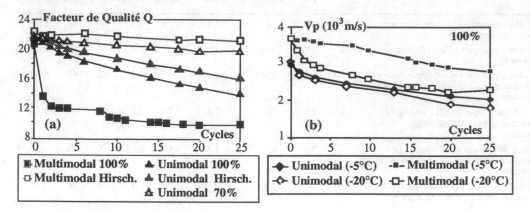

(i) Le phénomène de surpressions dans la glace et contre les parois solides développées par le front de glace lors de son passage dans les jonctions macropores/capillaires et durant sa progression dans les capillaires fins est prépondérant au sein des calcaires à réseaux poreux multimodaux et unimodaux étalés. La géométrie de ces réseaux en pores nodaux et capillaires associés favorise ce type de contrainte. Les contraintes maximales sont développées à saturation totale (l'ensemble du réseau, pores et capillaires, est saturé). L'endommagement se traduit par une fissuration importante mais ponctuelle marquée par les chutes brutales du facteur de qualité en atténuation : la fissuration d'un matériau et les phénomènes d'écoulements relatifs dans ces fissures sont les facteurs les plus influents sur l'atténuation des ondes ultrasoniques (Johnston *et al.*, 1979). Cet endommagement est confirmé par des essais de compression isotrope (Figure 6, faciès J) qui montrent une porosité de fissure croissante au cours des cycles gel/dégel (η_o initiale = 0,06%, $\eta_o \approx 1\%$ après 15 cycles gel/dégel). Par contre, aucune modification du milieu poreux n'est relevée par la porosimétrie au mercure : entre le réseau de fissures créées, le matériau reste sain. A saturation partielle (saturation Hirschwald, S_{48h}), les pores nodaux sont piégés d'air ; les contraintes qui se développent sont faibles et la présence d'air permet leur dissipation ; aucun endommagement n'est visible. La saturation Hirschwald de ces matériaux est faible (de l'ordre de 60%) ; elle est nettement inférieure au degré de saturation critique défini par Fagerlund (1971) au dessus duquel la dissipation des contraintes n'est plus possible et qui est de l'ordre de 80% pour ces matériaux (Remy, 1993). En conditions naturelles (saturation Hirschwald), ces matériaux peuvent donc être considérés comme non gélifs.

Figure 6. Essai de compression isotrope sur des matériaux ayant subi un nombre croissant de cycles gel/dégel (faciès J totalement saturé). Evolution de la porosité de fissure η_o.

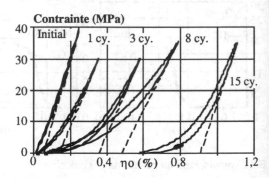

(ii) Les contraintes de type gonflement liées au phénomène de succion cryogénique sont les principales contraintes développées dans les calcaires à réseaux poreux unimodaux stricts. Les surpressions dans la glace lors de la progression du front de glace sont faibles du fait d'une taille de capillaire pratiquement constante. L'accumulation d'eau par aspiration vers les zones froides provoque un endommagement de type gonflement se traduisant par un élargissement progressif mais continu des capillaires. L'élargissement des vides n'a qu'une influence secondaire sur les propriétés ultrasoniques (O'Connell et Budiansky, 1977) : le facteur de qualité en atténuation diminue progressivement. Ce type d'endommagement se manifeste par une modification profonde du réseau poreux (Figure 7, faciès G2) : après 25 cycles, les courbes porosimétriques montrent une augmentation de la porosité totale résultant principalement d'une augmentation de la part connectée du réseau (porosité libre) qui est elle-même liée à un élargissement des capillaires (déplacement des courbes vers la droite). Le mécanisme de succion cryogénique est favorisé par la surface spécifique élevée du réseau connecté de ces matériaux ; celle-ci induit une teneur en eau adsorbée importante, jouant le rôle de canaux de circulation. Par contre, ce phénomène d'aspiration protège le matériau d'une fissuration à coeur (la porosité de fissure n'évolue pas). Cet endommagement apparaît que le matériau soit totalement saturé ou qu'il soit à saturation Hirschwald (S_{48h}). Le degré de saturation Hirschwald est élevé pour ces matériaux (> 90%) ; il est supérieur au degré de saturation critique. L'endommagement s'atténue à partir d'un degré de saturation de l'ordre de 70%. Ce mécanisme d'élargissement du réseau poreux peut être dangereux pour le matériau à long terme : l'augmentation du volume connecté conduit à une aspiration d'eau plus importante et à un accroissement de la succion cryogénique ; le phénomène devient "auto-entretenu" et peut conduire, à terme, à une destruction complète des matériaux se traduisant par un effritement généralisé.

Figure7. Courbes porosimétriques avant (N_t, totale et N_l, libre) et après les cycles gel/dégel (N_{tg} et N_{lg}) (faciès G2).

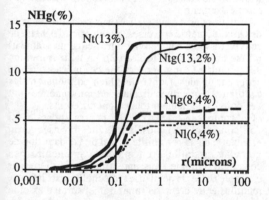

(iii) La distinction entre les deux comportements précédents est nette pour des cycles de gel brutaux (20°C/-20°C). Pour des cycles plus doux (20°C/-5°C), les calcaires à réseaux poreux unimodaux réagissent de la même manière ; par contre, le comportement des calcaires à réseaux poreux multimodaux est modifié. Ce type de gel ne permet pas une propagation notable du front de glace dans le réseau poreux ; les contraintes de type gonflement par succion cryogénique deviennent prépondérantes et les calcaires à réseaux poreux multimodaux totalement saturés tendent à se comporter comme les calcaires à réseaux unimodaux : l'évolution du

facteur de qualité devient plus progressive mais continue.

Ce modèle physique du comportement au gel des roches calcaires tenant compte de la géométrie des réseaux poreux a été validé par des observations et des mesures expérimentales sur des blocs d'enrochements constitués des matériaux calcaires précédemment décrits, installés depuis plusieurs années en protection de digues, de berges ou de piles de ponts et ayant subi les conditions de gel *in situ* (Remy, 1993).

4 CARACTERISATION DE LA SENSIBILITE AU GEL DE GRANULATS

La plupart des roches calcaires précédemment décrites sont exploitées en vue d'une utilisation en granulats d/D concassés dans les domaines routiers et bétons. La sensibilité au gel de ces matériaux granulaires est testée par les essais normalisés en vigueur en France et le comportement au gel de ces granulats est comparé au comportement des roches massives.

4.1 L'essai normalisé

L'essai de durabilité au gel sur granulats d/D fait l'objet de la norme NF P 18.593 (1990) qui repose sur l'évolution du coefficient Los Angeles d'un matériau avant et après avoir subi une succession de cycles gel/dégel. Le coefficient Los Angeles (L_A) définit la résistance à la fragmentation d'un échantillon de granulats et correspond au pourcentage en masse des éléments inférieurs à 1,6 mm produits en soumettant le matériau aux chocs répétés de boulets en acier normalisés.

L'état de saturation en eau initial du matériau repose sur une immersion totale sous vide partiel (4.10^{-2} bar). Les échantillons, placés dans un bac au fond duquel quelques millimètres d'eau subsistent, sont soumis à 25 cycles gel/dégel (+25/-25°C), d'une durée de 12 heures, avec un taux de refroidissement de 13°C/h (gel brutal). La sensibilité au gel des granulats, traduisant l'augmentation relative du coefficient L_A après gel (L_{Ag}) est exprimée par :

$$G = \frac{100(L_{Ag} - L_A)}{L_A} \qquad (2)$$

Les résultats d'essais normalisés menés sur des granulats issus de calcaires bioclastiques (faciès J) et micritiques à oolithiques (faciès G1 ; faciès S proche de G1 mais moins poreux, 2% ; faciès G indifférencié regroupant G1 et G2) pour différentes classes granulométriques sont résumés dans le tableau 2.

L'indice de gélivité est le plus faible pour les calcaires à faciès sublithographique (faibles porosités et réseaux poreux unimodaux étalés, faciès

S et G1). D'autre part, les faciès bioclastiques (J, réseau poreux multimodal) apparaissent les plus sensibles au gel ce qui est en contradiction avec les comportements observés sur roches massives.

Tableau 2.

Faciès d/D (mm)	S 6,3/10	G1 6,3/10	G 4/6,3	G 6,3/10	G 10/14	J 3/6	J 6/15
L_A (%)	24	24	27	28	34	44	39
L_{Ag} (%)	24	26	41	43	52	82	68
G (%)	≈ 0	8	52	54	51	86	75

4.2 Influence de l'état de saturation

L'immersion sous vide préconisée par la norme conduit à un degré de saturation proche de la saturation Hirschwald pour le faciès G1 (S = 82%) du fait d'un réseau poreux très mal connecté et constitué de pores de faibles rayons. Il en découle une faible sensibilité au gel pour ce type de faciès. Par contre, cette méthode de saturation conduit à des degrés de saturation très élevés pour les autres matériaux (90% pour les calcaires à réseaux poreux unimodaux stricts, faciès G ; 90 - 95% pour les calcaires à réseaux poreux multimodaux, faciès J). Dans ce dernier cas, l'essai normalisé reproduit les conditions rencontrées lors des essais de gel brutal sur roches massives très fortement ou totalement saturées présentés précédemment et conduit logiquement à des sensibilités au gel élevées (G > 50). Les calcaires à réseaux poreux multimodaux (faciès J) apparaissent particulièrement gélifs du fait d'un degré de saturation important couplé à un fort taux de refroidissement (gel brutal, 13°C/h).

Des essais de gel ont été réalisés sur des granulats constitués des faciès G indifférencié et J bioclastique à différents degrés de saturation obtenus soit par des vides partiels soit par des immersions à pression atmosphérique. Les indices de gélivité diminuent nettement avec une réduction du degré de saturation des échantillons (Figure 8). Ces indices sont inférieurs à la valeur seuil de 30, préconisée par les spécifications d'utilisation en structure routière, pour des degrés de saturation inférieurs ou égaux à 80% obtenus par simple immersion à pression atmosphérique. Ce seuil de saturation, au dessus duquel l'indice de gélivité augmente fortement, correspond au coefficient d'Hirschwald (S_{48h}) déterminé sur roches massives. En dessous de ce seuil, seule le réseau poreux connecté du matériau (porosité libre au mercure) est saturé ; le piégeage d'air limite la dégradation par le gel de ces matériaux.

Figure 8. Evolution de l'indice de gélivité avec le degré de saturation des échantillons de granulats au cours des cycles gel/dégel.

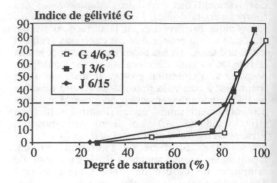

4.3 Effet d'échelle

Certains auteurs avancent un effet de la taille des granulats sur leur sensibilité au gel. Dans le cadre d'une utilisation en béton, Pigeon (1989) définit un concept de dimension critique de granulat liée à la notion de distance critique (Fagerlund, 1979) résultant d'une approche en terme de mécanisme de dégradation par le gel de type pressions hydrauliques. Un granulat est considéré gélif si sa dimension est supérieure à une dimension critique permettant l'expulsion de l'eau vers des pores piégés d'air ou vers l'extérieur et qui est fonction du taux de refroidissement.

Au cours des essais réalisés sur calcaires et pour des granulométries diverses, aucun effet d'échelle de ce type (associé à la gélivité intrinsèque du granulat) n'a été observé (voir Tableau 2). La taille minimum des granulats (3 mm) et l'échelle des mécanismes de propagation des fronts de glace (dimension des capillaires : quelques dizaines de micromètres) ont des ordres de grandeur totalement différents.

Par contre, un effet d'échelle pourra apparaître si l'on considère non plus la gélivité intrinsèque du granulat mais la sensibilité au gel de l'échantillon de granulats assimilé à un matériau granulaire peu consolidé (comme dans une structure routière, par exemple). Ce type d'effet d'échelle est lié aux relations entre degré de saturation et surface externe développée par unité de volume d'échantillons. La figure 9 présente l'évolution du degré de saturation Hirschwald avec une augmentation de la surface externe par unité de volume dans le cas du faciès J (calcaire bioclastique). Les plus faibles degrés de saturation correspondent aux plus faibles surfaces (éprouvettes de roches massives) : l'eau est principalement absorbée dans le réseau poreux du matériau. Une augmentation de la surface externe (éprouvettes fragmentées puis granulats de granulométrie de plus en plus fine) conduit à une augmentation du degré de saturation : un volume d'eau adsorbée sur la surface externe non négligeable vient s'ajouter au volume d'eau absorbée dans le réseau poreux.

Figure 9. Evolution du degré de saturation avec la surface développée par unité de volume de l'échantillon : éprouvettes massives, fragmentées, granulats de granulométrie de plus en plus fine (faciès J).

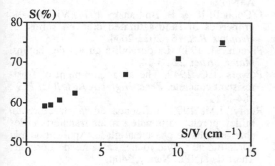

Dans le cas d'une structure routière, l'eau de surface des granulats constitue une zone de saturation partielle intergranulaire qui participe activement aux transferts dans la structure. A cette échelle, les mécanismes de gel prépondérant sont associés au phénomène de succion cryogénique et conduisent à des endommagements par gonflement. Une augmentation du volume d'eau adsorbée résultant d'une diminution de la taille des granulats favorisera ce mécanisme de succion cryogénique.

4.4 Signification de l'essai normalisé

Au vue des résultats précédents, quelques conclusions peuvent être tirées quant à la signification de l'essai normalisé de sensibilité au gel sur granulats calcaires (signification des conditions de gel et de saturation, validité de la mesure de l'endommagement).

D'après l'essai normalisé, le pôle non gélif est constitué par les calcaires peu poreux à réseaux poreux unimodaux étalés très peu connectés (faciès S et G1). Ce bon comportement est lié aux faibles degrés de saturation obtenus par la méthode préconisée par la norme, proches du degré de saturation Hirschwald. De plus, ces calcaires (faciès sublithographique) ont une résistance intrinsèque à leur matrice solide élevée du fait d'une structure cristalline micritique très dense ; le coefficient L_A initial de ces matériaux est faible (< 25) et son utilisation comme mesure de l'endommagement est significative.

Par contre, l'essai normalisé donne un caractère gélif aux granulats issus de calcaires à réseaux poreux complexes multimodaux et à porosité importante ($> 20\%$, faciès J) bien que ces matériaux soient non gélifs si l'on considère les mécanismes de gel et des conditions naturelles de saturation. La forte sensibilité au gel de ces granulats est liée à un degré de saturation important pendant l'essai, supérieur au coefficient d'Hirschwald, couplé à un taux de refroidissement élevé. De plus, ces matériaux ont une faible résistance initiale à la fragmentation ($L_A \approx 40$) du fait d'une porosité élevée et de faibles liaisons intercristallines ce qui limite la validité de la quantification de l'endommagement par la mesure du L_A.

Les matériaux calcaires à réseaux poreux unimodaux stricts sont gélifs vis-à-vis de la norme ($G > 50$) et vis-à-vis des essais sur roches massives (endommagement par le gel des éprouvettes à saturation Hirschwald - conditions naturelles). Cependant, l'essai normalisé risque d'avantager ces matériaux du fait du faible nombre de cycles réalisés (25). Nous avons vu précédemment que ces matériaux présentaient un endommagement progressif mais continu ; l'indice de gélivité mesuré après 25 cycles peut être ainsi minimisé.

La sensibilité au gel mesuré par la norme est relative à la gélivité intrinsèque du granulat ; reste à savoir si un granulat non gélif aura le même comportement une fois mis en oeuvre dans une structure routière ou dans un béton.

Dans le cas d'une structure routière, il faut considérer le volume de granulats comme un matériau granulaire où l'espace intergranulaire représente une échelle supplémentaire de porosité. Comme nous l'avons vu précédemment, la présence d'un volume d'eau adsorbée non négligeable dans cette porosité peut entraîner des endommagements de type gonflement de l'ensemble de la structure sans que les éléments unitaires ne soient eux-mêmes endommagés.

Dans le cas d'un béton, la non gélivité d'un granulat n'entraîne pas forcément la bonne tenue au gel du béton. A la suite d'un apport d'eau, l'eau qui sature d'abord les granulats peut être ensuite expulsée (perméabilité des granulats) vers la pâte de ciment et la réponse du béton aux sollicitations climatiques sera fonction de la faculté de la pâte de ciment à accomoder cet apport d'eau (Pigeon, 1989). L'essai de gélivité est dans ce cas insuffisant pour conclure ; des essais sur éprouvettes de bétons, similaires à ceux présentés précédemment sur roches massives, sont nécessaires.

5 CONCLUSION

Une analyse expérimentale du comportement au gel des roches calcaires couplée à une étude théorique des mécanismes de gel de l'eau dans un milieu poreux a permis de proposer un modèle de comportement au gel de ces matériaux qui est fonction de la géométrie de leurs milieux poreux. Le pôle non gélif est représenté par les calcaires à réseaux poreux unimodaux étalé et multimodaux (porosité piégée au mercure importante), le pôle gélif par les calcaires à réseaux poreux unimodaux stricts (porosité libre supérieure à la porosité piégée). La confrontation des résultats des essais normalisés de sensibilité au gel sur granulats à ce modèle montre que l'essai normalisé semble trop sévère vis-à-vis des matériaux calcaires. Les conditions d'humidité avant et pendant les cycles résultent en des degrés de saturation importants. La vitesse de refroidissement est également un facteur de sévérité : elle conduit à un éclatement des granulats.

Une adaptation de l'essai normalisé sur granulats calcaires devrait aller dans le sens d'une meilleure représentativité des conditions naturelles et notamment des degrés de saturation et des conditions de gel plus réalistes.

REMERCIEMENTS

Cette recherche a été réalisée dans le cadre du Programme National de Recherche/Développement en Génie Civil "MATERLOC - Calcaires". Les auteurs tiennent à remercier le comité de direction de ce programme et plus particulièrement M. A. Remillon.

REFERENCES

Aguirre-Puente, J. & J.J. Bernard 1978. Comportement au gel des matériaux de construction. In BTP (éd), *Le comportement thermique des matériaux de construction* : 29-72.

Aguirre-Puente, J., M. Fremond & J.M. Menot 1977. Gel dans les milieux poreux. *Proc. Int. Symp. Frost Action in Soils, Lulea* : 5-28, Rotterdam : Balkema.

Bellanger, M., F. Homand & J.M. Remy 1993. Water behaviour in limestones as a function of pores structure : application to frost resistance of some Lorraine limestones. *Eng. Geol.* 36 : 99-108.

Blachère, J.R. 1979. Le gel de l'eau dans les milieux poreux. *Proc. VIème Cong. Int. FFEN* : 295-303.

Bousquié, P. 1979. Texture et porosité des roches calcaires. Relations avec perméabilité, ascension capillaire, gélivité, conductivité thermique. Thèse de l'Université P. et M. Curie, Paris VI, 191p.

Everett, D.H. 1961. Thermodynamics of frost damage to porous solid. *Trans. Faraday Soc.* 57 : 1541-1551.

Fagerlund, G. 1979. Studies of the destruction mechanism at freezing of porous materials. *Proc. VIème Cong. Int. FFEN* : 168-196.

Gillot, J.E. 1980. Properties of aggregates affecting concrete in North America. *Quart. J. of Eng. Geol.* 13 : 289-303.

Johnston, D.H., M.N. Toksöz & A. Timur 1979. Attenuation of seismic waves in dry and saturated rocks : Mechanisms. *Geophysics* 44 : 691-712.

Jones, R.H. 1980. Frost heave of roads. *Quart. J. of Eng. Geol.* 13 : 77-86.

Li, Y. & N.C. Wardlaw 1986. Mechanisms of nonwetting phase trapping during imbibition at slow rates. *J. Colloid and Interface Sci.* 109 : 461-472.

Litvan, G.G. 1978. Freeze-thaw durability of porous building materials. In Sereda, P.J. & G.G. Litvan (eds), *Durability of building materials* : 455-463. STP ASTM.

Mamillan, M. 1984. Durabilité des pierres tendres. *CEBTP* rapp. 102 : 69p.

NF P 18-593 1990. Granulats : sensibilité au gel. AFNOR.

O'Connell, R. & B. Budiansky 1977. Viscoelastic properties of fluid-saturated cracked solids. *J. Geophys. Res.* 76 : 2022-2034.

Pigeon, M. 1989. La durabilité au gel des bétons. *Mater. Struct.* 22 : 3-14.

Powers, T.C. 1949. The air requirement of frost-resistant concrete. *Proc. Highway Res. Board* 29 : 184-211.

Remy, J.M. 1993. Influence de la structure du milieu poreux carbonaté sur les transferts et les changements de phase eau-glace. Application à la durabilité au gel de roches calcaires de Lorraine. Thèse de l'INPL, Nancy, 340p.

Remy, J.M., F. Homand & M. Bellanger 1994. Laboratories velocities and attenuation of P-waves in limestone during freezing. *Geophysics* 59 : 245-251.

RILEM 1980. Recommandations d'essais pour mesurer l'altération des pierres. *Mater. Struct.* 13 : 175-253.

Toksöz, M.N., D.H. Johnston & A. Timur 1979. Attenuation of seismic waves in dry and saturated rocks : Laboratory measurements. *Geophysics* 44 : 681-690.

Tourenq, C. 1970. La gélivité des roches. Application aux granulats. *Lab. Central P. et Ch.* Rapp. 6, 60p.

Tourenq, C. 1984. Le problème de l'utilisation des granulats hors spécifications. *Bull. Liaison Labo. P. et Ch.* n° sp. XIV : 181-183.

Walsh, J.B. 1965. The effect of cracks on the compressibility of rocks. *J. Geophys. Res.* 70 : 381-385.

Influence des paramètres géologiques sur la résistance mécanique des granulats dans la région de Bujumbura (Burundi)

The influence of geologic parameters on the mechanic resistance of the aggregates of the Bujumbura region (Burundi)

Egide Nzojibwami
Université du Burundi, Bujumbura, Burundi

RESUME: Dans la région de Bujumbura, trois types de roches peuvent être utilisées dans la confection des granulats: les roches basiques, les granites et les quartzites. Compte tenu de leurs bonnes caractéristiques mécaniques et de leur relative abondance, les roches basiques fournissent plus de 90% des granulats du Burundi. Toutefois, certaines roches apparemment saines se sont montrées quelquefois mécaniquement fragiles. Des études récentes, faites sur les granulats de la région, ont montré que la qualité des granulats est liée à des paramètres géologiques dont l'analyse n'entre pas dans les tests routiniers sur les granulats.

ABSTRACT: In the Bujumbura area, aggregates can be provided by three kind of rocks: mafic rocks, granites and quartzites. Considering their high strength and their good occurence, mafic rocks are to provide 90% of the Burundi aggregates. Meanwhile, apparently fresh rocks have some times been found mechanically weak. Recent studies on the Bujumbura area aggregates suggest that their mechanical characteristics are related to geological parameters which are not considered among current tests on aggregates.

INTRODUCTION

La ville de Bujumbura, capitale du Burundi (Afrique Centrale), comporte aujourd'hui plus de 350.000 habitants. A partir des années 1980, le pays a connu un important développement des infrastructures (routes, immobilier, barrages, communications,...) dont une grande partie a été réalisée dans la ville capitale et ses environs. Des carrières d'exploitation des matériaux de construction (sables, graviers, concassés et moellons) ont surgi spontanément autour de la ville pour répondre à ce besoin en croissance constante. La plupart de ces carrières sont exploitées artisanalement. Seules, quelques carrières sont exploitées industriellement par des compagnies de construction. Cette exploitation incontrôlée ne permet pas toujours de vérifier la qualité du matériau, ce qui conduit souvent à des disputes et des incompréhensions entre les différents partenaires de la construction. Le gouvernement, soucieux des problèmes

d'environnement, désapprouve cette pratique d'exploitation anarchique. Certaines exploitations industrielles ne respectent même pas les normes et les règles élémentaires de sécurité. Face à cette situation, les services gouvernementaux ont fait réaliser récemment des études préliminaires afin de quantifier les gisements et rationaliser leur exploitation. C'est dans ce cadre que nous avons étudié les différents paramètres géologiques et géotechniques qui conditionnent les granulats utilisés dans le pays.

CONSTITUTION GEOLOGIQUE DE LA REGION DE BUJUMBURA

La figure 1 présente la configuration géologique de la région de Bujumbura (Nzojibwami 1987). La zone étudiée fait entièrement partie de la branche occidentale du rift est-africain . Le fond du graben est occupé par le lac Tanganyika qui se prolonge au nord par la grande plaine de la

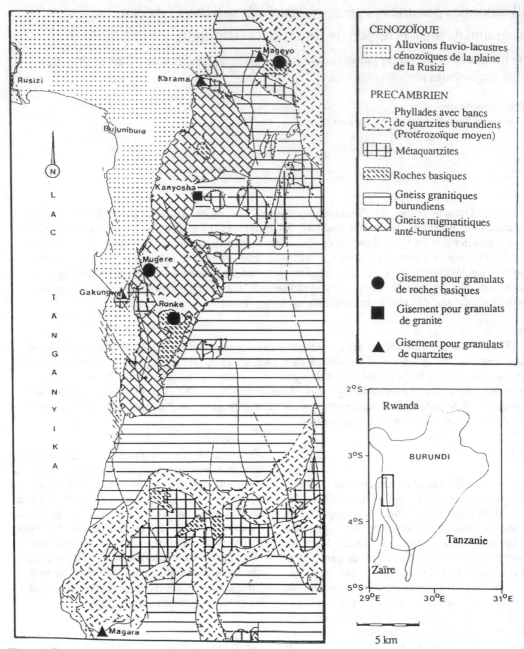

Figure 1: Carte géologique de la région de Bujumbura (Nzojibwami, 1987)

Rusizi, formée d'une épaisse série de sédiments fluvio-lacustres d'âge cénozoïque. Cette plaine est rapidement limitée à l'Est par des contreforts abrupts faits de roches cristallines précambriennes. Il s'agit d'un grand complexe gneissique qui entoure des ensembles de roches basiques (am-phibolites, dolérites, gabbros à amphiboles, gra-nulites basiques), de quartzites et de micaschi-stes. La région est profondément affectée par des phénomènes tectoniques (failles, schistosité, plissements) et métamorphiques (transforma-tions et réglages minéralogiques). Son architecture structurale est principalement marquée par une tectonique précambrienne de charriage (Theunissen (1988). De nombreuses cassures liées au rift sont également partout présentes (Tanganhydro group, 1992).

Le degré d'altération des roches est très important. Le manteau d'altération atteint souvent plusieurs dizaines de mètres (Nzojibwami et al. 1990a, 1990b). Compte tenu de leur faible altérabilité, les quartzites coiffent les sommets les plus élevés dans le paysage.

SELECTION DES GISEMENTS

La figure 1 montre certains gisements précambriens sélectionnés pour étude. Des gisements de laves basaltiques cénozoïques du Nord-Ouest du pays (figure 2) ont également été considérés dans cette étude. Les gisements ont été choisis sur base de leur exploitabilité en tenant compte des éléments ci-après.

1. La taille du gisement: la carrière doit pouvoir fournir suffisamment de matériaux pendant une durée d'exploitation industrielle d'environ vingt ans.

2. L'allure du gisement: certains gisements sont difficilement exploitables à cause de leur forme en filons verticaux étroits mais fortement allongés. On leur préfère des gisements de dimensions suffisamment importantes dans toutes les directions.

3. L'épaisseur des terrains de recouvrement: un grand nombre de massifs rocheux sont recouverts de plusieurs dizaines de mètres de terrains altérés, ce qui augmente considérablement les coûts d'exploitation.

4. L'environnement: certaines exploitations ne sont pas possibles parce qu'elles sont situées, soit dans des zones à forte densité de population ou à proximité d'infrastructures importantes, soit que les travaux risquent d'entraîner des glissements importants du terrain naturel.

5. L'accessibilité: Certains gisements sont inaccessibles à cause du relief tourmenté de la région.

6. La qualité du massif rocheux: les gisements sélectionnés d'après les cinq critères précédents ont été soumis à une analyse pétrographique approfondie pour déterminer leurs aptitudes pour la confection des granulats. Trois types de roches ont été retenues: les roches basiques, les granites et les quartzites. Cette question est étudiée en détail dans cet article.

METHODES D'ETUDE

Les études géologiques ont permis de délimiter les contours des gisements, d'analyser les joints (espacement, qualité, ouverture) et la stabilité des terrains.

Des observations pétrographiques ont été faites au département des sciences de la Terre de l'Université du Burundi. Elles ont permis de déterminer les facteurs microscopiques qui conditionnent la qualité des granulats. A défaut d'essais technologiques parfois très coûteux, les analyses au microscope polarisant donnent des indications qui permettent une estimation relativement précise du comportement mécanique des roches.

Des essais Los Angeles ont été réalisés au Laboratoire National du Bâtiment et des Travaux Publics (LNBTP) du Burundi. Les analyses pétrographiques ont été interprétées uniquement sur base de cet essai parce que c'est le test le plus courant et souvent le seul réalisé

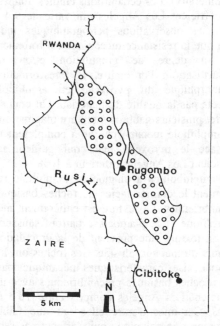

Figure 2: Position des massifs basaltiques du Nord-Ouest du Burundi

Tableau 1: Valeurs indicatives des caractéristiques physiques et mécaniques des roches basiques, granitoïdes et quartzitiques les plus courantes

	gabbro	basalte	granite	gneiss	rhyolite	quartzite
D	3	2.8-2.9	2.6-2-7	2.6-2.75	2.6-2.65	2.65-2.75
Dr	-	-	-	-	-	1280
Rc	13-23	13-45	24	30	34	36
Rt	150	-	170	150	250	280
LA	15	12	25-30	15-30	10-20	20
UM	-	0.7	0.54-1	0.6-0.8	0.80	0.40
CPA	0.4-0.5	0.45	0.5-0.6	0.55-0.62	0.45-0.55	O.5

D: Densité (t/m³) Dr: Dureté (kgfcm²) Rc: Résistance à la compression (kg:/mm²)
Rt: Résistance à la traction (kg/cm²) LA: Coefficient Los Angeles (%) UM: Usure sur meule CPA: Coefficient de polissage accéléré

pour classer les granulats dans le pays. Le tableau 1 donne, selon la littérature, les valeurs indicatives moyennes des caractéristiques physiques et mécaniques des trois types de roches considérées (Tourenq et Fourmaintraux 1971, Vanden Eynde, 1983). Par ailleurs, la plupart des caractéristiques mécaniques peuvent être obtenues sur base de l'analyse pétrographique, en considérant les propriétés mécaniques des minéraux constitutifs. Il faut noter toutefois que ces valeurs peuvent être altérées par différents facteurs qui sont analysés dans la suite de cet article.

LES ROCHES BASIQUES

Les gisements de roches basiques étudiées sont indiqués sur les figures 1 et 2. Trente-cinq échantillons ont été analysés en coupes minces au microscope polarisant et ont subi les tests Los Angeles. Le tableau 2 résume les relations entre le coefficient Los Angeles et la constitution pétrographique de la roche.

Le premier faciès pétrographique est celui des gabbros à texture magmatique ou doléritique. Ces roches sont formées de plagioclase et de pyroxène avec très peu d'amphibole qui résulte d'un début d'altération hydrothermale. Ces roches, les moins affectées par le métamorphisme, présentent les meilleures caractéristiques mécaniques avec des coefficients Los Angeles inférieurs à 10.5%.

Les gabbros à amphibole constituent le faciès le plus fréquent. La roche est faite de plagioclase partiellement saussuritisé (transformation du plagioclase en un minéral argileux, la saussurite) et d'amphibole résultant de l'hydratation complète du pyroxène. Sa texture semi-magmatique semi-métamorphique est dominée par des minéraux partiellement granulés (réduction granulométrique à l'echelle des minéraux). Les échantillons étudiés donnent un coefficient Los Angeles qui varie de 9.4 à 20%. Les observations pétrographiques indiquent que la résistance mécanique est fortement liée au degré de granulation et/ou de saussuritisation. Par contre, la transformation métamorphique du pyroxène en amphibole n'affecte pas la qualité du matériau. On connaît en effet plusieurs gabbros à texture magmatique où l'amphibole métamorphique a complètement remplacé le pyroxène mais qui gardent un coefficient Los Angeles inférieur à 10%.

Les amphibolites à plagioclase-amphibole représentent le dernier faciès de roches basiques précambriennes. La texture est entièrement métamorphique et le plagioclase partout saussuritisé. La texture fine provient de la granulation au cours du métamorphisme. Ces roches ont les plus mauvaises caractéristiques mécaniques parmi les roches basiques précambriennes avec un coefficient Los Angeles compris entre 18.8 et 34.3%. Certaines d'entre elles ont une faible orientation minéralogique qui n'affecte pas sensiblement le coefficient de forme des granulats.

Tableau 2: Relations entre les paramètres pétrographiques et le coefficient Los Angelès dans les roches basiques de la région de Bujumbura.

type de roche	composition minéralogique	texture	coefficient Los Angeles
1. gabbros	-plagioclase -pyroxènes -peu ou pas d'amphibole	-texture magmatique, doléritique -plagioclase clair -pas de granulation -peu ou pas d'altération hydrothermale	8.9 - 10.4
2. gabbros à amphiboles	-plagioclase -amphiboles	-texture magmatique avec minéraux partiellement granulés -plagioclase partiellement saussuritisé	9.4 - 20
3. amphibolites	-plagioclase -amphiboles	-texture métamorphique complètement granulée -plagioclase entièrement saussuritisé -vague orientation métamorphique	18.8 - 34.3
4. basaltes vacuolaires et/ou à olivine	-plagioclase en petites lattes -olivine -matrice très fine -vides vacuolaires	-magmatique volcanique	27 - 45

Les basaltes cénozoïques, vacuolaires et/ou à olivine du Nord-Ouest constituent un faciès distinct des précédents. Leur texture volcanique est formée de fines lattes allongées de plagioclase au milieu d'une matrice fine foncée. L'olivine (10 à 20% de la roche) est partout altérée en iddingsite. La résistance mécanique de ces matériaux est la plus mauvaise de la série de roches basiques avec un coefficient Los Angeles compris entre 27 et 45%.

De ces observations, il découle que les roches basiques fournissent généralement du granulat de très bonne qualité. Toutefois, le métamorphisme hydrothermal contribue à diminuer leur résistance mécanique. Les paramètres pétrographiques qui influencent la valeur du coefficient Los Angeles sont principalement la granulation

la saussuritisation du plagioclase La transformation hydrothermale du pyroxène en amphibole n'influence pas sensiblement la résistance mécanique de la roche. Les roches granulées dont la texture fine provient des effets tectonométamorphiques doivent être distinguées des roches fines à texture doléritique magmatique. Ces dernières ont de très bonnes propriétés mécaniques. L'orientation minéralogique est généralement faible même dans les roches les plus métamorphiques et n'influence donc pas le coefficient de forme.

La résistance mécanique des basaltes est fortement diminuée par les vides vacuolaires et l'altération de l'olivine en iddingsite. L'olivine est un minéral non souhaitable dans les roches à cause de son altération rapide. Les vides vacuolaires constituent des poches de résistance nulle dans la roche. Malgré leur apparente fraîcheur, les basaltes du Nord-Ouest ne conviennent pas pour la confection des granulats.

LES GRANITES ET LES GNEISS

De nombreux et importants massifs granitiques sont connus dans les environs de Bujumbura, mais ces roches sont généralement très altérées (figure 1).

Les trois massifs de granites ont sensiblement les mêmes caractéristiques pétrographiques. Tous sont constitués de cristaux de feldspath potassique, de plagioclase, de quartz, de biotite et de muscovite. Les roches sont généralement grenues et porphyriques. La taille moyenne des grains varie de 150 à 500 microns et celle des gros cristaux dépasse parfois 5 mm. Le quartz et les feldspaths constituent en proportions sub-égales 80% de la roche et les micas occupent les 20% restants.

Les coefficients Los Angeles mesurés dans ces granites varient de 25 à 40%. La dispersion des résultats dépend de la dimension des grains ainsi que du degré de fissuration et d'altération de la roche.

L'altération constitue le problème le plus important de ces granites. En climat tropical humide, le feldspath s'altère rapidement en kaolin. La biotite perd de la potasse interfoliaire et donne des minéraux argileux de type vermiculites. Au Burundi, les granites et les gneiss frais exposés aux conditions atmosphériques par des tracés routiers s'altèrent généralement très vite, souvent en moins de dix ans. C'est principalement pour cette raison que les granites ne sont pratiquement pas utilisés comme granulats dans le pays.

LES QUARTZITES

Six massifs quartzitiques ont été étudiés et 45 échantillons ont été soumis aux analyses pétrographiques et mécaniques. Les quartzites

Figure 3: Texture dentelée et ondulée dans les quartzites (largeur de la photo: 3.5 mm)

Figure 4: Micro-fissures dans un grain de quartz (largeur de la photo: 800 microns)

sont formés de quartz avec quelques paillettes de micas. Ce sont des roches métamorphiques grenues (600 à 800 microns) fortement recristallisées. Le coefficient Los Angeles varie entre 20 et 64%. On note que les roches en provenance du gisement de Magara au sud (figure 1) donnent les meilleures valeurs Los Angeles, inférieures à 30%. L'analyse pétrographique montre que les minéraux de ces roches ont des contours ondulés et dentelés imbriqués (figure 3) les uns dans les autres ce qui leur donne une bonne rugosité qui améliore la résistance mécanique. Les roches dont les minéraux ont des contacts rectilignes et angulaires sont moins résistantes.

Les quartzites de la région de Bujumbura donnent en général des coefficients Los Angeles moyens ou mauvais (supérieurs à 30%), malgré leur apparence de roche dure et fraîche. Les analyses pétrographiques montrent en effet divers éléments qui contribuent à la diminution des caractéristiques mécaniques.

Le principal facteur est la fissuration. En effet, un quartzite soumis à d'importants efforts tectoniques réagit par fissuration à cause de sa grande compétence. Cette fissuration se traduit aussi bien à l'échelle microscopique que par des fractures et des diaclases au niveau de l'affleurement. La figure 4 montre la micro-fissuration d'un grain de quartz suite aux efforts tectoniques qu'a connus la roche au cours de son histoire. Des massifs entiers avec des roches sans défaut apparent sont affectés par ces micro-fissures.

On remarque également que certains contacts entre minéraux sont tapissés d'un fin film de micas ou d'hydroxydes de fer surtout dans les roches les plus déformées par la tectonique, ce qui contribue à la diminution du coefficient Los Angeles.

Enfin, les quartzites présentent une forte abrasivité et usent beaucoup les outils d'exploitation et de concassage. C'est pour toutes ces raisons que ces roches se placent loin derrière les gabbros dans la fabrication des granulats au Burundi.

CONCLUSIONS

Dans la région de Bujumbura, trois types de roches peuvent être exploitées pour la fabrication des granulats: les roches basiques, les granites et les quartzites. L'analyse pétrographi-

que a permis d'analyser les paramètres géologiques qui influencent la résistance mécanique des granulats. Les roches basiques fournissent les meilleurs granulats compte tenu de leurs caractéristiques mécaniques généralement très bonnes. Toutefois, les phénomènes métamorphiques de saussuritisation du plagioclase et de granulation diminuent sensiblement la résistance des gabbros et des amphibolites précambriens. L'olivine et les vides vacuolaires présents dans les basaltes cénozoïques du Nord-Ouest du pays rendent ces roches impropres à la confection des granulats. Les granites sont rarement utilisés comme granulats compte tenu de leur forte et rapide altérabilité dans la région. A défaut de roches basiques, les quartzites présentent quelques fois des alternatives intéressantes. Il faut toutefois noter que ces roches sont très abrasives et que leur résistance mécanique est profon-dément altérée par les micro fissurations tectoni-ques et la nature des contacts entre les minéraux.

REFERENCES

Nzojibwami, E. 1987. Le Précambrien cristallin de la région de Bujumbura (Burundi). Thèse de doctorat, Université de Liège, Belgique.

Nzojibwami, E., Nahimana, L., Sakubu, S. et Sinzumunsi, A.. 1990a. Etude des pavés en pierres naturelles pour le revêtement des routes secondaires de Bujumbura. Rapport du Centre de Recherche Universitaire en Sciences de la Terre (CRUST), Université du Burundi.

Nzojibwami, E., Nahimana, L., Sakubu, S. et Sinzumunsi, A. 1990b. Recherche des matériaux de construction des routes dans les environs de Bujumbura; 1. Matéraux rocheux pour granulats. Rapport du Centre de Recherche Universitaire en Sciences de la Terre (CRUST), Université du Burundi.

Tanganydro group 1992. Sublacustrine hydrothermal seeps in Northern Lake Tanganyika, East African Rift: 1991 Tanganydro expedition. Bulletin des Centres de Recherche et d'exploration-production, Elf- Aquitaine, 16, 1: 55-81.

Theunisssen, K. 1988. Kibaran thrust fold belt (D1-2) and shear belt (D2). IGCP n° 255 Newsletter/Bulletin, 1:55-64.

Tourenq, C. et Fourmaintraux, D. 1971. Contribution de la pétrographie à l'étude des propriétés physiques et mécaniques des roches. Bulletin de liaison des Laboratoires routiers, n° 50.

Vanden Eynde P,. 1983. Caractéristiques géologiques et mécaniques des granulats. Académie royale des Sciences d'Outre-Mer, Mémoires in-8°, Nouvelle série, tome XVIII, fasc. 5, Bruxelles.

Dilational behaviour of crushed stone

Le comportement dilatant d'une roche broyée

Jørgen S. Steenfelt & Niels Foged
Danish Geotechnical Institute, Lyngby, Denmark

ABSTRACT: In a large scale triaxial set-up accommodating circular, cylindrical specimens of 500 by 500 mm size, the behaviour of a hyperitic crushed stone was investigated at the Danish Geotechnical Institute (DGI). The uses of crushed stone material imply large grain sizes. However, due to the inherent particulate nature crushed stone is most often considered to conform to the behaviour of other particulate materials such as sand. Despite very high strength in terms of angle of internal friction the crushed stone, however, showed very limited dilational potential. The implications for design of foundations subjected to high impact forces or dynamic loading where crushed stone is an integral part of the load transfer system are discussed in light of the test results and literature findings. The experience gained may prove valuable in analyses of scree deposits and rock fill properties in general by engineering geologists and geotechnical engineers.

RESUMÉ: Dans une mise en place triaxiale faite pour des échantillons de 500x500 mm de taille, les réactions d'une roche hyperitique broyée ont été examinées à l'Institut Danois de Géotechnique. L'utilisation de matériels de roches broyés est plus souvent considerée être conforme à la réaction d'autres matériels particuliers comme le sable. Malgré une force très haute en termes d'angle de frottement interne, la roche broyée a montré un potentiel de dilatation très limité. L'implication du design des fondations mis à l'epreuve de hautes forces d'impact ou des charges dynamiques où la roche broyée est une part inteégrale du système de distri butions des charges, est discuté à la lumière des résultats de tests et par les apports trouvés dans la literature. L'expérience obtenue peut se montrer de grande valeur dans les analyses de dépots d'éboulis de roches et des proprietés de remblais de roches en géneral fait par des ingénieurs-géologue et des géotechniciens.

1. INTRODUCTION

Crushed stone material is widely used for construction work, notably in dams, land reclamation and as bases for foundation constructions in shallow water.

The engineering properties of the crushed stone depend on the properties of the parent rock, the quarrying method and the resulting grain shape and size. For most of the applications rather large grain sizes are preferred from a cost benefit point of view. The largest grain sizes are encountered for dams and breakwater constructions, but even for foundation subbases mean grain sizes of $d_{50} = 10$—50 mm are common.

The mere size of the particles excludes the use of standard laboratory testing methods for acceptance and verification testing of the material. However, as the crushed stone material is a particulate media, the idea to test down scaled material suggests itself. This is akin to suggesting that the properties of the crushed stone may be compared to the properties of other particulate media as sand, ie as a frictional material.

Based on a comprehensive literature study of the properties of rockfill, it was, however, concluded that inference of crushed stone properties directly from scaled down testing might lead to erroneous conclusions.

The overall strength properties may be correct, but the implications are that the dilation potential for very coarse grained materials may be strongly reduced.

In order to furnish a design tool for verification and acceptance testing of crushed stone material for a particular foundation project, it was therefore decided to design a large scale triaxial test set-up at the Danish Geotechnical Institute (DGI). Basically, the set-up is an upscaled, more rugged version of the Danish triaxial apparatus (cf Jacobsen, 1970) with cylindrical specimens with a height/diameter ratio of 1 and smooth end platens. The new triaxial cell accommodates cylindrical specimens with 500 mm diameter.

In the paper the test set-up is briefly described and the test results for a quarried hyperite are presented and discussed in the light of available findings in the literature.

2. CRUSHED STONE MATERIAL

The crushed stone in question is a hyperite quarried at Kragerø in Norway. Hyperite is a type of gabbro with elongated, plagioclas mineral grains in a matrix of pyroxen with minor constituents of horblende, olivine, quartz and biotite.

The hyperite is greyish black with a high grain density (3.12 Mg/m^3) and a high compression strength, σ_c = 150–200 MPa, with a mean value of 170 MPa.

2.1 Grain size

Judged from the grain size curves for compacted and "virgin" material (Figure 1), compaction reduces the grain size. Thus, the material used in triaxial testing was sampled from a trial field compaction site in order to represent as closely as possible the properties of the in situ placed crushed stone.

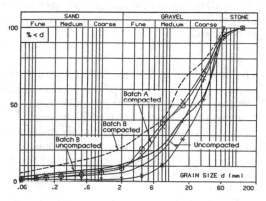

Fig. 1 Grain size distribution for crushed stone

At the field site, and subsequently in the laboratory, a tendency for segregation in the material was noticed. At the surface of a tipped heap, the larger particles segregated and formed a berm around the heap. Furthermore, when the material was worked with

hand tools, the finest particles had a clear tendency to filter through the underlying material.

The triaxial tests were carried out on the two different batches with grain size curves marked, Batch A and Batch B, respectively.

2.2 Grain density

The segregation tendency was also clear from grain density measurements as shown in Table 1.

Table 1 Grain density for crushed hyperite

Fraction size d [mm]	Grain density ρ_s [Mg/m^3]	Fraction part by weight [%]
> 8	3.15	64
2–8	3.09	27
< 2	3.02	9
0–64	3.12	100

2.3 Grain shape

A grain shape analysis of 100 grains in each of the fractions d < 16 mm, 16–32 mm, 32–64 mm and 12 grains > 64 mm showed no clear distinction between fractions. The grains may be characterised as equant to disc shaped as shown in Fig. 2.

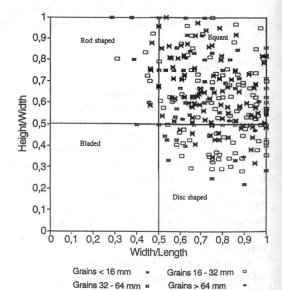

Fig. 2 Grain shape of 312 individual grains

2.4 Maximum and minimum void ratio

In order to characterise the obtainable densities the maximum and minimum densities, and the corresponding void ratios e_{min} and e_{max}, were determined in the laboratory. A 500 mm high, 480 mm diameter rigid steel cylinder was used for the density determinations (served as an oedometer as well). Loose tipping, drop hammer and vibratory hammer compaction were applied.

The values of maximum dry density depends very much on the method of compaction, and hence a high energy vibratory compaction was included to mimic the field compaction. However, the ASTM standard gives rise to lower densities as seen in Table 2.

Table 2 Laboratory determination of maximum and minimum densities and void ratios

Compaction or place-ment method	Dry density ρ_d [Mg/m^3]	Void ratio e
no compaction, loose tipping for e_{max}	1.82	0.72
low energy drop hammer under water	2.07	0.51
low energy drop hammer	2.14	0.46
medium energy vibrator hammer	2.24	0.40
high energy vibrator hammer	2.33	0.34
ASTM method* e_{max}	(1.83)	(0.71)
ASTM method* e_{min}	(2.15)	(0.45)

*Determined from another batch of Hyperite with a slightly larger mean grain size

2.5 Angle of repose

The angle of repose of loosely tipped, dry crushed stone material was determined to provide a rough estimate of the critical state angle φ_{crit}. Due to separation, where the larger particles tend to "run off", the angle cannot be measured with high accuracy. A best estimate from four measurements is $\varphi_{repose} = 40°$.

At the field site, however, angles up to 90° were observed for the compacted material. The slope angle deteriorated to 40-45° when a single larger particle was removed.

3. TRIAXIAL TEST SET-UP

Based on the specifications for the particular batch of crushed Hyperite to be tested and general handling and cost considerations a specimen size of 500 mm diameter was chosen for the triaxial cell.

According to the literature recommendations a specimen diameter some 6 to 10 times the size of the largest grain ensures that there is negligible influence on the test results from the specimen size. This criterion was met, as the test specifications required particles > 64 mm to be removed (maximum grain size 90 mm, with less than 5% by weight > 64 mm).

Figure 3 shows a schematic cross section of the triaxial test set-up. The test chamber is a rigid steel cylinder with a diameter of 800 mm allowing cell pressures up to 0.5 MPa and vertical stress up to 1 MPa (In 1994 an improved, more versatile Mark II version is completed allowing cyclic loading at 1 MPa cell pressure and axial stresses up to 5 MPa). The vertical stress is afforded by a hydraulic jack acting on a force transducer inside the cell on top of the upper end platen. Both the vertical load and the cell pressure are computer controlled through GDS pressure actuators.

The weight of the expelled pore water is measured by a force transducer allowing the volumetric strain to be calculated. The vertical displacement of the top platen is measured directly by an LVDT.

The overall control of the test, including pre-programming of test phases, automatic data acquisition, display and printing, is facilitated by an HP 332 computer. The control software is a slightly modified version of the general-purpose laboratory test software CUBIC, developed by DGI.

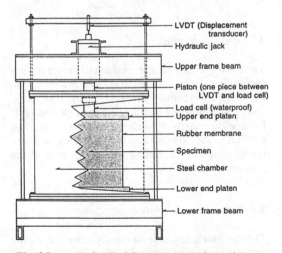

Fig. 3 Large-scale triaxial test set-up (schematic)

3.1 Specimen preparation

The specimen is enclosed in a cast rubber membrane, which widens from the 500 mm specimen diameter to a 600 mm collar at both top and bottom. The 3 mm thick membrane is manufactured in one piece at DGI using a special membrane casting form and a three component Ureol-Polyurethane resin.

This allows the specimen to dilate without loss of full support along the smooth end platens (stainless steel with silicone grease and a sheet of rubber membrane). On the inside the membrane is coated with thin teflon sheeting on silicone grease in order to protect the membrane from the very sharp corners of the crushed stone and to prevent friction between the membrane and the crushed stone.

The crushed stone specimen may be placed (with a pre-determined void ratio) by various methods:
- placed by hand
- compacted in 100 mm layers using a 15 kg drop hammer
- compacted/vibrated in layers using an electric high energy hammer.

In order to facilitate the high energy compaction the specimen is prepared within the cell using a purpose-built compaction mould as seen in Figure 4. The mould splits into six individual parts allowing disassembly after compaction where the specimen is supported by suction between the end platens and the membrane as shown in Figure 5.

The extremely sharp edges of the crushed stone caused minor leaks in the membrane despite the measures taken with the Teflon sheeting. However, in all cases sufficient suction could be applied to allow removal of the mould and subsequent sealing of the leaks.

3.2 Loading procedure

The triaxial tests were executed at constant cell pressure with constant strain axial loading to failure or to a maximum vertical stress of $\sigma_1 = 1000$ kPa. Unloading reloading cycles with a deviator stress difference $\Delta q = \Delta(\sigma_1 - \sigma_3) \doteq 200$ kPa were included to evaluate the deformation properties and to compensate for the limited volume capacity of the pressure actuator.

3.3 Evaluation of test data

The deviator stress is continuously corrected for the changes in cross-sectional area assuming a cylindrical

specimen. Measurements after the tests showed near-cylindrical specimens even after 10–15% vertical strain.

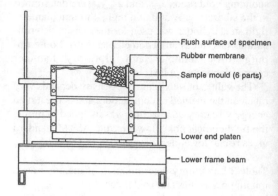

Fig. 4 Preparation of specimen in special mould

Fig. 5 Disassembly of compaction mould

Influence from membrane compliance was disregarded, as the tests were conducted at constant cell pressure, and as visual inspection indicated little change in membrane indentation from start to end of testing.

4. LABORATORY TEST RESULTS

4.1 Consolidation tests

For the purpose of evaluating the one-dimensional stiffness properties of the crushed stone (relevant for working stress conditions under the centre of a foundation) two consolidation tests were carried out in the steel cylinder used for the density determinations.

The loading system of the triaxial set-up was used.

To reduce friction the inside of the cylinder was coated by thin Teflon sheeting over silicone grease. Furthermore, specimen heights of 470 and 200 mm were chosen to elucidate the influence from possible arching or friction.. There was, however, no detectable difference in the stress-strain relations inbetween the two tests. Hence, the friction reduction was considered adequate.

The consolidation curve for the 200 mm specimen is shown in Figure 6.

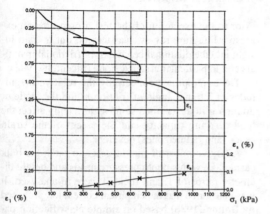

Fig. 6 Consolidation test data. The stress-strain curve includes unloading reloading cycles and creep phases

The stiffness properties were evaluated by primary loading to increasingly higher vertical stress levels, followed by a period of sustained loading (creep). At each stress level unloading/reloading cycles were carried out (normally two cycles at each stress level).

The secant stiffness for vertical loading up to $\sigma_l \approx$ 500 kPa was of the order $E_s = 110\text{--}120$ MPa, dropping to some 70–80 MPa for $500 < \sigma_l < 1000$ kPa.

The unloading reloading moduli were in excess of 1000 MPa. For vertical stresses up to $\sigma_1 \approx 1$ MPa the creep under sustained loading is proportional to the stress level with $\varepsilon_s \approx 0.1\%$ per log cycle of time at $\sigma_l = 1$ MPa.

4.2 TRIAXIAL TESTS

The triaxial specimens were prepared as described for the consolidation tests. However, to avoid membrane damage, the high energy compaction was not applied. Hence, the minimum void ratio obtained was $e = 0.4$. All specimens were fully saturated.

The tests were carried out as drained triaxial tests using a constant strain rate of 4% per hour, with a few unloading-reloading cycles. Failure, corresponding to maximum stress ratio, σ_l/σ_3, (equivalent to maximum deviator stress, q), was reached after 8-18% vertical strain.

The test results are summarised in Table 3 for both failure and characteristic states.

Table 3 Summary of test results at failure for drained triaxial tests on Kragerø hyperite

Test No.	Void ratio	Failure state						Characteristic state, $D = 0$				
		Cell pressure	Axial stress	Axial strain	Volume strain	Dilation rate	Angle of friction	Cell pressure	Axial stress	Axial strain	Volume strain	Character-istic angle
	e	σ'_{3f}	σ'_{1f}	ε_{1f}	$\varepsilon_{v,f}$	$D = -\dfrac{\partial \varepsilon_v}{\partial \varepsilon_1}$	φ'_{max}	σ'_3	σ'_1	ε_1	ε_v	φ'_{char}
		[kPa]	[kPa]	[kPa]			[°]	[kPa]	[kPa]			[°]
1	0.42	80	881	13.7	-0.8	0.24	56.5	80	605	3	0.8	50.0
2*	0.41	120	979	10.0	1.3	0	51.4	120	905	6	1.4	49.9
3	0.45	941	561	13.9	-2.3	0.31	59.7	38	238	1	0.2	46.3
4	0.40	41	685	9.2	-2.7	0.49	62.6	41	291	1	0.1	49.0
5	0.52	79	787	18.3	0.7	0.17	54.8	80	600	6.5	1.8	49.9
6	0.54	80	752	14.8	2.2	0.05	53.8	80	670	10	2.3	51.8
D0*	0.57	160	1072	10.7	2.9	-0.06	47.8	160	1052	11	2.9	47.4
D1	0.54	80	660	15.2	3.7	0	51.6	80	630	15	3.7	50.8
D2	0.44	80	706	18.3	0.6	0.11	52.8	79	600	8	1.3	50.0

Tests 1, 2, 3, 4, 5, 6 and D0, D1, D2 are carried out on two different batches of Hyperite with grain size curves marked Batch A and Batch B, respectively on Figure 1.
*vertical stress limit reached before fully developed failure

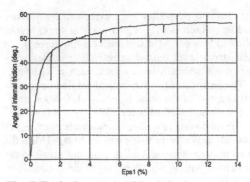

Fig. 7 Typical example of load-displacement curve, showing mobilised secant angle of friction $\varphi_{s,mob}$ versus axial strain ε_1

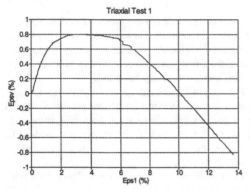

Fig. 8 Volume strain ε_v versus axial strain ε_1 for the test in Figure 7.

Irrespective of the dilation rate, $D = -\partial\varepsilon_v / \partial\varepsilon_1$, the strength mobilisation in the tests was as shown in Figure 7. There was no distinct peak and no reduction in strength towards a residual value or a critical state.

This is typical of tests with homogeneous strain conditions, which may be achieved in tests with smooth pressure heads and $H/D = 1$. No localization took place in the specimens, which remained cylindrical even at large vertical strain.

Hence, data for the characteristic state, marking the transition from compaction to dilation where the dilation rate is zero ($D = 0$), are included to evaluate the critical state behaviour. A large number of Danish triaxial tests indicate that the value of φ_{crit} to be entered in (1) is equal to φ_{char}, the mobilised angle of friction at the characteristic state.

Due to the coarse grained nature of the crushed stone, values of stresses are slightly fluctuating. Failure values are based on average values, typically corresponding to a standard variation of 0.1° of resulting secant angles of friction.

For all tests the rate of dilatancy, D, was constant during the failure phase as shown in Figure 8, even for tests subjected to 18% vertical strain.

The (initial) tangent stiffness E_t was of the order 13–50 MPa. The values compare well with the stiffness inferred from the oedometer tests, when the inevitable effect of initial bedding and the effects from Poisson's ratio, v are considered. The unloading/reloading stiffnesses are of the order 120–200 MPa, ie significantly smaller than found in the oedometer.

5. LITERATURE REVIEW

Prior to the execution of the test programme an extensive literature review was carried out to gain insight into the dilational behaviour of frictional materials and in particular crushed stone (Steenfelt, 1992).

The assumptions concerning the relative magnitudes of internal angle of friction φ and the angle of dilation ψ in the crushed stone significantly affect the load carrying capacity and the extent of the strain fields (and failure zones) developed.

A large number of triaxial tests on sand, the bulk carried out with smooth pressure heads and height diameter ratios of 1, and rockfill were re-analysed. It was concluded that the empirical correlation proposed by Bolton (1986) based on simple classification parameters and representative mean stress level, p' (in kPa), provided an excellent fit for the secant angle of friction, φ'_{max}

$$\varphi'_{max} - \varphi'_{crit} = 0.8\psi_{max}$$
$$= 3I_R° = 3° \left[I_D(10 - \ln p') - 1\right] \tag{1}$$

The fit of the angle of dilatancy ψ is more difficult as the angle only has a direct physical manifestation for sliding under plane strain conditions. Bolton's approximation in terms of the rate of dilatancy $D = -\partial\varepsilon_v / \partial\varepsilon_1$:

$$\left(-\frac{\partial\varepsilon_v}{\partial\varepsilon_1}\right)_{max} = 0.3°I_R \Rightarrow$$
$$\psi_{max} = 12.5° \left(-\frac{\partial\varepsilon_v}{\partial\varepsilon_1}\right)_{max} = 12.5°D \tag{2}$$

was found to underestimate ψ_{max} and hence φ'_{max} in equation (1).

Conversely, the general expression (3) valid for both plane strain and conditions of axial symmetry, usually applied for the calculation of the angle of dilatancy overestimated ψ_{max} and hence φ'_{max} in (1).

$$\sin \psi = \frac{\dfrac{\partial \varepsilon_v}{\partial \varepsilon_1}}{\dfrac{\partial \varepsilon_v}{\partial \varepsilon_1} - 2} = \frac{D}{D+2} \qquad (3)$$

With good approximation the expression for ψ_{max} in (3) may, for the range $0<\psi<20°$, be written as:

$$\psi_{max} \approx 22.5° \left(-\frac{\partial \varepsilon_v}{\partial \varepsilon_1} \right)_{max} = 22.5° D \qquad (4)$$

In the review it was recommended that the value of ψ derived from (1) be used in numerical calculations.

6. DISCUSSION OF TEST RESULTS

It appears from Table 3 that very high secant angles of friction were obtained with the Kragerø Hyperite. The empirical formula (1), with the critical angle φ_{crit} = 40° corresponding to the angle of repose (see Sec 2.5), significantly underestimates the angle of friction.

However, if a value of $\varphi_{crit} = \varphi_{char} \approx 50°$ (cf Table 3) is adopted a prediction within ±2° is achieved.

This still corresponds to very low values of the angle of dilatancy considering the high friction angles and the dense specimens. Moreover, as may be seen from Table 3, the transition from compression to dilatancy (the characteristic state) requires considerable axial strain. At the characteristic state it was found that $\varepsilon_v \approx 0.25 \, \varepsilon_1$ for all nine tests.

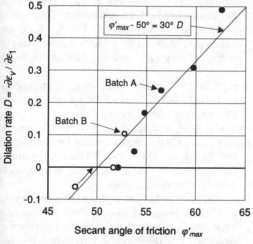

Fig. 9 Rate of dilatancy, D, at failure versus measured secant angle of friction, φ'_{max}

Only at high degrees of strength mobilisation a constant rate of dilatancy is achieved (cf Figure 8).

As shown in Figure 9 the measured secant angles of friction may be expressed by the approximation:

$$\varphi'_{max} - 50° \approx 30° \left(-\frac{\partial \varepsilon_v}{\partial \varepsilon_1} \right)_{max} = 30° D \qquad (5)$$

The recorded values of D are thus about 1/3 of the values anticipated using the expressions (1)–(2) and 3/4 of the values derived from (3). This is believed to reflect the properties of the crushed stone

Thus, it is concluded that the dilation potential of the crushed stone is significantly less than found for similar materials with smaller grain sizes. Furthermore, it requires high strain levels to mobilise both characteristic states and failure states.

6.1 Validation of findings

To validate the findings from the triaxial tests a series of tests was carried out on a standard test sand (Lund 1 with d_{50}=0.52 mm) in the large scale set-up.

The test results were compared with existing triaxial test data with 70 and 200 mm diameter specimens (smooth pressure heads and height /diameter ratio 1 as in the present tests) reported by Steenfelt (1992). The angles of friction as well as the rates of dilation were as previously determined by the small scale triaxial tests on the same sand. Thus, it was concluded that the high friction angles and low dilation ratios were attributable to the material and not to the test set-up.

From tests on sands with the same grain shape and mineralogy Steenfelt (1992) concluded that increasing mean grain size led to an increase in the critical angle of friction, φ_{crit}. Furthermore, it was found that the angle increased with increasing angularity. This tied in with a high value of φ_{crit} for the rockfill re-analyzed, and seems thus, to be confirmed with the findings for the present rockfill.

7. IMPLICATIONS IN DESIGN

By comparison with other available data on rockfill (f.inst Leps, 1970; Charles and Watts, 1980 and Indaratna et al 1993) the strength recorded for the Kragerø hyperite is in the upper range, but comparable to high strength basalt rockfill data.

However, most of the data are from specimens with a diameter < 0.3 m and with a height/diameter ratio of 2. This may explain the differences in observed dilational behaviour, where the reported values of the rate of dilation are higher than in the present investigation.

More importantly, the present investigation indicates that a significant vertical strain is required to initiate dilation.

This may have important implications in design. Over a large range of strength mobilisation, particularly for poorly compacted crushed stone, the material contracts. Thus, impact and cyclic loading may lead to significant pore pressures and hence reduction in effective stresses and reduced safety or even failure.

8. GEOLOGICAL IMPLICATIONS

In nature, the transport and deposition of the largest particles are controlled by friction along the boundary between the transporting agent, ice, water or air and the ground surface.

Deposits of scree or talus in mountaneous areas are potentially very labile, sedimentary deposits often placed at high slope angles.

The triggering mechanisms for surface sliding or deep-seated failures may be erosion at the foot of the talus, impact forces, falling debris, rising water level or seismic actions (Anderson and Richards, 1987).

The literature shows a large number of slides in scree and talus. In combination with development of high pore pressures in shear and intermixing with soil fines, such slides may develop to catastrophic avalanches or "Sturzstrom" (Hsü, 1978).

Man-made constructions, which interfere with such labile deposits, as in reservoirs or open cut mining, have also caused very extensive slides (Pariseau and Voight, 1979).

As most stability problems are principally concerned with the strength of the crushed rock material at low stress levels, it is important that this strength be measured together with the dilatancy potential.

Without proper quantification of the strength as a function of stress level it may be impossible to decide whether a high slope angle signals a competent material or an imminent slide.

9. CONCLUSIONS

There is a natural need for interplay between engineering geologists and geotechnical engineers in assessment of natural and man-made hazards.

The present paper illustrates one of the tools of the engineer in the interplay.

The large scale triaxial tests furnished valuable, qualitative and quantitative data for the evaluation of the strength and dilational behaviour of the crushed stone material.

With the data the design implications for various applications of the crushed stone may be assessed. This reduces the risks and allows economic optimation for man-made structures on crushed stone in particular and rockfill in general.

For engineering geological evaluation of the stability of natural slopes the large scale triaxial experience provides valuable information for quantification of the internal angle of friction and dilatancy potential.

10. ACKNOWLEDGEMENTS

The authors gratefully acknowledge financial support from the Danish Geotechnical Institute to perform the investigation and permission to publish the paper.

11. REFERENCES

Anderson, M.G., Richards, K.S. editors (1987). *Slope stability. Geotechnical Engineering and Geomorphology.* Chichester: John Wiley & Sons

Charles, J.A., Watts, K.S. (1980). The influence of confining pressure on the shear strength of compacted rockfill. Géotechnique 30, No. 4, 353-367

Hsü, K. J. (1978). Albert Heim: Observations on Landslides and Relevance to Modern Interpretations. *Developments in Geotechnical Engineering 14A.* 71-93. Amsterdam: Elsevier.

Indraratna, B., Wijewardena, L.S.S., Balusubramaniam, A.S. (1993). Large-scale triaxial testing of greywacke rockfill. Géotechnique 43 No. 1, 37-51.

Leps, T.M. (1970). Review of shearing strength of rockfill. *J. Soil Mech. Fdn Engng Div. Am. Soc Civ. Engrs* 96, SM4, 1159-1170.

Pariseau, W.G, Voight, B. (1979). Rockslides and Avalanches: Basic Principles and Perspectives in the Realm of Civil and Mining Operations *Development in Geotechnical Engineering 14B* 1-92. Amsterdam: Elsevier.

Steenfelt, J.S. (1992). Strength and dilatancy revisited. Miscellaneous Papers in Civil Engineering *35th Anniversary of the Danish Engineering Academy*, 155-186. Lyngby: Polyteknisk Forlag.

Geotechnical characteristics of some carbonate rocks from the Calabar Flank, southeastern Nigeria: Implications as construction materials

Caractéristiques géotechniques de quelques roches carbonatées du Calabar (Nigéria): Implications comme matériaux de construction

A. E. Edet & N. U. Essien
Department of Geology, University of Calabar, Nigeria

ABSTRACT: Carbonate rocks in adequate quantity and quality occur in the Calabar Flank (Nigeria). However, due generally to the susceptibility of most carbonate rocks to solution effect, these materials are not being harnessed for building purposes. In this paper some geotechnical characteristics of some samples have been evaluated through some tests (mineralogical, chemical, physical and mechanical) in order to access their suitability or otherwise as building stones. Results show that on the basis of a quali- tative scale two litho facies stromatolitic boundstone and bioclastic packstone are very suitable for construction purposes. The others which include bioclastic grainstone, sandy bioclastic grainstone, bioclastic pelloidal packstone and silty pelloidal mud- stone are fairly suitable as building construction materials.

RÉSUMÉ: Les roches carbonatés se trouvent en quantité et qualité suffisantes au Flanc de Calabar au sud-est du Nigéria. Cependant, en raison de la susceptibilité de la plupart des roches carbonates a l'effet de solution, ces matériaux sont exploités pour des fins de construction. Dans cet article, des caractéristiques géotechniques, de quelques échantillons ont été évaluées à travers examens (physiques, mécaniques, minéralogiques et chimiques) a fin d'évaluer à quel point ils sont propices, comme matériaux de construction. Les résultats attestent qu'à base d'une balance qualitative, deux lithofacies-pierres stromatolitiques et pierres bioclastiques-sont très appropriés comme matériaux de construction. Les autres dont la pierre sont fort bonnes pour la construction.

1 INTRODUCTION

The Calabar Flank (Fig 1) is a major sedi- mentary/tectonic basin with potentials for petroleum accummulation. As a result of this, it is expected that in the next few years oil prospecting companies will start the exploration and exploitation for hydro- carbon in the area. Exploration for oil in this basin is also necessitated by the fact that, the major hydrocarbon field in Nigeria (Niger Delta) is fast depleting.

For proper operation of these Petroleum companies, certain infrastructures (roads, houses etc) have to be put in place. It is on the basis of this that some carbonate rock samples are being qualitatively eva- luated as a preliminary measure for future use as building stones within the area. This is because carbonate rocks form about 50% of the entire rocks that make up the basin. This studies also form an aspect of on going research on the suitability of Nigerian carbonate rocks for construction purposes (see Teme 1991, Edet 1992)

2 GEOLOGICAL SETTING

The Calabar Flank essentially is a faulted continental margin basin (Edet & Nyong 1993). It consists of NW-SE trending crustal blocks of graben and horst structures, the Ikang Trough and Ituk High respectively. The Ikang Trough was an intracratonic mobile depression which accumulated mostly shales while the Ituk High, a relatively stable submarine platform, received predominantly limestones (Murat 1972; Petters 1982; Reijers and Petters 1987).

The stratigraphic sequence consist of conglomerates, sandstones, limestones, calcacerous sandstones, shales and marls ranging in age from ?Aptian to Maastrichtian.

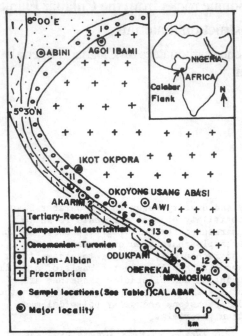

Fig. 1 Sample locations (Calabar Flank).

3 SAMPLE LOCALITIES AND ANALYSES

Samples were collected from localities as
shown in Fig 1. The tests/analyses carried
out include petrographical, mineralogical
and chemical and physio-mechanical (density,
porosity, uniaxial compressive strength).
In all cases the relevant International
standards were applied during the tests
and analyses.

4 LITHO-ENGINEERING DESCRIPTION AND PETROGRAPHY

4.1 Litho-engineering description

The carbonate rocks are collectively described as white to dark grey, fine to coarse
grained, massive to bedded, generally
closely to largely spaced joints but in
places the rock is fractured, moderately
weathered to fresh, pure, LIMESTONE, strong
to moderately strong.

4.2 Petrography

Petrographical studies indicate that the
carbonate rocks are mainly bioclastic
grainstone, (1,4,6,11,13), sandy bioclastic
grainstone(2,3), bioclastic pelloidal
packstone(5), silty pelloidal mudstone
(7,9,10,14), stromatolitic boundstone (8)
and bioclastic packstone. These are presented in Table 1

Table 1. Rock types, locations/numbers and
designations

Rock type	Locations/ numbers[a]	Desig- nation
Bioclastic grainstone	1,4,6,11,13	A
Sandy bioclastic grainstone	2,3	B
Bioclastic pelloidal packstone	5	C
Silty pelloidal mudstone	7,9,10,14	D
Bioclastic packstone	12	F

[a]See Fig 1

5 MINERALOGY AND CHEMICAL ANALYSES

5.1 Mineralogy

X-ray diffraction (XRD) analyses show that
calcite and quartz are the major minerals
that make up these rocks (Tabel 2). Others
include clay minerals (Table 3), dolomite,
microcline and muscovite.

Table 2 Calcite and quartz content of the
rocks

Designation(s)	Calcite %	Quartz %
A	60.7	6.8
B	86.7	4.9
C	83.7	13.9
D	52.1	13.4
E	56.1	1.0
F	56.7	2.2

From the above table (2), the silty
pelloidal packstone (D) contain relatively
low calcite and the sandy bioclastic grainstone (B) the highest. In addition, the
C (bioclastic pelloidal packstone) and D
contain the highest percentage of quartz.
Materials with high quartz content are
expected to be of higher quality interms
of strength.

This is because quartz has a retatively
high element-oxygen bond strength (2.4)
and a hardness of 7 compared to calcite
with a bond strength of 0.7 and hardness o
3.

Clay minerals consist mostly of kaolinit

illite and smectite as presented in Table 3.

Table 3 Clay mineral contents of the rocks

Designations	Kaolinite %	Illite %	Smectite %
A	50	30	5
B	33	50	0
C	50	30	0
D	50	28	10
E	85	5	5
F	60	5	30

From the above table (3), apart from B (sandy bioclastic grainstone) with 33% kaolinite and 50% illite, the rest (A,C, D,E,F) can serve as good construction materials by virture of the fact that they contain up 50% kaolinite. Kaolinites are known for their strong hydrogen bonding, little or no absorption of water molecules, low cation exchange capacity and low Atterberg limit values compared to the smectites and illites.

6.2 Chemical composition

Some chemical compositions of the carbonate rocks are shown in Tabel 4 below.

Table 4 Some chemical composition of the rocks

Designations	CaO %	SiO_2 %	$CaCo_3$ %
A	47.6	8.4	85.0
B	49.7	7.1	88.7
C	55.3	30.6	98.8
D	38.3	21.6	68.4
E	54.4	0.9	54.4
F	52.3	4.1	52.2

CaO ranges between 38.3 to 55.3 percent and constitute the dominant of the carbonate rocks. This assertion is supported by result of the calcimetric analysis which indicate a $CaCo_3$ content of 52.2 to 98.8% SiO_2 is relatively low in some suites. These properties of the rocks reflect its high degree of purity. Samples with high CaO is good for the production of lime which can be used as stabilizing agent.

6 PHYSIO-MECHANICAL PROPERTIES

6.1 Density

The mean density range between 2.34 to 2.83 Mg/m^3. These values are described as medium-high density. Variations could be attributed to cementing materials and fossil content.

6.2 Porosity

The mean porosity values are in the range 2.11 to 7.92%. Generally these are described as low to medium. Porosity has an inverse relationship with strength and hence durability.

6.3 Uniaxial compressive strength

Average uniaxial compressive strength values are between 20.73 and 37.62 MPa. These materials are described as moderately strong to strong. The physio-mechanical data are presented in Table 5 below.

Tabel 5 Some physio-mechanical data

Designations	Density Mg/m^3	Porosity %	Strength MPa
A	2.63	6.12	37.62
B	2.55	2.11	27.73
C	2.57	7.92	25.70
D	2.34	2.28	20.73
E	2.57	2.38	25.70
F	2.83	8.04	28.30

7 POTENTIAL AS BUILDING STONES

From the above sections, on the basis of the mineralogy-chemical-physical-mechanical characteristics, it is clear that there are certain lithofocies which can serve as good building stones. This observation is based on the contents of calcite, quartz, kaolinite illite, smectite, $CaCO_3$ and the values of density, porosity and uniaxial compressive strength. Using these parameters a rating scheme developed qualitatively for the carbonate rocks is presented on Table 6.

Table 6 Potential of some Calabar Flank carbonate rocks as building stones

Designations	1	2	3	4	5	6	7	8	9
A		M	L	M	M	M	H	M	M	M
B		H	L	M	M	L	H	M	H	M
C		H	M	M	M	L	H	M	M	M
D		M	M	M	M	M	M	M	H	L
E		M	L	H	L	M	M	M	H	M
F		M	L	H	L	H	M	H	M	M

1(Calcite) 2(Quartz) 3(Kaolite)
4(illite) 5(smectite) 6($CaCO_3$)
7(density) 8(porosity) 9(uniaxial compressive strength) L(low) M(Medium) H(High)

Carbonate rocks in suites A,B and C are fairly suitable while those of suites D E and F are described as suitable. Rocks of these groups (D,E,F) have been found to contain low to moderayely high calcite, quartz, kaolinite, illite, smectite and medium to high density and strength. Minor variations are due to non homogenuity of geologic materials. Generally, medium to high strength properties are essential for the suitability of these limestones as construction materials. Moderate quantities of $CaCO_3$ and clay are also required as binding materials. Weathering of these rocks may be accelerated under the conditions of humid tropical conditions where the study locality is located. This can be minimised by providing proper drainage and constructions carried out immediately as soon as excavations have been made.

8 CONCLUSIONS

The carbonate rocks exhibit satisfactory physical and mechanical properties, high quality in terms of its mineral and chemical composition and high content of non swelling clay type. All these combine to give the rocks the qualities of good construction materials. These rocks are however likely to undergo rapid disintegration when climatic (humid tropical) factor is taken into consideration.

ACKNOWLEDGEMENT

The German Academic Exchange Service (DAAD) for providing funds during period of analyses in Tuebingen, Germany.

REFERENCES

Edet, A., 1992. Physical properties and indirect estimation of microfractures using Nigerian carbonate rocks as examples Eng. Geol., 33: 71-80.

Edet, J.J. and Nyong, E.E., 1993. Depositional environments, sea level history and paleobiogeography of the late campania Maastrichtian on the Calabar Flank, SE Nigeria. Palaeogogr., Palaeoclimatol., Palaeoecol., 102: 161-175.

Murat, R.C., 1972. Stratigraphy and paleogeography of the cretaceous and lower Tertiary in southern Nigeria. In:T.F.J. Dessauvagie and A.J. Whiteman (Editors), African Geology. Ibadan Univ. Press, Ibadan, pp 251-226.

Petters, S.W., 1982. Central West African Cretaceous-Tertiary benthic foraminifera and Stratigraphy. Paleontolographica B, 179: 1-104.

Reijers, T.J.A. and Petters, S.W. 1987. Depositional environments and diagenesis of Albian carbonates on the Calabar Flank, S.E. Nigeria J Pet. Geol., 10(3): 283-294.

Teme, S.C., 1991. An evaluation of the engineering properties of some Nigerian limestones as construction materials for highway parements. Eng. Geol;, 31:315-326.

Geologic conditioning of aggregates to avoid ASR in concrete: A first evaluation of the Portuguese mainland

Conditionnement de l'utilisation des granulats pour éviter les ras dans le béton: Pré-evaluation du territoire Portugais

Henrique S. Silva
Laboratório Nacional de Engenharia Civil, Lisbon, Portugal

ABSTRACT: Concrete, which can be considered as a artificial stone, may deteriorate with consequences in the operationallity and security of structures. It depends on the mixture quality, namely the aggregate nature (some type of silica forms, like chert, silex, chalcedony, tectonized quartz, etc.), cements and water, and lifeservice factores. That is why a selection of aggregates based on an early evaluation of Country resources can help in avoiding or at least attenuate such concrete illnesses. The alkali-silica reaction (ASR) my be the most important cause of chemical deterioration of concrete. This paper deals with a methodology for evaluate the geological conditioning of aggregate use for concrete based on lithologic zonning of the Country, which was applied to Portuguese geologic resources.

RESUME: L'occurrence de réactions alcali-silice (RAS) au sein du béton contribue á son altération et provoque, généralement, une modification des propriétés structurales des ouvrages. Les granulats constituent la principale source d'éléments réactifs, nommément certains types de silice (le chert, le silex, la calcédoine, le quartz cataclastique, etc.) et, aussi, des minéraux avec une teneur élévée en alcali. Au Portugal Continental il y a des roches dont leur composition comporte ces éléments; ces roches sont souvent exploitées pour produire des granulats. Dans ce communication, ont décrit une méthode pour faire une évaluation globale des conditions lithologiques du pays dans le domaine de cette thématique. La methodologie utilizée et les zonements effectués sont présentés.

1 - INTRODUCTION

Among the main processes of concrete alteration there are included those resulting from the reaction between its components, with formation of expansive products (1,2).

The typical processes in the chemical alteration of concrete with formation of expansive products is based on the alkali-aggregate reactions (alkali-siliceous and alkali-carbonate reactions) and on the sulphate-alumina reactions (2,3).

Obviously, the main conditioning factors are the existence of reactive silica and alkalis, in adequate proportions, and an environment where the water content is higher than 80% and a sufficiently high pH, usually higher than 12 (4, 5, 6). The source of reactive silica is usually the aggregate, and the source of alkalis may be the cement and the additives or also the aggregate. The mixing water and the percolation water may also be a source of alkalis, in the first case with global distribution in the concrete mass and in the second case with a localised incidence, in the vicinity of seepage channels.

Of all the potentially reactive silica forms (Table 1), those of an amorphous, subcrystalline and cryptocrystalline nature, and those that are strongly tectonised, presenting an undulatory extinction, of which quartz can be pointed out, are the ones that usually produce more intense reactions in young concrete. The former are considered as having an assured reactive potentiality with alkalis, the process occurring at short term, as long as the other conditioning factors (pH, water content, temperature) are appropriate. The latter usually cause slow reactions, and their development reaches highest velocities only after tens of years (9,10).

In the Portuguese mainland there are some of the above mentioned potentially reactive silica forms, which make part of the mineralogical composition of rocks that provide aggregates. Moreover, some of these rocks have minerals with a high alkali content (sodium and potassium). In the following paragraphs a global lithological framework is given, taking into consideration these two important factors for the development of alkali-silica reactions in concrete (7, 8). Studies of this nature can be carried out on several scales:

i) on a national territory scale, giving access to a global knowledge, through the preparation of macrozonings;

ii) on a regional scale, introducing regional conditioning factors (sedimentologic, orogenic, tectonic, etc.), the knowledge being consubstantiated in microzonings;

iii) on a local scale, based on experimental criteria (mineralogical and petrographic analysis, alterability, expansibility, reactivity tests, etc.), leading to the characterization of the geologic formations which will be employed in the works.

The zonings presented make a contribution to the global knowledge, on a national scale, of fundamental aspects related with the nature of aggregates used in the manufacturing of concrete, in the light of the possible occurrence of alkali-silica reactions.

Table 1 – Potential sources of reactive silica forms and alkalis from Portuguese geological formations

POTENTIAL SOURCE OF REACTIVE SILICA FORMS	
Minerals	Opal, Opal CT (Cristobalite, Tridimite), Obsidian, Flint, Chert, Chalcedony, tectonized Quartz (TQ)
Rocks	Jasper, Lyddite, Phtanyte, Diatomite, siliceous Schist, Phyllite (with TQ, chert or silex), Quartzite (with TQ, chert or silex), Granitoid (with TQ), Volcanite (Rhyolite, Dacite, Andesite, Basalt), Limestone (with silex and chert), Dolomite (with silex and chert)
POTENTIAL SOURCE OF ALKALIS	
Minerals	Potassium: Sanidine, Orthoclase, Microcline, Leucite, Biotite & Muscovite; Sodium: Albite, Oligoclase, Nepheline, Sodalite
Rocks	Granitic, Sienitic & Dioritic families, feldspathic Hornfels, Leptynite, Arkose, Ectinite (Phyllite & Greywacke)

3370

2 - LITHOLOGIC FRAMEWORK OF THE PORTUGUESE MAINLAND

2.1 - Lithologic chart

In Portugal there are igneous, sedimentary and metamorphic rocks (2, 11). The predominant igneous rocks belong to the family of granites and gabbros (basalts mainly), and occupy about 30% of the Portuguese mainland. The predominant sedimentary rocks are of a carbonate nature (limestones, marly limestones, marls, dolomitic limestones and dolomite), which occur in about 10% of the territory and clastic (especially sandstones, conglomerates, sands and clays), which are distributed over about 20% of the Portuguese mainland. The metamorphic rocks occur in the other 40% of the territory and are predominantly composed of schists, almost always associated with metagreywackes, quartzites, marbles and rocks of a gneissic and amphibolic nature. Fig. 1 presents the simplified lithologic distribution of geologic formations in the Portuguese mainlan.

2.2 - Lithology and supply of potentially reactive silica

An analysis of the Portuguese Geologic Chart, in the scale of 1/500.000, 1/200.000 and 1/50.000 (13, 14) and of specific documents (15), reporting the lithologic and petrographic nature of geologic formations in the Portuguese mainland, made it possible to conclude as follows (2):

i) The above mentioned silica forms, to which a greater reactive potentiality is attributed, such as chert, flint, phtanite (including jasper quartz and jasper) and lyddite associated with palaeozoic formations, namely polimetamorphic, schist-greywacquic and quartzitic complexes; flint especially associated with mesozoic formations and of a carbonate nature, namely jurassic limestones; several silicifications (silcretes included), associated with formations of very different ages from palaeozoic siliceous schists, predominantly, to jurassic limestones with silicificated vegetable fossils and radiolarians and to cenozoic and quaternary arkosic, conglomeratic and sandy deposits.

ii) Other potentially reactive silica forms can be identified as quartz with a disturbed cristalline network, usually metastable, due to deformation and marked shearing, or finely granulated. The quartz with those characteristics is more often found in the metamorphic palaeozoic series (quartzites, leptynites, gneisses and ordovician and silurian granulites) and particularly in palingenetic granites and of early hercinian emplacement. Its reactive potentiality has been identified with the value of the undulatory extinction angle (7, 16, 17). In the most descriptive documents of geologic cartography, however, the values of the measured angles are not included, so that consideration of all the rocks in which the undulatory character of the extinction was recognised may therefore be too condemning.

2.3 - Lithology and the supply of alkalis

Considering the frequence of its occurrence, the minerals that due to their frequency may be source of alkalis are essentially feldspars, plagioclases (albit and oligoclase), feldspathoid and micas. In the Portuguese mainland the rocks with such minerals belong to the families of granites, syenites and ectinites (2). It is therefore essentially the use of igneous and metamorphic rocks as aggregates that may originate high contents of alkalis in the concrete. Some immature alluvial sands, resulting from the alteration of granite, may also contain a high percentage of feldspars, plagioclases and micas and also be an important source of alkalis. Besides, it is a common situation observed in works affected by alkali-silica reactions.

3 - ZONINGS OF LITHOLOGIC DISTRIBUTION

3.1 - Methods used

Zonings of the territory were made by taking into account the availability of outcrop materials for concrete aggregates which may be

3371

a source of the above mentioned potentially reactive silica forms and of alkalis. The origins of the silica forms were divided into two groups: rocks with amorphous, s u b - c r y s t a l l i n e a n d crypto-crystalline silica (chert, silex, phtanite, chalcedony and silicifications); and rocks with crystalline silica (deformed quartz and with cataclasis, recognised by the undulatory character of the extinction).

The analysis was made on the

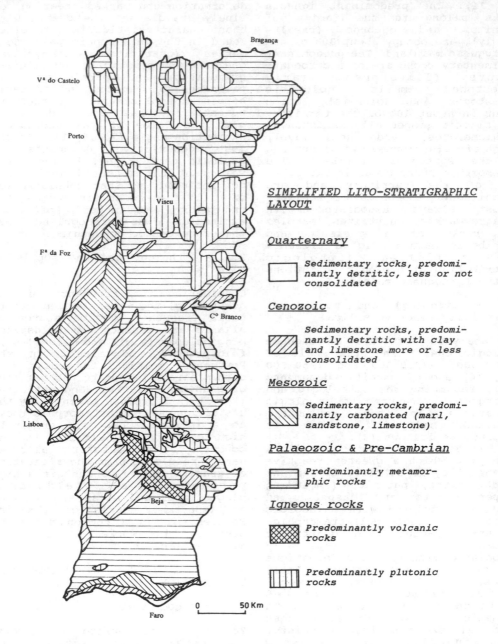

SIMPLIFIED LITO-STRATIGRAPHIC LAYOUT

Quarternary

Sedimentary rocks, predominantly detritic, less or not consolidated

Cenozoic

Sedimentary rocks, predominantly detritic with clay and limestone more or less consolidated

Mesozoic

Sedimentary rocks, predominantly carbonated (marl, sandstone, limestone)

Palaeozoic & Pre-Cambrian

Predominantly metamorphic rocks

Igneous rocks

Predominantly volcanic rocks

Predominantly plutonic rocks

0   50 Km

Fig. 1 - Simplified lithologic map of Portuguese mainland

rectangular mesh corresponding to the Portuguese Geological Chart in the scale of 1/50,000, each rectangle covering a mean of 640 km² (Fig. 2). The percentage of area occupied in each rectangle by the lithologic outcrop units was quantified on the basis of the mean area method (2). On this basis, three zonings were established which can gave an overall view of the Portuguese mainland, two of them concerning the occurrence of silica forms potentially reactive to alkalis and one related with the occurrence of alkali source formations. The identified areas where potentially reactive rocks outcrop were calculated and insert into a data base support. With this instrument it becames very easy in a regional scale to know where a

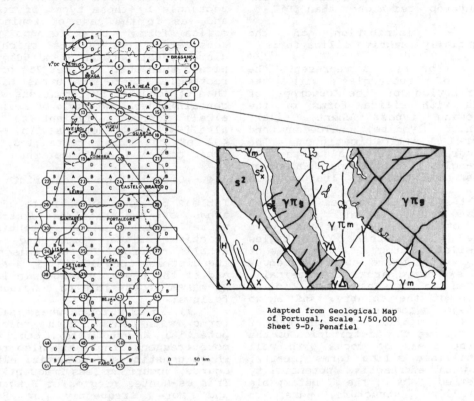

Adapted from Geological Map of Portugal, Scale 1/50,000 Sheet 9-D, Penafiel

▨ –	rocks with potentially reactive forms of silica
H –	conglomerate, arkose and schist
S^2 –	schist and greywacke with graphytic beds and lyddite
S_x^2 –	spotted slate
S_c^2 –	pellitic and psamitic hornfels
O –	schist, greywacke and quartzite
X –	schist and greywacke (flysch)
γm –	two-mica monzonitic granite
$\gamma \pi m$ –	two-mica monzonitic granite (essentially biotitic)
$\gamma \Delta$ –	granodiorite
$\gamma \pi g$ –	two-mica granite porphyry (with ondulatory extinction quartz)

Fig. 2 - Methodology used for evaluation of outcrop areas

3373

claimed quarry or borrow area can be dangerous or not and if further studies on reactivity research must be conducted locally, in a convergent process. Three groups of percentage of area of rectangles occupied by rocks with those species were considered: an outcrop area less than 5%, practically meaningless from a global point of view, which can be ignored; an outcrop area between 5 and 50%, and an outcrop area higher than 50%.

3.2 - Distribution of the potentially reactive silica forms

The map of Fig. 3 represents the mesh of rectangles with the distribution of the outcrops of rocks with silica forms of the following types: chert, silex, phtanite, lyddite, chalcedony and several silicifications. The histogram includes also the 22 rectangles with outcrops that occupy areas less than 5%. Its analysis makes it possible to verify that lyddites were mapped in 59 rectangles, while chert, phtanite and other silicifications were mapped in more than 30 rectangles and silex in about 20. The zones of the country where outcrops occupy more extensive areas are basically located in Trás-os-Montes, in NE region of the country, and Baixo Alentejo in the south.

In a similar way, the map of Fig. 4 shows the distribution of the outcrop areas of rocks with well crystallised silica forms (quartz) but with a reactive potentiality conferred by the metastable crystalline structure, due to deformation, and by cataclasis, which originates a high specific surface of minerals.

In this zoning the same three divisions of outcrop area referred to above were also considered. As can be assumed from the analysis of distribution, there are 67 rectangles where the outcrop areas of igneous and metamorphic rocks are higher than 5%, in which quartz presents an undulatory extinction. From the 67 rectangles, only in 21 the outcrop area of those rocks was higher than 50%.

3.3 - Distribution of rocks considered as a potential source of alkalis

Directing the analysis of the Portuguese Geologic Chart towards rocks which usually contain minerals with a high alkali content (granites, syenites and ectinites), there was prepared the zoning presented in Fig. 5. As regards the percentage of area occupied in the rectangle by those types of rocks, and, as in the case of zoning of silica forms, three groups were considered: less than 5%, which was disregarded; between 5 and 50%; and higher than 50%. In about 75% of the rectangles there are areas higher than 5% occupied by formations which contain minerals capable of being an alkali source sufficient to feed alkali-silica reactions. In about 40% of the total rectangles, the areas occupied are higher than 50%.

3.4 - Distribution by districts

Table 2 presents an estimate of the areas of the Mainland districts where rocks with potentially reactive silica forms and with alkalis outcrop. The comparison of the distributions of Fig. 3, 4 and 5 and of the areas indicated in Table 2, makes it possible to conclude as follows:

i) In the zones where silica forms with an ensured reactive potential (chert, silex, etc.) are more frequent, there are also rocks which constitute a potential alkali source, occurring less frequently in Trás-os-Montes (region of Bragança) and more frequently in Baixo Alentejo (region of Beja) and in Algarve (region of Faro). In the Western and Algarve borders however, the rocks with a high alkali content occur in very reduced areas (syenites and dolerites or basalts).

ii) The zones where rocks with quartz presenting an undulatory extinction and cataclasis occur more frequently are located in Minho (region of Braga and Viana do Castelo), Douro Litoral (region of Oporto) and Beira Alta (region of Guarda and Viseu); in these zones, rocks with high alkali contents can also be more frequently found. In most cases, these are rocks presenting the two conditions referred to above, namely the rocks of a granitic nature.

4 – CONCLUSIONS

The quality of concrete, in view of its resistance to alteration by alkali-silica reactions, comes from the nature of materials used in its manufacturing and the characteristics of the environment where it is to be used. The studies of the lithologic factors conditioning the use of aggregates must be object of concern as regards the definition and selection of the natural resources to be exploited and the project for the composition of the concrete to be used in the works.

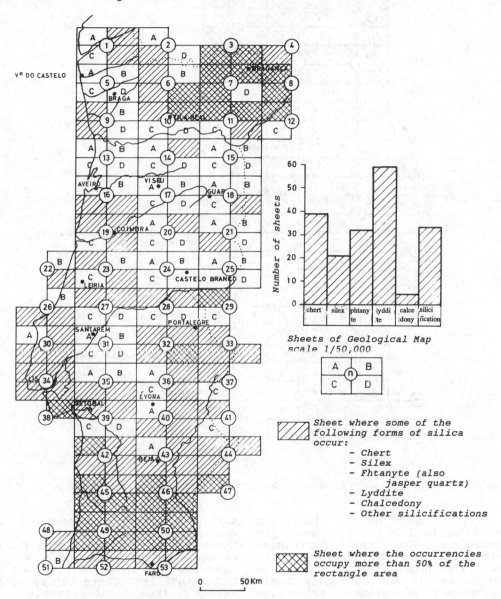

Sheets of Geological Map scale 1/50,000

A		B
	n	
C		D

▨ Sheet where some of the following forms of silica occur:
- Chert
- Silex
- Fhtanyte (also jasper quartz)
- Lyddite
- Chalcedony
- Other silicifications

⊠ Sheet where the occurrencies occupy more than 50% of the rectangle area

0   50 Km

Fig. 3 - Distribution of rocks with potential reactive silica forms others than quartz with ondulatory extinction

3375

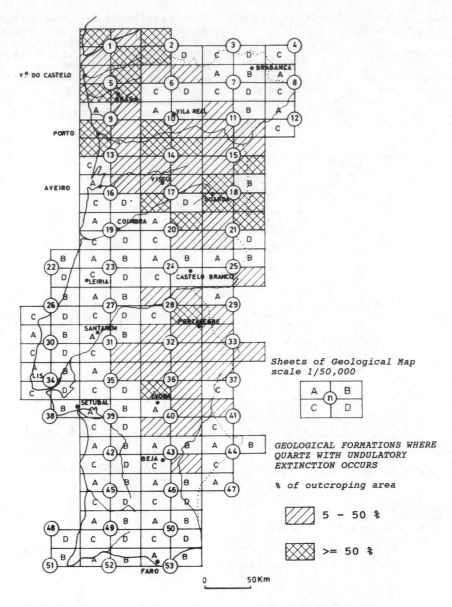

Fig. 4 - Distribution of rocks with tectonized and undulatory extinction forms of quartz

An analysis of the lithologic characteristics of geologic formations in the Portuguese mainland shows the existence of materials which, when are used as aggregates in an extremely basic environment, may cause alkali-silica reactions. The assumption that served as a basis for the identification of the reactive potentiality of tectonic quartz was the undulatory character of the extinction when observed in a polarising microscope. Normally, the reactive potentiality increases with the value of this angle, which was not quantified. There can therefore result from this zoning an aggravating conditioning for certain rocks, namely granitic, which constitute a source of the largest volume of aggregates used in the

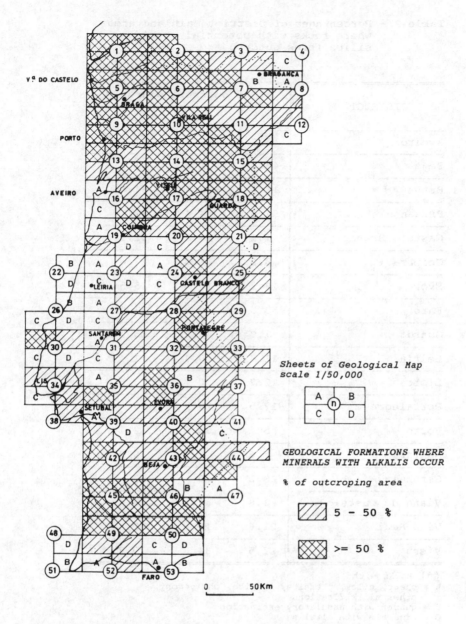

Sheets of Geological Map
scale 1/50,000

A	B
C	D

(n)

GEOLOGICAL FORMATIONS WHERE
MINERALS WITH ALKALIS OCCUR

% of outcroping area

5 - 50 %

>= 50 %

0   50 Km

Fig. 5 - Distribution of rocks whose minerals can be source of alkali elements

country. The usefulness of this study is due precisely to the global nature of the analysis prepared, but it obviously neither replaces nor invalidates studies and regional or local zonings based on criteria of analysis and petrographic and mineralogic quantification and on the appropriate tests.

The execution of studies appropriate for the evaluation of the reactive potentiality of this type of rocks in a certain site, may be the most effective way to confirm the situation resulting from the global zoning or, on the contrary, to desaggravate it.

Table 2 - Percentages of District mainland areas
where rocks with potential reactive
silica forms outcrop

DISTRICT	% of outcroping area(*)		
	A	B	C
Aveiro	7.1	18.4	41.4
Beja	48.8	1.6	47.9
Braga	11.4	33.3	64.5
Bragança	52.6	4.4	31.3
Castelo Branco	1.1	13.8	31.9
Coimbra	8.8	11.6	23.6
Évora	12.2	13	35.5
Faro	58.8	0	33.2
Guarda	1.9	45.9	65.7
Leiria	11.6	0.4	1.1
Lisboa	12.7	0	13.5
Portalegre	17.5	19.2	35.3
Porto	5	38.6	62.1
Santarém	7.6	2.1	9.9
Setúbal	28.2	0	32
Viana do Castelo	3.8	54.7	77
Vila Real	21.6	15	72.4
Viseu	17.5	34.3	79.7

(*) Rocks with:
A = chert, silex, phtanite, lyddite, chalcedony,
    other silicifications
B = quartz with undulatory extinction
C = minerals with alkalis

BIBLIOGRAPHIC REFERENCES

1 - Lauer, K. R., 1990; Classification of concrete damage caused by chemical attack; Materials and Structures, Nº 23, RILEM.
2 - Silva, H. S., 1992; Study on ageing of concrete and masonry dams. Physical and chemical alteration of materials (in Portuguese); Thesis, National Laboratory of Civil Engineering - LNEC, Lisbon.
3 - Silva, H. S. & Rodrigues, J. D., 1993; Relevance and evaluation methods of alkali-aggregate reactions in concrete and their occurrence in Portugal. The contribution of geologic knowledge (in Portuguese); GEOTECNIA Quarterly Journal Nº 67, SPG, Lisbon.
4 - Sousa Coutinho, A., 1993; Manufacture and properties of

concrete (in Portuguese); LNEC, Lisbon.

5 - CMCD, 1989; Alkali-aggregate reaction in concrete dams; ICOLD, Paris.

6 - Idorn, G. M., 1967; Durability of concrete structures in Denmark. A study of field behaviour and microscope features; Thesis, Technical University of Denmark, Copenhagen.

7 - Gillot, J. E., 1975; Alkali-aggregate reaction in concrete; Quarterly Journal of Engineering Geology, Vol. 9, Nº 3, London.

8 - Gogte, B. S., 1981; An evaluation of some common Indian rocks with speacial reference to alkali-aggregate reaction; Cement and Concrete Research, Vol 11, U.S.A..

9 - Dent Glasser, L. S. & Kataoka, N., 1981; The chjemistry of alkali-aggregate reaction; Cement and Concrete Research, Vol 11, U.S.A..

10 - Hyward, D. G., et al., 1988; Engineering and construction options for the management of slow-late alkali-aggregate reactive concrete; 16th ICOLD Congress, San Francisco.

11 - Teixeira, C. & Gonçalves, F., 1980; Introduction to the Geology of Portugal (in Portuguese); INIC - F: C. Gulbenkian, Lisbon.

12 - Soares da Silva, A. M., 1983; Lithologic chart of Portugal; in Environmental Atlas, CNA, Lisbon.

13 - SGP, 1972; Geologic Map of Portugal, in the scale of 1/50,000; DGGM - Portuguese Geological Service, Lisbon.

14 - Teixeira, C., 1972; Geologic Map of Portugal, in the scale of 1/500,000; DGGM - Portuguese Geological Service, Lisbon.

15 - Cotelo Neiva, J. M., et al., 1979; in Study and Classification of Rocks by Macroscopic Survey (in Portuguese); Ed. by J. Botelho da Costa, F. C. Gulbenkian, Lisbon.

16 - Dolar-Mantuani, L. M. M., 1981; Undulatory extinction in quartz used for identifying potentially reactive rocks; Int. Conference on Alkali-Aggregate Reaction in Concrete, Pretoria.

17 - Buck, A. D., 1986; Petrographic criteria for recognition of alkali-reactive strained quartz; 7th Int. Conference on Alkali-Aggregate Reaction in Concrete, Ottawa.

Classification des sols de régions tropicales par la méthode du bleu de méthylène

Classification of tropical soils by the methylene blue method

O.J. Pejon & L.V. Zuquette
Universidade de São Paulo, Brésil

RESUMÉ: Dans ce travail a éte étudiée l'efficacité de l'utilisation de la méthode d'adsorption du bleu de méthylène pour caractériser les sols des régions tropicales. Près de 130 échantillons ont été analysés, ramassés sur une superficie d'environ 2.900 Km^2, située dans la région centre-est de l'Etat de São Paulo-Brésil. Les résultats de capacité d'échange de cations et de la surface spécifique obtenus par adsorption du bleu de méthylène et par des méthodes traditionnelles présentent des coefficients de corrélation élevés et significatifs. Les indices d'adsorption du bleu de méthylène ont encore fourni une bonne indication de la composition minéralogique de la fraction argileuse des sols, comparés aux résultats obtenus par diffraction de rayons X. Le comportement géotechnique de ces sols a été encore analysé; les résultats par adsorption du bleu de méthylène comparés à la classification des sols tropicaux proposée par Nogami & Villibor (1981) témoignent d'une très bonne concordance.

ABSTRACT: In this paper we expose a study about the utilization of the methylene blue adsorption method to characterize soils of tropical regions. The tests were performed on the 130 samples of soils that were obtained in the surface area of 2,900 Km^2 in the center east region of the São Paulo State (Brazil). The results of cationic exchange capacity obtained by methylene blue adsorption and tradicional methods presented high correlation coefficient and good significance. The quantity of methylene blue adsorbed by the soils permited the evaluation of the groups of clay minerals from the soils and the results of the methylene blue method were similar to the results obtained by X-ray diffraction analysis. The use of the methylene blue adsorption method permited the assessment of the geotechnical behavior of the tropical soils and it presented good results when compared to tropical soil classification suggested by Nogami & Villibor (1981).

1 INTRODUCTION

La fraction argileuse présente dans les matériaux non consolidés est en grande partie responsable de leur comportement. La quantité de la fraction argileuse ainsi que ses propriétés physico-chimiques sont importantes.

Les méthodes employées par la Mécanique des Sols pour la caractérization de la fraction fine des sols s'avèrent inadequates pour les matériaux de régions tropicales comme l'ont démontré Nogami & Villibor (1981, 1985).

A partir des années 80 a été developpée spécialement en France, une méthodologie pour la caractérisation et la classification des sols bassée sur l'adsorption du bleu de méthylène

(Lan, 1977). Cette méthode présente des grands avantages grâce à sa facilité d'exécution et d'aplication. Dans la plupart des travaux il a été utilisé des échantillon de sols de régions à climat et évolution pédogénitique différents des régions tropicales.

Dans ce travail, on va étudier une série de matériaux de régions tropicales comprenant des types pédologiques variés et avec une génèse de lithologies diverses. A cet effet, il a été ramassé près de 130 échantillons dans la région de Piracicaba, Etat de São Paulo, Brésil (Figure 1). On cherché à établir une relation entre les valeurs de capacité d'echange de cations obtenus au moyen d'essai d'adsorption deu bleu de méthylène et au moyen de méthodes traditionnelles.

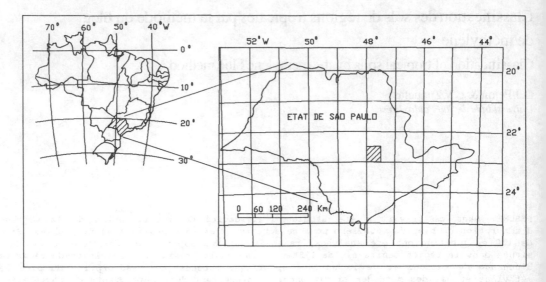

Fig.1 Région de origine des échantillons.

On à étudié également l'importance du pH, puisqu'il peut altérer les charges disponibles à la surface des minéraux argileux du groupe Kaolinite.

On a encore cherché à prévoir le comportement des matériaux non consolidés en comparant les résultats de la valeur d'adsorption du bleu de méthylène et de l'indice d'activité de la fraction argileuse avec la classification M.C.T. (Miniatura, Compactado, Tropical), developpée par Nogami & Villibor (1981) espécialement pour les matériaux tropicaux.

2 MÉTHODE D'ADSORPTION DU BLEU DE MÉTHYLÈNE

La méthode d'adsorption du bleu de méthylène permet la détermination de la capacité d'echange des cations (CEC) et la surface spécifique (SE) des minéraux argileux. Le bleu de méthylène est un colorant organique que présente la composition $C_{16} H_{18} N_3 S Cl 3H_2O$ et en solution aqueuse se dissocie en anions chlorure et cations bleu de méthylène ($C_{16} H_{18} N_3 S^+$) présentant la formule structurale décrite à la figure 2.
Selon Chen et al. (1974) le cation bleu de méthylène substitue les cations Na^{2+}, K^+, Ca^{2+}, Mg^{2+} adsorbés par les minéraux argileux provoquant un processus d'adsorption irréversible, se caractérisant par une forme de mesure de capacité d'echange de cations. La déterminacion de la surface spécifique (surface interne plus externe) est possible quand la surface des minéraux argileux est recoverte par une couche monomoléculaire de cations de bleu de méthylène, permettant, de cette manière, une fois les dimensions de cette molécule connues, de déterminer la suface totale recouverte (Hang & Brindley,1970). Ainsi, au moyen de la diffraction de rayons X ces auteurs sont arrivés à la conclusion que la molécule de bleu de méthylène peut être représentée par un solide ayant la forme d'un paralêlepipède dont les dimensions sont approximativement de 17,0 x 7,6 x 3,25 Å. Encore selon ces mêmes auteurs, la surface du minéral argileux est complétement recouverte quand il y a formation d'une couche monomoléculaire de bleu de méthylène sur la face ayant 130 $Å^2$.

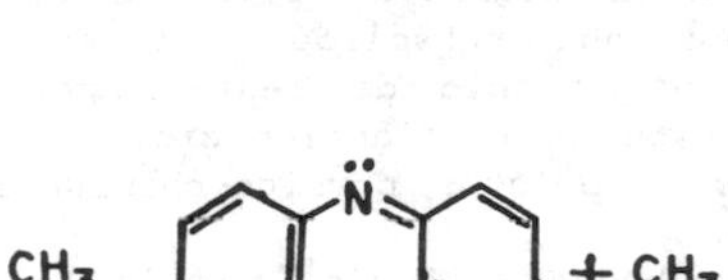

Fig.2 Structure chimique des cations bleu de méthylène.

3382

2.1 Determination de la quantité de bleu de méthylène adsorbé

Il a été utilisé pour cette détermination, la méthode appelé de "essai au bleu à la tache" (LAN 1977), en suivant dans les grandes lignes le procédé proposé par Beaulieu (1979). Quelques changements ont été apportés afin d'adapter cette méthode aux caractéristiques des sols tropicaux. Ainsi, la concentration du bleu de méthylène a été modifiée de 10 g/l pour 1,5 g/l, en raison de la prédominance de minéraux argileux du groupe Kaolinite dans les sols tropicaux, qui présentent une petite capacité de d'adsorption du bleu de méthylène. On a également changé la fraction granulométrique, utilisant des échantillons avec des particules inférieures à 2 mm, permettant, de cette manière, l'obtention d'un indice unique pour le sol étudié par l'essai, sans qu'il soit nécessaire de réaliser des courbes granulométriques.

La capacité d'echange des cations peut être obtenue à partir de l'expression suivante Chen et al. (1974):

$$CEC = \frac{V \times C \times 100}{P} \quad meq/100 \text{ g de sol} \quad (1)$$

où V est le volume de solution (cm^3), C est la concentration de la solution (normalité) et P est le poids de matériau sec (g).

La surface correspondant à 1 cm^3 de bleu de méthylène peut être obtenue à partir de l'expression suivante (Beaulieu 1979):

$$S = \frac{A \times N \times m}{Mbm \times 100} \quad m^2/cm^3 \quad (2)$$

où A est la aire active d'une molécule de Bleu (130×10^{-20} m^2), N est le nombre d'Avogadro ($6,02 \times 10^{23}$) et Mbm est la masse moléculaire du Bleu anhydre (319,9). Partant S = 3,67 m^2/cm^3

Par conséquent, la surface spécifique d'un matériau (SE):

$$SE = \frac{3,67 \times V}{P} \quad m^2/g \quad (3)$$

où V est le volume de solution utilisé (cm^3) et P est le poids de matériau sec (g).

3 CLASSIFICATION DES SOLS PAR LA MÉTHODOLOGIE M.C.T.

La classification M.C.T. est basée sur les indices et coefficients obtenus dans deux expériences de base, réalisées sur des éprouvettes de sols compactes de dimensions réduites qui sont: compactage selon le Mini-MCV et perte de poids par immersion dans l'eau (Nogami & Villibor 1981, Villibor & Nogami 1982). Cette Classification a été developé pour caracterization des sols tropicaux et présent très bonne rapport avec le comportement du sol, surtout pour utilization en construction des autoroutes. Selon Nogami & Villibor (1981) les sols peuvent être divisé en deux groupes: sols de comportement latèritique et sols de comportament non latèritique. Les principales caractéristiques de ces groupes son presenté dans la Tableau 1.

4 PRÉSENTATION ET DISCUSSION DES RÉSULTATS

Les sols des régions tropicales présentent, la plupart, des minéraux argileux du groupe des kaolinites et de grandes quantités d'oxyde et hydroxyde de fer et d'aluminium, matériaux peu recontrés dans des sols des régions où a été developpée la méthode du bleu de méthylène. Cuisset (1980) a vérifié qu'une variation d'adsorption du bleu de méthylène survenait avec le changement de pH pour les argiles kaolinites.

Prenant en considération ces faits, il est nécessaire de faire une estimation de l'efficacité de l'essai d'adsorption du bleu de méthylène pour les sols des régions tropicales. Comme il a été vu antérieurement la méthode d'adsorption du bleu de méthylène consiste en une mesure de capacité d'echange de cations des sols. Ainsi, pour vérifier sa validité pour les sols tropicaux, on a comparé les résultats de capacité d'echange de cations et de surface spécifique obtenues par le bleu de méthylène avec ceux obtenus par les méthodes conventionnelles.

On cherché encore à établir une relation entre la classification M.C.T., qui cherche à relationner le comportement des sols tropicaux avec leur minéralogie, et les résultats obtenus avec l'essai d'adsorption du bleu de méthylène.

4.1 Capacité d'échange de cations

Pour 78 échantillons, on a déterminé la capacité d'échange de cations au moyen de la méthode utilisée par l'Institut

Tableau 1. Propriétés typique des groupes selon la classification M.C.T. (Nogami & Villibor 1982).

CLASSES			N - SOLS DE COMPORTEMENT "NON LATERITIQUE"				L - SOLS DE COMPORTEMENT "LATERITIQUE"		
GROUPES			NA SABLES	NA' SABLEUX	NS' LIMONEUX	NG' ARGILEUX	LA SABLES	LA' SABLEUX	LG' ARGILEUX
DESIGNATION			sable, sable limoneux, limon	sable limoneux, sable argileux	limon, limon sableux et argileux	argile, argile sableuse, argile limoneuse	sable avec peu argile	sable argileux, argile sableuse	argile, argile sableuse
MINI-CBR sans immersion (%) (1)	tres haut	>30	haut a moyen	haut	moyen a haut	haut	haut	haut a tres haut	haut
	haut	12-30							
	moyen	4-12							
	bas	<4							
PERTE DE POIDS PAR IMMERSION (%)	haute	>70	moyenne a basse	basse	haute	haute	basse	basse	basse
	moyenne	40-70							
	basse	<40							
EXPANSION (%) (1)	haute	>3	basse	basse	haute	haute a moyenne	basse	basse	basse
CONTRACTION (%) (1)	moyne	0,5-3	basse a moyenne	basse a moyenne	moyenne	haute a moyenne	basse	basse a moyenne	moyenne a haute
	basse	<0,5							
PERMEABILITE log K(cm/s) (1)	haute	>(-3)	moyenne a haute	basse	moyenne a haute	basse a moyenne	moyenne a basse	basse	basse
	moyne	(-3) a (-6)							
	basse	<(-6)							
PLASTICITE	(%) IP LL		basse a NP	moyenne a NP	moyenne a haute	haute	NP a basse	basse a moyenne	moyenne a haute
	haute	>30 >70							
	moyenne	7-30 30-70							
	basse	<7 <30							

(1) Eprouvettes des sols compactes (densite optimale Proctor Normal)

d'Agronomie de Campinas (Camargo et al.1986) et par la méthode du bleu de méthylène. La corrélation entre les deux résultats a été analysée par la méthode des moindres carrés (Figure 3). On obtient un coefficient de correlation égal à 0,90 avec un niveau de confiance égal à 0,0005 pour 76 degrés de liberté, soit une corrélation hautement significative.

Ces résultats montrent clairement la corrélation existant entre les deux méthodes de détermination de capacité d'echange des cations, indiquant la validité d'utilisation de la méthode du bleu de méthylène, considérée beaucoup plus rapide et plus simple que la méthode chimique traditionnelle.

Malgré le coefficient de corrélation élevé, on note que pour quelques échantillons d'essais, la capacité d'echange de cations déterminée par l'intermédiaire de l'essai d'adsorption du bleu de méthylène, est bien inférieure à celle déterminée par la méthode traditionnelle. Cela se remarque principalement pour les échantillons ramassés à petite profondeur qui normalement ont une quantité plus grande de matières organiques.

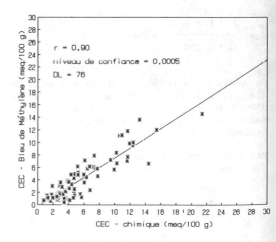

Fig.3 Correlation entre CEC - Bleu de Méthylène et CEC - méthode chimique traditionnel.

3384

Selon Casanova (1986) la matiére organique humidifiée n'adsorbe pas ou adsorbe très peu le bleu de méthylène, pouvant de ce fait justifier la différence rencontrée dans les résultats de capacité d'échange de cations pour ces échantillons. Encore selon ce auteur, les oxydes et hydroxydes de Fe et Al n'adsorbent pas le bleu de méthylène, prouvant donc que la capacité d'echange de cations déterminée ne concerne que les minéraux argileux, fait positif puisque l'objectif de l'essai est justement la caractérisation de la fraction argileuse des sols.

Une autre facteur que peut provoquer des erreurs dans la détermination de la capacité d'echange des cations et de la surface spécifique des matériaux non consolidés des régions tropicales est la variation des charges négatives qui apparait à la surface des minéraux argileux du groupe Kaolinite, pouvant même provoquer l'apparition de charges positives dans des valeurs de pH très basses. Pour cette raison, on a déterminé le pH de la suspension dans lequel l'essai a été réalisé, de même que le pH du sol dans l'eau et dans la solution de KCl, selon la méthodologie présentée par Camargo et al. (1986). Quand le pH dans le KCl est plus petit que le pH dans l'eau, il y a une prédominance de charges négatives, dans le cas contraire il y a prédominance de charges positives à la superficie des minéraux argileux (Demattê 1989).

En analysant les résultats de ces essais, on remarque que dans la majorité absolue des échantillons, il y a prédominance de charges négatives. On observe encore que dans la plupart des cas, le pH mesuré dans la suspension de sol et le bleu de méthylène, après la conclusion de l'essai, est au-dessus de la valeur du pH dans KCl.

Bien qu'il y ait une variation des charges superficielles avec le changement de pH, qui peut intervenir avec le volume de bleu de méthylène adsorbé et par conséquent dans la valeur de capacité d'echange de cations, cette variation est petite, de l'ordre de 1,0 à 1,5 meq/100g por une variation de pH de 2 à 8 (Cuisset 1980).

4.2 Surface Spécifique

La surface spécifique (surface interne et externe) a été déterminée par la méthode de l'ether monoethylique de l'éthylène glicol - EMEG (EMBRAPA-SNLCS 1979) et par la méthode d'adsorption du bleu de méthylène. A la figure 4 sont représentés les résultats de la corrélation des valeurs de surface specifique obtenues par les deux méthodes. Le coefficient obtenu, est de l'ordre de 0,87 avec un niveau de confiance de 0,005 pour 64 degrés de liberté, indique l'existence de corrélation entre les deux méthodes.

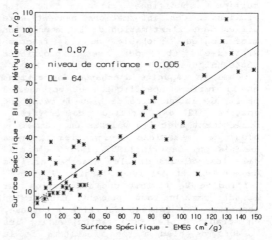

Fig.4 Correlation entre surface Spécifique - bleu de méthylène et surface spécifique - EMEG.

On observe encore à la figure 4 que les valeurs de surface spécifique obtenue par la méthode du bleu de méthylène sont, presque toujours, inférieures à celles obtenues avec l'éthylène glicol et peut être relationné aux taux d'oxide de fer et matière organique présents dans les matériaux étudiés et que peuvent affecter l'adsorption du bleu de méthylène.

4.3 Comparaison des résultats obtenus entre les essais d'adsorption du bleu de méthylène et de la classification M.C.T.

La classification M.C.T. s'est démontrée bien efficace pour caractériser les sols de régions tropicales, mais elle est laborieuse, car elle exige un temps relativement long pour l'exécution des essais. Ainsi, dans les cartographies géotechniques régionaux comprenant des surfaces étendues où le nombre d'echantillons soumis aux essais peut être élevé, il est nécessaire d'utiliser des techniques plus simples que permettent d'estimer les propriétés de la fraction fine des matériaux non consolidés de maniére rapide et sûre.

L'essai d'adsorption du bleu de méthylène fournit la capacité d'echange des cations des sols qui est représentée presque exclusivement par la fraction argileuse puisque les autres composants du sol ont une faible capacité d'adsorption du bleu de méthylène. Par conséquent, connaissant le poucentage de l'argile on peut calculer la capacité d'echange des cations et la surface spécifique de la fraction argileuse. Ces informations peuvent être utiles dans l'estimation de la composition minéralogique probable de la fraction argileuse et de ses propriétés géotechniques.

Comme la capacité d'échange des cations et la surface spécifique sont obtenues à partir de la quantité de bleu de méthylène adsorbé, il est pratique de travailler directement avec les valeurs en grammes de bleu de méthylène selon les indices définis par Lautrin (1987, 1989), c'est-à-dire les valeurs de bleu adsorbé par 100g de sol (V_B) et par 100g d'argile (A_{CB}).

L'indice V_B fournit une information sur le sol comme un tout, pouvant indiquer son comportement. Partant de cette hypothèse, on établit une corrélation entre les valeurs de bleu et les résultats de la classification M.C.T.. A cet effect, pour les 125 échantillons ont été réalisés les essais d'adsorption du bleu de méthylène et ceux de la classification M.C.T.. De tous les échantillons soumis à l'essai, 82 présentent un comportement latéritique, 28 un comportement non latéritique et 15 ont été écartés en raison de problèmes dans l'exécution des essais de classification M.C.T. pour des materiaux très sableux, donnant des résultats douteux.

Dans la Figure 5 sont représentées les valeurs de l'indice V_B en fonction du poucentage d'argile, avec la différenciation entre les matériaux de comportement latéritique et non latéritique, obtenues avec la classification M.C.T.. En analysant ce graphique, on remarque que approximativement 85% des matériaux à comportement latéritique présentent $V_B < 1,5$ et près de 88% des non latéritiques $V_B > 1,5$. Par conséquent, la valeur de bleu égal à 1,5 peut être considérée comme une ligne divisant entre les matériaux de comportement latéritique et non latéritique. De plus, à l'analyse du graphique de la Figure 5, on remarque que, au-dessous de la valeur $V_B = 1,0$ on ne rencontre aucun matériel à comportement non latéritique et avec $V_B > 2,5$ il y a seulement 3 échantillons à comportement latéritique entre les 82 analysés.

De cette manière, on peut conclure que, seulement avec le résultat de l'essai au bleu de méthylène, qui est simple et rapide d'exécution, on peut estimer le comportement latéritique et non

latéritique des materiaux analysés, avec probabilité de réussite de 85%. Quant aux échantillons qui présentent $V_B < 1,0$ ou $V_B > 2,5$, le degré de certitude quant au comportement du matériel est plus grand, atteignant presque 100%. Le plus grand degré d'incertitude concerne les échantillons qui se trouvent dans l'intervalle de $1,5 < V_B < 2,5$ car on rencontre des matériaux de comportement latéritique et non latéritique en nombre équivalent. Sur le total des échantillons soumis à l'essai près de 20% se trouvent dans cet intervalle et devront être analysés par d'autres méthodes pour déterminer leur comportement. Par conséquent l'essai au bleu de méthylène, identifiant près de 80% des matériaux, réduit substantiellement le nombre d'echantillon soumis à l'essai par d'autres méthodes ce qui est très important pour la cartographie geotechnique.

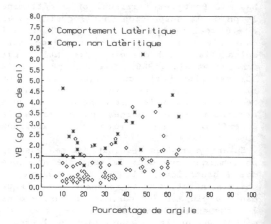

Fig.5 Valeur de Bleu en fonction de la pourcentage d'argile avec le comportement latéritique ou non latéritique obtenues pour la classification M.C.T.

D'autres informations peuvent encore abtenues à partir de l'essai au bleu de méthylène en élaborant un graphique de la valeur de bleu de méthylène adsorbé par 100g d'argile (indice A_{CB}) en fonction du pourcentage d'argile (particules plus petites que 0,002 mm) comme l'a proposé Lautrin (1989). On peut vérifier sur le graphique de la Figure 6 que les matériaux de comportament latéritique, dans leur grande majorité, présentent l'indice A_{CB} inférieur à 4 g/100g d'argile alors que les non latéritiques, supérieur à 5 g/100g d'argile.

On peut encore associer l'indice

d'activité de la fraction argileuse (A_{CB}) avec le coefficient c', qui fournit une indication sur la texture du matériel et est obtenu dans l'essai Mini-MCV (Figure 7). En analysant cette figure, on remarque que la ligne A-B-C obtenue empiriquement, sépare près de 90% des matériaux de comportement non latèritique et 85% de ceux de comportement latèritique. On observe que les matériaux plus sableux, avec c' inférieur à 0,7, même en présence d'argiles d'activité plus élevée, présentent un comportement latèritique, dû, probablement au petit pourcentage d'argile en relation au sol dans son ensemble. On remarque encore que, au-dessus de la valeur A_{CB} = 8 on ne trouve que 3 matériaux de comportement latèritique, c'est-à-dire, moins de 4% du total.

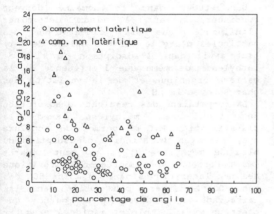

Fig.6 Indice d'activité de bleu (A_{CB}) en fonction de la pourcentage d'argile, avec le comportement latèritique obtenues pour la classification M.C.T.

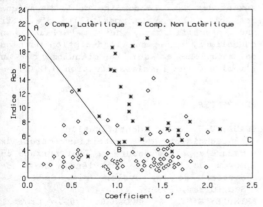

Fig.7 Relation entre l'indice d'activité de bleu (A_{CB}) e l'indice c' de la classification M.C.T.

Dans l'intervalle de A_{CB} entre 5 et 8 et à droite de la ligne A-B, il y a un mélange de matériaux de comportement latèritique et non latèritique, indiquant l'existence d'une région de transition, où la définition du comportement latèritique est difficile à déterminer. L'existence de cette région de transition, dans la classification M.C.T. a déja été proposée par Vertamatti (1990). On peut donc conclure que, l'essai d'adsorption de bleu de méthylène est une manière simple et rapide d'identification du comportement des matériaux non consolidés.

4.4 Relation entre la minéralogie de la fraction argileuse déterminée par rayons X et l'adsorption du bleu de méthylène

Les minéraux argileux sont les grands responsables de l'adsorption du bleu de méthylène ce qui permit de déterminer leur capacité d'echange de cations et leur surface spécifique. Les différents types d'argiles ont des capacités d'échange de cations distinctes, et donc, doivent adsorber d'autant plus de bleu de méthylène que leur capacité d'échange de cations est plus grande. Par conséquent, les matériaux argileux de grande capacité d'échange de cations présenteront des indices A_{CB} plus élevés.

Près de 53 échantillons ont été leur fraction fine analysée par diffraction de rayon X et l'on a cherché à déterminer les types de minéraux argileux prédomimants, en utilisant pour cela les informations et tableaux (Brindley & Brown 1980, Revut & Rode 1981, Gomes 1988, Santos 1989). Il a été fait une analyse semi-quantitative pour estimer la quantité de chaque type d'argile. Selon Gomes (1988), le pourcentage d'argile présent est proportionnel à la surface du pic caractéristique du mineral.

Avec l'analyse des résultats, on remarque que les échantillons de matériaux non consolidés qui contiennent, d'une maniére prédominante, de la kaolinite et gibbsite dans leur fraction argileuse, présentent l'indice A_{CB} < 3 et une capacité d'échange de cations inférieur à 9,4 meq/100g d'argile. Ces résultats concordent avec obtenus par Lautrin (1989). Ces matériaux présentent une fraction argileuse considérée inactive prouvée par le comportement latèritique obtenu dans la classification M.C.T..

Avec des valeurs de A_{CB} entre 3 et 5 (CEC entre 9,4 et 15,5 meq/100g d'argile) on remarque encore une prédominance des minéraux argileux du groupe des kaolinites, mais commencent à apparaître des minéraux du type Illite, Muscovite et Vermiculite alumineuse encore en petites proportions. Cette fraction fine peut encore être considérée comme peu active.

Quant aux valeurs de A_{CB} situées entre 5 et 8 (CEC de 15,5 et 25 meq/100g d'argile), on observe l'apparition de minéraux interstratifiés, ou du groupe d'argiles 2:1 en petite proportion ou encore dans quelques cas, la prédominance des micas et Illitas sur les Kaolinites dans la fraction argileuse. Ces matériaux présentent une fraction argileuse qui peut être considérée encore d'activité normale selon Lautrin (1989), mais par les comparaisons réalisées avec la classification M.C.T., on remarque que le comportement de ces matériaux peut être très variable.

En raison de la prédominance de matériaux non consolidés avec des argiles de faible activité existant dans les régions tropicales, peu de sols avec un indice $A_{CB} > 8$ ont été recontrés (8 échantillons). De ceux-ci, un seul indiquait, à la difraction de rayons X, la présence de grandes quantités d'argiles du groupe des Montmorillonites avec un indice $A_{CB} = 46,5$. Les autres se situaient dans le région de A_{CB} entre 5 et 8, indiquant la présence d'argiles actifs à très actifs (Lautrin 1989). Les diffractogrammes de rayons X montrent la présence de minéraux interstratifiés du type Biotite-Vermiculite ou Pyrophyllite-Vermiculite et en quelques cas des minéraux du groupe des smectites en petite proportions.

Solement dans le cas de une echantillon il existe un désaccord marquant entre le résultat de l'adsorption du bleu de méthylène et l'interprétation du diffractogramme de rayons X, puisque l'indice $A_{CB} = 15,5$ et la diffraction de rayon X a montré la présence de minéraux argileux du groupe Kaolinite dans la fraction argileuse.

5 CONCLUSION

Dans la détermination de capacité d'échange de cations, les résultats obtenus avec l'essai d'adsorption du bleu de méthylène ont été très positifs. Les valeurs de CEC obtenus avec le bleu de méthylène présentent une corrélation hautement significative avec la méthode traditionelle.

La détermination de la surface spécifique par la méthode du bleu de méthylène et par EMEG a montré une corrélation significative bien que les valeurs

obtenues avec le bleu de méthylène soient presque toujours inférieures à celles de EMEG. Pourtant, les valeurs de surface spécifique obtenues par les deux méthodes ont montré la même tendance de variation.

Les valeurs de pH déterminées dans l'eau et dans KCl sont montré que pratiquement tous les échantillons présentent un ΔpH négatif mettant en évidence la prédominance des charges négatives.

L'interprétation des diffractogrammes de rayons X de la fraction fine des matériaux étudiés, a montré qu'il existe une relation entre la minéralogie de la fraction argileuse et les niveaux d'adsorption du bleu de méthylène. Donc, les valeurs de l'indice d'activité des argiles A_{CB} obtenu avec l'essai au bleu de méthylène, on peut avoir une notion du groupe des minéraux argileux présent dans le sol.

Des études dans ce domaine doivent continuer, visant à mieux déterminer l'influence que la cimentation des particules d'argile par des oxides de fer, peut avoir dans l'adsorption du bleu de méthylène, de même que l'influence de la matière organique et de la variation des charges avec le pH.

La comparaison des résultats d'adsorption du bleu de méthylène avec la classification M.C.T. permet de conclure que, au moyen de l'indice d'adsorption du bleu de méthylène (V_B), on peut séparer les matériaux de comportement latéritique des non latéritiques dans au moins 80% des cas. Donc, il est plus utile dans les cartographies geotechniques parce que c'est un essai simple et rapide à exécuter pouvant réduire le nombre d'echantillon à être testés par la systématique M.C.T. ainsi que les autres essais d'exécution longue et laborieuse.

Avec les résultats déjà obtenus, associés aux études dans le domaine de la Mecanique des Sols, il y a la possibilité d'etablir une classification des materiaux non consolidés, basée sur l'adsorption du bleu de méthylène et sur la granulomètrie ou l'indice c' de la classification M.C.T..

RÉFÉRENCES

BEAULIEU J. 1979. Identification géotechnique de matériaux argileux naturels par la mesure de leur surface au moyeu du bleu de méthylène. Thèse de doctorat de 3o. cycle, Univ. de Paris-Sud, Orsay, 133 p.

BRINDLEY G.W. & BROWN G. 1980. Crystal strutures of clay minerals and their X-Ray identification. 3rd. ed., 305 p., Mineralogical Society, London.

CAMARGO O.A., MONIZ A.C., JORGE J.A., VALADARES J.M.A.S. 1986. Métodos de análise química, mineralogia e física de solos do Instituto Agronômico de Campinas. Secretaria de Agricultura e Agropecuária - Instituto Agronômico. 93 p.

CASANOVA F.J. 1986. O ensaio do azul de metileno na caracterização de solos lateríticos. Anais da 21ª Reunião Anual de Pavimentação. Salvador - BA. pp. 277-286.

CHEN T.J., SANTOS P.S., FERREIRA H.C., CALIL S.F., ZANDONADI A.R., CAMPOS L.V. (1974). Determinação da capacidade de troca de cátions e da área específica de algumas argilas e caulins cerâmicos brasileiros pelo azul de metileno e sua correlação com algumas propriedades tecnológicas. Cerâmica, 79:305-326.

CUISSET O. 1980. Propriétes électrocinétiques des particules argileuses. Application de la méthode électrophorétique aux problemes d'environnement et d'identification des sols. Rapport de recherche Labo Central P. et Ch., 96, sept, 101 p.

DEMATTÊ J.L.I. 1989. Curso de gênese e classificação de solos. Centro Acadêmico "Luiz de Queiroz" - Escola Superior de Agricultura "Luiz de Queiroz" USP. 210 p.

EMBRAPA-SNLCS - EMPRESA BRASILEIRA DE PESQUISA AGROPECUÁRIA e SERVIÇO NACIONAL DE LEVANTAMENTO E CONSERVAÇÃO DE SOLOS 1979. Manual de análise de solo. Rio de Janeiro, SNLCS.

GOMES C. F. 1988. Argilas: o que são e para que servem. Ed. Fundação Calouste Gulbenkian. Lisboa. 457 p..

HANG P.T. & BRINDLEY G.W. 1970. Méthylène blue absorption by clay minerals. Determination of surface areas and cation exchange capacities. Clays and Clay Minerals, 18:203-212.

LAUTRIN D. 1987. Une procédure rapide d'identification des argiles. Bull. Labo. P. et. Ch. 152:75-84.

LAUTRIN D. 1989. Utilisation pratique des paramètres dérivés de l'essai au bleu de méthylène dans les projets de génie civil. Bull. Labo. P. et Ch. 160:53-65.

NOGAMI J.S. & VILLIBOR D.F. 1981. Uma nova classificação de solos para finalidade rodoviárias. Anais do Simpósio Brasileiro de Solos Tropicais em Engenharia, COPPE/UFRJ, CNPq.

NOGAMI J.S. & VILLIBOR D.F. 1985. Additional considerations about a new geotechnical classification for tropical soils. First International Conference on Geomechanics in Tropical Lateritic and Saprolitic Soils. ABMS. Brasília. pp. 165-174.

REVUT I.B. & RODE A.A. 1981. Experimental methods of studying soil struture. Ed. Amerind Publishing Co. Pvt. Ltd., New Delhi. 530 p..

SANTOS P.S. 1989. Ciência e tecnologia de argilas. Ed. Edgard Blucher Ltda. 2nd ed., 499 p..

LAN, T.N. 1977. Un nouvel essai d'identification des sols: l'essai au bleu de méthylène. Bull. Liaison Labo. P. et Ch., 88:136-137.

VERTAMATTI E. 1990. Novos padrões geotécnicos para o tratamento de solos tropicais da Amazônia. Sexto Congresso Brasileiro de Geologia de Engenharia. ABGE. Salvador-BA. pp.289-294.

VILLIBOR D.F. & NOGAMI J.S. 1982. Comparações de solos: nova classificação. Revista do Departamento de Estradas de Rodagem do Estado de São Paulo, 127:42-48.

A new simple method for classification of lateritic and saprolitic soils

Une nouvelle et simple méthode pour la classification des sols latéritiques et saprolitiques

J.S. Nogami
Polytechnic School, University of São Paulo, Brazil

D.F. Villibor
Engineering School of São Carlos, University of São Paulo & Highway Department of State of São Paulo, Brazil

ABSTRACT: A simple and inexpensive procedure of soil identification is presented. This method recognizes the groups of the MCT geotechnical classification, which is developed especially for compacted tropical soils. The proposed procedure is based on measurements of: 1) the shrinkage of small discoid (20 mm diameter and 5 mm thick) moulded and dried specimens and 2) the consistency, using a mini-penetrometer, after soaking of these dried specimens.

RÉSUMÉ: Une méthode simple et peu coûteuse pour l'identification de sols est présentée, permettant de classer les sols selon la classification MCT, un système de classification développé exclusivement pour les sols tropicaux compactés. La méthode est basée sur la mesure de deux paramètres: 1) le retrait après séchage d'un petit échantillon moulé avec 20 mm de diamètre et 5 mm d'épaisseur; 2) la consistance du sol a l'aide d'un mini-pénétromètre après réabsorption d'eau par l'échantillon séché.

1 INTRODUCTION

In humid tropical areas, since there are many soils which present peculiar characteristics, the use of some traditional soil mechanics procedures is not advisable (Wooltorton, 1954, Nogami & Villibor, 1979; Novais-Ferreira, 1985; Geological Society Engineering Group, 1990). Such soils have been designated tropical soils by the Committee on Tropical Soils of ISSMFE (1985). Genetically, these tropical soils may be grouped in two large classes: lateritic soils and saprolitic soils. Lateritic soils are pedogenic and superficial soils, and saprolitic soils are residual and sub-superficial soils in natural conditions, with a clear macrostructure inherited from fresh rock..

One of the procedures which should not be used is the traditional soil characterization and geotechnical classification, proposed or presented in IAEG (1981) for mapping purposes and based on the liquid limit, plasticity index and grain-size given by a few sieves. Most of these classifications are based on Prof. Casagrande's proposal (1948). The limitations of the use of these classifications have great practical consequences, because they may indicate the wrong soils to be submitted to more detailed geotechnical studies or to preliminary planning.

To give a solution to such problems, the authors of this paper have been developing , for more than two decades, a new procedure of soil testing based mainly on mechanical and hydraulic properties, the so-called MCT method, from the use of small-sized (50mm diameter) or miniature specimens, obtained through compaction and to be used specifically for tropical soils (Nogami & Villibor, 1979). By incorporating the compaction procedure presented by Parsons (1976) and a new test - the loss of soil mass by immersion - in 1981, it was possible to develop a new geotechnical soil classification, which is also called MCT (Nogami & Villibor, 1981, 1985, 1986, Nogami et al, 1989).

This classification can be considered as a by-product of studies aiming at a better use of fine grained sandy lateritic soils as low cost pavement bases (Nogami & Villibor, 1991). Undoubtedly, the test, when used for classification purposes only, is too complex, lengthy and costlier than those necessary for the traditional geotechnical soil classification, as will be considered in the following section.

Owing to the possibilites of using the MCT classification for preliminary soil investigations and for general purpose geotechnical mapping, many attemps have been made to develop a more expedite visual-manual procedure, similar to that available for classifications based on Prof. Casagrande's proposal.

It should be pointed out that Prof. Casagrande's visual-manual procedure, now standardized in ASTM D 2488-84 (ASTM, 1990) cannot be adapted to identify MCT groups. So, many new procedures were tried, and the first one was presented about ten years ago (Nogami & Cozzolino,.88).

The objective of this paper is a fourth approximation, considerably simpler than the previous ones, and based on data obtained from small discoid specimens (20 mm diameter and 5 mm thick). The procedure have parts which is similar to that proposed by Nascimento et al (1964) for the study of lateritic soils.

2   THE MCT SOIL CLASSIFICATION

As the aim of the proposed test method is to classify soils according to the MCT soil classification, and this classification is not known outside Brazil, it was deemed advisable to present at least its basic characteristics. Further details can be found in Nogami & Villibor (1981, 1985, 1986).

The MCT soil classification uses the graph in Fig. 1, where coefficient c' is plotted in x axis, and index e' is plotted in y axis. Coefficient c' is the slope of the deformability curve obtained from compaction of specimens, according to the adaptation of the procedure presented by Parsons (1976), of the U.K. Transport and Road Reseach Laboratory, and is related to the determination of the "moisture condition" of earthwork materials Index e' is calculated as follows:

$$e' = [(20/d') + (Pi/100)]^{1/3} \qquad (1)$$

where: d' is the slope (in kg/m^3/moisture content in percent) of the dry side of the compaction curve obtained by using 12 blows of a light type rammer. and Pi is the loss of mass (in percent) by total immersion in water of the specimens produced by the above-mentioned compaction, and corresponding to a standardized moisture related to a fixed "moisture condition" value (10 for low dry density soils and 15 for high dry density soils).

To obtain such data, at least 1200 g of soil fraction which passes through a 2,00 mm sieve, as well as the moulding by compaction of at least 4 specimens, the obtainment of about 150 numerical data to be processed, and the plotting of about 15 curves are required. So, it is usually more difficult and costlier than obtaining the data necessary for the traditional soil classifications based on the liquid limit, plasticity index and grain-size. This makes the use of the MCT classification more difficult and costlier, especially when used for purposes other than road paving or construction of transportation ways.

To supplement the classification chart, there is one table with two parts: one related to the mechanical and hydraulic properties of the compacted soils of the soil groups, and another related to the desirability of their use in transportation ways, when compacted near the optimum of the AASHO standard compaction procedure (or ASTM D698-78), except in case they are used in bases, when the compaction procedure is the AASHO modified compaction procedure (or ASTM D1557-78). It is clear that the data considered in these tables are specifically based on soils, environments and construction procedures prevailing in the State of São Paulo, Brazil.

One of the most outstanding characteristics of the MCT soil classification is the separation of tropical soils in two large classes: one comprehending lateritic behavior soils, and the other comprehending non-lateritic behavior soils. These two classes of soils are separated by a dashed line in Fig. 1. Further, the names of the groups are related to grain-size or textural terms, but no traditional grain-size test is required for classification purposes. It should also be noted that the group with the same textural name (such as G' (clayey, from argila) and A' (sandy, from arena)) have very different properties and relative desirability for use in transportation ways, depending on whether their behavior is lateritic (L) cr non-lateritic (N).

3   TEST METHOD

Only a brief description of the test method will

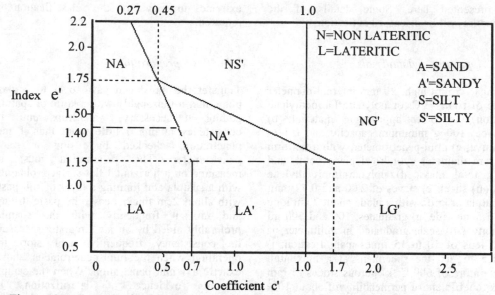

Fig. 1 - Graph of MCT soil classification. The Lateritic behaviour groups and Non-Lateritic behaviour groups are separated by a dashed line

Table 1- Properties and relative desirability in transportation ways of MCT soil classification groups.

BEHAVIOR		N = NON LATERITIC				L = LATERITIC		
GROUP	MCT	N A	N A'	N S'	N G'	L A	L A'	L G'
PROPER- TIES (1)	MINI-CBR NOT SOAKED	M, H	H	M, H	H	H	V, H	H
	SOAKED	M, H	M, H	L, M	L	H	H	H
V=VERY HIGH	EXPANSION	L	L	H	M, H	L	L	L
H =HIGH	SHRINKAGE	L	L, M	M	M, H	L	L, M	M, H
M=MEDIUM	PERMEABILITY (k)	M, H	L	L, M	L, M	L, M	L	L
L =LOW	SORPTIVITY (s)	H	L, M	H	M, H	L	L	L
RELATIVE DESIRA- BILITY AS: n = NOT SUITABLE	PAVEMENT BASE	n	4th	n	n	2nd	1st	3rd
	SELECT SUBGRADE	4th	5th	n	n	2nd	1st	3rd
	COMPACTED SUBGRADE	4th	5th	7th	6th	2nd	1st	3rd
	EMBANKMENT (CORE)	4th	5th	6th	7th	2nd	1st	3rd
	EMBANKMENT (SHELL)	n	3rd	n	n	n	2nd	1st
	EARTHROAD SURFACING	5th	3rd	n	n	4th	1st	2nd

CLASSIFICATION OBTAINED FROM TRADITIONAL INDEX PROPERTIES	USCS	SP SM	SM SC ML	SM,CL ML MH	MH CH	SP SC	SC	MH ML CH
	AASHO	A-2	A-2 A-4 A-7	A-4 A-5 A-7	A-6 A-7-5 A-7-6	A-2	A-2 A-4	A-6 A-7-5

(1) Specimens compacted near the maximum dry density and optimum moisture of standard compactive effort.

be presented here. Some details of the procedure will be discussed in Section 4.

3.1 Apparatus and materials

a) rings, 5 mm high, 20 mm internal diameter made of rigid PVC, or polytetrafluoroethylene (Teflon), or other appropriate material; b) balance, 100 g minimum capacity and 0.1 g. sensitive; c) mini-penetrometer, with a uniform 1.3 mm diameter cylindrical plane point, and 10 g total mass; d) polytetrafluoroethylene (Teflon) sheet; e) sieves of 2.00 and 0.42 mm; f) spatula or knife with a blade about 7 cm long and 1.5cm wide; g) graduates, 100 and 500 ml capacity ; h) scale graduated in millimeter; i) hand lens of 10 to 15 times magnification; j) device to dry the specimens at a maximum temperature of 60° C; k) porous stones (5 mm thick, coefficient of permeability of about 10^{-2} cm/s); l) filter paper, medium grade and sheets of coffee percolator paper; m) air circulator of the type used in micro-computers; n) porcelain mortar and a plain and a rubber-covered pestle of adequate size; o) ground glass plate similar to that used for determination of traditional soil indexes.

3.2 Preparation of test samples

'Air dry the sample and crush the aggregates in a mortar with a rubber-covered pestle. Pass the resulting material through a 2.00 mm sieve to obtain a passed fraction of about 100 ml; measure the volume of the retained fraction and of the passed fraction. After the proper characterization, discard the retained fraction. Then, pass the passed fraction through a 0.42 mm sieve to obtain about 30 ml of passed fraction. Measure the volumes of the retained and passed fractions. In this second sieving, it is advisable to use a plain porcelain pestle, pressing it gently so as not to destroy hard grains, but to break clay aggregates and weak grains.

3.3 Moistening

Moisten the 0.42 mm passed fraction, adding clean water until free water can be seen; leave it resting for at least 10 hours, except when

extreme urgency in the classification is required.

3.4 Paste preparation

Transfer the moistened paste over to a glass plate, mix it thoroughly with a knife or spatula, adding, if necessary, waterdrops until its consistency attains a little more than 1 mm penetration, checked by using a mini-penetrometer. The penetration must be measured on a horizontal flat surface, obtained with a spatula and limiting a layer of soil paste with about 2cm thick. Leave the paste drying, and mix it frequently with the spatula, preferably aided by an air circulator. Measure its consistency frequently, and stop the operation when the mini-penetrometer shows exactly a 1 mm penetration. When the sample presents evidence of lateritization, its consistency should be determined after kneading it at least 400 times. with the spatula. Each kneading involves a strong squeeze with the spatula of the paste at least 1mm thick, against the glass plate surface.That thickness is produced by successively gathering, mixing and smoothing the soil paste with a spatula or knife.

3.5 Moulding specimens

Take a portion of the paste, and mould a little ball of about 2cm diameter; press it with the thumb against a ring horizontally displayed on a Teflon sheet, or in case of very sticky soils, on a horizontal flat surface covered with a thin PVC film. When the ring is not made of Teflon, the use of petroleum jelly is necessary to prevent adherence of the soil to the border of the ring. The excess of soil is taken off with the straight edge of a piece of Teflon sheet, kept in an almost vertical position, and by doing short longitudinal movements against the upper border of the ring, so as to obtain a flat surface, but with knurling marks left by the movement of the piece of Teflon sheet. Such marks may aid in the indentification of the upper face of the specimen. Mould at least 3 specimens by applying the procedure described above.

3.6 Moulding little balls

Using the same paste used to mould the

cylindrical specimens, mould at least one little ball having the same internal diameter measure as the rings. (about 20 mm).

3.7 Drying specimens and balls

Transfer the rings with the respective test specimens and the balls to an oven, or any other appropriate device, and keep them there for at least 10 hours at a temperature of 60°C . When the soil is expected to have very high shrinkage, it is advisable to leave it drying in the open air overnight before transferring it to the oven. Weigh the dry weight of the ball to make it possible to determine its moisture content.

3.8 Shrinkage measurement

Measure the shrinkage directly, using the precision milimetric scale and the hand lens, considering the maximum distance between the outside border of the specimen and the inner border of the ring, keeping the scale radially. Repeat the measurement at least in 3 points. Do not consider statistically discrepant measurements. With some practice, it is possible to make measurements with 0.1 mm approximation. The largest causes of errors are parallaxes, imperfections of test specimens and ovalization of both the specimens and the rings used.

3.9 Soaking the specimens

Transfer the rings with the respective test specimens onto a saturated porous stone previously covered with coffee percolator paper, whose surface keeps at 10 mm negative head over the top. Measure the time of rise of the water front to reach the top of the specimen. Let it rest for about 2 hours, and afterwards take note of eventual changes observed on the top of the specimens, such as: no changes, fissuration, expansion, etc. Determine the consistency of the upper face of the specimen with the aid of a mini-penetrometer, keeping it in a vertical position, and measure penetration of the point near the confined border and in the center (Fig. 2). When penetration is about 2 mm at the upper face, invert carefully the rings with the specimens and repeat the

Fig. 2 - Rings, dried and soaked specimens and mini-penetrometer.

measurements on the newly exposed face. The values so obtained should be used for classification purposes. The measurements near the free border or fissure are not valid.

3.10 Additional determination

In special cases, some additional determinations may be necessary when the soils have shrinkage between 0.3 to 1mm and penetration near 2mm, but with values above and below that. In this case, two procedures can be followed:

1. Remould new cylindrical specimens, with paste prepared as indicated in 3.5, but add a little more paste and press it using a plane plate covered with filter paper, so that the density of the specimen becomes higher than in the regular procedure, and follow the test according to the preceding items.Many sandy lateritic behavior soils, may be identified with this procedure.

2. Prepare a new paste with the soil fraction which passes through the 0.075 mm sieve, in an amount only necessary to mould 3 cylindrical specimens, and then follow the regular procedure; classify the test specimens as lateritic or non-lateritic behavior soils, depending on whether penetration is larger or smaller than 2 mm.

3.11 Supplementary determinations

Many other determinations that may be useful to identify some particular types of soil in the same group, can be made, such as, for example: a) time of rise of moisture front to reach the top of the specimen in the capillary soaking; b) plasticity of the paste; c) crushing strength of dried balls; d) expansion rate after soaking, e) shape and size of blocks formed by craking in the expansion, f) peculiar forms of expansion such as spherical, cylindrical, etc.

4 CLASSIFICATION

4.1 *Calculation of c'*

Use the following formulas to calculate coefficient c':
a) Shrinkage Sk from 0.1 to 0.5 mm:

$$c' = (\log_{10} Sk + 1)/0.904 \qquad (2)$$

b) Shrinkage Sk >= 0.6 mm

$$c' = (\log_{10} Sk + 0.7)/0.5 \qquad (3)$$

Notes: 1) The samples giving shrinkage Sk lower than 0.1 mm cannot be properly classified by this method. 2) Equation (2) corresponds to the dashed straight line in Fig. 3 and equation (3) corresponds to the continuous straight line.

4.2 *Determination of the MCT group*

Use Table 2.

4.3 *Special cases*

Proceed in the special cases as follows:
1. Consider as class L ,if the results of section 3.10 gives penetration equal to or lower than 2mm;
2 Consider NA' as a second option of NS' group, in the following cases:
i) retained fraction in a 0.42mm sieve is more than 30 %;
ii) moisture content of balls obtained in item 3.7 is <= 30.%
3 Consider NA' as an additional option of NG' group when the moisture content of the balls obtained in item 3.7 is <= 35%

Table 2 - Determination of the groups of MCT classification.

COEFFICIENT c'	PENETRATION (mm)	MCT GROUP
<= 0.5	<= 3.0	LA
	3.1 to 3.9	NA
	>= 4.0	NA/NS'
0.6 to 0.9	<= 2.0	LA-LA'
	2.1 to 3.9	NA'/NS'
	>= 4.0	NS'/NA'
1.0 to 1.3	<= 2.0	LA'
	2.1 to 3.9	NA'
	>= 4	NS'
1.4 to 1,7	<=2.0	LA'-LG'
	2.1 to 3.9	NA'/NG'-NS'
	>= 4.0	NS'-NG'
>=1.8	<= 2.0	LG'
	2.1 to 3.9	NG'
	>= 4.0	NG'

Note: Signal / separates the second option group and signal - separates the same option group.

5 DISCUSSION OF RESULTS

5.1 *Results*

In the application of the proposed expedite method, of the 121 tested samples with the characteristics in Table.3, the following general results have been obtained:

Same group of original method 92%
Result of original method is
2nd option of the proposed method 7%
Discrepant results :.....................................1%

5.2 *The c' versus shrinkage Sk relationship*

The shrinkage Sk measured according to the described procedure was the simple property which presented the best correlation with the empirical c' coefficient of the MCT soil classification. The liquid limit was the second best correlation, but the squared R correlation coefficient was considerably lower(0.27 against 0.87). The experimental results obtained from 121 samples are represented in Fig. 3. Note that

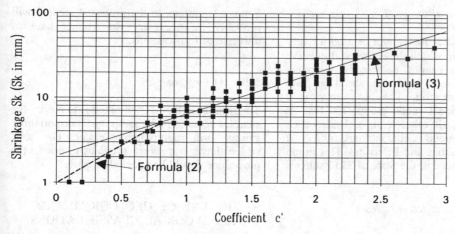

Fig. 3 - Relationship between shrinkage Sk and coefficient c'.

the above squared R was obtained in total samples, only one straight line(continuous line in Fig. 3) and logarithmic scale on Sk axis.

5.3 Consistency of dryed and soaked specimens and index e'

Many tests made on lateritic soils showed that one of the characteristics of their fines is the ability to irreversibly harden when dried. Instead, in non-lateritic tropical soils, the drying and soaking resulted in considerably weakening of the specimens. The results become more complex when the fines are mixed with silts and sands, for which reason an allowance was made to these soils, increasing the final penetration to 2 mm (for c' >= 6 mm) or to 3 mm(for c' <= 0.5mm). No quantitative correlation of consistency after soaking with index e' was developed.

5.4 Shape and size of specimens

The discoid shape of the specimens and their short diameter (20 mm) and height (5 mm) have many advantages related to: easiness in moulding, shrinkage measuring, drying and soaking rapidity, measuring penetration after soaking and small quantity of samples.

5.5 Sieving

The use of samples passing through a 0.42 mm

sieve is similar to that leading to its use in the determination of the index properties of non-tropical soils. For greater ease, the MCT soil classification considers only 2.00 mm of passed fraction so that it is necessary to determine the passed and retained fractions of said sieves. To simplify, the volumetric percentage of these fractions is adequate. In some special cases, it is advisable to use 0.075 mm of the passed fraction to enhance the lateritic or non-lateritic behavior.

5.6 Intensive kneading and pre-soaking

Intense kneading as recommended in the case of soils which have lateritic characteristics was necessary because the MCT classification is based on the behavior of intense compacted specimens. The too gentle manipulation is one of the most common errors in tests. Pre-soaking of the samples before moulding the paste was the adopted routine because some tropical soils show sensibility to this procedure.

5.7 Initial consistency of the soil paste to mould the specimens

The 1 mm penetration consistency has been chosen after many trials. It corresponds to a moisture content lower than the liquid limit, but considerably higher than the plastic limit.

5.8 Limitations of the MCT classification

The MCT classification cannot be used for very granular soils, in which the deformability curve by compaction cannot be clearly obtained. Most of these soils have zero shrinkage with the proposed method. The NA' group is a typical transitional soil group so that it confronts with all other groups. Under such condition, this group is characterized by a great variety of properties and behaviors. This group is not widely found in the State of São Paulo.

5.9 Universe of tested samples

The universe of tested samples has the peculiarities of the State of São Paulo, as specified in Tables 3 and 4 below:

Table 3 - Number of tested samples according to the MCT classes and groups.

CLASS	GROUP	SAMPLES	(%)
N		53	43.8
L		68	56.2
	NA	3	5.8
	NA'	6	5.0
	NG'	18	14.9
	NS'	26	21.5
	LA	7	5.8
	LA'	15	12.4
	LG'	46	38.0

Table 3 - Number of tested samples according to origin.

ORIGIN		SAM-- PLES	(%)
	LATERITIC	64	52,9
SAP- RO- LITE	granit,gneiss	26	21.5
	micaschist,phyllite.	10	8.3
	shale,siltstone	8	6.6
	diabase,basalt	2	1.7
	sandstone	1	0.8
	TRANSPORTED	10	8.3

Some adaptations to the proposed method will probably be necessary for another kind of universe of tropical soils.

5.10 Drying samples

The routine of the MCT method is to air dry the samples so as to make sampling, transportation, storage and manipulation easier. For soils which have not been previously dried, the procedure should be adapted properly.

6 THE USE OF GEOLOGICAL AND PEDOLOGICAL CLASSIFICATIONS

Before the development of the MCT method, the identification and preliminary characterization of tropical soils had to supplement the results of traditional index properties with general meaning pedological and geological terms (Nogami, 1978). Recently, the Report of Working Party of the Geological Society Engineering Group (1990) has recommended the use of the French pedological classification for geotechnical purposes.

From a pragmatical standpoint, the exclusive use of geological and pedological classifications presents many problems in tropical areas, mainly because:

a) Same geological or pedological classification group generally corresponds to profiles with layers having very different geotechnical behavior, in undisturbed or after compaction. The ranking of such soil layers may not be made following the available traditional index tests or geotechnical identifications. Mechanical or/and hydraulic tests may be used, but their cost is high so that they would only be appropriate in the final stage of geotechnical investigations.

b) Most geological maps and studies available in tropical areas are of a bedrok type, and do not consider pedogenic(mostly lateritic) and saprolitic layers, and pedological maps and studies do not consider properly the soils of C horizons, of transported or residual (saprolitic) origin. In many areas, the pedogenic superficial layer is the most important layer, and in others, the saprolitic layer can be more important. So, even the shape and thickness of the most important layers for geotechnical purposes are not obtainable from geological or pedological maps and studies.

c) The use of a more appropriate classifica tion, such as the French pedological will help the geotechnical investigationsis on tropical soils, but maps and studies based on such classification are not available in most tropical areas. Even in areas in which these data are available, to establish the proper ranking of the profile layers from a geotechnical standpoint is necessary for the preliminary stages of geotechnical studies, for mapping purposes and small works, the use of simple and low cost procedures such as the proposed identification method.

7 CONCLUSIONS

1 The proposed method becomes possible by the use of a relatively simple and low cost procedure the identification of tropical soils according the MCT group, which is a more appropriate means than the traditional ones to predict the behavior of compacted tropical soils for engineering works. So, it is particularly useful for geotechnical mapping purposes, small works and in the preliminary stages of geotechnical investigations.

2 The proposed method may also broaden the of use of less expensive design procedures developed by the use of the MCT method, which are now related mainly to low cost pavements, (more than 5.000 km rural roads and 3 million sq. m of streets built satisfactorily in Brazil), allowing the use of local soils, not considered traditionally suitable. Owing to the difficulties of the original MCT method, these procedures have been restricted to larger or specialized organizations, such as federal and state governments, universities, and few consulting and construction firms.

3 According to peculiarities mainly related to the universe of local or regional tropical soils, the proposed method may need some adaptation. Furthermore, the effectiveness of the proposed method and the associated MCT method may increase in the future, when properly adapted to a broader universe of tropical soils, environments and some kinds of engineering works, including for instance the planning of urban occupation, erosion control, foundations of small buildings, earth roads, small airports and earthdams, and others.

REFERENCES

ASTM .1990. Annual Book of Standards, Vol 04.08, Soils and Rock; Dimension Stone; Geosynthetics. American Society for Testing and Materials, Philadelphia, USA.

Casagrande, A. 1948. Classification and identification of soils, Transactions ASCE, Vol 113, 901/991, New York.

Committee on Tropical Soils of ISSMFE. 1985 Progress Report 1982-1985. ABMS. São Paulo, Brazil.

Geological Society Engineering Group Working Party .1990. Tropical residual soils. Quarterly Journal of Engineering Geology. Vol.23, No 1,

IAEG .1981. Rock and soil description and classification for engineering geological mapping. Commission on Engineering Geological Mapping. Bull. IAEG, N° 24 235/274.

Nascimento, U; Castro, E; Rodrigues, M. 1964. Sweeling and petrification of lateritic soils. Technical Paper 215. Laboratorio Nacional de Engenharia Civil. Lisbon.

Nogami, J.S. 1978. The adequate use of genetic grouping of lateritic and residual soils. Proc. III International Congress IAEG. Sec.II, vol 2, 153/159

Nogami, J.S.; Villibor, D.F. 1979. Soil characterization of mapping units for highway purposes in a tropical area. Bull. IAEG. No 19 196/199.

Nogami,J.S.; Villibor, D.F. 1981. Uma nova classificação de solos para finalidades rodoviárias. Anais Simpósio Brasileiro de Solos Tropicais em Engenharia, Vol 1, 30/41. COPPE/UFRJ, Rio de Janeiro.

Nogami, J.S.; Cozzolino, V.M.N. 1985. Identificação de solos tropicais: dificuldades e proposta de um método preliminar.Anais da XX Reunião Anual de Pavimentação-Fortaleza Vol 1, 115/134. ABPv, Rio de Janeiro.

Nogami,J.S.;Villibor,D.F. 1985. Additional considerations about a new geotechnical classification for tropical soils. Proc. First Int. Conf. on Geomechanics of Tropical Lateritic and Saprolitic Soils.Vol.1, 165/174. ABMS, São Paulo.

Nogami,J.S.;Villibor,D.F. 1986. The problem of classification of soils used as road construction materials in tropical areas.Proc. 5th International IAEG Congress. 5.4.7

1633/1699. Balkema, Rotterdam.

Nogami,J.S.; Cozzolino, V.M.N.;Villibor,D.F. 1989. Meaning of coefficients and index of MCT soil classification for tropical soils. Proc. of XII International Conference on Soil Mechanics and Foundation Engineering, Rio de Janeiro. Vol 1, 547/550. Balkema. Rotterdam.

Novais-Ferreira, H. 1985. General report of session 1, First International Conference on Geomechanics in Tropical Lateritic and Saprolitic Soils. Proceedings, vol 3, 139/170, ABMS.São Paulo.

Parsons, A.W. 1976. The rapid measurement of the moisture condition of earthwork material. LR 750.Transport and Road Research Laboratory. Crowthorne. UK.

Wooltorton, F.L.D. 1954. The scientific basis of road design. Arnold. London. UK.

Classification and assessment of the rock material for ash-retained dam

Classification et évaluation du matériau rocheux d'un barrage de retention de cendres

Zhang Guojing & Zhou Yingqing
The Reconnaissance & Design Institute of Sichuan Electric Power, Chengdu, People's Republic of China

ABSTRACT: Classification and assessment of rock as constructing materials for ash—retained dam has been a difficult area little of which has ever been studied. Application of fuzzy mathematic to the area is undoubtedly a new attempt. Incorporated with concrete project—Kaixian power plant, anthor present a set of rock classification criteria and build up a set of complete system of fuzzy clustering for rock classification and comprehensive assessment in this paper. The system was applied to rock classification and assessment, which avoid the disadvantage of an old assessment only by single factor and get excellent result.

RÉSUMÉ: Classification et commentaire, spécialement ceux sur la roche utilisée dans la construction de digue cimentée est une zone difficile qui est encore peu attaquée. L'application des indécises mathématiques dans cette zone est, san doute, un nouvel essai. En liant à un projet spécifique - centrale Kaixian, et selon la demande de conception, les caractéristiques de digues cimentées et à la référence de documents concernés et de principes, les auteurs ont établit un nouveau principe de classification de roches et un complexe système de rassemblement indécis et de commentaire compréhensif, qui évite la faiblesse de commentaire par unique facteurs de gagne dans la pratique des résultats excellents.

1 INTRODUCTION

Kaixian power plant, whose installed capacity is 100 thousands KW (two 50 thousands generating sets) is only thermal power plant in northeastern Sichuan. The ash—retained site of the plant was selected to situate at Chenjiagou where is 2-3 km away the north of the plant. Chenjiagou is a long and narrow gully with abundant rock materials. The gully distribute in northwestern direction and is 20-60 m in depth, 100-200 m in width. Through investigation and comparison, designer deside to take advantage of the rock masses around the site as the materials to build permeable rock —fill dam.

The dam was designed to be 27.5 m in height, 230 m in length with the width of 5 m at the dam top, 94. 4 m in the middle and 29.65 m in the two end of the dam' bottom. To build the dam need 210 thousands cubic meter of broken stones and 35 thousands cubic meter of rectanggular slab of stones. The rock masses for the dam must be satisfied with the condition that the dry unit weight(rd) may be equal or over 1.75 g/ cm³, porosity ratio(n) may be equal or below 35%, internal friction angel(φ) may be equal about 38 degrees, maximum compressive strength (Rd) must be over 80 MPa. Meanwhile, the content of the soft rock must be less than 10%.

The initial result of investigation show that it is difficult to classify the rock according to the experience in the past becauce of obvious difference in engineering geological and mechanical characteristic among the rocks. In addition to, some rock masses is good in whole, but some single property do not meet the design demand. Under these circumstances, the recoverable reserves of the qualified rocks in the site may not satisfy with the need for the dam according to the formerly classification, assessment and calculation. Selecting another or enlarging the quarry will undoubtedly lose the time limit for the project and raise the invest. For this reason designer made effort to develop a fuzzy clustering and comprehensive classification and assessment system (FCCA) for the rock materials.

2 GENERAL SITUATION OF THE ROCK QUARRY

The quarry with the overburden of 0.5-2.0 m and strippng ratio of 1/15, which is about 300 m in length and 40-65 m in width, located at the southwsetern of Chenjiagou gully and is about 100 m away from the center line of the dam. The rocks are mainly sandstone, siltstone and mudstone with the gently dip of 6-8 degrees and wide-ranging distribution. The mining slope is designed to be 1 : 0. 75-1 : 1. 00 .

3 ENGINEERING GEOLOGICAL AND MECHANICAL CHARACTERISTIC OF THE ROCKS

3. 1 *Engineering geological characteristic of the rocks*

In the quarry, the clay and silty clay with thickness of 0. 5-1. 0 m distribute over the rocks which are mainly purplish red sandy mudstone, mud—siltstone and yellow—grey sandstone. The sandstone and siltstone with single layer of 1. 3-3. 0 m in thickness exposed whole 12 layers from up to lower to make up of 40% within the mining depth. The joints in rocks grow mently with spacing of 0. 5-2. 0 m. The rocks with thickness of 1. 0-2. 0 m at top is badly erosin. The detailed engineer-

ing geological property of each layer of rock masses is showed in Tab. 1.

3. 2 *Mechanical property of the rocks*

In the basis of field logging and invesigation, sampling rocks were taken in each layer to test the physical and mechanical property of the rocks in laboratory. Then, the property parameter were calculated with statistics and listed in Tab. 1. The Tab. 1 show that the dry unti weight is 2. 20-2. 65 g/cm³, porosity ratio is 8-16% both of which meet the designing demand, but there is obvious difference in the internal friction angel (φ) and maximum compressive strength (Rd) in each layer of rock.

3. 3 *Quality classification of rock materials*

In the document and standard nowavilable in China, there are not a definite criteria for classification of rock

Tab. 1 Engineering geological and mechanical characteristic of rock

Rock Lager	Thickness (m)	Engineering geological property		Physical & mechanical property					
		N_1(m)	N_2(m)	r_a (g/cm³)	n (%)	c (kpa)	φ (deg.)	Rd (mpa)	Rw/Rd
A	2. 5	0. 8-1. 0	1. 0-1. 2	2. 13	8. 8	48	38	46. 4	0. 20
B	2. 0	1. 0	0. 6-0. 8	2. 17	9. 2	47	37. 8	51. 5	0. 22
C	1. 4	0. 7	0. 5-0. 8	2. 37	7. 0	50	41. 5	49. 3	0. 16
D	1. 5	0. 75	0. 6	2. 38	6. 0	54	44. 0	58. 0	0. 34
E	2. 2	1. 0	0. 8-1. 2	2. 36	7. 5	51	42. 5	49. 5	0. 18
F	1. 8	0. 6	0. 6	2. 42	6. 4	53. 3	45. 2	61. 3	0. 32
G	3. 0	0. 8	0. 8	2. 43	8. 0	55	48	52. 6	0. 38
H	1. 3	0. 6	0. 5	2. 41	8. 1	55. 6	45. 8	53. 7	0. 40
I	1. 0	1. 0	0. 9	2. 38	6. 0	57	47	64. 2	0. 41
J	2. 8	1. 5	1. 5	2. 64	5. 8	92. 6	53	82. 0	0. 48
K	2. 5	2. 0	1. 8	2. 66	5. 4	80	54	79. 8	0. 69
L	2. 0	2. 0	1. 2	2. 70	4. 8	81. 4	52. 2	92. 6	0. 50
M*	32. 0	0. 2	0. 2-0. 4	2. 12	6. 8	40	30	36. 0	0. 15

N_1, N_2 respectirely represent thickness of single lager of rock and joint spcing.
M* represent mudstone with add up thickness.

Tab. 2 Criteria of rocks for the materials of ash—retained dam

Level Factor	I		II		III	
	X	X*	X	X*	X	X*
N_1(m)	>1.5	1	0.8	0.53	<0.2	0.13
N_2(m)	>1.5	1	0.8	0.53	<0.2	0.13
r_d(g/cm)	>2.5	1	2.0	0.8	<1.75	0.70
n(%)	<6.0	0.4	10	0.67	>15	1
C(Kpa)	>70	1	50	0.71	<30	0.43
φ(Deg.)	>48	1	38	0.79	<30	0.63
Rd(MPa)	>100	1	80	0.8	<50	0.5
Rw/Rd	>0.45	1	0.25	0.56	<0.15	0.33

X=Original data   X*=alternated value of X

materials to the permeable rock—fill dam. In general, designers indicate the criteria in the light of the size and charactirestic of the dam and the property of the natural materials. If the materials meet the criteria, it is satisfied; or not. This simple classifying—assesssment method more often not one reduce the usable capacity of many rocks so as to cause vast waste. Author put forward a classification criteria (Tab. 2) in the paper acorrding to the designing demand and the property of the ash-retained dam of Kaixian power plant and in refrence to some documents and standards on the classification of rock materials. It is showed in Tab. 2 that the rock in level I with the best quality may be selected as rectanggular slab of stone, one in level II with the better quality do broken stone, one in level III with bad quality are soft rock which may be selected as filling materials becauce it is easy to be crushed.

4 FUZZY CLUSTERING AND COMPREHENSIVE ASSESSMENT OF ROCK

4. 1 *The principle of fuzzy clustering and comprehensive assessment*

4. 1. 1 *The principle and method of fuzzy clustering*

The fuzzy clustering is to study the relation among samples. One or two variables mqy generally be selected to weigh the similaraty between the samples. Based on the similaraty, the samples can successively be merged according to a certain law so as to reach the goal of classification.

In fuzzy clustering, it is very important to calculate the statistical index. In order to compare the similaraty and difference of various rocks, the paper adopt distance and angle index to which the artificial effect is little. The step of fuzzy clustering are showed as following,

a. data standardization; altring an original data with one in [0,1] in formula (1) so as to be convienentle to make analysis

$$X_{ik} = X_{ik}/X_{kmax} \tag{1}$$

where Xkmax represent the maximum of the factor Nok in the rock Nor,

b. calculation of the statistical index rij; in formula (2) and (3), calculating the index by which the similaraty between the classified rock is weighed so as to define the similaraty matrix R, distance index;

$$r_{ij} = [\sum_{k=1}^{n}(X_{ik} - X_{jk})^2/n]^{\frac{1}{2}} \tag{2}$$

angle index;

$$r_{ij} = \sum_{k=1}^{n}X_{ik} \times X_{jk}/(\sum_{k=1}^{n}X_{ik}^2 \times \sum_{k=1}^{n}X_{jk}^2)^{\frac{1}{2}} \tag{3}$$

where i,j=1,2,3·············n,n represent the layer of the rocks,

3403

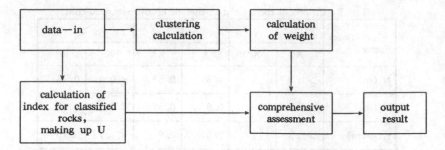

Fig. 1 Idea and process of fuzzy clustering and comprehensive assessment

c. calculation of a symmetry matrix; through the multiplication by itself, tuning the ordinary matrix calculated by formula (2) or (3) into a symmetry matrix which is essential for clustering, meanwhile, satisfied with $rii=1, rij=rji (i,j=1,2,3\cdots\cdots n)$ and $R*R=R$,

d. clustering; based on the practice, selecting a set of cutting matrix to classify the rocks.

4. 1. 2 *The principle and methed of fuzzy comprehensive assessment*

Fuzzy comprehensive assessment generally deal with the problem affected by several factors. It differentiate and determine the difference of the problem in whole degrees. When the influence of these factors to problome can not exactly made clear as well as the assessment to problome only based on single factors can not be made, this method is more suitable. The assessment and classification of the rock material are just this problem. The calculating steps for the fuzzy comprehensive assessment are showed as following;

a. defining tow limited vector

$$U=(U_1, U_2, U_3\cdots\cdots\cdots U_n)$$
$$V=(V_1, V_2, V_3\cdots\cdots\cdots V_n)$$

where U and V respectively represent factors and assessment result set,

b. assessment to each factors in U through V to get a fuzzy vector, many vectorsbeing

$$R_i = (r_{i1}, r_{i2}, r_{i3}, \cdots\cdots\cdots r_{in})$$

able to make up a fuzzy matrix

$$R = \begin{vmatrix} R_1 \\ R_2 \\ \cdot \\ \cdot \\ \cdot \\ R_n \end{vmatrix} = \begin{vmatrix} r_{11} & r_{12} & r_{13} & \cdots\cdots\cdots & r_{1m} \\ r_{21} & r_{22} & r_{23} & \cdots\cdots\cdots & r_{2m} \\ \cdot \\ \cdot & & \cdots\cdots\cdots\cdots \\ \cdot \\ r_{n1} & r_{n2} & r_{n3} & \cdots\cdots\cdots & r_{nm} \end{vmatrix}$$

c. distributing weight for each factor in the basis of the importance of each factor in the practical problem, all weight for each factor making up a fuzzy vector represented with A,

$$A = (a_1, a_2, a, \cdots\cdots\cdots a_n), a_i = 1 \qquad (4)$$

d. calculating B by fuzzy alternation

$$B = A \times R \qquad (5)$$

where B is assessment result being sub-set within V set, sum tatol of its sub—vectors should be one, otherwith, it must be dealed with standardization,

e. at last, finding out the level with maximum in every fuzzy sub-sets of assessment, so this layer of rock is corresponded with this level.

4. 2 *Idea and process of fuzzy clustering and comprehensive assessment*

It includ the steps as following the system of fuzzy clustering and comprehensive assessment for the rocks (Fig. 1).

a. data—in; input all original data into a computer describing the engineering geological and mechanical feature of rocks,

b. clustering analysis; after input of all original data, the computer dealling with the data, then carrying on a matrix alternation, at last outputint the classification result for the rocks in the form of clustering figure,

c. based on the classification, calculating each single index for each type of rocks to build up a U, and weight for every icdex according to original data and classification criteria in Tab. 2,

d. in the basis of the calculation in c, carrying on a comprehensive assessment according to the steps mentioned in 4. 1. 2, the system will make assessment for each type of rocks untill all rocks will be assessed.

e. output result; it include clustering figure, weigh value and comprehensive assessment

4. 3 Classification of rocks by fuzzy clustering

In formula (1), the original data r_d in Tab. 1 can be turned into

$$X_{ik} = (0.85, 0.87, 0.95, 0.94, 0.97, 0.97, 0.96, 0.95, 1.06, 1.06, 1.08, 0.85)$$
$$k = 1, 2, \cdots\cdots 13$$

With same formula, all original data can be dealed with until to get a data matrix with 13 * 8. Alternated value of original data X in Tab1 are all listed in Tab. 3.

After calculation mentioned above, the distance and angle index among each layer of rock are separately calculated in the formular (2), (3) to get tow matrix R_1 and R_2 with 13 * 8 steps, Then, alternation will be made to R_1 and R_2 until both matrix satisfy with $r_{ii} = 1, r_{ij} = r_{ji}$ (i, j = 1, 2, $\cdots$13) and R * R = R.

Through the calculation and alternation mentioned above, tow simililar clustering can be formed (Fig. 2).

Fig. 2 show that all 13 layer of rock are devided into four groups, among them, group1 include J, K, L, group2 does D, F, G, H, I, group3 does A, B, C, E and M belong to group4. Comparing with Tab. 1, it is discovered that each group of rock is similar in the engineering geological and mehanical characteristic. Based on the classification mentioned above, average value of property parameter of each group of rocks can be calculated and listed in Tab. 4.

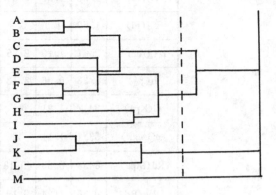

Fig. 2 Clustering figure

Tab.3 Alternated value of original data X in Tab1

Rock lager	Engineering geological property		Physical & mechanical property					
	N_1(m)	N_2(m)	r_d (g/cm³)	n (%)	C (kpa)	φ (deg.)	Rd. (mpa)	Rw/Rd
A	0.53	0.67	0.85	0.59	0.69	0.79	0.46	0.44
B	0.67	0.53	0.87	0.61	0.67	0.79	0.52	0.49
C	0.47	0.40	0.95	0.47	0.71	0.86	0.49	0.36
D	0.5	0.47	0.95	0.40	0.77	0.92	0.58	0.76
E	0.67	0.40	0.94	0.50	0.73	0.89	0.50	0.40
F	0.40	0.40	0.97	0.43	0.76	0.94	0.61	0.71
G	0.53	0.53	0.97	0.53	0.79	1.00	0.53	0.84
H	0.40	0.33	0.96	0.54	0.79	0.95	0.54	0.89
I	0.67	0.6	0.95	0.40	0.81	0.98	0.64	0.91
J	1.0	1.0	1.06	0.39	1.32	1.10	0.82	1.07
K	1.33	1.20	1.06	0.36	1.14	1.13	0.80	1.53
L	1.33	0.80	1.08	0.32	1.16	1.09	0.93	1.11
M	0.13	0.13	0.85	0.45	0.57	0.63	0.36	0.33

Tab. 4 Average value of property parameter of each group of rock

Level	1		2		3		4	
Factor	X	X*	X	X*	X	X*	X	X*
N_1(m)	1.83	1.22	0.88	0.59	0.75	0.5	0.2	0.13
N_2(m)	1.50	1.0	0.8	0.53	0.68	0.95	0.2	0.13
r_d(g/cm^3)	2.67	1.10	2.26	0.9	2.40	0.96	2.12	0.85
n(%)	5.33	0.36	8.13	0.54	6.90	0.46	6.80	0.45
C(Kpa)	84.7	1.21	49	0.7	55	0.79	40	0.57
φ(Deg.)	53.1	1.11	40	0.83	46	0.96	30	0.63
Rd(Mpa)	84.8	0.85	49.18	0.49	58	0.58	36	0.36
Rw/Rd	0.56	1.24	0.19	0.42	0.37	0.82	0.15	0.33

Tab. 5 Linear membership function attributing to level I, II, III of each factor

Factor	Membership function attributing to level I, II, III of each factor		
	Y_I	Y_{II}	Y_{III}
N_1	0　　　　　　　x<0.53 2.13(x−0.53)　0.53<x<1 1　　　　　　　x>1	0　　　　　　　x<0.13,x>1 2.5(x−0.13)　0.53<x<0.53 −2.13(x−1)　0.53<x<1	0　　　　　　　x>0.53 −2.5(x−0.53)　0.53>x>0.13 1　　　　　　　x<0.13
N_2	0　　　　　　　x<0.53 2.13(x−0.53)　0.53<x<1 1　　　　　　　x>1	0　　　　　　　x<0.13,x>1 2.5(x−0.13)　0.13<x<0.53 −2.13(x−1)　0.53<x<1	0　　　　　　　x>0.53 −2.5(x−0.53)　0.53<x<0.13 1　　　　　　　x<0.13
r_d	0　　　　　　　x<0.80 5(x−0.80)　0.80<x<1 1　　　　　　　>1	0　　　　　　　x<0.7,x>1 10(x−0.7)　0.7<x<0.80 −5(x−1)　0.8<x<1	0　　　　　　　x>0.80 −10(x−0.8)　0.80>x>0.70 1　　　　　　　x<0.70
n	0　　　　　　　x>0.67 −3.7(x−0.67)　0.67>x>0.4 1　　　　　　　x<0.4	0　　　　　　　x<0.4,x>1 3.7(x−0.4)　0.4<x<0.67 −3.03(x−1)　0.67<x<1	0　　　　　　　x<0.67 3.03(x−0.67)　0.67<x<1 1　　　　　　　x>1
C	0　　　　　　　x<0.71 3.45(x−0.71)　0.71<x<1 1　　　　　　　x>1	0　　　　　　　x<0.43,x>1 3.75(x−0.43)　0.43<x<0.71 −3.45(x−1)　0.71<x<1	0　　　　　　　x>0.71 −3.57(x−0.71)　0.71>x>0.43 1　　　　　　　x<0.43
φ	0　　　　　　　x<0.79 4.76(x−0.79)　0.79<x<1 1　　　　　　　x>1	0　　　　　　　x<0.63,x>1 6.25(x−0.63)　0.63<x<0.79 −4.76(x−1)　0.79<x<1	0　　　　　　　x>0.79 −6.25(x−0.79)　0.79>x>0.63 1　　　　　　　x<0.5
Rd	0　　　　　　　x<0.8 6(x−0.8)　0.8<x<1 1　　　　　　　x>1	0　　　　　　　x<0.5,x>1 3.33(x−0.5)　0.5<x<0.8 −5(x−1)　0.8<x<1	0　　　　　　　x>0.8 −3.33(x−0.8)　0.8>x>0.5 1　　　　　　　x<0.5
Rw/Rd	0　　　　　　　x<0.56 2.27(x−0.56)　0.56<x<1 1　　　　　　　x>1	0　　　　　　　x<0.33,x>1 4.35(x−0.56)　0.33<x<0.56 −2.27(x-1)　0.56<x<1	0　　　　　　　x>0.56 −4.35(x−0.56)　0.56>x>0.33 1　　　　　　　x<0.33

4. 4 *Fuzzy comprehensive assessment to the rocks*

4. 4. 1 *Build-up of membershiup function*

Application of fuzzy mathematic to assessment of rocks must, at first, assess each single factor, then, making comprehensive assessment. In the assessment to single factor, membership grade expressed by membership function is employed to define the fuzzy boundary of the problem. Otherwise, it is key to solving problem to accuratly build up the membership function. Three — phase linear relationship is emploied to build up the membership function. This method is a little deviation but easy to calculate and higher precision to assessment of rocks because of utiliztion of same law. U and V are respectively fuzzy set of effected factors and rock assessment, $U(N_1, N_2, r_d, n, C, \varphi, Rd, Rw/Rd)$, $V(I, II, III)$. Now, according to classification criteria, every linear membership function attributing to level I, II, III of each factor can be calculated (Tab. 5).

4. 4. 2 *Calculation of membership grade of factors to the rock with level* I, II, III

At first, all original data in Tab. 1 are divided by maximum boundary value in level I of rock classification, the result is listed Tab. 3, then the substituted into each membership function (Tab. 5) to ealculocte membership grade of every factor in accordance with each group of rock at level I. II. III. (Tab. 6). The membership grade is practicatcy the singleassessment of each factor to level I, II, III. Each factor can be of 3 value of membership grade, eight factors making up a 8×

3 matrix called fuzzy matrix and expressed by R. For exmple, initial data (after standardization) of 2nd group of rock is

$$(X_i) = (0.59, 0.53, 0.90, 0.54, 0.70, 0.83, 0.49, 0.42)$$

$$R = \begin{vmatrix} 0.13 & 0.87 & 0 \\ 0 & 1.0 & 0 \\ 0.5 & 0.5 & 0 \\ 0.48 & 0.52 & 0 \\ 0 & 0.96 & 0.04 \\ 0.19 & 0.81 & 0 \\ 0 & 0 & 1.0 \\ 0 & 0.61 & 0.39 \end{vmatrix}$$

4. 4. 3 *Calculation of the weight*

Eight factors mentioned above have different action and effect as well as importance in the assessment and classification of rocks, so they must be endowed with different weight Pi in the light of its importance. Then, the practical effected degree Ci of factor i can be calculated by formula (6)

$$Ci = Xi \times Pi \qquad (6)$$

where Xi=initial data (after standardization) of factor i (Tab. 2). based on Ci, the weight can be calculated by formular (7)

$$P^{\cdot}i = Ci/Si = Xi \times Pi/Si \qquad (7)$$

Tab. 6 Memtership grade of every factor

Group	1			2			3			4		
Membership Function	Y_I	Y_I	Y_{II}	Y_I	Y_I	Y_{II}	Y_I	Y_I	Y_{II}	Y_I	Y_I	Y_{II}
N_1	1	0	0	0.13	0.87	0	0	0.93	0.07	0	0	1
N_2	1	0	0	0	1	0	0	0.8	0.2	0	0	1
r_d	1	0	0	0.5	0.5	0	0.8	0.2	0	0.25	0.75	0
n	1	0	0	0.48	0.52	0	0.78	0.22	0	0.81	0.19	0
C	1	0	0	0	0.96	0.04	0.28	0.72	0	0	0.5	0.5
0	1	0	0	0.19	0.81	0	0.81	0.19	0	0	0	1
R_d	0.25	0.75	0	0	0	1	0	0.27	0.73	0	0	1
Rw/Rd	1	0	0	0	0.61	0.39	0.59	0.41	0	0	0	1

3407

where Si = boundary value of rock classification of a factor, it is a average value of three boundary value.

$$Si = (I + II + III)/3 \qquad (8)$$

Si of every factor may be expressed by a set.

Si = (0. 55, 0. 55, 0. 83, 0. 69, 0. 71, 0. 81, 0. 77, 0. 63)

in order to make comprehensive assessment, the weight of each factor must be standardizated with formula (9)

$$P^{*}i = (Ci/Si)/\sum_{i=1}^{8}(Ci/Si) \qquad (9)$$

Through calculation mentioned above, the weight of all factors can be calculated (Tab. 7) and expressed by Ai. For exmple, the weight of 1st group of rock is

Ai = (0. 07, 0. 11, 0. 09, 0. 03, 0. 10, 0. 21, 0. 25, 0. 14)

4. 4. 4 Comprehensive assessment

Ai (Tab. 7) and R (Tab. 6) are substituted into formula (8), the result of comprehensive assessment B can be calculated. For exmple, the 3rd group of rock is.

Tab. 7 Weight of all factors

Si	Pi	Weight Factor	Group			
			1	2	3	4
0. 55	0. 04	N_1	0. 07	0. 05	0. 05	0. 02
0. 55	0. 08	N_2	0. 11	0. 09	0. 08	0. 03
0. 83	0. 09	r_d	0. 09	0. 172	0. 01	0. 16
0. 69	0. 08	n	0. 03	0. 07	0. 06	0. 08
0. 71	0. 08	C	0. 10	0. 09	0. 10	0. 10
0. 81	0. 21	φ	0. 21	0. 26	0. 28	0. 27
0. 77	0. 32	Rd	0. 25	0. 24	0. 27	0. 26
0. 63	0. 10	Rw/Rd	0. 14	0. 08	0. 15	0. 08

Tab. 8 Comprehensive assessment result of all rock

Level	Group 1	2	3	4
I	0. 813*	0. 169	0. 398*	0. 105
II	0. 187	0. 513*	0. 385	0. 185
III	0	0. 318	0. 217	0. 71*

* represent maximum value

$$B = A \times R = (0. 05, 0. 08, 0. 01, 0. 06, 0. 1, 0. 28, 0. 27, 0. 15) \begin{vmatrix} 0 & 0. 93 & 0. 07 \\ 0 & 0. 8 & 0. 2 \\ 0. 8 & 0. 2 & 0 \\ 0. 78 & 0. 22 & 0 \\ 0. 28 & 0. 72 & 0 \\ 0. 81 & 0. 19 & 0 \\ 0 & 0. 27 & 0. 73 \\ 0. 59 & 0. 41 & 0 \end{vmatrix} = (0. 398, 0. 385, 0. 217)$$

With same emthod, four group of rock may all be assessed and the result are listed in Tab. 8

From Tab. 8, it may be seen that the maximum membership grade of the rock in group 1, 2, 3 and 4 are respectively 0. 813, 0. 513, 0. 398 and 0. 71 which respectively attribute to level I, II, I and III. Obviously, the rock of first, second and forth group attribute respectively to one in level I, II, III. The membership grade of third group rock to level I and II are respectively 0. 398 and 0. 385 both of which is of little difference. Third group of rock should be attributed to one in level I according to theoretical calculation, but is attributed to level II. Vdue to design demanol and close membership grade to level I to level

II. Based on the classification criteria mentioned above, first group rock can be acted as rectanggular slab stone, second and third group rock as broken stone and the forth group rock is only as filling materials.

5. CONCLUSION

1. Based on some documents and some of law on rock classification as well as the practice of Kaixian power plant. A new classification criteria of rock for ash-retained dam has been presented in this apper. The classification criteria work excellently in classification and assessment of rock for the ash-retained dam of

Kaixian power plant. It will also be inspected whether the criteria is valid for other project.

2. With fuzzy clustering and comprehensive assessment, the rock which is under dandard assessed by old method is qualified in whole and utilized in the ash-retained dam which has been in normal motion for five years. The practice prove the assessment result be suitable.

3. Weight distribution has been a difficult problem in fuzzy comprehensive assessment. Author more completly solve the problem in this paper. But, the initial weight of each factor Pi has been also suffered more artificial effect, it remain to be made farther improvement.

REFERENCE

Su Boqin, 1992. Analysis of collapse along Yongjiang River by fuzzy mathematic (Chinese). Proc. of 4th National Engineering Geology Conf. : 505—523

Huang Yunfei, 1992. Fuzzy expert system and its application to rock classification of hydraulic tunnel (Chinese). Site Investigation and Surveying Practice

Li Wenxiu, 1988. Fuzzy probability for the failure of slope in open-pit mines (Chinese). Journal of Civil Engineering.

Huang Zhenglai, 1986. Attempt to divide weathered zone of rock with acoustic wave in fuzzy sets theory (Chinese). Rock Mechanic and Rock Engineering

Zadeh, L. A. 1965. Fuzzy sets. Informa-tion and Contral; 338—353.

Nguyen, V. U. 1985. Some fuzzy set applioations in mining goomoohanice. Int. J. Rock Mooh. Sci. & Geomech; 399—379.

Mustafa, Gencer, 1985. Progressive failure in stratified and jointed rock mass. Rock Mech, and Rock Eng18; 267—292.

7th International IAEG Congress / 7ème Congrès International de AIGI, © 1994 Balkema, Rotterdam, ISBN 90 5410 503 8

Fractal characters of loess microstructure in deformation process and its engineering significance

Caractères fractales de microstructure de loess dans le processus de la déformation et sa signification

Hu Ruilin, Li Xiangquan, Guan Guolin & Ye Hao
The Institute of Hydrogeology & Engineering Geology, CAGS, People's Republic of China

ABSTRACT:With the Microstructure Image Processing System(MIPS) developed by the authors, seven kinds of quantitative data for loess microstructure have been achieved. Analysis result on them demonstrate that these factors behave in obvious fractal characters, which fractal dimensions are between 1 to 2.Under the load, the fractal structure of disturbed loess will change with pressure. This change process reflects the adjustment of loess microstructure to some extent. On the variation curve, the deformation stages and mechanical feature of soil can be determined.

RÉSUMÉ:En utilisant MIPS(Microstructure Image Processing System) mise au point par l'auteur, les 7 données quantitatives principales sur les facteurs de la microstructure de loess ont été obtenues. Les résultats d'analyse montrent que ces facteurs se caractérisent par la structure fractale évidente dont la valeur de dimension fractales est entre 1-2. Sous la pression, les changements de la structure fractale de loess remanié se sont produits.Le processus de ces changements réflète, a`des certains degrés, la loi d'ajustement de la microstructure de loess, à l'aide de la courbe de variation, on paut définir les phase de déformation et les caractéristiques mécaniques du sol.

Loess is a special soil widely distributed in Huanghe river drainage area .It is also a main environment of engineering geology activity in this region. It is popular that loess microstructure is a key to control engineering behaviors of construction foundation engineering and the root of engineering disaster such as foundation settlement, landslide, penetrative deformation of dam foundation and so on, so the microstructure in this area have being paid a good deal of attention by Chinese scientists with a great achievements. However, soil microstructure is a comprehensive product of various natural actions. It is hard to quantify the structure factors because of obvious heterogeneity and undefination so that the research level in microstructure still stay in qualitative analysis period. Furthermore, the theory of geotechnical engineering have to have being adopted the historical methods of continuous medium theories without ideal result for natural state. People have being had a strong desire to make a great progress about microstructure research to bring about great development in the theory of geotechnique and engineering practice. Fortunately the fractal theory improve a lot about undefination of rock structure, which gives a new way to study soil microstructure.

A new upsurge in microstructure research is coming up. It becomes a common idea that the fractal theories maybe become a breakthrough point for undefinablity of soil structure factors and the calculation of their fractal dimension, and then analyses their change characteristics on structure state of disturbed Loess in deformation processing. The research work is still going on with the support of the Chinese Geological Profession Development Foundation(CGPDF).

1.MICROSTRUCTURE IMAGE PROCESSING

As the first step to quantitative analysis on soil microstructure, some of effective measurement technology and method should be possessed so as to gain the quantitative information. Therefore we successfully developed Microstructure Image Processing System(MIPS), cooperated with the Hefei Industry University in 1990. The principle of this equipments shown as Fig.1:

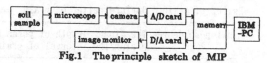

Fig.1 The principle sketch of MIP

Table 1 refers the major output items of MIPS. The quantitative structure information treated by this system is very abundant and beneficial to user to inquire into characters of soil engineering geology and models of quantified fractal structure from different aspects. Thereby we can determined its geotechnical engineering property. It is not difficult to understand that the MIPS is a new way to find out the mystery of soil structure and shows a great prospect.

Table 1 The main output items of MIPS

GRAIN	Coordinate verge, area, chord length, girth, shape coefficient direction angle, granular interval, grain diameter, specific surface

---

PORE	Coordinate verge, area, chord length, girth, shape coefficient direction angle, pore ratio, pore diameter, linking up ratio

---

CONNECTING BELT	Width, area

FRACTAL DIMENSIONED STRUCTURE FACTORS

STATISTIC VALUE OF STRUCTURE PARA.

---

2. FACTOR OF MICROSTRUCTURE AND ITS FRACTAL DIMENSION

2.1 Factors of microstructure

Modern concept about soil microstructure can be understood as a sum of grain shape, size, arrangement pattern in space, pore condition and granular link-up. Since the structure linking-up is hard to measure at present time, the Fabric of micromorphology seems more practical when people are using several kinds of the most common and directed methods (such as scanning electron microscope, optics microscope and so on) to study microstructure. The content what this paper involved is just limited as mentioned above. So we think microstructure character in narrow sense is including four main aspects: form of grain or aggregate, the arrangement pattern, contacting condition and pore property. The authors named these as Factors of Structure. Furthermore, we picked out seven microstructure factors from them (Table.2), which can be quantified and basically reflect the whole characters of soil microstructure.

2.2 Defination and Algorithm on the Fractal of Structure Factors

Fractal is one of the special types which is a common undefination phenomena existing in natural .It is characterized by obvious self-similarity, that is, the part is statistically similar to the entirety in shape, function, information and so on. The whole structure don't change if altering its geometric size of fractal object. The self-similarity can be described quantitatively in fractal dimension. As we know, the soil engineering character is owing to the comprehensive behaviors of soil structure unit. The character of structure unit is mostly controlled by grain aggregate or a single grain mineral. We can use fractal dimension of structure to express the undefination of soil microstructure because of obvious self-silimarity. Although most of structurefactors are definate in one structure

Table 2 The factor of microstructure and its meaning

NAME OF FACTOR	DEFINATION ON IMAGE	STRUCTURE PROPERTY
Grain size	Diameter of grain equal to its area	Particle size
Pore diameter	Diameter of pore equal to its area	Pore size
Soil density	Grain area in unit area	Density of soil or pore
Grain direction	The direction angle of the longest chord	Grain arrangement
Shape coefficient	The largest width in vertical to the longest chord/the longest chord	Shape of particle
Relief of grain	Surface relief of grain	Roughness of grain surface
Linking-up degree	The linary length of pores between grains/the interval of grain	Contacting degree & condition of pore development

unit, but they became dim and undefineted in the whole soil mass. For the long time, people used to use the way "average" to solve this problem, in fact, which has make great simplification to the characteristics of structure so as to lose a lot of important information. So the fractal dimension of structure factors obviously have dominant position reflecting the whole characteristics of structure factors. For the more importance, many characters(such as the surface reliefs of the grain and pore) are hard to be described with the historic statistics methods, but the fractal dimension can do that. Therefore we think that fractal geometry is a effective methods to handle undefinablity of factors of soil microstructure, and it will be make a great progress in the microstructure research with obvious practical advantages.

Nowadays, there are often 3 kinds of main mathematical algorithms for calculating fractal dimension. They are capacity dimension, box-counting dimension and relation dimension(Kenneth j.Falconer, 1991).However, how to define the concrete meanings of fractal morphology are different when people deal with the practical problems. In this paper, based on the actual situation, the author chiefly adopted two models:capacity dimension and box-counting dimension. the value of fractal dimension reflects the whole complex degree of structure factors. Table 3 shows the algorithm of fractal dimension which details will be presented in the authors' another paper because of this paper limited.

2.3 the Characteristics of Structure Factors in Disturbed Loess

The loess samples got from the Loess Plateau (Huangling Shanxi) have been image-processed in order to prove the fractal dimension of structure factors really existing , and then do data analysis based on the methods of calculation on Table 3. At last, relationship of ε and $N(\varepsilon)$ have been gained for each kind of structure factors. Their corresponding relationships are shown from Fig.2 to Fig.9. To my surprise, each factor perform in a wonderful regressive liner relationship, indicating much obvious fractal characteristic. This fact illustrated once again that applying fractal morphology methods to describing the features of soil microstructure is very effective.

In the light of slope got from liner part of each curve we have no trouble to gain the relevant value of each fractal dimension of structure factors. All the results are integrated in Table 4. From the table, we know that the values of structure fractal dimensions are between 1-2 for the disturbed loess homogenized enough. The fractal dimension value of grain density(Dpd) is the maximum, about 2.0005. The minimum is to the shape coefficient(Dci), ranging from 0.8701 to 0.9478. Relatively, the fractal dimension value of grain diameter is less complicate than pore. The fractal dimension value of grain is different from that of pore. The former is lower than the latter. The relief of pore wall (Dpr) is bigger than that of

Table 3 A sample illustration about the algorithm of fractal dimension of structure factors

STRUCTURE FACTORS	MATH MODEL	ε	$N(\varepsilon)$
Grain diameter		Grain size grade ranging from 1 to 256	The number of pore which diameter is bigger than ε
Pore diameter	$D=-\lim\dfrac{L_n N(\varepsilon)}{L_n(\varepsilon)}$	Pore size grade ranging from 1 to 256	The number of pore which diameter is bigger than ε
Density		Bord length of net of cutting-apart image	$N(\varepsilon)=\sum Pi(\varepsilon)$,here $Pi(\varepsilon)$refers ratio between grain area cut by ε and the net area
Relief of grain		Step length of ruler ranging from 1 to 256	The step numbers while measuring the bord of grain in ε
Grain direction	$D=\lim\dfrac{\sum P_i \cdot L_n P_i}{L_n(\varepsilon)}$	Chosen increase between 0 to π	$N(\varepsilon)=\ln(\sum Pi\ln(1/Pi))$, Here Pi is the probability of i area while cutting by ε
Shape coefficient of grain		Chosen intervals ranging from 0 to 1	

Note: The unit is a image point

Table 4.   The Fractal Dimension Average Structure Factors

Dps	Dbs	Ddi	Dpd	Dbd	Dpr	Dbr	Dci
1.21534	1.22892	1.04496	2.00054	2.00160	1.12920	1.17582	0.90436

grain(Dbr), caused chiefly by the cements between grains still existing and leading the
surface form of pore complicated. Generally,
the dimension of direction(Ddi) is about
1.04496, and that of pore density(Dbd) is
2.0016.

3. FRACTAL DIMENSION CHARACTERISTICS OF LOESS STRUCTURE FACTORS IN THE DEFORMATION PROCESS

In order to find out the change law of loess
microstructure under the different pressures, we specially made man-made disturbed
loess samples. The process of manufacture is
that we broken samples into pieces enough at
first and then put them into water to become
homogeneous mixture and sedimentate for 3 to 4
months in static state. After drawing out
water from sedimentation and drying it in the
air for a week, we get our experiment samples.
All of this actions are for the purpose of
eliminating the primary errors caused by loess
heterogeneity and keep these samples in the
same initial state and make the relevant

relationship between structure factors and
changeable pressure more clear. Those
samples were also taken from Huangling,
Shanxi. The method of experiment is common
compression test.chosen stress grades are
50,100,200,400,800,1600KPa.With image processing we got the relevant value D of structure
factors after different loads were exerted until deformation reached to be stable.

3.1 D--P relationship

Based on the values of fractal dimensions under different pressures, D--P relationship
diagrams of each structure factors have been
plotted(Fig.10--Fig.17).The major features are
as follow:
    -The changes of grain size and pore are
rather complex(Fig.10&Fig.11).In initial
stage, the values of fractal dimension fall
very fast and get lowest point when load pressure reaches to 400KPa.Then it gradually
rises, but at last relatively reduced again
after the pressure exceeds 800KPa. The fact

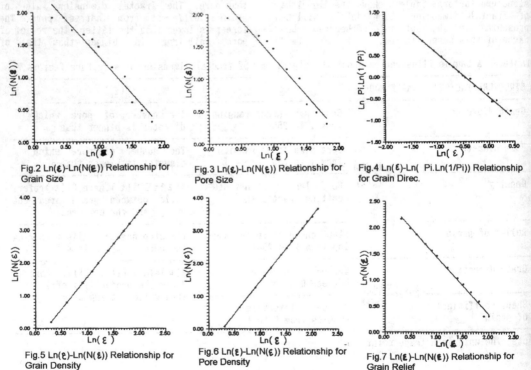

Fig.2 Ln(ε)-Ln(N(ε)) Relationship for Grain Size

Fig.3 Ln(ε)-Ln(N(ε)) Relationship for Pore Size

Fig.4 Ln(ε)-Ln(Pi.Ln(1/Pi)) Relationship for Grain Direc.

Fig.5 Ln(ε)-Ln(N(ε)) Relationship for Grain Density

Fig.6 Ln(ε)-Ln(N(ε)) Relationship for Pore Density

Fig.7 Ln(ε)-Ln(N(ε)) Relationship for Grain Relief

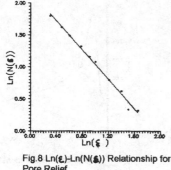

Fig.8 Ln(ε)-Ln(N(ε)) Relationship for Pore Relief

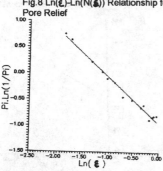

Fig.9 Ln(ε)-Ln(Pi.Ln(1/Pi)) Relationship for Grain Shape Coef.

tells us that the primary grain aggregate which connected with pore water will lose stability rapidly while exerting pressure, and form relatively firm grain or aggregate with the more homogeneous. This kind of avalanche action will not be terminated until the pressure approaches the preconsoildation pressure(about 400KPa). The change tendence of aperture and grain diameter have no difference during the whole process, but homogenization of pore seems more clear than grain in the primary, which tell us that the fractal dimension value is quite low, but in the latest is opposite.

-As shows in Fig.13 & Fig.14, the fractal dimension change of grain density is also rather complicated under the pressure. The value of fractal dimension will reduce during primary stage, and become rising gradually when the pressure gets to 200KPa, but the rise rate decrease progressively. It indicates the change course of grain distribution is from the relative heterogeneity , through homogeneity, to heterogeneity again. The reason for this appearance has no difference with that of grain diameter and aperture so as to bring about basically same result.

-Because the man-made disturbed samples is a sedimental product under ideal stable condition favourable to performing the optimum orientation of grains,they often appear perfect orientation. However, under the pressure the original direction will be disturbed and

cause the fractal dimension of grain orientation increased (Fig.12). There is a better linary relationship between the pressure and the dimension in double logarithm coordinates, that is:

$$Ddi=1.0211+0.00000825*P$$

Here, Ddi refers to the fractal dimension value of grain orientation.

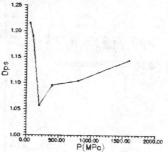

Fig.10 Dps--P Relationship

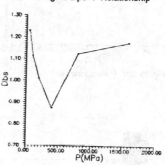

Fig.11 Dbs--P Relationship

-The surface reliefs of grain and pore in man-made disturbed loess will decrease clearly under the pressure affecting(Fig.15 & Fig.16),so the fractal dimension value of them will obviously reduce. Their change tendences can be described as follows:

$$Dpr=1.1343-0.0000105*P$$
$$Dbr=1.1687-0.000025*P$$

This effect will exert some influence on the connected force between grains.

-As Fig.17 shows, the grain shape coefficient is changing with the pressure increasing, and the value of fractal dimension adds up to the tendence of gradually complicating. The relationship between fractal dimension value of shape coefficient and the value of pressure is in liner relationship in the double logarithm coordinate as follow:

$$Dci=0.88897+0.000036*P$$

According to the change law of fractal dimension values of structure factors, we could divided structure adjustment of disturbed loess into three stages:

A.Structure Disturbed Stage: It emerge in the primary . The unstable grain or aggregate

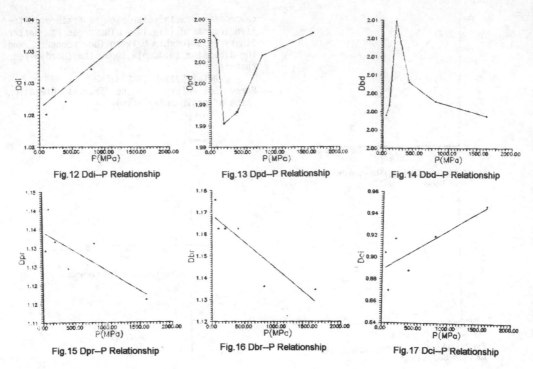

Fig.12 Ddi—P Relationship

Fig.13 Dpd—P Relationship

Fig.14 Dbd—P Relationship

Fig.15 Dpr—P Relationship

Fig.16 Dbr—P Relationship

Fig.17 Dci—P Relationship

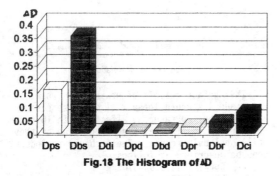

Fig.18 The Histogram of ΔD

come into being collapse into pieces.This stage doesn't stop until the value of pressure reaches preconsoildation pressure.

B.Structure Rebuilt Stage:It emerge in the middle-late time after loading and its main feature is grain amalgamate or integration.

C.Structure Firmed Stage:It appears in the latest time after loading, for example, 800KPa or 1600KPa. At this period the structure adjustment is becoming unclear and the change of fractal dimension tends to be stable.

These three stages mentioned above developed much perfect in the fractal dimension change curves of grain size and density factors.

3.2 ΔD

Fig.18 refers to the increases of each structure factors under the pressure(ΔD), in which the aperture is the most obvious. Its increase can reaches to 0.3536. The secondary is grain diameter, which increase is 0.1587. The rest is very small generally, no exceeding 0.08. So we think that aperture or grain diameter change is the major reason causing structure adjustment of disturbed loess under the pressure. Relatively, pore change more easily than grain.

4.CONCLUSION

A.Facts have proved that the Microstructure Image Processing System (MIPS)can be used as the effective measure to get quantitative information on soil microstructure.

B.Structure state of loess can be determined by 7 kinds of structure factors, that is grain diameter, aperture, grain orientation, grain density, grain shape coefficient, surface relief of grain and pore.

C.Each structure factors behave in obvious fractal characteristics and their values of fractal dimensions are between 1-2 . Fractal geometry is an effective method to quantify soil microstructure .

D.The structure state of man-made disturbed loess will change under the pressure affecting.Adjusting grain diameter, aperture and soil density perform in rather obvious three development stages, naming as the Structure Disturbed Stage,Structure Rebuilt stage, Structure Firmed Stage. The rest of structure factors will change chiefly in linary logarithm relationship with the pressure.

E.The sizes of pore and grain are the mostly sensitive factors to the structure adjustment. Relatively, the former is more changeable than the latter.

Characteristics of loess foundation and evaluation of its earthquake resistance behavior

Caractéristiques du loess comme fondation et évaluation de son comportement aux séismes

Duan Ruwen
Earthquake Research Institute of Lanzhou, SSB, Gansu, People's Republic of China

ABSTRACT: Based on dynamic test of loess engineering, properties and damage characteristics of earthquake in loess areas etc., the present earthquake-resistance behavior of the loess foundation in north-west China has been evaluated in this paper.

RESUME: Compte tenu du test dynamique sur le loess, de sa propriété de la construction et des caractéristiques de destruction du séisme dans la zone de loess, cet article a évalué la propriété anti-sismique de la fondation de loess dans le nord-ouest de la Chine.

1 PRESENT STATE OF LOESS FOUNDATION

Loess is a special kind of soil. Neo-loess (formed during Q3 and Q4) possess some special engineering properties, such as loose soil texture, macroscopic microstracture hydraulic collapse and lower strength when penetrated by water. As foundation of buildings, it has a great influence on earthquake resistant behavior. Cites have developed while the reforming and opening of China. Industrial and civil buildings have expanded to the third terrace. The depth of Q3 neo-loess formed on tertiary terrace by weathering is greater than 20 m. Loess area in Northwest of China is one of the areas where macroseism has occurred frequently. The prediction based on national seismic intensity regionalism map indicates that the fundamental scale of seismic magnitudes of some big and middle cities sited on Northwest loess area such as Xian, Lanzhou, Tianshui is 8. However, people now do not know much about dynamic characteristics of loess. Because research results are not written in the building earthquake resistance code, the dispose of loess foundation still adopts old methods when construct normal civil buildings. Most foundation are dug about 2 m deep, then 4 or 6 floors or even more floors civil building are constructed after simple treatment. In the fields of industrial construction, because of the quick development of village and town enterprises and their simplicity, non-standard and ignoring aseismic capability of factory buildings and foundation, hidden dangers earthquake resistance have been created.

2 RESULTS OF DYNAMIC CHARACTERISTICS TEST OF LOESS

Since Niggata earthquake, which happened in Japan in 1964, all of the world have paid more and more attention to the research of liquefaction of sandy soil. A large number of research projects and corresponding countermeasures have been proposed. However, seldom is known about loess dynamic characteristics and little research has been done. In 1978, during the microzoning research of earthquake resistance of construction field, we for the first time studied loess dynamic characteristics in Lanzhou. Then, we studied loess dynamic characteristics more deeply in Xian, Baoji, Tianshui and Xining, and proposed scientific bases for resolving seismic engineering geological problems and predicting seismic damage.

After ten years research, some achievements have been obtained. For example, loess seismic engineering property — that loess produce additional subsidence when loaded by earthquake i.e. earthquake subsidence character. Now people do not pay enough attention to earthquake subsidence of loess. In fact, earthquake damage in loess area is closely correlated to earthquake subsidence of loess, such as

3419

slope instability, soil glide, ground fissure and foundation subsidence. Based on our research, neo-loess will produce additional subsidence (i.e. earthquake subsidence) cannot be ignored, because the coefficients of those earthquake are greater than 10%. Soil samples for earthquake subsidence test were taken from undisturbed loess 5 to 10 m below ground. These samples were solidified under the imitated load of general civil building. After these samples were stably solidified, viberation experiment had been done and the relation of dynamic stress and remnant strain had been obtained. Figure 1 shows the record of earthquake subsidence test of different kinds of soil samples. Figure 2 shows loess remnant strain after 10 times viberation under dynamic stress (i.e. σ_d-ε_d relation figure). Based on the results of loess earthquake subsidence test, the amount of earthquake subsidence as to a loess sited for 20 m depth when the seismic intensities are 7, 8 and 9 was calculated by Zhang ZhenZhong in the use of the method of summation of individual layer in loess areas in Lanzhou, Xining and Xian. In Lanzhou, the maxium and minium amount of earthquake subsidence of some loess area are 25, 50, 130 cm and 1.1, 8.9, 57 cm when the seismic intensities are 7, 8 and 9 respectively. Based on calculation, in Lanzhou, the average amount of seismic subsidence are 20 to 50 cm and 90 cm when the seismic intensities are 8 and 9 respectively.

of earthquake subsidence. Although the induced factors are different for seismic subsidence and hydraulic collapse of loess, destroy of loess texture producing subsidence is the same final essence for both of them. They are effected by characteristics of loess such as density, void ratio, texture, load and preconsolidated degree. The exist of this property provides basis for predicting earthquake damage.

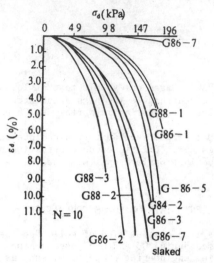

Fig.2 Remnant strain relation of loess after 10 times vibration under dynamic pressure

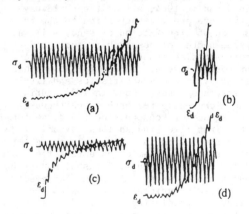

Fig.1 Result of earthquake subsidence test of various kinds of soil

Research results suggest that earthquake subsidence and hydraulic collapse of loess possess the same regularity, and all self-weight collapse loess possess the property

The characteristics of loess earthquake subsidence is that the amount of earthquake subsidence is great at the beginning of viberation. Earthquake subsidence mainly concentrates on first ten times viberation, it accounts for 80% of total amount of earthquake subsidence, it is different to the earthquake subsidence of sand and sandy loam, it will produce unstability, glide and subsidence in a short time and cause serious damage. The earthquake subsidence of loessial loam, sandy loam and soft soil are different from loess. For example, at the beginning of viberation of loessial loam and sandy loam, earthquake subsidence hardly occure, great residual deformation occure only after 10 times or 20 times vibration, then soil samples are failure, as shown in Fig.1a and Fig.1d. The amount of earthquake subsidence of soft soil is great at the very beginning, soil samples are broken after 4 or 5 times vibration, as shown in Fig.1b.

The results of research also suggest that loess earthquake subsidence whether happen or no is correlated to the index of loess physical properties. Earthquake subsidence will happen if the void ratio is greater than 1 and water content is 10% to 15% or greater. Earthquake subsidence might not happen if water content is less than shrinkage limit of contract of loess and void ratio less than 0.8.

Earthquake subsidence and hydrocompaction of loess are different. Loess with high water content has great earthquake subsidence and less hydraulic collapse, on the contrary, loess with low water content has greater hydraulic collapse and less earthquake subsidence. As G86-7 shown in Fig.2, when water content is small, earthquake subsidence will not happen, but earthquake subsidence will significantly increase when soil samples are slaked.

The strength of loess is low, the dynamic strength of hydraulic loess (water content greater than plastic limit) is even lower, and liquefaction may occure in some slaked loess.

3 CHARACTERISTICS OF SEISMIC DAMAGE IN LOESS AREA

Topography and landform of loess area in Northwest of China is complicated. Loess tableland, loess ridge, loess hillock and river terrace are extensive distributed. According to the survey of historical earthquake, the feature of earthquake damage is variant in different topography, landform and soil condition. Large amount of seismic landslide is major earthquake damage in the area of loess ridge, loess hillock and the edge of loess tableland. Sliding mass may cause more serious damage by burying villages and ponding up river, Earthquake subsidence is the major damage form in the area of loess tableland and terrace, where ground fissure are well developed. In addition, in the area of low-level terrace and where underground water level is low, the phenomena of sand boil and foundation subsidence is very common.

At 5:02 a.m. on January 23, 1989, an earthquake with magnitude 5.5 occurred in Gissar village sited at suburb of Dushanbe, Tajikistan, Republic of pre-USSR. In this earthquake, extensive liquefaction developed in the loess deposit of aeolian origin in the gently sloping hilly terrain and led to a series of catastrophic landslides accompanied by a large-scale mud flow. The most striking feature of the damage was a series of landslides and leading to a catastrophic mud flow. The landslide turned into a mud flow of vast scale and buried more than 100 houses in 5 meter depth of mud. About 220 villagers

died or missed in the debris. The reasons for such a liquefaction are thought to be the collapsible nature of highly porous loessial silt which had been wetted by irrigation water over the past years. The complete collapse of the loess structure due to the additional action of the seismic shaking seemed to lead to the catastrophic landslide. The time histories of acceleration recorded at Cymbulif closest to the epicenter are about 7 seconds, the main shaking lasted for about 4 seconds with a peak acceleration of 125 gal in the north-south direction. This example of earthquake damage is similar to the results of loess earthquake subsidence test. The amount of earthquake subsidence increase when loess is wetted. It also show that the amount of loess earthquake subsidence concentrate on first ten times vibration. In semiarid loess area in Northwest of China, because of the development of industry and agriculture, underground water level would rise if various kinds of engineering measure, urbanization, agriculture irrigation and industrial drainage works are mismanaged. The rise of underground water level will cause the increase of loess humidity and earthquake damage will also increase if earthquake occur.

Engineering characteristics of loess has great effect on earthquake damage as well. Analysing two maroseism in loess area by Liu Baichi discovered that both Tongwei earthquake with magnitude of 7.5 in 1718 and Tianshui earthquake with magnitude of 8 in 1654 caused slide group. 377 slide mass greater than 500 m distributed in Tongwei seismic area, which formed a ellipitic concentrated area of glide mass along NNW direction with a 80 km major axle. The gliding scale is great, gliding mass always connected to form skirt, almost everywhere had glide. However, there were only 59 glide mass in Tianshui seismic area. The distribution of all glide mass is similar to that of Tongwei seismic area along the same direction. But the area and scale is smaller and the major axle was only 40 km. The difference is caused by seismic engineering geological characteristics of different kinds of loess. On engineering geology map we can see that Tongwei is in the region of porous self-weight collapse loess, the strength of loess is low and aseismic capability is weak. However, Tianshui is in the region of self-weight non-collapse loess, the strength of loess is higher than that of Tongwei. Table 1 shows the soil strength test result of our research in these two regions in recent years. This property is consisted with the results

of our research on loess earthquake subsidence. Haiyuan earthquake which occurred
in 1920 also indicated that slide concentrated area did not always in extreme intensive seismic area, but in the area with
seismic intensity of 8 or 9. This also is
the result of the effect of soil condition.
This achievement is very significant in
evaluating earthquake damage.

Table 1. Geotechnical physical parameter
in the regions of Tongwei and Tianshui

Site	γ kN/m³	ω %	C kPa	ϕ 0
Tongwei	12.94–14.21	8– 16	14.7– 34.3	15–24
Tianshui	14.41–17.64	10– 25	39.2– 100	20–32

4 EVALUATION OF ASEISMIC CAPABILITY OF LOESS FOUNDATION

According to the survey of historical and
present earthquake, earthquake damage
possess the characteristics of repeatubi-
lity. Under this premise and based on
above achievement of research on loess
dynamic and seismic engineering geological
characteristics, we can evaluate the
aseismic capability of loess foundation
during future earthquake within a certain
loess area and give corresponding prediction of earthquake damage.

First thing of all to do is to investigate topography and landform maps engineering geological maps now can be obtained,
find out whether it is in the area of
self-weight collapse loess or in the area
of self-weight non-collapse, the depth of
loess and the characteristics of previous
earthquake subsidence loess based on the
result of the research on loess dynamic
characteristics. In addition, local
climate condition and the characteristics
and type of local building should be found
out. Finally, we can make corresponding
prediction based on three essential
seismic factors (i.e. seismic intensity,
distance to epicenter, characteristics of
frequency and sustained times), and
evluated seismic intensity of damage in
a certain region.

5 SUMMARY

Loess is a special kind of salty, not well
compacted soil formed by eolian sedimentation in arid and semiarid area. It is
mainly composed of by silt. It possesses
such properties as low density and humidity and granular cementing framework
structure. When loess is affected by
intensive shearing caused by seismic
dynamic stress, its texture will be quickly destructured, the flows like liquefaction will be produced, large amount of
seismic slide and additional subsidence in
loess area will occur and serious seismic
damage will be caused. In the results of
research on loess dynamic characteristics
discussed in above, the similarity of
earthquake subsidence and hydraulic
collapse affected coefficient and loess
engineering properity has important
practice significant for predicting and
evaluating seismic damage in loess area.
According to the test results of loess
earthquake subsidence and the calculation
by the method of summation of individual
layer, the amount of earthquake subsidence
in loess area when the seismic intensity
is 8 or 9 cannot be ignored. At present,
loess foundation situation in big and
middle cities in Northwest of China is
disadvantageous to earthquake resistance.
Department of structure and people should
pay attention to the consolidation and
conduction of loess foundation in these
regions, for the damage will be unusual
serious if macroseism occur in the regions.

REFERENCES

Duan Ruwen, Zhang Zhenzhong, 1990. Futher
Research on Loess Dynamic Character-
istics, Northwestern Seismological
Journal, 12(3).
Kenji Ishihara, Shigeyasu Okusa, et al.,
1990. Lique-faction-induced flow slide
in the collapsible loess deposit in So-
riet Tajik, Soils and Foundations, 3(4).
Liu Beichi, et al., 1984. Interpretation
of Aerial Photograph Taken in the
Regions of the 1718 Tongwei Earthquake
and the 1654 Tianshui Earthquake,
Journal of Earthquake Studies, No.1.
Wang Lanming, Zhang Zhenzhong, 1993. A
Method of Estimation the Quantity of
Earthquake Subsidence in Loess Deposits
during Earthquakes, Journal of Natural
Disasters, 2(3).
Zhang Zhenzhong, Duan Ruwen, Wang Lanming,
1990. Judgement and Prediction of
Probability of Loess Earthquake Subsid-
ence, Third National Conference of Soil
Dynamics Proceedings, Shanghai, Tongji
University Pub., May.

Les graveleux latéritiques dans les pays du Sahel: Cas des routes non revêtues

Lateritic gravels in the Sahel region: The case of unpaved roads

Idrissa Tockol
LNTPB, Niamey, Niger

Michel Massiéra & Paul A.Chiasson
Université de Moncton, N.B., Canada

M.Saliha Maïga
École des Mines et de la Géologie, Niamey, Niger

RÉSUMÉ: Les matériaux graveleux latéritiques sont habituellement utilisés pour la construction des routes non revêtues dans les pays du Sahel. Les auteurs ont constaté qu'il n'est pas possible de distinguer les graveleux latéritiques à partir des essais classiques d'identification (granulométries, limites d'Atterberg) et essais de compactage. Ils proposent d'utiliser des essais moins conventionnels pour la sélection de ces matériaux. Ces essais sont de nature mécanique (résistance maximale à la compression sèche) et surtout de nature chimique. A cet effet, un abaque complémentaire aux classifications classiques est présenté. Cet abaque permet de sélectionner les graveleux latéritiques en se basant sur la proportion de silice (SiO_2) vis-à-vis de celle des sesquioxydes ($Fe_2O_3 + Al_2O_3$).

ABSTRACT: Lateritic gravels are common road construction materials in the sahel countries of Africa. For these materials, creating a performance criteria, based on classical identification tests (Atterberg limits and gradation curves), and/or compaction tests, has proven to be unsuccessful. This is why the authors propose an unconventional testing protocol based on maximum dry compression strength and chemical content measurements. Using these tests, they created a graph which complements conventional classification systems. This graph underlines the relationship between the observed material performance and the chemical content of silica (SiO_2), and sesquioxides ($Fe_2O_3 + Al_2O_3$).

1 INTRODUCTION

Les graveleux latéritiques sont des sols résiduels qui se forment sous les climats tropicaux. En raison de leur abondance dans les pays en développement de la zone intertropicale, ces sols sont très utilisés en construction routière, soit comme couche de fondation, soit comme couche de base. Malgré leur disponibilité le long des tracés routiers et leur coût d'exploitation faible, les graveleux latéritiques n'ont pas une structure stable. Compte tenu du caractère évolutif de ces sols, il a été constaté que leur comportement sur le terrain est différent de celui prévu par les essais géotechniques classiques. Ces constats ont amené plusieurs chercheurs à s'intéresser aux problèmes des sols résiduels. Des recherches sur la genèse ont été effectuées par Dreyfus (1952), Magnien (1966), Little (1967),

Chatelin (1972), etc... Le comportement de ces matériaux a fait l'objet de plusieurs publications notamment Gidigasu (1971, 1972, 1975, 1985), Autret (1983, 1984), Malomo et al. (1983), Madu (1975), etc... Ces études des sols latéritiques ne sont pas arrivées à définir des essais géotechniques permettant de prévoir le comportement réel de ces matériaux sur la chaussée. La présente étude consiste à étudier deux aspects essentiels de dégradation des routes en terre. Ce sont la tôle ondulée et la perte en matériaux ("gravel loss") qui sont en effet les désordres les plus courants des routes en terre. A cet effet, des tronçons de routes non revêtues qui sont susceptibles ou non à ces deux types de dégradations ont été choisis. Des essais classiques et d'autres moins courants ont été réalisés sur les matériaux de carrière ayant servis à la construction de ces routes et sur les matériaux de chaussée. Cette étude a permis de dégager les facteurs qui permettent d'identifier les matériaux qui se sont mal ou bien comportés sur la route, et de faire une proposition de classification plus affinée des graveleux latéritiques. Cette classification doit être utilisée en complément de celles existantes.

2 IDENTIFICATION DES SITES

La région d'étude se situe en Afrique de l'Ouest. Deux pays de la CEAO (Communauté des États de l'Afrique de l'Ouest) ont été retenus pour cette recherche, il s'agit du Mali et du Niger. Le choix des tronçons a été guidé par certains critères tels:
- l'existence du dossier géotechnique;
- l'existence de tronçons soumis aux mêmes conditions de trafic, mais qui présentent des comportements différents;
- la proximité des sites par rapport aux laboratoires où se font les études;
- l'existence des données sur le trafic et l'entretien courant.

C'est ainsi que deux tronçons ont été choisis par pays: Au Niger, il s'agit des routes Say-Tapoa et Dembou-Farie. Au Mali, il s'agit des routes Bamako-Kangaba et Kati-Kolokani.

Sur ces différentes routes des sections qui se sont mal comportées (susceptibles au "gravel loss" et à la tôle ondulée) et des sections qui se sont bien comportées (faibles dégradations) ont été choisies par observations visuelles. Les différentes carrières qui ont servi à la construction de ces sections ont été identifiées à l'aide des dossiers géotechniques.

3 IDENTIFICATION PRÉLIMINAIRE DES SOLS

De nombreux échantillons de sols ont été prélevés sur la chaussée et en carrière afin de les soumettre aux essais classiques d'identification. Sur le terrain, des mesures de densité en place et de teneurs en eau, des vérifications des épaisseurs des couches existantes et les mesures des amplitudes et des longueurs d'onde de la tôle ondulée ont été effectuées. L'objectif visé était d'arriver à faire la différence entre les matériaux qui se sont mal ou bien comportés. Après identification, les échantillons de sols ont été classifiés selon les classifications AASHTO (American Association of State Highways and Transportation Officials) et USCS (Unified Soil Classification System).

4 RÉSULTATS DES ESSAIS PRÉLIMINAIRES

4.1 *Contrôles in situ*

Sur les deux sites retenus au Niger et au Mali, les longueurs d'onde mesurées sur des sections d'environ 200 m varient entre 500 et 700 mm. Quant aux amplitudes elles varient de 45 à 60 mm pour les tronçons du Niger et de 30 à 50 mm sur les tronçons du Mali. La vérification des épaisseurs des couches de base effectuée au Mali donne une épaisseur d'environ 150 mm pour les bons tronçons et 100 mm pour les mauvais. Le contrôle permanent des teneurs en eau effectué sur les sites retenus fait ressortir d'importantes variations, mais sans jamais atteindre la teneur en eau optimum et ceci même pendant la saison pluvieuse.

4.2 *Résultats des essais classiques*

Des analyses granulométriques et des limites d'Atterberg ont été effectuées sur les échantillons prélevés en carrière et sur la chaussée. Ces essais ont permis de vérifier l'état de dégradation des matériaux de la chaussée par rapport à ceux intacts prélevés en carrière. Ensuite, tous ces sols ont été classifiés selon les classifications AASHTO et USCS. Enfin, les résultats des matériaux des bons tronçons et des mauvais tronçons ont été comparés afin de déceler les facteurs qui les différencient (Tableaux 1 et 2).

4.3 *Observations*

4.3.1 *Analyses granulométriques*

Gidigasu (1971) au Ghana et Sandro (1985) au Brésil ont observé que la mesure de la granulométrie varie en fonction du mode de pré-traitement. Au cours de cette étude des analyses granulométriques par la voie sèche et par la voie humide ont été réalisées. Quand l'essai est effectué par la voie sèche, il y a séparation des particules fines des nodules avec un broyage des grains, alors que par la voie humide, il y a dissociation des grains sous l'action de l'eau et du brassage. Les résultats de cette étude démontrent une réelle difficulté à reproduire les analyses granulométriques, ce qui a pour conséquence de créer une dispersion, encore plus grande, des résultats au niveau d'une même carrière.

Tableau 1 Essais de laboratoire effectués au Niger, route Say-Tapoa, Points Kilométriques (PK) 3 et 27

Granulométrie et limites d'Atterberg	Zones de prélèvement			
	Carrière PK 3	Route PK 3	Carrière PK 26	Route PK 27
% < 0.80 mm	15	16	17	27
% < 0.50 mm	33	41	27	53
% < 2 mm	46	55	35	62
% > 5 mm	40	29	51	22
% > 10 mm	20	17	21	8
% > 20 mm	7	3	1	2
w_L en %	34.5	24	35	27
I_p en %	11.4	10	14	11
Classification:				
AASHTO	A2-6	A2-4	A2-4	A2-4
USCS	GC	GC	GC	GC

Tableau 2. Essais de laboratoire effectués au Mali, route Bamako-Kangaba, Points Kilométriques (PK) 18 et 51

Granulométrie et limites d'Atterberg	Zones de prélèvement			
	Carrière PK 18	Route PK 18	Carrière PK 51	Route PK 51
% < 0.80 mm	12	19	20	22
% < 0.50 mm	19	24	23	29
% > 2 mm	75	63	61	57
% > 5 mm	65	45	41	34
% > 10 mm	41	18	12	15
w_L en %	31	33	30	31
I_p en %	11	14	12	10
γ_d en t/m3	2.15	2.13	2.19	2.07
ω_{opt} en %	8.20	9.60	9.70	11.40
C B R à 95%	94	73	80	95
C B R à 98%	140	85	120	140
Classification:				
AASHTO	A2-6	A2-6	A2-6	A2-4
USCS	GC	GC	GC	GC

Les essais effectués sur les graveleux latéritiques de carrière et de chaussée montrent une dégradation plus ou moins grande selon la nature du matériau. Par contre, malgré la différence visuelle des matériaux, l'analyse granulométrique ne permet pas de faire la différence entre un bon et un mauvais matériaux pour la construction de routes en terre. Par exemple, la route Say-Tapoa (Niger) a été construite à la même époque. Le tronçon de route au Point Kilométrique (PK) 27 ne s'est pas dégradé alors que celui au PK 3 est très dégradé. Les deux matériaux graveleux latéritiques ont des

Tableau 3. Limites d'Atterberg effectuées à 1h et à 24h d'imbibition, route Say-Tapoa (Niger), carrières aux Points Kilométriques (PK) 3 et 27.

Carrière et durée d'imbibition		No. de l'échantillon				moy.	S.	c.v. en %
		1	2	3	4			
PK 3	w_L	30.9	30.0	30.6	29.3	30.2	0.71	2.3
Essai à 1h	I_p	14.5	12.7	14.4	11.3	13.2	1.53	11.6
PK 3	W_L	32.0	30.2	30.4	30.2	30.7	0.87	2.8
Essai à 24h	I_p	16.7	14.7	15.1	13.2	14.9	1.44	9.7
PK 27	W_L	32.6	29.7	30.7	30.7	30.9	1.21	3.3
Essai à 1h	I_p	13.6	12.1	12.0	15.1	13.2	1.46	11.1
PK 27	W_L	33.6	30.0	31.3	31.2	31.5	1.50	4.80
Essai à 24h	I_p	15.0	13.0	13.3	15.5	14.2	1.20	8.70

Légende:
- w_L = limite de liquidité, %
- I_p = indice de plasticité, %
- moy. = moyenne, %
- S. = écart type, %
- c.v. = coefficient de variation en % = (S./moy.) x 100

classifications AASHTO et USCS identiques et des limites d'Atterberg semblables (Tableau 1). Au Mali, sur la route Bamako-Kangaba, le tronçon au PK 18 s'est dégradé et présente de la tôle ondulée, alors que le tronçon au PK 51 se comporte bien. Les analyses granulométriques effectuées sur les graveleux latéritiques des chaussées donnent aussi des classifications presque identiques et des limites d'Atterberg semblables (Tableau 2).

4.3.2 Limites d'Atterberg

Selon certains chercheurs (Gidigasu 1972, Gidigasu et Yeboa 1972) le séchage au four crée une baisse de la limite de liquidité. Par contre Malamo et al. (1983) concluent que le séchage au four pendant 24 heures n'influe pas de façon significative sur la valeur des limites d'Atterberg.

Au cours de cette recherche, les essais effectués à 1 heure et à 24 heures d'imbibition ont permis de constater que la durée d'imbibition a une faible influence sur les valeurs des limites d'Atterberg (Tableau 3).

Par contre, il a été observé une bonne reproductibilité des essais des limites d'Atterberg au niveau d'une même carrière. Cependant, on ne distingue aucun facteur qui permet de faire la différence entre les matériaux qui se sont bien

comportés sur la route (carrière au PK 27) et ceux qui se sont mal comportés (carrière au PK 3).

4.3.3 Classification

Le but d'une classification est de prévoir le comportement mécanique du matériau sur la chaussée. En général, les classifications communément utilisées en génie civil et particulièrement en géotechnique sont basées sur la distribution des grains (granulométrie) et la plasticité du sol (limites d'Atterberg). Ces classifications, basées sur le principe de sols à granulométrie stable, sont inefficaces pour les sols résiduels qui sont évolutifs. En géotechnique routière, les classifications les plus utilisées sont les classifications USCS et AASHTO (HRB).

Au cours de cette étude, il a été constaté que presque tous les graveleux latéritiques analysés sont contenus dans une même classe, notamment A2-4 pour la classification AASHTO et GC pour la classification USCS (Tableaux 1 et 2). Ceci confirme le problème d'identification de ces matériaux. Ces classifications ne nous permettent donc pas de différencier les bons matériaux des mauvais. Il est évident que si les essais d'identification n'arrivent pas à déceler la différence entre les bons et les mauvais matériaux, le

Tableau 4. Analyse de résultats des essais Proctor, route Say-Tapoa (Niger), carrières aux Points Kilométriques (PK) 3 et 27.

Carrière et mode de compactage		No. de l'échantillon				moy.	S.	c.v. en %
		1	2	3	4			
PK 3	γ_d en t/m³	2.16	2.07	2.05	2.09	2.09	0.48	2.30
compactage A	ω_{opt} en %	8.8	10.3	10.5	9.4	9.8	0.79	8.1
PK 3	γ_d en t/m³	2.14	2.04	2.07	2.05	2.08	0.45	2.20
compactage B	ω_{opt} en %	10.4	10.8	11.2	10.6	10.8	0.34	3.10
PK 27	γ_d en t/m³	2.09	2.06	2.09	2.02	2.07	0.33	1.60
compactage A	ω_{opt} en %	9.5	10.5	10.0	11.4	10.4	0.81	7.80
PK 27	γ_d en t/m³	2.07	2.05	2.04	2.01	2.04	0.25	1.20
compactage B	ω_{opt} en %	8.7	10.3	9.2	11.8	10.0	1.37	13.7

Légende: Compactage A: Compactage à l'énergie Proctor modifiée où les 5 points de compactage sont réalisés sur un même échantillon

Compactage B: Compactage à l'énergie Proctor modifiée où chaque point de compactage est réalisé sur échantillon renouvelé

moy.:   moyenne
S: écart type
c.v.:   coefficient de variation en %

classifications basées fondamentalement sur ces essais ne pourront le faire.

5 ESSAIS MÉCANIQUES

Des essais de compactage, avec et sans renouvellement des graveleux latéritiques, ont été effectués afin de vérifier l'impact du mode de compactage sur les valeurs des densités sèches maximales et des teneurs en eau optimum. Les sols ont été, d'une part, compactés selon la norme ASTM D1557, c'est à dire en exécutant les 5 points de l'essai sur un même échantillon et, d'autre part, selon la méthode préconisée par Peltier (1969), c'est à dire en renouvelant l'échantillon pour chaque point de compactage. Les résultats obtenus sont présentés au Tableau 4.

Malomo et al. (1983) ont conclu que l'effet de recompacter un échantillon de latérite plusieurs fois n'affecte pas les résultats de l'essai Proctor. Pour notre part, il a été effectivement constaté que le fait d'exécuter le compactage sur un même échantillon ou sur échantillon renouvelé influe peu sur les valeurs des densités sèches obtenues (Figure 1). A l'exception d'un essai, la variation des résultats obtenus lors du compactage est relativement faible au sein d'une même carrière (Figure 2). Cependant, on ne constate pas de différence entre les résultats de compactage des bons et des mauvais matériaux graveleux latéritiques.

Novais-Ferreira et Correira (1965) ont défini un indice de dureté à partir de l'essai d'abrasion Los Angeles. Cet indice correspond au rapport du module granulométrique Mg avant l'essai et celui après l'essai. Dans cette étude, un indice de dureté modifié IH_c a été déterminé à partir de l'essai de compactage. Le module granulométrique Mg est défini comme la somme des poids de particules passant aux tamis 1" (25 mm), 3/4" (19 mm), 1/2" (12.5 mm), 3/8" (10 mm), N.4 (4.75 mm), N.10 (2 mm), N.40 (0.425 mm) et N.200 (0.075 mm). L'indice de dureté est le rapport des modules granulométriques avant et après compactage. Cet indice de dureté a été déterminé pour les différents graveleux latéritiques des carrières utilisées pour la construction des routes en terre (Figure 3). Il est facile de constater que les valeurs d'indice de dureté sont relativement dispersées. En général, plus le matériau contient des nodules durs, plus l'indice de dureté est élevé. Le seuil d'acceptabilité serait, à l'exception d'un résultat, de 0.80.

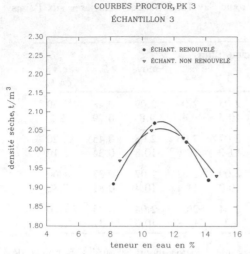

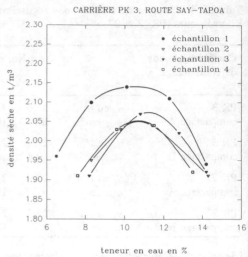

Figure 1. Courbes de compactage à l'énergie Proctor modifiée des échantillons compactés avec et sans renouvellement de matériau.

Figure 2. Courbes de compactage à l'énergie Proctor modifiée sur quatre échantillons d'une même carrière.

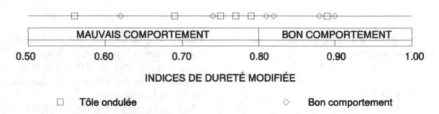

Figure 3. Indices de dureté modifiée sur les matériaux prélevés au Niger (route Say-Tapoa: PK 3 et PK 27) et au Mali (routes Bamako-Kangaba, PK 18 et Kati-Kolokani, PK 25 et PK 33).

Des essais de compactage, à énergie variable, en utilisant le marteau normalisé de masse 4.54 kg une hauteur de chute de 457 mm et 5 couches par échantillon, ont été réalisés en utilisant respectivement 55, 25 et 10 coups par couches. Ces essais ont été réalisés sur des graveleux latéritiques provenant du même échantillon de la carrière PK 3. Ces essais ont montré que les graveleux latéritiques se dégradent sous faible énergie de compactage (Figure 4). De plus, après les premières dégradations, le matériau évolue très peu et les valeurs des indices de dureté sont très proches à 55, 25 et 10 coups par couches. Ceci confirme les constatations faites lors des essais de compactage réalisés avec et sans renouvellement de matériau.

6 REMARQUES

Les essais classiques d'identification et les essais mécaniques courants, ne nous permettent pas de faire la différence entre les graveleux latéritiques qui se sont bien ou mal comportés sur la route en terre. Aucun facteur ne nous permet de dire si tel matériau est susceptibles ou non au phénomène de la tôle ondulée et de la perte en matériau. Pour pouvoir sélectionner des graveleux latéritiques pour les routes non revêtues, il est nécessaire d'avoir d'autres types d'essais qui permettent de faire une différence entre un matériau qui se comporte bien et celui qui se dégrade. C'est l'objet des autres études.

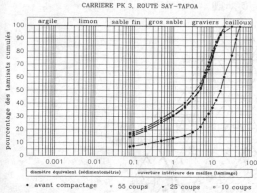

Figure 4. Courbes granulométriques avant et après compactage à énergie variable (granulométrie par voie sèche).

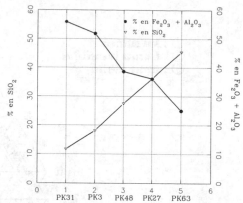

Figure 5. Variation du pourcentage de silice (SiO_2) et de sesquioxydes ($Fe_2O_3 + Al_2O_3$) en fonction des carrières situées au différents points kilométriques (PK).

7 AUTRES ÉTUDES DU COMPORTEMENT DES LATÉRITES

7.1 *Essais de résistance maximale à la compression sèche*

L'évolution de la cohésion et de la portance des matériaux utilisés comme épaulement non revêtu se fait, en Australie, par l'essai MDSC ("Maximum Dry Compressive Strength"). Cet essai de résistance maximale à la compression sèche est utilisé pour vérifier si certains matériaux non cohérents présentent une cohésion suffisante pour être utilisée pour certains projets (Cocks et Hamory, 1988).

Cet essai a été adapté en vue de faciliter son utilisation pour la construction des routes non revêtues. Les éprouvettes sont confectionnées dans des moules Proctor standard et à l'énergie du Proctor modifié. La confection se fait à des teneurs en eau qui correspondent à l'optimum Proctor et légèrement en dessus et en dessous de l'optimum Proctor (+- 2%). Après démoulage, les éprouvettes mûrissent pendant 6 heures à la température ambiante et pendant 12 heures au four à 105°C. Les teneurs en eau qui ont été mesurées après le mûrissement sont pratiquement les mêmes. Les essais n'ont été effectués que sur les graveleux latéritiques provenant d'une seule carrière.

L'ensemble des résultats indiquent que la résistance à la compression sèche est supérieure à 1700 kPa, résistance minimale exigée par la norme australienne. Les résistances les plus élevées ont été obtenues pour des échantillons avec des teneurs en eau initiale correspondant à l'optimum -2%. Cet essai, dans lequel les échantillons sont dans un état proche des conditions sur le terrain, présente des possibilités intéressantes.

7.2 *Minéralogie*

Des essais d'analyses chimiques ont été réalisés sur cinq échantillons provenant de différents tronçons étudiés. En effet, selon différentes études (Madu, 1980), il existe, pour les latérites, une bonne corrélation entre le pourcentage de silice (SiO_2) et celui des sesquioxydes ($Fe_2O_3 + Al_2O_3$).

Lors de cette recherche, les mêmes observations ont été effectuées. Lorsque le pourcentage de silice croit, celui des sesquioxydes baisse (Figure 5). De plus, il a été observé qu'il existe une relation entre le comportement des graveleux latéritiques sous trafic et la variation de ces pourcentages. Cet aspect est très important. Plus le matériau contient des silices, plus il se comporte mieux sous trafic. Ces observations ont permis de mettre au point un abaque complémentaire aux classifications des

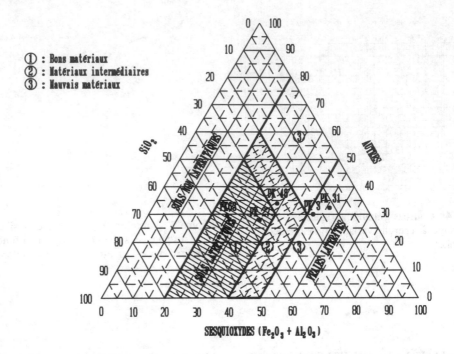

Figure 6. Projet d'abaque complémentaire aux classifications classiques dans le cas des graveleux latéritiques.

graveleux latéritiques utilisées dans la construction des routes non revêtues (Figure 6). Cet abaque permet de sélectionner les graveleux latéritiques en se basant sur la proportion de silice vis-à-vis de celle des sesquioxydes. Il doit être confirmé en effectuant d'autres vérifications sur le terrain.

8 CONCLUSIONS

Cette étude a permis de constater qu'aucun des essais classiques ne permet de distinguer les graveleux latéritiques qui se comportent bien de ceux se comportant mal sous trafic. Il a aussi été observé une grande dispersion des résultats d'analyses granulométriques sur un même échantillon. Les résultats de cet essai dépendent beaucoup des traitements initiaux et de l'habileté de l'opérateur, d'où la difficulté de reproduire l'essai. Les limites d'Atterberg et les essais de compactage Proctor ont une bonne reproductibilité. La variation de l'énergie de compactage a très peu d'incidence sur la destruction des grains du matériau.

La majorité des sols, y compris ceux prélevés sur la chaussée, se retrouvent dans la même classe selon les classifications AASHTO et USCS. Il est donc nécessaire d'affiner ces classifications pour qu'elles soient utiles dans la sélection des graveleux latéritiques utilisés pour la construction des routes non revêtues.

Les résultats des analyses chimiques ont permis de constater qu'il existe une relation entre le pourcentage de silice (SiO_2) et le comportement satisfaisant des graveleux latéritiques sous trafic. Les matériaux ayant un fort pourcentage de silice sont moins susceptibles au phénomène de la tôle ondulée et de la perte en matériaux.

Une meilleure maîtrise du comportement des graveleux latéritiques peut être obtenue:
- En améliorant les méthodes d'analyses granulométriques afin d'avoir une bonne reproductibilité de cet essai;
- En introduisant la mesure de l'indice de dureté modifié dans les études géotechniques;
- En poursuivant les essais d'analyses chimiques en vue d'établir des relations qui peuvent permettre d'utiliser ces essais

comme critère de sélection des graveleux latéritiques pour les routes non revêtues;
- En approfondissant la notion de résistance maximale à la compression sèche.

9 REMERCIEMENTS

Ces travaux ont été réalisés grâce à une subvention du Centre de recherches pour le développement international, Ottawa, Canada. Les institutions qui ont collaborées au projet sont respectivement, en Afrique, l'École des mines et de la géologie de Niamey, le Laboratoire national des Travaux Publics et Bâtiment du Niger, le CNREX LBTP du Mali et, au Canada, l'Université de Moncton.

10 RÉFÉRENCES

Autret, P. 1983. *Latérites et graveleux latéritiques*. ISTED, Paris, 38 p.

Autret, P. 1984. *Latérites*. Colloque international, Routes et développement, Paris, vol. 4, pp. 135-151.

Chatelin, Y. 1972. *Les sols ferralitiques T1, Historique, Développement des connaissances et formation des concepts actuels*. ORSTOM, Paris.

Cocks, G.C., Hamory, G. 1988. *Road construction using lateritic gravel in Western Australia*. Proceedings of the second International Conference on Geomechanics in Tropical Soils, Singapour, Vol. I, pp. 369-403.

Dreyfus, J. 1952. *Les latérites, généralités, leur utilisation en technique routière*. Revue générale des routes et aérodromes, no. 245.

Gidigasu, M.D. 1971. *The importance of soil profiles to the engineering studies of lateritic soils in Ghana*. Fifth Regional Conference for Africa on Soil Mechanics and Foundation Engineering, Luanda, pp. 55-60.

Gidigasu, M.D. 1972. *Mode of formation and geotechnical characteristics of laterite materials of Ghana in relation to soil forming factor*. Building and Road Research Institute, Ghana, 71 p.

Gidigasu, M.D., Yeboa, S.L. 1972. *Significance of pretesting preparations in evaluating index properties of laterite materials*. Transportation Research Record, no. 405, pp. 105-117.

Gidigasu, M.D. 1975. *Behaviour of laterite materials in Roadway Structure*. Building and Road Research Institute, Ghana, 37 p.

Gidigasu, M.D. 1985. *Sampling and testing residual soils in Ghana*. Sampling and Testing Residual Soil, Review of International Practice, pp. 65-74.

Little, A.L. 1967. *The use of tropically-weathered soils in the construction of earth dams*. Third Asian Regional Conference on Soil Mechanics and Foundation Engineering, Haïfa, pp. 35-41.

Madu, R.M. 1975. *Some nigerian soils - Their characteristics and relative road building properties on a group basis*. Sixth Regional Conference for Africa on Soil Mechanics and Foundation Engineering, Durban, pp. 121-129.

Madu, R.M. 1980. *The use of the chemical and physicochemical properties of laterites in their identification*. Seventh Regional Conference for Africa on Soil Mechanics and Foundation Engineering, Accra, vol. I, pp. 105-117.

Maignien, R. 1966. *Compte rendu de recherches sur les latérites*. ORSTOM-UNESCO, Paris.

Malomo, S. Obademi, M.O., Odedina, P.O., Adebo, O.A. 1983. *An investigation of the peculiar characteristics of laterite soils from Southern Asia*. Bulletin de l'A.I.G.I., Paris, no. 28, pp. 197-206.

Novais-Ferreira, H., Correira, J.A. 1965. *The hardness of lateritic concretions and its influence in the performance of soil mechanic tests*. Sixth International Conference on Soil Mechanics and Foundation Engineering, Montreal.

Peltier, R. 1969. *Manuel du laboratoire routier*. Edition Dunod, Paris, pp. 19-152.

Sandro, S.S. 1985. *Sampling and testing residual soil in Brazil*. Sampling and Testing Residual Soil, Review of International Practice, pp. 31-50.

Geotechnical characterization of weathered material in tropical regions

Caractérisation géotechnique du matériel d'altération dans les régions tropicales

E.G.Collares
University of São Paulo, São Carlos, Brazil

R.Lorandi
Civil Engineering Department, University Federal of São Carlos, Brazil

ABSTRACT: The weathered zone in tropical areas is generally thick and heterogeneous. These zones are composed of several levels of unconsolidated materials with different physical properties. The purpose of this work is to characterize these different levels leading to a correct evaluation of the geotechnical properties of the terrain, like subsidy to regional geotechnical mapping. Through the utilization of the alteration profiles and its correlation with the landforms, it is possible to represent the vertical variations and to delimit weathered zones with homogeneous distribution. The profiles are obtained after field data is collectd and the several levels are measured and described. The utilization of these profiles avoids the omission or the overestimation of materials that are fundamental in the assessment of the terrain. The landform analysis is leading through the use of field data and aerial photographic interpretations. This combination provides the space-time correlation of the profiles with some other characteristics of the physical environment making it possible to delineate the weathered units. The material properties are analyzed using the concepts of the lateritic and saprolitc soils, peculiar to tropical environments. The characterization encompasses the MCT Tropical Soil Classification and a expeditious textural classification, combining the field observations with lab experiments.

RÉSUMÉ: Le manteau d'altération dans les aires tropicalles c'est en général épais et héterogène. Ce manteau est composé de divers niveaux de matériaux non consolidés avec diférents propriétés physiques. L'intention de ce travail est celui de caractériser ces diférents niveaux en conduisant à une correcte évaluation géotechnique du terrain. Au milieu de l'utilisation des profils d'altération et leur rapport avec les "landforms" a été possible représenter les variations verticales et délimiter les zones de décomposition avec de distribution homogène. Les profils ont été obtenus d'après la collecte de donnés de champ, où divers niveaux ont été mesurés et décrits. L'utilisation de ces profils a évité l'omission ou la super-évaluation de matériaux lesquels sont fondamentaux dans l'évaluation du terrain. L'analyse des "landforms" fut éffectuée au milieu de l'utilisation de donnés de champ et de l'interprétation de photographies aériennes. Cette combinaison permit la corrélation espace-temps des profils avec des caractéristiques du milieu physique, en se rendant possible délimiter les unités décomposées. Les propriétés des matériaux ont été évaluées en utilisant des concepts relatifs aux sols latéritiques et saprolitiques propres des milieux tropicaux. La caractérisation renferme la classification MCT de sols tropicaux et une classification de la texture combinant les observations de champ avec des épreuves de laboratoire.

1 INTRODUCTION

The proposal application area presented in this work refers to the Quadricule of Bragança Paulista (SP, Brazil), limited by the meridians 46º30' - 46º45' W and by the latitudes 22º45' - 23º00' and located to the north of São Paulo city.

This region presents granite-gneissical cristaline rocks as rocky substrate. It constitutes, originally, of granitoids with an ample textural and compositional variation, which locally occur metamorphosed and/or mylonitized.

Due to an intense weathering action on these rocks, usual condition in tropical zones, a thick and heterogenous weathering mantle generally occurs, overlaying the unaltered rocky substrate.

Because of this it became extremely important to have an accurate and complete study of the unconsolidated materials leading to a correct geotechnical evaluation of the terrain.

The mechanism to be used should, therefore, show the variation of the alteration material, either in plant or in vertical profile and allow a previous knowledge of the geotechnical behaviour of the principal horizons.

The tools used for such a procedure were: alteration profiles; soil compartmentation by landforms and laboratory tests in accordance with the singular characteristics of tropical soils.

2 SOIL EVALUATION BY ALTERATION PROFILES

According to Deere & Patton (1971), alteration profile is defined as a sequence of material layers with different physical properties which developed "in situ", through the physical or chemical alteration processes.

By, adequating this to the tropical conditions, and also including the colluvial soils, each profile to be described must lead, mainly, to the distinction among the reworked, lateritic and saprolitic soils.

The data collecting process concentrates on a criterious field survey, since the preceding data, such as drill holes, are rare in poor countries like Brazil. In this survey, as a support, it was used a chart (Figure 1), which envolves three main types of data: the base data; data referring to the unconsolidated material an data referring to the rocky substrate.

The base data refer to some general characteristics of the point to be described. It is worth to mention, however, the items "Relief position", which sets the point at the slope, i.e., locates the point at the summit, half slope and low slope or valley; and the "rockiness" which indicates either the occurence of boulders or not.

For the unconsolidated materials, besides the characteristics and properties of the different materials, it is important to describe their space distribution, which designs the level which is being described. The data referring to the rocky material are the same used in a traditional geological survey.

Having done all the survey, each described point will be represented by an alteration profile, which, correlated to

the landforms, will allow a delimitation of homogenous areas, where each one of them will be represented by a unic typical alteration profile.

3 COMPARTIMENTATION BY LANDFORMS

The compartmentation of tropical soils by landforms, was proposed by Zuquette (1991). This compartmentation allows the space--time correlation of the weathered materials with other characteristics of the physical environment, leading, therefore, to the directioning of the extrapolations and delimiting the homogenous occurrences.

The term landform was accepted by Zuquette (1991) as being the physical environment elements which have a defined composition, as the variations of the physical and visual characteristics, such as: topographical form, drainage model and morphology. Thence, one must map the landforms considering the similar characteristics acoording to morphology and morphography, and which reflect the same evolution conditions (genesis).

The compartmentation envoled a preliminary photo interpretation, a field survey and finally a definitive photo interpretation, combining the alteration profiles with the relief elements.

In the preliminary photo interpretation a preliminary compartmentation of the landforms was obtained, by using as attributes the slope geometrical forms, declivity, lithology, weathering mantle absence, density and form of the drainage channels, relief dissection degree and vegetation.

In the field survey there was a checking of the preceding delimitations and lateral variations of the different horizons within their delimitations.

In the final photo interpretation it was studied the combination of all the described profiles with the other physical environment characteristics, encompassed in the different landforms, as to permit the delimitation of the units of unconsolidated materials, then presented by a unic typical alteration profile.

4 THE CHARACTERIZATION OF THE MATERIALS

The characterization of the unconsolidated materials involved two working stages: field stage and laboratory tests.

In the field survey, a description of the different materials was carried out according to the genesis and alteration

Figure 1. Diagram with the physical environment characteristics used in the field survey.

DATE:		FIELD CARD	Nº
		GEOTECHNICAL SURVEY AND ALTERATION PROFILE	
		EXECUTION: Collares, E.G.	

BASIC DATA		UNCONSOLIDATED MATERIAL
Point nº	Altitude	Surface Level
		Texture
Localization		Color
		Origen
Relief Position		Alteration Stage
		Thickness
Vegetal Covering		Constitution
		Lithology
Declivity		Compactness
		Consistency
Erosion		Permeability
		Lateral Continuity
Rockiness		LEVEL 2
		Texture
Current occupation		Color
		Origen
Water level		Alteration stage
		Thickness
Obs.		Constitution
		Lithology
ROCKY MATERIAL		Compactness
		Consistency
Homogeinidity degree		Permeability
		Lateral continuity
Lithology		LEVEL 3
		Texture
Texture		Color
		Origen
Mineralogy		Alteration stage
		Thickness
Alteration degree		Constitution
		Lithology
Frissure degree		Compactness
		Consistency
Structure		Permeability
		Lateral continuity
SKETCH		

degree as well as the tactile-visual identification of texture and the mica content in the soil levels and petrographical and mineralogical description in the less weathered levels.

At this stage, threre was a preocupation of correlating, still in the field, the materials which were being observed with the others previously described. This procedure made it possible to reduce and direct the samplings.

The laboratory stage involved Screen Analysis tests, Mini-MCV and Mass Loss by Immersion.

The screen analysis test involved sieving and sedimentation and was carried out in molds standardized by ABNT (Associação Brasileira de Normas Técnicas). The test results led to a textural classification of the materials, fitted in such a way as to distinguish similar materials but susceptible of distinct geotechnical behaviour. The criteria used in the classification are presented in Collares & Lorandi (1993).

The Mini-MCV and Mass Loss by Immersion tests were used for the MCT Classification of tropical soils, proposed by Nogami & Villibor (1985).

5 UNITS OF UNCONSOLIDATED MATERIALS

The resulting map of the used criteria shows 29 units of unconsolidated materials (Figuere 2), each of them represented by a typical profile and by a chart involving the principal characteristics of the unit.
    The typical profiles of alteration, representing each unit, show one or more distinguished levels, that means they are the levels of higher representativity for the area, as in terms of sand distribution, as in terms of geotechnical importance for the unit.

Figure 2. Map of unconsolidated materials (scale 1:50,000) of the Quadricule of Bragança Paulista (SP - Brazil).

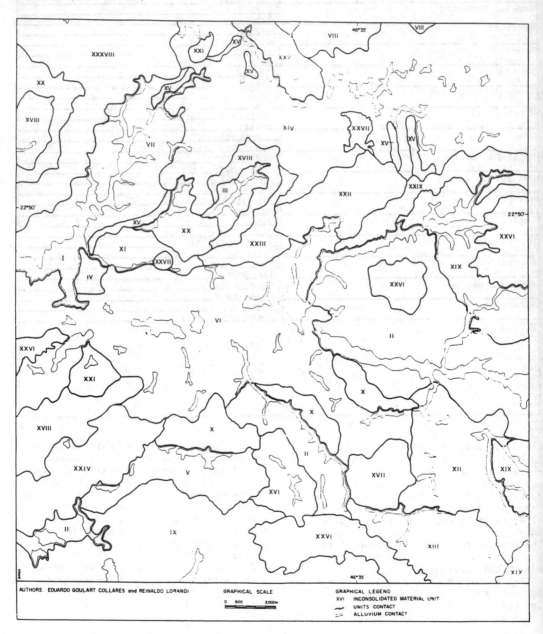

Figure 3 shows the informations contained in profiles and the Figure 4 classifies the informations and describe the symbols utilized.

Figure 3. Key used as an example in the interpretation of the alteration profiles.

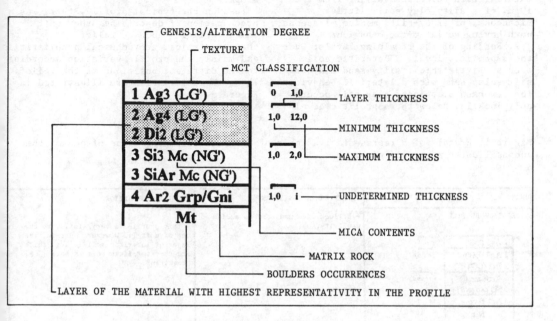

Figure 4. Information legends from the alteration profiles.

GENESIS/ALTERATION DEGREE	TEXTURE	MATERIAL COMPOSITION
1 – Reworked Soil	Ag – CLAY	MATRIX ROCK
2 – Lateritic Residual Soil	Ag1 – Clay with little sand (or silt)	Gni – Undifferentiated gneiss
3 – Saprolitic Residual Soil	Ag2 – Silt-sandy clay	GnX – Biotite/Gneiss and Biotite/
4 – Saprolitic	Ag3 – Sand-silty clay	Schist intercalations
5 – Altered Rock	Ag4 – Sandy clay	Gri – Undiffrentiated granitoids
6 – Undamaged Rock		Grp – Porphyritical granitoids
AL – Alluvium	Si – SILT	Grq – Inequigranular granites
	Si1 – Silt	Mi – Mylonites
	Si2 – Silt with little clay (or sand)	Mig – Migmatitic granites
	Si3 – Sandy silt	MGn – Mylonitized gneiss
		MGr – Mylonitized granites
	Ar – SAND	Peg – Pegmatites
	Ar1 – Sand with little silt (or clay)	
	Ar2 – Silty sand	MICA CONTENT
		MC – High mica content
	Ag/Si – CLAY AND SILT	Mc – Medium mica content
	AgSi – Silt clay	mc – low mica content
	SiAg – Clayey silt	
MCT CLASSIFICATION		BOULDERS OCCURRENCES
(LG') – Clayey Lateritic	Ag/Ar – CLAY AND SAND	MT – High boulders occurrence
(NG') – Clayey Nonlateritic	AgAr – Sandy clay	Mt – Scattered boulders
(NS') – Silty Nonlateritic	ArAg – Clayey sand	
(NA') – Sandy Nonlateritic		
	Si/Ar – SILT AND SAND	
	SiAr – Sandy silt	
	ArSi – Silty sand	
	Di – WELL DISTRIBUTED GRANULOMETRY	
	Di1 – Silty-sand clay	
	Di2 – Silty-clay sand	
	Di3 – Clayey-silt sand	

In the example of Figure 5, the profile reading of Unit XII must be done in the following order:
1. Reading of the distinguished level: the unit is composed, mainly, of a residual material with thickness varying from 2 to 10 m, of a silty clay soil, little micaceous, or sandy-silt, medium micaceous; both having no lateritic behaviour.
2. Reading of the remaining levels: Over the saprolitic level, a lateritic soil occurs, little thick, silty-sand clay or silty-clay sand with a lateritic behaviour. The reworked silty-sand clay soil occurs only locally. Below the saprolitic level the silty-sand saprolite appears, resulting from a porphiritic granite alteration.
3. Observations related to the occurrence of boulders: there is the occurrence of boulders scattered along the slope.

Besides the typical profile, there was a preocupation of describing other unit characteristics such as relief characteristics, geotechnical peculiarities and mainly the profile variation according to the different positions at the relief. These characteristics are illustrated in Figure 5.

Figure 5. Example of a representative table of the main characteristics of one of the unconsolidated material units.

UNITY XII ALTERATION PROFILE 1 Ag3 (LG') 0 1,0 2 Ag3 (LG') 0,5 3,0 2 Di2 (LG') 3 ArSi mc (NS') 2,0 10,0 3 Si3 Mc (NS') 4 Ar2 Grp 3,0 ∣ Mt	RELIEF CHARACTERISTICS - Hillock with moderate to medium declivity.	PROFILE DESCRIPTION PECULIARITIES: - This profile is very similar to the one of unit II, however, on this, the layer of lateritic soil is less thick and the saprolitic soil is more proeminent.
	TESTS BP382/01 - residual lateritic, silty--clay sand (LG'). BP382/02 - saprolitic, sandy silt (NS').	DESCRIPTION: - The lateritic soil is thicker in the plain areas, being able to reach over 3 m deep. - In the areas with a little higher declivity, the saprolitic soil prevails, medium micaceous, overlayed by a thin layer of lateritic soil. - The saprolitic soil is generally deep (sometimes over 10 m) and rarely is the saprolitic soil observed.
PRINCIPAL CHARACTERISTICS - Hillock with moderate to medium declivity. - Predominance of saprolitic soil in thick layers, sometimes superior to 10 m. - Boulders scattered at low slope.	GEOTECHNICAL PECULIARITIES - At the points 391 and 393 occur ravines and small slope slides, due to the occupation.	- Scattered boulders only occur at the low slope.

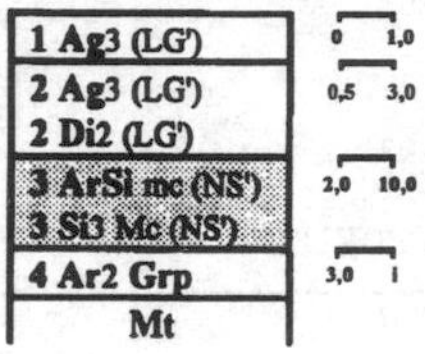

6 CONCLUSIONS

The use of alteration profiles proved to be efficient, leading to represent all the variations occured in the materials due to the weathering action. This method avoided the omission or the overestimation of some of the materials.

The way of characterizing the soils avoided a high number of samplings, besides evidentiating peculiary characteristics to the soils of tropical origen, as the lateritic or saprolitic character of them.

The procedure adopted for the compartimentation of the units led to a global analysis of the area physical environment, considering that a diversity of attributes were considered in the grouping of the homogenous occurrences.

In this way, the resulting map of unconsolidated materials describes not only the weathered mantle, but also a series of local physical environment geotechnical characteristics.

This procedure facilitated, a great deal, the crossover of the necessary information for the obtention of the General Geotechnical Zoning Map for the area.

ACKNOWLEDGEMENTS

The authors are indebted to PADCT/CAPES, that supplied the financial support to carry out of this work.

REFERENCES

Collares, E.G. & R.Lorandi 1993. A caracterização dos materiais inconsolidados na compartimentação geotécnica da região de Bragança Paulista (SP). In I Simpósio Brasileiro de Cartografia Geotécnica. São Paulo, Brazil (in press).
Deere, D.V. & F.D.Patton 1971. Slope stability soils. In Conference on Soil Mechanics and Foundation Engineering, 4º San Juan. Proceeding. American Society of Civil Engineering, V.1, p.87–170. New York, USA.
Nogami, J.S. & D.F.Villibor 1985. Additional considerations about a new geotechnical classification for tropical soils. In First International Conference on Geomechanics in Tropical Lateritic and Saprolitic Soils. Annais. V.1, p.165–174. Brasilia, Brazil.
Zuquette, L.V. 1991. Mapeamento Geotécnico: Ribeirão Preto. Cientific report, FAPESP. V.2, 269p. São Paulo, Brazil.

7th International IAEG Congress / 7ème Congrès International de AIGI, © 1994 Balkema, Rotterdam, ISBN 90 5410 503 8

Application of the cluster statistics in MCT data soil analysis
Application de l'analyse statistique des groupements aux données MCT des sols

A. R. L. Gonçalves & R. Lorandi
Civil Engineering Department, São Carlos, Brazil

J. M. M. Gonçalves
Geoscience and Environment Program, Rio Claro, Brazil

ABSTRACT: The MCT (Miniature, Compacted, Tropical) geotechnical tropical soils classification has been used in Brazil not only by the transportation ways geotechnicians but also in other geotechnical soil studies, like in Engineering Geological Mapping. For instance, it has been select 16 representative soil horizons, from 14 profiles of the urban area of São Carlos city-Brazil. The Cluster Analysis has been applied and it was determined the "distance and correlation coeficients" for these samples. By this analysis it was found the efficiency of this sampling, reproductibility of the Mini MCV testing and any error in the procedure. This overall procedure allow us to decide on the necessity of reanalysis of the samples for more representative final results.

RÉSUMÉ: La classification geotechnique MCT pour sols tropicaux a été utilisée au Brésil, non seulement dans les milieux géotechniques concernants à la construction de voies, mais aussi dans les autres études, comme dans la Géologie de l'Ingénieur. Pour cela ont été choisis 16 horizons représentatifs de sols originaires de 14 profils de l'aire urbaine de la ville de São Carlos-Brésil. Le programe CLUSTER d'analyse de groupements fut appliqué et on a déterminé les "coefficients de distance et corrélation" pour échantillons. Au moyen de ces analyses on a déterminé l'efficacité pour ces échantillonnages, la possibilité de répétition de la technique d'essai Mini-MCV et quelque erreur d'exécution. Ce conjoint de procès a porté à la résolution d'une nouvelle analyse des échantillons pour y trouver un resultat final plus réprésentatif.

1 INTRODUCTION

The purpose of the investigation was to use the MCT mechanical test in samples of lateritic soils from the urban area of São Carlos city-Brazil with the MCV (Moisture, Condition, Value) procedure.

The "Cluster Analysis" was used with the data obtained in order to know the samples similarities. By this analysis it could obtained the distance and correlation coeficients for the soils samples.

The area of interest is located in the central region of São Paulo State-Brazil, about 22°00' of latitude south and 47°55' longitude west.

2 METODOLOGY, RESULTS AND DISCUSSION

Based on the data of the MCT mechanical test, Nogami & Villibor (1985a) and Cozzolino & Nogami (1993), with the Mini-MCV procedure

results, Parsons (1976), it was selected 16 representative lateritic soil horizons, from 14 profiles of the area.

It was created a start matrix where the n lines represented the number of compacting shocks and the p columns the densities d_1, d_2 and d_3, corresponding to the same soil moisture. This configure the three repeated of the analysis for the same soil sample. To create this matrix and change it to a binary system it was used two computer programs.

The "cluster" program perform a analysis to determine the "distance and correlation coefficients" of the soils samples.

To obtain the "correlation coefficients" it was generate 137 dendograns. The "distance coeficient" was obtained using B horizons from 5 different soil profiles, and indicate the similarities among the average density of the clayey, silty and sandy lateritic soils (table 1).

Table 1. Average densities obtained over the Mini—MCV mechanical test, in same horizons of the differents soils profiles

| Samples | Soil Profiles | g/cm^3 | | | | |
		d_1	d_2	d_3	d_4	d_5
(1)	A_2	1.898	1.877	1.881	1.861	1.814
(2)	A_{10}	1.758	1.802	1.810	1.770	1.712
(3)	A_{11}	1.803	1.802	1.789	1.761	1.769
(4)	A_{12}	1.646	1.645	1.677	1.656	1.605
(5)	A_{14}	1.892	1.862	1.949	1.973	1.932

2.1 Correlation coeficients

Studing the 137 dendograns it has been found a good correlation, 98% to 99% among the three repeated MCT tests for the same soil sample. However, some data showed significant bad correlation. This bad correlation shows that something is wrong in the test procedure, therefore the data must be reworked. To perceive exactly the confiability level of the repeated tests it was done a statistical analysis for the 137 dendograns (table 2).

Table 2. Percentages of the samples for correlation coeficients

Correlation coeficient (%)	Percentage of the samples
0 - 9	0.364%
10 - 19	0.000%
20 - 29	0.000%
30 - 39	0.364%
40 - 49	0.000%
50 - 59	0.000%
60 - 69	0.000%
70 - 79	0.730%
80 - 89	2.554%
90 - 99	88.321%
100	8.029%

Table 2 shows for 88.321% of the samples the correlation coeficient given by the three repeated tests was from 90% to 99%. For 8.029% of the samples it has been found a correlation coefficient of 100% and for 2.554% of the samples it was found 80% to 89%. For only 1.458% of the samples a correlation coefficient of 80% was obtained. Studing the dendograns in more details it has been found that lower correlation levels when the test is repeated only once. When the analysis is repeated twice, a correlation of 80% or

better is obtained. Therefore, it is necessary to repeat the Mini—MCV analysis at least three times.

Table 3 shows the correlation coefficient between 90% and 100% in more detail. It car be observed that 79.927% of the samples had a correlation coefficient of 98% to 99,9%.

Table 3. Correlation coefficients between 90% and 100%.

Correlation coeficient (%)	Percentage of the samples
90 - 91	0.000%
92 - 92,9	0.730%
93 - 93,9	1.094%
94 - 94,9	1.459%
95 - 95,9	1.094%
96 - 97,9	3.649%
98 - 99,9	79.927%
100	8.029%

2.2 Distance Coefficient

The "distance coefficient" was obtained using the same horizon B from different soil profiles. It was used the medium valu of the densities obtained in the Mini—MCV procedure. The baseline used was the numbe: of compacting socks 12 (Table 1). It has been obtained the dendogran ds1 (table 4) showing the following:

Sample 1 (MCT classification: silty non lateritic soil) show a 7.92% similarity level with sample 5 (MCT classification: sandy non lateritic soil).

Sample 5 (MCT classification: sandy non lateritic soil) show 12.29% similarity level with sample 3 (MCT classification: clayey lateritic soil) and no similarity with sample 2 (MCT classification: sandy non lateritic soil).

Sample 3 showed the best similarity leve

in this dendogran (ds1), 19.18% similarity
level with sample 4 (MCT classification:
clayey lateritic soil).

Table 4. Dendogran ds1 showing the Distance Coefficient

```
.1839       .1555      .1271       .0987       .0703      .0419
  .1981       .1697      .1413       .1129      .0845      .0561      .0277
+----+----+----+----+----+----+----+----+----+----+----+
                                .-------------------- - 1------ .0792
                                I
                      .---------------------------------------- - 5------ .1229
                      I
                      I                     .----- - 2------ .0340
                      I                     I
  .------------------------------------------------------------- - 3------ .1918
  I
.------------------------------------------------------------------- - 4
+----+----+----+----+----+----+----+----+----+----+----+
.1981       .1697      .1413       .1129      .0845      .0561      .0277
  .1839       .1555      .1271       .0987      .0703      .0419
```

3 CONCLUSIONS

The low similarity levels among the samples
showed that the sampling was efficient and
sufficient for the proposed search.

The lateritic process in tropical soils
may have influenced the similarity level
between the soil samples and need a more
detailed study.

The granulometric fraction in tropical
soil may be a potencial active factor in
the similarity level of the soil samples
and need a more detailed study.

The study of the samples 3 and 4, of
clayey lateritic soils, show the necessity
of more sampling for the same kind of soil.

The number of repeated tests for the MCT
analysis using the Mini-MCV procedure must
be at least three.

ACKNOWLEDGEMENTS

The authors are indebted to CNPq - Conselho Nacional de Desenvolvimento Científico e Tecnológico, for the financial
support of part of this work.

REFERENCES

Cozzolino, V.M.N. & Nogami, J.S. 1993. The
  MCT Geotechnical Classification for
  Tropical Soils. Solos e Rochas 16:77-91.
Nogami, J.S. & Villibor, D.F. 1985a.
  Additional considerations about a new
  Geotechnical Classification for Tropical
  Soils. In:First Intern. Conf. Geomec. in
  Tropical Lateritic and Saprolitic Soils.
  1:165-174.
Parsons, A.M. 1976. The rapid measurement
  of the Moisture Condition of Earthwork
Material. L.R. 750. Transport and Road
Research Laboratory. Crowthorne, UK.

A demonstration to the definition of red clay

Une démonstration pour la définition de l'argile latéritique

Yu-Hua Wang

Architectural Design Institute of Guizhou Province, Guiyang, People's Republic of China

ABSTRACT

In the tropical and subtropical areas of China various types of parental rocks have experienced decomposition, dissolution / erosion, leaching and concentration as a result of geochemical laterization in the Quaternary inter-glacial moist / hot environment, hence leading to substantial changes in their occurrence, composition, texture and color. This may be a reasonable explanation of the mechanism of formation of red clay, which is widely spread in southern China. The laterite distributed on carbonate rocks is referred to as red clay and the transported and re-accumulated laterite as secondary red clay. Laterite is dotted about in northern China while it is widely distributed in southern China, covering an area of 2 million km2. Its subtype - - red clay is distributed in the area south of thirty - three degrees north latitude in China, covering an area of 1080000 km². The red clay is composed mainly of kaolinite minerals, with clay particles accounting for 50~80%; liquid limit, 50~120; porosity ratio, 1.0~2.2, saturation degree, >95%; and thickness, 5~15m. As observed in its sections, red clay is hard in the upper part and soft in the lower part. Fissures are well developed in hard plastic red clay, exhibiting fragmental texture. Red clay can be used as natural foundation for buildings, and also can be used as building materials for reservoir dams and roads. Research on red clay is of great significance both in architectural engineering and in economic profits.

RÉSUMÉ

Dans les régions tropicales et subtropicales de Chine, les roches mères ont subit décomposition, dissolution, érosion, lessivage, et concentration, changé d'état, composants, texture et couleur, transformé en l'argile latéritique distribué largement en Chine du sud. Tout cela à pour cause de l'environnement humide-chaud, de l'interglaciation quaternaire, sous l'effet de la latérisation géochimique. La latérite distribuée dans la roche carbonaté est référé comme argile latéritique, et la latérite transportée et reaccumulée comme argile latéritique secondaire.

L'argile latéritique, est dispersé dans le Nord, s'étendant largement dans le Sud sur une superficie de 2 millions km². L'argile latéritique secondaire est répartie dans les régions sud de latitude 33° N en Chine, sur une superficie de 1 080 000 km². L'argile latéritique contient principalement kaolinite, avec des particules d'argile 50-80 %, le limite de liquidité 50-120, la porosité 1.0-2.2, la saturation >95%, et l'épaisseur 5-15 m. L'observation de la section de l'argile latéritique montre que son dessus est dur, et l'inférieur est mou. Les fissures se développent en l'argile latéritique plastique dure, avec une texture fragmentaire. L'argile latéritique peu être utilisée comme la base naturelle de bâtiments, et aussi comme le matériel architectural pour les barrages et les routes. La recherche sur l'argile latéritique a une signification assez grand pour la construction architecturale et le profit économique.

COMPARISON OF THE DEFINITIONS OF RED CLAY

In the tropical and subtropical areas of China various types of parental rocks once experienced the geo-logical processes of laterization, leading to their decomposition, dissolution / erosion, leaching, migration and concentration in the Quaternary interglacial moist / hot environments. Consequently, great changes took place in the shape, composition, texture and colour of the rockbodies. This may account for the formation of the so – called " laterite " which is now widely spread in southern China. Laterite is a more widely spread and more characteristic Quaternary deposit than loess. Comprehensive studies of it are of great significance in practice.

In the early 1960' s many engineering geologist engaged in civil and industrial construction successfully discovered the red clay which can be used as natural foundation for buildings since it is characterized by high dispersion, high plasticity and high foundation strength in terms of the grain composition, physical and mechanical properties, horizontal and vertical distribution characters and excellent foundation performance of eluvial – slope deposits widely spread on carbonate rocks of the Guizhou – Yunnan Plateau. After more than ten years of study till the 1970' s, red clay as a specific type of foundation soil was formally defined and listed in "Standards of Foundation Design" (TJ – 7 – 74) and "Standards of Engineering Geological Survey"(TJ – 21 – 77).

Since the 1970' s, research on red clay has been carried out on a nation – wide scale. As a result, a better understanding has been gained of its assignment, distribution characteristics, age of formation, geological process of pedogenesis, relations with the parental rocks and secondary variations, as well as characteristic indicators for its classification.

New advances in the study of red clay in recent years help us to realize the necessity of reconfirming its definition. At the Conference on Engineering Geology held in 1981 in Xiamen, the author published the paper " The Status Quo of Red Clay Research" in which the definition of red clay was well documented. At the same time, the author appealed for revising its definition.

In 1982, in order to revise "TJ – 7 – 74 Standards", red clay was put on the list of specific research projects. Through more than three years of investigation in twelve provinces and regions we gathered 179 pieces of opinions and suggestions, 2383 indicator data, 123 loading test data and 2260 physico – mechanics data from 47 units engaged in surveying, designing and construction. The data available provide the basis for solving the problems involved in the revision of the definition of red clay, including its assignment, distribution, forming age, geological processes of pedogenesis, relations with its parental rocks, secondary variations, specific indicators, etc. The definition and de-

scription of red clay proposed by the author were approved at two meetings for examination in 1984 and 1985, respectively. The newly comprised " Design Standards of Construction Foundation" (GBJ – 7 – 89) has embraced the revised definition of red clay. As compared with the original standards, the newly comprised ones have involved the following aspects of revision:

(1) Carbonate rocks which were considered as the parental rocks of red clay are replaced by carbonate series rocks.

(2) The geological process of pedogenesis of red clay is a geological process of laterization instead of a weathering process.

(3) More genetic types of red clay including those of slope diluvial, alluvial – diluvial and ice – borne origins (i. e., secondary red clay) have been recognized instead of one type(the eluvial – slope type).

(4) The distribution of red clay is closely related with its pedogenetic history. More importantly, in the practice of construction it must be taken into consideration that in which areas the red clay foundation design standards should be implemented.

(5) As for the classification indicators of red clay, which indicator is better, its status or character? Studies have shown that the indicator of soil character is better since it shows no variation with the outer conditions. The indicator of soil status is very sensitive to the outer conditions. Among the classification indicators, liquid limit is the most sensitive indicator.

(6) The porosity of red clay tends to decrease from west to east. In the west where the Yunnan – Guizhou Plateau lies the porosity is generally greater than 1. 0 while in the east it is relatively small.

(7) In regard to the natural content of water in red clay, only in the hard plastic part is it close to the plastic limit, but in the other parts it is greater than the plastic limit.

(8) The saturation degree of red clay is generally greater than 95%.

In order to validate the new definition of red clay a special discussion will be given in the following section

ASSIGNMENT OF RED CLAY

Red clay is widely spread in the Quaternary strata of China, covering an area of distribution much greater than that of loess in China. All deposits including those of eluvial – slope and transport – accumulation origins

Table 1. Comparison of red clay and laterite developed
on granitoids in Guangdong Province

Soil type	W(%)	e	W_L(%)	W_p(%)	Ip
Laterite derived on granite	26	0.8	35	20	15
red clay	32	1.00	55	29	26

were formed under the climatic conditions favoring laterization is called laterite. In laterite – distributed areas the deposits which were formed as a result of laterization of carbonate series parental rocks, followed by transport and accumulation, are called red clay and secondary red clay. Accordingly, it is considered that:

(1) Red clay is the special part of laterite, which processes special characters and properties. For example, we can compare red clay with laterite, which are both developed on granitoids in Guangdong Province (Table 1).

(2) Compositionally, red clay is the portion of laterite which is most dispersed and highest in clay particles. (Table 2).

(3) In regard to the indicator of soil character, red clay is a portion of laterite which has the highest plastic – liquid limit so that it is characterized by the highest plasticity(Table 2).

(4) Red clay is generally high in saturation degree, up to 95%, but the saturation degree of laterite varies between 80 - 90%

(5) As compared with laterite, red clay is more extracted.

Generally speaking, red clay is embraced in laterite, but there is no reason to regard red clay as laterite. For example, laterite developed on red layers in Jiangxi and Hunan provinces and on volcanic rocks in the coastal areas of Southeast China, as well as on vast sandy shales contains only 30 - 50% clay particles, with the porosity ranging from 0. 6 - 1. 0, as is the case for the laterite developed in Shandong, Shanxi, Hebei and Liaoning provinces. As for the laterite developed on basalts in Hainan Island and the Leizhou Peninsula, its liquid limit ranges from 30 - 60%, and its porosity varies from 0. 7 - 1. 1. A considerable amount of such laterite is subclay, and the rock / soil interface characteristics are significantly different. Laterite distributed in carbonate rock interbeds should also be called red clay because its characteristic indicators are extremely close to those of red clay developed on carbonate rocks (Table 2).

DISTRIBUTION OF RED CLAY

During the Quaternary period laterization – climatic changes directly controlled the distribution of laterite, and the laterization climate and the distribution of carbonate rock series also controlled the distribution of red clay. Why are there signs of laterite distribution in northern China, but almost no signs of red clay distribution there? That is because the wet – warm climate conditions in favor of laterization were prevailing during the Middle – Early Pleistocene interglacial period in northern China, while during the Late Pleistocene and later periods the wet – warm climate conditions favoring laterization shifted towards the area south of the Yangtze River. As a result, the early laterite in northern China has been eroded or covered. This may account for the little or dispersed occurrence of laterite in northern China but the extensive occurrence of laterite and red clay in southern China(Fig. 1). Statistical data show that the distribution area of laterite is about 2000000 km² while that of red clay is about 1080000

Table 2. Comparason red clay and laterite

Soil type	W_L (%)	W_p (%)	W(%)	e	Content of clay particles(%)
Red clay	45~120	20~60	20~75	0.90~2.20	40~80
Red clay interbedded with basalt	45~80	20~55	30~70	0.80~2.10	
Red clay interbedded with sandy shales	45~80	20~50	25~70	0.80~2.0	
Laterite	20~60	20~30	20~30	0.6~1.1	30~50

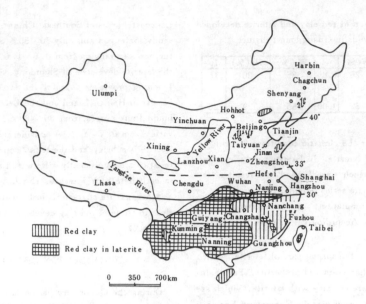

Fig.1. Geological sketch map showing the distribution of laterite and red clay in China.

Table 3. Comparison of red clays form different location

Location	Southern China								Northern China
Province	Yunnan	Guizhou	Guangxi	Sichuan	Hunan	Hubei	Guangdong	Anhui	Shanddong
WL(%)	50~80	50~120	45~92	45~83	45~82	40~80	45~90	23~60	23~60
Wp(%)	20~50	20~60	20~48	20~40	20~50	20~45	17~50	20~50	17~60
e	0.80~2.00	0.80~2.20	0.80~1.70	0.80~1.80	0.80~1.80	0.80~1.80	0.80~1.60	0.80~1.20	0.60~0.90

km², of which brown red and brownish – red clay covers an area of 740000 km² and yellow and yellowish – brown clay covers an area of 340000 km².

The following features are presented in regard to the distribution of laterite and red clay.

(1) In southern China the entire Quaternary strata are controlled by such special topographical features as being high in the west and low in the east as a result of lifting of the Tibet Plateau. Therefore, there exist significant differences in the distribution of laterite and red clay in going from west to east. No laterite and red clay are recognized in the region of the Qinghai – Tibet Plateau except the valley area of southern Tibet because of no wet – warm climatic environment there. The Yunnan – Guizhou Plateau is an area where red clay is most widespread and the percentage of its distribution area is the highest in China. As a consequence of the widespread of carbonate rocks on both sides of the Nanling Range and in hilly lands and islands along

the coast of China, laterite is of extensive occurrence and red clay is of subordinate distribution.

(2) Due to ever increasing differences in climate conditions between the north and the south of China since the Late Pleistocene, red clay developed on parental rocks of the same type but in different locations would show great differences in nature and dispersion. An analysis of liquid limit indicators and porosities of red clay samples from various provinces and regions indicates a tendency of decreasing from west to east. Such differences are more remarkable in the north of China with respect to the liquid limit indicator and porosity of red clay(Table 3).

(3) In southern China's provinces the various indicators of red clay vary regularly in the westeast direction and there are also some differences in the depth and extent of laterization. In the red clay – distributed zone including eastern Yunnan, southern Sichuan, northern Guangxi, western Hunan, western Hubei, etc.

with Guizhou Province as its center red clay usually contains 6 - 15% Fe_3O_2. For this reason, the red clay is mostly yellow and brownish – yellow in colour. In other areas red clay is much higher in Fe_2O_3 up to 15 – 25% as a result of a high extent of desilication and Fe – Al enrichment.

FORMING AGE OF RED CLAY

The data available show that the establishment of the Quaternary time scale is based mainly on paleontology paleoclimate, paleogeomorphology and crustal tectonic movement data, as well as on paleoantropology and ancient cultural remains. Such is the case for the history and age of formation of red clay. The precise determination of the forming age of red clay requires modern sophisticated techniques. Comparatively speaking, palaeomagnetic dating and sporo – pollen analysis are the most reliable techniques for this purpose.

(1) Extensive occurrence of Quaternary glaciation in the various provinces and regions of southern China was followed by glacial and interglacial phases, as remarkably reflected by cumulus phases and topographical features. Let's take Guizhou Province for example (Table 4), which can be compared with the eastern regions of China with respect to its glacial and interglacial history and stratigraphic ages, as well as to the

developing position, geomorphological unit of pedogenesis and general thickness of red clay. Red clays of different ages are different in distribution area and continuity. Early Pleistocene red clay is only recognized in local places, Middle Pleistocene red clay is more widely distributed, mainly on hilocks and ridges, and Late Pleistocene red clay is most widely distributed either in karst depressions or at the slope of hilly land.

(2) The Himalayan Movement is the foundation of modern topographic development in China. From Lower Tertiary through Cenozoic the northward compression of the South Asian subcontinent resulted in the formation of the Tibet Plateau. At the same time, in going from west to east, a three – step topographic pattern was created. The intermittence, recycle and difference of the crust uplifting movement since the Cenozoic led to the formation of greatly undulated ancient planes of denudation. The division of stratigraphic ages and the determination of geological history, i. e., the so – called " geocultural period", by geomorphologists are just based on these ancient planes of denudation. The pedogenetic history of red clay can be compared with the " geocultural period". According to the geocultural data from western Hubei and Guizhou provinces(Table 5), for example, it is reliable enough to determine the evolutionary history of red clay in terms of plant and animal fossils accumulated in karst

Table 4. Relationship between glacial and interglacial periods
and development of red clay in Guizhou Province

Stratigraphy	Glacial and internglacial		Red clay	
	Southern China	Guizhou	Occurrence	Geomorphological unit
Q₃	Dali glacial period Lusgan – Dali interglacial period	Luowan glacial period Xiaoliang – Luowan interglacial period	15~20m above river water face	Karst depressions sur – and valleys; red clay measuring 5~20m in thickness
Q₂	Lushan glacial period Dagu – Lushan interglacial period Dagu glacial period	Xiaoliang glacial period Guankou – Xiaoliang interglacial period Guankou glacial period	30~60m above river water surface	Amcient planes of denudation, red clay measuring 3~8 in thickness
	Boyang – Dagu inter - glacial period Boyang glacial period	Longjingquan – Guankou interglacial period Longjingquan glacial period	80~150m above river water surface	Ancient planes of denudation, red clay measuring 0~5m in thickness
Q₁	Dongcheng – Koyang interglacial period Dongcheng glaical period Early glacial period – Don - gcheng interglacial period Early glacial period		150~250m above river water surface	Moderate – low hills, no red clay recoglized

3449

Table 5. Relationship between the geocultrual and the development
of red clay in Guizhou

Age	Geocultural period		Red clay	
	Western	Hubei Guizhou	Acient plane of denudation	Thickness of red clay
Q₃	Sanxiao period	Xiagu period	Karst depression Karst vallley	5~15m
Q₂—Upper Tertiary	Shanpen period	Shanpen period	2~3 grade ancient planes of denudationon	0~3m
Lower Tertiary	Oxi period	Gaoyuan period or Daloushan	The highest plane of denudation	No red clay

Table 6. Acidity – alkalinity data for the earthy
bodies in the various parts of China

Province or region	North of the Qinling Range	Hubei, Hunan, Anbui, Jiagxi, Zhejiang	Yunnan, Guizhou, Guangdong, Guangxi, Fujian
PH	7~10	5~8	3~6
Salt staturation degree(%)	70~100	30~70	<30

Table 7. The cation – exchange capability data
for the various parts of China

Province of region	Inner Mongolia, Xinjiang, Northeast China	Hebei, Shaanxi, Shanxi, Shandong	Yunnan, Guizhou, Hunan, Jiangxi, Zhejiang	Guangdong, Guangxi
Cation – exchange capability (me / 100g)	>20	10~20	5~15	<10

caves which are correspondingly compared with the ancient planes of denudation.

(3) As it was developed in different geomorphological locations the Pleistocene red clay has experienced varying degree of erosion and denudation. Consequently, a great difference is noticed in the preservation status of red clay from one location to another. Since it was developed on the highest ancient planes of denudation, the Lower Pleistocene red clay has been, for the most part, denudated and hence is of rare occurrence. Red clay recognized now in karst depressions and on killocks and ridges is mostly Middle to Late Pleistocene in age.

RELATIONSHIP BETWEEN RED CLAY AND PARENTAL ROCKS AND GEOLOGICAL PROCESSES OF PEDOGENESIS

As described above, the formation of laterite and red clay is predominantly dependent on wet – hot climate environments. Throughout the Quaternary period the wet – warm climate conditions are changeable from place to place. Studies show that wet – warm climate conditions were strongly variable before and during the Late Pleistocene and slightly variable after the Pleistocene. Therefore, the current acidity – alkalinity and cation – exchange capability of the earthy bodies can, to some extent, reflect the regularities governing the

Table 8. Chemical composition of red clay

Province or region	Chemical compont(%)						
	SiO_2	Al_2O_3	Fe_2O_3	CaO	MgO	K_2O	Na_2O
Yunnan	40	24	19	0.30	0.40	0.63	0.16
Guizhou	40	23	13	0.22	3.00	0.70	0.30
Southern Sichuan	50	27	27	0.52	0.77	0.20	0.30
Northern Guangxi	60	17	17	0.56	0.65	/	/

Table 9. Clay particle contents and clay mineral compositions of laterite and red clay

Province or region	Inner Mongolia Xinjiang	Northeast China	Hebei, Shandong	Jiangsu, Anhui	Hubei, Hunan, Jiangxi	Yunnan, Guizhou, Guangdong, Guangxi
Clay particle content(%)	17~23	18~40	16~45	30~60	30~70	40~80
clay mineral composition	Hydromica	Hydromica, montmorillonite	Hydromica, vermiculite	Hydromica, vermiculite, kaolinite	Kaolinite, hydromica	Kaolinite

fluctuation of climate in a certain period of the geological history. The acidity – alkalinity and cation – exchange capability data obtained recently from the earthy bodies in various parts of southern and northern China are presented in Tables 6 and 7. In the various provinces and regions to the south of the Yangtze River the laterite – and red clay – distributed areasare characterized as being acidic in nature and low cation – exchange capability, suggesting that the mediums described above are strongly depleted in K, Na, Ca and Mg, but highly enriched in Si, Al and Fe which account for 90% or more. This is a clear indication of strong laterization (Table 8).

Similarly as described above, obvius variations are noticed in thecontents of clay particles and the composition of clay minerals inlaterite and red clay(Table 9). Laterite and red clay are predominatedby kaolinite minerals, which are the product of laterization. In the vertical profile it is observed that the closer to the parental rocks, the more predominant the primary minerals and the less the secondary minerals will be.

Under the wet – hot climate conditions during the Quaternary interglacial period there took place a complete process of laterization from the parental rocks to red clay. In this process the following components had experienced considerable variations: the transport amounts of Ca and Mg are within the range of 97 ~ 100%; and the enrichment amounts of Si, Fe and Al are 40~70%, 7~25% and 10~25%, respectively.

In fact, the history of laterization can be summarized as follows:

At the first stage of laterization from parental rocks to soil two aspects of variation took place: in the one hand the parental rocks were broken and decomposed and the primary clay minerals were partly accumulated and partly carried away by flowing water, and on the other hand the water – soluble minerals were dissolved to form saline solutions, which were partly lost and partly retained in – situ, followed by oxidation, hydrolysis, etc. to form secondary minerals.

At the second stage of formation of red clay the various components involved in pedogenesis would further evolve under wet – hot climate conditions. In case the annual average temperature was 17~25℃, the annual accumulated temperature 5000 ~ 9500℃ and the annual precipitation 1000~1500mm, the chemical components K, Na, Ca and Mg, as well as silicates tended to decrease further whereas the oxides of Fe and Al tended to be further enriched. Owing to the differences in latitude and geomorphological features, as well as in wet – hot climate conditions, the depth of laterization would also vary from place to place.

Table 10. Physicomechanics indices of laterite
and red clay form Guangdong Province

Parental rock and soil type		Index									
		W (%)	ρ (%)	Sr (%)	e	WL (%)	Wp (%)	Ip	It	C (kpa)	φ (degree)
Laterite	Granite	26	1.86	86	0.80	35	20	15	0.74	22	30
	Basalt	31	1.84	88	1.00	42	20	22	0.73	57	19
Red clay		32	1.86	92	1.10	55	29	26	0.58	43	22

Table 11. Comparison of red clay, secondary red clay
and other kinds of clay

Soil type	W_L(%)	Wp(%)	W(%)	e
Red clay	50~120	20~60	20~70	1.00~2.20
Secondary red clay	45~100	20~50	20~60	0.80~1.80
Common clayey soil	32~54	17~31	19~35	0.35~1.05

Because of these differences in laterization the laterite and red clay developed on different types of parental rocks would be different in clay particle contents and physicomechanics indices. Red clay developed, for example, on parental granitic rocks and sandstones has a content of clay particles ranging from 30~40%, but that on parental basalts and limestones has a content of clay particles varying from 35~50% and from 50~80%, respectively. Also, great differences are noticed in physicomechanics index for red clay from the north – central part and for laterite developed on granitic rocks in the southeastern part and on basalts in the southern part of Guangdong Province (Table lO).

It is worth to point out that laterite developed on some basalts seems to be one kind of red clay in consideration of its porosity greater than unity and its liquid limit greater than 50%. Careful examination of the geological section, however, provided evidence for the existence of loam which is never associated with red clay. Moreover, significant differences are noticed between laterite and red clay with respect to their respective rock / soil interface characteristics.

SECONDARY RED CLAY

Red clay accumulated since the Pleistocene as a result of glaciation and water flow action, i. e, of eluvial origin, displays signs of variation in grain composition and physicomechanics properties. Such red clay is generally characterized as being coarse in grain size, light in colour and soft in structure, and hence is called secondary red clay.

Differences in paleogeographical and geomorphological environments as well as in the mode and intensity of water flow action may be a reasonable explanation of the great differences in properties for secondary red clay. On the Yunnan – Guizhou Plateau the karst geomorphological features represented by karst valleys are poorly developed because of the unnormal river – erosion patterns, i. e., the relief is gentle in the upper reaches of the rivers but greatly undunated in the lower reaches. For this reason, the karst valleys and karst depressions are relatively small in size and secondary red clay is of limited occurrence. Comparatively speaking, in the provinces and regions of eastern China secondary red clay is widely spread. Secondary red clays in different areas are different in grain composition and liquid limit index. Even so are the secondary red clays of different origins. In the Guiyang Basin, for example, red clay of alluvial – diluvial origin is distributed in the shallow depressions and contains more or less gravels about 2. 0 mm in size, breccias, gravel sands, detritus, etc. However, secondary red clay in the Duyun Basin, southern Guizhou Province contains more glacial boulder clay, although its clayey fraction is extremely similar to that of red clay.

The data for secondary red clay and red clay are presented in Table ll. Comparing secondary red clay with primary red clay, we can notice some differences between them, but as viewed from their plasticity and

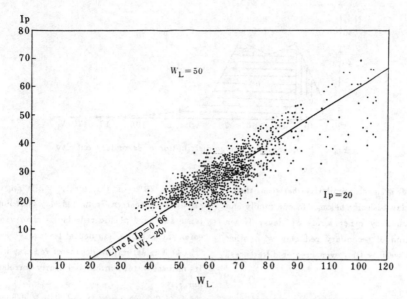

Fig 2. Diagram of the plasticity of red clay

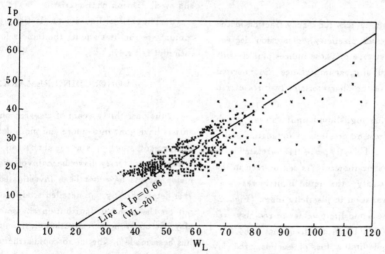

Fig 3. Diagran of the plasticity of secondary red clay

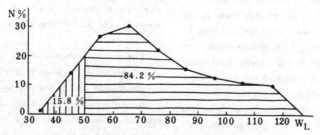

Fig 4. Frequency percentage distribution of red clay

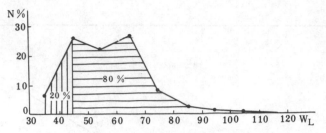

Fig 5. Frequeney percentage distribution of seeondary red clay

dispersion there is much similarity between them. Secondary red clay basically belongs to one family and cannot be matched by other kinds of clays. If we compare this kind of secondary red clay with other kinds of clays, we may find even greater differences. (Table ll).

CHARACTERISTIC VALUE INDICES OF RED CLAY

The characteristic indices for clayey fraction in the foundation soil include plasticity, dispersion degree, saturation degree, state, etc. These indices plus classification indices are all of great importance. Summarized in the following are the characteristic features of red clay.

(1) Red clay has high liquid limit values, which makes it distinct from other kinds of foundation soils. The liquid limit of red clay is perfectly correlated with other physicomechanics indices. As can be seen in the diagram of its plasticity, the liquid limit increases against no obvious variation in plasticity index (Figs. 2 nd 3). In case the liquid limit values of red clay are within the range of 50 − 120 %, the frequency is 84. 2 % and in case the liquid limit values of secondary red clay are within the range of 45 - 85 %, the frequency is 80 % (Figs. 4 and 5).

(2) Non other foundation soils than red clay have a high porosity. The porosity of red clay tends to decrease gradually from the west to the east of China. In comparison to the clayey soil of low dispersion, red clay is dense in structure, although its porosity is relatively high. Nevertheless, the porosity of red clay is not a decisive index for the foundation strength.

(3) Red clay layers are characterized as being hard in the upper parts and soft in the lower parts. Hard plastic red clay on the Yunnan − Guizhou Plateau accounts

for 60 - 70 % of the total, while in the various provinces of eastern China red clay is of limited occurrence and hard plastic red clay is all developed in fissures and exhibits fragmental texture.

(4) The saturation degree of red clay is over 95 %. Its natural water content is linearly correlated with its porosity.

(5) The strength of red clay is dependent on its mode of occurrence, the developing extent of fissures and swell − shrink characteristics.

(6) The ratio of liquid over plasticity and the water content are the best among the indices for the classification of red clays.

CONCLUDING REMARKS

In the past thirty years of research on red clay scientists have gone into more and more details of it. At the present stage of study, almost all the problems concerning red clay have been involved, but several major aspects have not been investigated in considerable detail. Firstly, no detailed study has been carried out on the regional distribution characteristics of red clay. Secondly, no investigation on the qualitative basis has been made on the microscopic structure of red clay as well as its macroscopic structure resulting from the development of fissures. Finally, further investigations should be made on the engineering performances of red clay, for example, the effects of structure, lagging strength, the long − term effect of foundation deformation, etc.

The repeated revision of the definition of red clay objectively reflects a history of how red clay has been recognized and understood step by step by scientists. It is impossible to establish a permanent definition and repeated consideration of the problems concerning red clay is needed so as to make it more scientific, more complete and more objective the definition to red clay.

REFERENCES

Wang Yuehua, 1979, Examination of the homogenity of red clay foundation: Guizhou Architecture, No. 2.

Wang Yuehua, 1981, A contribution to the status quo of research on red clay: The Proceedings of the First Symposium on Engineering Reconnaissance, Beijing, Architecture Press.

Wang Yuehua, 1987, On fissure red clay: The Proceedings of the Third Symposium on Engineering Reconnaissance sance, Beijing, Architecture Press.

Wang Yuehua, 1989, Distribution characteristics of laterite and red clay in China: Advances in Engineering Geology of China, Chengdu, Chengdu Science and Technology University Press.

Un mode de caractérisation génétique des sols résiduels compatible avec les données de l'expérimentation géotechnique: L'approche normative

Genetic characterisation system for residual soils compatible with the geotechnical experimentation data

G. E. Ekodeck
Institut de Recherches Géologiques et Minières, Yaoundé, Cameroun

RESUME : Les données issues des modes traditionnels de caractérisation génétique des latérites et des sols résiduels n'ont, contre toute attente, pas de signification en géotechnique. Nous proposons, exemples illustratifs à l'appui, une méthode normative de caractérisation génétique dont les résultats se montrent compatibles avec ceux de l'expérimentation géotechnique.

ABSTRACT : The data obtained from the traditional genetic characterization systems for laterites and residual soils, are hopelessly not significantly reliable to the geotechnical ones. Hereby is offered and illustrated a normative genetic characterization method, the results of which appear to be compatible with those from the geotechnical experimentation.

1 INTRODUCTION

Sous les conditions thermodynamiques de l'interface continentale, les processus d'édification des minéraux qui prévalent dans la séquence évolutive des matériaux de la lithosphère sont : l'allitisation, la monosiallitisation et la bisiallitisation, (Pedro et Delmas 1971).

Ils sont favorisés, en milieu d'altération ouvert dont le pH est compris entre 6 et 9, par l'hydrolyse des minéraux primaires et par l'évacuation en solution (lixiviation) des éléments ferromagnésiens et calcoalcalins, et de la silice, et en milieux fermé, ou confiné, par l'accumulation de ces éléments, l'évaporation et la saturation ionique (Paquet 1969). On peut observer également dans ces milieux, des phénomènes d'accumulation relative ou absolue d'éléments peu mobiles sous forme d'oxydes et/ou d'hydroxydes (fer ferrique, aluminium, manganèse) dont la teneur conditionne le cuirassement (Maignien 1966).

La caractérisation courante de l'altération s'effectue par le biais de bilans qui sont chimicominéralogiques ou hydrogéochimiques. Les premiers, concernant la phase résiduelle de l'altération, intéressent les géotechniciens. Ils nécessitent une connaissance suffisante des compositions minéralogique et chimique qui permet le choix d'un invariant par lequel peuvent être effectuées des comparaisons, qualitatives ou chiffrées, entre l'état initial, celui de la roche saine et homogène, et les diverses étapes de sa transformation.

Les principales démarches sont isovolume (Millot et Bonifas 1955), isoéléments (cf. Leneuf 1959), ou isominérales (cf. Lelong 1969). D'autres démarches n'ont pas d'invariant (cf Joachin et Kandiah 1949, ou Wackermann 1975).

La caractérisation géotechnique des sols résiduels a porté sur la détermination et la recherche de l'amélioration de leurs propriétés physiques et mécaniques.

En considérant que leur comportement est lié à l'ampleur de l'altération, de nombreux auteurs ont tenté de chercher des corrélations significatives entre les données de la nomenclature génétique et les propriétés géotechniques. On peut citer les travaux de Gidigasu (1974), Tuncer et Lohnes (1977), etc.. Si tous ces auteurs ont pratiquement abouti au constat d'inadéquation, cela est dû au fait que la quantification des propriétés géotechniques concerne chaque échantillon de matériau, considéré de façon globale et individuelle, alors que la caractérisation génétique s'effectue de façon sélective et comparative.

La méthode normative en expérimentation

depuis les années 1980, notamment dans le domaine géotechnique (Ekodeck 1984, 1989, 1990) est un moyen globalisant et sans invariant, qui s'affranchit de la référence à la roche mère.

Nous en présentons d'abord les grandes lignes, ainsi que le degré d'affinité de ses résultats minéralogiques avec la réalité. Nous exposerons ensuite l'étude d'un cas illustratif pris au Cameroun.

2 LA METHODE NORMATIVE (EKODECK 1989).

2.1 Critères de choix

L'exécution d'analyses chimiques sur les roches entraîne la rupture des liaisons cristallines naturelles. L'information quantitative obtenue n'a plus rien à voir avec la structure réelle de la roche. Pour l'exploiter en relation avec la composition minéralogique, on élabore des "restructurations". Parmi elles, De La Roche (1978) distingue : les normatives, qui reconstituent une composition minéralogique virtuelle, les paramétriques, qui analysent les relations entre les variables chimiques et les évènements envisageables, et les statistiques, qui définissent l'échantillon moyen après analyse multivariable.

Nous avons choisi la voie normative qui pour nous, donne de la réalité naturelle une image plus éloquente.

2.2 Objectif, principe et démarche

Inspirée dans sa conception de la méthode américaine CIPW (Cross et al. 1903), cette méthode, part de l'élaboration d'une composition minéralogique virtuelle à partir de la combinaison théorique des molécules d'une douzaine de constituants chimiques, et débouche sur la détermination de paramètres altérologiques qui permettent des caractérisations ponctuelles, des études évolutives et/ou prospectives de géologie appliquée au Génie Civil ou à la recherche minière.

Pour sa mise en œuvre optimale, les résultats des analyses chimiques pondérales des éléments majeurs doivent faire apparaître l'eau de constitution ou à défaut, la perte au feu qui peut y être assimilée. Le fer doit aussi être exprimé sous ses deux formes courantes.

La gamme minéralogique complète comporte trente deux étalons, triés sur les données de nombreux auteurs (Cross et al. 1903, Smith et al. 1965, Dixon et al. 1977, etc.), et qu'on peut rencontrer

dans les produits d'altération des roches. On distingue, les minéraux accessoires non siliciques, les silicates anhydres primaires ou résiduels, les hydrosilicates ou silicates supergènes, les oxyhydroxydes, la silice.

La démarche comporte principalement, la caractérisation taxonomique des échantillons sur la base de leur richesse en silice et en alumine, la reconstitution des silicates et des oxyhydroxydes, et l'évaluation des paramètres altérologiques.

2.3 Les paramètres altérologiques

Ils caractérisent les degrés d'enrichissement, d'appauvrissement et de fermeture du milieu d'altération.

2.3.1 Paramètres d'enrichissement du milieu d'altération

Ils expriment l'importance relative des oxyhydroxydes libres d'aluminium "ial" et de fer ferrique "ifl", et le degré de cuirassement (indice d'induration potentielle "iip").

$$ial = \frac{(Gib + Cor)\ 100}{Gib + Cor + Goe + Hém + Co}$$

$$ifl = \frac{(Goe + Hém)\ 100}{Gib + Cor + Goe + Hém + Co}$$

Gib : Gibbsite; Cor : Corindon; Goe : Goethite; Hém : Hématite; Co : coefficient permettant d'éviter pour ces paramètres, des valeurs indéterminées ou infinies; il s'exprime comme suit :

$$Co = \%aT - (\%aAl_2O_3 + \%aFe_2O_3);$$

%aT : pourcentage analytique total des éléments majeurs
%aAl$_2$O$_3$: teneur initiale en alumine
%aFe$_2$O$_3$: teneur initiale en fer ferrique

Ces deux paramètres sont complémentaires et doivent être évalués ensemble.

$$iip = \frac{(Ox + Hydrox,\ Fe^{3+}.\ Al^{3+}.\ Mn^{2+})\ 100}{Tous\ les\ minéraux}$$

Ox : Oxydes; Hydrox : Hydroxydes.

On peut définir un indice d'induration potentielle due au fer "iip.fe" :

$$iip.fe = \frac{(Ox + Hydrox, Fe^{3+})\ 100}{Tous\ les\ minéraux}$$

On peut opérer de façon similaire pour l'aluminium "iip.al" et pour le manganèse "iip.mn".

2.3.2 Paramètres d'appauvrissement du milieu d'altération

Ils se réfèrent à l'ordre d'hydrolyse des silicates primaires et néogénétiques, à la libération des éléments mobiles et aux néogenèses minérales conséquentes.

Il s'agit de l'indice de lixiviation potentielle "ilp" et du degré virtuel d'altération de la roche "dvar".

$ilp = N_1/D_1$

N_1 = (Gib)100 + (Kand)75 + (Sil.al° Ca-alc)50 + (Sil.al° FeMg)25, et D_1 = Gib + Cor + Sil.

Gib : Gibbsite; Cor : Corindon; Kand : Kandites; Sil : Silicates; Sil.al° : Silicates d'altération; Ca-alc : calco-alcalins; FeMg : ferromagnésiens.

Les coefficients de pondération que l'on trouve dans l'expression de ce paramètre tiennent compte de l'ordre d'apparition des minéraux secondaires conforme à la séquence de Goldich (1938). Quatre stades théoriques d'évolution en expliquent les valeurs attachées aux différents groupes de minéraux de néogenèse et peuvent être mis en rapport avec la nomenclature de Pedro et Delmas (1971) :
stade 1. les minéraux primaires ferromagnésiens sont tous devenus des hydrosilicates ferromagnésiens : les silicates sont altérés à 25 %;
   stade 2. les minéraux primaires calco-alcalins à leur tour, sont tous devenus des hydrosilicates calcoalcalins : les silicates sont altérés à 50 %;
   (ces deux premiers stades caractérisent la bisiallitisation);
   stade 3. tous les hydrosilicates ferromagnésiens et calcoalcalins ont cédé la place aux kandites (monosiallitisation) : les silicates sont altérés à 75 %;
   stade 4. toutes les kandites ont cédé la place à la Gibbsite (allitisation) : les silicates sont altérés à 100 %.

$dvar = N_2/D_2$

avec : N_2 = N_1 + (Si.am)100, et D_2 = D_1 + Qtz + Si.am
   Qtz : Quartz; Si.am : Silice amorphe.

Sa détermination, similaire à celle de

l'"ilp", prend en compte le Quartz et le produit de sa dissolution par l'eau, la Silice amorphe, pour mettre l'accent sur l'évolution globale de la roche. Les coefficients de pondération ont une signification similaire à ceux de l'"ilp".

2.3.3 Paramètre de fermeture du système d'altération : L'indice de confinement potentiel "icp"

$$icp = \frac{(Silicates - Kandites)\ 100}{Gibbsite + Corindon + Silicates}$$

Il se réfère au fait qu'en milieu fermé, les phyllites de types 2/1/1 et 2/1 se forment préférentiellement, alors qu'en milieu ouvert, on assiste à la formation des kandites (phyllites de type 1/1) et des hydroxydes d'aluminium (phyllites de type 0/1 par extension).

2.4 Les diagrammes d'interprétation et de caractérisation altérologiques

Les diagrammes mis au point en appui à la méthode normative altérologique, en vue de l'exploitation des données peuvent être répartis en deux types :

   1. Type F = f(z)
   2. Type Fo = $g(\sum_{i=1}^{n} Fi)$

Les facteurs F, Fo et Fi sont normatifs.

2.4.1 Diagrammes type F = f(z)

Ils permettent d'examiner l'évolution d'un facteur F en fonction de la profondeur z. Le facteur F peut être minéralogique ou paramétrique.

2.4.2 Diagrammes type Fo = $g(\sum_{i=1}^{n} Fi)$

Ils permettent d'examiner l'évolution d'un facteur Fo en fonction d'un ou de plusieurs autres facteurs Fi. Parmi ceux confectionnés dans cet ordre d'évaluation, nous avons retenu les diagrammes triangulaires, conçus sur leur modèle classique, et les diagrammes carrés.

1. Les diagrammes triangulaires, signalés pour mémoire, permettent d'apprécier l'action des principaux facteurs pédogénétiques, en établissant les prédominances (études ponctuelles), ou les tendances évolutives (études séquentielles).

On distingue, le diagramme Lixiviation – Induration – Confinement (L-I-C), et le diagramme Altération – Induration – Confinement (A-I-C).

2. Les diagrammes carrés permettent de caractériser un échantillon ou une suite d'échantillons prélevés dans un profil en combinant les facteurs d'appauvrissement et d'enrichissement du milieu d'altération.

On distingue, le diagramme Lixiviation – Oxyhydroxydes libres de fer ferrique – Oxyhydroxydes libres d'aluminium, d'où on peut extraire des informations qualitatives d'intérêt agropédologique, et le diagramme Altération – Oxyhydroxydes libres de fer ferrique – Oxyhydroxydes libres d'aluminium d'où on peut déduire un certain nombre de caractères, notamment d'intérêt géotechnique.

Dans ces diagrammes, chaque échantillon de roche analysé est représenté par deux points, et chaque profil d'altération par deux courbes.

3 FIABILITE DE LA METHODE NORMATIVE

Il s'agit d'établir le degré d'affinité entre les compositions minéralogiques réelles et virtuelles.

3.1 Mode d'appréciation du degré d'affinité

Nous avons procédé par traitement statistique des couples de données en déterminant l'équation de leur droite de régression, ainsi que leur coefficient de corrélation r. Ainsi, deux caractères d'une population de N individus, dont l'un normatif (x) et l'autre réel (y), sont liés de façon significative lorsque le coefficient de corrélation r, en valeur absolue, est compris entre 0,7 (correspondance acceptable) et 1,0 (correspondance parfaite), cette liaison étant directe ou inverse suivant que r est positif ou négatif.

3.2 Affinité entre la méthode pétrochimique combinée et la méthode normative

La méthode pétrochimique combinée, proposée par Lelong (1967) pour la détermination minéralogique quantitative à partir des résultats de l'analyse chimique, roche totale, et des teneurs en SiO_2 et en Al_2O_3 du "complexe d'altération", est réputée convenable à l'étude des mélanges Quartz – Kaolinite – Gibbsite – Phyllites

micacées, caractéristiques des sols tropicaux.

Notre méthode normative a été appliquée aux échantillons de Lelong (1967) prélevés dans des profils d'altération de roches allant des granites aux diorites. Les comparaisons effectuées pour différents minéraux et groupes de minéraux caractéristiques ont permis d'en établir les équations des droites de régression :

Quartz : $y = 0,93x + 1,4$; $r = 0,97$;
(N = 50)
Feldspaths : $y = 1,06x - 0,1$; $r = 0,98$;
(N = 50)
Ensemble Quartz-Feldspaths :
(N = 100) $y = 0,93x + 0,8$; $r = 0,98$;
Kaolinite : $y = 0,80x + 2,1$; $r = 0,86$;
(N = 50)
Gibbsite : $y = 0,90x + 1,9$; $r = 0,96$;
(N = 50)
Groupe Goethite-Anatase :
(N = 50) $y = 0,65x + 1,2$; $r = 0,94$;
Pour tous ces minéraux réunis (fig. 1) :
(N = 250) $y = 0,92x + 0,8$; $r = 0,94$;

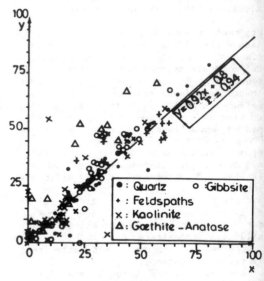

Figure 1 : Corrélation entre les données minéralogiques de la méthode pétrochimique combinée (en ordonnées) et celles de la méthode normative (en abscisses).

La corrélation entre les deux méthodes est excellente. La validité reconnue à la méthode pétrochimique combinée de Lelong (1967) peut donc s'étendre à notre méthode normative.

3.3 Affinité entre les teneurs pondérales réelles et normatives de minéraux repères

Les références prises sont, pour les sols sur roches acides, celles de Leprun (1979) (profils NAY 1 sur granite et THN 1 sur migmatite), et pour les sols sur roches basiques, celles de Pion (1979) (profil KOL 3 sur gabbro). Le Quartz et les Feldspaths sont les seules espèces minérales pour lesquelles les données pondérales réelles ont été présentées par les auteurs cités. Les comparaisons entre ces données et leurs correspondants normatifs, de la roche mère aux horizons superficiels donnent les résultats suivants:

1. Roches mères acides :
Quartz : $y = 0,80x + 0,6$; $r = 0,88$; (N = 18)
Feldspaths : $y = 0,88x + 5,0$; $r = 0,98$; (N = 18)
Ensemble Quartz-Feldspaths : (N = 36) $y = 0,77x + 4,6$; $r = 0,95$;

2. Roches mères basiques :
Quartz : $y = 0,39x - 2,3$; $r = 0,59$; (N = 23)
Plagioclases : $y = 0,86x + 2,5$; $r = 0,87$; (N = 23)
Ensemble Quartz-Plagioclases : (N = 46) $y = 0,59x - 0,2$; $r = 0,60$;

En dépit du nombre limité des échantillons considérés, les corrélations sont excellentes pour les minéraux des roches acides, et très bonnes pour les Feldspaths des roches basiques. Elles sont moins bonnes pour le Quartz des roches basiques dont la teneur est surestimée en valeurs normatives. Ceci provient du fait que, le fer ayant été totalement exprimé par les auteurs en fer ferrique, nous ne pouvions pas reconstituer les silicates ferreux normatifs. Ainsi, la silice non combinée enrichissait le stock de Quartz normatif, et le fer ferrique, celui des oxyhydroxydes correspondants.

3.4 Constat d'ensemble.

La méthode normative se prête valablement à l'étude des variations minéralogiques relatives des produits d'altération. Par conséquent, elle peut constituer un mode rationnel de caractérisation qualitative et quantitative de l'altération.

4 CARACTERISATION NORMATIVE ET EXPERIMENTATION GEOTECHNIQUE : ETUDE DE CAS

4.1 Présentation générale de la région étudiée

La région de Yaoundé, limitée par les parallèles 3° et 5° Nord et par les méridiens 11° et 13° Est, est située sur le plateau du Sud Cameroun, dont l'altitude moyenne est de 750 m (fig. 2).

Le climat actuel y est de type équatorial, à quatre saisons, deux humides et deux sèches, avec des moyennes annuelles, pour la température, de 24°C, et pour les précipitations, de 1600 mm par an. C'est donc une région chaude et humide, dont les caractères climatiques ne sont que la partie actuellement observable d'une alternance polycyclique de périodes arides (comme actuellement) et de périodes pluviales qui, ayant pris place au Sud de la barrière saharienne au cours des temps tertiaires et quaternaires (Pias 1970), a favorisé l'action des phénomènes d'altération des roches du socle.

Parmi les principaux groupes lithologiques de la région de Yaoundé (fig. 2), on distingue les migmatites et les gneiss, intrudés par des plutonites de diverses natures. Leur paragenèse minérale est essentiellement constituée par du Quartz, du Feldspath potassique, du Disthène, des Grenats, de la Biotite, et des accessoires (Rutile, Zircon, Apatite, Monazite,

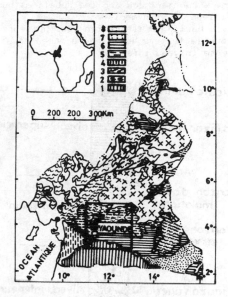

Figure 2 : Esquisse géologique du Cameroun (Bessoles et Trompette 1980, modifié), et situation de la région de Yaoundé.

(1 : Complexe du Ntem; 2-3-4 : Complexe de base; 5-6 : séries intermédiaires; 7 : Précambrien récent; 8 : Couverture volcanique et sédimentaire).

Graphite, Opaques). Avec les micaschistes, les amphibolites, les schistes quartzeux et les quartzites micacés, ils constituent les formations de la "zone mobile d'Afrique Centrale" (Bessoles et Trompette, 1980).

Les formations d'altération dont l'épaisseur varie, suivant les sites, entre 2 et 25 mètres, montrent, de haut en bas, les horizons suivants (fig. 3) :

1. La pellicule humifère;

2. L'horizon meuble, argilosableux (niveau superficiel);

3. L'horizon d'accumulation d'oxydes et d'hydroxydes de fer ferrique et d'aluminium, surmontant des argiles rubéfiées ou bariolées (niveau médian);

4. L'horizon d'altération, qui présente encore la texture de la roche mère sous-jacente (niveau inférieur).

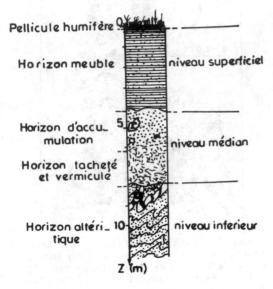

Figure 3 : Organisation verticale typique d'une séquence régionale d'altération.

(r : fragment de roche du socle; q : petit filon de quartz).

4.2 Modes d'analyse et d'exploitation des données analytiques

4.2.1 Méthodes analytiques

Les échantillons de roches du socle et ceux prélevés avec ou sans remaniement dans les différents niveaux d'altération ont fait l'objet d'analyses classiques. Les résultats pris en compte dans le cadre du présent essai sont :

1. Sur le plan géologique, ceux des analyses pétrographiques et minéralogiques et ceux des analyses chimiques des éléments majeurs.

2. Sur le plan géotechnique, ceux des analyses physiques et des essais de mécanique, (essais de compression simple pour les roches du socle, et essais de cisaillement rectiligne allant de la dessiccation à la saturation pour les produits d'altération des différents niveaux pour mieux appréhender l'influence de l'eau sur leurs propriétés mécaniques).

4.2.2 Méthodes d'exploitation des résultats

Sur le plan géochimique, les résultats des analyses chimiques ont été utilisés dans la restructuration normative.

Sur le plan rhéologique, la méthode utilisée est celle de Dayre et al. (1978), fondée sur l'analyse de la courbe effort-déformation au cours d'un essai. Par la discrimination des phases de serrage "s", d'élasticité "e" et de plasticité "p", elle permet d'aboutir à une classification cohérente des matériaux naturels à partir d'une représentation triangulaire du modèle classique, utilisant la relation "s" + "e" + "p" = 100. L'intérêt d'une telle classification réside dans le fait qu'elle s'applique indifféremment aux roches et aux sols.

Les données normatives, génétiques, ont été rapprochées de celles issues de l'expérimentation géotechnique par corrélation statistique.

4.3 Résultats analytiques

Les résultats globaux, présentés sous forme de tableaux, portent sur les données statistiques (moy : moyenne, e.t : écart-type, c.v : coefficient de variation, N : nombre d'individus de la population), et se rapportent aux caractères géotechniques, rhéologiques et normatifs.

4.3.1 Identification géotechnique

Les caractères relevés ont trait aux poids volumiques (pvg poids volumique des grains; pvs poids volumique sec), aux rapports volumétriques (c compacité; e indice des vides), à la granulométrie (a argiles; l limons; s sables; g graviers), et aux limites d'Atterberg (W_L limite de liquidité; W_p limite de plasticité).

Tableau 1a. Niveau superficiel

caractères	unités	moy	e.t	c.v	N
pvg	g/cm³	2,65	0,04	0,02	25
pvs	g/cm³	1,54	0,05	0,03	25
c		0,58	0,03	0,05	25
e		0,72	0,07	0,10	25
a	%	47,50	18,10	0,38	25
l	%	8,90	4,10	0,46	25
s	%	40,70	17,40	0,43	25
g	%	2,90	3,20	1,10	25
W_1	%	51,00	17,40	0,34	25
W_2	%	25,20	9,00	0,36	25

Tableau 1b. Niveau médian

caractères	unités	moy	e.t	c.v	N
pvg	g/cm³	2,79	0,07	0,03	13
pvs	g/cm³	1,76	0,10	0,06	13
c		0,53	0,03	0,05	13
e		0,69	0,09	0,15	13
a	%	21,50	12,80	0,60	13
l	%	4,80	2,70	0,56	13
s	%	24,70	13,20	0,53	13
g	%	49,00	21,40	0,44	13
W_1	%	67,60	11,70	0,17	13
W_2	%	33,80	6,80	0,20	13

Tableau 1c. Niveau inférieur

caractères	unités	moy	e.t	c.v	N
pvg	g/cm³	2,60	0,04	0,02	28
pvs	g/cm³	1,58	0,30	0,19	28
c		0,61	0,12	0,20	28
e		0,69	0,27	0,39	28
a	%	11,90	8,60	0,72	28
l	%	9,90	6,60	0,67	28
s	%	65,90	17,20	0,26	28
g	%	12,30	14,80	1,21	28

Tableau 1d. Roches du socle
(Tous les faciès sont confondus)

caractères	unités	moy	e.t	c.v	N
pvg	g/cm³	2,93	0,23	0,08	30
pvs	g/cm³	2,81	0,20	0,07	30
c		0,96	0,02	0,02	30
e		0,04	0,03	0,63	30

Les poids volumiques ont des valeurs plus élevées dans le socle que dans les niveaux d'altération. La compacité se comporte de façon similaire, à l'inverse de l'indice des vides. La démarcation du niveau médian par rapport aux autres niveaux d'altération est essentiellement due à l'accumulation du fer ferrique.

La granulométrie des produits d'altération montre la partition des caractères des différents niveaux des profils d'altération et leur évolution (fig. 4). En faisant abstraction des graviers qui marquent surtout l'intervention du fer sous forme de concrétions, cette évolution, sous l'influence de l'altération, part des sables (base des profils) aux argiles limono-sableuses (sommet des profils), en passant par des sables argilo-limoneux.

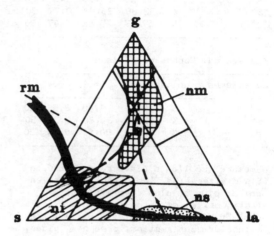

Figure 4 : Circonscription et évolution des caractères granulométriques des matériaux des profils d'altération.

(Pôles du diagramme : g : graveleux; s : sableux; la : limono-argileux; sens des abréviations et des signes : rm : roche mère; ni : niveau inférieur; nm : niveau médian; ns : niveau superficiel; points encadrés : valeurs moyennes).

4.3.2. Analyse rhéologique

Les caractères relevés de l'analyse de la déformation sont : le serrage "s", l'élasticité "e" et la plasticité "p").

Tableau 2a. Niveau superficiel

caractères	unités	moy	e.t	c.v	N
"s"	%	15,72	20,86	1,33	45
"e"	%	20,57	18,12	0,88	45
"p"	%	63,71	34,06	0,53	45

Tableau 2b. Niveau médian

caractères	unités	moy	e.t	c.v	N
"s"	%	9,75	6,95	0,71	36
"e"	%	59,79	22,58	0,38	36
"p"	%	30,46	21,68	0,71	36

Tableau 2c. Niveau inférieur

caractères	unités	moy	e.t	c.v	N
"s"	%	7,65	10,00	1,31	30
"e"	%	32,88	25,92	0,79	30
"p"	%	59,47	27,66	0,47	30

Tableau 2d. Roches du socle

caractères	unités	moy	e.t	c.v	N
"s"	%	3,56	9,08	2,55	30
"e"	%	63,67	24,20	0,38	30
"p"	%	32,83	20,91	0,64	30

A mesure que l'on s'élève dans les profils d'altération, les matériaux deviennent de plus en plus compressibles et plastiques, et de moins en moins élastiques. En effet l'altération, par les phénomènes de soustraction, crée des vides, fragilise le milieu et induit la formation des argiles. Le niveau médian, siège de l'accumulation des oxyhydroxydes de fer ferrique, se démarque de cette tendance générale, par un regain d'élasticité et par une diminution conséquente de la plasticité ainsi que le montrent les diagrammes "s"-"e"-"p", notamment lorsqu'on fait varier les conditions hydriques (fig. 5).

Figure 5 : Variation du comportement rhéologique des matériaux des profil d'altération.

a : Aperçu global sous l'effet des phénomènes d'altération.

b : Aperçu relatif aux différents niveaux d'altération, en fonction du degré de saturation Sr.

(Pôles des diagrammes : "s" : serrage; "e" : élasticité, "p" : plasticité; pour le sens des abréviations et signes, cf. fig. 4; aux extrémités de chacune des flèches de la fig. 5b : 1 : Sr = 0%; 2 : Sr = 100%).

4.3.3 Traitement géochimique

Les caractères normatifs obtenus sont les paramètres d'évolution du milieu d'altération par appauvrissement (ilp indice de lixiviation potentielle, dvar degré virtuel d'altération de la roche) et par enrichissement (iip indice d'induration potentielle, ial importance de l'aluminium libre, ifl importance du fer libre).

Tableau 3a. Niveau superficiel

caractères	unités	moy	e.t	c.v	N
ilp	%	64,31	23,41	0,36	18
dvar	%	32,66	13,50	0,41	18
iip	%	16,01	9,79	0,61	18
ial	%	9,18	7,68	0,84	18
ifl	%	8,45	4,88	0,58	18

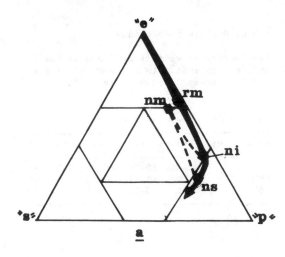

Tableau 3b. Niveau médian

caractères	unités	moy	e.t	c.v	N
ilp	%	69,07	7,95	0,12	13
dvar	%	49,61	14,91	0,30	13
iip	%	28,31	14,60	0,52	13
ial	%	2,47	3,02	1,22	13
ifl	%	30,43	16,16	0,53	13

Tableau 3c. Niveau inférieur

caractères	unités	moy	e.t	c.v	N
ilp	%	54,64	26,76	0,49	23
dvar	%	29,8	19,33	0,65	23
iip	%	15,73	11,03	0,70	23
ial	%	7,42	6,97	0,94	23
ifl	%	9,65	7,00	0,73	23

Tableau 3d. Roches du socle

caractères	unités	moy	e.t	c.v	N
ilp	%	12,00	14,10	1,18	18
dvar	%	6,50	6,88	1,06	18
iip	%	3,76	4,21	1,12	18
ial	%	3,56	4,68	1,31	18
ifl	%	0,72	1,91	2,65	18

Les enseignements apportés par ces paramètres montrent que les formations du socle sont dans l'ensemble très peu lixiviées et très peu altérées. L'évolution de la lixiviation et celle de l'altération montrent un passage relativement brutal, de la roche saine aux produits d'altération.

L'individualisation des oxyhydroxydes, ferrugineux notamment, est tout aussi brutale. Sa manifestation paroxysmale s'effectue dans le niveau médian des profils. C'est ce qui confère à ce niveau l'illusion d'une lixiviation et d'une altération plus élevées que dans les autres niveaux (phénomène de dilution). Sinon, les valeurs optimales des paramètres correspondants se situeraient à mi-chemin entre celles des niveaux inférieur et superficiel.

Les diagrammes carrés illustrent on ne peut plus clairement tous ces phénomènes.

Le diagramme "Lixiviation - Oxyhydroxydes libres d'aluminium et de fer ferrique" (fig. 6) situe l'évolution supergène des profils d'altération de la région dans la zone de lixiviation moyenne à forte, soit à la frontière entre la bisiallitisation et la monosiallitisation. C'est le domaine de la prédominance de la formation de la Kaolinite. Du point de

vue agropédologique, la désaturation en bases de la fraction fine des sols y est faible à moyenne.

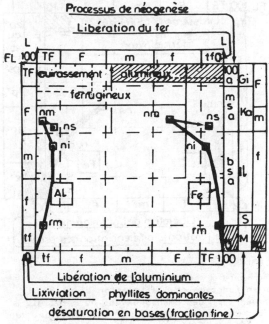

Figure 6 : Evolution des profils en fonction de la lixiviation et de la libération des oxyhydroxydes de Fe et Al.

(degrés de lixiviation, de libération, de désaturation : tf : très faible; f : faible; m : moyen; F : fort; TF : très fort; processus de néogenèse : a : allitisation; msa : monosiallitisation; bsa : bisiallitisation; phyllites : Gi : Gibbsite; Ka : Kaolinite; Il : Illite; S : Smectites et Chlorites; M : Micas et Hydromicas; pour les autres abréviations et signes, cf. fig. 4).

Le diagramme "Altération - Oxyhydroxydes libres d'aluminium et de fer ferrique" (fig. 7) montre que la pédogenèse se situe dans la zone d'altération faible à moyenne. Les horizons des profils y sont peu différenciés. Du point de vue rhéologique, le comportement prédominant de ces matériaux est élastoplastique. Ils sont utilisables dans les travaux du Génie Civil, moyennant l'élimination de la fraction fine.

4.4 Corrélations statistiques significatives : présentation, commentaire et interprétation

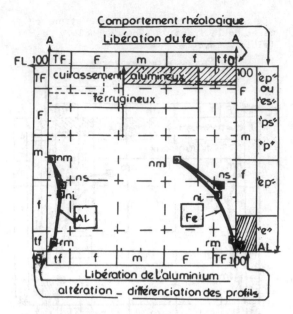

Figure 7 : Evolution globale des profils en fonction de l'altération et de la libération des oxyhydroxydes de Fe et Al.

(degrés d'altération, de libération, de différenciation : tf : très faible; f : faible; m : moyen; F : fort; TF : très fort; comportement rhéologique : "ep" : elasto-plastique; "es" : élasto-compressible; "ps" : plasto-compressible; "p" : plastique; "ep" : élasto-plastique; "e" : élastique; pour les autres abréviations et signes, cf. fig. 4).

4.4.1. Corrélations entre propriétés physiques et paramètres normatifs

Elles ont été établies à l'intérieur du groupe lithologique donné constitué par les gneiss et les migmatites.

Matériaux du niveau superficiel (N = 25)

a	$= 0,66$ilp	$+ 3,60$;	$r =$	$0,81$
a	$= 1,08$dvar	$+ 11,60$;	$r =$	$0,75$
W_L	$= 0,55$ilp	$+ 12,30$;	$r =$	$0,74$
W_L	$= 0,98$dvar	$+ 18,30$;	$r =$	$0,71$
IP	$= 0,31$ilp	$+ 4,70$;	$r =$	$0,76$
IP	$= 0,52$dvar	$+ 7,80$;	$r =$	$0,73$
pvg	$= 0,23.10^{-2}$dvar	$+ 2,58$;	$r =$	$0,75$

N.B. : IP, indice de plasticité $= W_L - W_P$.

Matériaux du niveau médian (N = 15)

c	$= 0,21.10^{-2}$iip	$+ 0,60$;	$r =$	$0,88$
e	$= - 0,51.10^{-2}$iip	$+ 0,70$;	$r = -$	$0,86$
pvs	$= 0,64.10^{-2}$iip	$+ 1,57$;	$r =$	$0,91$
W_L	$= 1,36$ilp	$- 27,90$;	$r =$	$0,95$
IP	$= 0,80$ilp	$- 22,30$;	$r =$	$0,95$

Matériaux du niveau inférieur (N = 28)

c	$= - 0,45.10^{-2}$ilp	$+ 0,88$;	$r = -$	$0,94$
e	$= 1,03.10^{-2}$ilp	$+ 0,77$;	$r =$	$0,91$
pvs	$= - 1,2.10^{-2}$ilp	$+ 2,30$;	$r = -$	$0,97$

Roches du socle (Migmatites et Gneiss) (N = 8)

c	$= - 0,5.10^{-2}$ilp	$+ 0,97$;	$r = -$	$0,71$
c	$= - 1,03.10^{-2}$dvar	$+ 0,98$;	$r = -$	$0,88$
e	$= 0,6.10^{-2}$ilp	$+ 0,03$;	$r =$	$0,72$
e	$= 1,5.10^{-2}$dvar	$+ 0,02$;	$r =$	$0,84$

Si l'indice de lixiviation potentielle ilp et le degré virtuel d'altération de la roche dvar ne sont réellement significatifs que pour la roche mère et les niveaux inférieur et superficiel des profils, pour le niveau médian, c'est l'indice d'induration potentielle iip.

L'évolution de la compacité c, de l'indice des vides e, et même du poids volumique sec pvs est fonction de celle de la lixiviation et/ou de l'altération, dans les matériaux susceptibles de garder leur volume global invariable. C'est le cas des roches du socle et des matériaux de l'horizon altéritique. Les corrélations significatives établies à cet effet viennent à l'appui des bilans isovolumétriques de Millot et Bonifas (1955). L'absence totale de relation de ce genre dans les matériaux du niveau superficiel des profils atteste les remaniements de divers ordres dont ils sont l'objet. Par contre, pour ces matériaux qui sont relativement les plus évolués des profils, des corrélations significatives ont été établies entre leur degré d'évolution (indice de lixiviation potentielle ilp, degré virtuel d'altération de la roche dvar) et celles parmi leurs propriétés physiques, qui sont directement liées à la plasticité (teneur en argile a, limite de liquidité W_L, Indice de plasticité IP), ou sur lesquelles les remaniements n'ont pas d'influence (poids volumique des grains pvg).

Les relations établies pour les matériaux du niveau médian des profils montrent la dualité de composition et de comportement de ces derniers. L'immobilisation des oxyhydroxydes essentiellement ferrugineux confère une trame rigide à ces matériaux dont le volume reste pratiquement invariable, d'où les relations qui s'établissent entre l'indice d'induration potentielle iip d'une part et, d'autre part la compacité c, l'indice des vides e, et le poids volumique sec pvs. De plus, ces matériaux ont une matrice sablo-argileuse dont les propriétés plastiques sont liées à leur degré d'évolution, d'où les relations établies entre l'indice de lixiviation potentielle ilp d'une part, la limite de liquidité W_L et l'indice de plasticité IP d'autre part.

4.4.2 Corrélations entre propriétés de la déformation et paramètres normatifs

Elles ont été établies à partir d'une série de profils sur migmatites et gneiss, abstraction étant faite du niveau médian de ces profils. La séquence considérée est donc la suivante de bas en haut : Roche mère - niveau inférieur - niveau superficiel.

Evaluation à teneur en eau nulle (N = 32)
"s" = $0,67.10^{-2}$ilp + 2,92;   r = 0,78
"s" = $1,25.10^{-2}$dvar + 1,35;   r = 0,77

Evaluation à teneur en eau variable
(N = 65); (Teneur en eau naturelle allant de 0 à 31 %)
"e" = $- 0,58.10^{-2}$ilp + 61,05; r = $- 0,69$

Evaluation à teneur en eau de saturation
(N = 29)
"s" = $0,24.10^{-2}$ilp - 0,20;   r = 0,84
"s" = $0,43.10^{-2}$dvar - 0,57;   r = 0,81
"e" = $- 1,53.10^{-2}$dvar + 60,74; r = $- 0,70$

Les corrélations établies font apparaître que la compressibilité varie en raison directe du degré d'évolution, et l'élasticité en raison inverse quelles que soient les conditions hydriques envisagées. Des corrélations moins significatives (r < 0,7) ont par ailleurs indiqué comme autres tendances, pour les échantillons séchés à l'étuve, une réduction du domaine de comportement élastique au fur et à mesure qu'augmente le degré d'évolution du matériau et, pour des teneurs en eau non nulles, une augmentation du domaine de comportement plastique directement proportionnelle au degré d'évolution.

5 CONCLUSION

Les matériaux de la lithosphère, soumis aux conditions thermodynamiques qui règnent à l'interface continentale, subissent une évolution qui affecte leurs propriétés originelles, notamment leurs propriétés physiques et mécaniques. Les altérologues et les ingénieurs des travaux publics en conviennent. Si certains aspects fondamentaux de cette évolution, qu'on retrouve dans le langage courant, sont restés essentiellement qualitatifs et ont parfois fait l'objet d'appréciations altérologiques quantitatives pratiquement inutilisables par l'ingénieur, la

restructuration normative aboutit à une quantification plus appropriée de ces notions par le biais de paramètres phénoménologiques. Le principal inconvénient de cette méthode réside dans son caractère artificiel, qui, du reste, est inhérent à toutes les méthodes de restructuration.

Les relations significatives établies entre les propriétés physiques et rhéologiques d'une part, et les paramètres altérologiques d'autre part, restant en harmonie avec les faits d'observation, permettent d'entrevoir, malgré le nombre limité des cas pris en compte, les possibilités d'une quantification rationnelle de l'évolution des propriétés géotechniques des matériaux de la lithosphère en fonction de certains de leurs caractères génétiques.

REFERENCES

Bessoles, B., R.Trompette 1980. Géologie d'Afrique. La Chaîne panafricaine : "Zone mobile d'Afrique Centrale (partie sud) et zone mobile soudanaise" Mém. BRGM 92:101p.

Cross, W., J.P.Iddings, L.V.Pirsson, H.S. Washington 1903. Quantitative classification of igneous rocks. Univ. of Chicago Press.

Dayre, M., D.Fabre, J.Letourneur, P. Antoine, Y.Orengo 1978. Eléments pour une classification géotechnique des terrains. Proc. 3rd Int. Congr. I.A.E. G. II 2:131-139.

De La Roche, H. 1978. La chimie des roche présentée et interprété d'après la structure de leur faciès minéral dans l'espace des variables chimiques. Fonctions spécifiques et diagrammes qui s'en déduisent. Application aux roches ignées. Chem.Geol. 21:63-87.

Dixon, J.B., J.B.Weed, J.A.Kittrick, M.H. Milford, J.L.White 1977. Minerals in soil environments. Soil Science Society of America, Madison, Wisconsin, USA, 948 p.

Ekodeck, G.E. 1984. L'altération des roches métamorphiques du Sud Cameroun et ses aspects géotechniques. Thèse Doct. d'Etat ès Sci. Nat., IRIGM, Univ. Sci. et Médic. Grenoble I France 368p.

Ekodeck, G.E. 1989. L'algorithme généralisé de restructuration normative altérologique et les diagrammes correspondants d'interprétation et de caractérisation. Doc. inéd., dact., 116p.

Ekodeck, G.E. 1990. Essai de corrélation entre paramètres altérologiques normatifs et caractères géotechniques des sols structurés : cas des produits d'altération des roches de la région de Yaoundé, Cameroun. Actes du Sympos. Internat. sur les sols structurés, A.I. G.I., 9-13 Avril, Yamoussoukro, Côte d'Ivoire, p.1-7.

Gidigasu, M.D. 1974. The degree of weathering in the identification of laterite materials for engineering purposes Eng.Geol. 8(3):213-266.

Goldich, S.S. 1938. A study in rock weathering. J.Geol. 46:17-23.

Joachin, A.W.R., S.Kandiah 1941. The composition of some local soil concentrations and clays. Tropical Agric. 96:6-65.

Lelong, F. 1967. Détermination quantitative par voie chimique des constituants minéraux de produits argileux d'altération tropicale. Bull.Gr.Fr.Arg. 19:49-67.

Lelong, F. 1969. Nature et genèse des produits d'altération des roches cristallines sous climat tropical humide (Guyane française). Thèse Fac.Sci.Univ. Nancy I, France. 182p. et Mém. Sc. Terre 14, 188p.

Leneuf, N. 1959. L'altération des granites calcoalcalins et des granodiorites en Côte d'Ivoire forestière, et les sols qui en sont dérivés. Thèse Doct. d'Etat ès Sci.Nat. Fac.Sci.Univ. Paris, France. 210 p.

Leprun, J.C. 1979. Les cuirasses ferrugineuses des pays cristallin de l'Afrique Occidentale sèche. Genèse-Transformations-Dégradation. Thèse Doct. Fac. Sci. Univ.L.P., Stasbourg France et Mém.Sci.Géol., 58:224p.

Maignien, R. 1966. Compte rendu de recherches sur les latérites. Rech.sur les Ress.Nat. IV UNESCO Paris France 155p.

Millot, G., M.Bonifas 1955 : Transformations isovolumétriques dans les phénomènes de latéritisation et de bauxitisation. Bull.Serv. Carte Géol.Als.Lorr. France 8:3-10

Paquet, H. 1969. Evolution géochimique des minéraux argileux dans les altérations et les sols des climats méditerranéens tropicaux à saisons contrastées. Thèse Doct. d'Etat ès Sci.Nat. Fac.Sci.Univ. Strasbourg France et Mém. Serv. Carte Géol.Als.Lorr. 30:212p.

Pedro, G., A.G.Delmas 1971. Sur l'altération expérimentale par lessivage à l'eau et la mise en évidence de trois grands domaines d'évolution géochimique. C.R.A.S., Paris France 273 D:1543-1546.

Pias, J. 1970. Les formations sédimentaires tertiaires et quaternaires de la cuvette tchadienne et les sols qui en dérivent. Mém. ORSTOM Paris France 43:407p.

Pion, J.C. 1979 : Altération des massifs cristallins basiques en zone tropicale sèche. Etude de quelques toposéquences en Haute Volta. Thèse Doct., Fac. Sci. Univ., L.P. Strasbourg France, et Mém. des Sci.Géol. 57:220 p.

Smith, A.W., A.S.Beward, L.G.Berry, B. Post, S.Weissmann & M.B.Lutz 1965. Index to powder diffraction file A.S.T.M. Ed. Philadelphia USA, 581p.

Tuncer, E.R., Lohnes, R.A. 1977. An engineering classification for certain basalt derived lateritic soils. Eng.Geol. XI 4:319-339.

Wackermann, J.M. 1975. L'altération des massifs cristallins basiques en zone tropicale semi-humide. Etude minéralogique et géochimique des arènes du Sénégal oriental. Conséquences pour la cartographie et la prospection. Thèse Doct. d'Etat ès Sci.Nat., Univ. L.P., Strasbourg France. 373 p.

Parameters for bearing capacity and settlement analyses of residual soils

Paramètres pour l'analyse de la capacité portante et du tassement de sols résiduels

M. M. Zhao & K. W. Lo
National University of Singapore, Singapore

ABSTRACT: The problem of predicting the bearing capacity and settlement of foundations in partially saturated, mixed granular and cohesive soil formations is a common one. However, in view of their marked heterogeneity, routine methods of sampling and laboratory testing, and subsequent determination of soil parameters for foundation design, are not amenable to these soils. In these circumstances, a study was initiated to determine appropriate site investigation procedures for the residual soils of Singapore, which fall into this category of materials. In addition, correlations were established between the soil parameters for foundation design and results of standard field testing obtained from several construction projects involving such formations. A semi-empirical approach is proposed for the design of foundations in these soils, for which the veracity of the correlations adopted was confirmed by back-analysis from measured responses of as-constructed shallow and piled foundations subject to applied loading.

RESUMÉ: La prédiction de la capacité de charge et du tassement de fondations dans une formation de sol, saturée partiellement, granulaire mixte et cohésive, est un problème courant. Cependant, à cause de sa hétérogénéité marquée, des méthodes ordinaires de prélèvement d'échantillon des essais aux laboratoires et la détérmination suivante des paramètres du sol pour la conception de fondation ne sont pas convenable pour ce genre de sols. Dans ces circonstances, une étude a été effectuée pour déterminer les procédures d'examen sur place appropriées pour les sols résiduels à Singapour, qui correspond à la même catégorie de sol décrit ci-dessus. En outre, les corrélations ont été établiées entre les paramètres de sol pour la conception de fondations et les résultats des essais sur place venant des projets de construction sous les mêmes conditions de formation. Une approche semi-empirique est proposée pour la conception de fondation dans les sols résiduels, pour lesquels des corrélations adoptées ont été confirmées par les analyse des résultats obtenus des fondations sous constraints.

1 INTRODUCTION

Highly heterogeneous mixed granular and cohesive deposits, which may in addition be partially saturated, are commonly encountered worldwide, and variously referred to as residual, saprolithic and indurated soils. Although extensive investigations have been carried out by researchers in this field (Stroud 1974; American Society of Civil Engineers 1982; Committee on Tropical Soils of ISSMFE 1985; Brand and Phillipson 1985; Publications Committee of 2-ICOTS 1988), no definitive study seems to have been conducted on the choice of parameters that would be appropriate to the design of foundations in such soils. In view of the fact that

these soils are generally not amenable to routine methods of site investigation which are mainly applicable to either purely granular or cohesive soils, a study was conducted into the special boring, sampling and testing requirements for determining the necessary parameters for designing foundations in the Singapore residual soils. The study also included field measurements to verify the proposed design parameters.

2 CLASSIFICATION OF SOILS

The soils of the present category that are pertinent to Singapore are the Old Alluvium formation and residual soils of the sedimentary and granitic rock formations which cover the majority of the island republic (Figure 1), hereinafter referred to collectively as "residual soils" except where further differentiation is found necessary. The classification test results for the respective soil types are provided by Figures 2 and 3. Accordingly, the Old Alluvium formation consists of lightly-cemented Pleistocene sediments, which predominantly consist of poorly-graded clayey, coarse sand, or less often sandy clay, occasionally containing stringers of pebbles up to about 40mm in equivalent diameter, clay lenticles and sand pockets. The lateral variation of material types is generally rapid and frequent in this formation. As to the granitic residual soils, these may be broadly differentiated into silty clay or clayey silt, followed by coarser-grained silty sand or sandy silt, deposits. The degree of compaction and grain size of these deposits generally increase with depth, as a reflection of their decreasing degree of decomposition. The sedimentary residual soils are visually alike to, and comprise similar materials as, the granitic residual soils, although the distribution of the former materials tends to be less uniform than the latter, due to the fact that it follows the layering of the parent rock types, that is shaley mudstone, siltstone and sandstone, and these rocks had, in turn, been subjected to considerable tectonic activity in the

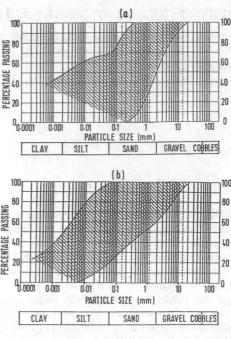

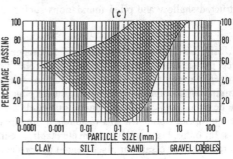

Figure 2   Particle size distribution
(a) Old Alluvium   (b) sedimentary residual soil
(c) granitic residual soil

In view of the elevated levels at which the residual soil deposits can occur, a zone of partial saturation can exist to depths of some 30m.

3 SITE INVESTIGATION

3.1 *Boring and sampling*

Those residual soils whose mechanical behaviour is essentially cohesive, for instance even in cases where significant amounts of granular material are embedded in a silty clay matrix with little intergranular contact, exhibit firm, stiff or hard

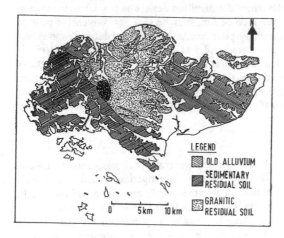

Figure 1   Residual soils of Singapore

geological past, thereby resulting in extensive folding and faulting. Similarly as in the case of the granitic residual soils, however, the sedimentary residual soils tend to exhibit an increasing degree of compaction with depth due to a corresponding decrease in the degree of decomposition.

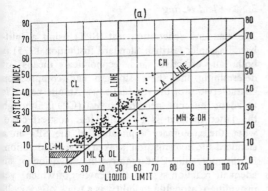

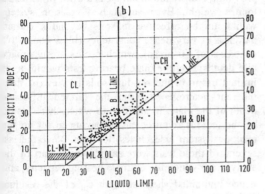

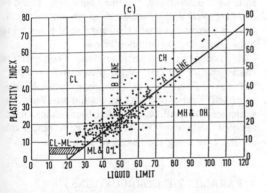

Figure 3   Plasticity charts
(a) Old Alluvium   (b) sedimentary residual soil
(c) granitic residual soil

consistencies, whereas for mainly granular behaviour where significant intergranular rolling and sliding occur, the corresponding soils may be more appropriately described as loose, medium dense or dense (vide British Standards Institution 1986 for definition of consistencies). Rotary methods of boring are generally recommended in the residual soils. Wash boring would tend to both loosen the soil

by the action of pumping and cause softening to occur due to the ingress of water from the drilling fluid. Percussive drilling, on the other hand, would result in excessive remoulding of cohesive soil due to the impact of the claycutter, or loosening of granular material due to application of the shell.

In firm to stiff deposits, a hollow stem auger advanced by means of wire-line drilling using a bit of hardened steel, and itself serving as a casing, would provide a suitable means of excavating a 100mm diameter borehole. In this case, jacked-in 50mm internal diameter thin-walled steel tube samplers would be able to maintain an acceptable level of sample disturbance, as indicated by Table 1. Otherwise, a 150mm diameter steel casing may be used in conjunction with tricone roller drill bits having hardened steel cutting teeth, for constructing boreholes in these deposits. In this instance, due care should be given to the choice of drilling fluid as a trade-off between ease of drilling and the need to minimise ground softening by water ingress. Furthermore, the use of standard 100mm internal diameter open-drive steel tube samplers would be adequate to keep soil disturbance within tolerable limits.

For generally very stiff to hard soils, on the other hand, a tricone roller drill bit with tungsten carbide cutting teeth, followed by 100mm diameter steel casing, may be employed in advancing the borehole, in conjunction with which 63mm internal diameter triple tube sampling would provide a reasonable quality of sample taken.

For 50mm diameter jack-in samplers, the area ratio should be limited to 10% to ensure good-quality samples. In the case of the 100mm diameter open-drive sampler, the area ratio could be increased to a

Table 1. Undrained shear strength c_U (kN/m²) for varying specimen size and preparation method

Borehole	Depth (m)	Sample diameter (mm)			
		38	50	100	150
A	1.0	10	32.5	50	55
	2.0	4	60	60	68
	3.0	40	85	-	80
B	0.8	-	82.7	81.4	-
	2.3	-	86.5	86.5	-
	6.3	-	106.8	108.1	-

Note:
1. Samples from borehole A obtained by open-drive method
2. Samples from borehole B obtained by direct jacking-in.

3471

maximum of 25%, while for the 63 mm diameter triple tube sampler, an intermediate area ratio would be appropriate.

3.2 Laboratory and field testing

3.2.1 Laboratory testing

To determine the shear strength characteristics of firm to stiff residual soils, either 50mm jacked-in tube samples or 100mm open-drive samples may be extruded and tested directly in triaxial compression. Table 1 shows a comparative study of sample disturbance associated with various sampling and test specimen preparation methods. Accordingly, the undrained shear strength c_U of test specimens of various preparations was determined in the triaxial test apparatus. Thus, test specimens of 100mm and 150mm diameter were extruded directly from corresponding open-drive sampling tubes, as well as single 50mm diameter and triple 38mm diameter specimens from 100mm diameter tubes. Other test specimens were extruded directly from 50mm diameter jacked-in samplers. All specimens were then subjected to unconsolidated, undrained triaxial compression in accordance with BS 1377 (British Standards Institution 1975). As a result, it was found that the values of c_U obtained from 38mm diameter specimens were significantly lower than those of the 50mm diameter and larger specimen sizes, due to sample disturbance. On the other hand, the corresponding values for 50mm diameter jacked-in samples approximated those of 100mm diameter open-drive samples, beyond which size there was little variation in the undrained shear strength of soil specimens. For very stiff to hard residual soils, 63mm diameter triple-tube samples extruded directly for triaxial compression testing were found to provide reasonable values of c_U.

In the case of compression characteristics, however, it was found that even for soil samples extruded directly into a 152mm diameter hydraulic oedometer (Rowe and Barden 1966), tests carried out with or without flooding of the soil specimen overestimated the coefficient of compressibility m_v by a factor of up to 4. On the other hand, stress-path testing of 63mm diameter triple tube samples in the triaxial compression apparatus gave values of Young's modulus which compared well with those determined by field pressuremeter testing as outlined in the following discussion.

3.2.2 Field testing

The standard penetration test (SPT) using a split-barrel sampler (British Standard Institution 1975) is both an effective and economical means of determining the shear strength and compression characteristics of the residual soils. An alternative is to adopt the cone penetration test (CPT), although the latter method is generally less cost-effective when the customary application of the SPT in a standard site investigation is taken into consideration. Furthermore, the CPT requires the provision of an adequate reaction system, the heaviest of which, yet providing reasonable mobility, is a truck of some 20 tonnes weight. Even so, such a reaction system would not be effective in soils beyond a certain hardness or very dense consistency (blow count "N" in excess of 50 or so), which occurs at customary pile depths. Routine plate bearing tests would be of limited usage in the residual soils in view of their inherent heterogeneity and the limited influence depth of the plate, in addition to which there would be the tedious requirement of having to set up the reaction system at each test location. Nevertheless, several of these tests were carried out as part of the study, from which the ultimate bearing capacity, and hence undrained shear strength c_U, of the residual soils were determined, and the values of c_U found to compare well with corresponding laboratory result used in correlation studies with standard penetration blow count "N", as outlined below.

In order to determine the compression characteristics of these soils, pressuremeter testing has been found to be the most cost-effective approach. As the following discussion will show, the settlement of shallow and piled foundations in the residual soils may be predicted with reasonable accuracy by using the results of such tests.

4 PARAMETRIC CORRELATIONS

The measurement data upon which the following correlations have been based have been summarised in a separate publication (Lee 1987), and may be referred to therein.

Figure 4 contains a plot relating the SPT blow count "N" to the CPT cone point resistance q_C, for which the empirical relationship

$$q_C \ (kg/cm^2) = 2N \qquad (1)$$

may be found to be applicable to the residual soil as a whole.

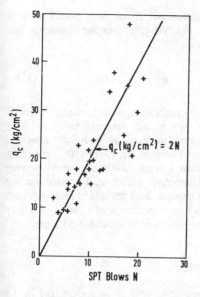

Figure 4   Correlation between cone point resistance and blow count

As shown in Figure 5, for the Old Alluvium formation, the blow count "N" may also be correlated with the pressuremeter deformation modulus E_M according to the relationship

$$E_M \ (MN/m^2) \ = \ 1{\cdot}75N \qquad (2),$$

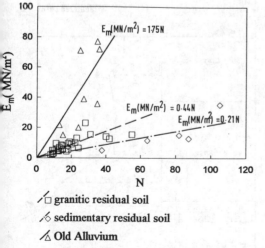

granitic residual soil

sedimentary residual soil

Old Alluvium

Figure 5   Correlation between deformation modulus and blow count

while for the residual soil of the sedimentary formation,

$$E_M \ (MN/m^2) \ = \ 0{\cdot}21N \qquad (3),$$

and for the residual soil of the granitic formation,

$$E_M \ (MN/m^2) \ = \ 0{\cdot}44N \qquad (4).$$

In the case of the pressuremeter limit pressure p_L, the corresponding correlations with blow count "N" may be determined from Figure 6 as

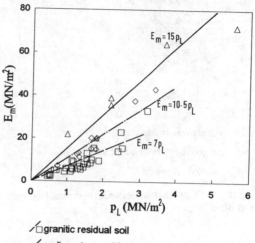

granitic residual soil

sedimentary residual soil

Old Alluvium

Figure 6   Correlation between deformation modulus and limit pressure

$$E_M = 15p_L \qquad (5)$$

for the Old Alluvium,

$$E_M = 10{\cdot}5p_L \qquad (6)$$

for sedimentary residual soil, and

$$E_M = 7p_L \qquad (7)$$

for granitic residual soil.

Stress path tests have also been performed on the residual soils, in the laboratory, which generally confirm the above deformation moduli obtained by field testing. Furthermore, the undrained cohesion c_u

of the residual soils, as determined by the preceding recommended laboratory testing procedures, may be correlated with the standard penetration blow count "N" according to the relationships indicated in Figure 7, whereby, for the Old Alluvium,

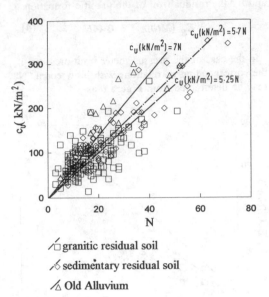

granitic residual soil

sedimentary residual soil

Old Alluvium

Figure 7   Correlation between undrained cohesion and blow count

$$c_U(kN/m^2)=7N \qquad (8),$$

while for the sedimentary residual soil,

$$c_U(kN/m^2)=5\cdot7N \qquad (9),$$

and for the granitic residual soil,

$$c_U(kN/m^2)=5\cdot25N \qquad (10).$$

5 FOUNDATION DESIGN

As in the case of the above parametric correlations, the measurement data upon which the following discussion is based have been summarised in Lee's (1987) publication.

5.1 *Shallow foundations*

5.1.1 *Bearing capacity*

Based on results of plate loading tests mentioned in

the earlier discussion, it has been found that, in determining the ultimate bearing capacity q_U, whereby

$$q_U=c_UN_C+\gamma D \qquad (11),$$

in which N_C = bearing capacity factor for effect of cohesion; γ = total unit weight of soil; and D = embedment depth of plate, an undrained cohesion given by either Equation (8), (9) or (10), depending on the particular ground conditions, would provide a reasonably accurate prediction. For safe bearing capacity, the trend of load-settlement plots obtained from measurements indicate that a factor of safety of 3 would be advisable to avoid local failure.

5.1.2 *Settlement*

The settlement of strip footings constructed in the Old Alluvium and sedimentary residual soil, as well as of bearing plates in the Old Alluvium, have been determined by precise levelling in order to verify the parametric correlations of the foregoing discussion. The strip footings were subjected to loading by arch structures, which were themselves buried in backfill soil (vide Figure 8). The analytical expression for

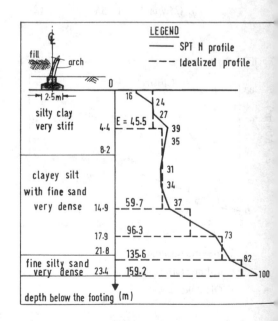

Figure 8   Idealization of shallow foundation in Old Alluvium

3474

settlement "w" adopted for comparison with measurement results was derived by Giroud (1972) as

$$w = P_{Hm}(qB/E) \qquad (12)$$

where q = average normal contact stress exerted by foundation on ground surface; L = length of rectangular foundation; B = width of rectangular foundation; H = thickness of the underlying soil layer; E = elastic modulus of underlying soil layer; and P_{Hm} = dimensionless coefficient depending on Poisson's ratio ν_s, L/B and H/B.

In order to obtain the Young's modulus "E" of Equation (12), the approach suggested by Wilun and Starzewski (1975) was adopted. Accordingly, the oedometer modulus E_{oed} corresponding to the pressuremeter deformation modulus E_M was first determined via the expression

$$E_{oed} = \beta E_m \qquad (13),$$

where β = coefficient relating to soil structure which is dependent on soil type and E_M/pL, from which the Young's modulus of the soil may be deduced as

$$E = \delta E_{oed} \qquad (14),$$

where δ is a function of the Poisson's ratio ν_S for the soil. A similar approach was adopted for the bearing plates.

The analytical settlement plots of Figure 9 have thereby been obtained, and may be compared with the measurement results of corresponding sites on the same figure. Accordingly, there appears to be reasonably good agreement between the two sets of results for bearing plates in Old Alluvium and strip footings in sedimentary residual soil, thereby justifying the proposed site investigation and testing procedures for determining the compression characteristics of the corresponding soils. In the case of strip footings in the Old Alluvium, which was generally not as free-draining as the more granular sedimentary residual soil, it was found that a drained Young's modulus of E'=0·33E_M was applicable.

5.2 Deep foundations

5.2.1 Bearing capacity

At various sites of the Old Alluvium, and sedimentary and granitic residual soils, bored piles have been fully-instrumented to determine their

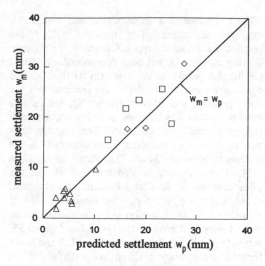

☐ strip footing in Old Alluvium

◇ strip footing in sedimentary residual soil

△ bearing plate in Old Alluvium

Figure 9 Comparison of settlements

respective adhesion factors α, and pile base softening factors ω, when subjected to load testing. In addition, using these bearing capacity factors and corresponding correlation coefficients for undrained shear strength indicated by Equations (8)-(10), the ultimate bearing capacity of piles determined from routine load tests at other sites, where the residual soils also occurred, were confirmed.

In all cases, the basis for analysing the ultimate bearing capacity Q_U of piles was as follows:

$$Q_U = \alpha c_{US} A_S + 9 \omega c_{UB} A_B \qquad (15)$$

where α = adhesion factor; c_{US} = average undrained cohesion along pile shaft; A_S = surface area of pile shaft; ω = pile base softening factor ; c_{UB} = undrained cohesion within influence zone of pile base; and A_B = cross-sectional area of pile base .

Following Meyerhof's (1951) suggestion, Equation (15) may be re-formulated as

$$Q_U(kN) = k_S N_S A_S + k_B N_B A_B \qquad (16)$$

where k_S = ultimate skin friction coefficient; N_S = average blow count along pile shaft; k_B = ultimate end-bearing coefficient; N_B = blow count within influence zone of pile base; and the respective pile areas A_S and A_B are in m².

Table 2 contains a summary of test results for instrumented bored piles installed within the respective residual soil types. Generally, the ultimate bearing capacity Q_U has been determined via Chin's (1970) plot, while the ultimate skin friction Q_{US} has been deduced from the flat portion of the corresponding plot of skin friction Q_s versus pile head settlement ρ, which was found to occur after a relatively small pile settlement. The ultimate end-bearing load Q_{UB} has then been obtained as the difference between Q_U and Q_{US}. Based on Equation (16), it may thereby be deduced that for the test piles installed in the Old Alluvium, the empirical coefficient $k_S \approx 5$, while $k_B \approx 34$. Hence, in view of Equation (15), the adhesion and pile base softening factors may be determined as $\alpha \approx 0.7$ and $\omega \approx 0.55$, respectively. By similar reasoning, $\alpha \approx 0.7$ and $\omega \approx 1$ for the sedimentary residual soil, while for the granitic residual soil, $\alpha \approx 0.5$ and $\omega \approx 0.6$.

Table 2. Bearing support components from pile instrumentation

Soil type	Pile length (m)	d (m)	Q_B (t)	Q_S (t)
O	10.15	0.45	30	250
O	10.1	0.45	38	335
O	20	0.56	35	433
O	20	0.56	41	504
O	10.8	0.53	5	305
O	7.65	0.46	44	416
S4	12.4	0.46	61	265
S4	7.6	0.53	41	311

Note: O = Old Alluvium and S4 = sedimentary residual soil.

With regard to the safe bearing capacity, in view of the tendency of the ultimate skin friction to be mobilised at a relatively small settlement, a partial factor of safety of 1 would be appropriate to this component of bearing support, while a study of the load-settlement measurement plot has indicated that a partial factor of safety of 3 against ultimate end-bearing failure would be advisable to avoid local failure in end-bearing.

5.2.2 Settlement

The settlements of pile foundations installed in the Old Alluvium and granitic residual soil were measured by means of dial gauges attached to the pile head, during load testing in the standard as well as fully-instrumented configurations, in order to provide an independent check to the foregoing settlement analysis of shallow foundations and bearing plates, on the proposed semi-empirical approach for determining the compression characteristics of the residual soils. An example of a fully-instrumented test configuration is shown in Figure 10. The analysis of pile settlement was based on the derivation of Poulos and Davis (1980), whereby the pile head settlement ρ may be determined as

$$\rho = PI_0R_KR_bR_v/(E_sd) \tag{17}$$

where P = applied axial load; I_0 = settlement-influence factor for incompressible pile in semi-infinite soil mass with Poisson's ratio $\nu_S = 0.5$; R_K = correction factor for pile compressibility; R_b = correction factor for stiffness of bearing stratum; R_v = correction factor for soil Poisson's ratio ν_S; E_s = average Young's modulus of soil along pile shaft; and d = pile diameter.

The analytical plots of settlement in Figure 11 were thus obtained and may be compared with the corresponding measurement results. Accordingly, reasonably good agreement was obtained between the two sets of settlement results, thereby providing

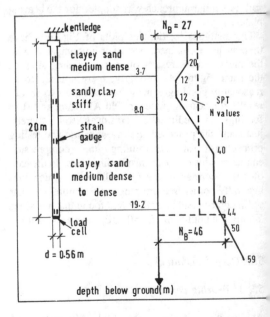

Figure 10 Idealization of bored pile in Old Alluviu

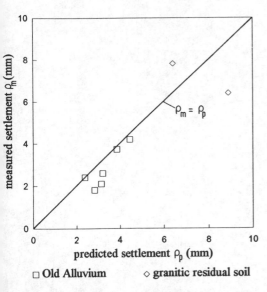

Figure 11 Comparison of pile settlements

further justification to the adoption of the proposed semi-empirical approach for determining the settlement of foundations in the residual soils. As in the foregoing case of strip footings in the Old Alluvium, however, the relatively low permeability of this formation vis-a-vis the other two types of residual soils had to be taken into account, in that the virtually undrained conditions under which pile loading took place meant that an undrained Young's modulus given by $E_U = 1.5E_M$ was applicable to the Old Alluvium.

5 RECOMMENDATIONS AND CONCLUSIONS

In the light of the overall findings of the preceding study, the following recommendations and conclusions can be made:

5.1 Site investigation

Rotary drilling techniques should be adopted in advancing boreholes in the residual soils. In firm to stiff deposits, either 50mm jacked-in thin-walled samplers may be employed in 100mm diameter boreholes, or standard 100mm open-drive samplers in 150mm diameter boreholes, whereas in very stiff to hard deposits, 63mm triple tube sampling in 100mm boreholes would be advisable. The shear strength of these soils may be determined by triaxial compression testing of directly-extruded samples,

but laboratory testing to determine their compression characteristics is unlikely to be viable. The most cost-effective field testing method for the residual soils is the standard penetration test which may be correlated to the shear strength and compression characteristics of these soils (the same conclusion was made by Stroud 1974 for insensitive clays and soft rocks). Supplementary pressuremeter testing could also prove to be useful in the investigation of a particular site.

6.2 Design parameters

As a preliminary measure, the following proposals may be adopted, bearing in mind that these recommendations may have to be modified slightly as more case studies become available. Notwithstanding this possibility of modification, the suggested approach to evaluating foundation response would still be appropriate.

Shallow foundations: Based on the results of plate loading tests, for the determination of ultimate bearing capacity, an undrained cohesion c_U given by Equation (8), (9) or (10), depending on soil type, would be appropriate. For safe bearing capacity, a factor of safety of 3 would be prudent. In the case of settlement computation, the correlations indicated by Equations (2)-(4) may be adopted, together with a corresponding conversion factor of 1, 1·33 and 1·33 applied to the pressuremeter deformation modulus E_M to obtain the Young's modulus E, for the Old Alluvium, sedimentary residual soil and granitic residual soil, respectively. In the case of the Old Alluvium, however, further consideration has to be given to its relatively low permeability by the use of $E_U = 1.5E_M$ or $E' = 0.33E_M$, as applicable.

Deep foundations: For ultimate skin friction, the adhesion factor $\alpha = 0.7$, 0·7 and 0·5 for the Old Alluvium, sedimentary residual soil and granitic residual soil, respectively. In determining the ultimate end-bearing, corresponding pile base softening factors $\omega = 0.55$, 1 and 0·6 would be appropriate. The respective partial safety factors for ultimate skin friction and end-bearing may be taken to be 1 and 2.5, respectively. For the settlement analysis of piles in these soils, similar considerations as for the shallow foundations would apply.

ACKNOWLEDGEMENT

The collation of data for this paper was carried out by Mr. Lee Chong Hee as part of his M.Sc. dissertation requirement, the use of which is gratefully acknowledged.

REFERENCES

American Society of Civil Engineers (ASCE) 1982. *Engineering and construction in tropical residual soils.* New York: ASCE.

Brand, E.W. & H.B. Phillipson 1985. *Sampling and testing of residual soils - A review of international practice.* Hong Kong: Scorpion Press.

British Standards Institution (BSI) 1986. *Foundations: BS8004.* London: BSI.

British Standards Institution (BSI) 1975. *Methods of test for soils for civil engineering purposes: BS1377.* London: BSI.

Chin, F.K. 1970. Estimation of the ultimate load of piles not carried to failure. *Proc. 2nd South East Asian Conf. on Soil Engg.:* 81-90. Singapore.

Committee on Tropical Soils of ISSMFE (1982-1985) 1985. *Peculiarities of geotechnical behaviour of tropical lateritic and saprolitic soils.* Sao Paulo: Brazilian Society for Soil Mechanics, Edcile services Graficos Editora Ltada.

Giroud, J.P. 1972. Settlement of rectangular foundation on soil layer. *Journ. Soil Mech. Fdn, Div., ASCE* 98(SM1): 149-154.

Lee, C.H. 1987. *Settlement and bearing capacity of foundation in singapore residual soils and Old Alluvium.* Dissertation submitted in partial fulfilment of degree of M.Sc. from National University of Singapore.

Meyerhof, G.G. 1951. The ultimate bearing capacity of foundations. *Geotechnique* 2(4): 301-332.

Poulos, H.G. & E.H. Davis 1980. *Pile foundation analysis and design.* New York: John Wiley. Publications Committee of 2 ICOTS 1988. *Proc. of the 2nd Inter. Conf. on Geomichanics in Tropical Soils*: 625pp. Rotterdam: Balkema.

Rowe, R.W. & L. Barden 1966. A new consolidation cell. *Geotechnique* 16(2): 162-170.

Stroud, M.A. 1974. The standard penetration test in insensitive clays and soft rock. *Proc. 1st Europ. Symp. Penetr. Test.:* 367-375. Stockholm.

Wilun, Z and K. Starzewski. 1975. *Soil mechanics in foundation engineering, vol. 1.* London: Surrey University Press.

Géologie de l'ingénieur des massifs des sols saprolitiques au climat tropical
Engineering geology of the saprolite soils in tropical climates

E. L. Pastore
Instituto de Pesquisas Tecnológicas de São Paulo, Brésil

P.T.Cruz
Escola Politécnica da Universidade de São Paulo, Brésil

J.O.Campos
Instituto de Geociências, UNESP, Brésil

RESUME: Cet article présente les caractéristiques et propriétés de geologie de l'ingenieur des massifs de sols saprolitiques aux régions tropicales, souvents engagés par des travaux publics et miniéres. En prenant la même demarche des études des massifs rocheux, on present en separé les propriétés du material, des structures residuélles et du massif (ensemble material et structures residuélles).

ABSTRACT: This paper presents the engineering geology properties of the saprolite soil masses in tropical regions, related to civil and mineral works. These properties are discussed according to the same criteria used for rock masses.

1 INTRODUCTION

Les massifs de sols saprolitiques trouvés couramment aux régions tropicales sont un des plus importantes matériaux engagés pour des travaux publiques et miniéres, à cause de leurs larges surface d'ocurrence et épaisseur. Ces massifs se trouvent situés au sud de l'Asie, sud-est de l'Amérique du Nord, à l'Amérique Centrale, au Caraibe et à l'Amerique du Sud (Sowers, 1963), et sont composés par des produits de décomposition des massifs rocheux, qui mantiennent preservées nettement les structures et discontinuités originalles. Les massifs de sols saprolitiques sont couramment recouverts par des couches de sols residuels ou transportés, latéritiqués, matériaux fort différentes au point de vue morphologie et comportement géotechnique.

Parmi les nombreux articles qu'on trouve sur les massifs de sols tropicaux, les plus anciens d'après Deere et Patton (1971) traitent de l'identification des couches dans les profil d'altération, comme ces de Derby (1846) et Branner (1986) au Brésil, Blackelder (1925) en Californie (États Unis) et Brok (1943) à Hong Kong. Du point de vue propriétés géotechniques on a les articles de Deere et Patton (1971), Sowers (1967) et Vargas (1974). Plus récemment on peut mentionné les études de Vaughan et Kwan (1984) et Vaughan (1990) qui s'occupent fondamentalement des questions relatifs à l'identification géotechinique, contraintes naturelles et analyses des propriétés de résistance au cisaillement et compressibilité des massifs de sols tropicaux residuels. Ces chercheurs, cependant, font réference aux couches les plus superficieles du massif qui ont été déjà latérisées. Cook et Newill (1988), d'autre part, ont devellopés une méthodologie pour la description des sols residuels tropicaux sur le terrain, en faisant di fférence entre le sol et le massif de sol, compte tenu de ses composition, structures residuélles et comportement géotechnique. En tenant compte que les massifs de sols saprolítiques héritent les anisotropies des massifs rocheux, les théories appliquées en Mécanique des Roches ont été aussi appliquées aux massifs de sols, comme outils áuxiliares des études géotechniques. À ce propos, des chercheurs, outres que Cook et Newill (1988), présentent des études, tel que Vargas (1963), de Mello (1972) et Cruz (1989).

De cette façon, on présente, dans cet article, des caractéristiques et des propriétés géotechiniques des sols saprolitiques, callés sur des resultats obtenus par plusieurs chercheurs, ainsi que les recherches en laboratoire efféctuèes par Pastore (1992).

2 LES MASSIFS DE SOLS TROPICAUX

Les massifs de sols tropicaux sont formés fondamentalement par deux couches, d'aprés Pastore (1992):

1. Une couche supérieure de sol latéritique
Cette couche peut être composée par sol residuel, c'est-à-dire, ce qui n'a pas été transporté par les agents géologiques, ou par des matériaux transportés, tel que les sols colluvionaires et depôts de talus, toujours affectés par les transformations pédologiques comme la laterisation.

Dans les dépôts de talus et colluvions, telles transformations touchent la matrice de sol qui enveloppe les cailloux de roches.

La composition granulométrique de cette couche et son épaisseur sont très variables, compte tenu du relief et des roches du substratrum. Cette couche est équivalente à la couche B pédologique, ne présentant pas des structures de la roche d'origine, mais présentant des structures identifiées par la Pédologie. Les concepts ici mentionnés sont d'accord avec ceux établis par le Comite des

Sols Tropicaux de l'Association Internationale de Mécanique de Sols et de Génie de Fondations (1985), selon lcqucl, sol latéritique, dans le sens géotechnique, doit avoir les caracteristiques suivantes:

. appartenir aux couches pedologíques A et B des profils bien drainés développés sous climat tropical humide et avoir de fractions d'argile constituées essentiellement par des kaolinites et oxydes de fer et d'aluminium hydratés. Ces éléments sont groupés en structures poreuses composées par des agregats trés stables. Les couleurs prédominantes de ces sols sont le rouge et le jaune.

2 . Une couche inférieure de sol saprolitique

Cette couche est constituée essentiellement par les sols résiduel qui ont comme des caractéristiques principales, la présence de la structure residuélle de la roche d' origine, pouvant avoir jusqu'a 10% de blocs de roche, d'après Deere et Patton (1971).

Les discontinuités du massif rocheu, telles comme fractures,joints et failles se trouvent préservées, malgré la profonde décomposition.

L'épaisseur et la composition granulométrique de cette couche sont aussi trés variables, en fonction du relief et des roches d'origine. Les compositions granulométriques les plus ordinaires sont les sables limoneux, peut argileux et les limons argileux, peut sablés. Les minéraux, les plus ordinaires, trouvés dans les sols sont la kaolinite et la mica. Les caractéristiques minéralogiques et quelques propriétés de la fraction limoneuse que compose la plus part de ce type de sol ont été étudié par Pastore (1982).

Les couleurs plus ordinaires de cette couche sont le blanc, le creme et le jaune. Les concepts ici mentionés sont d'accord aves ceux du Comite des Sols Tropicaux de la Association Internationale de Mécaniques de Sols et de Génie de Fondation (1985), selon lequel, sols saprolitiques sont:

. sol d'après le concept géotechnique
. montre nettement les structures residuélles qui permettent identifier, sans difficultés, la roche d'origine et, il est typiquement résiduel. Dans cet article on traite essentiellement des caractéristiques et des propriétés géologique-géotechniques de la couche de sol saprolitique.

3 CARACTERISTIQUES ET PROPRIE TES GEOLOGIQUE GEOTECHNIQUES

On a dit auparavant, que les caractéristiques et les propriétés géologique-géotechniques des massifs de sols saprolitiques seront présentées en suite de façon individuelles pour le matériel sol, les discontinuités resi duélles et le massif, suivant la tendance mondiale de les étudier à l'imitation des massifs rocheux. Malgré la manque des résultats, ce qui ne permet pas de faire une analyse statistique, quelques conclusions sont, cependant, possibles d'être etablies

3.1 Sols saprolitiques

On présente ici, en tenant compte des essais de laboratoire, des résultats concernants au caractéristiques de granulometrie, plasticité et minéralogie, ainsi que les propriétés géotechniques tels que la perméabilité, compressibilité et résistance au cisaillement de sols saprolitiques naturels, sans aucune interférence de discontinuites, couramment préservés dans les massifs.

3.1.1 Caractéristiques granulométriques, de plasticité et minéralogiques.

La composition granulometrique, la plasticité et la minéralogie des sols saprolitiques sont étroitement liées à la texture et à la minéralogie des roches d'origine et à la décomposi

tion. Dans le cas particulier de la granulometrie, Vaughan (1990) remarque que les resultats des essais dependent de l'énergie utilisée, parce que, on a dans le sol la présence de grains frágiles, tel que les feldspaths alterés. Compte tenu que les roches peuvent avoir des granulations fines jusqu'a grossières, on peut attendre que la granulométrie des sols saprolitiques, originaires de telles roches, seront trés variables.

Malgré tout, on peut dire, d'après les resultats d'essais granulometriques, présentés sur le tableau 1, que les sols saprolitiques originaires des roches de granulation moyenne à grossière, tel que les granites, les migmatites et les gnaisses, sont composés surtout par des sables limoneux et limons sableux. Déjà les sols originaires des roches de granulations fine, tel que les basaltes et quelques types de granites, sont composés surtout par des limons argileux et des argiles limoneux. Des coubes granulométriques de plusieurs types de sols, compilé par Pinto et al (1993), présentent pratiquement les mêmes predominance.

En ce qui concerne les limites d'Atterberg, quelques valeurs sont aussi donnés sur le tableau 1, oú on peut verifier une trés importante variation. L'emploie de ces limites en sols tropicaux ont été beaucoup contestés par plusieurs chercheurs, selon lesquels la dispersion des resultats est liée à la grande sensibilité que ces sols montrent quand remaniés. Dans ce sens on trouve Vaughan et al (1988) que en faisant l'analyse à ce sujet proposent un'autre technique d'identification de sols tropicaux basé sûr la comparasion entre l'indice des vides natural et desagregé. Nogami et Villibor (1980) proposent une nouvelle classification apellée MCT pour les sols tropicaux, utilisant les essais de compactation et d'immersion dans l'eau.

Du point de vue mineralogique, les sols saprolitiques presentent la composition suivante, d'áprés Nogami (1978): fraction sable:

minéralogie complexe avec des mineraux en différentes dégres d'alteration , fraction limon: principallement kaolinite, mica et quartz et fraction argile: kaolinite, haloysite, atapulgite et smectites. Vaughan (1990) fait refférence à la kaolinite, haloysite, atapulgite et smectites comme mineraux de la fraction fine des sols residuels et au quartz et feldspaths, en plusieurs etats d'alteration, comme components principaux de la fraction grossière.

Pastore (1982) étudiant la fraction limoneux des sols saprolitiques a trouvé la kaolinite, la mica et le quartz comme principaux mineraux de cette fraction. C'est trés courant dans la même , la kaolinite avoir la forme "d'accordeon" , ce qui donne aux sols sapro-litiques, provenant des granites des propriétés trés particulier , specialement ceux liés au gonflement.

3.1.2 Propriétés Géotechniques

Pour une anályse plus correct des propriétés géotechnique des sols saprolitiques, ces sols

Tableau 1 . Granulometrie, limites et classification des sols saprolítiques (Pastore, 1992)

roches d'origine	argile < 0.002	limon < 0.074 mm	sable < 2.0 mm	LL	LP	USCS (1)	MCT (2)	chercheur
magmatiques et metamórfiques	10	30	60	30 - 65	0 - 25	ML-MH	NA'	Sowers (1963)
migmatites	0 5	5 20	95 75	29 - 40	NP	--	NA '	Campos (1974)
gnaiss	5	30	65	42 - 62	12 - 25	ML-MH	NA '	Mori (1982)
migmatites	19 16	32 32	49 52	53 48.5	20.5 32.5	MH CL	NA '	CESP (1986)
gnaiss	35	47	18	20 -75	0 -38	ML-MH	NS '	Vargas (1974)
granodiorite	50	20	30	67	26	MH	NS '	Milititsky e Nudelman (1981)
gnaiss	15	47	38	42-62	12-25	ML-MH	NS '	Mori (1982)
granite	6	58	36	52	19	MH	NS '	Pastore (1982)
basalte	35	45	20	-	-	-	-	Sardinha e outros (1981)
basalte	50	30	20	-	-	-	-	Leme (1985)
basalte	53	20	27	28-68	7-31	CL-MH	NG '	Vargas (1974)

(1) Classification Universelle des Sols (2) Classification des Sols Tropicaux

doivent être subdivisés par deux groups, selon la presençe ou pas des foliations resi duélles:

1 . Sols saprolitiques homogénes: y compris

les sols originaires des roches magmatiques, sedimentaires et metamórphiques sans stratification ou foliation trés evident, tels comme les grés, les argilites, les granites, les calcaires, les basaltes et quelques types de gnaisses et migmatites, oú le comportement est basiquement isotropie.

2 . Sols saprolitiques hétérogènes: y compris les sols originaires des roches sedimentaires et metamórphiques avec stratification ou foliation evidents, tels comme les ardoises, les phylites, les schistes, quelques types de migmatites et gnaisses et les argile feuilletées, oú le comportement géotechinique est basiquement anisotropie.

3.1.2.1 - Sols saprolitiques homogénes

1 . Permeabilté

Des resultats de coeficients de permeabilité de quelques sols saprolitiques homogénes ont été compilés par Pinto et al (1993), callés sur des essais de laboratoire, oú les valeurs sont presentés sûr le tableau 2.

Ces resultats montrent que pour les éprou-vettes avec texture homogéne il y a un écarte trés importante de valeurs de permeabilité, ce qui, d'aprés les chercheurs auparavant mencionés, elle est due à la variation de l'índice des vides et porosité des sols essaiés. Il faut ici mencioner, cependant, que, d'aprés Costa Filho (1985), la determination du co-eficient de permeabilité de sols saprolitiques, sûr des éprouvettes de laboratoire, est trés douteux, car c'est presque impossible de tenir en compte les interferences des descon-tinuites residuélles couramment trouvées dans les massifs.

Encore, d'aprés Richardson et Brandt (1984, apud Costa Filho, 1985), à cette cause, en Hong Kong (Chine), on ne fait que trés ra-rement des essais au laboratoire avec des éprouvettes de sols saprolitiques pour de-terminer le coeficiente de permeabilité.

2 .Compressibílté

En ce qui concernent les valeurs de com-pressibilité, la correlation couramment utilisée pour les sols sedimentaires, entre la limite de liquidité et l'índice de compression, semblet ils n' avoir pas de bons resultats pour les sols saprolitiques, remarque fait par La

Tableau 2 . Coefficients de permeabilité des sols saprolítiques homogénes (Pinto et al, 1993)

sol saprolítique	Kv (cm/s)	Kh (cm/s)
grés	2.5×10^{-4}	-
basalte	2.8×10^{-4}	2.6×10^{-3}
granite	7.0×10^{-4}	3.5×10^{-3}

cerda et al (1985). D'aprés ces cheurcheurs, une correlation un peut plus confiable, il y a entre l'índice de compression et l'índice des vides natural. Ils montrent en plus des resul-tats d'essais oedométriques pour plusieurs types de sols saprolitiques, remarquant, cependant, que la determination de la com-pressibilité en laboratoire ont toujours des problems à cause de troubles de la estructure de sols quand du prelévement. Par conse-quence, les resultats les plus confiables seriont ceux qu'on determinent en place.

3 . Résistance au cisaillement du sol satu-ré

Pinto et al (1993) presentent les caracteris-tiques de resistence au cisaillement de plusi-eurs types de sols saprolitiques callés sur des resultats d'essais de laboratoire. Les valeurs moyennes de la cohésion et d'angle de frottement interne obtenus pour les sols homogénes saturés, sont presentés sur le tableau 3.

On n'y a pas beaucoup des resultats d'essais pour ceux types de sols mais on a rajouté sur le tableau 3 les resultats des sols de basaltes, obtenus par Humes (1989), et aussi leurs de Lumb (1962), pour les granites.

3.1.2.2 Sols saprolitiques hétérogènes

1 . Permeabilité

D'après les resultats d'essais présentés par plusieurs cheurcheurs, entre eux Lumb (1962), Sowers (1963), Maciel (1991) et CESP-LECEC (1986), realisés au laboratoire, parallelement et perpendiculairement à foliation residuélle, dans les cas des sols bandés, on a des valeurs presentés sur le tableau 4 (Pastore, 1992).

L'anisotropie verifiée est plus evident en sols avec une grande quantité de mica, tel que les sols essaiés par Sowers (1963) et CESP-LE-CEC (1986).

2 . Résistance au cisaillement

Les valeurs de cohésion et de l'angle de frottement interne ont été determinés aussi par plusieurs chercheurs, en special à la boîte de cisaillement, parallelement et perpendiculairement à foliation residuélles de sols saprolitiques.

Tels resultats montrent une résistance plus faible vis-à-vis le sens parallele, comme on peut observés sur le tableau 5.

3.2 Descontinuites residuélles

Les descontinuites residuélles les plus importantes pour l'etude de massifs de sols saprolitiques, compte tenu de l'anisotropie causés, sont les structures plâtes.

Telles structures sont basiquement les foliations, stratifications, fractures, joints et failles originaires du massif rocheu.

La résistance au cisaillement des descontinuites residuélles, ainsi que celles du massif rocheu, dependrent du type de material de remplissage, de l'ondulation et de la rugosité des parois. Deere et Patton (1971) et Pastore (1992) presentent des courbes intrinséques hipotéthiques pour les massifs alterés et les massifs de sols saprolitiques, respectivement, avec des descontinuites residuélles.

3.3 Massif saprolitique (sol plus descontinuités)

On comprendre par proprietés des massifs de sols saprolitiques, les proprietés qu'on obtient avec des essais et des recherches en place, qui donnent comme resultats le comportement de l'ensemble sol, structures residuélles et d'autres formes complexes de structures, éventuellement presentes au terrain.

Les proprietés plus usuellement determinés er place sont l'índice (N) S.P.T. (Standart Penetration Test), les coefficients de perméabilité et, la compressibilité.

Pastore (1992) a etudié au laboratoire l'influence d'une descontinuité residuélle dans la résistance au cisaillement de sol saprolitique de migmatite, et Pastore et Cruz (1993) presentent des études sûr la resistence au cisaillement de massifs de sols saprolitiques.

1 . Índice (N) S.P.T.

D'après plusieurs resultats de forages compilés par Pastore (1992), on peut vérifier que les valeurs (N) S.P.T. le plus courants pour les massifs de sols saprolitiques de migmatites/gnaisses sont trés variables, dans un écart de 5 a 20, c'est-à-dire, compacité variable de molle à moyenne, d'après la classification de Terzaghi et Peck (1967).

2 . Permeabilité

Quelques valeurs de coefficient de permeabilité determinés en place par plusieurs types d'essais, ont été compilés par Costa e Vargas (1985). Pour les massifs de sols saprolitiques de gnaiss il y a un écart variable de 1 a 5×10^{-4} cm/s et pour les migmatites il y en a de 3×10^{-3} a 1×10^{-5} cm/s.

3 . Compressibilité

Tableau 3 . Paramétrés de résistance des sols saprolítiques homogénes saturés

sol saprolí-tique	c (kpa)	φ (degrés)	c' (kpa)	φ ' (degrés)	chercheur
argilite/siltite	77	15	43	29	Pinto et al (1993)
grés Caiuá	1.25	21	1.25	36	Pinto et al (1993)
granite fin	0	25.5-34	-	-	Lumb (1962)
granite grossier	0	36 - 38	-	-	Lumb (1962)
granite	40	7	10	31.5	Pinto et al (1993)
basalte	27	11.5	19	30	idem
basalte 1.1<e<1.5	48	12	33	27	idem
basalte 1.5<e<1.9	9	11	7	33	idem
basalte IP<20	7	12	6	33	idem
basalte 20<IP<60	37	11	22	30	idem
basalte IP>60	65	13.5	62	20.5	idem
basalte	-	-	35-55	22 30	Humes (1989)

Tableau 4 . Coefficients de permeabilité des sols saprolítiques heterogénes (Pastore, 1992)

sols saprolítiques	coefficients de permeabilité (cm/s)	
	perpendiculaire	parallele à la structure
sable limoneux (gnaisses)	$5x10^{-5}$ - $4x10^{-3}$	$1x10^{-4}$ - $1x10^{-5}$
limon sableux (migmatites)	$1x10^{-5}$ - $5x10^{-5}$	$2x10^{-5}$ - $2x10^{-4}$

Des essais sur place realisés par Sandroni (1981) sur de massifs de sols saprolitiques de gnaiss, ont presentés le resultats suivantes, en KPa:

Pressiomètre   40.000 - 80.000

Essai à la plaque

40 cm   70.000

80 cm   47.000

160 cm   84.000

4 . Résistance au cisaillement

La résistance au cisaillement des massifs de sols saprolitiques est determinée couramment par les essais au laboratoire sur des éprouve-ttes taillées à partir des blocs indeformés prélevés au chantier.

Pastore (1992) etudiant l'influence de des

continuites residuélles sûr la resistance au cisaillement de sols saprolitiques de migmati-tes a realisé des essais triaxiaux (R_{sat}), en changeant la position du plan de la desconti-nuité par rapport à l'axe de la contrainte principale σ_1, tel que Hoek (1983) pour les roches.

La figure 1 présente la courbe contrainte effective à la rupture (σ'_{1r}) - angle (β), défini entre l'axe de telle contrainte et le plan de la descontinuité. On doit tiré l'attention sur les points de la courbe de la figure 1, lesquels ont été liés en tenant compte les índices des vides de chaque éprouvette. Ceux índices definent deux groupes distincts. On vérifie, qu'il y a un chute maximun de la

Tableau 5 . Choésion et angle de frottement interne de sols saprolitiques à la boite de cisaillement (Pastore, 1992)

Roches d'origine	types de sol	résistance au cisaillement						cheucheur
		parallele		perpendiculaire		residuélle		
		c' KPa	φ' (degres)	c' KPa	φ' (degres)	c'$_r$	φ'$_r$ KPa (degres)	
quartzite à oxyde de fer	limon sableux	20	37	50	44	-	-	Sandroni (1985)
quartzite à mica	sableu limoneux	40	22	45	27	-	-	Sandroni (1985)
migmatite à mica	limon à mica	40	20	52	23	-	-	Campos (1974)
		30	21	49	22	-	-	
schiste	limon sableux	78	28	100	27	-	-	Durci et Vargas (1983) apud Maciel (1991)
filite à mica	limon	10	29	60	41	-	-	idem
filite	-	0	18	0	24	-	-	De Fries (1971) apud Deere et Patton (1971)
migmatite à mica	limon sableux	8	22	-	-	3	17	CESP - LECEC (1986)
		9	19	-	-	0	18	
		35	22	60	26	24	20	
migmatite	limon sableux	15	26	-	-	4	23	CESP - LECEC (1986)
		13	40	-	-	0	33	
		0	36	-	-	0	33	
		27	24	-	-	27	23	
gneiss	limon sableux	45	35	38	36	-	-	Campos (1989)
		27	28	27	29	-	-	
gneiss	sable limoneux	Sans aucune différénce. Ne presentent pas des chiffres.						Maciel (1991)

résistance du sol quand β = 45°, ce qui montre avoir une interférence de la structure residuélle, selon sa position par rapport à la contrainte σ'1r.

Pastore et Cruz (1992) comparent des resultats en roche, presentés par d'autres cheurcheurs, en proposant un coefficient d'anisotropie.

4 CONCLUSIONS

Les massifs de sols saprolitiques au climat tropical presentent, d'une façon generale, caractéristiques et propriétés géotechiniques suffisamment complexes à cause de l'herité des structures et particullarités des massifs rocheux. En plus, la mineralogie conséquente de la profonde decomposition liée à une trés importartante variation des índices des vides augmentent la dispersion des resultats des essais tant sur place qu'au laboratoire.

Compte tenu des resultats des essais presentés par plusieurs cheurcheurs ont verifie que l'influence des structures residuélles sont trés evidentes dans les propriétés des materiaux auparavants mencionés, ce qui mettre en doute les resultats des essais qui n'ont prenent en compte telles traites.

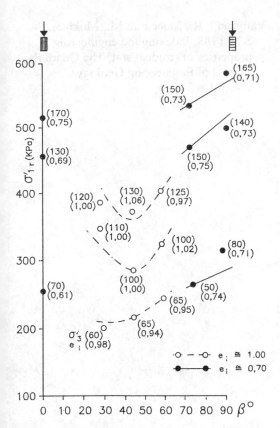

Figure 1 Graphique σ'$_{1r}$ - β (Pastore, 1992)

5 REMERCIEMENTS

Les auteurs remercient à la FAPESP - Fundação de Amparo a Pesquisa do Estado de São Paulo, pour les dépenses de voyage au 7 th Congréss.

6 REFERENCES

Cesp/Lecec.1986.Caconde:vertedouro suplementar. Ensaios geotécnicos especiais sobre amostras indeformadas da fundação. Relatório LEC-SO-01/86)

Committee on tropical soils of the ISSMFE. 1985.Progress Report (1982-1985) theme 2, topic2.3, p.67.

Cook,J.R. & Newill, D.1990.The field description and identification of tropical residual soils. In International Conference on Geomechanics in tropical soils, 2,

Costa Filho, L.M. et al. 1989.Engineering properties and design assesment of tropical soils: fabrics and engineering properties. In International Conference on soil mechanics and foundation engineering, 12, ISSMFE

Costa, L.M. & Vargas, E.A.Coord.).1985.Mechanical and hydraulic properties of tropical lateritic and saprolitic soils: hydraulic properties. In Committee on tropical soils of ISSMEF. Progress report (1982-1985). theme 2, topic 2.3, p.67

Cruz, P.T. 1989. Raciocínios de mecânica das rochas aplicados a saprólitos e solos saprolíticos. In Colóquio de solos tropicais e subtropicais e suas aplicações em engenharia civil, 2.UFRG p.121

Deere, D.V. & Patton, E.D. 1971.Slope stability in residual soils. State of the art paper. In Panamerican Conference on soil mechanics and foundation engineering, 4 ASCE.v.1, p.87

Hoek, E. 1983. Strength of jointed rock masses. Geotechnique, v.33 ,n.3, p.187

Humes, C. 1990.Uma síntese da resistência ao cisalhamento de solos naturais superficiais e saprolíticos de basalto. In Congresso Brasileiro de Mecanica dos Solos , 9, v.1, p.487

Lacerda, W.A. et al (Coord.) 1985. Mechanical and hydraulic properties of tropical lateritic and saprolitic soils: compressibility properties of lateritic and saprolitic soils. In Committee on tropical soils of the ISSMEF. Progress Report (1982-1985). theme 2, topic2.4,p.85

Lumb, P.1962. The properties of decomposed granite. Geotechnique, v.12 n.3,p.226

Maciel, I.C.Q.1991.Aspectos micro-estruturais e propriedades geome-cânicas de um perfil de solo residual de gnaisse facoidal.Tése PUC-RJ, 183 p.

Mello, V.F.B.1972.Engineering residual soils.Southeast Asian Conference on Soil Engineering, 3

Nogami, J.S. & Villibor, D.F. 1980. Caracterização e classificação gerais de solos para pavimentação: limitação do método tradicional, apresentação de uma nova sistemática. In Reunião anual de pavimentação, 15. ABPV.v.1, p.38

Nogami, J.S.1978. The adequate use of genetic grouping of lateritic and residual soils. In International Congress of the International Association of Engineering Geology, 3. v.2, p.153

Pastore, E.L. & Cruz, P.T. 1993. Resistência ao cisalhamento de maciços de solos saprolíticos. Publicação IPT 2052

Pastore, E.L.1982. Contribuição ao estudo dos solos saprolíticos compactados derivados de rochas ácidas, com destaque às obras viárias. Tése de mestrado EESC-USP, 193 p.

Pastore, E.L.1992. Maciços de solos saprolíticos como fundação de barragens de concreto gravidade. Tése de doutorado EESC-USP, 290 p.

Pinto, C.S., Gobbara, W., Peres, J.E.E., Nader, J.J.1993.Propriedades dos solos residuais.Solos do interior de São Paulo. ABMS/EESC-USP, p.95

Sowers, G.F. 1963. Engineering properties of residual soils derived from igneous and metamorphic rocks. In Panamerican Conference on soil mechanics and foundation engineering. 2. ISRM, p.39

Terzaghi, K. & Peck, R. 1967. Fundamentals of soil mechanics. John Wiley, 700p.

Vargas, M.1974.Engineering properties of residual soils from south central region of Brazil. In Internatinal Congress IAEG, 2

Vaughan, P.R.1990.Keynote paper: characterising the mechanical properties of in-situ residual soils. InInternational Conference on Geomechanics in tropical soils, 2, Balkema, v.2, p.469

Vaughan,P.R.&Kwan,C.W.1984.Weathering, structure and in situ stress in residual soils. Geotechnique, v.34, n.1, p.43

Vaughan,P.R., Maccarini,M., Mokhtar, S.M.1988. Indexing the engineering properties of residual soil. The Quarterly Journal of Engineering Geology, v.21,n.1,p.29

The loess sediment types of Slovakia, their fabric and properties

Les sédiments loessiques de Slovaquie, leur fabrique et propriétés physico-mécaniques

A. Klukanova

Dionyz Stur Institute of Geology, Bratislava, Slovakia

ABSTRACT: Based on investigation of the Slovak Carpathian loess soils it appears that the loess mass is composed of various primary rock fragments and minerals, products of their secondary changes and different organic substances. These components were incorporated in loess soils under various conditions. They are diversely distributed both horizontally and vertically depending on the terrain relief, the climate, the degree of dissection, the water table level and other factors. The fabric of loess not only reflects the sediment's genesis, but also affects its engineering geological properties. Four basic groups of loess sediments have been characterized. These are: typical loess, sandy loess, clayey loess and loess-like sediments. They differ from each other by their fabric, microstructure, genesis, physical and mechanical properties, tendency to collapse etc. Collapse tendency is one of the most significant properties of loess deposits. It is caused by a sudden volume change under the influence of humidity, stress or loading. To explain the mechanism of loess collapse, laboratory modelling of the process has been carried out in a triaxial cell, monitoring the microstructural changes.

EXTRAIT: Dans le cadre des études des limons loess de Carpathes Slovaqnes on a constaté que le loess se composé de divers fragments de roches primaires et de mineraux et aussi de produits de leurs altérations secondaires et encore de substances organiques diverses. Le loess est distribué diversement - horizontalement et aussi verticalement en dépendance de la morphologie du terrain, du climat, de degré de disséction, de niveau de l'eau et des autres facteurs. La structure de loess ne reflete pas la genese du sediment, mais elle influence aussi les propriétés de ingénieurs-géologiques. Nous caractérisons les quatre groupes des sédiments loess. Ce sont: les loess typiques, les loess sablonneux, les loess argileux et les sédiments semblables au loess. Ils se different par sa structure, microstructure, genese, ses propriétés physico-mécaniques, par son tendance vers collapsus. Le collapsus est un des propriétés les plus significantes des sediments loess et repose dans le changement brusque de volume sous l'effet de l'humidité, de stress et de surcharge. Pour expliquer le mécanisme de collapsus de loess, ce processus était reproduit dans la cellule triaxielle et les changements microstructuraux etaient observe.

1 INTRODUCTION

The Slovak Carpathians display considerable geomorphological variability, climatic and morphologic zonality. Characteristic feature include the diversity of the Quaternary geodynamic processes. These factors have a considerable influence upon the formation of loess cover of the Carpathian system, as they preconditioned and determined the particularities and the evolution of hyper-geneous processes. They influenced the contents of Fe and Mn hydroxides, carbona-tes, chlorides, sulfates, as well as chemical composition of pore solutions, organic matter and the origin of secondary minerals. All these factors were active to a different degree and in certain cases even determined the formation and origin of various loess lithotypes. We have distinguished following four principal types of loess sediments in Slovakia: typical loess, sandy loess, clayey loess and loess-like sediments. However this classification may not apply only for the Carpathian region, but it could be used at a larger scale.

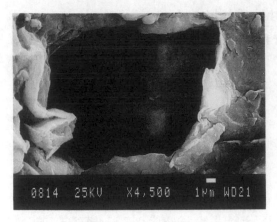

Fig. 1. One part of disturbed basical structural element. a – core, b – clay film

Fig. 3. Skeletal microstructure of typical
loess. Locality Mnesice brick plant near
Nove Mesto nad Vahom.

The microstructural analysis of undisturbed samples has been made using scanning
electron microscope (SEM) JEOL JSM 840. The
samples were always examined at two vertical
planar surfaces, one being parallel and the
other perpendicular to the bedding. The
results obtained, along with physical and
mechanical testing, X-ray, DT-analysis, and
other methods show that loess possesses a
complicated fabric.

In spite of the fact that the organization
of the individual structural elements in
loess (Fig.1) is complicated, it generally
shares a common principles. The basic structural unit of loess soils is form by a core
represented by quartz, carbonate, or more
rarely feldspar. The core is coated by a
thin clay film composed mainly of illite,
montmorillonite and mixed structures, sporadic iron and manganese hydroxides and

finely dispersed quartz and calcite particles.

The contacts between the basic structural
elements of loess soils are mediated through
clay bridges and buttresses (Fig.2). Carbonates and salts also occur to a lesser extent. Under natural conditions the above
structural elements of loess soils contribute to larger structural forms: in other
words, the structure ranges from micro- to
macroaggregates. Macroaggregates are unstable structural element, being easily broken
down.

2 TYPICAL LOESS

Typical loess (Fig.3) is characterized by a
skeletal microstructure (sensu classification of Grabowska-Olszewska et al, 1984),
which determines its granularity. Silt is a
dominant fraction. It is homogeneous with
isotropic, non-oriented fabric. The basic
structural element is represented by silty,
rarely sandy grains coated with several
superimposed clay films. The clay minerals
have phase contacts. They are medium aggregated. The presence of phase contacts adds
considerably to stiffness of sediments in
dry state. As mentioned above, the contacts
between basic structural elements formed by
clay bridges and buttresses. These are
characterized by considerable instability at
the contact with water. Intergranular pores
represent the main type of pores, enabling
the saturation of soil with water. This type
of soil tends to collapse. The problem
connected with soil collapse will be discussed below. The statistic summary of
physical properties of loess sediments are
given in table 1.

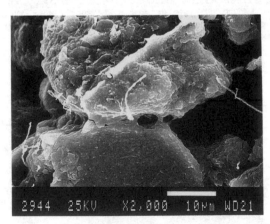

Fig. 2. The dominant contacts between two
grains are clay bridges and buttresses.

3 SANDY LOESS

Sandy loess (Fig.4) differs from typical loess mainly by its granularity – sandy grains with clay coating being dominant. Its microstructure is skeletal too.

Fig. 4. Microstructure of loess like sediment from Hajnacka locality. The skeletal microstructure is formed mainly by sandy grains.

The pores are intergranular, mutually permeable. The clay fraction content is lower than in typical loess and many sandy grains are not completely coated with clay films. The contact types are of the same nature as in typical loess. Although there are negligible microstructural differences only, we can state that the sandy loess tends to collapse mainly in the case of non complete coating of clay films. Sandy loess may also be of aeolian origin but its provenance could be different. Physical properties of sandy loess are in table 2.

4 CLAYEY LOESS

Clayey loess (Fig.5) ranges from matrix to matrix-skeletal microstructure, depending on its clay mineral content. The former contains more clay minerals and its fabric is heterogeneous whereas the later is characterized by lower content of clays and its fabric may be either homogeneous or heterogeneous. The properties of clayey loess are different in a complex, and it tends to alter to form various soil derivatives (e.g. delluvial loess, loess-like sediments, etc.). Its primary fabric is isotropic but owing to epigenetic alterations, it possesses frequently anisotropic fabric. This way we can explain considerable variability of

Fig. 5. Skeletal-matrix microstructure of clayey loess from Safarikovo locality.

some of its properties. The clay minerals are well aggregated, and its contacts are of the coagulation type. It occurs in matrix form. The coagulation type of contacts are less stable and thus the clayey loess does not form, the contrary to typical loess, vertical walls. The contacts between non-clay grains represent the cover types. The presence of the clay matrix with well aggregated clay minerals prevents the collapse. Structural elements consist of complicated aggregates, retaining water easily owing to their great surface energy, however, water weakens the bonds between the structural elements. Presumably this clayey loess will tend to swell. The pores represent predominantly the intermicroaggregate type. It is able to retain water for a long time which explains its relative high natural moisture. The cover types of contacts between non-clay grains and the presence of mostly intermicroaggregate type of pores cause, in contrast to typical loess, lower permeability. The microstructural analysis has shown, that the clayey loess investigated is also of aeolian origin, but it was subsequently loamified. Besides of climatic conditions the most significant factors were the character of source area and the transports distance. Physical properties of clayey loess are in table 2.

Table 1. Summary statistic of physical properties of typical loess from

	Units	Sum	Average	Minimum	Maximum	Variance	Skewness	Kurtosis
testing moisture	%	11	5.391	4.205	6.577	0.605	0.117	-0.445
bulk density of moisture soil	kg.m^{-3}	13	1.453	1.399	1.508	2.773	-2.867	-1.132
bulk density of dry soil	kg.m^{-3}	13	1.379	1.337	1.421	2.137	-0.188	-0.848
specific density	kg.m^{-3}	13	2.722	2.710	2.733	5.865	-0.293	-1.141
liquid limit	%	11	31.144	30.771	31.517	0.190	-0.358	-1.021
plasticity limit	%	10	24.573	24.241	24.906	0.170	7.041	-1.248
plasticity index	%	12	6.775	6.045	7.506	0.373	-0.250	-1.059
consistency index	%	12	3.802	3.345	4.259	0.233	0.171	-1.244
pore content	%	13	49.334	47.814	50.854	0.776	0.308	-0.884
saturation degree	%	13	15.064	10.095	20.034	2.536	6.098	-0.930
carbonate content	%	9	17.284	15.857	18.712	0.728	-0.257	-0.684
organic matter content	%	10	1.637	1.268	2.004	0.187	-0.217	-1.104

Table 2. Physical properties of loess sediments

Locality	Loess type	dept [m]	Grain-size fraction % — < 0.002 mm	0.002 - 0.063 mm	> 0.063 mm	natural moisture W_N %	liquid limit W_L %	plasticity limit W_P %	plasticity index I_P %	consistency index I_c	BULK DENSITY of dry soil ρ_d kg.m^{-3}	specific density ρ kg.m^{-3}	pore content n %	saturation degree S_N %	carbonate content O_u %	organic matter content O_m %
Mnesice	typical l.	20.0	14.0	72.0	14.0	12.38	31.14	24.57	6.77	2.55	1379	2721	49.33	15.06	17.28	1.63
Vyskovce	sandy loess	1.5	7.6	40.0	52.4	4.29	23.80	20.68	3.12	6.25		2730			5.58	
Vyskovce	sandy loess	2.5	9.4	35.3	55.3	3.35	24.71	21.26	3.45	6.19		2727			11.57	
Hajnacka	sandy loess	2.0	8.7	28.1	63.2	5.17	25.98	19.70	6.28	3.31		2741			17.56	
Jesenske	clayey l.	1.5	28.1	48.8	23.1	6.54	39.71	22.36	17.4	1.91		2744			0.7	2.61
Safarikovo	clayey l.	2.0	37.6	51.3	11.1	11.47	55.72	30.13	25.6	1.73		2739			1.12	4.09

5 LOESS-LIKE SEDIMENTS

The loess-like sediments (Fig.6 and 7) represent a large group of soils with diverse mineral composition and granularity. Their microstructures are: skeletal, matrix, honeycomb and laminar with homogeneous, heterogeneous, isotropic and anisotropic fabric. With respect to various types of microstructure they have distinct microstructural features and relations between them, influencing their properties. In our opinion the variability of loess-like sediments depends from both, different genesis and postgenetic processes. Differences are controlled, among others factors to mention, by the type of source area, climatic conditions, transport durations, etc. Physical properties of loess-like sediments are in table 3.

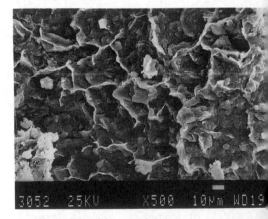

Fig. 6. The honey-comb type of microstructure of loess-like sediments from locality Krcava (12 m).

Table 3. Physical properties of loess-like sediments

Locality	Depth [m]	Grain-size fraction <0.002mm	0.002-0.063mm	<0.063mm	natural moisture W_N %	liquid limit W_L %	plasticity limit W_P %	plasticity index I_P %	consistency index I_c	BULK DENSITY of dry soil ρ_d kg·m⁻³	of moisture soil ρ_n kg·m⁻³	specific density ρ_s kg·m⁻³	pore content n %	saturation degree S_N %	carbonate content O_u %	organic matter content O_m %
#Krcava	0.5	45	48	7	12.1	53.3	20.5	32.8		1680		2690	34	90.4	1.01	0.1
#Ipelsky Sokolec	1.5	22	62	16		36.1	18.7	17.4								
#Ipelsky Sokolec	2.5	30	56	14		40.8	19.9	20.9								
#Fabianka	1.0	22	55	23		36.2	18.5	17.7								
#Rimavska Sec	1.5	16	63	21		33.3	20.6	12.7				2674			2.52	1.38
*Kysta	4.0	36	49	15	21.0	58.9	24.2	34.7	1.09	1678	2031	2654	36.7	95.9	0.2	0.39
*Kysta	5.0	39	48	13	23.1	64.1	25.0	39.1	1.05	1612	1984	2671	39.6	93.9	0.3	0.52
*Cejkov	3.0	10	62	28	28.7	27.4	16.1	11.3	1.03	1816	2103	2645	31.3	91.7	1.9	0.42
*Cejkov	5.0	18	49	33	28.8	38.7	17.9	20.8	1.1	1826	2114	2675	31.7	90.9	0.9	0.19
*Borsa	4.0	28	42	30	19.1	56.1	21.3	34.8	1.08	1759	2095	2660	33.9			
*Orechova	8.0	45	48	7	22.5	67.1	27.4	39.7	1.12	1620	1980	2650	38.9	93.3	0.7	0.16
*Orechova	7.0	40	52	8	20.0	59.5	27.3	32.2	1.23	1625	1950	2624	38.1	85.3	0.8	0.13
*Orechova	9.0	34	51	15	17.6	38.3	18.8	19.5	1.06	1710	2020	2690	36.4	83.0	1.4	0.1
*Krcava	10.0	24	66	10	19.5	43.4	19.6	23.8	1.0	1710	2040	2660	35.9	92.8	0.8	0.07
*Krcava	12.0	53	37	10	27.2	77.4	27.2	50.2	1.0	1560	1980	2630	40.8	100.0	0.8	0.5
*Vojnatina	1.0	6	31	63	20.4	41.2	21.1	20.1	1.03	1660	1999	2689	38.3	88.5	0.0	0.3
*Secovska Polianka	4.0	30	59	11	20.6	43.0	18.6	24.4	0.92	1730	2090	2650	34.6	100.0	0.6	0.3
*Secovska Polianka	10.0	41	49	10	20.4	51.8	21.5	30.3	1.04	1750	2110	2630	33.3	100.0	1.0	0.07

monolith sample, * sample of drill core

6524 25KV X500 10µm WD19

Fig. 7. The matrix type of microstructure of loess-like sediment from locality Orechova (8 m).

5 LOESS COLLAPSE

In terms of loess collapse we understand sudden reduction of volume due to increase in moisture and loading. The collapse in many cases leads to the failure of the foundation and break downs of complete buildings because of uneven and excessive subsidence. This is why we regard loess a problematic foundation soil.

The problems of collapse in general are even more grave in Slovakia, where the loess sediments spread in the area of almost 7000 km².

It follows from the mentioned above, that it is important to monitor and evaluate the collapse mechanisms of soils, surficial collapsing and sagging manifestations and changes in the geologic environments and the damage to it, to foresee its temporal and spatial effects and to take measures to minimize these effects to an acceptable level.

6.1 The conditions and mechanisms of collapsing

A number of national as well as foreign authors have referred to the conditions and

mechanisms of collapsing. According to the Czechoslovak standard the collapse may occur if one of the following conditions has been identified:
- the soil is of aeolian genesis
- the content of silty fraction is more than 60%
- the content of clayey fraction is below 15%
- the degree of saturation is less than 0.6 and the liquid limit is less than 32%
- porosity is greater than 40 %
- actual natural moisture is lower than 13%
- Collapse soils are those, whose collapse due to saturation surpass 1 %.

The conditions of collapse indicate that the causes of this process vary. They are influenced by the genetic features of soil, post-genetic processes, compactibility, hydrogeologic conditions, mineral composition etc. According to microstructural analysis (Klukanová, 1988) and the results of Sajgalik and Modlitba (1983) we can divide the process of collapse into three phases which, however, take place simultaneously.

Phase 1 -
- initial stage of destruction of the original microstructure due to increasing humidity and external pressure,
- clay films, bridges and buttresses start to break (Fig.8),
- aggregates, microaggregates disintegrate,
- intensity of dissolution of carbonates and their migration in the soil increases (Fig.9 and 10).

Phase 2
- disintegration of microstructure continues
- clay particles are water-transported in the soil,
- content of carbonates is decreases
- others fabric elements compress,
- total volume of soil decreases.

Phase 3
- a new microstructure develops after collapsing (Fig.11) - basical relations among structural elements changed,
- basic structural units disintegrate, they are no more coated by clay films and have no mutual contacts by clay bridges.
- clay films of silty grains were destroyed or removed completely,
- clay particles were aggregated, and a coherent matrix formed in some places (Fig. 12),
- the soil acquired heterogeneous structure in contrast to the original homogeneous,
- percentage of particular pore sizes changed and the general porosity has been reduced.

The proportion of individual size fractions of pores has been changed. We have found out (Klukanová, 1988), that in a collapsed soil there is a smaller proportion (in per cent)

of pores sized below 0.005 mm. It is very interesting to see that the amount of pores below 0.005 mm and macropores did not change (Fig.13). The number of pores of the size between 0.005 mm and 0.01 mm increased considerably, especially the fraction around 10 ɱm. The total amount of pores changed in average from 45 per cent to 37 per cent during the test. Our results indicate, that leading role in the collapse process plays the translocation of clay films and bridges.

There are also indications, that collapse process depends from the fabric of loess. Collapse loess is composed of silty and sandy grains, coated by clay films, whereby the contacts are mediated through clay bridges and buttresses. We presume, that the most collapsible is a soil with the exact ratio between the amount of clay minerals and sandy, or silty grains in such a manner that the grains are coated by clay films and connected by clay bridges. No other forms of clay minerals are present in such fabric. Any deviation from the balanced ratio between the fractions (another form of clay minerals present in the fabric) will result in lower degree of collapsing.

We intend to verify this presumption by the direct monitoring of the collapsing factor, because it could lessen the general investment. According to my opinion any soil with the said fabric would be collapsible should the moisture and/or strain in the foundation soil increase. The degree of collapsing will be influenced by porosity (especially pores around 0.01 mm in diameter), the contents of carbonates, oxides, metallic hydroxides (mostly Mn and Fe) and soluble salts. It follows that collapsing is a very complicated process, dependent upon many factors with considerable variability, resulting in uneven collapsing in areal as well as vertical sense.

To explain the mechanism of loess collapse we have constructed the model of loess collapse in triaxial cell and consequently monitored of changes in fabrics by scanning electron microscope (Klukanova, Sajgalik, 1993). Figures 8 - 13 demonstrate the results of this test.

7 CONCLUSION

The formation of the essential fabric of loesses including the organization and the mutual interaction of its components (be it in solid, liquid or gaseous phase) is controlled by various geochemical processes. In particular, the processes, which took place during the formation of the loess material (glacial crushing, frost weathering), during its transport and those geochemical pro-

Fig. 8. Disintegrated clay film coated the quartz grain.

Fig. 11. The formation of a new microstructure after collapse.

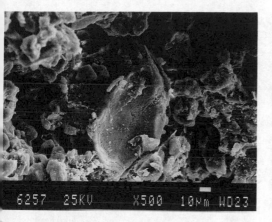

Fig. 9. Dissolution of carbonates in the soil.

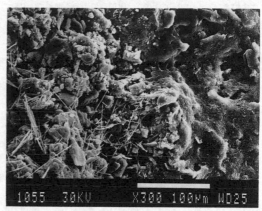

Fig. 12. A coherent matrix formed in some place.

Fig. 10. Dissolution of carbonates in the soil.

Fig. 13. A macropore did not change during collapsibility test.

cesses, which acted during the deposition of the sediments, are fundamentally important. During the lithogenesis, which generally follows loess deposition, other processes, induced by site conditions, become important. Since the site conditions in the Western Carpathians due to their relief vary, the process of lithogenesis resulted in the development of four types of loess. They are: typical loess, sandy loess, clayey loesses and loess-like sediments. The laboratory test showed that each of these types is distinguished not only by a varied fabric, but also by variable geotechnical properties.

On the basis of microstructural analysis the process of collapse can be divided into three phases:

Phase 1 – the original microstructure is destroyed due to increasing humidity and external pressure. Clay films, bridges and buttresses are broken, aggregates and micro-aggregates disintegrate, and there is more extensive dissolution of carbonates and mobilization of Fe, Mn – oxides.

Phase 2 – Involves the disintegration of microstructure. Clay particles are water-transported in the soil, volume of carbonates decreases and other fabric elements are compressed. Total volume of the loess decreases.

Phase 3 – During this phase there is the formation of a new microstructure after collapsing is completed. Basic relations among structural elements are changed, clay films of sandy or silty grains are destroyed or removed completely. Clay particles aggregate and in some places form a coherent matrix. The less acquires a heterogeneous structure in contrast to its original homogeneity. The percentages of particular pore sizes changes and in general porosity is reduced.

Loess collapse is controlled by its fabric and the extent of collapse by porosity, mainly by pores in the size of about 0.01 mm, and carbonate content.

REFERENCES

Grabowska-Olszewska, B., Osipov, V.I., Sokolov, V.N. 1984. Atlas of the microstructure of soils. Panstowe wyd.nauk. Warsawa. 414 p.

Klukanova, A. 1988: Microstructure of loess sediments and their alterations due to their deformational properties. Manuscript – thesis. University Comeniana, Faculty of Natural Science, Department of Engineering Geology. 109 p.

Klukanova,A.,Sajgalik,J.,1993: Changes in loess fabric caused by collapse: An Ex perimental Study. Quaternary International. Pergamon Press.

Sajgalik, J., Modlitba, I. 1983: Sprase Podunajskej niziny a ich vlastnosti. Veda, vyd. SAV, Bratislava, 204.

The swelling behaviour of some tropical soils

Le comportement expansif de quelques sols tropicaux

B.S.Bueno, D.C.de Lima & D.L.Cardoso

Federal University of Viçosa, Minas Gerais, Brazil

ABSTRACT: Damages caused to highway works by swelling soils in tropical areas where a dry season is followed by a long wet period can be of considerable magnitude. To identify and remedy the behaviour of some of these soils an experimental laboratory program was carried out at the Federal University of Viçosa. Eighteen samples from the Viçosa county were tested, 04 from alluvial soils, 14 of residual soils from granite-gneiss rocks, whose behaviour was compared with the one of a commercial bentonite. Results of laboratory routine tests such as clay contents, Atterberg limits and LNEC swelling index allowed prediction of soil swelling potentials based on a proposed swelling scale. Simultaneously, a lime stabilization study using lime contents in the range of 0% to 4% was carried out aiming to control soil expansion. The study showed that very small lime content (0 to 1%) can increase soil expansion, but larger amounts (1 to 2%) clearly drop the swelling potential.

RESUMÉ: Les dégâts sur les autoroutes et les chaussées dûs a l'expansion des sols tropicaux où la saison sèche est suivi par une longue période humide peuvent être assez importants. Un programme expérimentel a été mis au point à l'Université Fédéral de Viçosa pour identifier ces sols et remédier leur comportement. Dix-huit échantions de la ville de Viçosa on éte essayé où quatre des sols alluvioraires et quatorze des sols reésiduels des roches granite-gnaises. Leurs comportements ont été comparés avec ce d'un échantion de bentonite. Les résultats des essais typiques de caractérisation tell que le teneur en argile, les limites d'Atterberg et l'indice de l'expansion LNEC ont permit la prevision de potenciels d'expansion. Pendant tout ça, l'estabilization au chaux a été étudiée avec des teneurs en chaux entre 0% et 4% avec le but de contrôler l'expansion. Les études ont montré que l'expansion augment pour des petits teneurs en chaux (0 % a 2%) et pour le teneurs en chaux plus grands, le potenciel d'expansion se réduit.

1. INTRODUCTION

Damages caused by swelling soils to engineering works in tropical countries can be of considerable magnitude, although there is no available data to give a quantitative support to this statement. In a two season climate where a long wet term, dominated by heavy rains, is followed by a very dry period, some of the superficial soils experiment severe moisture losses which induce large soil structure contraction and lead to relatively high volume change. When wet these dry soils swell. This class of materials is named expansive or swelling soils.

The large moisture content loss to which some tropical soils are exposed can cause intense tension cracking which reaches relatively deep levels. Under these weather conditions even moderate expansive soil may be detrimental to lightweight geotechnical works such as highway pavements and earth slopes.

To investigate the swelling behaviour of some tropical granitic-gneiss lateritic, saprolitic and alluvial soils, occuring at the county of Vicosa, an experimental laboratory testing program was carried out at University of Vicosa, Minas Gerais State, Brazil. The main results of this research program are presented and dicussed in this paper. Comparisons are also made with previous research works of FONSECA and FERREIRA(1981), who tested granitic-gneiss residuals soils from the State of Rio de Janeiro, and GARCIA (1973) and SIMOES and COSTA FILHO (1981) who worked with the most expansible known Brazilian soil, the massape, from the State of Bahia.

2. THE EXPERIMENTAL PROGRAM

The laboratory test program included sieve analysis, Atterberg limits, LNEC swelling index, swelling potential and pressure determinations via oedometer tests, according to ASTM D4546, and soil expansion measured at samples placed in a triaxial chamber and subjected to different confining pressures. Parallel to this a lime stabilization study was carried out to investigate the reduction of LNEC swelling index of all tested samples. Lime contents in the range of 0.5% to 4% were used.

3. THE TESTED SOILS

To represent the most widespread soil occurrences in Viçosa county, eighteen soil samples were selected: 04 of alluvial and 14 of residual origin (02 saprolitic and 12 lateritic), whose behaviour could be compared with the one of a commercial bentonite. Table 01 introduces results of sieve analysis and Atterberg limits of all tested soils.

Table 01: Basic data of tested soils

SOIL	CLAY (%)	SILT (%)	SAND (%)	LL (%)	PI (%)	LNEC (%)
BEN	100	0	0	221	155	76.7
GUI-A	45	15	40	54	26	32.7
GUI-C	25	17	58	43	15	26.7
HV	60	28	12	78	34	17.5
LEC	68	17	15	71	34	15.5
IPA	58	25	17	78	39	18.4
ETA	40	19	41	63	30	05.8
EA-B	35	37	38	64	28	07.2
EA-C	10	38	52	36	17	11.5
FNB-S	25	15	60	45	15	20.8
FNB-1	59	11	30	77	17	20.6
FNB-3	55	17	28	79	36	17.4
SG	50	25	25	77	38	23.0
NV	33	20	47	59	20	12.9
VS	03	17	80	29	10	08.1
GM	6	05	34	78	33	13.2
IC	05	19	76	26	13	05.1
IV	19	36	45	56	20	14.3
MC	51	16	33	76	36	10.2

CARDOSO et alli (1993) separated the tested soils in four major groups according to their pedological classification as:

a) podzols (HV, LEC, SG, MC and IPA). In these strong yellowish clay soils kaolinite and goethite were the main clay minerals present;

b) latossols (ETA, EA-B, NV, FNB-S, FNB-1, FNB-3, SGIC and IV). Although with similar mineralogy to the previous group such soils displayed hematite, which is responsible for their redish colour, small clay content and therefore low plasticity;

c) saprolitic (EA-C and VS). Besides kaolinite and gibbsite, such soil also presented quartz, K-feldspar and a significative amout of mica in both silt and clay fractions;

d) vertissols (BEN, GUI-A and GUI-C). These montmorilonitic soils presented the largest LNEC swelling indexes among all tested soil even when clay content were relatively low (GUI-C).

4. TEST RESULTS

4.1 Natural soils `

Identification of expansive soils has usually been made comparing soil Atterberg limits (LL and PI) with values given by swelling scales, see, for example, CHEN (1988). However, as shown by CARDOSO (1994), all proposed scales were grounded on local experiences and therefore should not be used without restrictions at any place. This emphasizes the importance of local studies and weakens the idea of a global swelling scale able to identify and classify the expansive soils irrespective to soil composition, origin and climate conditions.

To confirm the above statement, results of LNEC swelling index (CASTRO, 1964) were ploted versus clay content for the tested soils and showed considerable dipersion with the class of soil and exhibited low coeficient of determination (r^2 = 0,4032). Slightly better results were obtained when clay contents were replaced by PI results (r^2 = 0.7583).

When test results from soils of different places are compared to those from Vicosa, see Figure 1 which plots PI versus LNEC swelling index, for example, not only considering the tested soil data but also data from research works of GARCIA (1979) and FONSECA and FERREIRA (1981), it can be seen that there is a general trend suggesting the increase of LNEC swelling index with the increase of PI for montmorilonitic soils (data from GARCIA, 1979) and samples GUI-A, GUI-C and FNB-S, from Table 01). Although not belonging to the vertissol group, sample FUB-S presented a of large expansion which was tributed to its considerable amount organic mater.

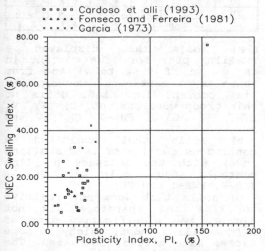

Figure 1 - LNEC swelling index versus Plasticity index

Such good agreement occurred despite of four of GARCIA's samples have been stabilized with lime. On the other hand, data from CARDOSO et alli (1993) and from FONSECA and FERREIRA(1981) are scatter. This may be tributed to the different classes of tested soil, ranging from homogeneous transported aluvial clayey soils to micaceous saprolitic materials.

When PI is plotted versus expansion pressure, Figure 2, it can be clearly noticed two classes of soils: (1) gneiss derived soils with large clay content and massape samples display expansion pressure above 100 kPa; (2) less clayey samples which show values of expansion pressure smaller than 50 kPa. The data are rather scatter and do not suggest a reliable relationship between PI and expansion pressure.

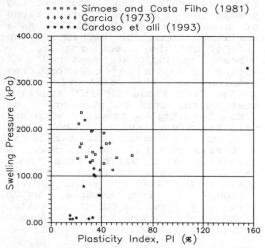

Figure 2 - LNEC swelling index versus expansion pressure

However when the analysis is restrained to the tested soils (CARDOSO, 1994), it is possible to propose a swelling scale to identify and classify expansive soils and to estimate swelling potential (PSE) as a function of clay content (C), plastic index (PI) and LNEC swelling index. Such prediction gives an r^2 =0,9694 and can be identified as:

$$PSE = -0.07C + 0.2244PI + 0.1114LNEC - 4.0411 \qquad (1)$$

Table 2: Proposed swelling scale (Cardoso et alli, 1993)

CLAY CONTENT (%)	PI (%)	LNEC(%)	SWELLING LEVEL	EXPECTED PSE (%)
>50	>30	>15	HIGH	>3
25-50	20-35	12-15	MEDIUM	1-3
<30	<20	<12	LOW	<1

This scale can correctly predict the swelling behaviour of 86% of FONSECA and FERREIRA's samples, 100% of GARCIA's and only 56% of SIMOES and COSTA FILHO's. The good prediction of FONSECA and FERREIRA's data can be tributed to the fact that they worked with soils with with genetic origin similar to the actually tested soils. Although working with massape, most of GARCIA's data refer to lime stabilized samples which display swelling behaviour of the same order of magnitude as those of the tested soils. This fact can explain the good swelling predictions. The poor predictions of SIMOES and COSTA FILHO's data swelling behaviour stresses the need of defining local scales.

Free expansions measured on either undisturbed or compacted samples in the oedometer test differ from those of LNEC swelling test which refer to oven dry and compacted samples prepared with the soil fraction passing trough sieve 40. In both cases the sample is laterally confined and the resulted expansion also differ from those measured at the triaxial cell, as shown by MONTEIRO and BUENO (1994).

Figure 3, for example, which shows sample GUI-C triaxial expansions under three different confining pressures.

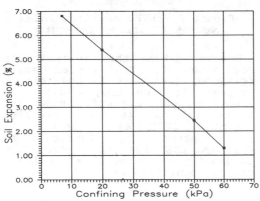

Figure 3 - Soil expansion versus confining pressure

These results are generally higher than those of oedometer test and the expansion measurement in the triaxial cell in many cases may represent a better way to simulate field conditions, specially when these conditions differ from one of total lateral confinement.

4.2 Soils stabilized with lime

Table 03 shows LNEC swelling data when lime contents up to 4% were added to the tested soils. Three different soil behaviours can be distinguished from the analysis of these data:

Table 3 - LNEC swelling index for lime stabilized soils (lime contents of 0.5, 1, 2 and 4%)

SOIL	LNEC SELLING INDEX (%) FOR DIFFERENT LIME CONTENTS				
	0	0.5	1	2	4
BEN	76.7	81.0	75.5	77.1	72.7
GUI-C	32.7	335.8	30.9	27.7	26.9
GUI-A	26.7	25.9	21.7	19.4	12.1
HV	17.5	19.5	18.5	13.7	11.4
LEC	15.5	17.1	17.9	14.9	17.3
IPA	18.4	14.5	18.8	8.2	8.5
ETA	5.8	9.1	8.0	2.0	0.9
EA-B	7.2	9.3	9.0	2.3	2.33
EA-C	11.5	10.4	11.4	11.0	11.1
FNB-S	20.8	13.3	11.1	7.6	7.1
FNB-1	20.6	16.1	14.8	6.7	7.0
FNB-3	17.4	18.7	16.1	6.7	7.2
SG	23.0	25.7	18.7	16.1	14.3
NV	12.9	12.1	11.5	4.6	3.4
VS	8.1	11.2	13.9	12.6	13.5
GM	13.2	13.5	12.0	5.2	4.7
IC	5.1	5.5	4.5	5.7	5.3
IV	14.3	16.8	15.6	12.6	11.6
MC	10.2	13.1	9.6	4.7	2.3

a) soils that displayed a swelling peak for lime content in the range of 0.5% to 2% and from there on a continuous decrease as lime content increased. Soils of this group were BEN, GUI-C, SG, HV, ETA, IPA, EA-B, FNB-3, GM, IV and MG.

b) soils that exhibited a continuous decline of LNEC swelling index with the increase of lime contents. Such soils were GUI-A, FNB-S, FNB-1 and NV;

c) soils that were not reactive to lime and therefore did not display any significative drop of LNEC swelling index, which, in some cases, have already increased. The soils included in this group were LEC, VS, EA-C, and IC.

The reason why most of the tested soils showed a LNEC swelling index peak for lime contents in the range of 0,5 to 2% can be tributed to chemical changes occuring in the soil difuse double layer. It is known that the addition of lime to a fine-grained soil gives a surplus of ions Ca^{2+} which tend to replace trivalent ions such as Al^{3+} and Fe^{3+}. This replacement increases the particle net negative charge which results in an increase of soil pH. Besides that the double layer difuses and thickens giving rise an increase of soil expansion. Further addition of lime will tend to neutralize negative charges, dropping double layer thickness and therefore reducing soil expansion/contraction and plasticity. In the following step, and as a result of pH increase, lime cimentitious action develops bonding soil particles. The required amount of lime to entirely satisfy clay exchange needs varies with soil constituents, temperaure, etc., but in general ranges between 2 to 10%.

5.CONCLUSIONS

From the analysis of the experimental data the following conclusions can be reached:

a) reasonable predictions of soil swelling behaviour can be made via swelling scales defined in terms of Atterberg limits and LNEC swelling index;

b) available swelling scales are based on local experiences and should not be used without restrictions at any place since soil expansion is strongly dependent on soil constituency and genesis and weather conditions;

c) lime can be used to control soil expansion although in some cases (low lime content, soils displaying large percentages of montmorilonite and saprolites with large percentge of botite) it shows small benefits or even increases soil expansion.

6. BIBLIOGRAFY

CARDOSO, D.L., A contribution to identification and satabilization of expasive soils from Vicosa county, MSc thesis, Department of Civil Engineering, Federal University of Vicosa, Brazil, 1994, 194p

CARDOSO, D.L., BUENO, B.S., LIMA, D.C. and FONTES, M.P.F., On the identification of swelling soils, Paper submitted to Geotecnia, Geotechnical Portuguese Society, Lisbon. (in Portuguese)

CASTRO, E., Soil swelling test, LNEC- National Laboratory of Civil Engineering, 1964, 11p (Technical paper 235).

CHEN, F.H., Foundation on expansive soils, New York, Development in Geotechnical Engineering, 1988, vol. 14, 463p.

FONSECA, A.M.C.C. and FERREIRA, C.S.M., Metodology to determine an erodibility index of soils, Proceedings of the Symposyum on Engineering of Brazilian Tropical Soils, Rio de Janeiro, 1981, vol. 1, pp. 646-667. (in Portuguese).

GARCIA, J.J.M., Suelos expansivos su etabilizacion con cal, MSc thesis, COPPE-UFRJ, March 1973, 100p.

MONTEIRO, G.C. and BUENO, B.S., Meassurement of soil expansion in the triaxial cell, CNPq Research Report, 1994, 35p. (in Portuguese)

SIMOES, P.R.M. and COSTA FILHO, L.M., Mineralogical, chemical and Geotechnical characteristics of expansive soils from Reconcavo Baiano, Proceedings of the Symposyum on Engineering of Brazilian Tropical Soils", Rio de Janeiro, 1981, vol. 1, pp. 569-588. (in Portuguese).

Tropical environment and geotechnical behavior of cenozoic deposits

Environnement tropical et comportement géotechnique de dépôts cénozoiques

S. Bongiovanni
UNESP, Universidade Estadual Paulista, Campus de Assis, São Paulo, Brazil

Jayme de O. Campos
UNESP, Universidade Estadual Paulista, Campus de Rio Claro, São Paulo, Brazil

F.X. de T. Salomão
IPT, Instituto de Pesquisas Tecnológicas, São Paulo, Brazil

ABSTRACT: The cenozoic deposits cover extensive areas in the São Paulo State, Brazil, and are related to the stepped levels in the landscape, binded to flattened regions. These deposits are used in engineering works like embankments, road pavements, construction materials and even buildings foundations. To the geotechnical purposes, the sandy materials with weak cementation have a special interest, but in the natural state they present collapse characteristics. They have been called fine sandy soils, sandy lateritic soils, sandy lateritic soils, modern sediments and superficial deposits. They occur in the middle south-west of the São Paulo State, in which the rocks are mainly of the mesozoic sedimentary and magmatic types, from which the cenozoic deposits came from. They are well developed pedological types, with a tendency to the latosols and podzols formation. They are represented by 6 pedological classes and are grouped in 3 geotechnical unities (G.U.).

RESUMÉ: Les dépôts cénozoiques couvrent grandes surfaces de l'État de São Paulo, Brésil, et se sont rapportés aux niveaux échelonnés dans le paysage, liés aux régions applainées. Ces dépôts sont utilisés en Génie Civil dans les constructions des remblais, bases de routes, matériaux de constructions et même comme fondations des édifices. En ce que concerne aux propos de la géotechnique, les matériaux sableux avec du faible ciment ont un intérêt spécial, mais dans l'état naturel ils sont collapsibles. Ils ont été només sols fins sableux, sols sableux latéritiques, sédiments modernes et dépôts superficiels modernes. Ses occorrences sont aux sud-ouest de l'État, où les roches sont nettement des types sédimentaires et magmatiques du Cretacé, desquelles les sédiments sont originés. Les sols locaux sont bien évolués, du point de vue pédogénétique, avec la tendence aux dévelopement des latosols et des podzols. Ils sont représentés par 6 classes pédogénétiques et 3 unités géotechniques (U.G.).

1 INTRODUCTION

The most difficulties in Civil Engineering for the identification, classification and behavior of soils originating in the tropical environment are the problems that can be expected by the unusual results when the more conservative methods of testing for the common soils are employed. The more appropriate characteristics concerning the first ones are their sensibility to drying and to the reshuffe, the swelling potencial and the self - stabilization Cruz (1987). This author affirms that the soil structure in a typical profile is the result of a weathering process, mainly chemical associated to the washing and laterization, and the soil structure has a great relasionship with the bondings between the soil components, consequently with the chemical evolution or the weathering changes; the past stress history of the soil is less important to the actual structure and the geotechnical characteristics. This peculiarity represents the main difference between the soils that were submited to a stress history, like the common sediments, and those formed in the tropical environments. They have a macro-structure similar to lumps, that means, clayey or sandy particles of different resistances or stabilities the first ones weakly stable because of the iron and aluminium oxides cement, and the another ones stables because they are granular and resistants, that results in high permeability owing the macropores. The collapsing phenomenon is intimately connected to the saturation degree et to the loading to which the soil is submited. The stable or meta-stable state depends therefore on the stabilily or the weakness bondings; when the soil goes through from the meta-stable to the stable condition, it happens normally a structure collapsus, and the permeability owing the macropores is reduced to compatible values with the new structure. At the same time, the shear resistance and the compressibility are influenced by this new structure(Campos,1993).

2 GEOLOGICAL ENVIRONMENT

The studied area ranges almost 300 km² and is located in the southwest of the state (Fig. 1). The pluviometric rate is of 1.250 mm/year, with a montly variation of 30 to 240 mm, the higher concentration during the summer (December/March). From June to September it rains only 15% of the annual rate. In this period there is a water deficit in the soil and the rivers

reach their lowest levels.

According with the koeppen's classification, there are two sudvidions in this area: Cwa temperate climate with dry winter and hot and rainy summer; and Cfa temperate, without dry season and with ho summer. The annual temperature average is around 22°C; the hottest month aveage temperature (January) is 24°C and 25°C, and the coldest one (July) is around 17°C and 18°C.

The geologic substratum of the area consists of mesozoic sedimentary and magmatic rocks pertaining to the São Bento Group (Serra Geral Formation - basalts and sandstones), Bauru Group (siltstones, sandstones and conglomerates), besides the cenozoic deposits.

The radiometric measurements show that volcanism came out around 140 and 120 million years at lower Cretaceous. At the end of the volcanisms events and the tectonic movements at the Cretaceous, it happe-

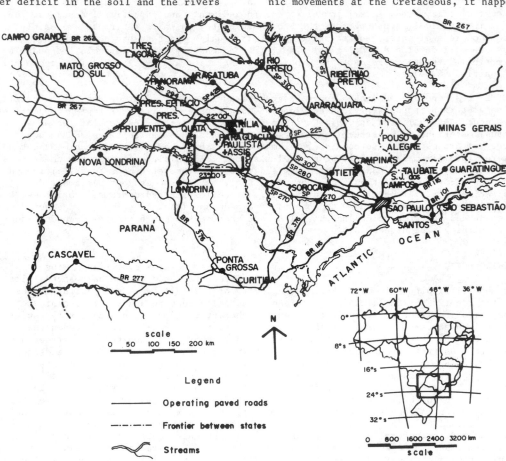

Fig. 1 - Map of the studied region.

ned the Bauru Group sedimentation, whose the main lithology is a fine to very fine sandstone with a argillaceous or calciferous/argillaceous cement, being worst to badly select.

Landim et al (1974) and Soares & Landim (1976) related the origin of the cenozoic deposits at two distinct geomorphologic surfaces: the South American Surface of the Terciary period, and the old surfaces of the Quaternary period defined by King (1956). These authors consider that the cenozoic deposits are composed of unstructured sands placed upon the lower sedimentation stage or directly over older formations, gravel deposits or flint-stones lines, constituted by quartzite or limonite fragments. (Fúlfaro, 1979) refers to the cenozoic deposits as non consolidated sediments similar to soils, with flint-stones lewels at the base, recovering others older lithological types of the cenozoic, mesozoic and paleozoic periods, including rocks of the basement complex. To the geomorphology, the area is developed over the province named western plateau (Moraes Rego, 1930, apud Ab' Saber, 1954; de Martonne, 1954; Almeida, 1964), being the relief characterized by wide ridges with undulant convex summits that compound big hills advancing towards the confluents of the Paraná River. They also appears small plateaus that are considered as effects of the local flattening or are witnesses of a primitive surface in which were stablished all the consequent drainage of the province (Almeida 1964, op.cit). Therefore, the relief evolution of the western plateau would be related fo erosive processes.

Ponçano et al (1981) divided the western plateau in 4 geomorlogical zones, being the studied region in the zone of undivided areas, not included in any of the raised areas of the Marilia, Catanduva and Monte Alto Plateau.

Hence, the cenozoic deposits correspond to stepped levels in the landscape, binded to the flattened stage. They are thin deposits that covers both the basement complex and the inland uplands, and its origin is related to climatic or tectonic factors. Generally in this area are predominant the well developed pedological soils, characterized by upper horizons with deeply weathering of the original minerals, greatly influenced by the climate, with a tendency to form latosols or podzols in the profiles.

The soils are represented by the following pedological classes: dark red clayed texture; dard red latosol with a medium texture; violet latosol; structured violet soil; yellow red podzol with a

sandy texture/medium abruptic; yellow red podzol with a sandy texture/medium non-abruptic. The climatic characteristics of the are (sub-tropical, hot, humid) is propicious to the developement of ferralitic weathering processes by the hydrolisis of the minerals from the pedological substratum, with the silica and bases release and iron and aluminium sesquioxides concentration. It must be also considered the influence of the relief and the geological substratum on the developement of the soils that can be observed by the physical and chemical weathering of the mineral components, as well through the processes related to their movement as a solute across the soil, the water carrying a basic role in this process.

3 METHODOLOGY

The cenozoic materials tested at the laboratory were selected in a field work in which were bored 7 pits located over the pedological classes. They were obtained disturbed and undisturbed samples.

The tests achieved were: characcterization; grain size analysis; atterberg limits - Casagrande and cone methods; compaction with the miniature equipement; normal compaction and loss of soil mass by water imersion; tropical soils classification - MCT methodology; comparison among the geotechnical classification - unified and MCT, and the pedological classification and special tests; consolidation/collapsing with permeability measures.

3.1 Grain size analysis - ABNT - NBR 7.181

The results show that the cenozoic deposits reported have quite homogenous grain size, with a wide predominance of fine sand, being the values between 60% and 82%. The exception are the structured violet soil which present 72% of clay and 18% of fine sand, and the violet latosol with 50% of clay and 42% of fine sand (Fig. 2).

3.2 Atterberg limits - Casagrande and cone methods - ABNT - NBR 6.459-71

For the liquid limit tests, the Casagrande's equipement is a routine in the soil mechanics brazilian laboratories. The utilization of the cone penetrometer aiemd at the comparison between the results and its easiness or difficulties of execution. As a result, the test for the Atterberg limits by the cone method (BS 1.377) is more easily applicable with a good resolution and may became a routine test.

3.3 Compaction with the miniature equipement (mini MCV), normal compaction and loss of soil mass by water imersion

These tests were executed aiming the determination of the parameters used in the MCT geotechnical classification - miniature, tropical (Nogami & Villibor, 1981, 1982, 1985; Nogami, 1985) - of tropical soils for road pavement purposes.

3.4 Tropical soil classification - MCT methodology

This classification, proposed by Nogami & Villibor (1981), utilizes the results of tests on reduced dimension sampling (5 cm), accomplished in accordance with the MCV determination. It is convenient only for the soils passing totally through the 2mm sieve or when the influence of the retained fraction in the sieve is despicable or foreseeable, and is not applicable to essencially granular soils.

The soils are divided in 2 large classes: soils with a lateritic behaviour (L) and soils with a non - lateritic behaviour (N). This division is not based on geologic or pedological criteria, but essencially in technological considerations.

Nogami (1985) considers the soil lateritic when: a) it belongs to A and B horizons of well drained profiles, developed under tropical humid climate conditions; b) its clayey fraction is essencially constituted of the kaolinitic group minerals with aluminium and iron hydrated oxides, being these components grouped in highly porous and instable structures.

According the Geological Society Engineering Group Working Party Report: Tropical Residual Soils (1990), the soils that result from the sandstones weathering, in seasonal and dry tropical environmental conditions with kaolinites and hydrated iron and aluminium oxides, are classified as ferrisols or ferruginous soils, which have erosive and dispersive characteristics and collapse potencial. On the other hand, the basalts weathering in the same environmental conditions, results in fersiallitic soils with the halloysites predominant.

3.5 Comparison between the unified soil classification, the MCT classification and the pedological classification

Six pedological classes, representative of the soils from the south west of the São Paulo State, were found in the studied area. The unified classification considers these soils as sandy clays, clayey sands, sands and silt clayey sands. In the MCT classification, all these soils are located in the group of lateritic behaviour (L) spreaded in 3 areas: clayey lateritic (LG'), sands lateritic (LA), and sands/sandy lateritic (LA/LA'). This last area was called like that, because in the MCT classification diagram the dot corresponding to these soils was placed just at the interface between the two areas. Comparing the two geotechnical classifications, their results are too much near. Table 1 shows the tests results for the textural and mechanics characterizations of the cenozoic deposits studied.

3.6 Consolidation/collapsing tests with permeability measures

The collapsing soils have a structure that can reasonably resists to a loading when partially saturated but undergo a sudden volume decrease due to the collapse structure when saturated.

Collapse settlement usually results from loss or reduction of the bonding between the soil particles due to the presence of water, and often occurs in intensed leached residual soils formed from quartz-rich rocks. Another mechanism of rupture includes a loss stabilizing influence of the surface tension in menisci water at particles contacts in partially saturated soils, and loss of strength of the particles themselves when wetted (Geological Society Engineering Group Working Party Report: Tropical Residual Soils, 1990). They are considered collapsing soils those with a essencially sandy or clayed nature, non saturated, characterized by a meta-stable structure of the porous type, in which the silt and sand grains and clay lumps are hold in their positions by a weak cement (clays, iron oxides or carbonates), or by the capillary menisci (Lopes, 1987).

Vargas (1973) affirms that the collapsing of the porous soils seems to be negligible at small applied pressures; it increases with the applied pressures to a maximum, and them decreases to a minimum at the higher applied pressure; in a case studied by this author, the critical pressure above which no collapsing was observed, was around 5 kgf/cm^2. But at

TABLE 1. Tests results

PEDOLOGICAL CLASSE	SHAFT	DEPTH (m)	GRAIN SIZE Medium Sand (%)	Fine Sand (%)	Silt (%)	Clay (%)	CLAY MINERALOGY (%)* Kaolinity	Gibbsite	Chlorite	Goetite	LIMITS LL LP IP (%)	LL (%) CONE	PLASTICITY	COL. ACTIVITY	PROCTOR NORMAL h_ot (%)	γmax (kg/m³)	MCT CLASS.	UNIFIED SOIL CLASSIFICATION
Dark red latosols with clayed texture	1	1.00-1.30	4	60	3	33	19.44	2.17	-	-	24 16 8	31	Medium	Low	11.7	1902	LG'	SC
		2.00-2.30	4	61	3	32					26 16 10	34						
Dark red latosols with a clayed medium texture	2	1.05-1.35	1	82	2	15	10.06	-	-	-	16 14 2	20	Low	Low	10.2	1968	LA'	SM
		2.00-2.30	2	75	7	16					17 14 3	22						
	6	1.00-1.30	2	80	3	15	-	-	-	-	19 15 4	20					-	SC-SM
		2.00-2.30	4	77	3	16					21 15 6	25						
Structured violet soil	3	0.60-0.90	1	18	9	72	51.1	1.40	14.83	-	60 27 33	62	High	Low	29.9	1493	LG'	CH
		2.00-2.30	-	-	-	-					66 35 31	68						
Red yellow podzol with sandy/fine abruptic texture	4	0.60	2	80	8	10	12.07	-	-	-	18 14 4	25	Low to Medium	Low	12.6	1858	LA'/LA'	SC-SM
		1.70-2.00	2	73	6	19					25 17 8	28						SC
Red yellow podzol with a sandy/medium and not abruptic texture	5	1.00-1.30	4	68	4	24	17.79	-	-	-	26 15 11	31	Medium	Low	12.2	1916	LA'	SC
		2.00-2.30	4	60	11	25					29 18 11	37						
Violet latosol	7	0.70-1.00	2	42	6	50	-	-	-	-	46 24 22	51	High	Low	-	-	LG'	CL

* % clay regarding to total sample
Plasticity, low < 7; medium 7 to 17; high > 17

3507

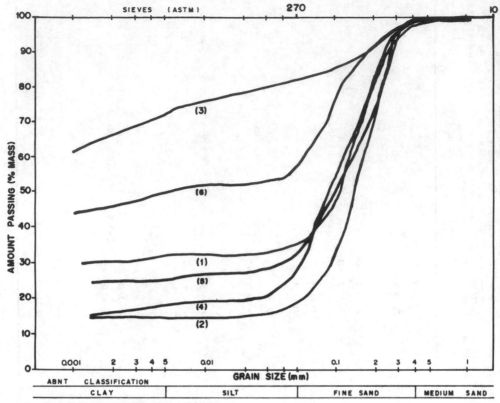

SIEVES (ASTM)

ABNT CLASSIFICATION

| CLAY | SILT | FINE SAND | MEDIUM SAND |

1 - Dark Red Latosol With Clayed Texture
2 - Dark Red Latosol With a Clayed Medium Texture
3 - Structured Violet Soil
4 - Red Yellow Podzol With Sandy / Medium Abruptic Texture
5 - Red Yellow Podzol With Sandy / Medium Non Abruptic Texture
6 - Violet Latosol

Fig. 2 - Cenozoic deposits - grain size distribution.

such a high pressure, the settlement of the soil due to its compressibility, at the natural moisture content, is intolerable for pratical purposes.

In the consolidation tests with the cenozoic deposits carried out, were utilized samples at the natural moisture content (not flooded), and another ones flooded at the beginning, in which were obtained the coefficient of permeability (k) values, in all the pressure levels. The loading at the upper part of the samples during the flooded consolidation tests, produced sudden changes in the soil volume, that is, a voids index variation by the expelling of the contained water. Among the six types of soils found only the structured violet soil can be considered as a non-collapsing soil, that is, it doesn't occur the abrupt diminution of the voids index when flooded and loaded.

4 GEOTECHNICAL UNITIES

The six soil types concerned were grouped in geotechnical unities related to their textural characteristics (grain size and plasticity), clay activity, and, from the compaction tests, the optimum water content and the dry specific density. The clay mineralogy held by x-ray diffracion, pointed outh clays of low activity (kaolinite, gibbisite, chlorite, goethite).

The unities assembly was possible by the utilization a methodology suggested by Salomon (1984), which permits in a relatively simple way the physical environment interpretation, and to obtain' the useful geotechnical data of the cenozoic material aiming its utilization in engineering works.

They are considered 3 geotechnical unities in the area:

G.U. 1: dark red latosol with clayey texture; dark red latosol whit a clayey medium texture; yellow red podzol with a sandy texture and medium non abruptic texture; G.U. 2: structured violet soil with medium texture; yellow red podzol with a sandy and a medium abruptic texture; G.U. 3: structured violet soil; violet latosol.

The G.U. 1 and the G.U. 2 unities group excellent soils for the embankments, foundations, being also used in local buindings constructions, mixed with cement as a plaster, instead the tradicional sand. This kind of use is more economic, mainly for the needfully families houses.

The G.U. 3 soils are considered of bad quality for foundations and inadivisable to embankments, because they have high-plastic properties, that difficult the compaction procedures.

5 CONCLUSIONS

In resume, the main aspects relating to the studied cenozoic deposits are:
a) They are consequence from the weathering of the Serra Geral Formation basalts and from the Adamantina and Marilia Formations sandstones in a tropical environment;
b) They are insaturated soils, with a porous structure and a grain disposition lump shaped and unstable;
c) They are latosols and podzols pertaining to the ferruginous, ferrisols or fersiallitic groups soils;
d) There are a good correlation between the topography, the geomorphology, the parent rocks and these soil origin;
e) They can be also colluvion deposits, instead residual soils.

This environment understanding (geology, geomorphology, pedology), and the knowledge of cenozoic deposits behaviour, became possible to evaluate its applicability in engineering works, as road base pavements, embankments, foundations and construction material.

The accomplishement of a relashionship between the unified soil classification and the geotechnical classification for the tropical soils - the MCT methodology - can be useful to stablish a expectation of the cenozoic deposits utilization.

GRATEFULNESS

The authors express their gratitude to the FAPESP - Fundação de Amparo à Pesquisa do Estado de São Paulo, Brasil and to the CNPq - Conselho de Desenvolvimento Científico e Tecnológico, Brasil, for the financial support to this work.

REFERENCES

Ab' Saber, A.N. 1954. A geomorfologia do Estado de São Paulo. In: Aspectos geográficos da terra bandeirante. IBGE, Cons. Nac. de Geog. Rio de Janeiro, p. 1-97.

Almeida, F.F.M. de. 1964. Fundamentos geológicos do relevo paulista. In: Geologia do Estado de São Paulo. IGG, bol. 41: 169-263.

ABNT - Assoc. Bras. de Normas Técn. 1981. Análise granulométrica de solos. NBR-7181.

Bongiovanni, S. 1990. Uma abordagem de geologia de engenharia do cenozóico da região de Paraguaçu Paulista, SP, Brasil. Dissertação de Mestrado. UNESP/IGCE, Rio Claro, 102 p.

British Standards Institution. 1975 a. Methods of tests for soils for civil engineering purposes. BS 1377. BSI, London.

Campos, J. de O. 1993. Sols tropicaux brésiliens traités avec la chaux et le carbonate de sodium. Symp. Intern. Sols Indurés-Roches Tendres, Athène, Grèce. Comptes Rendus Balkema publ. v.2, p. 1233-1245.

Cruz, P.T. da. 1987. Solos residuais: algumas hipóteses de formulações teóricas de comportamentos. Sem. em Geot. de Solos Tropicais, ABMS, Brasília. Anais, p. 79-111.

Fúlfaro, V.J. 1979. O cenozóico da Bacia do Paraná. In: Simpósio Regional de Geologia, SBG. Atas, São Paulo, p. 231-241.

Geological Society Engineering Group Working Party Report: Tropical Residual Soils. Quart. Journ. of Engin. Geol. v.23 1:101 p.

King, L.C. 1956. A geomorfologia do Brasil Oriental. Rev. Bras. de Geog. Rio de Janeiro. Ano XVIII nº 2:147-265.

Landim, P.M.B.; Soares, P.C.; Fúlfaro, V.J. 1974. Cenozoic deposits in South-Central Brazil and the engineering geology. In: 2nd IAEG Intern. Cong. São Paulo. Proceed. theme III v.VI 11.1-11.7.

Lopes, J.A.U. 1987. Terra roxa-Pr: um caso notável de problemas em fundações rasas provocado por fenômenos associados a colapsos de solos. In: 5º CBGE Cong. Bras. de Geol. de Eng. ABGE, São Paulo. Anais tema 2, v.2:359-375.

Nogami, J.S. & Villibor, D.F. 1981. Uma nova classificação de solos para finalidades rodoviárias. In: Simp. Bras. de Solos Tropicais em Eng. ABMS/CNPq, Rio de Janeiro. Atas COPPE/UFRJ, p.30-41.

Nogami, J.S. & Villibor, D.F. 1982. Algumas comparações entre uma nova classificação de solos e as tradicionais, princi

palmente para finalidades rodoviárias.
In: 7º Cobramsef - Cong. Bras. de Mec.
dos Solos e Eng. de Fundações. ABMS,
Olinda/Recife. Anais. v.5, Tema: solos e
rochas (2ª parte: 160-173).

Nogami, J.S. & Villibor, D.F. 1985. Charac
terization, identificacion and classification of tropical lateritic and saprolitic soils for geotechnical purposes:
preliminary remarks. In: Progress report: pecularities of geotechnical behaviour of tropical lateritic and saprolitic soils. Committee on tropical soils,
ISSMFE, Brasilia, proceed, theme 1, topic 1.1:3-9.

Pastore, E.L.; Ignatius, S.G.; Salomão, F.
X. de T.; Campos, J. de O.; Bongiovanni,
S. 1990. Correlação entre as classifica-
ções pedológicas e geotécnicas de alguns
solos do interior do Estado de São Paulo.
In: 6º CBGE - Cong. Bras. de Geol. de
Eng. 9º Cobramsef - Cong. Bras. de Mec.
dos solos e Eng. de Fund. ABGE/ABMS, Sal
vador. Anais, v.2:261-270.

Ponçano, W.L.; Carneiro, C.D.R.; Bistrichi,
C.A.; Almeida, F.F.M. de.; Prandini, F.
L. 1981. Mapa geomorfológico do Estado
de São Paulo, esc. 1:500.000. Public.
SICCT/Pró-Minério/IPT, 94 p.

Salomão, F.X. de T. 1984. Interpretação
geopedológica aplicada a estudos de geologia de engenharia. Dissertação de Mestrado. IG/USP, SP.

Soares, P.C. & Landim, P.M.B. 1976. Depósi
tos cenozóicos na região centro-sul do
Brasil. Not. geomorf. Campinas, 16(31):
17-39.

Vargas, M. 1973. Structurally unstable
soils in Southern Brazil. 8th Intern.
Conf. on Soil Mechan. and Found. Engin.
ISSMFE. Moscow. Proceed. p. 239-246.

Fabric and mineral characteristics of some collapsible soils from northern Egypt

Structure et caractéristiques minérales de quelques sols collapsables du nordest de l'Egypte

M. Darwish – *Faculty of Science, Geological Department, Cairo University, Egypt*

M. Aboushook – *Faculty of Engineering, Al-Azhar University, Nasr City, Cairo, Egypt*

S. Mansour – *Building Research Institute, Dokki, Cairo, Egypt*

ABSTRACT: This work is concerned with the study of some Egyptian collapsible soils, concerning their geological textural and mineralogical aspects, in relation to their engineering behavior. The examined samples represent sites of the new urban areas and cities in northern part of Egypt. Experimental techniques have been applied in order to delineate, their fabrics, mineral compositions as well as the mechanical behavior. The detailed investigations of the tested samples indicated that partial arrangement, skeleton grains, cement and pore filling matters and the interparticle voids had important role in determining the mechanical behavior of such soil types. The collapse increases with the increase in gypsum and clay content. Lower values of collapsing is recorded with the increase of textural maturity (well sorted, close- packed and higher grain roundness). The voids (inter- and intragranular spaces) play an important role in collapse beahavior. It is found when the voids occupy less than 15% of the total volume, samples show low values of collapse.

RESUME : Ce travail porte sur l'étude de quelques affaissements de terrains en Egypte. Il cherche à établir un lien entre leurs caractèristiques géologiques, physiques, minéralogiques et leur comportement mécanique. Les échantillons examinés sont reprèsentatifs des nouveaux sites urbains de la partie nord de l'Egypte. Cet article fait le bilan des expériences menèe pour décrire leur structures, leur compositions minérales ainsi que leur comportement mécanique. L'étude détaillé des échantillons a indiqué que les facteurs principaux qui dèterminent le comportement mécanique de ce type de terrains sont : l'arrangement interne, la forme des grains, le ciment, la matrice et les espaces interparticulaires. On montre ègalement que l'affaissement augmente avec la quantité de gypse et d'argile. Plus les valeurs d'affaissement enregistrées sont faibles, plus le parametre de maturité structurale est élevè (i.e. grains bien ronds, bien compactés et bien classés). Les espaces inter et intragranulaires jouent en outre, un rôle important dans le comportement de l'affaissement. Il a été prouvé que, quand les vides constituent moins de 15 % du volume totale des échantillons, les valeurs de l'affaissement sont plus basses.

1 INTRODUCTION

Collapsible soil deposits cover many locations in desert areas in Egypt particularly in arid and semi-arid areas. New communities and cities have been constructed in these areas. In Egypt, studies, concerning the behavior of the collapsible soils, were carried out by different researchers of mentioned: Radwan and Gaber 1980; El- Sohby et al. 1985& 1988; El- Saadany 1981; Sabry 1987; Sharkis 1990; Lotfy 1990 and Masour et al. 1993. The purpose of this paper is to present fabric and mineral compositions of some Egyptian collapsible soils in relation to their engineering behavior.

Natural slightly cemented soil samples were obtained from five virgin sites in Egypt. The location of these sites are: New Ameriyah city, New Borg El- Arab city , 10th Ramadan city, El- Oboor city and 6th October city (fig. 1). Block chunks from several open pits were extracted at depths ranging from 2.0 m to 6.0 m from ground surface.

To identify soil characteristics, physical properties tests and oedometer tests were performed (Table 1). The physical properties included determination of natural water content (w), natural bulk density (ɣ) and particle size distribution. Each sample was subjected to wetting test using an oedometer apparatus. The collapse potential (Cp) is calculated

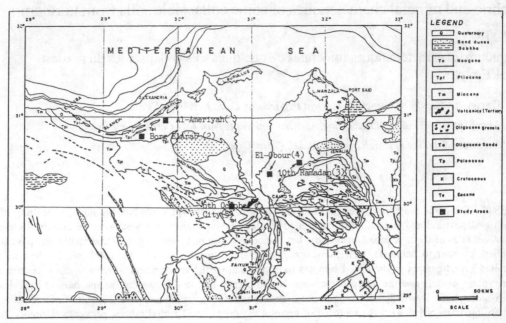

Fig. 1   Geological & Location map of the study areas (After EGSMA 1982)

Table 1. Physical and collapsing properties of the representative samples from the three main provinces

	Properties/Area*	1	2	3	4	5
Physical	γ_b kN/m²	15	16	18	19	16-20
	w %	2.5	1.5	2	1.5	2-4
	w_L %	32	32	36	28	NP-37
	w_P %	17.5	20	25	14	NP-22
	I_P %	14.5	12	11	14	NP-15
	Sand %	20	63	88	55	60-100
	Silt %	45	28	3	19	5-16
	Clay %	35	9	9	26	0-8
Collapse	C_P** %	6.5	10	9	4	5-9

* Province I (1: New Ameriyah city; 2: New Borg El- Arab city)
  Province II (3: 10th Ramadan city; 4: El- Oboor city)
  Province III (5: 6th October city)

**C_P: Collapse potential = H/H. It was measured for all tested samples at an inundation pressure of 200 kN/m² in the oedometer.

at 200 kN/m² wetting pressure, according to Jennings and Knight formula (1975):

$C_p = H_c/H_o$ where,

C_p: Collapse potential %

H_p: Change in the height upon wetting

H_o: initial height

For determination soil texture, a detailed and extensive lithologic and petrographic study has been done. Special attention has been paid for the 6th October samples as an applicable example for the Egyptian collapsible soils. Thin sections have been prepared after impregnation from natural undisturbed samples, before and after wetting test in the oedemeter. The procedure outlined by Adams et al.(1984) was mainly undertaken for preparing the impregnated thin sections and forr the microscopic examination. The study was carried out in order to delineate the main particle shape(s), arrangement, position and the amount of cementing agents. The type and status of the visual pore spaces are considered.

2 GEOLOGICAL SETTING

The surface features of Egypt are shared unequally between geographical elements, the deserts and the Nile valley and Delta. Collapsing soils are observed in many localities situated towards the deserts around The Nile valley. Such deserts are characterized by a complete absence of a drainage system. Previous studies referred to some locations of collapsing soils and their geological setting, e.g. El- Sohby et al. 1985; El- Saadany 1981 and Mansour 1992.

Collapsing soils are observed in many areas in northern part of Egypt towards the western desert. It is mainly characterized by a complete absence of a drainage system. The most of collapsing soils in Egypt were deposited in shallow water depths in a loose structure of high void ratio. Rivers, flood streams and rainfall are responsible for these formations. The absence of geological changes on the sediments in addition to the moderately deep water helped in keeping this loose structure without changing. These sediments which are generally fine granular materials are composed of sand and cementing material. The cementing material could be composed of one or more of clay, iron oxides, carbonates, gypsum....etc. The engineering behavior of these types of soil differs from one location to another.

The geological features and occurrence of the studied collapsing soils can be classified as follows under three main provinces(Fig. 1):

2.1 Province I

It is the inter-ridges areas parallel to the north coast of the Mediterranean. These areas include the area between New Borg El- Arab city and extends southward along Cairo- Alex. desert road passing through New Ameriyah city and they represent the northern part of the western desert. This region was subjected to various orogenic movements which created different dopocenters from the Paleozoic and Mesozoic eras and thick sequence of fluviatile to shallow marine sediments (Said 1962). The soil formation in these areas consists of calcareous clayey sandy silt, and it is described as shaley clayey silt (loess). Wind, sea and rain contributed in the formation of these deposits. They are situated between the parallel ridges consisting mainly of coastal oolitic limestone dated to Plio- Pleistocene(Shukri et al. 1956; Said 1962). The collapsible soils were formed in the area between these ridges or behind them, as a result of the interaction of mixed fresh water and marine conditions with parent rocks.

2.2 Province II

It is the inter dune areas. These areas occupy the east of the Nil Delta along Cairo- Ismailia desert road and Cairo- Belbies desert road where the 10th Ramadan city and El- Oboor city are newly constructed. The soil formation in these areas is mainly calcareous silty clayey sand and weakly indurated sandstone. It was formed under mixing conditions of very shallow marine embayments and fluviatile systems during the Plio- Pleistocene age. The cementing materials that came by the stream rivers banded the coarse grained particles together and the collapsing soil deposits were formed.

2.3 Province III

It is the Six October plateau. This plateau lies between Cairo- Alexandria desert road and Cairo El-

Fayoum road in which the 6th October city exists. The collapsing soil formation in this area consists of calcareous slightly cemented sand, weakly cemented sandstone with intermittent lenticular shale interbeds. The sand grains (mainly quartz) are almost coated with The sand particles are covered with thin layer of clay films and or ferruginated matter which when dry, constituted the cementing material binding the coarse grained particles. Such soil sequence is related to Oligocene age which followed in some areas by volcanic eruptions and flows (dominated by basaltic types). Sedimentologically these Oligocene sediments are of continental origine dominated by old river systems(channel and flood plains). The collapsing soils were deposited in this plateau by old rivers and drainage nets from older sandstone sediments and according to some geological factors, this type of collapsing sandy product was formed.

3 FABRIC & MINERAL COMPOSITION

The pattern of solid particl arrangement (soil skeleton), soil matrix and cement, and associated pore arrangement (void) are described through the detailed microscopic examination through polarizing light, following the standard techniques of physical and microchemical tests for mineral identification of Tickell 1967; Coquette & Pray 1970 and adams et al. 1984.

3.1 Description of natural samples from province I & II

Detailed micro- texture analysis of undisturbed samples from New Ameriyah city (location 1), New Borg EL- Arab city (location 2) , 10th Ramadan city (location 3) and El- Oboor city (location 4) are described in Table 2. From this table, it can be observed that the examined samples could be classified into two main types which are quartz arenite existing at locations 3 & 4 and quartz litharenite existing at locations 1 & 2. From the geological point of view, these soils are transported type. The soil of location 3 is transported by fluvial action while the rest are of near shore mixed marine and continental deposits

3.2 Description of natural samples from province III (6th October city)

This important province is studied in detailed as a model analogue for the collapsible soils in many desert areas in Egypt and neighboring countries. More than 15 samples have been selected and studied in detailed. Hereunder, the results can be presented under three main petrographic types that represent the essential varieties met within the 6th October plateau. The detailed description and characterization of these three types before and after wetting test in the oedometer are presented in Table 3. From this table, it can be concluded that:

Type 1 is characterized by : a) Most of the grains are composed of quartz and some feldspars, with subangular and subrounded borders and moderately sorted in natural condition. The altered feldspars increase the clay matrix content. Feldspar act in different way than the quartz grains as they have cleavage planes which may be affected by any vertical pressure; b) Clay matrix and iron oxides are randomly disturbed, that means the sample is heterogeneous; c) Inundation of the sample resulted in removing the clay matrix; d) Increase of mechanical pressure up to failure resulted in disintegration of detrital grains.

Type 2 is characterized by : a) Most of the grains are composed of quartz (about 90%) with subangular shape and moderately to well sorted and partially open packed in natural condition; b) The clays present as thin films coating the quartz grains. Also iron oxides are randomly disturbed as binder; c) Inundation of sample resulted in removing the clay matrix accompanied with slight decrease in voids; d) Inundation of samples and increase in vertical pressure up to failure resulted in decrease in sorting (well sorted).

Type 3 is characterized by : a) Most of the grains are composed of quartz (about 90%) with subrounded shape and well sorted and closed packed in natural condition; b) Clay matrix is randomly disturbed and iron oxides are present as cement; c) Inundation of sample resulted in slight increase in pore spaces due to the decrease in clay content; d) Inundation of sample and increase in vertical pressure up to failure resulted in removing the clay matrix completely and decrease in sorting (well sorted), i.e. decrease in maturity.

Table 2 : Textural and petrographic aspects of soil samples from Provines I & II

Location	Condition	Description			Voids	Rock name, Maturrity and Origin
		1. Grains	2. Matrix	3. Cement		
10th of Ramadan(1)	Undisturbed before wetting	About 90% of framework. Medium to fine quartz sand, subrounded to angular borders, Moderately Sorted and Partially Open Packed. Most of the grains are in Point contact. Rare mica detritus are scattered.	Clay matrix is rare	Iron Oxides and silica, (<10)	About 25%. Intergranular small size mesopores of irregular shape.	Quartz arenite Mature mineralgically and texturally. Transported, aquaeous, fluvial deposit
New Ameriyah(2)	Undisturbed	About 65% of framework. Medium quartz sand, rounded to subroundedborders. Moderately sorted and open packed. Shell debris and carbonate detritus. Quartz > rock fragments.	Clay (10% to 15%)	Little Carbonates	About 35% Small size mesopores to micropores with irregular shape.	Quartz Litharenite. Immature mineralogically and texturally. Transported aquaeous mixed marine and conti-netal coastal cone.
Borg El Arab (3)	Undisterbed	About 70%. Fine quartz sand and rock fragments, shell detritus having angular borders. Poorly sorted and open to partially closed Packed. Tock Fragments > quartz grains.	Clay (10% to 15%)	Little Carbonates	About 15% Intergranular Micro pores with irregular shape.	Type Quartz Litharenite. Immature Mineralogically and texturally. Transported, aqualous, mixed marine and continental.
El Obcur City(4)	Undisterbed	About 95 of franework. Coarse and very fine quartz sand, little glauconite and rare phosphatic grains. Moderately sorted and open packed.	Rare clay	Calcite, dolomite and iron oxides, intergranular filling	< 10% Intergranular and micro fractures. Micropores with irregular shape.	Type Dolomitic calcareous quartz arenite. Submature mineralogically and texturally. Transported aquaeous, mixed contaunebtal land marine deposit.

3515

Table 3 Texture analysis of representive type samples (province III)(5)

Type	Condition	Framework			Voids	Name and Maturity
		1. Grains	2. Matrix	3. Cement		
1	Undisturbed Before Wetting	Represent about 85% of framework Medium to fine quartz sand and some altered feldspars. The borders of grains are subangular and sudrounded, Moderately sorted and partially open packed. This sample is hetrogeneous in clay matrix content	Composed Mainly of clay(<5%)	Composed mainly of iron oxides in the from of fine crystalline (about 10%)	Represent about 20% The voids are inter grnular and in partially open status. Micropore and irregular in shape.	Subarkose Immature mineralogically and submature texturally. This sample is hetrogeneous in clay matrix content.
1	Oedometer After Wetting	Represent about 85% of framework Medium to fine quartz sand and, some altered feldspare, The borders of grains are subangular and sudrounded, Moderately sorted and partially open packed. This sample is hetrogeneous in distribution and content of feldspars and cement	Clay is less than 5%	Composed mainly of iron oxides (about 10%).	Represent about 25%.The voids are intergranular and in open status, Micropore and irregular in shape.	Subarkose Immature mineralogically and submature texturally.
2	Undisturbed	Represent about 95% of framework Medium to fine quartz sand having rounded to sudbrounded borders. Moderately to well sorted and partially open packed This sample is hetrogeneous in matrix content.	Clay is less than 5%.	Composed mainly of iron oxides (5%) as binder.	Represent about 20%. The majority are intergranular with subordinate intragranular, mostly connected, fine and small mesopores size with irregular shape.	Ferruginous quartz arenite Mature mineralogically and submature texturally.
2	Oedometer	Represent about 95% of framework Medium to fine quartz sand having rounded to sudbrounded borders. Moderately to well sorted and partially open packed. This sample is hetrogenous in matrix and cement distribution	Clay is undetectable	Iron oxides are undetectable.	Represent about 25%.The voids are inter grnular and partially filled with very fine grains.	Ferruginous quartz arenite Mature mineralogically and submature texturally.
3	Undisturbed	Represent about 90% of framework Medium quartz sand having subrounded borders. Moderately to well sorted and closed packed. This sample is hetrogenous in matrix content	Composed mainly of clay (about 7%)	Iron Oxides are undetectable	Represent about 15%. They are intergranular and partially filled with clay matrix. Fine to small mesopores with irregular shape.	Argillaceous quartz arenite. Mature mineralogically and submature texturally.
3	Oedometer	Represent about 90% of framework Medium to fine quartz sand having subrounded and subangular borders poorly sorted and open to packed	Clay batch Coating the borders (<5%)	Iron oxides are undetectable	Represent about 20% Intergranular voids are in open status.	Argillaceous quartz arenite Mature mineralogically and submature texturally.

4 DISCUSSIONS & CONCLUSIONS

The information obtained from several investigators (Dudley 1970; Barden et al. 1973; Sowers 1979; Clemence & Finbarr 1981 and Clough et al. 1981) indicated that the cementing agents in collapsing soils could be one or more of the clay, silt, calcium carbonate, iron oxides, gypsum and soluble salts. Mc Gown and Collins 1975 indicated that the chemical or clay bonding of the collapsible soils together with environmental and other factors such as amount of pore spaces are important factors in engineering behavior of these soils. Sowers 1979 stated that the most wide spread cementation due to clay and calcium carbonate bonds are usually strong but may weakened by water. On the other hand, bonds by various iron oxides and colloidal silica cementation are relatively insensitive to softening by water. The shape of the coarse grained particles and their packing and pore spaces and their sizes play an important role on collapse behavior of soil.

Based on thin section analysis, Clough 1981 stated that relatively weaker cement sand exhibit less cementation bons and their particles are more rounded and less density packed than strongly cement sands.

The collapsibility behavior of the selected soil samples from the new urban areas in northern Egypt were carried out in detailed geological and engineering framework. The interaction between the textural, fabric, mineralogical directly control the mechanical aspects of the studied samples in their natural state, both before and after wetting. The following main points will be concluded and discussed:

4.1 Particle arrangement

Quartz grains were predominant in all tested samples and varied from 80 % to 90 % of framework. Most of the sizes of the grains were medium to fine and had angular to rounded borders. The higher values of collapse percent ((C_p) were recorded for samples of locations 2; 3 and 5 in which the particles are subrounded borders, poorly sorted and partially open packed. The presence of some composite type of quartz grains in sample of location 5 seems to be a structural defect which weakens the strength of the sample and increases the deformability. The lower values of collapse percent were observed for samples of locations 1 and 4 in which the grains are more well sorted , closed packed and rounded to subrouned.

4.2 Matrix and cement content

Clay is the basic element of the soil matrix and varied from 5 to 10 % of frame work. It is clear that the collapse increases with the increase in clay content . On wetting, the clay acts as a lubricant agent facilitating the rearrangement of the coarser soil particles into a denser packing leading to a soil collapse. Iron oxides are the main cementing agents and are found with minor amounts to 10 %. Gypsum and silica are observed in some samples with less amounts as cementing agent.

4.3 Pore arrangement

Intergranular voids that occupy 10 to 30 % of the total volume were identified in the tested samples. Most of pores are of irregular shape and their sizes varied from large mespores (500u) to micropores (<63u). The samples of higher values of collapsibility have 25 to 30 percent mespores voids of their total volume. On the other hand in the samples of lower values of collapsibility , the voids occupy 10 to 15 percent of the total volume. These voids are in small mesopores sizes. Intragranular voids as shown in sample of location 5 seem to be a structural defect which weakens the strength of the sample and increases the deformability under any pressure. That is mean that the interparticle voids play an important role in collapse behavior. this is agreement with Larionov 1965 and El- Sohby et al. 1988.

From the previous discussion, it can be concluded that:

1. Investigation of the peculiarities of the fabric of samples may determine to a great extent the mechanical properties of collapsing soils.

2. The tested samples consist basically of skeleton grains and matrix. materials including clay and/or cement binder agents.

3. The mechanical properties of the tested samples are determined by the skeleton grains and the pore spaces (voids).

4. wetting the collapsible soil samples reduces greatly the cementation effect and disintegrates the grains and reduces their sorting.

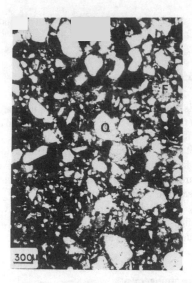

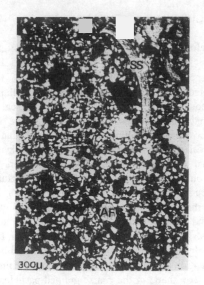

Fig. a   Quartz litharnite texture of rounded to subrounded quartz grains (Q). Limestone detritus (brown color) and fossil remains (f). Notice the open intergranular pores (blue) and the increase of quartz grains than the other lithic fragments.

"Plane Polarized Light (PPL) for New Ameriyah location (1)".

Fig. b   Quartz litharenite with abundant fossil remains, sea shells (SS)& Algal fragments (AF). Carbonate detritus (dark brown color). Quartz grains in subordinate frequency being fine to very fine and mostly angular. Notice the open intergranular pores.

"PPL for Borg el Arab location (2) ".

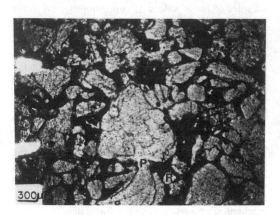

Fig. c Moderately sorted quartz arenite, with intergranular open pores (R). Notice the points of contacts between grains (P).

"PPL for 10th Ramadan location (3)".

Fig. d The same as Fig. c but crossed Nicole (XPL). Notice the composite polycrystalline quartz grain (Q) in the middle of the photo. Bar scale in microns.

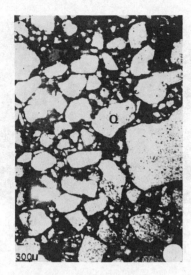

Fig. e Submature quartz- arenite with micro to fine crystalline calcite and dolomite cement (dark brown color) coating the quartz grains (Q). Notice the little intergranular pores (blue color) and some irregular microfracture in quartz grains.

"PPL for El Oboor location (4)".

Fig. f Same as Fig. e but under (XPL)

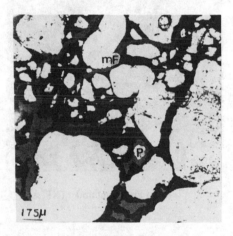

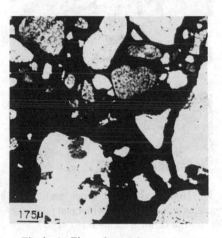

Fig. g Highly ferruginated quartz arenite spot which reflects an irregular micro- fractures pattern (mF). The intergranular pores are filled with iron oxides with isolated preserved open pores (P)

"PPL for 6th October location (3), type 1 ".

Fig. h As Fig. e, but under (XPL).

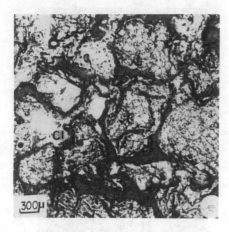

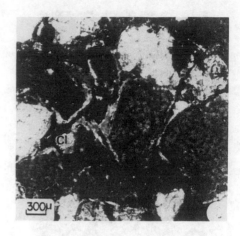

Fig. i Argillaceous quartz arenite with clays (C1) which partially coat the quartz grains and fill some of the intergranular pores. Quartz grains are rounded to subrounded and moderate to well sorted.
  "PPL for 6th October location (3), type 2 ".

Fig. j As in Fig. i, but under (XPL).

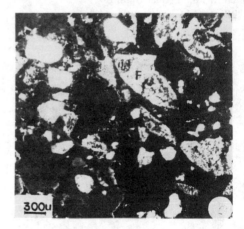

Fig. k Highly altered feldspar grains (F) in a ferruginous subarkose type with open pores.
  "PPL for 6th October location (3), type 3 ".

Fig. l As in Fig. j, but under (XPL).

ACKNOWLEDGMENTS

The present authors are gratefully indebted to Prof. Dr. M. Serif , professor and director, General Organization for Housing, Building and Planning Research (GOHBPR) and Prof. Dr. M. Mashour, professor of soil mechanics & foundation engineering, Structural Design Department, Zagazig University, Egypt for their critical discussions and encouragement during the preparation of the present work.

REFERENCES
.

Adams, A.E.,Mackenzie, W.S. & C. Guilford 1984. Atlas of sedimentary rocks under the microscope. Longman Group Limited, Essex, ngland.

Barden, L.; Mc Gown, A. & K. Collins 1973. The ollapse mechanism in partially saturated soils. ngineering Geology J. July, 49-60.

Clemence, S.P.& A.Q. Finbarr 1981. Design considerations for collapsible soils. J of
   Geotech. Eng.Div. Proc. of ASCE, March.

Clough, G.W. 1981. Cement sands under static loading. JSMFE Div., ASCE,107: GT6: 799-817.

Choquette, P.W. & L.C. Pray 1970. Geological nomelactur and classification of porosity in sedimentary carbonates. Bull. Am. Ass. Petrol. Geology, 54:207-250.

Dudley, J.H. 1970. Review of collapsible soils. J. of SMFE div., ASCE, 96, 3: 925-947.

EGSMA. 1982, Egyptian Geological Survey and Mining Authority, The Geological Map of Greater Cairo.

El- Sohby, M.; Mazen, S. & A. Elleboudy 1985. Comparative study of problematic soils in two areas around Cairo. Proc. of the Symp.on Evironmental Geotechnics and Problematic soils and Rocks, Bangkok: 515-526.

El- Sohby, M.; Elleboudy, A. & O. Mazen 1988. Stress- strain characteristics of two calcareous arid soils. Proc of Int. Conf. on Calcareous Sediments, Perth: 17-24.

Jennings, J.E. & K. Knight 1975. A guide to construction on or with materials exhibiting aditional settlement due to collapse of grain sructure. The 6th Regional Conference for Africa on SMFE, Durban, South Africa.

Larionov, A.K. 1965. Structural characteristics of loess soils for evaluating their constructional properties. Proc. of The 6th ICSMFE, Montreal, Canada, 1:64-68.

Lotfy, H.A. 1990. Evaluation of oedometer test for prediction performance of collapsible soils. Civil Eng. Res. Mag., Civil Eng. Dept., Faculty of Eng., Al- Azhar Univ., 12, 13:14-29.

Mansour, S.; Aboushook,M.; Mashour, M. & M.Sherif 1993. A comparative study of the geotechnical properties of some collapsible soils in the world and in Egypt. Al- Azhar Eng. 3rd. International Conf., Cairo, 3:141-148.

Mansour, S. 1992. A study of the geotechnical characteristics of some Egyptian collapsing soils. Ph.D. Thesis, Structural Design Dept., Faculty of Engineering, Zagazig University, Egypt.

Radwan, A.& A. Gabr 1980. Geotechnical studies at New Ameriyah city. Civil Eng. Res. Mag., Al-Azhar Univ., 2:1-21.

El-Saadany, M. 1989. Stabilization of Egyptian collapsible soils with hydrate lime. Al- Azhar Eng. Conf. AEC'89,4.

Sabry, M. 1987. Prediction of collapsibility of sandy soil. Scientific Eng. Bull., Faculty of Eng.., Cairo Univ.,4.

Said, R. 1962. The geology of Egypt. El Sevier Pub. Company, Amsterdam.

Sharkis, E. 1989. Geotechnical study of soils at 6th October city. M. Sc. Thesis, Faculty of Eng. Cairo Univ.

Shukri, N.; Philip, G. & R. Said 1956. The geology of the Mediterranean coast between Rosetta and Bardia. Part Ii, Pleistocene Sediments, Geomorphology and Microfabric, Bull. desert Inst. Egypt,37:2.

Sowers, G. 1979. Introductory soil mechanics and foundation. Published by Collier Macmillan.

Tickell, F. 1967. The techniques of sedimentary mineralogy. Developments in Sedimentology, Amsterdm, 4:

Engineering-geological properties of the laterites: An example of the humid subtropics of Georgia

Propriétés de géologie de l'ingénieur des latérites: Un exemple des régions subtropicales humides de la Georgie

L. I. Varazashvili
Georgian Technical University, Tbilisi, Georgia

E. D. Tsereteli
Sakgeologia, Tbilisi, Georgia

ABSTRACT: Within humid subtropics of West Georgia typical laterites and lateritic eluvial formations are of substantial occurance. They form peculiar engineering-geological conditions and obtain the role of a specific element as the basis for construction and for development of special agricultural crops.

RESUME: Dans les limites des regions subtropicales humides de la Georgie les latérites typiques et les formations éluviales ressemblantes aux latérites sontttrès repandues,lesquelles forment une ambiance geologique particulière et jouent le rôle d'un élément specifique comme une base de construction et créent les conditions du developpement des cultures agricoles speciales.

Study of the nature of facies formation, matter composition and physico-mechanical properties of laterites of Georgia has definite factual and methodological importance for the cognition of general regularities of the formation of engineering-geological conditions of eluvial crust weathering in the Earth's humid and warm climatic zone.

The essence of laterization process is that under frequent and intensive atmospheric precipitation and increased temperatures deep physico-chemical weathering of iron and aluminium bearing rocks takes place. Silica SiO_2 and absorbed bases of Ca, Mg, Na and K are removed from rocks,while colloidal Fe_2O_3 and Al_2O_3 are accumulated there. Their low molecular ratio serves as an obvious indication of intensive laterization.

Within subtropics of West Georgia the typical laterites expand to the absolute height of 400-500 m, becoming the most important element as the basis for construction and for intensive development of the territory.

Facies and geochemical changes of laterite rocks, their colour, thickness, weathering degree, structural composition and engineering-geological properties are determined by mineral-petrographic composition of the original rocks undergoing active hypergenesis and by relief conditions.

Laterites of West Georgia vary in age from the Kimmerian up to the Quaternary, inclusive. However, up to the present, paleogeographic conditions of laterization processes went on in almost the same climatic conditions: annual temperature – never lower than 13-15°, and annual precipitation – 2000-3000 mm at humidification factor of 4,1. They are formed on volcanogenic rocks of the Middle and Upper Eocene, on the Cretaceous-Paleocene carbonaceous rocks, on the Oligocene-Neogene clays, on the Pliocene molasses and on the Quaternary coastal-marine alluvial deposits. The thickness of laterite rocks varies in wide range. The highest thickness (up to 30-

60 m and in some sections of Aska-
na bentonites – up to 150 m) is
achieved in volcanogenic formations, and the lowest (5-15 m) in
the deposits of marine and river
terraces, composed basically of the
rocks rich in quartz and other minerals stable to hypergenesis.

On alkaline volcanogenic formations laterites of red, brownish-
red and yellowish-brown colour enriched in aluminum oxide, iron and
hydrochlorites are generated; in
the zone of active hypergenesis
almost uniform ferritization is
observed (average content of iron
oxide – 11-15%). As a result, gradual thinning of aluminum oxide
and thickening of silica is taking
place. For these rocks acid reaction medium (P_H – 4,7-5,0) and active kaolinization are characteristic. Lithologically they are
presented by loams and clays having porous, granular cloggy, loose
structure. High porosity (65-70%)
and humidity (42-47%) are determined by step-like aggregation.
Physically bound water is sorbed
inside aggregates, as well as on
their surfaces, which stipulates
high values of all kinds of humidity. Maximum molecular moisture
capacity equals 32% and 33%, plasticity limit is 41% and 44%, yield
point – 59% and 65% and absolute
moisture capacity is 65% and 68%,
respectively. The latters frequently exceed porosity in case of insignificant swelling (in average
3,6% for loams and 4,9% for clays).
This happens because average density of adsorbed water in heavily
latericized soils is 1,85 g/cm^2.
Consequently, the volume occupied
with this water is 1,85 times less
than that occupied with gravity
water. This explains as well the
increase of the value of their
unit weights. Plasticity increases
to 29. The amount of clay fractions
(0,002 mm) in heavily latericized
loams average 33% and in clays –
39%.

When laterization is reduced the
amount of clay fractions in mechanical composition is increased up
to 43-44% in average. Thermograms
of sludge fractions of typical latericized soils point that kaolinite and halloysite predominate in
them. From colloids hydromica with

montmorillonite admixture is encountered. In highly latericized
rocks dust-like fractions are up
to 32 and 22%. They basically have
high filtration coefficient, solid
consistency, high values of unit
weight (2,78-2,83 g/cm^2), low density of rocks (1,13-1,63 g/cm^3),
increased strength indices (ℓ 15÷20;
C = 0,2000÷0,950 kgf/cm^2) and deformations, the final value of
which changes within the limits of
31-186 mm/m . In water saturated
medium $\not$ decreases only by 3-5°,
while adhesion decreases by 0,05-
0,45 kgf/cm^2.

In natural conditions these rocks
stand high angles in slopes (46-60°)
at slope height of 5-15 m; they are
water-resistant, do not soak and
despite of favourable climatic conditions, the development of landslide processes is limited.

In laterites generated on Upper-
Pliocene-Quaternary marine terrace
deposits kaolinite argillaceous
formations dominate (55-60°). At
the same time thickness of hypergenesis notably increases from young
to old terraces and argillaceous
composition increases respectively.
In these deposits desilication is
noticed on profile depth where silicium content from the upper layer
of the horizon on the depth of 15-
19 m decreases from 58÷ 60% to 52%.
Aluminum and iron oxides composing
18% and 9%, respectively, stay in-
varied on the entire depth of section. From 3,0 m appreciable seli-
nization of soils is marked, where
composition of carbonates achieves
1% and that of easily soluble salts
– 4%. These conditions often favour
the formation of subsurface ortstei
horizon.

On Middle Quaternary terraces
yellowish and brownish tints of
clay and loam with thickness from
3,5 to 8-10 m prevail. In clays the
composition of argillaceous fractions is 38-63% and of dust fractions – 52-60%. Argillaceous fraction everywhere has hydromicaceous
composition with, in some places,
gypsum and calcite admixture. In
chemical composition the basic mass
consists of silicium and aluminum
oxides, the amount of which in nature is 60 and 12%, respectively.
The amount of iron oxide is a
little higher. The content of finely soluble salt per 100 g of rock

does not exceed 0,3 g/l.

Plasticity number of rocks is chiefly within the limits of 19-2o and does not exceed 24, which is somehow connected with peculiarities of chemical composition. Yield point of clays mainly equals 36-49, and plasticity point is 22-25. The consistency of rocks is solid, seldom- semisolid and in single cases - plastic. In the main, clays are of low and average water resistance, though relatively water resistant varieties are often encountered, too. In some places areals of water resistant clays occurance are noticed, which apparently is connected with increased natural humidity of rocks up to 28-29%.

Internal friction angle of clays changes in wide range, chiefly from17-19 to 21-22°. Coefficient of internal friction varies from 0,28 to 0,44 and adhesion from 0,29 to 0,84 kg/cm^2. After water saturation internal friction angle decreases by 2-7°, achieving 12-14°, coefficient of internal friction decreases to 0,22-0,37 and adhesion - to 0,1-0,5 kg/cm^2. Modulus of clay settling under 5 kg/cm^2 load makes 40,1-90,4 mm per meter. In loams clay fraction content is from 17 to 20%, with prevalence of 22-23%, dust-like fraction content is 38-68% with prevalence of 60-62%. The remaining part is medium-grain sand.

Argillaceous materials in rock is almost everywhere of hydromica composition. In some cases the admixture of gypsum is also encountered. The basic mass of rock consists of calcium oxide up to 68-76% and aluminum oxide up to 14-16%. The content of easily soluble salts per 100 g of loam is 0,06-0,3 g/l.

Plasticity number is 8-14 with prevalence of 11-12, yield point changes from 28 to 42 and plasticity limit - from 17 to 21. Consistency of loams is mainly solid or semisolid, seldom - soft.

Loams are characterized with weak water resistance. Total soaking of rocks occurs during one hour. They practically are nonswelling or weakly swelling.

Internal friction angle of loams is 18-20°, coefficient of internal friction is 0,23-0,39 kg/cm^2. After water saturation these values decrease respectively. Internal friction angle - by 3-5°, down to 16-20°, coefficient of internal friction - to 0,23-0,37 and adhesion to 0,06-0,27 kg/cm^2.

Under 5 kg/cm^2 load settling modulus makes up to 125 mm per m.

In elluvial clays filtration coefficient is 0,03 m per day while in loams with skeleton material inclusion it varies from 0,25 to 0,64 m per day.

Laterites of molassal deposits of 2-5 m thickness are presented with clays and loams, and compared with hypergenecized volcanogenic deposits of Eocene are characterized with less natural humidity (average value 25÷ 36 %), porosity (46-50%) and increased density (1,75-1,79 g/cm^3). In clays yield point is 33-65, plasticity limit is 21-48, plasticity number is 17-34, swelling is 4,1-10,6%, internal friction angle is 17-21° and adhesion is 0,5-1,2 kg/cm^2. Loams here have natural humidity 25-27%, porosity 43-46%, yield point - 43-44, plasticity limit 2,8-30, plasticity number 14-15, swelling 2,8-21,2%, internal friction angle 21-22°, adhesion - 0,45-0,75kg/cm^2.

The second rock mass of 1-2 m thickness is presented with clays and loams with water and rock wastes up to 15-20%. Their physico-mechanical properties are: natural humidity - 18-42%, volume weight of damp ground - 1,66-1,96 kg/cm^3, volume weight of gritty ground - 1,14-1,63 g/cm^3, porosity - 40-59%, in aleurite sand content is 8-22%, dust content is 12-55%, clay content is 26-80%, yield point 35-86, plasticity limit - 21-48, plasticity number - 14-48, swelling 2,8-12,7%, internal friction angle - 16-23°, adhesion - 0,27-1,1 kg/cm^2.

According to geologomorphological conditions of bedding the both above described masses are potential mediums for landslides and here eluvium actually performs

the role of basic deforming horizon. This is favoured also by the
presence of plastic and soft-plas-
tic clays and loams in eluvium
lows, high soil swelling and low
shift coefficients. In water medium they become soaked, in most
cases they are characterized with
swelling where swelling pressure
is measured within 1-2 kgf/cm^2.
Medium is neutral (pH 6,8-7,0).
Strength factors are also low when
ultimate values of internal friction angle and adhesion reduce

from 14-23^0 and 0,400-1,300 kgf/cm^2

to 5-19^0 and 0,150-0,600 kgf/cm^2
in water -saturated condition. In
this connection, the slopes comp-
posed of such rocks are landslide
resistant, while construction
foundations are subjected to considerable deformation.

Particularly low strength and
deformation properties are characteristic for laterites generated
on poorly lithified argillaceous
Oligocene-Miocene rocks presented
with high hydrophilic montmorillonite and montmorillonite-hydromi-
caceous, heavily swelling and compressible argillaceous formations
easily yielding to landslide processes with landslide coefficient
0,7-0,9. In the process of construction these rocks need performing draining measures.

Highly hypergene materials differ distinctly from lower eluvium
and original base in colour, humidity and density.

Regional-geological factors affect the engineering-geological
features of laterite rocks formed
under the effect of zonal-geologi-
cal factors.

Laterite rocks generated on different original basis and elements
of relief, in the same climatic
conditions may be characterized
with different composition, conditions and properties.

The highest hypergene variation
is characteristic for laterites
generated on tufogenic volcanogenes.
The transformation of original
rocks into laterite weathering
crust happens at considerable
change of chemical composition,
therefore, geochemical zonation is
more clearly traced in the profile
of these formations than in the
traditionally existing weathering
zones.

REFERENCES

Emelianova, E.P. 1972. Comparative
  method of estimation of stability
  and prediction of landslides.
  Moscow:Nedra.
Javakhishvili, E.A. 1977. Weathering of Baios and Cretaceous volcanogenic deposits in connection
  of landslide formation: Proc.
  Hydrogeology and Engineering Geology. Tbilisi.
Javakhishvili, E.A. 1976. Rock
  weathering as the evidence of
  the prognosis of landslide show
  inmountain-folded regions. Materials of scientific-technical
  Conference, Tbilisi.
Lomtadze, V.D. Engineering geology,
  engineering geodynamics.Lenin-
  grad:Nedra, 1977.
Problems of engineering-geological
  study of crust weathering processes. 1971. Materials of Laza-
  revskoe conference. MGU.
Tsereteli, D.V. 1966. Pleistocene
  deposits of Georgia. Tbilisi:
  Metsniereba.
Tsereteli, E.D. and M.E.Oniani.1969
  Engineering-geological conditions
  and reguliarities of development
  of modern exodynamic processes
  on the territory of the Georgian
  SSR. Grusgeologia, v.1, Tbilisi.
Tsereteli, E.D. and D.D.Tsereteli.
  1985. Geological conditions of
  mudflow development in Georgia.
  Tbilisi:Metsniereba.

Effect of deterioration on the chemical and physical properties of Göreme tuffs

L'effet de détérioration sur les propriétés chimiques et physiques des tufs de Göreme

T.Topal & V.Doyuran
Geological Engineering Department, Middle East Technical University, Ankara, Turkey

ABSTRACT: The tuffs and unique morphological structures, the so-called "fairy chimneys" of the Göreme area have been chemically and physically deteriorated due to atmospheric effects. Studies towards conservation of these historical and touristic values should be based on the understanding of the processes which cause deterioration of the tuffs. In this study, the deterioration phenomena have been investigated through petrographical and geochemical methods. These involved thin section studies, X-ray powder diffraction method, and major and trace element analyses to determine the degree and depth of weathering. The effect of deterioration on the physical properties of the tuffs (unit weight, porosity and water absoption) has also been investigated. An attempt is made to correlate the chemical and physical parameters to understand the net effect of deterioration phenomena acting on the tuffs. The deterioration of Göreme tuffs produces chemically weathered zones restricted mainly to the uppermost 2 cm of the zone of alteration. However, no significant change on the physical properties of the tuffs through weathering zones exists.

RESUMÉ: Les tufs et les structures morphologiques uniques, soi-disant "cheminées magiques" de la région de Göreme ont été détériorés physiquement et chimiquement à cause des effets atmospheriques. Les études vers le préservation de ces valeurs touristiques et historiques doivent se baser sur la compréhension des processus qui crént la détérioration des tufs. Dans cette étude, les phenomènes de détérioration ont été étudiés par les méthodes pétrographiques et géochimiques. Ces méthodes concérnet les études petrographiques en lames minces, la méthode de diffraction de rayons X en poudre et les analyses des éléments majeurs et traces pour déterminer le dégre et la profondeur de l'altération. L'effet de la détérioration sur les caractéres physiques des tufs (poids unitaire, porosité et absorbtion d'eau) a été aussi étudié. Un essai est fait pour comprendre l'effet du phenomène de détérioration en faisant les corrélations entre les parametres physiques et chimiques. La détérioration des tufs de Göreme produit chimiquement des zones d'altération confinées principalement 2 cm le plus haut de la zone d'altération. Cependant il existe no changement particulière à travers des zones d'altération dans les caractères physiques des tufs.

1 INTRODUCTION

The Cappadocia region is an important touristic site of Turkey. The region is very popular with geomorphologically distinct features the so called "fairy chimneys" and the relicts of ancient civilizations. The fairy chimneys are formed within the tuffs by the natural processes of weathering and differential erosion. The area within Nevşehir-Ürgüp-Avanos triangle, is one of the well known areas where the fairy chimneys were dwelled and contain valuable wall paintings inside belonging to Byzantine times (Figure 1). Weathering and erosion, however, are still active and these natural processes along with

man-made activities threaten the future of the chimneys. Although new fairy chimneys are being formed, the old ones need to be protected due to their historical heritage. Studies towards conservation of these historical and touristic values have already been started. Nevertheless, they should be based on the understanding of the processes which cause deterioration of the tuffs.

In this paper, present state of weathering of the fairy chimneys within the aformentioned triangle is investigated. Emphasis is given to the change, if any, of chemical and physical properties of the tuffs with depth.

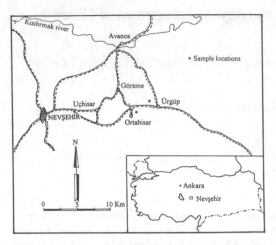

Figure 1. Location map of Nevşehir-Ürgüp-Avanos area

2 SITE GEOLOGY

The Cappadocia region is generally underlain by thick and extensive deposits of volcano-sedimentary sequence (Miocene-Pliocene) of the Ürgüp Formation (Pasquare, 1968). Although the formation comprizes a number of well-distinguished members, in the study area only the Kavak and Tahar members contain the fairy chimneys. Both members are characterized by non-welded tuffs. Due to the fact that the most of the dwellings and wall paintings are found within the fairy chimneys of the Kavak member, our study is focused on this member.

The Kavak member represents the product of the first intermittent volcanic activity which produced the Ürgüp Formation. Thus, this member constitutes the basal portion of the volcano-sedimentary sequence. The chaotic arrangement of pumice fragments within the unit suggests that it is an ash flow tuff deposited in a lacustrine environment. It covers the existing paleotopography and shows laharic and/or tuffitic character. It is dirty white to pink, and contains phenocrysts of plagioclase, quartz, biotite and opaque minerals. Various rock fragments and pumice are also commonly observed. In the matrix volcanic glass shards are rather common (M.E.T.U., 1987; Türkmenoğlu et al., 1991). The Kavak member is about 120 m thick and dips 10 to 15 degrees N and NE (Erdoğan, 1986).

Some of the fairy chimneys formed within the Kavak member contain the so-called "cap rock". Petrographical studies revealed that the cap rocks are made of welded-tuffs.

3 METHODS OF INVESTIGATION

The fairy chimneys which are partially covered by lichens and cut by joints show clear effects of weathering especially in Göreme Open Air Museum. Since the weathering products are also protected in those places, field studies are concentrated to find such typical locations in the study area in order to determine the state and depth of weathering.

Two block samples of the Kavak member were collected from two different locations. The Ortahisar sample has a lichen cover and the Ürgüp sample contains joint with intense discoloration (Figure 1). Since the depth of weathering in the tuffs of the Göreme Open Air Museum is found to be about 8 cm (Türkmenoğlu et al.,1991), the sizes of block samples were selected to fully incorporate the weathered zone.

For the petrographic studies, thin sections were systematically prepared from the two block samples to provide a continuous view of the tuffs from both lichen covered and iron stained surfaces to a depth of 23 cm. For the chemical analyses, 20 samples were obtained by scratching on the two block samples in the laboratory from the surface to a depth of 23 cm.The scratched samples were then powdered to pass through 200 mesh.

Petrographic and mineralogical analyses were done by a polorizing optical microscope. Relative abundance of clay minerals was determined by X-Ray diffraction (XRD) on less than two-micron-size fractions on 10 unoriented powder samples corresponding to each block (total 20 samples). The types of clay minerals, however, were determined by XRD on the same size samples as before, but on oriented samples corresponding to each block (total 10 samples) . The oriented samples were tested after air-drying, glycolation, and heating. A RIGAKU diffractometer with CuKα radiation was used in the measurements. The measurements were performed at the General Directorate of Mineral Research and Exploration (M.T.A.) in Turkey. Ten powdered samples from each block (total 20) corresponding to the same depth of samples for XRD, were used for the major and trace element determinations by X-Ray Fluorescence (XRF). These analyses were carried out at the laboratory of Leeds University in U.K.

For the determinations of physical properties namely, the unit weight, porosity, and water absorption, core samples having 25 mm diameter were recovered from the block samples. The tests were carried out according to I.S.R.M. (1981) RILEM (1980), and TSE 699 (1978).

4 MINERALOGICAL PROPERTIES OF WEATHERED TUFF

Mineralogical and petrographical investigation of the tuff samples revealed that the samples consist predominantly of quartz, plagioclase feldspar, biotite, and opaque minerals. Pumice being the most abundant one, and volcanic rock fragments (mostly andesite and basalt) can be easily identified both with naked eye and under the microscope. The minerals and the rock fragments are embedded within a tuffaceous matrix composed of volcanic glass.

In the Ortahisar sample, extensive mechanical but very little chemical weathering of the minerals are observed within the upper 2 cm of the sample. Plagioclase feldspars are highly fractured at this zone. The cleavage planes of the feldspars and biotites are widened due to weathering. According to M.E.T.U. (1987), lichens play a very important role in deterioration of tuffs within this zone. Biotites and rock fragments, however, are slightly discolored forming iron oxide staining at this depth. The effect of mechanical weathering decreases with depth and ceases almost at a depth of 8 cm whereas no chemical weathering can be observed after 2 cm depth.

In the Ürgüp sample, extensive mechanical and slight chemical weathering of the minerals are also noted within the first 2 cm of the sample nearer to the joint surface. In this zone, plagioclase feldspars are highly fractured and widened. Biotites are highly discolored forming iron oxide staining so that the outer boundary of the minerals can hardly be traced. On the other hand, the rock fragments are slightly discolored. The effect of mechanical weathering decreases with depth and ceases almost at a depth of 10 cm whereas chemical weathering in the form of discoloration in biotites and rock fragments, extends laterally about 17 cm from the joint surface with decreasing intensity. In the field, however, discolored zone of up to 20 cm was observed.

In both samples, the tuff matrix is partly altered to smectitic clay. The smectitic clay is not only found at the weathered zone but also away from it. So, the alteration of volcanic glass to smectitic clay has probably started after the deposition of the tuffs even before the formation of the fairy chimneys.

X-Ray Diffraction (XRD) analyses of 20 unoriented (powder) samples from the source of alteration to fresh parts of Ortahisar and Ürgüp samples revealed that very small amount of clay minerals exists within the tuff samples other than quartz, feldspar, and mica minerals. XRD analyses of the samples indicated no significant change in the clay content of the samples with depth. This conclusion is achieved from the presence of very small peak area (also not changing with depth) of the clay minerals.

XRD analyses of 10 oriented samples after air-drying, glycolation, and heating at 300°C revealed mainly "smectite" type of clay minerals. A very small amount of mica (probably illite) is also observed. Since smectite is the major and the first formed clay mineral within the rock, this suggests that no considerable leaching processes had taken place within the system.

5 CHEMICAL PROPERTIES OF WEATHERED TUFF

Chemical weathering of rocks may result changes in initial elemental concentrations by leaching and enrichment (Borchardt et al., 1971; Borchardt and Harward, 1971). In order to determine the depth of chemical weathering, major and trace element concentrations in percent and ppm, respectively, from the source of alteration (lichen and joint surface) at 10 different depths from each block sample (total 20 samples) were determined by means of XRF (Table 1-4). Because chemical weathering is more pronounced near the source of alteration, much closely spaced samples were collected from the highly altered zone. The depths of samples from the source of alteration were 0-0.5 cm, 0.5-1 cm, 1-1.5 cm, 1.5-2 cm, 3-4 cm, 5-6 cm, 8-9 cm, 11-12 cm, 15-16 cm, and 22-23 cm.

Since the weathering products derived from the fresh tuffs are of major interest, analytical data are normalized by dividing each elemental compositions of every sample by the same elemental composition of the fresh tuff sample. For this purpose, the Ortahisar sample obtained at a depth of 22-23 cm is considered to be a reference sample. So, all the analyses belonging to other depth intervals are normalized to this sample. This permits easy comparison between the two different tuffs and between different weathering zones of each tuff. The relative contents (normalized values) of each major and trace element of both Ortahisar and Ürgüp samples, are plotted against depth in Figures 2 through 7. Since the sampling points correspond to a range of depths (e.g. 3-4 cm), an average depth is used during plotting of graphs. Considering that the weathering proceeds horizontally in the fairy chimneys, the depth of analysis of every element is shown in abscissa of the graphs.

Table 1. Major element analyses of Ortahisar samples.

Depth (cm)	0-0.5	0.5-1	1-1.5	1.5-2	3-4	5-6	8-9	11-12	16-17	22-23
SiO_2	68.43	72.66	73.42	73.81	73.35	73.24	73.40	73.52	73.19	73.49
TiO_2	0.18	0.17	0.17	0.17	0.18	0.19	0.18	0.19	0.18	0.19
Al_2O_3	12.26	12.34	13.01	13.11	13.24	13.38	13.31	13.22	13.31	13.35
Fe_2O_3	1.30	1.16	1.12	1.14	1.15	1.14	1.11	1.17	1.15	1.16
MnO	0.05	0.06	0.05	0.05	0.05	0.05	0.05	0.05	0.05	0.05
MgO	0.77	0.69	0.53	0.42	0.38	0.39	0.33	0.38	0.33	0.36
CaO	2.43	1.87	1.75	1.74	1.79	1.82	1.78	1.77	1.80	1.77
Na_2O	1.94	2.11	2.24	2.19	2.25	2.33	2.28	2.31	2.39	2.29
K_2O	3.51	3.64	3.83	3.90	3.95	3.93	3.93	3.98	3.94	3.88
P_2O_5	0.05	0.03	0.02	0.02	0.02	0.03	0.02	0.02	0.03	0.02
L.o.I.	9.48	5.15	4.10	3.99	4.05	3.93	3.59	3.52	3.81	3.57
Total	100.39	99.88	100.25	100.56	100.42	100.42	99.99	100.13	100.18	100.15

Table 2. Trace element analyses of Ortahisar samples.

Depth (cm)	0-0.5	0.5-1	1-1.5	1.5-2	3-4	5-6	8-9	11-12	16-17	22-23
Sc	8	2	2	1	3	0	3	0	2	0
V	34	30	26	22	24	22	24	20	23	24
Cr	27	19	22	17	17	16	9	20	15	16
Co	9	9	7	9	11	17	8	7	4	8
Ni	9	6	4	3	5	6	3	4	4	5
Cu	7	7	12	12	16	26	9	18	4	11
Zn	37	30	33	32	31	31	32	36	31	32
Rb	123	133	137	135	140	137	141	139	139	139
Sr	225	236	231	221	225	229	223	219	234	224
Y	12	13	14	14	12	14	13	13	13	12
Zr	82	84	86	82	88	87	87	87	90	84
Nb	12	13	13	13	12	13	13	13	14	13
Ba	868	956	948	898	853	841	839	842	843	838
Pb	17	14	16	17	17	16	16	17	16	16
Th	21	23	23	23	24	23	22	23	23	21
U	3	4	5	4	5	5	5	5	5	5

Table 3. Major element analyses of Ürgüp samples.

Depth (cm)	0-0.5	0.5-1	1-1.5	1.5-2	3-4	5-6	8-9	11-12	16-17	22-23
SiO_2	70.03	70.47	70.72	71.33	71.45	71.33	71.42	72.18	72.38	72.29
TiO_2	0.17	0.18	0.18	0.19	0.18	0.19	0.19	0.19	0.19	0.21
Al_2O_3	12.98	13.27	13.57	13.74	13.72	13.63	13.81	13.82	13.73	13.87
Fe_2O_3	4.98	4.20	3.10	2.89	2.78	2.36	1.68	1.74	1.57	1.44
MnO	0.05	0.05	0.06	0.06	0.07	0.10	0.09	0.08	0.12	0.06
MgO	0.52	0.45	0.46	0.40	0.40	0.40	0.40	0.40	0.38	0.42
CaO	2.01	1.96	1.96	1.95	1.93	1.87	1.88	1.83	1.84	1.87
Na_2O	2.18	2.13	2.58	2.27	2.34	2.38	2.31	2.33	2.35	2.44
K_2O	3.40	3.52	3.70	3.71	3.80	3.80	3.92	3.96	4.00	3.98
P_2O_5	0.02	0.02	0.02	0.02	0.02	0.02	0.02	0.02	0.03	0.02
L.o.I.	4.18	3.83	3.91	3.78	3.84	3.92	3.77	3.83	3.93	3.98
Total	100.53	100.08	100.27	100.33	100.53	99.99	99.50	100.38	100.53	100.57

Table 4. Trace element analyses of Ürgüp samples.

Depth (cm)	0-0.5	0.5-1	1-1.5	1.5-2	3-4	5-6	8-9	11-12	16-17	22-23
Sc	0	0	0	0	0	1	2	2	2	1
V	26	23	18	18	17	25	21	21	15	21
Cr	16	20	22	21	11	10	15	13	19	17
Co	12	12	10	12	8	10	8	9	5	6
Ni	3	3	4	3	3	3	4	3	13	5
Cu	6	5	4	4	6	4	4	4	5	8
Zn	41	38	37	36	36	35	33	34	34	33
Rb	117	121	127	128	129	132	136	135	137	137
Sr	244	235	237	232	228	231	233	228	229	234
Y	11	9	9	11	11	12	11	12	11	11
Zr	81	81	88	88	88	89	93	86	91	89
Nb	12	14	13	14	14	14	14	15	14	14
Ba	856	810	818	819	853	911	905	894	979	864
Pb	21	18	17	18	18	19	19	17	20	19
Th	23	23	23	24	25	23	23	23	25	22
U	4	5	4	5	5	6	6	4	4	6

The relative content variations of major elements (oxides) of Ortahisar samples (solid lines) with depth indicate that significant enrichment of MgO, CaO, P_2O_5, and L.o.I occurs within upper 2 cm of the weathering zone. The other elements do not change significantly (Figure 2). In the case of Ürgüp samples (dashed lines), however, a significant Fe_2O_3 and very little MgO enrichments within the first 2 cm of the joint surface are observed. Fe_2O_3 enrichment extends almost laterally to 17 cm depth (Figure 3). This depth is also determined from the discoloration of the sample observed in the field and under the microscope. Such reddish color is attributed to Fe_2O_3 development. Similar reddish color deteriorating the wall paintings also exists in the Chapel at the north of Elmalı Church in Göreme

Open Air Museum. MnO in these samples, however, fluctuates with depth.

If we compare both Ortahisar and Ürgüp samples, the Ortahisar sample has more enrichment in MgO, P_2O_3, and L.o.I. On the other hand, the Ürgüp sample has more Fe_2O_3 enrichment (Figure 4). Excepting Fe_2O_3, all the variations within the samples are confined almost to the upper 2 cm of the weathering zones.

The relative content variations of trace elements of Ortahisar samples indicate that there is an enrichment of V, Ni, and Zr mostly within the upper 2 cm of the weathering zone (Figure 5). In the case of Ürgüp samples, no significant change is observed. However, Ni shows fluctuations (Figure 6). In Figures 5-7, Sc is not plotted because Sc does not exist at a depth of 22-23 cm of Ortahisar sample selected as standard for normalization. Nevertheless, a close examination of the variations of Sc with depth in Ortahisar sample (Table 2) clearly shows an enrichment within upper half a centimeter of the weathering zone.

A comparison of the relative content variation of trace elements in both samples reveals that only V, Ni and Zr of Ortahisar samples show variations (Figure 7). However, these variations are so small when compared with major element analyses. For this reason, major element analyses may be used to determine the depth of weathering of the tuffs. Evaluation of these analyses indicates that the depth of weathering in Ortahisar and Ürgüp tuffs is about 2 cm from the source of alteration. However, discoloration may extend laterally upto 17 cm in the case of Ürgüp tuffs.

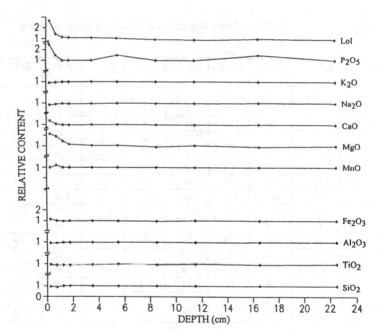

Figure 2. Relative concentration variations of major elements in Ortahisar samples.

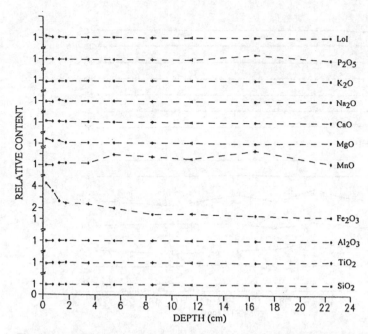

Figure 3. Relative concentration variations of major elements in Ürgüp samples.

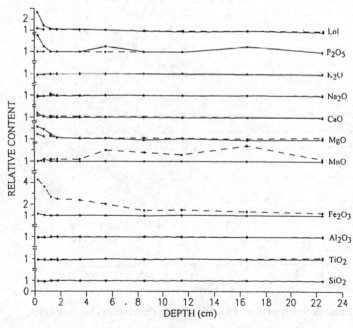

Figure 4. Comparison of major element concentrations of Ortahisar (solid lines) and Ürgüp (dashed lines) samples.

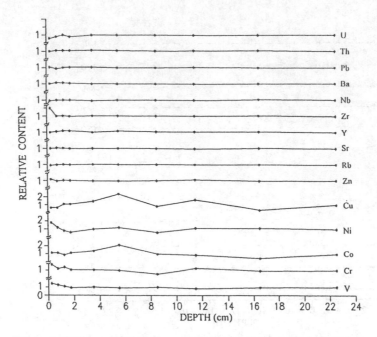

Figure 5. Relative concentration variations of trace elements in Ortahisar samples.

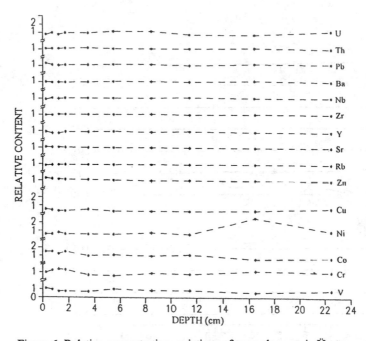

Figure 6. Relative concentration variations of trace elements in Ürgüp samples.

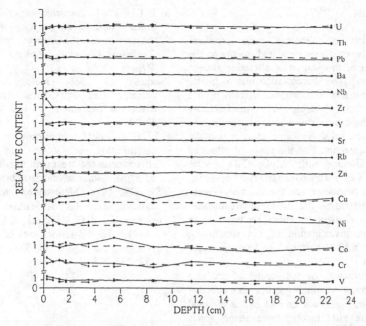

Figure 7. Comparison of trace element concentrations of Ortahisar (solid lines) and Ürgüp (dashed lines) samples.

6 PHYSICAL PROPERTIES OF WEATHERED TUFF

The effect of deterioration on the physical properties of tuffs is investigated by determining the variations of dry and saturated unit weight, effective porosity, and water absorption (by weight) through weathering zones. The core samples of 25 mm diameter are extracted from the block samples and the tests are performed at 4 different depths. The test results are given in Tables 5 and 6.

Table 5. Physical properties of Ortahisar samples.

Depth (cm)	0-3	4-7	8-11	20-23
Dry Unit Weight, kN/m³	13.5	13.6	13.5	13.6
Sat. Unit Weight, kN/m³	17.6	17.6	17.6	17.6
Effective Porosity, %	41.1	40.8	40.8	40.5
Water Absorption,%	20.9	20.6	20.7	20.8

Table 6. Physical properties of Ürgüp samples.

Depth (cm)	0-3	4-7	8-11	20-23
Dry Unit Weight, kN/m³	13.7	13.6	13.6	13.5
Sat. Unit Weight, kN/m³	17.6	17.6	17.6	17.5
Effective Porosity, %	39.9	39.9	39.9	40.1
Water Absorption,%	21.7	21.9	22.0	21.8

Based on the test results, no significant change on the physical properties from source of alteration to fresh part of the tuffs is observed. Comparison of the test results between the two blocks also yields similar findings. Therefore, the effect of deterioration on the physical properties of the tuffs is not significant. Although the samples are taken from the area where the weathering products are well preserved, it is also possible that the some portion of the weathered material may be removed from the surfaces of the fairy chimneys.

7 CONCLUSION

The deterioration of Göreme tuffs produces chemically weathered zones restricted mainly to the uppermost 2 cm of the source of alteration (lichen and joint surfaces). The discoloration due to biotite and rock fragment alteration extends laterally upto a depth of 17 cm. Feldspars are mechanically weathered and the depth of weathering is confined to upper 8 cm of the tuffs. Volcanic glass is partially altered to form smectite. For the study of effect of weathering of Göreme tuffs, MgO, CaO, P_2O_5, Fe_2O_3, and L.o.I as major elements, and Sc, V, Ni, and Zr as trace elements are the most significant parameters to determine the depth of weathering. However, major element variations are much more significant. Thus, investigation of the weathering phenomenon of tuffs should be based on the major element analyses.

An overall evaluation of the results of analyses reveals that chemical weathering is not very effective in the region. No significant change in the physical properties of the tuffs through weathering zones exists. Since upper 2 cm of the tuff surface shows signs on alteration, this depth should be considered during the conservation studies.

ACKNOWLEDGEMENT

This study is financially supported by METU Research Fund Project (AFP). The authors gratefully acknowledge Dr. Marjorie Wilson of Leeds University for XRF analyses and also to Dr. Jerf Asutay and his team of MTA for XRD analyses.

REFERENCES

Borchardt, G.A., Harward, M.E., & Knox,E.G.1971. Trace element concentration in amorphous clays of volcanic ash soils in Oregon.*Clays and Clay Minerals*. 19: 375-382.

Borchardt, G.A., & Harward, M.E. 1971. Trace element correlation of volcanic ash. *Soil Sci. Soc. Am. Proc*. 35: 626-630.

Erdoğan, M. 1986. *Nevşehir-Ürgüp yöresi tüflerinin malzeme jeolojisi açısından araştırılması*. Ph.D thesis, İTÜ Maden Fak.

I.S.R.M. 1981. *Rock characterization, testing and monitoring*. International Society For Rock Mechanics Suggested Methods. Oxford: Pegamon Press.

M.E.T.U. 1987. *Investigation of the mechanism of stone deterioration for the purpose of conservation of Göreme tuffs*. Middle East Technical University, Ankara, Turkey, Progress Report

Pasquare, G. 1968. Geology of the Cenozoic volcanic area of Central Anatolia. *Acedemia Nazionale Dei Lincei*, Roma. 9.Serie 8. Fasc 3: 57-201.

R.I.L.E.M. 1980.Recommended tests to measure the deterioration of stone and to assess the effectiveness of treatment methods. Commission 25-PEM, *Material and Structures*. 13: 175-253.

TSE 699. 1978. *Methods of testing for natural building stones*.Turkish Standards Institution:15 p.

Türkmenoğlu, A.G., Göktürk, E.H., & Caner, E.N. 1991. The deterioration of tuffs from the Cappadocia region of Turkey. *Archaeometry*. 33: 231-238.

Measuring the polishing aptitude of ornamental stones

Mesure de l'aptitude au polissage de roches ornementales

V. Poizat
Université de Liège, Laboratoires de Géologie Appliquée, Belgium

Ch. Schroeder
Université de Liège, Laboratoires de Géologie de l'Ingénieur, d'Hydrogéologie et de Prospection Géophysique, Belgium

ABSTRACT: The polishing aptitude of a rock can be quantitatively estimated by measuring its micro-roughness following a given intensity of polishing. Such a principle is taken into account in the notion of Polished Stone Value. This basic principle is applied to ornamental rocks. A new global parameter is defined, including a "local hardness contrast". The introduction of the grain size influence leads to offer a simple method to predict the polishing aptitude and visualize its evolution with compositional variations.

RESUME : L'aptitude au polissage d'une roche peut être estimée quantitativement par la mesure de sa microrugosité après un polissage donné. Une telle approche est reprise dans le Coefficient de Polissage Accéléré. Cette notion est appliquée aux roches ornementales. Un nouveau facteur global est défini, faisant intervenir un "contraste de dureté local". L'introduction de l'influence de la granularité conduit à présenter une méthode simple d'évaluation de l'aptitude d'une roche au polissage et de l'évolution de celle-ci en fonction de la variation de la composition minéralogique.

Most of the time, people involved in the ornamental stones industry carry out polishing tests on a few samples in order to appreciate whether a rock can be considered as a marble, commercially speaking.

This method is indeed fundamental, not only because it is easy to perform but also because it brings a very simple answer: "yes"... or "no".

Yet, it only gives punctual informations on a finite number of samples. Hence, it should not be applied to heterogeneous deposits.

This paper describes a complementary technique which does not lead to the same kind of answer but enables one to quantify the polish of a given rock and its variations within a heterogeneous deposit.

The last chapter gives a summary of the direct application of this technique on the marble deposits of Sinkat (eastern Sudan).

ROUGHNESS AND PSV

The polishing aptitude of a rock can be quantitatively estimated by measuring its micro-roughness following a given intensity of polishing.

Such a principle of an intrinsic micro-roughness is intituively taken into account in the notion of Polished Stone Value (PSV). As a matter of fact, this parameter is supposed to represent the material's micro-roughness through the bias of another parameter, i.e. the energy lost by a pendulum which sweeps over the rock's polished surface.

The rougher the rock, the higher the PSV.

There is a direct link between the APC and the mineralogical composition, as shown by Fourmaintraux and Tourenq:

$$PSV = a\,Dmp + b\,Cd + c$$

where Dmp is the mean hardness of the rock,
   Cd its hardness contrast,
   a, b, and c three coefficients.
In the general case, the coefficients' values are :
   a = 0.015; b = 0.07; c = 30.

$$Dmp = \sum_i C_i Dv_i \qquad Cd = \sum_i C_i |Dv_i - Dv_b|$$

where C_i is the proportion of mineral i,
   Dv_i its hardness,
   Dv_b the major phase's hardness.

The starting idea of this study was to correlate the PSV with an objective roughness measure made by some mechanical way in a bid to check the value of PSV as a roughness parameter.
Twelve samples were considered. Amongst them, two main types of rocks were present : calc-silicates bearing marbles (from Sudan) and various granitic rocks (from Norway).
Micro-roughness measurements were performed on a rugosimeter on 1.5 centimeter profiles. Altogether, a sharp mineralogical "mapping" of these profiles was performed by observation through a microscope.
The well known Ra parameter was considered as the most representative roughness parameter.

$$R_a = \frac{1}{L} \int_0^L y(x)dx$$

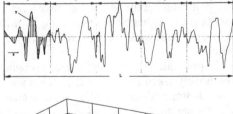

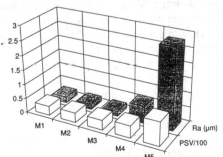

Fig.1: PSV/Ra correlation for calc-silicates bearing marbles.

Obviously (see Fig. 1), the PSV does not fit as a quantitative measure of micro-roughness. Indeed, it seems not sensitive enough, although the general trend does tally with the one shown by our measures.

ROUGHNESS AND PC

Another correlation was attempted between Ra and a new parameter named PC (Polishing Coefficient). The latter was supposed to follow the same sort of equation as the PSV, but with other coefficients a', b', and c' (of course, it does not represent the pendulum's loss of energy anymore).
These were to be determined by the classic orthogonalized matrix method, in which PC was considered as the theoritical value of Ra.
The coefficients' values are:
a' = $5.75 \ 10^{-4}$; b' = $3.608 \ 10^{-3}$; c'= $-5.43 \ 10^{-3}$.
The correlation according to these values is illustrated in Fig. 2.
There is an undeniable connection between the two parameters but some points of the curve are totally out of the trend.

The conclusion of this second attempt is that there are some other parameters which do affect micro-roughness, despite the essential influence of mineralogy.

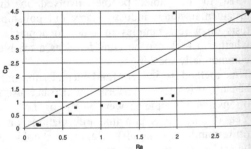

Fig.2: CP/Ra correlation.

INFLUENCE OF MINERALOGY

This influence is absolutely obvious on Abbott' diagrams as shown on Fig. 3. Abbott's diagram represent the distribution of altitude (cumulate percents on measured profiles).

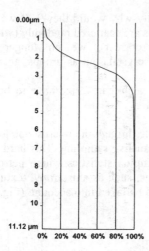

Fig.3: Abbott's diagram relative to green marble.

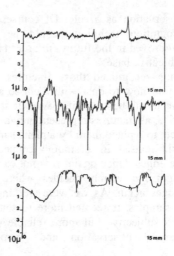

Fig. 4: Various profiles.

The given example is related to an epidote-diopside-calcite marble. In an ideal case, the profile -and the Abbott's curve- would appear as a kind of stairs with three different altitudes corresponding to the three different minerals.

As this is definitely not an ideal case, one can only notice the existence of three distinct slopes. It is therefore quite clear that mineralogical composition has a major influence upon roughness.

Abbott's diagrams should be dealt with some caution: they do show the heterogeneity of a given rock and allow us to guess the number of forming minerals in the best of cases. On the other hand, they give no possibilities of identifying them or quantifying their proportions.

At least two other parameters are susceptible to show a distinct influence upon micro-roughness: fractures and grain size.

FRACTURES

At first, when viewing the rocks' profiles (Fig. 4), one can easily understand that fractures, which appear as the steepest "valleys", can be of great influence.

Another study could be made in a near future in order to determine this influence and perhaps to establish the connection with IC (*Indice de continuité*)...

The IC (TOURENQ, C., DENIS, A., FOURMAINTRAUX, D., 1971: Propagation des ondes et discontinuités des roches. Symp. Soc. Int. Mec. Roches, Conn.I. 1., Nancy) can be determined through a simple measurement of the sonic velocity on a small sample. As one already knows the rock's mineralogical composition, one can easily calculate its theoretical sonic velocity. The IC is then:

$$IC = 100 \frac{V_{measured}}{V_{calculated}}$$

Nevertheless, that sort of work was beyond the scope of the current study and it was decided not to bother about fractures.

These fractures would then be simply "erased" from the profiles and the Ra parameters re-calculated. Unfortunately, it is rather difficult to be sure of what is a fracture and what is not...

Once again, after a short glance at the profiles, one can admit that fractures are the most contrasted areas, topographically speaking. The idea is then to virtually flatten the profiles by the following method: the altitude of each point would be replaced with the mean of the three altitudes of the given point and its direct neighbours.

It is noticeable that after three "flattenings", what has been observed as fractures under the microscope figures as the only steep "valleys" to remain. It seems sensible to consider the previous observation as a rule. Of course, this simple recognition technique could be and should be improved in the future, in order to get rid of its subjective base.

After having recognized those fractures, one still has to find how to deal with them so as to "erase" them in the best possible way. It seems sound practice, as fractures represent only a very small surface, to replace them by a plane surface which would consist in a smooth connection between the edges. Once again, it is not easy to find an algorithm which decides where the fracture's edges begin. As we were working on only a few samples, it seemed more reasonable to do it in a subjective but more reliable way: through careful observation and "common sense".

The new Ra parameters can then be calculated and fortunately, the difference between the old ones and the new ones ranges between 0 and 10%.

This means that the use of subjective principles such as "common sense" has little influence if any on the measured Ra.

GRAIN SIZE

As previously mentioned, the other parameter (and actually the one which appears the most interesting) is grain size.

Still, the definition of what is called "grain" has to be set clearly. In order to be able to propose the right definition, one has to imagine what the effect of grain (whatever it is) size is.

Let us consider a couple of two-phases rocks formed by 50% of quartz (hard) and 50% of calcite (soft). Rock #1 is made of 5 millimeter crystals alternating in a perfect manner, while rock #2 is made of 1 millimeter crystals, showing a perfect alternation as well.

Once again, our common sense tells us rock #2 is rougher than rock #1, as its profile looks more "comb-like". In fact, the value of the corresponding Cd should be higher (their Dmp of course remain the same).

The conclusion would be therefore that roughness increases as crystals size decreases. Unfortunately, crystals do not alternate perfectly

(and actually why would they?). If we do admit that crystals are scattered randomly (which is not absolutely correct), we no longer consider "grains" as crystals but as groups, aggregates of crystals.

Hence, the aggregates' size have to be sampled and studied.

It is possible through the microscope to establish a very exhaustive sampling. This hard work can lead to complex statistics where petrographical observations, such as anisotropy, texture, shape, etc., should be taken into account. (Fig. 5).

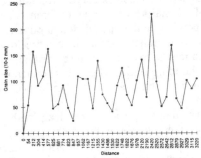

Fig. 5: Size distribution along profiles in various types of marble.

This kind of approach would be far too brain-racking and would bring hardly any improvement to a technique which aims to be mainly practical.

Inspite of the deep theoretical interest of petrographical descriptions, only the mean length of the grains (L_i) of each mineral i on the measured profile was retained.

PCg

In a view to justify one's feelings and ideas about the two theoretical profiles and their application to real ones, one should remember what roughness does really mean.

3540

Roughness is the alternation of altitudes at a microscopic scale. Intrinsic roughness would be, then, the remaining altitude alternation after a given intensity of polishing. One can admit that the harder a mineral is, the "less polished" it will be, and therefore the higher it will remain, compared with other minerals.

Roughness does not come from plane horizontal surfaces, even if they are not at the same altitude. It comes from the junctions between these surfaces. The smaller the grains are, the more junctions there are.

Beside the Dmp which is independent to such phenomenon, the Cd has to be altered so as to take grain size into account.

For each mineral i, the Cd_i which was baptised "local hardness contrast" corresponds to the equation:

$$Cd_i = C_i \left| Dv_i - Dmp \right| \frac{L}{L_i}$$

Two innovations are contained in the latter equation.

The first lies in the use of Dmp. The use of Dv_b makes sense if the rock is, for example, made of 95% of the same mineral, but it soon becomes unrealistic for more equal proportions of minerals. For instance, what would be its meaning for a rock made of 55% of mineral 1 and 45% of mineral 2?

Instead of holding the considered grain as surrounded by the main mineral in the rock, it sounds wiser to take account of every minerals and of their respective proportions by using Dmp instead of Dv_b.

The second innovation is the introduction of a L/L_i factor (L = profiling length = 15 millimeters; L_i = average grains size in millimeter).

This factor represents the influence of grain size. Multiplicating $1/L_i$ by L allows us to get rid of the scale factor.

The same numerical method was used in order to get the sharpest correlation between Ra and the new parameter PCg (Polishing Coefficient with Grain size taken into account).

The result is illustrated on Fig. 6 and follows the equation:

$$Ra \approx PCg = a'' \, Dmp + b'' \, Cd_g + c''$$

where $Cd_g = \sum_i Cd_i$

$a'' = 1.40 \; 10^{-3}; \quad b'' = 5.66 \; 10^{-5}; \quad c'' = -0.367.$

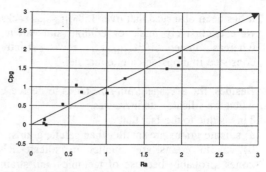

Fig. 6: Ra/PCg correlation.

Although the number of samples needs to be increased, it is obvious that the correlation is quite good. The gap between Ra and PCg is so thin that it might be due only to experimental errors.

At this stage, the microscopist is able to assess and quantify the polishing aptitude of a rock through the mere knowledge of its mineralogical composition and its granulometry.

The cut-off Ra still has to be chosen by the polisher, according to his experience and technical skills.

An interesting feature of the above formula is the possibility to visualise the evolution of the polishing aptitude with compositional variations. The Sudanese marbles case is particularly demonstrative.

MARBLE DEPOSITS OF SINKAT (SUDAN)

The geology of Sinkat is very complex and has had a great influence upon the area's marble deposits.

The marble's history can be briefly summarized as follow:

- chemical precipitation of limestone flecked with silica aggregates;
- metamorphism generating diopside;
- chemical reactions (skarn) with cross-cutting dykes with epidote and accessory garnet near the dykes.

The result is a calcite-diopside massive deposit (with various proportions of both minerals depending on deposition conditions) showing some epidote-rich zones scattered all over the deposit, near the dykes.

It has been observed that over 15%silicates rocks were exclusively made of epidote and calcite, whereas under 10%silicates rocks strictly consisted in diopside-calcite marble.

Besides, the average grain size tends to decrease when the silicate rate increases.
This is due to the fact that:
a) silicate grains are smaller than calcite grains;
b) calcite grains are smaller in silicate-rich zones, probably because of tectonics and strain concentrations.
The three roughness parameters' evolution is represented in Fig. 7.

At first sight, the three curves show a siimilar pattern. All the parameters grow towards a maximum value which is reached around the 50%silicates point.
It is obvious though that PSV is far more inert than the other two.

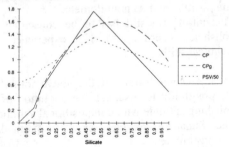

Fig. 7: Roughness parameters in the case of Sudanese marble.
Moreover, the grain size effect moves the maximum value from 50%silicates to about 60%silicates.
The sharp alteration of slope which affects the curves between the 10% and 15% of silicates is due to the change of the silicate's nature (petrological observation).
Because of the artificial and empirical nature of PCg, values under 5%silicates fall under zero. This has of course no physical meaning. The PCg should therefore be applied only to rocks with 5%silicates and over.

Such a diagram can be very useful to the professional who wants to estimate the polishing aptitude, not only of a rock but of a whole deposit. However, the polisher still has to assess his own technical skills so as to have an accurate idea of the maximum PCg he can reach. This can be achieved with a single testing session on a serie of samples (with various known PCg).

CONCLUSIONS

The above theory offers a simple method for quantifying a rock's polishing aptitude, providing an acute microscopic examination.
The true innovation is undoubtedly the introduction of the grain size influence.
Furthermore, an interesting feature is probably the link built between geology and the practical aspects of polishing.

This is definitely not an analytical study (which often leads to complex equations where physical meaning is ambiguous) but a simple method which finds its roots in common sense and experimental data.

ACKNOWLEDGMENT

This work was achieved within the scope of a thesis entitled "The marble deposits of Sinkat (Eastern Sudan)" and directed by professor F. Dimanche, Head of the Applied Geology Department of the University of Liège (Belgium).
It was made possible thanks to the very efficient support of Mr V. Brancaleoni, director of the Groupe des Carrières de Sprimont and partner of the belgo-sudanese association which extracts marble in Sinkat. We wish to thank Dr Imad Eldin Hassan and Mr Issam Hassan for their assistance.
We are also deeply indebted to Mr A. Magnée and G. Monfort, engineers at the Metalurgical Research Centre (CRM) of Liège, for their precious help.

Déformations de grès consécutives à leur consolidation avec un silicate d'éthyle

Sandstone deformation after its consolidation with an ethyl silicate

C. Félix
Ecole Polytechnique Fédérale, Lausanne, Suisse

ABSTRACT: When sandstones are consolidated with an ethyl silicate product, they are subject to intense shrinkage, particularly if they contain clay minerals. Shrinkage takes place during the consolidant hardening: it is not irreversible as water absorption is in fact sufficient to cancel its effects because of hydraulic swelling. Therefore the shrinkage phenomenon and its measurement cannot be ignored when tests are made to control the efficiency of sandstone consolidation with an ethyl silicate product. At the present time, the link between shrinkage and long-range consolidation is not well understood, so that the consolidation of some sandstone facies with such a product appears to be a chancy process.

RÉSUMÉ : Les grès consolidés avec le silicate d'éthyle subissent de fortes contractions volumiques se produisant pendant le processus même de durcissement du produit. Cette déformation, particulièrement importante dans les facies contenant des minéraux argileux, n'est toutefois pas irréversible: l'absorption d'eau l'annule en effet par gonflement hydraulique. Les essais destinés à évaluer l'efficacité d'un produit de traitement de consolidation doivent donc comporter des mesures de ces variations dimensionnelles caractéristiques des grès. Le processus de durcissement du silicate d'éthyle, et conséquemment de consolidation de la pierre traitée, restant par ailleurs mal connu, la réussite d'un traitement de ce type semble aujourd'hui relever surtout du hasard.

1 INTRODUCTION

L'éthique actuelle de la restauration des monuments anciens met l'accent sur la conservation des matériaux originels. Le but premier de la restauration étant de ralentir la dégradation des matériaux et non de remettre à neuf les monuments, des interventions techniques comme la consolidation, le nettoyage et les traitements superficiels (l'hydrofugation par exemple) des maçonneries en pierre naturelle revêtent dès lors une importance considérable.

L'évaluation de l'efficacité des produits, polymériques notamment, utilisés pour ces interventions n'obéit malheureusement à aucune méthodologie d'essai ayant, au niveau international, l'accord des spécialistes (Laurenzi-Tabasso 1993): la recherche d'une solution est donc laissée à la sagacité des expérimentateurs, à l'habileté des vendeurs, et à la chance.

Dans ce travail, on examine une situation réelle, celle de la restauration du château de Prangins (Vaud, Suisse), où la consolidation des grès avec un ester de silice a été définitivement abandonnée en cours de réalisation. Après l'apparition de fissures dans les zones consolidées, et seulement dans celles-ci, il était en effet difficile de ne pas invoquer une relation de cause (consolidation) à effet (fissuration), et cela malgré les résultats favorables fournis préalablement par des essais menés en laboratoire et in situ (point 4 ci-dessous). Les nouveaux essais réalisés pour essayer d'expliquer l'apparition de cette fissuration (point 6 ci-dessous), et voir si on pouvait y remédier, ont montré que la contraction importante des grès en voie de consolidation, c'est-à-dire leur diminution de volume pendant le processus même de leur consolidation, pouvait être à l'origine de la fissuration observée au voisinage immédiat des interfaces grès-mortier. Ce sont les principaux résultats des essais qui ont suivi l'apparition des fissures qui sont rapportés dans cette note.

2 LA SILICATISATION DES PIERRES

On désigne généralement par silicatisation d'une pierre naturelle son durcissement par imprégnation de silice amorphe (liant) provenant de la transformation de composés inorganiques ou organiques du silicium qui ont été absorbés dans ses pores et capillaires. Parmi les composés organo-siliciés, dont certains sont utilisés comme consolidants depuis le XIXe siècle déjà (Grissom and Weiss 1981), le silicate d'éthyle (ester éthyle de l'acide orthosilicique) [Si(OC$_2$H$_5$)$_4$] (alkoxysilane) est aujourd'hui le produit le plus utilisé pour la silicatisation des pierres naturelles.

Au contact de l'humidité de l'air, ce produit s'hydrolyse lentement pour former un gel de silice amorphe (polymère) et de l'éthanol qui s'évapore.

$$Si(OC_2H_5)_4 + 4\,H_2O \; \text{-->} \;\; 4\,C_2H_5OH + Si(OH)_4$$

$$Si(OH)_4 \; \text{-->} \;\; SiO_2 \;+\; 2\,H_2O$$

Ces réactions entraînent une importante réduction de volume lors du passage de la solution au gel. Plusieurs semaines de réaction sont nécessaires pour assurer le "séchage" complet du gel (polycondensation): de 2 à 4 semaines, si les conditions ambiantes idéales sont atteintes, soit entre 10 et 20° C de température et plus de 40 % d'humidité relative.

Le silicate d'éthyle que nous avons utilisé pour tous les essais dont il est question dans ce travail se présente dans le commerce sous forme d'un liquide prêt à l'emploi contenant 75 % en poids de tétraéthylsilicate et 25 % en poids de solvant (alcool). Très fluide (0.6 centipoise à 20° C), ce consolidant peut être appliqué sur la pierre par pulvérisation ou à la brosse, et possède un excellent pouvoir de pénétration par capillarité sur un support sec.

Si l'on en croit l'abondante littérature à son sujet, il est utilisé avec satisfaction tant pour la consolidation des grès que celle des calcaires. Cet optimisme doit cependant être tempéré par la remarque suivante, faite par L. Sattler et R. Snethlage (1988, p. 953), que nous croyons toujours d'actualité, au moins dans sa seconde partie: "In fact until recently the efficiency of stone treatments with SAE (silicic acid esters) was not controlled by an appropriate laboratory test. Additionally, the understanding of the consolidating processes on the façades was rather poor".

3 CONSOLIDATION DES GRÈS DU CHÂTEAU DE PRANGINS

Le siège romand du Musée national suisse devant être installé au château de Prangins (construit en grès local entre 1732 et 1739 sur un site moyenâgeux déjà fortifié), celui-ci fait actuellement l'objet de nombreux travaux de restauration et d'aménagement. Le grès tendre originel (grès molassique chattien, Molasse horizontale - désigné par Pr dans les tableaux et figures) est conservé autant que faire se peut, mais d'assez nombreux blocs fort altérés doivent être remplacés. Deux grès appartenant à la Molasse du Plateau suisse sont utilisés à cet effet: le grès dur de Schmerikon (Aquitanien de la Molasse plissée, extrémité est du Lac de Zurich - désigné par Schm dans les Tableaux et Figures) pour les éléments en saillie, et le grès de la Mercerie (grès dur marneux de l'Aquitanien lausannois, Molasse horizontale - désigné par Me dans le Tableau 1) pour les parties en élévation (fonds de mur). Les 3 facies gréseux ainsi associés ont des propriétés physiques et mécaniques différentes (Tableau 1). C'est pourquoi un traitement de consolidation, avec un silicate d'éthyle, a été envisagé pour l'ensemble des arénites (2500 m^2), afin de consolider les plus altérées et les plus faibles, et avec l'espoir de redonner ainsi à l'ensemble une certaine homogénéité.

Après des essais de laboratoire favorables (Félix 1992), une partie de l'aile est du château a donc été traitée, pour servir en quelque sorte de témoin. Malheureusement, 4 à 5 semaines après l'application du consolidant par un restaurateur, les architectes responsables du chantier ont noté l'apparition de nombreuses et très longues fissures dans le grès et dans le mortier de jointoiement consolidés, toujours au voisinage

Tableau 1. Principales propriétés physiques et mécaniques des grès feldspathiques à ciment carbonaté (10 à 20 %) et contenant des minéraux argileux gonflants (2 à 3 %) utilisés à Prangins (Pr: grès originel, Schm et Me: grès de substitution, voir le texte). GVr désigne un grès bigarré des Vosges (France), facies à Voltzia, contenant des smectites. GM 1 désigne le grès de Birkach (Allemagne), orthoquartzite contenant de la kaolinite (2 à 3 %). VD est un grès tendre témoin de la Molasse horizontale suisse (Villarlod) ayant la même composition minéralogique que le grès de Prangins Pr.

Caractéristiques	Unités	Lit	Pr	Schm	Me	GVr	GM 1	VD
Masse volumique apparente	kg/m^3		2380	2399	2375	2078	2036	2270
Masse volumique réelle	kg/m^3		2699	2658	2697	2651	2650	2670
Porosité totale	% volume		11.9	9.6	12.2	21.9	22.8	15.9
Absorption d'eau, p. atm.	% volume		8.3	6.6	8.2	14.2	13.2	11.7
Coefficient d'Hirschwald	%		70	69	68	65	58	74
Coefficients de capillarité	mg/(cm^2.min^1/2)	I-	36.3	5.8	14.5	73	61	42.8
		II	-	6.0	17.0	93	79	44.0
Coefficients de dilatation hydraulique	mm/m	I-	1.19	0.59	1.51	0.78	0.25	1.91
		II	-	0.69	1.47	0.43	0.27	2.03
Résistances Etat sec à la		I-	58.2	109.7	72.3	71.4	41.9	51.0
	N/mm^2	II	49.9	95.0	59.1	54.7	-	48.6
compression Etat saturé		I-	27.0	51.3	28.5	49.7	35.4	22.1
		II	21.6	54.6	23.4	40.3	-	20.3
Coefficients de ramollissement hydraulique	%	I-	46	47	40	70	84	43
		II	43	57	41	74	-	42
Vitesses du son, état sec	m /s	I-	1942	2998	2634	2185	3015	1667
		II	2235	3010	2840	2448	2983	1930

des interfaces de ces 2 matériaux traités, tandis que les zones non traitées restaient intactes de toute trace de fissuration (les murs traités et non traités étaient entièrement à l'abri des intempéries, notamment de la pluie: des protections en plastique, installées depuis plus d'une année sur les échafaudages, isolaient en effet les maçonneries des agressions de l'environnement extérieur).

4 ESSAIS AVANT LA CONSOLIDATION

La fonction essentielle d'un consolidant est de rendre à la partie altérée de la pierre des propriétés mécaniques aussi proches que possible de celles de la partie saine, et cela de la manière la plus homogène possible par la formation d'un nouveau liant. Un bon consolidant doit donc (Pouchol 1992):
- pénétrer assez profondément dans la pierre pour assurer une bonne liaison, sans aucune discontinuité majeure, entre la partie consolidée et la pierre restée saine;
- ne pas modifier de façon trop importante la porosité de la pierre et les propriétés physiques qui en découlent, comme la perméabilité à la vapeur d'eau;
- ne pas former de produits secondaires nuisibles, tels des sels;
- ne pas modifier l'aspect et la couleur du support.
    Des essais ont été réalisés en laboratoire:
- sur des pierres altérées provenant du château de Prangins, et dont une partie a été consolidée en laboratoire pour des essais;
- sur des pierres saines, traitées et non traitées, provenant de carrières;
- sur des carottes prélevées dans les maçonneries du château après traitement de consolidation in situ.
    En tenant compte de la diversité des facies des grès de la maçonnerie, ces essais ont montré que (Félix 1992):

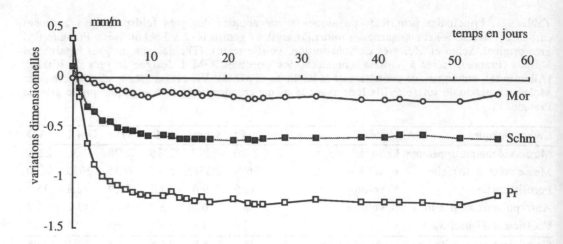

Fig. 1 Déformations linéaires de 2 des 3 grès du château de Prangins et du mortier de jointoiement (Mor) pendant l'absorption du silicate d'éthyle (dilatation au temps 0) et sa période de durcissement (contraction)

- le traitement avec le silicate d'éthyle ne modifiait pas de façon sensible la couleur de la pierre sur de grandes surfaces;
- suivant l'état de dégradation de la pierre, la profondeur de pénétration du produit était de l'ordre de 2 à 4 cm pour une absorption de consolidant d'environ 3.0 kg/m² (contrôles des profondeurs de pénétration par des mesures de vitesses du son et des essais de coloration);
- la consolidation était vraiment efficace entre 1 et 3 cm de profondeur, rendant alors au grès une cohésion comparable mécaniquement à ce qu'elle était à l'origine (contrôles par des essais de flexion biaxiale sur des disques -diamètre: 12 cm, épaisseur: 1 cm- provenant des carottages effectués après traitement in situ du grès);
- le traitement diminuait de façon importante (environ 3 fois) l'absorption d'eau par le grès (contrôles par des essais à la pipe de Karsten), mais sans en changer le régime de séchage (contrôles par des essais d'évaporation).

C'est sur la base de ces résultats d'essais favorables que la consolidation fut donc décidée pour les 2500 m² de façades, la quantité de consolidant à utiliser ne devant pas dépasser 3.0 kg/m² et ne pas être inférieure à 2.5 kg/m².

C'est après que 200 m² de grès divers eurent été traités par le restaurateur que la consolidation fut stoppée en raison de l'apparition de fissures ... qui n'avaient été ni prévues, ni détectées pendant la longue phase des essais préliminaires !

5 VARIATIONS DIMENSIONNELLES DES GRÈS TRAITÉS AVEC LE SILICATE

Les variations volumétriques des grès ont des origines diverses. L'absorption et l'adsorption d'eau, de même que les variations thermiques (en présence ou non d'eau) en sont les causes les mieux connues; elles constituent même un facteur important d'altération menant à des morphologies symptomatiques, notamment caractérisées par des fissurations (Delgado 1991; Félix 1991; Hamès et al. 1987; Hudec 1980).

Le retrait volumique dû à la polymérisation de résines injectées dans les pierres est capable de produire des effets comparables, engendrant dans celles-ci des contraintes qui sont directement proportionnelles aux déformations linéaires (Tran et Descalle 1985). Sachant que le durcissement par hydrolyse du silicate d'éthyle s'accompagne d'une diminution importante du volume de la matière solide, il nous a paru indispensable de connaître les variations dimensionnelles des pierres traitées avec ce type de consolidant (Figures 1 et 4).

6 MESURES EXPÉRIMENTALES

Les mesures ont porté sur différents types de grès, dont ceux utilisés au château de Prangins (Tableau 1), et sur le mortier de jointoiement des

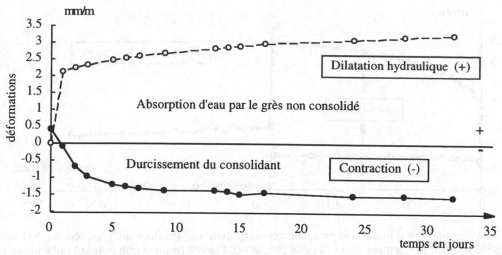

Fig. 2 Comparaison des déformations d'un grès tendre témoin (VD), a) pendant l'absorption d'eau - grès non consolidé (dilatation hydraulique), b) pendant le durcissement du consolidant absorbé par le grès (contraction)

maçonneries du château (composition du mortier: 12 parts de sable sec 2 mm, 1 part de ciment HTS, 0.25 part de ciment blanc Lafarge, 1.5 part de chaux hydraulique, 2 parts de chaux hydratée; module d'élasticité statique: 1731 N/mm^2, soit 1/3 à 1/6 du module des grès utilisés dans la maçonnerie; contrainte moyenne à la rupture: 3.6 N/mm^2).

Pour chaque faciès, 6 prismes de 1.5 x 1.5 x 10.0 cm, séchés en étuve à 60° C jusqu'à poids constant, ont été imprégnés par capillarité (pendant environ 4 heures) jusqu'à saturation de consolidant liquide (en volume, cette saturation correspond approximativement au coefficient d'Hirschwald (Tableau 1); après durcissement, le gel de silice représente généralement moins de 10 % du volume de la porosité totale). Les déformations linéaires provoquées par l'absorption du silicate d'éthyle immédiatement après l'imprégnation, et celles engendrées pendant tout le temps nécessaire au durcissement complet de ce consolidant ont été mesurées suivant le grand axe des prismes, perpendiculairement au litage sédimentaire, sur des bâtis identiques à ceux utilisés pour la détermination du coefficient de dilatation hydraulique isotherme par absorption d'eau (Félix 1991). Après consolidation des pierres, on a également mesuré les déformations linéaires provoquées par l'absorption d'eau isotherme (dilatation hydraulique) et par le séchage par

évaporation des éprouvettes traitées ainsi saturées en eau (Figure 4). Toutes les mesures ont été effectuées à 21 ± 1° C, dans un local où l'humidité relative pouvait varier entre 50 et 60 %.

Quels que soient l'intensité des déformations enregistrées et les processus en cause, on peut constater 2 phénomènes principaux (Figures 1, 2 et 4):

1. un gonflement de la pierre pendant la phase d'absorption, par capillarité, du produit liquide consolidant ; le mortier et le grès quartzite ("orthoquartzite") sont toutefois pratiquement insensibles, de ce point de vue, à l'absorption de consolidant;

2. une contraction de la pierre lors du durcissement du produit (une fois la contraction maximale atteinte, les variations dimensionnelles enregistrées pendant la phase de durcissement sont uniquement fonction des fluctuations du taux d'humidité relative dans le local de mesure - adsorption d'eau).

7 COMPARAISON DES DÉFORMATIONS DUES À LA CONSOLIDATION ET À L'ABSORPTION D'EAU

Notre but n'est pas, présentement, d'étudier les causes et les processus menant aux contractions

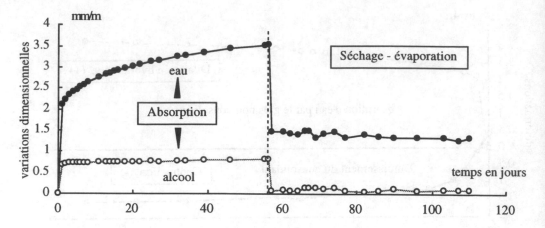

Fig. 3 Grès non consolidé. Comparaison des déformations linéaires du grès témoin VD non consolidé pendant l'absorption d'eau (liquide polaire) et d'alcool (liquide non polaire) (dilatation), et pendant le séchage par évaporation (contraction). A noter, en particulier, les déformations résiduelles

des grès traités avec le silicate d'éthyle. Une comparaison s'impose toutefois avec les effets gonflants bien connus produits sur les mêmes facies par l'absorption, et l'adsorption, d'eau isotherme (Félix 1983; Hamès et al. 1987; Hudec 1980).

On sait en effet que la dilatation hydraulique des pierres dépend de leurs structures, de la nature du liquide absorbé (Figure 3), et de leur minéralogie, et plus particulièrement encore de la présence et de la nature des minéraux argileux (Figures 1 et 2). La lecture des graphes des Figures 1 à 4 montre que:

Tableau 2. Récapitulatif des déformations des grès exprimées en mm/m.

	VD	Pr	Schm	GVr	GM 1
λ	2.00	1.19	0.64	1.00	0.26
δ	2.00	1.65	0.70	1.30	0.40
μ	-	2.40	0.88	1.00	0.42

- les grès non consolidés, réputés sensibles au gonflement hydraulique, se contractent fortement lors du durcissement du silicate d'éthyle (Figure 1): les valeurs des contractions dues à la consolidation δ, exprimées en mm/m, sont d'ailleurs proches, et même généralement supérieures aux valeurs des coefficients de

dilatation hydraulique des pierres non consolidées λ (Tableau 1);
- le traitement consolidant avec le silicate d'éthyle n'empêche pas les importantes dilatations provoquées par l'absorption d'eau (dilatations hydrauliques μ), celles-ci étant au mieux du même ordre de grandeur pour les pierres traitées et non traitées (Tableau 1, Figure 4): la contraction volumique due à la consolidation n'est donc pas irréversible;
- lors du premier séchage après le traitement de consolidation (Figure 4), la déformation résiduelle est proche de zéro, sauf dans le cas de GVr où elle reste négative et relativement importante (± 0.5 mm/m). Nous ignorons totalement de quelle manière la contraction due à la consolidation est en relation avec le gain de résistance mécanique dû à l'effet du consolidant: des essais ultérieurs devraient montrer si l'action de l'eau sur les pierres traitées, au moins celles contenant des minéraux argileux gonflants, n'annihile pas rapidement ce gain de résistance mécanique.

Les constatations faites à propos des grès sont également valables pour les calcaires (on a examiné le cas de quelques calcaires biodétritiques et oolithiques poreux), bien que ceux-ci soient beaucoup moins sensibles que les grès aux déformations hydrauliques et à celles dues au processus de consolidation avec le silicate d'éthyle. Ceci n'est sans doute qu'une conséquence révélatrice du rôle (délétère) joué

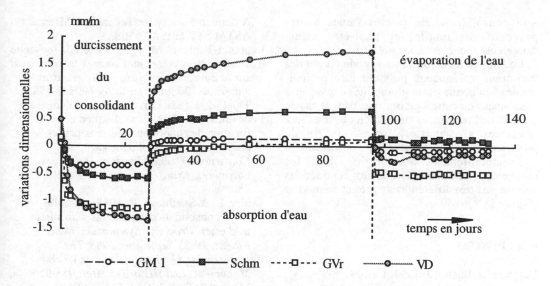

Fig. 4 Grès consolidés. Comparaison des déformations linéaires de grès de facies différents (voir le Tableau 1), a) pendant l'absorption (dilatation au temps 0) et le durcissement du consolidant (contraction), b) pendant l'absorption d'eau par les mêmes pierres consolidées (dilatation hydraulique), c) pendant leur séchage par évaporation (contraction)

par les minéraux argileux (gonflants ?) tant dansle processus de dilatation hydraulique que de contraction par durcissement du silicate d'éthyle.

8 CONCLUSIONS

Les faits sont clairs.

Lorsque des grès, et même des mortiers et des calcaires, sont traités avec le silicate d'éthyle, il y a d'abord dilatation du matériau pendant l'absorption du produit liquide, puis contraction volumique pendant la phase de durcissement du consolidant. La contraction est importante dans les facies gréseux contenant des minéraux argileux (ou certains minéraux argileux ?), pouvant atteindre dans les cas étudiés 2 mm/m, mais elle n'est pas irréversible: un cycle d'absorption d'eau suffit en effet pour l'annihiler rapidement et complètement, même si pendant le séchage certaines pierres conservent une déformation résiduelle non négligeable.

La consolidation n'empêche donc pas la dilatation hydraulique, qui est même quantitativement au moins aussi forte après qu'avant le traitement. La dilatation hydraulique étant un facteur important de détérioration des

grès, cette constatation ne peut qu'amener à se poser des questions sur l'efficacité réelle d'une consolidation de grès avec un silicate d'éthyle.

La contraction volumique étant un phénomène important accompagnant le durcissement du silicate d'éthyle (bien qu'on n'en connaisse pas le rôle dans le processus de consolidation), il n'est pas logique de l'ignorer, donc de ne pas la mesurer, notamment dans les essais destinés à juger de la validité d'un traitement avec ce type de produit.

Les mesures faites dans le cadre de cette étude posent plus de questions qu'elles n'apportent de réponses ou de certitudes, la question la plus importante étant: Quel type de lien le gel de silice établit-il dans la pierre consolidée? en particulier, s'il y a des minéraux argileux interstitiels ou matriciels, ce qui est le cas pour nombre de roches détritiques ou en voie d'altération? Avec, comme corollaire immédiat: Comment mesurer efficacement l'effet consolidant?

Les réponses aux questions précédentes n'existant pas encore, sur le plan pratique, cela implique que:
- l'utilisation d'un silicate d'éthyle comme consolidant sans protection hydrofuge est sans doute un leurre pour de nombreux grès;

- la consolidation de pierres saines à titre préventif est inutile, et peut-être même dangereuse, pour les facies gréseux.

Le fait que les déformations produites par des variations volumiques puissent être préjudiciables à la bonne tenue (durabilité) d'une pierre nous amène en outre à penser que, dans le cas du château de Prangins, la fissuration observée aux interfaces grès-mortier, juste à la fin du processus de durcissement du consolidant, a vraisemblablement son origine dans les mouvements différentiels dus aux contractions volumiques très différentes du grès et du mortier (Figure 1).

RÉFÉRENCES

Delgado Rodrigues, J. 1991. Causes, mechanisms and measurement of damage in stone monuments. *Proceedings European Symposium, Bologna, Italy, 13-16 June 1989, Science, Technology and European Cultural Heritage*: 124-137. Baer, N.S., Sabbioni, C. & Sors, A.I. (eds). London: Butterworth-Heinemann Ltd.

Félix, C. 1983. Sandstone linear swelling due to isothermal water sorption. *Proc. First Int. Conference Esslingen, 6-8 September 1983, Technische Akademie Esslingen, Germany, "Materials Science and Restoration"*: 305-310. Ed. Lack + Chemie.

Félix, C. 1991. Essais physiques et critères de sélection pour la restauration des grès. *Mines et carrières, les techniques* 73: 87-90.

Félix, C. 1992. Consolidation des grès du château de Prangins avec l'ester de silice Wacker OH. Résultats des essais et mesures en laboratoire et in situ. EPF-Lausanne, Rapport 21/92/LCP, 420 pages.

Grissom, C.A. & Weiss, N.R. 1981. Alkoxysilanes in the conservation of art and architecture : 1861-1981. *Art and archaeology technical abstracts* 18: 150-198.

Hamès, V., Lautridou, J.-P., Ozer, A. et Pissart, A. 1987. Variations dilatométriques de roches soumises à des cycles "humidification-séchage". *Géographie physique et Quaternaire* XLI: 345-354.

Hudec, P.P. 1980. Durability of carbonate rocks as function of their thermal expansion, water sorption, and mineralogy. In P.J. Sereda & G.G. Litvan (eds), *Durability of building materials and components*,

American Society for Testing and Materials, ASTM STP 691: 497-508.

Laurenzi-Tabasso, M. 1993. Materials for stone conservation. *Actes du Congrès International sur la conservation de la pierre et autres matériaux, 29 juin-1er juillet 1993*: 54-58. Thiel M.-J. (éd). Paris: Unesco- Rilem.

Pouchol, J.-M. 1992. Consolidation et hydrofugation des matériaux pierreux par les résines silicones. *Colloque européen, Construction, réhabilitation, Apport des polymères, Lyon, 8-10 septembre 1992*: 65-78.

Sattler, L. & Snethlage, R. 1988. Durability of stone consolidation treatments with silicic acid ester. *Proc. Int. Symposium IAEG, Athens, 19-23 September 1988, The Engineering Geology of Ancient Works, Monuments and Historical Sites*: 953-956. Marinos P.G. & Koukis G.C. (eds). Rotterdam: Balkema.

Tran, Q.K. et Descalle, P. 1985. Etude des variations dimensionnelles de pierres calcaires imprégnées en fonction des propriétés élastiques des résines radiodurcissables utilisées. *Actes Ve congrès international sur l'altération et la conservation de la pierre*: 809-815. C. Félix (éd). Lausanne : Presses Polytechniques Romandes.

La méthode des éléments distincts (DEM): Un outil pour évaluer les risques d'instabilités des monuments en maçonnerie – Applications à des cas égyptiens

The distinct element method (DEM): A tool for the evaluation of risk instability of blocky stone monuments – Application to Egyptian monuments

T. Verdel
Laboratoire de Mécanique des Terrains, Ecole des Mines de Nancy, Ineris, France

ABSTRACT : To identify and to study some of the phenomena which lead to the degradation of historical monuments, the use of numerical modelling, as usually carried out in Geotechnic, can be very useful. But, seeing that ancient buildings or monuments are much different from new constructions because of their blocky (stone) structure, only suitable numerical methods may be used. Unlike the Finite Element Method which may be unsuitable for the analysis of such masonry structures, the Distinct Element Method allows the representation of hundreds joints and blocks and is able to simulate their non-linear behaviour very easily. This is why this method was used to model some famous egyptian monuments. It specially helped us to understand their present stability conditions or to determine their possible future global mechanical behaviour.

RESUME : Pour identifier et étudier certains des phénomènes qui conduisent à la dégradation des monuments, l'utilisation des modèles numériques, tels qu'ils sont employés en Géotechnique, peut s'avérer très utile. Mais, les monuments historiques étant des constructions très différentes des édifices actuels en raison de leur structure généralement composée de blocs, seules des méthodes numériques adaptées peuvent être utilisées. Contrairement à la méthode des éléments finis qui est peu destinée à la modélisation de ce genre de structure, la méthode des éléments distincts permet de représenter des centaines de joints ou de blocs et elle est capable de simuler leur comportement non-linéaire très facilement. C'est pourquoi cette méthode a été utilisée pour modéliser quelques monuments égyptiens célèbres. Elle nous a spécialement aidé à comprendre les conditions de stabilité actuelles de ces monuments ainsi qu'à déterminer leur possible évolution mécanique globale.

1 INTRODUCTION

Pour faire face aux nombreuses et complexes menaces qui pèsent sur la conservation des monuments, il est nécessaire de mobiliser "toutes les sciences et toutes les techniques qui peuvent contribuer à l'étude et à la sauvegarde du patrimoine monumental" comme le rappelle la Charte de Venise dans son article 2 (ICOMOS, 1966).

Avec le souci de s'intéresser non seulement aux effets du mal mais surtout à ses causes, la géotechnique est une discipline qui peut permettre de mieux comprendre certains phénomènes de dégradation, grâce en particulier aux outils de modélisation dont elle dispose. De part son objet d'étude habituel, la géotechnique peut ainsi intervenir sur trois catégories de monuments :

1. Le monument délimité par son environnement naturel et notamment les tombes et autres excavations souterraines, dont la stabilité dépend du comportement mécanique du milieu, souvent rocheux, au sein duquel l'ouvrage historique est situé.

2. L'édifice monumental en rapport avec le milieu naturel, qu'il soit fondé sur un sol déformable, situé au pied ou au sommet d'un talus, ou d'une falaise éventuellement rocheuse et fracturée. Dans ce cas, la stabilité du monument sera étroitement liée à celle de son environnement.

3. L'édifice monumental en lui-même, dont la bonne conservation dépend moins de son rapport avec le milieu naturel que de sa propre structure, souvent complexe, et en général constituée d'un très grand nombre de blocs de pierre.

En raison de la spécificité structurale des monuments historiques qui fait d'eux des milieux généralement discontinus (qu'ils soient creusés dans des mileux rocheux fracturés ou constitués d'un assemblage de blocs de pierre en maçonnerie), leur modélisation requiert l'usage d'une méthode particulièrement adaptée à la représentation des discontinuités. C'est pourquoi nous avons fait appel à la méthode des éléments distincts dont le fonctionnement nous a paru tout à fait adapté à ce problème.

2 LA METHODE DES ELEMENTS DISTINCTS

La Méthode des Eléments Distincts (MED), proposée par Cundall (1971) a été développée dans le but d'étudier le comportement mécanique des milieux discontinus soumis à un chargement statique ou dynamique. Le milieu étudié est représenté par un assemblage de blocs (distincts), rigides ou déformables, qui interagissent par leurs côtés et leurs sommets (via l'introduction de raideurs normales et tangentielles). Les discontinuités (aussi appelées "joints") sont alors interprétées comme les limites d'interaction entre ces blocs. Si les blocs sont choisis déformables, ils sont discrétisés en éléments continus comme dans les méthodes plus classiques.

D'autre part, la MED repose sur une résolution de type explicite où les équations de mouvement sont directement résolues pour chaque point nodal (noeuds du maillage ou centre de gravité des blocs s'ils sont rigides) en fonction de la variable "temps". Ainsi, elle permet aisément de modéliser les grands déplacements et les grandes rotations et permet d'introduire facilement des lois de comportement non-linéaires pour les blocs (déformables) comme pour les discontinuités. De plus, cette méthode autorise le détachement complet d'un bloc et permet d'identifier les nouveaux contacts susceptibles de s'établir entre blocs, au fur et à mesure du déroulement du calcul.

Les figures 1 et 2 illustrent succinctement le principe de fonctionnement de la MED

3 LA MODELISATION FACE AUX SPECIFICITES DES MONUMENTS

Les monuments historiques présentent malheureusement des spécificités qui s'accommodent mal d'une modélisation. L'âge, la géométrie et la valeur des monuments en sont particulièrement responsables.

En effet, dès lors que l'on connaît mal l'histoire du monument étudié, on connait mal, a fortiori son histoire mécanique (y-a-t-il eu des effondrements, des reconstructions, des ensevelissements puis des fouilles archéologiques ?), ce qui interdit, a priori, toute possibilité de calage des calculs.

De même, la géométrie d'un monument historique

Figure 1. Principe de l'interaction entre blocs dans la Méthode des Eléments Distincts.

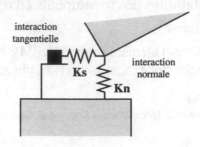

Figure 2. Principe de la résolution explicite par cycles dans la Méthode des Eléments Distincts.

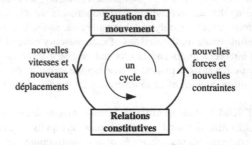

est souvent bien plus complexe que celle d'un édifice moderne, ce qui rend difficile une représentation géométrique fidèle sur l'écran d'un ordinateur.

Enfin, la valeur des monuments interdit souvent qu'on y prélève des échantillons, qu'on y installe des instruments de mesure pour des raisons de conservation ou simplement d'esthétique, ce qui nous oblige, pour la modélisation, à faire de nombreuses hypothèses et interdit également tout calage des calculs (Verdel, 1993).

Disposant de peu de données ou d'informations peu précises, l'étude des monuments historiques par la modélisation numérique va donc entrer dans la catégorie des problèmes à données limitées.

Dans ce cadre, et selon la méthodologie développée par Starfield et Cundall (1988) pour cette catégorie de problème, la modélisation géotechnique, et en particulier la modélisation numérique, pourra être utilisée comme un outil, certes de calcul, mais surtout comme un outil de simplification et de représentation faisant l'interface entre l'objet étudié et le scientifique, de façon à rendre mieux perceptible la réalité mécanique de l'objet. Elle sera aussi utilisée comme un outil d'investigation permettant de simuler des phénomènes, des sollicitations, des états géométriques, de façon à vérifier des hypothèses ou à en

Figure 3. Plan de la tombe des babouins (avec relevé des fissures), exemple de monument délimité par son environnement naturel.

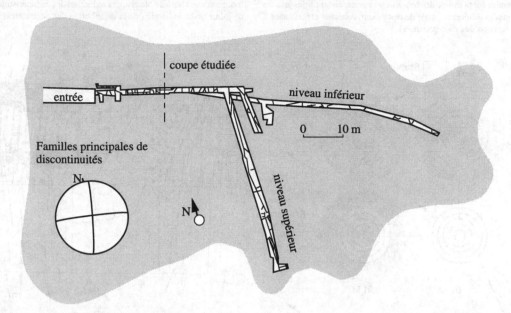

suciter de nouvelles. Son but sera donc essentiellement de compléter ou de rendre objective l'interprétation intuitive et synthétique que l'on peut faire d'un ensemble d'observations, le plus souvent visuelles.

4 EXEMPLES EGYPTIENS

Dans le cadre d'une collaboration que notre laboratoire mène depuis 1985 avec la faculté d'ingénieurs de l'Université du Caire, sur l'étude et la protection du patrimoine égyptien, nous avons été amené à modéliser un certain nombre de monuments d'importance significative dont l'état de conservation méritait une investigation scientifique approfondie. Nous en avons retenu ici trois exemples pour illustrer l'apport des modèles numériques à l'étude géotechnique des monuments historiques.

Le premier est la tombe des Babouins de Saqqarah, située dans l'ancienne nécropole de Memphis à 20 km au sud du Caire. Il s'agit d'une tombe excavée dans le désert libyen, donc d'un ouvrage délimité par son milieu naturel (rocheux et fracturé en l'occurence).

Pour la catégorie des édifices en rapport avec leur milieu naturel, nous avons retenu l'exemple des deux colosses de Memnon, imposantes statues d'environ 800 tonnes chacune, dressées sur la vallée limoneuse, à l'ouest de Louxor, en Haute-Egypte.

Enfin, le troisième exemple est celui du IXe Pylône du temple de Karnak près de Louxor. C'est un édifice à l'origine constitué d'environ 230 000 blocs qui représentaient une masse globale d'environ 30 000 tonnes. Ce pylône, dont la stabilité dépend essentiellement de l'équilibre de ses milliers de blocs, entre dans la troisième catégorie des ouvrages historiques, à l'étude desquels, la géotechnique peut appliquer ses méthodes de modélisation numérique.

Chacun de ces ouvrages présente des instabilités observées visuellement dont nous avons cherché à identifier l'origine et à expliquer les mécanismes mais, pour chacun d'eux, la géométrie et la distribution des matériaux constituent les seules données fiables dont nous disposons. Les caractéristiques mécaniques des matériaux ne sont généralement pas connues et il n'existe aucun dispositif de surveillance permettant de connaître l'évolution mécanique récente de la structure de ces ouvrages. Dépourvu de données fiables et de moyens de calage, on cherchera donc essentiellement, grâce à la modélisation numérique, à valider des phénomènes pressentis, à en évaluer les effets et les conséquences possibles.

5 CAS DE LA TOMBE DES BABOUINS

Excavé à faible profondeur (toit à environ 2 m de profondeur) dans des roches marno-calcaires fracturées en blocs, la tombe des babouins est constituée de

Figure 4. Ensemble des résultats des différentes simulations menées sur la tombe des babouins et correspondant à des jeux de données différents (angle de frottement, cohésion et résistance à la traction des discontinuités).

Figure 5 : L'un des modèles réalisés pour la tombe des babouins montrant une chute de blocs après réduction des caractéristiques de joints (avec indication de la distribution des contraintes).

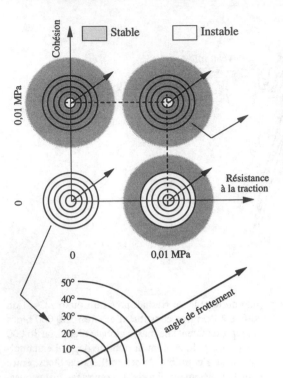

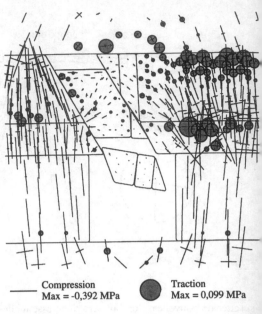

2 galeries d'environ 100 m de longueur en partie superposées (figure 3). Des chutes de blocs récentes y ont été observées et certains endroits de la tombe menacent ruine. Une première modélisation numérique menée en éléments finis nous a amenés à formuler des réserves (compte tenu du facteur correctif introduit à cause de la fracturation naturelle) quant à la stabilité de cette tombe, mais c'est par l'utilisation de la MED que nous avons pu (en tenant compte des discontinuités existantes) mieux comprendre les mécanismes d'instabilités observés. Comme le montre la figure 4, la stabilité de la tombe est fortement liée aux propriétés mécaniques des discontinuités et la figure 5 montre l'un des résultats obtenus avec le code UDEC (Itasca, 1991) illustrant ce propos. Il semble en effet que, selon les valeurs retenues pour la cohésion, l'angle de frottement et la résistance à la traction des discontinuités, une chute de blocs au toit de la galerie soit possible. De telles instabilités ayant été effectivement observées dans la tombe, on peut donc s'attendre à ce que ces paramètres aient des valeurs faibles (pour lesquelles le calcul prédit effectivement une

chute de blocs). Malgré la simplicité des modèles, il est possible de conclure dans ce cas que les chutes de blocs ne trouvent pas leur origine dans la charge appliquée par les terrains sus-jacents comme c'est souvent le cas pour les excavations à grande profondeur, mais dans la réduction des caractéristiques mécaniques des discontinuités au cours du temps. Cette réduction peut notamment être expliquée par des changements de conditions environnementales depuis l'ouverture de la tombe en 1965.

6 CAS DES COLOSSES DE MEMNON

Les colosses de Memnon sont deux grandes statues d'environ 15 m de hauteur (au dessus du sol) et pesant environ 800 tonnes chacune. Elles sont situées en Haute Egypte, près de Louxor et reposent dans la vallée, sur une couche de limon (figure 6). Ce sont les seuls colosses de l'Antiquité égyptienne (ils datent de 1500 ans avant J.-C. environ) que l'on ait retrouvé debout. Ils étaient constitués, à l'origine, de deux blocs : l'un pour le socle et l'autre pour la statue. L'une d'entre elles s'est effondrée en 27 avant J.-C. à la suite d'un tremblement de terre et a été reconstruite en petits blocs un siècle après.

Figure 6. Colosses de Memnon (fissures du socle et inclinaison), exemple d'un monument dont la stabilité est étroitement liée à son environnement.

Figure 7. Modèle tridimensionnel des colosses de Memnon en éléments finis (code CESAR et modèle continu élastique).

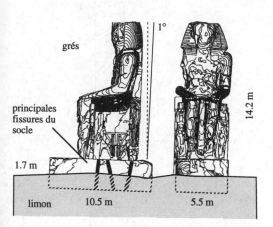

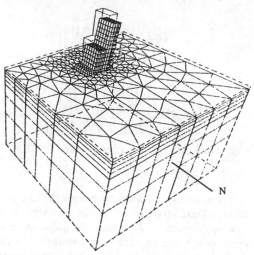

Les deux colosses présentent actuellement des fissures similaires dans leur socle ainsi qu'une inclinaison vers l'arrière d'environ 1 degré, nettement observable à l'oeil nu (figure 6).

Les facteurs présumés responsables de cet état sont : la flexion du socle sous la très forte charge que représente la statue, l'incapacité portante du sol, le rôle passé des crues du Nil et le rôle des éventuels séismes. Ce qui place ce cas dans la catégorie des monuments dont la stabilité dépend essentiellement de ses interactions avec l'environnement naturel.

Pour évaluer l'importance de chacun de ces facteurs, nous avons donc mobilisé et couplé plusieurs méthodes de modélisation afin de mieux découpler les phénomènes dont ils sont à l'origine.

Dans une première étape, nous avons fait appel au modèle analytique de la poutre simple, à laquelle nous avons assimilé le socle des colosses. Cela nous a permis de prévoir une flexion possible du socle sous l'effet de la charge imposée par la masse énorme de chacune des statues. Les tractions induites à la base du socle (sous l'effet de la flexion) présentent un facteur de sécurité minimal de 2 vis à vis de la rupture théorique du grès dont il est constitué. Compte tenu des défauts structuraux qui peuvent affecter la matrice rocheuse de la pierre, un tel facteur ne constitue pas une sécurité suffisante et l'on peut donc expliquer la fissuration du socle, effectivement localisée aux endroits les plus sollicités.

Selon une autre approche analytique, la capacité portante du sol calculée par les formules de Caquot et

Kerisel (1966) présente un facteur de sécurité inférieur à 2 vis à vis de la charge moyenne appliquée sur le sol par chacun des colosses, notamment en cas de crue du Nil. Ce résultat nous a permis d'expliquer, en raison du faible facteur de sécurité, l'existence d'un tassement différentiel important (donc d'un basculement important) induit par l'apparition de déformations plastiques dans le sol.

Dans un troisième temps, nous avons réalisé un modèle tridimensionnel continu et entièrement élastique, en éléments finis, de l'un des colosses (code CESAR, LCPC, 1990). Nous avons pour cela utilisé des caractéristiques mécaniques plausibles pour chacun des matériaux en présence. Ce modèle nous a permis de confirmer l'existence de fortes tractions à la base du socle ainsi qu'une flexion du socle et un léger basculement vers l'arrière (figure 7).

Avec des modèles bidimensionnels, plus souples d'emploi et riches de nombreuses fonctions, nous avons poursuivi une investigation encore sommaire. Ainsi, en utilisant la méthode des différences finies (pour un modèle toujours continu) avec le code FLAC (Itasca, 1991), nous avons montré que les déformations plastiques, susceptibles de se produire sous l'effet des sollicitations importantes, semblent en partie responsables du basculement vers l'arrière des colosses. De même, la simulation d'une crue généralise l'entrée en plasticité théorique du sol (que les contraintes effectives gouvernent dans ce cas) et conduit à une augmentation du tassement différentiel, donc du basculement (voir figure 10).

Figure 8. Contraintes dans le socle après modélisation bidimensionnelle discontinue tenant compte des fissures du socle (code UDEC)

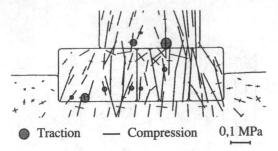

● Traction — Compression 0,1 MPa

Figure 9. Déplacements induits par la sollicitation dynamique, à l'équilibre, 10 secondes après la fin de la sollicitation (déplacement maximal = 20 cm).

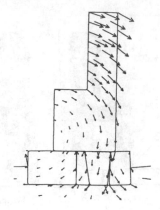

Enfin, avec les modèles discontinus (MED avec le code UDEC, Itasca, 1991), le réalisme de la représentation a pu être amélioré par la prise en considération des interfaces "socle-statue" et "socle-sol" ainsi que des discontinuités effectivement observées dans le socle. Après calcul, la distribution des contraintes est changée et certains blocs du socle sont nettement plus sollicités que d'autres. Les tractions apparaissant à la base de la statue pourraient expliquer certains désordres qu'on y observe (figure 8). D'autre part, le basculement prédit par cette étape de calcul se rapproche nettement de la valeur observée.

Quand on ajoute à cette étape la simulation d'une sollicitation sismique selon la méthode développée par Lemos (1987) avec UDEC (onde de cisaillement d'amplitude 4 MPa, de fréquence 5Hz, de durée limitée à 10 secondes pour raisons pratiques et amortissement proportionnel à la masse) on mesure le rôle important qu'ont pu jouer les secousses sismiques survenues par le passé dans la région de Louxor puisqu'on assiste à une augmentation du basculement d'environ 90 % au terme d'une telle simulation (figures 9 et 10). Cela nous donne une valeur très proche de la valeur observée.

Ainsi, par la mise en oeuvre d'une méthode de modélisation "évolutive", permettant dans un premier temps d'expliquer la fissuration du socle et une part du basculement des colosses puis, dans un deuxième temps, d'évaluer le comportement des colosses après la rupture de leur socle, il a été possible de reconstituer, au moins dans ses principes, les principales phases de l'évolution mécanique de ces monuments. Cette méthode nous a permis notamment de mettre l'accent sur la nécessité de renforcer le socle des colosses (qui assure en partie la stabilité de l'ensemble) et sur l'intérêt qu'il y aurait à les instrumenter en mesurant notamment leur basculement au cours du temps (de

Figure 10. Valeurs du basculement arrière obtenues par les différents modèles.

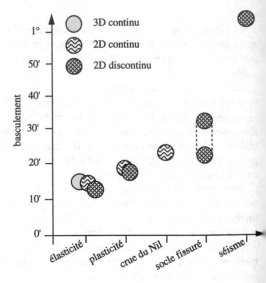

façon à quantifier l'influence des variations de la nappe ainsi que d'éventuelles secousses sismiques ou vibrations diverses).

Du point de vue des méthodes utilisées, l'apport de la MED est évident. Comme le montre la figure 10, cette méthode nous a permis de retrouver une valeur de basculement nettement plus proche de la valeur observée (simulation dynamique mise à part) grâce à la prise en compte des discontinuités effectivement observées sur les colosses (en simulant un séisme la valeur est sensiblement égale à la valeur observée).

Figure 11. IXe Pylône du grand temple de Karnak (avec indication de la coupe modélisée), exemple de monument dont la stabilité dépend avant tout de sa propre structure.

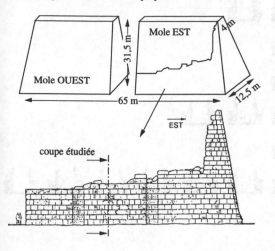

Figure 12. Recherche d'un remplissage de talatats équivalent

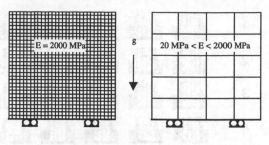

7 CAS DU IXe PYLONE DU TEMPLE DE KARNAK

Le IXe pylône du Temple de Karnak en Egypte est un portail de pierres constitué de deux môles en vis-à-vis. Il fut édifié sous le règne de Horemheb, au cours de la XVIIIe dynastie, à Louxor. C'est une structure colossale (figure 11) qui représentait, à l'origine, un volume total (fondations comprises) d'environ 17000 m^3 soit une masse approximative de 30600 tonnes dont 25 % en fondations (ces valeurs sont des estimations). La structure de cet édifice est composée de parements en gros blocs de grès et d'un remplissage formé par l'empilement plus ou moins régulier d'un très grand nombre de petits blocs, les talatats (au nombre de 230000 environ, à l'origine).

Le problème posé pour ce pylône est atypique. En effet, les talatats proviennent du démontage d'autres temples que les archéologues cherchent à reconstituer. Ils sont par ailleurs recouverts d'inscriptions. Il s'agit donc de trouver une méthode qui permettent d'extraire les talatats du coeur du Pylône en minimisant les perturbations aux parements. Une fois cette méthode définie, il faut aussi prévoir les instabilités auxquelles la structure peut être sujette et proposer les moyens de s'en prémunir. Nous avons donc essayé de modéliser ce problème qui rentre dans la catégorie des cas où la stabilité du monument dépend avant toute de sa propre structure (monument constitué d'un très grand nombre de blocs).

Etant donné le caractère discontinu du milieu, nous avons fait appel à la MED en deux dimensions (code UDEC, Itasca, 1991), sur la base de données mesurées par Martinet (1992) pour les matériaux et de données plausibles pour les joints. Etant donné la forme allongée de la structure, un calcul bidimensionnel (en déformations planes) peut être justifié car la section étudiée (indiquée sur la figure 11) se répète sur une assez grande longueur (cette option exclue donc de l'analyse, la partie haute du pylône)

La difficulté de représenter tous les blocs de la section étudiée (1000 environ), nous a amené à substituer au remplissage de talatats, 20 gros blocs ayant une déformabilité équivalente (modification du module d'Young) et qui permettent néanmoins de différencier les méthodes d'évidement possibles (figure 12).

Puis, nous avons défini différentes méthodes d'évidement possibles ou réalisables en pratique moyennant d'éventuels dispositifs de soutènement provisoire (figure 13). Ces évidements peuvent être classés en deux catégories selon la direction principale d'extraction : verticale ou horizontale. Mais, pour être sûr de pouvoir comparer toutes les simulations, il fallait d'une part, que les perturbations des parements soient comparables donc significatives et d'autre part, que ces perturbations ne conduisent pas à un effondrement généralisé du pylône. Une fois encore nous avons spéculé sur les données du problème de manière à mettre au point un modèle de base "discriminant" vis à vis des méthodes d'évidement à simuler. Cette étape a été menée à bien par une série de simulations d'un évidement total et instantané réalisées pour différents jeux de paramètres (module d'Young des blocs et angle de frottement des joints modifiés).

Ayant ainsi optimisé un jeu de paramètres, les différentes simulations d'évidement ont été progressivement réalisées et nous avons identifié l'avantage d'une méthode d'extraction à caractère horizontal commençant par le centre (simulation n° 9, figure 13 et figures 14 et 15).

Enfin, à partir de cette méthode, de nouvelles spé-

Figure 13. Les différentes méthodes d'évidement modélisées pour l'étude du IXe Pylone.

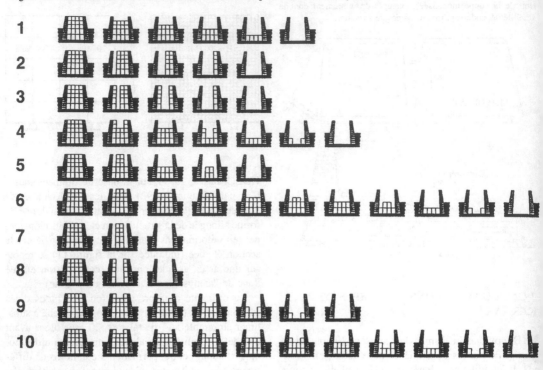

Figure 14. Synthèse des déplacements horizontaux obtenus aux parements pour l'ensemble des simulations.

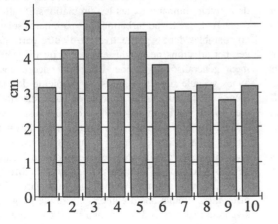

Figure 15. Meilleure méthode d'évidement obtenue par les simulations.

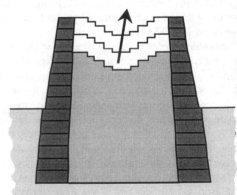

Figure 16. Effondrement possible du pylône à la fin de l'évidement (pour la méthode d'évidement retenue précédemment). La fermeture des joints est indiquée par les petites "jauges" noires dont la hauteur augmente en certains points "bas" de la structure.

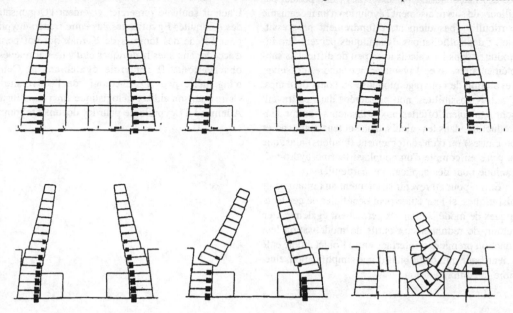

culations sur les données nous ont amenés à envisager plusieurs mécanismes d'effondrement possibles :

1. Une chute des blocs de parements les plus hauts dès le début de l'évidement dans le cas où les joints seraient très glissants (mécanisme néanmoins peu probable).

2. Un effondrement généralisé des parements à la fin de l'évidement par écrasement des blocs les plus bas sous la charge des blocs supérieurs (figure 16).

Ainsi, principalement utilisée comme un outil de simulation permettant de comparer différents scénarios d'intervention possibles, la simulation nous a ici apporté des réponses difficilement accessibles par d'autres moyens d'investigations. Dans ce cas particulier, la MED est certainement la seule méthode actuellement fiable permettant de faire ce genre d'investigations. Elle permet, d'autre part, de prévoir certains mécanismes d'instabilités contre lesquels on pourra au préalable se prémunir.

9 CONCLUSION

Les monuments historiques présentent des spécificités que le géotechnicien ignore dans son domaine d'étude habituel. En tenant compte de ces spécificités et notamment de l'impossibilité d'appréhender les monuments dans leur complexité et leur nature historique, les outils de modélisation numérique permettent, en tant qu'outils heuristiques, d'améliorer sensiblement l'investigation scientifique qui leur est habituellement réservée. L'utilisation combinée de plusieurs méthodes de modélisation permet, en particulier, de mieux différencier et identifier les phénomènes conduisant aux instabilités qui menacent les monuments. Par leur capacité de représentation, ces outils facilitent la compréhension et la lecture des mécanismes (particulièrement lorsque plusieurs méthodes sont utilisées simultanément). Ils aident ainsi à envisager les moyens de mieux les étudier (par des méthodes expérimentales par exemple) et de mieux lutter contre leurs conséquences.

Dans ce schéma, la méthode des éléments distincts se distingue par sa capacité à tenir compte de l'existence de discontinuités qui caractérisent souvent les monuments anciens construits en maçonnerie ou fissurés ou les monuments souterrains excavés dans des milieux rocheux fracturés. Améliorant ainsi le réalisme de la représentation faite de l'objet, elle conduit à des résultats auxquels les méthodes classiques n'ont généralement pas accès.

Hormis la prise en compte des discontinuités, cette méthode présente l'avantage de tenir compte de la non-linéarité du comportement mécanique des matériaux et des joints (une des caractéristiques principales des maçonneries, par exemple) ; elle permet par ailleurs de suivre aisément l'évolution d'un mécanisme particulier (basculement, effondrement progressif, etc.) ; des sollicitations dynamiques pervent être introduites dans les calculs avec peu de difficultés supplémentaires, avec l'avantage, ici encore, de suivre l'évolution de l'ouvrage modélisé au cours du temps. D'autres possibilités, non présentées dans ce travail, sont également offertes aux utilisateurs comme l'introduction, dans les calculs, d'un soutènement (cadres ou cintres) ou d'un renforcement (boulons) ainsi que la prise en compte d'un couplage thermo-hydro-mécanique pour des applications particulières.

Enfin, pour en revenir strictement aux monuments historiques, si leur étude peut bénéficier de ces techniques de modélisation, ils permettent également, en retour, de redonner aux outils de modélisation leur fonction première d'interface entre l'objet étudié et le scientifique qui, par le biais d'une simplification, améliore la compréhension.

REMERCIEMENTS

L'auteur souhaite remercier vivement l'Organisation des Antiquités Egyptiennes, le centre franco-égyptien pour l'étude des temples de Karnak qui ont permis d'accéder aux sites historiques et d'y mener relevés et observations. Il remercie également le Centre d'Ingéniérie pour l'Archéologie de l'Université du Caire pour son aide et sa logistique ainsi que l'Institut Allemand d'Egyptologie pour les documents fournis.

REFERENCES

Caquot A. et Kerisel J. 1966. *Traité de Mécanique des Sols*. 4e édition. Paris : Gauthier - Villars.

ICOMOS 1966. *Charte Internationale sur la conservation et la restauration des monuments et des sites*. Paris.

Itasca Consulting Group. 1991. *UDEC 1.8 and FLAC 3.0 manuals*. Minneapolis.

LCPC (Lab. Central des Ponts et Chaussées) 1990. *Manuel de CESAR 2.0*. Paris

Lemos J.V. 1987. *A Distinct Element Model for Dynamic Analysis of Jointed Rock with Application to Dam Foundations and Fault Motion*. PhD Thesis, University of Minnesota, Minneapolis

Starfield A.M. & Cundall P.A. 1988. *Toward a methodology for rock mechanics modelling*. In International Journal of Rock Mechanics and Mining Sciences, n° 25-3 : 99-106. London : Pergamon Press.

Verdel T. 1993. *Géotechnique et Monuments Historiques, méthodes de modélisation appliquées à des cas égyptiens*. Thèse de Doctorat. Nancy : INPL-Ecole des Mines.

Verdel T., Piguet J.P., Helal H. et Abdallah T. 1993. *Etude de l'évidement du IXe Pylône du temple de Karnak par la méthode des éléments distincts - Recherche d'une méthodologie*. Revue Française de Géotechnique, n°65 : 57-66. Paris

Martinet Gilles 1992. *Grès et Mortiers du Temple d'Amon à Karnak (Haute-Egypte). Etude des altérations, aide à la restauration*. Paris : LCPC

Degradation of building stone of some Romanesque churches in Moravia

Dégradation de la pierre de quelques églises romanes de Moravia

M. Šamalíková & J. Locker
Institute of Geotechnics, Brno Technical University, Czech Republic

ABSTRACT: The article deals with the petrographical description of building stone types which have been used during the Romanesque period for construction of some churches in Moravia. These churches are under the government protection at present, and therefore the attention is paid to their restauration. The degradation of the durbachite from the basilica in Třebíč is an example of the frost weathering. The article presents the results of the study of feldspars weathering separated from strong and decomposed blocks of this basilica. The degradation is documented by SEM and EMA.

RESUMÉ: L'article est une description pétrographique de la pierre de construction, dont on se servait á l'époque romane pour la construction de petites églises en Moravie. Aujourd'hui, la plupart d'entre elles sont protégées par l'office du patrimoine et leur restauration exige souvent une approche particuliére.
La désagrégation de la durbachite á la basilique de Třebíč est un exemple de la dégradation de la pierre de construction. Dans l'article, on présente les résultats d'une étude sur la désagrégation des feldspaths, séparés des blocs exposés au vent et décomposés des parois de pierre. Les résultats montrent que la pierre utilisée se dégrade surtout sous l'effet de la gelée. Cette désagrégation est documentée á l'aide du SEM.

INTRODUCTION

They are several churches of the Romanesque period in Moravia. Two of the most significants are the St. Prokopus Basilica of Třebíč and the Porta Coeli of Předklášteří. Both basilicas represent structures from the period of transition style of Romanesque - Gotic periods.

A St. Prokopus Basilica in Třebíč was built, base upon historical sources, from 1101 up to the first half of the 13th century. It was the property of the Benedictin monastery. During the 15th century it was damaged due to Hussite wars. Its partial reconstructions have been taking with some breaks since the beginning of the 17th century. The original crypt 27 m long and 8 m wide belongs to the most valuable parts of the monastery from the historical point of wiev. This rests on 16 Romanesque columns (Fig. 1) and represents a part of the original church. It is built prevailingly of durbachite (syenite) (Fig. 2) having the ornamentation of Anthracolithic system sandstone.

The second basilica is the Porta Coeli - The Gate to Heaven - in Předklášteří near Tišnov. It was originally a part of the nunnery founded before the year 1250 in an uninhabited area. That is why the whole complex was provided with fortification with gates and keeps. The church gate represents the gate to heaven and is

Fig.1 General view of the crypt in Třebíč

Fig. 2 St. Prokopus Basilica of Třebíč

made of sandstone. It is one of the Central European gems (Fig. 3).

Both churches are under the government protection at present. Because of their historical significance the attention to the degradation and future reconstruction has been paid last year.

STUDY OF THE DEGRADATION OF THE DURBACHITE IN TŘEBÍČ

The durbachite (granite-syenite) of the Třebíč massif is coarse-grained, porphyric (Fig. 4) with phenocrysts of potash feldspar and, locally, with small amphibole prism. Enclosures of older rocks of some centimeters in size are conspicuous. This high-quality material appears to have been recovered, in the course of the amelioration of the monasterial estates from boulders spread in the fields in the vicinity (Fig. 5).

The weathering of the durbachite has been studied mostly on the feldspars. The grains were separated from the strong and the weathered building stone blocks and analysed by the aid of X-ray diffraction analysis and finally

Fig. 3 Basilica Porta Coeli in Předklášteří

Fig. 4 Durbachite masonry, St. Prokopus
Basilica

Fig. 5 Outcrops of the durbachite in the vicinity
of Třebíč

0.1794   0.2154   0.3253   0.6410 nm

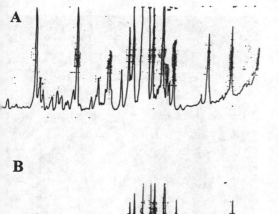

Fig. 6 X-ray diffraction curves of the feldspars
(with biotite and amphibole), A strong
durbachite, B weathered durbachite

documented and determined by the SEM and
EMA.

Fig 6 shows the X-ray diffraction curves of
strong and weathered potash feldspar. The result
shows that it is no substantial difference between
both samples.

The difference however, has been shown from
the study by the SEM and EMA. Meanwhile the
feldspars from strong durbachite show very
typical surface with clean undisturbed cleavage
plains (Fig. 7) without clay minerals or other
secondary mincrals, thc feldspars from thc
weathered durbachite show significant fissures
according to the cleavage plains and some newly
formed minerals between individual parts.
According to the result of the EMA the gypsum
(Fig. 8, 9) and some iron oxyhydroxides are
present.

Also chloritization of the biotite is typical.

DEGRADATION OF THE ANTHRACOLITHIC
SYSTEM SANDSTONES

These rocks are present mostly in Porta Coeli in
Předklášteří (Fig. 10). In Třebíč they are present
only in ornaments of the columns capitals (Fig.
11).

The anthracolithic system sandstones are of
two different colours: yellow-brown and red-
brown to red.

Fig. 7 Undisturbed cleavage plain of the feldspar from strong durbachite, SEM enl. 2680

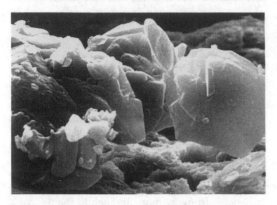

Fig. 8 Gypsum in the fissure of the cleavage plain of the feldspar from weathered durbachite, SEM enl. 2300

Fig.9 Another form of gypsum on the feldspar from weathered durbachite, SEM enl. 2300

Fig. 10 Hematite sandstone ornaments, Basilica Porta Coeli

Fig. 11 Limonite sandstone ornaments in St.Prokopus Basilica of Třebíč

3564

They with the limonite are very weak and their frostresistance is low.

The red sandstones with the hematite are very strong and their frostresistance is good.

STUDY OF THE CRUSTS

In the both cases we paid a special attention to the surface crusts of the building and ornamental rocks.

In the case of St. Prokopus Basilica of Třebíč only in recent time when the good atmosphere of our highland was changed and influenced by chemical air pollution some newly formed components appeared. They form on the surface only a black layer of undetermined mineral composition (Fig. 12 and 13).

In the case of the Basilica Porta Coeli in Předklášteří the red sandstone with hematite are until present very good and unweathered. But the surface is much worse compared to that in Třebíč. The degradation takes place in the sinne of desquamation of layers saturated with the new aggresive secondary minerals. We determined gypsum as the most frequent mineral (Fig. 14, 15).

CONCLUSION

From the results obtained we could conclude that

Fig. 12 Character of the crust, strong durbachite, feldspare cleavage plain, SEM enl. 2680

Fig. 14 Gypsum particles, the crust of sandstone, Porta Coeli, EMA

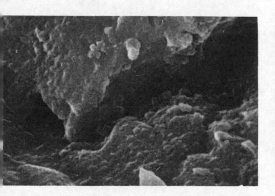

Fig. 13 Character of the crust, weathered durbachite, fissure of the feldspar cleavage plain, SEM enl. 6100

Fig. 15 Gypsum particles, the crust of sandstone, Porta Coeli, EMA

in the case of the St. Prokopus Basilica of Třebíč the main reason for the degradation of the durbachite is the frost weathering of the feldspars, which took place more rapidly on the blocks of the original slightly weathered rocks boulders.

In the case of the Basilica Porta Coeli in Předklášteří the red sandstone with the hematite is according to the frost action relatively stable. It seems to be the degradation is caused by forming the gypsum in the 2 up to 3 mm thick layer which is flaking off.

These results are not final and the study will continue in the future.

Natural weathering on stone slabs

Météorisation sur façades de pierre

María Beatriz P. de De Maio & Marisa Domínguez
National Institute of Industrial Technology, Argentina

ABSTRACT: Observations and analysis are presented, carried out upon stone facings made of calcareous and granite. They were submitted to a programmed natural weathering application, durability laboratory tests being performed at the same time, so as to establish a possible relationship between both scales.

Evaluations that are reported belong to the first stage of research, during which different steps were outlined, as follows: arrangement of several walls (concrete blocks, common bricks and ceramic hollow bricks); characterization of stone facings; selection of fixing methods; joint types; selection of different waterproofing and/or protection products; among durability laboratory tests other determinations were carried out, such as freezing and thawing, action of sulfuric, nitric atmospheres with acid pH, analysis of staining process.

Partial results are concerned with early four months of natural weather monitoring foreseen as a long term projection, as well as with those obtained from laboratory experimentation.

RESUME: On se presente des observations et des analyses sur façades de pierre, de marbre et de granit. Elles furent soumises a un programme d'action naturelle à l'intempérie (meteorisation).

Des essais de laboratoire de durabilité ont été fait en mème temps pour établir une possible rélation entre les deux échelles. Les informations sur les deux échelles appartient à la première étape de l'investigation pendant laquelle on a suivie les pas suivants: construction de differents murs avec du béton, des briques communes et des briques creuses céramiques, forme de façades de pierres, élection de methodes de fixation de types d'ajustage, selection de differents produits de protection et / ou imperméabilisation.

Parmis les essais de laboratoire de durabilité, on a fait d'autres déterminations, comme congelation et décongelation, action des atmosphères sulfuriques, nitriques avec pH acide, et analyse d'un procès de tacheté (staining).

Les résultats partiels devient des premiers quatre mois d'observation comme une partie d'un projet de grande portée.

1 INTRODUCTION

The plan is to develop a long reach research work along which the idea is to analyse results of laboratory testing on calcareus and granite slabs, so being able to choose and eventually adjust those which actually show damage produced by natural weathering.

Although there exist previous studies about marble and granite durability related to ambient pollution (De Maio & Dominguez, 1993), there are no antecedents referring exclusively to damage upon these stones due to ambient factors, associated with climate characteristics.

For such evaluation, a site was selected, away from big cities' smog and car pollution's influence. At this place, several walls of different constructive characteristics were built: concrete blocks, ceramic hollow bricks and common bricks) as considered those most commonly used by constructors. Such walls were orientated facing South and North so as to analyse the performance of stone slabs under site's conditions of temperature, humidity, insolation, etc.

In order to obtain as much data as possible, different methods were used for joints placement and materials, leaving for future stages the evaluation of different protection / waterproofing products.

Partial results that are presented belong to early four months of the natural weathering program.

2   WORKING PLAN

It was organized in stages as follow:

I   Selection of site, distribution of walls and construction system.

II   Selection of materials to be used

a)   Petrographic and physicochemical characterization of calcareous and granite slabs to be used

b)   Joints material: characteristics

III   Selection of laboratory durability methods to apply.

IV   Planning of natural weathering monitoring. Confection of control form

V   Partial evaluations program: periodicity

3   STAGE DEVELOPMENT

3.1. *Stage I*

As it was said, this stage consisted of building walls following the selected scheme.

According to it, walls were built up upon a lean concrete subfloor with concrete chaining on base and crown, both being laterally reinforced with concrete. Walls were 400 cm long and 220 cm high. They were divided by longitudinal slight walls, 50 cm width, each one corresponding to concrete blocks, ceramic hollow bricks or common bricks respectively, being each slight wall separed from the other by 10 cm The orientation was towards North, counterfaces facing South.

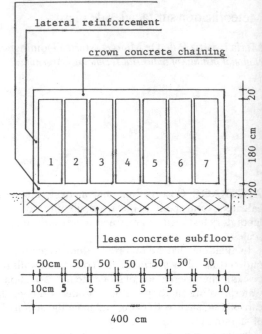

Stone slabs were placed over each slight wall, size 11x23x2 cm (height, width, thickness) and selected according to different fixation alternatives and joint types.

3.2. *Stage 2:*

a) *Materials selection*

Granite slabs were Sierra Chica (R-1), Rojo Dragón (R-2) and Gris Mara (R-3) types; while calcareous slabs consisted in Dolomite (D-1), Travertine (T-2) and Rosa Quilpo Marble (M-3).

Table 1. Physicomechanical properties of stones

Sample	Density kg/m³	Absorption %	Porosity %	Flexural strength MPa
R-1	2680	0.2	0.4	10.1
R-2	2670	0.3	0.5	11.0
R-3	2700	0.4	1.2	11.9
D-1	2650	1.2	1.9	11.7
T-2	2430	1.8	3.1	12.6
M-3	2720	0.2	0.3	9.8

Mineralogical characteristics were determined by analysing thin-sections using an optical microscope, with the following results:

Sample R-1: Sierra Chica Granite
Texture: very coarse granular
Composition: 30% quartz, 35% microcline, 10% biotite, 5% amphibole and opaque minerals, 20% chlorite, sericite and clay.

Sample R-2: Rojo Dragón Granite
Texture: coarse granular
Composition: 25% quartz, 45% microcline, 8% plagioclase, 5% biotite, 2% opaque minerals, 15% muscovite, chlorite, sericite and clay.

Sample R-3: Gris Mara Granite
Texture: Fine to medium granular
Composition: 25% quartz, 35% microline, 20% plagioclase, 5% biotite and opaque minerals, 10% muscovite, zoicite and clay.

Sample D-1: Dolomite
Texture: Fine granular crystalline
Composition: 60% dolomite, 60% calcite, 12% iron ore and clay, 8% quartz.

Sample D-2: Travertine
Texture: Granular and fibrous crystalline, vesicular
Composition: 93% calcite, 7% clay and iron ore.

Sample D-3: Rosa Quilpo Marble
Texture: Coarse granoblastic
Composition: 85% calcite, 10% quartz and plagioclase, 5% amphibole and apatite.

b) *Fixation methods*
One of the methods used mortar formed by a mixture of sand, lime, cement (3:1:1) with enough water to reach necessary fluidity for its straining between slabs and masonry.

Next method was using an air chamber between slab and masonry with brass staples as a fixation

c) *Joints material*
Two kinds were used: cement base sealer and polyurethane base sealer.

3.3 *Stage III*
Durability laboratory testing:
They were carried out according to international standards, and in their absence other materials' methodologies were adapted.

a) *Freezing and thawing*:
According to procedure UNI 9724 Part. 8. Tarnished cubic samples were tested, side=7 cm, having been submitted to 20 cycles of 3 hs. at 35 ºC into water + 3 hs. at -15 ºC to air.
The evaluation of the performance of the different

kinds of stones show granites with no visible damage, being appreciated just a slight raise of microfissures.

Dolomite and Rosa Quilpo marble did not show changes either, while Travertine appeared slightly granular close to vesicules_although there is no development of fissuration .

It may be concluded that tested samples are not considered easily frozen.

b) *Weather resistance*
ASTM C 217-90 was taken as a guide, as establishing attack procedure with sulfuric acid, for the evaluation of the resistance of slates to meteorization. Nitric and ascetic acid were also used.

The procedure consisted on submerging stone samples of 10x5x2 cm with mirror polished surfaces. At the start of the test dry weight was determined, as well as superficial hardness, observating them with a stereomicroscope. Testing took 7 days using solutions of sulfuric and nitric acid 1% ,and 0.5 N for ascetic acid. Last one's degree of disolution was taken from that established by the Environmental Protection Agency (EPA), which resembles it to the environmental natural agressive agent. In each case, solutions were renewed each two days while being inmersed.

Once it took place, register was repeated for dry weight. superficial hardness and optical detail observation.

To determine, with as much precision as possible, the degree of attack suffered by each sample in accordance with the acid used, attack depth was measured using a roughnessmeter which reproduces as hills and valleys the alteration of the superficial level.

This registration of the surfaces state technique was adopted because methodology applied to slates (ASTM C 217-90) is not sufficiently precise.

The comparative evaluations done before and after testing show that no significant changes are produced about samples weight; while the observation of attack surfaces show higher resistance among granites against remarkable variation among calcareous.

Table 2. *Alteration observed on samples attacked by acid solutions.*
Observations - Attack by sulfuric acid

R-1 Slight loss of colour; attack on micas. No loss of polish on quartz and feldespates.

3569

R-2 Slight loss of color; certain degree of alteration on ferromagnesian minerals.

R-3 Higher loss of colour due to alteration of min. Fe / Mg varying towards greenish color.

Apparition of white spots.

D-1 Loss of polish and colour (tending to white).

T-2 General loss of polish and color. Tarnish and white deposits at pore areas.

M-3 Whitening of surface. Higher loss of polish on calcite and lower on silicates.

Observations - Attack by nitric acid

R-1 Slight loss of colour with higher intensity on min. Fe / Mg, with some tarnish. On feldespates internal microfissuration.

R-2 No changes in general; excepting for min. Fe/Mg with slight loss of polish.

R-3 Slight loss of color .

D-1 General loss of surface polish.

T-2 General loss of polish.

M-3 Moderate loss of polish. Apparition of oxide spots.

Observations - Attack by ascetic acid

R-1 Slight loss of colour although polish remains.

R-2 No remarkable changes.

R-3 Apparition of small white stains on some feldespates. Scarce min. Fe / Mg present slight loss of polish.

D-1 Higher loss of polish on yellowish areas in relation with white ones. White superficial deposits.

T-2 Slight loss of color. Total loss of polish. Vesicules deepen.

M-3 Total loss of polish. Formation of an irregular surface due to differential attack on biggest calcite crystals, marking twins and contacts. Sectors with silicates are in relief. Maximum roughness surface.

General conclusions

On granites, ferromagnesian minerals are the most affected, especially micas, producing dasquamation and loss of colour. Texture is the most influential property, not like is absorption.

General performance to acids action gives similar response.

In calcareous stones varieties the most important characteristic is their chemical composition; this is visible on tarnished , wrinkled, colorless and porous surfaces.

Sulfuric acid produces superficial gypsum deposits with consequent color whitening.

About nitric acid, the inmediate consequence is polish loss; while ascetic is the most aggresive, producing surfaces with deeper attack (roughness).

From the mineralogical point of view, samples with higher contents of calcite mineral are more attacked than those who have silicate compound.

c) Staining test

Not existing standarized testing about staining problems on stone surfaces, different procedures were put into practice, as mentioned at the consulted bibliography, as much as results obtained from one's own experimentation.

Granite: Stains from oil derivates in red varieties (R-1 and R-2). Effects from staining were shown by a darkening of polished surface not interesting polish degree.

On grey variety (R-3) grease and oil were used, quite common a problem due to absorption of this variety. The produced effect darkened the stained area, and in cases when the staining product had some dye, it was transmitted to the granite surface.

Calcareous: The formation of calcareous incrustation is a very common problem on marble surfaces. To be able to reproduce this phenomena at the laboratory, a marble slab (M-3) was exposed to air, with one of its faces covered by white cement (traditionally used for its fixation) resting its opposite face polished and free to the air. The slab was submerged up to a thirth of its longitude in vertical position into a container with water containing salts that had 2 to 4% of iron in the solid residue. Consecutively, cycles of 16 hs. of inmersion were fulfilled (under noted conditions) and of 8 hs. of drying at 30 ºC. After 4 months, it was observed a yellowish deposit over the polished surface that was submerged, while the portion that was exposed to air remained clean; not so the mortar over its counterface, which also showed a yellowish staining.

The analysis of this results allows to believe that capilar raise took place through the sector of the slab that was submerged, dragging salts with iron oxide, leaving a deposit of dust beneath water's level and then staining mortar, not so the area of the slab that was exposed to the air, because the same was always submitted to water evaporation.

This mechanism is visualizad through contents of iron because it is just supplied by the water, taking into account that neither marble nor mortar have iron contents.

This procedure could appear as the one observed

on slabs exposed to weather inclemency, which after rain washing, solubilize limes belonging to fixation mortars, producing white "drippings" that coincidently appear from joints between slabs.

3.4 Observations on natural weathering

As a consequence of the scarce period of time described slabs have been exposed to weather inclemency, no relevant evaluation can be made, safe for speaking about staining produced by humidity as shown by granitic slabs fixed with mortar, particularly walls facing South (lower insolation) in coincidence with cement base joints.

The time period that was considered is of four months during the winter, at which maximum global radiation arrived to 10984 Kjoule/m², with a duration of 9.8 hs. Mean temperature was of 9 ºC, with precipitation that went from 700 to 1000 mm. Relative humidity was between 70 and 95 %.

4 MONITORING PLANNING

At the same time laboratory testing went on, monitoring of slabs under weather conditions took place. For that purpose, forms were put into practice, which registered changes and / or results of "in situ" studies.

4.1 Visual observation on slabs

- Colour: Changes. Evaluation will be done by comparison to Munsell Colour Charts (ASTM, 1968)
- Polish: Changes.
- Stain: Type, origin, extension, distribution.
- Deposits of dust
- Efflorescences: Tipe, origin, quantity.
- Disintegrations: film disjunctions, contour scaling, swelling deformation, grain disgregations, pitting, alveolar erosions.
- Ruptures: cracks, fissures, spalling.
- Superficial hardness
- Depth of meteorization : using replica support films of slabs' surfaces.
- Mineralogical alterations
Observation frequency: monthly

4.2 Joints observation:
- Watertightness testing
Testing frequency: three-month

4.3 Protection products monitoring
Frequency: monthly

5 CONCLUSIONS

It is considered that from an year on natural weathering will offer significant results, so allowing to carry out comparison with laboratory testing.

This way, most adequate laboratory testing could be selected for evaluating stone slabs, taking into consideration not only intrinsical facts but also those who depend on their location, orientation, thermic changes, low temperatures, rainfall, etc.

REFERENCES

ASTM C 217 - 1990. Weather resistance of slate De Maio, M. B. & Domínguez, M. - 1993 - Durability analysis of ornamental rocks under natural weather conditions. Proc. of Int. RILEM / UNESCO Cong. 344-350, Paris

EPA -1980 - Test methods for evaluating Solid Waste. SW 846. Section 7

UNI 9724 Part. 8 - 1990 - Prove di gelitivitá (in elaborazione).

Weathering of building carbonate rocks under SO_2 polluted atmospheres

La météorisation de roches carbonatées sous des atmosphères polluées de SO_2

C. M. Grossi & R. M. Esbert
Department of Geology, University of Oviedo, Spain

A. J. Lewry & R. N. Butlin
Building Research Establishment, Watford, UK

ABSTRACT: The behaviour of two Spanish monumental carbonate rocks under SO_2 polluted atmospheres was compared. The rocks differed in composition (dolomite and limestone, respectively), porosity, pore size distribution and hydric properties. Both stones were exposed to clean and SO_2 polluted atmospheres in an Atmospheric Flow Chamber (AFC). During the exposure period, half of the samples were subjected to simulated rain (wetting and drying cycles) and the other half remained dry; dry deposition on wet and dry surfaces was simulated. After exposure, extracted reaction products (soluble salts) and "run-off" solutions were analyzed using ion chromatography. The stone surface was principally examined by means Fourier Transform Infrared Spectroscopy (FTIR) and Scanning Electron Microscopy (SEM). The degree of reaction was determined by calculating the sulphate content of the rocks. The physico-chemical properties of these rocks control their weathering behaviour and reactivity.

RESUMÉ: Le comportment de deux roches carbonatées espagnoles sous des atmosphères polluées en SO_2 a été comparé. Les roches différaient en composition (dolomite et calcaire respectivement), en porosité, en réseau poreux et en propietés hydriques. Les deux pierres ont été exposées à des atmosphères propres et polluées en SO_2 dans une chambre d'écoulement atmosphérique. Pendant la période d'exposition, la moité des échantillons a été exposée à une pluie simulée (cycles d'arrosage et de séchage) et l'autre a été gardée sèche; un depôt sec sur les surfaces humides et sèches a été simulé. Après exposition, les produits extraits de la réaction (sels solubles) et les solutions de ruissellement ont été analysés à l'aide de chromatographie ionique. La surface de la pierre a été principalement éxaminée par microscopie électronique à balayage et par spectroscopie infra-rouge. Le degré de réaction a été déterminé par la calcul du contenu de sulfate des roches. Les propiétés physico-chimiques de ces roches controlent leur réactivité et dégradation.

1 INTRODUCTION

In this paper, the behaviour of two Spanish monumental carbonate rocks under SO_2 polluted atmospheres was compared; the reactivity was related to their composition, porosity and some physical properties (water absortion and capillarity).

The rocks have been used in two important Spanish Cathedrals: Oviedo and Burgos. So, they have been subjected to pollutant action in both cities although under different climatic conditions: Atlantic climate (Oviedo) and Continental climate (Burgos).

Laspra dolomite besides being used in the Cathedral of Oviedo (Asturias - NW Spain) (Esbert and Marcos, 1983), has been utilized as building or restoration material in other Asturian monuments such as the Pre-Romanesque "Santa María del Naranco" (Esbert at al, 1992).

Hontoria Limestone has not only been used in the Cathedral of Burgos but also in the important Cathedral of León; both cities are located in Castilla-León (Spanish North Meseta). Moreover, it has been very utilized in many buildings of Burgos Region and it can be found in some buildings of Madrid, País Vasco, Asturias... (Marcos et al, 1993).

Extensive black crust formation has been observed when both these materials have been exposed to urban environments. (Esbert and Marcos, 1984; Marcos, 1992).

2 PETROPHYSICS

2.1 Petrography

Laspra Dolomite: The blocks examined come from the Cathedral of Oviedo and were removed during ancient restorations.

Laspra is a lacustrine, white, micritic, soft and homogeneous dolomite from the Eocene basin situated round Oviedo in the Central Part of Asturias. It is principally composed of dolomite (90-100%) and quartz (0-5%). Occasionally it contains some argillaceous minerals (0-4,5%). Barite and some opaques can be found as accessory minerals.

Hontoria limestone: The material under study comes from the quarries of Hontoria de la Cantera (Burgos).

This rock is a Turonian (Upper Cretaceous) bioesparite (Folk, 1962). It is a white, coarse and heterogranular limestone with a great number of voids.

The composition of the rock is almost entirely calcite (100% determined by XR-diffractometry); but, occasionally, it can contain terrigenous components up to 2%, principally quartz. The grains are mainly fossil remains sealed by a sparitic cement (Marcos et al, 1993).

2.2 Porosity and hydric properties

Porosity and pore distribution were determined by mercury porosimetry. Obtained results are presented in Table I and Fig. 1.

Two hydric properties were determined according to C.N.R.-I.C.R. standards: free water absorption by total immersion (Doc NORMAL 7/81) and capillarity (Doc NORMAL 11/85). Table II shows the values of the physical properties: absorption capacity and capillarity coefficient. In Figs. 2 and 3, the water absorption Kinetics of both materials was compared.

Table II. Hydric properties

Material	C.I.(%)	C.A. $(Kg/m^2.min^{1/2})x10^{-2}$
Laspra	12	56
Hontoria	6	80

C.I.= absorption capacity (Normal 7/81, 1981) (free water absorption)
C.A.= capillarity coefficient (Normal, 11/85, 1985); units according to Mamillan (1981) (capillary suction normal to the bedding)

Then, *Laspra dolomite* is a rock with a very high open porosity ($\approx$ 30%), very fine pore opening radii (0.01-1 μm), a negligible macroporosity and relatively slow water absorption kinetics.

Hontoria limestone has a high porosity ($\approx$ 20%) with a large proportion of pore opening radii between 1 and 10 μm plus very high values in water absorption kinetics.

Table I. Porosity and pore radius distribution (intact rock)

Material	Specific surf area (m²/g)	open porosity (%)	Macrop. (%)	Microp. (%)	Radius mode (μm)	Radius distribution (μm)
Laspra	4.03	30.7	--	30.7	0.4	0.01-1
Hontoria	0.27	20.3	4.2	16.1	1.7	1-7.5

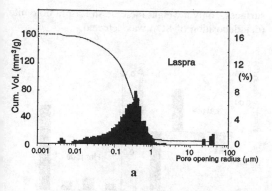
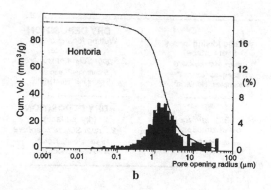

Fig.1 Pore distribution of selected carbonate stones: a) Laspra; b) Hontoria

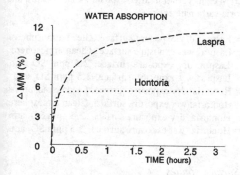

Fig.2 Free water absorption in Laspra and Hontoria carbonated rocks.

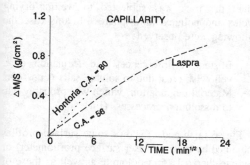

Fig.3 Water capillary suction in Laspra and Hontoria carbonated rocks.

If both materials are compared, it can be observed how the percentage of water absorption was higher in Laspra dolomite than in Hontoria limestone due to Laspra's higher open porosity; its absorption kinetics, however, was slower due to smaller pore opening radii.

Hammecker et al (1992) studied these materials and found that Laspra dolomite adsorbed approximately 100 times more water than Hontoria limestone. This property is closely related to the rock pore distribution.

3 EXPERIMENTAL

Dry deposition has been simulated in an Atmospheric Flow Chamber (AFC) with a sulphur dioxide (SO_2) polluted atmosphere. Laspra dolomite and Hontoria limestone samples (5x3x1 cm^3; exposed surface: 15 cm^2) were placed in the climatic chamber (Fig.4). The climatic chamber consists of four exposure chamber units with controlled humidity and temperature and the ability to simulate polluted atmospheres. Detail of these apparatus can be found in previous papers (Lewry et al, 1992 and 1993a).

Two different atmospheres have been simulated; half the samples were exposed to clean air whilst the rest were subjected to a polluted atmosphere of 2.5 ppm SO_2 (equivalent atmospheric concentration = 20 μg.m^{-3} - Lewry et al, 1992).

The four exposure chamber compartments were divided into two sets of two; each set had a wet and dry chamber. During the exposure period, half the samples were subjected to simulated rain; the other half remained dry. In this way, dry deposition on both wet and dry surfaces has been simulated.

surfaces, only a weight increase in Laspra dolomite (dry deposition of SO_2) was detected.

	DRY DEPOSITION
Wetting /drying cycles (wet surface) *clean atmosphere* 5 samples "Laspra" 5 samples "Hontoria"	Wetting /drying cycles (wet surface) *2.5 ppm SO_2 atmosphere* 5 samples "Laspra" 5 samples "Hontoria"
(dry surface) clean atmosphere 5 samples "Laspra" 5 samples "Hontoria"	**DRY DEPOSITION** (dry surface) *2.5 ppm SO_2 atmosphere* 5 samples "Laspra" 5 samples "Hontoria"

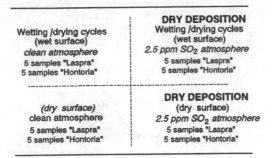

Fig. 4.- Schematic representation of tests carried out on Laspra and Hontoria rocks into the climatic chamber.

In the wet chambers, wetting/drying cycles have been carried out; the cycles have consisted of 8 hours wetting followed by 16 hours drying. The aqueous "run-off" from the surface of each sample was collected weekly for analysis. The exposure period was 30 days and the samples were wetted for 22 days; temperature was constant at $19\pm3°C$ and relative humidity was 84%. This test simulates a year's natural exposure in an area with an annual average rainfall of 800 mm.

4 ANALYSES. RESULTS

After 30 days exposure, the following studies have been carried out:

4.1 Weight variation
4.2 surface changes
4.3 Cation (Ca^{2+}, Mg^{2+}, Na^+ and K^+) and anion (SO_4^{2-}, NO_3^- and Cl^-) content of extracted reaction products (soluble salts) and run-off solutions.

4.1 Weight variation

The weight variation at the end of the test is plotted in Fig. 5

In this figure it can be observed how a larger weight variation occurred when the surface of the rocks was wet and exposed to 2.5 ppm SO_2 polluted atmospheres. *Laspra dolomite* (L4) showed a weight loss due to material reaction and deterioration. *Hontoria dolomite* (H4) had a weight increase due to dry deposition of SO_2 on its surface and an apparent lack of material loss from the surface.

In the case of polluted atmospheres and dry stone

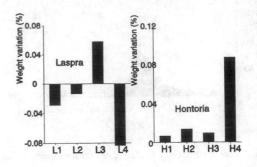

Fig. 5. Weight variation after exposure tests. Laspra and Hontoria carbonate rocks.

L1 = Laspra: dry exposure surface. Clean atmosphere.
L2 = Laspra: wet exposure surface. Clean atmosphere.
L3 = Laspra: dry exposure surface. 2.5 ppm SO_2 atm.
L4 = Laspra: wet exposure surface. 2.5 ppm SO_2 atm.
H1 = Hontoria: dry exposure surface. Clean atmosphere.
H2 = Hontoria: wet exposure surface. Clean atmosphere.
H3 = Hontoria: dry exposure surface. 2.5 ppm SO_2 atm.
H4 = Hontoria: wet exposure surface. 2.5 ppm SO_2 atm.

4.2 Surface studies

After the 30 day exposure period to different conditions in the AFC, Hontoria limestone had no variation in appearance and no deterioration was observed. Laspra dolomite had noticeable damage when the rock was subjected to wetting/drying cycles and polluted atmospheres. In this case the following were observed:

. Blisters, efflorescences and formation of a yellowish crust due to soluble salts (Fig. 6b).
. Material degradation, which was probably due to dissolution processes (Fig. 6c).

Fig. 6 (a, b and c) shows some surface profiles (obtained by means a laser surface profilometer) of non-weathered Laspra dolomite as well as those of samples subjected to polluted atmospheres and simulated rain. Surface roughness increased with material degradation; this resulted in the formation of new reaction sites.

Morphological studies carried out by Scanning Electron Microscopy (SEM) are presented in Fig 7. Energy Dispersive analysis of X-Ray (EDX) was used to perform a study of the chemical composition. The SEM/EDX results in Fig. 7 correspond to the surface profiles showed in Fig. 6.

Dolomite crystals and their composition can be observed in Fig. 7a which is a non weathered Laspra dolomite sample.

Fig. 7b shows the presence of sulphur and calcium corresponding to crystals with a typical gypsum morphology; these are typical of blisters and crust formation.

Degradation due to dissolution can be observed in Fig 7c; dolomite (calcium, magnesium carbonate) and sulphur-calcium compounds were detected by EDX.

The newly formed material on the Laspra dolomite surface appeared to be gypsum; this was deduced from its morphology and chemical composition. No magnesium sulphate was detected by SEM-EDX examination.

4.2.2 Fourier Transform Infra-Red spectroscopy

Surface material from Laspra and Hontoria rocks was analyzed by this technique. The analyzed material belonged to non weathered samples as well as samples subjected to dry deposition of SO₂ both on dry and wet surfaces.

When the rock surface was dry, no reaction products could be detected.

In the case of wet surfaces, reaction products were detected in Laspra dolomite and in Hontoria limestone.

The reaction product detected on Laspra dolomite surface was calcium sulphate dihydrate (gypsum), whilst the main product on Hontoria limestone surface was calcium sulphite hemihydrate (although sulphate was also detected) (Fig. 8).

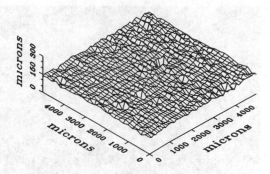

Fig 6a. Non weathered Laspra dolomite.

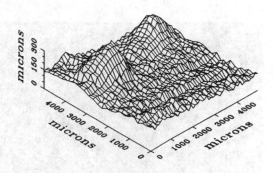

Fig 6b. Blisters on a Laspra dolomite wet surface subjected to a 2.5 ppm SO₂ atmosphere.

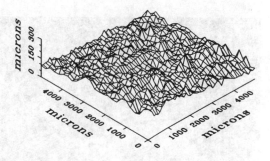

Fig 6c. Degradation by dissolution in Laspra dolomite.

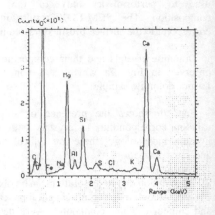

Fig 7a. Non weathered Laspra surface: dolomite crystals (left-SEM) and their composition (right-EDX).

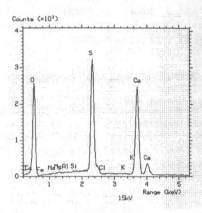

Fig 7b. Yellowish crust and blisters: gypsum crystals (left-SEM) and chemical composition (right-EDX).

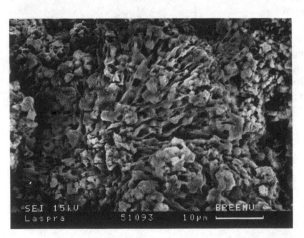
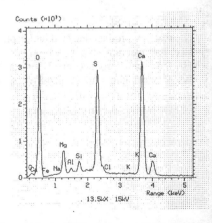

Fig 7c. Degradation by dissolution: the resultant high porosity can be observed in the SEM micrograph (left). The EDX analysis (right) shows the presence of dolomite and sulphur-calcium compounds.

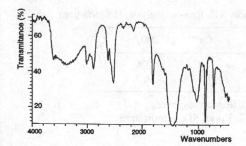

Fig.8a FTIR spectrum of non weathered Laspra dolomite surface.

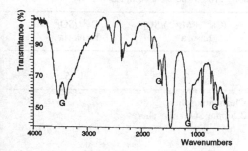

Fig. 8b FTIR spectrum of Laspra dolomite surface exposed to a 2.5 ppm SO₂ atmosphere and wetting/drying cycles.

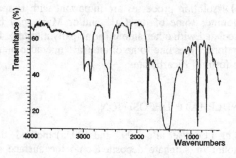

Fig.8c FTIR spectrum of non weathered Hontoria limestone surface.

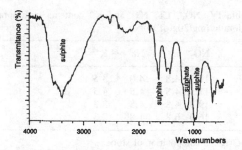

Fig. 8d FTIR spectrum of Hontoria limestone surface exposed to a 2.5 ppm SO₂ atmosphere and wetting/drying cycles.

4.3 Chemical Analyses

They were carried out both on extracted salts and on run-off collected during the exposure period.

4.3.1 Analysis of soluble salts

After exposure, the soluble salts were extracted from ground powdered samples with water. The solutions obtained were analyzed by ion chromatography.

The variation of SO_4^{2-}, Ca^{2+} and Mg^{2+} content in Laspra and Hontoria rocks is shown in Table III. In addition to the newly formed sulphates, NO_3^-, Cl^-, Na^+ and K^+ were detected in Laspra dolomite. Their variations are presented in Table IV.

Table III. SO_4^{2-}, Ca^{2+} and Mg^{2+} content in extractions

| | Laspra (mg) | | | Hontoria (mg) | |
	SO_4^{2-}	Ca^{2+}	Mg^{2+}	SO_4^{2-}	Ca^{2+}
1[1]	9.4	28.7	13.6	5.5	19.6
2	20.0	29.5	14.5	10.8	20.4
3	32.9	29.6	8.75	6.1	19.6
4	82.1	35.7	12.6	87.4	49.1

[1] Original composition of stone

1 = Dry exposure surface. Clean atmosphere.
2 = Wet exposure surface. Clean atmosphere.
3 = Dry exposure surface. 2.5 ppm SO₂ atmosphere.
4 = Wet exposure surface. 2.5 ppm SO₂ atmosphere.

Table IV. NO_3^-, Cl^-, Na^+ and K^+ contents in Laspra dolomite (mg/20g of stone)

	NO_3^-	Cl^-	Na^+	K^+
1[(1)]	17.3	4.8	4.0	3.5
2	17.5	4.5	4.5	4.5
3	19.3	4.8	4.5	3.3
4	16.0	0.8	1.8	2.3

[(1)] Original composition of stone

In Table VI, mole balances of cations against anions are presented.

Table VI. Mole balance in extractions

	Laspra cations/anions	Laspra $(Ca^{2+}+Mg^{2+})/SO_4^{2-}$	Hontoria Ca^{2+}/SO_4^{2-}
1	1.81	13.1	8.6
2	1.84	6.4	4.5
3	1.37	3.2	7.7
4	1.76	1.6	1.34

It can be seen from these tables that when a wet Laspra dolomite surface is exposed to SO_2, sulphate formation is an important factor in degradation. Visible deterioration due to dissolution was supported by the removal of nitrates and other impurities by leaching.

In the case of Hontoria limestone dry deposition of SO_2 only occurs when its surface is wet.

4.3.2 run-off analysis

This was carried out on samples subjected to wetting/drying cycles. The solutions, collected weekly, were analyzed by ion chromatography. The amount of the ions in the run-off from weathering samples are presented in Tables VI (Laspra) and VII (Hontoria).

Table VI. Run-off analysis. Laspra (mg)

	Ca^{2+}	Mg^{2+}	Na^+	K^+	SO_4^+	NO_3^-	Cl^-
1	0.6	0.04	0.08	0.04	2.4	0.2	0.2
2	13.2	6	2.6	1.4	47.6	13.6	4

Table VII. Run-off analysis. Hontoria (mg)

	Ca^{2+}	SO_4^{2-}
1	1.7	3.6
2	7.9	16.5

1 Clean atmosphere
2 2.5 ppm SO_2 atmosphere

In Table VIII, mole balance of Ca^{2+} and Mg^{2+} against SO_4^{2-} in run-off after exposure of stone to SO_2 atmospheres is presented.

Table VIII. Mole balance in run-off

	$(Ca^{2+}+Mg^{2+})/SO_4^{2-}$ Laspra	Ca^{2+}/SO_4^{2-} Hontoria
1		
2	1.2	1.15

1 Clean atmosphere
2 2.5 ppm SO_2 atmosphere

Tables VI, VII and VIII show a reaction between SO_2 and carbonate material to form sulphates or sulphites which, subsequently, are dissolved and removed by washing in run-off. These salts are, principally, of calcium and magnesium in Laspra dolomite and calcium in Hontoria limestone.

Dissolution processes are important with Laspra dolomite; some of the Ca^{2+} and/or Mg^{2+} may be associated with other anions (principally nitrates). In Hontoria limestone some of the Ca^{2+} may also be in the form of bicarbonate.

5 SULPHATE DEPOSITION

Using Lewry et al's (1992 and 1993a) method, the amount of sulphate deposited onto the surface of stone was calculated. This determination was made assuming sulphate is in the form of $CaSO_4$ and $MgSO_4$ in Laspra dolomite and $CaSO_4$ in Hontoria limestone.

Table IX shows the amount of sulphate formed during a 30 day exposure period, on the stone surface.

Table IX. Estimated sulphate deposited onto surfaces of Laspra and Hontoria rocks (mg).

	Laspra	Hontoria
2.5 ppm SO₂ Dry surface	31.5	0.9
2.5 ppm SO₂ Wet surface	185	128

The amount of sulphate deposited onto a dry surface of these carbonate rocks appeared to be in accordance with their specific surface area. Thus, the amount was greater with Laspra than with Hontoria (Table X).

Table X. Sulphate deposition onto a dry surface and specific surface area

	Laspra	Hontoria
composition	Dolomite (95%)	Calcite (100%)
spec. surface area (m²/g)	4.03	0.27
est. depos. sulphate (mg)	31.5	0.9
stone weight gain (mg)	17.6	2.6

Both stones belong to the stone reactive group mentioned by Furlan and Girardet, but Laspra dolomite is much more reactive than Hontoria limestone. These results agreed with those produced by other authors such as Mirwald and Zallmanzig (1986), who point out that the characteristics of the pore system are more important than chemical composition in the weathering behaviour of pure carbonate stones.

From the results of this work it can be inferred that the pore distribution, in spite of the porosity value, is the main rock characteristic controlling dry deposition of SO_2.

Thus, Hontoria limestone, in comparison with other carbonate rocks such as Laspra dolomite or Portland limestone, has a low capacity to adsorb water vapour (Hammecker et al, 1992; Valdeón et al, 1992). In the same way this may indicate a low

capacity to adsorb reactive gases from the air.

The initial product of the reaction, calcium sulphite hemihydrate, can only be oxidized to gypsum at high humidity conditions (Mirwald and Zallmanzing,1986; Götürk et al, 1993). Calcium sulphite is the major reaction products of Hontoria limestone due to the rock's petrophysics. Hontoria stone does not retain water within its pores.

Calcium sulphate is the only reaction product remaining on the Laspra dolomite surface. This is due to its great capacity to retain water because of its very small pore opening radii (Hammecker et al, 1993). Also, Laspra contains, unlike Hontoria limestone, a significant number of impurities which may favour the oxidation reaction.

Finally, Dry deposition of SO_2 is always higher on wet surfaces. It can be from the following reasons:

. Water is a important transport medium for the reaction products. It dissolves reaction soluble salts resulting in renewal of surface reaction sites (Lewry et al, 1993b).

. The presence of liquid water on exposed surfaces accelerates SO_2 deposition due to its solubility in water which, furthermore, can penetrate porous materials (Spiker et al, 1992; Johnson et al, 1989).

. Surface moisture reduces the surface resistance to dry deposition (Everett et al, 1988). According to Spiker et al (1993), materials with acid buffering capacity, such as wet carbonate minerals, usually have low surface resistance to dry deposition of SO_2.

6 CONCLUSIONS

Both Laspra dolomite and Hontoria limestone are reactive under SO_2 polluted atmospheres due to their composition (95% dolomite and 100% calcite, respectively) and physical properties.

The reactivity appears to rely on the pore system characteristics. In this way, the higher reactivity of Laspra dolomite is attributed to its smaller pore opening radii and greater specific surface area which favours dry deposition of SO_2 (see Tables I and X). Moreover, the high content of water inside the rock favours chemical reactions.

7 ACKNOWLEDGMENTS.

The authors wish to acknowledge FICYT (Fundación para el Fomento en Asturias de la Investigación Científica Aplicada y la Tecnología) for the financial support of this research. Also to CICYT (Comisión Interministerial de Ciencia y Tecnología Proj. PAT01-1093-CO3-01: "Modelos de interacción piedra-ambiente para el diagnóstico del deterioro de la piedra monumental".

8 REFERENCES

Esbert, R.M. & R.M. Marcos 1983. *Las piedras de la Catedral de Oviedo y su deterioración*. Oviedo: Colegio Oficial de Aparejadores y Arquitectos Técnicos de Asturias.

Esbert, R.M. & R.M. Marcos 1984. Incidencia de los factores ambientales en los mecanismos de alteración de las piedras de la Catedral de Oviedo. *I Congreso Español de Geología. Tomo I:* 635-645.

Esbert, R.M., J.C. García-Ramos, A. Nistal, J. Ordaz, M. Valenzuela, F.J. Alonso & C. Suárez de Centi 1992. El proceso digital de imágenes aplicado a la conservación de la piedra monumental. Un ejemplo: Santa María del Naranco. *Revista de Arqueología, n. 139:* 7-11.

Everett, L.H., T.M. Band, P.A.T. Burman, R.N. Butlin, M.J. Cooke, R.U. Cooke, A.T. Coote, J. Eynon, R.C. Haines, G.O. Lloyd, M.I. Manning, C.A. Price, W.B. Whalley, R.B. Wilsob, G.C. Wood & T.J.S. Yates 1988. *The effects of acid deposition on buildings and buildings materials in the United Kingdom.* Department of Environment.

Folk, R.L 1962: Spectral subdivison of limestone types. Classification of carbonate rocks. *Symp. Am. Ass. Pet. Geol., 1:* 62-84. W.E. Ham.

Furlan, V. & F. Girardet 1992. Pollution atmosphérique et réactivité des pierres. *7th Int. Congress on Deterioration and Conservation of Stone:* 153-161. Lisbon.

Götürk, H. M. Volkan & S. Kahveci 1993. Sulfation mechanism of travertines: effect of SO_2 concentration, relative humidity and temperature. *Conservation of Stone and Other Materials, Vol. 1:* 83-90. RILEM.

Hammecker, C., R.M. Esbert & D. Jeannette 1992. Geometry modifications of porous network in carbonate rocks by ethyl silicate treatments. *7th Int. Congress on Deterioration and Conservation of Stone:* 1035-1062. Lisbon.

Johnson, J.B., S.J. Haneef, C. Dickinson, G.E. Thompson & Wood G.C. 1989. Laboratory exposure chamber studies: pollutant and acid rain presentation rates and stone degradation. *Science, Technology and European Cultural Heritage:* 474-477. Commission of the European Communities.

Lewry, A.J., D.J. Bigland & R.N. Butlin 1992.- A chamber study of the effects of sulphur dioxide on calcareous stones. *7th Int. Congress on Deterioration and Conservation of Stone:* 641-650. Lisbon.

Lewry, A.J., D.J. Bigland & R.N. Butlin 1993a. The effects of sulphur dioxide on calcareous stones: a chamber study. Submitted for publication to *Atm. Environment*

Lewry, A.J., J. Asiedu-Dompreh, D.J. Bigland & R.N. Butlin, R.N. 1993b. The effect of humidity on the dry deposition of sulphur dioxide onto calcareous stones. Submitted for pub. to *Construction and Building Materials*.

Mamillam, M. 1981. Connaisances actuelle des problèmes des remontées d'eau par capillarité dans les murs. *The Conservation of Stone II:* 59-72. Bologne.

Marcos, R.M. 1992. *Tratamientos de Conservación aplicados a rocas carbonatadas: La Catedral de León.* Tesis Doctoral. Departamento de Geología. Area de Petrología y Geoquímica. Universidad de Oviedo.

Marcos, R.M., R.M. Esbert, F.J. Alonso & F. Díaz-Pache 1993. Características que condicionan el comportamiento de la caliza de Hontoria (Burgos) como piedra de edificación. *Boletín Geológico y Minero, Vol. 104, n.5:* 123-133.

Mirwald P.W. & J. Zallmanzig 1986. The influence of air pollution on natural stone. *Glodschmidt informient. Building Protection, 1/86 No. 64:* 10-19.

Doc. NORMAL 7/81 1981. *Assorbimanteo d'acqua per immersione totale. Capacità d'imbibizione.* Rome: C.N.R.-I.C.R.

Doc. NORMAL 11/85 1985. *Assorbimento d'acqua per capillarità. Coefficiente d'assorbimento capillare.* Rome: C.N.R.-I.C.R.

Spiker, E.C., V.J. Comer, R.P. Hosker & S.I. Sherwood 1992. Dry deposition of SO_2 on limestone and marble: role of humidity. *7th Int. Congress on Deterioration and Conservation of Stone:* 641-650. Lisbon.

Valdeón, L., R.M. Esbert & C.M. Grossi 1992. Hydric properties of some Spanish building stones: a petrophysical interpretation. *Mat. Issues in Art and Archaeology III* 911-916.

Basalts from Três Irmãos Hydroelectric Dam, São Paulo State, Brazil:
A review of methodology for alterability evaluation

Roches basaltiques du barrage de Três Irmãos à São Paulo, Brésil: Une révision de la méthodologie pour l'évaluation de l'altérabilité

E.B. Frazão
Instituto de Pesquisas Tecnológicas do Estado de São Paulo, IPT, Brazil

A.B. Paraguassu
Escola de Engenharia de São Carlos, Universidade de São Paulo, Brazil

ABSTRACT: An experimental study on the alterability of basalts from Três Irmãos Hydroeletric Dam foundations was performed by means of alteration test (dry and wetting cycles) and Los Angeles abrasion test carried out after the samples have been submitted to alteration tests. The distribution of weight loss due to alteration and abrasion, with time, and the grain size distribution presented after those tests allowed the evaluation of disaggregation of the studied samples. Several criteria as suggested by different authors, have been used to define alterability indexes. The alterability indexes adopted allowed to state properly the grade of alterability of the basalt samples.

RESUMÉ: À partir des essais cycliques d'altération par saturation et séchage et d'abrasivité Los Angeles nous avons effetuée une recherche sur l'altérabilité de basaltes provenant des fondations du Barrage Hydroeletrique de Três Irmãos, São Paulo. La repartition des valeurs de la perte de masse dûs aux essais d'altération et d'abrasivité en fonction du temps (des cycles), et aussi la repartition granulometrique, ont permi d'évaluer la desagregation des échantillons. Les indices d'altérabilité adoptés ont réussi a établir les degrés d'altérabilité des basaltes etudiés.

1 INTRODUCTION

Several studies on alterability of different types of rocks are described in the international literature. In Brazil, the majority of the studies for geotechnical purposes consider, specially, basalt rocks. It is motivated by their large occurence in the Paraná basin, where the most important hydroeletric power plants are placed, with high consumption of basalts as construction material. Other reason is the high variability in the degree of alteration presented by the basalt rocks and the consequent variability in mechanical behaviour.

2 SAMPLES

Blocks of basalts extracted from the massif during the excavation were left under environmental conditions during six months for the evaluation of their alterability by macroscopic inspection.

Four basalt types were selected for this study; two of them presenting low alterability and two high alterability, as indicated by macroscopic aspects.

The characteristics of the basalt samples were: sample A: dark gray compact basalt; sample B: dark red compact basalt; sample C: grayish green microvesicular basalt; sample D:

grayish green micro-vesicular basalt.

The samples present clay mineral of hydrothermal origin with the following content: 5% (sample A), 15% (sample B), 20% (samples C and D).

3 TESTS

Alteration tests were performed through dry and wetting cycles and by imersion in ethilene glycol. The quantification of the expected disaggregation was done by weight loss.

All the samples submitted to alteration tests through dry and wetting cycles were tested by Los Angeles abrasion, in order to evaluate the expected decay in strength.

These alteration tests were performed through 12h wetting and 24h drying, with 60 cycles for sample A, 54 for sample B, 36 for sample C and 20 for sample D.

Other physical tests carried out, to get complementary informations, were: longitudinal wave velocity measurements (in dry and saturated states), dry density, apparent porosity and water absorption and grain size analysis.

4 ALTERABILITY EVALUATION CRITERIA

The evaluation of the alterability of studied samples was performed by the following criteria:

4.1 *Weight loss index, Iwl (Frazão, 1993):*

$$Iwl = \frac{wo - wf}{wo} \times 100$$

where: wo = initial weight of the sample; wf = final weight.

This criterion was adopted after alteration and Los Angeles abrasion tests.

4.2 *Farjallat's alterability index, Kdt (Farjallat, 1971):*

$$Kdt = 1 - \frac{(200 - F)}{(200 - I)}$$

where: 200 = sum of the maximum possible percentage weight loss (due to both alteration and Los Angeles abrasion tests) F = sum of percentage of weight loss due to alteration and Los Angeles tests at the end of the considered time; I = sum of percentage of weight loss due to alteration and Los Angeles tests at the beginning of the considered time.

4.3 *Yoshida's alterability index, Rf (Yoshida, 1972):*

$$Rf = \frac{(100 - P) R}{100}$$

where: 100 = maximum limit of weight loss; P = weight loss due to alteration test, in a selected time interval; R = strength in a given degree of alteration.

4.4 *Fineness modulus index, Ifm (Frazão, op.cit):*

$$Ifm = \frac{fmo - fme}{fmo}$$

where: fmo = fineness modulus at the beginning of the considered time; fme = fineness modulus at the end of the considered time.

4.5 *Physical indexes, through the comparison of samples with lower and higher degreees of alteration, (after Frazão, op.cit.):*

1) apparent porosity index, In:

$$In = \frac{nw - no}{nw}$$

where: nw = apparent porosity of the sample presenting higher degree of alteration; no = apparent porosity of the sample presenting lower degree of alteration.

2) apparent absorption index, Id:

$$Ia = \frac{aw - ao}{aw}$$

where: aw = apparent absorption of the sample presenting higher degree of alteration; ao = apparent absorption of the sample presenting lower degree of alteration.

3) ultrasound velocity index, Ivp:

$$Ivp = \frac{Vpo - Vpw}{Vpo}$$

where: Vpo = P wave velocity in the sample presenting lower degree of alteration; Vpw = P wave velocity in the sample presenting higher degree of alteration.

5 RESULTS

FIGURE 1 and 2 show the results of alteration tests by dry and wetting cycling and Los Angeles abrasion tests carried out on the samples submitted to those tests.

It can be observed that the samples A and B show lower dis-aggregation rate both after alteration and Los Angeles abrasion tests, but samples C and D show higher disaggregation rate than A and B. It can be noted that sample A and sample B show similar rate and C and D different rates between them.

FIGURE 3 shows the evolution of fineness modulus index with time of the four samples. The behaviour is similar to that observed in the previous Figures.

FIGURES 4 and 5 show the evolution of Farjallat's al-terability index (Kdt) and Yoshida's alterability index (Rf), respectively. Note that samples A and B show again low dis-aggregation, in similar rate, and samples C and D show higher disaggregation, but with different rate between them.

TABLE 1 presents the results of physical and physico-mechanical tests carried out on the four samples.

The analysis of the data set of TABLE 1 shows that sample A represents the lowest alterable with basalt among the samples tested.

With the data of TABLE 1, the physical and physico-mechanical indexes were calculated, consi-dering the A as the sound sample (or the lowest alterated). These results are shown in TABLE 2. It can be observed that in this Table, that the indexes values are in

agreement with other indexes as formerly presented.

Finally, it can be noted that all the alterability indexes were suitable to distinguish the alterability of the four samples studied.

6 CONCLUSION

According to the results of this study it can be concluded that:
   a) the alteration tests per-formed, mainly the drying and wetting test, were suitable to point out the alterability of the samples;
   b) the Los Angeles abrasion test was efficient to verify the mechanical decay due to the alteration process;
   c) the alterability indexes adopted were suitable to represent the alterability of the samples;
   d) according to the alterability indexes adopted the following sequence of alte-rability of the samples can be stated: A < B < C < D.

REFERENCES

FARJALLAT, J.E.S. 1971. *Estudos ex-perimentais sobre degradação de rochas basálticas*. São Paulo. (Phd Thesis, Geosciences Insti-tute, University of São Paulo).

YOSHIDA, R. 1972. *Contribuição ao conhecimento das características tecnológicas de materiais rocho-sos*. São Paulo. (Phd Thesis, Geosciences Institute, University of São Paulo).

FRAZÃO, E.B. 1993. *Metodologia para avaliação da alterabilidade de rochas a partir de estudo expe-rimental em amostras de basaltos da U.H.E. de Três Irmãos*. Estado de São Paulo. São Carlos. (Phd Thesis, São Carlos Engeneering School, University of São Paulo).

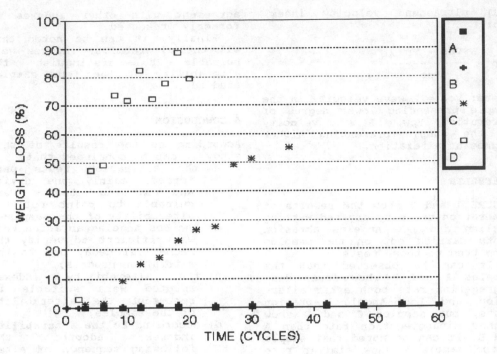

Figure 1. Evolution of weight loss due to alteration tests for A, B, C and D basalt samples.

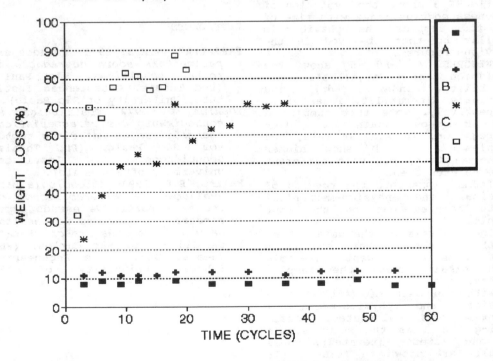

Figure 2. Evolution of weight loss due to Los Angeles abrasion tests for A, B, C and D basalt samples.

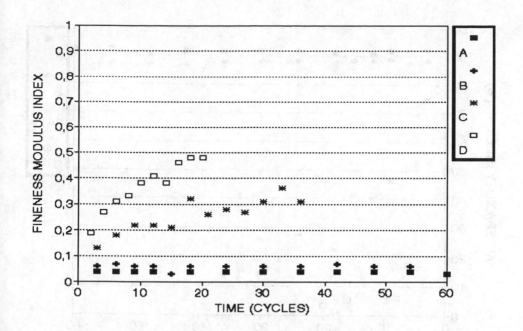

Figure 3. Evolution of fineness modulus index for the A, B, C and D basalt samples.

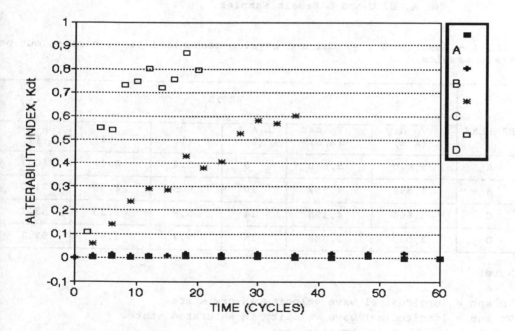

Figure 4. Evolution of Farjallat's alterability index for A, B, C and D basalt samples.

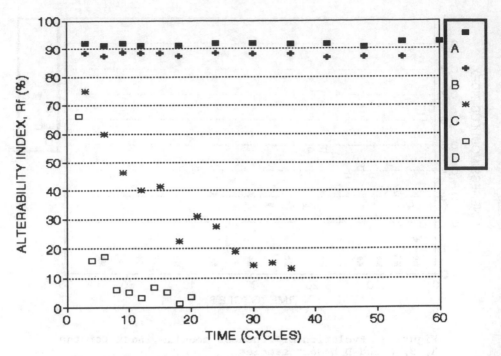

Figure 5. Evolution of Yoshida's alterability index
for A, B, C and D basalt samples.

TABLE 1 - Results of physical and physico-mechanical tests carried out on
basalt samples.

	TESTS					
SAMPLES	Vp dry (m/s)	Vp sat (m/s)	L.A.A. (%)	n (%)	a (%)	E.G. (%)
A	5,470	5,680	8	3.4	0.42	10.7
B	4,698	4,957	10	6.9	2.50	31.0
C	3,864	4,395	14	8.9	3.33	53.1
D	3,020	3,888	15	9.7	3.65	92.7

Note:

Vp sec = longitudinal wave velocity in dry state.
Vp sat = longitudinal wave velocity in saturated state.
L.A.A. = Los Angeles abrasion loss.
n   = apparent porosity.
a   = apparent absorption.
E.G.   = ethylene glycol loss.

TABLE 2 - Results of physical and physico-mechanical indexes for the B, C and D basalt samples relative A basalt sample.

PROPERTIES	ALTERATION INDEX VALUES
apparent porosity, In	$InD = (nD - nA)/nD = 0.65$ $InC = (nC - nA)/nC = 0.61$ $InB = (nB - nA)/nB = 0.51$
apparent absorption, Ia	$IaD = (aD - aA)/aD = 0.88$ $IaC = (aC - aA)/aC = 0.87$ $IaB = (aB - aA)/aB = 0.83$
Los Angeles abrasion loss, Ila	$IlaD = (laD - laA)/laD = 0.47$ $IlaC = (laC - laA)/laC = 0.43$ $IlaB = (laB - laA)/laB = 0.20$
Ethylene glycol loss, Ieg	$IegD = (egD - egA)/egD = 0.88$ $IegC = (egC - egA)/egC = 0.80$ $IegB = (egB - egA)/egB = 0.65$
ultra sound velocity, Vp	dry state $IvpD = (VpA - VpD)/VpA = 0.45$ $IvpC = (VpA - VpC)/VpA = 0.29$ $IvpB = (VpA - VpB)/VpA = 0.14$ saturated state $IvpD = (VpA - VpD)/VpA = 0.13$ $IvpC = (VpA - VpC)/VpA = 0.23$ $IvpB = (VpA - VpB)/VpA = 0.32$

Weatherability of calcareous rocks: Quantitative preliminary approach
Altérabilité des roches calcaires: Évaluation quantitative préliminaire

L. Aires-Barros, C. Gomes da Silva, C. Figueiredo & P. Rego Figueiredo
Laboratory of Mineralogy and Petrology of Instituto Superior Técnico (LAMPIST), Technical University of Lisbon, Portugal

ABSTRACT: Three types of limestones and three types of calcitic marbles were submitted to thermal fatigue weathering essays (ageing laboratory tests).

Before the beginning of the laboratory tests, at each of their pauses ("1 and 5 laboratory years") and at their end the rock samples were weighed, their exposed surfaces studied by reflected light microscopy, the textural characteristics of each of such surfaces have been converted, by means of a video camera, into grey level images susceptible of being processed using image analysis techniques; pH of immersion waters was measured and such waters were chemically analysed in terms of their major cations.

Taking into account data obtained from these ageing laboratory tests two weatherability indexes were calculated, one of them being based on weight loss and chemical leaching suffered by the samples and the other based on the evolution of textural characteristics of the exposed surfaces.

The evolution of such indexes with time is discussed, and the rocks are seriated in terms of those indexes (weatherability seriation).

RÉSUMÉ: Trois types de calcaires et trois types de marbres calcitiques ont été soumis à des essais de fatigue termique.

Avant les essais, à chaque pause et à la fin on a pesé les échantillons et etudié leurs surfaces exposées à l'aide de la microscopie de lumière réfléchie. À l'aide d'une caméra video on a fait l'acquisition d'images en niveaux de gris des caractéristiques texturelles de ces surfaces susceptibles d'être traitées par les techniques d'analyse d'images. On a mesuré le pH des eaux d'immersion et dosé leurs cations majeurs.

À partir des résultats de ces essais, on a calculé deux indices d'altérabilité, l'un en fonction de la perte de poids et du lessivage soufferts par les échantillons essayés; l'autre basé sur l'evolution texturelle des caractéristiques des surfaces exposées à l'attaque laboratorial accéléré.

L'evolution de ces indices avec le temps est discutée et les roches sont sériées en fonction de ces indices (sériation selon l'altérabilité).

1 THERMAL FATIGUE TEST

The thermal fatigue weathering tests used at LAMPIST (Laboratory of Mineralogy and Petrology of Instituto Superior Técnico) consist of submitting disc-type rock specimens with a polished surface to repeated cycles, each one with a period of insolation (at about 70° C) followed by a period of immersion in distilled water (AIRES-BARROS, et al. 1970).

Each test lasted 10 "laboratory years" with pauses at 1 and 5 "years".

Each "laboratory day" consists of a 15 min. (10 min. of insolation + 5 min. of immersion) cycle and so each "laboratory year" corresponds to 365.25 "laboratory days", i.e., approximately 91.5 hours.

Before the beginning of the laboratory ageing tests and at each of their pauses rock samples were weighed, their exposed surfaces studied by reflected light microscopy, the textural characteristics of each of such surfaces having been converted, by means of a video camera, into grey level images susceptible of being processed using image analysis techniques; pH of immersion waters was measured and such waters were chemically analysed in terms of their major cations.

2 ROCK SAMPLES STUDIED

2.1 Geological setting

The calcitic marbles referred to in this paper were collected at São Sebastião, Rosal and Encostinha quarries, under exploitation in the surroundings of Borba (NE Alentejo/Portugal) in the Southern part of Estremoz anticline's NE flank.

The three types of limestones tested belong to bio-clastic formations of upper Jurassic (Azul de Sintra) and Cretaceous (Encarnadão and Amarelo de Negrais) ages (West Lisboa-Sintra area).

Azul de Sintra type was exploited in the region of Sintra (Alto da Biscaia quarry) while rock types usually known as Encarnadão and Amarelo de Negrais are extracted from Negrais-Pero Pinheiro quarries.

Both of these groups of rocks have been used along History as ornamental rocks applied in a lot of monuments all over the country. Nowadays they continue to be important ornamental rocks used in large buildings and monuments.

2.2 Petrography

Rock material from Rosal quarry is rose-coloured, has medium granoblastic texture and is crossed by thick dark green to greyish veins. Microscopic observation reveals the rose-coloured part to be formed by abundant twinned calcite and some quartz, while veins are mainly formed by phylossilicates, fine grained calcite and opaque minerals.

The marble exploited at Encostinha quarry has a colour between salmon and cream. Its texture is medium granoblastic. A few veins cross it. Under the microscope the granoblstic mattrix shows to be mainly constituted by abundant twinned calcite and some quartz. The veinlets are formed by phylossilicates, turmaline, quartz, opaque minerals and fine-grained calcite.

Crystalline limestone from São Sebastião quarry is a medium granoblastic whitish rock crossed by thin and scarce reddish veinlets. Microscopic observation reveals its groundmass to be formed by abbondant twinned calcite and some quartz, while veins are formed by fine-grained calcite and iron oxides.

Azul de Sintra type is a blue-greyish crystalline limestone with a granoblastic texture.

The so-called Encarnadão and Amarelo de Negrais types are respectivelly rose and yellow microscrystalline limestones. In thin sections they can be classified as biosparite-biomicrosparite, the yellow one being lighty marly.

2.3. Chemical composition

Chemical analysis of sound samples of rocks submitted to ageing test are presented in Table I.

Table 1. Chemical analysis of samples.

	Fe_2O_3 (%)	CaO (%)	MgO (%)
A	0.27	50.85	0.52
B	0.21	51.64	0.40
C	0.08	52.60	0.54
D	0.12	56.33	0.58
E	0.09	54.39	0.31
F	0.97	48.48	0.38

General legend:
A - Rosal Marble   B - Encostinha Marble
C - S. Sebastião Marble   D - Azul Sintra
E - Encarnadão Negrais   F - Amarelo de Negrais

3 RESULTS AND WEATHERABILITY INDEXES

The main results derived from ageing accelerated tests described (chemical leaching and weight lost) are referred in Table 2 and 3.

Table 2. Chemical analysis of immersion waters.

	Years	Fe_2O_3 (%)	CaO (%)	MgO (%)
	1	0.000	0.017	0.001
A	5	0.000	0.013	0.001
	10	0.000	0.009	0.000
	1	0.000	0.003	0.000
B	5	0.000	0.006	0.000
	10	0.000	0.008	0.000
	1	0.000	0.002	0.000
C	5	0.000	0.010	0.000
	10	0.000	0.010	0.000
	1	0.000	0.014	0.000
D	5	0.000	0.018	0.000
	10	0.000	0.020	0.000
	1	0.000	0.022	0.000
E	5	0.000	0.009	0.000
	10	0.000	0.009	0.009
	1	0.000	0.004	0.001
F	5	0.000	0.009	0.001
	10	0.000	0.008	0.001

Table 3. Sample weights and pH of immersion waters as a function of laboratory time.

	Years	Weight (g)	pH
	0	41.20	
A	1	41.18	7.35
	5	41.15	7.72
	10	41.03	7.95
	0	43.45	
B	1	43.42	7.90
	5	43.39	8.00
	10	43.36	8.20
	0	42.33	
C	1	42.32	8.16
	5	42.29	8.16
	10	42.26	8.16
	0	41.21	
D	1	41.17	7.88
	5	41.15	7.97
	10	41.12	7.98
	0	40.40	
E	1	40.38	8.11
	5	40.35	8.14
	10	40.32	8.15
	0	38.98	
F	1	38.95	7.98
	5	38.91	8.29
	10	38.87	8.30

Table 4. Weatherability indexes.

	years	K_{geoq}	K_{text}
	1	5	3.38
A	5	12	5.86
	10	17	8.32
	1	7	5.91
B	5	14	8.76
	10	21	8.93
	1	2	2.72
C	5	9	6.99
	10	17	9.71
	1	7	3.86
D	5	13	5.43
	10	20	5.95
	1	5	3.70
E	5	12	4.60
	10	19	5.39
	1	8	1.75
F	5	18	4.70
	10	28	9.91

By looking closely at the values of K_{geoq} one can see that, according to their susceptibility to weathering, the rocks under study can be seriated as follows:

K_{geoq}. Rosal $= K_{geoq}$. S. Sebastião $< K_{geoq}$. Encarnadão $< K_{geoq}$. Azul $< K_{geoq}$. Encarnadão $< K_{geoq}$. Amarelo

From these data two types of weatherability indexes have been derived and are discussed:

3.1 Weatherability index based on chemical leaching

AIRES-BARROS et al (1970) defined an alterability index using the following formula:

$$K = (1 + gj) K_{min}$$

where K is the weatherability index; K_{min} is a factor regarding mineralogical influence - it can be considered to have the same value as the weight loss, ΔP; gj is a factor regarding textural, permeability and porosity influence; it can be assumed that its value is the same as the total chemical leaching.

The application of the above formula to our weathering tests originated the values in table 4. So that no confusion will arise when referring to this index it was designated K_{geoq}.

3.2 Weatherability index based on textural evolution (pores, fractures, joints, and so on) of the rock sample surfaces.

AIRES-BARROS et al. (1991) defined a weatherability image analysis index (designated K_{text} in this paper) by means of the following equation:

$$K_{text} = I_{Rto} = \frac{1}{(1-A_{to})} \times \frac{\sum_{j=1}^{n-1} \Delta_j A}{t_{n-1} - t_o} \times 100$$

where I_{Rto} is the image analysis weatherability index of the R rock (relative) to t_o; n is the number of laboratorial observations; $t_{n-1} - t_o$ is the laboratory test time (lab. years); $\Delta_j A = A_j - A_{j-1}$ is the relative "voids" are increment at j time interval; $\Delta_j t = t_j - t_{j-1}$ is the j time interval.

Using this index we have an overall average weatherability rate conditional to that part of rock initially considered sound.

This weatherability index proposed needs acurate measurements of the effects observable on rock samples. The image acquisition of voids (pores, fractures and "enlarged" joints) was performed on the lower gray levels of the video camera image, using a "robust" thresholding algorithm ("black top hat") as they show a good contrast against the background. Electronic and optical noises induced by image acquisition process produce poor definition of voids. In order to have a more clear definition of voids, a morphological filter ("conditional opening transform") was performed on the previously defined binary images. At this step, compared observations of binary and stereomicroscope images showed "reasonable" correspondence.

K_{text} values obtained (table4) enable to seriate the tested rocks as follows:

K_{text} Encarnadão < K_{text} Azul < K_{text} Encostinha < K_{text} S. Sebastião < K_{text} Amarelo

4. FINAL REMARKS

It should be emphasized that it is possible to extend the weatherability index proposed by AIRES-BARROS (1970) to igneous, sedimentary and metamorphic rocks.

The above mentioned index can be adapted in such a way as to be used by image analysis techniques.

It is also possible to characterize quantitatively the weatherability grade using ageing laboratory tests, accompanying such ageing by measurements based upon image analysis and adapting the quality weatherability index defined.

The studies carried out enabled us to establish a "bridge" with similar and parallel studies carried out on building rocks used in monuments. In order to obviate the problems arisen one can notice that the use of profilometer image analysis techniques as well as the comparison between the effects of the ageing laboratory test and the effects of decay observed in ornamental stones pannels (v.g. Estrela Basilica) can be very useful.

The research carried out shows both the theoretical and the pratical interest of the weatherability studies on one hand. On the other hand it can propitiate and generate future studies, using image analysis techniques and, above all, developing non-destructive analysis techniques.

REFERENCES

AIRES-BARROS, L. (1970) Note préliminaire sur un indice d'altérabilité - 1st Int. Cong. of IAEG, vol. 1, pp. 573-577. Paris

AIRES-BARROS, L. et al. (1975) Dry and wet laboratory tests and thermal fatigue of rocks. Eng. Geol., vol. 9, pp. 249-265

AIRES-BARROS, L. et al. (1991) Experimental correlation between alterability indexes obtained by Laboratorial ageing tests and by image analysis Coll. Int. La détérioration des matériaux de construction, pp. 199-208. La Rochelle

AKNOWLEDGEMENTS

This study was partly financed by STRDA/C/CEN/665/92 project (JNICT, Portugal)

7th International IAEG Congress / 7ème Congrès International de AIGI, © 1994 Balkema, Rotterdam, ISBN 90 5410 503 8

Physical properties and petrographic characteristics of some Bateig stone varieties

Propriétés physiques et pétrographiques de quelques variétés de 'Pierre Bateig'

S. Ordóñez & M. Louis
Universidad de Alicante, Spain

M. A. García del Cura
Universidad de Alicante & Instituto de Geología Económica, Facultad de Ciencias Geológicas, Madrid, Spain

R. Fort, M. C. López de Azcona & F. Mingarro
Instituto de Geología Económica, Facultad de Ciencias Geológicas, UCM, Madrid, Spain

ABSTRACT: Bateig stone is an allochemical calcareous rock (mainly biosparite - biomicrite) worked in Novelda, named Novelda stone too (Alicante province, Spain) that belongs to a transgressive unit of Middle - Late Miocene age. This stone is a highly workable material that often was used in monumental Spanish architecture. Using criterium publising by the M.I.T., Bateig stone may be classified as a Medium to Low Limestone. Data from Mercury Porosimeter has been used to stimate a DDS (Durability Dimensional Stimation). The DDS equation is supported on the calculus of the pressure of salt crystallization in porous mediums. The DDS values obtained, for different varieties of Bateig stone let us establish that the microporosity, size pores less than 0,1 μm, is the main factor responsible for durability of materials.

RESUMÉ: La "Pierre Bateig" est une roche carbonatée à allochèmes (biosparite-biomicrite) extraite en Novelda, elle est appellée aussi "Pierre de Novelda" (Province de Alicante, Espagne); elle appartient à une unité transgresive du Miocène Moyen- Supérieur. Cette roche est très facile à travailler et souvent a été utilisée dans la architecture monumentale espagnole. En utilisant le critérium publié par le M.I.T. la "Pierre Bateig" peut être classifiée comme "Moyen à Low Limestone". Les données du Porosimètre de Mercure ont été utilisées à la fin d'établir une DDS (Durability Dimensional Stimation). L'équation de la DDS est basée dans le calcule de la pression de cristallisation des sels dans un moyen poreux. Les valeurs obtenues pour les différentes variétés de la Pierre Bateig permettent d'établir que la microporosité, pores plus petits de 0,1 μm, est le principal facteur responsable de la durabilité des matériaux.

1 INTRODUCTION

Bateig stone is an allochemical calcareous rock (mainly biosparite - biomicrite) extracted in Elda municipal district (Alicante) from the early XX century, however is knew in scientical literature as Novelda stone too (Alicante), because the manufacture is located in Novelda. Bateig is the name of hill where the stone are mined. Similar stones were mined in the past in other geographical location in Novelda municipal district, Casas del Señor,...

The Bateig stone is a highly workable material that often was used in monumental Spanish architecture since XIXth-XXth century, mainly in Madrid (Linares Palace, Príncipe Pío Railway Station, Ministry of Agriculture, Palace of Comunications, Almudena Cathedral, Central Telephone Company Office...) and Valencia (General Post Office, some elements of City Hall...), and Alicante (some modernist houses,) ; in Novelda we find monuments in Bateig stone since XIIIth century: La Mola Castle (XIIIth century), San Pedro Church (XVIth century), City Hall (XVIIth century) and Modernist Museum-House (1904). To day this stone is used in modern building and restoration of monuments (Almudena Cathedral of Madrid, Leon Cathedral, Salamanca monuments,...) too. This stone was used in the restoration of the Old Facade of the University of Alcalá de Henares (Madrid province) dated in 1925 (García de Miguel at al, 1992)

The present day commercial varieties of Bateig stone are:

Bateig Blanco (White Bateig), Bateig Azul (Bleu Bateig), Bateig Llano (Layer Bateig), Bateig Diamante (Diamond Bateig) and Bateig Fantasia (Fantasy Bateig).

The active quarries of Bateig stone are located in the west part of Bateig hill (551 m a.s.l.). There are two main zones of quarries, one zone just located in the side of the hill, from there are extracted White Bateig, Diamond Bateig, Bleu Bateig and locally Fantasy Bateig; and another located in the foot of the hill, from there are extracted Layer Bateig. At the moment,

geological relations among diferents varieties ar not well understood. The production c commercial products are 9.000 cubic meters pe year, and the annual extracted material are up t 30.000 cubic meters. Dimension stone ar extracted using diamond wire cutting machine The total lengt of active faces are more than km, and the average heigh less than 10 m sometimes with benches dipping in agre stratification. The potencial resources are up t 13 millions of cubic meters. The indentifie reseves are probably less than 5 millions of cubi meters.

2 GEOLOGY

Bateig stone outcrop belongs to a transgressiv unit of Middle - Late Miocene age in the Betic Balearic domain (External Zone of Bet Cordillera- Prebetic Zone), that represents north-eastern prolongation of the external part the Betic thrust and fold belt. This transgresiv unit is developed in a basin formed during th Early and Middle Miocene, coetaneously with th westward drift of the Internal Zone of the Bet Cordillera. This foreland basin , with an activ sector (foredeep), located in front of the ne formed reliefs, received huge olistostrom masses. (Sanz de Galdeano & Vera, 1991). Th Trias (Keuper facies) has played an importa role. The diapiric processes affect the Bate stone and condition their outcrops and folds.

From sedimentological point of view Bate stone primary sediments belong to a continet shelf with no continous deposition marked b erosional unconformities, strongly agitated wate that are ideals for benthic, nektonic and planton organismes at the same time causing stror reworking and movement of sediments. Sediments displays abundant organic shells (mainly foraminifera), detrital quartz and neoformed silicates (attapulgite, glauconite and cristobalite). Sedimentary structures such as extensive bioturbation, burrows and flaser bedding are abundant, mainly in the variety named "Fantasy".

3. PETROGRAPHY

The original sediments of Bateig stone displays laminated primary structures, however was deeply bioturbated and primary structures are destroyed, only some varieties as Layer Bateig preserve primary laminations. The characteristic features of Fantasy Bateig, as pointed out before, may be related with burrowed laminated sediments, while other varieties display homogeneous aspec probably related with deepest reworking and burrowing of sediments. In the outcrops the "Blue Bateig" variety display a sharp contact with "White Bateig" variety that apparently cut bedding planes, and probably may be interpreted as conected with a later diagenetic process of oxidant water infiltration that flux trough primary and secondary (faults and extensional joints) high permeability layers, and flux the moderate reduced formation waters. The "Diamond Bateig" layers are allways located in the lower part of stratigraphic section just under "Blue Bateig" outcrops.

Some textural and mineralogical microscopical features may be used as differential character of the five varieties studied in this paper (table 1).

The White Bateig (W), and the Diamond Bateig (D) varieties have up to 5 % of siliceous cement and authigenic layered silicates: smectites

and palligorskite. Fantasy Bateig (F) has mainly detritic layered silicates: micas. Glauconite is widespread in Bateig stone varities: small grains and internal moulds of carbonate microfossil test mainly Foraminifera.(García del Cura et al. 1994).

The **White Bateig** is a biomicrite that consist mainly of a 15 % of terrigenous components, mainly angular microcrystalline quartz, dolostone extraclast and minor feldspars, muscovite, tourmaline,.. Two modes (0.06 - 0.12 & 0,12 - 025 mm) has been identified in terrigenous fraction. The 56 % of components are **fossils** (foraminifera: *Rotalidae, Textularidae, Globigerinidae*,... ; mollusca and briozoa). The 19% are orthochems (15% micrite matrix and 4% sparite cement). As authigenic constituents has been identified opalo C, phylosilicates and glauconite. Some sparite cements and even fossils display a grain disminution type of recrystallization. The under optic microscopy porosity (inter- & intra granular) is up to 10%.

The **Blue Bateig** is a biomicrite that consist mainly of a 15 % of terrigenous components, mainly monocrystalline and polycrystalline quartz, dolostone extraclasts, K- Feldspar, phyllosilicates, opaque minerals, tourmaline, muscovite, rock fragments (metacuarcite and slates). It has been identified clay galls and also clay filling shells. The 60% of components are fossils (foraminifera: *Textularidae, Rotalidae, Globigerina s.p., Globigerinoides s.p., Globorotalia s.p.*,... briozoa, echinodermata). The 15 % are orthochems (10 % micritic matrix and 5 % sparry cement). Also has been identified locally a siliceous fibrous cement. The under optic microscopy porosity is lower than the 10 %.

The **Layer Bateig** is a sandy biosparite with

Table 1. Summary of petrographical features of the Bateig stone.

Components	W	B	L	D	F
% Terrigen.	15	15	20	6	15
% Fossils	56	60	55	69	65
% Ortochems	19(m)	15(m)	5(s)	20(m)	12
% Porosity	10	10	20	5	8
% Calcite	85	86	78	78	85
% Quartz	7	8	12	5	6
% Philosil.	5	2	3	5	6
% Dolomite	2	2	2	2	2
% K Felds.	1	2(w)	5 (w)	1	1
Accesor.	mu, t	t, mu	t, mu	t, mu	mu, z

m = micrite; s = sparite; mu = muscovite; t = turmaline; w = weathered; z = zircon.

Table 2. Physical properties of bateig stone.

CHARACT.	W	B	L	D	F
Type (MIA)	Medium	Medium	Low	Medium	Low
Porosity 7-0,005 μm	18,32	12,57	14,60	17,79	
Porosity 200-7 μm	0,35	0,17	1,33	17,79	
Mineral.	cal, q,	cal, q,	cal, q,	cal, q,	cal, q,
Chemistry $CaCO_3$ $MgCO_3$	>85 >1	>86 >1	>78 >1	>87 >1	>85 >1
Density	2,26	2,31	2,13	2,20	2,10
W. absort.	4,86	4,32	6,60	5,79	6,10
Compress. strength	35,0	31,27	22,05	35,41	30,07
Modulus rupture	6,08	5,00	*	*	*
Sonic velocity	3392	3545	3328	3569	*

some laminations that consist of 20 % of terrigenous components, mainly monocrystalline quartz with anhydrite inclusions, probably derived from idiomorphic authigenic quartz of Keuper facies (Trias), besides wheatered K - Feldspar, schist and clay balls. The 55 % of components are fossils (foraminifera: *Rotalidae, Textularidae, Globigerina s.p.,..*; briozoa fragments, mollusc shells, echinoderma plates. The 5 % are ortochems, mainly sparite cement, sometimes has been identified some calcitic rim cements and locally some grain disminution processes over skeletic carbonates. Besides siliceous interparticula and intraparticula cements has been described too.

The **Diamond Bateig** is a biomicrite that consist of 6 % of terrigenous components mainly monocrystalline quartz, K - Feldspar, schist and clay balls. Terrigenous components display bimodal size distribution. The fossils components size display low sorting distribution: *Rotalidae, Globigerina s.p., Turborotalia s.p. and Heterostegina s.p.* has been identified. The main ortochem component is micrite. The under optic microscopy porosity is lower than 5 %.

The **Fantasy Bateig** is a biomicrite - biosparite that display a highest variability in the micrite/sparite ratio and percent of terrigenous, from one thin section to another. It is difficult to define the average of components, but may described as a mix of the previous described varieties.
Amphistegina s.p. and scarce *Globigerinidae* are present.

4 MECHANICAL PROPERTIES AND CHARACTERISTICS OF STONES.

A summary of physical properties and characteristics of Bateig varieties using criterium published by the Masonry Institute of America (M.I.T.), see Harbem & Purdy (1991), is contained in the table 2. The White, Blue and Diamond Bateig varieties may be classified as Medium Density Limestones, and Layer and Fantasy Bateig belong to Low Density Limestone type.

The sonic velocity of Bateig stone varieties is also contained in table 2. We tested using Lineal Regresion Analysis to compare sonic velocity (m/sec) values of Bateig stones varieties with compressive strength (Mpa) ones. The equation of the line is:

$$UCS(Mpa) = 0,0262 V(m/sec) - 57,56$$

, the correlation coefficient, $r = 0,71$, and data number is up to 24.

The color measured using L*a*b* color system, recommended by CIE, over unpulished samples has been listed in the table 3. Color values reflected there are the average of 36 measures obtained from three samples for each Bateig type.

5. PORE SIZE DISTRIBUTION AND DURABILITY.

Pore size and pore volume distribution was obtained by Mercury Porosimeter data, Autoscan - 60. Mercury Porosimentry offers the possiblity to obtain data about porosity, pore size (from 213 μm to 0,0018 μm) and normalized volume vs diameter.

Table 3. Color measured using l*a*b* color system.

	W	B	L	D	F
L*	83,9	79,9	81,8	85,5	-
a*	,5	-0,1	1,3	0,7	-
b*	8,0	5,7	12,0	9,1	-

L* Black to White.
a* - green + red.
b* - blue + yellow

The data from Mercury Porosimeter may be used to stimate the stone durability. Thus may predict the behaviour of a stone during salts crystallization an/or hydratation and frost weathering.

The theoretic problem of the pressures of crystallization of salts in porous mediums was studied by Fittzer and Shethlage (1982) by analogy with the thermodynamics of the freezing of dissolutions in porous mediums, developed by Everet (1961). According to these authors the crystals of the saline phases show a preference to grow in large pores. When the crystallization of the saline phases in the large pores ceases, for having filled the available space, crystallization will then continue in the connected small pores.

When the phase minerals precipitate in the pore, the pressure of crystallization can be expressed, according to Fittzer and Shethlage (1982), as:

$$p = 2\sigma \left(\frac{1}{r} - \frac{1}{R} \right)$$

, where p, is the pressure of crystallization; σ is the ionic interfacial tension salt-solution ; r and R are the radia of the coarse and small pores considered, respectively.

Following the method employed Rossi-Manaresi and Tucci (1989), we m estimate the pressures of crystallization that develope in the rocky materials. For this we h considered five intervals, or classes, of pores function of the radius expressed in μm : class r < 0.01; class 2, 0.01 < r < 0.1; class 3, < r < 1; class 4, 1 < r < 10; class 5, 10 < 100. As centers of the classes of distribut we have considered respectively the value 0.003, 0.03, 0.32, 3.16, and 31.6 μm.

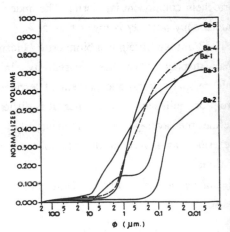

Fig. 1 Normalized Volume vs Diameter Cur of Bateig Stones varieties. Ba-1, White Bate Ba-2, Blue Bateig; Ba-3, Layer Bateig; Ba Diamond Bateig; Ba-5, Fantasy Bateig.

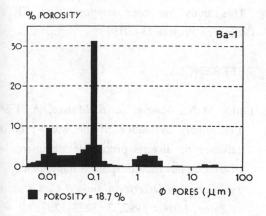

POROSITY = 18.7 %

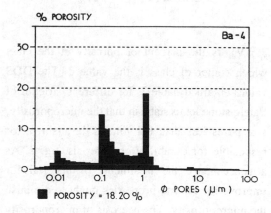

POROSITY = 12.7 %

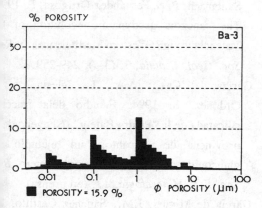

POROSITY = 15.9 %

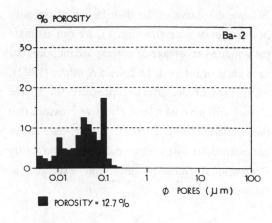

POROSITY = 18.20 %

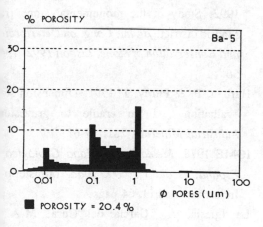

POROSITY = 20.4 %

Fig. 2 Histograms of pore size diameter of Bateig stones.

Using the curve of the distribution of porosity in function of pore size (fig. 1), we can estimate the volume percentage of correspondant pores for each class or interval, La Iglesia & others (1992).

The sum of the effective pressures of all the classes will give us a total effective pressure that the material will support. From the above considerations we may define a Durability Dimensional Stimator, that may expresed as follow:

$$DDS = \Sigma \left[4 \left(\frac{1}{d} - \frac{1}{D} \right) \% V_d \right]$$

, $\% V_d$, is the percent of porosity of the class which center of class is the value d. The DDS values obtained, table 2, for different varieties of Bateig stone let us stablish that the microporosity, size pores less than 0,1 μm, is the main factor resposible for durability of materials, fig. 2. As a consecuence the treatments of material for improve the natural properties ought to diminish the microporosity. The necrosis of microporosity may obtained also with natural climatic conditions exposure during long time, as has been point out by La Iglesia & others (1991). The DDS values obtained for samples studied in this paper is contained in the table 4.

Table 4

1	2	3	4
W	64,64	18,7	35,0
B	85,82	12,16	31,27
L	62,88	15,90	22,05
D	51,05	18,20	35,41
F	22,10	20,40	30,07

(1)Durability Dimensional Stimation (m^{-1}).
(2)Percent of porosity.
(3)Uniaxial comprenssive strength (Mpa)

This study has been supported by CICY (Research Project 93-0019)

REFERENCES

Bello, M.A., Martin, L. & Martin, A. 199 Preliminary evaluation of the water-proofir efficacy of diverse protective treatments diverse stones used in spanish monuments. 7 Int. Cong. on Deterioration and Conservati of Stone, Lisbon 1992, 3, 1273-1277.

Fonboté, J.M., Guimerá, J., Roca, E., Sabat, Santanach, P. & Fernández-Ortigosa, F. 199 The cenozoic geodynamic evolution of Valencia trough (Western Mediterranean). Re Soc. Geol. España, 3, (3-4), 249-259.

García del Cura, M.A., La Iglesia, A. Ordóñez, S. 1994. Estudio dela fracci silicatada de la "Piedra Bateig" (Neógeno de provincia de Alicante): un ejemplo glauconitización en medio marino somero. B Soc. Esp. Mineralogía, 17, (en prensa).

Garcia de Miguel, J.M., Sanchez Castillo, Puche Riart, O. & Gonzalez Aguado, M. 1992. Study of the monumental stone fr Madrid district. 7th Int. Cong. on Deteriorati and Conservation of Stone, Lisbon 1992, 1, 56.

Harben, P. & Purdy, J. 1991. Dimension sto evaluation . From cradle to gravesto Industrial Minerals, 281: 47-61.

IGME 1978. Memoria del Mapa Geológico España, escala 1:50.000 Elda. Ministerio Industria. Madrid. 64 pags.

La Iglesia, A., García del Cura, M.A. Ordóñez, S. 1992. Aproximación a la físi química e la alteración de los materiales pétr de la Catedral de Toledo. Congreso Rehabilitación del Patrimonio Arquitectónic

Edificación, Canarias, I, 344-350.

Louis Cereceda, M., Alonso Pascual, J., Martinez Pastor, V. & Alcaide Romero, J.S. 1992. Caractérisitiques du grès naturel de Bateig très utilisé dans l'architecture. *7th Int. Cong. on Deterioration and Conservation of Stone,Lisbon 1992*, 3, 1205-1211.

Rossi- Manaresi, R. & Tucci, A. 1989. Pore structure and salt crystallization: "salt decay" of Agrigento biocalcarenite and "case bardening" in sandstone. *Proc. 1st Int. Symposium Bari "The conservation of monuments in the Mediterranean Basin"*, 97-100 p.

Fittzer,B and Shethlage,R. 1982. Ueber Zusammenhange zwischen salzkristallizations druck und Porenradienverteilung. *G.P. New Letter*, 3, 13-24.

Sohnee, O. 1982. Electrolyte crystal-aqueous solution interfacial tensions from crystallization data. *Jour. Crystal Growth* 57, 101-108.

Sanz de Galdeano, C., & Vera, J.A. 1991. Una propuesta de clasificación de las cuencas neógenas béticas. *Acta Geologica Hispánica*, 26, 205-227.

Engineering geological problems related to the preservation of some medieval castles in Slovakia

Problèmes géologiques et géotechniques en rapport avec la préservation de quelques châteaux médiévaux en Slovaquie

A. Hyánková & J. Vlčko
Department of Engineering Geology, Comenius University, Bratislava, Slovakia

ABSTRACT: Research work and practical studies aimed at toward the protection and conservation of historic sites have shown important role of geosciences, namely engineering geology. The paper presents some results of two-year engineering geological research programme dealing with restoration of some medieval castles in Slovakia.

RÉSUMÉ: Les travaux de recherche et les connaissances pratiques dirigées à la protection et conservation des sites historiques ont montré la tâche des sciences géologiques, surtout de la géologie de l'ingénieur. L'article apporte quelques résultats da la résolution du programme de géologie de l'ingénieur de deux ans dirigé à la protection des châteaux médiévaux en Slovaquie.

1 INTRODUCTION

A great number of medieval castle ruins (dating back mostly to the 13th century) located at steep ranging cliffs belongs to the typical landscape of Slovakia. Since the beginning of the 18th century the castles were loosing their primary defence function, many of them were destroyed by fire or during military actions and later abandoned. At present the majority of the castles have the character of ruins and they are in a rather poor state. Some of them endanger the safety of settlements (living places) and the traffic below, others are located in dense forests, in places difficult of access. The latter being part of nature reserves represent a hazard mostly to individuals.

Recently, a two-year's programme has been initiated and sponsored by the Ministry of Environment, in which about 40 castles or castle ruins (from among 110) are going to be investigated under the auspices of the Ministry in order to prevent their gradual disintegration due to both natural and man-made influence. The paper presents some results of the engineering geological survey aimed at the restoration of some historic sites in Slovakia.

2 WORKS AND TECHNIQUES EMPLOYED

The preliminary stage of investigation carried out on a large regional basis, within a very short period of time, with low financial costs requires a cost-effective approach. The experience gained from previous studies and practical works aimed toward the protection and conservation of historic sites proved that each historic site represents a completely different object of interest from geological, geotechnical and historical point of view. That was the real reason why no standardized methods and techniques were applied and the survey was strictly focussed at these priorities:
- to undertake an engineering geological study of all rock slopes and to determine in the rock mass and in the castle walls those sections which represent a potential source of instability,
- to design the appropriate remedial works in designated sections,
- to prepare graphic outputs that can be easily

understood by specialists from branches dealing with historic building conservation, landscape architecture, etc.

The engineering geological survey was conducted largely by visual techniques without drilling and laboratory tests. The main accent was laid on engineering geological mapping and consequently engineering geological maps at a scale of 1: 500 were prepared from each site. Further activities were focussed on measuring the joint orientation, on location of past and active slope failures, rock falls, etc., on observation of weathering processes in rock mass and rock material and on karst processes. Special attention was paid to the character of anthropogeneous sediments and their thickness, which is quite different from the thickness found there at the time of construction. These sediments act now as a source of active pressure on the castle ruins and when accompanied by other natural and man-made factors they may cause damage.

Use was made of terrestrial photogrammetry for surveying and calculating joint orientation patterns in places difficult of access as well as in places which have been identified as potential sources of instability. Finally typical profiles from stereopairs were drawn.

3 ENGINEERING GEOLOGICAL FACTORS INFLUENCING CASTLE ROCK STABILITY

The majority of castle ruins is located on steep ranging slopes in areas of flysh and neovolcanic uplands and core middle mountains at a height of 350 to 700 m a.s.l. (Fig.1). Due to intensive neotectonic movement of the West Carpathians in the periods of late Neogene and in the Quaternary a complicated geologic-tectonic structure developed. This development resulted in the formation of a highly dissected relief and a diverse climatic zonality of the territory. The effect of the climatic conditions at the above mentioned heights is quite evident, manifesting sudden daily and seasonal temperature fluctuations ranging from -20 to +20 C, the daily temperature fluctuations range from 10 to 15 C. The mean annual amount of precipitation varies between 600 to 800 mm. In this respect, the

prolonged snow cover (100 to 140 days a year) is of great importance. The vertical climatic zonality and the variety of lithology control not only the exogeneous processes like weathering, erosion, slope movements, etc. but also superficial sediment formation and groundwater conditions.

The presence of overthrusted tectonic units is a typical feature of the West Carpathians and this has a substantial influence on the character and the state of the castle rocks. A very frequent case is when soft rocks mostly claystones and marlstones are overlain by rigid limestone-dolomitic complexes. Overthrust lines act, when affected by shear forces, as an important plain of weakness and potential shear plains. Such geological structure forms favourable conditions for deep-seated creep deformations inhibiting ridge loosening, block failures (block rifts and block field) followed by rock falls and rock toppling. This geological structure was observed on several sites of interest. As an example we present the Plavecký hrad Castle, located 60 km NE of the capital in the Malé Karpaty Mts. The castle rock is formed by Triassic limestones with a lense of dolomites (Veterník nappe formation). As a consequence of Neogene differential movements and retrograde displacement, the carbonate rocks were displaced (overthrusted?) on flyshoid Paleogene sandstone and conglomerates. Due to uplift and horizontal stresses release the existing fault system was rejuvenated (Fig.2) and with the karst and weathering processes open tension cracks appeared. In places where the fault system and jointing has been intensively developed the castle walls fell down. We may assume that the ridge loosening and associated phenomena are the direct cause of the castle walls damage (Fig.3).

Despite completely different engineering geological conditions the situation found at the Lietava Castle in Sulovské vrchy Mts. is similar (Fig.4). The castle is located on a massive carbonate conglomerates castle rock, which form a rather bizzare landscape with a number of cliffs. The basal conglomerates found in the bedrock are underlain by Cretaceous claystones and marlstones. This again led to deep-seated creep movement and a subsequently to a differentiated displacement of conglomerate

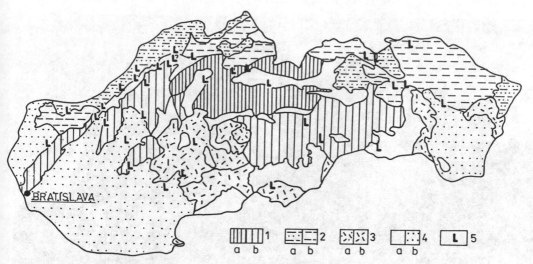

Fig.1 Location of investigated sites in the map of engineering geological zoning (after Matula 1965):1-region of the core mountains, 2-region of Carpathian flysh, 3-region of the Neogene volcanites, 4- region of the Neogene tectonic depressions, 5-medieval castles

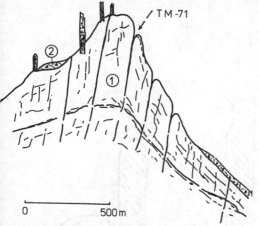

Fig.2 Open tension cracks at Plavecký hrad Castle rock. 1 - Triassic limestones, 2 - slope sediments, TM-71 - dilatometer

Fig.3 Ridge loosening of Plavecký hrad Castle rock

cliffs along subvertical tectonic planes running deep into rock mass took place. The castle ruin shows a high degree of destruction and the displacement of several cliffs is quite evident from open cracks separating individual castle objects (Fig.5). On the basis of terrestrial photogrammetry the orientation of faults and joints (dip direction/dip) exhibited on places difficult of access was calculated and profiles showing places of potential instability were

drawn by computer (Fig.6). In order to determine the actual activity of creep movements we decided to install dilatometers (TM-71) in tension cracks which enable monitoring and a facilitate the subsequent decision concerning remedial works.

A very complicated geological structure developed in the Klippen-Belt zone of the West-Carpathians. It is characterized by intensive folding and extreme tangential

Fig.4 Lietava Castle

Fig.5 Open cracks in the walls of Lietava Castle as the result of a creep movement

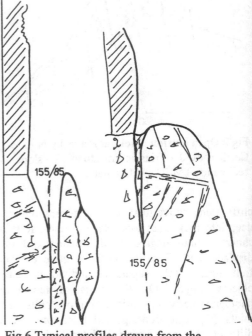

Fig.6 Typical profiles drawn from the photogrammetric survey from Lietava Castle

Fig.7 Lednica Castle. A typical sketch drawn photogrammetric survey. Construction material: 1-sandstone, travertine;2-limestones, claystones;3- sandstones; 4- limestones, marlstones, travertines, sandstones; 5- tension cracks; 6- faults; 7- block slide; 8- boundaries between different construction materials; 9 - Jurassic limestones; 10- Cretaceous marls and marlstones; 11-slope sediments

compression and it reminds of a "tectonic megabreccia" (Matula 1965). Due to disharmonic folding rigid Jurassic limestones-sandstone strata are sunken into plastic Cretaceous marlstones in the form of blocks and lenses which in the present relief rise actively and form landscape dominants, on the top of which castles were built in the past.

The Lednica castle in Biele Karpaty Mts. represents a typical example of the fact that the physical state of the object is substantially influenced by the complex geotectonic structure of the bedrock. The geology of the Lednica castle Klippe (Fig.7) represented on a sketch drawn on the basis of photogrammetric survey showing that important tectonic lines found in the bedrock continue in the castle walls in the form of open cracks and visible diturbances. It may be assumed that the process of crack formation is the result of a creep movement which is necessary to monitor so that

appropriate remedial works may be designed.

Failures of quite different origin brought about the destruction of the Čachtice castle in Čachtice Karpaty Mts.. The castle rock is formed by Triassic dolomites (Hauptdolomite) in which the main discontinuities are the bedding plains. A constant dynamic process of loosening and damages associated with the collapse can be observed on the castle ruins. The main reason for this process is an active earth pressure of anthropogeneous fills of great thickness occurring at critical places. Calculation proved that the coefficient of earth pressure at rest reaches 80% of limiting state of equilibrium.

The western part of the castle ruin manifests signs of destruction illustrated in Fig.8. For restoration purposes we have delineated in the western castle walls zoning units in terms of their physical state, where the highly endangered parts (1) are affected by active earth pressure of anthropogeneous fills and by the presence of

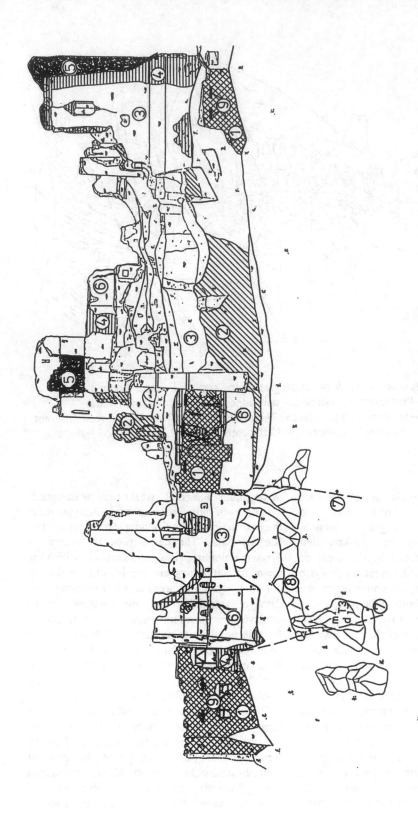

Fig.8 Zoning sketch of western walls at Čachtice Castle. Failures in the walls: 1- major; 2- moderate; 3- minor; 4- walls reconstructed in 1992; 5- fallen down walls; 6- open cracks; 7- faults; 8- Triassic dolomites; 9- supporting wooden beams

bedding planes (7) which have resulted in the formation of open cracks found in the castle walls (6). The stability of the castle walls (1) has been temporarily secured by means of supporting wooden beams. The zoning of the physical state of the castle objects has been carried out on all investigated castles in a similar way as in the case of Čachtice castle - either with the use of photogrammetric photography or in a ground plan.

4 CONCLUSION

In this paper we have focussed our attention on the description of engineering geological and geotectonical factors causing the destruction of a number of castles and the presentation of the results of their engineering geological inventory. Naturally, there exist several other factors influencing the destruction processes: climatic conditions (influencing weathering of construction materials, mortar, etc.), natural or induced seismicity, the vegetation cover and first of all poor maintenance and the adverse influence of man.

REFERENCES

Matula, M. 1965. Regional engineering geology of Czechoslovak Carpathians. Maps. Slovak Cartography Bratislava

Comparison of weathering resistance of some building stones based on treatments simulating different external conditions

Comparaison de la résistance à l'altération de quelques matériaux de construction basée sur des traitements simulant différentes conditions externes

P.E. Ihalainen & R. P.J.Uusinoka
Tampere University of Technology, Finland

ABSTRACT: The most significant factors determining the weathering of natural rock material proved to be the water saturation of the samples and the chemical composition of the pore water. The action of hydrolysis caused by the acidity of the pore water combined with repeated freeze and thaw in 100% relative humidity proved to be the most significant factor in the strength decrease of the materials studied. The decrease in the strength of the rock material was most reliably illustrated by the changes in tensile strength, measured by the changes in the modulus of rupture and the point load index. The superiority of granites to marbles was clearly indicated by the results of all the treatments. As to the strength parameters, certain differences were also observed among the marbles.

RESUMÉ: La saturation en eau des échantillos et la composition chimique de l'eau dans les pores de la roche se sont relevés comme les facteurs les plus significatifs déterminant l'altération du matériau rocheux naturel. L'action de l'hydrolyse, causée par l'acidité de l'eau dans les pores, combinée avec des cycles répétitifs du gel et dégel dans l'humidité relative de 100 %, s'est montrée comme le facteur le plus significatif dans la réduction de la durabilité des matériaux étudiés. L'illustration la plus fiable de la réduction de la durabilité du matériau rocheux a été obtenue par le changement de la résistance à la rupture par traction, mesuré à l'aide des changements dans la résistance à la flexion et dans l'index de charge ponctuelle. La supériorité des granites par rapport aux marbres a été clairement démontrée par les résultats de tous les traitements. Quant aux paramètres de durabilité, quelques différences ont aussi été observées parmi les marbres.

1 INTRODUCTION

As weathering proceeds in the thin slabs and panels of stone cladding even the slightests changes in the strength of the building stone material have a crucial effect on the durability of the cladding. The degree of weathering whose effects can be measured through the changes in strength of the test materials depends both on the external conditions and the resistance of the particular rocks. Variations in weathering resistance can be determined by submitting a series of different rocks to a series of artificial treatments in different conditions. The differences in effectiveness among these conditions can be observed in turn by the changes in strength of a particular rock after each

treatment including repeated variations of temperature in periods of 6 hours, known as the Icelandic temperature variation cycle (Wiman 1963, Martini 1967).

2 METHODS OF STUDY

The effect of some weathering processes with their influence on the strength of the rock material was examined by exposing small-dimensioned samples to artificial weathering in a special climatic cabinet. Besides the temperature variation, some samples were also exposed to the influence of solutions accelerating the weathering. The samples were subjected to four different types of weathering conditions shown in Table 1. The variation of temperature and the duration of the freeze-thaw cycles are presented in Fig. 1. In case of samples 3 - 4 (rows 3 - 4 in Table 1) each cycle was terminated by immersing the samples in solutions listed in Table 1. Depending on the material the samples were examined after 25, 50, 100, 400 and 800 freeze-thaw cycles to determine their strength. The solutions and concentrations of liquids with an accelerating influence on weathering were chosen to simulate the climatic stresses to which the building stones are exposed on the industrialized coastal area. Similar methods have been introduced e.g. by Aires-Barros et al. (1975) and Ihalainen (1993).

The changes in strength of the specimens were determined by means of the uniaxial compressive strength, modulus of rupture and point load index. The uniaxial compressive strength was measured according to norm DIN 52 105, except the dimensions of the cubic samples. The modulus of rupture was determined according to norm DIN 52 112 and point load index as well as anisotropy indices according to specifications of ISRM (1985). To determine the water absorption of the materials, the specimens were first evacuated to a pressure of 0.10 mbar, then immersed in water and finally weighed 24 hours after the submersion.

Table 1. Conditions of the artificial weathering treatment. RH = relative humidity

Sample set Nr.	Temperature variation	Liquids used in treatment of samples	Theoretical type of weathering treatment
1	-25°C - +18°C	H_2O (occasional condensed moisture)	Cooling and warming (physical)
2	-13°C - +13°C	H_2O (100 % RH)	Frost weathering (physical)
3	-13°C - +13°C	Sea salt - solution, 2% (100 % RH)	Crystal growth (salt + frost weathering) + solution (physical + chemical)
4	-13°C - +13°C	Diluted H_2SO_4, pH=1 (100 % RH)	Frost weathering + leaching + hydrolysis (physical + chemical)

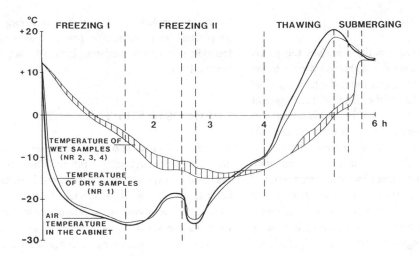

Fig 1. The variation of temperature in the climatic chamber and in the specimens

3 MATERIAL STUDIED

The small-dimensioned samples submitted to the treatments consisted of Finnish granites with both a homogeneous (Kuru Grey) and a porphyric (Viitasaari Red) type, and of two Italian Carrara marbles.

Kuru Grey is a medium grained, mostly homogeneous granite with hypidiomorphic texture, typical of Finnish granites. Principal minerals are quartz, potassium feldspar, plagioclase, biotite and some epidote. The specimens herewith treated consist of relatively fresh and compact granitic material with slight alteration in a few feldspar grains. Some microcracking also occurs in these grains.

Viitasaari Red is a medium grained, occasionally porphyric granite consisting of potassium feldspar, quartz, plagioclase and biotite. Some quartz-feldspar material occurs as fine-grained mortar among the coarser-grained crystals. There is hardly any alteration in the minerals of the rock, but the amount of microcracks in feldspars and quartz is occasionally abundant. The appearance of the Viitasaari granite is reddish with scattered porphyric grains of potassium feldspar.

The marbles consist practically of calcite only. Among the two samples used in the tests certain textural differences were observed in the shape of the grains: the one (A) has a relatively loose texture with more rounded grains as compared with the other (B). The latter has a locked-in texture with significantly more angular grains.

The samples used in the experiments were sawn by a diamant saw into cubic or beam-like forms from stone slabs calibrated to constant thicknesses in a stone processing plant. The only significant exception from the specifications of the norm DIN 52 105 was the shorter edge length of 32 mm in the cubic samples, used for determining the uniaxial compressive strength. The dimensions and the shape of all the other specimens were set according to the specifications of each norm. The number of samples for determination of each strength parameter was 5 - 15.

4 CONCLUSIONS AND DISCUSSION ON THE RESULTS

The results of the different strength determinations after each treatment are presented in Tables 2 - 4.

Tables 2 - 4. Results of the determinations of the uniaxial compressive strength (Table 2), the modulus of rupture (Table 3) and the point load index (Table 4) of the three types of building stones (Kuru Grey and Viitasaari Red granites as well as Carrara marble, the latter being of type A - see Tables 7 and 8). All the strength parameters are in MPa. See Table 1 for the descriptions of different treatments (rows 1 - 4).

Table 2	Compressive strength			
Kuru	1	2	3	4
Unweathered	223	223	223	223
400 cycles	238	224	234	231
800 cycles	215	216	221	224
Viitasaari	1	2	3	4
Unweathered	160	160	160	160
400 cycles	192	159	151	171
800 cycles	172	179	156	173
Carrara	1	2	3	4
Unweathered	79	79	79	79
400 cycles	78	72	76	0
800 cycles	81	77	88	0

Table 3	Modulus of rupture			
Kuru	1	2	3	4
Unweathered	24.4	24.4	24.4	24.4
400 cycles	21.5	21.8	24.9	23.0
800 cycles	20.3	22.3	20.3	17.5
Viitasaari	1	2	3	4
Unweathered	12.6	12.6	12.6	12.6
400 cycles	11.3	11.4	11.6	12.0
800 cycles	-	-	-	-
Carrara	1	2	3	4
Unweathered	4.8	4.8	4.8	4.8
400 cycles	3.6	2.5	2.9	0
800 cycles	1.7	1.1	1.4	0

Table 4	Point load index			
Kuru	1	2	3	4
Unweathered	12.7	12.7	12.7	12.7
400 cycles	12.1	11.7	12.8	12.4
800 cycles	11.0	11.6	11.7	10.6
Viitasaari	1	2	3	4
Unweathered	8.4	8.4	8.4	8.4
400 cycles	6.8	8.4	8.0	7.1
800 cycles	7.6	7.2	7.2	7.4
Carrara	1	2	3	4
Unweathered	1.8	1.8	1.8	1.8
400 cycles	1.4	1.4	1.4	0
800 cycles	1.3	1.2	1.3	0

Table 5. Anisotropy indices of unweathered as well as weathered rock materials

Kuru	1	2	3	4
Unweathered	1.24	1.24	1.24	1.24
400 cycles	1.24	1.28	1.05	1.10
Viitasaari	1	2	3	4
Unweathered	1.27	1.27	1.27	1.27
400 cycles	1.39	1.03	1.11	1.09
Carrara	1	2	3	4
Unweathered	1.13	1.13	1.13	1.13
400 cycles	1.11	1.11	1.22	1.01

The most effective weathering of the samples was obtained by using treatments based on the combination of repeated freeze-thaw cycles and diluted sulphuric acid. In most cases all the three strength parameters of the specimens are lower in sample set 4 than in samples treated with other methods to simulate weathering. The results in Tables 2 - 4 indicate that the influence of diluted sulphuric acid significantly accelerates the decay of a natural stone, even if the average temperature is low.

The weathering resistance of the granitic rocks proved to be significantly better than that of the carbonate rocks. No real difference between the durability of the homogeneous and porphyritic granites could be observed. A number of strength parameters of the rock materials shows a slight increase in the initial phases of each treatment. This phenomenon is most often due to the decrease of anisotropy (Table 5) among the samples, also observed e.g. by Papadopoulos & Marinos (1992) in their studies on the weathering of the Athenian schist. Depending on the orientation of feldspar crystals in the practically homogeneous rock mass microcracking tends to open in all these directions of the afore-mentioned orientation making any preferred direction of rupture less important. The influence of the changes in anisotropy seems to coincide best with the changes in strength parameters of the porphyric granite. The increase of microcracking in the test samples was indicated by the increased water absorption (Table 6).

Table 6. Water absorption (weight %) of samples after the 400 freeze-thaw cycles with the treatments described in Table 1. (Columns 1-4, see Table 1; 0 = fresh, untreated material; 3* = like Nr. 3, but soaked in pure water during several days)

Sample	0	1	2	3	3*	4
Kuru	.21	.22	.22	.21	.22	.26
Viitasaari	.14	.14	.16	.16	.19	.22
Carrara A	.17	.23	.29	.25	.27	.56

By comparing the results of water absorption tests with the data in Tables 2 - 4 one can observe that the increase of water absorption seems to give a relatively good illustration of the progression of the weathering, except in connection with the specimens exposed to salt water. Salts tend to crystallize in the pores of the rock material and thus prevent the penetration of water into the pores (Table 6) (Cooke 1979).

The methods determining the tensile strength seem to be the most useful tools for indicating the start of weathering. Unlike the values of the compressive strength, those of the modulus of rupture and point load index have mostly decreased during the initial phase of weathering (Tables 2 - 4). This is due to the microcracking whose increase acts as a natural start of the tensional rupture caused by the strain. As regards the compressive stress, the slight increase in microcracking means no rupture at first hand. The cracks will sooner be closed elastically thus promoting the compressive strength to become more evenly distributed throughout the sample tested. This phenomenon can be inferred from inspecting the strength parameters of the Carrara marble A. The values of the uniaxial compressive strength remains fairly good, sometimes showing a slight increase, until a complete crumbling of specimens takes place. As to the values of the modulus of rupture and point load index, a gradual decrease is observed along with the progress of the weathering (see Fig 2).

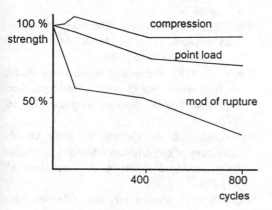

Fig 2. Behaviour of the relative strength of the Carrara marble A in the different tests after the treatment Nr.2 described in Table 1.

Differences in the change of strength, caused by weathering were also determined among the two types (A and B, see the description in previous chapter) of the Carrara marble. Comparison of the strength parameters between these samples were made after one hundred freeze-thaw cycles. See the results in Table 7. Clear differences between the strengths of these two samples were observed. The greater strength of the sample B was due to its locked-in texture with angular mineral grains, while the loose texture with more rounded grains in sample A remarkably reduced its strength parameters.

Heating both the marbles up to 105°C also significantly changed some strength properties of the specimens (Table 8). These changes took place in a lower temperature than expected. See the results obtained e.g. by Griggs et al. 1960.

Table 7. Variation of the strength parameters (MPa) of the two types of the Carrara marble (A and B) after 100 freeze-thaw cycles. 0 = fresh, untreated material; columns 1 - 4 according to Table 1.

Compr. strength	0	1	2	3	4
Carrara A	79	84	93	70	0
Carrara B	85	73	80	80	75
Mod. of rupture	0	1	2	3	4
Carrara A	4.8	3.4	2.8	2.4	1.9
Carrara B	12.6	12.1	12.4	12.6	13.1
Point load index	0	1	2	3	4
Carrara A	1.8	1.6	1.7	1.6	1.4
Carrara B	3.1	2.9	3.0	3.0	2.9

Table 8. The effect of heating on the strength parameters (MPa) of the two types of the Carrara marble (A and B).

Strength parameter	Unheated	Heated 105°C
Compression (A)	79	79
Modulus of rupt (A)	11.22	4.76
Point load index(A)	1.97	1.77
Compression (B)	79	85
Modulus of rupt (B)	17.60	12.55
Point load index(B)	4.37	3.11

The described methods to simulate weathering did not prove to be very effective in characterizing the progression of weathering in the case of firm, nearly non-porous plutonic rock materials. The absolute change of the measured parametres was quite small. It is obvious that both the composition of pore water and the rate of saturation of the pore net have a significant influence on the effect of rock weathering. This can be seen when comparing the strength parameters of specimens treated with procedures described in Table 1. Similar observations has been made e.g. by Ihalainen (1993).

The crystallization of sodium chloride in the pores of the specimens was not very efficient, which is due to the Icelandic type of temperature variation as well as the high relative humidity in the climatic cabinet. See also Winkler & Singer (1972). However, some salt crystallization in the pores of the rock specimens proved to take place. The salt crystallized in the pores of the specimens apparently decreased the total pore volume of the material. Some weight loss and increase in water absorption capacity was observed among the salt-treated specimens after soaking them in pure water for several days (Table 6) It is obvious that there is a relation between the total porosity and some strength parameters of rock materials..

The samples subjected to treatment Nr. 3 seem to show greater strength values together with greater standard deviations of the results as compared with those subjected to treatment Nr. 2. According to the results in Tables 2 - 4, the strengthening effect of salt crystals is most clearly observed in the homogeneous materials such as the Kuru Grey granite and the Carrara marble. Because of the test conditions it was not possible to determine reliably the effectiveness of salt weathering in comparison to frost weathering. This mutual effectiveness has also been studied e.g. by Cooke (1979).

It is rather difficult to simulate in the laboratory the changes in the strength parameters of building stones caused by weathering, since the effectiveness of the climatic and environmental factors affecting the rock surface show a great variation. The effectiveness of weathering depends e.g. on the factors of local macro- and microclimate. However, it can be estimated that 400 artificial freeze-thaw cycles equal the conditions of stone cladding exposed on outer walls for a period of 10 - 20 years at most in the climatic conditions prevailing e.g. in southern Finland. On the southwestern coast of Finland the air temperature, measured 2 meters above the ground surface, sinks annually below 0°C about 85 times. An unweathered, firm silicate rock is so resistant against weathering that the necessary series of laboratory experiments to determine the changes in strength inevitably takes a long time.

REFERENCES

Aires-Barros, L., Graca, R.C. & Velez, A. 1975. Dry and wet laboratory tests and thermal fatigue of rocks. *Engineering Geology* 9: 249-265.

Cooke, R.U. 1979. Laboratory simulation of salt weathering processes in arid environments. *Earth Surface Processes* 4: 347-359.

DIN 52 105 1988. Prüfung von Naturstein. Druckversuch. Deutsche Normen.

DIN 52 112 1988. Biegeversuch. Deutsche Normen.

Griggs, D.T., Turner, F.J. & Heard H.C. 1960. Deformation of rocks at 500°C - 800°C. *Rock deformation*. Geological society of America, Memoir 79.

Ihalainen, P. 1993. Methods applied in determining the variations of strength and structure of plutonic rock material exposed to artificial weathering treatment. *Bull. Soc. Geol Finland* 65: 68-76.

ISRM 1985. Suggested method for determining point load strength. *J. Rock Mech. Min. Sci. & Geomech. Abstr.* 22: 51 - 60.

Martini, A. 1967. Preliminary experimental studies on frost weathering of certain rock types from the West Sudetes. *Biuletyn Peryglacjalny* 16: 147 - 194.

Papadopoulos, Z. & Marinos, P. 1992. On the anisotropy of the Athenian schist and its relation to weathering. *Bull. Int. Assoc. Eng. Geol.* 45: 111-116.

Wiman, S. 1963. A preliminary study of experimental frost weathering. *Geogr. Ann.* 45: 113 - 121.

Winkler, E.M. & Singer, P.C., 1972. Crystallization pressure of salts in stone and concrete. *Geol. Soc. America Bull.* 83: 3509-3513.

Natural defects in decorative and dimensional stones – A critical study

Les défauts naturels des pierres décoratives – Un étude critique

Dharmapuram Venkat Reddy
Geology Section, Department of Civil Engineering, Karnataka Regional Engineering College, Surathkal, Srinivasnagar, Mangalore, DK, India

ABSTRACT : Stone is the basic building material of the earth's crust, used by man from pre-historic times. In India the use of stones for construction and architecture is as old as civilization. The Indian natural stone industry has the unique distinction of possessing all varieties of decorative and dimensional stones. Geologically, India possess extensive natural stone deposits. The success of the commercial stone industry solely depends upon the availability of defect-free raw materials. In selected instances, enterpreneurs with commercial zeal took leases of quarries, exploited them initially and abandoned them mid-way due to the bad quality of stone deposits. All available natural stone deposits are not suitable for decorative and dimensional requirements. Natural defects in the commercial stones adversely affect the quality. Detection of defects in decorative and dimensional stones plays a vital role in the quality assessment. Macro-defects such as colour variation, textural characters, structural discontinuities, inclusions, intrusives, contact zones etc., are identified by careful visual examination of entire stone deposits. Micro-fractures, stress minerals, alterations of minerals in rocks will make the stone deposit worthless. These defects are identified by systematic micro-petrographic analysis of the entire stone deposits. Chemical analysis is also required for identification of certain inclusions in marbles and limestones. Quality assessment of decorative and dimensional stone is an important factor in the natural stone industry.

Imperfections naturelles en pierres orniees et dimensionnels - Une etude critique

RÉSUMÉ: La pierre est le matériel de construction basique de la croûte de la terre, utilisée par l'homme des temps préhistoriques. En Inde l'usage des pierres pour construction et architecture est si âgé que la civilisation.
L'industrie indienne des pierres naturelles a la distinction unique d'avoir de toutes les variétés de pierres ornementales. En regardant la géologie de l'Inde, il y a beaucoup de dépôts de pierre naturelle. Le succès de l'industrie de pierre commercielle dépende pleinement sur les matières premières qui sont libres de défaut.
En cas choisis, les entrepreneurs avec le zèle commercial ont pris le bail des mines, ils les sont exploités au commencement et ils les ont abandonné dû a qualité mauvaise des dépôts de pierre. Tous les dépôts de pierre naturelles ne sont pas convenables pour les conditions ornementales et dimensionnels.
Les défauts naturelles en les pierres commerciales affectent contrairement la qualité. La découverte des défauts en pierres ornementales et dimensionnelles represents un rôle essentiel pour imposer la qualité. Les défauts grandes comme le changement de couleur, les caractères textuelles, les discontinuités de structure, les inclusions, les intrusions, les zones de contacts etc. On identifié par examen visuel des dépôts complétés de pierre. Les fractures secondaires, la modification des minérales rendront le dépôt de pierre sans valeur. L'analyse chimique est aussi exigé pour identifier certaines inclusions en marbres et les pierres calcaires.
L'évaluation de la qualité de la pierre ornementale et des dimensions est le facteur important dans l'industrie de pierre naturelle.

1 INTRODUCTION

The Indian natural stone industry has the unique distinction of possessing all varieties of stones which only a few countries in the world possess. Indian granites now-a-days are very popular in Japan and there is a vast scope for them in the future. Importers demand only defectless stone blocks. Supply of defective stones by an individual or organisation leads to the cancellation of assignment with the exporting firm. Scientific knowledge of natural stones is very vital before utilisation of natural stones for commercial purposes. In this paper an attempt has been made to study critically the natural defects in decorative and dimensional stones.

2 NATURAL DEFECTS

The following are natural defects in commercial stone deposits. They are as follows:
- i) Colour variation
- ii) Textural & Structural characters
- iii) Structural discontinuities
- iv) Micro-discontinuities
- v) Intrusives
- vi) Inclusions
- vii) Contact zones
- viii) Alterations

2.1 Colour variations

The colour of decorative and dimensional stones challenges the structural engineers while selecting a coloured stone for a particular architectural design. A pleasing colour of the decorative stone will immediately gain customer acceptance. However, irregular colour variation in commercial stones constitutes a major defect. These variations are due to the predominance of a few minerals, accessory minerals, textural characters, macro and micro-macro structures, inclusions, intrusives, alterations, differential weathering, etc.

2.1.1 Colour variations in Igneous Rocks

Igneous rocks usually show strong colour pigments. Granites, Syanite gabbro, dolerite, basalts, etc show strong colour. These colours will be fast for many years even with adverse climatic conditions.

In Igneous rocks, colour variations are caused due to colour indices of certain accessory and mafic minerals. In some varieties, the presence of biotite, hornblende, ferromagnesian minerals will mask the strong colour pigment of quartz and orthoclase. In a few igneous rocks the presence of excessive pyrite, biotite, causes a brown tinge. However in a few rocks. the presence of garnet, epidote, sphère, tourmaline will bring down the colour combinations. These minerals will not take uniform polish. Leucocratic and mesocratic igneous rocks show lighter colours due to predominance of light coloured silicate minerals. Presence of darker coloured mafic minerals, even in small percentages, also causes colour variations. Melanocratic rocks with the presence of lighter coloured silicate and non-silicate minerals also shows undesired colour variations. Such rocks are usually not prefered in the commercial stone industry.

Granites generally occur as sheet rocks or boulders. Exploitation of sheet rocks involves a long process of quarrying. Sheet granite deposits in Karnataka and Andhra Pradesh show very little colour variation. In Karnataka granite outcrops occupied about 8947 sq.km. (Vasudev, 1991). However, boulder deposits of granites in Karnataka and Andhra Pradesh show much colour variation. It is generally observed that the field boulder deposits expose much surface area to solar radiation. Hence boulder deposits exploitation requires careful identification of colour variation.

2.1.2 Colour variation in Sedimentary Rocks

The most abundant minerals in sedimentary rocks are quartz, feldspars, carbonates, clays, iron oxides etc. In sedimentary rocks, colour variation depends upon the stability of the colour pigment of detrital and non-detrital minerals. Iron-rich sedimentary rocks are unstable under severe weathering conditions. These rocks show undesired colour coatings on surfaces.

The Bander series of the Vindhyan system is considered as a store house of excellent pleasingly coloured sand stones in India. In historical buildings such as the Fathepur-Sikkri, the Buddhist stupas of Sanchi, Sarnat and Barhut these rocks have been utilised. In a few deposits, iron-rich cementing materials exhibit unusual colour variations. Lime stones of the Cuddapah, Kurnool and Vindhyan system are widely utilised for architectural purposes in the country. Limestone slabs of varied colours are available in Batemcherla, Tandur, Kurnool, Guntur, Anantapur, Ranga Reddy, Nalgonda districts of Andhra Pradesh. The Bhima series lime stones occur mostly in the Gulbarga district of Karnataka and in a small part in the Ranga Reddy district in Andhra Pradesh. Under decorative stones, these rocks are marketed as Shabad stones. A few of these deposits also show pale olive green, pink and purple shades. The lower Bhander stage lime stones in Rajasthan are mostly quarried from Suket, Satalkheri, Kumot, Morak, Hirakhetri. These rocks as marketed as 'Kota' stones.

Limestone deposits also show great variation in colours. Lime stones occasionally consist of ferrous carbonates. These compounds will not show any colour variation. However, under prolonged atmospheric weathering conditions, ferrous compounds gradually dissolve in pore water and oxidise to form a yellow coating on the stone surfaces. In some instances yellowish limonitic clots are seen in exterior decorative lime stones. The exporter, importer or user will not find it possible to detect ferrous compounds by the usual identification methods. In such situations chemical analysis will yield possible clues for the identification of ferrous compounds in lime stones.

Carbonaceous lime stones show dark grey to black colour. These rocks show the bleaching action of atmospheric agents. These rocks gradually lose their original colour and show dull colour combinations.

Carbonaceous rocks which consist of organic shells, show irregular colour patterns. In some instances, these enhance the beauty of the stone. In a few cases undesired colour variations are seen. In a few sedimentary rocks, the presence of dendrites causes bluish fern like coatings. These are composed of iron-manganese oxides. These are generally deposited along the joint and bedding planes of lime stone deposits. These rocks are located as defective stones in the commercial stone industry. However in rare situations, the presence of dendrites enhances the beauty of the stone deposit. In spite of this, few architects prefer these rocks for exterior decorations.

Generally sedimentary rocks will not show colour fast pigments. Colour variations in sedimentary deposits are common.

2.1.3 Colour variation in metamorphic rocks

Metamorphic rocks are well distributed in the Archaean gneissic complex in India. Generally metamorphic rocks show stable colour pigments. Low grade metamorphic rocks show light colours. However high grade metamorphic rocks exhibit strong colours. In schist and gneiss rocks, ferro-magnesian minerals are arranged more or less in parallel to sub-parallel bands. In some instances colour indices of these rocks are masked by bands. The presence of excess of mica-flakes in a few metamorphic rocks shows an undersired silvergrey colour variation.

Granulites, marbles, quartzites etc show colour combinations. The presence of some impurities and traces of ferrous carbonates causes a brown or yellow tinge on the surfaces of the marbles. These impurities are detected by chemical analysis.

2.2 Textural and Structural Characters

Textural characters mostly depends on constituent minerals and the mode of formation of the rock. The textural characters of a stone deposit determine the quality of the stone. Textural characters enhance the beauty of the stone deposit. However textural variations will reduce the quality of the stone deposit.

2.2.1 Textural variations in Igneous Rocks

In stone trade parlance, dolerite (black granite) has high commercial value due to its uniform textural and colour characters. For the purpose of export, dolerites are commercially exploited in Warangal, Khammam, Nalgonda and other districts in Andhra Pradesh. In this state, presently 230 dolerite quarries are in operation. In Karnataka, export oriented quarries are in operation in Belur, Hassan, Chamarajnagar, Kolegal, Deodurg, Malavalli, T.Narasipur, Gundlupet, Nanjangud, Kankapura, Averemale, Hoskote, Karenhalli, Doddaballapur, Yelburga, Bangarpet, Kolar, Chellakurthi, Davangere, Buntwal, Puttur and other talukas. In Tamilnadu, dolerite, amphi-bolites (black granites) reserves are estimated to be about 6-7 million cum. Export oriented quarries are in operation in Gindivanam, Kollakurichi, Thirukollur, Villipuram talukas of South Arcot district. Dolerite quarries are in operation in Dharmapuri, Kuppam districts.

Porphyritic dolerite shows unusual

textural characters. In a few dolerites, quartz phenocryst causes greyish white patches on the black surface. Quartz minerals will not take uniform polish along with other minerals. Quartz porphyry dolerites are treated as waste materials in the natural stone industry (Fig.1).

Fig.1 : PORPHYRITIC DOLERITE

Porphyritic granites, syenites, diorites, felsites show attractive textural and colour combinations. These rocks are preferred by some architects. However, a few rocks show very large crystal grains. This dissimilar variation in grain size brings down the quality of the stones.

Vesicular basalts are not in demand for export. In some conditions, vesicules are filled with secondary minerals. Basalt rocks are abundantly available.

2.2.2 Textural and Structural Variations in Sedimentary Rocks

The textural characters of sedimentary rocks depends upon the grain size parameters of the detrital and non-detrital and cementing materials. In sedimentary rocks, the size, shape, roundness, fabric and packing of the mineral grains plays a vital role. Compacted or consolidated sand stone shows high packing density of mineral grains. Unconsolidated or poorly compacted sand stone shows poor compactness of mineral grains. Large size cobble, pebble, detrital grains in conglomarate, breccia show unequal irregular, angular and sub-angular grains. In the natural stone industry these rocks are not of much value.

Textural characters in sedimentary rocks either enhance its beauty or bring down the quality. The quarry owner has to study the textural characters of sedimentary rocks before exploiting them for commercial applications.

Structures are large scale features in sedimentary rocks. Primary structures such as ripple marks, cross bedding, graded bedding etc on clastic and carbonate rocks shown dissimilar structural variations. Styolites, stromatolites, biogenic structures etc also show undesired structural variations in limestones. These will bring down the quality of the commercial stone deposit.

2.2.3 Textural and Structural Variations in Metamorphic Rocks

Metamorphic rocks show granular, banded, linear, foliation and platy cleavages. Schists and gneiss rocks show parallel to sub-parallel platy minerals. In some rocks excess banding will mark the colour index of the rock. Generally architects prefer less banding in rocks. In a few rocks excessive clotting of greyish black hornblende, biotite minerals show undesired textural characters. Gneisses shown deformed structures. This texture includes regular handed nature, irregular bands, folded bands and nabulas pattern with clots of patches of dark minerals strew-in a leucocratic ground mass (Balasubramanyam et al, 1991. Some architects prefer these rocks to marbles. These rocks are stable under adverse climatical and environmental conditions. Nebulas gneisses are well distributed in the Archaean gneissic complex. In some instances Nebulas gneisses form substitutes for marbles. For instance the Taj Mahal whcih is one of the prime tourist attractions in India, was constructed with white marble by the Mughal Emperor Shahjahan in the seventeenth century at Agra. With the passage of time, the white marble has turned pale and vugose (Tripati et al, 1984). With the setting up of an oil refinery at Mathura and threat to the monument due to the refinery emission has further increased.

2.3 Structural Discontinuities

In nature no rock is perfectly continuous. Natural discontinuities are formed in rocks due to genetical and tectonic conditions. Rocks will be broken by fractures, joints, faults, folds, unconformities, etc. These are commonly known as structural discontinuities. Prolonged weathering and inherent geological disturbances will form undesired cavities and cracks. However, the intensity and nature of discontinuities depends upon the origin, nature, homogeneity and hetrogeneity of the rock mass. Mega-discontinuities in natural stones will make valuable deposits worthless. These can be identified by systematic structurall mapping of the stone deposit. Regional structural trends have to be established before the exploitation of stone deposit for commercial utilization. In our coutry, the majority of stone deposit leased areas are limited due to the local government's leasing rules and regulations. For instance, in Karnataka State, the leasing pattern granted is less than one hectare. In Andhra Pradesh State, the majority of leases of stone deposits is less than 2 hectares. However, the area extent available for leased deposit is not sufficient for carying out the systematic study of structural discontinuities. In such a situation, combined leased area structural mapping is required.

2.4 Micro-discontinuities

Micro-discontinuities are not directly visible on the stone deposit. In some instances, a few micro-fractures are seen on the polished stone surface. In another situation micro-fractures are developed in the stone deposit after being exposed to solar radiation. These deposits insitu conditions will not show any signs of discontinuities. However minerals which are already in stress-conditions, suddenly being released from insitu conditions, develop micro-discontinities. The many valuable stone deposits are later rejected in the natural stone industry. However, it is not possible to detect micro-fractures by surface structural mapping. Systematic micro-petrographic analyses will be useful for determination of stress minerals in commercial stone deposits.

2.5 Intrusives

Intrusives constitute major defects in natural stones. The major intrusives are quartz, feldspar, pegmatite, epidote, veins, reefs and basic intrusives.

2.5.1 Quartz veins and Rocks

These are observed traversing into country rocks. For instance quartz veins has intruded into granitic outcrops in Ranga Reddy Pally, Ranga Reddy district, Andhra Pradesh. The characteristic trend of quartz veins is to run parallel to the

major structural trend of the area which is in the northwest-southeast direction (Venkat Reddy, 1983). In a few outcrops intrusive quartz reefs displaced the mineral constitutents. In some field conditions, the workability pattern of the deposit, is required to be altered. However it depends upon the width and length of the quartz intrusive (Fig.2).

Fig.2 : INTRUSION OF QUARTZ REEF

2.5.2 Feldspar vein and reefs

These are also equally bring down the quality of the commercial stone deposit. In a few outcrops, quartz and feldspar veins cut into each other. Quartz veins disturb the feldspar veins and embay them. Feldspar veins are also observed in granitic outcrops in Andhra pradesh and Karnataka.

2.5.3 Pegmatite Rocks

Pegmatite rocks show gradual changes of mineral composition in the country rock. Granite pegmatite is composed of feldspar, quartz, muscovite, tourmaline, topaz, apaltite, rare earths which gradually change the quality of the deposit.

2.5.4 Aplite Veins

These are common in granites and plutonic rocks. Aplite veins are observed in the grey and pink granites of Hyderabad, Ranga Reddy and Mehabubnagar districts, Andhra Pradesh. Aplite intrusives do not penetrate enough into country rocks. These intrusives also lower the value of the stone deposit.

2.5.5 Epidote Veins

Epidote vein intrusives are common in basic rocks. Epidote veins occur comparatively less in granites. Epidote vein intrusions have occured parallel to the strike direction of joints in the granitic outcrops of Ranga Reddy Pally (Venkat Reddy, 1985). Epidote group minerals shows brownish green to greyish black colour. Epidote vein intrusions also lower the value of the commercial stone deposit.

2.5.6 Dykes

Dyke intrusion into natural stone deposits adversely affects the quality. Small dimension tabular dyke intrusions influence the walls of original commercial stone deposit. Small dykes are observed in the granitic outcrops of Andhra Pradesh, Karnataka and Tamilnadu. Large dimensional dolarite dykes are utilised in the decorative and dimensional stone industry. Dykes are commonly called "black granite" in trade paralance. These are commercially exploited in South Indian States.

2.5.7 Lineations

Platy or spherical minerals are arranged to form in a directional way to show lineation. For instance greenish black hornblende minerals are arranged in a parallel to sub parallel manner in schists and gneissic rocks. Extensive lineation bands on the schist and gneiss rocks show undesired colour variations. These rocks are of little value in the commercial stone industry.

2.6 Inclusions

Inclusions in commercial stones are considered as major defects. Inclusions in the natural stone deposit appears more or less as irregular patches. A few stone deposits show angular, sub-angular and rounded inclusions. The size of inclusions may range from a fraction of centimeter to several hundred meter dimensions. Inclusive bodies show very sharp contacts with enclosing rocks. The common inclusions in are xenolith, segregation and schlierens. Inclusions in commercial stone deposits adversely affect the quality. These stones are treated as defective stones in the dimensional stone industry. In some instances, the workability of commercial stone deposits will be affected by these inclusions.

Less predominant minerals in commercial stones will not take uniform polish and show colour variations. Less predominant minerals are calcite, epidote, diotite, chlorite, micas, iron-oxides, etc. In a few situations, the commercial stone deposit workability also changes. Accessory minerals are common in natural rocks.

2.7 Contact Zones

Contacts of two divergent origin rocks are always considered as weak planes. The contact of intrusive rocks shows the following characters - colour, textrual, mineralogical and chemical variations. These changes depend upon the nature of the country rock and invaded magma. Contact zones are more pronounced in metamorphic and igneous rocks. Contact zones in natural stone deposits adversely affect the quarrying operations. Contact zone boundaries of the natural stone deposit show dissimilar colour, textural and mineralogical characters. Quarry owners have to demarcate contact zones before exploitation of the stone deposit for commercial utilization.

2.8 Alterations

Alteration of minerals in rocks changes the physical and chemical properties. Altered minerals distort the binding capacity and this leads to decrease in strength. Alterations are more pronounced along the bedding planes, joints, fractures, contact zones and other structural discontinuities. In many instances, altered minerals spread over larger areas in rocks making the natural stones unsuitable for commercial utilization. Iron-richminerals form brown to reddish yellow pits on the rocks. Altered minerals such as feldspar, amphibole, mica, etc result in the formation of dull colour and lustre The polished surfaces of decorative stones often show flaky, brown coloured pits. In rare instances altered minerals will enhance the beauty of the commercial stone. Micro-petrographic analysis will help to determine the extent of the altered rocks are treated as defective stones. Quarry owners have to determine the percentage of altered minerals in leased natural stone deposits before exploitation for commercial applications.

3. CONCLUSIONS

1. Detection of defects in decorative and dimensional stones is a very important factor in commercial stone industry.
2. Macro-defects such as colour variation, textural variation and macro-structural discontinuities can be detected by careful visual examination of insitu natural stone deposit.
3. Micro-discontinuities in commercial stone deposit adversely affect the quality. These are identified by systematic micro-petrographic analysis of the entire stone deposit.
4. Every enterpreneur is interested in the exploitation of quality stones from the quarries. However due to natural or technical reasons bad quality stone blocks have been obtained.
5. The quarry owner and exporter should be quality conscious about export oriented natural stones and keep up the national pride.

ACKNOWLEDGEMENTS

I am sincerely grateful to my wife Smt. Vasumathi, who has always given her support and encouragement.

REFERENCES

Balasubramanyam, M.N., Sarvotham, H. and Gopal Reddy (1991). Gneisses and granite of Andhra Pradesh as dimensional stones - A review. In Proc. National Seminar on Dimensional Stones and Quarrying Processing and Marketing, Mining, Engineering Association, Hyderabad Chapter, pp. 27-30.

Tripati et. al 1984. Results of laboratory Experiments with Pollutants on marble rocks of Taj Mahal, Indian Minerals, Vol. 38, No.3, pp. 1-8.

Vasudev 1991. Ornamental granites of Karnataka Resources prospects and problems and appraisal, Kaigarika Varthe bulletin vol.9, No.9 and 10, pp. 32-36.

Venkat Reddy, D. and Pavanguru, R.(1998). Building and Ornamental stones in India. Journal of National Building Organization, Government of India, Vol. XXXIV No.1, pp. 1-8.

Venkat Reddy, D.(1985). Geological and hydrogeological studies around Salarnagar Project Area, Pargi Taluk, Ranga Reddy District, Andhra Pradesh, India. Unpublished Ph.D. Thesis submitted to Geology Department, Osmania University, Hyderabad, A.P.

Engineering-geological aspects of reconstruction of Charles Bridge in Prague

Aspects géologiques et géotechniques de la reconstruction du pont Charles à Prague

P. Pospíšil & J. Locker
Institute of Geotechnics, Technical University of Brno, Czech Republic

M. Gregerová & P. Sulovský
Department of Mineralogy, Petrography and Geochemistry, Masaryk University, Brno, Czech Republic

ABSTRACT: The paper deals with the building stone of Charles Bridge in Prague - the Czech Republic. It generally describes causes of building stone erosion (fracturing) by the tension in the bridge structure resulting from the reconstruction during the end of 1960's and the begining of 1970's due to the application of unsuitable building materials. The paper also presents very detailed analysis both of the present stage and the causes of building stone degradation caused by aggressive environment in Prague.

RÉSUMÉ: L'exposé traite de la pierre de construction du Pont Charles á Prague. Il analyse globalement les causes de la dégradation de la pierre de construction sous l'influence de la pression résultant de l'emploi de matériaux impropres lors de la restauration de la fin des années 60 et du début des années 70. Puis, il examine en détail l'évaluation de la dégradation de la pierre de construction, causée par l'environnement agressif de Prague.

HISTORY OF THE CHARLES BRIDGE

The foundation stone of the bridge was laid personally by the King Charles IV on 9th July 1357. The bridge was originally called "Stone Bridge". It was built as the second stone bridge across the Vltava river in Prague in the place of formerly built Judita's Bridge which was destroyed by the flood in 1342.

The new stone bridge was built bigger and as building material sandstone was mostly used instead of marlite. The bridge is 515.76 m long, the width is on average 9.40 m. It has 16 arches with the span from 16.62 to 23.38 m and 17 piers from 6.30 to 10.84 m thick.

Charles Bridge is at the first look a solid homogeneous stone structure. But when observed from a closer distance many varieties of building stones with different textures, structures, mineral composition and stage of degradation are distinguished.

Because the Charles Bridge was damaged many times by floods and by man during the wars we can find also differences between the parts of the bridge which were additionally built during the different periods. Four times it was partly destroyed by the floods in 1432, 1496, 1784 and 1890.

Arches and piers which were destroyed during each of the floods had to be repaired, but the used building stones were many times differed from the original ones. Also the construction technologies were different during the centuries when the reconstructions were carried out. It included different types of mortars (lime and cement).

This commonly results in a great material diversity of the bridge.

Very problematic was the last recontruction (1966-1975) when the filling materials in the bridge were removed and replaced partly with porous concrete and partly with the rigid concrete slab without dilatation. The parapet walls were fixed to this concrete slab and the bridge deck was built as granite paving put in cement mortar.

PRESENT STAGE

The changes in the bridge structure caused that the bridge became very rigid. But the different properties and behaviour between the original stone structure and the newly used elements resulted in the creation of tension in the bridge structure. Partly the parapet walls and mainly some of the arches were fractured parallel to the lengthwise axis of the bridge. The fractures are not only along the edges of ashlars, some of which are cut in some places.

The stage of ashlar erosion which depends on many factors is different from place to place and from arch to arch. The type of surface erosion and its depth is present in Tab. 1. The depth of sub-surface erosion was verified by geochemical methods.

were oriented along to the lengthwise axis of the borehole cores. In the depth they were oriented across to the lengthwise axis.

Four types of sandstones were identified of 21 samples taken from Charles Bridge. The dominant type of the sandstone is silicarenite.

Silicarenite

It is of gray-brown, white-gray and rusty-brown colour.

The structure is massive, cellular, often with colour streaks. The texture is half-rounded

Tab 1. Categorization of surface erosion of building stone (examples)

	Category A	Category B	Category C	Category D	Category E
Arch No. 4	20%	5%	3%	62%	10%
Arch No. 6		95%		5%	
Arch No. 9		5%	90%		5%

category A - dry, non-weathered surface
category B - wet, non-weathered surface
category C - wet to dry with coats, slightly weathered
category D - wet to dry with or without coats, surface corroded up to 4 mm
category E - wet to dry with or without coats, surface corroded over 4 mm

NEW INVESTIGATION

The stage of erosion of building stones and fractures in the structure initiated a new investigation in January 1994 as preliminary investigation before the presumed reconstruction of the bridge.

For the purposes of petrographical and geochemical study 21 core samples with the diameter 20 and 35 mm were taken. They were situated on different types of rocks, on the ashlars of different age of application into the bridge structure and on the ashlars of different stage of degradation. The lengths of the cores were approximately 220 mm. Besides these samples also surface samples of efflorescences and crusts were taken.

PETROGRAPHIC CHARACTERISTIC OF BUILDING STONES OF CHARLES BRIDGE

In order to enable the observation of the changes in the mineralogical and chemical composition of sandstones the thin-sections near the surface

psammitic or subangular psammitic. The granularity corresponds to that of medium-grained psamites, rare coarse-grained psamites.

Mineral composition: quartz (92 - 98 %), ± K-feldspars, rock fragments of quartzite and metaquartzite, clay minerals, Fe-oxyhydroxides muscovite, biotite, tourmaline, ± zircon, ± apatite, ± chlorite, ± glauconite.

The binding material is of coating and touch character. The polymict binding material is formed of clay minerals, Fe-oxyhydroxides and silt particles of quartz. Sometimes it is possible to distinguish the matrix and cement. Gypsum rarely occurs in the pores near the surface. Three samples of eleven silcarenites were strongly porous and friable.

Green (glauconite) sandstone

It is gray-brown and greenish in colour. The structure is massive, macroscopically compact. The texture is non-uniform, half-rounded psammitic, medium grained.

3628

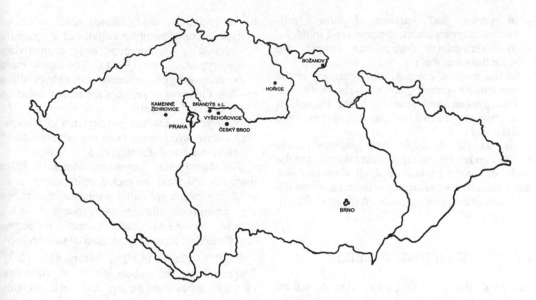

Fig.1 The localities of building stones of Charles Bridge (marked ●)

Mineral composition: quartz (90 - 93 %), K-feldspar, glauconite, clay minerals.
Binding material is prevailingly formed of silt, locally occurs coating of clay minerals. Clastic psammitic grains have a close connection, therefore empty pores are rare. Subsurface zone is strongly friable.

Arcose sandstone to polymict conglomerate

It is light brown to gray-brown in colour. The structure is massiv and cellular. The texture is non-uniformly subangular to half-rounded coarse psammitic to finely psefitic. Mineral composition: quartz (85 - 87 %), K-feldspars (localy up to 15 %), rock fragments of quartz porphyr, sericite schist, metaquarzit and claystone, clay minerals, muscovite, biotite, calcit, Fe-oxyhydroxides, gypsum.
The binding material is mostly of coating to basal character. It is possible to find aleuritic and psammitic matrices locally. Minerals in the binding material are represented by Fe-oxyhydroxides, calcit and rarely gypsum.

Sandstone with Fe-oxyhydroxide cement

This sandstone is rusty-brown up to dark brown. The structure is massive and very porous. The sandstone is strongly friable.
The texture is non-uniform, half-rounded psammitic. Its granularity corresponds to that of coarse grained psamites.
Mineral composition: quartz (more as 95 %), rock fragments of quartzite and metaquartzite, Fc-oxyhydroxides.
For this sandstone the coating cement of Fe-oxyhydroxides on the grains of quartz and siliceous rocks is typical. Fine psammitic matrix prevailingly of quartz grains is observed locally.
The binding material near the surface is preserved only in fragments.

PETROGRAPHIC SUMMARY

Samples of sub-surface zone differ from the samples of deeper levels (15-20 cm) of the ashlar as follows:
- in colour
- in compression strength

- in porosity and character of pores (sub-surface samples are more porous and friable)
- in desintegration (sub-surface samples are scale desintegrated)
- in the mineral composition (feldspars in the sub-surface samples are strongly kaolinized; also gypsum and scarcely halite by optical microscopy in the sub-surface pores were observed

It is on the base of petrographical study recommended to use the silicarenite to quartzite with minimum contant of alkali elements, iron and calcium as reconstruction material. From the rock structure point of view massive rocks are preferred.

GEOCHEMICAL CHARACTERISTIC

The rock samples of drill cores were studied on the base of polished thin sections. Rock slices, cut along the core axis, document the surficial, most weathered layer of the stone (4 - 5 cm thick). For comparison, thin sections from deeper levels (from the depth of 10 - 15 cm) of the examined blocks were prepared and observed, too. A part of the outermost portion of each drillcore (first 3 mm from the surface) was cut away parallel to the surface, mounted in epoxy resin and carefully lapped until the section's surface revealed grains of secondary minerals covering the rock surface and filling rock pores in the thin surficial layer. The sections were polished and examined with electron microprobe, together with the polished thin sections. Specimens of stone surface with the most developed efflorescences were examined in scanning electron microprobe and identified by EDX spectra.

The rock relatively uninflicted by action of environmental factors (i.e. from inside the ashlars) and rock of the surface layer differ significantly in several parameters:
a) the rock porosity is much higher in the sub-surface layer, than inside the stone. The outer zone is usually friable, strongly weathered, with individual clastic grains easily coming off the surface. In some blocks, the depth porosity profile starts with few millimetres thick, tightly packed layer encrusted with neogenic minerals, followed by a zone of extremely high porosity (3 - 15 mm thick).

b) the surficial crust is formed either solely of gypsum, or of complex sulfates of the jarosite - alunite group, interpenetrating or overlying the gypsum zone (Fig.1). The third most common neogenic mineral - amorphous silica - fills the pores or replaces the matrix rather in the less surficial parts.
c) outer zones of stones are depleted in feldspar minerals, which have been intensively altered (kaolinized) and disintegrated.

The commonest secondary minerals differ mutually and from the rock-forming minerals in solubility (water solubility product of quartz = - 2.6, amorphous silica = -4, gypsum = -4.12, alunite = -83.4) and thermal properties (coefficient of thermal expansion of sulfates = 30 - 40.10^{-6}, quartz 14.10^{-6}, feldspars 12.10^{-6}) The mentioned differences in mineral properties in their depth distribution and rock porosity variations lead to scaling of the stones, i.e. separation of rock in thin flakes parallel to the original stone surface. Exfoliation of the surficial layer may in some case be invoked by wrong bedding of the stone block in the are construction.

The most frequently occurring precursor of secondary pore fillings is colloid silica resulting from decomposition of various minerals, above all feldspars. The precipitation of gelous silica and/or replacement of clayeous sandstone cement by opal CT may seem to be the least harmful process. Nevertheless it causes extensive closing of pore spaces, which prevents the moisture to leave the stone.

Its occurrence is not so conspicuous as that of another abundant compound - calcium sulfate. According to the equilibrium calculations, it should precipitate in ambient conditions as dihydrate - gypsum (Fig.2,5). The electro microprobe study of outer zones of stones used in Charles Bridge revealed also the presence of other two mineral forms - anhydrite (anhydrous calcium sulfate) and bassanite (calcium sulfate hemihydrate - plaster of Paris). Their presence may indicate high content of other salt dissolved in pore solutions, e.g. KCl or NaCl which decrease the formation temperature of anhydrite from 66°C to almost 30°C. This explanation is supported also by the quite common presence of potassium sulfate hydrate syngenite $K_2SO_4.CaSO_4.H_2O$, which formed

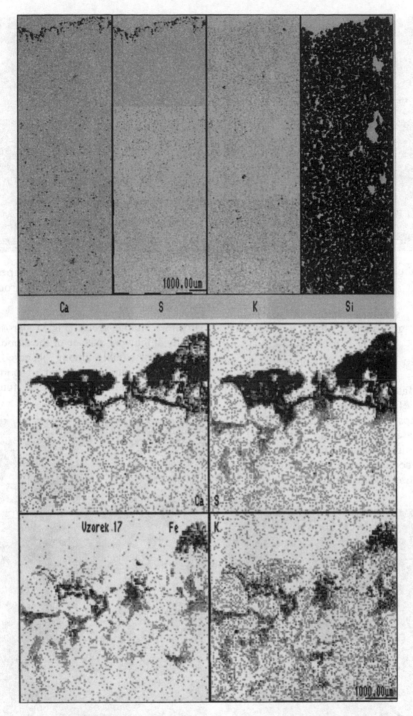

Fig. 1 The areal distribution of elements in the polished thin section
from surface to deeper levels (whole above, detail below)

Fig. 2 The tabular crystals of gypsum

Fig. 4 The aggregates of spicular-shaped crystals of halotrichite-pickeringite-bílinite group

from pore solutions high in potassium. According to Van't Hoff, if K_2SO_4 concentration exceeds 3,3 milimole, gypsum transforms to syngenite; the process can be reversible.

Other sulfate minerals detected in surficial zones and efflorescences belong to several mineral groups with wide compositional range - simple sulfates of sodium, potassium and ammonium, alums, basic sulfates of monovalent (potassium, sodium, ammonium) and trivalent (iron prevailing over aluminium) ions of the

jarosite - alunite group (Fig.3). Jarosite occurs usually at the very surface of the stone, outlining as the least soluble among observed neogenic minerals the current position of pore solution thickening front. Jarosite crust functions as moisture trap. Underneath it, continuing

Fig. 3 The aggregates of whisker-forming crystals of halotrichite (left) and izometric druse of jarosite (right)

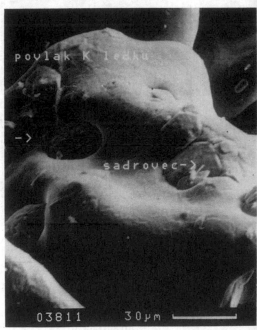

Fig. 5 The tabular crystals of gypsum coated with K-alum

dissolution of smaller feldspar grains, mica minerals and iron oxides/oxyhydroxides present in the groundmass leads to pronounced decrease in rock consistency. As soon as this sealing crust exfoliates due to action of volume changes caused by mineral hydration or heating, the highly porous subsurface zone suffers by further pore opening and grains loosening by wind or rain.

In places most exposed to the action of polluted atmosphere, efflorescences are sometimes composed also of whisker-forming alums of the halotrichite group (Fig.3), containing divalent magnesium and iron.

The potential supply of elements composing the neogenic minerals can be:
a) chemical solutions leaking from the bridge body;
b) primary rock-forming minerals;
c) atmospheric aerosol;
d) water taken up from Vltava river.

Alkalies, calcium, and sometimes also aluminium released from altered feldspars and micas serve as source to secondary minerals precipitating in pores of the surficial layer. Another supply of calcium are grains of carbonate (calcite, dolomite). Iron is supplied either by remobilization of primary limonitic cement, iron disulfides, or, together with magnesium, from decomposition of mafic minerals (biotite, glaukonite). Anions necessary to compose the secondary minerals forming efflorescences on the rock surface, filling the pores, replacing the cement or pervading the rock along weakened layer boundaries are for the most part introduced from outside - by acid rain, aerosols, road salting. Nevertheless, certain part of sulfate and chlorine ions may have been present in original fresh rocks (average S content in North Bohemian Cretaceous sandstones is 0.024%, Cl - 10 ppm).

"Health" disposition of the building stone is in close relation to the climate and its variations. The most important environmental factor controlling the weathering processes is water. The rock takes in water and releases it. Water acts as leachant as well as transport medium in the form of ionic and/or colloid solutions, supports microbial processes, and concurs in physical weathering processes (freeze-thaw effects etc.).

The present composition of the atmosphere in Prague varies strongly in dependence on climatic conditions. In the historical centre, sulfur dioxide is nevertheless present in concentrations constantly exceeding 180 microgrammes per cubic metre. Peak values are even higher - maintaining during periods of climatic inversion 500 - 800 $\mu g.m^{-3}$ (Červený et al. 1984). Concentration of nitrous gases averages 100 $\mu g.m^{-3}$. Total volume of emissions in Prague amounted in 1980 60.500 t of SO_2 and 23.060 t of NO_x. These gases can enter the stone either directly, or with rain water. The precipitation total in the environs of the Charles Bridge is around 500 litres per square meter a year. The yearly average composition of rainwater in the area analogous to Prague is as follows: pH value = 3.8, NO_3 - 3.81 mg/l, SO_3 - 18.3 mg/l, Cl^- 1.05 mg/l, NH_4 - 2.55 mg/l. The acid pore solutions dissolve less resistant components of the matrix and clastic minerals sandstone, becoming enriched in alkalies, calcium, alumina, ferric oxide, and silica. Increase in acidity of pore solutions enhances decomposition of clayeous sandstone cement and some clastic minerals. Alteration of feldspar grains (kaolinization) starts at pH values = 4 - 5, leading to release of colloid silica and alkali metals or earths. Evaporation of pore solutions at the stone surface in dry weather supports the capillary movement towards the surface and solution thickening. Concentration increase at the thickening front leads to precipitation of less soluble minerals, above all jarosite.

Temperature is another very important factor participating in physical weathering, and it substantially influences the types and rates of reactions in chemical weathering. The surface layers of stone blocks on the sunny side often reach much higher temperatures than usually expected (over 50°C); progressing staining of the rock surfaces by mineral efflorescences, microorganisms, soot and other solid particles of atmospheric aerosol promotes the temperature rise, enhancing thereby the reaction rates.

The occurrence of varied sulfates, which must have formed under different conditions, can be explained by distinct changes in the rock environment (composition of pore solutions, great daily and seasonal variations of

Table 2: Overview of neogenic mineral species found in stones of the Charles Bridge

sylvine	KCl
halite	$NaCl$
potassium nitrate	KNO_3
gypsum	$CaSO_4.2H_2O$
bassanite	$2CaSO_4.H_2O$
anhydrite	$CaSO_4$
opal	$SiO_2 . nH_2O$
scawtite	$Ca_6(Si_3O_9)_2 . CaCO_3 . 2H_2O$
jarosite	$KFe_3[(OH)_6/(SO4)_2]$
syngenite	$K_2Ca[SO_4]_2 . H_2O$
amarillite	$NaFe(SO_4)_2.6H_2O$
halotrichite - pickeringite - bilinite	$(Fe,Mg,2Na)(Fe, Al)_2[SO_4]_4 . nH_2O$
hydrated glaserite	$K_3Na[SO_4]_2 . n H_2O$
ungemachite	$K_3Na_9Fe[OH/(SO_4)_2]_3 . 9H_2O$
kalinite with admixture of amonnium analogue	$K_x(NH_4)_{1-x}Al[SO_4]_2 . 11 H_2O$

temperature etc.). Owing to the fact the bridge is a very complex structure, we can find there places conveying various physical conditions, depending on the block setting in the bridge construction, i.e. its orientation to the sun, whether their surface is sheltered from rain, and on the distance from river level and from paths draining rainwater collected at the road surface). The spectrum of mineral species depends also on the supply of elements available for neogenic minerals formation (controlled by the presence of rather easily decomposable minerals in the rock, and also on the accessibility to the acid rainwater intake), which varies from stone to stone.

Globally, temporal and spatial variations of physico-chemical conditions of weathering and formation of secondary minerals lead to their uneven distribution. While gypsum crusts prefer to form rather in shaded places where the pore solutions thicken slower than in sunlit areas, in suitably sheltered niches highly soluble mineral species occur, too: rock salt, sodium/potassium chloride, potassium nitrate. The source of soluble secondary minerals can be sought not only in the atmosphere, but also in road salting in broader environs (salts sprinkled over nearby roads and street may in dry spells be transported by wind and deposited on stone surface).

We have detected formation of neogenic minerals at the surface of ashlars known to be replaced in the bridge structure during the reconstruction in 1970's, indicating the weathering process is still running at rather high rate. It invokes the necessity to exchange the worst damaged building stones as soon as possible. It is utterly necessary to avoid using rocks containing minerals that may serve as source of secondary minerals formation, above all carbonates, feldspars and iron minerals. In addition to it, road salting should not be applied on the Charles Bridge and its vicinity.

REFERENCES

Červený, J. et al. 1984. Podnebí a vodní režim ČSSR. SZN, Prague.

Lang, M. 1989. Destruktivní působení novotvořených minerálů hornin lícního zdiva Karlova mostu v Praze. Geologický průzkum 10: 289-291.

Novotný J., Poche E. (1946): Karlův most Praha. 1946.

Rybařík, V. 1988. Středočeské stavební a sochařské pískovce. Geologický průzkum

Ornamental granite 'Cinzento Claro de Pedras Salgadas': Its exploration and exploitation

Le granite ornemental 'Cinzento Claro de Pedras Salgadas': Exploration et exploitation

L.M.O.Sousa & C.A.C.Pires

Geology Department of Trás-os-Montes e Alto Douro University, Vila Real, Portugal

ABSTRACT: Fracturation data is presented in this paper. It pertains to an area where ornamental granite is exploited. Regional faulting is formed by three main sets: N10°-30°E, N40°-50°W and N60°-80°E. They are materialised by shear stripes. The mapping of these faults is of primordial importance since they conditionate the quarry jointing. Space frequency between the joints of the same family were measured. Variable dip joints were identified. These joints turn hardly difficult block extraction.

RÉSUMÉ: On presente quelques donnés sur l'étude de la fracturation d'un lieu où s'explore un granit ornemental. La fracturation regional est formée par trois familles de failles: N10°-30°E, N40°-50°W et N60°-80°E, qui sont visibles dans le terrain par des couloirs de cisaillement. La cartographie de ces failles est trés important parce qu'elles affectent la fracturation dans les carrières. On a mesuré l'écartements des joints de la même famillie, dans les carrières située prés des couloirs de fracturation. Les joints subhorizontaux, avec des inclinations variables, prejudicient beaucoup l'extraction des blocs de ce granit.

1 INTRODUCTION

In last years the rock industry has suffered important changes, due to the development and improvement of cut and polishing techniques of ornamental rocks. The gangsaw can work with large blocks (3x2x2m), reducing the cost of the final product.

The necessity of high-quality raw material demands the execution of exploration campaigns to identify the most favourable places for the setting of extractive units.

In Portugal, concerning the exploration of granitic ornamental rocks, several works have been made (Ramos, 1990; Lopes & Dias, 1991; Pires, 1992; Gonçalves & Lopes, 1992; Moreira, 1992). These works should be continued so as to identify new varieties of ornamental rocks and evaluate the potentialities of the ones which are already being exploited.

In the district of Vila Real (North of Portugal) a homogeneous leucocratic granite is exploited, and its physico-mechanical properties are presented in Table 1. The commercial name of this granite is "Cinzento Claro de Pedras Salgadas". There are 15 quarries producing blocks for cut and for paving-stones.

The average production of blocks for cutting is rather low and that is due to the low producing capacity (5-20%) as well as to the small area of the quarries. Only a few of them have a monthly production over 50m³.

The transformation of the smalleast blocks into paving-stones reduces the exploitation costs and also contributes for decrease the average size of the waste pile materials.

Nava et al. (1989) refer the more important features that have to be studied in order to decide about the quarries locations. In this paper only fracturation has been analysed as is the most important factor conditioning the extraction of large blocks for future cut.

Table 1. Physico-mechanical properties of the ornamental granite (Moura et al.,1983-85).

Compression breaking load (Mpa)	101.5
Compression after freezing test (Mpa)	83.8
Bending strength (Mpa)	16.7
Volumetric weigth (Kg/m3)	2618
Water absorption at N.P. conditions (%)	0.21
Apparent porosity (%)	0.56
Thermal linear expansion coefficient (x10^6/°C)	9.0
Abrasion test (mm)	0.2
Impact test: minimum fall high (cm)	65

2 STUDIED AREA

2.1 Geographic location and geologic setting

The studied area is located in the district of Vila Real (North of Portugal) at the NE of Pedras Salgadas, in the granitic massif of Vila Pouca de Aguiar (post tectonic in relation to F3). This massif is parallel to the Luarca-Verin-Chaves-Vila Real-S.Pedro Sul-Penacova fault (Vila Real Fault) towards the NNE-SSW direction, that has controlled its emplacement (Fig. 1).

By the East and West sides the post-tectonic granites are in contact with metamorphic rocks belonging to the Peritransmontano (Early Palae-ozoic) and with sin-tectonic granites (Noronha & Ribeiro, 1993).

2.2 Geology

Based upon grain size, quantity of biotite and presence of potassic feldspar, three types of granites were identified:

- porfiroid medium grained biotitic granite (granite of Pedras Salgadas)
- medium to coarse grained biotitic granite (granite of Bragado)
- fine grained two mica granite

Quaternary deposits are associated to the streamlets and there are small spots of me-tasedimentar rocks, made up essentially by horn-felses (Fig. 2).

The Granite of Pedras Salgadas, where quar-ries are located, is: homogeneous, leucocratic, with a few megacrysts of potassic feldspar. The equigranular matrix is formed by feldspars, globular quartz and biotite. The modal propor-tions (vol per cent) of quartz, plagioclase, potassic feldspar, biotite and muscovite are respectively: 35.6%, 31.8%, 27.3%, 4.4% and 0.8%.

2.3 Regional fracturation

The proximity of Vila Real Fault, located 1000m East of Monte da Grulha (small area where the majority of the quarries are located) can be de-tected in the fracturation that affects the Granites of Pedras Salgadas and Bragado. The over abun-dance of N10°-30°E faults (Fig. 3), parallel to the Vila Real Fault is notorious. These faults control the jointing of the Monte da Grulha and therefore the setting of quarries in this place. This set of faults is materialised in field by shear corridors. The development of alteration stripes (subparallel bands of approximately 50m wide) has frequently take place along these faults. The quarries are located in the compartments in be-tween these stripes.

Another important set of faults has N40°-50°W direction. It's less important than the last one.

The third sct has N60°-80°W direction, is less frequently observed and produces alteration in stripes with just a few meters wide.

Some other faults have N40°-60°E direction but its appearence is only sporadic.

On the crossing points of these faults the in-tensity of fracturation and alteration are obvi-ously more intense.

In the exploitation of ornamental granite it is crucial the stuydy of these sets of faults. In fact, its identification would have prevent from situ-ations as those found on the C, I, U, S and Q quarries. Naturally, those quarries are either abandoned (V, Q e S) or producing only small blocks which are cut on the place (C e I).

3636

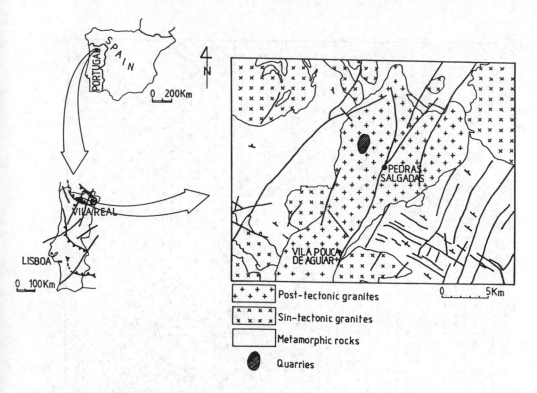

+ + +	Post-tectonic granites
x x x x	Sin-tectonic granites
	Metamorphic rocks
⬮	Quarries

Figure 1 · Geographic location and geologic setting.

3 FRACTURATION IN QUARRIES

The processes related to the emplacement, cooling and consolidation of the massifs originate the primary joints (Dutartre, 1982). These joints together with the ones resulting from tectonic actions, form the final jointing pattern.

The most important fractures seen at the quarries are sub-vertical joints. The sub-horizontal ones are less frequent.

3.1 Sub-vertical joints

The sub-vertical joints are usually closed, sowing little or any surface alteration. They often gather in sets of parallel joint stripes.

The pattern and density of fracturation varies from quarry to quarry and in some of them the principal set of faults of the massif (NNE-SSW) doesn't appear with the expected frequency. This

fact had already been referred by Ramos et al. (1984; 1985), with respect to one quarry of the Monte da Grulha and to others located in the several different places of Portugal where igneous rocks are exploited.

We can see that fracturation pattern in each quarry is an accordance to the nearest fault pattern. Good examples of this are the quarries D, N, A and O (Fig. 4). The cartography of the most important faults around the place of a future quarry gives us very important information about the local jointing.

Although the fracturation pattern is different for each quarry, the totality of the joints on the quarries reveals the existence of three sets of joints, whose directions are very similar to the three sets of regional faults that affect the Monte da Grulha.

As being already referred the joints are grouped in fracturation stripes which must be avoided as they will turn impracticable the

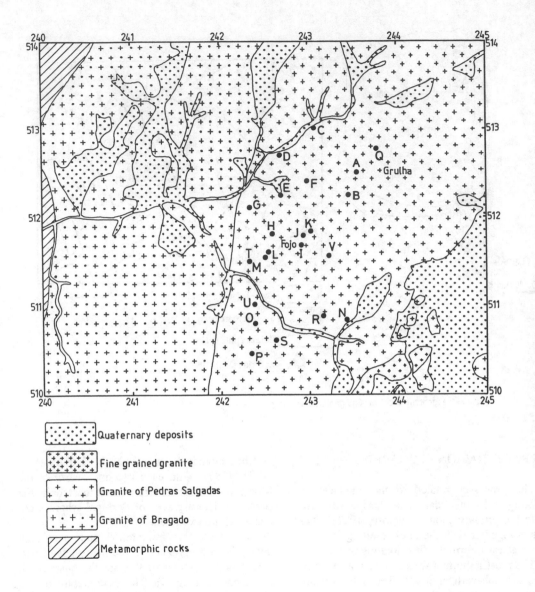

Quaternary deposits

Fine grained granite

Granite of Pedras Salgadas

Granite of Bragado

Metamorphic rocks

Figure 2 · Geology of studied area.

granite exploitation. Some quarries are located on the limit fracturation stripes. In these cases the spacing between the joints has been measured. Quarries O, D and N are very typical examples of this fact. The H quarry is farther from the fracturation stripe reason why the spacing is considerably larger (Fig. 5).

During the exploitation the measuring of the spacing between the several joints belonging to the same set could gives us valuable information about the position of a fracturation stripe. The increasing or decreasing in the joint spacing will show us the central area of the fracturation stripe. Therefore it is possible to "guide" the exploitation on the trail of less fractured zones.

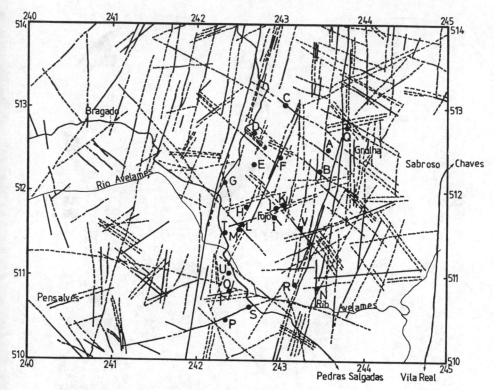

Figure 3. Regional fracturation and quarry locations.

3.2 Sub-horizontal joints

These joints are more or less parallel to the topographic surface. With small dips they do not cause great damages to the granite exploitation. Where well spaced its presence makes the blocks extraction easier.

Other types of joints, characterised by variable dips, associated to steep topographic zones, and tending to follow the local topographic surface They become, in some places, almost parallel to vertical faults (Fig. 6). This type of joints causes great damages during the exploitation because they turn difficult to obtain blocks with the help of the natural fracturation.

The location of these joints, near to the places where steep dips can be found, could be explained by the erosion process (Fleischmann, 1991) based on the motion of blocks caused by alpine and tardihercynic faults (Molina et al., 1989).

The quarries must then be installed on places where the topography is as regular as possible.

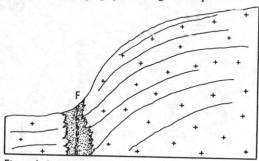

Figure 6. Sub-horizontal variable dip joints.

4 CONCLUSIONS

The regional fracturation study puts in evidence three sets of faults: N10°-30°E, N40°-60°W and N60°-80°E, which are materialised on the ground by fracturation stripes.

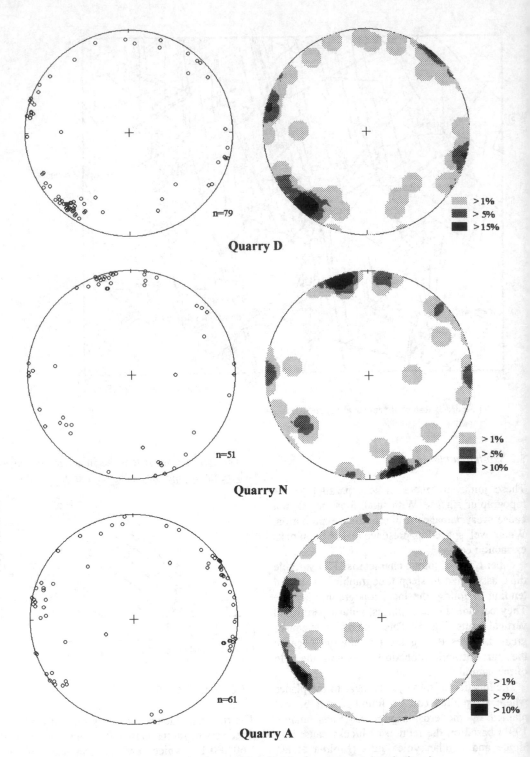

n=79

Quarry D

n=51

Quarry N

n=61

Quarry A

> 1%
> 5%
> 15%

> 1%
> 5%
> 10%

> 1%
> 5%
> 10%

Figure 4 - Joints in the quarries (Schimdt net, lower hemisphere)

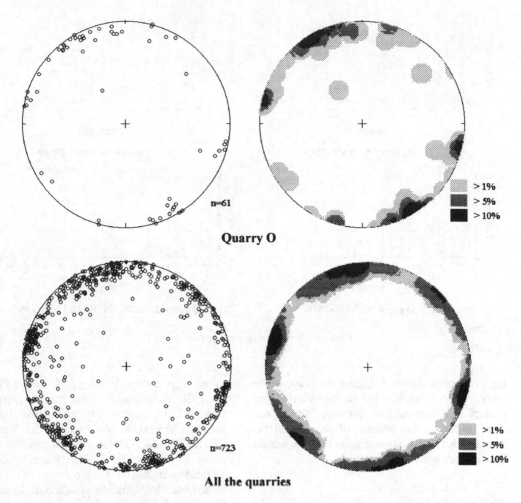

Quarry O

All the quarries

> 1%
> 5%
> 10%

Figure 4 (cont.) - Joints in the quarries.

The mapping of these faults is of great importance since they conditionate the quarry jointing.

The joint space frequency will give valuable information about the proximity of a fracturation strip and, therefore, "guide" the exploitation to the massif zones with less density of fracturation.

Sub-horizontal joints were identified with variable dip and associated to places where the topography is steep. In order to avoid these kinds of joints the future quarries must be installed where the topography is the most regular.

A careful study, before the quarry installation

and also during its activity, appears to be of primordial importance because, only this way the less fracturation zones of the massif can be identified. Planning the quarry activity in the time and space, with the establishment of the extend limits of the exploitation areas, buildings, block park, waste pile and accesses, will allow the exploitation in a more rational way, increasing the production and reducing the costs. As the quarries are all very close and there is an high percentage of the extracted material that is thrown into the waste piles, the necessity of a global planing becomes more and more urgent. Only so a continuous exploitation will be possi-

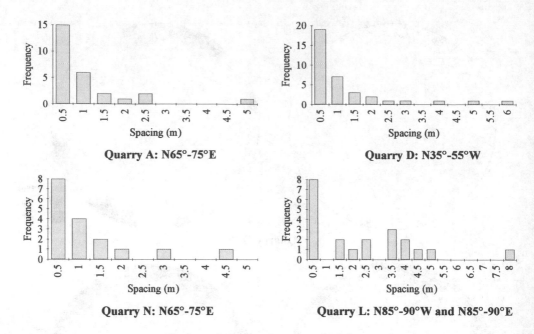

Quarry A: N65°-75°E

Quarry D: N35°-55°W

Quarry N: N65°-75°E

Quarry L: N85°-90°W and N85°-90°E

Figure 5 - Space frequency between joints .

ble in the near future. Common accesses, waste piles and rock-breaker, for all the quarries, are examples of actions that can contribute to put some order on exploitation of the natural resources, that although existing in large quantities they can always came to an end.

REFERENCES

Dutartre, P. 1982. Ètude de la fracturation du granite de la Margeride (région de Saint-Albain-Sur-Limagnole. Lòzere). Geometrie, cinématique, densité des fractures. *Documents du BRGM* 41.

Fleischmann, K.H. 1991 Interaction between jointing and topography. A case study at Mt Ascutney, Vermont, USA. *Journal of Structural Geology* 13(3):357-361.

Gonçalves, F. & Lopes, L. 1992. Aspectos das potencialidades em rochas ornamentais do Alentejo. *A Pedra* 43:7-22.

Lopes, A.T. & Dias, J.M. 1991. Activação da zona de produção de granitos ornamentais de Guarda-Pinhel. *A Pedra* 42:39-45.

Molina, E., Moreno-Ventas, I. & López-Plaza, M. 1988. Proposicion de um modelo genetico sobre ciertos montes isla graniticos cupuloformes ("Bornahrdt inseelbergs") del Sistema Central Español. *Geologia de los granitoides e rocas asociadas del Macizo Hespérico*:479-487. Editorial Rueda. Madrid.

Moreira, A. 1992. Maciço granítico de Monção-Definição de áreas com potencialidades para a produção de granito ornamental. *Bol Minas* 29(4):339-365.

Moura, A.C. et al. 1983-85. *Catálogo de Rochas Ornamentais Portuguesas*.

Nava, P.M. et al. 1989. Metodologia de investigación de rocas ornamentales: granitos *Boletín Geológico y Minero* 100(3):129-149.

Noronha, F. & Ribeiro, M.A. 1993. *Mineralizações em contexto granítico*. Excursão pós congreso. IX Semana de Geoquímica e II Congresso de Geoquímica dos Países de Lingua Portuguesa.

Pires, C.A.C. 1992. *Inventariação, Caracterização e Avaliação das Potencialidades em Rochas Ornamentais do Distrito de Vila Real*. Relatório Interno. UTAD.

Ramos, J.M.F. 1990. Potencialidades dos granitos aflorantes na região da folha 14 D: Aguiar da Beira. *Estudos, Notas e Trabalhos* 32:37.63.

Ramos, J.M.F., Moura, A.C. & Grade, J. 1984. Exploração de rochas ígneas para pedra ornamental. Alguns aspectos de natureza geológica e estrutural. *Geonovas* 7:47-55.

---- 1985. Análise sumária da fracturação em rochas ígneas ornamentais. *Boletim da Sociedade Geológica de Portugal* 24:313-327. Lisboa.

Engineering geology for solving the stability problems of the artificial cavities in Budapest

Géologie de l'ingénieur pour la solution des carrières souterraines à Budapest

M. Gálos & P. Kertész
Department of Engineering Geology, Technical University Budapest, Hungary

I. Nagy
FÖMTERV, Budapest I, Hungary

ABSTRACT: The coarse limestone from the Miocene age is an important building material of the actual Budapest town. Its quarries are typical morphological elements of the town and its region. The aim now and in the future of the engineering geological tests and investigations is a complex evaluation of the stone material as well as of the construction erected on it.

RESUME: Le calcaire grossier du Miocene est une importante matière à construction de l'actuelle ville de Budapest. Ses carrières présentent des éléments characteristics de morphologie dans la ville et autour d'elle. Le but actuel et future des recerces est l'investigation en géologie de l'évaluation complex de la pierre à bâtir et des constructions situées sur ses gisements.

1. BUILDING STONES AROUND BUDAPEST

The first urban settlement on the actual territory of Budapest was developed in the roman imperial periode; after the dissolution of the empire the town Aquincum was abandoned. In the 13th century, after a break of quite 1000 years the new center of the country was founded here, needing a lot of civil and official buildings. The overwhelming part of the building materials consisted of limestones in the antiquity as well as in the medieval time.

The triple reason of the roman town marking were the presence of strategy, thermal water and the suitable building stone: in the surroundings of the roman town the from Italy well known fresh-water limestone (lapis triburtinus, travertine) quarries gave the building stones generally marble and some other building were imported from greater distances.

The new (already hungarian) building periode active since the 13th century used also these limestones, which were extracted mostly from the covering layers on the castle hill (Várhegy). In this limestone complex natural, until now known caves exist; the activities, forming cellars from the caves were stone quarrying; the extracted blocks were used for the local constructions. The engineering geology is continuously dealing with these cavities,

upon which the metropolitan life, the metropolitan trafic is to observe.

Toward the end of the Middle Ages the possibilities of the fresh-water limestone quarrying were decreasing, and a new type of limestone appeared, which was already used from the milleium in the surroundings of actual Budapest, a coarse limestone from the Miocene. This limestone was already used by the romans (f.i. westward from Budapest in Sóskút), the quarrying florished in the Middle Age and due to its good cuttability and to other favourable properties it was transported to great distances. A part of the old quarries was later included in the territory of the town, and these quarries cause a lot of problems due to increasing traffic and the development of the new constructions, needing here also a continuous activity in engineering geology. This way, the miocene limestone should be analysed as a building stone as well as a void volume of the old quarries with stability problems of its surroundings.

2. THE COARSE MIOCENE LIMESTONE AND ITS USE IN THE HISTORY

The coarse limestone is a very important

building stone f.i. in France, in the neighbourhood of Hungary we find it at the Austrian-Hungarian border; this way it is the most important building material of many austrian towns, included Vienna. In the historical past the quarries belonged to Hungary, but no custom regulations hindered the transport to Austria. This limestone, as well as the coarse limestone near Budapest sedimented in the Miocene, but near Budapest the Sarmatian, near Vienna the Tortonian was the geological age. The coarse limestone of the Badenian (the so called Leithakalk, Leithalimestone) has better property parameters, the durability is more advantegeous, than the sarmatian one.

The formation of it in the southern part of the Budapest region is connected with a slowly subsiding basin in the Oligocene and low Miocene. The southern part of it was inundated by sea water. The basin lost in the Sarmatian its connection with the sea and the water became brackish, sedimentation and subsidence went together. The predominant product of the age is the coarse limestone.

The Pleistocene was characterized by a rise of the surrounding territory and by the further subsidence of the Pest plane, caused probably by tectonics and compactation. The closing periode of the development history here is denudation and accumulation in the Holocene. The actual morphology is a product of all these activities.

The coarse limestone forming independent units may occur on the surface or covered by younger sediments of unimportant thickness.

The coarse limestone of the Sarmatian is a characteristicly porous, generally ooidic limestone. The ooids appear like sand grains, therefore names as sandstone ore calcareous sandstone are also frequently used. Quatrz-quarzite fragments, fossils and bioclasts are ruling, their incrustation forms the ooids, the particles are linked together by calcitic bonds (Fig. 1.) The petrophysical properties depend on the petrological character: the porosity and bonding properties separe stronger and weaker types.

The fossils are in this Budapest region bioclasts, from single celled animals or shells and gastropods. The incrustation by calcium-carbonate occure subsequently, and the calcite contact of the crystals assure the bonds with cemented or structural character. The detrital grains may be of sandy size, but the washed gravels of small or medium sizes occur also frequently, as transition membres to sandstones or conglomerates. Clay particles occur even in the typical limestones, its presence of some percents increases the water sensibility. Besides these syngenetic clays the occurence of tuf layers or the mixture with tuf materials is also common, due to a volcanic activity and assuring special petrophysical properties.

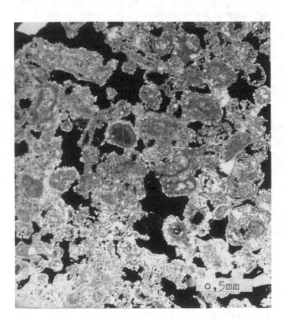 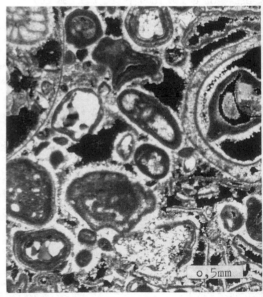

Fig.1. Typical fabric of the carse limestone

The 20-50 percentage of the porosity causes a relatively low apparent density, low resistance but good cuttability and heat insulation. During the use of this limestone these properties were searched for, but the durability was not always satisfactory. The apparent density of 1,5-2,5 Mg/m³ shows generally the variability of strength and durability. The uniaxial compressive strength may increase to about 100 MPa, the minimal values are at the limit of the bloc formability. The often used limestone types have a compressive strength of 20-60 MPa. In some layers even types of higher resistance were occuring, but these types were not exploited for masonry aims but more as ornamental or structural stones.

The properties of this limestone are water-dependent and the rock is sensitive to air pollution. The cohesive energy of the internal bonds decreases with the water content, the strength parameters become also worsening, up to 30-40 %, but decreases with 60-70 % are also known, altering in situ layers as well as building blocks. The well chosen building stones had the opportunity to remain sound during centuries when water insulation and water protection operated well, but the important increase of the air pollution caused often an important deterioration. Its basic phenomenon is the calcite-gypsum transformation, which results in the development of an impermeable crust, after closing the pores. The accumulation of dust behind the crust is starter of a deep desintegration. Therefore the restoration, cleaning and treatment of urban masonry and other stone elements is a continuous task.

The coarse limestone of these stone elements is a loyal companion of the hungarian history of architecture, in the beginning not more than a local building material, later material of prefabricated elements for the whole country.

The old roman quarries, reopened after centuries of unactivity, were not enough for the needs of a developing country. The church of Zsámbék (13th century) is one of the best examples, its quarry lies only few hundreds meters from it. The church was saved through the various war periods, an earthquake damaged it in the 18th century (Fig.2.). The turkish wars and occupation periode (16.-18th century) was followed by an important building activity, houses and military buildings even in Buda were more or less destructed. The old, as quarries used cellars at the Budapest (Buda) castle were already parts of the buildings, the near fresh-water lime-stones occurences (f.i. on the Gellérthegy -Gellért-hill) exhausted, the role of the coarse limestones increased rapidly. In the buildings of the Baroque periode and of the 19th century the coarse limestone took over the leadership (Fig.3.).

Fig.2. The church of Zsámbék (13th century)

Fig.3. The Hungarian Parlament House (End of the 19th century)

3. EXTRACTION OF THE COARSE LIMESTONE EXCAVATION OF THE "CELLARS" OF KŐBÁNYA

The surface and surface-near occurences of the miocene coarse limestone border the actual Budapest in a semicercle from SE-S-SW (Fig.4.). On the right-hand (Western) Bank of the Danube river the quarrying has a multicenturies old history, however the exact timing of it is not known enough exactly. At the lefthand (Eastern) bank it was the 19th century in which the real metropol was developed, for these constructions it was suitable to find quarries where the Danube crossing was not necessary.

However at the righ-bank region there were quarries whose overburden was to thick for economic exploitation in open pits, the new quarries at the left bank in the district Kőbánya (hungarian name: Quarry) of Budapest needed generally underground operations, due to a missing morphology and to a thick overburden.

In the first time they lay on open fields, in the case of suitable stone types the cavities were excavated until the quality change or the water-level was reached. The thickness of the overburden has been determined by local practice. After good experiences cavities were later deepened only for technological aims (f.i. brewing) too, with a secondary using of building blocks.

The cavities, formerly stable, suffered various impacts (f.i. traffic, change of the water system, loading, inundation) whose damages required stabilization works. The first step is always the complete localization of them, many of the cellars beeing unknown or not passable. Hence a continuous engineering geological survey is necessary in this town district of Kőbánya, whose typical operation system will be shown in the following analysis.

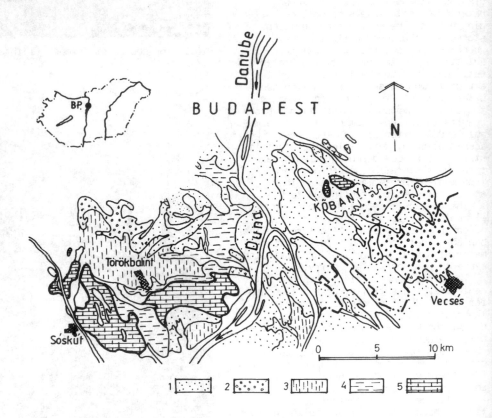

Fig.4. Geological map showing the occurences of the coarse limestone around Budapest
1. sand, sand with gravel, silt, 2. gravel, sandy gravel, 3. loess, loess with sand, sandy loess, 4. silt, mud, clay, 5. Miocene (Sarmatian) coarse limestone

3648

4. ENGINEERING GEOLOGICAL SURVEY IN A CELLAR SYSTEM

The quarrying, becoming really intensive at the and of the 19th century used a special extraction method. From an outcrop they deepened a working open pit as a starting field for the cellarlike underground quarries, where the building stone was marketable and the cellar remained as a useful space. The cellar systems of this region were established this way, differently used later with the development of the town. These cellars are mostly independent on the actual settlements, they may exist under actual roads or building endangering their normal using . Fig.5. shows a characteristic example: the cellar system in the area of Maglódi út (street)- Maláta utca (street).

The layers, becoming more and more thin of the coarse limestone are from littoral sedimentation. The thickness of the over- burden is under 10 m, and it disappears at the end of the cellar, and remains under 2-3 m. Under the cover the not really con- solidated layers or banks of the littoral limestone are occuring. The thick-banky limestone starts in a depth of 5-6 m. The cellar was excavated in this rock type, with an overburden of thin layer types of the limestone.

The rock bodies in the cellars surro- undings are shown in Fig.6. The drilling Nr. M8 furnished the samples for rock mechanical evaluation from the overburden and from the rocky surrounding up to the footwall.

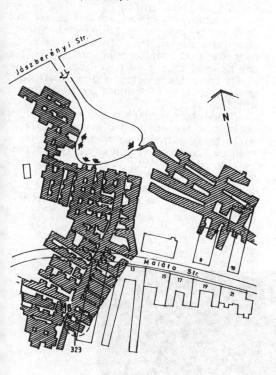

Fig.5. Ground-plan of a typical cellar- system (Budapest-Kőbánya

A failure happened in cellar Nr.323 of the system in 1993, whose cause were de- termined by the means of engineering geo- logy as well as the possibilities of an intervention. The local observations and measurements were complemented by a set of laboratory tests in petrophysics and rock mechanics.

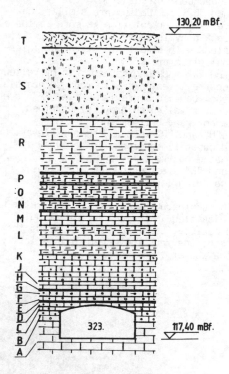

Fig.6. Rock varieties by cellar-arm. Nr.323

The variation of the rocks are very important:

"A" rock body is the level of the pre- viously excavated coarse limestone. White-yellow, fine grained, with gravels, many fossils, macroporous. The carbonateous sand is well graded, homogeneous, the ooid grains have a maximal size of 0,2-0,4 mm.

"B" rock body is a greenish-yellow, weaker, gravelly coarse limestone. At the upper and lower boundary of the layer a clear discontinuity is occuring. The quartz gravels are rounded, of 2-3 mm size.

"C" rock body is a yellowish-pinkish ooidic coarse limestone with many bioclasts. The light crosslamination characterizes a more intensive transportation.

"D" rock body is similar to rock body "A", it consist of well graded calcareous grains of sandy size.

"E" rock body is a thin layer of medium grey, fine grained coarse limestone. The sharp, lower and upper boundaries are incrustated by a cement-film of subsequent dissolution.

"F" and "G" rock bodies are similar, however their orientation is more laminated.

"H" rock body is a medium grey coarse limestone with clasts of determined internal structure.

"J" and "K" rock bodies consists of yellow coarse limestones with rounded gravels.

"L" rock body is a greyish-white, "M" rock body a yellowish grey, finegrained, muddy coarse limestone.

"M", "N" and "P" rock bodies are clayey, decayed finegrained, intensively laminated coarse limestones.

"R" and "S" are the younger, detrital covering layers.

The southern ens of the cellar system was excavated in the coarse limestone of good quality, becoming thiner towards the surface, the roof is therefore immediately in the weaker thin layers or in their vicinity.

The stability of the thin layers forming the roof is disappearing after the excavation, the interbeds figure as discontinuity surfaces. The rocky complex should be considered as consisting of a set of rock bodies, divided by separating discontinuities. The layers, themselves may be evaluated as continuous rock blocks, forming the rock bodies.

The stability of a hanging layer in the thin-banked complex may be determined by the internal section of the cellar, the thickness of the layer in question as well as the quality of the rock type and the active load. The failure is a long time-needing process, forming a natural vaulting by succesive deterioration of the layers, when the limestone in the overburden is not enough thick, the failure goes quite to the surface. A typical failure is shown in Fig.7.

The stability of cellars and cellar systems may be assured by lining with concrete form blocks. This lining may take over the load bearing role, and the cellar may be used this way as required.

In the cellars out of using or without any possibility of using a complete termination of it.

This latter was the decision for the cellars at the end of the system Nr.323, while their maintenance in the thin-bedded limestone (where water filtration is also possible) didn't seem useful. The end of this cellar arm is at a great distance from the entry, its using is this way difficult. The filling was its final operation.

Fig.7. Characteristic roof-failure in cellars

Effects of acid rain on granitic building stone

Effet de la pluie acide sur des pierres des bâtiments en granite

Shoichi Kobayashi, Takabumi Sakamoto & Satoru Kakitani
Faculty of Science, Okayama University of Science, Japan

ABSTRACT: By comparing the process of alteration of granitic building stone chemically and mineralogically, with results from an artificial chemical weathering, we are able to predict the effects of acid rain. The Deva gate of Takamatsu Saijo Inari Temple, which is located in Okayama City, southwestern Japan, was reconstructed using so-called Kitagi granite in 1952. Recently, a part of the inside wall of the Deva gate has been exfoliating due to alteration such as chemical weathering by the environment. Smectite and chlorite are products of the alteration, and appear to be formed from plagioclase and biotite in the granite, respectively. The trend in mineral and chemical changes on the surface of plagioclase and biotite during the alteration in the Deva gate agree with the results from artificial chemical weathering of granite using a HNO_3 solution at pH 4. The results strongly suggest that the degradation of granite is caused by acid rain.

RÉSUMÉ: Le processus d'altération de la pierre des bâtiments en granit est chimiquement et minéralement comparable, par ses résultats, à celui d'une altération chimique artificielle et ceci permet de prévoir les effets de la pluie acide. La porte de Déva du Takamatsu Saijo Inari Temple, qui se trouve dans la ville d'Okayama, au sud-ouest du Japon, fut reconstruite en 1952 avec le granit dit "de Kitagi". Récemment, est apparu sur une partie de la face interue de la porte, un phénomène d'exfoliation causé par une altération chimique survenue en environnement naturel. La smectite et la chlorite, qui sont les produits d'altération semblent étre respectivement formés à partir de la plagioclase et de la biotite dans le granit. Dans le phénomène de dégradation de la porte de Déva, la nature des transformations chimiques et minérales sur la surface de la plagioclase et de la biotite concorde avec les résultats d'une altération chimique artificielle du granit obtenue en utilisant une solution HNO_3 de pH 4. Les résultats laissent fortement suggérer que la dégradation du granit est causée par une pluie acide.

1 INTRODUCTION

Granite, limestone, and sandstone are commonly used as construction materials for buildings and monuments. The effect of acid rain on constructions has recently become the center of a scientific and political issue in Japan as a result of various reports concerning the degradation of concrete and some types of stone, which constitute buildings, monuments, and rock-cliff Buddha sculptures (e.g. Seki and Sakai, 1987). Alterations of the construction materials by acid rain have been reported and reviewed by Fassina (1978), Gauri and Holdren (1981), Sharp et al. (1982), Charola (1987), Ishi (1992) and so on. As seen from these papers, stone constructions most susceptible to acid rain are those built out of calcareous stone such as limestone, calcareous sandstones and so on, whereas granitic constructions are not easily affected. Granitic rock seems to be highly resistant to weathering due to its physical and chemical properties such as compactness, very low porosity and limited dissolution rate in aqueous solution.

Kobayashi et al. (1993) strongly pointed out, however, that it may be possible for granite to be altered by acid rain based on the results from the artificial chemical weathering using a HNO_3 solution at pH 4 of granite from Kitagi Island (Kitagi granite), Okayama Prefecture (Fig. 1). In

Fig. 1. A map showing the localities of Takamatsu Saijo Inari Temple and Kitagi Island.

Fig. 2. Deva gate of Takamatsu Saijo Inari Temple, Okayama, Japan.

this paper, we will go on to make the first report of environmental damage of a granitic construction in Japan from observations of the Deva gate at Takamatsu Saijo Inari Temple, which is located in the northwestern part of Okayama City, southwestern Japan (Fig. 1). The geological aspects of the degradation of the gate, will be described in comparison with results obtained from the artificial chemical weathering.

2 DESCRIPTION OF THE DEGRADATION OF THE DEVA GATE

The Deva gate is commonly the main entrance to a Buddhist temple in Japan. The two Deva gods, usually shown with scowling face and muscular body, are generally enshrined in niches on both sides of the wooden Deva gate as guardian deities

Table 1. Chemical composition of granite and its composed minerals.

Granite			Plagioclase			Alkali feldspar		Biotite	
SiO_2	74.27	wt%	SiO_2	65.78	wt%	65.30	wt%	35.43	wt%
TiO_2	0.08		TiO_2	0.00		0.00		2.76	
Al_2O_3	12.93		Al_2O_3	21.38		18.49		15.19	
Fe_2O_3	1.24		Cr_2O_3	0.00		0.00		0.00	
MnO	0.03		FeO	0.00		0.00		29.52	
MgO	0.15		MgO	0.42		0.22		3.41	
CaO	1.35		MnO	0.00		0.00		0.58	
Na_2O	3.61		CaO	2.42		0.02		0.14	
K_2O	4.17		Na_2O	9.38		0.90		0.59	
P_2O_5	0.02		K_2O	0.27		15.69		8.17	
Total	97.85		Total	99.65		100.62		95.79	
			Atomic ratio[1]						
Y	18.2	ppm	Si	11.576 } 16.01		11.973 } 15.97		5.646 } 8.00	
Nb	7.5		Al	4.434		3.996		2.354	
Rb	144.3		Al					0.499	
Sr	129.8		Ti					0.331	
Zr	88.7		Fe					3.934 } 5.65	
Zn	68.5		Mn					0.078	
Cr	0.0		Mg	0.111		0.061		0.810	
Pb	25.1		Ca	0.455 } 3.83		0.004 } 4.05		0.024 } 1.87	
Ba	635.9		Na	3.200		0.320		0.182	
Cl	34.1		K	0.061		3.671		1.661	

[1] O=32 for plagioclase and alkali feldspar, and O=22 for biotite.

of the temple. At Takamatsu Saijo Inari Temple, the old wooden Deva gate was destroyed by fire, and reconstructed using Kitagi granite in 1952. The size of the gate is 15 m in both length and breadth, and 10 m in height (Fig. 2). The Kitagi granite, reported by Tainosho et al. (1979), is of medium grain size and belongs to the so-called Sanyo zone. The major mineral proportions are quartz 37.5%, plagioclase 23.4%, alkali-feldspar 30.3% and biotite 8.7% with minor amount of opaques. The chemical composition of the granite and its main constituent minerals are shown in Table 1. Thirty years after the reconstruction, the inside and outside walls of the Deva gate were coated with a waterproof substance as some stains of dissolved mortar began to appear on the surface of the wall. The waterproof substance

consists of acrylic resin varnish with small amounts of solvent such as toluene and xylene. The substance has a feature that it is impermeable to water but vapor-permeable. Recently, a part of the inside wall, which is not exposed to rain directly, has been subject to deterioration. The granite has disintegrated into loose, small fragments and individual minerals. This phenomenon is similar to weathering by the environment of granitic rock found in nature. The most altered part of the inside wall has exfoliated with the waterproof substance as shown in the photograph Fig. 3. In addition to this, some icicle-like calcite

Fig. 3. Inner wall of the Deva gate.

Fig. 4. Icicle-like calcite formations under eaves of the Deva gate.

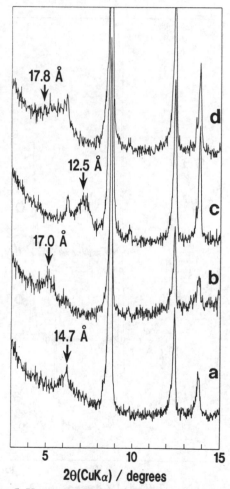

Fig. 5. X-ray powder patterns of weathered products after various treatments. a: Untreated, b: Ethylene glycol saturated, c: Ammonium nitrate treated, d: Glycerol saturated after magnesium acetate treatment.

formations are hanging from the eaves of the gate (Fig. 4).

3 ALTERED PRODUCTS ON THE GRANITE SURFACE OF THE DEVA GATE

X-ray diffraction patterns of the clay fraction separated from a sample of the weathered granite surface of the Deva gate by hydraulic elutriation are shown in Fig. 5. A reflection of 14.7 Å suggests the presence of clay minerals in the weathered granite (Fig. 5a). The 14.7 Å reflection shifts to 17.0 Å by ethylene glycol saturation

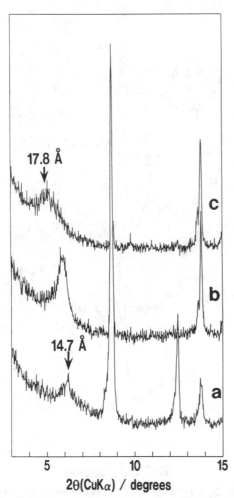

Fig. 6. X-ray powder patterns of weathered products after HCl treatment. a: Untreated, b: Magnesium acetate treated, c: Glycerol saturated after magnesium acetate treatment.

(Fig. 5b), 12.5 Å by ammonium nitrate treatment (Fig. 5c), and 17.8 Å by glycerol saturation after magnesium acetate treatment (Fig. 5d). Though the 14.7 Å reflection still remains even after these treatments, the reflection disappears after HCl treatment as shown in Fig. 6. The results reveal the formation of smectite and chlorite. Accurate chemical analyses of the altered products are impossible because the original minerals are mixed in a collected clay fraction. However, spot analyses using a SEM equipped with an EDS suggest the existence of Si, Al and a small amount of Fe on the altered plagioclase surface.

4 INDIVIDUAL MINERALS OF THE GRANITE FROM THE DEVA GATE

The surface, and cracks in the granite of the inside wall of the Deva gate, exhibit extreme weathering. Fig. 7a shows a photomicrograph of plagioclase on the altered granite surface of the gate. Many fractures formed by dissolution have developed in the plagioclase crystals. An altered product having a low refractive index replaces a part of the plagioclase. Biotites in the altered granite appear pale brown in the interior but deep brown at the border. In the most altered section of granite, chlorite almost completely replaces the biotite (Fig. 7b). This fact agrees with previous reports that biotite is replaced by chlorite in the early stages of a natural weathering (Stephen, 1952; Kato, 1964, 1965). XRD examination and microscopic observation indicate that plagioclase and biotite are altered remarkably and most likely the major sources of smectite and chlorite formed by weathering. Consequently, the alteration process of plagioclase and biotite in the granite is discussed in detail as follows.

The chemistry shows a characteristic change on the surface of each mineral with the progress of alteration. Fig. 8 shows the ratio (Na + Ca + K)/ Total Al against the ratio Si/Total Al of altered plagioclase with the same ratios of the artificial weathered plagioclases reported previously (Kobayashi et al. 1993). In the artificial weathering, the ratios (Na + Ca + K)/Total Al and Si/Total Al decrease linearly with an increase in duration. Both ratios of altered plagioclase from the Deva gate appear on the same line. Therefore the chemical change during the alteration of plagioclase on the Deva gate shows a similar tendency to that on a leached surface of plagioclase by a HNO3 solution of pH 4. As a result, by com-

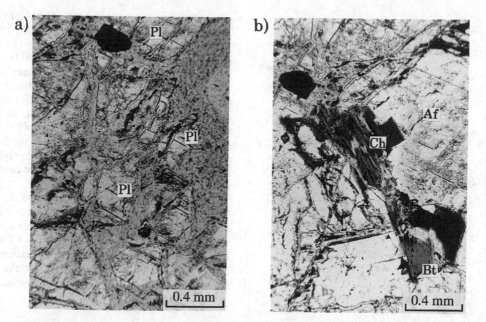

a)

b)

Fig. 7. Photomicrographs showing a) plagioclase and b) chlorite replaces biotite in altered granite from the inside wall of the Deva gate. Pl=plagioclase, Ch=chlorite, Bi=biotite and Af=alkali-feldspar.

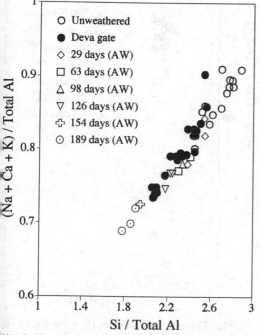

Fig. 8. The chemistry of plagioclase in the altered granite from the Deva gate and the leached surface of plagioclase by a HNO3 solution of pH 4. AW= artificial chemical weathering.

parison with the data from the artificial weathering, it can be stated that the process of alteration on the inside of the Deva gate is the same as the weathering by an acid solution with desilicification and dealkalitization. Therefore, it may be inferred that plagioclase is progressively replaced with a kind of clay mineral from the effect of an acid solution on the Deva gate.

The relation between the ratio $(Fe + Mg + Mn + Cr)/Al^{VI}$ in the octahedral sheet and the ratio $(Si + Ti)/Al^{IV}$ in the tetrahedral sheet of biotite on the Deva gate is shown in Fig. 9 with that from the artificial weathering. The former ratio remarkably decreases, whereas the later one remains unchanged. The former ratio in the core of biotite crystals is nearly constant. The ratio, however, is plotted over a wide range from 3.3 to 6.4 in the altered rim. The direction of the change is in agreement with that for the artificial weathering. The most altered biotite corresponds chemically to a leached surface of biotite for 189 days by a HNO3 solution at pH 4. Fig. 10 shows the relation between the ratio $(K + Na + Ca)/Total Al$ in the interlayer and the ratio $(Si + Ti)/Al^{IV}$ in the tetrahedral sheet of biotite. The change in the former ratio shows the same trend as the ratio $(Fe + Mg + Mn + Cr)/Al^{VI}$. Nagasawa (1972) reviewed that the alkali and Mg content of biotite decreases with chloritization in its early stage un-

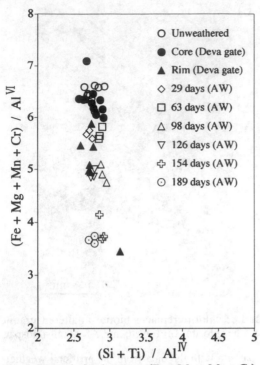

Fig. 9. The relation between (Fe + Mg + Mn + Cr) /Al^{VI} and (Si + Ti)/Al^{IV} of biotite in the altered granite from the Deva gate and a leached surface of biotite by a HNO3 solution of pH 4. AW= artificial chemical weathering.

der a natural environment. Therefore, the chemical changes of biotite observed in the present study are in agreement with this.

The trend in mineralogical and chemical changes on the granite surface by alteration agrees well with the results from the artificial chemical weathering using a HNO3 solution at pH 4. Therefore, it is considered that the characteristic variations on the mineral surface are produced chemically by an acid solution.

5 ORIGIN OF THE ACID SOLUTION

In Japan, no acid rain had been reported until now. Recently, however, various acidities of rain waters have been observed not only in large cities but also in small villages among the mountains, especially in southwestern Japan. In the case of the Okayama region, dissolved chemical species in rain water have been measured over a period of

several years by Ichikawa et al. (1984) and Toyosawa et al. (1985, 1986, 1987, 1989, 1990). The results are summarized in Fig. 11. The concentration of dissolved ions, SO_4^{2-}, NO_3^-, Cl^-, NH_4^+, increases with year, and pH values range from 4.7 to 4.2 showing a decreasing trend with year. The concentration of nitrogen, sulfur oxides and other components will be highest, and pH values will be lowest at the beginning of the rain. In Fig. 11, each value is an average of the first 1 mm of rain water to fall. Rain water in equilibrium with atmospheric CO_2, the natural cause of acidity, has a pH of 5.65 on calculation.

As mentioned above, it can be stated that the degradation of granite of the Deva gate is related to the effect of acid solution. We have to consider what the origin of the acid solution is. On the Deva gate, some icicle-like calcite formations are observed under the eaves separate from the degradation of granite (Fig. 4). It is recognized by

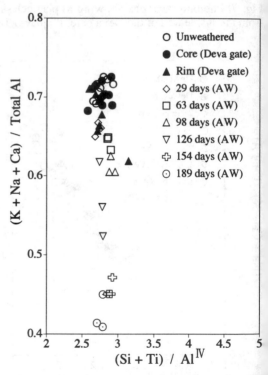

Fig. 10. The relation between (K + Na + Ca)/Total Al and (Si + Ti)/Al^{IV} of biotite in the altered granite from the Deva gate and a leached surface of biotite by a HNO3 solution of pH 4. AW= artificial chemical weathering.

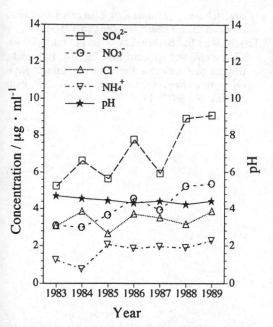

Fig. 11. Dissolved ions and pH of the rain water in southern Okayama Prefecture (Ichikawa et al. 1984, Toyosawa et al. 1985, 1986, 1987, 1989, 1990).

many authors (e.g. Ishi, 1992) that such a formation is caused by acid rain. Therefore, degradation of the granitic construction can be convincingly attributed to the effect of acid rain.

The process of degradation may be inferred as follows : Acid rain, including SO_4^{2-}, NO_3^-, and Cl^-, penetrates into the granitic building stone from the roof of the gate. On the inside of the wall, the water gathers between the granite and waterproof substance and interacts with the granite. The acidity of the rain water increases with evaporation of H_2O from micro pores in the waterproof substance. Finally, the granitic building stone exfoliates with the waterproof substance.

6 CONCLUSIONS

We have reported the first case in Japan where the alteration of granitic building stone implies the process of weathering by an acid rain. Smectite and chlorite were recognized as alteration products in the weathered granite. Plagioclase changes partially to smectite, and biotite is likely to be altered to chlorite. The trend in chemical

and mineral changes on the surface of plagioclase and biotite during the alteration in the Deva gate agree with the results from artificial chemical weathering. Therefore, it is concluded that the characteristic variations on the mineral surface are produced chemically by an acid solution such as an acid rain.

In addition to the acid rain, aerosols of SO_2 and NO_x, produced by the combustion of fossil fuels, are potent for alteration of stone. During periods of dryness, they accumulate as particulate matter on stone surfaces and are activated by subsequent wetness. McGee (1989) reviewed that the degradation of limestone constructions has accelerated over the last 20 to 25 years, coinciding with increased industrialization and the use of high sulfur fuels. The features described above suggest that if we do not carry out a policy to protect granitic building stone against acid rain, even granitic rock may be damaged easily.

ACKNOWLEDGMENTS: We would like to thank chief manager Mr. T. Nozaki, and Mr. K. Ikeyama, head of the preservation section, of Takamatsu Saijo Inari Temple for giving us the opportunity to study some specimens from the Deva gate, and other useful information. We are grateful to Mr. L. E. Anthony of Okayama University of Science for correcting and proofreading the paper, and Miss S. Kiba of Okayama Univ. Sci. (Present address: Toppan Moore Co., LTD) for help with the experiments. We also wish to thank the Analytical Center of Okayama Univ. Sci. for the use of their facilities.

REFERENCES

Charola, A.E. 1987. Acid rain effects on stone monuments. *J. Chem. Educ.* 64: 436-437.

Fassina, V. 1978. A survey on air pollution and deterioration of stonework in Venice. *Atmospheric Environ.* 12: 2205-2211.

Gauri, K.L. & G.C. Holdren, Jr 1981. Pollutant effects on stone monuments. *Environ. Sci. Technol.* 15: 386-390.

Ichikawa, S., S. Toyosawa and K. Ishii 1984. Sansei-u ni kansuru chousa kekka ni tsuite [Results from investigation of acid rain]. *Annual Rep. Okayama Pre. Inst. Environ. Sci. Public Health.* No. 8: 237-242.

Ishi, H. 1992. *Sansei-u [Acid rain].* Iwanami Shoten, publishers Tokyo: p.242.

Kato, Y. 1964. Mineralogical study of weathering products of granodiorite at Shinshiro City (Ⅱ). *Soil Sci. and Plant Nutriation.* 10: 264-269.

Kato, Y. 1965. Mineralogical study of weathering products of granodiorite at Shinshiro City (Ⅲ). *Soil Sci. and Plant Nutriation.* 11: 30-40.

Kobayashi, S., T. Sakamoto & S. Kakitani 1993. Artificial alteration of granite under earth surface conditions. *J. Clay Sci. Soc. Japan.* 33: 81-91.

McGee, E. S. 1989. Mineralogical characterization of the Shelburne marble and the Salem limestone; test stones used to study the effects of acid rain. *U. S. Geol. Surv. Bull.* No.B 1889: 1-25.

Nagasawa, K. 1972. Alteration of micas during weathering. *J. Mineral. Soc. Japan.* 10: 528-539.

Seki, Y. & H. Sakai 1987. Salt crystallization decay of the rock-cliff Budda sculpture and water-rock interaction at Daifukuji Temple, Tateyama, Chiba, central Japan. *J. Japan. Assoc. Min. Petr. Econ. Geol.* 82: 230-238.

Sharp, A.D., S.T. Trudgill, R.U. Cooke, C. A. Price, R.W. Crabtree, A.M. Pickles & D.I. Smith 1982. Weathering of the balustrade on St. Paul's Cathedral, London. *Earth Surface Process. Landforms.* 7: 387-389.

Stephen, I. 1952. A study of rock weathering with reference to the soils of the Malvern Hills. *J. Soil Sci.* 3: 20-33.

Tainosho Y., H. Honma & K. Tazaki 1979. Mineral chemistry in granitic rocks of East Chugoku, Southwest Japan. *Memoirs Geol. Soc. Japan.* No. 17: 99-112.

Toyosawa, S., S. Ichikawa, K. Ishii & K. Morita 1985. Shissei taiki osen ni kansuru chousa kenkyu [The pollution of rain water. Ⅰ]. *Annual Rep. Okayama Pre. Inst. Environ. Sci. Public Health.* No.9: 90-98.

Toyosawa, S., S. Ichikawa & K. Ishii 1986. Shissei taiki osen ni kansuru chousa kenkyu [The pollution of rain water. Ⅱ]. *Annual Rep. Okayama Pre. Inst. Environ. Sci. Public Health.* No.10: 63-69.

Toyosawa, S., S. Ichikawa & K. Ishii 1987. The pollution of rain water. Ⅲ. *Annual Rep. Okayama Pre. Inst. Environ. Sci. Public Health.* No.11: 55-59.

Toyosawa, S., S. Ichikawa & K. Ishii 1989. Den-en chiiki to mizushima rinkai kougyo chiiki ni okeru usui seibun no hikaku [Characteristics of acid rain in rural districts and Mizushima industrial area]. *Annual Rep. Okayama Pre. Inst. Environ. Sci. Public Health.* No.13: 31-43.

Toyosawa, S., S. Ichikawa & K. Ishii 1990. Characteristics of acid rain in east and south district of Okayama Prefecture. *Annual Rep. Okayama Pre. Inst. Environ. Sci. Public Health.* No.14: 8-11.

7th International IAEG Congress / 7ème Congrès International de AIGI, © 1994 Balkema, Rotterdam, ISBN 90 5410 503 8

Intrinsic factors influencing the decay of the granite as a building stone

Caractéristiques intrinsiques qu'influencent l'altération du granite comme pierre de construction

R. M. Esbert, A. Pérez-Ortiz, J. Ordaz & F. J. Alonso
Department of Geology, University of Oviedo, Spain

ABSTRACT: The granite used in the construction of a dolmen in Axeitos (NW of Spain) is taken as a methodological example of the influence that some characteristics of this kind of rock at different scales (e.g. fracture network and alignment of biotites) have on its physical properties (especially on sonic waves velocities). The anisotropic behaviour of the granite also conditions its forms of deterioration when it is subjected to salt crystallization tests.

RESUMÉ: On étude l'influence que certaines caractéristiques des roches granitiques, a différentes échelles (p. ex. le réseau des fractures et l'alignement des biotites) ont sur leurs propriétées physiques (essentialment sur les vitesse de propagation d'ondes elastiques). Pour cette étude on a consideré le granite utilisé pour la construction du dolmen de Axeitos (NO-Espagne). Le comportament anisotropique du granite determine differents formes d'altération lorsque les éprovettes sont soumises a l'essai de cristallisation de sels.

1 INTRODUCTION

Although granitic rocks have been widely used as a building material from early times, some aspects related to its mechanisms of deterioration have not been investigated in depth. However, some recent studies have been focused on the decay of the granite in monuments (e.g. Casal et al., 1992; Robert et al., 1992; Vicente Hernández et al., 1993).

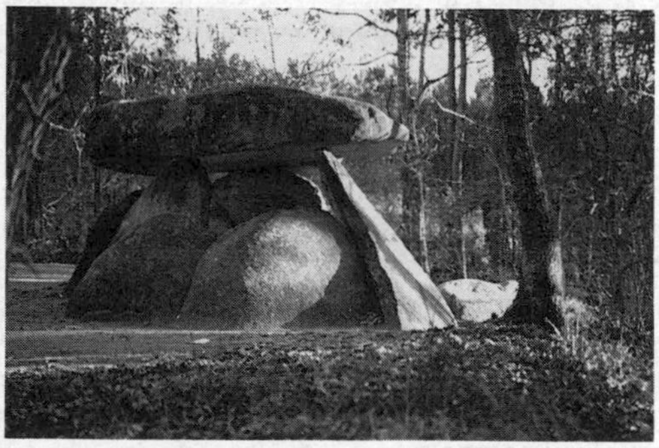

Fig. 1.- General view of the dolmen of Axeitos (La Coruña, Spain).

The aim of this paper is to point out the incidence that certain structural and textural characteristics of the granite have on its physical properties, as well as on the development of some deterioration forms typical of this kind of rock.

For this purpose, the granite used in the construction of a megalithic monument, the dolmen of Axeitos, located in the area of La Coruña (Galice, NW of Spain) (Pérez-Ortiz et al., 1994), has been selected in order to analyze this incidence (fig. 1).

From the geological point of view, this is a late hercynic granite, located in the western margin of the leucocrat exernal facies of the Caldas de Reis granitic pluton (Cuesta, 1991).

In the field, the Axeitos granite shows an important fracture or joint system that correlates with the fissure network observed at "intact rock" scale (volume of rock free of main discontinuities). These fractographic features are responsible for the anisotropic behaviour of the granite, influencing the values of some significant properties (sonic waves velocity, capillary suction and water vapour permeability), as well as some specific forms of decay.

2 CHARACTERISTICS OF THE AXEITOS GRANITE

2.1 Rock mass scale

In the rock mass, the Axeitos granite shows a blocky aspect, with a deep fracture system and a clear orientation of the biotites. In order to study the fracture system in the outcrop, two cuts or courses, C1 and C2, perpendicular to the two principal families of fractures have been marked (Fig. 2). The directions of both cuts are N40° and N140°, respectively.

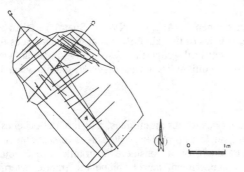

Fig. 2.- Graphic representation of the fractures in the field. The two cuts C_1 y C_2 can be observed. They are perpendicular to the two main families of fractures.

The orientation, spacing and persistence of the discontinuities have been measured and mapped along the two courses (ISRM, 1978). The spherical projection of the joint orientations proves the presence of two conjugated, subperpendicular families of discontinuities (Fig. 3).

Other fractographic parameters at rock mass scale, such as "fracture surface density" and "fracture volumetric density" (Gervais et al., 1993), have also been quantified.

The fracture surface density (B_A) is defined as the length of the fracture per unit surface (m/m²). The fracture volumetric density (S_V) is the surface of the fracture per unit volume (m²/m³).

Those parameters have been determined by means of stereological procedures. The equations used are:

$$B_A = \frac{\pi}{2} P_L \quad ; \quad S_V = \frac{4}{\pi} B_A$$

where:

P_L = Number of intersections of the fracture

boundaries with the grid lines, divided by the length of the lines.

The measures have been carried out with the grid in position 0° and 45°, respectively. The results obtained are shown in Table 1, where the values of P_L, necessary for calculating the fracture surface, are also indicated.

Table 1. Density fracture parameters

ORIENTATION GRID	PARAMETERS		
	P_L	B_A	S_V
0°	1.33	2.09	2.66
45°	1.51	2.36	3.03
Average Values	1.42	2.23	2.85

For a calculated fracture volumetric density of 2.85 m²/m³, and considering a uniform spatial distribution of the fractures, it can be deduced that the block size free of main discontinuities will be around 0,125 m³ (a cube of 0.5 m side) (See Gervais et al., ibid.).

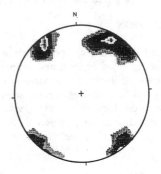

Fig. 3.- Spherical projection of the fracture system. Two orthogonal poles are observed.

It is also macroscopically remarkable the preferred orientation of biotites. Four stations were selected to represent this orientation and 225 biotites traces measured. The average orientation of the biotites is N130°, that is concomitant with one of the fracture families (Fig. 4).

The extraction of the quarry blocks of Axeitos granite was carried out systematically (Gundersen et al., 1987), taking into account the above mentioned structural anisotropy.

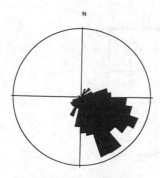

Fig.4.- Diagram of the biotite orientation. The preferred orientation is N130°.

2.2 Intact rock scale

The most noticeable feature of Axeitos granite, at intact rock scale, is its well developed fissure network, that is related to the joint system observed in the rock mass. It is also observed, at this scale, the alignment of biotites, coinciding with one of the preferred directions of fissuring.

The rock is an heterogranular granite, with a grain size ranging from 0.1 to 8.0 mm. It is formed by quartz, 43%, potasic feldspar (orthose and microcline), 29%; plagioclase (An_{4-20}), 23%; biotite, 3.8% and accesory minerals (apatite, muscovite, zircon and others), 1.2%. According to the classification of Streckeisen (1976), the rock corresponds to granite sensu stricto.

The microfissuring network is formed by inter-, intra- and transgranular cracks, mainly affecting quartz grains and, to some extent, feldspars and biotites (Fig. 5). Microfissures help the granite porosity (around 2%), rather high for this type of rock (Montoto and Esbert, 1976). Potasic

Fig. 5.- Detail of the internal microcracking of the quartz grains. Note two principal orthogonal directions. (Pol. CNx32).

Fig. 6.- Grains of plagioclase and feldspar physically and chemically altered. (Pol CNx32).

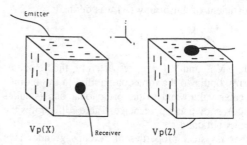

Fig. 7.- Scheme of the position of the transducers, for the meassuring of Vp, in relation to the orientation of the specimens (X and Z)

feldspars, plagioclases and biotites are chemically altered to sericite and chlorite in a variable degree (Fig. 6).

According to the "petrographycal index of deterioration" (Ordaz et al., 1978), the Axeitos granite shows a general index Id=2.8, that can be considered as high.

Laboratory ultrasonic techniques have been applied in order to characterize the granite physically. Velocities of longitudinal wave propagation (Vp) have been measured in three perpendicular directions on two types of samples: prismatic blocks (30x20x10cm) and cubic specimens (5x5x5cm) (Fig. 7). The equipment used was a New Sonicviewer, mod. 5217-0 (Oyo Co.). The voltage used for the tests was 200 V; the adapted frequency was 54 KHz for the emitter and 40 Khz for the receiver

Measurements were carried out at different moisture contents. The humidity of the blocks extracted from the quarry corresponds with its

Table 2. Values of V_p for the Axeitos granite.

SAMPLES	MOISTURE CONTENT	V_p(m/s)			A(%)
		X	Y	Z	
Blocks (30x20x10 cm)	Quarry Humidity	2200±440	3260±500	3500±490	59
Specimens (5x5x5 cm)	Dry	2250±230	2600±220	3000±220	33
	Water Saturated	3990±380	4700±740	5100±280	28

X= Direction orthogonal to the alignment of biotites
Z= Direction parallel to the alignment of biotites

natural water content, whereas the specimens were tested under dry and water saturated conditions. A coefficient of anisotropy (A) was defined:

$$A = \frac{V_{p(z)} - V_{p(x)}}{V_{p(x)}} \cdot 100$$

where $V_{p(z)}$ and $V_{p(x)}$ are, respectively, the velocities correspondig to the direction parallel to the biotites (greater value) and to the orthogonal one (smaller value), shown as a percentage. Table 2 illustrates the results obtained.

Other physical properties of Axeitos granite have also been determined (table 3). The results verify that the distribution of the microfractographic network and the alignment of biotites are the main responsible for the physical anisotropies detected.

Table 3. Physical properties of Axeitos granite

PHYSICAL PROPERTIES		AVERAGE VALUES
Open porosity $(n_0)^*$		2.0 %
Coefficient of water saturation $(W_s)^*$		0.8 %
Coefficient of capillarity $(C)^{***}$	Direction X	2.5×10^{-4} g/cm^2·s$^{1/2}$
	Direction Z	2.9×10^{-4} g/cm^2·s$^{1/2}$
Coefficient of water vapour permeability $(K_V)^{***}$	Direction X	0.25×10^{-4} g/cm^2·24h.
	Direction Z	0.66×10^{-4} g/cm^2·24h.

*RILEM, 1980; **NORMAL, 1981; ***NORMAL, 1985;

So, the Vp shows different values in the three directions, regardless of the specimen size and their moisture content. The higher values correspond to the direction of microfissuring that is concomitant with the orientation of biotites. The coefficient of anisotropy is higher in the prismatic blocks than in the cubic specimens. It is also pointed out the influence of the orientations on the capillarity and water vapour permeability, being higher the values measured in direction Z.

3. INDUCED ALTERATION: CORRELATIONS WITH THE INTRINSIC CHARACTERISTICS

3.1. Procedure

Salt crystallization cycles were carried out on specimens of Axeitos granite in order to study the correlations among the directions of textural anistropy, the evolution of the induced microfissuring and the forms of decay generated throughout the test.

Two series of specimens, cubic (5cm side) and prismatic (5x15x25cm), were tested according to the X and Z directions. An aqueous solution of decahydrated sodium sulphate ($Na_2SO_4.10H_2O$) at 14% was used. The absorption was by capillary suction. The procedure followed was the one proposed by RILEM (1980). The number of cycles was 50.

The evaluation of damage was made taking into account the observation of superficial modifications of the specimens, as well as weight variations and values of ultrasonic waves velocities after every five cycles.

3.2. Results

The results obtained are shown in Table 4. The damage induced by the salt crystallization test changes according to the shape and orientation of the specimens.

a) Cubic specimens. The specimens X-oriented in the sense of the rise of the saline solution tend to deteriorate by granular disaggregation. The first minerals to disaggregate are the biotites, followed by plagioclases and K-feldspars; the last ones in detaching are the quartz grains. The final result is an arenization that is preferably located in a band, of one centimeter wide, in the central part of the specimens (Fig. 8).

In the specimens Z-oriented in the sense of the rise of the saline solution, the degradation is mainly due to fissurating. Fissures accumulate also in the central part of the specimens, coinciding with the limit of the dry section of the specimen and the one impregnated with the saline solution. In the 50th cycle the specimens eventually split.

In general, the losses of material are greater in direction Z than in the X one, due to the detachment of the material involved in the processes of fissuring and final rupture.

The values of Vp decrease along the cycles with respect to the initial ones (non-tested specimens). This decrease is apparent from the 35th cycle, and it is greater in the specimens Z-oriented (about 100%) than in the X ones (30%).

b) Prismatic specimens. Prismatic specimens were also tested according the above mentioned directions. There was no significant damage in X orientation, only slight pitting. In the Z orientation, detachment of grains (mainly of the assemblage plagioclase-K feldspar) were observed from the 12th cycle, as well as a progressive increase of thefissurating, until scaling and traces

of plaques took place. A staining by lixiviation of iron from biotites was also evident after the test. The loss of material is much greater in the prisms Z oriented than in the X ones. However, the decrease of Vp is negligible (see Table 4).

A)

B)

Fig. 8. Macroscopic aspect of specimens of Axcitos granite tested by salt crystallization cycles. A) Specimen oriented X after the 50th cycle. Note the disaggregation in the central part. B) Specimen Z oriented, split at 50th cycle.

Table 4. Results of the salt crystallization test.

SPECIMENS	ORIENTATION	WEIGHT LOSS (%)	Vp DECREASE (%)	DAMAGE
Cubic	X	1	30	Disaggregation Small fissures
	Z	5	100	Fissures Splitting
Prismatic	X	0.25	0.25	Slight Pitting
	Z	5	5	Fissures Scaling Traces of plaques

3663

4. CONCLUSIONS

The structural characteristics that granitic rocks show at rock mass scale are intrinsic factors that can influence some of its physical properties at intact rock scale, and also its decay as a building stone.

So, Axeitos granite shows an anisotropic behaviour that can be clearly related to the preferred orientation of its discontinuities at different scales (fractures, fissures and cracks) and with the alignment of the biotites. This behaviour is evidenced through the variation of some physical properties, such as velocity of sonic waves, capillary suction and water vapour permeability. Ultrasonic techniques are especially useful for detecting the above mentioned anisotropies.

It has been experimentally shown, by means of an accelerated ageing test (salt crystallization), that the orientation of these discontinuities condition or favour the appearance of some typical deterioration forms shown by this granite: granular disaggregation, scaling and deplaquation.

The anisotropic behaviour of some granites, as the Axeitos one, will be kept in mind in order to understand or interpret the evolution of its decay as building stone. The different states of decay shown by exposed dimensional stones in monuments can be partially conditioned by the relation between the orientation of the stones and their structural and textural anisotropies.

These aniotropies will favour also, to a certain extent, the selective action of the external (environmental) agents (water, pollutants, soluble salts, etc.).

ACKNOWLEDGEMENTS

We thank to the European Economic Community for its financial support of the research project: "Conservation of Granitic Rocks. Application to Megalithic Monuments" (Contract No. STEP-CT90-0110). Also to the "Comisión Interministerial de Ciencia y Tecnología" (CICYT, Spain) Project: PAT 91-1093-C03-01.

REFERENCES

Casal, M., J. Delgado Rodrigues & B. Silva 1992. Construction materials and decay problems of Salomé church in Santiago de Compostela. Proceed. 7th Int. Cong. on Deterioration and Conservation of Stone, Lisbon, 1:3-10.

Cuesta, A. 1991. Petrología granítica del plutón de Caldas de Reis (Pontevedra, España): Estructura, mineralogía, geoquímica y petrogénesis. Edic. Laboratorio Xeoloxico de Laxe-Universidad de Oviedo, Serie Terra Nova, nº 5, 417 p.

Gervais, F., J. Riss & S. Gentier 1993. Caractérisation stéréologique de la géometrie d'un massif rocheux fracturé: aplication aux carrières de Comblanchien (Côte d'Or, France). Bull. Soc. Géol. France, 164(3):459-471.

Gundersen, H.J.G. & E.B. Jensen 1987. The efficiency of systematic sampling in stereology and its prediction. Journal of Microscopy, 147(3): 229-263.

ISRM 1978. Suggested methods for the quantitative description of discontinuities. Rock Mech. Min. Sci. & Geomech. Abstr. 15:319-368.

Montoto, M. & R.M. Esbert 1976. Alteración de granitos: Evolución a rocas blandas y degradación de propiedades geomécanicas. Memorias del Simposio Nacional sobre Rocas Blandas, Madrid, 1-13.

NORMAL 1981. Doc. 7/81: Assorbimento d'acqua per immersione totale e capacitá d'imbibizione. CNR-ICR, Roma, 5p.

NORMAL 1985. Doc 11/85: Assorbimento d'acqua per capillaritá. Coefficiente di assorbimento capillare. CNR-ICR, Roma, 7p.

NORMAL 1985. Doc. 21/85: Permeabilitá al vapor d'acqua. CNR-ICR, Roma, 5p.

Ordaz, J., R.M. Esbert & L.M. Suarez del Río 1978. A proposed petrographical index to define mineral and rock deterioration in granitic rocks. Proc. Symp. on Deterioration and Protection of Stone Monuments, Paris, 1: 2-6.

Pérez-Ortiz, A.; Ordaz, J.; Esbert, R.M.; F.J. Alonso 1994. Microfissuring evolution of the granite from the Axeitos dolmen along the salt crystallization test. IIIrd Int. Symp. on the Conservation of Monuments in the Mediterranean Basin, Venice (in press).

RILEM 1980. Essais recommandées pour l'altération des pierres et évaluer l'efficacité des méthodes de traitment. Matériaux et Constructions, 13 (75): 216-220.

Robert, M., V. Verges-Belmin; A.M. Jaunet, M. Hervio & P.H. Bromblet 1993. Mise en evidence de deux microsystemes d'alteration chimique et physique) dans les monuments en granite (Bretagne). Proceed. 7th Int. Cong. on Deterioration and Conservation of Stone, Lisbon, 1:129-135.

Streckeisen, A. 1976. To each plutonic rock its proper name. Earth Sci. Rev., 12:1-33.

Vicente Hernández, M.A., E. Molina Ballesteros & V. Rives Arnau (eds.) 1993. Alteración de granitos y rocas afines, empleados como materiales de construcción. Consejo Superior de Investigaciones Científicas, Madrid, 190 p.

Natural stones geotechnical characterization

Caractérisation géotechnique des pierres naturelles

A. Mouraz Miranda

Instituto Superior Técnico, Technical University of Lisbon, Portugal

ABSTRACT: Natural Stones are quarry or erratic produced, as blocks and are then sawn, cut or splitted. They are intended for building cladding, flooring, decorative furniture and memorials. Industry and trade recognizes each stone by a particular Denomination. Petrography, colour and structure are specific characters of each Natural Stone. Technical and mechanical properties of each stone have different importance as envisaged use is concerned.

RÉSUMÉ: Les pierres naturelles sont produites soit en carrière, soit d´erratiques. Les blocs produits sont sciés, coupés ou cassés. Les pierres sont destinées au révêment de façade, de plancher, aux objects decoratives ou le funéraire. L´industrie reconnâit chaque pierre par une non ou désignation. La petrographie, la couleur et la structure sont les charactéres de chaque Pierre Naturelle. Les propriéties mecaniques et techniques de chaque pierre ont une importance differentte pour chaque utilisation.

1 INTRODUCTION

The terminology proposed at ISO and CEN level to designate marbles, granites, slates and other stones to be used for cladding, flooring, decorative works and memorials is Natural Stones. The origin is the German word Naturstein. English(US) Dimensional Stone is superseded. Spanish Rocas Ornamentales is misleading. Natural Stones are not only decorative, for architectural purposes, but are essential elements of construction requiring structural dimensioning.

The industry of Natural Stones is thousand years old. Europe has been traditionally the major producer and end user. SME´s are dominant in the industry and trade.

Tradition of the mettiér is to distinguish each Natural Stone by a peculiar Name. The origin of the Name is linked to a geographical location, colour or structural feature. Most of the Names are not easily explained. Few examples of Names are: Baltic Brown, Rosso Levante, Comblanchien, Rosa Aurora, Rosa Porriño. Name or Denomination of a Natural Stone is a feature to be noted on Engeneering Geological Characterization.

2 MAIN FEATURES FOR NATURAL STONES CHARACTERIZATION

The standardization work of CEN Technical Committee TC246 is of foremost importance to understand the main characteristics of Natural Stones to be used.

In the forefront and at any production, processing or application stage is the type of stone. This feature corresponds to a NAME or DENOMINATION.

For each Name a certain geographic location is attached, e.g. Rosa Aurora is produced in the municipality of Estremoz in Portugal.

A petrographic character is also bounded to a Name, e.g. Rosa Aurora, is a calcitic crystalline limestone, medium to fine grained, also described as a marble.

The colour is a very important feature and serves as a distinction and is normally the base of acceptance or rejection of a stone.

The rock structure at mesoscopic level gives each Natural Stone its decorative character. It is part of the individuality of each stone, e.g. the orbicular texture of Baltic Brown.

Laboratorial tested properties envisaged for Natural Stones vary in accordance with the intended usage or application.

Mechanical properties to be determined are intended for structural dimensioning. The most important ones as far as claddings are concerned are:
(a) the flexural or bending strength;
(b) the resistance to the fixation elements or anker resistance.

Facade claddings are the most expensive items within the industry and trade, and are the ones requiring specialised finishing and drilling of anchoring holes.

Weathering and freezing resistance are relevant for application in external pavings and façades in climates with icy cold winters or in humid foggy polluted urban areas.

The gelivity freezing-defreezing test is an old standard. MORANDINI (1991) gives a new insight for the use of this test.

Weathering tests are more difficult to ascertain. A group of tests is based on salt action. Salts interact differentially with one or a very limited number of minerals. Test results despite quantitative in nature are difficult to translate to usage and application purposes. A second group of weathering tests is based on cyclic ageing actions. Dry and wet tests are within this group (see AIRES-BARROS and MOURAZ MIRANDA, 1989). All freezing and weathering tests have two major drawbacks:

(a) they are time consuming;
(b) the provided information is not absolute and should be treated intercomparatevely.

Mechanical abrasion resistance is a flooring selection parameter like slipperiness. Both are important for safety on use. Safety is a relevant concern.

Bulk volumic mass (density) and water absortion are industrial relevant parameters which are used for dimensioning. Gravimetric tests with weighting are used.

3 IN SITU AND ROUGH PRODUCTS EVALUATION

For the Natural Stones industry, blocks (Rohblocke) and slabs (Rohplatten) are rough products not intended for final application.

It is a geotechnical task to evaluate a producing site for blocks and the slabs. Blocks are quarried at a rock mass (fr. masse minérale) or from "érratics". Evaluation should be carried on type of stone to which a Denomination or a Name is attached. The mineable dimensions of blocks is a determinant factor for economic purposes. Petrographic, colour and mesoscopic features are determinant. "Defects" as described by the industry, must be pointed and if possible mapped. Defects can be schlierens, nodules, small fractures and cracks, weathered zones or decoloured parts in the material. Very small blocks are useless for processing (approx. below 1 cubic meter). No standard exists for in situ acceptance or rejection.As a rule of thumb a mineable or recoverable block should possess a minimum of 2cubic meters of the consistent quality. All other material should be treated as waste.

At quarry level either on front or surface observation different types of stone must clearly be separated or mapped. All produced blocks must be marked with the natural layer (SENZO).Some materials are sawn or cut normally to the natural layer,others are sawn or cut in counterlayer.

Geotechnical evaluation should clearly mark differently both aspects.

For certain applications requiring large amounts of material the available or mineable quantity of stone must be assessed.Very few producing sites have been properly evaluated for mineable reserves. Due to the fact that most companies involved on block production are SME´s with a limited financial capability the overburden cost removal is of critical importance, therefore, geotechnical evaluation of the overburden is required for thickness and type.

Rock mass fracturing is quoted to be a decisive item on quarry evaluation. It is our point of view it should be linked to the type of stone (in the commercial sense and value) plus the average dimensions of the recoverable block with the major dimension parallel to the natural layer.

Rough blocks must be squared to a paralelipipedic shape with the major dimension (length) paralel to the natural layer.

Blocks must be checked for the Natural Stone type, rejecting all blocks with "defects". Small cracks, which must be clearly marked in the block, are acceptable only if they are supposed not to introduce severe damage to sawing or cutting operations.

Rough slabs are the product of a sawing, cutting or splitting process of a Natural Stone block. Each individual slab must be checked for type according to the NAME principle, elsewhere explained. Colour variations must be within an acceptable range.No tests are envisaged presently for colour. Veining and any other structural rock feature is normally part of a Natural Stone.

Dimensions are indicated in centimeters for length, width and heigth. Thickness is indicated in millimeters

4 CONCLUSIONS

For a geotechnical or engineering geologist evaluation of Natural Stones the following items are relevant:

(a) At visual or non-testing stage:
(a.1) Type of stone, in accordance with a given Name or Denomination;
(a.2) The petrographic character or classification;
(a.3) Colour or colour range variation;
(a.4) Structure, particularly veining.

(b) Physical and resistance parameters to be stated for each Natural stone, either from direct sample testing or used from prevïons data-banks:
(b.1) Bending or flexural strength;
(b.2) Anchorage resistance;
(b.3) Bulk volumic mass (density);
(b.4) Water absorption;
(b.5) Abrasion resistance;
(b.6) Slipperiness;
(b.7) Weathering and freezing resistance.

REFERENCES

AIRES-BARROS; MOURAZ MIRANDA A.
(1989) Weathering and weatherability of rocks
and its significance in geotechnics
in *WEATHERING: its Products and Deposits- vol.
III - Products deposits; Geotechnics,*
ed. Teophraustus Publications, SA Athens

CEN/TC 246/ WG3 (1993)- Natural Products -
Products: Draft of Names for Blocks, Slabs,
Claddings, Tiles, Stairs and Cubic Works

MORANDINI, A. FRISA (1991) - La
Caractterissazione Tecnica dei materiali Lapidi:
Proposte per una normativa Europeia
concernemente la prova di gelivitá
MARMO MACCHINE, pg. 120-130

On the importance of a careful characterisation of decoration stones aimed to durability assessment: An example of faked aesthetic equivalence in stones used in public buildings lining in Turin

L'importance de la caractérisation soigneuse de pierres de construction pour l'évaluation de leur durabilité: Un exemple relatif au revêtement d'un bâtiment publique à Turin

R. Mancini & A. Morandini Frisa – *Centro di Studio per i Problemi Minerari & Politecnico di Torino, Italy*

P. Marini – *Dipartimento Georisorse e Territorio, Politecnico di Torino, Italy*

ABSTRACT: Decoration stones are selected because of their aspect, but aesthetically equivalent stones can be very different in mineralogical composition, in structure and therefore in durability, and became quite different upon weathering. The example is provided and analysed of two decoration stones, visually identical and sometimes marketed under the same name, widely used in building facing in Turin, the one being very durable, the other subject to fast degradation. It makes clear the importance of more strict controls and accurate characterisations that became mandatory only lately, and whose methodologies are still the subject of intensive studies. Differences in behaviour of the two stones when set in place and exposed to weathering are described, and the petrographic, chemical, mechanical and technological tests that should be carried out in order to avoid inappropriate use are exposed and discussed. In the concluding remarks, criteria for aesthetic behaviour forecasting and a scheme of lab tests program tailored to this aim are presented.

RESUME: Les pierres ornamentales sont selectionnées pour leur aspect, mais des pierres très semblables extérieurement souvent ont des différentes compositions minéralogiques, structures et durabilitées. On a analisé deux pierres très pareilles, quelquefois présentées aves le mème nom, beaucoup utilisées in Torino comme pierres de décoration; la première est très résistante, l'autre est characterisée par une degradation très rapide. Ça mit en évidence l'importance d'une précise caractérisation, dont les méthodologies sont l'objet de nombreux études. On analise les différences entre les deux pierres après leur mise en ouvre et on discute les essais pétrographiques, chimiques, méchanique et téchnologiques mises à point pour individualiser les i utilisations incorrectes. On présente aussi un schéma de tests en laboratoire programmée au but.

1 FOREWORD

The problem of choosing a matching stone is important in repairs and in designing building facing.

Colour is the most important factor, but resistance to weathering is important too, since a durable match has to be obtained.

Investigations and testing of durability are by far more difficult than visual aspect check, and tend to be disregarded, due to the fact that facing stones has usually no structural function, hence weathering does not cause a danger as far as stability is concerned.

Even so, durability check pays when facing stones has to be chosen for important buildings. In this report we present an example of the durability differences of two visually similar stones, and of the laboratory testing methods that could be employed in

order to assess these differences and to avoid inappropriate use of that stones.

The two stones compared, both showing a pleasing green colour, have been used in many buildings in Torino. Trade names are, respectively, Verde Vogogna and Verde Roja; the first is cheaper and sold sometimes as Verde Roja. Durability, in weather conditions of Torino, is markedly different, as seen in figures 1 and 2, showing the present appearance of two buildings where the two stones types have been used, after the same duration of exposition (40 years) to weathering.

Though being the loss of mechanical strength a scarcely relevant feature when comparing materials whose function is not structural our testing campaign rely mostly on strength loss as indicator of weathering; the reason is that mechanical testing provides quantitative parameters upon which a

Fig. 1 - Green Vogogna stone after 40 years natural exposure.

Fig. 2 - Green Roja stone after 40 years natural exposure.

comparison can be set, and moreover, provides criteria to assess the necessity of replacement of the facing slab: mechanically degraded slabs, indeed are unsafe.

Being the material to be tested usually in the form of slabs, the most significant parameter used to define the soundness of the architectural items is the bending strength.

2 MACROSCOPIC, PHYSICAL AND MINERALOGICAL FEATURES OF THE TWO STONES

Freshly quarried and polished samples of the two stones have been compared; a synthesis of the results of the comparison is shown in table 1.

Trade name		Green Vogogna	Green Roja
Macroscopic properties	Colour	green	green
	Grain size	fine grained	fine grained
	Polishability	good	good
Physical properties	Specific gravity (kg/m³)	2890	2770
	Sonic sound (m/s)	2800 (normal)	5450 (normal)
		5300 (parallel)	6300 (parallel)
	Water absorption (%)	0.293	0.117
Petrographic data	Name	chloritic schist	metapelite
	Mineralogic composition	50% quartz	85% sericite
		20% epidote	10% quartz
		15% chlorite	2% carbonates
		10% white mica	accessories: tourmalin
		5% carbonates	graphite
		accessories: titanite, rutile, apatite	
Chemical composition data	carbonates (%)	4.95	1.9
	sulphates (%)	//	//

Table 1 - Macroscopic and physical properties of the stone in the fresh state.

It can be seen that the two stones do not belong to the same petrographic type. The most relevant difference , to our purposes, is the higher contents of carbonates in the Vogogna stone, suggesting a lower resistance to acidic fog (which actually has been experimentally demonstrated).

Water absorption is higher in the Vogogna stone, and again suggests higher vulnerability.

Sonic speed difference from parallel and normal to schistosity tests, being lower in the Roja stone than in the Vogogna suggests a lower anisotropy of the former.

For both stones, slabs are cut parallel to the schistosity in order to obtain maximum bending strength.

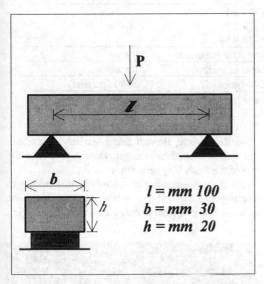

Fig. 3 - Bending test according to UNI norm.

3 MECHANICAL AND ELASTIC PROPERTIES

Comparison has been made on probes obtained from freshly produced slabs. The results are shown in table 2 (bending test has been performed according to the Italian UNI norm).

The formula to obtain bending strength is:

$$\sigma_t = (3 \times P \times l)/(2 \times b \times h^2) \quad [1]$$

A preliminary discussion about the meaning of the bending test reccomanded by UNI Norm is the subject of a precedent study (Mancini et alii, 1994).

Verde Roja stone is stronger than Vogogna, but less hard, according to the Knoop test (due to the lower quartz contents). NCB conical indenter test provides a different hardness ranking (indeed the penetrator probes the stone at a larger scale, and the influence of fine quartz grains is not felt in the NCB test, while is pre-eminent in the Knoop test).

Properties	Green Vogogna	Green Roja
Dynamic elastic modulus (MPa)	35200	71100
Bending strength (MPa)	32.2	81.1
Knoop micro - hardness (MPa)	2119	1248
NCB cone indenter	8.1	9.3

Table 2 .- Mechanical and elastic properties of the stone in the fresh state.

4 WEATHERING EFFECTS OBSERVED

The experimental material for our study has been provided by slabs removed for repair works from the two already quoted buildings. References has been made, in order to quantitatively describe the effects of weathering, to the properties measured on the fresh materials, as described in tables 1 and 2 (hardness tests have been obviously omitted, being the surface of slabs no longer polished).

Three slabs have been tested of Vogogna stone, having been exposed with different orientation and used for different architectural functions; a single slab of the Roja stone, having been in this case the exposure to weathering of all slabs quite similar.

Visual appearance changed markedly from the one of the fresh stone, much more conspicuously in Vogogna than in Roja stone; in the former loss of material in the form of detaches scales is observed, and mechanical tests results, here under reviewed, are favourably biased by the fact that the probes cannot be obtained from the outer layer of the slabs.

The results of the comparison tests are summarised in table 3.

Water absorption shows a significant increase only in the Vogogna stone; however, referring to the single parameter, the materials seems to be still in the range of a sound stone. Roja stone is not significantly modified by weathering, as far as absorption is considered.

Sonic speed changes are scarcely meaningful for both stones.

Chemical changes are apparent when the surface layer is considered. In the Vogogna stone, carbonates disappear and sulphates practically absent in the fresh material, show up, while Roja stone is apparently unaffected by chemical weathering agents.

Trade name		Green Vogogna	Green Roja
Use		1) wall facing 2) window vertical frame 3) sill	wall facing
Exposure orientation		1) SSW 2) ESE 3)NNE	S
Duration		40 y	40 y
Measured degradation indicator	Water absorption increase	1) 10.2 % 2) +17.4 % 3) + 41.6 %	negligible
	Bending strength variation	1) -30.1 % 2) -19.2 % 3) -27.9 %	-30 %
Sulphates		fairly light in surface	traces
Carbonates		absent in surface	traces
Sonic speed (m/s)		5200 (parallel)	5980 (parallel)

Table 3 - Properties of weathered stones by natural exposure to weathering agents

Strength is strongly affected in both materials. To be reminded, data obtained from Vogogna material are favourably biased for the already explained reason, and, when the strength of the slab instead of the strength of the probe is considered, for the lack of consideration of the loss of material.

The Roja stone, though being securely weakened, still retains a bending strength value higher than the one of the fresh Vogogna stone.

Roja stone weakening is apparently due mainly to freezing-thawing effect, while in the Vogogna material also chemical agents play an important role.

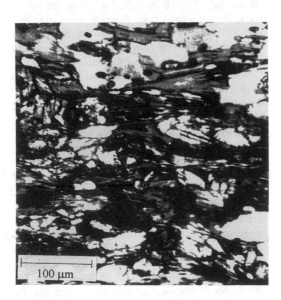

Fig. 4 - Thin section of Green Vogogna.

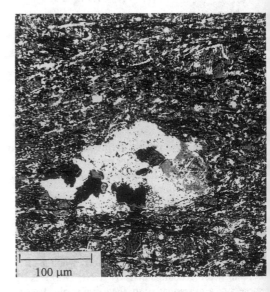

Fig. 5 - Thin section of Green Roja.

| Trade name | Green Vogogna | | Green Roja | |
Type of ageing test	Bending strength	Water absorption variation	Bending strength	Water absorption variation
Acid mist exposure				
48 h	- 9.3 %	+ 20.51 %	- 0.5 %	+ 6.3 %
96 h	- 10.8 %	+ 12.2 %	- 7 %	+ 13.2 %
144 h	- 14.4 %	- 14.2 %	- 3 %	+ 9.6 %
Combined thawing - freezing (25 cycles) plus acid mist				
48 h	- 27.3 %	//	- 4.2 %	- 21.3 %
96 h	- 27.7 %	+ 4.2 %	- 8.4 %	+ 63.3 %
144 h	- 28.8 %	+2.6 %	- 11.4 %	+ 31.2 %

Table 4 - Results of accelerated ageing test.

5 ACCELERATED AGEING TESTS

Both Vogogna and Roja stones are quoted as not frost sensitive when tested according to the UNI norm, based on freezing-thawing cycles and compression strength upon treatment.

Prescribed test conditions are apparently too mild.

We devised an harder accelerated ageing testing procedure, where exposure to acid mist (pH 2) in an ageing cell and freezing-thawing cycles are combined, and the mechanical degradation is checked through the bending strength reduction.

Results are summarised in table 4

The erratic behaviour of water absorption deserves some comment. Actually, carbonates/sulphates transformation entails a volume increase, hence a reduction of water absorption is in principle expected, due to pores filling; sulphates, however, are soluble and are slowly leached away, which partly occurs in the test conditions (in natural ageing sulphate leaching is more complete).

The effect of volume increase is more or less compensated by leaching in the different tests. As to the strength, Vogogna stone is severely affected by acidic mist exposure, and shows a strong synergism between freezing-thawing and acid mist effects. The damage level produced by artificial accelerated ageing is quite similar to the one observed in natural, 40 years, exposure.

Roja stone is apparently less sensitive; acid mist alone has practically no effect and combined exposure produces less damage than observed in natural, 40 years exposure. A still harder ageing test should be advised in order to reproduce the effects of the worst weathering conditions actually met by the facing stones in an urban environment.

6 CONCLUSIONS

The research on artificial ageing test for facing stones is still underway, a series of warnings however emerges from the cases examined:

Inconspicuous differences are important in dictating the long term behaviour: being co-operation between different weathering agents the rule, a small change in the contents of a sensitive mineral (in the examined case, the carbonates) can open the way to a combined degradation mechanism. On the contrary, the abundance of inert minerals (in the examined case, quartz) does not represent a remedy.

Artificial ageing tests are necessary, it is difficult , however, to calibrate them in such a way that the actual weathering effects are correctly reproduced in any stone type.

Loss of mechanical strength is always to be expected as a result of long exposure. Materials showing a low strength in the fresh state should be more securely checked, it seems obvious, than the stronger ones (in the studied case, the Vogogna stone).

Acceptance rules should not be based on "percent decrease of strength" upon ageing tests, rather upon absolute lower limits of the strength value.

Anyhow, the choice of a stone for a long term use should not be done on the basis of the only visual appearance of the fresh material.

REFERENCES

Ente Italiano di Unificazione: Norma UNI 9724 parti 1-8, *Edilizia - Materiali lapidei - Metodo di prova*

Mancini Renato, Marini Paola 1994 - Confronto fra due metodi di prova di resistenza a trazione indiretta per le rocce. *Renc. Jeunes Cherch. Geol. Appl., Lausanne 21 avril 1994* (printing).

Mancini R., Marini P., Morandini Frisa A. 1994 - Evoluzione e suscettibilità delle pietre ornamentali all'alterazione. *III Int. Symp. on the Conservation of Monuments in the Mediterranean Basin, Venice 22-25 June 1994* (printing).

Marini P. 1993 - Degrado artificiale in rocce ornamentali. *Giornata Informativa sulla Ricerca nel campo delle Materie prime*, Roma.

Environmental implications of surface treatment of building stones and ornamental rocks

Implications environnementaux du traitement des façades des bâtiments et des roches ornementales

Asher Shadmon

Stone Technology International – Consultants, Jerusalem, Israel

ABSTRACT : Loss of strength of stone and problematic "dressed" surfaces contribute to failures and accidents. Environmental aspects of stone facades and pavements, and their geotechnical factors are observed; flexure tests are studied more closely as well as fundamentals of stone surface treatment. Polishing of stone is the more complex of primary surface "dressing" and is reviewed at some length as the process is closely related to friction and slip resistance properties and information sources are not readily accessible.

RESUME : La perte de la resistance de la pierre et les surfacages problématiques, contribuent à défauts et à des accidents. Les aspects environmentaux des façades et des pavements de pierre et leur facteurs géotechniques sont considerées: des essais de flexion ainsi que les données fondamentaux du traitement de surface de pierre sent etudiés de manière plus proche. Le polissage de pierre étant le plus complexe des surfacages primaires, est passé en reveu plus longuement étant donné pour les procès est lié intimement aux proprietés de friction et de résistance au glissance et que l'accès aux sources d'information n'est pas immediat.

Environmental symptoms in stone facades have always existed. Only in the last few years have failures accelerated due to pollution, etc., and not less important, the change of stone working technologies.

Solid stone, at one times predominant in buildings, was replaced by thinner veneers earlier this century. In mid-century the thickness began to decrease further and many buildings are presently being clad with granite, limestone and slates with veneer less than 10 mm thick.

Thicker stone units can accommodate cracks, spalls, fractures, discontinuities and other macro and micro blemishes or failures typical of products of nature with its wide range of properties.

Geotechnical knowledge of dimension stone properties has not kept pace with the fabrication and other cost saving cladding techniques. The manifold increases of stone varieties from worldwide new sources lack the track record of existing stone products which have withstood the test of time.

Knowledge of the geotechnical parameters of stone and cladding techniques is essential to prevent turning the one time symptoms into failures. Witness to this are major failures reported in the last decade from places as far apart as Chicago and Helsinki.

Stone is brittle rather than ductile, loses strength when exposed to the elements, with strength varying in different directions. To boot, strength is affected by surface finish and by the content of moisture.

Attention to geotechnical properties starts in the quarry. Bedding planes, splitting directions, stylolites, relics of previous bedding planes are a few examples that represent the genesis, depositional and other modes of formation and generation.

Added to this are tectonic influences, residual stresses from cooling magmas, rock bursts. All these add to possible if not probable causes for failure caused by environmental and other external influences, including manmade.

The knowledge of these factors at the deposits helps to prevent failures provided that notice is taken during transformation of the rock into dimensional stone and extended during the application and use.

Although quarries take into account such features when removing blocks from the quarry face, this empirical knowledge is not always transferred to the fabricator who transforms the blocks into slabs.

Slabs in turn are converted into products by finishers who are now, perhaps quantitatively, the largest handlers of dimension stone readied for application. Diminished attention can be expected during the surface finish process, the last stage before the stone is used.

The finish of dimension stone involves various surface treatments including dressing, abrasion, thermal treatment and polishing. Although the testing of physical properties of stone has become a standard procedure, the implications of the environment effect on the surfaces, important to recent experiences in failures, requires more quantitative attention.

Since much of these failures can be due to inadequate flexural strength, more notice is taken of the latter, especially when updating standards and specifications by ASTM and ISO basing its recommendations on the work by CEN, the European Committee for Standardization.

Besides revision of flexural strength testing usages, the effect of surface treatments including polishing, and slip resistance of floors, paving and other frictional parameters, are attended to.

There is confusion about the terminology used for tensile strength measurements. The Modulus of Rupture. i.e. the resistance of a stone slab to bending or flexure is presently discussed in CEN/TC 246/WG1 and WG2, respectively the "Terminology; classification" and "Test methods" CEN working groups.

In the meantime "bending test" is applied to the unconfined centre point knife-edge loading or three point loading test; testing of "flexural strength" of natural stone is used for the four-point, also known as the quarter point, unconfined loading test. This test which uses steel rollers for loading rather than a knife-edge gives more information than the former and enables statistic interpretation of the break caused.

FLEXURAL STRENGTH

In flexural strength testing a load is applied at the quarter points of the stone specimen preferably from the batch that will be used for a construction. The load is increased gradually until the test specimen fails by fracture. The maximum load is recorded for the calculation of the flexural strength.

The testing may be carried out for various thicknesses and surface finishes. The test is repeated wet and dry, parallel and perpendicular to the bedding or rift.

The anisotropy observed in the latter can be significant and e.g. in Rosa Porrina pink granite the strength difference relative to the rift may be

anything between 5% and 35%. The quotation of unqualified values of technical data without full details of the test variations, given by the stone supplier, is more often than not the lower one.

The flexural test differs from the bending test in that the latter does not take into account the production thickness of the stone or the exterior finish used in the final application. Also, the failure occurs at the weakest point of the test piece and not directly under the applied load as in the rupture test.

This enables not only the comparison of differences of flexural strength between the various dimension stone but also serves to compare the strength of stones of the same type.

For these reasons the flexural test is important although actually the compressive strength is the strongest physical property of a stone.

A major cause in failure is the wide temperature ranges to which buildings are exposes, e.g. in Chicago, which can expect up to 80 freeze/thaw cycle per year. A dark coloured stone wall can reach some 70°C whereas extreme cold exterior wall design is based on -23°C.

Presently, an environmental exposure test is applied by using boosted cycles, each simulating multiple cycles, and thus accelerating environmental effect although there are not means to equate accelerated cycles to one year of actual weathering.

An approximation is achieved by placing the stone specimen, exposed-face down, into a container partially filled with water, and into an environmental chamber with temperature cycles alternating e.g. from -28°C to 70°C.

At various intervals, up to a few hundred cycles, specimens are tested for flexural strength loss. The number of cycles depends on a pre-determined equivalent figure of one year exposure in a specific environment obtained by measuring the flexural strength of non-exposed stone specimen. It has been shown that the loss of strength in stone levels off after 250-300 cycles.

LOSS OF STRENGTH

The evaluation of testing of the stone to be used gives a measure of its potential strength loss which can be of significant amounts and is a reason why high safety factors are required in the stone specifications. The high variation in stone properties, being a natural product, requires higher safety factors than those established for manufactured building materials like concrete and steel.

Limestone may need a safety factor of 8 times, marble 5 times for flexure and 10 times for a concentrated load, granite 4 times for connections (e.g. anchors) and 3 times for flexure etc. A minimum safety factor can be determined by a specially developed equation based on flexural strength.

Depending on environmental factors, length of exposure and dimensions, marble may lose up to 70% of its flexural strength, thermally finished granite up to 45%, limestone up to 30%, and polished granite up to 5%.

Investigations over many years of the influence of moisture content on the compressive strength of stone have not produced generic norms for the effect of moisture content; generally with the increase in moisture content the compressive strength decreases.

From the sporadic quantitative investigations available, sandstone of the Ecco Series in the South African Karroo systems with a 15% porosity, may lose up to 50% of its compressive strength when saturated. (Colback et al, 1965).

Besides the effect of environment on strength of stone, the impact on various treatments of stone surfaces has been studied sparsely. Polishing processes and thermal treatment are but a few of the "dressings" applied to stone surfaces.

Thermal finishes like flame-textured stone causes near surface damage, especially micro fracturing unless carried out under strict supervision and weight losses during the flaming is checked. Shocks by the flaming process can damage 2 to 3 millimeters of surface. Thermal treatment by plasma heating is claimed to reduce the damage effect. Testing the flexural strength by freeze/thaw cycles will quantify the damage caused. The thicker the slab the lesser the damage to be expected.

Various arbitrary measurements have been suggested for establishing when a stone is considered damaged, e.g. by measuring the compressive strength of a specimen in the dry state after undergoing the freeze/thaw cycles, so as to compare the compressive strength with samples not exposed to frost. The stone is considered damaged if the drop in compressive strength is more than 20%.

POLISHING

Polishing, as distinct of grinding, is not easily reproducible as the processes react differently with the workpiece and is dependent on environmental factors like temperature and humidity in addition to the various physical,

mechanical and chemical parameters of the applied polishing medium.

Firstly, we have to see what happens when we polish stone. Polishing of stone is a complex process that has been evolved with a varied degree of success as an art and has received little attention from applied science as have other material. There is no universal theory explaining conclusively the mechanism of polishing surfaces in general - different surfaces react in an individual manner to particular processes; the effect on each material has to be judged by its merits. That polishing has been practised from times immemorial has been born out by archaeological finds, whether in the form of early stone artifacts or gemstones.

Until the end of the last century, the polishing process was considered to be a continuation of the grinding process on a smaller scale. One of the earlier modern studies (Rayleigh, 1901), postulated that polishing is not essentially a continuation of grinding. For polishing, a flexible backing is essential where the polishing particles can sink in, the polished material being worn away almost molecularly contour-wise from the prominences, without the formation of pits as in grinding. Rayleigh's observations were made on glass.

Beilby (1921), who included mineral crystals in his experiments, came to the conclusion that polishing was the result of surface flow. There is still considerable controversy about the Beilby layer. At best, it does not seem to apply to all stones or crystals. Bowden (1937), showed that the surface flow occurs, provided that the melting point of the polished agent is higher than that of the material, except for materials with a very high melting point.

Rabinowicz, et al (1935) showed that the weight-loss by polishing corresponds to the material removed to obtain the smoother surfaces, i.e. - no evening out - polishing is not the result of a smearing action. Samuels (1967) located, by etching a repolished surface, the grooves under the layer by preferential action of the etchant on the underlying distorted metal, thus showing that polishing was a form of deformation.

When the melting point of the surface is reached by high speed polishing, rate of removal may become less and the surface becomes wavy and not polished - (e.g. in Wood's metal).

Although chemical aspects of polishing have been noted in the 19th Century, only in the last few decades W.M. Preston (1930) and W.H. Hampton (1930) showed that chemical reactions take place in glass polishing.

Kaller (1956) postulated that the role of a

polishing medium is to supply a material with a high chemical activity. Polishing materials, especially artificially produced ones, are generally submitted to heat treatment before use, thus producing particles with a large number of internal defects in their crystalline structure.

In the polishing of rock, the problem can be broadly divided, for practical purposes, into those concerned with high silica content (quartz at one end) and those with high calcium contents (calcite at the other end). Some of the quartz polishing problems may have certain affinities with those of glass polishing. Rocks with high calcium content which are able to take a polish, are technologically designated as marbles. Some marbles may also have a fairly high silica content. The theory of polishing of such marbles, or for that matter "granite", is more complex than that of the polishing of high calcium marbles (Shadmon, 1994).

It is likely that for the foreseeable future stone polishing will remain an art rather than a reproducible technology. To satisfy environmental demands both from an aesthetic and public safety angle the best that can be expected is to control the quality of the surfacing process, including the polishing of stone.

No recognized standard for stone polishing can be expected without ability to measure the results of polishing. This is especially important for tiles and paving as public awareness of slip control has come into prominence. Improved polishing methods have created a demand for highly polished floors. The need for testing of resistance to slipping is now part of the essential requirements of the EEC Directive for Building Products.

FRICTION

The friction force is dependent on the roughness of the sliding surface. With very smooth surfaces the friction tends to be high because the real area contact grows excessively. With very rough surfaces the friction is high because of the need to lift one surface over the asperities of the other. In the intermediate range of roughness, the friction is at a minimum and almost independent of roughness.

Exceptionally, a rough surface gives higher friction than a smoother one when a rough body slides on a much softer one. The aspirities of the rough surface digs into the softer material and when sliding occurs, a much larger part of the soft surface is sheared.

Some basic laws of friction are relevant to the studying of the polishing process:

1. In any situation where the resultant of the tangential forces is smaller than some force parameter specific to that particular situation, the friction force will be equal and opposite to the resultant of the applied forces and no tangential motion will occur.

2. When tangential motion occurs, the friction force always acts in a direction opposite to that of the relative surfaces. Macroscopically observable variables include applied load, size of the region of contact and sliding velocity.

(a) Friction force f is proportional to the (normal) force F of between the surfaces L.

$$F - fl \ \tan \theta \ = f \ (\text{or} \ \mathcal{M})$$

(b) the friction force is independent of the apparent area of contact

(c) the friction force is dependent of sliding velocity

Exceptions to (a) and (b) occur with very heard (diamond) or very soft materials (teflon). Also there is the possibility of a surface with a thin hard surface layer and a softer substrate which is reached through the broken-through surface layer at high loads.

Deviations of (b) are sometimes noted in very smooth and very clean surfaces. Strong interaction between the surfaces takes place and the friction becomes independent of the load but proportional to the apparent area of contact (which has become the real area of contact).

The static friction coefficient is a function of time and contact.

The kinetic friction coefficient is a function of velocity throughout the range of velocities and only in limited velocity ranges may be taken to be a constant independent of the sliding velocity. At speeds the friction usually decreases.

SLIP RESISTANCE

Corresponding accepted safety factors (see p) as for strength are non-existent for slip resistance in stone, lately receiving professional attention due to its inclusion in the Americans With Disabilities Act.

Although tests exist to evaluate resistance to wear of flooring, paving, etc. no such tests have been developed for slip resistance on stone surface per se and use is made of tests which deal with a wide range of surface coverings.

Friction is related to polishing. However these properties differ in that the aim of polishing is to obtain a smooth surface which shows at the same time a high reflectivity, not essential in evaluating frictional force. Various aspects of friction accompany the polishing process from its beginning till after its termination. A smooth surface is not essentially a "frictionless" surface. Bailey and Pratt (1955) state that atomically smooth surfaces as in mica, produced by a cleavage, show a very high friction.

This is not surprising when considering that the actual area of contact between two surfaces increases with smoothness.

SURFACE MEASUREMENTS

There are several instruments available to conduct a surface analysis of smooth surfaces. For polishing evaluation the Talysurf gives an assessment of the surface topography or rather roughness of a flat surface whereas the gloss meter gives an idea of the reflectivity of a surface. Both these instruments require dry clean surfaces.

The Friction Meter which measures sliding friction proportional to the movement of sliding gives an evaluation of the friction value and can also be used on a wet surface. The higher the static coefficient of friction the safer the floor surface.

Measuring instrument types include static, dynamic and a combination of the two which simulates a slip situation by using a pendulum released from a standard height. After impact of the pendulum at the lowest point of its swing on the stone surface to be tested, the edge of the test price is dragged across the test surface and the retardation of the pendulum is a measure of the coefficient of friction with a rather indicative value. In the static instrument, the force to move a loaded pad of a known area is used to calculate the coefficient of friction. The static/dynamic method measures the resistance to continued movement of a self powered unit across the test surface (e.g. the "Tortus" unit).

Values presently generally accepted are:

0.63 - 1.00	Very safe
0.42 - 0.63	Safe
0.29 - 0.42	Secure
0.21 - 0.29	Insecure
0.00 - 0.21	Dangerous

Typical values for granite in various conditions are as follows:

Polished.dry	0.40 - 0.55
Polished.wet	0.15 - 0.25
Honed.dry	0.30 - 0.40
Honed.wet	0.20 - 0.30
Textured.dry	0.55
Textured.wet	0.35

The values recommended in the U.S.A. are a static coefficient of friction of 0.5 of 0.6 for accessible routes and 0.8 for ramps. These figures were arrive at by measuring the friction between a show sole material and a flooring surface by three separate tests: The NSB-Brungrader Tester, the PTI Drag Sled Tester and the Horizontal Pull Slip Meter. The latter is used for the ASTM C-1028 Static Coefficient of Friction Test. There is little comparison available between results of these various tests.

Polished stone surfaces require periodical checks as wear takes its toll and may change a polished surface into a honed one. Another measurement for friction coefficients is described in ASTM D-2047.

Results of similar experimental tests in Europe have shown that pavings with relatively similar friction coefficients may lose their slip resistance under certain conditions of use.

Slip resistance in stone does not necessarily depend on the smoothness of a flat surface as coarse and not so flat surfaces can be slippery when wet or worn. Dressed limestones can be quite slippery and pavings require repointing with a point chisel in heavy traffic areas. Also, granite can lose much of its slip resistance in high traffic wear.

Unless the results of the various slip measurement devices, and there are some 60 of them, can be standardized and agreement reached upon what constitutes a slip-resistant stone surface, it looks unlikely that a standard coefficient of friction will be determined in the foreseeable future.

REFERENCES

Bailey, A.I. and J.S. Courtney-Pratt. 1955 -
The Area of Real Contact and the Shear Strength of Mono-Molecular Layers of a Boundary Lubricant: Proceedings Royal Society A227, London

Beilby, G.T., 1921 - "Aggregation and Flow of Solids"; Macmillan, London

Bowden,, F.P. and T.P. Hughes, 1937 -
Polishing Surface Flow and the Formation of the Beilby Layer; Proceedings Royal Society, London

Colback, P.S.B. and B.L. Wiid, 1965 - The Influence of Moisture Content on the Compressive Strength of Rock, proceedings of the 3rd Canadian Symposium on Rock Mechanics, Toronto.

Hampton, W.M., 1930 - Journal Society Glass Technology 14, London

Kaller, A., 1956 - The Glass Polishing Process; Natur Wissenschaft 43.

Preston, F.W., 1930 - Chemical & Physico-Chemical Reactions in the Grinding and Polishing of Glass; Journal Society Glass Technology 14

Rabinowicz, E., 1935 - "Friction and Wear of Material"; John Wiley and Sons, New York

Rayleigh, Baron (John William Stoult), 1901 - "Polish"; Proceedings Royal Institute 16, London

Samuels, L.E., 1967 - Metallographic Polishing by Mechanical Methods; Sir Isaac Pitman and Sons Ltd., Melbourne

Shadmon, A. 1994 - The Polishing of Stone (in preparation)

5 Case histories in surface workings

Études de cas dans les travaux de surface

Inverse analysis for seepage parameters of the rock mass in Laxiwa project

Analyse inverse des paramètres de percolation du massif rocheux du projet Laxiwa

Zhou Zhifang & Gu Changcun
Department of Engineering Geology, Hohai University, Nanjing, People's Republic of China

ABSTRACT: In this paper, a reverse analytical model for the evaluation of the hydrogeological parameters of fractured rock mass is introduced, which is based on the combination of finite analytical method (FAM) and simplex algorithm. Considering the geological and hydrogeological conditions of rock mass and by means of Snow's theory of tensor, a practical case illustrates that the method presented here is very useful.

1 THE ANALYSIS OF SEEPAGE CHARACTERISTICS OF THE ROCK MASS IN DAM SITE

1.1 *Geological conditions*

Shimen dam site of Laxiwa hydroelectric project is located in the upstream mouth of Longyang Gorge in the Yellow River which flows in about E—W direction at the dam site. The river valley in the site is narrow, symmetrical and of "V"—shape. the river bank is high and steep, with the height of 600~800m, and the average slope of 40 ~ 65 degrees. The strate in the site is mainly composed of a metamorphic rock of Longyang Qun (T_1) and Xiaxi Qun (T_{1pn2}) in the period of Triassic system down series, a granite of Secondary Era in the period of Yinzhi, and a little of ebullition, slide drift and diluvium of Quaternary system (Q_n). The metamorphic rock mainly consists of metamorphic sandstone of different thickness, limestone, sandstone and slate which are bedded and developed in touch with the intrusive mass of the granite. Therefore, the strata possess complicated textures with more interlayer sand fractured zones. The granite, in close touch with wall rock and in the state of bathylite, is distributed in the area from No. 1 Suspension Bridge to Chouqun Coombe, which is of medium- coarse texture, massive structure, high durability and stiff compact character (Fig. 1).

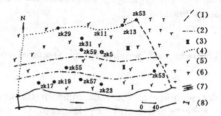

(1) boundary of lithological line;
(2) line of subarea; (3) subarea;
(4) boundary of the area; (5) granite;
(6) metamorphic rock; (7) river;
(8) axis of the dam.
Fig. 1 Outline of the area

Three groups or main tectonic fractures are developed in granite mass, with the strikes of NNE, NNW and NWW, and with the densities of 0.990, 0.682 and 0.094 per meter respectively. Laxiwa hydroelectric project which is the biggest one of those constructed on the Yellow River is lying on the granite mass. Therefore, the study of the tensor of hydraulic conductivity (Simplified as THC) which determines the magnitude and the direction of permeability of the rock mass becomes very important in estimating the uplift in foundation, the water pressure (dimension and direction) acting on the underground power house, and the stability of the high slope.

1.2 *Hydrogeological conditions*

The type of underground water in the fractured

zones in the site may mainly be described as fracture phreatic water. The underground water storing and moving in fractures of the rock mass, is supplemented with rain and discharged to the river. the underground water in the granite embeds deeply, with the embedment of $197 \sim 177m$, in the left bank from the boreholes No. 15 to No. 11, and with the embedment of $229 \sim 160m$, in the right bank from the boreholes No. 32 to No. 8. The overall gradient of the underground water is slight, i. e., $13 \sim 18$ degrees. The annual water table fluctuates very little. The high water level appears from July to October, in concordance with the rainning season. The characters of underground water in metamorphic rock are the same as those in the granite except the embedment of underground water is deeper. In conclusion the hydrological characters in this region are dry climate, little precipitation (average 255mm per year), steep geography and insufficient alimentative water.

1.3 *Characters of permeability of fractured rock mass*

The permeability of the rock mass in the dam site is determined mainly by its tectonic fractures, and also depends on lithological character, weathing action, decompression and the morphological units of area where the rock mass is. Therefore, the permeability is obviously characterized by structural actions, developed fractures, interlayers and fractured zones. In metamorphic rock, the permeability is much larger than that in granite. the data observed during the same period from the boreholes show that the gradient of the phreatic curve near the bank slope is slighter in metamorphic rock than that in granite. Meanwhile, the data of water loading in boreholes show that the permeability in the rock mass obviously increases along the direction from the bank slope to the river valley, due to more severe weathing action and decompression action near the river valley. Also, the horizontal zoning of permeability can clearly be seen in some galleries. The fluctuations of the phreatic curve differ from place to place. This also denotes horizontal zoning.

Hence in general speaking, the permeability in the rock mass is inhomogenous. On the other hand, in the individual rock mass, the magni-

tude and the direction of the permeability mainly depend on the size and the distribution of texture planes of fractures. It is well known that the quantity of tectonic fractures in main direction which controlling the seepage action in the rock mass are more than those in other direction so that seepage in the rock mass bears a typical anisotropic feature. The dimension and the direction of anisotropic property are concerned with geometrical parameters of texture planes.

In order to obtain parameters of seepage, it is necessary to collect and analyse the geometrical parameters of texture planes in situ.

2 COLLECTION IN SITU AND STATISTICS OF DATA FOR INVERSE ANALYSIS

2.1 *Collection and statistics of data from texture planes controlling seepage in rock mass*

1. Statistics of strikes of texture planes Because of geographical and morphological characters, the statistics of geometrical parameters of texture planes should be carried out in some galleries. the methods of data collecting and its statistical analysis can be introduced as follows. In left bank, the galleries PD35, PD1, PD35, PD15, PD5-1, PD5-2 and PD11 are chosen. In each gallery, several sample points are selected and exactly measured, each of which has a statistical area about $2 \times 2m$. In these galleries, 462 fractures have been measured in the strike of the texture planes so that the distribution column can be drawn out. the column obviously denotes that strikes of texture planes in that area can be divided into three groups. the statistical values of strikes of three groups are listed in Table I by using of the mode method of the diagram.

Table I

Group of strikes	NNE	NNW	NWW
Value of statistics	33.08	353.57	304.50

2. The statistics of dips and dip angles of texture planes

When the strikes of texture planes are collected, the dips and the dip angles can also be collected at these areas. In totally, 50.65% in the NNE-direction of texture planes, 36.15% in the NNW- direction, and 13.20% in the NNW — direction are gotten. After analyse and statistics, the average, variance, the type of distributed function, and the density function of the dips and dip angles in each group of texture planes can be listed in Table II.

Table I

strike of texture planes	NNE	NNW	NWW
dip	NW	SE	SW
mean value of dip angle μ	84.69	89.26	83.38
variance of dip angle σ	26.15	21.87	27.63
type of distribution	normal distribution		
density function	$f(x) = \dfrac{1}{\sqrt{2\pi}\sigma} e^{-\frac{(x-\mu)^2}{2\sigma^2}}$		

2.2 *The data collection and processing of water level in observation boreholeg*

12 boreholes for long-term observation of underground water levels are distributed in left bank slope. these boreholes which are initially used for exploration, are 0.06m in diameter, and 200m in depth. the data about borehole core, the packer test, and water levels for each borehole have been collected in detail.

Since different horehole is finished at different time, data of water levels observed in different borehole are not contemp/oraneous. The data of water levels observed are sufficient, i. e., the longest series is of 5 years, the shortest, several months. But in order to get the synchronous data in different boreholes, it is necessary to extrapolate and to interpolate the curve of data. The observed value of water levels on May 18, 1989, is chosen as the standard value in inverse computation. the water level of some boreholes which are lack of contemporeneous data of water levels must be filled up according to the variation of water levels in other boreholes about the same period (Fig. 2). Therefore, the observed values of water levels of each borehole at the same instant in the area investigated are listed in Table III.

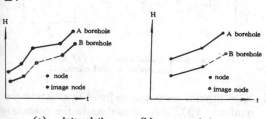

(a) interpolating   (b) extrapolating

Fig. 2 Correct of underground water level in boreholes

3 THE METHOD OF INVERSE COMPUTATION

3.1 *The mathematical model of seepage in the fractured rock mass*

The Laxiwa dam site is located on the plateau in north-west China, and there is a little precipitation every year, specially the period between October and next May. The water level in boreholes observed on May 18, 1989 is chosen as the standard water level. Because the motive factors (seepage velocity, water level or water pressure of underground water) are basically stable in the seepage field under natural condition, the movement of underground water in the fractured rock mass can satisfy the seepage control equation:

$$\nabla \cdot (\widetilde{K}\nabla H) = 0 \tag{1}$$

in which, ∇——Hamiltonian operator, $\widetilde{K}$——the tensor of hydraulic conductivity, H——any hydraulic head in the vadose field.

Table III

No.	borehole	measured water level (m)	remarks
1	ZK17	2233.34	
2	ZK19	2233.37	measured on: May 18, 1989
3	ZK57	2233.85	
4	ZK23	2233.91	
5	ZK55	2234.42	
6	ZK59	2234.60	
7	ZK5	2240.09	
8	ZK31	2278.79	
9	ZK29	2288.30	
10	ZK11	2284.45	
11	ZK13	2261.73	
12	ZK53	2232.48	

Let OX, OY and OZ axes be north, east and vertical upwards respectively, then THC can be expressed as:

$$\widetilde{K} = \sum_{i=1}^{n} \cdot K_{ei} \cdot \begin{bmatrix} K_{11} & K_{12} & K_{13} \\ K_{21} & K_{22} & K_{23} \\ K_{31} & K_{32} & K_{33} \end{bmatrix} \tag{2}$$

in which, K_{ei}——the coefficient of permeability generalized from fracture group No. i; $K_{11} = 1 - \cos\beta_i \sin^2\gamma_i$; $K_{12} = -\sin\beta_i \sin^2\gamma_i \cos\beta_i$; $K_{13} = -\cos\beta_i \sin\gamma_i \cos\gamma_i$; $K_{21} = K_{12}$; $K_{22} = 1 - \sin^2\beta_i \sin^2\gamma_i$; $K_{23} = -\sin\beta_i \sin\gamma_i \cos\gamma_i$; $K_{31} = K_{13}$; $K_{32} = K_{23}$; $K_{33} = 1 - \sin^2\gamma_i$; α_i, β_i, γ_i——the strike, dip, and dip angle of texture planes of fissur group No. i respectively; n—— groups of texture planes in the rock mass.

In equation (2), $\tilde{K}$ is a symmetrical tensor. Because the three groups of texture planes which control the movement of underground water in the Laxiwa dam site belong to high angle fractures (Table II), i. e., $\gamma_i = \pi/2$ ($i = 1, 2, 3$), the seepage problem in the rock mass can be simplified as 2—dimensional problem. Meanwhile, the strike is related to the dip by $\alpha_i = \beta_i - \pi/2$, then

$$\tilde{K} = \sum_{i=1}^{n} K_{ai} \begin{bmatrix} \cos^2\alpha_i & \sin\alpha_i\cos\alpha_i \\ \sin\alpha_i\cos\alpha_i & \sin^2\alpha_i \end{bmatrix} \quad (3)$$

therefore, equation (1) becomes

$$\frac{\partial}{\partial x}(K_{xx}h\frac{\partial H}{\partial x}+K_{xy}h\frac{\partial H}{\partial y}) +\frac{\partial}{\partial y}(K_{yx}h\frac{\partial H}{\partial x}+K_{yy}h\frac{\partial H}{\partial y}) =0 \quad (4)$$

in which, h is the thickness of the aquifer in the fractured rock mass, i. e., $h = H - z$; z is the altitude of impervious base.

The matrix in Equation (3) is also symmtrical, therefore there must be an orthogonal matrix Q which satisfies the equation (5).

$$\tilde{K} \sim Q^{-1}\tilde{K}Q = \text{diag } (K_1, K_2) \quad (5)$$

then

$$K_1 = (K_{xx}+K_{yy}+\sqrt{(K_{xx}-K_{yy})^2+4K_{xy}^2}) /2 \quad (6)$$
$$K_2 = (K_{xx}+K_{yy}-\sqrt{(K_{xx}-K_{yy})^2+4K_{xy}^2}) /2 \quad (7)$$
$$\text{tg } (2\theta) = 2K_{xy}/ (K_{xx}-K_{yy}) \quad (8)$$

in which, K_1 and K_2——Principal coefficients of permeability, θ——principal direction of OZ axis.

The new coordinate system $(X' OY')$, can be obtained by properly revolving of the original coordinate axes

$$\begin{aligned} x' &= x \cdot \cos\theta + y \cdot \sin\theta \\ y' &= -y \cdot \sin\theta + y \cdot \cos\theta \end{aligned} \quad (9)$$

Equation (4) can be simplified as

$$\frac{\partial}{\partial x'}(K_1h\frac{\partial H}{\partial x'}) +\frac{\partial}{\partial y'}(K_2h\frac{\partial H}{\partial y'}) =0 \quad (10)$$

To the control equation (1) or (10), the boundary condition can be given in the seepage field.

The Yellow River is considered as a certain water level boundary. The touching line between the granite and the metamorphic rock can be acted as a given boundary.

Because the east boundary has little effect on the seepage field and the area is of symmtry, the direction perpendicular to the river which is nearly parallel to the direction of streamline can act as the boudary of streamline.

Therefore

$$H|_{\Gamma_1} = H_0$$

$$\frac{\partial H}{\partial n}|_{\Gamma_2} = 0 \quad (11)$$

in which, Γ_1—— boundary of first type, Γ_2——boundary of streamline, H_0—— given hydraulic head, n——the outer normal direc-

tion of boundary of stream line.

Hence, the mathematical model for seepage may be composed of equation (10) and equation (11).

3. 2 Method for inverse calculation

On the basis of the data from several observation boreholes in seepage field, it is possible to gain the general solution from calculating THC by the inverse analysis. And according to an important character there is a possibility to solve the normal problem and the inverse problem can be simplified to a series of normal problems, i. e., a group of parameters first given is taken as initial values from which any hydraulic head can be calculated by numerical method. the errors of parameters between the values measured and calculated can gradually be corrected and the final hydraulic head is gradually converged until the calculated value of hydraulic head is perfectly in accordance with the measured one. It is shown that method of solution is feasible when characters of geometrical parameters of texture planes are considered.

The finite element method of local coordinate is used in solving the equations (10), (11), of mathematical model. In this method, area is divided into several rectangles and discretization is made for the control equation (5). To each rectangular unit (Fig. 3), the hydraulic head at the central point p ($\xi = 0$, $\eta = 0$) may be expressed as:

$$H_p = \sum_{i=1}^{8} C_i H_i \quad (12)$$

in which, C_i is the coefficient of finite element method, concerned with the coordinate of 8 boundary points of each unit.

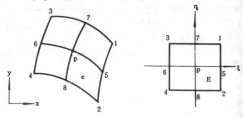

(a)　x—y any quadrilateral　　(b) η—ζ square
Fig. 3　Conversion of local coordinate

Each interior point can be listed to satisfy the equation (12). And according to the boundary conditions [equations (11)], linear equation groups enveloped can be listed. Numerical solutions of hydraulic heads in the normal problem can be gained by using of the SOR iteration method. An article shown that solution of nunerical method is more exact than that of the finite unit method.

3. 3 The detail steps of computation

In this paper, the inverse calculation of THC is not directly to compute each component in the matrix, but first to get the coefficients K_{ei} (i= 1, 2···, n) of permeability generalized from each fracture, then to solve THC of the whole area by equation (2). the advantages are as follows:
1) Fractures which control the seepage in the rock mass are generally classified as three groups so that the unknowns can be reduced in inverse analyses, (six independent components of THC in the 3-D problem and three in the 2-D problem);
2) The occurrence of each fracture mentioned above has been known in situ, i. e. , each element in the matrix of equation (2) must uniquely be determined so that the limited condition of THC can increase in practice. It has been proved that when n<3 or n=3, THC solved by inverse analyses in unique.
The detail steps of solution are:
1) to choose suitable coordinate system;
2) to collect and arrange geometrical parameters of texture planes controlling the seepage;
3) to set up the hydrogeological model in the area in order to determine the relevant seepage mathematical model of fractures;
4) to evaluate the each component value in the matrix of equation (2);
5) to collect the data of underground water level in boreholes in the area and correct them necessarily;
6) to suppose the initial value of coefficient of permeability $(K_{ei}, i=1,2, \cdots ,n)$ of each fracture;
7) to set objective function by using of the method of least squares; i. e. ,

$$E (K_{ei}^k) = \sum_{j=1}^{M} \omega (j) \sqrt{(H_j^c - H_j^0)^2} \qquad (13)$$

in which, E——independent variable of objective function; K_{ei}^k——coefficient of seepage of texture plane group No. i in subarea No. k; ω——weight function; H^c——calculated hydraulic head; H^0——measured hydraulic head; M——total number of observation boreholes in the area.
8) to get numerical solution of normal problem

by the finite element method under thd condition of parameters first supposed;
9) to gain the optimal solution of objective function by using of the optimal method of simplex algorithm;
10) to get the numerical solution of THC in each subarea by using the optimal solution and equation (2).

4 THE EXAMPLE AND RESULTS

According to the characteristic of fracture phreatic water in the left bank of Laxiwa dam site, the seepage field can be divided into three subareas I , II , III (Fig. 1). It is supposed that medium of each subarea is anisotropic and the permeability on the boundary between two subareas is inhomogeneous. Because there are three groups of fractures which have controlled the underground water in each subarea (n = 3), there are new unknown components in the whole area (k=3).
Eight observation boreholes are located in three subareas (Fig. 1). On the southern boundary, the water levels of each node are obtained by estimating from the levels in the profile of the Yellow River observed on May 18, 1989, and the gradient of the riverbed. On the eastern and northern boundary, the water levels of each node are obtained by linear interpolation according to the data observed from boreholes. And the western boundary of streamline are treated as follows:
First, a column of image nodes may be reflected along the outer normal line of the boundary (Fig. 4). If the column M-1 the image nodes out of the area may bo column M+1; then, according to the property on the boundary of streamline.

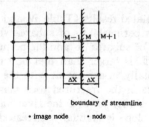

boundary of streamline
• image node   • node

Fig. 4   Treatment of boundary

$$\frac{\partial H}{\partial n}\Big|_{\Gamma_2} = 0 \qquad (14)$$

We have
$$H_{M+1} = M_{M-1} \qquad (15)$$
The calculating processes are programmed in

FORTRAN language and the processes are complished by microcomputer IBM- PC. The area is sorted into 342 nodes and 272 units. When the initial values are supposed, if the objective function $E = 0.01253$, the results calculated are listed in Table IV and Table V. The comparison of calculated values and measured values are listed in Table VI.

Table IV   Coefficient of permeability of each group of fractures in different areas

subarea / strike	I	II	III
NNE	0.2531	0.1687	0.0016
NNW	0.1752	0.0851	0.0012
NWW	0.3812	0.1591	0.0124

Table V   The magnitude and direction of principal coefficient of permeability in each area

subarea	principal coefficient of permeability		principal direction of permeability
	K_1	K_2	$\cos\theta$
I	0.5146	0.2963	0.41967
II	0.2563	0.1575	0.06826
III	0.0132	0.0025	0.77389

Table VI

No.	No. of borehole	No. of node	calculated water level (m)	measured water level (m)	$|H^c - H^0|$
1	ZK17	38	2233.34	2233.501	0.161
2	ZK19	40	2233.57	2233.382	0.012
3	ZK57	62	2233.85	2233.326	0.524
4	ZK23	64	2233.91	2233.025	0.885
5	ZK55	93	2234.42	2234.557	0.137
6	ZK59	163	2234.60	2240.507	5.907
7	ZK5	200	2240.09	2240.121	0.031
8	ZK31	248	2278.79	2278.045	0.745

The calculated results (Table IV and Table V) show that permeability of subarea I is larger than that of subarea II and the permeability of subarea II is larger than that of subarea III. This is totally in agreement with the characters of seepage in the fractured rock mass, i. e., the permeability towards the river vallery from the bank slope is gradually increased. At each point in a subarea, the permeability in different directions may be different, i. e., anisotropy. And the magnitudes of permeability of each group of fractures in different area may also be different. This is mainly caused by that each group of fractures is located in different morphological units (different subarea) which are endured different decompression actions

Futhermore, Table VI shows that errors between the calculated hydraulic head H^c and measured head H^0 are within a satisfied range which meets the requirement of engineering. But the error ($\Delta H = H^c - H^0$) in the borehole ZK59 is not so satisfied. the reason is that borehole drilled in the gallery PD 25 is located under the water level, and the discharge corridor is thus formed.

5 CONCLUSION

From the calculated results in inverse analysis of hydrogeological parameters in the fractured rock mass in Laxiwa dam site, conclusions can be drawn as follows:
1) The character of seepage in granite is obviously inhomogeneous and anisotropic;
2) It is feasible to solve the seepage problem of fractured medium by the finite analytical method;
3) According to the water level data of limited boreholes in a given area, THC of anisotropic fractured rock mass can be obtained by inverse and the optimal method. When the groups of texture planes controlling the seepage are less than three, the number of unknown for gaining the parameters can also be reduced.

In this paper, is introduced the study of the distribution of each hydraulic head in seepage field, when the discharges in this field are unknown. Therefore, the parameters gained are relative values. It is known that the magnitude of each hydraulic head in seepage field is concerned with the permeability. Hence, it is very useful for the calculations of following parameters including the magnitude and the distribution of uplift in the dam foundation, the magnitude and the direction of water pressure acting on underground openings, and the stability of the high slope affected by underground and surface water. Besides, the absolute value of parameters of each point in the area can easily be gained by using the data of water loading tests in boreholes and data of boundary discharge.

REFEREMCBS

Zhou Z. F. 1991. Environmental hydraulics. P. 1311-1316. Rotterdam: Balkema.

Vír – Dam on the Svratka river: Improvement of structural stability

Barrage de Vír sur la rivière Svratka: Récupération de la stabilité structurale

J.Verfel

Zakládání staveb Praha, a.s., Foundation Engineers and Contractors, Czech Republic

ABSTRACT: During the years 1947–57 was completed the construction of "Vír - (Vortex)" dam on the Svratka river. It is a gravity type of concrete dam, with the maximum crown height of 76.5 meters above the bottom of structure's foundation. Bedrock foundation base of the site comprises   gneiss and mica schist. At the bottom of river gorge was detected a fault of 2–5 m width, running at oblique direction towards a centerline of the valley trough. Unfavorable overturning T/N ratios casted doubts about the stability of dam's blocks No.5 to No.8. The following article describes the method of stability improvement.

RESUMÉ: Entre 1947 et 1957 a été construit le barrage de Vír sur la rivière Svratka. Ce barrage est construit en béton, de type gravité. La hauteur maximum du barrage est 76,5 m au-dessus des fondations. Le sous-sol rocheux est formé de gneiss et de mica-schistes. Au centre de la vallée, en travers de son axe, se trouve une faille de 2 à 5 m de largeur. Suite aux circonstances adverses T/N, la stabilité des blocs 5 à 8 a été mis en doute. Dans cet article, est décrite la solution adoptée pour récupérer cette stabilité.

During the years 1947–57, under the directions of former Czechoslovakian government, was completed the project of Vír dam on the Svratka river. It is gravity type of concrete dam, shaped in the arch of 305.7 meters radius. The maximum crown height of is 76.5 meters above the bottom of structure' foundation base.

Bedrock foundation at the dam's vicinity is formed by gneiss and mica schist. Gneiss layers interlaced with mica schist lenses can be found on the right bank of the trough and they are weathered within first 1–2 meters of depth. Along the left bank are mica schists with veins of aplite, weeathered into extent of 4–5 meters of depth. At the bottom of trough was detected a fault of 2–5 m width, running askew the centerline of the trough valley (refer to Fig.1).

For unknown reasons the bottoms of blocks No.5 thru 8 were set on the base sloped 11–12° down the waterstream. That was one of the causes of unfavorable T/N ratio, which is recommended as 0.65. Actual ratios for block No.5 is 0.84, block No.6 is 0.95, block No.7 is 1.02 and block No.8 is 0.96.

This unfavorable condition of T/N ratio led to doubts about the stability of blocks No.8 and toward the design of secondary system to back up the block's stability. After extensive discussions about the most suitable back up system (VUIS, HDP, prof. Wünsche, at al), the approach proposed by prof. Wünsche was selected. This method utilised propping the dam blocks on the support offerred by the passive resistance of dam's foundation subbase on the dam's air side. The activation of forces was accomplished by pressing in the wedges on rollers between the heel of the dam and a spreader plate and by prestressing of the tieback anchors (refer to Fig. 2). Anchors for blocks No.5 and No. 6 were fabricaded on the site by stranding together 703 wires, each of 2.8 mm in diameter. Application of double coat of hot asphalt, tar paper wrap and final asphalt coat guarded against corrosion. This protection proved later to be unsufficient. During the check up of prestress level in 1966 a severe corrosion of the majority of the wires was found.

Results of measurements taken during the prestressing and tieback anchors activation, while pressing in the wedges, were evaluated in 1956 by ing. Čermák. At that time eleven strain gages were built in at the base of support system. The evaluation has concluded, that only 20 to 30 percent of

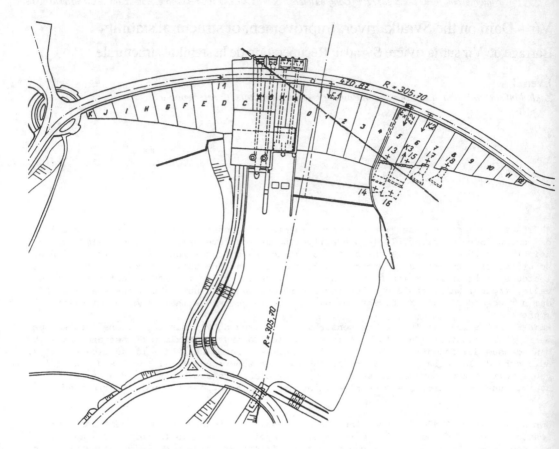

Fig. 1 - Site plan of the Vir dam. 1 - fault, I1 to I8 - inclinometric borings, E1 to
          E2 - strain gages, K1 to K3 - dilatometric borings

theoretical prestress force was transferred into the base of support system. During the supplemental prestressing the prestress loss of 7 to 12 percent was detected after 14 days and 18 percent after 18 consecutive months.

Underneath the block No.5, where anchor borehole passed thru the fault (refer to Fig.2), the attempt to prestress anchor III to 4o MN failed. The tieback anchor slipped out by the multiple of decimeters.

Fracture of some anchors in block No.6 and slippage of anchor III within the block No.5 were reason to propose lowering of water level in reservoir by 7 meters in 1974.

Another method for improving the dam´s structural stability was the subject of the projct submitted by Aquatis Brno Co., where the foundation subbase of prof. Wünsche´s spreader plates was reinforced by 1o8 mm diameter pipes. To execute this particular scheme, it would be necessary to drill boreholes of 152 mm min. diameter, that can

be made only by a drilling rig weighing more than 10 metric tons. Mobility of this rig upon the spreader plates, tilted 31 or 36 degrees from horizontal plane, would be extremely difficult. Therefore we have recommended to reinforce the foundation subbase by 59 mm diameter borings, drilled with the diamond bits by the drilling rig Diamec D 252. The boreholes were pressure injected with cementitious suspension (water/cement 1:1) and afterwards filled with grout of 1.92 t/m^3 density, reinforced by insertion of steel bar of 32 mm diameter (refer to Fig.2).

During the eighties and in the beginning of nineties were performed numerous shear tests and stability calculations for blocks 5 thru 8.

Actual work to improve the stability of the above mentioned blocks commenced in 1990. There was not, however, united consensus about the construction method to use. To clarify our picture, we have studied numerous reports about the result of measure-

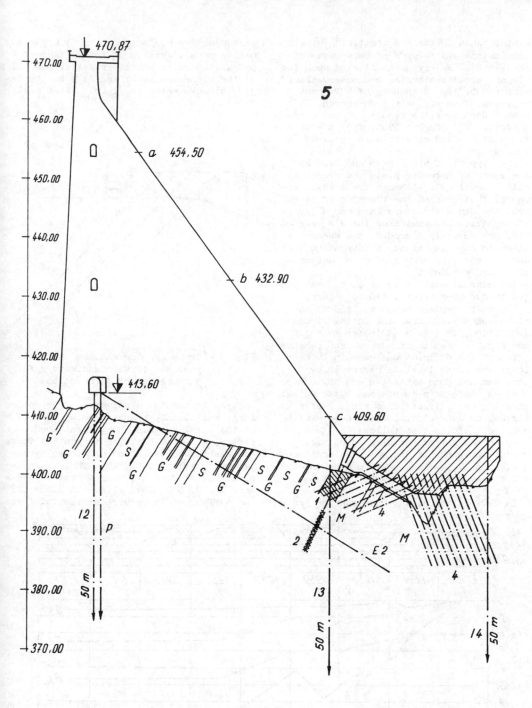

Fig. 2 – Cross section thru the block No.5, showing reinforcing borings and safety de-
vice of prof. Wünsche. a, b, c – marks for trigonometric measurements, I2, I3,
I4 – inclinometric borings, P – pendulum, E2 – strain gage, 2 – tieback anchor,
G – gneiss, S – mica schist, M – migmatite, 1 – fault

ments, taken during the monitoring, their conclusions and records of observation of the Vír dam. We have concentrated upon the "Report about monitoring and observations", prepared by ing. Štěpánský of ŘVR Praha, representative of their Brno branch. This report describes the results of dam´s monitoring up to August 1965. From the contents of studied materials we conclude the following:

- The support of block No.5 and No.6 by the means of spreader plate and wedges is still functional. Although the tieback anchors, that enabled the transfer of loads from the dam´s structure into sound layers of migmatite and gneiss on the air side of dam, are severely corroded, the shear strength capacity of dam´s foundation was not exceeded (no displacement or deformations are evident).
- The displacement of 20 to 30 mm, calculated by our specialists, did not ocurr.
- The only displacement, recorded at the elevation 409.40 above the bottom of dam´s foundation, is 1 to 2 mm (refer to Fig.2 and Fig.3 – results of survey measurements). Although the decision was made in 1966 to lower the water level in reservoir by 7 m, this was not done until the end of 1974.

Since there were doubts about the actual carrying capacity of 32 mm steel (refer to Fig.2), on June 25th, 1993 we have performed the field test as follows:

Up on the surface of cleaned up mica schist layer at the dam´s site a small concrete pad 4.24 m long, 0.91 m wide and 0.9 m deep was casted, in order to test our theory.

In between the mica schist and the pad

were placed 4 hydraulic jacks (1.0 MN capacity each). Section of the surface of pad´s foundation subbase was inclined (57 % of the plane was level and 43 % on the degree slope – refer to Fig.4). After small

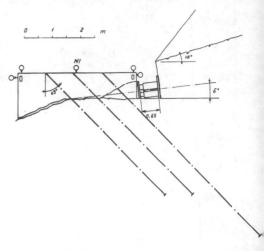

Fig. 4 – Cross section showing the testing pad. M – 16 Mahr indicators for measurement of deformations

initial horizontal displacement the entire force was already transferred by mere 57 % of the level surface. Horizontal shear stress was 0.927 MPa. The tangential force was exerted by jacks, normal force by the block´s gravity (106.7 kN). Thus, with the plan size of 3.88 m², normal contact pressure was 27.5 kPa. To slide the block off the surface would require horizontal force

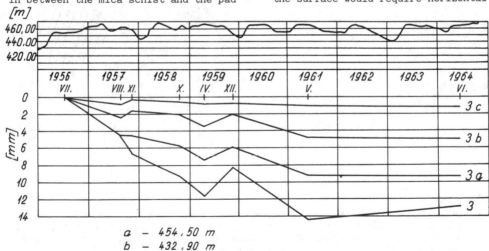

a – 454,50 m
b – 432,90 m
c – 409,40 m

Fig. 3 – Summary of survey measurement trigonometrical results

of 2.054 kN, while horizontal shear stress would reach 0.53 MPa. This compares okay to 0.927 MPa even with the reduced area of the bottom of footing pad.

After this first phase of test we have drilled 3 boreholes, 6 and 8 meters deep and inclined at 45 degrees thru the pad and the subbase foundation layers (refer to Fig.4).

Next we have filled the boreholes with a cementitious grout of 1.94 t/m^3 density and reinforced by insertion of 3 steel bar of 32 mm diameter, that corresponded to re-inforcement of spreader plate. Next we have gradualy increased the tangential force to 3.171 MN, where we recorded 2.4 mm displace-ment.

During gradual decrease of the force we have recovered and recorded following dis-placement:

3.171 MN	–	2.400 mm
2.115 MN	–	2.235 mm
1.272 MN	–	1.910 mm
0.881 MN	–	1.185 mm
0.000 MN	–	0.37o mm

Because during the second phase of test some elastic deformations have occurred, we have implemented the third phase of the test. There, up to a force of 3.282 MN elastic deformations occurred (2.77 mm). Forces above this level caused faster grow-ing deformations and their slower recovery. Maximum force, that we reached during this test was 4.138 MN accompanied by the dis-placement of 9 mm.

On the basis of previous findings we came to the conclusion, that the securing of the dam´s blocks stability according to the me-thod proposed by prof. Wünsche is feasible – in spite of severe corrosion of tieback anchors of blocks No.5 and No.6. The wed-ges, that had activated the spreader pla-tes are so slender, that they are self-clamping.We have therefore filled the gaps between spreader plates and dam´s blocks with concrete and anchored the spreader plates with 32 mm steel rods (refer to Fig. 5). We have chosen to do this work in the drought year, where the water level in re-servoir was lowered by as much as 25 meters.

Our main attention was concentrated on the securing of the heel of structure within the passive zone area (refer to Fig. 5 and Fig.2). By the pressure grouting and reinforcing of the sound gneiss (migmatite) with the preload of 7 m min. height of concrete block the stability of questioned blocks should be secured.

We are reasonably sure, that future mea-

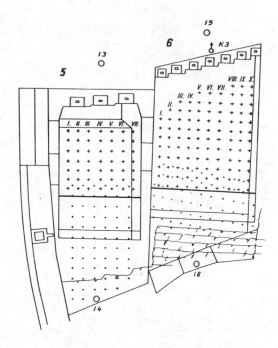

Fig. 5 – Plan layout of spreader plates and ballast block behind dam´s blocks No.5 and No.6

surement in precise inclinometric borings and strain gages will prove, that future displacements will be not be only small (around 1 mm), but also elastic, as well.

REFERENCES

Štěpánský, M. 1965. Report about monitoring and observations. ŘVT, Brno.
Doležalová, M. 1992. Numerical analysis of Vír dam. Bulletin 23 CSAV, Brno.

Geological conditions and permeability tests at Kouris dam (Cyprus)

Caractéristiques géologiques et essais de perméabilité au barrage de Kouris (Chipre)

P. Michaelides

Geological Survey Department, Cyprus

ABSTRACT: The Kouris dam is situated within the middle miocene "Pakhna Formation" of the circum Troodos sedimentary succession. The lithologies present consist of alternating, subhorizontal series of calcareous sandstone, chalk and marl (with their local varieties), more than 200 m in thickness. They represent the northern limb of a wide syncline whose axis passes through the Limassol Salt Lake. This fact leads to the development of a relatively dense rock - fracturing characterised by two differently oriented systems of well formed, open discontinuities (joints) which control the secondary permeability of the foundation stratum of the dam. The latter was tested during the design of the dam by a great number of Lugeon (packer) tests and found to be directly dependent on the type of the lithologies present. The sandstones (especially of the "Khalassa Sandstone" Unit) were found to be the most permeable needed to be treated by contact, consolidation and 3 row curtain grouting.

RESUMÉ: Le barrage de Kouris est situé aumilieu de miocene "Pakhna Formation" du circum Troodos succession sedimentaire. Les lithologies presentes consistent en series alternées subhorizontales de gres de chaux de la craie et terre englaissée (avec leur varietées locales) plus que 200 m enepaisseur. Elles representent le membre du Nord d' une large syncline l'axe delaquelle passe à travers du Lac Salé de Limassol. Ce fait meme aussi au développement d' un relativement dense roche fracturant caracterisé de deux differemment systemes orientés d' interruptions ouvertes bien formées qui controllent la permeabilité secondaire de la fondation stratum du barrage. Ce dernier a été testé pendant le dessin du barrage d'un grand nombre d'essais de Lugeon (packer) et on trouvé qu' il est fortement dependent du type de lithologies presentes. Les grés (specialement ceux de "Khalassa bloc de grés" se sont trouvés d'être le plus permeables qui ont besoin d'être traités de contact, consolidation et trois langs injection de rideaux.

1 INTRODUCTION
1.1 General description of the dam

The Kouris dam is the largest dam in Cyprus constructed between 1984-88. It has a central clay core zoned earthfill embankment with gravel shells and a height of 113 m, a total volume of 9,4 MCM (= Million cubic metres) and a crest length of approx. 550 m. The reservoir provides a water storage volume of 115 MCM and extends at top water level (247 m elev.) about 5 km upstream covering a surface area of ca. 3,6 km^2. The catchment area is 310 km^2.

The design of the dam is shown in Fig.1 and the engineering data are given in Table 1.

The dam is the keystone of the Southern Conveyor Project which is the main water supply system of the island providing to the southern and eastern regions about 26 MCM/year domestic and 32 MCM/year irrigation water. It covers some 40% of the entire area of the

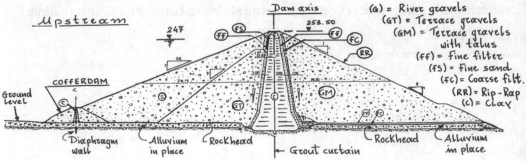

(G) = River gravels
(GT) = Terrace gravels
(GM) = Terrace gravels
with talus
(FF) = fine filter
(FS) = fine sand
(FC) = Coarse filt.
(RR) = Rip-Rap
(c) = Clay

Fig. 1. Typical embankment cross-section (original) of Kouris dam

island or 70% of the area controlled by the Cyprus Government. It operates with a 1,20 m dia. pipeline and is 180 km in length.

1.2 Location of the dam

The dam lies on the Kouris river some 15 km northwest of Limassol town (Fig. 2). This river is one of the largests on the island with approx. 40 km in length, but its flow is only seasonal. The computed annual discharges at Kouris dam (1917-80) are shown in Fig. 2.

The river rises in the central region of the Troodos Ophiolite and flows south through younger sedimentary formations (mainly carbonates) of the Lefkara Formation (Palaeocene-Oligocene) and Pakhna Formation (middle Miocene).

1.3 Historical background of the project

The site was originally recommended by the Water Development Department (WDD) of the Cyprus Government within the "Cyprus Water Planning Project" of 1968. The preliminary site and fill material investigations were carried out in 1970-71 in cooperation with the Geological Survey Department (GSD). The Feasibility Study was completed in 1979.

The detailed geological site investigation was carried out in 1981-82 by the WDD and GSD in collaboration with the consulting engineers. The latter (SOGREAH, France and Hydroconsult, Cyprus)

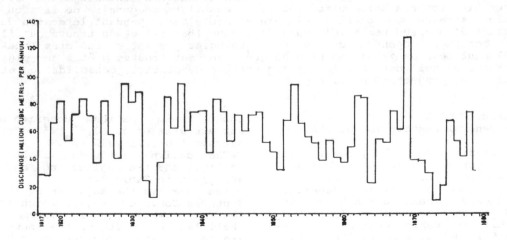

Fig. 2. Computed annual discharges 1917 - 80 at Kouris dam (WDD)

3698

Table 1. Main data of Kouris dam

A. THE EMBANKMENT	
1. Type of dam ...	Earthfill with clay core and gravel shells
2. Max. height above lowest foundation.......	113 m
3. Max. height above river bed..............	102 m
4. Crest elevation......	253,50 m
5. Crest length............	550 m
6. Crest width.............	14 m
7. Core volume.........	900 000 m
8. Shell volume.........	7,6 MCM
9. Total volume of earthfill..........	9,4 MCM
B. THE RESERVOIR	
10. Capacity..............	115 MCM
11. Surface area	3,6 km^2
12. Top water level	247 m
C. HYDROLOGY OF THE AREA	
13. Drainage area	310 km^2
14. Length of drainage basin	29 km
15. Max. discharge.......	400 m^3/s
16. Flood peak discharge	2600 m^3/s
D. THE SPILLWAY	
17. Type	side channel
18. Crest elevation	247 m
19. Discharge Capacity..	1920 m^3/s
20. Total concrete vol.	60500 m^3/s
E. THE DIVERSION CHANNEL	
21. Length	633 m
22. Diameter...............	4,2 m
23. Max. disch. capacity..	300 m^3/s
F. GROUTING WORKS	
24. Total drilling	40.000 m
25. Weight of cement + bentonite grout..	1 550 000 Kg
G. DESIGN,CONSTRUCTION,SUPERVISION	
26. Final design: SOGREAH (France) (1981-82) + Hydroconsult (Cyprus)	
27. Construction period:... 1984-88	
28. Main Contractor:.... IMPREGILO (Italy) with J+P (Cyprus)	
29. Supervision:.. SOGREAH (France) + Hydroconsult + WDD (Cyprus)	

prepared the Final Design of the dam which was overviewed by an international panel of experts acting as advisers to the WDD.

The contract for the dam construction was awarded in July 1984 to the joint venture of IMPREGILO S.p.A. of Italy and J+P of Cyprus. The works commenced on the 1st September 1984 under the supervision of the parties mentioned above.

2 GEOLOGICAL AND PERMEABILITY CONDITIONS AT THE DAM SITE
2.1 General

Both the dam site and reservoir area rest within the Pakhna Formation of the middle Miocene which overlies the Palaeocene - Oligocene Lefkara Formation (Fig. 3).

The Pakhna Formation was developed in relatively deep sea where normally mudstones, siltstones, marls and chalks were deposited. The calcareous sandstones (calcarenites) in opposite represent periodic influxes of turbidites and gravity flows into the basin. These sediments form in the project area a conformable sequence with a thickness in excess of 200 m. Starting with the oldest the sequence present is composed by the following units:-
   a) the Lophos Beds,
   b) the Khalassa Sandstone (or Lower Calcarenite),
   c) the Khalassa Marl, and
   d) the Kandou Sandstone (or Upper Calcarenite).

The names are after villages located very closed to the type localities.

The more recent superficial deposits form a discontinuous mantle on the above units. These are
   a) the alluvial deposits covering the river valley floor,
   b) the terrace deposits covering patches of the dam site and reservoir area, and
   c) the slope talus and scree (colluvium).

Fig. 4 shows the stratigraphic conditions described in more detail below.

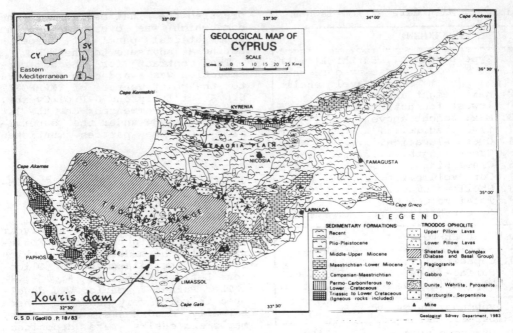

Fig. 3. Geological map of Cyprus (GSD) with location of Kouris dam

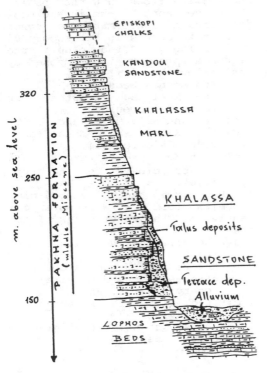

Fig. 4. Stratigraphy at Kouris dam

2.2 The Lophos Beds

These beds represent a pelagic carbonate sedimentation with strong variations in $CaCO_3$ - supply being reflected in alternating thin beds of chalk, marl and mudstone (in places with thin "soft" bands). Calcareous sandstones and coarser calcirudites are in opposite the products of periodic influxes of turbidites and debris flow into the (lower-middle miocene) sedimentation basin. This sequence developed just below the foundation of the dam crops out in the dam reservoir. The total thickness is in excess of 100 m.

The water pressure (Lugeon) tests carried out in these beds revealed a very low (secondary) permeability which is typical for marly formations. In the few calcareous bands however the permeability can reach values of up to 10 lugeons (lit/min/m at 10 kg/cm^2 pressure).

2.3 The Khalassa Sandstone

This unit (called also "Lower Calcarenite") lies conformably on the Lophos Beds. The boundary is transitional and marked by an increase in the number of sandstone layers. It is taken at the first distinct, i.e. coarse grained to fine conglomeratic, (ca. 1 m) thick, continuous sandstone band. The thickness of this unit at dam site is about 100 m.

The Khalassa Sandstone can be subdivided into three secondary groups:-

a) The Lower Group which is composed mainly of yellowish - cream, coarse grained to fine conglomeratic, thick (up to 2 m) bands of calcareous sandstones interbedded with subordinate beds of marl - chalky marl. The thickness is about 35 m,

b) The Middle Group which consists of chalky marl - marly chalk interbedded with thin bands of hard (overconsolidated) marl and thinly bedded - laminated (slightly lignitic) siltstone and mudstone. Thin layers of yellowish sandstone (in places poorly cemented) are also present. The thickness of this group is about 20 m, and

c) The Upper Group which is similar to the Lower Group but, with an increase in the number of fine conglomerates and with few bands of sandy limestone. Its thickness is about 45 m.

The in situ permeability tests carried out during the design stage of the dam (1981-82) showed variable but in general medium to high lugeon values. The high values can be attributed to several factors, as for example the primary permeability of the calcareous sandstones. However, the secondary permeability of the calcareous sandstones as well as of the interbedded lithologies controlled by the number, type and spacing of the discontinuities, is the most significant factor which governs the overall permeability of the rock-mass. Details about the discontinuities are given below.

In view of the above it is not surprising that the Khalassa Sandstone was found to be the most permeable stratigraphic unit at the dam site. The most permeable lithological horizons established are those connected with densely fractured, coarse grained, or only slightly cemented calcarenites. In opposite, marls and chalky marls as well as siltstones and mudstones are less permeable. It was also found, that the permeability decreases (not always gradually) with depth as a result of sealing due to overloading. In opposite, in the near surface weathering zone, especially where the overburden is not thick and unloading through erosion tends to open discontinuities, the lugeon values obtained are at maximum.

A characteristic permeability profile in the Khalassa Sandstone Unit is given in Fig. 5.

2.4 The Khalassa Marl

This unit overlies the Khalassa Sandstone conformably and with a gradual change which indicates a cessation of turbidite influxes. The main lithological types present are greenish-yellowish laminated marls, whitish chalky marls and marly chalks and, thin beds of fine grained calcareous sandstone. The thickness of the individual beds ranges mostly from 5 to 40 cm while the total thickness of the unit is about 70 m.

The permeability of this unit at dam site is not sufficiently investigated due to its position above top water level (247 m elev.). In the few tests carried out it was found out that it is low reaching in places values up to 10 lugeons.

2.5 The Kandou Sandstone

This unit (named also "Upper Calcarenite") is lithologically and depositionally similar to the Khalassa Sandstone. It lies conformably on the Khalassa Marl and has a thickness of about 50 m. Topografically it is above top water level.

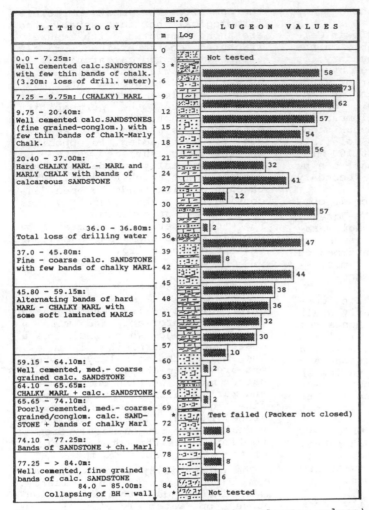

Fig. 5. Lithology vs Permeability (lugeon values) in Khalassa Sandstone, Borehole BH.20

2.6 The Superficial Deposits

These are essentially of Late Pleistocene to recent age and can be subdivided into three groups, namely
a) the alluvial deposits,
b) the terrace deposits, and
c) the slope talus and scree.

2.6.1 The Alluvial Deposits

The alluvial deposits cover the river valley floor at dam site with a thickness of about 6 m. Downstream of the dam site they increase gradually reaching at the river delta some 20 m. Lithologically they are strongly heterogenous composed of greenish - grey, subrounded to rounded, hard/sound gravels and boulders of igneous rocks (e.g. harzburgite, dunite, wehrlite, pyroxenite, gabbro, lava but mostly diabase) of the Troodos Ophiolite as well as whitish, subrounded and tabular gravels and boulders from sediments (e.g. chalk, limestone, sandstone and chert) of the Lefkara and Pakhna Formation. The overall predominant fraction originates from diabase. The ratio of igneous to sedimentary clasts is about 3:2.

These clasts are embedded in a greyish, usually coarse grained sandy matrix, which is (very) loose and contains only minor amounts (up to 5%) of silt and clay. The latter appears mostly in isolated thin pockets.

The permeability of the alluvial deposits at dam site was investigated by lefranc tests in some of the exploratory boreholes. The purpose was also to determine the occurrence of fines (silt and clay) and, therefore the inhomogeneity of the deposits. The results are shown in Table 2.

Table 2. Results of lefranc permeability tests in the alluvial deposits at the dam site

BH. No	K (cm/sec)
O.BH 23	$7,34 \times 10^{-1}$
O.BH 24	$0,52 \times 10^{-1}$
O.BH 28	$1,65 \times 10^{-1}$
O.BH 29	$2,11 \times 10^{-1}$
O.BH 30	$0,42 \times 10^{-1}$
O.BH 31	$0,54 \times 10^{-1}$
O.BH 33	$0,52 \times 10^{-1}$

Based on the classification of Cassagrande and Fadum these alluvial deposits are "clean sandy gravels" with "good permeability". Therefore, considering their limited (6 m) thickness they were removed below the clay core of the embankment.

2.6.2 The Terrace Deposits

These represent old alluvial deposits covering patches of the dam site and reservoir at different elevations. They are lithologically similar to the alluvial deposits, but moderately cemented (by $CaCO_3$). Their thickness varies from 5 to 9 m. Locally they are masked by reddish topsoil (terra rosa) or whitish talus deposits ("havara"). In general, they are strongly inhomogeneous due to lenses and layers of variable materials. Therefore, they were removed below the embankment clay core in order to avoid differential settlements.

2.6.3 The slope talus and scree

These occurred fairly extensively on the dam abutments and reservoir slopes and represent in situ developed accumulations of whitish-brown, not uniform, in general slightly to moderately cemented (by $CaCO_3$) silt, locally named "havara", with variable amount of sand and, in places, large angular stones of up to 3 m in diameter originating from old rock falls from the adjacent bedrock (mainly chalk and sandstone). The thickness reaches in gentle side gullies a max. of 16 m. They were also removed from below the embankment.

3 KARSTIC FEATURES

The only karstic features found during the investigation were two small (20 and 50cm long) solution cavities at 9 and 13 m depth in borehole O.BH 21A, left abutment. Also, during dam construction (i.e.the excavation of the spillway channel on the right abutment) similar features were found on top of the Khalassa Sandstone.

4 GEOLOGICAL STRUCTURE AND ROCK - FRACTURING

The structure in and around the Kouris dam site is quite simple. The beds strike N70°E and dip 6°-8° to the south (downstream). Their apparent dip is approx. 1°-2° from the right towards the left abutment due to the orientation of the dam axis against the dip direction of the beds. The two abutments of the dam are geologically identical.

The broader area could be the northern limb of a syncline whose axis passes possibly through the Limassol Salt Lake. Normal faulting with graben formation may have affected the syncline in the Late Miocene. Only one (minor) fault was encountered at the dam site striking diagonally across the valley (i.e. N 100° E) and dipping 75° NNE. Its throw is only 1 m. But, during the construction of the dam two relatively extensive displacements were also found indicating old slope destabilisations most probably created by gravity forces. Morel (1969) described two geological factors which have controlled the geomorphological evolution of the area,

a) the nature of the rock undergoing erosion, and
b) intermittent uplift.

Thus , river cliffs of almost vertical attitude in the reservoir area are characteristic of the hard Khalassa Sandstone Unit, while the more erodible Khalassa Marl gives rise to gentler slopes. The intermittent uplift of the area is also evidenced by the occurrence of old river terraces and hanging valleys.

The rock fracturing developed is characterised by two different systems of joints,

a) those striking ENE - WSW and
b) those striking NNW - SSE.

These intersect each other at steep angles (75°-90°) and produce, since they are perpendicular to the bedding planes, tabular or rectangular rock blocks. The ENE - WSW striking joints coincide with the strike of the bedding planes, therefore they are also called "strike joints" while the NNW - SSE striking joints are almost parallel to the general dip direction of the beds, named therefore "dip joints". The most predominant system is of the ENE-WSW striking joints.

The spacing of all joints is directly proportional to the bed - thickness. In the hard, competent chalks and calcarenites they have a closer spacing (20 - 60 m), especially in the coarse grained-fine conglomeratic or poorly cemented lithologies. In the less competent but more plastic and deformable marls the spacing of joints varies between wider intervals (usually greater than 0,5 m). Both types close gradually with depth. Their surfaces are usually smooth in the marls and more or less rough in the sandstones and chalks. Also, their opening (width) varies from few millimetres (e.g. "hair" cracks) in the marls to several centimetres in the coarse grained sandstones. In these open discontinuities, drilling water was often lost. Polished surfaces with evidence of slickensiding were not found.

Joints closed to the ground surface are open or partly filled with loose sand or silt with clay. Also , calcarenitic fragments occur as infilling material. In most cases, these joints are moderately stained by brown Fe- or black Mn-oxide indicating water circulation. In permeable calcareous horizons this staining can be also found at depth. Closed to sheared zones in the reservoir area they show a variable (from mm to several cms) displacement and a dense spacing creating therefore a highly fractured rock-mass.

These joints, as mentioned above, govern the whole of the rock-mass and, control directly the secondary permeability of the beds. The latter, consequently, is much more significant than the primary permeability which is, in fact, restricted to the only coarse grained - fine conglomeratic calcareous sandstones.

5 DESIGN OF GROUT CURTAIN

Based on the geological and permeability conditions at dam site and reservoir area and, taking into account the probable water losses calculated by the consultants it was decided that the foundation stratum of the dam should be carefully grouted. The grouting pattern decided (Fig. 6, 7, 8) consists of

a) contact grouting,
b) blanket (or consolidation) grouting, and
c) curtain (or main) grouting.

The contact grouting was to strengthen the immediate foundation stratum below the clay core zone and in particular, to prevent any migration of core (clay) material through the underlying open rock - discontinuities. It is made of primary and secondary, very shallow (1 m deep), perpendicular to the slope holes drilled both upstream and downstream of the dam axis.

The blanket (or consolidation) grouting was to consolidate the foundation rock below the clay core. It consists of 6 m deep grout holes at a 4 m spacing. They are located in the central zone of the clay core trench (primary and secondary holes) and in the outer

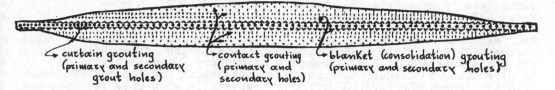

Fig. 6. Plan of grouting (WDD)

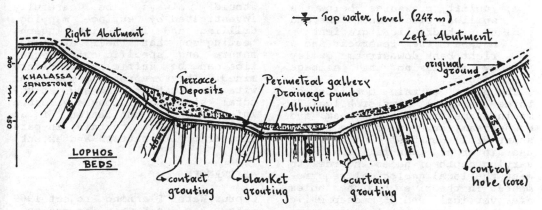

Fig. 7. Longitudinal profile of grouting (WDD)

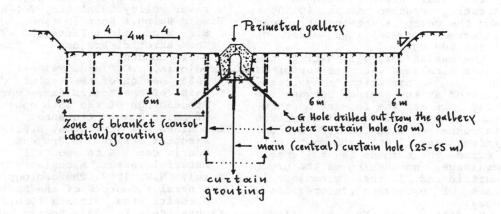

Fig. 8. Section of the setting out of grout holes (WDD)

zone of it (only primary holes).

The curtain or main grouting was designed along the whole length of the dam axis. Its extent is determined by the actual geological and permeability conditions, in particular the position and nature of the highly permeable Khalassa Sandstone contained between the almost watertight Lophos Beds and Khalassa Marls. In fact, it was designed to cover the whole thickness of the calcareous unit below the dam abutments (up to 65 m in

3705

depth) and to penetrate the Lophos Beds below the river bed to a depth of 25 m. On the abutments it was extended (through galleries) further 100 m into the bedrock. On the right abutment it had a change in direction in order to cover also the foundation stratum of the spillway blocks with the purpose to reduce

a) uplift pressures below the spillway blocks and slabs,and
b) the hydraulic gradient between the reservoir and a right bank downstream gulley by increasing the seepage length.

The grout curtain consists of three parallel rows of holes, 76 mm in diameter. The upstream and downstream, outer rows have a constant drill depth of 20 m while the internal (central) row has a varying depth of holes in order to meet the local geological requirements. In the river bed the holes are vertical and 25 m deep while on the dam abutments they are perpendicular to the slope and up to 65 m long. They are separated into primary and secondary holes at a spacing of 8 m.

The total drilling length for grouting reached about 40.000 m and the cement + bentonite grout injected was of the order of 1.550.000 kg.

The design for eliminating the water leakages was also supported by a "perimetral" gallery, located at the contact between the clay core and the foundation stratum. This (reinforced) structure is connected with the 100 m long (right and left) abutment galleries. The purpose is to enable drainage, monitoring of the grout curtain and further grouting in case of excessive (future) leakages.

Kridiotis and Ysiquel (1985) calculated on the basis of several hydraulic assumptions, considering also the site geological + permeability conditions, that "... the rate of loss could be 57 lit/s or 1,8 MCM per annum if the reservoir were to remain at elev. 252 m throughout the year. This rate of loss is relatively low and acceptable for the Southern Conveyor operation."

6 CONCLUSIONS

Rock discontinuities, e.g. joints, in gently bedded calcareous sandstones often affect seriously the secondary permeability of the foundation stratum of dams, especially when they are open and (through the bedding planes) interconnected. These conditions should always be carefully investigated by surface mapping, drilling and in - situ (lugeon) testing so that their attitude, nature and specific characteristics can be sufficiently determined. They represent, together with the stratigraphic and structural site conditions a valuable information for the design of an effective grout curtain, in particular of its pattern and extent.

REFERENCES

Cyprus Water Planning Project 1969 Eng. Rep. on Kouris dam region. Nicosia (WDD).
Eleptheriou, S.A. & Petrides, G.P. 1972. Kouris proposed dam (GSD).
Greitzer,Y.& Constantinou,C. 1969. Rep. on the Hydrogeol. of Kouris river valley.Tahal Ltd, Tel-Aviv
Howard Humph.& Sons in assoc. with Sir Macdonald & Partners, 1972. The Akrotiri Project - Kouris dam. London.
Kridiotis, C. and Ysiquel, A. 1985. Design of the grout curtain of the Kouris dam, Cyprus. Proceedings of the 15th congress of Large Dams. Lausanne.
Loucaides,P.& Stylianou, N.P.1972. Akrotiri Irrigation Project - Kouris dam - Site and Fill material Invest. Nicosia (WDD).
Morel, S.W. 1969. The geology and mineral resources of the Apsiou-Akrotiri area. Nicosia (GSD).
Michaelides, P. 1982. Kouris Dam - Geological dam site investigations 1981-82. Nicosia (GSD).
Petrides, G.P. 1968. Reconnaisance Geolog. Invest. of Kouris dam sites & reservoirs.Nicosia (GSD)
Sogreah Consult.Eng.& Hydroconsult 1981. Feasibility Study Kouris Dam - Southern Conveyor Project-Rev. of exist. data. Grenoble.
Sogreah Consult.Eng.& Hydroconsult 1982.Feasibility Study Kouris Dam - Southern Conveyor Project, Geological Report. Grenoble.

7th International IAEG Congress / 7ème Congrès International de AIGI, © 1994 Balkema, Rotterdam, ISBN 90 5410 503 8

Investigations for the Tawau Dam, Sabah, Malaysia

Prospection géotechnique au site du barrage de Tawau, Sabah, Malaisie

B.K.Tan
Universiti Kebangsaan Malaysia, Bangi, Malaysia

ABSTRACT: A case study for the investigations of the proposed Tawau damsite in Sabah, Malaysia, and vicinities is presented. The proposed damsite is located on Sg. Tawau some 10 km upstream from Tawau. The damsite area is underlain by Tertiary and Quaternary volcanics such as andesites, basalts and pyroclastics. These rocks are weathered to various degrees, ranging from grades II to VI. Sub-horizontal fissures are common features of the volcanic rocks, thus resulting in high permeability values of as much as 10-3 cm/sec. Possible construction materials include fresh basalt located about 2 km upstream from the damsite (limited quantity) and fresh microdiorite from the Kukusan Hill located some 8 km south of the damsite (enormous supply). Other geological constraints of relevance to the proposed project include: seismicity and recent or active faults. There is cause for concern from seismicity and faulting as the proposed damsite appears to be located close to two faults, and the Tawau area has experienced earthquakes in recent times, the latest event occurring at the end of 1991.

I INTRODUCTION

The Tawau Dam was planned for water supply to the Tawau area and vicinities. Various site investigations have been conducted for this project, including engineering geology, construction materials, foundation investigations, etc. This paper presents some results from these investigations. Details of the results of the investigations are contained in the Consultants' Report, Jurutera Rakanan Sabah (1990).

2 GEOLOGIC SETTING

The geology of the Tawau area has been mapped by Liaw (1979) on a scale of 1:50,000. The stratigraphy is summarised in Table 1. In essence, only volcanic rocks have been reported in the area, and they comprise basalts, andesites, pyro-clastics and microdiorite. Quaternary sediments cover the coastal plains, Figure 1. The

Quaternary and Tertiary rocks overlie slightly older Tertiary (Oligocene to Miocene) Kalumpang Formation (sedimentary rocks: shale, sandstone, tuff, rare limestone & chert) which occurs at much greater depths but does not outcrop in the Tawau area. Surface mapping in the area revealed that most of the rocks exposed near the damsite consist of pyroclastics, ie. blocks of andesite ranging from several cm to 1 m embedded in a fine-grained matrix. Occasional sections of andesite and agglomerates are also encountered. Basalt showing columnar jointing and forming high sub-vertical cliffs along the river bank was encountered at several localities upstream from the damsite. Otherwise, most of the basalt appears as boulders up to 1-2 m length on or within the dark brown residual soils of the basalt. Borehole data also confirms that the basalt is underlain by pyroclastics. As a matter of fact, the borehole data confirms the

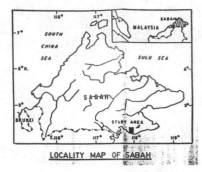

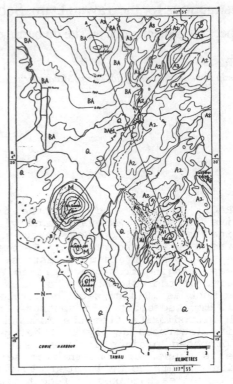

Fig.1 Geologic map of the Tawau area (Liaw, `79)
(Q = Quaternary sediments, BA = Basalt,
A = Andesites/Pyroclastics, M =
Microdiorite).

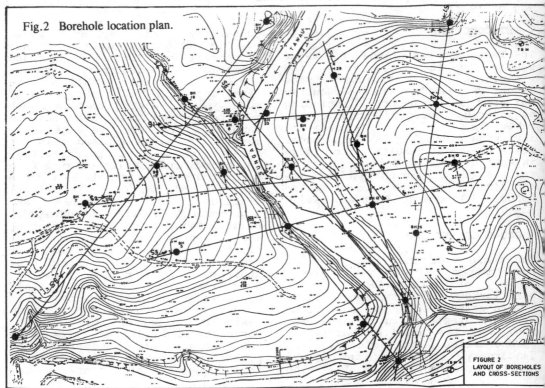

Fig.2 Borehole location plan.

FIGURE 2
LAYOUT OF BOREHOLES
AND CROSS-SECTIONS

predominance of the pyroclastics at the damsite and its immediate vicinities.

Away from the damsite and located west of the Tawau river are three small hills aligned roughly north-south. These hills, namely: Mt. Gemok, Middle Hill & Kukusan, consist of micro-diorite. The micro-diorite is well-exposed in all the three hills, especially at the Kukusan hill where a huge quarry is currently in operation. Quaternary sediments compri-sing sands, gravel beds, tuffs, clay, etc. are exposed at several road cuts in the Tawau area.

3 SUBSURFACE GEOLOGY

Some 22 boreholes have been drilled in the damsite area. Figure 2 shows the borehole locations with associated directions of cross-sections (S1-S8). Example cross-sections, S2 & S4, located along the dam axis and perpendicular to the axis are shown in Fig. 3 & 4 respectively.

The borehole data shows that the damsite proper is underlain by predominantly pyroclastic rocks, with some thinner intercalations or interlayering of andesite. Basalt is encountered only further upstream, ie. some 2-3 km north of the damsite proper.

Residual soils developed on the pyroclastics or andesite comprise mainly silty clay with S.P.T., N values of < 10 at the surface, but with N values increasing consi-derably with depth up to N > 50. The thickness of the residual soils varies from 0.6m - 22.5m at the damsite proper. One particular borehole shows 30.5m of residual soils and has yet to reach bedrock! The excessive thickness of the residual soils over pyroclastics/andesite in some of the boreholes could be due to the greater degree of weathering or deeper weathering along major fracture planes or faults.

The proven thickness of the pyroclastics layers ("bedrock") ranges from 0.3m - 48.5m. In fact the thickness of the pyro-clastics can exceed the 48.5m proven since the particular borehole termi-nated within pyroclastics. Many of the boreholes show the pyroclastics to be of several tens of metres thick.

On the other hand, thickness of the andesite layers is much less, ranging from 0.4m - 17.5m, mostly < 5m thick. The thickness of basalt proven ranges from 3.0 - 19.4m, but data is limited.

The pyroclastics and andesite are weathered to various degrees, ranging from grades II - VI, mostly grades IV - VI (ie. soil-like), with very few or limited sections of better grades of II - III (more rock-like). Thus, in general, the damsite proper is underlain mostly by highly weathered (IV) to completely weathered (V) materials, ie. poor foundation materials. How-ever, some bands or zones of better/harder materials (grades II, III) are found intermittently. An examination of the rock cores taken from the boreholes also confirms the "poor" or highly weathered nature of most of the rocks at the damsite proper. On the other hand, the basalt is of much better grade - mostly grade II, with some III. Some fresh (grade I) basalt is exposed on the western bank (cliff) of the Tawau river about 2 km upstream from damsite.

Numerous occurrences of "core loss" sections in the boreholes indicate existence of fractures or fissures within the volcanic rocks. Since the rocks comprise lava flows and pyroclastics, the fissures are mostly subhorizontal. Multiple subhorizontal fissures are also common. The fissures or "core loss" sections range in thickness from 0.1m - 2.8m, but mostly < 1m thick.

In view of the common occurrence of fissures mentioned above, the permeability values of the "bedrock" (based on either Constant Head tests for soils or Packer tests for rocks) are rather high. For the residual soils, the permeability values range from 2 x 10^{-2} cm/sec to (3-9) x 10^{-7} cm/sec, with most values around 10^{-6} - 10^{-7} cm/sec. In the case of the rocks, the permeability values range from 1 x 10^{-3} cm/sec to 2 x 10^{-7} cm/sec, with most values around 10^{-4} cm/sec. Thus, in general, the rocks show higher permeabilities than the residual soils (due to the existence of fissures in the rocks). However, low permeability values in the rocks are also encountered where fissures are

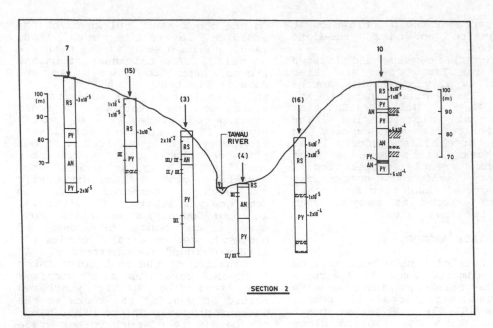

Fig.3 S2, Cross-section along dam axis.

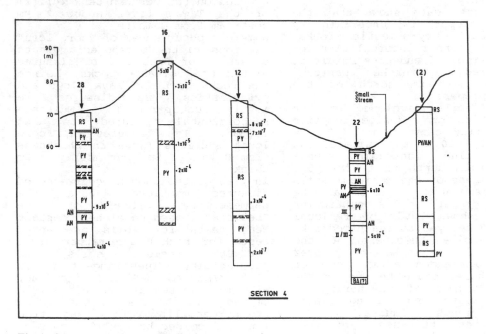

Fig.4 S4, cross-section perpendicular to dam axis.

absent or limited. The high permeabilities caused by the fissures would mean that grouting of the dam foundation would be necessary, and that grout intake would be expected to be rather high.

The cross-section, S2, located along the dam axis (Fig. 3) shows thick residual soils (RS) of upto ~ 20m. Again, pyroclastics predominantes over andesite, and the rocks are mostly of poor weathering grades (IV - VI) with only very little sections of grades II/III. Permeabilities are mostly around 10^{-4} - 10^{-5} cm/sec. (high). In terms of "core loss" sections, this cross-section would seem to indicate that the right abutment is better than the left abutment (more "core loss" sections or fissures) - thus, one reason for locating the diversion tunnel in the right abutment.

Cross-section, S4, (Fig. 4) located perpendicular to the dam axis and along the left abutment, shows the great thickness of residual soils of upto ~ 20m. Weathering grades of the pyroclastics are also poor (IV - VI), with only a few sections showing better grades of II/III. "Core loss" sections (fissures) are frequent, thus, high permeabilities of mostly 10^{-4} - 10^{-5} cm/sec. The great thickness of residual soils along section 4 could be due to a) faulting along this alignment, or b) colluvial/ slope deposits. However, there is no conclusive evidence as to which is the case although faulting at the damsite area is strongly indicated from the study of aerial photographs.

4 CONSTRUCTION MATERIALS

The construction materials investigated include i) fresh rocks (basalt, microdiorite), ii) weathered rocks (pyroclastics, andesite), & iii) soils (clay core materials & sand).

4.1 Fresh rocks (basalt & microdiorite)

Basalt occurs as fresh/sound rocks ~ 2-3 km upstream from damsite. The proven thickness of basalt is 19.4m at one locality, but only ~ 15m is fresh rock. The rock also gets thinner further north or northwest (3-3.5m only). Thus, the amount of fresh basalt is limited, estimated at ~ 50,000 m3 only, and insufficient for the proposed Tawau dam (rockfill dam). Limited laboratory test data on the basalt cores show point load strength index, Is (50), ranging from 2.5 - 7.9 MPa, ie. high strength rock. Microdiorite occurs as fresh/sound rock at three hills located west of the Tawau river, namely Mt. Gemok, Middle Hill & Kukusan Hill. Mt. Gemok is the largest of the three and nearest to the damsite (~ 5 km away). Unfortunately, it is a forest reserve area, and hence it is not available for quarry purposes. The Middle Hill is ~ 10 km from the damsite, has several natural rock exposures or rock faces that would have made good quarry sites. Unfortunately, the hill is completely surrounded by villages and new housing projects. In addition, there are two water reservoirs on top of the hill. Hence, it appears that this hill cannot be used for quarry. The Kukusan Hill is the smallest of the three hills and is located furthest from the damsite (~ 12 km away). This hill currently has a huge quarry operating, producing fresh/sound microdiorite. The quantity available or rock rserrve is also enormous - more than sufficient to meet the requirements of the Tawau Dam.

4.2 Weathered rocks (pyroclastics & andesite).

As shown by the borehole data, the pyroclastics & andesite are mostly highly weathered to completely weathered (grades IV - V), with some limited bands of less weathered (grades II - III) rocks. Hence, the pyroclastics & andesite can only be used as "weathered rockfill". In view of the type of dam finally selected/designed, ie. rockfill dam, it would seem that such weathered materials cannot be utilised. Also, laboratory test data shows the pyroclastics/andesites to have low strengths: Unconfined compressive strengths ranging from 657 - 11770 kN/m2, mostly < 5000 kN/m2; point load strength index, Is (50), of mostly < 1 MPa.

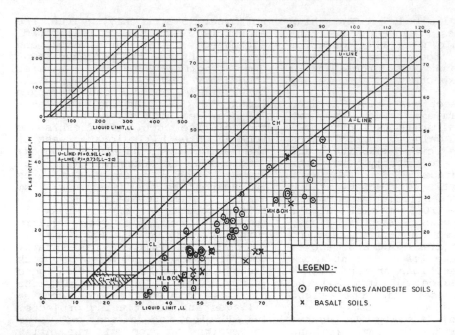

Fig.5 Plasticity chart.

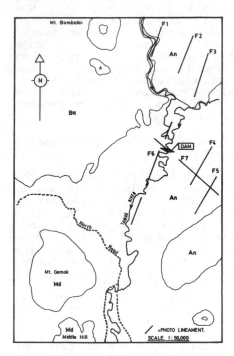

Fig.6 Lineaments of Tawau area based on aerial photographs.

4.3 Soils (clay core materials & sands)

Prior to the selection of the type of dam, a possible option was an earthdam with clay core. As such, soils were investigated for possible use as clay core. However, the results indicate that the residual soils of the pyroclastics/ andesite in the damsite area comprise mostly silty soils (Fig. 5). The grain size analyses also indicate that the fines contents are mostly silt with little or no clay. Hence, it is not possible to use these residual soils for the clay core. The residual soils from the basalt contain slightly more clay, but are also generally silty in nature - also the quantities of clay useable are again limited. Sand is needed for the filter zone (in the case of earthdam) and also for concrete structures. Since the damsite area and vicinities are underlain by volcanic rocks which contain little or no quartz, little sand is derived from these rocks (unlike granite or granitic area which produces abundant quartz sand). Thus, the source of sand is also limited.

3712

Table 1, Stratigraphy of the Tawau area (Liaw, `79)

AGE	ROCK TYPES
QUATERNARY	olivine basalt flow hypersthene olivine basalt flow hypersthene basalt flow 'Quaternary Sediments' (and associated dacitic tuff)
PLIOCENE	quartz andesite flow hornblende andesite pyroclastic rocks and lava hypersthene andesite flow hypersthene microdiorite intrusion

A stratigraphic sequencs of the various rock units in the study area.
(modified after Kirk, 1962)

5 OTHER GEOLOGIC CONSIDERATIONS

Other geologic factors considered include: faults/lineaments and earthquakes.

5.1 Faults/lineaments

Study of aerial photographs (1:50,000) reveals several northeast trending lineaments and one northwest trending lineament, Fig. 6. All the lineaments (F1 to F7) are located within andesite/pyroclastics terrain. It appears that the Tawau damsite is located close to two faults, namely F6 & F7. These are young & "active" (see 5.2 below) faults and are of concern. However, field evidence is difficult to come by, as is usually the case in tropical areas (dense vegetation and thick soil cover often hide or obliterate the evidence for faulting). Examination of another set of aerial photographs on 1:25,000 scale reveals similar lineaments, in particular F6 & F7, with F7 intersecting the damsite. The presence of faults at the damsite proper would possibly be revealed during the construction stage during the excavation of the foundations, spillway & diversion tunnel.

5.2 Earthquakes

Sabah as a whole is located close to the western "Pacific Rim of Fire" and has been subjected to earthquakes from time to time. The Semporna Peninsula, where Tawau is located, is situated at the southwestern section of the "inactive" Sulu Trench and the Sulu Volcanic Arc (Lim, 1985). Earthquakes or seismic activities in Sabah have been reviewed by Lim (1983, 1985) and recorded by the Malaysian Meteorological Service (1985), and the record shows several earthquake events for the Tawau area. The latest event reported actually occurred recently, ie. at the end of 1991. Thus, for the Tawau Dam, it is necessary to consider and incorporate earthquake loadings in the design of the dam. Furthermore, Lim (1985) concluded that the earthquakes in the Dent-Semporna seismic zone are closely related to major northeast and northwest trending faults and that the increase in seismic activity in this zone in recent years indicates that it is under high tectonic stress and that fault activities may have been renewed along the northeast and northwest trending faults (i.e. active faults are involved).

3713

6 CONCLUSIONS

The Tawau damsite is underlain predominantly by pyroclastics, with some andesite. These rocks are weathered down to grades IV - VI (soil-like), with very few or limited sections of grades II - III (rock-like). Sub-horizontal fissures are common in the pyroclastics/andesite, resulting in high permeabilities eg. 10-4 cm/sec for the rocks, thus necessitating considerable foundation grouting. The pyroclastics/andesite also have low unconfined compressive and point-load strengths. Potential fresh rocks for rockfill can be obtained from the Kukusan Hill which produces microdiorite. Soils from basalt and pyroclastics/andesite are mostly silty and unsuitable for clay core purpose. Major faults (active) and earthquake events are of concern and need to be incorporated in the design of the dam.

REFERENCES

Jurutera Rakanan Sabah (1990). Tawau Water Supply Extension Scheme Stage II, Design Report, Engineering Geology, Oct. 1990.

Kirk, H.J.C. (1962). The geology and mineral resources of the Semporna Peninsula, North Borneo. Geological Survey Dept., British Territories in Borneo, Memoir 14, 1962, 178 pp.

Liaw, K.K. (1979). Petrology and geochemistry of the volcanic suite of North-Tawau area, Semporna Peninsula, Eastern Sabah, Malaysia. B.Sc Hons. thesis, Dept. of Geology, Universiti Malaya, 35 pp (unpublished).

Lim, P.S. (1985). Seismic activities in Sabah and their relationship to regional tectonics. Geological Survey Malaysia Annual Report for 1985, pp 465-480.

Lim, P.S. (1983). History of earthquake activities in Sabah, 1897-1983. Geological Survey Malaysia Annual Report for 1983, pp 350-357.

Malaysian Meteorological Service (1985). Macroseismic study of Malaysia. Malaysian Meteorological Service, Feb. 1985, 74 pp.

7th International IAEG Congress / 7ème Congrès International de AIGI, © 1994 Balkema, Rotterdam, ISBN 90 5410 503 8

Hydrofracture test and grouting treatment design of the Hydroelectric Project of Zimapán, Hgo., Mexico

Essais de hydrofracturation et projet de traitement par injections dans la central Hydroélectrique de Zimapán, Mexique

A. Foyo & C. Tomillo
University of Cantabria, Santander, Spain

J. Ibarra Maycotte & Carlos González Cruz
Comisión Federal de Electricidad, México D.F., Mexico

ABSTRACT

The geology of the Zimapán Dam foundation have had a definitive influence to design the grouting treatment. The dam foundation is made of a sequence of dolomitic limestones and dolomitic breccia of the late Jurassic, Mesozoic Era, where the carstic phenomena is the main structural feature. To know the hydraulic characteristics of the foundation, some permeability test have had carried out, as well as two hydrofracture test under high pressure conditions. From the analysis of the results, the grouting treatment was design.

RÉSUMÉ

La géologie de la fondation du barrage de Zimapán a eut une grande influence dans le projet du traitement par injections. La fondation du barrage est sur une formation de calcaires dolomitiques et brèches dolomitiques du Jurassique supérieur, Mésozoique, où le phénomène carstique est très important. Quelques essais de perméabilité et deux essais de hydrofracturation, ont été exécutés pour la connaissance des caractéristiques hydrauliques de la fondation. A partir de l'analyse des résultats, le traitement par injections a été projeté.

1. INTRODUCTION

The Zimapán Dam site is located 300 m down stream of the River Tula and River San Juán confluence, over the River Moctezuma, border between Hidalgo and Querétaro provinces, in México. The site is called "The Infiernillo Canyon".

From the hydrological point of view, the project is located in the Moctezuma River Basin, inside of the Pánuco Hydrological System, in the Gulf of México Region.

The project consist in a Large Arch Dam 207 m high, 122 m length in the top zone and only 46 m length in the bottom zone. The Zimapán Dam, close a reservoir with a capacity of 1.360×10^6 m^3 with a surface about 22.9 Km^2.

1.1 Geology

Between the physiographic provinces of "Sierra Madre Oriental" and the "Eje Neovolcánico Transmexicano", the Zimapán Project is situated. Fig. 1.

The stratigraphycal sequence from the late Jurassic, they include "Las Trancas" Formation, late Jurassic and early Cretaceous, made by a sequence of sandstones and limestones at the bottom and claystones and limestones at the top of the formation.

The main geological formation of the region is "El Doctor" Formation, early Cretaceous, 1.500-2.000 m thickness and made by a sequence of dolomitic limestones and dolomitic breccia with a important development of the carstic features.

The first tectonic events that affect the region, they were the Pre-Laramic normal faults of the late Jurassic and early Cretaceous period.

As consequence of Laramic Orogenesis, Paleocene and early

Eocene, the tectonic structure is characterized by large folds and thrust faults.

The Post-Laramic events, Paleocene and Oligocene, they could be divided in two stages. A first stage with distension faults and a second stage with compressional structures, like sweet folds.

The results showed that the maximun acceleration on dam site it was 0.014 g, equivalent at a Richter magnitude of 5.0 earthquake 67 km away. Furthermore, the seismic characteristics of the main structural discontinuities they have analised. The most dangerous structures could be the "El Doctor"

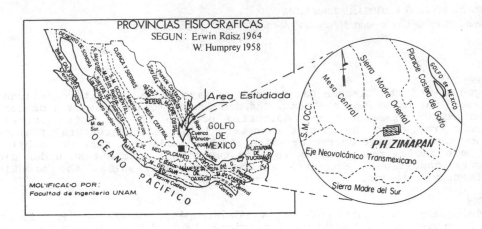

Fig. 1.

Situation plan of the Zimapán Hydroelectric Project in México and detail of it situation between the "Sierra Madre Oriental" and the "Eje Neovolcánico Transmexicano" physiographic provinces.

Finally, normal faults in relation with the basaltic eruptions, they were the last tectonic events.

The main volcanic formation is called "Las Espinas" formation, middle Tertiary, and made of three levels. The lower level is made of volcanic rocks of andesitic composition, the middle level is a mixture of volcanic and sedimentary rocks and the upper level is made of andesitic breccia and rhyolitic rocks.

The thickness average of formation is about 400 metres.

1.2 Seismic analysis

To analise the seismic coefficient of the project a risk study was carried out from the determination of the epicenters situation in an area of 100 km around of dam site and about the events that they occur between 1948 and 1984.

fault and the "La Florida" fault.

The possibility of the reactivation of these geological structures, is practically null. Consequently, the seismic risk coefficient on Zimapán Dam Site, it will be about 0.10 g.

2 THE GEOLOGY OF DAM SITE

The Infiernillo Canyon cut a sequence of limestones and dolomites of the "El Doctor" formation, which they have a sweet dip to the rock mass in the left abutment and to the river course in the right abutment.

The good quality of the rock they were determined from the RQD index and the compressional seismic wave velocity, obtained from the investigation boreholes.

The main large structural discontinuity is "El Doctor" thrust fault, situated at the bottom of the "El Doctor" formation.

3716

2.1 Carstic features on dam site

The discontinuities that affect to the rock mass on dam site, they have a subvertical disposition.

The structural conditions of the rock mass, they have a narrow relation with the carstic features. Therefore, the characteristics of the carstic cavities they have been carefully analised. Fig. 2.

Consequently, the carstic register included from the characteristics of the fissures filling, topographic situation, magnitude to the kind of the structure where the carstic feature it have been developed.

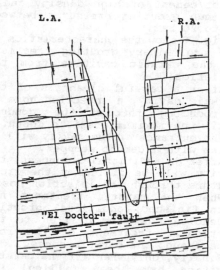

Fig. 2
The Infiernillo Canyon.
→ Flow directions through the joint and strata discontinuities.
L.A. Left Abutment
R.A. Right Abutment

According to the results obtained from the geological and structural characteristics of the carstic features, they conducted to the next conclusion:

a) In the left abutment the carstic cavities have a bigger development than the right abutment, possibly as consequence of the structural dispossition of the strata, with a dip direction toward inside the rock mass.

b) The average of the magnitude of the carstic cavities they were little, maximun about 50 cm.

c) In the hight levels of the axis of dam site, the concentration of the carstic features is bigger, according with the major percentage of the pure limestone in front of the dolomitic levels in the rock mass.

d) Alike, the lower levels have the most important carstic development, consequently with a bigger concentration of the fractured dolomitic layers.

e) The best part of the carstic cavities on dam site, they have not clayey filling.

Consequently, the geological evidences show the existence of two separates carstic levels, caused by the regional tectonic movements, so that by the existence of the large clayey horizontal layers.

Finally, in both margins of dam site, the limestones and dolomites have had a hight RQD index and a compressional velocity up to 5.500 m/s. The maximum permeability in Lugeon Units, it has been about 2.1 UL.

3 GROUTING TREATMENT

The Zimapán Dam Site, such as all the fractured rock masses, it has a considerable number of structural discontinuities as bedding layers, little faults, joints or carstic cavities.

These geological structures, they could be the ways to the hydraulic conductivity and some problems, from the seepage through the foundation, until the uplift over the bottom or the abutments of the dam.

To avoid these risks, the construction of a prolongation of the dam in the foundation, it can be necessary. This extension of the dam is called the "impervious grout curtain".

In the Zimapán Dam Site, a important net of grouting, drainage and inspection galleries, they were carried out in the foundation.

With about a section of 10 m², its favourable distribution must avoid a too long grouting holes, as well as the main criterion will be that the holes cut of the most possible number of structural and geological discontinuities.

With this criterion, a total of 1.822 linear metres of galleries were builded, with a distribution in four levels for each margin and a vertical spacing of 48 metres. Fig. 3.

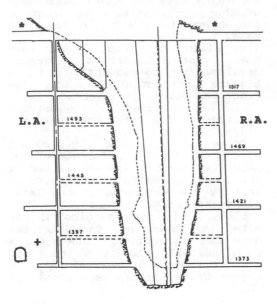

Fig. 3
The Zimapán galleries diagram
* Access Tunnel
L.A. Left Abutment
R.A. Right Abutment
+ Deviation Tunnel

3.1 Grouting treatment of carstic cavities

Although with a little section, the carstic cavities could be a important way to water flow. Their grouting treatment must be prior.

Usually, is practically imposible to find and cut the carstic cavities, only by means of the investigation boreholes, whereas will be easier across the galleries.

The galleries must cut the main carstic features developed in the vertical discontinuities and the vertical boreholes carried out from these galleries, they must cut the carstic cavities situated between two of them.

The proposed method to the carstic treatment from the galleries, should include, after the excavation of the galleries, a careful geological study of the carstic structures, because only from the knowledge of

the exact distribution of the discontinuities affecting by the carst phenomena, a correct grouting treatment will be possible.

In the Zimapán Dam Site we would like to remember the importance of the carstic register, which it included for each one of the carstic cavities, from the magnitude, the characteristics of void filling, the topographic situation to the kind of the geological structure where the carst was developed.

With this criteria, the grouting treatment of the carstic cavities, it was made from the bottom of the valley to the crest of the dam, using to void filling a mixture water/cement of high density and a maximun grouting pressures between 10 to 20 bar.

Fig. 4 show the characteristics of the preliminary grouting treatment of the carstic cavities from the galleries.

After a careful cleanness of the carstic cavity, a injection hose is introducing through the conduct until the maximum possible depth.

The entrance of the cavity will be close with a cement stopper.

After an additional cleanness, the cavity will be filled by the grout mixture through the injection hose.

When the carstic treatment has been finished, additional grouting boreholes could be drilled to the particular treatment of the rock mass around the carstic zones.

Finally, only when all the carstic cavities have been filling, the grouting treatment of the courtain it will be carried out.

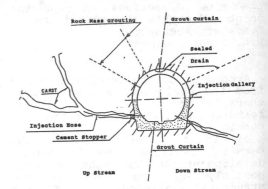

Fig. 4
Grouting treatment of carstic cavities.

3.2 Impervious Grout Curtain

To design of the impervious grout curtain, the principle of the Grouting Intensity Number, GIN (Lombardi and Deere 1993), it was taking into account. Fig. 5.

The Grouting Intensity Number, GIN, is defined as the result of the final grouting pressure in bar, multiplied by the maximun cement absorption in l/m, GIN = V x P.

The GIN principle is founded in the next considerations:

1. There are a lot of difficulties to establish a correct correlation between the water absorptions or Lugeon values, and the grout absorptions.

2. The grouting process will be conditioned by the rock mass characteristics.

3. The risk of hydrofracturing can be avoid, because the grouting pressure will depend of the grouting absorption in each stage of the treatment.

With these considerations, for a maximum grouting pressure of 50 bar the corresponding theoretical limit for grout absorption, it will be 40 l/m with a maximum volume of 200 l/m. Consequently the constant value for the GIN will be 2.000 bar * l/m.

To design the Zimapán grout curtain with a maximun of security, a "practical limit" to grout absorption of 20 l/m, it will be taking into account.

3.3 Drainage Curtain

Obviously, is not possible to seal the total of the differents discontinuities that affecting to rock mass with the impervious grout curtain. Consequently, to permit that the possible water seepage through the grout curtain it has a way to go out, an additional drainage curtain will be necessary.

Like this, from the drainage galleries, some boreholes with a diameter about 100 mm they will be drilled.

These boreholes must be drilled with a rotary method and without covering. Furthermore, using water pressures about 2 to 3 bar, the holes must be perfectly washing.

3.4 Inspection Galleries.

To have a correct control and monitoring of the stress states that affecting to the Zimapán Dam Site,

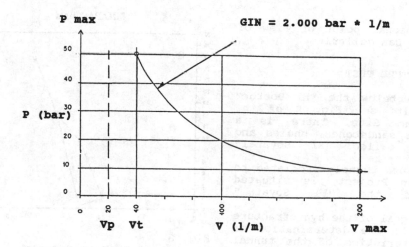

GIN = 2.000 bar * l/m

Fig. 5
Evolution of the Grout Intensity Number, GIN, for the Zimapán grout curtain.
P   Grouting Pressure
Vp Practical value of grout absorption
Vt Theoretical value of grout absorption
V   Volume of grout

four inserted levels of inspection galleries will be builded with a section about 10 m².

Fig. 6 show the tridimensional distribution of the complex sketch of the impervious, drainage and inspection galleries.

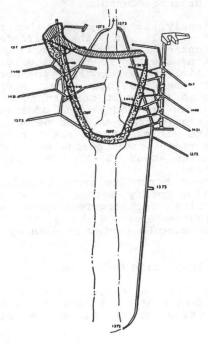

Fig. 6
A tridimensional point of view of the Zimapán Dam galleries.

4 HYDROFRACTURE TEST

About 40 m below the "El Doctor" thurst fault, see Fig. 2 of the Zimapán Dam Site, there is a sequence of sandstones, shales and limestones called "El Soyatal" formation.

The pressure tunnel associated to the Zimapán Project, is situated inside of the "El Soyatal" formation.

The main goal of the hydrofracture test will be the determination of the characteristics of the tunnel covering for a pressure conduction.

Furthermore, the results obtained could be used to know the geological characteristics of the lower zone of dam foundation.

The test consist in the control of the seepage flow through the rock mass using three increase and

decrease pressure cycles to define three critical points and, consequently, three Critical Pressures during the test.

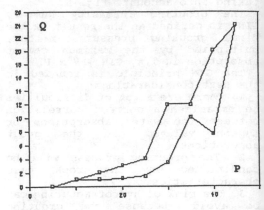

Fig. 7
Borehole S-1. First Cycle.
Pressure-Flow Graphic.
P Pressures (bar)
Q Flow (l/mi)

Figure 7 show that the hydraulic fracturing appear about the ascending pressure level of 45 bar with a water absorption of 24 l/mi

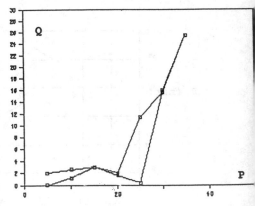

Fig. 8
Borehole S-1. Second Cycle.
P Pressures (bar)
Q Flow (l/mi)

In the second cycle and about the pressure level of 30 bar the induced fractures start to close and consequently, the water absorption decrease. Fig. 8.

Finally, during the third cycle, in the pressure level of 30 bar, the induced fractures were open again and, possibly, in the descending pressure level of 25 bar, the fractures were practically closed. Fig. 9.

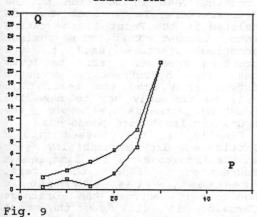

ZIMAPAN DAM

Fig. 9
Borehole S-1. Third Cycle.
P Pressures (bar)
Q Flow (1/mi)

From the analysis of the results obtained in the Hydrofracture Test carried out in the argillaceous limestones of the " El Soyatal" formation, we can conclude that to produce the hydraulic fracturing, a minimun pressure of 45 bar it will be necessary.

5. CONCLUSIONS

In the Zimapán Hydroelectric Project, the geological characteristics of the Dam Site, they have had a great influence for the impervious and drainage grout curtain design.

Furthermore, the carstic features affecting at the limestones of the "El Doctor" formation, they have needed a special treatment, only possible, from the exactly knowledge of the structural characteristics of the rock mass.

Like this, a total of 2.106 structural discontinuities, 1.177 joints to the left abutment and 929 to the right abutment, they have been cuantified through eigth

gallery levels in both margins. Table A.

Table A. Carstic degree of fractures

	Levels	1	2	3
1	1565-1529	386	36	9%
2	1529-1505	331	28	8%
3	1505-1481	150	14	9%
4	1481-1457	251	27	11%
5	1457-1433	208	28	13%
6	1433-1409	355	30	8%
7	1409-1385	168	16	9%
8	1385-1362	257	12	4%
	TOTAL	2108	190	9%

1. Number of fractures
2. Carstic fractures
3. Percentage of carstic fractures

Finally, from the analysis of the results of more than 300 Lugeon Permeability Test and about 12.500 grouting stretch along of more than 250 injection boreholes carried out in the "El Doctor" formation, and more than 150 not standard Lugeon Permeability Test and two Hydrofracture Test, we can conclude that:

A) The main structural discontinuities affecting to the limestones of the Zimapán Dam Site, they are subvertical fractures. Only a little percentage of their, 9 %, have got a development of carstic features.

B) The general permeability, Lugeon permeability, of the rock mass when is not affecting by carst, is little, with a maximun of 2.10 Lugeon Units.

With these two characteristics, if we don't take into account the little and controlled carstic zones, the rock mass of the Zimapán Dam foundation could be classified as a low permeability and low deformability rock mass.

After the special grouting

3721

treatment carried out into the carstic zones of the foundation, the design of the impervious grout curtain it was carried out by means of the GIN principle.

From a theoretical point of view, the employment GIN principle should permit the treatment of the best part of the wider fissures, using low grouting pressures through the primary holes.

Then, through the secondary holes, should be possible the treatment of the rest of the most important fissures. As consecuence of a foreseeable smaller grout absorption, the grouting pressures will be bigger.

Finally, the grouting treatment through the tertiary holes, it should affect only to a little number of closed fissures. The final grouting pressures will be the much bigger.

From these criteria, the results obtained in the grouting treatment of the Zimapán Dam foundation, can be considered satisfactory. Fig. 10.

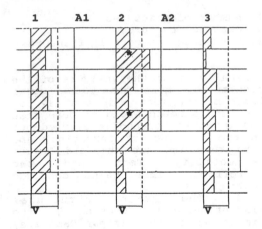

Fig. 10
Grout curtain
1, 2, 3 : Primary, Secondary and Tertiary holes.
A1 and A2 : Additional holes.
V "Practical Limit Volume".
* Absorption > V.

Nevertheless, as well as in the primary and secondary holes the grout absorption capacity or "Groutability", it should be the consequence of the natural characteristics of the geological discontinuities, it is hard to believe that the high pressures induced across the tertiary holes to introduce through the closed discontinuities a little amount of the grout mixture, they can avoid that the grouting treatment have been carried out under hydraulic fracturing conditions.

Furthermore, the GIN value, should be determined by means of previous Grouting Test in addition to the Hydrofracture Test, such as the related in the Point 4, and taking into account that the maximun grouting pressure used to the grouting treatment, must be lower than the hydrofracture pressure determinated from the test. This will be the only way to make the grouting treatment, avoid the hydraulic fracturing phenomena.

Finally, in rock masses with a little secondary permeability, such as the limestones of the Zimapán Dam foundation, after the carst treatment, the "Hydraulic Monitoring" across the Drainage Curtain, it will be the best instrument for the knowledge and control of the seepage and uplift during the life time of the dam.

REFERENCES

Ewert, F.K. 1985. Rock Grouting with Emphasis on Dam Sites. Springer Verlag. Berlin.
Foyo, A. 1993. Permeability, Groutability and Hydraulic Monitoring of Large Dams Foundations. EUROCK'93. Lisboa.
Foyo, A. & Tomillo, C. 1993. Permeability and Groutability of the Valparaíso Dam foundation. NW of Spain. Int. Conference on Grouting. Salzburg. Austria.
Gómez Laá, G. & alt. 1985. In search of a deterministic Hydraulic Monitoring model of concrete dam foundation. XV ICOLD. Lausanne.
Lugeon, M. 1933. Barrages et Géologie. Librairie de l'Université. Lausanne.
Lombardi, G. & Deere, D 1993. Grouting design and control using the GIN principle. Water Power & Dam Construction.

L'étude du matériel argileux et du mélange argilo-graveleux pour la construction du noyau du barrage de Fierza

The study of the clay and of the mixture clay-gravel for the construction of the core of the dam Fierza

N. Goro
ISP No. 3, Tirana, Albanie

N. Konomi
Universiteti Politeknik, Fakulteti i Gjeologjisë dhe i Minierave, Tirana, Albanie

RESUME: Dans cet article l'on traite des principales qualités physico-mécaniques et minéralogiques du matériel argileux. Il en résulte que les argiles à elles seules ne sont pas convenable pour la construction d'un barrage haut, parceque les déformations qu'elles subisent mettant en question sa stabilité. Les auteurs donne la méthode suivie pour diminuer les déformations et augmenter la résistance des argiles en leur ajoutant du matériel graveleux.

ABSTRAKT: In this paper the main physical-mecanical and mineralogical properties of clays are treated. It was found that the clays ane not suitable in construction of a high dam, because their deformations put in to doubt its stability. The auther presents the ways he has followed in order to decrease the deformations and increase the stability of the clays through the mixture of clays with gravel.

INTRODUCTION

Fierza se trouve au nord-ést de l'Albanie, dans une gorge étroite de roches granitiques et diabasique qui ont été soumises à une activité téctonique intense et à de multiples facteurs d'érosions.

L'un des principeux problèmes à résoudre était l'étude et le choix du matériel qui servirait à construire le barrage.

Quelques caracteristiques de ce barrage (fig. 1) seraient: la hauteur maximale 167m, la longueur (à la couronne) 400m, la largeur (à la couronne) 12m, la largeur (à la base) 700m. Le volume total du matériel qui a servi à construire le barrage est de 8.000.000m³ dont: 1.240.000m3 de mélange d'argile et de graviers, 4.900.000m³ de calcaire concassé, 1.260.000m³ de graviers (alluvions), 375.000m³ de filter et 225.000m³ de materiaux d'escavatitions.

Tenant compte du volume important de ce barrage, il était nécessaire de trouver une carriere ayant suffisament de matériel pour que ses propriétés physico-mécaniques soient les mêmes pour tout le barrage.

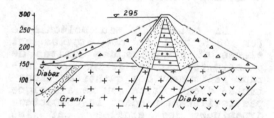

Fig. 1. Shoèma du barrage de Fierza.

1. PROPRIETETES PHYSICO-MECANIQUE DES ARGILES

Les analyses mineralogique du matériel servant à la construction du noyou d'argile montrent que le mineral principal argileux est la kaolinite mélangée à un peu d'illite, tandis que le ciment est ferrugineux et alumineux. Les principales propriétés physique de ce matériel sont présentées sur le tableau nr.

1. L'analyse granulométrique (fig. 2) montre que la fraction argileuse est importante.

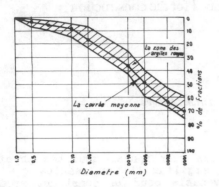

Fig. 2. Composition granulométrique des argiles rouges.

La composition minéralogique et chimique des argiles, et surtout la presence des minéraux de fer sous forme d'hydroxydes, limite au maximum leur gonflement en présence de l'eau.

de compactage allant de 50 t/m³ à 200 t/m³ (table 2).

Table 2. Les resultats de compactage

Energie (t-m/m³)	γ_{ak} (kg/m³)	W_{op} (%)
62.5	1510-1520	25.3-29.0
100	1520-1540	26.0-28.6
200	1570-1580	26

Il en résulte que l'augmentation de l'énergie de compactage (fig. 3) est accompagnée d'une faible augmentation du masse volumique séche ce qui s'explique par la teneur en eau moleculaire qui ne diminue pas facilement, et par la predominatio de la fraction argileuse qui constitue 50% de la masse étudiée.

L'étude des argiles sous l'effet des charges statiques (en compresion) a été faite à l'aide des échantillons soumis à des charges uniformes, appliquées graduellement. Les efforts maximale ont été peis 30 bars. Cette valeur est en même tempsla valeur maximale de la charge exercée à la base du barrage.

Table 1. Les propriétés physiques .

Sort de sol	W_n	Diamétre du grains (mm)				W_1	W_p	I_p	γ_s	W_{mol}
		>0.05	-0.005	<0.005	<0.002					
	%	%	%	%	%	%	%	%	g/cm	%
Argil	28.7	16.5	32.0	51.5	35.5	54.0	34.0	20.0	2.79	29.0

La teneur en eau moléculaire (en 63.5 bar) de ces argiles est élevée et s'aproche des limites de plasticité, car les grains des minéraux retiennent des quantités importantes d'eau. Cette eau ne s'éloigne pas facilement pendant le compactage dynamique des argiles, car les grains des minéraux ne s'approchent pas facilement l'un de l'autre, ce qui rend difficile ce compactage.

Les propriétés mécaniques et hydrodynamiques du sol sont liées à son compactage.

Les grains,étant plus prés l'un de l'autre à cause du compactage, le coefficient de l'infiltration diminue. Il en est de même du glissement des grains l'un par raport à l'autre, ce qui rend le sol plus résistent.

Les argiles de Çernice ont été soumises à une étude detaillée au laboratoire et sur place, l'énergie

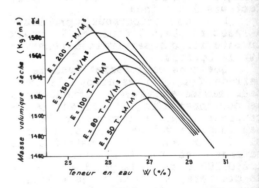

Fig. 3. Les courbes de compactation des argiles rouges en fonction de l'énergie.

L'étude des courbes de compression montre que pour des charges de 1 à 3 bars la compaction est limitée. Le coefficient de compression dans ce cas varie de 0.0045 à 0.0065 cm²/kg.

Dans le cas où les charges extérieures deviennent supérieures aux forces de liasson structurale des argiles. Ces dernières se rompent et la déformation augment. Il en résult une augmentation du tassement qui pour des contraintes de 30 bars atteint des 12% (fig. 4). Ces déformations importantes mettent en question la stabilité du barrage.

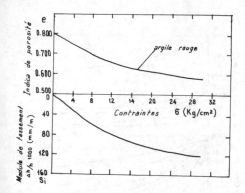

Fig. 4. La courbe de compression des argiles rouges (échantillon preparé avec dmax et Wop pour une énergie de 62.5 t-m/m³).

La résistance au cisaillement des argiles de Cernica a été étudiée dans des conditions de cisaillement plan (directement) et dans des conditions de contraintes volumiques (triaxiales). Dans le prémier cas les échantillons sont preparés saturés et consolidés dans les conditions où se trouvent les sols dans le barrage. D'après l'interprétation statistique des donnés obténues, sont précisée les valeurs moyennes des paramètres de la résistance au cisaillement (tableau 3).

Cette résistance augmente quand on passe vers les conditions d'essais d'argiles saturées et consolidées. Cela montre que toutes les contraintes sont passé au contacts des grains.

Les essais ont été fait aussi dans des conditions de contraintes volumiques avec la mesure des préssions de pore. À cause des petits valeurs de perméabilité cette argile present une certaine pression de pore qui est mésurée pendant les essais.

D'après les conditions de construction et d'éxploitation de barrage, les essais ont été fait CU avec du drenage et CD.

Pendant les éssais volumiques l'angle de frottement (Ø) augmente de 24° en éssais (CU) à 27° à l'éssais (CD). Cela nous pérmet de dire que les argiles de Cernica ont une importante qui peut éxplique d'aprés les conditions physique et chimique, la composition mineralogique, la structure des agregats, les liasons colloïdales etc.

2. LES PROPRIETETES MECANIQUE DU MELANGE ARGILE-GRAVIER

L'amélioration des propriétés de déformation des argiles de Cernica est fait artificiellement en augmentant les fractions à gros grains qui ont été prises dans les carrières des graviers à Gri et Bushtrica.

Ansi en laboratoire on a réalisé le compactage de mélanges argile-gravier pour de différentes énergies de compactage et pour des divers raports variables des fractions gravieres. Par éxemple pour la même énergie de compactage 62.5 t-m/m³ on voit que l'augmentation du contenu en graviers s'accompagne d'une croissance du masse sec (dmax)

Table 3. La résistance au cisaillement

Sort de sol	Methode d'essais					
	Wop		Ws		Ws	
	Ø (°)	C (bar)	Ø (°)	C (bar)	Ø (°)	C (bar)
Argile rouge	23	0.79	18	0.56	23	0.50

D'après les résultats ci-dessus on voit que pour le même sol, quand la téneur en eau tend vers la téneur en eau saturée, la résistance au cisaillement diminue.

et d'une diminution de la teneur en eau optimale.

L'intervalle de la teneur en eau dans laquelle le sol est compacté le mieux en mélange, se distingue clairement au sommet des courbes (fig. 5). qui ont une configuration plus plate pour les argiles et tendent vers une configuration pointue quand la proportion de graviers atteint les 50%. Dans ce cas, les sols ayant les particule plus proche l'une de l'autre ont une résistence plus grande contre les forces de cisaillement et en même temps une deformation et une pérméabilité plus faible par rapport à d'autres états. Les paramètres obtenues dans ces états servent pour des études plus détaillés.

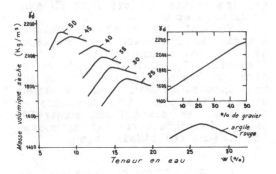

Fig. 5. Les courbes de compactage du mélange argile rouge-gravier sous l'énergie de 62.5 t-m/m³.

Sur les résultats présantés sur la fig. 5, on voit que si l'on passe du compactage d'argile vers le mélange de 25, 30, 35, 40 et 45% de graviers, le masse volumique sèche augmente avec l'augmentation du pourcentage de gravier. Quand le pourcentage de graviers atteinttes 50%, les variations du compactage sont très faible tandisque les valeurs de la téneur en eau optimales diminuent. L'augmentation de l'énergie de compactage de 62.5 à 150 t-m/m³ influe tre peu sur les paramètres de compactage.

Sous l'infuence des charges statiques en passant de l'argile vers le mélange, argile-gravier le compactage a tendance à diminuer. Cela se voit dans les résultats donnés ci-dassous:

-Les valeurs du coefficient de compactage (a-cm²/kg) qui pour des valeurs de contraintes normales dei 3 bars pour l'argile sont de 0.0045-0.0065 cm²/kg, et pour le mélange argilo-gravier de 0.003 cm²/kg. Sur

la fig. 6 on distingue très clairement où le déformation diminue avec l'augmentation des graviers.

En particulier, ce phénomène se voit où les contraintes arrivent à 30 bars, ou on a obtenu un tassement de 12% pour l'argile, de 7.9% pour le mélange argile avec 25% de graviers. L'augmentation du pourcentage des graviers de 25 à 50 %, s'accompagne d'une diminution du tassement jusqu'a 4.3% (table 4).

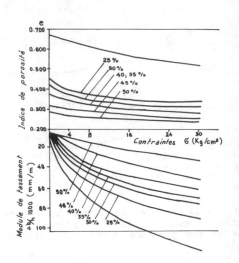

Fig. 6. Les courbes de compresion du mélange argile rouge-gravier (échantillon preparé avec dmax et Wop pour une énergie de 62.5 t-m/m³).

Cela fait que les argiles qui étaient defavorables pour le noyau de barrage à cause de leur grande deformation, sont devenues très favorables, et d'une bonne qualité quand on a crée un mélange argile-graveleux.

Aussi l'augmentation du gravier influe positivement sur la dureté de sol. L'étude de la résistance au cisaillement est éffectuée en mésurant la presion de pore pour des contraintes totales, ce qui corespond aux conditions de sol en barrage (fig. 7).

Les résultats obténus quand le mélange contient 30 et 35% de gravier, prouvent que: l'augmentation du masse volumique s'accompagne d'une augmentation de l'angle de frottement interne de 25° à 27° et

d'une diminution de cohesion de 0.86
à 0.75 bar.

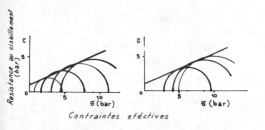

Fig. 7. La résistance au cisaillement du
mélange argile rouge-gravier (30 et 35% de
rgravier; éssais triaxial, θ=300mm,
h=750mm).

Dans les mélanges d'argile
rouge avec 25, 30, 35% de gravier,
le coefficient de filtration rest
présque le même (10⁻⁸ cm/sec).
    L'augmentation du pourcentage
de gravier dans le mélange jusqu'à
40%, s'accompagne d'un augmentation
du coefficient de filtration, de
l'ordre de 10⁻⁷ cm/sec, tandisque
pour 45 et 50% de gravier, cette
valeur arrive jusqu'à 10⁻⁵ cm/sec.
    Les résultats obténus au
laboratoire, sont verifiés in situ.
Ainsi ont été choisi les résultats
les plus convenables (de compactage,
résistence de deformation et de
filtration) du mélange argile-
graveleux (30-40%).
    Ces résultats ont servi pour
faire les calculs du projet de
barrage de Fierza et pour assurer
toutes les conditions techniques
pendant sa construction.

(à la base du barrage pour une
contrainte normale de 30 bars),
comme il a été prévu pendant les
résultats éxperimentaux.

CONCLUSIONS

    1. Le sol de la carrière de
Cernica est constitue d'argiles.
Etant donné leur structure ces
argiles résistent bien au tassement
(à la compaction) et aux contraintes
de cisaillement. Sous l'éffet des
charges importantes, les tassements
atteignent jusqu'à 12%, ce qui
compromet la stabilité du barrage.
Ces argiles sont inconvenables à la
construction du nuyau du barrage.
    2. Le mélange argilo-graveleux
(30-40% de graviers) améliore les
propriété physico-mécaniques des
argiles, qui restent pratiquement
imperméable. L'augmentation du
pourcentage du gravier a pour
conséquence:
    - L'augmentation du poid
volumique du mélange et la
diminution de la teneur en eau
optimale, quand la quantié de
gravier atteint le chiffre de 45%.
    - La diminution du tassement du
mélanges pour 50% de gravier, cette
diminution est de 3 fois celle des
argiles pures.
    - L'augmentation de la
résistence face aux contraintes de
cisaillement.
    - La diminution de la
perméabilité, de l'ordre de 10⁻⁷-10⁻⁸
cm/sec, jusqu'à une teneur de 40% de
gravier. Au dela de 40%, la
perméabilité n'est plus négligeable.

Table 4. Les paramètres mecanique et hydraulique du mélange argile-gravier

Energie de compactage	Gravier plus de 5mm	Wop	dmax	Coeffic. de filtration	Modul de tassement	Essai triaxial ø	
t-m/m3	%	%	kg/m3	cm/sek	mm/m	°	C bar
	0	27.8	1550	2.12X10⁻⁹	120.46	--	--
	25	16.6	1830	1.03X10⁻⁸	79.18	--	--
	30	14.7	1920	2.23X10⁻⁸	71.16	25	0.86
62.5	35	13.5	1990	2.64X10⁻⁸	64.96	27	0.75
	40	12.0	2060	2.58X10⁻⁷	62.86	--	--
	45	9.3	2120	1.03X10⁻⁶	48.82	--	--
	50	7.8	2150	2.98X10⁻⁵	43.46	--	--

    Sans prendre en consideration
la consolidation pendant la
construction du barrage, deux ans
après la constrution totale, les
tassements ont diminués de 25 mm/ml

C'est pourqua il est recomandé que
le noyau du barrage contienne de 30
à 40% de gravier.

3. Il existe une bonne corrélation entre les resultats obtenus en laboratoire et ceux obtenus sur le terrain, pendant la construction du barrage.

REFERENCES

1. BISHOP, A., HENKEL, D. - 1961 - Opredjelenie svoistv gruntov b trehosni ispitanijah.
2. BOZO, L., GORO, N. - 1984 - Mekanika e dherave dhe e shkëmbit. SH. B. L. U. Tiranë.
3. British standart of mecanik soils, 1975.
4. CAQUOT, J. - 1966 - Traite de mécanique des sols. Paris.
5. COINE, A. - 1960 - Soils dams.
6. Essais triaxeaux. - 1970 - DUNOD, Paris.
7. GORO, N., Leka, L., XHAXHIU,T. - 1976 Raport mbi analizat laboratorike te materialeve te ndërtimit të bërthamës së digës së Fierzës. N.GJ.GJ., Tiranë.
8. GORO, N. 1976 - Raport mbi provat e rezistencës në rrëshqitje në shpatet e HC Fierzë.
9. GORO, N. - 1984, 1993 - Udhëzues metodik mbi provat laboratorike. N.GJ.GJ., Tiranë.
10. -GORO, N. -1990 - Udhëzues metodik mbi studimin e materialeve te ndërtimit. N.GJ.GJ., Tiranë.
11. JACK, W. - 1975 - Compacted fill.
12. Le compatage des sols. -1970- DUNOD, Paris.
13. VERDENY, J. - 1975 - Mécanique des sols. Paris.

Ecological repairs of the Orava dam reservoir shore banks

La réparation du barrage d'Orava et l'étude de la zone inférieure des rives

Otto Horský
Geoinza, Madrid, Spain

Lubomír Lincer
Geotest a.s., Brno, Czech Republic

Jaromír Nešvara
Nešvara IGGT, Ostrava, Czech Republic

ABSTRACT: In the year 1990 the Orava dam, one of the greatest dam reservoirs of Slovakia, was drained off after 37 years periods of its existence. The drainage was used for shore line maintenance and enabled investigation the abraded and slided banks in all its length, as far as the silted reservoir bottom. The article give some experiences of watered and drained valley slope, formed by soft neogene sediments. The experiences relate to do fact that the zone under the sliding mass has lower abrasion resistance.

RESUME: En 1990 le barrage d'Orava (une des barrages les plus grandes de la Slovaquie) a été vidé au bout de 37 ans pour la période de six mois. Pendant cette période, on a effectué non seulement les travaux de reconstruction mais aussi les recherches des parties inférieures des rives. Dans l'article, nous donnons les résultats de cette recherche. Il s'agit de la portée de transformation des pentes(qui sont constituées de la terre argileuse) au cours de l'abrasion et de la facon supposée de l' assainissement.

1. INTRODUCTION

At the beginning of 1990 the Orava dam reservoir was started to be drained off so that its operating mechanisms could be repaired since the dam had been operating for 37 years. The dam was drained off for about half a year, from 4th April, 1990 till 30th September, 1990. The drained off dam was used for its shore line maintenance as well as for its bottom research to elucidate the problems of bottom silting, shore slopes deformations, etc.

We have been studying the causes of landslide occurences on the Orava dam reservoir shore slopes since 1967. During thaws in the spring of 1967 shore slope landslides endangered some recreation areas and a port and drew our attention to the necessity to study the dam influence on its surroundings and to protect the shore banks against surf, water level fluctuation, etc. The study results of the shore bank changes

up to 1982, the factors and conditions of the reshaping process were published by O. Horský (1990).

The dam draining has enabled the research works (especially geophysical) to be extended to the reservoir bottom and to get new experience.

2. THE SITUATION, PARAMETERS AND WATER REGIME OF THE DAM.

The Orava dam reservoir is situated in the northwest part of Slovakia on the border on Poland. The reservoir was built by constructing a concrete dam 44 m high in the depression created by the Orava river (just bellow the confluence of the Čierna Orava and the Biela Orava) in 1953. At its maximum capacity the water covers an area of 32.8 km². The shoreline measures about 70 km.

The purpose of the Orava dam is to retain water of flooded rivers and on the contrary, when there are low water levels, to raise the rate of flow of the Orava and the Váh (the Orava empties into the Váh) to use water power in industry and agriculture.

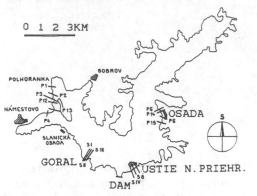

Fig.1 Dam reservoir situation

The total volume of the reservoir at the flood level of 603.0 m a.s.l. is 345.9 million m³ of water (in the Adriatic system), the constant volume being only 27.3 million m³, the reservoir itself being 298.1/278.0 and the channel storage being 20.4/40.6 million m³ of water. When there are floods, the dam can retain 500 - 600 m³/s of water.

Consequently, the characteristics of the dam regime are as follows:

a/ the water level fluctuates up to 16.8 m during a year between the levels of 585.5 - 602.8 m over the sea (Figure 2). The lowest difference between the minimum and maximum water level was recorded in 1966 (-3.1 m) and in 1974 (-3.5 m), other years being from 5.0 up to 16.5 m. In February and March the water level is the lowest and from April to July the reservoir is being refilled. From August to November the water level is the most constant.

b/ the dam regime is not constant from a long- termed point of view. It is influenced by both climate changes and the changeable use of water power etc. We have divided the 37 years'operation of the dam into several periods according to its water levels and their rate and frequency. Histograms of the frequency of water levels and their durations in those periods are shown at Figure 2.

At this picture we can see that during the first years after being filled with water the dam did not have its constant regime. In the

histogram of frequency of that period no water level is more constant. The most frequent water levels occupy the range of 8 m. On the contrary, in the following years 1963 - 1967 the dam regime became rather constant, there were no significant changes in maximum water levels within a year and the most frequent water level reached the frequency of 6.5 % at the level of 601.5 - 601.6 m a.s.l. The range of water levels lowered to 2 m at the levels of 600 and 602 m over the sea with the frequency of over 2 %. In the following period the water regime of the dam does not differ from the previous one in the range of water levels and the frequency of over 2 %. The level of maximum frequency and duration is not that significant, though.

Since 1976 there has been a considerable change in the dam regime (probably caused by the Liptovska Mara Dam on the Váh, which was put into operation). From 1976 to 1984 the annual maximum water levels mostly ran between the levels of 598.8 and 601.6 m a.s.l. and in comparison with the previous period, the most frequent water level of this period dropped 1.5 m down (599.6 - 599.7 m a.s.l.) level. In 1985 the water level was not falling down so much as in previous years (the lowest level was 592.0 m a.s.l.), presumably because of mild winter, especially in February and March. The number of days with the highest frequency of water levels within a year (up to 16 %) went up. In 1986 the water level in the reservoir kept at the level of 598.8 - 598.9 m a.s.l. with small breaks from 12th September to 25th November (57 days according to the records). In 1987, on the contrary, the water level stayed at the level of 600.6 - 600.7 m a.s.l. for 58 days. It was in the period from 20th July to 4th November, that the water level fluctuated between the levels of 600.48 and 600.83 m a.s.l.

c/ the dropping and raising tendency of the reservoir water level of the dam can be sudden and quick. The operating regulations of the dam, when considering its shore banks and possible streaming pressures, allows the maximum decrease of its water level to 1.2 m a day. The actual maximum decrease in a day runs from 30 - 60 cm. At lower water levels in the dam, however, there has been a decrease of 100 cm and a raising tendency up to 190cm a day measured.

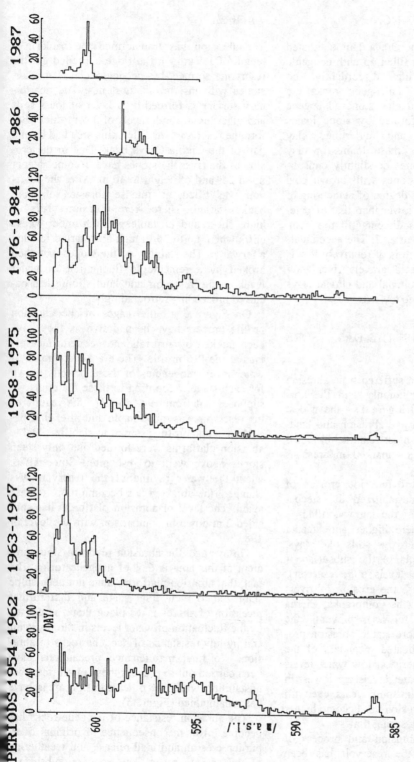

Fig. 2 Histograms of the frequency of water levels and their durations in separate periods.

3. GEOLOGICAL SITUATION

The immersed area of the Orava dam is situated in so called Orava basin filled up with neogenic clayey sediments lying discordantly on paleogene flysh rocks. Paleogene rocks are represented by sand and clay stones. Neogene sediments are mainly formed by slope loams sometimes calcareous and underlined clay stones. In some places these loams or clay stones contain sand stones or slightly consolidated calcareous sand stones with brown coal inserts up to 40 cm. The degree of cementing of neogene rocks is rather lower than that of paleogene rocks. Neogene sediments fill the basin up to the level of 620.0 m a.s.l. The accumulating base of erosion which is of relatively low is represented by fluvioglacial gravels, river gravels of higher terraces, deluvial and eluvial clays changing into loamy scree towards.

4. THE CHANGING OF THE DAM SHORE SLOPES.

The areas that have most suffered from abrasion were Osada, Ústie, Polhoranka and the area bellow the Goral hotel (Figure 1) - the widest range and intensity of wave abrasion and landsliding process and the most remarkable recess of shore slopes has been marked in areas as follows:

a/ in the areas, where the side erosion of original streams had destabilized the steeper slopes of the valley before the dam was filled,

b/ in the areas, where higher and steeper slopes are formed by erosive soils like clays, sands and neogene soils in the subsurface (within the reach of water levels in the reservoir),

c/ in the areas, where the erosion of rivers emptying into the dam (the Polhoranka, Čierna Orava, Biela Orava, Jelšava), prevents the slopes from creating more stable abrasion profiles of shores. Complicated meanders of the beds of those rivers emerge at low water levels in the reservoir. The water levels are lowest in February and March, in some years even till June. That means the erosion has become higher because of floods during spring thaws.

The most intense abrasion and process of shore - reshaping of the reservoir has been marked in the locality of Osada.

4.1 Osada.

The abrasion has transformed the bank in a length of 1.5 km - a flat ridge corbelled out the reservoir, formed by neogenic clays or claystones with inserts of silt stones. The quarternary surface is formed by a layer of loess soils and glacifluvial sandgravels of 3-5 m thick. Before the reservoir was filled, this area had been a part of the Čierna Orava bank. Due to the erosion of the river the slopes have become steeper up to 24° and of height to 40 m. After the reservoir was filled, an intense abrasion of these banks started. There were abrasion cliffs of high, sheer and unstable walls formed. These cliffs where up to 15 m high and started to slide afterwards. The stability of the slopes was then broken by a remarkable fluctuation of water levels in the reservoir and land sliding has occurred proceeding retrogradingly.

On Figure 3 all the stages of the abrasion profile process have been drawn as they have been marked by airprints and geodetic measurings at the P 6 profile. There are histograms of water level frequencies of these periods drawn for each stage. According to these materials we suppose, that from 1962 to 1975 the regime of the reservoir was quite stable and after that period the slope profiles started to stabilize. Stable abrasion platforms were formed and only steep slopes above went to reshaping. Since 1976, when the water regime of the reservoir was changed, the abrasion has become more intense again. The level of abrasion platforms has subsided 2 m down in comparison with the level of 1983.

On average, the abrasion platform slope gradient of this area is 6.5° of neogene rocks. The fact, that abrasion platforms have not been slope stabilized to their final stage and that further recession of shores takes place there, is caused by the fluctuation of water levels in the reservoir and by low resistance of neogene rocks to abrasion. No final maintenance precautions have been carried out to prevent these platforms from unstability, while the other abrasion areas have been maintained (Figure 5).

The abrasion resistance of neogene rocks has become lower in consequence of original overburden erosion and load release and weathering of the surface rocks. The zone of subsurface

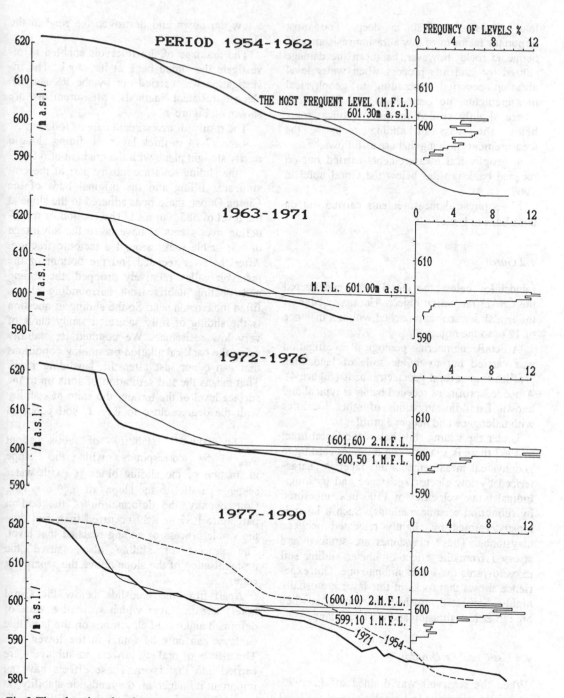

Fig.3 The abrasion development of the bank at the P6 (Osada).

load release is 18-20 m deep. The most important factor to of low abrasion resistance of neogenic rocks, however, has been the damage caused by landslide process when water level abrasion occurred. According to geophysical measurements, neogene clays and claystones where slightly eroded in quite a thick zone, below the mass of sliding rocks - the measurements proving that are as follows:

a/ geophysical measurements carried out on receded bank profiles below the Goral hotel in 1967;

b/ geophysical measurements carried out on landslides at Ústie.

4.2 Goral

Landslides below the Goral hotel endangered the ČSAD recreation object. The upper edge of the landslides scar approached within a distance of 12 m to the object.

A detail engineering geological investigation was carried out to divide soils of landslided profiles into several geoelectric beds (Figure 4). A bedded profile of receded banks is typical and known from investigations of other localities with paleogene and neogene profiles.

Under the sliding soils below the Goral hotel in 1967 there is a layer 6-7m thick. According to geophysical measurements this layer is characterized by low electric resistance and by longitudinal wave velocities of 1100 m/s (measured by refraction seismic methods). Such a layer is mostly formed by slightly relocated neogene claystones. Those claystones are strained and stressed from the effect of higher sliding soil mass towards the lower sliding zone. Our experience shows that rocks of this layer are usually highly dislocated, crushed and mixed, so they are rather less abrasion and wash resistant.

4.3 Ústie nad Priehradou

When the reservoir was drained off in 1990, especially during its last quick phase at the level down of 590.0 m a.s.l., and landslides of the original bank at Ústie occurred.

Due to landslide process a head new scar formed above the separating edge on the slope at the distance of 18 m. The sliding block subside a few dm down and destroyed the road to the church.

The drainage of the reservoir enabled to investigate the slided bank at its length. The investigation was carried out by the RS and RP electric resistant methods. Measurements are shown on Figure 4.

The results of investigation are as follows:

- the 12-14 m thick layer, is sliding along a nearly straight plane with the gradient of 6.5°.

- the sliding soils are mostly part of the erosion scar filling and the original beds of the Čierna Orava; these beds adhered to the slope at the level of 585.0 m a.s.l. The erosion of meandering river seems to have taken the advantage of less stable rocks around a tectonic fracture. After the river receded and the bottom of the reservoir valley relatively dropped, the sliding and floating debris from surrounding slopes filled the erosion scar. So the sliding in question is the sliding of fully saturated sandy clays of very low resistance. We counted its stability using the back calculation presuming conditions that can occur just after the lansliding itself. That means the full saturation of soils up to the surface level of the terrain, the state of stability with the degree close to F = 1, soil cohesion $c' = 0$.

The counted stability of soils about $\varphi_{res} = 13.5°$ corresponds with the slope inclination of the sliding block α (with water seepage parallel to the slope) of 6.5° $\alpha = \varphi/2$. Consequently, the deformation of the bottom part up to RS7 or RS9 occurred first, and then the whole process of sliding reached that level. The process of sliding also caused the destabilisation of the slope above the separating edge.

Apart from the landslide below the Goral hotel, here the layer which we call a zone of deformed and mixed claystones on the landslide sublayer can only be found in the lower part. The effects of braking powers on sublayers are carried into that layer. These effects have an important influence on the landslide stability (at a very low resistance of wash-out filling sediments of the original erosion scar in the upper part of the slope).

We suppose, that a similar erosion of neogenic rocks occurs even in the process of high abrasion block sliding in the area of Osada.

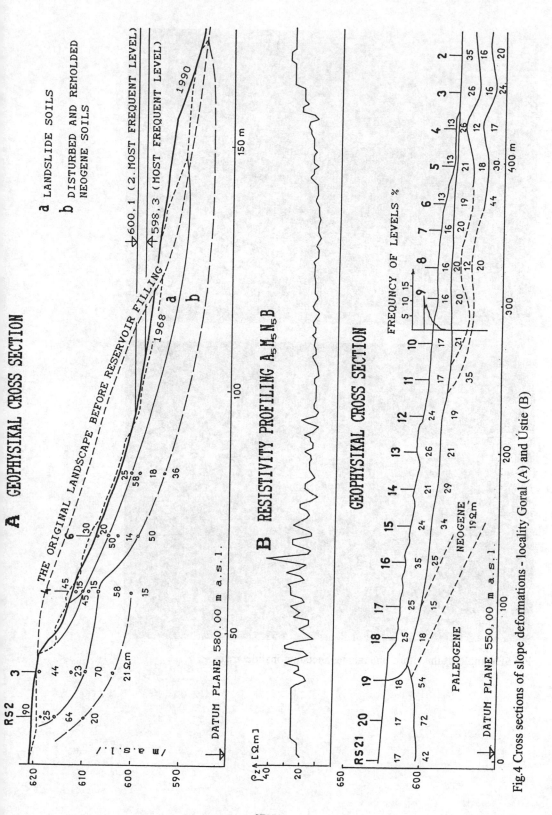

Fig.4 Cross sections of slope deformations - locality Goral (A) and Ústie (B)

3735

Fig.5 Example of the shore line mantenance in nonslide area.

We have analyzed the development of the abrasion-slide process in the P6 profile. We have taken the slope profiles of 1953 and 1962 as a starting point of slope development. That means before the time reservoir was filled and then the slope sliding occurred According to the known shape of the slided block and place of the separating scar we have been able to count the depth of the damaged zone in its sublayer. As for the slided slope profile below the Goral hotel, its thickness was 7/9 of the thickness of soils above the slided block, before it started to recede. We have used the same design conditions for the P 6 profile. We have found out, that the base of such an eroded layer corresponds with the present surface of the abrasion platform (with the part in question). In case that our conclusions were correct, the layer of eroded and easier to abrade rocks has been moved away by water level effects of the reservoir. Sublayer rocks of this layer will be more resistant to abrasion and the conditions of this part of the abrasion profile will become stabilized. Then it will make it possible to carry out the maintenance works even in this part of the dam banks.

5. CONCLUSIONS

In 1990 the Orava dam reservoir - one of the biggest reservoirs in Slovakia - was drained off for about half a year after 37 years of operating. The Orava dam is situated in a basin of neogenic clayey sediments lying on paleogenic flysh rocks. The drainage of the reservoir was used for a lot of repairing and maintaining works on the banks as well as for further searching works on water subsurface parts of its banks. It also helped to get new experience which will later enable us to make better prognoses on the development of bank abrasions and following maintenance.

REFERENCES

Bláha, P. 1993. *Geofyzikální metody při výzkumu svahových deformací.* Geotest, Brno and NIS ČR - středisko GEOFOND Praha.

Horský, O. 1990. The causes of morphological changes at the water edges of Orava-Reservoir. *Proc. 6th IAEG*, p.2859-2868. Rotterdam, Balkema.

Lincer, L. and Nešvara, J. and Bláha, P. 1991. *Závěrečná zpráva o inženýrskogeologickém průzkumu v zátopové oblasti údolní nádrže Orava.* Geotest, Brno.

Carsticity in the Zimapan Hydroproject Gorge

Phénomènes carstiques dans la zone du barrage de Zimapan

Vicente Páez J. & Carlos González C.
Mexican Federal Power Comission (CFE), Mexico

ABSTRACT: It is described the geological characteristics of the carbonated rocks which outcrop in the gorge, emphasizing into the development and distribution of the carst phenomena affecting the rock mass.
Quantitatively the carst features represent a very small portion of the rock mass, developed by different phreatic levels during several historical geology stages in the site; the carst development took advantage of different planes of weaknesses or discontinuities of the rock mass in order to establish its cycle.
The carst distribution analysis, the kind of fillings, and their relation with the stratigraphic, structural and tectonic characteristics have been of great usefulness in the impervious grouting and drainage curtains design of the dam, and so for their genesis explanation.

RÉSUMÉ: Les caractéristiques géologiques des roches carbonatées qui affleurent aux zones d'appui du barrage sont décrites en donnant une attention particulière á la formation et distribution du carst qui est le principal facteur de perméabilité du massif rocheux.
Le phénomène carst occupe un petit volume du massif rocheux et a pour origine le gradient hydraulique engendré par les différents niveaux phréatiques durant l'historie géologique du site. La formation du carst s'est effectuée à travers les plans faibles du massif rocheux creusant ainsi les conduites où s'écoulent l'eau (et autres solutions) de la surface jusqu'aux niveaux inférieurs.
L'analyse et l'étude statistique de la distribution du carst, des différents matériaux de remplissage et de leurs relations avec la stratigraphie, la structure et la tectonique du massif sont d'un grand intérêt pour le projet du rideaux d'injection et de drainage du barrage et permettent de reconstituer l'histoire de son site.

1.- Introduction

Tha Zimapan damsite is located in the Infiernillo Canyon, approximately 400 m downstream of the San Juan and Tula rivers coalescence, where is named Moctezuma River, just at the Queretaro and Hidalgo states border (Figure 1). This canyon has a deepness of approximately 500 m (between 1362 and 1850 m.a.s.l.) and the works are located between 1362 and 1565 m.a.s.l.

The structural and lithological characteristics of the rock mass keep a closed relation with the genesis and development of the carst; then, it was necessary the recollection of the detailed geological information in all the outcrops, excavations and drillings into the gorge.

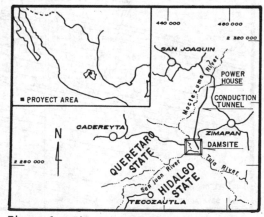

Figure 1.- Zimapan Damsite Location

2.- Geological characteristics of the gorge

In the dam courtain abutments is outcroping the El Doctor Limestone (Aptian-Cenomanian), which is constituted by two lithological units: the first one is placed between 1565 and 1500 m.a.s.l., and consists of dolomitic limestone and dolomites with light and dark gray colours; the second one is placed from 1500 m.a.s.l. to all down to the riverbed, and is composed exclusively of dark gray dolomites.

The stratification has a general strike N60°E, dipping 10° to 12° NW (toward left abutment). The Laramidic compressive stresses originated the El Doctor overthrust (with regional character), which emplace in tectonic contact the El Doctor Limestone with the Soyatal Formation, placed 35 m of average deepness below the dam foundation (Figure 2).

At local levels, the Laramidic stresses took advantage of stratification planes or crossed between them, developing clayey horizons (milonites).

3.- Fracturing

The discontinuities in the rock mass are predominantly subverticals, their mapping in each of the galleries, adits and outcrops in the abutment trenchs has been permitted their classification, obeying their orientation, in four structural families, as follows:

ALPHA:N22.5°-67.5°E   (average N45°E)
BETA: N22.5°-67.5°W   (average N45°W)
GAMMA:N0°-22.5°W y N0°-22.5°E   (average N-S)
DELTA:N67.5°-90°E y N67.5°-90°W (average E-W)

In section, the gorge was divided in 24 m intervals. In each of these intervals, it was summarized the total of discontinuites per family; the same was done for those

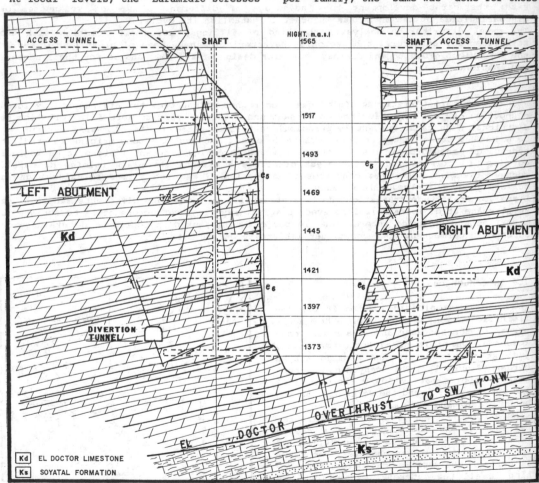

Figure 2.- Geological section of the Gorge

structures with carst development (Table 1).

The results were represented in circular rossettes which indicate orientations and quantities in percentage for each family (Figure 3).

In the Figure 4 (fracturing-carsticity) are presented these diagrams, reaching the following results:

- Between the 1565 and 1433 m.a.s.l. the BETA fracturing rules.

- There is a transition zone between 1433 and 1409, with smooth predominancy of the ALFA family. This zone presents an important overthrusting at the 1418.

- Below the 1409 is defined a clear predominancy of the ALFA family.

- In the diagrams is observed that the percentage of structures with carst, commonly match with the predominant family.

- It was quantified a total of 2105 structures, 1176 for the left abutment and 929 for the other; 193 (9%) of them have carstic structures (Figure 3). The 44% of the structures with carsticity corresponds with the BETA family; 23% to the ALFA one; 19% to DELTA; and 14% to the GAMMA family.

4.- Quantity and distribution of carst.

Based on the number of registered voids in the gorge, it was done the Table 2, which analysis shows:

In both abutments:
- The carst is notably higher, in number and magnitud into the left abutment (291 voids) than in the right abutment (152 voids), without considering the access tunnels.

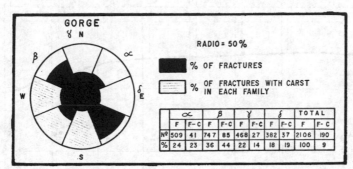

	$\propto$		β		γ		δ		TOTAL	
	F	F-C	F	F-C	F	F-C	F	F-C	F	F-C
Nº	509	41	747	85	468	27	382	37	2106	190
%	24	23	36	44	22	14	18	19	100	9

RADIO = 50%
% OF FRACTURES
% OF FRACTURES WITH CARST IN EACH FAMILY

Figure 3.- Rossette of percentages for fracturing-carst

TABLE 1.Statistics of fractures and fractures with carst, in both right abutment and left abutmen of the gorge

		LEFT ABUTMENT										RIGHT ABUTMENT									
L E V E L		ALPHA		BETA		GAMMA		DELTA		TOTAL		ALPHA		BETA		GAMMA		DELTA		TOTAL	
		f	f-c	f	f-c	f	f-c	f	f-c	f	f-c	f	f-c	f	f-c	f	f-c	f	f-c	f	f-c
1565-1529	#	55	09	80	7	64	5	35	5	234	26	19	--	58	4	47	4	28	2	152	10
	%	24	16	34	9	27	8	15	14	100	11	13	--	38	7	31	9	18	7	100	7
1529-1505	#	26	3	88	7	45	3	28	2	187	15	27	--	71	10	29	1	17	2	144	13
	%	14	12	47	8	24	7	15	7	100	8	19	--	49	14	20	3	12	12	100	9
1505-1481	#	9	2	47	3	14	--	12	4	82	9	9	--	25	4	20	1	14	--	68	5
	%	11	22	57	6	17	--	15	33	100	11	13	--	37	16	29	5	21	--	100	7
1481-1457	#	21	--	74	13	36	3	17	--	148	16	19	1	52	9	19	--	13	1	103	11
	%	14	--	50	9	24	2	12	--	100	11	18	5	50	17	18	--	14	8	100	11
1457-1433	#	18	2	45	14	20	1	28	3	111	20	24	--	41	5	13	--	19	3	97	8
	%	16	11	41	31	18	5	25	11	100	16	25	--	42	12	13	--	20	16	100	8
1433-1409	#	71	7	55	6	51	3	37	6	214	23	35	3	37	--	32	--	37	4	141	7
	%	33	10	26	11	24	6	17	16	100	11	25	9	26	--	23	--	26	11	100	5
1409-1385	#	34	7	10	2	26	3	16	3	86	15	25	--	17	--	17	1	23	--	82	1
	%	39	21	12	20	30	12	19	19	100	17	30	--	21	--	21	6	28	--	108	1
1385-1362	#	57	3	19	--	12	1	27	--	115	4	60	3	28	1	23	1	31	2	142	7
	%	50	5	17	--	10	8	23	--	100	3	42	5	20	4	16	4	22	6	100	5

f=fracture   f-c=fracture with carst   #=number of fractures   %=percentage of fractures and fractures with carst

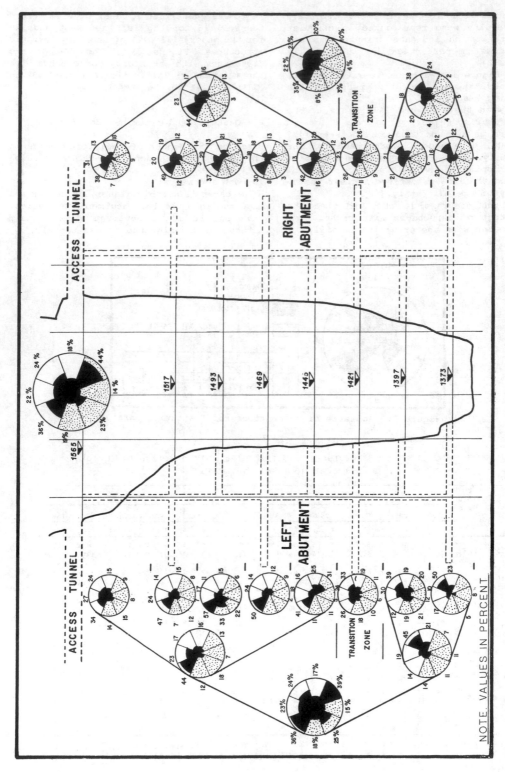

Figure 4.— Diagrams of fracturing-carsticity

NOTE. VALUES IN PERCENT

3742

Table 2.- Voids acounted in the gorge

	NUMBER OF VOIDS	
LEVELS	LEFT ABUTMENT	RIGHT ABUTMENT
1565-1529	64	44
1529-1505	45	35
1505-1481	22	14
1481-1457	44	20
1457-1433	35	14
1433-1409	48	8
1409-1385	25	3
1385-1362	8	4
TOTAL	291	152

- Between the 1565-1505 levels exists a high portion of voids. This interval is composed of dolomitic rocks, with higher dissolution potential than the dolomite.
- From 1385 level down to the river bed and below it, the carst is scarce; in general, the carsticity at lower levels is poor, the reason could be that the freatic level is recent and with very low gradient.
In right abutment:
- Generally, the carst is scarce (there is no presence between 1493-1385 levels of the canyon wall). It is explained that the larger carst development in left abutment (contrasting with the right one) is due to the presence of a terrace with smooth topography favouring the reception and percolation of the meteoric water; more over, the dipping of the strata is toward the interior of the left margin (Figure 2), controlling and inducing more and more the infiltration into the massif core (Figure 5).

The classification of the voids according to their size (Table 3) evidence what more than the 80% of them are smaller than the 50 cm; it means that in general the carst is rather small and that a voluminous development obey probably to a special geological conditions (f.r.: coalescence of a fat clayey horizon with an important fracturing zone).

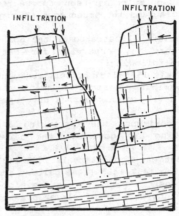

Figure 5.-Scheme of the gorge showing the paths which follows the infiltration into fractures and stratification, near the canyon walls.

Table 3.- Classification of the voids according to their size

RIGHT ABUTMENT

SIZE (cm)	# VOID G-1517	# VOID G-1493	# VOID G-1469	# VOID G-1445	# VOID G-1421	# VOID G-1397	# VOID G-1373	# VOID ABUT.	# VOID S.	# VOID A.T.	TOTAL
> 10	1	2	3	6	2	1	-	-	51	97	163
10-50	3	-	3	1	2	-	3	26	20	707	765
50-100	-	-	-	-	-	-	3	21	1	46	71
> 100	-	-	1	-	-	-	-	1	1	18	21

LEFT ABUTMENT

SIZE (cm)	# VOID G-1517	# VOID G-1493	# VOID G-1469	# VOID G-1445	# VOID G-1421	# VOID G-1397	# VOID G-1373	# VOID ABUT.	# VOID S.	# VOID A.T.	TOTAL
> 10	-	-	21	9	19	1	-	43	18	-	111
10-50	9	-	5	13	16	8	4	63	20	29	167
50-100	4	-	-	1	2	3	-	6	6	-	22
> 100	6	1	-	1	1	1	-	2	8	1	21

G= Gallery   ABUT.= Abutment   S=Shaft   A.T.= Access Tunnel

The carst distribution depend greatly on water circulation into the massif, the same having a structural control (Figure 5), evidently there is a structure-carst relationship, and it is fundamental in the definition of the geological guides for the carst detection.

5.- Carst in structures

In general, in the voids is possible to observe the presence of the responsible structure which favoured the carst development. This association has been represented by circular diagrams in percentages (Figure 6).

The diagrams are illustrated in the gorge section (Figure 7), observing for both abutments the following:
- The intervals 1565-1481 of left abutment and the 1565-1505 of right abutment are correlated, having mainly strata developed carst.
- To lower horizons, the fracturing and normal faulting are governing the carst development.
- In left abutment the intervals 1505-1481 (shaft) and the 1457-1433 (1445 gallery and abutment) present important carst development in the intersection of clayey stratas and fractures.

The water flow into the massif (vadose or gravitational, and the freatic or pressurized) determine the geometry of the voids. It is very scarce the carst of the gorge developed in freatic conditions; the voids of this kind generally are of semicircular geometry and subhorizontal elipsoidal in stratas or subvertical elipsoidal in normal fractures (Figure 8). Almost all the carst of the gorge was originated in gravitational conditions (Figure 5) with abrasion-dilution phenomena, developing lenticular shapes in stratas and fractures (Figure 9).

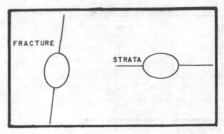

Figure 8.- Carst in freatic conditions (Llopis Lladó, 1970)

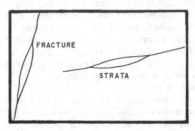

Figure 9.- Carst in vadose conditions (Llopis Lladó, 1970).

Outstand the carst development in the intersection of fractures with clay strata, wich make the function of an impervious barrier avoiding the direct flow of water to lower horizons, and commanding the flow over the clayey plane following its maximum inclination; at the same time the fluid erode and disolve the rock until it find a new conduit downslope (Figure 10).

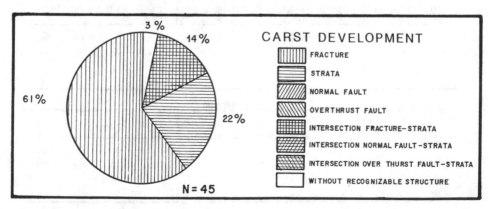

Figure 6.- Diagram of carst in structures

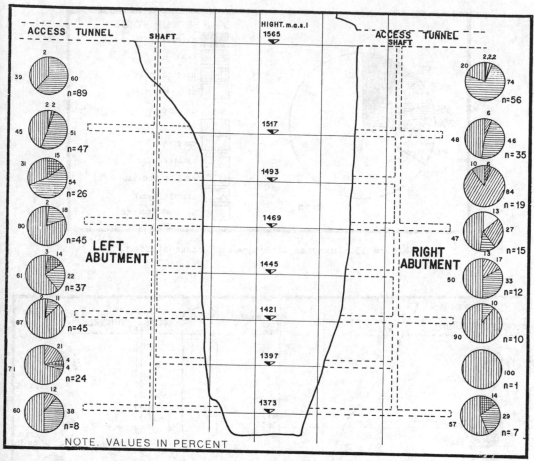

Figure 7.- Association of carsticity with structures

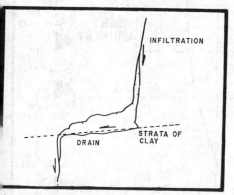

Figure 10.- Carst in clayey strata.

hese case is very well observed in 1517 and 445 galleries of the left abutment.

6.- Types of fillings.

Very often the voids present a peculiar filling. These characteristics were plotled in circular diagrams (Figure 11), and in the gorge section (Figure 12).
In both abutments:

- The carstic voids with calcite coated walls without any other filling (semi-empty) are predominants.

- There were identified two horizons for the clayey fillings; the firstone is restricted to the interval 1505 - 1457 in the left abutment (in the shaft and abutment), correlated with the 1565 - 1505 interval in right abutment.

- The second level is located from 1433 elevation to the bottom, being more noticeable in 1421 el, which coincides with thrusting clayey horizon (1418).

The microcrystaline calcite filling are concentrated specially in the interval 1565-

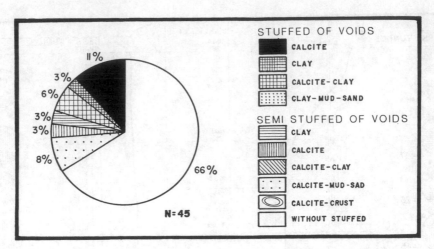

Figure 11.- Diagram with types of carst stuffed

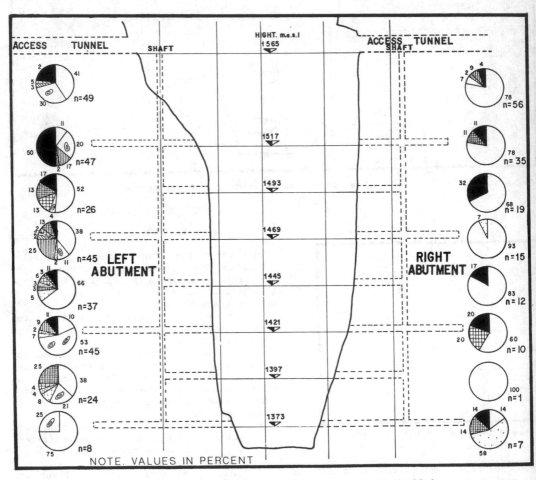

Figure 12.- Diagrams with types of carst stuffed

1505 (limestone and dolomite limestone zone).

-The presence of fillings is due to the subterranean flows wich deposite the transported materials when they reach a flat freatic level or the base erosion level, where they lose energy (Figure 13).

them, a zone were the water has been follow a descendent path in order to reach this level.

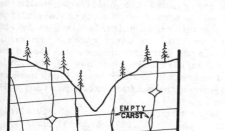

Figure 13.- Relation between carst fillings and freatic level

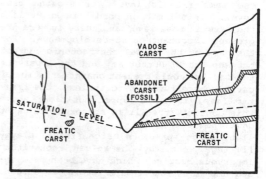

Figure 14.- Freatic and vadose conditions for carst development

7.- Origin of carst

The carst characteristics described in this paper are indicatives of the proccess which originated and favoured their development into the gorge. The carbonated rocks (limestons and dolomites) which outcrop in the site are characterized by a mineralogy with high solubility properties.

The number of detected holes or voids is high, nevertheless their dimmensions are not so large indicating that the length of time of water circulation was not so large, too. This supply has been done by means of the two following mechanisms:

- Lateral supply of the freatic level. The carst of this type was developed in freatic conditions, it means that the cavity was full of water, which moved following a hydraulic gradient, which in the case of the gorge seems to be very low, resulting in subhorizontal caverns, developed very near of the level or into it (Figure 14).

- Surficial supply. The carst development of this type was created in vadose conditions, it means that the void never was full of water. The water was infiltrated by the fracturing doing an abrasive work in vertical direction, searching for the freatic level or the superficial base erosion level (the river) (Figure 14). The vadose carst prevail in all the gorge, this mechanism was eased due to the lowering of the freatic level from upper horizons to its present level. The rock mass which used to cover the freatic level has constituted, since

The variations of the freatic level have been related to regional tectonic uplifts of the ground. The fluctuations what it have had the uplifts (variations in velocity) have produced freatic levels which were stables temporarily, favouring the deposit of the water running materials reaching these levels or the erosion base level. Evidences of the matter are present in the horizons with clayey fillings identified into the gorge at the 1505-1457 and 1433-1385 levels. The first interval represents an ancient stable freatic level, it was abandoned due to a rapid lowering of the water level associated with a rapid too, terrain uplift.

The lower levels down to gallery 1421 of the left abutment (second interval mentioned above) are characterized by a relative abundance of clayey and granular fillings, the larger carstic conduits present evident abrassion and development in subhorizontal sense. These characteristics seem to indicate that the velocity of the terrain uplift was slower or maybe ceased; them , the work of the river has been slower and the freatic level has experienced a gradual lowering; consequentely, the development of carst has been give preferentely in freatic conditions, with low hydraulic gradient.

The volcanic rocks, what in remote times covered almost all the paleotopography of the limestones, were capable of store water yielding due to infiltration with the consecuent carst development in the limestone, these filtrations carried in volcanic material which filled the voids reached the freatic level (f.i., the red clay in the gorge).

Generally, the carst development takes place into the rock massif until a determined level, which could be a geological contact with an impervous formation what make a barrier to the water circulation; or the same freatic level, or erosion base level. This limit or border is known like the carstic base level and there is were are present important carst developments. For the case of the gorge, the tectonic contact between the El Doctor Fm. and the impervous Soyatal Fm. could present characteristics of carstic base level; nevertheless, the freatic level is in an upper position, above this contact, then it is not expected carst development in it.

The gorge presents significant clayey fillings what, by themselves, constitute impervous barriers which have been functioned like local carst base levels, developing important voids.

8.- Conclusions

The statistic study of the structural, dimmensional and filling type characteristics prooved to be a good tool in order to yield a generalized model of the carst affecting the gorge. Having the following: the left abutment has more carsticity than the right one, it is due to the smooth topography between 1565-1520 levels, and so for the strata dipping toward the interior of the massif.

The carsticity have the majority of its developments in fracturing, so, in the 1565-1433 interval along the NW fracturing; in the 1433-1409, along the NW too, but with tendency on NE fracturing; and the 1409 all down the river, along the NE one. Nevertheless, the stratification plays a major role, too.

Any discontinuity (fault or fracture) associated with clayeys horizons could develope carst in their intersection with high probability.

It is not expected the presence of carst below the 1370 level (freatic level), because this is the present carst base level at the site.

The majority of the carst voids in the gorge does not have any filling.

The clayey and granular filling are present in two levels: The 1505-1457, representing an ancient freatic level temporarily stable, which were abandoned by a drastic lowering of the freatic level (now, fossil or dead carst) by a rapid terrain uplift.

The 1433-1385 level, mainly at levels down the 1421, indicate a gradual regional upli-

ft, which have influenced in the same way like the variation of freatic level.

The results obtained in this study give us a guide to the action to follow in the grout courtain works. In the defined structural levels, the carst development are more abundant in the more frequent structures; those imply what the drillings should be oriented in accord to the morphology of the carst conduits more suitables in each level.

More over, for the extra grouting holes in intervals of lot of consume, it should be considered that more tham 80% of the voids have dimmensions smaller than 50 cm, measure which govern the maximum extra grouting hole distance.

Nowadays, the impervous courtain is almost finished and for 3187 grouted intervals, only 102 presented high consumes due to carst (3.2%).

Finally, the information which it is collecting in the grouting works will permit to define the relation between the massif carst percentage and the high consume percentage, which has been very useful in the retro-planning of the next grouting boreholes, for this and future similar projects.

Bibliography

Castany G.; 1971. Tratado práctico de las aguas subterráneas. Ed. Omega.

Clayton, K. M.; 1985. Limestone geomorphe logy. Ed. Longman Group.

Freeze, R. Allan and Cherry, John A.; 1979 Groundwater. Ed. Prentice-Hall, Inc.

Ham, William E; 1962. Classification of carbonate rocks. Symposium of the America Association of Petroleum Geologists.

Lascano Sahagún, Carlos; 1984. Las formas cársticas del área de la Florida en l Sierra Gorda de Querétaro. Instituto d Geografía (U.N.A.M.)

Llopis Lladó, Noel; 1970. Fundamentos de hidrología cárstica. Ed. Blume.

Schneider, Bernard: Informes técnicos 1990 1993.

Wahlstrom, Ernest E; 1974. Dams, dam foun dations, and reservoir sites. Ed. Elsevie Scientific Publishing Company.

Problème posé par la présence d'une galerie de mine à proximité d'un projet de barrage

Problems caused by the presence of an underground mine next to site of a dam

B. Couturier
BRL Ingénierie, Nîmes, France

ABSTRACT : The case studied above is a good example of the possible interferences due to the geological context between an existing underground structure (gallery) and a planned above ground project (dam). The location of this dam, apparently quite favorable, suffered in fact some latent defects which could only be evidenced through progressive and well planned reconnaissance works.

RÉSUMÉ : L'exemple traité montre les relations qui pourraient s'établir entre un ouvrage souterrain existant (galerie) et un projet nouveau de surface (barrage) par le biais du contexte géologique. Celui-ci apparemment assez favorable, révèle en fait un certain nombre de vices cachés que seuls des reconnaissances précises et progressives ont permis de mettre en évidence.

1. INTRODUCTION

Le site de barrage se trouve au Sud de la France, dans le département de l'Hérault, sur la commune de Camplong. L'ouvrage envisagé, à vocation touristique, aurait environ 12 m de hauteur et formerait un plan d'eau de 4,5 ha créant un pôle d'attraction principal pour la vallée de l'Espaze.

2. CADRE GEOLOGIQUE

L'ensemble de ce projet se développe dans un bassin d'âge Stéphanien qui forme un sillon houiller, profond et étroit, à cheval sur une zone majeure de dislocations tardi-hercyniennes.

Ce bassin, connu depuis le Moyen-âge, fait aujourd'hui encore l'objet d'une exploitation à découvert, le minerai étant acheminé vers le Bousquet d'Orb par d'anciennes galeries, à 4 km au nord-est de la zone du projet.
Une de ces galeries longe la future retenue et se trouve à une distance horizontale de 40 m de la crête du barrage en rive gauche et à une profondeur de 35 m sous le versant (fig. 1).

Signalons aussi qu'à l'amont de la retenue la rivière se perd partiellement dans le bed-rock et qu'un revêtement de son cours a été nécessaire. Une preuve de ces infiltrations existe dans la galerie d'évacuation du minerai où l'on peut observer une venue d'eau de plusieurs litres par seconde qui jaillit du rocher à l'intersection d'une fissure verticale et d'un joint de stratification.

3. GEOLOGIE DE L'AMENAGEMENT

Le substratum rocheux est constitué de grès-quartzite en bancs décimétriques à métriques avec des passées schisteuses minces, plus rarement charbonneuses. En thalweg, on observe un large développement des alluvions sablo-graveleuses.

D'un point de vue structural, les couches de grès et schistes sont pentées de rive droite vers rive gauche, d'environ 35° et ont pratiquement la direction de la vallée. A l'aval du site, elles sont recoupées par une grande faille d'orientation Est-Ouest, sur le trajet de laquelle les terrains semblent assez désorganisés.

La possibilité d'avoir ce même type d'accidents transverses sous les alluvions de la cuvette et jouant le rôle de drains vers les anciennes galeries de mines, nous a conduits à faire réaliser un certain nombre d'investigations pour évaluer ce risque.

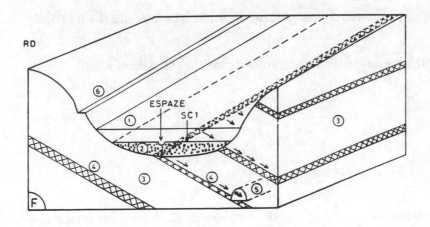

RD

ESPAZE

SC1

F

① SURFACE DU PLAN D'EAU

② ALLUVIONS

③ SUBSTRATUM PENTÉ RD vers RG DE 35°

④ NIVEAUX PERMÉABLES DANS LES SCHISTES

⑤ ANCIENNE GALERIE DE MINE

⑥ ROUTE

Ⓕ PLAN DE FAILLE TRANSVERSAL (GÉOPHYSIQUE)

──▶ CHEMINEMENT DES EAUX

●──▶ VENUE D'EAU EN GALERIE

SC1 SONDAGE

**Fig.1– Schéma théorique des risques de pertes
au niveau de la retenue**

4. TRAVAUX DE RECONNAISSANCE

Ils se sont déroulés en deux phases distinctes.

4.1. Travaux exécutés en première phase

Les reconnaissances comprenaient :

- des mesures de sismique réfraction (7 dispositfs de 55 m avec 12 capteurs espacés de 5 m, amont–aval sur la longueur de la cuvette) au droit desquels étaient juxtaposées des mesures électriques (15 dispositifs de 7 électrodes espacées de 5 m) pour enregistrer les résistivités apparentes) ;

- deux sondages carottés de 30 m de profondeur, inclinés de 45° vers l'amont, recoupant les anomalies détectées par la géophysique, et dans lesquels ont été réalisés des essais Lugeon à 5 bars afin de tester la perméabilité de ces zones particulières.

L'analyse des mesures sismiques montre dans toute la cuvette des vitesses de fond du substratum de l'ordre de 3400 à 3900 m/s témoignant d'une roche relativement saine.

Dans cet ensemble homogène, il existe 2 zones qui font exception, à savoir une zone aval à environ 90 m à l'amont du futur barrage et une zone amont en queue de retenue où les vitesses de fond sont de l'ordre de 2800 m/s et dont l'origine pourrait être un accident tectonique.

L'examen des mesures de résistivité apparente du rocher en profondeur confirme la présence des 2 zones faibles avec des valeurs inférieures à 500 ohms x mètre encadrées par des valeurs plus fortes.

Le dépouillement des sondages réalisés sur ces deux zones anormales confirme ces résultats. En effet, sur la zone aval, le sondage SC1 recoupe dans la série schisteuse très désorganisée (RQD de 29 % sur toute la hauteur) des zones très fracturées, brisées. Ce sont des veines de charbon de 1,2 et 1,1 m d'épaisseur, avec des RQD nuls ou faibles, qui témoignent d'une roche de très mauvaise qualité vers 25 m de profondeur.

En ce qui concerne la zone amont, on retrouve des dispositions un peu similaires mais avec des terrains dans l'ensemble moins fracturés et oxydés. Les zones de charbon sont par contre toujours aussi fracturées et un RQD inférieur à 5 % confirme que ces bancs sont brisés.

Les essais Lugeon réalisés sur la zone aval ont montré une très forte perméabilité pour l'horizon charbonneux à 20–25 m avec des valeurs supérieures à 20 l/m/mn pour des pressions légèrement supérieures à 1b, confirmant que la capacité d'absorption du terrain est ici importante.

Rappelons que les couches de terrain sont pentées de rive droite vers rive gauche et que leur direction correspond à celle de la vallée. Cela signifie qu'une couche profonde brisée et plus perméable, du type de celle touchée sur le sondage, peut exister sous les alluvions perméables de la cuvette. On constate sur la figure 1 que sa "tranche" sera baignée sur toute sa longueur qui est en direction avec celle du plan d'eau.

De plus, la fracturation globale des terrains montre bien que l'on doit se trouver aussi dans une zone faillée transversale à la vallée et parallèle aux grands accidents régionaux existants (plan F de la fig. 1)

4.2. Travaux exécutés en deuxième phase

Devant ces résultats, une prospection géophysique plus large de l'ensemble du fond de la cuvette a été décidée.

Un balayage complet de la future retenue a été effectué par méthode électromagnétique. Cette prospection comprenait 500 stations de mesures plus des calages électriques afin de repérer les zones à faible résistivité apparente, liées à des couches charbonneuses sous–alluviales.

Les résultats des mesures électromagnétiques ont mis en évidence 4 axes de faibles résistivités (400 ohms x mètre), témoignant de la présence de zones charbonneuses ou plus fissurées, dont 2 particulièrement développés et alignés selon la direction des couches.

De plus l'examen du tracé des valeurs de forte résistivité (100 ohms x mètre) montre un décrochement transverse de la vallée que l'on peut interpréter comme une zone de discontinuité tectonique.

5. CONSEQUENCES POUR LE PROJET

Les résultats des investigations faites sur la cuvette du barrage vont donc dans le même sens, à savoir la confirmation de la présence de zones perméables (charbon) ou fracturées (failles) en cuvette pouvant drainer les eaux de la future retenue vers les galeries de mines sous–jacentes.

La répartition plus ou moins discontinue de ces anamolies du substratum masquées par 4 à 5 m de terrains de couverture, conduit à envisager un étanchement artificiel de la cuvette qui représente une surface importante d'environ 4.5 ha.

Dans un tel contexte, l'étanchement de la retenue peut s'envisager de plusieurs façons :

– injections traditionnelles,
– mise en place d'un tapis de matériaux argileux compactés sur toute la surface,
– mise en place d'une membrane étanche.

Les injections ne sont pas adaptées au problème posé car la localisation des zones perméables reste beaucoup trop aléatoire et imprécise. Il est en effet impossible de décaper les alluvions (volume trop important) pour les rechercher. Il pourra donc toujours subsister une zone non traitée ou mal traitée avec les conséquences que cela entraîne.

Pour la réalisation d'un tapis étanche, il apparaît qu'il n'y a pas de matériaux compactables dans les alentours immédiats de l'aménagement.

De ce fait, seule la solution membrane reste possible. Pour un tel projet, l'examen de cette solution montre qu'elle est très contraignante (préparation du support, fixation de la membrane, liaison avec les berges et le barrage, installation des plages artificielles, etc.) et que son coût est élevé puisqu'il double pratiquement le montant des travaux.
En conséquence, cette solution onéreuse ne sera proposée qu'en dernier recours si, après mise en eau et examen du comportement de l'ouvrage sur une période suffisante, les débits de fuite s'avéraient trop importants et incompatibles avec sa gestion.

6. CONCLUSION

Bien que de dimensions modestes, ce barrage à vocation touristique présente des risques de fuite de eaux de la retenue qui ne sont pas négligeables et conduisent à prendre un certain nombre de précautions. L'évaluation réelle du risque a été possible après des étapes successives de travaux de reconnaissances appropriés et relativement lourds compte tenu de l'importance du projet.

Cependant, ils ont permis d'imaginer un modèle géologique du terrain suffisamment réaliste pour confirmer l'intéraction qu'il pouvait y avoir entre les données naturelles (lithologie et structure) et artificielles (anciennes galeries de mines), tout en recherchant une solution pour un éventuel étanchement.

REFERENCES

COUTURIER B. (1987)
Les études géologiques dans les projets de barrages.
Thèse doctorat d'Etat, Université de Grenoble I, 350 p.

COUTURIER B. (1990)
Barrage de Camplong sur l'Espaze.
Etude géologique préliminaire. CNARBRL, rapport inédit.

COUTURIER B. (1993)
Barrage de Camplong sur l'Espaze.
Reconnaissance géologique complémentaire. CNARBRL, rapport inédit.

GAUTHIER F.B. (1993)
Commune de Camplong.
Etude électromagnétique.
CEBTP, rapport inédit.

MARINIER P. et BERTRAND Y. (1992)
Barrage de Camplong (Hérault).
Reconnaissance géophysique. SEGG, rapport inédit.

Geological-geotechnical mapping and characterization on rock mass of River Setubal Dam, Minas Gerais, Brazil

Zonage et cartographie géotechnique et caractérisation du massif rocheux du barrage de Setubal, Minas Gerais, Brésil

Silvana R. Liporaci & Nahor N. Souza Jr
Engineering Geological Department, E.E. São Carlos, University of São Paulo, Brazil

ABSTRACT: This hydraulic scheme is located on the Setubal River, at Jequitinhonha River Valley. The lithology is varied. Carbonatic quartz-schist is dominant, with pockets and intercalations of other rock types.
The schistosity is the principal structural feature, followed by tectonic cleavage and four joint sets.
The characteristics showed through geological-geotechnical mapping, directioned alterations and definitions ol the spillway excavation design, in the light of the rock mass behaviour, in order to achieve striketdip on spillway slopes.
The rock mass quality was evaluated according to the degree of fracturing. The intact core length, according to RQD, was not considered.
According to mapping executed on this geologically and structurally complex rock mass, it was observed that the survey method and probabilistic study of discontinuities proposed by Priest & Hudson may lead to unreal evaluation of a rock mass quality, due to the generalization of a local information

RÉSUMÉ: Le développement des sources hydrauliques de la rivière Setubal est située dans la Vallée de la Jequitinhonha. La lithologie est variée. Le quartz-schiste carbonatique est dominant et présente des portions et des intercalations d'autres roches.
La schistosité est la principale caractéristique structurelle suivie par le clivage tectonique et quatre conjoints de diaclases.
Les caractéristiques montrées à travers les soulevés géologique-géotechniques, ont directioné les changements et les nouvelles définitions du projet de creusement de l'evacuateur selon la condition réelle du massif rocheux, pour obtenir la meilleure définition de la pente de l'evacuateur.
Le massif rocheux a eté évalué selon le degré de fracturation. La longueur de l'echantillon intact n'a pas eté considerée.
Selon la relevée executée sur ce massif rocheux, géologiquement et structurellement complexe, il a été observé que la méthode de soulevé et d'étude probabilistique des discontinuités, proposée par Priest & Hudson peut fournir des évaluations irréelles sur la qualité du massif rocheux dans un projet, à cause de géneraliser les données obtenues localement.

1- INTRODUCTION

The hydraulic exploitation is located on the Setubal River, at Jequitinhonha River Valley, in the northern region of State of Minas Gerais, Brazil.
The rock masses that lie along the Setubal Dam belong to the Precambrian, they are part of São Francisco Super-Group of the Salinas Unit. The lithology is quite diverse with the predomination of carbonatic-quartz-schist and quartzite in which the most salient are quatz-sericite-graphite-schist and sericitic-calciferous quartzite.

2 - SYNTHESIS OF THE GEOLOGICAL AND GEOTECHNICAL CHARACTERISTICS OF THE ROCK MASS

The excavations of the temporary deviation gallery were developed in the quartz-micaschist with graphite and eventually some portions of quartzitic rock. Both the rock present sound appearance (D1/D2) and they present some fractures (F1/F2) with the exception of few segments where it is very or highly weathering (D4/D5) and medium to highly fractured (F3/F4).

Only two planes of schistosity were found to be open with appreciable continuity containing decomposed filling material with small angle of friction.

In the emergency deviation canal, common and rock excavations were conducted with slopes of 3V:2V and 5V:1H respectively. From the geotechnical point of view, the excavations were made mainly in saprolite and desarticulated decomposed rock (D3 to D5 and F3 to F5). The excavation in the coherent rock (D1 to D3 and F1 to F3) happened only in the middle portion of the canal.

The right wall of the canal was appreciably influenced by the dip of the schistosity, particularly in the reaches of occurrence of saprolite and decomposed rock (D3/D4/D5), acquiring the final inclination of approximately 45°.

In the water intake, the excavations were conducted in first and second category material, represented by colluvium, residual soil, saprolite (rock D4/D5), eventually some portions of D2/D3 rock at the base of the slope.

The inclination of 63° (2V:1H) of the right slope generated during the rainy period generalized ruptures along the schistosity besides the numerous wedge failures due to the interaction of shistosity with the main plane of fracture.

From the structural point of view, the schistosity (1) (N 20° E/45° NW) is without doubt the main discontinuity with medium concordant attitude with the direction of flow and reclining toward the left shoulder.

The tectonic cleavage (6) occurs cutting the surface of schistosity, exhibiting medium attitude (N 30° E/80° to 90° SE) sometimes toward NW.

Beside these, there exist four more sets of discontinuities(joints and/or fractures) of regional nature which were identified in the field with medium attitude: (2) - (N 60° E/65° NW); (3) - (N 15° W/60° NE);(4) - (N 40° W/55° SW);(5) - (N 45° E /60° SE).

3 - METHODOLOGY FOR SURVEY OF THE ROCK MASS

During the stages of preliminary and basic designs, a survey of the discontinuities revealed in the rock out croppings existing along the river banks and in the grottos was made.

During the construction and execution of the project, geological and geotechnical mapping of the excavation surfaces (slopes and foundations) was made with the purpose of determining and recording the "in situ" characteristics of the surfaces for installation of hydraulic structures. This allowed a verification of the parameters used in project and provide for necessary change in the rock mass conditions instead of those initially assumed (preliminary and basic design)

The study of geological and geotechnical characteristics recording by mapping provide the necessary basis for the changes and the definition of the design of installation and excavation for the spillway on the right abutment in order to fit the real conditions for the behaviour of the rock mass.

The mapping works (scale 1:100) for the slopes were checked with the topographic survey of the excavated surfaces, beside the normal staking, included labeling the absolute levels on the slope faces at 5m vertical intervals at 10m horizontal intervals. In the foundations, the staking was 5m each (Figures 1 and 2).

The mapping of final excavated surfaces consisted of delimitation of the geologically and geotechnically uniform areas considering the characteristics like: Weathering (D1= sound rock until D5= extremely weathering rock); Consistency (C1= very consistent until C5 consistency of the soil or crumbly); Fracturing (F1= occasionally fractured , smaller than 1 fracture per meter, until F5 = extremely fractured or fragmented , bigger than 20 fractures per meter) Considering litology types: micaschist(MX);quartzite (QT);quartz-schist(Qx).Components Minerals: quartz (qz); graphite (gf) and calcite (ca). Considering soil levels: colluvium (CO); residual soil; (SR); saprolite (SA); and others characteristics Shown in figure 1.

Different discontinuities (schistosity, tectonic cleavage, Joint/ fracture, etc.) and their principal characteristics such as: attitude, weathering, opening, filling, roughness, etc., were catalogued according to the recommendations of ISRM (1981).

4 - INDICES EMPLOYED FOR THE MEASUREMENT OF THE INTENSITY OF DISCONTINUITIES

The form of the index used to describe the intensity of discontinuity, frequently depends upon the nature of the study and upon the technique of investigation.

Thus the results of the inspection of a line of observation or the quality of the rock obtained during the field investigation can be described by the method proposed by Deere: the Rock Quality Designation (RQD) is given by:

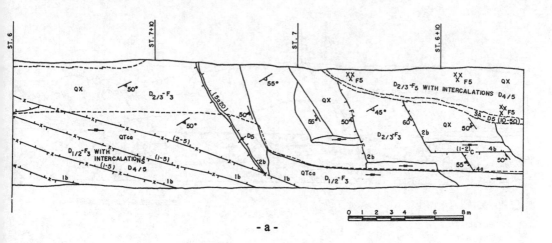

- a -

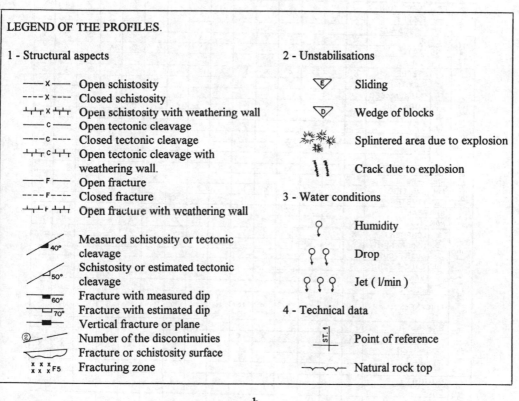

LEGEND OF THE PROFILES.

1 - Structural aspects

—x— Open schistosity
----x---- Closed schistosity
⊥⊥τx⊥⊥τ Open schistosity with weathering wall
—c— Open tectonic cleavage
----c---- Closed tectonic cleavage
⊥⊥τc⊥⊥τ Open tectonic cleavage with weathering wall.
—F— Open fracture
----F---- Closed fracture
⊥⊥τF⊥⊥τ Open fracture with weathering wall

⟋40° Measured schistosity or tectonic cleavage
⟋50° Schistosity or estimated tectonic cleavage
▬■60° Fracture with measured dip
▬⊐70° Fracture with estimated dip
▬■ Vertical fracture or plane
②⟋ Number of the discontinuities
⌣ Fracture or schistosity surface
x x x F5 / x x x Fracturing zone

2 - Unstabilisations

▽E Sliding
▽D Wedge of blocks
※※※ Splintered area due to explosion
〗〗 Crack due to explosion

3 - Water conditions

♀ Humidity
♀ ♀ Drop
♀ ♀ ♀ Jet (l/min)

4 - Technical data

⌐ST.1 Point of reference
〜〜〜 Natural rock top

- b -

Figure 1 - Geological-geotechnical mapping of the right wall of the deviation canal: a - profile; b - legend of the profiles (Fig. 1a. and Fig. 2a.)

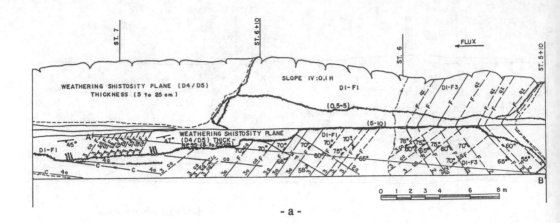

- a -

DISCONTINUITY Nº	I	Ia	Ib	2	2a	2b	.3	3a	3b	4	4a	4b	5	5a	5b
TRACE CLASS — FRACTURE / JOINT				X	X	X	X	X	X				X	X	X
TRACE CLASS — SHISTOSITY	X	X	X												
TRACE CLASS — TECT. CLEAVAGE										X	X	X			
TRACE — PLANE	X	X		X	X		X	X		X	X		X	X	X
TRACE — UNDULATE			X	X	X	X	X	X	X	X	X	X		X	X
TRACE — IRREGULAR															
SURFACE / TYPE — STEPPED	X	X	X	X	X	X	X	X	X	X	X	X	X	X	X
SURFACE / TYPE — SMOOTH				X						X					
SURFACE / TYPE — SLICKENSIDED															
SURFACE / WEATHERING — OXIDATION		X			X	X		X	X		X	X		X	
SURFACE / WEATHERING — D1	X			X			X	X		X			X		
SURFACE / WEATHERING — D2					X	X		X					X		
SURFACE / WEATHERING — D3			X	X	X						X	X	X	X	X
SURFACE / WEATHERING — D4			X						X		X	X			X
SURFACE / WEATHERING — D5									X						X
APERTURE — CLOSED	X			X			X			X			X		
APERTURE — ROCK - ROCK							X			X			X		
APERTURE — MILLIMETRIC		X	X		X	X		X	X		X	X	X	X	X
APERTURE — CENTIMETRIC			X			X			X			X	X	X	
APERTURE — GENERAL			X	X	X	X									
APERTURE — PARTIAL	X	X						X	X		X	X	X	X	
FILLING / COH. — COHERENT	X	X		X	X	X		X	X		X	X	X	X	
FILLING / COH. — UNCOHERENT			X												
FILLING / TYPE — QUARTZOUS				X	X	X									
FILLING / TYPE — CALCIFEROUS	X			X	X		X								
FILLING / TYPE — SANDY		X	X					X	X		X	X	X	X	
FILLING / TYPE — CLAYEY			X									X			X
FILLING — THICKNESS (mm)	1 to 2	1 to 5	1 to 30	1 to 10	1 to 10	1 to 10	1 to 2	1 to 5	1 to 40	0,5 to 1,0	0,5 to 5	1 to 20		1 to 5	1 to 20
FILLING — EXPANSIVE															
FILLING / COLOR — CLEAR	X			X	X	X				X					
FILLING / COLOR — GREEN			X						X		X	X	X		
FILLING / COLOR — RED		X	X			X		X	X		X	X			X
FILLING / COLOR — DARK															
WATER — HUMIDITY															
WATER — DROP															
WATER — JET			X												
ATTITUDE	N 350° to 50° / 05° to 50° NW			N 40° to 75° / 35° to 75° NW			N 280° to 315° / 35° to 88° NE or SW			N 10° to 60° / 80° to 90° NW or SE			N 10° to 50° / 2° to 60° SE		

- b -

Figure 2 - Geological-geotechnical mapping of the floor of the deviation gallery: a - scanline A-B; b classification of the discontinuites

$$RQD = 100 \sum_{i=1}^{n} x_i / L, \text{ where :}$$

x_i = is the length of the ith reach ≥ 0.1 m

n = is the number of intact reaches ≥ 0.1 m

L = is the length of scanline or borehole along which the RQD value is required.

Priest & Hudson (1976, 1979) proposed that the study of the survey of the existing discontinuities at least along two perpendicular lines traced on the rock mass (outcroppings and the excavated slopes) should give, according to statistical theory, same probability distribution or density as that of the rock mass as a whole.

For these authors, according to standard theory, if each small segment of the mapping line has equal or smaller probability of containing a point of intersection of a discontinuity, these points follow Poisson distribution and the associated spacing follow a negative exponential distribution expressed as :

$$f(x) = \lambda e^{-\lambda x}$$

$f(x)$ = frequency of a discontinuity spacing x

x = discontinuity spacing

λ = average number of discontinuities per metre.

These authors also propose a formula for the calculation of RQDt* index as:

$$RQDt^* = 100 e^{-\lambda x}(\lambda t + 1)$$

This general expression for RQDt*(theoretical)can be used for other values of the length of intact reaches than the conventional value of Deere ($t = 0.1$m).

5 - CORRELATION BETWEEN DEGREE OF FRACTURING AND THE INDICES OF DEERE - PRIEST & HUDSON

For the evaluation of the discontinuities in the Setubal rock mass, the degree of fracturing was utilized. It is the number of discontinuities existing in each one meter length of the bore or mapped surface, independently of the spacing existing between them and also irrespective of the size of the intact reaches in view of the fact that the spacing of schistosity and of tectonic cleavage is of the order of millimeters.

It was found that the degree of fracturing obtained from the borehole, a given location is analogous to

that obtained on the surface. This fact was also suggested by Deere with reference to RQD.

It was also observed that the degree of fracturing increases proportionately with its degree of weathering in view of the fact that sealed discontinuities in the sound rock are not quite perceptible.

In the rock with degree of weathering D3, D4, D5, these discontinuities are well-defined by decomposition or even open.

For each degree of fracturing, average number of fractures per metre and the average spacing between the fractures were considered. The table 1 Shows degree of fracturing adopted for Setubal.

The Figure 3 shows the curve of distribution of discontinuity spacings of Setubal rock mass adapted to the Theory of Priest Hudson (1976).

In reality, this curve reflects more the classification adopted for the rock mass indicating that there is a correlation between the degree of fracturing and the Theory of Priest & Hudson.

On the basis of the line A-B (Figure 2), traced on the geological and geotechnical mapping of the floor of the deviation gallery between the stakes 5 + 10 and 7, theory of Priest & Hudson is applied. The total length of the line is 32m such that 21m have degree of fracturing (F3), thus total estimated number of fractures for the reach is 109 and average number of discontinuities per meter (λ) is: 3,4

The Figure 4 shows that the discontinuities located along A-B follow a negative exponential distribution. In other words, the discontinuities with small spacings are more common thus allowing the use of the Priest & Hudson model.

The Figure 5 shows that 95,4% of the rock mass has the probability of containing intact reachs ($t \geq 0.1$m) and that 14,7% of the rock mass may contain intact reaches of 1.0m length or greater which confirms the field observations.

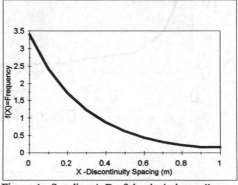

Figure 4 - Scanline A-B of the deviation gallery

Table 1 - Shows degree of fracturing adopted for Setubal Project

DEGREE OF FRACTURING	N° AVERAGE OF FRACTURES BY METRE (λ)	SPACING AVERAGE BETWEEN FRACTURES (x)	$f(x) = \lambda e^{-\lambda x}$
F1=occasionally fractured - 1 fract/m	1 fract/m	1,000m	0.368
F2=little fractured - 2 to 5 fract/m	2 fract/m 3.5 fract/m	0.500m 0.286m	0.736 1.288
F3=averagelly fractured - 6 to 10 fract/m	8 fract/m	0.125m	2.943
F4=much fractured - 11 to 20 fract/m	15.5 fract/m	0.065m	5.702
F5=extremelly fractured > 20 fract/m	25 fract/m 33 fract/m 50 fract/m 100 fract/m	0.040m 0.030m 0.020m 0.010m	9.197 12.140 18.394 36.788

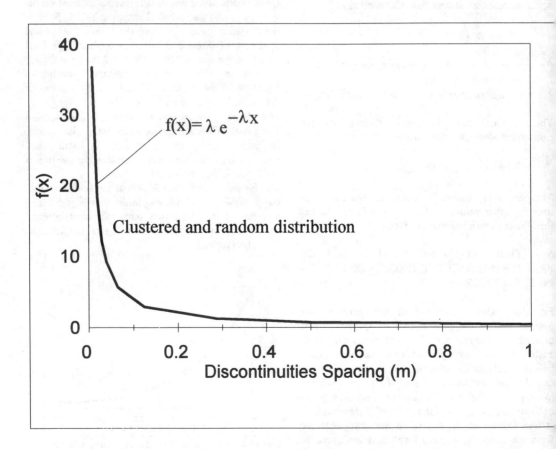

Figure 3 - Discontinuites spacing distribution based on degree of fracturing adopted for Setubal Project

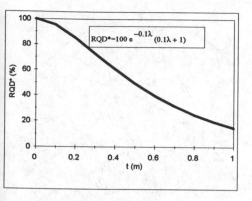

$$RQD^* = 100 \, e^{-0.1\lambda} \, (0.1\lambda + 1)$$

Figure 5 - RQD_t^* calculation of the scanline A-B

6 - STUDIES OF STABILITY OF THE SLOPE OF THE RIGHT WALL OF THE SPILLWAY

The Setubal Dam design requires a spillway of 240m length and of 40m height excavated in soil (saprolite) and sound rock. Figure 6 shows the study of spillway slope stability.

As the excavation of other hydraulic structures proceeded and the geological and geotechnical services of the surfaces conducted, it was observed that:

The surface of schistosity (1) (N 20° E / 45°NW) is commonly plane, rough and continuous (up to 100m), notably in the weathering rock portions (D3 to D5) where these surfaces, in general, are found to be open and oxidized or covered with decomposed material of small angle of friction.

In the zones of contact among different lithologies, the schistosity also presents itself open and weathering. In sound rock the schistosity is generally closed and cemented by SiO_2 or $CaCO_3$, in this case, when the schistosity is open, they present milliletric widths and rock-to-rock contacts.

Tectonic cleavage (6) (N 30° E/ 80°- 90° SE) occurs intersecting the surface of schistosity. It is nearly imperceptible in the sound rock. In the portions of weathering rock (D3 to D5), these surfaces become more evident, consisting persistent and with spacing frequently of millimetric size which open easily.

From the four discontinuity sets (joints/fractures) identified in the field (N 60° E/65°NW; N 15°W/60° NE; N 45° E / 60°SE and N 40° W / 55°SW) the first two were found to be more critical with respect to stability cuts. Separately or combined with other discontinuities, they are potentially unstable as regards plane rupture or wedging.

The set (2) is, as a rule, characterized by surfaces of plane compartmentation and spacing of centimeter to meter size which occur passing through varied strata of different lithologies mainly of quartzite rock.

In the sound rock, these fractures are generally blocked or cemented by quartz (SiO_2) whereas in the altered rock, they are open, oxidized and of easy landslide.

The other notable set (3) is characterized by irregular surfaces, persistent (3 to 30m) with metric spacings either blocked or sealed by $CaCO_3$ or open with oxidized or weathering walls.

The set (5) occurs with less frequency if compared with the last two, exhibiting, however, expressive planes (persistence of up to 70m) with millimetric openings with characteristic of undulating, rough and weathering walls.

The set (4) occurs with less frequency than the last three sets exhibiting less expressive planes with openings of millimeter to centimeter sizes when weathering and having rough characteristics.

The shistosity allied to tectonic cleavage and to the other sets of detected discontinuities lead to various wedge failures and slipping of blocks and slabs, observed in the excavated slopes of emergency deviation canal , deviation gallery and water in take (Figures 1 and 2).

On the basis of the mapping for geological and geotechnical characterization of the slopes of excavation for the above mentioned hydraulic structures, a study was conducted to predict the best attitude and the inclination of the right slope of the spillway aiming to minimize the conditions of instability of slopes excavated in the soil, saprolite and sound rock.

The various discontinuities described above were dealt with in the equatorial diagram of equal area (Shimidt-Lambert) where the poles as well as the large circles, for the sets located in the field, were plotted.

Through this study, it was possible to examine the potential instabilities caused by plane rupture, wedge failure or by toppling. The slope in saprolite, attitude N 20° E / 45°NW was considered with an angle of friction of 28° (obtained from tests) and the slope in sound rock, attitude N 20° E / 85° SE with an angle of friction of 45° (by literature) (Figure 6 a and b).

7 - CONCLUSIONS

The use of degree of fracturing as indices to measure the intensity of the occurrence of discontinuities in a rock mass is shown to be feasible. It was found that

3759

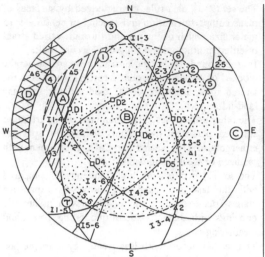

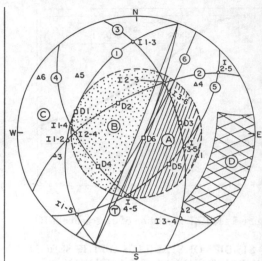

DIAGRAM I - Attitude of the slope:
N 20° E / 45° NW
Strength Parameters of the Saprolite:
Cohesion = C = 0,   Angle of friction = ∅ = 28°

Potential Instabilities:
-Plane Failure:None -The plane 1 touches the zone A
-Wedge Failure:None - The interceptions of the planes 1-2 and 1-4 touch the zone A.
-Toppling Failure: Yes, plane 6.

DIAGRAM II - Attitude of the slope:
N 20° E / 85° SE
Strength Parameters of the joint rock-rock:
Cohesion = C = 0,   Angle of friction = ∅ = 45°

Potential Instabilities:
-Plane Failure:Planes 3-5. The plane 6 touches the face of the slope.
-Wedge Failure:Combination of the planes 3-5, and 3-6. The planes 2-3 touch the face of the slope. The planes 4-5 are intercepting themselves in the limit of the angle of friction
-Toppling Failure:None

LEGEND OF THE DIAGRAMS:

  The great representative circle from each set of discontinuities

  Interception point between two planes of the discontinuities

  The great representative circle from slope face

D₁ Dip direction of each set of discontinuities

▲₂ Representative pole of each set of discontinuities

  Zone A - Unstable - surface of the discontinuity intercepting unfavorably the face of the slope, with dip greater than the angle of friction.

  Zone B - Stable - discontinuity with dip greater than the angle of friction (∅), but not intercepting the face of the slope.

  Zone C - Stable - discontinuity with dip smaller than the angle of friction (∅), without intercepting the face of slope.

  Zone D - Unstable - with possibilities of toppling failure

Figure 6 - Study of stability of the right slope of the spillway: diagram I and II

the degree of fracturing can be correlated with the two traditional indices proposed by Deere and Priest & Hudson.

As regard the theory of Priest & Hudson, one should take exception about the statement that the survey of the existing discontinuities in just two lines can provide the quality of the rock mass as a whole.

It was observed in the field that this procedure can either overestimate the quality of the rock mass or otherwise penalize it, this is due to the fact that there may be intercalations of various lithologies having different susceptibilities to fracture and weathering.

8 - ACKNOWLEDGMENT

Acknowledgment to Engineering geological Dep of the CEMIG- Companhia Energética de Minas Gerais (Energetic Company of Minas Gerais) and Association SPEC-DAM

9 - REFERENCES

Goodman, R.E. 1989. Introduction to rock mechanics - Second edition, John Wiley & Sons, Inc.

Hasui, Y. & Mioto, J.A. 1992. Geologia estrutural aplicada - ABGE (Associação Brasileira de Geologia de Engenharia & S.A. Industrias Votorantim - VOTORANTIN.

Hobbs, N.E.; Means, W.D. & Willians, P.F. 1989. An outline of structural geology - John Wiley & Sons, inc.

Hock, E. & Bray, J.W. 1977. Rock Slope Engineering - Revised second edition. The Institution of Mining and Metalurgy, London.

ISRM - International Society for Rock Mechanics 1981. Comission on testing methods for rock characterization testing and monitoring, E.T. Brown - Royal School of Mines, Imperial College of Science and Technology, London, England - by Pergamon.

LNEC - Laboratório Nacional de Engenharia Civil 1988. Estudo das descontinuidades e sua influência no comportamento das rochas e maciços rochosos. 1o. relatório: as descontinuidades nos maciços rochosos, características geométricas e influência na deformabilidade dos maciços. Ministério das Obras Públicas, Transportes e Comunicações, Lisboa.

Loczy, L. & Ladeira, E.A. 1976. Geologia estrutural e introdução à geotectônica - Ed. Edgard Blücher Ltda.

Priest, S.D. & Hudson, J.A. 1976. Discontinuity spacings in rock - in: Int. J. Rock Mech. Min. Sci. & Geomech. Abstr. vol. 13, p. 135-148. Pergamon Press, England.

Relatório 1991. Mapeamento geológico e geotécnico - Projeto executivo do Aproveitamento de Setúbal - Consórcio SPEC-DAM.

Geological and geotechnical features of Lend dam site (South east of Iran)

Les caractéristiques géologiques et géotechniques du site du barrage de Lend (au sud-est de l'Iran)

S.M.R.Emami
*Graduate School of Science and Engineering,
Saga University, Japan*

Y.Iwao
*Department of Civil Engineering, Saga University,
Japan*

J.Ghayoumian
Graduate School of, Kumamoto University, Japan

P.Ghazanfary
Jahad Engineering Services Company, Tehran, Iran

ABSTRACT: In this article the engineering geological study of the Lend dam site will be presented. To find out about the subsurface engineering geological features of the dam site, field investigation and laboratory tests were performed. It was clarified that the right abutment consists of stratified sandstone with high bearing capacity which due to joint sets are permeable. The left abutment is formed by shaly layers which in the surficial part are weathered and exhibit low geotechnical characteristics. The foundation corresponds to both type of the rocks . Based on the field investigation, laboratory and in-situ testing, the necessary comments were proposed.

RÉSUMÉ: Dans cet article on présente les études du site de barrage Lend du point de vue de la géologie de l'ingénieur. Pour trouver des conditions géologiques du barrage on a réalisé des recherches sur le site ainsi que des tests en laboratoire. On a montré que la côte droite de la vallée est formée des couches de roche sableuse d'une grande résistance. Ces roches sont perméables à travers des fissures. La côte gauche est formée par des roches argilleuses brisées lequelles présentent des caractéristiques géotechniques défavorables. Le barrage sera construit sur les deux types de roches. Sur la base des recherches du terrain on a proposé des tests sur le site ainsi que des tests de laboratoires suivis des commentaires nécessaires.

1. INTRODUCTION

The Lend dam, 21m high, 2148m crest length and reservoir volume of $233 \times 10^6 m^6$, is planned to be constructed at northern part of Chabahar city, southeastern part of Iran. The aim of this dam is to irregrate the downstream valley, to improve the water provision to neighboring village and to avoid eventual flooding.

From geological point of view the study area is located in Makran zone. This zone covers a wide region including southern part of Balochestan area of Iran and Pakistan and a portion of Hormozgan prefecture. The main geological features of the zone are thick sediments of upper Cretaceous (Colored Melange), which are covered with flysh sediments of Eocene age with thickness of more than 5000m (Tehrani, et al 1984). In this paper, the detail of the Lend dam site including geological features of the area and engineering geological aspects of the subsurface sediments in the dam site are briefly presented.

2. GEOLOGICAL CHARACTERISTICS Of THE DAM SITE

The dam site was studied on the surface with detail engineering geological mapping. The subsurface condition were investigated by drilling 14 boreholes whose depths varied between 10 and 30m and several adits.

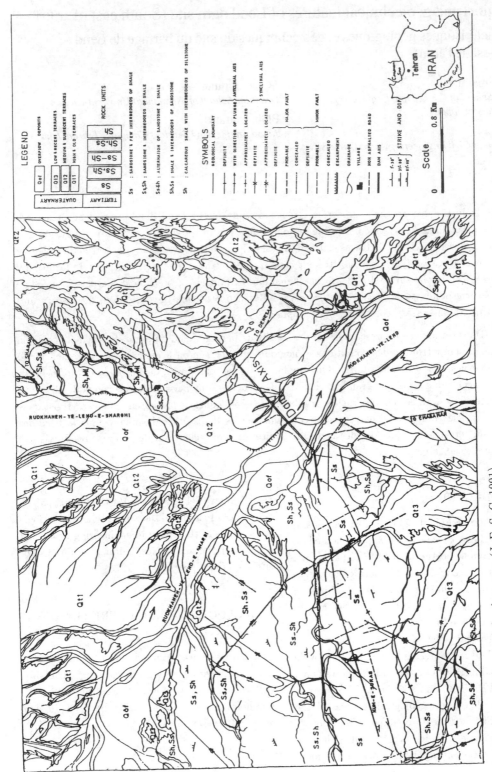

Fig. 1 Geological map of the dam area (J. E. S. C. 1991).

2.1 Lithology

Rock making up the right abutment is a Neogene stratified sandstone, the sandstone presents different bedding thicknesses. Based on the bedding thickness the sandstone unit is classified into four groups as follows:

1. Thickly bedded sandstone (Ss1): This type includes layers with average bedding thickness of 65cm. The outcrop of the layers can be observed at the dam axis. The layers shows least weathering rate.

2. Medium bedded sandstone (Ss2): The sandstone layers comprise a sequence of sandstone layers with average bedding thickness of 3.0 cm and thinly bedded shaly layers. In general, this group shows moderate weathering rate.

3. Thinly bedded sandstone (Ss34): The outcrop of this type can be observed at right abutment. The sandstone is a coarse grained sandstone with an average bedding thickness of 12 cm. The layers are characterized by E - W trends and N dips.

4. Very thinly bedded sandstone (Ss4): This type of sandstone is limited to the right abutment. The layers bedding thickness are 3.5cm. The left abutment is formed by shale and siltstone layers.

Due to physical conditions, the layers are impermeable. The apparent thickness of the layers is about 150m.

X-ray diffraction analysis clarified that Quartz, Kaolinite and Feldspars are the main rock forming minerals. The shaly layers changes to calcareous shale toward the east.

The foundation rocks correspond to both sandstone and shaly layers with maximum 8.0m of alluvial deposits. Based on morphology and geological condition , two type of quaternary deposits are observable in the dam site vicinity. The first type is old quaternary marine terraces (Qt1) which are not extended to the dam axis. The second type is new quaternary marine terraces (Qt2) with an average height of 5 to 20 m from river bed. The thickness of beds in the dam site reaches to 1.5 - 6.0 m. A geological map for the dam site is given in Fig. 1.

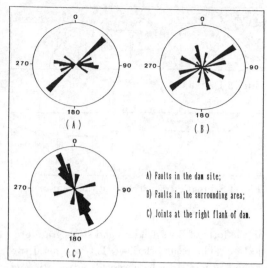

A) Faults in the dam site;

B) Faults in the surrounding area;

C) Joints at the right flank of dam.

Fig. 2 Rose diagram of fractures(J. E. S. C. 1991)

2.2 Structural geology

In the vicinity of the dam site several minor faults can be seen which their trends agree with the general structural model of the Makran zone. The main trends are west - east, sorthwest - southeast and northeast - southwest trends.

Rose diagram for the faults in the vicinity of the dam site and for a major regional fault called Ghasre Ghand fault are presented in Fig. 2. It is obvious that the trend of the faults in the vicinity of the dam site agree with the trend of Ghasre Ghand fault which is regional fault with trend of NE - SW.

There are several type of local faults in the dam site which in some case 20m displacement could be measured. In the right abutment a local fault caused displacement of sandstone layers at the flank of an anticlinal.

The ridge of the dam forms a local anticlinal, where the anticlinal axis plunges toward the dam foundation. The anticlinal has created several sets of joints which the tentional joints is the most important. In the left abutment, the shaly beds form several local folds which mostly are asymmetric folds.

Investigation and measurement of the joint system was performed and the results were prepared

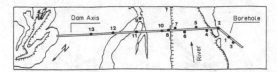

Fig. 3 Dam axis.

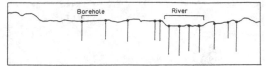

Fig. 4 Schmatic cross section of the dam axis.

in the form of Rose diagram and stereogram projection. The results clarified that in general two groups of joints can be observed, systematic and nonsystematic joints. The systematic joints are as a result of tectonical forces and mostly are related to the local folding in the area. The nonsystematic joints are surficial and are attributed to the weathering phenomenon in the sandstone layers and shrinkage crack in the shaly beds.

3. ROCK SLOPE STABILITY IN THE DAM SITE

The effect of joints and other planes of weakness of different orintations on slope instability phenomenon in the dam site was investigated. Analysis of rock slope stability in the dam site were based on the results of the field discontinuties investigation and their analysis using stereogram.

The results approved that the right abutment and the western part of reservoir are the only locations which due to geomorphological, geological and structural features show potential for instability. In both areas the lithology are inhomogenous and due to concentration of weak zones the conditions for plane and wedge types of failures are provided.

It is worth mentioning that even in the case of instability, the movement will not threaten the dam. It is because in the case of the first location, the movement will be toward outside of the dam, and in the case of the instability of the sec-

Table 1 Engineering properties of rock in the dam site.

Property (Sandstone)	Max.	Min.	Ave.
γ_d (dry density)	2.7	2.5	2.6
γ_{sat} (g/cm^3)	2.8	2.6	2.65
Water absorbtion (%)	3.2	1.1	2.1
Prosity (%)	8.2	2.9	5.4
Un. com. str. (Mpa)			
dry	123	29	81
saturated	82	21	51
Young modulus (ε)			
dry (* 10^4 Mpa)	3	1.73	2.15
saturated	2.3	0.95	1.67
Poisson's ratio			
saturated	0.42	0.19	0.35
Shear modulus			
dry (* 10^4 Mpa)	0.17	0.13	0.15
saturated	1.1	0.4	0.62

Property (Shale)	Max.	Min.	Ave.
γ_d (g/cm^3)	2.59	2.02	2.36
γ_{sat} (g/cm^3)	2.67	2.3	2.53
Water absorbtion (%)	11.5	1.4	5.4
Prosity (%)	28.2	28.05	16.1

ond location, the volume of moved materials will be insignificant.

4. IN-SITE AND LABORATORY TESTS

To find out about the engineering geological characterics of rock mass in the dam site, field procedures including permeability test and rock classification and laboratory tests were performed. Figurs. 3 and 4 illustrates dam axis and a schematic cross section of the dam axis including the drilled boreholes and Fig. 5 presents a schematic view of Lend river in the dam site including the locations of the boreholes.

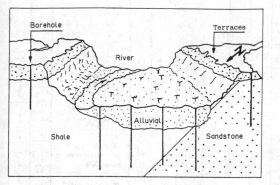

Fig. 5 The schmatic view of the Lend river in the dam site.

4.1 Laboratory Tests

The core samples were tested for physical and mechanical compressive strength, modulus of elasticity and poisson's ratio. The resultes are given in Table 1.

4.2 Permeability

Single packer permeability test were carried out at 1.0 m interval and under 3, 5 and 7 atmosphere pressures in the all of the drilled borholes in the dam foundation. In all cases, the diagram of flow as a function of pressure was prepared. The results of the permeability tests are presented in Table 2. It can be concluded that sandstone due to several joint and surficial weathering is slightly permeable and the shaly layers is almost impermeable.

4.3 Rock mass classification.

Based on the results of the geological investiga-

Table 2 Variation of permeability with depth at the dam site (Lugeon).

Depth (m)	BH5	BH6	BH7	BH8	BH9	BH10	BH11	BH12	BH13	BH14
~ 10	2. 13	6.3	0	---	95.6	58.1	2.8	0.8	9.4	3.2
10 ~ 20	0.1	0.6	0	6. 11	23.8	1.1	0.1	0.6	3.3	1.9
20 ~ 30	---	1.9	0	---	---	---	---	---	---	---
Ave. / BH	0.9	3. 10	0	3.6	50.4	35.8	1.8	0.7	6.8	2.7

Depth (m)	BH1	BH2	BH3	BH4
~ 10	44.8	92.6	27.2	10.9
10 ~ 20	15.1	55.2	75.5	43.2
20 ~ 30	---	44.3	46.3	80
Ave. / BH	32.4	60.7	52.6	34.3

Table 3 Rock mass classification according to R.M.R.

Parameter	Left flank		Rating		Right flank		Rating	
	D. *	S. *	D.	S.	D.	S.	D.	S.
Un. com. str. (Mpa)	3~10	1~3	1	0	80~90	44~50	8	5
R. Q. D.	15		3		60		13	
Joints spacing (cm)	5		8		31		22	
Joint condition	smooth ~ rough		6		smooth ~ rough		6	
Ground water cond.	dry ~ moist		7		completely dry		10	
Joint orientation	fair~favorable		-3		favorable		-2	
Total of ratio	-------		22	21	-------		57	54

Type of rock	Condition	Rating	Class	Description
Sandstone	dry	57	III	fair rock
	saturated	54	III	fair rock
Shale	dry	22	IV	poor rock
	saturated	21	IV	poor rock

D. : Dry, S. : Saturation

tion and performed index tests, the rock mass was rated and classified using Bienaiawski's classification system (Hoek et al 1980). The results of the classification is exhibited in Table 3. R.M.R. values of 21 - 22 for shale and 54 - 57 for sandstone classify the rock mass as poor rock and fair rock respectively.

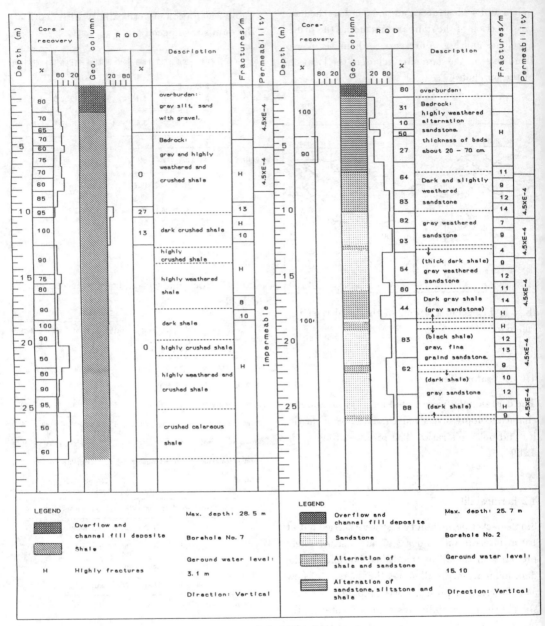

Fig. 6 Well logs No. 2 and 7 of the right and left flank of dam.

5. RESULTS

The two type of rocks in the dam site present different characteristics. Sandstone shows relatively high compressive stength, while preparation of core samples from shaly rocks was impossible.

On the other hand shaly rocks shows to be almost impermeable in site while sandstone rocks because of several joint set are slightly permeable.

The well logs for two of the boreholes are given in Fig 6. Boreholes number 2 and 6 have been

Table 4 Variation of R. Q. D. with depth at the dam site.

Depth (m)	BH5	BH6	BH7	BH8	BH9	BH10	BH11	BH12	BH13	BH14
~ 10	0	0	2	0	20	0	6	0	2	0
10 ~ 20	12	0	3	10	25	17	25	45	23	19
20 ~ 30	33	6	0	--	--	--	35	--	--	31
Ave. / BH	15	2	2	5	22	8	22	23	12	17

Depth (m)	BH1	BH2	BH3	BH4
~ 10	61	43	51	11
10 ~ 20	89	60	45	44
20 ~ 30	--	75	70	46
Ave. / BH	75	59	55	33

drilled in sandstone and shaly layers respectivly.

Core recovery from test boring advanced into sandstone ranged from approximately 90 to 100%, and that of shaly layers ranged from 60 to 90%. Rock Quality Designation ranged from 10 to 88% for sandstone and 0 to 27% for shaly layers. Table 4 shows the variation of R. Q. D. with depth. Both type of rocks have enough bearing capacity for a rockfill dam and the dam will not faces differential settlement.

6. CONCLUDING REMARKS

Field investigation indicates that the dam reservoir area is not expected to have any major slope instability problems. The dam site comprises two type of rocks with different geotechnical characteristics. The sandstone layers in the right abutment showed to be slightly permeable which shoud be improved. The shaly rocks are impermeable revealed poor geotechnical behavior compared to the sandstone. As the dam is a rockfill dam both type of the rocks have enough bearing capacity and the dam will not face differential settlement. The following comments are made to improve the stability of the dam:

1. Due to existence of local faults in the right abutment, drilling of several boreholes is proposed to identify fractured zone.

2. The drilled borehole can be used for grouting for water tightness of sandstone layers.

3. As shaly layers in the surficial part are weathered with weak geotechnical characteristics, removal of weathered part is proposed.

4. Removal of alluvial deposits in the dam foundation also is commented.

REFERECES

Bell, F. G. 1983. Fundamental of engineering geology, Butterworths Co.

Brown, E. T.: Rock characterization testing and monitoring I.S.R.M.

Hoek, E. & E. T. Brown 1980. Underground ex cavations in rock.

Jahad Engineering Services Company, 1991. Geological and geotechnical report of the first stage of Lend project, Tehran, Iran.

Tehrani, K. & A. Darvish Zadeh 1984. Geology of Iran, Minstry of Education.

Dam site investigation in southern central Arabian Shield, Saudi Arabia

Étude de sites de barrages dans le centre-sud de l'Arabie saoudite

B. H. Sadagah, O. M. Tayeb, A. S. Al-Sarrani & M. T. Hussein
Faculty of Earth Sciences, King Abdulaziz University, Jeddah, Saudi Arabia

ABSTRACT: Saving water of rainfall and groundwater is one of the prime objectives in the five year development plans for the agricultural and civil societies in the highlands in the south west of Saudi Arabia. Preliminary investigation of two sites and their abutments was conducted. This study include geological, structural, geotechnical and hydrogeological aspects of the proposed sites across the valley. Rock mass classification and engineering geology map were prepared. The study aims to evaluate the foundation of the abutment of each location, finally to choose the best location suitable to be site for construction of he dam. Furthermore, vertical and horizontal permeability of the valley deopsits was determined, in addition to grain size analysis of the wadi deposits conducted in locations along the valley. Special consideration for the effect of the presence of discontinuities on the rock masses was taken. Assembly of these information was put together to choose the suitable site for dam.

RÉSUMÉ: La conservation des eaux de précipitation et des eaux souterraines est parmi les objectives essentielles des plans de développement à cinq années pour les societés agricoles et civiles à la montagne de sud-ouest de l'Arabie Saoudite. Les examens preliminares de deux sites de barrage et de leur aboutements ont été effectués. Cette étude inclut des aspects géologiques, structurels, géotechniques, et hydrogéologiques pour les sites proposés. Une classification du massif rocheux et une carte de géologie de l'ingénieur ont été préparées. Cette étude a pour but d'examiner l'abutement de chaque site et finalement de choisir le site le plus acceptable pour le barrage. La perméabilité verticale et horizontale des depôts de vallée a été déterminée, en addition de la granulométrie. Des considerations speciales pour l'effects de la presence des discontinuités du massif rocheux ont été adoptées. L'assemblement de toutes ces informations a aidé le choix du site de barrage.

1 INTRODUCTION

The purpose of this study is to do a preliminary investigation of the abutments of two proposed dam sites. This investigation include a geological, structural, geotechnical and hydrogeological aspects of the sites, in order to evaluate the foundation conditions of the abutments at each location and to choose the best site for dam construction.

1.1 *Location*

The study area is located in the south central part of the Arabian Shield in Saudi Arabia about 60 km north east of Al-Baha city in the Al-Aqiq quadrangle (Greenwood, 1975) between Latitudes 20° and 20° 30′N, and Longitudes 41° 30′ and 42° E (Figure1).

The studied sites are located in wadi Quhlah which is a branch of major wadi in the area, see Figure 2. Wadi Quhlah vary in width and gently sloping eastward.

1.2 *Topography and climate*

The area is generally surrounded by moderately to high mountains, however the relief in wadi Quhlah vary from low to moderate.

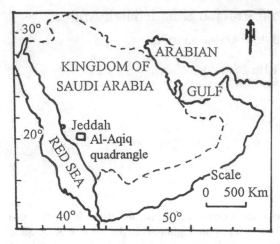

Figure 1. Location of the Al-Aqiq quadrangle.

Temperature in the summer is about 30°c in the morning, and cool during night. The weather is mainly dry and sunny, rotary storms are frequent in the area. In the winter the temperature is about 15°c, rain is frequent and weather is generally coal. Accordingly, this wadi is usually dry in summer and flows in winters.

Vegetation is mainly bushes and small trees.

2 GEOLOGY

The geology of the Kingdom of Saudi Arabia consists of two main units: (i) the basement igneous and metamorphic rocks called the Arabian Shield located along the western coast of the kingdom. The rocks are of the Precambrian era which are mainly Precambrian igneous and metavolcanic rocks, (ii) the sedimentary rocks cover lies at the upper part of the Arabian Shield and expanded eastward and called the Arabian shelf. The age of this unit is from Cambrian to Recent.

2.1 *Geology of Al–Aqiq quadrangle*

Al-Aqiq quadrangle underlined by two assemblages of Precambrian rocks: (1) metabasalt, greywacke, and chert of Baish and Bahah Groups, and (2) metamorphosed andesite to rhyolite and associated coarse clastic rocks of the Jeddah and Ablah Groups. The Baish, Bahah, and Jeddah Groups were folded,

metamorphosed to greenschist facies and intruded by gabbroic to quartz dioritic plutons during the Aqiq orogeny. The Ablah Group rests unconformably on older dioritic and layered rocks. Details are given by Greenwood (1975).

2.2 *Brief geology of the studied area*

The rock units in the study area, as shown in Figure 3, are formed of rhyolite, dolomitic limestone and fine grained sandstone. Rhyolitic rocks are fine grained dark grey intrusion, younger than the limestones and sandstones, Greenwood (1975) presented some details. The valley deposits are consists mainly of sand and gravel. Silt is located on the terraces near the wadi banks.

2.3 *Structure*

Field mapping (Figure 3) show that the area include the following structural units: (1) a number of small scale normal and thrust faults. These faults are of almost E-W or N-S directions and located at the narrowest part of the wadi, some faults in the upstream has different directions. (2) three folds of successive anticline and syncline in rhyolite and dolomitic limestone, plunging generally toward north. (3) the rock masses are intersected by mainly three major joints sets of different attitudes at different locations in the area, plus randomly oriented joints. (4) a shear zone of small scale located in the dolomite limestone about 1 meter width and 5 meters length. (5) a number of andesitic dikes are wide spread in the area. They vary in width from 20 cm to 2 meters, and in length more than 2 meters.

The major joint sets prevailing in the study area are shown below.

3 GEOTECHNICAL PROPERTIES

Due to aim of the study, two proposed sites for locating the dam were choosen. Accordingly, the site investigation program was designed toward the identification of the rock masses properties at those two sites and at the upstream area, in addition to the wadi deposits properties along the valley. The detailed field investigations and measurements followed by laboratory tests were

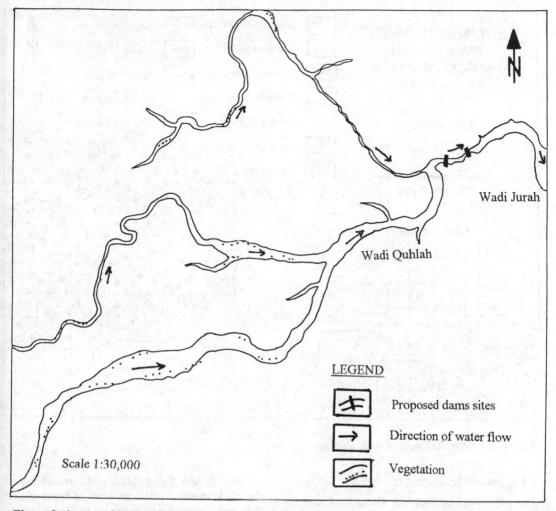

Figure 2. A general layout of the tributaries of wadi Quhlah.

carried out by Tayeb (1993) and Al-Sarrani (1993), and supervised by the first author.

3.1 *Rock masses properties*

The upstream part of the study area was divided primarily into a number of stations. The stations were chosen according to field observations of the change in lithology and engineering parameters. While the proposed dam sites was divided according to the change in the engineering properties of the rhyolitic rock masses, the properties of the rock masses are given as follows:

3.1.1 *State of weathering*

The state of weathering identified by observing the degree of change in color and disintegration on the rock fabric, according to the procedure recommended by the Geological Society of London (1977). The results given in Table 1 show that the state of weathering of the existed rock masses is ranging between W2-W4 in the upstream area (slightly to highly weathered), while mainly W2 (slightly weathered) in the proposed dam sites.

3773

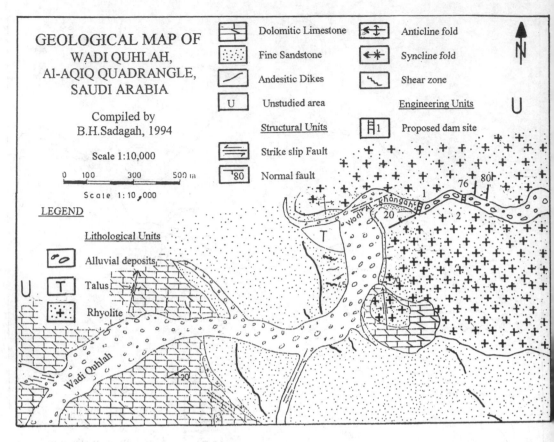

Figure 3. Geological map of wadi Quhlah.

3.1.2 Uniaxial compressive strength, UCS

The uniaxial compressive strength of various rock masses was estimated by using (i) Schmidt hammer (L-type). The recommendations of the ISRM Commission (1978) has been followed to identify the rebound number, R, for the rock masses (see Table 1), and (ii) the point load test on lumps was conducted to calculate the Is_{50} (see Table 1).

Rock cores of NX size has been drilled out of representative rock blocks of the existing rock types, in order to determine the UCS and elastic moduli from the stress-strain curve for each core sample. The average readings of 3 to 7 tested core samples at every station is given in Table 1.

3.1.3 Rock mass discontinuities

The spacing between the adjacent discontinuities largely control the size of the rock blocks, which in turn govern the stability and permeability of the rock masses. The intensity of fracturing of the rock masses, as introduced by Deere (1964) as Rock Quality Designation (RQD), was measured along a scanline in the field in three dimensions X, Y and Z. Where X-direction is parallel to the wadi axis, and Y-direction is perpendicular to the wadi axis while Z-direction is vertical measurement along the exposed rock face slope. The length of the scanline was 50 times the average spacing between the joints planes, as suggested by Priest and Hudson (1981). The average 3-D RQD readings given in Table 1 show that the joints planes in the upstream area are generally closely spaced, according to the terminology given by the Geological Society of London (1977). This is due to the bedding nature of the sedimentary rocks. However, were igneous rocks are located, the joints spacing was widely spaced and

Table 1. Some of the geotechnical properties of the existed rock masses.

Station No.	Rock type	Weathering state	Av. R No.	Av. Is$_{50}$, MPa	Av.UCS, MPa	Av. RQD	Av. Joint Roughness Coefficient	Aperture, mm
1a	rhyolite	moderately	29.8	2.19	30.6	64.0	11.0	2-6
1b	rhyolite	slightly	60.2	24.50	342.8	94.2	8.2	0-2
2a	limestone	slightly	71.7	16.20	226.4	98.3	10.6	0-2
2b	rhyolite	slight.-mod.	56.6	9.50	133.0	64.9	5.8	0-2
3	sandstone	slightly	35.7	5.44	76.5	62.7	6.6	0-2
4a	rhyolite	slight.-mod.	41.5	15.10	210.6	25.0	3.8	2-6
4b	sandstone	slightly	58.3	12.24	213.5	38.3	5.8	2-6
5	limestone	slightly	61.8	8.80	122.0	93.0	8.6	2-6
6	limestone	mod.-highly	46.1	13.50	146.0	19.0	4.6	2-6
7	limestone	slightly	53.6	9.00	125.7	96.3	10.6	6-20
8	limestone	slightly	52.5	6.50	87.1	58.3	12.2	2-6
9	limestone	slightly	49.7	8.10	166.3	76.0	13.4	6-20
10	sandstone	moderately	51.5	6.70	94.0	95.5	8.6	2-6
11	sandstone	slight.-mod.	44.8	12.30	172.0	90.0	10.2	0-2
13	sandstone	slight.-mod.	29.8	30.60	30.0	64.0	5.8	2-6
14	rhyolite	slight.-mod.	67.4	24.50	340.0	57.5	7.2	0-2
15	rhyolite	slight.-mod.	67.4	11.00	256.8	57.5	7.4	1.5
16	rhyolite	slightly	67.5	9.20	232.8	80.7	7.4	3
17	rhyolite	slightly	68.2	7.60	201.7	99.7	7.8	5
18	rhyolite	slightly	61.6	7.10	213.7	100.0	7.0	2
19	rhyolite	slightly	62.5	11.30	270.0	99.4	6.2	2.5
20	rhyolite	slightly	66.4	10.20	239.0	99.6	7.8	3.5
21	rhyolite	slightly	64.9	10.60	246.0	100.0	6.2	2
22	rhyolite	slightly	66.1	9.82	232.6	84.7	5.0	4
23	rhyolite	slightly	59.8	8.10	342.8	97.0	8.2	5

moderately wide spaced in the proposed dam site 1 and 2, respectively.

The aperture of the different rock masses show that the igneous rocks has slightly open aperture than the sedimentary rocks. In general, most of the apertures in the area are tight (see Table 1).

Joint Roughness Coefficient (JRC) was measured by using the roughness gauge, where it was placed along the joint surface. Five profiles were traced along each joint plane then using Barton and Chouby (1977) chart to find the corresponding JRC value for each profile. Average of the five JRC readings were calculated and given in Table 1.

The roughness angles were measured along each JRC profile. The average of the roughness angles of each station is given in Table 2.

Fracture spacing of the discontinuities, related to different joints sets, is moderately to closely spaced (Geological Society of London, 1977) in the upstream area, whereas it is generally widely spaced to extreme wide spaced in the proposed dams sites (see Table 2). This is mainly due to the presence of the igneous rocks.

3.1.4 Rock mass classifications

The geomechanics classifications of Barton et al. (1974), and Bieniawski (1974) were applied, in order to identify the quality of the rock masses in the study area. Table 2 shows the quality of the rock masses in the upstream area and the proposed dam sites.

Table 2. Discontinuities geotechnical properties and quality of the   rock masses according to the geomechanics classifications.

Station No.	Rock type	Fracture spacing, cm	Roughness angle, i°	R system rating	Rock Quality	Q system rating	Rock Quality	Rock mass zone
1a	rhyolite	17	16	56	fair	0.63	v. poor	xiv
1b	rhyolite	26	21	75	good	9.71	fair	xv
2a	limestone	40	25	85	v. good	16.58	good	ix
2b	rhyolite	9	12	53	fair	16.58	good	xi
3	sandstone	4	12	78	good	5.70	fair	i
4a	rhyolite	4	9	39	poor	0.00	ex. Poor	xvi
4b	sandstone	3	14	53	fair	2.75	poor	ii
5	limestone	37	18	83	v. good	14.55	poor	v
6	limestone	3	10	37	poor	0.35	v. poor	viii
7	limestone	33	22	83	v. good	2.38	poor	v
8	limestone	19	25	65	good	2.23	poor	vii
9	limestone	11	27	63	good	0.72	v. poor	vi
10	sandstone	31	17	82	v. good	3.61	poor	iii
11	sandstone	14	23	62	good	0.93	v. poor	iv
13	sandstone	15	9	64	good	4.72	fair	i
14	rhyolite	9	18	58	fair	12.70	good	xi
15	rhyolite	9	18	78	good	14.38	good	x
16	rhyolite	14	18	77	good	20.16	good	x
17	rhyolite	60	35	82	v. good	32.38	good	xvii
18	rhyolite	200	16	97	v. good	150.00	ex. good	xviii
19	rhyolite	26	15	85	v. good	74.33	v. good	xii
20	rhyolite	41	19	85	v. good	33.10	good	xiii
21	rhyolite	21	14	80	good	15.83	good	x
22	rhyolite	17	14	72	good	21.18	good	x
23	rhyolite	17	16	82	v. good	12.15	good	xii

3.1.5 *The engineering geological map*

The description of the existed rock masses and discontinuities characteristics given in Tables 1 and 2 were studied thoroughly, in addition to the ratings of the rock mass classifications. The engineering geological map shown in Figure 4 was then constructed. The map show that the rock masses are divided into 18 zones of different rock quality. The best rock quality according to the zoning is given number 18.

3.2 *Wadi deposits*

The wadi deposits are formed of unconsolidated detrital material. It represents a heterogeneous assemblage of cobbles, gravels, sand, silt and clay. The distribution of these granulometric facies varies in the wadi both in the vertical and horizontal directions. Generally these deposits exhibit a net bimodality, poor sorting and both negative and positive skewness. Figures 5 and 6 represents an example of a grain size distribution curve for the wadi deposits.

3.2.1 *Vertical and horizontal permeability measurements*

Samples were obtained from seven sites within the study area (Figure 4) at depths of 50, 70, 80 and 90 cm from the ground surface. Permeability measurements were carried on these samples using a constant head permeameter. The results of these measurements are shown on Table 3.

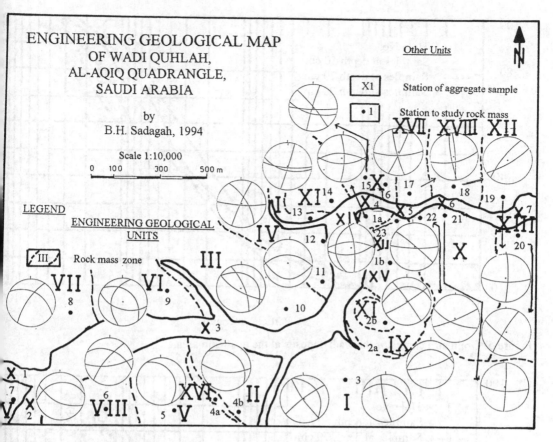

ENGINEERING GEOLOGICAL MAP
OF WADI QUHLAH,
AL-AQIQ QUADRANGLE,
SAUDI ARABIA

by
B.H. Sadagah, 1994

Scale 1:10,000
0   100   300   500 m

LEGEND
ENGINEERING GEOLOGICAL
UNITS

Other Units

X1   Station of aggregate sample
• 1   Station to study rock mass

III   Rock mass zone

Figure 4. Engineering geological map of the rock mass zonation in the study area.

Table 3. Values of vertical permeability (K_z) and horizontal permeability (K_x).

Station No.	Horizontal permeability, K_x cm/sec	Vertical permeability, K_z cm/sec
1	0.71	0.25
2	0.18	0.09
3	0.58	0.57
4	24.88	0.22
5	0.37	0.29
6	0.65	0.49
7	1.02	0.51

Using

$$K_x = \frac{z_1 k_1 + z_2 k_2 + ... z_n k_n}{z_1 + z_2 + ... z_n} \quad (1)$$

or the calculation of the horizontal permeability

(K_x) and

$$K_z = \frac{z_1 + z_2 + ... + z_n}{\frac{z_1}{k_1} + \frac{z_2}{k_2} + ... + \frac{z_n}{k_n}} \quad (2)$$

for the calculation of the vertical permeability (K_z). For the details of the above equation refer to Bouwer (1978).

The vertical permeability in these wadi deposits varies from 0.09 to 0.57 cm/sec and the horizontal permeability ranges from 0.18 to 24.88 cm/sec.

4 Discussion and conclusions

The above study show that the area of study in need to conserve rain water. The geological, geotechnical, and hydrogeological conditions of the rock masses in the upstream and dam sites

3777

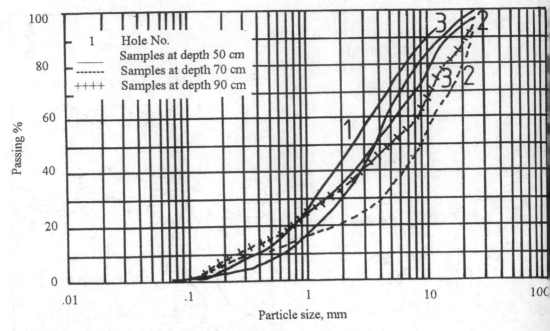

Figure 5. Grain size curves of the wadi deposits at the upstream area.

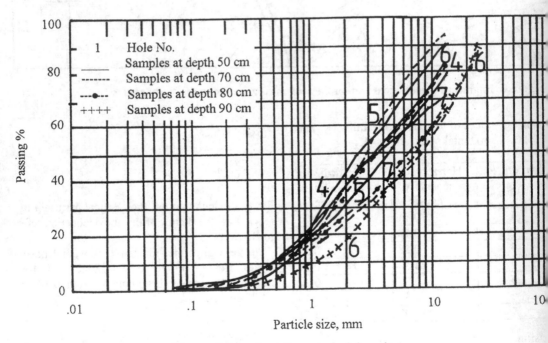

Figure 6. Grain size curves of the wadi deposits at the proposed dam sites.

are suitable for constructing the dam. Chosen the best site for dam construction was based on the geomechanics classifications ratings, in addition to the other discontinuities characteristics. The intensity of fracturing plays the optimum role in choosing the best dam site. The intensity of fracturing and joints aperture should be as low as possible. The RQD value could be misleading, therefore in order to identify the percentage of different quality regions of the rock masses in the selected dam sites, the newly introduced method by Sadagah and Şen (1992) to recognize the distribution function of the block size (BSDF%) is recommended to be used. According to the above study, site number 2 was chosen as the best site for locating the dam due to the above reasons.

REFERENCES

Al-Sarrani, A. S. 1993. *Site investigation for dam site.* A graduation project. Faculty of Earth Sciences, King Abdulaziz University, Jeddah, Saudi Arabia:98p.

Barton, N., R., Lien and J., Lunde 1974. Engineering classification of rock masses for the design of tunnel support. *Norwegian Geotechnical Institute Publication.* 106: 48.

Barton, N. R. and V. , Choubey 1977. The shear strength of rock joints in theory and practice. *Rock Mechanics.* 10:1-54.

Bieniawski, Z.T. 1974. Rock mass classification in rock engineering. *Proceedings of the Symposium on Exploration for Rock Engineering, Johannesberg* 1·97-106..

Bouwer, H. 1978. *Groundwater hydrogeology.* Mc Graw Hill.

Deere, D.U. 1964. Technical description of rock cores for engineering purposes. *Rock Mechanics and Engineering Geology.* 1:16-22.

Geological Society of London, 1977. The description of rock masses for engineering purposes. Report by the Geological Society Engineering Group Working Party. *Quaternary Journal of Engineering Geology.* 10:355-388.

Greenwood, R.W. 1975. *Geology of Al-Aqiq quadrangle,* Kingdom of Saudi Arabia. Ministry of Petroleum and Mineral Resources. GM23: 15p.

International Society for Rock Mechanics Commission on standardization of laboratory and field tests 1978. Suggested methods for the quantitative description of discontinuities in rock masses. *International Journal of Rock Mechanics and Mining Sciences & Geomechanics Abstracts.* 15:319-368.

Priest, S.D. and J.A. Hudson 1981. Estimation of discontinuity spacing and trace length using scanline surveys. *International Journal of Rock Mechanics Mining Sciences & Geomechanics Abstracts.* 18:183-197.

Sadagah, B.H. and Z. Şen 1992. Block size distribution and quality classification in naturally fractured rocks. *Bulletin of the Association of Engineering Geologists.* 29:in press.

Tayeb, O.M. 1993. *Geotechnical properties for rocks and soils in wadi quhlah.* A graduation project. Faculty of Earth Sciences, King Abdulaziz University, Jeddah, Saudi Arabia: 82p.

Proserpina dam (Mérida, Spain): An enduring example of Roman engineering

Le barrage de Proserpina (Mérida, Espagne): Une construction romaine durable

M.Arenillas
Universidad Politécnica, Madrid, Spain

J.Martín
Confederación Hidrográfica del Guadiana,
Mérida, Spain

R.Cortés
Universidad Politécnica, Valencia, Spain

C.Díaz-Guerra
Ingeniería 75, S.A., Madrid, Spain

ABSTRACT: The Proserpina Roman dam was built to supply water to the Colonia Augusta Emerita (Mérida) which dates back to the year 25 B.C. Therefore the dam is around two thousand years old. Recently several studies and works have been carried out aimed at the restoration of this construction. When the silt which had partially filled the reservoir was removed, seven metres of the lower part of the structure, which had previously been covered, were exposed. Consequently it can be said that the Proserpina dam, which rises more than 21 metres above its foundations, is one of the most important works of Roman hydraulic engineering remaining in Spain.

RESUME: Le barrage romain de Proserpina fut construit afin de pourvoir à l'approvisionnement en eau de la colonie Emerita Augusta, fondée en l'an 25 a.C. Il a donc une ancienneté de l'ordre de 2.000 ans. Récemment, différents travaux et études ont été menés à bien afin de rehabiliter cette construction. Aprés avoir retiré les limons qui remplissaient en partie le réservoir, on a découvert les sept mètres inférieurs de la structure, méconnus jusqu'à ce jour, et qui n'avaient pas pu être observés amparavant à cause des couches de sédiments. Par conséquent, le barrage de Proserpina avec plus de 21 métres de hauteur sur fondations, s'avère être un des ouvrages de génie hydraulique romain les plus importants conservé en Espagne.

1 INTRODUCTION

The Proserpina Roman dam, some 5 km to the NNW of Mérida, forms a small reservoir in the stream of Las Pardillas, a tributary of the Aljucén river which in turn empties into the Guadiana (fig. 1).

The dam, located on the Mérida granite, exploits a slight narrowing of the valley opened by the intersection of two fractures. The granites supporting the dam display marked mineralogical variations from one area to another. In the central zone we find a granite of two micas, medium grained with very evident feldspar crystals in sizes ranging from milimetres to centimetres and rose coloured. Superficially they appear relatively altered and are generally very rounded in form. The fracturation is not very pronounced, predominating in the directions N10°-20° and N100°. Frequently these fractures are filled with seams of quartz.

The reservoir receives water from its own catchment basin (8.5 km²) and from an adjoining one (15.2 km²), whose flow is diverted along a canal which is also of Roman origin. Thus the territory supplying water to the dam covers a total area of 23.7 km² (Table 1).

Table 1. Principal statistics of the construction

Present Name: Presa de Proserpina	
River: Arroyo de las Pardillas	
Township: Mérida	
Basin:	
Local	8.5 km²
Outside	5.2 km²
Reservoir:	
Volume	5.0 hm³
Area	70.0 ha
Dam:	
Height above foundations	21.6 m
Crest length	427.8 m
Date of construction: 1st century B.C.-2nd century A.D.	

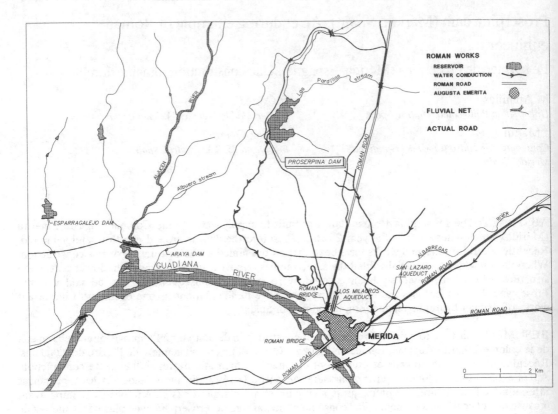

Fig. 1. The surroundings of Mérida

The dam has been functioning up to the present day with normal interruptions due to damage incurred by the passage of time. During this long period the reservoir's water has been used for various different purposes.

Originally the reservoir formed part of the water supply system for the 'Colonia Augusta Emerita' (Mérida), founded in the year 25 B.C. Water reached the city along an 8.5 kilometre long canal whose numerous remains are still visible. The most remarkable are those of the aqueduct of "Los Milagros" built across the river Albarregas in the vecinity of Mérida. However, other applications of the water cannot be overlooked. In Roman times water was also used to power mills, some of which were located inside the city. At the foot of the dam remains of another of these mills can be found.

Likewise there is evidence of its use for irrigation purposes. From the 17th century on, the water was mainly used in a wool wash-house, situated immediately downstream from the dam. In the last decades the Proserpina reservoir has provided an important source of recreation for the inhabitants of Mérida and nearby villages, which as a result has caused some problems in the upkeep of the reservoir.

There is no doubt that such a dam, which has been in almost continuous service for two thousand years, must have sustained serious damage and subsequent repairs.

The earliest reference to this issue dates back to 1633 when Moreno de Vargas tells of serious repair work carried out on the dam in 1617; more reconstruction work of a less important nature took place in 1698, 1791 and 1865.

At the start of this century the serious state of deterioration of the dam and the canal from the adjoining basin provoked successive projects by the hydraulic administration (División Hidráulica del Guadiana). The first one in 1910 was drafted by the engineer Pascual de Luxán and although it was never put into operation it served as a basis for the following plans of José de Castro in 1931 and for those of Raúl Celestino in 1936, which were finally put into practice. The resulting

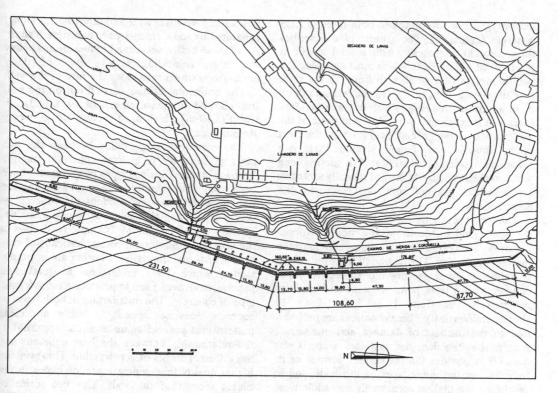

Fig. 2. Proserpina dam: ground plan

reconstruction work took place in the forties and involved the starting up of the dam's supply canal and repair work on the dam itself. Later two series of cement injections were carried out on the body of the dam (1944-45 and 1970).

However the failure of the deepest outlets, together with the eutrophication of the reservoir's water and a rapid deterioration in the surrounding area caused the Confederación Hidrográfica del Guadiana in 1991 to initiate a series of activities aimed at rehabilitating the dam and the reservoir.

After pumping out the water stored in the reservoir, a new outlet was built in a lower area on the left side of the dam (called 'La Sangradera'). Then silt which covered the lower part of the reservoir was removed. Moreover studies were begun to determine the measures needed to carry out structural repairs and archeological restoration on the dam in addition to a general improvement in the conditions of the surrounding area. The removal of the sediments revealed a new, previously unknown part of the structure whose characteristics will be mentioned later.

At the same time as the silt was being removed,

drilling and scraping work was carried out in order to determine the structure's characteristics and the condition of the foundations. To achieve this, initially nine holes were drilled from the crest of the structure (one of them at an angle) and another four were drilled upstream, two of which were horizontal and two at an angle of 45°. In the same way various scrapings were made near the downstream face of the wall and in the downstream embankment. Subsequently the Servicio Geológico de la Dirección General de Obras Hidráulicas has carried out a new series of drillings in the masonry wall and the downstream embankment to complete the investigation.

2 CHARACTERISTICS OF THE DAM AND THE FOUNDATIONS

The Proserpina dam consists of a masonry wall joined downstream to an earth embankment, which is very wide in the central section of the structure and relatively smaller in the two side sections. The wall consists of two granite faces made up of ashlar work, dressed stone or masonry, with a core between both faces of lime concrete (calicanto). The ground plan describes three straight alignments, with a total length at the crest of 427.80 m, to which must be added the twelve metres of a small wall built on the right side fifty years ago (fig. 2). On the left side there is, in addition, an auxiliary wall of some hundred metres in length which is used to seal some of the sections where the ground level is below the crest of the dam. In one of these sections there is a rectangular opening of almost five metres long and two metres high which until recently served as an overflow channel and is known as 'La Sangradera'.

The lower seven metres of the upstream face of the wall (which constitute the newly exposed masonry) are vertical and the remaining fourteen metres are sloping. The incline of this area, which is achieved by staggering the successive stone courses, varies from point to point and oscilates between values from 1 in 2, to 1 in 10 (horizontal/vertical). These variations are probably due to the amount of damage and number of repairs that the dam has sustained since it was built. It is obvious that at certain periods of its long history the outer layer of this wall, and in particular the highest area, easily accessible from the crest, has been used as a 'quarry' for other works in the surrounding area.

There are nine buttresses joined to this wall which are distributed at irregular intervals along the central section of the dam. Eight of these originate from the lower section of the wall (which has recently been exposed) and only one (the nearest to the right extreme) remains outside this central area. The first eight are vertical at the oldest section of the wall and from this level upwards they extend to the crest at a slightly less steep angle than that of the wall. This is also achieved by staggering the successive courses. In the lower section the eight buttresses finish off in a semicircular form at around four and a half metres away from the wall. In the upper part which is staggered, all the buttresses are rectangular in form.

The downstream face is covered by the embankment which almost reaches the crest. In spite of this, by drilling and scraping it has been established that the downstream wall is vertical at different points. Probably this is the case along the whole length of the structure. The aforementioned explorations have also led to the detection of sixteen buttresses in the central section of the dam. These are vertical masonry structures 1.4 metres wide and three metres along, separated by some six metres between each shaft. All of these buttresses extend to two metres below the crest, coinciding with a step of thirty centimetres which runs along the face of the wall.

The horizontal drilling carried out on the wall has allowed a measurement of its foundation width (5.90 m) which must correspond to twenty Roman feet.

The core (calicanto) filling between the two faces of the wall is a 'lime concrete' (mainly Roman) in which various types of aggregates have been used. The samples extracted by drilling the wall show that in some sections the concrete is very compact and thus relatively non-permeable. The aggregates used in this concrete (with sizes in the order of centimetres) consist basically of very hard schists and quartzites. In other areas granite stones, which have undergone a weathering process, have been used to produce a more fragile type of concrete. The distribution of both types of concrete does not appear to follow any clear pattern thus preventing an accurate appraisal of the relationship between the lime concrete and the different stages of construction. However the higher quality lime concrete predominates in the oldest section of the wall. The last series of explorations carried out have demonstrated that the cement injections performed in 1944-45 and in 1970 only succeeded in improving the condition of the Roman concrete in localised areas of the wall.

The dam has two water intake towers ('bocines'), joined to the downstream face of the wall, thus contained within the embankment. The water intakes are situated at three different levels. The two lowest ones correspond to the tower located in the deepest part of the dam while the third one is connected to the tower nearest the left side. Both 'bocines' have undergone some serious modifications and so many repairs that it is difficult to determine their original forms.

The main tower displays an irregular but practically square cross-section whose sides oscilate between 5.5 and 6.0 m approximately. Its present inside depth is 18 m.

The lower water intake of this 'bocín' consists of two lead pipes undoubtedly of Roman origin. These measure 22 cm in diameter inside and are 1.5 cm thick and are located at about three metres above the foundations. The intermediate intake is situated almost four metres higher up and, judging by its characteristics, dates from more recent times, possibly from the repairs carried out in 1698. The inlet is formed by a circular hole 25 c

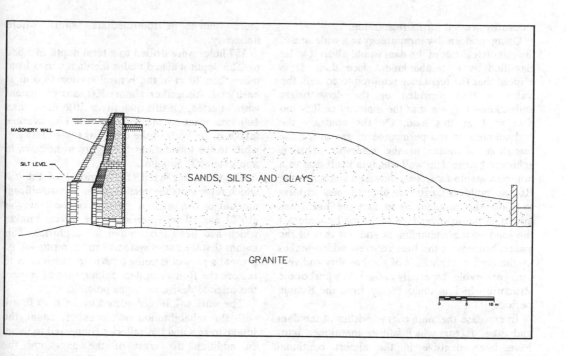

MASONERY WALL

SILT LEVEL

SANDS, SILTS AND CLAYS

GRANITE

Fig. 3. Proserpina dam: cross plan

in diametre cut into a slab of granite which has been embedded rather carelessly in the oldest part of the wall.

The shortest tower has an intake consisting of a cast iron tube of 40 cm in diameter, installed in the forties to replace the Roman one. This intake is especially interesting as it is the only one at a sufficient level to allow water to reach Mérida from the reservoir along the aqueduct of 'Los Milagros'. The tower is square in cross-section, 7 m wide on the outside and 11.5 m high.

Although today it is in disuse, the upper intake was used until only a few years ago, unlike the middle and lower ones which had not functioned for centuries as they were obstructed by the silt which partially filled the reservoir.

The downstream embankment is some 30 m wide at the crest, sloping down-river at a gradient of around 2 in 1 (horizontal/vertical). In the central section of the dam the foot of the embankment is supported by a wall of granite of around 4 m in height (fig. 3).

The embankment which has had to be reconstructed on various occasions, shows two clearly distinguishable areas: the lower part is formed by silts similar to those extracted from the reservoir, while the higher levels are made up of

slightly clayey sands which are the product of alterations in the granite. The whole embankment is in general quite permeable.

The foundations of the dam rest on a granite base on the verge of a small marshy pool. The idea of increasing this natural reservoir was probably what inspired the Roman builders to place the dam here. As a result of the form of the verge, the wall is not straight in alignment.

The drilling explorations carried out have demonstrated the poor condition of the foundations at some points of contact between the structure and the terrain due to the weathering of the granite on which the dam is resting. Most of these alterations must have existed before the dam was built as, when the foundations were laid, only the outer layer of the weathered granite had been removed. Therefore the structure was built on altered rock.

3 RESTORATION WORK

The emptying of the reservoir, the removal of the silt and the subsequent explorations carried out have allowed a quite detailed analysis of the structure and have also helped to determine measures needed for its restoration.

Using modern-day terminology in a wide sense, the characteristics of the dam would allow it to be classified as an embankment face dam. It is evident that the resistance required to contain the water is that provided by the downstream embankment, given that the masonry wall is too slim to bear this load. On the contrary the embankment is too permeable to guarantee the retention of water in the reservoir. This is achieved by the dam wall which in itself acts as a non-permeable face. Two thousand years ago the Roman builders must have applied these criteria to solve the problem as in fact the buttresses joined to the up-river face of the wall counteract the load of the embankment and not that of the water. Moreover, the lime concrete which makes up the core of the wall is of good quality and very non-permeable, especially in the lower part of the structure which is, undoubtedly, from the Roman period.

In any case the main characteristics of the dam and other factors which will be mentioned later have been decisive in the almost continual functioning of the structure over a period of more or less two millennia. However, its present state viewed from present-day criteria demands that very specific restoration work be carried out.

On the one hand it is evident that the foundations require reinforcement to be able to afford adequate support to the structure, particularly in the section corresponding to the base of the wall.

On the other hand, the removal of the silt which had filled the reservoir (approximately 800,000 m³) has also meant the removal of a protective barrier against leakages and uplift, which must be replaced by another protective system. Something similar occurs with the deficiences detected in the lime concrete which constitutes the core of the wall, and whose homogeneity and non-permeability must be assured as far as possible.

All of these problems have led to the development of a grouting plan, carried out recently, which has succeeded in water-proofing the core of the dam wall, strengthening the structure's contact with the terrain and forming a non-permeable curtain along the whole length of the dam.

The cement was injected from the crest of the dam through holes drilled at three metre intervals in the first phase. After checking the results of the first phase the work was completed by injecting more cement at intermediate points, where necessary.

157 holes were drilled to a total depth of 3,880 m. The depth reached under the dam varies from more than 20 m in the central section to 6 m at each end. Altogether almost 300 tons of cement were injected. Of this quantity a little more than 100 tons went to the foundations. The average intake in the body of the dam was some 100 kg/m while in the foundations this figure went down to approximately 55 kg/m.

On the other hand, the removal of the silt which represents the recuperation of a significant amount of the original capacity of the reservoir (16%) has allowed access to the lowest intakes which had been inoperative for centuries. This means that the outlet systems can be improved. At present a project is being drawn up which aims to restore the Roman outlets, taking care to respect the original design as far as possible.

The work will be completed in the near future with the rehabilitation of the supply canal, the intake towers and the galleries connected to them. In addition the crest of the dam and the embankment will be remodelled and the surrounding area tidied up.

4 THE PAST, PRESENT AND FUTURE OF THE PROSERPINA DAM

The Proserpina dam must have been built, or at least started, around the year 25 B.C. practically at the same time as the 'Colonia Augusta Emerita' was founded.

According to information presently available the Roman engineers must have designed a dam of similar dimensions to the existing one although it is possible that it was built in two stages. Without going into detail about these possible stages of construction, which have been analised elsewhere, it remains clear that the dam was already in full operation from the first half of the second century A.D. (in the times of Trajan or Hadrian) and was supplying water to Mérida along the Los Milagros aqueduct. The C1-method, used to establish the date of a wooden plug found in the silt of the reservoir, confirm these estimations.

Throughout its history the dam has been adapted to the needs of each period and must have seen moments of greater and lesser activity. But the unquestionable fact is that it has survived to the present day serving its original purpose that of a reservoir. As a result the Proserpina dam (and that of Cornalvo, also in Mérida and dating from the same period) can be regarded as exceptional structures and perhaps unique examples of Roman hydraulic engineering.

The reasons for its survival to the present day seem clear. On one hand the almost systematic use of the structure has ensured that it has been constantly maintained without being allowed to suffer irreversable damage. On the other hand the fine characteristics of the downstream embankment, the load-bearing capacity of the granite foundations, sufficient in spite of the alterations, and the practical non-existence of serious floods (the dam is located on a tributary with a very small catchment basin) have guaranteed the stability of the structure since its construction.

In spite of everything, the leakages through the dam wall must have been continuous although not very serious, considering the general high quality of the lime concrete core. In addition the embankment, sandy in the upper levels and silted in the lower levels, must always have had good drainage towards downstream, thus avoiding substantial increases in the load the dam wall had to bear. Due to this factor and to the contribution of the buttresses on the upstream side, the wall has been able to counteract the pressure of the embankment without suffering serious damage when the reservoir was empty. This kind of pressure is usually what causes breakages in dams of this type. With a full reservoir the embankment must have been vital, given that the wall, even with its downstream buttresses, is too slim to support the pressure of the water without breaking.

In view of these circumstances and with the help of the restoration work taking place it seems clear that the future of the ancient Proserpina dam is assured for a long time to come.

REFERENCES

Alcaraz Calvo, A., Arenillas Parra, M. y Martín Morales, J. 1993. La estructura y la cimentación de la presa de Proserpina. *IV Jornadas Españolas de Presas*. Comité Nacional Español de Grandes Presas. p. 719-733. Murcia.

Alvarez Martínez, J.M. 1977. En torno al acueducto de Los Milagros de Mérida. *Publicaciones Eventuales*, nº 27. p. 49-60. Barcelona.

Alvarez Martínez, J.M. 1981. En torno a algunos aspectos de la fundación de Augusta Emerita. *Revista de Estudios Extremeños*, p. 155-161. Mérida.

Alvarez Sáenz de Buruaga, J. 1976. La fundación de Mérida. *Simposium Internacional Conmemorativo del Bimilenario de Mérida*. p. 19-32. Mérida.

Arenillas Parra, M., Martín Morales, J. y Alcaraz Calvo, A. 1992. Nuevos datos sobre la presa de Proserpina. *Revista de Obras Públicas*. 3311: 65-69. Madrid.

Arenillas Parra, M., Díaz-Guerra Jaén, C. y Cortés Gimeno, R. 1992. *La presa romana de Proserpina* (Mérida). Confederación Hidrográfica del Guadiana. Mérida. (Unp).

Cantó, A.Mª. 1980. Sobre la cronología augústea del acueducto de Los Milagros de Mérida. *Homenaje a Sáenz de Buruaga*. p. 157-176. Mérida.

Celestino Gómez, R. 1980. Los sistemas romanos de abastecimiento de agua a Mérida. Estudio comparativo para una posible cronología. *Revista de Obras Públicas*, Madrid.

Confederación Hidrográfica del Guadiana-Ingeniería 75, S.A. 1992. *Estudio de actuaciones para la rehabilitación de la presa romana de Proserpina*, Mérida. (Unp).

Fernández Casado, C. 1972. *Acueductos romanos en España*. Madrid.

Fernández Casado, C. 1983. *Ingeniería hidráulica romana*. p. 125-135. Madrid.

Jiménez Martín, A. 1976. Los acueductos de Emerita. *Simposium Internacional Conmemorativo del Bimilenario de Mérida*. p. 111-125. Mérida.

Moreno de Vargas, B. 1633. *Historia de la ciudad de Mérida*, 2nd reed (1974). p. 77-90. Badajoz.

Creep modelling of an endangered slope adjacent to Wusheh Dam, Taiwan

Modèle de fluage relatif à un versant instable au réservoir de Wusheh, Taiwan

C.W.Yu & J.C.Chern
Geotechnical Research Center, Sinotech Engineering Consultants, Inc., Taiwan

ABSTRACT: Creep of a heavily weathered and altered slaty rock mass in a steep slope near Wusheh Reservoir is considered to be the major cause of slope instability. Drilled samples obtained in the creep zone were tested to evaluate the creep characteristics of the rock mass. A visco-elastoplastic creep model was incorporated in a computer code to analyze the time-dependent movement of the slope, and the analytical results were compared with the field monitoring data. Various remedial support measures were studied for their effectiveness in stabilizing the endangered slope.

RÉSUMÉ: La présence de roches schisteuses très altérées sur une pente fortement inclinée au réservoir de Wusheh est considérée comme le principal facteur d'instabilité et de lent glissement de ce versant. Les échantillons obtenus par forage dans cette zone ont été testés en laboratoire pour évaluer les propriétés de glissement des roches. Un modèle visco-elastoplastiques de filuage a été incorporé dans un programme informatique destiné à analyser le mouvement du versant en fonction du temps. Les résultats ont été comparés avec les données mesurées sur le terrain. L'efficacité de plusieurs méthodes de stabilisation de versant a été testée.

1 INTRODUCTION

Wusheh Reservoir is located at about 4 km the up stream of the conjunction of Wanda and Drawshui creeks, which serves as a part of Sun Moon Lake hydraulic scheme in central Taiwan. Since its completion in late 1950's, the slope along a road connecting the reservoir and the power plant became a constant problem for the reservoir administration. The locations of the endangered slopes are shown in Figure 1.

Since 1960's, several stages of stabilization work including installing tendons, surfacial shotcrete protection, retaining wall, etc. have been implemented. But these remedial measures failed to stabilize the endangered slopes. Continued movement in endangered slope no.3 has been monitored since the completion of the latest stabilizing work in 1987.

2 SITE DESCRIPTION

2.1 Geology

The rock formation in the reservoir area belongs to the early Miocene strata. It consists mainly of dark grey slate, which is more or less phyllitic and the slaty cleavage is well developed. The rock mass forming the endangered slope is highly weathered, especially in the shallow depth due to the high relief of the slope. The rock mass can be rated as very poor in surfacial zone to fair to good rock mass in the deeper strata according to Bieniawski's rock mass classification system.

2.2 Behavior of the endangered slope

The last slope stabilization work was carried out during 1985 to 1987. It included additional pre-stressed tendons and surfacial shotcrete protection. A monitoring system including inclinometers and load cells were also istalled to assess the effectiveness of these remedial measures and the long term stability of the slope. The layout of the remedial works and the instruments are shown in Figure 2.

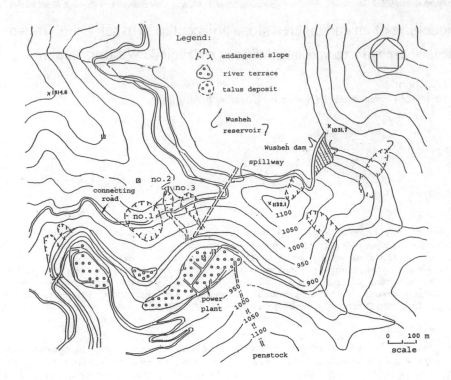

Fig.1 Locations of the endangered slopes.

endangered slope no.3

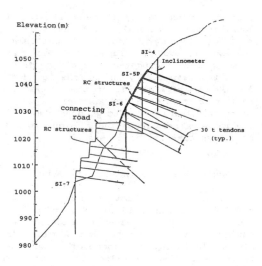

Fig.2 Layout of the remedial works and the
instrumentations.

Since the completion of the stabilization work in 1987, no significant movements in slopes no.1 and no.2 were observed. However, no.3 slope still underwent continued creeping movement. The movements of the slope as measured by inclinometers are illustrated in Figure 3. Over the years, creeping movement in the surfacial layer with total value over some 30 cm has been recorded and the local rainfall appears to play an important part in accelerating the slope movement. In the current investigation project , geophysical survey, geologic drilling, sampling and rock mechanics testing were carried out. From the results of slope stability analysis using the rock mechanics properties obtained, the slope still maitains a proper margin of safety (Sinotech, 1992). Based on the results of these investigations, it was concluded that the mechanism of the long term slope movement is associated with creep behavior in the highly weathered surfacial layer and rainfall.

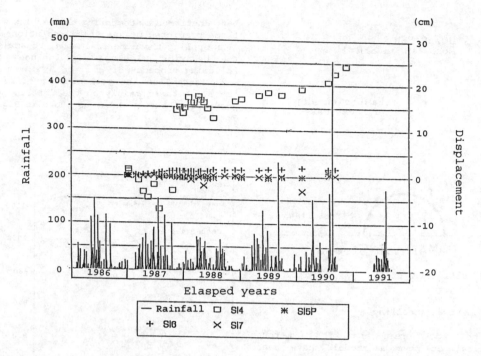

Fig.3   Inclinometer measurement for endangered slope no.3

3 CREEP MODELLING

3.1 Sampling and testing

To model the creep characteristics of the rock mass, sampling and testing on rock samples recovered from drilling were firstly conducted. This provided the basis for selecting the proper creep model for further analysis. The HX size specimens selected from borehole drilling for testing were highly fractured, clay containing slaty materials. The specimens were firstly subjected to isotropic confining pressures corresponding to the in-situ state of stress, and sustaining deviatoric stresses were then applied. The deviator stress started from a low value and increased incrementally until sample failed. Typical test results are shown in Figure 4.

From the test results, it is apparent that the rate of creep is closely associated with the stress state. When the stress/ strengh ratio (SSR) as defined in Figure 5 is less than about 0.5, only elastic deformation would occur. As the SSR is greater than 0.5 and less than 0.8, only primary creep can be

observed. For SSR greater than 0.8, secondary creep becomes apparent.

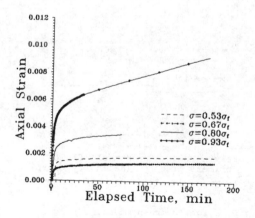

Fig.4   Laboratory creep test results.

3791

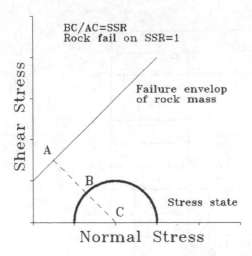

Fig.5   Definition of SSR

3.2 Numerical modelling

For the rock creep analysis, numerous empirical or physical models have been discussed for representing the creep behavior of rocks, e.g., Lama et al(1978), Goodman(1989). Based on the evidence of creep test results obtained in the study, Burger's 4-component viscoelastic model appears to be well-suited for describing the rock creep behavior. The model is schematically illustrated in Figure 6. It is a combination of Kelvin solid and Maxwell fluid, which control the primary and secondary creep respectively.

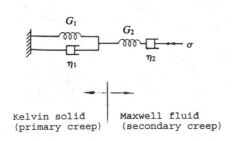

Kelvin solid | Maxwell fluid
(primary creep) | (secondary creep)

Fig.6   Burger's creep model.

The viscoelastic model was incorporated into an elastoplastic computer code, FLAC version 3.22 (Itasca, 1993), and the validity of the newly developed code has

been confirmed by comparing the results with those predicted by analytical solutions.

In appling the numerical model to analyze the behavior of slope, it includes the following procedures:

(1) establish the geological profile of the slope based mainly on the drilling and geophysical survey data, as shown in Figure 7;

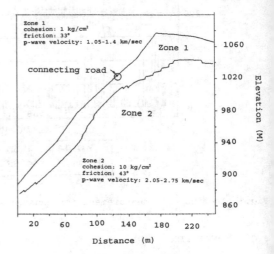

Fig.7   Geological profile for numerical modelling.

(2) evaluate the mechanical properties of the rock masses, such as strength and deformation characteristics, according to test results and empirical approaches;

(3) obtain the in-situ state of stresses of the slope by using elastoplastic model in which the existing supports were incorporated, see Figure 8;

(4) assign creep parameters of rock mass based on in-situ stress calculated and guided by the laboratory creep test results;

(5) calibrate the creep parameters based on the field monitoring data;

(6) predict the future performance of the slope under various remedial support schemes.

4 CASE STUDY

Using the procedures described above, prediction analyses were carried out to examine the effectiveness of various stabilization schemes. These included:

(1) installing 20 m long, 30 ton capacity

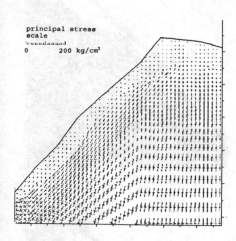

Fig.8 Principal stress states of the endangered slope.

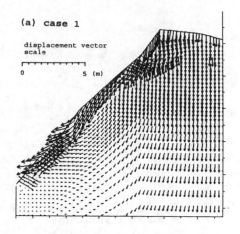

(a) case 1

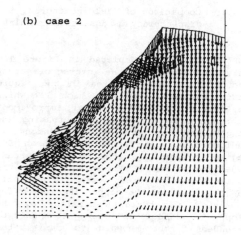

(b) case 2

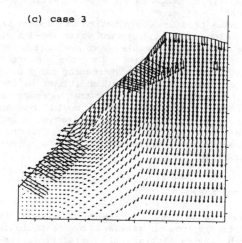

(c) case 3

Fig.9 Predicted movements from analytical results for 3 cases.

tendons;

(2) installing 30 m long, 30 ton capacity tendons; and

(3) installing 30 m long, 60 ton capacity tendons.

The predicted slope movements for the three cases from 1987 till the end of 1990 are shown in Figure 9. The horizontal slope movements at the ground surface of the inclinometer (SI-4) are compared for the 3 cases and illustrated in Figure 10.

In case (1), which corresponds to the actual support installed in 1987, the slope movements were used to calibrate the creep parameters. The trend and magnitude of creep movement can be modelled by the method quite well. It may also be seen that the support work could not prevent the slope from creeping movement. The reason may be due to the fact that the tendons were not anchored in the stable deep strata. In case (2), the length of the tendons were incresed to 30 m, an improvement in the slope stability may be noted. But it still can not prevent the slope from long term creep movement. In case (3), the capacity of the tendons was increased to 60 tons. The results show slope movement at early stage due to adjustment in stresses and stabilizing trend with increasing time.

The effectiveness of the support works is mainly due to improved stress conditions in the slope. This may be illustrated by the

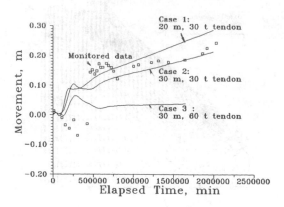

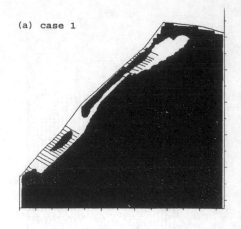

(a) case 1

Fig.10 Comparison of field monitored
displacement and analytical results.

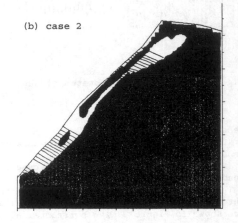

(b) case 2

SSR distributions depicted in Figure 11, which corresponds to the inverse of safety factor. The unshaded area in the figure indicates the region with high SSR which would have high creep potential. Improvement in SSR can be obtained by increasing the length and capacity of the tendons as indicated by the reduction of high SSR region in case (3). However, the current modelling schemes still can not prevent the entire slope from creep movement especially in lower part of the slope. Therefore, for improving the overall stability of the slope, the tendons in lower part should be extended further upward to resist the downward movement in the middle part of the slope.

Aside from the contribution of tendons to the slope stability, ground water level also plays an important role. Deep drainage holes have been proposed in the present stabilization work. By decreasing the ground water level, the stress conditions in the slope would be improved by the increase in effective stresses. This would be an effective way of reducing the potential of creep movement. However, this aspect was not taken into consideration in the creep modelling.

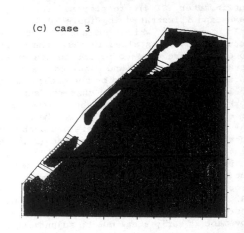

(c) case 3

5 CONCLUSION

Long term creep movement of slope is difficult to be assessed by traditional stability analysis, and, therefore it is

Fig.11 SSR distributions for different
scheme of slope improvement work.

difficult to estimate the support required to stabilize it. The model described in this paper provides a method to estimate the long term creep behavior and to assess the effectiveness of supporting measures quantatively.

From the results of laboratory tests, it appears that creep behavior of rock mass is closely associated with its state of stress. At lower stress/strength ratio, only elastic responce is expected. As the stress/strength ratio increases, primary creep or secondary creep can develop. However, the laboratory test data can not be used directly in estimating the creep parameters of the rock mass due to the complexity of the rock mass in the field, representativeness of the sample tested, etc. They are mainly used to divide the regions of different creep characteristics. Field monitoring data are always required to calibrate the creep parameters in order to assess the over-all performance properly.

6. REFERENCES

Lama, R.D.,Vutukuri, V.S. 1978. Handbook on mechanical properties of rocks. Vol. 3, pp209-323.
Goodman, R.E. 1989. Introduction to rock mechanics, second edition. John Wiley & Sons.
Itasca Consulting Group, Inc. 1993. Fast Lagragian Analysis of Continua, version 3.22.
Sinotech Engineering Consultants Inc. 1992. Report on safety evaluation of Wusheh Dam and appurtenant structures.

Joint opening and head distribution in the foundation rock of the Albigna Gravity Dam

Ouverture des joints et distribution de la charge hydraulique au massif de fondation du barrage poids d'Albigna

K. Kovári & St. Bergamin
Swiss Federal Institute of Technology, Zurich, Switzerland

ABSTRACT: The interaction between joint water pressure and joint opening in the rock foundation of dams may play a crucial role both for seepage loss and stability analysis. A comprehensive research project was carried out in the rock foundation of the Albigna gravity dam focusing on the monitoring of head distribution along a number of long boreholes with the Piezodex sliding piezometer probe and on the measurement of strain profiles with the Sliding Micrometer. The observation of the deformational behavior and the piezometric pressure distribution as a function of seasonal reservoir water level fluctuation over a period of 6 years yielded a wealth of interesting information. In the granitic rock mass two large "active joints" could clearly be identified which open and close due to variation of the reservoir water level. At a given critical water level the opening of the joints becomes specially pronounced leading to a higher water conductivity. A back analysis of the measured potential field for several reservoir water levels indicates the correlation between the changing conductivity of the two particular joints and their opening and closing.

RÉSUMÉ: Les interactions entre la pression d'eau agissant dans les joints des rochers de fondation de grands barrages et l'ouverture de ces joints peuvent jouer un rôle décisif à la fois pour les fuites d'eau et pour l'analyse de stabilité. Un projet de recherche étendu, portant sur le rocher de fondation du barrage-poids d'Albigna, a pour objectif l'auscultation de la distribution de la charge sur toute la longueur des trous de forage utilisant la sonde piézométrique coulissante Piezodex et la détermination de profils de déformation à l'aide du Micromètre Coulissant. L'observation du comportement des déformations ainsi que de la distribution de la pression piézométrique en fonction du marnage pendant une période de 6 ans livre beaucoup d'informations intéressantes. Deux grands «joints actifs», dont le degré d'ouverture varie en fonction du niveau d'eau, ont pu être identifiés clairement. À un niveau d'eau critique donné, l'ouverture de ces joints devient particulièrement marquée, conduisant à une conductivité plus élevée. Un calcul de vérification du champ potentiel mesuré pour différents niveaux d'eau montre une bonne corrélation entre le changement de la conductivité de ces deux joints particuliers et leur ouverture et fermeture.

1 INTRODUCTION

Flow through rock joints plays a central role in a number of rock mechanical problems. The water pressure in fractures can have a decisive effect on the assessment of slope or dam stability (uplift, hydrostatic pressure, seepage forces) and the amount and rate of water flow in the rock mass is a key factor in the evaluation of the suitability of the disposal of radioactive waste or underground storage of oil and gas in unlined caverns.

The water flows through geologic media in the pores and/or along discontinuities (fissures, fractures etc.) in the rock. Because the discontinuities can have a controlling effect on the seepage, the flow trough rock joints is of great practical importance. Unfortunately, rock joints are very difficult to characterise quantitatively in terms of geometry and hydraulic properties, which is in contradiction to the potential accuracy of the numerical models presently used to describe fractured rock masses. Models, depending on the level of sophistication and dimensionality (2D, 3D), require more or less detailed data regarding the hydraulic boundary conditions and permeability as well as the position, direction, extent, distance and width of the joints.

In order to contribute to the clarification of the nature of the seepage through a discontinuous rock mass and to test the applicability of the presently used models, we embarked in 1987 on an extensive measuring campaign in the rock foundation of the Albigna dam (canton Graubünden, Switzerland) which is still ongoing (Arn, 1989; Kovári et al., 1989). The field measurements were complemented by numerical modelling of the results. As the pressure and strain distributions in the foundation of the dam had to be collected at a great number of measuring points, the availability of portable measuring devices was essential for the field program. In the following the principles of the measuring systems used are briefly described.

2 PIEZODEX MEASURING SYSTEM

The distribution of piezometric head along boreholes in rock is measured by the Piezodex system which has no hydraulic coupling between the probe and the liquid in the measuring interval in contrast to earlier developed systems (Kovári & Köppel, 1987).

In order to define a series of measurement intervals, a chain of packers is introduced into the borehole which are all interconnected by a central line and inflated simultaneously with gas or a volumetrically constant mortar under a pressure as required (Fig. 1). Nitrogen is used as the inflation gas because of its very low diffusion through the rubber membrane. The inflation pressure is selected in accordance with the anticipated maximum value of liquid pressure along the borehole. A standard pressure tank is employed for supplying and maintaining the pressure throughout the observation period. Packers of this

type can be reused elsewhere after concluding the measurement program. The position of the individual packers or the location and length of the measurement intervals are chosen in accordance with the fissure system derived from the rock core and can be arranged in irregular intervals. The number of measuring intervals is limited by the maximum length of the packer chain (approximately 120 m), by the minimum length of a single packer (1.0 m) and by the design-related minimum length of a measuring interval (0.50 m).

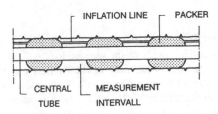

Fig. 1 Interconnected chain of packers defining individual measurement intervals.

To measure the water pressure without any hydraulic coupling between the probe and the liquid, a pressure transmitter is installed in each measuring interval (Fig. 2). It is the purpose of these cells to transmit the liquid pressure to the probe's force sensor during the short period of a reading. The water exerts the pressure via an ultra thin, soft and elastic membrane onto the pressure transmitter in the form of a piston (Fig. 3a). When the probe is in measuring position, the piston is raised slightly (0.1 - 0.2 mm) from its normal position against the membrane (Fig. 3b).

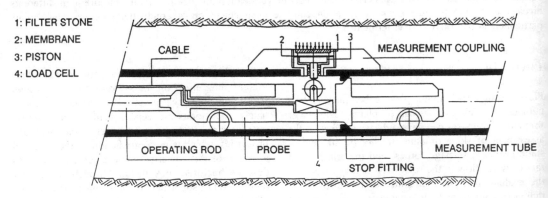

Fig. 2 Piezodex: The sliding probe in measuring position slightly lifting the built in pressure transmitter which is loaded by water pressure (Kovári & Köppel, 1987).

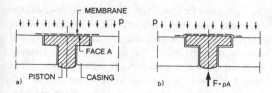

Fig. 3 Pressure transmitter with membrane:
a) Piston in normal position, pressed there by the water pressure.
b) Piston pressed slightly outward by the measured force F exerted on it by the probe in measuring position.

During this phase the water pressure acts directly on the low-friction controlled piston. In this way the pressure is obtained by way of a force measurement. The force is given by the product of the water pressure and the area of the membrane-protected piston surface. By selecting various piston dimensions, it is possible to account for different sensitivities or measuring ranges with the same force sensor. To achieve a high precision of measurement (± 150 - 500 mm water column depending on measuring range), the axis of the force sensor must comply with that of the piston. It was possible to solve this problem of coaxial transmission of force very simply through high-precision positioning of the probe using the cone-and-ball principle, similar to a Sliding Micrometer (Kovári et al., 1979). Conical stops are fitted to the central tube and spherical stops to the Piezodex probe.

A detailed description of the measuring system with measurement examples has been published elsewhere (Kovári & Köppel, 1987).

3 THE SLIDING MICROMETER

If the continuous axial strain profile i.e. strain distribution is measured along a borehole in the rock foundation of a dam, any sharp peaks mark the presence of so-called active discontinuities that open or close as the result of fluctuations of the level of reservoir water.

The sliding micrometer proved to be a suitable instrument for taking precise measurements of the strain profile along a borehole. This is achieved by inserting and grouting in of a hard PVC tube into the borehole. The inside of the tube is equipped in intervals of 1.00 m with annular coupling elements with conical stops which serve to hold the spherical heads of the probe during reading (Fig. 4a). When the two

stops touch, the position of the ball center is uniquely defined in relation to the cone. To ensure that the probe glides in the tube from one measuring station to the next, the ball (probe) and cone (measuring marker) surfaces are notched and designed to let the micrometer to be turned alternately into reading and sliding positions by rotating the probe ± π/4 (Fig. 4b/c).

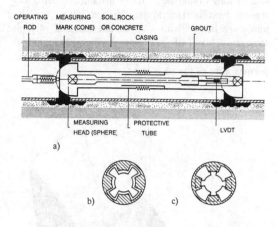

Fig. 4 Sliding Micrometer (Kovári et al., 1979):
a) Schematic view of the instrument
b) Sliding position
c) Measuring position

The movement of neighboring measuring markers toward each other as a result of deformations in the rock is signalled by a change in their spacing, i.e. by the change of two readings. The strain is obtained relative to the length of 1.00 m. With this measuring system, a change of distance between two neighboring marks can be determined with a precision of ± 3 μm. To conduct a complete series of readings, the probe is first secured and the reading taken at the mouth of the borehole. The probe is then moved on until it reaches the bottom of the borehole. Measurements are also taken in reverse direction for control purposes.

A detailed description of the measurement system can be found elsewhere (Kovári et al., 1979).

4 THE ALBIGNA GRAVITY DAM

The Albigna Gravity Dam is situated in the southeastern part of Switzerland at an elevation of 2000 m above sea level. It has a crest length of 759 m and consists of 20 m wide blocks. The largest block in the middle of the valley is 115 m tall. There are 5 m wide hollow joints between the blocks acting as uplift pressure relief openings (the basic design is that of a hollow-joint gravity dam). The storage capacity of the reservoir is 70 million cubic meter of water.

The dam is based on a prominent rock ridge consisting of a sound, coarse-grained granite of high strength. The rock foundation exhibits three regular sets of joints (spacing: 1 to 10 m) and a fourth, less regular near horizontal set of joints (Fig. 5). The joints are healed with a crystalline material of high strength. This is of great importance with regard to the deformation and strength properties as well as permeability of the rock mass.

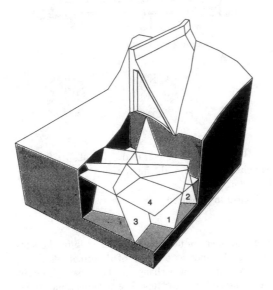

Fig. 5 Position of the main fissure systems in relation to the dam wall (Kovári et al., 1989).

Although the rock foundation was in the design phase considered very favorable, already during the first reservoir filling in summer 1960 some unexpected problems occurred. In continuation of a healed rock joint a clearly visible crack developed in one of the blocks. Over the following years, some other unexpected problems developed, involving increased water loss and a slight increase of permanent deformations. At this point the decision was made to determine the deformation characteristics in the rock foundation as a result of reservoir water level fluctuation. For this purpose six boreholes with a total length of approx. 400 m were drilled and instrumented. The measurements in the boreholes revealed

the existence of a few active fissures (i.e. fissures that open and close as a result of the fluctuation of the reservoir level) thus providing an explanation for the unexpected behavior of the structure. After clearing away the alluvial deposits on the upstreamside of the dam, a 300 m long scarp of joint S was indeed discovered at the foot of the wall. Later, a second joint (L) was discovered through sliding micrometer measurements (i.e. distinct peaks in the strain characteristic). The two joints S and L were directly responsible for the seepage under the central area of the dam (blocks 14 and 16). To prevent water from entering these major joints, a large Neoprene cover was fitted to the foot of the dam on the upstreamside (Fig. 6). The justification of the measures was confirmed by later control measurements (Kovári & Peter, 1983). In course of the investigations detailed information were gathered about the rock mass in the foundation of the dam and its hydraulic characteristics. In order to best utilize resources the drilling and instrumentation of the boreholes was concentrated in one crosssection perpendicular to the crest of the dam. The crosssection selected for observations was located in the hollow-joint between two dam blocks the behavior of which was known from strain measurements in two Sliding Micrometer boreholes with 95 measuring sections since 1980. In spring of 1987, five new boreholes of a total length of 260 m were drilled and instrumented with 39 Piezodex measuring stations.

These new observations were not related to the difficulties mentioned above. They served for a research project aiming at the empirical and theoretical investigations of the percolation of water in jointed rock masses. During 1987 the emphasis of the reading program was on observing the water pressure and strain during the reservoir filling (one measurement per 10 m increase of the reservoir level). Afterwards attention focused on readings at maximum and minimum water levels.

5 MEASUREMENT RESULTS

5.1 Piezodex Measurements

Each of our readings is defined by a specific location, time and exact reservoir level and therefore the data can be presented in a variety of relationships (Kovári et al., 1989; Bergamin & Kovári, 1993). Figure 7 presents the reservoir levels jointly with the hydraulic potentials measured at location PD.2.8 (indicated by a star). The potentials were calculated from the value of the measured pressure and the elevation of the

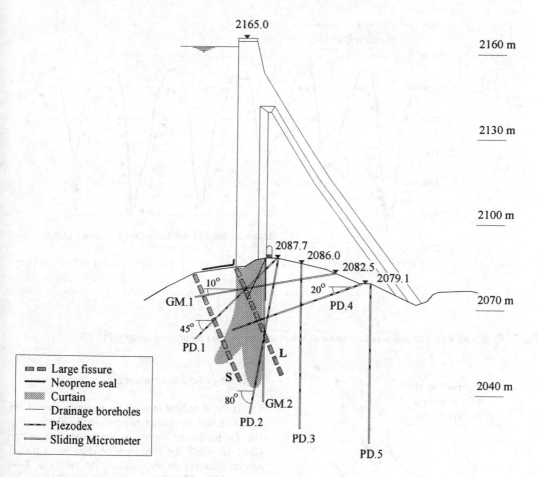

Fig. 6 Crosssection of joint 14 with Piezodex and Sliding Micrometer boreholes, the position of the large active fissures S, L and the fitted neoprene seal.

measuring points. The time intervals between measurements varied; for the study of the long term behavior, from June 1989 on, we limited ourselves - with the exception of location PD.2.8 - to readings of pressure and strain at minimal and maximal reservoir levels.

If the time of measurement is omitted, the relationship between the reservoir level and observed pressure can be presented as in Figure 8. It is apparent from the figure that the slope of the reservoir level vs. pressure curve markedly changes above the reservoir level 2135 m. A plausible explanation for the above average increase of the measured values is to assume an abrupt change of permeability in the rock mass due to opening of joints. Furthermore, the relationship demonstrates a hysteretic pressure response to the cyclic change of the reservoir level (Fig. 8), i.e.

higher pressures are measured at decreasing than at increasing reservoir levels.

Although the piezometric measurements are obtained for a specific point, they have to be attributed to the full length of the zone between the packers isolating the measuring interval. The depth distribution of the pressure at the maximum reservoir level (2160 m) is presented in Figure 9. The two continuos lines limiting the measurements represent the hydrostatic pressures along the borehole as they would exist due to a water column in an borehole without packers and open to the atmoshere (i.e. not fitted with packers) and due to the full water column of the reservoir. Figure 9 shows that (1) considerable pressure differences exist between individual stations, and (2) pressures close to maximum prevail in the deepest section of the borehole. In order to demonstrate

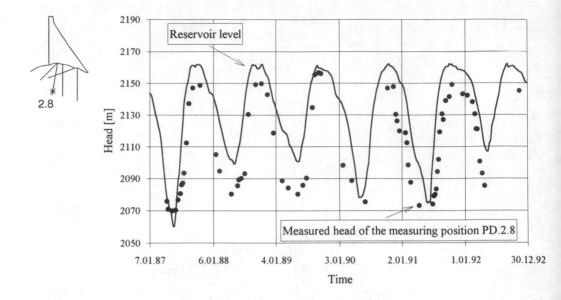

2.8

Fig. 7 Reservoir level and measured pressure as a function of time for measuring section PD.2.8.

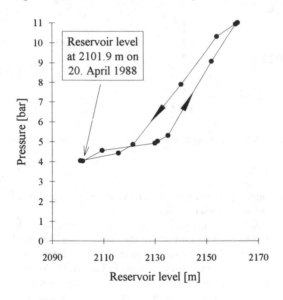

Fig. 8 Pressure development in measuring section PD.2.8 as a function of reservoir level (April 88 to May 89).

the good repeatability of the measurements, we plotted the range of pressure values observed at maximum reservoir levels at various times. The differences between the measurements are insignificant.

5.2 Sliding Micrometer Measurements

The strain in mm/m in a plane perpendicular to the dam axis was measured along two boreholes (Fig. 10). Of particular interest is the measuring station GM.1.46 where the extension reached more than 3 mm in response to increase of the reservoir level from 2103 to 2161.70 m above s.l.. In the borehole GM.2 extensions in the rock mass can be observed up to a depth of 50 m. The two peaks at 5 and 25 m indicate the presence of two active joints intersecting the borehole. However, for the interpretation of the measurements it is important to realize that the measured strain is not identical with the width of the observed joints. Firstly, only the strain component parallel to the borehole is captured during measurements; secondly, it is uncertain whether the particular section is intersected by one or more joints; thirdly, the initial width of the joint at the time of the in situ calibration of the instrument is unknown.

Figure 11 shows the extension in the measuring interval GM.1.46 (active joint S) as a function of the reservoir level. Two facts should be noticed in this figure: (1) the sharp turn in the curve at a reservoir level of 2130 m and (2) the greater extension at decreasing than at increasing reservoir levels.

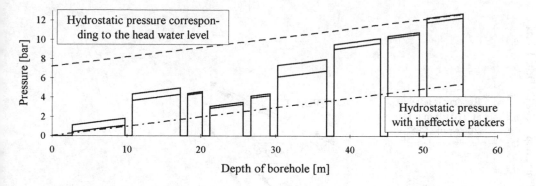

Fig. 9 In borehole PD.2 measured pressures as a function of borehole depth with range of readings at a reservoir level of 2160 m (bottom line: hydrostatic pressure in the borehole with water and open to atmosphere; top dashed line: hydrostatic pressure corresponding to the head water level).

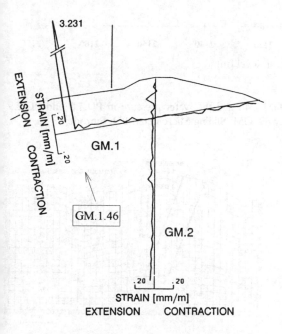

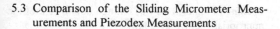

Fig. 10 Strain profiles along two boreholes in the measurement plane with a rise in the reservoir level from 2103 m to 2161.7 m (Kovári et al., 1989).

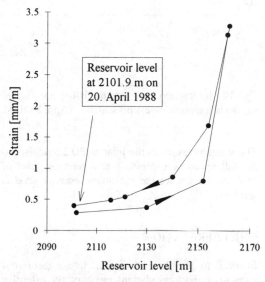

Fig. 11 Strain development at the active joint S (measuring position GM.1.46) due to reservoir level fluctuation (April 88 to May 89).

5.3 Comparison of the Sliding Micrometer Measurements and Piezodex Measurements

For purposes of comparison the pressure and strain measurements from the piezodex boreholes PD.1 and PD.2 and sliding micrometer section GM.1, which were taken during the same time period, are presen-

ted in a dimensionless form in Figure 12. All the sections considered are in the range of the active joint S. The straight line at PD.1.7 indicates that the joint here is permanently open and reflects the hydrostatic pressure corresponding to the reservoir level. The dimensionless pressure and strain curves PD.2.9 and GM.1.46 show in agreement a sharp turn when the reservoir level reaches 2130 m. It is worth pointing out that the rate of extension of the joint at GM.1.46 is slower than the rate of pressure increase at PD.2.9.

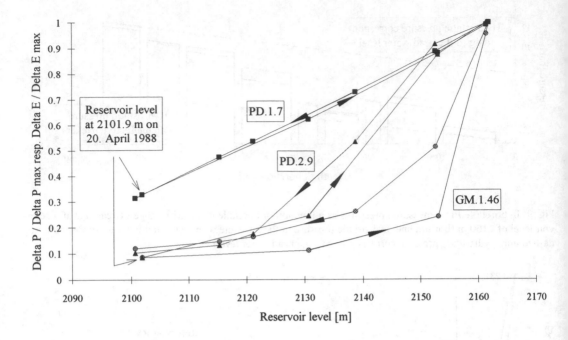

Fig. 12 Interaction between the fissure opening in GM.1.46 and the water pressure in PD.1.7 and PD.2.9 shown in a dimensionless presentation (April 88 to May 89; GM: Sliding Micrometer, PD: Piezodex).

The water pressure in the joint at PD.2.9 adjusts to the full hydrostatic pressure at a reservoir level of 2150 m, i.e. before the maximum reservoir level is reached.

6 INTERPRETATION

In order to put the observations into a theoretical framework, backcalculations involving the extensive database were carried out with a finite element model (Kovári et al., 1989). The objective of the modelling was to numerically reproduce the measured potentials corresponding to specific reservoir levels for various model assumptions.

6.1 Description of the models

For the simulations the AQUA-ROCK finite element model developed by Th. Arn (1989) was used. The model allows the simulation of the three-dimensional flow in a porous medium and in one- or two-dimensional rock mass structures. However, the model is not designed to handle coupled hydro-mechanical interactions, such as increase of permeability due to

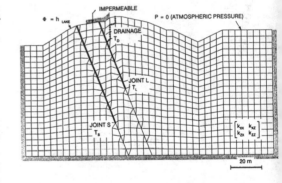

Fig. 13 Geometry, boundary conditions and network division of the finite element model inclusive of two large active fissures S and L (Kovári et al., 1989).

opening of joints. This was omitted due to lack of the appropriate constitutive laws.

In accordance with the observational program which assumed a two-dimensional flow in a plane, the backcalculations were done for a slice of unit thickness perpendicular to the axis of the dam. In the other two dimensions (length and depth) the slice was extended sufficiently far to assume no flow boundary conditions and negligible influence on the potentials in the range of the piezodex boreholes. The flow through the surface of the rock mass on the upstream side of the dam was defined to be driven by the potential corresponding to the reservoir level. An eight meters broad strip of the surface adjacent to the reservoir side of the dam was considered impermeable along with parts of the rock-concrete contact which was sealed with a neoprene seal. The geometry, boundary conditions and flownet (800 right angle elements) used in the calculations, are shown in Figure 13 along with the two major joints S and L.

The main types of models (Arn, 1989) used were selected according to the description of the rock mass. The dam foundation was modelled as:

1.- a homogeneous isotropic continuum.

2.- a rock mass with two major zones of evenly distributed bundle of fractures extending into infinite (Snow, 1965) representing joints S and L and a series of minor low permeability fractures perpendicular to the main zones. The increase of the fracture width at full reservoir levels is simulated by increasing the permeability coefficient of the main fracture zone.

3.- a homogeneous isotropic medium with two major joints and drainage D. The transmissivity of the upstream joint S could be varied, which allowed to consider the permeability variation due to opening/closing of the joint S (indicated by the sliding micrometer measurements) in response to the reservoir level fluctuation.

In addition, for each type of model, parameter variation studies were performed using three different reservoir levels (2162 m, 2135 m, 2101 m).

6.2 Results

The simulation results are plotted jointly with the water pressure measurements (presented in the trapezoid form) along the piezodex boreholes in Figures 14 - 16. This form of presentation has the advantage that the differences can be directly located. In Figures 14 - 16 the water pressures calculated with each type of model are compared to the measured values at maximum reservoir levels.

The result of the direct comparison can be summarized as follows:

Fig. 16 Homogenous, isotropic continuum with the large fissures S and L in comparison with the measurements with the reservoir level at 2162 m (Kovári et al., 1989).

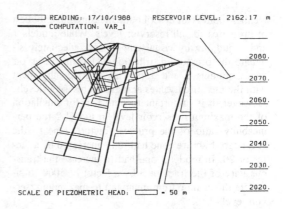

READING: 17/10/1988 RESERVOIR LEVEL: 2162.17 m
COMPUTATION: VAR_1

SCALE OF PIEZOMETRIC HEAD: ⊏⊐ = 50 m

Fig. 14 Homogeneous, isotropic medium (main model 1) in comparison with the measurements with the reservoir level at 2162 m (Kovári et al., 1989).

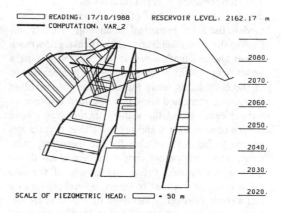

READING: 17/10/1988 RESERVOIR LEVEL: 2162.17 m
COMPUTATION: VAR_2

SCALE OF PIEZOMETRIC HEAD: ⊏⊐ = 50 m

Fig. 15 Continuum with infinitely persistent fissure systems (main model 2) in comparison with the measurements with the reservoir level at 2162 m (Kovári et al., 1989).

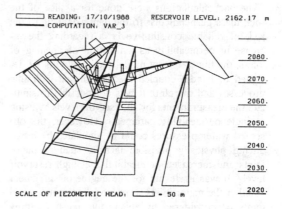

READING: 17/10/1988 RESERVOIR LEVEL: 2162.17 m
COMPUTATION: VAR_3

SCALE OF PIEZOMETRIC HEAD: ⊏⊐ = 50 m

Neither model can describe the measured distribution of pressures for all reservoir levels. While models 1 and 2, judging by visual comparison, adequately simulate the pressure distribution at minimal reservoir levels, models 2 and 3 provide a good agreement with the measured values at maximal reservoir levels. However, it is important to note that for simulation of the maximum reservoir level in model 2 the permeability ratio of the primary fracture zone to the secondary fracture zone had to be increased by a factor of 20. In model 3 one had to increase the transmissivity of the fracture S by a factor of 5000 to simulate the change from minimum to maximum reservoir level.

7 CONCLUSIONS

The interpretation is based on both an extensive field data base and the results of three types of rock mass models used for numerical simulation of the water flow in the dam foundation. Of particular importance is that the calculations could be checked against a broad range of the reservoir level fluctuations at one of the boundaries while the conditions at the other boundaries remained constant. The field instruments - the piezodex and the sliding micrometers - have proven their accuracy and precision over 6 years and proved to be most suitable for the measuring campaign. The observations confirmed beyond any doubt the importance of opening and closing of the two major joints in response to reservoir level changes - a fact already previously identified by sliding micrometer measurements - for explanation of the water seepage in the foundation of the Albigna dam. However, it were the later piezodex measurements that proved the direct hydraulic consequences of the deformations of the joints, i.e. the change of permeability due to activation of the joints.

The backcalculations were done for a slice of the rock mass perpendicular to the dam axis. Due to the lack of reliable constitutive laws regarding the increase in permeability as a function of opening of joints, the finite element model used could not be designed to handle coupled mechanical and hydraulic processes and therefore it was not possible to simulate the measurements for all reservoir levels by using a single permeability parameter. However, the observed water pressures could be simulated by introducing physically plausible factors, such as major joints and increased permeabilities at high reservoir levels. It was irrelevant for the degree of agreement whether the major joints were modelled as individual joints or considered by anisotropic tensors (Snow type of fractures bundle extending into infinite); already a simple extension of the homogenous, isotropic porous media model provided satisfactory results.

8 ACKNOWLEDGMENTS

The research project was partly financed by the Commission for Promotion of Scientific Research, Swiss Economics Department, and the National Co-operative for the Storage of Radioactive Waste (Nagra). The project was supported in addition by the Engineering Office for Structural Works of the City of Zürich (IBA). The authors would like to thank persons involved for giving this research project their valuable assistance.

9 REFERENCES

Arn, Th. 1989. Numerische Erfassung der Strömungsvorgänge im geklüfteten Fels. Dissertation, ETH Zürich, Institut für Bauplanung und Baubetrieb, Fels- und Untertagebau, Mitteilung 1/89.

Bergamin, St. & Kovári, K. 1993. Staumauer Albigna: Resultate der Piezodex- und Gleitmikrometermessungen. ETH Zürich, Institut für Geotechnik, Professur für Untertagbau (not published).

Kovári, K., Amstad, Ch. & Köppel, K. 1979. New developments in the instrumentation of underground openings. 4th Rapid Excavation and Tunneling Conference. Atlanta, USA, June 79.

Kovári, K., Arn, Th. & Gmünder, Ch. 1989. Groundwater flow through fissured rock: field investigations and interpretation in the Albigna dam area, Graubünden, Switzerland. Nagra Technical Report 90-10. Baden, Switzerland, November 1989.

Kovári, K. & Peter, G. 1983. Continous strain monitoring in the rock foundation of a large gravity dam. Rock Mechanics and Rock Engineering, Vol. 16 (3), p. 157 - 171.

Kovári, K. & Köppel, K. 1987. Head distribution monitoring with the sliding piezometer system "Piezodex". Field Measurements in Geomechanics. Kobe, Japan, April 87.

Snow, D.T. 1965. A parallel plate model of fractured media, Ph. D. Thesis, University of California, Berkeley.

Performance of the Foz do Areia and Segredo concrete face rockfill dams

Comportement des masqués en béton des barrages en enrochement de Foz do Areia et Segredo

Pedro L. Marques Filho
Intertechne Consultores Associados, Curitiba, Brazil

Paulo Levis
Companhia Paranaense de Energia, COPEL, Curitiba, Brazil

ABSTRACT:The concrete face rockfill dams of Foz do Areia (160 m high, comissioned 1980) and Segredo (145 m high, comissioned 1992) were built with basaltic rocks. Behaviour was excelent. However, the measured deformations were quite large, when compared with similar structures built with other rock types (such as quartzites in Cethana dam, hornfels and diorite in Alto Anchicaya dam, sandstones and siltstones in Mangrove Creek dam) or with compacted gravel dams (Salvajina and Golillas dams). This paper compares the behaviour of these different rockfill materials and the design and construction features that guarantee similar and adequate performances, in spite of distinct deformation characteristics. It also analyses the foundation conditions of Foz do Areia and Segredo dams, and presents a more general criteria of plinth design extended to poorer foundation conditions.

RÉSUMÉ: Les barrages en enrochement avec masque amont en béton de Foz do Areia (160 m d'hauteur, commissionée en 1980) et de Segredo (145 m d'hauteur, commissionée en 1993) ont un remblai de roches basaltiques. Leur tenues ont etées excelentes. Cépendent, les enrochements basaltiques son plus deformables, en comparaison avec des enrochements construits avec d'autres tipes de roches (par example des quartzites dans la barrage de Cethana, des cornubianites et dioritcs en Alto Anchicaya, des arenites e siltites en Mangrove Creek) ou avec des remblais de gravier utilisées dans les barrages de Golillas et Salvajina. Dans ce rapport, on fait une comparaison entre ces differents tipes d'enrochements, et une analise des mesures de projet et construction qui garantisent une bonne tenue, meme en utilizant des enrochements differents. Dans ce rapport, on analise, aussi, les conditions de fondations rencontrées en Foz do Areia et Segredo, ainsi qu'un critère general de projet pour des fondations de qualité inferieur.

1 INTRODUCTION

The behaviour of concrete face rockfill dams (CFRDs) depends mainly of three basic features:

1. the deformability of the rockfill;

2. design and construction measures adopted for the plinth, perimetric joint and transition materials to counterbalance deformability;

3. foundation quality in terms of permeability and erodibility.

Concerning the two first ones, Foz do Areia and Segredo are good examples of dams built with rockfill materials of low modulus of compressibility, where adequate design and construction cares assured a good performance.

Concerning the third feature, both dams were built on good foundations, but in Foz do Areia the occurrence of weathered and pervious seams required extensive treatments. Present trend is to accept poorer foundation conditions, and some comments concerning a more general criteria for the design of the cut-off slab, based on a foundation classification, is also presented.

2 BASIC FEATURES OF FOZ DO AREIA AND SEGREDO DAMS

Foz do Areia (160 m high, comissioned 1980) and Segredo (145 m high, comissioned 1992) were built by Companhia Paranaense de Energia, COPEL, in the Iguaçu river, in southern Brazil.

Detailed descriptions of these dams are presented in Pinto et al., 1982, 1985 and 1993 and Marques Filho et al., 1985.

Both dams were built with basaltic rocks, the first one with massive basalts, amygdaloidal basalts and basaltic breccias, and the second one almost entirely with very sound massive basalts.

3 ROCKFILL DEFORMATIONS AND GENERAL PERFORMANCE

3.1 Deformations measured at Foz do Areia and Segredo dams

Deformations measured at Foz do Areia and Segredo dams were considerably larger than those observed in rockfill dams built with other types of rock or with compacted gravel. In Foz do Areia, maximum settlements measured 4.0 m (construction plus impounding), and the deflections of the concrete slab reached 0.78 m. In Segredo, corresponding values measured 2.20 m and 0.35m, respectively.

In less deformable CFRDs of comparable height, such as Alto Anchicaya (Colombia) (Materon 1985) and Cethana (Australia) (Fitzpatrick et al 1985), the first one built with hornfels and diorite, the second one with quartzite, and the gravelfill dams Golillas (Amaya and Marulanda 1985) and Salvajina (Hacelas et al. 1985) (Colombia), the maximum slab deflections measured, respectively, 0.160 m, 0.174 m, 0.160m and 0.160 m. Maximum settlements in Cethana, Golillas and Salvajina measured approximately 0.45, 0.39 and 0.30 m.

3.2 Leakage

The performance of CFRDs has been historically evaluated in terms of leakage. In Foz do Areia, the leakage immediately after reservoir filling was 240 l/s, subsequently reduced to 50 l/s. In Segredo, an initial peak of about 400 l/s decreased very rapidly to 50 l/s

(Pinto et al. 1993). In Cethana, the infiltrations measured 7 l/s and in Salvajina about 60 l/s, so that the values obtained in Foz do Areia and Segredo may be considered within normal limits, in view of the much greater size of these dams.

In Alto Anchicaya, in spite of the better rockfill properties, the initial leakage measured 1,800 l/s, later on reduced to 180 l/s. In Golillas, where general deformations were very small, the inicial leakage peaked 1,080 l/s, subsequently reduced to 650 l/s. In both dams, the reservoir had to be emptied to permit remedial treatments.

3.3 Rock strenght, rockfill granulometry and compressibility moduli

Main features of Foz do Areia and Segredo dams concerning rockfill properties, deformations and behaviour are listed in table 1, together with data from other modern CFRDs.

The deformation characteristics of a rockfill are related with granulometric distribution after compaction, strenght of the rock, shape of the rockfill particles, thickness of the layers, energy of compaction, and use (or not) of water during compaction.

Since the last three factors are more os less standardized in modern CFRDs, the main cause of distinct deformational behaviour usually rely on the first three parameters.

The mechanism of deformation is related with the breakage and crushing of contact points between rock fragments, and fracturing and crushing of the rock fragments themselves. This effect is obviously reduced when particles have a round shape, so that gravelfills deform very little.

The effect of rock strenght is obvious. However, both in prototypes and laboratory tests, this effect is somewhat hidden by the superposition of the granulometric features.

In a poorly graded rockfill, the stresses at the contact points are very high and there are voids enough between the particles so that the breakage effect is not considerably reduced during construction. In this case, even the more resistant rock types may result in deformable rockfills.

The deformation characteristics of a rockfill are usually expressed in terms of modulus of

Table 1. Construction and behaviour parameters of CFRDs

N°	Dam	Comp.	H	Rock Type	L	γe	γr	e	Cu	Ev	Et	D	O	S	T	Leakage (l/s)
1	Foz do Areia (Brazil)	1980	160	Massive Bas./Bas. Breccia	0.8	2.12	2.81	0.33	6	37/56	88	775	25	55	25	236/60
	Foz do Areia (Brazil)		160	Massive Bas./Bas. Breccia	1.6	1.98	2.51	0.27	14	26/29						
2	Segredo (Brazil)	1993	145	Massive Basalt	0.8	2.12	2.91	0.38	7	62	145	35	-	-	-	400/50
	Segredo (Brazil)		145	Massive Basalt	1.6	2.01	2.88	0.44	10	37						
3	Wilmot (Aust)	1970	35	Graywacke	1	2.20	2.70	0.23		115	160	25				<1
4	Cethana (Aust)	1971	110	Quartzite	0.9	2.10	2.65	0.26	21	145	310	174	12		7.5	7
5	Paloona (Aust)	1971	38	Clayey Chert	1	2.00	2.65	0.33		75	115	40	0.5	5.5		<1
6	Serpentine (Aust)	1972	39	Quartzite	1	2.10	2.65	0.26		115	95	25	1.8	5.3		
7	Mackintosh (Aust)	1981	75	Graywacke	1	2.20	2.70	0.23		40	95	155	4.8	2	2.8	14
8	Murchison (Aust)	1982	94	Rhiolite	1	2.30	2.70	0.17		225	650	44	12	9.6	7	2
9	Tullabardine (Aust)	1982	26	Graywacke	1	2.20	2.70	0.23		90	170	7		0.7	0.3	4
10	Bastyan (Aust)	1983	75	Graywacke	1	2.20	2.70	0.23		160	300	58	4.8	21.5		7
11	Lower Pieman (Aust)	1986	122	Dolerite	1	2.30	2.80	0.24		160/80	224		7	70		
12	Winneke (Aust)	1980	85	Sound Siltstone	0.9	2.05	2.65	0.29	59	51/59			9	19	24	58
	Winneke (Aust)			Weathered Siltstone	0.5	2.05	2.60	0.27								
13	Alto Anchicaya (Col)	1974	140	Hornfels/Diorite	0.6	2.29	2.80	0.22	14	145		160	125	100	15	1800/180
14	Mangrove Creek (Aust)	1981	80	Sandstone/Siltstone	0.45	2.21	2.70	0.22	400	60						14.2
	Mangrove Creek (Aust)			Weathered Sard./Silt.	0.45	2.06	2.37	0.31		100						
15	Shiroro (Nigeria)	1984	125	Granite	1	2.23	2.67	0.20	38	76		90	30	>50	21	1800/100
16	Fortuna (Panama)	1982	60	Andesite	1					33		40				
17	Khao Laem (Tailand)	1984	130	Limestone	1	2.10	2.70	0.29		50/40		125	5	8		53
18	Kotmale (Sri Lanka)	1984	90	Charnoquite	1	2.20	2.80	0.27		66			2	20	5	
19	Golillas (Col)	1984	125	Gravel	0.6	2.25	2.65	0.18		162	160					1080/650
20	Salvajina (Col)	1984	148	Gravel	0.6	2.24	2.80	0.25	50	393	40	40	9.1	19.5	15.4	60
	Salvajina (Col)			Soft Sand./Silt.	0.9	2.26	2.74	0.21		54						

H - Height of dam (m)
L - Layer thickness (m)
γ e - Unit weight of the rockfill (kN/m³)
γ r - Unit weight of the rock (kN/m³)

e - Void ratio
Cu - Coefficient of uniformity
Ev - Constructive modulus (MPa)
Et - Transversal modulus (MPa)

Joint movements (mm):
O - Opening
S - Settlement
T - Tangential

compressibility. Settlement measurements during construction give "constructive or vertical moduli" (Ev), and settlements during impounding give "water-load or transversal moduli" (Et). Representative values measured in Foz do Areia, Segredo and other dams are shown in table 1 and figure 1, where the much poorer parameters obtained in the two first ones are quite evident.

rocks, and do not improve very much even when changing the blasting methods.

The water load moduli are greater than the construtive moduli, figure 1. They obviously depend, like the construtive moduli, of the type of rock. These factors affect directly the deformation of the concrete slab during impounding, and naturally the deflections measured in Foz do Areia and Segredo dams were greater than those found in other dams of higher compressive modulus, as shown in table 1 and figure 2.

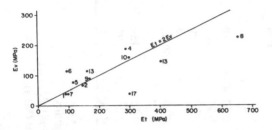

Figure 1. Ev x Et

The gradation of a rockfill is evaluated in terms of void ratio and coefficient of uniformity. Void ratios close to 0.20-0.25 indicate a rockfill of low deformability, and a coefficient of uniformity greater than 15 indicates low void ratios.

When using weaker rocks, the crushing effects during compaction may result in a rockfill material that is so dense and massive that it's deformation features may result better then the ones found in a rockfill made up of stronger rocks, as in Mangrove Creek dam, Australia (Mackenzie and MacDonald 1985).

3.4 Parameters and deformations measured at Foz do Areia and Segredo dams

The basaltic rocks used in Foz do Areia and Segredo dams had very high compressive strenghts, around 150 MPa. However, the granulometric parameters after compaction resulted quite poor, void ratios around 0.30 - 0.40, coefficients of uniformity around 6 - 8, table 1.

The poorer granulometric parameters of basaltic rockfills seem to be typical of these

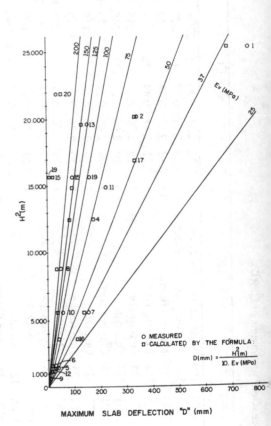

Figure 2. Slab deflections x height of dam

3.5 Joint movements

However, the deflection of the concrete slab, important as it may be, is not the critical feature concerning leakage.

Leakage in CFRDs are mainly caused by excessive movements along the perimetrical joint, which may locally damage the waterstops.Other causes are, of course, leakage through the foundation or through cracks or other types of defects in the concrete slab itself, somewhat less common and not considered in the present analysis.

Figure 3 shows how the movements of the perimetrical joint occurr, and the components that are usually measured by means of joint-meters: settlement normal to the plane of the concrete slab, opening normal to the perimetrical joint and shear paralel to the perimetrical joint.Actually, the movement that produces separation between the plinth and the slab occurs aproximately normal to the dam axis, and the opening and shear movements shown in the above mentioned figure are in fact components of this movement. However, concerning the waterstops, the shear component is real, and represents one of the main causes of waterstop damage, when it locally exceeds the capacity of adaptation of the perimetric joint waterstops.

Table 1 and figure 4 depict joint movements measured in some modern concrete faced dams. These results show that the movements measured in Foz do Areia, considering the height of this dam, are quite similar or only slightly greater, then the ones found in other less deformable dams of good behaviour. The movements measured in Segredo are not shown, because they were impaired by a large flood that occurred a few weaks before starting final impouding. However, the smaller slab deflection indicate that they are probably similar or smaller then the ones measured in Foz do Areia.

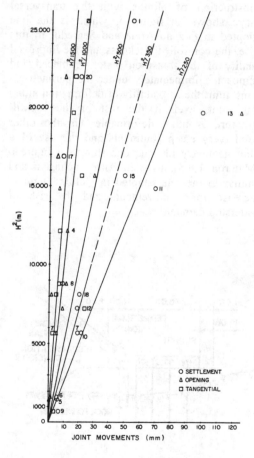

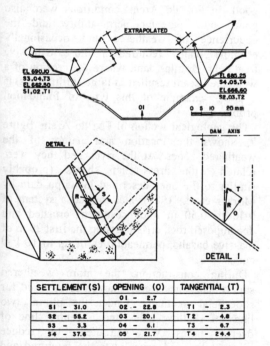

SETTLEMENT (S)	OPENING (O)	TANGENTIAL (T)
	O1 – 2.7	
S1 – 31.0	O2 – 22.8	T1 – 2.3
S2 – 55.2	O3 – 20.1	T2 – 4.8
S3 – 3.3	O4 – 6.1	T3 – 6.7
S4 – 37.6	O5 – 21.7	T4 – 24.4

Figure 3. Foz do Areia Dam - Joint Movements (mm)

Figure 4. Joint Movements x Height of Dam

As shown in figure 4, the movements of the perimetrical joint tend to increase with the height of the dam and with the deformability of the rockfill, but in a moderate, somewhat irregular and almost unpredictable way. In fact, the protopype measurements indicate that the magnitude of these movements depend, in a larger extent, on the shape of the valley abutments, the geometry of the plinth, the use of less deformable material under the perimetrical joint, and on workmanship of compaction work in this area. Foz.do Areia and Segredo are good examples of this fact.

The geometry. of the cut-off slab, both in longitudinal and transversal sections, for instance, is a critical feature. When the valley slopes are not too abrupt, they allow the construction of plinths with the transversal shape shown in figure 5, which is the one adopted in Foz do Areia and Segredo. In this case, the controlled thickness and the improved quality of the transition materials placed and compacted immediately under the perimetrical joint, limit the possibility of deformation along this joint, even if the rest of the rockfill structure is more deformable. On the other hand, very steep abutments and the use of a plint geometry like the one shown in figure 6 (Materon 1985), used in Alto Anchicaya and similar to the one adopted in Golillas, tend to increase these movements and the risk of waterstop damage.

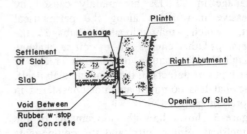

Figure 6. Alto Anchicaya (Materon 1985). Vertical section of the plinth (paralel to dam axis)

4 LEAKAGE THROUGH FOUNDATION

For the cut-off slab design, the requirements concerning support capacity are usually not critical, since stresses are very low. The significant features are permeability and erodibility of the foundation, under the very large hydraulic gradients that occur in this area.

In Segredo, foundation was exceptionally good. In Foz do Areia, conditions were also good, but greater permeability and the occurrence of weathered and occasionally erodible seams required a more extensive treatment. Golillas dam, where the erosion of a weathered seam resulted in large infiltrations, is a good example of this type of foundation problem.

The geological section of Foz do Areia, figure 7, shows the location and nature of the weathered zones. At the river bed they were related with stress-relief shears (probably similar to the ones described in Itaipu dam by Moraes et al., 1982), and formed a system of 0.30 to 0.50 m thick seams of crushed and decomposed rock criss-crossing the first 20 m of massive basalts, permeabilities of up to $10^{(-2)}$ cm/s.

During construction, the more weathered superficial zones were removed or exposed for superficial treatment. Under the plinth, two lines of shallow holes and a central line of deeper holes were grouted, to reduce permeability and consolidate the fractured and weathered zones. Grout courtains required about 33.600 linear meters of grout holes, and 695.000 kg of cement.

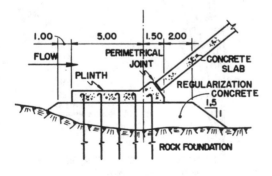

Figure 5. Segredo Dam - Vertical section of the plinth (normal to dam axis)

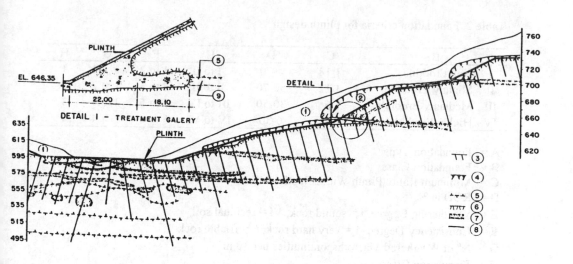

Figure 7. Foz do Areia Dam. Geological section along the plinth (river bed and left abutment)

(1) Soil and decomposed rock
(2) Hard, slightly weathered rock
(3) Grout curtain exploratory holes
(4) Surface of the hard, slightly weathered rock
(5) Flow contact
(6) Excavation surface
(7) Pervious zone
(8) Stress relief fractured horizons
(9) Weathered basaltic breccia

In Segredo, these numbers reduced to 5.250 m of grout holes, and 72.000 kg of cement.

Present trend in CFRDs, after the experience of Salvajina, (Sierra 1989), is to accept poorer foundations, when deep excavations are required to reach sound, non-erodible rock. In this case, the width of plinth has to be increased to reduce the hydraulic gradients to acceptable gradients for the type of foundation.

In Segredo, were rock was exceptionally good and soil cover very thin, the excavations were designed chiefly to shape the desired plinth geometry, practically without concern with rock quality.

The same criteria was generally used in Foz do Areia, but complicated by locally poorer rock conditions. Since the foundation was predominantly of good quality, a 1:20 plinth width: depth of water relation was used in the dimensioning of the cut-off slab. However, at river bed, where the weathered seams outcroped, the width was locally enlarged by using mass concrete under the plinth, and the number of grout lines increased.

More recently, one of the authors was involved in the design and construcion of a concrete faced gravelfill dam, in Argentina, Pichi Picun Leufu (Machado et al. 1993). In this dam, to avoid deepening the excavations, poorer rocks were accepted as foundation, and a classification criteria was devised to define the ratio between plinth width and water height based on foundation parameters. This criteria, presented in table 2, was based on a similar one proposed by Sierra, 1989, after the experience of Salvajina, but for Pichi Picun Leufu it was extended to include foundations of medium hard to hard rock criss-crossed by weathered seams, in part as a result of the experience of Foz do Areia.

Table 2. Foundation criteria for plinth design

A	B	C	D	E	F	G	H
I	Non Erodible	1/18	> 70	I to II	1 to 2	< 1	1
II	Slightly Erodible	1/12	50-70	II to III	2 to 3	1 to 2	2
III	Medianly Erodible	1/6	30-50	III to IV	3 to 5	2 to 4	3
IV	Highly Erodible	1/3	0-30	IV to VI	5 to 6	> 4	4

A = Foundation Type
B = Foundation Class
C = Minimum Ratio: Plinth Width/Depth of Water
D = RQD in %
E = Weathering Degree: I = sound rock; VI = residual soil
F = Consistency Degree: 1 = very hard rock; 6 = friable rock
G = N° of Weathered Macrodiscontinuities per 10 m
H = Excavation Classes:

1 = requires blasting
2 = requires heavy rippers; some blasting
3 = can be excavated with light rippers
4 = can be excavated with dozer blade

REFERENCES

Amaya, F. & A. Marulanda 1985, Golillas dam-design, construction and performance, ASCE Geotechnical Engineering Division, *Detroit Symposium on CFRD.*

Hacelas, J., C. A. Ramirez & G. Regalado 1985, Construction and performance of Salvajina dam, ASCE Geotechnical Engineering Division, *Detroit Symposium on CFRD.*

Machado, B. P., P. L. Marques Filho, B. Materon & E. Bechis 1993, Pichi Picun Leufu - The first modern CFRD in Argentina, Chinese Society for Hydro-electric Engineering, ICOLD, *Beijing Symposium on CFRD.*

Mackenzie, P. R. & L. A. McDonald 1985, Use of soft rock for rockfill, ASCE Geotechnical Engineering Division, *Detroit Symposium on CFRD.*

Marques Filho, P. L., E. Maurer & N. B. Toniatti 1985, Deformation characteristics of Foz do Areia dam as revealed by a simple instrumentation system, *Trans. 15 th*

Congress on Large Dams, Q 56-R.21, Lausanne, Switzerland.

Materon, B. 1985, Alto-Anchicaya dam-design, construction and performance, ASCE Geotechnical Engineering Division, *Detroit Symposium on CFRD.*

Pinto, N. L. de S., B. Materon & P. L. Marques Filho 1982, Design and performance of Foz do Areia concrete membrane as related to basalt properties, *Trans. 14 th Congress on Large Dams*, Q 55-R.51, Rio de Janeiro, Brazil.

Pinto, N. L. de S., P. L. Marques Filho & E. Maurer 1985, Foz do Areia dam-design, construction and behaviour, ASCE Geotechnical Engineering Division, *Detroit Symposium on CFRD.*

Pinto, N. L. de S., S. Blinder & N. B. Toniatti 1993, Foz do Areia and Segredo CFRD dams - 12 yrs evolution, Chinese Society for Hidro-electric Engineering, *Beijing Symposium on CFRD.*

Sierra, J. M. 1989, Concrete face dam foundations, *De Mello Volume*, São Paulo, Brazil, Edgar Bluchter Ed.

Analysis of Porto Primavera Hydroelectric Power Plant foundations performance during the second stage diversion

Appréciation du comportement des fondations du barrage Porto Primavera pendant la deuxième phase de déviation du fleuve Paraná

G. Re & M.C.A.B. Guimarães
Themag Engenharia Ltda, São Paulo, Brazil

L.B. Monteiro
CESP Cia. Energética de São Paulo, Brazil

ABSTRACT: Since May 1993 Porto Primavera dam structures have been subjected to part of the hydraulic load applied as a consequence of the completion of the last phase of the river diversion. This stage of works was of exceptional interest because of the expected differential behaviour of the foundations, caused by distinct geological and geomechanical characteristics.

This paper relates the main observations carried out during this first partial reservoir filling discussing particulary the foundation reactions in terms of deformations and pore pressure development.

RESUMÉ: Depuis Mai de 1993 les structures du barrage de Porto Primavera ont reçu les premières charges hydrauliques partiales à cause de la première mise en eau de l'oeuvre. Ce chargement partial fut conséquence de l'exécution de la dernière phase de deviation du fleuve Paraná.

Le premier remplissage du réservoir d'un barrage est toujours important en vue de suivre le comportement du barrage proprement dit et de sa fondation. En plus, les caractéristiques géomécaniques du massif rocheux sous quelques blocs de la structure faisaient attendre une réaction differentielle soit sous le point de vue des déformations, comme sous celui des souspressions.

On décrit ici les resultats des observations faites au cours des travaux de déviation du fleuve Paraná jusqu'á la stabilisation totale des mesures.

1 INTRODUCTION

Porto Primavera Hydroelectric Power plant, owned by CESP-Cia. Energética de São Paulo - , is under advanced stage of construction on the Paraná river on the borderline of the States of São Paulo and Mato Grosso do Sul, in Southeastern Brazil, about 700 km West from the City of São Paulo.

The layout of the project comprises an homogeneous 10.2km long earthfill dam on the right bank with maximum height of 32m. The concrete structures are located mainly on the riverbed and include 18 Intake-Powerhouse gravity units, with 560m of total lenght and 62m of maximum height, and 16 Spillway blocks 315m long with tainter gates

(fig.1).

The left abutment of the dam is intersected by the two navigation lock facilities necessary for the fluvial traffic during both construction and operation periods.

At site the works began in 1978 but, due to Federal Government economic policy, the construction schedule suffered several delays. The operation of the first generating unit is only foreseen for 1996.

In view of the form of the concrete blocks and the hydraulic loads to be applied it had been stated during the evolution of the structural design that foundation requirements would not be severe neither in terms of shear strength nor support capacity needs. For such reasons the excavation levels

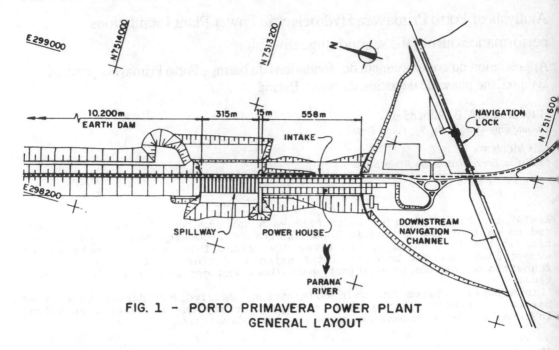

FIG. 1 - PORTO PRIMAVERA POWER PLANT GENERAL LAYOUT

for the structures were defined permitting the permanence in the foundation of lower quality rock (III or III* classes - see Table 1) and the foundation treatment design was simplified consisting only of an upstream grout and drainage system.

The geologic conditions of the foundations change along the axis of the concrete structures from left abutment to right becoming poorer in geotechnical standpoint under the last Intake-Powerhouse units and under Spillway blocks. So differential behaviour of the foundation was expected caused by those distinct geological and geomechanical characteristics of the rock mass. Several structural blocks were carefully instrumented in order to detect such occasional differential behavior during the first loading of the structures.

The second stage diversion of the Paraná river occurred from March to August 1993. It caused the flooding of the area close to the conkrete structures and the passing of the flowing water by the Spillway. The main observations carried out in that period are the object of this paper and are presented in the following items as well as the measures adopted to solve some improperties related to excessive inflow in part of the structure.

2 MAIN GEOLOGIC FEATURES

The site of Porto Primavera Powerplant is geologically located in the center of the Paraná basin in which both sedimentary and magmatic rocks can be found. The litholooies that directly affect the dam foundations are:

1. Colluvial and alluvial deposits (cenozoic)

2. Sandstones of Caiuá Formation (cretacic)

3. Basaltic lava flows of Serra Geral Formation (juro-cretacic)

All the concrete sructures are supported by the basaltic rock mass but there is an important structural discontinuity in the flows distribution along the dam axis. At the left side (from block U-0: through block U-16) basaltic lava flows present large thickness and great lateral continuity. They are called "macroflows" there and reach an average thickness of about 25m.

The physical and geomechanical characteristics of such rock mass seem to be equivalent to those found on other dam sites in Southern Brazil. No particular problems were identified in such lithology.

From unit U-16 towards right bank the structural arrangement of the rock mass changes. Lava flows present both reduced thickness (from a few decimeters up to some meters) and lateral continuity, up to some tens of meters. Here they were called "microflows".

The region where the contact, between the "macro" and "microflows" occurs presents a singular lithology, named earlier "cathaclastic basalt" with clayey composition and peculiar physical and geomechanical characteristics. This basaltic lithology presents low dry density, low modulus of deformability, very high porosity and is extremely sensitive to moisture changes. Due to its low dry density that lithology was named "light-weight basalt".

The occurrence of "light-weight basalt" at Porto Primavera site was object of several reports discussing the genesis of such formation and its implications on the structural design (see Tressoldi, 1986, Ferraz, 1986 and Marques, 1987 and 1987a). It can be noted, however, that this lithology has only a secondary importance as support of these structures because it was almost totally removed from the foundation.

Despite the almost complete removal of the "light-weight basalt" from the structure foundations, the region of its occurrence continued to be an important structural discontinuity between the "macro" and "microflows" regions.

In terms of geomechanical characterization, "macroflows" could be classified as class I/II rock mass while "microflows" could be considered as a III/III* class rock mass (see table 1). This different characterization led to the expectation of distinct foundation behavior.

Due to the relatively low load (up to 2.5MPa) applied by the structures, both rock mass types were considered able to support the concrete blocks with deformations within allowable limits, but expected settlements of Spillway blocks were about two times higher than those for the Powerhouse blocks.

On the other hand the more frequent presence of discontinuities in "microflow" rock mass (mainly lithologic contacts between lava flows) not only influenced its deformability characteristics but also contributed to turn it into a much more permeable foundation.

Table 1. CESP-IPT geomechanics classification of basaltic rock masses in Porto Primavera project

CLASS	FRACT.DEG	WEATH.DEG
I	<2	
II	2/10	SOUND TO
III	11/20	FEW ALT.
III*	>20	
IV	>20	VERY ALT.
V	CAHOTIC	EXTR.ALT.

Under the shear strength standpoint, despite the great number of discontinuities present, mainly in "microflow" rock mass, only one set of fractures was considered with some concern to stability analysis, much more due to its continuity and permeability than to its shear strength characteristics.

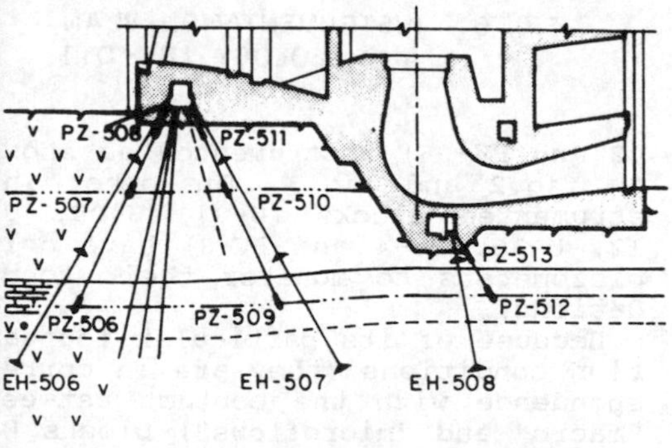

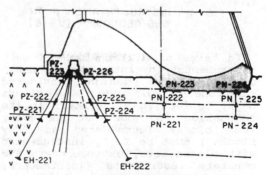

FIG. 2 - INSTRUMENTATION PLAN
AND GEOLOGY (U-06)

FIG. 3 - INSTRUMENTATION PLAN
AND GEOLOGY (BV-02)

This set of discontinuities (named "el.196 disc.") was intersected by the upstream grout curtain but not by the deep drain holes, in order to try to reduce the inflow to the drainage system.

Fig.2 and fig.3 show the cross section of typical concrete bloks as well as the main geologic features present in the foundations.

3 FOUNDATION MONITORING SYSTEM

The instrumentation to be installed in the concrete blocks was designed in order to achieve the following objectives:

a - to verify the foundation behaviour during construction phase

b - to monitor the response of both structures and foundations during the reservoir filling

c - to do the surveillance of the structures safety during the operation period.

In view of the geologic features affecting the basaltic rock mass, intrumentation included mainly piezometers (both standpipe and pneumatic types) and multiple rod extensometers. In some blocks, with particular purposes, a few total pressure cells and concrete strainmeters were also instmlled to try monitor the stresses in the concrete-rock contact.

As regards the piezometers, emphasis was placed on instrumenting the more permeable features of the rock mass in addition to the concrete-rock contact. The distribution of such devices under the structures pursued a scheme the purpose of which was the determination of the magnitude and distribution of the seepage flow trough the rock discontinuities mainly in upstream to downstream direction.

As regards movements and deformations of the rock mass, multiple rod extensometers were installed deep in the foundation crossing the more fractured zones and anchored in good quality rock deep enough to be considered not influenked by the stresses bulb.

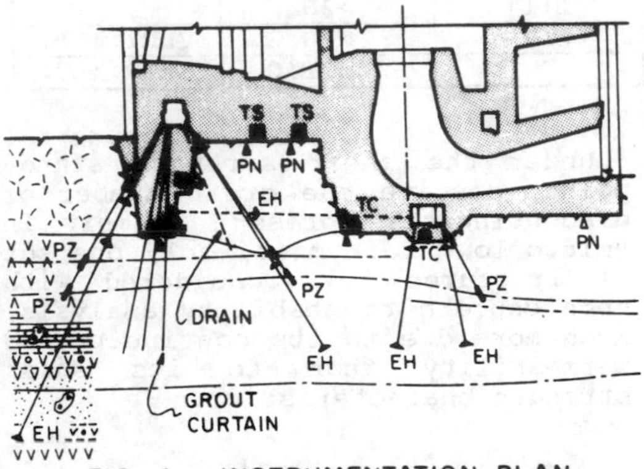

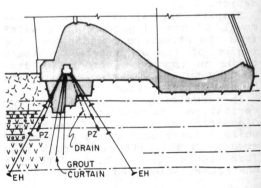

FIG. 4 - INSTRUMENTATION PLAN
AND GEOLOGY (U-18)

FIG. 5 - INSTRUMENTATION PLAN
AND GEOLOGY (BV-01)

In terms of distribution of such instrumentation into the concrete structures the design considered that one out of every two blocks ought to receive piezometers in its foundation and one out of every four blocks (considered as "key blocks") must be fully instrumented with piezometers, extensometers and concrete instruments (jointmeters, strainmeters, thermometers etc.).

It result was four Intake-Powerhouse blocks (U-02, U-06, U-10 and U-14) and two Spillway blocks (BV-

02 and BV-06) instrumented as shown in fig.2 and fig.3. The other instrumented blocks (U-04, U-08, U-12, U-16, BV-4 and BV-8) have only piezometers to monitor their foundations.

Because of its particular foundation conditions (they are in correspondence with the contact between "macro" and "microflows") blocks U-18 and BV-1 were also instrumented with a complete set of instruments, as can be seen in fig.4 and fig.5.

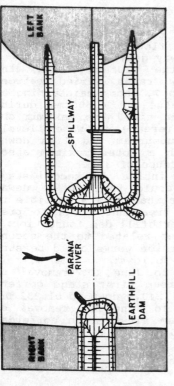

A — FIRST STAGE - FIRST STEP COFFERDAMS

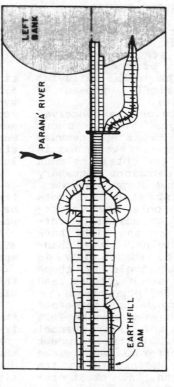

B — FIRST STAGE - SECOND STEP COFFERDAM

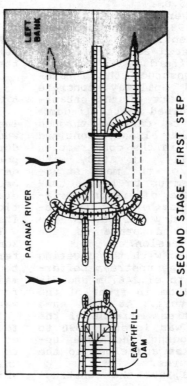

C — SECOND STAGE - FIRST STEP

D — SECOND STAGE - SECOND STEP

FIG. 6 — PORTO PRIMAVERA POWER PLANT
PARANÁ RIVER DIVERSION SCHEME

4 THE RIVER DIVERSION SCHEME

The Paraná river diversion scheme designed to permit the construction of the concrete structures in the riverbed and part of the earthfill embankment in the right alluvial plain can be briefly exposed as follows (see fig.6).

The river diversion was conceived in two main stages. The first one was performed in two steps because of the great volume of work and in view to the need for obtaining rock in volumes and dimensions necessary to build the second step cofferdam.

During the first step left cofferdam comprised only the Intake-Powerhouse area. An auxiliary cofferdam was built on the right bank narrowing the original river channel from 1300m to about 1100m in width. The crest levels of these cofferdams were at el.246.5m and 245.0m (upstream and downstream).

First stage second step cofferdam enlarged the work area on the left bank contracting the river channel more than 700m. Upstream crest level of this cofferdam was raised up to el.250.0m, meanwhile in its downstream strecth the crest remained at el.245.0m.

These works were carried out respectively in 1981 (first step) and 1982 (second step).

In the dewatered area inside the first stage cofferdam it was then possible to construct the Intake-Powerhouse and Spillway concrete structures and part of the enbankment of the riverbed earthfill dam.

The works were carried through 1993 due to the already mentined delays occurred in the construction schedule. Finally, in 1993, with the completion of the main structures of the dam up to levels compatible with the diversion operations it was possible, at the beginning of the dry period, to initiate the second stage of the Paraná river diversion.

The works began with the lowering of the first stage upstream cofferdam first down to el.246.5m and finally to el.236.0m in front of the Spillway structure. At the same time the downstream strecht of the same cofferdam was lowered down to el.232.0m, flooding the area upstrem the Intakes and close to the Spillway structures.

From March 11,1993 the construction of the second stage cofferdam took place which ended in June 4, with the total closure of the river, now flowing trough six lowered Spillway gates.

During the closure works, the river flow rates varied between 5,500 up to 7,300cms maintaining an average value of 6,500cms during most of the time. The maximum observed difference in water levels was 2.9m upstream and 1.1m downstream with a total maximum absolute difference of 3.4m.

The area inside the second stage cofferdam could then to be dewatered after what it was possible to begin the treatment works to prepare the earthfill dam foundation.

Main dates related to the second stage diversion works could be summarized as follows:

April 2 to June 10: removal of the downstream first stage cofferdam from el.243.0 down to el.232.0m

April 6 to June 28: removal of the upstream first stage cofferdam from el.246.5m down to el.236.0m

May 15 and May 22: flooding of the area close to the structures downstream and upstream respectively

June 17 to July 4: construction of the dumped rockfill of the second stage cofferdam

August 31 to September 11: dewatering of the area and beginning of the excavation and treatment works

5 OBSERVED BEHAVIOUR

The most important characteristics whose alterations were monitored during diversion phase were, in importance order, the pore(pressure developed in the rock mass, the seepage flow rate collected by the drainage system and the deformations suffered by the stuctures and its foundations.

It was observed that the piezometric head was among the parameters that presented the most intense changes. The intensity varied mainly depending on the type of foundation where the observed block was.

In general way Intake-Powerhouse foundation presented more increase of piezometric pressures from left to right side (from U-01 to U-18 units) not only due to the permeability characteristics of the foundation but also because of the existence of the downstream cofferdam of the Powerhouse.

The increase of the pore pressure varied from 1 or 2m (U-02) up to 8m (U-18) in the concrete-rock contact downstream of the grout curtain. In the rock mass discontinuities such increase was of the same order. Pore pressure variations were much more intense upstream of the grout curtain reaching values up to 18m, almost reaching at the same elevation of the river water level.

In the Spillway foundation (in which "microflow" predominate) the increase of piezometric levels were more uniform and intense. It reached up to 11 to 15m downstream and up to 20m upstream of the grout curtain.

In order to visualize the uplift pressure variations in the foundations fig.7 and fig.8 show the registered readings respectively, in unit U-6 (with average intensity variations) and in BV-2 block (with high intensity variations). The piezometric surface referred to the el.196m discontinuity before and after the river diversion can be seen in fig.9 in which the more intense pore pressure increase under Spillway blocks is clear.

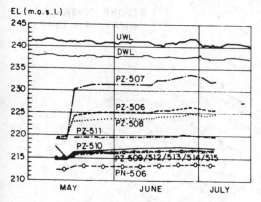

FIG. 7 - PIEZOMETRIC LEVELS U-06

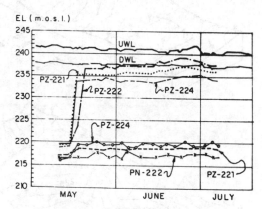

FIG. 8 - PIEZOMETRIC LEVELS BV-02

As regards the water inflow to the drainage system, fig.10 shows clearly the different behaviour between Intake-Powerhouse and Spillway structures. As it can be seen through readings evolution, seepage increased suddenly flooding the area close to the structures. Spillway blocks had a total inflow of about 100l/min before river diversion, value that abruptly raised to 2400l/min. Right Lateral Wall blocks also presented an important growth in seepage from about 350l/min up to 900l/min. Intake blocks were dry before diversion and after that only presented a total flow of about 400l/min distributed from units U-10 to U-18. It is clear that downstream water level must be the main responsible for such intense seepage through structure foundation.

Table 2 shows (in l/min) the total absolute values of seepage inflow collected before and after river diversion by the drainage system in all structures

Table 2 Flow rate in l/min before and after diversion

STRUCTURE	BEFORE	AFTER
ASSEMBLY HALL	680	761
INTAKE	0	401
CENTRAL WALL	237	514
SPILLWAYS	108	2431
RIGHT WALL	343	961
TOTAL	1368	5068

From Table 2 it is possible to conclude that seepage flow increased up four times with the diversion, the Spillway and the Right Transition Wall being responsible for 70% of the total value.

Deformations registered by the deep multiple rod extensometers during the diversion phase were very moderate and in agreement to

3821

those expected. Their pattern was very uniform a little distension occurring in upstream oriented instruments and shortening in those oriented towards downstream.

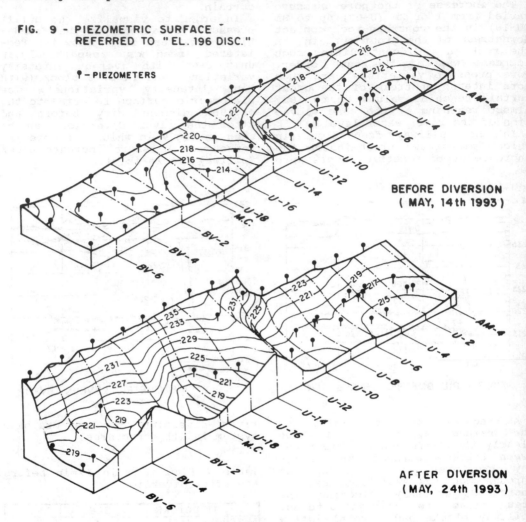

FIG. 9 - PIEZOMETRIC SURFACE REFERRED TO "EL. 196 DISC."

♀ – PIEZOMETERS

BEFORE DIVERSION (MAY, 14th 1993)

AFTER DIVERSION (MAY, 24th 1993)

Table 3 presents both horizontal and vertical displacements occurred from May through July showing that all structural blocks (with exception of block Ta-2) moved 0.3 up to 1.8mm towards downstream. Regarding vertical displacements the picture is not so homogeneous probably due to the different weight of the blocks (some of them were still under construction) and due to the non uniform distribution of the pore pressures in the foundation. It can be seen as a consequence that vertical displacements indicated settlement of some blocks (0.05 up to 0.30mm)(while others raised 0.05 up to 0.50mm.

Table 3 Displacements registered in structure foundations (micra)

BLOCK	HORIZ.	VERT.
AM-03	370	50
TA-02	-10	-50
TA-06	290	110
TA-10	1170	-290
TA-14	1110	-180
TA-18	1760	-280
BV-01	540	-470
BV-02	380	60
BV-06	570	210

Obs.: (+) to downstr. and compress.

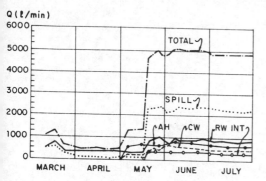

**FIG. 10 - FLOW RATES THROUGH
DRAINAGE SYSTEM**

6 CONCLUSIONS

The immediate reaction of almost all instruments to loading(changes due to the river diversion certified the good performance of the instrumentation system as well as indicated that foundation rock mass behaved satisfactorily under support standpoint but presented permeability characteristics locally higher than expected.

Regarding the observed displacements, back calculations$performed with the deformations registered in Spillway blocks indicated an average deformability modulus of about 12GPa, perfectly in agreement with the value adopted for such kind of foundation in Finite Elements Analysis (10GPa for class III rock mass). Total expected settlement of about 0.8mm was not reached but the maximum reading of the deepest extensometer was 0.6mm, value that certified the good approach of the performed estimates.

As regards seepage estimates, it must be noted that in Porto Primavera project the adopted value for the design inflow was deduced from an extensive survey made on concrete dams in operation on the same kind of rock mass foundation. This criterion was adopted because of the difficulty in well representing in a 2D seepage analysis the actual three dimensional situation of the seepage in an highly fractured rock mass through the drainage system. The adopted value of 1.5l/min per meter of drainage curtain was ex-

ceeded more than four times in the Spillway and in the Right Transition Wall blocks. It can be remarked however that despite the high value for the total flow rate in this part of the structures (about 3600 l/min) 75% of such value is due to only 17 drain holes.

Back analysis computations made after diversion attested that the rock mass presents in the worse case an average permeability coefficient of 5x10E-2cm/s. Such value can be considered very high and an investigation program is under way in order to analyze the influence of each drain in the foundation uplift pressure distribution.

Under the foundation treatment standpoint it can be noted that either the grout curtain or the drainage system both presented good performance.In fact, drain action was effective to control the uplift pressures and the grout curtain created a real barrier in the upstream to downstream direction (fig.9).

Concluding, it can be stated that the structures and foundations general behaviour regarding uplift pressures and deformations was as expected.

Some concerns exist regarding the seepage inflow in part of the structures, mainly in the right part where structural blocks take place on "microflow" rock mass.

Due to the preliminary results of an investigation program still under way however, in view of the pattern of,the,uplift pressure distribution in the foundation, it seems possible to reduce the number of effective drains, closing those which have more expressive water contribution, without increasing the uplift diagrams.

The end of such investigations and the conclusions about the relationship between all parameters involved in the problem such as flow rates, uplift pressures, upstream and downstream water levels etc. could allow the establishment of oxeration rules for the drainage system in order to obtain a reasonable equilibrium between the necessary stability safety and the desirable operation economy.

ACKNOWLEDGEMENTS

The authors wish to thank CESP for its permission for publishing the data contained in this paper. They are also grateful to THEMAG for the collaboration in its preparation.

REFERENCES

Ferraz, J.L. et al. 1986.
  Caracterização geomecânica de litologia basáltica de baixa densidade nas fundações da UHE Porto Primavera. II Simp.Sulam. Mec.das Rochas. Porto Alegre (BR)
Marques, J.D et al. 9987
  Análise de feições estruturais do maciço rochoso da barragem de Porto Primavera. V Congr. Bras. Geol. Eng.. Sao Paulo (BR)
Marques, J.D. et al. 1987
  Considerações genéticas sobre o basalto pouco denso de Porto Primavera. Idem Idem
Tressoldi, M. et al. 1986.
  Ocorrências de basalto de baixa densidade na Usina de Porto Primavera e aspectos de interesse ao Projeto. II Simp.Sulam. de Mec. das Rochas. Porto Alegre (BR)

7th International IAEG Congress / 7ème Congrès International de AIGI, © 1994 Balkema, Rotterdam, ISBN 90 5410 503 8

Geotechnical problems and treatment in the foundations of Sardar Sarovar (Narmada) project, Gujarat (India)

Problèmes géologiques et géotechniques et traitement des fondations du barrage de Sardar Sarovar (Narmada), Gujarat (Inde)

D. I. Pancholi, S. I. Patel & H. M. Joshi
Sardar Sarovar Narmada Nigam Limited, Gandhinagar, India

ABSTRACT : The construction for 1210 m long and 128.5 m high concrete gravity dam across the river Narmada (Sardar Sarovar Project) is in progress. The construction of dam will involve placement of about $7*10^6$ m^3 of pre-cooled concrete to be laid at 55° F.

The preconstruction stage sub-surface exploration was carried out during 1963 to 1978 involving drilling of 129 drill holes aggregating to 8073 running metres and the construction stage sub-surface exploration involved drilling of 216 holes aggregating to 7017 running metres. The surface geological mapping, covering an area of 5,04,000 m^2 was also carried out on 1:100 scale. These explorations and progressive excavations for fault zone treatment and construction sluices revealed about 12 m wide reverse fault in the river bed, weak sedimentary formations below foundations on right bank and weak red bole layer below foundations on left bank. Also, on the right bank, limestone was encountered below the sandstone layers.

The dam foundations are mainly occupied by Deccan Trap intruded by dolerite dykes. The sheared contacts of dyke, steeply inclined fault plane/shear zones, low dipping shear planes, weathered flow contacts, weathered joints and closely jointed weathered rockmass have been exposed during the construction stage. These weak features required special treatment in consultation with design engineers.

The paper gives details of various weak features (surface and sub-surface) and the treatment provided to stabilize the concrete blocks of 128.5 m high dam against sliding or settlement as well as seepage through the foundations.

RÉSUMÉ: On est en train de construire un barrage-poids de grande hauteur en béton sur la rivière Narmada (Projet Sardar Sarovar). La construction du barrage comprendra le placement de $7x10^6$ m^3 de béton préaleblement réfroidi et placé à 55°F. La phase de reconnaissance géologique a été conduite pendant 1963-1978 comprenant 129 sondages avec un total de 8073 m pour l'étude du project et 216 sondages avec un total de 7017 m pendant le suivi de construction. La cartographie géologique de surface sur 5 km² a été aussi faite à l'échelle 1:100. Ces travaux et l'excavation progressive pour le traitement de la zone de faille et des tranchées de drainage ont découvert une faille inverse de 12 m de largeur dans le lit de la rivière, des formations de roches tendres sous la fondation dans la rive droite et une couche tendre rouge sous la fondation dans la rive gauche. Dans la rive droite on aussi trouvé des calcaires sous les couches de grès.

La fondation du barrage est formé par les "Deccan Traps" avec des intrusions doleritiques. Des cisaillements au contact des diques, des plans de faille très inclinés, de surfaces de cisaillement a angle bas, des zones d'altération au long des joints, ont été exposés pendant la construction. Ces zones de faiblesse ont exigé un traitement particulier. Cette communication décrit les détails de ces zones de faiblesse (en surface et en sous-sol) et les mesures de traitement pour la stabilization des massifs de béton de 128.5 m d'hauteur soit contre le glissement où l'affaissement, soit concernant les fuites d'eau sous la fondation.

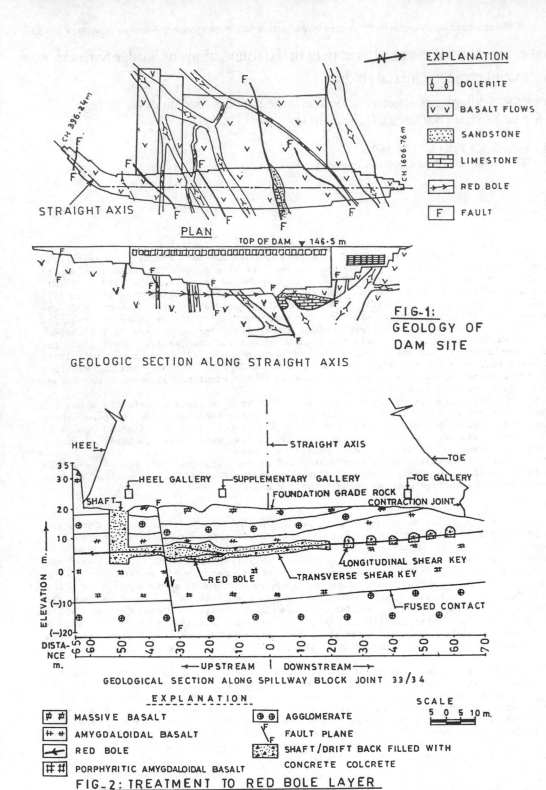

EXPLANATION

	DOLERITE
	BASALT FLOWS
	SANDSTONE
	LIMESTONE
	RED BOLE
F	FAULT

PLAN

FIG-1: GEOLOGY OF DAM SITE

GEOLOGIC SECTION ALONG STRAIGHT AXIS

STRAIGHT AXIS

TOP OF DAM ▽ 146·5 m

HEEL

STRAIGHT AXIS

TOE

HEEL GALLERY — SUPPLEMENTARY GALLERY — TOE GALLERY

FOUNDATION GRADE ROCK

SHAFT

CONTRACTION JOINT

LONGITUDINAL SHEAR KEY

RED BOLE — TRANSVERSE SHEAR KEY

FUSED CONTACT

ELEVATION m.

DISTANCE m.

←— UPSTREAM | DOWNSTREAM —→

GEOLOGICAL SECTION ALONG SPILLWAY BLOCK JOINT 33/34

EXPLANATION

	MASSIVE BASALT		AGGLOMERATE
	AMYGDALOIDAL BASALT		FAULT PLANE
	RED BOLE		SHAFT/DRIFT BACK FILLED WITH CONCRETE COLCRETE
	PORPHYRITIC AMYGDALOIDAL BASALT		

SCALE
5 0 5 10 m.

FIG-2: TREATMENT TO RED BOLE LAYER

3826

INTRODUCTION

The concrete gravity dam across river Narmada is in advance construction stage. This will create a reservoir with live storage of 5800 Mm^3 submerging an area of 370 Km^2. This dam will generate 1450 MW power and irrigate 1792 m Ha land.

The geotechnical investigations were commenced in 1963 and continued during pre-construction and construction stage. These includes (i) surface geological mapping, diamond core drilling, test pits and trenches, 90 cm dia calyx holes, shafts and adits, (ii) laboratory tests to determine geomechanical properties (iii) insitu tests to determine the deformability and shear parameters of various rocks, fault zones, red bole, flow contacts, bedding planes, and (iv) the isotope tracer studies to verify the nature of limestone occurring 30 to 40 m below the foundations on the right bank and also exposed just upstream of the dam seat.

The foundation trench along dam alignment, varying in depth from 5 to 33 m to reach desirable foundation grade has been excavated in advance. This helped in identifying foundation defects, study their characteristics and evaluating their properties for the design of the structure as well as for the foundation treatment.

GEOLOGY

The dam is situated in Deccan Trap province comprising numerous basalt flows of Cretaceous-Eocene age. The porphyritic, dense or amygdaloidal flows of basalt are underlain by sedimentary sequence of Bagh beds of Cretaceous age. The 7 to 56 m thick basalt flows having rolling dip of 5 to 15 are separated by hard agglomerate. The Bagh sedimentaries are consisting of quartzitic sandstone, argillaceous sandstone, shale, pebbly sandstone and limestone. The contacts of sedimentaries and some of the basalt flows are sheared. The Bagh sedimentaries and basalts are intruded by ENE-WSW aligned dolerite and thin trap dykes. The thickness of dolerite dykes varies from 5 to 45 m. The contacts of dyke are tight, but at places, calcified and at some places, sheared.

The fractures, shears and fault pattern in the area trend in NNE-WSW direction (parallel to Narmada-Tapi lineament). In addition, NE-SW, NNW-SSE (the west coast lineament) trends are also observed.

A steeply dipping, almost E-W trending, 12 m wide river bed fault, identified in the initial stages of investigations has, brought the sedimentaries on the right bank in juxtaposition with the basalt on the left bank (Fig-1).

GEOTECHNICAL PROBLEMS AND THEIR TREATMENT

(1) Sub-surface problems and treatment :

Based on the investigations and observations, the foundation defects identified and treated during construction stage are briefly described below.

Treatment to Red bole layer

During the excavation of trench for the treatment of river bed fault, the red bole was exposed on left side of trench at EL (-)10.8 m. It is delineated in the foundations of left bank spillway blocks 28 to 42. This red bole layer separates two amygdaloidal basalt flows which are intruded by two, 17 to 28 m wide dolerite dykes trending in ENE-WSW direction. Shearing across the dyke was observed in lateral continuity of red bole.

The red bole layer is 5 to 40 cm thick, red/green colour, and consisted of sheared rock powder and rock fragment with 6 to 10 % clay. Insitu tests in open trench as well as in a drift on left bank have indicated angle of internal friction (0) 17 and no cohesion. The red bole layer is having rolling dip of 5 to 15 towards upstream. The sub horizontal shearing in dolerite dykes, in continuity of red bole layer contains 1 to 3 cm thick sheared material.

On account of its orientation and low shear parameters, this red bole layer, occurring at shallow depth (10 to 30 m) below foundations, posed threat to the stability against sliding of spillway blocks 28 to 42, resting over this layer. The underground treatment, carried out through approach shafts consisted of excavation of 3 m wide drifts in grid pattern at right angle to each other leaving 4.5*8.5 m rock pillar between them. The drifts are excavated in such a way that the red bole layer is intercepted at mid height of the drifts and later backfilled with concrete to achieve adequate factor of safety against sliding along this weak feature (Fig-2).

In blocks 28 and 29, a branch of red bole/shear joint with considerable extent from upstream to downstream and located 2.5 to 3 m above the main red bole layer has also been treated separately.similarly the branching- two or three off shoots is also common in drifts in blocks 34 to 36 where height of drifts increased to 4.5 m due to removal of such branches alongwith main red bole layer during excavation. Total excavation involved for red bole treatment is 47590 Cu.m. in a total running length of 4858 m.

Treatment to sedimentaries

On the right side of the river,due to displacement along river channel fault, sedimentary rocks, comprising of quartzitic sandstone, argillaceous sandstone, pebbly sandstone, shale and limestone of varying thickness,occurred at shallow depths below basalts in the foundations of spillway blocks 44 to 51.The sedimentaries trend in ENE-WSW direction with gentle dip of 8 towards right bank and apparent dip of 7 in the d/s direction. Argillaceous sandstone is hard but highly sheared and fractured in the proximity of river bed fault. The contacts of lower and upper argillaceous sandstone are sheared,associated with gougy material and containing clayey material with sheared rock pieces. Its thickness varies from a centimeter to 20 cm. Series of insitu tests conducted in drift excavated along contact of sedimentaries and also in sedimentary rocks indicated low value of shear parameters for argillaceous sandstones and its contacts (0 varying from 11 to 26 and no cohesion). The shear parameters of other rock types are found to be fairly high (0 more than 40). Highly sheared tuff layer, overlying upper argillaceous sandstone, is of pinching and swelling in nature and at places, varying in thickness from 1 to 3.8 m. The 1 to 30 cm thick contacts of this tuff layer are sheared and containing gougy material with rock fragments.

The under ground treatment carried out through approach shafts consists of removal of argillaceous sandstone layers and their contacts by excavating 3 m wide and 3.6 to 6 m high drifts in a grid pattern, leaving rock pillar of 8.5*8.5 m (Fig-3). These drifts are backfilled with concrete/colcrete. Excavations were made in such a way that crown of upper drifts is in sound basalt and of lower drifts in quartzitic sandstone. Tuff layer was removed from the crown of upper argillaceous sandstone treatment drifts. In blocks 49-50,the quartzitic sandstone inbetween argillaceous sandstone layers is less than a metre thick, hence, some of the lower and upper drifts are combined. The maximum height of the drifts is 12.5 m.

The total quantity of rock excavation involved is 88870 Cu.m.in a total running length of 5186 m.

To ensure proper contact between colcrete and rock at crown, consolidation cum contact grouting has been done from foundation rock. The holes are spaced 2 m c/c in grid pattern.This grouting is aimed to fill any voids remaining due to concrete shrinkage and/or inadequate filling of irregularities in the rock surface, as well as open joints.

Limestone :- The limestone forms the part of sedimentaries encountered below the dam seat. Similar limestone is exposed on the east of Mokhadi fault, 500 m upstream of the dam axis. The sub-surface explorations have established the thickness of limestone varying from 62 m in the upstream to 33 m in the downstream. The water losses and poor core recovery has indicated the presence of open joints.

To know the nature of limestone, isotope tracer studies were carried out. For this, entire limestone section near the dam vicinity was divided into three parts, viz (1) upstream of Mokhadi fault (2) below the dam seat,and (3) 225 m d/s of dam toe.

These studies indicated that the limestones are siliceous in nature and have no solution channels/cavities. However, to reduce the water loss through limestone, the depth of consolidation grouting is extended two meters in the limestone and the depth of curtain grouting is extended below the limestone through the gallery of the dam on the right bank.

(2) Surface problems and treatment :

The foundation defects, identified during the construction stage are river bed fault, steeply inclined minor faults/shear zones, low dipping shear planes, weathered flow contacts,weathered joints and closely jointed rockmass. The treatment provided to all these weak features is described below.

Faults

Total eight faults of local extent have

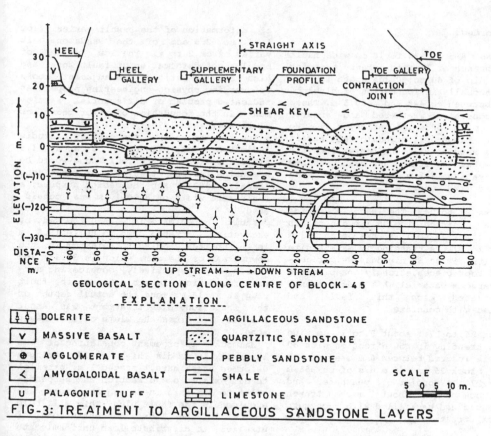

Geological section along centre of block-45

EXPLANATION

⋔	DOLERITE	⊟	ARGILLACEOUS SANDSTONE
V	MASSIVE BASALT	⊡	QUARTZITIC SANDSTONE
⊕	AGGLOMERATE	⊟∘	PEBBLY SANDSTONE
<	AMYGDALOIDAL BASALT	⊟	SHALE
U	PALAGONITE TUFF	▦	LIMESTONE

SCALE
5 0 5 10 m.

FIG-3: TREATMENT TO ARGILLACEOUS SANDSTONE LAYERS

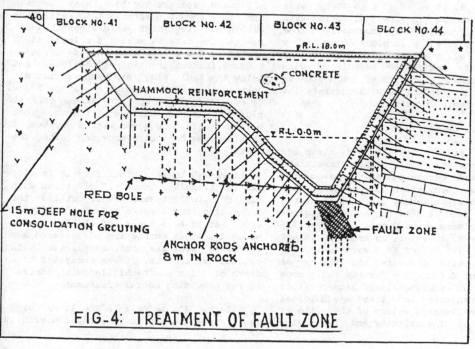

FIG-4: TREATMENT OF FAULT ZONE

3829

been recorded.

1. Blocks 4-5 : A 4 to 10 cm wide fault and its branch of limited extent trending N90 with dip of 60 to 80 due south is exposed diagonally in foundation of blocks 4 and 5 between ch 446 and 460.5 m. These faults consist of weathered and sheared material with brownish sticky clay. Two different flows of amygdaloidal basalt occur as hanging wall and footwall of the fault. Dental treatment has been provided by excavating 1.5 m deep trench and backfilled with concrete.

2. Blocks 21 to 24 : The 10 to 20 cm wide fault is observed from 36.1 m d/s of ch 726 m in block 21 to 11 m u/s of ch 793 m in block 24. It is trending in N40-50 direction with 60 to 80 dip due NW. It consists of rock pieces, strongly consolidated by calcareous material. A 1 m deep trench is excavated along this fault and backfilled with concrete.

3. Block 25 to 27 : About 5 to 10 cm wide fault is trending in N50 direction with 80 due NW is located between 6 m u/s of ch 810 in block 25 and 46 m u/s of ch 856 m in block 27. It consist of weathered and sheared rock pieces. About 1 m deep trench is excavated and backfilled with concrete for dental treatment.

4. Block 30 to 34 : A 5 to 20 cm wide fault, trending in N60-75 direction with steep dip towards NW is encountered from 65 m d/s of ch 934 m in block 30 to 63.9 m u/s of ch 1021 m in block 34. It consists of sheared rock pieces associated with calcification. A trench of 1 m depth is excavated and backfilled with concrete for providing dental treatment.

5. Block 41 to 44 :

River bed fault : The 12 m wide river bed fault, consisting of highly fractured, sheared and weathered basalt, associated with about 15 cm thick clay gouge present as intermittent lenses along fault zone. The fault zone is skew to the dam alignment cutting across the foundation of spillway blocks 41 to 44. It is trending in N80 direction with dip of 60 towards NW.

Insitu tests indicated low values of modulus of deformation for the fault zone material and correspondingly higher values for sedimentaries and fresh amygdaloidal basalt. The average values of the ratio of the modulus of elasticity and the modulus of deformation of the basalt varies from 1.87 to 2.5 and of the sandstone, it varies from 2 to 4. The low values of modulus of deformation of fault zone and high modulus ratio of the adjacent rocks of varying physico-engineering properties indicated problem of differential settlement in the foundation of river bed block 41 to 44 (Mehta 1990).

The two dimensional photo-elastic studies indicated that plug depth of about 1.5 times the width of the fault zone would be adequate for the treatment of fault zone (Desai, 1983). For the design purpose, a fault zone width of 8 to 10 m was considered.

Fault zone material was removed by excavating a trench down to EL (-)16.5 m in the upstream and to EL (-)6.5 m in the downstream of the dam, which is 34.5 m and 24.5 m deep respectively, considering normal foundation levels outside the fault plug at EL 18 m. Thus the actual depth of the fault zone plug treatment was about 2.6 to 3.4 times the width of the fault zone.

Hammock reinforcement consisting of two layer of 36 mm dia high yield strength deformed steel bars, spaced 1 m, parallel to dam axis in one direction and perpendicular to dam axis in other direction, was provided in the fault plug. This will help to avoid the local concentration of stresses to distribute load uniformly to safeguard against any local weak pockets and to prevent differential settlement within the plug. High yield strength deformed anchor bars 36 mm dia, 8 m long and in a grid of 3*3 m were also provided (Fig-4).

Consolidation grouting of the foundation below the fault plug has been carried out down to 15 m depth through a grid of 3*3 m. To ensure positive contact between concrete plug and the rock surface of trench, grout buttons duly connected with a system of grout pipes have been provided along the contacts.

6. Blocks 45 to 47 : About 2 to 15 cm wide fault, trending in N40-70 direction with dip of 55 to 70 towards NW is cutting the foundation from 6 m u/s of ch 1255.6 m in block 45 to 62.4 m u/s of ch 1319.1 m in block 47. It is consisting of sheared and weathered rock fragments associated with calcite and zeolite. It was excavated to a depth of 1.2 m and backfilled with concrete for providing dental treatment.

7. Blocks 45 to 48 : A 2to 15 cm wide fault, trending N40 to 80 direction with 50

to 70 dip due SE is recorded between 66.6 m d/s of ch 1258.3 m in block 45 and 57.6 m u/s of ch 1348.1 m in block 48. It consist of sheared, weathered rock fragments and zeolite. About 0.5 to 3 m deep trench was excavated for providing dental treatment.

8. Blocks 48 to 55 : About 2 to 40 cm wide fault trending in N35-70 direction with dip of 50 to 80 towards SE is observed between 69.2 m d/s of ch 1330.2 m in block 48 and 38.5 m u/s of ch 1484.6 m in block 55. It is consisting of sheared and weathered rock fragments, calcic material and little clay at places. A trench was excavated to a depth of 1 to 2.5 m and backfilled with concrete for providing dental treatment.

Sheared contacts of dolerite dykes :

1. Blocks 28 to 32 : A dyke is running oblique to the foundation in N50 to 75 direction. Its northern contact (dip 80 due NW) is moderately consolidated and associated with 5 to 15 cm wide shear zone. The shear zone is running along as well as close to the contact of dyke. It is consisting of rock pieces, associated with calcite and sheared material. Very close jointing is pronounced near the contact in a width of 0.3 to 2 m. A trench of 0.7 to 1.5 m depth was excavated and backfilled with concrete for dental treatment along southern contact of dolerite.

2. Blocks 34 to 38 : About 40 m wide dolerite dyke trending in N55 to 65 direction, dipping 55 to 65 towards NW is obliquely traversing the foundation of blocks. The contacts of dyke with basalt are sheared. The width of the shear zone was found tobe 20 to 30 cm and width of associated shattered zone is about 5 to 6m. It consists of soft and sheared material and is devoid of clay. The sheared contacts of dyke with basalt were provided dental treatment by excavating 6 to 7 m wide and about 2.5 to 3.5 m deep trench and backfilling with concrete.

3. Block 56 to 60 : A 45 m wide dolerite dyke is recorded diagonally in foundations of spillway blocks 56 to 60. It is trending in N50 to 70 direction with dip of 55 to 80 due SE. Its southern contact is sheared, associated with zeolite and calcite. A 3 to 40 cm wide shear zone consisting of sheared/weathered rock pieces with calcite and clayey material, is trending in N60 to 65 direction with dip of 50 to 75 due SE and running along and close to the southern contact. About 1 to 3.5 m deep trench was excavated for providing dental treatment along this contact.

The northern contact in blocks 58 to 60, is marked by 3 to 40 cm wide shear zone, trending in N65 to 75 direction with dip of 70 to 80 due SE. It consists of sheared and weathered rock pieces, associated with weathered calcite, zeolite and clayey material. The cumulative width of the shear zone, inclu-ding slightly weathered rockmass of basalt and sub parallel minor shears, is 0.7 to 3.7 m. About 1.8 to 3.5 m wide and 1.5 to 3.5 m deep trench was excavated along the contact for providing dental treatment. In addition, reinforce-ment raft along shear zone at EL 91 m has been provided.

Red bole layer

Block 34 : About 5 to 50 cm thick, up-stream dipping red bole layer was treated by providing underground grid of concrete shear keys. Near toe, this red bole consi-sting of sheared rock pieces associated with clay was provided open concrete shear plug treatment by excavating 9 m deep open trench down to EL 7.95 m i.e. below the red bole layer and backfilled with concrete

Block 37 : 13 m deep trench, extending from d/s 46 to 68 m was excavated at the toe of the block as a part of the red bole layer treatment and backfilled with concrete. This layer is sheared and having rolling dip of about 20 towards u/s. Thickness of this material varies from 30 cm to 1 m. Clayey material, associated with shear zone is about 1 to 8 cm thick.

Low dipping shear zones in block 46 and 47

During the course of underground treatment to argillaceous sandstone layers, a down-stream dipping low angle shear zone of limited extent is observed in approach shafts and in some of the drifts. Its con-tinuity below foundation of block 47 was also confirmed by exploratory drift. The stability of blocks 46 and 47 was checked in design against sliding and adequate tre-atment is provided.

Block 46 : The shear zone 10 to 30 cm thick, trending in N40 to 60 direction, and dipping 10 to 35 due NW, is recorded beyond ch 1294 m and from 43 to 557 m d/s and also exposed on the u/s cut face of the foundation trench. It is comprised of sheared rock fragments associated with clay and calcite. The stability of block 46 resting over this weak plane was checked considering internal friction angle as 25 and open concrete shear key at the toe portion of the foundation was designed. Accordingly, a 16 m deep trench has been excavated at the toe of the block extending from d/s 47.5 to 77.4 m between ch 1275 and 1301.8 m. This trench was excavated to complete the remaining treatment of the argillaceous sandstone and tuff layers alongwith the treatment of the shear zone in the overlying basalt. This trench has been backfilled down to EL (-)1 m. Further, to avoid differential settlement, reinforcement raft has been provided at EL 6.7 m.

Block 47 : In block 47, this shear zone is consisting of sheared rock pieces associated with 1 to 6 cm thick clay gouge and recorded in the foundation between 43 m d/s and 61 m u/s. At places, the cumulative width of the shear zone, including fractured rockmass is about 80 cm. It is wavy in disposition (strike vary between N40E to N60W) and dipping 10 to 35 due S. Treatment to this shear zone in the basalt has been provided in the central part of the block, extending from 20 m u/s to 38.5 m d/s. The shear zone is exposed on the right face of the trench between EL 4.4 and 7.2 m. Deepest level in the trench is at EL 4.3 m. The trench has gone down to 12 m in depth from general foundation level and backfilled with concrete. Problem of differential settlement was anticipated in this area due to removal of 5 to 10 m good basalt rock cover overlying the weak sedimentaries and tuff layers. A reinforcement has been provided in the trench area at EL 7.4 m.

Shear zone in block 48 : A low angled shear zone of limited extent, trending in N55 to 75 with dip of 15 to 30 due NW was encountered from 11 m u/s to 68 m d/s in foundation and was recorded on the transverse cut face of the foundation near the block joint 47/48 near ch 1330 m, exposing in the block from 69.2 m d/s to 3 m u/s. In the u/s, it merged with the upper contact of amygdaloidal basalt. It is 1 to 5 cm thick consisting of chlorite/zeolite and associated with sheared rock fragments and clayey material.

Another sub horizontal weak plane - flow contact between pyroclastic rock and underlying amygdaloidal basalt, is located above and close to the above mentioned shear zone. The contact is tight, but at places, slickensides and thin film of clay are observed.

The stability analysis of the foundation of block 48 against sliding along weak planes was checked. Excavation for foundation was done on partial removal of the weak planes from the foundation.

Weathered, jointed rock zones :

Block 3 : The foundation between ch 438 and 442 m, and from 15 to 18m d/s is occupied by moderately weathered basalt. The foundation in this part was lowered down to 1 to 2m depth from general foundation level, but no improvement was observed in rock condition. To prevent uneven settlement, after covering the foundation near toe, 1.5 m thick (EL 124.5 to 126 m) R.C.C. raft, between ch 434 and 443.14 m from 5.5 to 17 m d/s was constructed using 36 mm diameter steel bar spaced at 30 cm c/c bothways.

Block 15 : The rock beyond 21 m d/s is profusely jointed, moderately weathered and traversed by sub horizontal joint sets. A raft of 5*5 m, located between 17.5 and 27.5 m d/s, using 36 mm diameter high yield strength deformed steel bars placed at 40 cm c/c bothways has been constructed. This raft was provided for even distribution of load and to take care of high stresses.

Block 16-17 : A 0.1 to 1.5 m thick weathered zone is present, 5 to 6 m below the general foundation level. A trench was excavated in the central portion of the block for insitu tests and later on considered as a part of the foundation preparation. The ratio of deformation modulus and modulus of elasticity is 1.37. The weathered zone was removed from this portion during the excavation. The remaining part of the foundtion is lying over weathered rock zone having 5 to 6 m rock cover. For uniform distribution of load and to prevent differential settlement, if any, two tier reinforcement of 33 mm diameter tor steel bars, spaced at 0.6 m c/c, was provided in entire foundation.

Block 37 : A weathered and highly jointed doleritic patch, encountered between ch 1076 and 1095 m extending from straight axis to u/s 25 m was treated locally down to 1.5 m depth.

Block 57 : The u/s portion of the foundation around EL 85 m is occupied by moderately weathered to fresh and moderately to profusely jointed porphyritic basalt. steep rock face has been found between ch 1518 and 1523 m and from 38 m u/s to 10 m d/s (EL 68.5 to 85 m). This cut face is traversed by number of low persistence stained joints. Three sub horizontal high persistence weathered joints (2 to 8 cm thick) have been recorded, dipping inside as well as towards free side of the cut face. Near EL 69 m, sub horizontal weathered flow contact, dipping inside the cut face, is recorded between 18.5 and 39.5 m u/s. A load distribution arrangement has been provided as treatment. Considering elevation difference of about 18 m at the cut face, contact grout holes were taken to seal the shrinkage gap between concrete and steep rock face.

CONCLUSION

The Deccan Basalt is generally competent rock for the dam foundation. The varied geomechanical properties of adverse features in basalt flows posed various foundtion problems and needed special measures. The treatment to these problems was evolved by presenting geological details to the design engineers.

The foundation problems due to the presence of sedimentary rocks, red bole layer, river bed fault, indicates that the extensive geological and geotechnical investigations are necessary for the quantitative assessment and evaluation of the foundations of the dams.

REFERENCES

Desai N.B. and JAVA G.L. (1983): Treatment of geological fault in foundation of Sardar Sarovar (Narmada) Dam. CBIP 52nd Annual section V, 13-31.

Mehta P.N. and INDRA PRAKASH (1990): Geotechnical problems and treatment of foundation of major dams on Deccan Traps in the Narmada Valley/Gujarat/Western India. Proceedings Sixth International IAEG Congress 1921-1927.

A type of epigenetic time-dependent structure and its engineering geologic significances

Une structure épigénétique dépendant du temps et son importance du point de vue de la géologie de l'ingénieur

Lansheng Wang & Tianbin Li
Chengdu Institute of Technology, People's Republic of China

ABSTRACT:During geological investigations for Tongjiezi damsite on Dadu river in China, a series of specific structures were discovered, which was obviously different from both general geologic structures and surperficial structures.A comprehensive study was conducted. It is concluded that this is a typical kind of epigenetic time-dependent structure.Its engineering geological significances are quite different from other structures.

RÉSUMÉ: Pendant l'investigation géologique de la fondation du barrage de Tongjiezi sur le fleuve Dadu, une série de discontinuités structurales sont révélées dans les couches de basalte permien presque horizontales. Leur caractère diffère beaucoup de la structure générale. L'étude synthétique comprenant des essais mécaniques l'analògie géoméchanique la simulation numérique, la détermination de l'effet de Kaiser du géostress historique, datant l'activité des failles été fait.
On arrive à la conclusion que ces structures ont été dérivées de la déformation avec une dépendance du temps un peu profond sous la surface. Quand la décharge de la surcharge à causé l'érosion régionale, la détente de l'énergie de la contrainte des masses basaltiques compétentes a été produite du pliocène au pléistocène. Leur signification du point de vue de la géologie de l'ingénieur est aussi discutée dans cette communication.

1 INTRODUCTION

A specific structure which is different from the general structure and the superficial discontinuities,has been found in Tongjiezi damsite area on Dadu River of China, during geological explorations and foundation excavation in 1986. The damsite lies on a gentle anticline with less than 8° dip angle.The anticline's axis is roughly parallel to the river with direction of NNE (Fig. 1).The centric part of anticline is mainly composed of the basalt(P_β),which belongs to the Permian and covered by the sandstone and shale(P_2s),Triassic's sandstone and mudstone(T1f) as well as the limestone(T2t) on both right and left banks. This specific structure was discovered at first in the basalt rockmass of the damsite. The

structures are composed of a set of interstratal and innerstratal fractures(C,Lc) along the basalt strata, and a series of small conjugate thrust faults(F3,F6,etc.) and fissures(Fig.2).Major geological characters of them are described as follows.

1.Interstratal and innerstratal fractures are a set of shear-tensile dislocation fissures,which made rockmass relax.A series of cavities can be found along some part of fractures,which result in the water-loss during borehole hydraulic pressure test. Surfaces of fractures in general are undulating and accompanied with some lightly sliding traces,which show the dislocation direction faces to the axis of the anticline. According to the exploration datas,the maximum developing depth of fractures is at

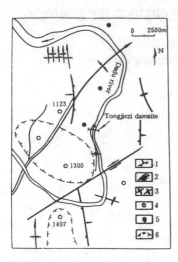

Fig.1 Regional geologic map in the
area of Tongjiezi damsite
1.thrust fault; 2.strike-slip fault
3.fold; 4.planation surface-1300m;
5.planation surface-900m; 6.distri-
bution range of planation-1300m.

least over 150-200m.
2. Conjugate thrust faults distribute sparsely and asymmetrically.
The dip angles of them are about
25-30°. The separation distance of
faults is about 5-10m (maximum one
is 16m)and decreases gradualy with
depth. During the excavation some
ground water with H_2S smell overflowed from the faults,which shows
that the developing depth of faults
can be hundreds of meters.It's very
interesting that in the deep valley
of the river some gravels are interlocked by the faults and some have
been brocken(Fig.3a) or scored with
teeth traces(Fig.3b).This case indicates that the faults had removed
or reslided during the forming of
the deep valley. According to the
teeth traces,the dislocation direction of faults are also facing to
the axis of the anticline.But it's
true that all faults in this area
are not active based on the regional tectonic analysis. Same kinds
of faults and fissures also can be
found in the overlying strata.
  Thus it can be seen that this
specific structure system is obviously different from the general
tectonic structures. Although their
mechanical characteristics are
similar to the superficial discontinuities under the wide valley,the
developing range and growing depth
of them are much larger than the
latter and it's not controled by the
present valley shape. In summary,
this structure system is different
from others and has no   detail
systematical research up to now.
We name this structure   system
the epigenetic time-dependent
structure system. Folllowing comprehensive studies are conducted,
including studying the geostress
situation, dating the formation
times and activity of them,the
formation mechanism of them and
their engineering geologic significances.

2 STUDY ON GEOSTRESS SITUATION BY KAISER EFFECT TEST

The maximum geostress in rockmass
during the present geological period can be determined by means of
Kaiser effect test of rock samples.
According to AE characteristic
curves in Fig.4,which resulted from
the Kaiser effect test of samples
in 6 directions, the 3-dimensional

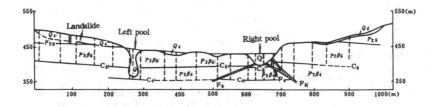

Fig.2 Geologic profile of Tongjiezi damsite area

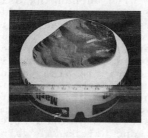

Fig. 3 Gravels interlocked in fault(a),teeth traces on surface of gravel(b)

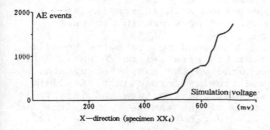

Fig.4 Curve of AE events-simulation voltage

geostress tensors(see table.1) are confirmed.

Tab.1 Results of 3d geostresses

principal stress	magni. (MPa)	direction azimuth	dip angle
σ_1	38.02	294.4	-13.9
σ_2	15.25	188.5	-47.3
σ_3	-0.28	216.1	39.3

This results show that the principal compresive geostress is in the direction of NWW.(290°± of azimuth),which is different from the principal ones(EW,NW etc.) happened during the geologic history for the tectonic systems. An explanation is that this geostress tensors got from the Kaiser effect test in this area is the residual stress stored in the rockmass, which made the specific structure system form and develop.

Based on back analysis of finite element method, the optimum fit between the results of the test and the geostress back analysis is got, when the horizontal boundary stress increased to 30 MPa,and the vertical stress was about 8 MPa. It means that this specific struture was formed in this case when the overlying strata was about 300m in height.

3 DATING FORMATION AND ACTIVITY OF THE FRACTURES BY SEM AND TL

It's said in general that the Scanning Electrical Microscope method can only defines the relative age of the present activity of faults(Y.Kanaore,1982).The research has shown that it is impossible to destroy all quartz grains in the fracture zone during the resliding of faults. The evidence is that the quartz grains in the same fracture zone show different surface textures. Even if they have been damaged,the original surface textures also can be found(Fig.5). So the forming and removing age of the faults are able to be dated by the statistics analysis on the surface textures of quartzes grains.

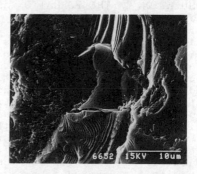

Fig.5 Different surface textures on a quartz-grain in the fault zone

According to above mentioned, the results of the statistics analysis on surface texture of quartz grains (Fig.6) are obtained using the classification system (Y.Kanaori,1982). This result shows that the faults in the specific strcture system was formed in Pliocene to lower Pleistocene(N2-Q1), and reslided during middle-upper Pleistocene(Q2-Q3). The same analysis for interstratal fracture zone of C5 shows they had been formed and removed at the same times.

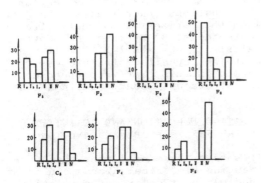

Fig.6 Adjustment statistic histogram

Analysis results of thermoluminescence(TL) absolute-age test for the faults also show that the last stronger resliding of them was at 2.2-2.3 Ma before,or in the middle-upper Pleistocene(Q2-Q3).The same analysis for the interstratal fractures of C5 got the resliding time at 2.5-3.0 Ma before.

4 SIMULATION ON THE FORMATION AND EVOLUTION OF THIS STRUCTURES

Summarizing the characteristics mentioned above,it could be concluded that this structure system is formed by the releasing of high residual geostress stored in the crust during elevating,denudating and unloading. The geomechanical model on evolution process of them can be summarized as energy storing --denudating and unloading--deforming and failing.In order to study and verify the forming mechanism and evolution process of this

structure, a series of simulation researchs have been performed.

4.1 Rock mechanic test on unloading state

Triaxial test with equal confined pressure($\sigma_1 = \sigma_3$) is adopted to study the deformation and fracture characteristics of basult rock samples on unloading state.The unloading process is designed in two ways with different stress path. In the first one(UL-1) the confined pressure decrease with unchanging axial pressure (σ_1).In the second one(UL-2) axial pressure increases with decreasing confined pressure.Meanwhile, the general test (LL) of increasing axial pressure with unchanging confined pressure is also taken to compare with these two special tests.The axial direction of samples are conformed to the maximum principal geostress mentioned above. Several conclusions are gained as follows.
  1. On unloading state,not only tensile fractures but also shear ones appear in the rock samples, whose fracture shapes are similar to the fractures and faults in the field. Angles between shear-tensile fissures and axis of sample(α) decreased with increasing of the confined pressure in fracture period.
2. On unloading state,the tensile fissures parallel to σ_1 easily occured on the sample's surface, and finally formed into ravelling slices. The fracture and failure types of the sample varied from tensile to tensile-shear style with increasing of confined pressure at fracture and failure period. Some part of the shear fracture are tracing along tensile fissures.
  3. Rock failed more intensely on unloading state ,especially in UL-1 test,comparing with the general test. This phenomenon can also be found in the axis part of the anticline in field,where the rockmass are strongly crushed.
  4. The characteristics of rock deformation are greatly affected by the stress path(fig 7).On unloading state,ditatancy or expansion clearly appeared in the sample.

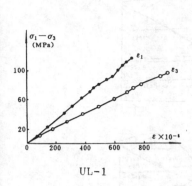

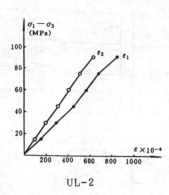

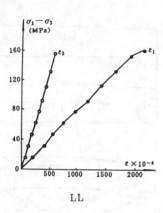

UL-1 UL-2 LL

Fig.7 Strain-stress curves of the basalt sample

4.2 Geomechanical model test using equivalent material

The geomechanical model is prepared on the basis of the law of similarity. The mechanical modal is shown in Fig.8,where expresses

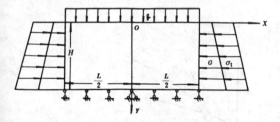

Fig.8 mechanical model for simulation research

residual geostress with the value of 30 MPa(0.086 MPa in model); q the weight of overburden and G the lateral stress caused by gravity pressure of rockmass.The model test was performed under the decreasing of q, and divided into 5 stages, of which the unloading processes from 1st to 4th stage (Fig.9a-c) simulated the regional denudating and the last one simulated deep erosion of river(Fig.9d). This test shows some very important informations as follows.

1. The model began with the centeral part upheaving and bulging when the q decreased to a critical value(Fig.9a,q corresponding to

overburden of 300m) and then gradually upheaved with decreasing of q. Finally, the model became a gentle anticline with dip angle of 8° on both sides,(Fig.9c),which was very similar to the original one in the field.

2.Fractures and faults occured and develop accompanied with the evolution of the upheaval. All of them had the tensile or shear-tensile mechanic characters,and gradually grew from the surface to the deep part. The dislocation directions of them were facing to the center.

3. The interstrtal fractures occured along the weak discontinuities of C4,C5 and the surface of basalt strata.they are previous to the occurance of interstrtal fractures. Most of the interstratal fractures occured close to the interstratal ones.(Fig.9)

4. The forming and evolution process of congurate faults was a progressive failure process tracing to the tensile fractures. Some of the faults were resliding during deep excavation.(Fig.9d).

It can be seen that this model test is successful in reappearing the formation and evolution process of this specific structures.

4.3 Numerical simulation

The mechanical model and unloading stages which are same as that of geomechanical simulation are used in numerical simulation.(see Fig.8). By means of the rheological finite element method,the numerical

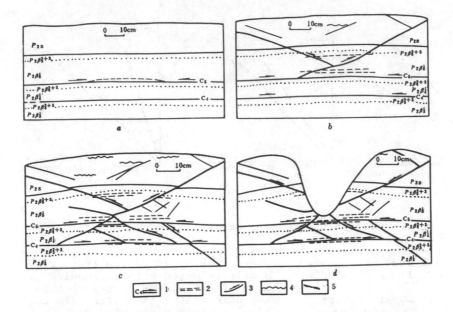

Fig.9 Evolution process of deformation and fracture in geomechanic model
1.interstratal fracture,2.innerstratal fracture,3."fault" and dislocation
direction,4.tensile fissure,5.shear fissure.

caulation have been done.The main
results are as follows.

1. As the unloading went on, the
upheaval deformation of the rock
strata occured gradually and became
obvious . There were some dis-
coordination of the uplift deformation near the weak discontinuities
(C4,C5). The upheaving deformation
of rock strata above the weak discontinuities was obviously more than
that below the weak discontinuities
(Fig.10) . The tangential displacement of the weak discontinuities
increased smoothly with unloading,
and the normal displacement changed
from compression to tension. The
features mentioned above are in
accordance with the phenomena
observed in the geomechanical model
test.

2. The rockmass deformation is
characterised by time-dependence in
the formation of the specific fractures. Furthermore, on different unloading stages, the property of the
time-dependent deformation is also
different.

3. The stress state of rockmasses
is affected by unloading process.
With the development of unloading,

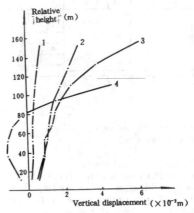

1,2,3,4--unloading steps

Fig.10 Relative vertical displacement curves at different height in
the middle section of the model.

the maximum principal stresses(near
horizontal in direction) went up
(Fig.11),and the minimum principal
stresses (near vertical) went down,
which made a high stress difference
in the rockmass.Meanwhile, because

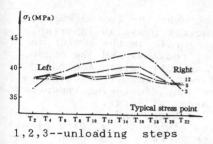

1,2,3--unloading steps

Fig.11 Variance of maximum principal stress with unloading,0,1,2,3 represent the unloading stages.

of the difference of lithological characters, the stress distribution is extremely different. The stress in the hard basalt are much higher than those in the relatively soft rockmass. Thus there existe a high stress gradient from the hard rock-masses to the soft ones. Compared with the boundary stresses, the maximum principal stress in the hard rockmasses increased by 40%, very obvious in the middle parts (Fig.11),while the maximum princi-pal stress in the relatively soft rockmasses decreased about 10%. It can be seen that the distribution and variation of stresses are fully correspondent with the deformation and faillure of the geomechanical model.
4. The high elastic strain energy was stored in the rockmasses before the formation of specific fracture system,and it was gradually released with the unloading of rockmasses. The situation of the element failure shows that the tensile failure areas are mainly located at the top part of the model and near the weak dis-continuities(C4,C5), the shear failure areas developed downwards as the unloading continued. These results also coincide with the results of the geomechanical model test.

5 DISCUSSIONS AND CONCLUSIONS

5.1 Formation mechanism of the epigenetic structure

Based on a series of comprehensive study and analysis mentioned above, it is concluded that the epigenetic

structure in research area origina-ted from time-dependent deformation at shallow depth under the original ground surface of Pliocene to early Pleistocene epoch,when unloading of overburden due to regional erosion caused the releasing of stored resi-dual strain energy in the competent basalt rockmass.In order to emphasize the time-dependence of formation processes, we defined this type of specific fracture structure as the epigenetic time-dependent structure. Its evolution process and mechanical mechanism can be summarized as four periods(Fig.12),namely,formation of energy-stored mass(Fig.12a),unload-ing and rebounding deformation(Fig. 12b),unloading failure and epigene-tic structure formation(Fig.12c), superficial modification(Fig.12d).

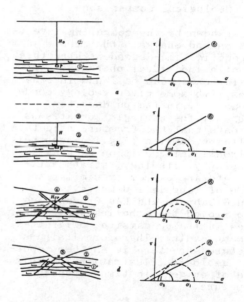

Fig.12 The evolution process and mechanical mechanism of epigenetic time-dependent structures.
(1)energy-stored mass;(2)over-burden;(3)denudated overburden;
(4)broad valley;(5)river valley;
(6)strength envelope for rockmass;
(7)strength envelope for disconti-nuity

5.2 Development conditions of the epigenetic structure

The research results have shown that the epigenetic time-dependent structure developed in a special geologic environments that always possess the following basic conditions.

1. Lithologic condition. Have the rockmass which stored high strain energy, i.e., the competent rockmass with block structure or layered and quasi-bedded structure such as magmatic rock, canbonate rock etc.

2. Structure condition. Be repeatedly subject to process of structural stress field in the geologic history, but without great deformation and failure of rockmasses.

3. Topographical condition. Be also subjected to regional erosion and unloading due to rising of crust in a large area during present geologic process.

5.3 Geological comparisons

It is shown by the comparing investigation and analysis that the epigenetic time-dependent structure is not a individual phenomenon. The similar cases can be found in other sites with some given geology conditions, i.g., in Itaipu damsite of Brazil. The geologic comparisons indicate that the fracture structure of Itaipu damsite (Fig.13) and their environmental condition such as topography, lithology, strata attitude, etc. are extremely similar to those of Tongjiezi damsite.

John G. Cabrera and Las Cruces have also proved that the low angle shear zones in Itaipu damsite originated from unloading. Therefore, Itaipu's fracture structure is another important geologic illustration for this kind of epigenetic time-dependent structure.

5.4 Engineering geological significances

developing near the ground surface, the structure plays an important controlling role in the evolution of surface rockmass and the human engineering activities. According to the research so far, this kind of structure has following important significances at least.

1. The fractures and the low angle faults are always the sliding planes of surface rockmasses movement, controlling the rockmass stability. The instability of the landslide groups with sliding-cracking model on the left bank of Tongjiezi damsite and rock slope mass on its right bank during the excavation of the construction are under the control of the discontinuities.

2. The occurance of the epigenetic structure makes the rockmass crush and loosen intensely as a result of unloading. In the expensive foundation treatment at Itaipu damsite, approximately 25% of the foundation area, containing shear zones and high fractured rockmass, was excavated and replaced by concrete. In the case of Tongjiezi site, the fracture zone of rockmass can be divided into highly -fractured, midium fractured and disturbed zone from the center of gentle anticline to its two limbs. Rockmass has much more holes and higher permeability to groundwater in the highly-fractured zone than the other zones. So, it is necessary for engineering geologists to find out the zoning laws and resonably plan the buildings.

3. It has been proved that the formation and evolution of the epigeneti structure is a process of stress relief. Thus the present geostresses o rockmass are clearly low as a whole. Whereas, it is possible there exsist local geostress concentration . For example, the very weak earthquake due to resevior filling had happened in construction of Tongjiezi damesite.

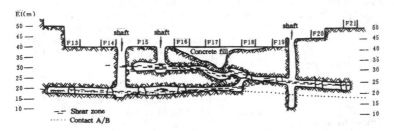

Fig.13 Distribution of shear zones in Itaipu damsite of Brazil.

But situation of the area is quite
different from that of the area with
some active faults .
4. what is worth studying in further
is that some of the epigenetic structures may be closed structures which
will become the groundwater-rich
belts,even storing structures or movement channels of gas and oil.

ACKNOWLEDGMENTS

Thanks prof.Zhang Zhuoyuan for his
help.Assoc.Prof.Xu Jin and Mr.Zhao
Qihua are also acknowledged for their
cooperation.Inst.of Geology,Academia
Sinica provided the financial aid to
the study.

REFERENCES

Li.T.B.,1988, Study on the mechanism
    of epigenetic time-dependent
    deformation and failure of rock
    mass,(in Chinese), M.S.thesis of
    Chengdu College of Geology.
John G. Cabrera & Las Cruces,1986,
    Buckling and shearing of basalt
    flows beneath deep valleys. Proc.
    5th cong.IAEG. Buenos Aires.
Y.Ranaori,1982, Fracturing model
    analysis and relative age dating
    of faults by surface textures of
    granite grains from fault gouges.
    Eng. Geol.,Vol.19, No.3.pp261-281.

Geotechnical problems related to the construction of the Taka lake reservoir site in Greece

Problèmes géotechniques relatifs à l'étanchéité du lac Taka en Grèce

D. Rozos & N. Nikolaou
Institute of Geology and Mineral Exploration, Athens, Greece

ABSTRACT: Taka lake area, consist of alluvial deposits and diluvial fluvio-terrestrial formations of a great thickness, while its margins are composed of sediments from the Olonos-Pindos isopic zone (mainly limestones). The extensive karstification of the later and the climatic conditions (high evaporation), result in the complete drying of the lake during the summer period. As the irrigation necessities of the area ask for a permanent reservoir with a strong capacity (about 20-30 million m^3), a full investigation of the geomechanical characteristics of the area was necessary, in order to design the most suitable method of impounding. After a close examination of all the engineering geological problems referring to the quaternary deposits, dredging of the central and south parts of the lake in combination with the construction of a low embankment, was considered to be the most suitable solution for the impounding of this reservoir.

RÉSUMÉ: Le substratum du lac Taka, est constitué par des formations sédimentaires alluviales et diluviales d'origine fluvio-terrestriale à grand épaisseur, tandis que les marges du lac sont constituées des sédiments (surtout des calcaires) attribués à la zone isopic Olonos-Pindos. La karsification étendue des calcaires et les conditions climatiques (intense évaporation), ont comme résultat l'assèchement complet du lac, pendant l'été. Comme les besoins de la région pour l'irrigation demandant un reservoir d'eau permanent avec une capacité élevée (a peu près 20-30 millions m^3), une investigation des charactéristiques géomecaniques de la région a été nécessaire, a fin de projecter la méthode la plus convenable pour le stockage de l'eau. Après l'examen aprofondi des problèmes géotechniques liés aux formations quaternaires, on a considéré que la meilleure solution était le dragage de la partie central et meridional du lac combiné avec un barrage de terre de petite hauteur.

1 INTRODUCTION

The local authorities of Arcadia province in their efforts to overcome irrigation problems in the Arcadic plateau, have decided to construct a permanent reservoir, having a desirable capacity of 20-30 million m^3, in the area of the seasonal lake of Taka (Fig.1).

Previous investigations in the area (Anon,1963, Anon,1967) resulted in the proposal to construct a circular concrete dam all around the lake, but due to the lack of information, mainly concerning the thickness of the deposits in the area of the lake and also their geomechanical characteristics, the study was not completed.

To this end, a geotechnical investigation was carried out (Andronopoulos *et al*,1988), which was focused on: i) the determination of the lithostratigraphy, the study of the physical properties and the mechanical characteristics as well as the evaluation of the permeability parameters of the Quaternary deposits which cover the Taka lake area, and ii) the definition of the palaeogeographic relief of the karstic limestone bedrock (Fig. 2). For that purpose, the investigation programme which was carried out during two successive summers, due to the climatic and hydrologic conditions prevailing in the area, was included:

a) The drilling of eight sampling boreholes (to a maximum depth of 60m). In the boreholes, the

Fig.1. Location map of Taka lake site in Peloponnesus, Greece.

positions of which are shown in Fig.2, in situ tests (for example, SPT and permeability tests) were carried out and piezometers were installed in order to obtain information about the level of the groundwater table.

b) Drilling of two observation boreholes, to be used for the pumping out tests.

c) The caring out of a geophysical survey (Lazou, 1988, Lazou,1989) to study the lithostratigraphy and subsurface distribution of the Quaternary deposits and to estimate their thickness. From this survey, which included 88 geoelectrical soundings, useful information on the shape of the palaeogeographic relief was obtained. The geophysical survey results were also confirmed, to the depth of 60m, by the results of the drilling.

d) Laboratory tests on disturbed and undisturbed samples.

2 MORPHOLOGY AND GEOLOGICAL SETTING OF THE WIDER AREA

The area that includes Taka lake, with an average altitude of 660 m, is situated in the southern part of the Arcadic plateau, which is located in the Central Peloponnesus (Fig.1). This area, of about 20 Km², is connected with the rest of the Arcadic plateau only eastwards, while in all the other directions it is enclosed by hills and mountains producing a dense hydrographic net and reaching an altitude up to 1500 m.

This morphology of the area is the result of the tectonic evolution, as well as the action of weathering-erosion processes (Georgulis, 1984, Lekkas, 1978). The latter have contributed to the infilling of the tectonic basin with Quaternary deposits.

Alpine sediments form the bedrock of this basin as well as the surrounding slopes. These are Cretaceous limestones of the Olonos-Pindos geotectonic zone, which in the southern part of the area overthrust flysch formations of the Gavrovon-Tripolis geotectonic zone (Fig.3).

The Cretaceous limestones are thin to thickly bedded and characterised by high fracturing and intense karstification. Around the lake there are swallow holes with a high flow capacity, since they are connected with a network of subterranean passages. In addition, a number of the swallow holes have been artificially enlarged in some places, in order to facilitate the drainage of the lake (Fig.3).

Thus, apart from the high evaporation in the area (which has been calculated at about 30-40%), the dense karstification of the bedrock causes the flooded area of the lake, which extends over 5-6 Km² and with a three meters maximum water height (near the southern part), to dry out during the summer (Figs 4, 5).

3 ENGINEERING GEOLOGICAL CONSIDERATIONS OF THE SITE

The Quaternary deposits, which cover the lake area, are fluvio-terrestrial origin and they were found to have a considerable thickness.

From the boreholes, which came to a depth ranging from 25 to 60 m, only one (T-10) drilled at a distance of 200 m from the NE boundaries of the basin, reached the limestone bedrock at a depth of 25 m. Further information concerning the thickness of the deposits in question (geophysical survey), revealed that the sedimentation basin is progressively downfaulted (Lazou, 1988), from N-NW to S-SE (Fig. 6).

This morphology, which is due to gravity faults of NE to SW direction, has resulted in a considerable thickness of the Quaternary deposits in the SE part of the basin (more than 200 m).

As it was found from the lithostratigraphic examination of the lake area (logging of boreholes, geophysical survey, etc), the various facies of the Quaternary deposits were not continuous but showed lateral and vertical transitions.

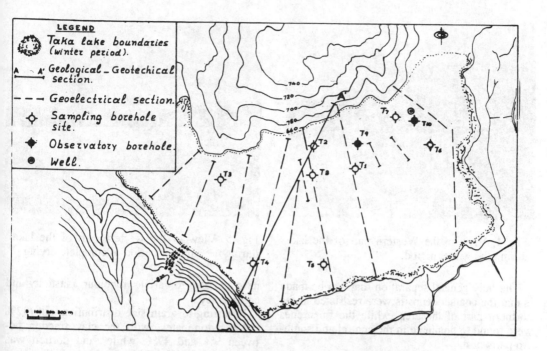

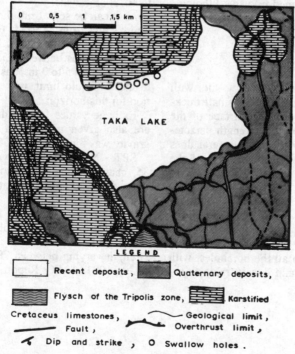

Fig. 2. Topographic map of Taka lake site, showing the location of the site investigation works and the A-A' section.

TAKA LAKE

LEGEND

Recent deposits, Quaternary deposits,

Flysch of the Tripolis zone, Karstified

Cretaceus limestones, Geological limit,

Fault, Overthrust limit,

Dip and strike, O Swallow holes.

Fig. 3. Simplified geological construction of the studied area.

Fig. 4. View of the Western part of the lake, during the winter period.

Fig. 5. View of the western part of the lake during the summer period of complete drying.

The only general separation that can be made is that the coarser deposits were restricted to the northern part of the area, while the finer ones were found to dominate in the central and southern parts of it.

From an engineering point of view, as the evaluation of the results of the in situ and laboratory tests indicates, the above fluvio-terrestrial deposits can be distinguished into five lithological facies, as follows:

3.1 Recent deposits

This unity, which consists of clayey silts with some plant remnants, acquires very small thickness, and it appears only in the surface of the lake. It is characterised by low strength parametres and high permeability, but in general does not affect the geomechanical behaviour of the Quaternary deposits as a whole.

3.2 Clayey silts

This unity dominates the Quaternary formations, since they were found in all the boreholes, with its various horizons to gain considerable thick-

ness and also to variate in colour, plasticity and consistency.

Referring to grain size distribution, silt (32% to71%) dominates, with the clay fraction between 9% and 47%, while sand portion was found to be 1% to 25%. In the upper horizons (5 to 8 metres in depth) at the southern part of the lake, it also contains a small percentage (up to 10%) of gravels.

As it can be seen from the Table 1, Atterberg limits varies from 34.9% to 64.3% the liquid and 14.8% to 26.9% the plastic limit. Moisture content usually lies between 15.0% and 43.2%, but in the depth of 5 to 7 m, has values almost equal to that of liquid limit, revealing a liquid condition for this horizon.

Extreme values for the dry and bulk density are also given in Table 1, while the specific gravity was found 2.67 - 2.70.

SPT - N values were found between 11 and 35, characterising the various horizons as medium dense to dense formations (IAEG Report, 1981).

Tests carried out for the determination of the mechanical properties (Table 1) indicate a firm to stiff soil formation, with uniaxial compressive strength varying between 78 and 484 KPa (BS 5930, 1981; IAEG Report, 1979, Hoek *et al*.

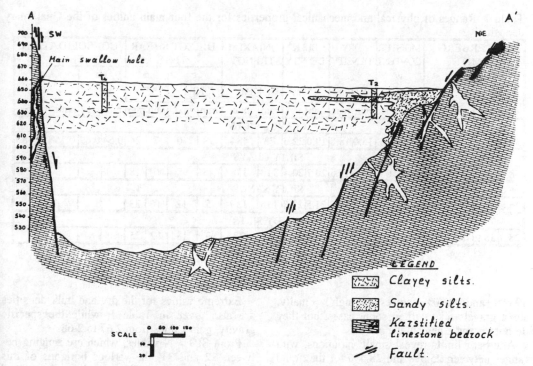

Fig. 6. Geological - geotechnical section A - A' showig the thickness of the quaternary deposits and the palaiogeographic relief of the downfaulted karstic limestone bedrock.

1981). Cohesion values ranges from 38 to 70 KPa, while the angle of internal friction was estimated between 7° to 24°.

Compression index values are generally considered low, ranging between 0.158 and 0.180, with void ratio from.0.640 to 0.686.

Despite the rather wide range of the various strength parameters and physical characteristics of this unity, due to the varying clay and sand percentage, but also to the presence of some gravels in certain cases, their mean values still characterise formations with satisfactory geomechanical behaviour.

3.3 Silty clays

This unity consists of very coherent plastic clay deposits (42% to 81% of clay), with 10% to 44% of silt and 1% to 16% sand , which were found in various horizons at the Northwest and South parts of the lake. In the northestern part of the lake and at the depth of 25 to 27 m, gravels with percentage of 8% to 14% also take part in the grain size distribution of the relevant horizon.

Atterberg limits and espesially liquid limit show rather large variations, with ranges 48.3% to 80.0% for the liquid and 21.5% to 29.9% for the plastic limit. In relation to these values, moisture content was found to be very low (14.9%-16%) indicating a rather dry condition for this unity.

The range of the values for dry and bulk density are also given in Table 1, while the specific gravity was found 2.68 - 2.70.

From the SPT - N values, which are raging from 23 to 45, the horizons of this unity are characterized as medium dense to dense formations.

Tests for the determination of the mechanical properties (Table 1) reveal stiff soil formations, with uniaxial compressive strength of about 310 to338 KPa. Cohesion values ranges from 45 to 98 KPa, while angle of internal friction was estimated between 6° to 18°.

3.4 Silty sands

The sand (40% - 52%) dominates in this unity, while the silt and clay portions do not exceed

Table 1. Ranges of physical and mechanical properties for the four main unities of the Quaternary deposits.

ATTERBERG LIMITS				MOISTURE CONTENT		DRY DENSITY		BULK DENSITY		UNIAXIAL STRENGT		DIRECT SHEAR TEST				CONSOLIDATION TEST			
w_L %		w_P %		w %		KN/m^3		KN/m^3		q_u KPa		c KPa		φ °		eo		Cc	
min	max	min	max	min	max	min	max	min	max	min	max	min	max	min	max	min	max	min	max
CLAYEY SILTS																			
34.9	64.3	14.8	26.9	15.0	43.2	15.0	18.8	19.0	22.0	78	484	38	70	7	24	0.64	0.69	0.158	0.180
SILTY CLAYS																			
48.3	80.0	21.5	29.9	14.9	16.0	17.3	17.5	20.1	20.4	310	338	45	98	6	18				
SILTY SANDS																			
25.6	28.4	14.7	19.9	12.7	21.0	17.8	18.2	21.5	21.8	135	139	5	12	16	33				
SANDY SILTS																			
28.9	38.9	14.1	24.0	18.7	31.5	16.5	16.9	20.0	20.3	76	78	9	18	12	25				

39.0% and 22.0% correspondingly. Finally, some gravels take part in some cases, but they do not exceed a percentage of 9%.

Atterberg limits reveal small variations, with ranges between 25.6% and 28.4% for the liquid and 14.7% and 19.9% for the plastic limit, while moisture content has values around the latter (12.7%-21.0%)

Extreme values for the dry and bulk density are also given in Table 1, while the specific gravity was found 2.65 - 2.67.

Standard Penetration Test (SPT) N values are generally ranging between 23 and 29 indicating medium dense formation.

Tests carried out for the determination of the mechanical properties (Table 1) indicate a firm soil formation, with uniaxial compressive strength of 139 KPa). Finally, the shear strength parametres were found to be, 5-12 Kpa the cohesion and 16°-33° the angle of internal friction.

3.5 Sandy silts

This unity was mainly found at the margins of the lake and mainly consists of silt (45% - 66%) with sand up to 36%, while the percentage of the clay, 14% - 21%, is small.

Atterberg limits, as expected, show rather medium variations, the ranges being 28.9% to 38.9% for the liquid, and 14.1% to 24.0% for the plastic limit. Water content usually varies between 18.7% and 31.5%, namely between liquit and plastic limits.

Extreme values for the dry and bulk densities are also given in Table 1, while the specific gravity gain values from 2.67 to 2.68.

From SPT - N values, which are ranging between 22 and 31, the various horizons of this unity are characterised as medium dense and rarely dense.

Tests carried out for the determination of the uniaxial compressive strength (Table 1), revealed values of about 78 KPa, indicating a soft soil formation. Cohesion values ranges from 9 to 18 KPa, while angle of internal friction was estimated between 12° to 25°.

From the above discussion, it is clear that the lake deposits and especially clayey silts, which dominate, present a good geomechanical behaviour. Thus, they can either be easily excavated, even with steep slopes, without failures, or used as embankment material, for the increase of the reservoir capacity.

4 HYDROGEOLOGICAL CONDITIONS

Since the main problem of the area is the dense karstification of the limestone bedrock, selection of the type and the size of the reservoir will depend on both the geomechanical behaviour and the permeability characteristics of the Quaternary deposits, overlying the limestone bedrock.

From the piezometres installed in the area were found that the groundwater table fluctuated from 0.8m (at the centre of the lake area) to

3.94m towards the margins.

Consequently, a gradual drawdown of the water table from the centre of the lake, mainly to the limestone fringes and less to the SE site, was defined. The highest drawdown was noticed at the northern part of the lake, where the coarse grained deposits dominate.

This configuration of the groundwater table apparently is connected to the thickness of the Quaternary deposits, the grain size distribution of the different facies, as well as the distance from the limestone fringes. Thus, the coefficient of permeability (k) obtained by means of pumping-in tests (Maag tests), revealed formations of medium to low permeability (k between 10^{-5} to 10^{-7} m/s). Also, from pumping-out tests the permeability(k) and transmissivity (T) were found equal to 10^{-6} m/s and 2 m^2/day respectively. For these last tests, the Boulder method was used, since it was considered to be the most appropriate for the examination of this semi-confined type of aquifer (Kruseman et al, 1979, Anon, 1981).

From the evaluation of these permeability characteristics, a lateral water leakage to the karstified limestone bedrock was revealed. This leakage seamed to be higher near the northern part of the lake, where the thickness of the deposits is smaller (25 to 40 m), and the coarser facies prevail. Consequently, the possible dredging of the Quaternary deposits, in the frame of the creation of a new tight reservoir, should be planned away from this part of the lake in order to avoid undesired leakages.

5 CONCLUSIONS

Summarising the results of the above considerations, the lithostratigraphic conditions and the geomechanical behaviour of the Quaternary deposits of Taka lake area, which are going to be the foundation ground of the planning reservoir, are given.

Grain size distribution curves and Atterberg limits conferm the separation of the Quaternary deposits into four main unities, apart from the surficial soil.

Moisture content values were usually found around the plastic limit, but in some cases they were as high as that of the liquid limit indicating full saturation of the relevant horizons. This is due to the permeability of the materials but also

to the percentage and type of clay mineral present. The values of the rest physical characteristics (dry and bulk densities, specific gravity)do not show significant differentiation from each other.

The uniaxial compressive strength values reveal that the Quaternary deposits in question are characterised as very firm to stiff soils and only these of the sandy-silt unity were found to be soft soil formations. Also, from shear strength parameters, it is concluded that there is a general decrease of cohesion values and a corresponding increase of friction angle, from silty clays gradually to silty sands.

From the examination of the shear strength parameters it is obvious that there is a decrease in the values of cohesion and a corresponding increase of those for the angle of friction, from the lower unity (finer) to the upper unity facies (coarser).

The compression index values for clayey silts, which dominate in the quaternary deposits, are low, indicating that no problems from consolidation settlement are to be expected.

Referring to hydrogeological conditions of the examined deposits, they show medium to low values for the coefficient of permeability (k) and Transissivity (T) (between 10^{-5} and 10^{-7} m/s and 2m^2/day, respectively).

From the above considerations of the geomechanical properties and hydrogeological condition of the Quaternary cover of the progressively downfaulted karstic limestones, they can be considered as foundation ground with satisfactory charactcristics for the development of a reservoir, but also as good embankment construction material.

Based on the above results of the lithostratigraphy and the geotechnical behaviour of the deposits in Taka area, dredging of the central and south parts of the lake in combination with the construction of a low embancment, was found to be the most suitable solution for the formation of the reservoir.

6 REFERENCES

Andronopoulos, B., Rozos, D., Nikolaou N. 1988. Geotechnical investigation of lake Taka, Arkadia P. Unpublished Report, IGME (in Greek). Athens.

Anon 1963. Primary study of S complex and of

artificial lake Rachiani. *Report E 1/15 Elektrocunsult.* Unpublished Report,. Athens.

Anon 1967. *Study of Nestani, Piana, Silymna, Taka, Kandila and Alea project. Report E 1/24 Elektrocunsult.* Unpublished Report,. Athens.

Anon 1981. *Ground water manual. A water resources technical publication. United States Department of the Interior.* USA.

BS. 5930 1981. *Code of practice for site investigation.* British Standards Institution.

Georgulis, I. P. 1984. Geological and hydrogeological investigations at Mandinia county. *PhD thesis, University of Athens (in Greek).* Athens.

Hoek, E., Bray, J.W. 1981. *Rock slope engineering.* London:The Institute of Mineral and Metallurgy.

IAEG 1979. Report of the Commission of engineering geology mapping. Classification of rocks and soils for Engineering Geological Mapping. Part I: Rock and Soil Materials. *Bulletin of IAEG.* 19: 364-371. Krefeld.

IAEG 1981. Report by the Commission on engineering geology mapping . Rock and Soil description and classification for Engineering Geological Mapping. *Bulletin of I.A.E.G.* 24: 235-274. Aachen, Essen.

Kruseman, G.P., De Ridder, N.A. 1979. *Analysis and evaluation of pumping test data.* The Netherlands: International Institude for Land Reclamation and Improvement.

Lazou, A. 1988. *Geoelectrical investigation at Taka lake area, Arcadia P.* Unpublished Report G-7455, IGME (in Greek). Athens.

Lazou, A. 1989. *Geoelectrical investigation at Taka lake area, Arcadia P.* Unpublished Report G-1403, IGME (in Greek). Athens.

Lekkas, S, P. 1978. Contribution to the geological structure of the area SE of Tripoli. *PhD thesis, University of Athens (in Greek).* Athens.

Lessons from a landslide occurred after the filling of a small reservoir

Expérience acquise à partir d'un glissement de terrain à la suite du remplissage d'un petit réservoir

G. Guidicini
Engevix Engenharia S.A

J. Lousa
CELG, Centrais Eletricas de Goias S.A.

ABSTRACT: It is usually difficult to justify costs of geological site investigations when we are dealing with projects other than large ones. Geological information available for small hydroelectric power plants is in general very poor and restricted to the "essential", as if problems should only occur on large projects. Problems are not strictly related to the size of the project itself, but rely upon factors such as the physiography of the site, the geological structure and hydrogeological behaviour. Nevertheless it is always easier to justify investigation costs for large projects than for small ones.

This paper describes the case of an unpredicted slide on a slope close to the penstocks of a small hydroelectric power plant in Northern Goiás, Central Brazil, which almost caused serious damage. Geological investigations at the design stage were restricted to the foundation of the main structures.

Once the origin of the slide was identified, it became clear that more emphasis should have been given to the geological investigations, in order to achieve a better understanding of the global hydrogeological behaviour of the site.

RESUME: Il est difficile de justifier le coût des prospections géologiques quand il s'agit d'ouvrages de petites dimensions. Les informations géologiques disponibles pour les petites usines hydro-electriques sont limitées a l'essentiel, comme si les problèmes se limitaient aux gros ouvrages. En fait, les problèmes n'ont aucun rapport avec la taille de l'ouvrage, mais sont dus à facteurs tels que la physiographie du site, sa structure géologique et son comportement hydro-géologique. Malgré tout, il est toujour plus facile de justifier le coût des prospections pour les grands ouvrages que pour les petits.

Cet article décrit le cas du glissement imprévu d'un talus voisin de la conduite forcée d'une petite usine hydro-electrique située dans le Nord de l'Etat de Goiás (Brésil central) que aurait pu causer de serieux dégâts. Les prospections géologiques, à l'occasion du projet, étaient reduites à la fondation des structures principales.

Identifiée la cause du glissement, il était clair que l'on aurait dû approfondir les prospections de façon à mieux comprendre le comportement hydro-géologique du site.

1 - INTRODUCTION

Sao Domingos hydroelectric power plant is located along Sao Domingos river, in the State of Goiás, at a short distance from the small homonymous town, at 12° 55'S and 47° 50'W of Greenwich. The owner is CELG-CENTRAIS ELETRICAS DE GOIAS S.A. and it was built between 1985 and 1987. ENGEVIX ENGENHARIA S.A. was in charge of the design of this 12 MW power plant, with two Francis turbines that have a discharge of 19,0 m³/s each, and a nominal head of 44,6m.

The site geology is very complex. Low degree metamorphic rocks, such as phyllites and schists are strongly disturbed by intrusive bodies of basic rocks represented by tonalites. The normal sequence of phyllites is cut into wedges of decametric dimension, tilted by shearing stresses caused by the intrusions. Along the area between the intake and the powerhouse, phyllites and tonalites appear alternatively. The strike of phyllites is approximately 30° divergent from the axis of the penstocks and its dip is almost vertical.

Weathering has an important role at the site. A normal weathering profile presents a clear differentiation between phyllites and tonalites. During investigations by core drillings, large and abrupt steps in the bedrock depth have been found, caused by sudden changes of lithology. Phyllites usually exhibit one to three meters of residual soils while tonalites can reach twelve or more meters, side by side. A normal sequence of soil profile is constituted by an upper horizon of mature soil that, in the case of tonalites, can reach two or three meters. Below this horizon, saprolitic soils occur, gradually increasing in strength and evolving into weathered rock. This differential behaviour in respect to weathering is justified by the high susceptibility of tonalites to surface degradation agents and to its low content of silica.

2 - SOME DATA ON THE PROJECT

The layout was developed in order to make good use of a natural drop of about 30 meters along the river valley, for a horizontal distance of about 600 meters. An earth dam, with a maximum height of 20 meters, is located just upstream of a stretch of the river with rapids. A headrace channel along the right abutment brings the water to an intake structure located just upstream of a small earth dike, at a distance of about 250 meters from the powerhouse. Photo 1 shows a general view of the site, particularly of the intake area. The spillway is just behind the flap of the plane, at the left end of the earth dam.

Figure 1 presents a sketch of part of the layout related to the headrace channel. The earth dike close to the intake structure involves two low pressure penstocks, with a diameter of 2,4 m each. The conduits are envelopped by a concrete shell. The dike is 12 meters high and its foundation is made up of saprolitic soils originated from phyllites. No grouting curtain was made

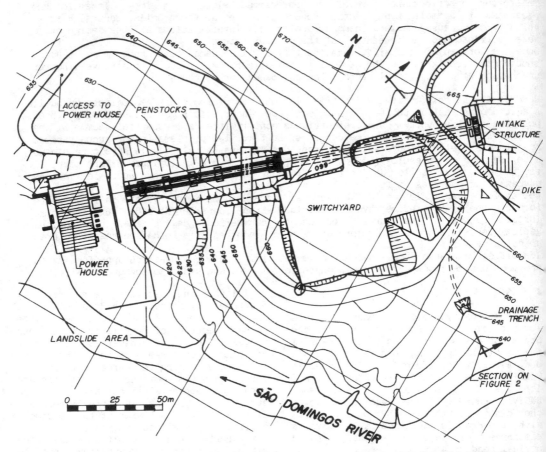

FIGURE 1 - ADDUCTION CIRCUIT AND THE LANDSLIDE LOCALIZATION

in the dike foundation, since it was considered that injections in saprolitic soils would not be effective.

The internal drainage system of the dike consists of a traditional sandy blanket, closed toward downstream but open toward south, i.e. toward the bottom of the valley. Theoretically, seepage should be collected and led through the drainage blanket and discharged along the right side of the river valley. Figure 1 indicates the position of the drainage trench in plan and figure 2 presents a vertical section parallel to the axis of the dike and along the drainage trench.

Downstream of the dike, the penstocks follow horizontally for another 100 meters, into a trench open through cut-and-cover technique and then follow the slope down to the powerhouse site. In this area, the foundation of the anchorage and supporting blocks for the penstocks required the removal of surface colluvial and residual soils, as well as of some talus bodies lying along the slope.

This slope reminds us an amphitheatre, because of its round configuration and presents a mean declivity of 23°, in spite of higher inclination angles at the upper parts. As previously mentioned, some talus bodies were clearly detected in this area, since they represent a normal mechanism of local hillslope evolution.

Figure 3 presents an aereal view of the site and the arrow indicates the landslide position.

3 - DESCRIPTION OF THE LANDSLIDE

The landslide occurred in the middle of the hillside, close to the left penstock. A total volume of about 1000 m³ of material detached from the original position along the local horizon of superficial soils and moved toward the site of the powerhouse. The horizontal displacement was of some 30 meters and

Figure 3 - General view. The arrow indicates the landslide area.

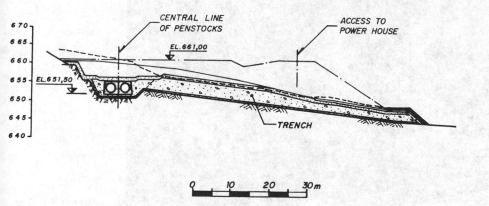

FIGURE 2 – *SECTION ALONG THE DRAINAGE TRENCH OF DIKE*

the vertical one of 15 meters. Figures 1 and 3 are intended to show the position of the area affected by the occurrence. The left penstock was slightly affected by the flank of the depleted mass, without causing damage. A cable gallery was also affected and about 50 meters of small drainage trenches were obstructed or destroyed. Figure 4 gives a clear idea of the extension and magnitude of the landslide.

The displaced material was constituted by colluvial and saprolitic soils both from phyllites and tonalites and also by some loose material left along the hillside from the construction. The main scarp of the landslide exhibited an irregular surface containing weathered rock and saprolitic soils. Schistosity is clear in upstream/downstream direction, dipping almost vertically. Along the top of the landslide multiple small ressurgences of clean water could be observed rising from an intensely weathered rock body, possibly a fault gouge. The total flow of water was estimated in two to four litres per second.

As referred above, the hillside along the amphyteatre was originally covered by low permeability colluvial and residual soils and also by some talus bodies. Along the penstocks superficial soils had been removed in order to supply adequate foundation conditions to these structures, exposing the underlying more permeable saprolitic soils and weathered rocks.

Around the area affected by the landslide, at some distance, tension cracks could be observed, indicating that new regressive movements should be expected. These cracks advanced up to the position of the access road to the powerhouse.

4 - THE ORIGIN OF THE LANDSLIDE

The origin of the landslide is clearly related to the filling of the reservoir. The lake was formed in a very short time, starting in December, 1st, 1989 and reaching the expected maximum level 8 days later. The first movements along the penstock slope were observed in January, 27th, 1990, as tension cracks. The next day, the landslide occurred. The time lag of 50 days between these events can be considered as the period of maturation of

Figure 4 - The landslide seen from downstream

the conditions necessary to the process outbreak.

The mechanics of the occurrence probably registers a first phase of gradual seepage of water from the headrace channel, inside the reservoir, along the favorable attitude of the rock mass, toward the penstocks hillside. The strike of phyllites, almost coincident with the direction of the penstocks, and its vertical dip, favour the movement of water in that direction. A second phase probably consisted in the storage of water and growing of internal pressure retained by the layer of superficial soils, considerably less pervious than the underlying rock mass (in spite of the existence of a strip of terrain, along the hillside, where soils had been removed). At last, internal water pressure must have overcome the mechanical resistance of the soils, inducing the landslide.

Based on the geometry and the duration of the problem, the mean permeability of the envolved stretch of the hillside resulted as 2×10^{-2} cm/s.

At this time, the internal drainage system of the small dike remained absolutely dry. At the south side of the dike, along the outlet of the trench, no water was observed. The real behaviour of water did not follow the design assumptions, ignoring the drainage system. As a matter of fact, water from the headrace channel flowed toward the structured and permeable deep rock mass, emerging along the penstock hillside, since the attitude of the rock layers favoured this route.

Figure 5 clearly shows that the crown of the area affected by the landslide was located some 25 meters below the water level of the upstream reservoir, at a horizontal distance of about 150 meters.

5 - ON THE QUALITY OF GEOLOGICAL INVESTIGATIONS

Once the origin of the slide was identified, it became evident that more emphasis should have been given to the geological investigations, in order to achieve a better understanding of the global hydrogeological behaviour of the site.

The geological investigations undertaken to support the design decisions included core drill holes with modified Lugeon pressure tests, percussive drilling with infiltration tests, trenches and inspection pits and also permeability tests, besides surface mapping and conventional investigation for construction materials, through field works and laboratory tests.

17 NX double barrel core drillings were executed, being a total of 320 meters, with 30 pressure water tests and 43 infiltration tests.

Another 10 drill holes had been executed previously, corresponding to 170 meters of poor recovery and without water tests. The mean depth of core drillings was 18 meters. Percussive holes were 11, totalizing 80 meters. A significant part of core drillings were executed in the spillway area, located at the left side of the main earth dam, due to

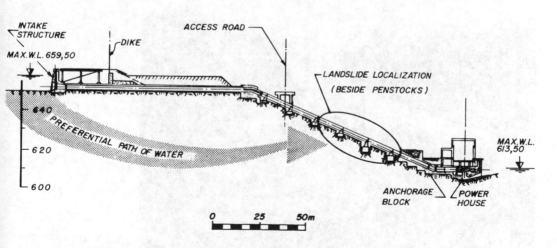

FIGURE 5 - SECTION THROUGH THE ADDUCTION CIRCUIT

difficulties to provide reasonable foundation conditions for the concrete structures, since weathering was particularly intense and deep in that area.

Site investigations were restricted to the "essential", and drill holes were strictly executed with the purpose to define the quality of the foundations of the structures, rarely exceeding 20 meters in depth.

6 - FINAL CONSIDERATIONS

The repair of the slope was carried out with a retaining wall of gabions, consisting of four tiers of stepped-front units. Above the gabions, the hillslope was rebuilt with compacted soil, as can be seen in photos 4 and 5. A drainage blanket was placed above the zones of depletion and accumulation of the landslide, leading all the seepage water to a main collecting pipe, as seen in Figure 6. The flow was about 60 l/min, practically constant since the reconstruction of the slope.

As a final analysis of the design behaviour, one can say that design values of the internal drainage system of the small dike close to the intake structure were not confirmed in the prototype. Due to the high permeability conditions of the foundation and to a peculiar physiographic situation, a flow net was established toward the penstock area. The lateral drainage trench of the small dike was useless. This flow net was particularly intense and caused a

landslide close to the left penstock. Fortunately, the destructive action of the unexpected landslide was limited and required small expenses to correct.

The erroneous assumptions were due to a lack of information regarding the global hydrogeological behaviour of the site. No investigations were oriented toward the global hydrogeological characterization of the site.For this to have been done, the depth of drilling should have been increased, according to the dimension of the geological framework. Due to the size of the project, it would be quite difficult to justify an investment on deep drill holes.

The unexpected slide in Sao Domingos power plant fortunately did not result in heavy damages, but it reminds us of the need for an appraisal of geological questions in a global context, even in small projects.

REFERENCES

IAEG Commission on Landslides. Suggested nomenclature for landslides. Bulletin of IAEG, 41, pg. 13- 16. 1990.

AKNOWLEDGEMENTS

The authors aknowledge CELG - CENTRAIS ELETRICAS DE GOIAS S.A. and ENGEVIX ENGENHARIA S.A. for their permission to present this paper to the IAEG 7th Congress.

Figure 6 - The landslide area after repair

Geology and geotechnics of the rock mass of the Caldeirão dam site (Portugal)

Géologie et géotechnique du massif de l'endroit du barrage de Caldeirão (Portugal)

J.M.Cotelo Neiva, Celso Lima & Nadir Plasencia
Electricidade de Portugal, Oporto, Portugal

ABSTRACT: Seismic refraction, rock weathering, faulting and spacing of joint sets helped to define the zoning of the schist-granitic rock mass, which was completed by in situ and laboratory mechanical tests. This zoning and a detailed geological study helped to define the dam site and to choose the ground improvement and the devices to observe the behaviour of the foundation.

SOMMAIRE: La réfraction sismique, la météorisation des roches, les failles et l'espacement des diaclases ont permis de définir un zonage du massif schisto-granitique, ce qui a été complété par des essais mécaniques. Ce zonage et une reconnaissance géologique détaillée ont permis de définir une localisation précise pour le barrage et de prévoir le traitement du massif rocheux, ainsi que le système d'observation du futur comportement des fondations.

1 INTRODUCTION

The own flows of the Caldeirão brook are relatively small, about 21 million m3 in an average year. Therefore, a water diversion of 95 million m3 from the Mondego is added to that brook in an average year. All the water is dammed in that brook at 920 m from its mouth in the River Mondego. Profiting by a fall of 193 m, a pressure tunnel and penstock carry the water to a powerhouse which is located nearby and downstream from the Mizarela bridge on the Mondego (Fig. 1).

The Caldeirão dam was placed to create a reservoir which, at its full storage capacity at level 702, has a volume of 5.5 million m3.

The Caldeirão multipurpose development generates 44 million kWh in an average year, the city and surrounding regions of Guarda being also supplied with 1 million m3 of water per year.

The geology, the geotechnics and the foundation treatment of this dam, as well as some devices for observation of the behaviour of this foundation, will be discussed.

2 LOCATION

The dam site is situated 35 m upstream of the old bridge and the Caldeirão gorge and near the Mocho Real mirador, in the county of Guarda and 5.2 km WSW of this town.

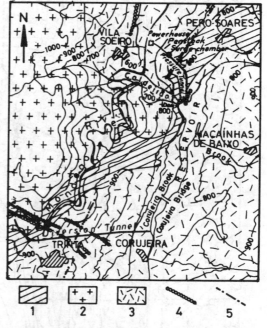

Fig. 1 General layout of the Caldeirão development. 1-mica-schist and hornfels, 2-medium-grained granite, 3-coarse-grained porphyritic granite, 4-dolerite dyke, 5-quartz vein.

The Caldeirão brook, which is a tributary of the right bank of the River Mondego, flows along a plateau with incised meanders in granite and schists, and also in longer segments along the NE-SW, N-S, E-W and, more rarely, WNW-ESE faults.

The dam is placed immediately upstream of the Caldeirão gorge at the granite-schist contact and in a vertical E-W fault. In a geomorphological section along the old Mocho Real bridge, the V-valley has a thalweg of 125° included angle (slopes 35° on the left bank and 20° on the right bank), above level 685; but that angle is of 25° with slopes 79° and 76°, respectively, on those banks, below level 675. 35 m upstream from this place, a section along the reference surface of the arch dam is almost symmetrical with 127° included angle, although the left bank in the granite is convex-concave and the right bank in the hornfels is convex.

Pre-Ordovician graphitous hornfels and mica-schist, which locally alternate with metagraywacke, outcrop in the E and centre of the dam region (Fig. 2). They are very folded, with WNW-ESE folds and N75°-90°, 80°NNE-90°-80°SSE schistosity, which are mainly obliterated by later N-S folds, with 75°E dip and N0°-15°E, 65°-90°E schistosity. This schistosity had a torsion during the Hercynian granitic emplacement, and it passed locally to N30°E, 65°-85°ESE. The joint sets of schists are shown in Table 1.

In the W, a fine to medium-grained biotite-muscovite granite magma intruded the schists concordantly with the main Hercynian folding and only produced a low degree of contact metamorphism. In the SW, a medium-grained porphyritic biotite-muscovite granite magma intruded that granite. In the NE and SE, the

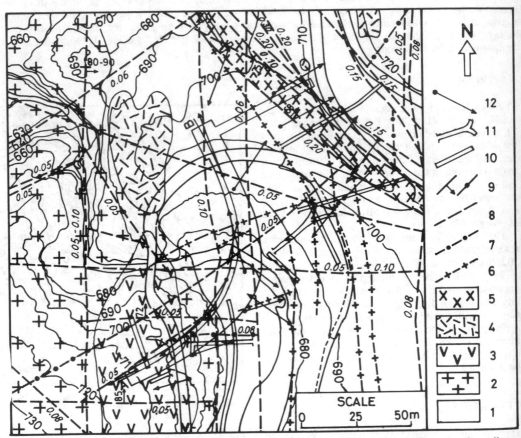

Fig. 2 Geological map of the Caldeirão dam site. 1-mica-schist, hornfels and metagraywacke; 2-medium-grained granite; 3-medium-grained porphyritic granite; 4-coarse-grained porphyritic granite; 5-dolorite vein; 6-pegmatite vein; 7-quartz vein; 8-fault; 9-dip (sloped, vertical); 10-trench; 11- gallery; 12-rotary drill. A-B - Reference surface of the dam.

coarse-grained porphyritic biotite-muscovite granite magma intruded the schists. The porphyritic granites produced a mica-schist and hornfels contact metamorphic aureole. The joints of granites are shown in Table 2.

Aplite-pegmatite veins are mainly orientated N5°W-N15°E, 65°-90°E. Others are orientated N80°-90°W, 80°N-90°-80°S, and others are N30°-40°W, subvertical. They mainly cut schists, but they produced tourmalinization and muscovitization at the direct contact.

Some quartz veins cut the schists and granites; they are generally narrow, subvertical and orientated: a) N80°W-N80°E, 60°-70°N, the most common; b) N0°-15°E, 65°-80°E; c) N43°-50°W, vertical; d) N60°E, 80°NW. Some contain wolframite and sulphides.

A 9.0-10.50 m-thick dolerite vein, N50°W, 68°-70°NE, along a fault, cuts mica-schists. Some quartz veins cut this dolerite vein.

The faults shown on the geological map (Fig. 2) are given in sets in Table 3. The fault spaces are filled with gouge and, locally, with quartz vein and seldom with dolerite vein (Neiva, 1986).

The valley of the Caldeirão brook has a late alluvium.

A geological section along the reference surface of the dam is given in Fig. 3.

In a circle with a radius of 30 km and with the centre on the dam site, several earthquakes occurred at historical times. The most important had a magnitude ranging between 3.5 and 5.5, and the nearest epicenter was located 30 km from the dam.

5 MECHANICAL SURVEYING

In 1965, the mechanical exploration comprised galleries and boreholes for a gravity dam (Almeida, 1966). In 1985-1986, trenches and new boreholes were carried out for an arch dam, Fig. 2 (Neiva, 1986).

The galleries and boreholes showed that the rock mass of the right bank towards NE is more altered (weathering reaches a depth from 2 to 19 m) than in the N (weathering is up to a depth of 5 m) and on the left bank (weathering is up to a depth of 1-2 m).

The depth of weathering is smaller in granite and hornfels than in mica-schist and dolerite.

6 PERMEABILITY TESTS

Lugeon tests were carried out in boreholes at 0.5, 1, 0.5 MPa pressures (Tecnasol, 1986, Teixeira Duarte, 1985). The loss of water and the small absorption

Table 1. Joint sets in hornfels and mica-schist

Sets	Strike	Dip	Spacing (m)	Size	Frequency (%)
A	N5 W - N15°E	64 - 90° E	0.40 - 1.50	great	27
B	N65 - 90° W	80°NNE-90°- 75° SSW	0.30 - 1.20	great	24
C	N5 - 25° E	14 - 30° WNW	0.20 - 0.50	medium	24
D	N65 - 80° W	8 - 15° NNE	0.40 - 1.20	small	10
E	N60 - 75° E	73 - 90° NNW	0.30 - 1.30	small	9
F	N70 - 80° E	35 - 45° SSE	-	small	3
G	N22 - 28° E	60 - 68° ESE	-	small	3

Table 2. Joint sets in granites

Sets	Strike	Dip	Spacing (m)	Size	Frequency (%)
a	N80 - 90° W	80° N - 90° - 80°S	0.20 - 1.50	great	41
b	N0 - 20° E	65 - 90° E	0.25- 1.40	great	23
c	N10 - 15° E	15 - 40° WNW	0.10 - 2.00	medium	14
d	N45 - 50° E	65 - 90° SE	0.10 - 0.60	small	9
e	N5 - 15 ° W	33 - 50° ENE	0.25 - 0.55	small	7
f	N55 - 65° W	vertical	-	small	3
g	N15 - 25° W	vertical	-	small	3

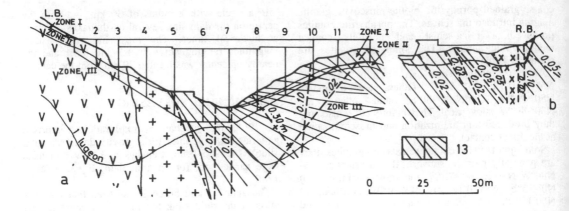

Fig. 3 Geological and geotechnical section along the reference surface of the dam: a. of the arch dam, b. of the small gravity dam. Legend as in Fig. 2. 13-schistosity.

will not be considered. In granites and schists, whether the flows are in very narrow joints or joint filling with altered materials (laminar flow - L) or in open and relatively clean joints (turbulent flow - T) and intermediate cases (LT flow), occur in about the same proportions, but the L regimen predominates in the dolerite vein.

The tests showed that the data on the elastic recovery index of joints are mainly negative, but there are several positive and 0 results. However, the rocks present mainly micro-joints.

The 1 U.L. curve is shown in the geological-geotechnical section (Fig. 3).

Table 3. Fault sets

Sets	Strike	Dip	Thickness (m)
1	N30-50°W	65°NE-90°-83°SW	0.10 - 0.60
2	N5°W-N20°E	63°E-90°-63°W	0.05 - 5.00
3	N85-90°W	77°N-90°-77°S	0.02 - 0.10
4	N55-68°E	Vertical	0.03 - 0.27
5	N62-75°W	75°NE-90°-83°SW	0.05 - 0.40

7 GEOPHYSICAL EXPLORATION AND GEOLOGY

Electric resistivity and seismic refraction of surveying were carried out (LNEC, 1984).

The map of apparent resistivity of the right bank in hornfels and mica-schist shows that: a) the highest values of electrical resistivity predominate in NW and W in the hornfels and at the direct contact with

the granite; b) the lowest values belong to the altered dolerite vein; c) a fault was found and there is also some information on two others.

An attempt was made to relate the seismic profiles to the weathering degrees and rock discontinuities. On the left bank, in the W2,F2 granite, the velocity of the compressional waves is 2,000-3,000 m/s in a thickness of 1 m and 4 m at low and high levels, respectively, but, in the W1,F1-2 granite, it is > 4,500 m/s immediately beneath this surface; in the W2,F2-3 schists, it presents a V1=2,000 m/s in a thickness of about 2 m from level 686 to near the valley floor; immediately underneath and on the valley floor, in the W1,F1-2 schists, it has a V1≥4,000 m/s. On the right bank, the W3,F3-4 schist has V1=700-1,500 m/s and corresponds to a developed zone of increasing thickness up to 2 m or 4 m from near the valley floor to level 680 and 701, respectively; below this zone, there is another zone of W2,F2-3 schist, with a V1 varying between 1,900 and 2,500 or 3,000 m/s and with an increasing thickness of 0.5 m and 3 m on the valley floor and at high levels, respectively; and underneath the last zone, another with W1,F2 presents a V1≈4,000 m/s.

8 MECHANICAL TESTS AND GEOLOGY

The mechanical tests were carried out (LNEC, 1967). They are related to the geology.

Tensile strength and deformability moduli, determined in samples extracted from galleries and boreholes, are related to the weathering and jointing degrees (Table 4).

In mica-schist and W1,F2 hornfels, the values of deformability modulus are 14.0-43.4 GPa,

Table 4. Uniaxial strength and deformability moduli of rocks

		Uniaxial strength (MPa)	Deformab. moduli (GPa)
Granite	W1, F1-2	133.3-167.8	40.0-51.8
	W1-2, F1-2	112.0-153.0	45.2-49.6
	W2-3, F2-3	-	28.3-35.0
	W2-3, F2	-	24.5-29.8
Aplite	W2, F2	81.0	-
Mica-schist and Hornfels	W1, F2	49.6-103.2	44.0-53.0
	W1, F2-3	35.0	16.6-21.1
	W2-3, F2-3	15.0-20.0	14.0-16.0
	W4. F3-4	2.3	6.5-9.0
Dolerite	W2, F2-3	183.5	55.0
	W2-3, F3	-	37.0

parallel to the schistosity. In prisms immersed in water for 45 hours, nonlinear elastic behaviour with secant elasticity moduli of 30.0 GPa and Poisson's coefficient of≈0.20 were determined in the W1-2,F1-2 granite; the same behaviour and 20.0 GPa and≈0.15, respectively, were obtained in the W1-2,F2 schists.

The strenght and deformability moduli ellipsoides are orthotropic in mica-schists and hornfels with almost cyclical sections parallel to the schistosity. In the granite, they are scalene; the ellipsoid of strength is positive with the axis perpendicular to the N10°-15°E,15°-40°WNW joints; and the ellipsoid of deformability moduli is negative with ZZ axis orientated N60°W.

Joint shear tests in cores gave the following data: 0.07-0.2 MPa for cohesion and 30°-45° for friction angle in W2-3,F2-3 to W1,F2 granite; and 0.16-0.38 MPa and φ =29°-40° in W1,F2 hornfels passing to mica-schist; and 0.21 MPa and φ =52° in W3,F3 hornfels passing to mica-schist.

Large flat jack deformability tests were carried out "in situ" in mica-schist passing to W1,F2 hornfels. On loaded surfaces parallel to the schistosity, 6.0-27.0 GPa were obtained; the directions of maximum deformation form angles of ≈ 22° with the perpendicular to the schistosity. In creep tests, this deformation and primary recovery occurred in less than 10 hours, and the sudden decreases in the deformability modulus were generally normal.

In joint shear tests "in situ" in W1-2,F2 mica-schist passing to hornfels, the cohesion and friction angle were 1.1-3.0 MPa and 48°-59° in tests perpendicular to the schistosity; 1.3-1.4 MPa and 62°-64° in those parallel to the schistosity; and 0.6-0.8 MPa and 30°-56° in the schistosity plane.

9 GEOTECHNICAL ZONING

Rock mass weathering and joint spacing degrees in the surface, trenches, galleries and boreholes, the seismic wave velocities, the tensile strength and the deformability moduli, determined in samples from cores, helped to define the zoning of the rock mass (Table 5).

The limit between zones II and III almost coincides for all the used factors; however, in the right bank, the limit between zones I and II, given by the seismic wave velocities, is from 0.5 to 1 m above that based on the degrees of weathering and joint spacing, which is attributed to the rock mass decompression (Fig. 3).

Table 5. Geotechnical zoning of the granitic rock mass

	Zone	W	F	Seismic Waves Velocities (Vl) (10³ m/s)	Uniax. strenght (MPa)		Defor. Moduli (GPa)		Thick. (m)
					Granite	Hornfels and mica-schist	Granite	Hornfels and mica-schist	
Left bank	I	W3	F3	800	-	-	< 25	10 - 15	0.2 - 2.0
	II	W2	F2	2000 - 3000	81	35	35 - 40	17 - 21	0.5 - 4.0
	III	W1	F1-2	> 4500	133-168	50 - 103	45 - 52	44 - 53	> 30
Valley floor	III	W1	F2	>4000	-	50 - 103	-	44 - 53	> 30
Right bank1	I	W3	F3-4	700 - 1500	-	8 - 13	-	10 - 15	0.2 - 6.5
	II	W2	F2-3	1900 - 3000	-	35	-	17 - 21	0.5 - 2.5
	III	W1	F2-3	4000	-	50 - 103	-	44 - 53	> 28

W-weathering   F-joint spacing

10 GEOTECHNICAL DISCUSSION

An arch dam is easily adaptable to the geomorphological and geotechnical characteristics of the site placed in good granite, hornfels and mica-schists with good mechanical characteristics (zone II). On the right bank, this dam is extended to a small gravity dam which is partially on a poorer foundation.

The orientation both of joints and faults, although the latter have a low fault spaces, and of the quartz veins indicated that the consolidation grouting should be carried out by cement injections after washing out. Each fault and each vein should be treated by fans of holes with washing out before the cement injection. All the foundation treatments must be applied down to a fixed depth, which will be a function of the bulbs of pressure of the dam, hydrostatic pressure, thermal variation of seasons and a probable earthquake.

The depth of weathering of the thick dolerite vein, underneath the gravity dam, must be considered.

In order to reduce the permeability, the orientations of joints, faults and veins must be taken into account to define the grout curtain, which must go 10 m below the 1 U.L. permeability curve.

A good drainage system should be oriented to cut joint sets, faults and veins, in order to avoid possible uplift pressures beneath the dam blocks (Neiva, 1986).

It will be convenient to consider an earthquake of magnitude 5.5 for the dynamic calculations of the dam stability.

11 FOUNDATION AFTER EXCAVATION

After the excavation of the arch dam foundation, W1,F2 granite and W1,F2 hornfels and mica-schist were found on the left bank and valley floor, but the hornfels and mica-schist showed more joints and both were W1,F2-3 on the right bank.

In the foundation of the small gravity dam, the mica-schist is W1-2,F2-3 up to the dolerite which is W5,F5; and the mica-schist between two faults is W3-4,F4 and it is W2,F2-3 towards the SE fault.

All the faults have narrow fault spaces (0.01-0.10 m), and they were detected in the surface geological survey (Fig. 3).

12 THE DAM

The concrete composite dam has the following characteristics: a double curvature arch dam, 39 m high and with a 122 m-long crest, a right concrete abutment and a small gravity dam (Figs. 2, 3).

13 CONSOLIDATION GROUTING AND GROUT CURTAIN

In the arch dam, a fault special treatment was not applied because the faults are only a few centimetres thick in depth, and the opened joints are not significant.

The stresses induced by the dam on the rock mass were determined, in order to design a deep consolidation by grouting. A total of 36 fans of 11 holes each were defined. These holes, which irradiate from the general drainage gallery and with a maximum space of 5 m, comprise a zone with pressure increases higher than 0.3 MPa. As the rock foundation is very good and of small permeability, uneven and even boreholes were drilled in alternated profiles.

In all treatments, the washings out were done under pressure between holes. They were injected with blast furnace cement (1/4 and 1/2) under the grouting pressure which increases with depth, although a minimum of 0.75 MPa was always reached. In order to avoid cement injections at the dam joints, a copper slice was placed parallel to the concrete-rock contact in each joint (EDP, 1988).

The grout curtain was projected with holes whose orientation is tangent to the reference surface of the dam. These holes were orientated normally to the banks, crossing each joint set with angles of about 60°. Primary holes, separated 5 m, were drilled (Fig. 4a-1). Secondary holes were necessary in the zones of greater absorption (from the left abutment to the J4 joint and from the J10 joint to the J11 joint), and each was placed between primary holes (Fig. 4a-2). Forty primary holes and nine secondary holes were drilled, in a total of 1,311 m of drilling.

To improve the superficial ground foundation upstream of the main curtain, a complementary curtain with 10-15 m-long holes, dipping 450 towards upstream and each placed at the entrance of a primary hole, was carried out. However this complementary curtain was only performed at zones with high absorptions of cement (Fig. 4a-3, 4c-3).

In order to close the main curtain, 8 holes were drilled on the left bank from the crest, as shown at Fig. 4a.

As the dolerite vein is altered up to a depth of 18-24 m, an excavation up to 5 m was carried out under the gravity dam, where a reinforced beam was built, making a bridge between the two mica-schist walls of the vein and supporting the upper loads. A general consolidation grouting was carried out with shorter fans of holes and cement injections in the dolerite vein. The primary grout curtain was done as it would be for the arch dam; the secondary curtain was reduced (Fig. 4b). In order to close the curtain at the right bank, a fan of 6 holes in a vertical plane was carried out from the crest (Fig. 4b).

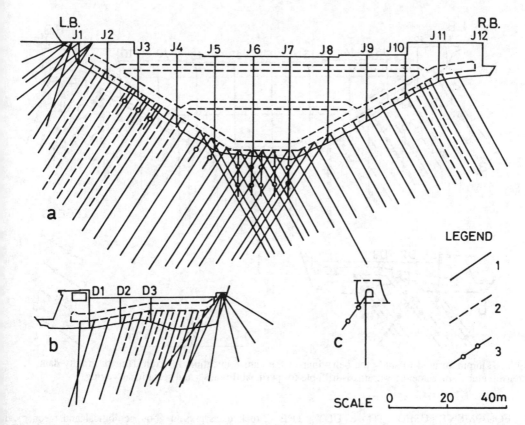

Fig. 4. Grout curtain: a. of the arch dam, b. of the small gravity dam, c. radial section perpendicular to the foundation. Drillholes for impermeability: 1-primary, 2-secondary, 3-complementary.

A drainage system was carried out, in order to avoid possible uplift pressures beneath the arch dam blocks. Three 30 m-long holes were drilled in each block and perpendicular to the foundation surface, except in the J5-J6 and J7-J8 blocks where four holes were drilled in each one (Fig. 5a-3).

In the small gravity dam, two holes of dip 51° were drilled in each block. A hole dipping 51° was also drilled in the D1-D2 block (Fig. 5b-3).

14 BOREHOLES TO STUDY THE BEHAVIOUR OF THE DAM FOUNDATION WITH TIME

Before the foundation treatment eleven vertical holes were drilled in the general drainage gallery in the arch dam (EDP, 1988), each one being 28 m long and in the middle of each block (Fig. 5a-4). They cut the W1,F1-2 granite with RQD of 99% and the W1,F2 to W1,F2-3 hornfels and mica-schist with RQD of 98.5%. The permeability tests showed < 1 lugeon in the last segments of holes 3 to 8, 10 and 11 while they showed > 1 lugeon in holes 1, 2 and 9.

The seismic surveying among boreholes, as well as between them and the foundation surface, showed the S wave velocity of > 4,500 m/s between holes 2 and 9, the same velocity 2 m below the surface between holes 2 and 1, and also 1.5 m below between holes 9 and 11.

After the foundation treatment was applied, the permeability became 0 U.L. in eight boreholes, in the last 20 m of the borehole on the valley floor and also in the last 11 m and 23 m of boreholes 1 and 3, respectively, of the left bank.

The seismic surveying was repeated. Unfortunately at this moment we have not those results, but we expect some improvements on the superficial ground.

Permeability tests and seismic surveying should be carried out some months after the reservoir is full and also each four years. If any behaviour changes in the ground foundation are detected by the equipment placed therein, those permeability tests and seismic surveying must be carried out and the data compared with the previous results.

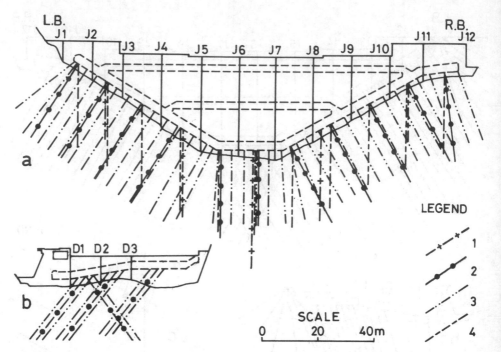

LEGEND

SCALE
0   20   40m

Fig.5 . Equipment used to study the behaviour of the dam foundation. a. Arch dam; b. gravity dam. 1. extensometer, 2-piezometer, 3-drain, 4-drillhole for permeability tests and seismic surveying.

15 EQUIPMENT USED TO STUDY THE BEHAVIOUR OF THE DAM FOUNDATION

In addition to the 11 vertical boreholes for permeability tests and for the seismic surveying, there are also 35 boreholes used as drains in the arch dam, and 7 drains in the gravity dam.

Eleven piezometers, each being 30 m long, were placed in the middle of each block and perpendicular to the arch dam foundation contact (Fig. 5a-2). Furthermore, 5 foundation extensometers were also placed (Fig. 5a-1).

In the small gravity dam, 4 piezometers were placed, 2 in the D1-D2 block and one in each of the other two blocks (Fig. 5b-2).

16 CONCLUSIONS

1. Some years ago, it was intended to build a gravity dam, but the geomorphology and the geological structures which were very near and upstream of the Caldeirão gorge enabled to build an arch dam extended to a small gravity dam.

2. The dam site is situated at the contact of Hercynian granites with hornfels and pre-Ordovician mica-schist. All are cut by pegmatite veins, quartz veins and a dolerite vein (Fig. 2). The rocks and the rock mass present good geological and mechanical characteristics. The foundation was defined after the zoning of the rock mass (Fig. 3).

3. In the Lugeon permeability tests, the (L), (T) and intermediate (LT) cases occur in the same proportions, except in dolerite where the (L) predominates.

4. The arch dam was founded in zone III, in W1,F1-2 granite and hornfels, passing to W1,F2 and W1,F2-3 mica-schist, presenting a few faults with a thickness of 0.01-0.10 (Fig. 3). The foundation of the small gravity dam is situated on the W2,F2-3 mica-schist and on the W5,F5 dolerite vein, which implied the construction of a reinforced beam making a bridge between the two mica-schist walls of the vein.

5. It was not necessary to apply a consolidation at the surface of the rock mass. However, it was applied, in depth, in fans of holes drilled from the general drainage gallery. Their holes were injected with blast furnace cement, under pressure. In the foundation of the small gravity dam, it was applied a general consolidation with shorter holes in the altered dolerite vein.

6. The grout curtain was projected based on the orientations of the discontinuity sets of rock mass. It was applied with cement injections in primary, secondary and complementary holes (Fig. 4).

7. The drainage system of the arch dam foundation consists of 35 holes, each being 30 m long. Eleven piezometers and 5 extensometers were placed at the foundation (Fig. 5a). In the small gravity dam, 7 drains were drilled and 4 piezometers were placed (Fig. 5b).

8. A device of eleven vertical holes, each being 28 m long (Fig. 5a), enabled to carry out the seismic tomography and the permeability tests of the foundation before and after its treatment. We expect some improvement on the superficial ground. These studies should be repeated periodically, and also when both there is a large increase in the flow of the drainage system and there are significant changes in the data given by the instruments placed in the foundation. So, changes in the foundation can be detected and treated adequately.

REFERENCES

Almeida, F. M. 1966. Escalão de Vila Soeiro. Estudo geológico. Unpublished report.

EDP 1988. Aproveitamento hidroeléctrico do Caldeirão - Anteprojecto. Unpublished report.

LNEC 1967. Estudo das fundações da barragem do Caldeirão. Unpublished report.

LNEC 1984. Prospecção geofísica em locais da barragem do rio Mondego e afluentes (1º relatório) - Local de Vila Soeiro. Unpublished report.

Neiva, J. M. C. 1986. Geologia e geotecnia do local da barragem do Caldeirão. Unpublished report.

TECNASOL, Lda. 1986. Barragem do Caldeirão. Reconhecimento por sondagens. Unpublished report.

Tcixcira Duartc, Lda. 1965. Escalão de Vila Soeiro. Barragem do Caldeirão - Reconhecimento por sondagens. Unpublished report.

Teixeira Duarte, Lda. 1985. Inventário do Alto Mondego. Barragem de Vila Soeiro. Reconhecimento geotécnico. Unpublished report.

The zoning of granitic rock mass of the Alvarenga dam site (Portugal)
Le zonage du massif granitique de la région du barrage d'Alvarenga (Portugal)

J.M.Cotelo Neiva
Electricidade de Portugal, Oporto, Portugal

ABSTRACT: A 155 m-high arch dam will be constructed at Alvarenga. Geological and geophysical surveyings of the dam site, as well as mechanical tests on the granitic rock mass, were carefully carried out. Two zonings were defined: one is based on rock weathering, degree of jointing, decompression and RQD, while the other is based on longitudinal seismic wave velocities. The two zonings are compared and discussed. They partially overlap. A final zoning is defined and completed with data on permeability and mechanical properties of the granitic rock mass. Ground improvement and a system to study the behaviour of the foundation with time are mentioned.

SOMMAIRE:On a procédé à une étude géologique, géophysique et à celle des propriétés mécaniques du massif et du granite de l'endroit où on construira le barrage-voûte de 155 mètres de hauteur. Ceci permettra d'établir un zonage fondé sur la météorisation du granite, l'espacement des diaclases, la décompression, le RQD et un autre zonage fondé sur la vitesse des ondes sismiques longitudinales. Les 2 zonages sont comparés et discutés; un zonage final est établi, complété par des valeurs de pérméabilité et de propriétés mécaniques du massif granitique. Des considérations relatives au traitement du terrain et à celui des systèmes d'observation du comportement des fondations y sont exposées.

1 INTRODUCTION

A dam with a height of 155 m above the foundation will be constructed at Alvarenga (Arouca country, northern Portugal) in the valley of the River Paiva, on a granitic outcrop. The crest of this arch dam will reach a total development of 430 m at level 284, and it will create a reservoir of 843×10^6 m³ at level 283.

The geomorphology and geology of the dam site and the mechanical properties of the rocks helped to define two zonings in the rock mass. They partially overlap. One of them is completed by the other and by the mechanical properties. The conditions of dam foundation, the ground improvement, the drainage system and the study of the evolution of the foundation with time will be discussed.

2 LOCATION AND GEOMORPHOLOGY

The dam will be situated at 120 m SW of the church of the Alvarenga village (Arouca country), in a narrow 600 m -long segment of the valley of the River Paiva.

Near the end of the Pliocene, the River Paiva, tributary of the left bank of the River Douro, had several meanders cutting a plateau. Tectonic movements caused a continental uplift and a rejuvenation at the mouth of the Douro. There was a rejuvenation of the relief and downcutting of the Douro and its tributaries. Nowadays, the Paiva has steep banks and incised meanders.

On the dam site, the valley has a thalweg 105° include angle with slopes of 41° on the left bank and 34° on the right bank, and it was excavated along the NNW-SSE granite joints and a narrow fault.

The detailed geomorphology of the area suggests that, up to level 215, the river would have downcut due to slow decreasing of the base level of the Paiva, because the slopes of the left and right banks are currently 37° and 33°, respectively. From this level to level 170, the downcutting was quicker, nowadays presenting slopes of 61° and 40°, respectively; and up to level 145, it was slower with slopes of 43° and 48° respectively. Between levels 145 and 140, the valley is almost flat and of 36 m wide; the erosion uplifted granitic plates, limited by subhorizontal decompression joints and subvertical tension and traction joints in the valley floor.

3 GEOLOGY

The dam site is situated on the Alvarenga granitic outcrop whose geological surveying was carried out on the scale of 1/500, but show reduced in Fig. 1.

The medium-grained two-mica Hercynian granite is locally porphyritic, and it shows a WNW-ESE lineation of micas and rarer of feldspar megacrysts. It contains biotite-rich xenoliths.

In the dam site, mainly at high levels, the granite is more decompressed and altered in the left back than in the right bank, due to jointing and insolation.

The prismatic structures are common in granite; however, at high levels, there are also concentric structures, mainly on the left bank. The subvertical and subhorizontal joints of this granite are given in Table 1. E joint set cuts B joint set; D joint set

contain some wolframite, arsenopyrite, pyrite, sphalerite, galena and chalcopyrite.

In the dam site and in its surroundings, several faults cut the granite and the sets are show in Table 4. The slickensides with striae of slipping are common.

4 SEISMICITY

The epicentres from 1130 to 1975 with intensities between III and IV and magnitudes ranging between 3.4 and 5.5, were plotted in a circumference with a radius of 50 km and with its centre placed on the dam site. The epicentre of the earthquake with a magnitude of 5.5 was located 30 km from the dam site and, for a return period of 1000 years, the

Table 1. Main joint sets of granite

Set	Strike	Dip	Spacing (m)	Size	Frequency
A	N5E-N30°W	75°E-90°-70°W	0.30-3.50	great	great
B	N65-100°W	50°S-90°-70°N	0.10-4.00	medium	great
C	N0-25°E	75°E-90°-75°W	0.15-2.00	great	medium
D	N5-30°W	4-42°ENE	0.15-2.00	medium	great
E	N75-105°E	3-25° S	0.20-2.00	medium	great
F	N30-50°E	18-60° NW	0.10-1.50	small	medium
G*	N5-25°E	7-45°WNW	0.30-4.00	medium	medium
H**	N4-38°W	5-45°WSW	0.38-2.00	medium	medium
I**	N45-60°W	25-63°NE	0.20-0.50	small	small

* Only in the left bank. ** Only in the right bank

locally cuts B and C joint set; A joint set cuts joints of all other sets and, locally, displaced them.

The aplite and quartz veins, which are mentioned in the Table 2 and 3 respectively, cut the granite.

maximum calculated acceleration is 93 cm/s^2 ,and the maximum velocity is 8.6 cm/s. It will be convenient to predict an earthquake with a magnitude of 5.5 for the dam dynamic analysis.

Table 2. Aplite vein sets

Set	Strike	Dip	Thick.(m)	Freq.
1	N80-90°E	62-90°N	0.30-0.70	medium
2	N57-70°W	60-64°NE	0.20-2.00	medium
3	N27°E	Subver.	0.08	small
4	N27°W	80°ENE	0.1	small

In the direct contact with the quartz veins, the granite shows partial sericitization of its feldspars and enrichment in sulphides. The quartz veins

Table 3 . Main vein sets of quartz

Set	Strike	Dip	Thick.(m)	Freq.
a	N2-20° W	85-85°ENE	0.02-0.08	great
b	N5-30°E	82-90°WNW	0.01-0.30	great
c	N35-45°E	56-90°NW	0.02-2.00	great
d	N18-30°W	30-90°WSW	0.01-0.08	med.
e	N67-70°W	23- 48°SW	0.05-1.00	med.
f	N-S	19°W	0.20-0.60	small
g	N40-45°W	38-63°NE	0.01-0.20	small

Table 4. Fault sets

Set	Strike	Dip	Thick.(m)
I	N0-18°W	79-90°E	0.02 - 0.25
II	N3-22°E	82°E-90°-80°W	0.05 - 0.50
III	N30-48°E	68-80°NW	0.07 - 1.20
IV	N10-30°W	12°-42°ENE	0.05 - 0.70
V	N57-62°W	64-70°NE	0.05 - 0.10
VI	N80-90°E	68-90°N	0.05 - 0.10
VII	N3°W	20°W	0.03 - 0.15

5 MECHANICAL SURVEYING

Initially, the TE-290, TE-260, TE-225 and TD-290, TD-266, TD-230, TD-200 trenches were excavated. The three TE are on the left bank, and the four TD are on the right bank. Later on, the small TE-240, TE-210-250 and TE-190-195 trenches were also excavated in order to clarify some geological doubts (Fig.1). The paths that join the trenches were also useful for the detailed geological surveying (Neiva, 1962, 1992).

Fair to altered granite (W2-3, F2-3) predominates in the trenches of the left bank, but the granite is fair (W2, F2-3) in the TE-225 trench .

On the right bank, fair granite (W2, F2-3) predominates (61%)in the TD-290 trench , but there is also fair to altered granite (W2-3, F2-3), although there is, upstream and in the middle of the trench, some (30%) very altered granite (W4, F3-4). In the TD-260 trench , the fair granite (W2, F2-3) occupies 74% and W2-3, F3 occupies 17%. At lower levels, the trenches cut the fair granite (W2, F2-3).

Seven galleries were excavated in each bank. In the galleries of the right bank, 3 and 10 m of fair granite (W2, F2-3) were found in GD1 (at level 268) and GD240, respectively. They are followed by: 6.1 m of fair granite (W2, F2) in GD1 and GD240; 12.5 m of fair to good granite (W2-1, F2) in GD1; 15 m of W1-2, F2-3; afterwards 8 m of good granite (W1,F1-2) in GD240.

The first 6 m, 5.5 m and 4 m of GD2 (level 214), GD210 and GD180 galleries, respectively, were excavated in fair granite (W2, F2-3), followed by 9 m, 8 m, 6 m, respectively, of good granite (W1-2, F2-3) and, afterwards, followed by 2.5 m, 25 m and 28 m, respectively, of good to fair granite (W1, F2). However, near some quartz veinlets, but mainly near the F8 fault , containing a thick quartz vein, the granite is altered into W2-3, F2-3 in a zone of 1 to 2 m thickness of those 25 m and 28 m.

In both GD3 (level 177) and GD150 galleries , the first 2 m are of good to fair granite (W1-2, F2-3), followed by 17.5 m and 33 m, respectively, of good granite (W1, F2).

The macroscopic decompression of granite is significant, e.g. it takes at least up to 22 m in the GD1 (level 268) gallery, it extends up to the F8 fault in the GD210 and GD 180 galleries (at 32m and 21 m respectively), and it reaches about 5 m in the GD 150 gallery.

In the GE6 (level 276) and GE8 (level 256) galleries, in the left bank, the first 8m and 17.5 m, respectively, are of fair to altered granite (W2-3, F2-3) followed by 8.5 m and 15 m, respectively, of fair granite (W2,F3). Afterwards in the GE8 gallery there are also 3 m of good granite (W1, F2-3), followed by 11 m of W1, F2. This gallery cuts the F5 fault.

In the GE5 (level 233), GE200 and GE4 (level 172) galleries , the first 13.5 m, 7 m and 7 m, respectively, are of fair granite (W2, F2-3), followed by 25 m, 15.5 m and 7 m, respectively, of good to fair granite (W1-2. F2-3) and, afterwards, by 19 m, 25 m and 15.5 m, respectively, of good granite (W1,F2).

In the GE150 gallery the first 7 m were excavated in good to fair granite (W1-2, F2-3), followed by good granite, being 7.5 m of (W1, F2-3) and, afterwards, 28 m of (W1,F2).

A total of 42 boreholes were drilled, as follows: 22 in the left bank, 15 in the right bank and 5 in the valley floor (Fig.1). They helped to locate faults, aplite, pegmatite and quartz veins in depth and to study faulting, joints and weathering of rock mass from the surface inwards.

Weathering, joints, faulting, decompression and RQD of drillcores and galleries (Neiva, 1992) helped to distinguish three geotechnics zones in the granitic rock mass (Fig.2), refered in the item 9.

6 PERMEABILITY TESTS

Lugeon tests were carried out in 5 m-long segments of S1 to S15 drillholes at 0.5-1.0-0.5 pressures (Sondagens Ródio 1962), and from SE1 to SE15, SIE1 and SIE2, from SD1 to SD7 and from SID1 to SID3A drillholes at 0.25-0.5-1.0-0.5 MPa pressures (Tecnasol, 1986). The data is given in Table 5.

In the left bank, it was recognised that: a) open joints predominate at high and mean levels of the valley; b) at low levels, the whole of C, D, I and < C cases (12 to 13 % each) is slightly higher than N cases (42 %) of closed joints; c) from high levels to the valley floor there is a clear increase on closed joints, reaching 92 % in the valley floor.

In the right bank, the observations are the following: a) at high levels, D and I cases exceed the PM and C cases , showing the predominance of joints filled with weathering materials; b) at mean levels, there is a predominance of cases of null absorption (52%) corresponding to closed joints,

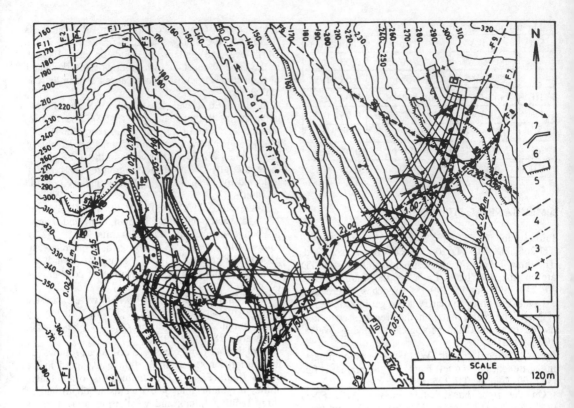

Fig.1 Geological map of the Alvarenga dam site. 1-granite, 2-aplite vein, 3-quartz vein, 4-fault, 5-trench, 6-gallery, 7-borehole. AB-reference surface of the dam.

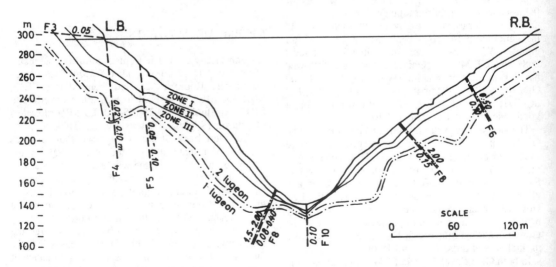

Fig.2 Geological and geotechnical section along the reference surface of the dam. Legend as in Fig.1 from 1 to 4.

while the open joints (PM and C) reach 26%; c) from high levels to the valley floor, there is an increase in the cases of null absorption.

Table 5. Permeability tests

Cases	Left bank	Valley floor	Right bank
PM	28%	1%	15%
C	18%	2%	12%
D	7%	-	15%
I	11%	3%	11%
< C	5%	-	5%
> D	5%	2%	4%
N	26%	92%	38%

All the permeability tests in holes drilled in both banks and in the valley floor (Table 5) show that: a) the left bank is the richest in clean and open joints; b) the right bank is the richest in joints filled with decompressed filling and also in closed and impermeable joints: c) at the valley floor the granitic rock mass is nearly impermeable.

The permeability tests defined the approximate locations of the 1 and 2 U.L. curves.

7 GEOPHYSICAL EXPLORATION AND GEOLOGY

Electric resistivity and seismic refraction surveyings were carried out by S.R.G. (1961) and LNEC (1982).

In the left bank the apparent resistivity profiles were carried out along the 325, 295,285, 275 and 255 contour lines. Above level 295 and downstream of the reference surface of the dam, there is a zone of low resistivity, corresponding to a weathered granite which extends to a thin zone of 10-20 m upstream of that surface; at high levels and in the abutment, medium resistivities were found. In the right bank, resistivity profiles were carried out along the 290, 280, 270, 260, 250 and 240 contour lines. Above level 235, the resistivity ranges between 7900 and 12000 Ωm; a 8 to 15 m-wide valley with low resistivity, oriented N60°E above level 265 and upstream the reference surface of the dam, corresponds to a fault in the geological map.

Seismic profiles were carried out in places of apparent electric resistivity profiles. Although the interpretation of dromochrones is difficult due to the abrupt topography of both banks and to several degrees of weathering and decompression of granite, three zones could be distinguished in the granitic rock mass at high levels of the left bank, which was confirmed by seismic fans between boreholes and between them and galleries. Two zones were also distinguished above the level 250 of the right bank.

LNEC (1985) carried out seismic profiles in the floor of the GE8 (256), GE200, GE4 (172) and GE156 galleries of the left bank, and of the GD1 (268), GD240, GD210, GD180 and GD150 galleries of the right bank and seismic fans between these galleries. From the velocity of seismic longitudinal waves it was inferred in the left bank: a) in depth, the rock mass has generally V_1 =3600-6000 m/s, but at higher levels, like at the G8 (256) gallery , there is an upper, altered and decompresse zone with a thickness of 10 m with V_1 ~1200 m/s, which overlies good granite of V_1 =5400 m/s; b) between GE200, GE8 (256) and GE5 (233) galleries, the granitic rock mass is more heterogeneous with values of V_1 ~1990-5080 m/s, generally increasing from the surface inwards of the rock mass and from higher to lower levels. In the right bank: a) V_1 generally ranges between ~3500 and 6000 m/s, except in the zone nearer the surface, corresponding to granitic rock mass with good mechanical characteristics, whose values increase from high levels to the valley floor; b) there are some exceptions like: in the GD1 gallery , which contains a weathering zone with a thickness of 3 m, with V_1 < 2100 m/s, which overlays a rock mass of V_1 < 4000 m/s; and also in the 240 gallery , which has 5 m-thick zone at the end, with V_1 ~ 2300m/s followed by rock mass with V_1 ~6000m/s .

However, the mentioned thicknesses do not always correspond to the observed degrees of weathering and decompression of granite in galleries and drillcores.

8 MECHANICAL TESTS AND GEOLOGY

LNEC (1965, 1970, 1985) carried out mechanical tests in areas selected by EDP.

In chambers excavated in the GE8 (level 256), GE200 and GE150 galleries, in the left bank, and in the GD240 and GD150 galleries, in the right bank, large flat jack tests (LFJ) were carried out in slots parallel to stresses introduced by the dam in the rock mass. The average values of E are given in Table 6.

If the tests carried out in mainly altered and faulted rock of the GE150 gallery are not considered, the deformability modulus increases when the level decreases. Generally, the least weathered granite presents the least jointing degrees and the highest value of deformability modulus.

In galleries at the same level in both banks, e.g. about level 270 (GE6 and GD1) and level 175 (GE4 and GD3), the results of deformability modulus are respectively: upper vertical (left bank (LB) 9.4 and 30.8 GPa, right bank (RB) 16.3 and 22.5 GPa);

lower vertical (LB 16.1 and 23.5 GPa, RB 2.7 and 40.8 GPa); left horizontal (LB 1.5 and 28.4 GPa, RB 14.3 and 41.6 GPa); right horizontal (LB 2.2 and 49.4 GPa, RB 21.31 and 50.8 GPa). Therefore, generally the deformability moduli increase with decrease in level inwards both banks, but the highest value was found in the right bank.

and W1 granites, and they range between 28.9 and 48.3 GPa. Rupture strength was determined in 54 of these samples, and it ranges between 93 and 168 MPa, but the 125-145 MPa values are more common. There is a positive linear correlation between the elasticity moduli and the values of rupture strength.

Table 6. LFJ tests in chambers of galleries

Galleries	Levels	Granite	Stress orientation	Cut parallel to joint	Deformability Modulus (GPa)
GE8	256	W2,F2-3	N52°W,13°NW	N38°E,77°SE	8.8
GE200	200	W1,F1-2	N36°W,22°WNW	N54°E,68°SE	15.8
GE150	150	W3,F4	N22°W,42°WNW	N68°E,48°SSE	3.8
GD240	240	W1,F3	N10°E,15°ENE	N80°W,72°SSW	9.4
DG150	150	W1,F3	N10°W,40°WNW	N80°E,50°SSE	10

The granitic rock mass might be improved with cement injections preceded by washing out, as tested in the GE6 gallery. The results were generally good, but they were very good for closing vertical and subvertical joints (Table 7).

Furthermore, the anisotropy of the granite for the elasticity modulus and rupture strength was studied by the least-squares method using values determined in 9 directions. The results of the elasticity modulus and rupture strength are significantly higher in the good W1 granite than the fair W2 granite.

Table 7. Results of cement injections in the gallery GE6

	Deformability Moduli	
	Before Injection (GPa)	After Injection (GPa)
Upper vertical	9.6	24.3
Lower vertical	16.4	23.8
Left horizontal	1.5	33
Right horizontal	2.2	45

Table 8. Joint Shear tests in galleries

Gallery	Joint	Thick. (mm)	Cohes. (MPa)	ϕ (°)
GE4	N10°W,71°ENE	3(A)	0.02	28
GE4	N10°W,71°ENE	6(A)	0.03	25
GE5	N - S, 87°E	11(B)	0.03	38
GD180	N18°W,82°ENE	15(C)	0.02	23
GD180	N30°W,4°WSW	10(D)	0.06	38

A-with clay, B-with clay (1mm) and quart (10mm), C-with clay (5mm) and quartz (10mm) D-coarse-grained altered material.

The rock mass presents elastic-plastic behaviour shown by the creep tests, but the plasticity has a small influence.

In subvertical joints, nearly parallel to the valley, and also in subhorizontal joints as well, shear tests were carried out in situ in galleries. Cohesion is rather low, and the friction angle is very dependent on the thickness and composition of the joint filling (Table8).

Joint shear tests in granite drillcores were also carried out in the laboratory. The drillcores are from several galleries of both banks. The data is given in Table 9.

In laboratory, the elasticity moduli were determined in 104 samples of drillcores from W2

Table 9. Laboratory joint shear tests

	Cohesion (MPa)	ϕ (°)
Subhoriontal joints	0.12-0.16	44-46
	0.12-0.13	39
Subvertical joints	0.12-0.15	40-42

The elasticity modulus and rupture streng values are represented by triorthogonal ellipsoid with the axis XX along the direction WNW-ESE subvertical joint set and parallel the granite lineation, the axis YY along th direction of NNE-SSW subvertical joint set and the axis ZZ is nearly vertical.

9 GEOTECHNICAL ZONING OF THE GRANITIC ROCK MASS

A first zoning was based on weathering, joint spacing, decompression and RQD of the granitic rock mass, while a second zoning was based on the velocities of longitudinal seismic waves. Three zones were distinguished in each zoning.

The two zonings were compared. In the left bank, from the valley floor to level 220, the zones coincide. However, above level 220, the limit between zones I and II of the first zoning is 1-3 m below the corresponding one of the second zoning. In the right bank, above level 260, there is a coincidence of two zonings; between levels 260 and 150, the lower limits of zones I and II of the first zoning are 1-3 m and 1-5 m, respectively, below the corresponding limits of the second zoning. Therefore, the first zoning was chosen as the best representative of the zoning of rock-mass, and it was improved with the data on the velocities of longitudinal seismic waves. The characterisation of the three zones was completed with the average data on specific gravity, Poisson's coefficient and deformability moduli (Table 10). The permeability only gave a small contribution to distinguish the three zones; it showed that decompression reached a significant depth.

The three zones and the permeability curves are

of infiltrated water, which has been more intensive between subvertical faults and the subhorizontal fault of the upper part of the left bank, and it contributed to a higher degree of granite weathering and located the 1 and 2 U.L. curves more irregularly and deeply.

10 FOUNDATION TREATMENT

The geomorphology, geology and geotechnic of the Alvarenga dam site indicate that this is favourable to build a concrete arch dam with a height of 155 m and a crest of 430 m.

The dam foundation may be located in zone II, but the concrete blocks must be properly incised in the foundation because sliding may occur on the right bank related to the N45°-60°W, 25° - 63°NE joint set as well on the left bank related to the N5°-30°W, 4°-42°ENE joint set.

The subhorizontal fault F3 is exceeded by the excavation of the foundation, but other faults must be treated.

Each fault and quartz vein should be treated in the surface using a trench filled with concrete down to a depth which would be a function of the hydrostatic pressure. In a greater depth, the treatment should be applied with fans of holes, which were washed out before cement injections, under an adequate pressure.

The significant number of joint sets and their orientations and the data of the permeability tests suggest consolidation grouting of the granitic rock mass in order to fill the opened joints with cement

Table 10. Geotechnical zoning of the granitic rock mass

Zone	W	F	Decompres.	R Q D (%)	Perm. (UL)	Vl (m/s)	Specific Gravity	v	Def.Mod. (GPa)
I	W2 - 4	F3 - 4	great	< 80	> 11	300 - 3500	2.5-2.6	0.23	< 4
II	W2	F3	med. to low	> 80	> 11	3800 - 4500	2.6-2.7	0.28	4 - 9
III	W1	F2 - 3	absent	90	< 11	> 5000	2.7	0.26	> 9

shown in the geological and geotechnical section along the reference surface of the dam (Fig.2).

The geotechnical zoning is related to the local geomorphology and geology of the segment of the valley. The zones converge in the valley floor, but they are more opened than the valley. Weathering and decompression of granite are related to the increase in thickness of joints and to the decrease in spacings among them as a function of increase in height, and they can explain the increase in thickness of the zones with increase in level.

In the right bank and in the part of the left bank at the hanging wall of faults, there is an accumulation

under an adequate pressure. The cement injections should be applied in holes separated no more than 5 m and orientated to cut all joint sets.

Consolidation grouting and fans used for fault treatments must be carried out reaching a depth defined by the bulbs of pressure calculated as a function of dam load, hydrostatic pressure, thermal variation of seasons and probable prediction of an earthquake of magnitude 5.5.

The orientations of joint sets, faults, pegmatite veins and quartz veins must be taken into account to define the grouting curtain. This curtain must go 10 m below the 1 lugeon permeability curve.

An adequate drainage system should have its holes oriented to cut the tectonic structure of the foundation in order to avoid possible uplift pressures beneath the dam blocks.

A 40 m - long hole should be drilled vertically in the middle of each dam block from the main drainage gallery before ground improvement is carried out. These drillholes will help not only to carry out seismic surveying and to know the tomography of the foundation as a function of velocity of seismic waves, but also to better study the permeability of the granitic rock mass. After consolidation grouting and grout curtain have been carried out, the seismic tomography and the permeability tests must be repeated in these holes in order to compare these results with the previous data, to evaluate the general improvement of the granitic rock mass and to find out whether it is necessary to apply any complementary treatment.

Similar studies should be made six months after the reservoir is filled, to be repeated two years later and also every five years, in order to locate any foundation portion which should be treated.

11 CONCLUSIONS

1. In the dam site, aplite veins, quartz veins and faults cut the medium-grained, locally porphyritic, two-mica Hercynian granite (Fig.1). The granite presents an enrichment of sulphides and the feldspars are partially sericitized at its direct contact with quartz veins.

2. A detailed geological surveying was carried out in the surface and in the depth. The mechanical properties of the rock mass have also been determined. The anisotropies of the deformability moduli and rupture strength are represented by ellipsoids with axes oriented along the joint sets.

3. An earthquake of magnitude 5.5 should be considered for the dynamic calculation of the dam.

4. Lugeon permeability tests have given the following information: a) the clean or open joints are commoner in the left bank; b) the joints filled with decompressed material, which can be washed out under pressure, and the impermeable closed joints occur at a greater percentage in the right bank; c) there is an increase in the occurrence of closed joints from high levels to the valley floor; d) both permeability curves of 1 and 2 lugeon have been drawn, and they are very dependent on the subhorizontal joints in the valley floor and in the infiltrated water between subvertical and subhorizontal faults in the middle and upper parts of the left bank.

5. Two zonings were defined in the rock mass: the first is based in the weathering degrees, faulting, decompression and RQD, while the second zoning is based in the velocities of longitudinal seismic waves. The two zonings overlap in the left bank, from the valley floor to level 220, and also in the right bank above level 260. In the right bank, between levels 260 and 150, the lower limits of zones I and II of the first zoning are 1-3 m and 1-5 m, respectively, below the limits of the same zones of the second zoning. On the left bank, above level 220, the first zoning presents the lower limits of zones I and II, 1-3 m below the respective lower limits of the second zoning. The first zoning was chosen as the most representative; improved with data on the velocities of longitudinal seismic waves, it was also completed with mechanical properties of rocks and rock mass (Table 10 and Fig.2).

6. The geomorphology, geology and geotechnic of rock mass of the dam site are adequate for a concrete arch dam with a height of 155 m and a crest of 430 m. The dam foundation may be located in the zone II, but the concrete blocks must be properly incised in the foundation in order to avoid sliding. A good foundation treatment must be applied down to a fixed depth, which will be a function of the bulbs of pressure of the dam blocks, hydrostatic pressure, thermal variations of seasons and a probable earthquake of magnitude 5.5 .

7. The grout curtain must be defined in order to cut the orientations of joint sets, faults and aplite-pegmatite veins and quartz veins, and it must be 10 m below the 1 U.L. permeability curve.

8. The holes of the drainage system should be oriented to cut the tectonic structures of the foundation.

9. Vertical hole, 40 m long, should be drilled in the middle of each dam block from the general drainage gallery, in order that permeability tests and seismic tomography may be carried out before and after the foundation treatment, six months after the reservoir is filled and every five years, and also when the piezometers and extensometers placed at the foundation present significant anomalies, or there is an exceptional flow in the drainage system. So any foundation portion that requires treatment will be located.

REFERENCES

LNEC 1965. Determinação das propriedade mecânicas do terreno da fundação da barragem d Alvarenga. Unpublished report.

LNEC 1970. Determinação das propriedade mecânicas do terreno de Alvarenga. Unpublishe report.

LNEC 1982. Prospecção geofísica no local c barragem de Alvarenga. Unpublished report.

LNEC 1985. Estudo das fundações da barragem de Alvarenga. Ensaios complementares. Unpublished report.

Neiva, J. M. C. 1958. Reconhecimento geológico preliminar do local da barragem de Alvarenga. Unpublished report.

Neiva, J. M. C. 1962. Geologia do aproveitramento hidroeléctrico de Alvarenga. Unpublished report.

Neiva, J. M. C. 1992. Geologia e geotecnia do local da barragem de Alvarenga.Unpublished report.

S.R.G. 1961. Estudo geotécnico por método sísmico de las fundaciones de la presa de Alvarenga y de la zona de implantacíon de las obras subterráneas de la central y condiciones. Unpublished report.

Sondagens Ródio, Lda 1962. Barragem de Alvarenga. Relatório das sondagens geológicas e ensaios de permeabilidade. Unpublished report.

TECNASOL, Lda 1986. Barragem de Alvarenga. Reconhecimento da fundação (trabalhos executados em 1981/82). Unpublished report.

7th International IAEG Congress / 7ème Congrès International de AIGI, © 1994 Balkema, Rotterdam, ISBN 90 5410 503 8

Geological evaluation of foundation rock-mass of Jamrani dam project, district Nainital, Uttar Pradesh, India

L'évaluation géologique des roches du fondement du Jamrani barrage à Nainital, Uttar Pradesh, Inde

V. K. Sharma
Geological Survey of India, Lucknow, India

ABSTRACT: The geological evaluation of the foundation rock-mass of 130m high Jamrani dam in outer Himalaya was made utilizing the data of exploratory core drilling, water percolation tests (WPT) and in-situ deformation and shear tests. The Lower Siwalik rock (Middle Miocene to Pliocene) at the dam site comprise alternate sequence of sandstone, siltstone, clayshale and sandrock. The statistical evaluation of the exploratory drill core data like-frequency distribution of rock quality designation (RQD), mean RQD, average joint frequency and variation curves of Lugeon values provide a technique for assessing the quality of foundation rock-mass in quantitative and qualitative terms. In-situ determination of deformation and shear parameters of various lithounits of the foundation rock has provided additional data as regards the foundation constituents which highlight the problem of differential settlement requiring suitable design consideration for evolving corrective measures.

RÉSUMÉ: L'évaluation géologique des roches de la fondation du barrage du Jamrani, de 130m d'hauteur et situé dans l'Himalaya externe a été fait par forages, essais de percolation d'eau, et d'autres essais. Les roches à la base de la formation Siwalik d'âge Miocène moyen à Pliocène compris une alternance de grès, des limons et des argiles. La qualité des roches de la fondation était établie par l'évaluation statistique des donnés des forages dont la distribution de la qualité de la roche (RQD), valeur moyenne de RQD, la fréquence des joints et les variations des valeurs Lugeon. Les déterminations en place de la valeur de la déformation des roches et d'autres paramètres ont fournit des nouvelles données sur les constituants de la fondation et ont montré la nécessité de traitement spécial pour améliorer la qualité de la fondation.

1 INTRODUCTION

The Jamrani dam project, located within outer Himalaya, envisages construction of a 130m high roller compacted concrete (RCC) dam across river Gola about 10 Km upstream of Kathgodam township in district Nainital, U.P. for storage of 208.6 million cubic meter of water for irrigation, industrial and domestic purposes. An ogee type of spillway crest with 3 bays of 11 m each separated by two 3.5 m wide piers would pass a routed flood discharge of 3630 cumec. A stilling basin with a dentated end sill is envisaged as an energy dissipation device.

1.2 The lithological characters, structural details such as dist-ribution and orientation of discontinuties, water percolation test data, deformation and shear characteristics of different lithounits present in the

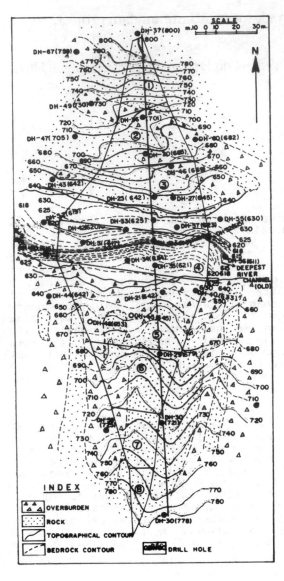

Fig.1 Bedrock contour plan of the dam seat area.

foundation are the main parameters for assessing the rock-mass. The foundation rock of Jamrani dam consist mainly an alternating sequence of sandstone, siltstone, clayshale and sandrock. Exploratory drilling by 42 drill holes for a cummulative length of 3300 m was completed and water percolation tests in each drill hole was carried out in

3m section in bedrock. In addition the in-situ deformation and shear tests were carried out on the various lithounits of the bedrock. These data have been evaluated for outlining the infirmities in the foundation rock and identification of the specific foundation problem.

2 GEOLOGY OF THE DAM SITE

The Jamrani dam site is located on Lower Siwalik rocks (Middle Miocene to Pliocene) of the outer Himalaya, characterised by a succession of sandstone, alternating with bands of siltstone and clayshale. Rock types occuring at the site could be categorised as : (i) Fine to medium- grained sandstone (FMS), occasionally micaceous hard and compact well-cemented and massive at places with calcareous cement. (ii) Sandrock (SR) with predominent pebbles of quartzite ranging in size upto a few cms. The mineral grains are poorly interlocked or cemented and the rock is friable. (iii) Clayshale (CS) and (iv) Siltstone (ST) are grey-brown to chocolate in colour with argillaceous or feebly calcareous cement. They are soft and friable, often plastic with many shrinkage cracks and occur as thin seams to thicker bands displaying facies variation along their strike direction. Some CS and ST bands contains angular intra-formational fragments referred as Sedimentary breccia (SB) which often show mottled structure, appear to have been deposited intraformationally possibly indicating a small break in deposition and hence may not be correlated with tectonic breccia. The rocks occuring at the dam site strike NW-SE and dip 35° - 40° in NE (upstream direction. These rocks are transected by five sets of prominent joints (Table - 1)

3 EVALUATION OF FOUNDATION GEOLOGY

An evaluation of the geological

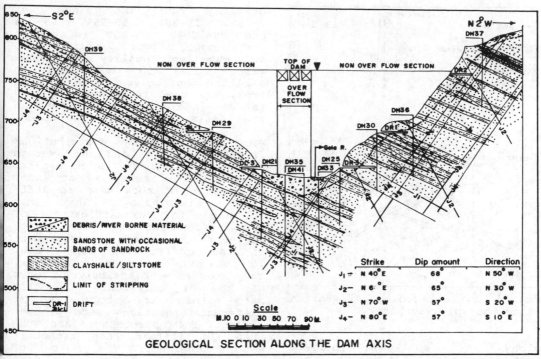

	Strike	Dip amount	Direction
J_1 —	N 40° E	68°	N 50° W
J_2 —	N 6° E	65°	N 30° W
J_3 —	N 70° W	57°	S 20° W
J_4 —	N 80° E	57°	S 10° E

DEBRIS/RIVER BORNE MATERIAL

SANDSTONE WITH OCCASIONAL BANDS OF SANDROCK

CLAYSHALE / SILTSTONE

LIMIT OF STRIPPING

DRIFT

Scale
M.10 0 10 30 50 70 90 M.

GEOLOGICAL SECTION ALONG THE DAM AXIS

Fig.2

setting, attitude of features such as weaker clayseams and bands, shear zones etc. demands portrayal of foundation grade geological plan, bedrock countour plan and quality of rock mass.

3.2   Sub-surface exploration by means of vertical drill holes were completed in the dam seat area to define the depth to fresh rock, foundation geology and the quality of the bedrock. The explorations established the presence of 7 m to 27 m thick overburden in the river bed comprising boulders and pebbles embedded in sandy matrix. The bedrock countour plan of the dam seat area was prepared based on the results of geological logging of individual drill holes and deciphering the fresh rock elevation. No visible sign of weathering is noticed in the drill cores except some limonitic leaching along joint planes at depth. The bedrock contour plan (Fig. 1) shows

that the deepest level at which the fresh rock is available is 611.13 m and at the maximum scoured depth, the width of old-channel varies from 2m to 8m. The bedrock slope of spillway block portion on right abutment is gentler than that of the left abutment.

3.3   The bedrock includes bands of fine to medium-grained sandstone (FMS), Sandrock (SR) clayshale (CS), siltstone (ST) and sedimentary breccia (SB). The ST, CS, SB and SR form the weaker bedrock members and occupy about 35% of the bedrock. The sub-surface geological conditions interpreted on the basis of drilling explorations along the dam axis (Fig. 2) reveal that even though quite a few bands show lateral facies variations, some of clayshale/siltstone bands in the river section may be correlated with those occurring on the left abutment. A typical geological section

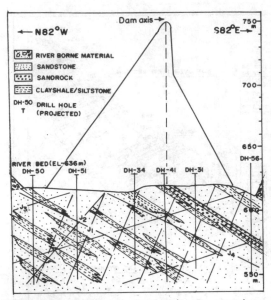

Fig. 3 Typical geological section along the spillway block.

along the spillway indicate that various lithounits are obliquely disposed (Fig. 3) with respect to the dam axis and dip upstream.

Table 1.

Joint	Strike	Dip amount & direction
J-1	N 40°E - S 40°W	68° NW
J-2	N 60°E - S 60°W	65° NW
J-3	N 70°W - S 70°E	58° SW
J-4	N 80°E - S 80°W	57° SE
J-5	N 50°W - S 50°E	35°-47° NE (Bedding joint)

4 ROCK QUALITY OF FOUNDATION

The qualitative assessment of foundation rockmass was made based on RQD determinations in drill cores recovered in the exploratory bore-holes. The rock quality designation (RQD) as defined by Deere (1964) has been determined in each hole by counting only the combined length of pieces of 10 cm or more in length, as percentage of the length of the individual

drilling run. The range of 0-25%, 25-30%, 50-75%, 75-90% and 90-100% of RQD refers to very poor, poor, fair, good and excellent quality of rock - mass. The RQD determined in all the boreholes in dam site area, was used to generate qualitative index for rock mass on the basis of study of cores.

4.1 The frequency distribution/ frequency polygon and the average RQD value ($\overline{X}$) in a particular rock unit was taken as an index of quality of rock mass in different lithounits. This may be used for classification of rock masses and may give design engineer an overall assessment of the quality of rock mass. To assess the quality index, the frequency distribution curve of RQD of different lithounits met within the boreholes at different elevations were plotted against RQD percentage and the mean RQD ($\overline{X}$) in each lithounit was determined. The value of mean RQD ($\overline{X}$) and frequency distribution of FMS, SR, CS and ST were separately analysed and plotted. The mean RQD ($\overline{X}$) and frequency polygon of FMS, SR, CS and ST in river bed section left abutment and right abutment have been shown (Fig. 4). In general, FMS has good to very good rock quality (as per Deere 1964) and larger area of frequency polygon as conpared to other constituent bands of SR, CS and ST of the foundation rocks.

5 JOINT FREQUENCY

The performance of discontinuties/ joints affect the foundations to a great extent. The physical characters of rock and the nature, spacing of joints control the behaviour of rock - mass as a whole espeçially when subjected to load. Conversely, the stress disposition within the rock-mass will be controlled by disposition of the planes of discontinuties/joints.

5.1 Exploratory core drilling data were used to assess the joints in the rock -mass. The frequency of joints has been

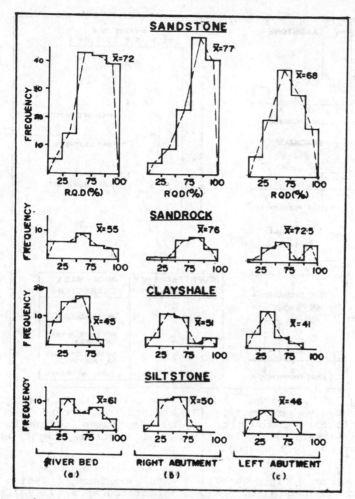

Fig. 4 Diagram showing the comparision of
frequency polygon of the rock quality
designation and the mean RQD value
in different lithounits in (a) river
bed section (b) right abutment & (c)
left abutment.

determined in each hole drilled
in dam seat area by counting
number of joints per metre of
each drilling run. The joints
frequency per metre (Jt/m) as
deduced from drill data were
categorised into four types
– wide (Massive), moderately
wide (Blocky), moderately close
(Fractured) and close (sheared
or crushed) depending on joint
frequency in the range of 0-
1 Jt/m, 1-2 Jt/m, 2-3 Jt/m and
more than 3 Jt/m respectively.
Quality of rock-mass, depends

largely on the closeness of
the joint spacing. Hobbs (1975)
has shown that fracture spacing
within the range of 10 to 50
cm are critical and small variations in fracture spacing result
in the exceptionally large changes
in the rock-mass factor.

5.3 The means frequency of
joints per meter in different
rock types met within the foundation were calculated for assessing
the quality of rock mass in
different types of lithounits

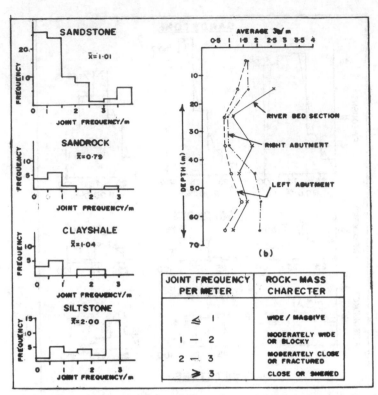

Fig. 5 Diagram showing the comparision of joint frequency per metre (Jt/m) and the mean Jt/m in different lithounits. (b) depthwise distribution of average Jt/m in river bed section, right abutment and left abutment.

at the dam site. In general the mean joint frequency of FMS, SR, CS and ST are 1.00, 0.79, 1.04 and 1.95 Jt/m respectively (Fig.5). The siltstone, one of the constitutent bands within the foundation has shown a moderately wide frequency of joints. The FMS is massive to blocky as its joint frequency value hovers around 1.00. The values of average Joint frequency plotted against the centre of vertical depth segments of 10 m each (Fig.5b) indicate variations of rock-mass characters and that in general average Jt/m decrease with depth.

6 WATER PERCOLATION TEST

The rock-mass was explored by vertical coring drill holes and the water percolation tests (WPT) were conducted in these drill holes in 3m sections using packer

with maximum testing pressure calculated as 0.171 Kg/Cm^2 for every metre of rock cover and 0.12 Kg/Cm^2 for every metre of every metre of overburden cover. The water loss has been calculated in terms of "Lugeon" (Water loss of one litre per metre of drill hole under a pressure of 1 MPa maintained for 10 min. in the test sections of drill holes and is approximately equivalent to the permeability of 1×10^{-7} m/sec which is selected by the inspection of five values obtained from each of the five cyclic runs of 10 min. duration).

6.2 The range of Lugeon value in left abutment, river section and in right abutment areas are shown in Fig.6. A reduction of the order of 13% to 89% in the values of water loss has been noticed in the rock-mass

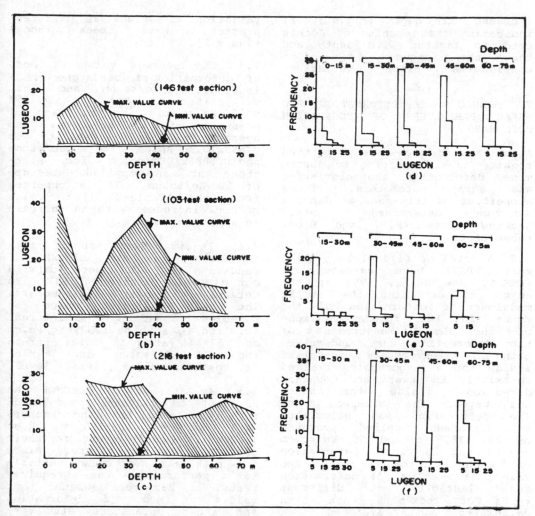

Fig. 6 Variation curves showing max. and min. Lugeon value with depth in set of drill holes in (a) left abutment (b) right abutment and (c) river bed section and the frequency distribution of Lugeon value in different depth segments (d) left abutment (e) right abutment and (f0 river bed section.

below a depth of 30m indicating water tightness at depth (Sharma, 1991,92).

6.3 The variation curves of Lugeon value verses depth were plotted bassed on WPT results in the drill holes on right abutment, left abutment and in river bed section (Fig. 6,a,b,c). The area lying between the maximum and minimum value curves represent the range of variation in the water loss value which has been conceived as the qualitative measure of rock-mass permeability. It is deduced from the curves that the permeability of rock-mass at the right abutment is of higher order than that on the left abutment and in river bed section. The frequency of Lugeon value in different depth segments (15-30m, 30-40m, 45-60m, etc.) and in the set of drill holes on right abutment, left abutment and in river bed section indicate that frequency of Lugeon value in the range of 0-5 Lugeon increase with depth (Fig.6,d,e,f) indicating that opening of Joints tend to tighten with depth and that the frequency of dis-continuities offer variation.

7 MODULOUS OF DEFORMATION AND SHEAR PARAMETERS OF FOUNDATION ROCK-MASS

Assessment of rock-mass properties require in-situ testing to determine deformation characteristics and shear parameters. These properties of rock-mass at Jamrani dam were determined by plate jacking tests (PJT) and Block shear tests.

7.2 A total of five plate jacking test (PJT) were carried out (CSMRS, New Delhi, 199) in the adits to determine the static modulous of deformation of the rocks in the foundation zone of the dam. Measurement of the deformation were undertaken using extensometers installed inside the 6 m borehole, drilled co-axially in upward and downward directions inside the adits and beyond the slumped zone. The deformation was calculated at different applied pressure of 24, 36, 48 and 60 Kg/sq.cm considering the total deformation of each loading cycle. The values of modulous of deformation and elasticity at different types of rock-mass have been calculated and averaged out from the five tests and the

variation of values at different pressure have been shown (Table 2).

7.3 The average value of mod. of deformation of sandstone (FMS) is 0.5371×10^5 Kg/sq.cm and that of siltstone (ST) is 0.4558×10^5 Kg/sq.cm at an applied pressure of 60 Kg/sq.cm. The average mod. of elasticity is 0.593×10^5 Kg/sq.cm for sandstone and 0.5086×10^5 Kg/sq.cm for silt-stone at an applied pressure of 60 Kg/sq.cm. It is observed from the Table-2 that value of mod. increases with an increase in applied pressure.

7.4 In-situ block shear tests consisting of six blocks of sandstone in adit were carried out (UPIRI, Roorkee, 1992) to determine shear parameters. The test indicated value of cohesion 'C' and angle of internal friction (ϕ) to be 1.525 Kg/sq.cm and 67°54' at peak shear stress and 0.0 Kg/sq.cm and 62°18' at residual shear stress level.

7.5 An attempt to perform the PJ test on clayshale and sand-rock was made but the anchors could not be installed in these lithounits possibly due to their soft and friable nature. Since the rock was not found suitable for performing the required tests, it has been assumed that values of mod. of deformation and shear parameters are very low to negligible.

Table 2. Average modulous value of foundation rocks at the dam site.

Pressure (Kg/Sq.cm)	Average Ed x 10^5 (Kg/Sq.cm)	Average Ee x 10^5 (Kg/Sq.cm)	Ee/Ed.
(i) Sandstone (FMS)			
24	0.4670	0.5145	1.10
36	0.4854	0.5345	1.09
48	0.5083	0.5486	1.07
60	0.5371	0.5930	1.10
(ii) Siltstone (ST)			
24	0.4049	0.4054	1.16
36	0.4190	0.4906	1.17
48	0.4412	0.4990	1.13
60	0.4558	0.5086	1.12

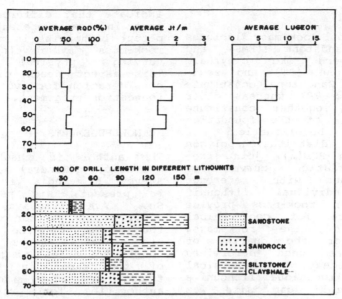

Fig. 7 Depthwise distribution of average RQD, average Lugeon value vis-a-vis drill length in different lithounits.

8 EVALUATION OF FOUNDATION PROBLEMS

The dam would mainly be founded on the fine to medium-grained hard and compact sandstone and relatively weaker bands of clay-shale, siltstone and sandrock. The clayshale, siltstone sandrock and a few clayseams, which together constitute about 30-35% of the foundation rock-mass, have been interpreted from the drill cores. The true thickness of these vary from 0.5m to 9.0m but at places show pinching and swelling. The direction of the dam axis of nearly 42° askew to the strike of the formations dipping upstream. The overall attitude of the bedding of the lithounits is NW-SE with an average dip of 40° in northeasterly direction. Four persistent joint sets (Table-1) recognised in the dam found-ation area, are obliquely disposed with respect to the dam axis, the joint set-3 dip towards downstream. However, its traces as shown in the spillway block portion (Fig.3) is oriented only in apparent direction. The depthwise distribution of average value of RQD, Joint frequency and the water percolation test results in the various lithounits indicate that the rock-mass properties tend to improve below 30-35m (Fig.7), depth from the general bedrock level. This information may be of use for the design of consolidation and curtain grouting programme. The main problems facing the dam foundations are likely to be :-

1. The variation in rock-mass character such as rock quality and joint frequency of the consti-tuent lithological bands.

2. The presence of incompetent bedrock members and further reduction of strength due to joint frequency.

3. Variation in the modulous of deformation values of the constituent lithological bands viz. FMS, SR, CS and ST.

An appreciable variation in rock-mass properties and deformation characters of the constituents of the foundation rock-mass lead to highlight the problem of unequal settlement in the foundation if loaded by the proposed high dam.

CONCLUSIONS

The foundation rock-mass includes competent sandstone (FMS) and weaker members like clayshale (CS), siltstone (ST) and sandrock (SR). Sandstone constitute about 65% to 70% whereas other members put together constitute about 30-35% of the foundation rock-mass. Determination of frequency distribution/polygon of RQD, mean RQD($\overline{X}$), joint frequency, variation curves of lugeon values with reference to the individual lithounit of foundation rock-mass provide the technique for geotechnical evaluation of the rock-mass. An index of the quality of rock-mass has been determined and an assessment of the physical characters of individual lithounits at the Jamrani dam site was made. It has been found that the competent FMS which has 1 joint per metre on an average has larger area of RQD frequency polygon and mean RQD ($\overline{X}$) in the range of 68-77%, considering the weaker members such as ST, CS and SR the mean RQD ranges are 46-61%, 41-51% and 55-76%, respectively. ST indicates relatively higher joint frequency (2 Jt/m). The values of the deformation and shear parameters of the competent lithounits are also found to be higher than that of weaker lithounits which is quite in conformity with the deductions made from statistical analysis of drill core data. This gives strength to the technique adopted and it is felt that a similar approach may be utilised even in the cases where foundation rock-mass contains only one lithological unit with considerable structural variations.

The variation curves of the Lugeon values giving an idea about the water tightness of the foundation rock-mass at the Jamrani dam site indicate that the joints are getting tightened below 30-35 m depth from bedrock level.

Overall analysis and study of data presented in the paper indicate that differential settlement in the competent and incompetent lithounits in the foundation rock-mass may develop due to their varying physical properties. This aspect needs careful design consideration for evolving remedial foundation treatment.

ACKNOWLEDGEMENTS

The author is thankful to the Director General, Geological Survey of India for permission to present this paper and to Sh. A.K.Srivastava, Director, G.S.I., Lucknow for going through the manuscript. The author extend his grateful thanks to Dr. A.K. Tangri for translating the abstract of the paper in french. The thanks are also due to the project authorities for providing assistence during data collection.

REFERENCES

Central soil and material research station (CSMRS), New Delhi report 1992. Report on plate jacking tests at Jamrani project (U.P)

Deers, D.U. 1964. Tehnical description of rock cores for engineering purposes. Rock Mechanics and engineering geology. Vol.1, no.1,17-22.

Hobbs, N.B. 1975. Factors affecting the prediction of settlement of structures on rock with particular reference to the chalk and Trias. Review paper: session IV, Rocks, Britishes Geotechnical society, 579-610.

Sharma, V.K. 1991. Progress report on the pre-construction stage geological investigations of Jamrani dam project, River Gola District Nainital, Uttar Pradesh, unpublished progress reports Geological survey of India, field season 1989-90, 1990-91 and 1991-92.

U.P.Irrigation Institute, Roorkee, 1991. Block Shear test in drift no. DR-1 at Jamrani dam site, Kathgodam (Nainital) Technical memorandum 62 (MT-49).

Geometrical and mechanical characterization of the joint sets of a dam foundation

Caractérisation géometrique et mécanique des familles de diaclases d'une fondation de barrage

José Muralha & Nuno Grossmann
Laboratório Nacional de Engenharia Civil (LNEC), Lisbon, Portugal

ABSTRACT: The paper describes some of the studies and tests performed by LNEC to determine the joint network and the mechanical parameters of the main joint sets of the rock mass upon which the Fridão dam will be founded. The study of the joint network used a set of techniques developed at the LNEC, and was based on the description of more than 1800 joints that were detected during the result of the field survey conducted in 40 zones of 7 adits of both banks. The study, first, looked at each one of the 40 zones, then, at each one of the adits, thereafter, at each bank, and, finally, at the whole rock mass. After the study of the joint network, 54 laboratory shear tests were performed on joints belonging to the 3 main sets. The results were the 4 ultimate shear stresses and the shear stiffness. A statistical description of all these parameters is presented.

RESUMÉ: La communication décrit quelques études et essais réalisés par le LNEC pour déterminer le système de diaclases et les paramètres mécaniques des familles les plus représentatives du massif rocheux sur lequel le barrage de Fridão sera fondé. L'étude du système de diaclases a été faite selon des techniques développées au LNEC. Elle s'est basée sur la description de plus de 1800 diaclases rencontrées dans 40 zones à l'intérieur de 7 galeries creusées dans les deux berges. L'étude a été conduite, successivement, pour chacune des 40 zones, pour chaque galerie, pour chaque berge et, finalement, pour tout le massif. Après cette étude, 54 essais de cisaillement en laboratoire on été réalisés sur des diaclases des 3 familles les plus représentatives. Les résultats concernent les 4 contraintes de cisaillement et la rigidité au cisaillement. Une description statistique de tous ces paramètres est présentée.

1 INTRODUCTION

The development plans of EDP (Electricidade de Portugal), the Portuguese electric power company, foresee the construction of the Fridão dam, a 94 m high concrete arch dam on the river Tâmega, a right bank tributary of the river Douro, in northern Portugal (Fig. 1), mainly for the production of hydroelectric energy. During the elaboration of the design, EDP requested LNEC to perform the necessary in-situ and laboratory tests for the characterization of the metamorphic rock mass on which the dam and its appurtenant works will be built.

The map of the Serviços Geológicos de Portugal, the official Portuguese geological department, (Sheet 10-A, Celorico de Basto, Pereira 1989) shows that the Fridão dam will be located in the Canadelo unit (Lower Devonian), which comprises alternating phyllites, grey schists, and metamorphosed siltstones.

The rock mechanics investigation for the Fridão dam (LNEC 1989) included:

- in situ, the study of the rock mass joint fabric, seismic refraction tests along the dam foundation and the deviation tunnel, 30 BHD dilatometer tests, and 7 LFJ large flat jack tests (using 2 flat jacks per test);

- in the laboratory, 39 uniaxial compression tests and 54 joint shear tests.

This paper shall only deal with the methodology used for the geometrical and mechanical characterization of the joint sets of the dam foundation.

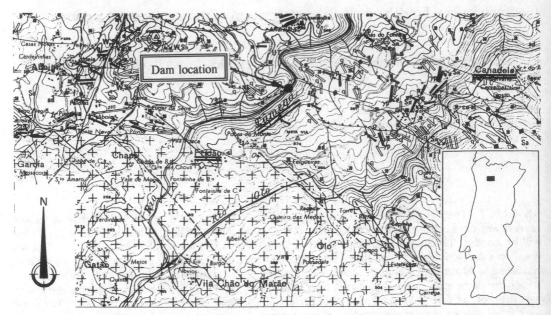

Fig. 1. Fridão dam location

2 GEOMETRICAL CHARACTERIZATION OF THE JOINT SETS

The study of the joint network of the dam foundation rock mass was performed using a set of techniques developed at LNEC. In this chapter, some general remarks about the methodologies used by LNEC to evaluate the relevant parameters of the jointing and their particular application to the Fridão dam, will be presented.

As for many other rock mass characteristics, the determination of the jointing parameters is also performed with the help of some kind of sampling. This sampling should try to be uniform under 2 different, but equally important, viewpoints:

- in order to detect any heterogeneity, eventually occurring in the rock mass, the sampling locations should cover the whole foundation rock mass in a uniform manner; in other words, each point of the rock mass should have the same probability of being considered as a sampling location;

— in order to assess, at every sampling zone, the rock mass discontinuity system, the sampling at each location should cover all attitudes in a uniform way.

Due to physical limitations in the choice of the sampling locations, which nearly always exist, very often it is not possible to obtain a uniform sampling of the jointing. These limitations obviously increase when the access to the interior of the rock mass is scarce or even non-existent.

For the correction of the sampling bias related to the lack of uniformity in the covering of all attitudes at a given sampling location, the notion of sampling quality has been proposed (Grossmann 1984). For a given attitude with the strike σ and the dip δ, the sampling quality $P(\sigma,\delta)$ corresponds to the probability that the considered sampling would detect 1 discontinuity surface with that attitude, if the attitudes of all discontinuity surfaces of the rock mass would follow a uniform distribution.

For a sampling on a rectangular observation surface and assuming that the areas of the discontinuity surfaces follow a Bessel function distribution (Grossmann and Muralha 1987), the sampling quality is

$$P(\sigma,\delta) = \frac{2\sin\varepsilon}{\pi^2}\left[1 + \frac{\frac{\pi}{2}(a\,|\cos\omega|+b\,|\sin\omega|)-(a+b)}{\frac{\pi^2 ab}{4\sqrt{\pi\overline{A}}}+(a+b)}\right] \quad (1)$$

where:
- a and b are the lengths of the sides of the rectangle;
- ε is the angle between the attitude (σ,δ) and the plane containing the rectangle;
- ω is the angle between a projection on the plane of the rectangle of a normal to the attitude (σ,δ) and a side with the length a of the rectangle;
- $\overline{A}$ is the mean area of the discontinuity surfaces.

As regards the lack of uniformity in the covering of the whole rock mass, there are virtually as many different cases occurring in practice, as rock masses to be studied, and, therefore, each problem has to be analysed by itself.

For a dam foundation, the discontinuity surface sampling should try to sweep the whole valley cross-section, i.e., the upper, middle and lower parts of both banks. In each of these zones, the sampling should look not only at the direct dam foundation area, but also at the adjacent downstream and upstream areas, covering the rock mass volumes around the pressure bulb and the grouting and drainage curtains. Furthermore, if adits and/or boreholes are available, the sampling should also cover the evolution of the jointing from the surface to the interior of the rock mass. This means that the jointing study for an ordinary dam foundation can easily comprise up to 50 sampling locations.

In order to study the homogeneity of the jointing characteristics of the rock mass, the information collected at each sampling location, should provide a full picture of the jointing occurring in its neighbourhood, i.e. it should be representative of the surrounding volume. Due to the variability of the jointing parameters, this implies that, at least, around 30 discontinuity surfaces should be sampled with every observation surface.

For dam foundations, the joint sampling quite often exceeds 1000 discontinuity surfaces, which dictates that over 5000 different pieces of information have to be collected during the field survey, usually under difficult conditions. The discontinuity surface sampling is, therefore, as a rule, a demanding and time consuming job that tends to be put, with obvious risks, on the shoulders of young and unexperienced persons.

The sampling, at each location, should always include all discontinuity surfaces intersecting the chosen observation surface and special care should be taken to avoid the frequent tendency to disregard some discontinuity surfaces just because they have a small intersection (trace length). Another common flaw is to just record "another discontinuity surface like the previous one", instead of measuring and registering its characteristics, only to try to make things a little bit easier. There is also the temptation of trying sometimes to classify in advance some given discontinuity surfaces into a given set; this should always be made at a later stage of the study and never during the field survey.

Usually, the characteristics to be registered for each discontinuity surface are the geologic type, the attitude, the intersection with the chosen observation surface (which may not be the full visible trace length, as sometimes the discontinuity surface continues beyond the limits of the observation surface), the mean aperture and the type of its filling (if there is one). In some cases, other features, as the depth of occurrence (if the observation surface is a borehole wall or core) and the relation with the discontinuity surfaces that are cut by or end at it, are also recorded.

Generally, shallower zones of a rock mass present a higher weathering, a higher jointing density and a larger aperture of the joints than deeper zones. Moreover, due to toppling phenomena, some joint attitudes near the surface can be misleading. As the outermost weathered zone of the rock mass will be removed in order to allow the dam to be founded on a sounder socle, the sampling of the shallower zones of the rock mass should, usually, be avoided and preference given to observation surfaces inside adits. However, as in many cases explosives are used to excavate the adits, care should be taken where new fragmentation has been created and blasting induced joints should always be disregarded.

The most common observation surfaces for a joint sampling correspond to geometrically well defined domains: prismatic surfaces (stretches of an adit, with its walls, roof and/or floor), cylindrical surfaces (boreholes or scanlines) and plane surfaces (circles or rectangles). The determination of the attitude of joints along boreholes is only possible if the core is oriented (using, for instance, integral sampling or TV cameras). The sampling in adit stretches is only easy to perform if the cross-section is not too large. So, in most cases, plane observation surfaces are used.

The application of these general ideas to the particular case of the Fridão dam originated that from the existing 12 adits (6 in each bank) not all were selected for the joint network study. On the left bank, adits GE1.1 in the upper part of the slope and GE6.1 in the lower were chosen, whilst in the middle section adits GE3.1 and GE4.1 were selected (Fig. 2). This option was taken because the geophysical tests had revealed a certain heterogeneity in that zone and, furthermore, those 2 adits were not very long. On the right bank, 3 adits were picked: GD6.1 in the lower part, GD4.1 in the middle and GD2.1 in the upper part of the slope (Fig. 3). The adits corresponded to those where the BHD dilatometer and LFJ large flat jack deformability tests were performed. Figures 2 and 3 also reveal that, for both banks, the location of the 40 (20+20) selected observation zones on the walls of the adits rendered a fairly uniform and representative sampling.

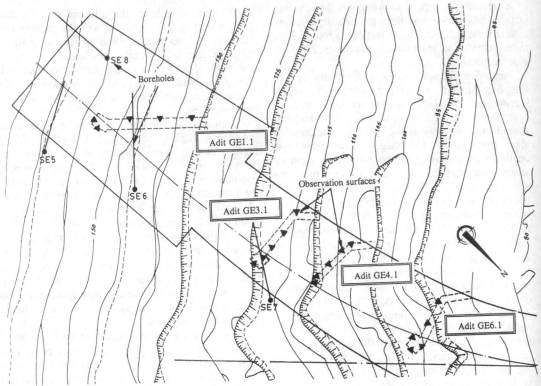

Fig. 2. Left bank. Location of the observation surfaces

At the Fridão dam, all observation surfaces were rectangles. The size of the majority of these surfaces (36 out of 40) was 3.75 m² (2.5m × 1.5m), but 3 zones in adit GE4.1 (1.5m × 1.5m) and 1 in adit GD4.1 (2.5m × 1.0m) presented different contours. Therefore, the joint survey covered a total area of 144.25 m² (70.5 m² on the left bank and 73.75 m² on the right).

In comparison with the whole volume of the dam foundation, the rock mass volume sampled for the jointing study was extremely small (less than 0,1%). Even so, this particular study provided the description of more than 1800 joints, for a foundation area of about 8000 m².

The method followed for the definition of the discontinuity system of the rock mass was the one developed at LNEC (Grossmann 1977):

- to start with the study of small zones of the rock mass (e.g. the individual sampling locations, or groups of only a few of them) for which it may be assumed, without reasonable doubt, that they are homogeneous as regards the jointing; at this stage, the discontinuity subsets of each zone are defined and so no discontinuity surface shall be discarded, even if it does not belong to significant subsets;

- to define larger zones in the rock mass for which, as regards the jointing, it still may be assumed that they are homogeneous; this assumption shall be based on the analysis of the results obtained for the small zones;

- to study those larger zones of the rock mass, the corresponding discontinuity sets being obtained by grouping the discontinuity subsets of the respective small zones and not directly from the data collected at the sampling locations; the advantage of this procedure is that no conditions of equal representativeness, such as a similar sample size or sampling quality, have to be imposed to the different sampling locations; still, at this stage, no discontinuity surface shall be discarded, even if it does not belong to the more significant discontinuity sets;

- to repeat, if necessary, the two previous steps using now the larger zones as small zones and defining even larger homogeneous rock mass volumes;

- having, finally, obtained the discontinuity set of the whole rock mass, or of its homogeneous parts, to study the characteristics (attitude, intensity area, aperture, etc.) of all occurring discontinuit

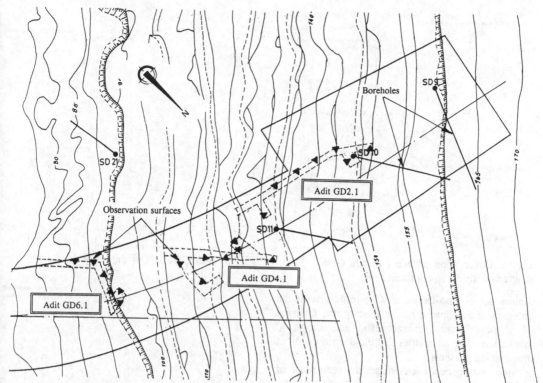

Fig. 3. Right bank. Location of the observation surfaces

sets; based on these characteristics and only at this stage, the most significant discontinuity sets will be identified and retained and the less notable shall be discarded.

The determination of the discontinuity sets and its relevant parameters is preferably made by automatic means, both for obtaining the results more quickly and avoiding mistakes or subjective decisions. The use of a computer presents, furthermore, the advantage that the correction of the non-uniformity of the jointing sampling, as regards the covering of the attitudes, can be automatically implemented (Grossmann 1983).

For the current case, the first step of the discontinuity system determination was performed for all 40 observation surfaces; then, the resulting subsets were grouped, in a second stage, for each adit; in a third step, the resulting joint sets were grouped for both banks; finally, the sets obtained for each bank were grouped for the whole rock mass. Figures 4 to 6 show an example of this step-by-step technique. These stereographic projections are part of the graphical output of the computer codes developed at LNEC. Figure 4 illustrates the results of the analysis for a single observation surface (Zone 1

in adit GD6.1); Figure 5 the results from adit GD6.1 and Figure 6 the outcome for the right bank.

The different geometrical parameters of the joint sets have to be described in a statistical manner, as they usually show a large variability.

The distribution of the attitudes of the discontinuity surfaces of a discontinuity set can be modelled by a bivariate normal distribution on the tangent plane at the mean attitude. This distribution requires 5 different parameters for its definition, namely, the strike σ and the dip δ of the mean attitude, the maximum and the minimum standard deviations, σ_M and σ_m respectively, and the angle ω_M that identifies the orientation of the maximum dispersion (Grossmann 1985).

The intensity I of a discontinuity set describes the degree of jointing that the whole lot of the discontinuity surfaces of that set have induced in the rock mass, independently of the individual extent of each discontinuity surface. So, the intensity is quantified by the sum of the areas of the discontinuity surfaces of the set which occur in a unit volume of the rock mass, and its units should be m^2/m^3. Though under certain conditions (a discon-

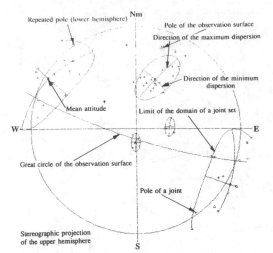

Fig. 4. Observation surface 1 of adit GD6.1. Pole diagram with the discontinuity sets

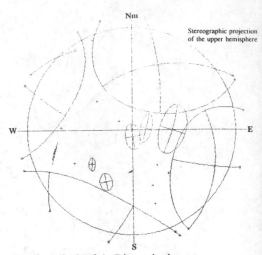

Fig. 5. Adit GD6.1. Discontinuity sets

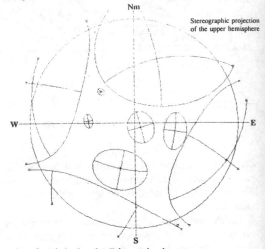

Fig. 6. Right bank. Discontinuity sets

tinuity set of parallel joints with infinite area) the intensity is the inverse of the spacing, the intensity is a more general characteristic and allows to overcome some difficulties that the determination of the spacing presents.

In a homogeneous rock mass, the occurrence of the discontinuity surfaces of a joint set can, usually, be modelled by a Poisson process, which only requires one parameter to be entirely defined: the intensity. Using the maximum likelihood principle, it is possible to establish a relation between the intensity of the discontinuity set, the geometric characteristics of the chosen observation surface and the sum i_T of the lengths of the intersections of the discontinuity surfaces of the considered set with that observation surface, i.e. the trace lengths inside the surveyed zone (Grossmann 1988).

For a plane observation surface, the intensity of a set can be determined by

$$ I = \frac{i_T}{S \sin \alpha} \qquad (2) $$

where S is the area of the observation surface and α is the angle between the mean attitude of the given discontinuity set and the observation surface.

The persistence of a discontinuity set can be described by the distribution of the areas of its discontinuity surfaces. The distribution of those areas can be modelled by a Bessel function distribution (Grossmann and Muralha 1987), that requires only one parameter, the mean area $\overline{A}$, to be fully characterized.

For a rectangle, the expression that renders the mean area of a discontinuity set, given the sum of

the lengths of the N intersections sampled inside that observation surface, is

$$ \overline{A} = \frac{\pi}{4\left(\dfrac{N}{i_T} - \dfrac{a|\cos \omega| + b|\sin \omega|}{a\,b} \right)^2} \qquad (3) $$

where the geometrical parameters a, b and ω have already been defined for equation (1).

For the Fridão dam rock mass, the results of the study revealed that the most important discontinuity set presents a mean attitude of N66°W-74°NE, parallel to the schistosity and, thus, shall hereafter be referred to as X. This joint set is responsible for around 50% of the total jointing in the whole rock mass. The mean attitude of set X is approximately perpendicular to the direction of the future dam impulses on the right bank and parallel on the left.

This relative position, associated with a high rock mass anisotropy, could, theoretically, lead to some difficulties. However, the schistosity is rather incipient and the anisotropy coefficient (around 2) of the rock deformability under uniaxial compression (LNEC 1989) does not anticipate any inconvenience.

With a 25% contribution to the jointing, the discontinuity set with a mean attitude N7°E-83°NE is the second most relevant. It is approximately perpendicular to set X and will be referred to as V, since it is almost vertical.

Only these 2 joint sets occur in the whole rock mass. However, but with a lower degree of relevance, several sub-horizontal discontinuity sets are also detected on each bank. On the left bank, the joint set with the mean attitude N76°E-34°NW accounts for about 25% of the jointing in that slope. On the right bank, the horizontal joints are even less important and the joint set with the mean attitude N16°E-22°SE is responsible for only 7% of the jointing in that slope. These sets will be, globally, referred to as H.

The parameters of the statistical distributions of the attitude, intensity and area of joint sets X, V and H are presented in Tables 1 and 2.

Table 1. Parameters of the attitude distribution for the most relevant joint sets

Joint set	Strike (Nm °E)	Dip (°)	Standard deviation Max.(°)	Standard deviation Min.(°)	Angle (°)
X	294	74 NE	23	18	61
V	17	83 SE	20	15	72
H (left bank)	256	34 NW	20	14	55
H (right bank)	16	22 SE	19	12	-10

Table 2. Parameters of the intensity and area distributions for the most relevant joint sets

Joint set	Intensity (m^2/m^3)	Mean Area (m^2)
X	5.4	1.4
V	2.8	1.0
H (left bank)	2.6	2.2
H (right bank)	0.9	1.5

3 MECHANICAL CHARACTERIZATION OF THE JOINT SETS

The evaluation of the mechanical characteristics of the main joint sets was based on the analysis of the results of 54 laboratory shear tests (18 on discontinuities from the sub-vertical set X, which is parallel to the rock mass schistosity, 20 from the sub-vertical set V and 16 from the sub-horizontal sets globally referred to as H).

The shear tests were performed according to the following basic procedure:
- applying a stress normal to the joint surface;
- submitting the upper part of the joint to a shear stress, without allowing the lower part to move, while keeping the normal stress constant;
- recording the shear displacement and the respective shear stress until a displacement of about 6 mm is reached;
- bringing the discontinuity back to its initial position (usually a mated one);
- repeating this cycle for different normal stresses.

Generally, the tests are performed under 4 increasing normal stresses σ_n, taking into account the expected pressures or impulses that the dam will transfer to the foundation. In this case, the applied stresses were 0.5, 1.0, 1.5 and 2.0 MPa and the areas of the tested discontinuities around 200 cm^2.

The direct results of the tests are 4 graphs (one for each different normal stress) relating the shear stresses and the shear displacements. Based on the analysis of these graphs a behaviour model can be established. The option for the adopted model must be taken considering the use that shall be given to it and the fit between the experimental data and the model.

The study of the graphs that came out of the 54 tests revealed some important trends: i) the diagrams are monotonous increasing curves with a horizontal asymptote, ii) no well defined peak shear stress occurs, iii) the shear stress increases relatively quickly and reaches a value close to the maximum shear stress for shear displacements around 1 mm. As a result, an elasto-plastic model was employed. It is formed by two linear branches: an inclined initial one, starting at the origin of the axes and with a slope defined as the shear stiffness K_s, and a horizontal second one, with an ultimate shear stress τ independent from the displacement. Concerning the relation between the shear stiffness and the normal stress, still, 3 distinct elasto-plastic models could be envisaged:
- K_s and τ are independent from the normal stress and the 4 bi-linear diagrams are determined separately from each graph;
- K_s is constant for all normal stresses;
- the shear displacement at the intersection of the 2 linear branches of all 4 diagrams remains constant.

After fitting these 3 models to the experimental data, the model with a shear stiffness independent

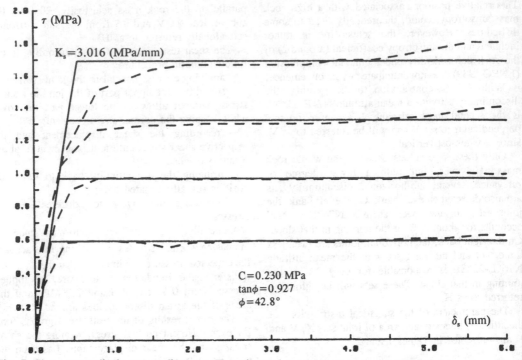

Fig. 7. Shear stress-displacement diagrams. Experimental results and elasto-plastic model

from the normal stress was chosen. The parameters K_s and τ were calculated minimizing the sum of the squared distances between the model and the points on the 4 graphs, with the constraint that the resulting shear stresses must relate linearly with the corresponding normal stresses, thus complying with a linear Coulomb failure model (Fig. 7).

In practice, this procedure follows an iterative scheme (Muralha 1990). First, all the shear stresses have to divided by the maximum shear stress and the shear displacements divided by the maximum shear displacement. This operation must be performed previously because the distances are computed perpendicularly to the model branches and so all values must be rendered non-dimensional. Otherwise, the parameters would depend on the adopted stress and displacement units. Next, all experimental pairs (δ_s, τ) from the 4 graphs are sorted into two groups: those with a displacement lower than a certain cut-off value and the remaining. Both these groups should include pairs of values from the 4 graphs that result from the application of the 4 different normal stresses. In order to make this possible, for all 4 different normal stresses, the shear stress should be measured at pre-defined shear displacement values. Moreover, these values should cover with a particular detail the sector where the

shear stiffness is to be defined (displacements up to 1 mm). With the (δ_s, τ) pairs of the first group, the shear stiffness is determined using the following expression

$$K_s = \frac{\sqrt{\left(\sum \delta_s^2 - \sum \tau^2\right)^2 + 4\left(\sum \delta_s \tau\right)^2} - \left(\sum \delta_s^2 - \sum \tau^2\right)}{2\sum \delta_s \tau} \quad (4)$$

The second group defines, through a simple linear regression, the parameters of the linear Coulomb failure envelope (apparent cohesion C, friction coefficient $tg\ \phi$ or friction angle ϕ). Finally, the sum of the squared distances between experimental data and the model is computed. Using different values for the cut-off displacement allows to determine the best fit for the experimental data.

Figure 8 presents the cumulative distribution functions of the shear stiffnesses determined from all tests and grouped according to the discontinuity sets X, V and H. Some statistical parameters of those distributions, as the arithmetic mean, the median, the standard deviation and the coefficient of variation, are also presented. The main conclusions that can be drawn are:

- the shear stiffness is a highly variable parameter, even for joints of the same discontinuity set (coefficients of variation Cv from 37 up to 73%);

3896

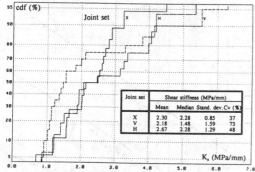

Fig. 8. Shear stiffness cumulative distribution functions of the joint sets

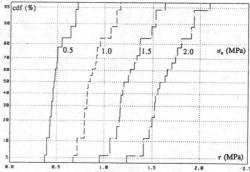

Fig. 9. Shear strength for constant normal stresses cumulative distribution functions of joint set X

- the shear stiffnesses of the joint sets show a positive skewness, and so the medians are smaller than the arithmetic means; this fact, which has also been registered in other studies, leads to suggest that in practical cases, for safety reasons, the medians should be preferred as a central tendency parameter.

As concerns the strength of the discontinuity sets, the analysis began by the plotting of the cumulative histograms and the evaluation of some simple statistical parameters for the 4 normal stresses of the tests. An example of the histograms is presented in Figure 9 and the computed statistics (Tables 3 to 5) were the arithmetic mean $\bar{\tau}$, the standard deviation s_τ, the coefficient of variation Cv and the lower and upper limits of the 95% confidence interval for the mean, $L_{95\%}$ and $U_{95\%}$, respectively.

Since the large majority of the test results of the shear strength for each discontinuity followed a linear Coulomb envelope, it could be foreseen that the mean shear strength of the sets would also relate linearly with the normal stress. What could not be expected were the linear relations, with correlation coefficients greater than 0.999, between the shear strength standard deviations and the normal stresses that all 3 discontinuity sets disclosed. These last relations can be seen in Figure 10, along with the corresponding straight lines.

The following conclusions referring to the shear strength of the 3 sets could be drawn:
- under the 4 normal stresses applied, discontinuity set V presents the lowest shear strength and set H the highest;
- set H presents a considerably higher scattering than the other two, which can also be noticed in the confidence interval around the mean;
- the coefficients of variation of sets X and H seem to decrease asymptotically with the normal stress, whilst those of joint set V appear to present a constant value (around 18.5%) for all 4 normal stresses.

The shear strength is a parameter often required to evaluate the safety of a given failure scenario. So, it is convenient that design values are available.

The analysis of the cumulative histograms of the shear strength for the 4 normal stresses of each

Table 3. Statistical parameters for the shear strength of joint set X

σ_n (MPa)	$\bar{\tau}$ (MPa)	s_τ (MPa)	Cv (%)	$L_{95\%}$ (MPa)	$U_{95\%}$ (MPa)
0.5	0.500	0.096	19.2	0.453	0.547
1.0	0.881	0.130	14.8	0.817	0.945
1.5	1.262	0.169	13.4	1.178	1.345
2.0	1.643	0.210	12.8	1.539	1.747

Table 4. Statistical parameters for the shear strength of joint set V

σ_n (MPa)	$\bar{\tau}$ (MPa)	s_τ (MPa)	Cv (%)	$L_{95\%}$ (MPa)	$U_{95\%}$ (MPa)
0.5	0.416	0.078	18.8	0.380	0.453
1.0	0.760	0.138	18.2	0.696	0.825
1.5	1.105	0.204	18.5	1.010	1.200
2.0	1.449	0.272	18.7	1.322	1.576

Table 5. Statistical parameters for the shear strength of joint set H

σ_n (MPa)	$\bar{\tau}$ (MPa)	s_τ (MPa)	Cv (%)	$L_{95\%}$ (MPa)	$U_{95\%}$ (MPa)
0.5	0.555	0.126	22.7	0.489	0.622
1.0	0.957	0.190	19.8	0.856	1.058
1.5	1.359	0.259	19.0	1.222	1.496
2.0	1.757	0.332	18.7	1.589	1.935

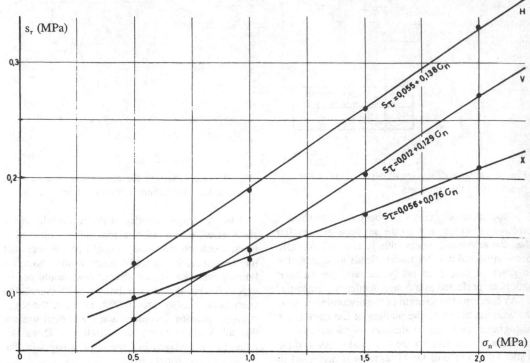

Fig. 10. Linear relations between the shear strength standard deviation and the normal stress for the joint sets

discontinuity set and the fact that, in all cases, the medians were relatively close to the respective means, made it possible to establish that, for a given normal stress, the shear strengths of the sets follow Gaussian distributions. This conclusion could be validated since it was found that all 12 distributions passed the Kolmogorov fit test with confidence levels above 0.99.

It was therefore possible to evaluate the percentiles of the shear strength for all normal stresses, which, due to the above mentioned linear relations between the shear strength parameters and the normal stress, also present linear relations with the normal stress. For the case of discontinuity set V, Figure 11 presents the apparent cohesion and the friction angles of the design Coulomb envelopes for the 1st, 5th and 10th percentiles, along with the normal distribution curves that generated them.

This analysis showed that the widespread engineering practice that decreases empirically the shear strength maintaining the friction angle and disregarding the apparent cohesion, is less conservative than considering the 10th percentile for the shear strength, except for very low normal stresses. This means that the common engineering routine is not as safe as it is thought, specially for high normal stresses.

This simple statistical approach of the mechanical characteristics of the discontinuity sets for the case of the Fridão dam, established a practical procedure that allowed to move from the usual arithmetic mean values to a better perception, in terms of safety, of the problems related with the joint's strength and deformability.

ACKNOWLEDGMENT

The authors express their gratitude to EDP - Electricidade de Portugal for allowing to publish results of the rock mechanics studies for the Fridão dam.

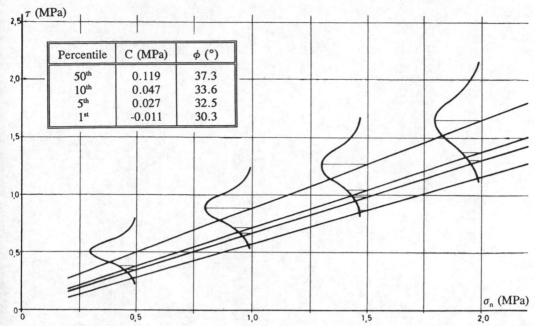

Percentile	C (MPa)	ϕ (°)
50th	0.119	37.3
10th	0.047	33.6
5th	0.027	32.5
1st	-0.011	30.3

Fig. 11. Design Coulomb envelopes for joint set V

REFERENCES

Grossmann, N.F. 1977. Contribuição para o Estudo da Compartimentação dos Maciços Rochosos (Contribution to the Study of the Jointing of the Rock Masses), Research Officer thesis. Lisbon: LNEC.

Grossmann, N.F. 1983. A Numerical Method for the Definition of Discontinuity Sets, in Proceedings of the 5th International Congress on Rock Mechanics, Melbourne, Australia, Vol. 1, pp B17-B21. Rotterdam: A.A. Balkema.

Grossmann, N.F. 1984. The Sampling Quality, a Basic Notion of the Discontinuity Sampling Theory, in Brown, E.T. and Hudson, J.A., Eds. — ISRM Symposium Design and Performance of Underground Excavations, Cambridge, UK, pp 207-212. London: British Geotechnical Society.

Grossmann, N.F. 1985. The Bivariate Normal Distribution on the Tangent Plane at the Mean Attitude, in Stephansson, O., Ed. — Proceedings of the International Symposium Fundamentals of Rock Joints, Björkliden, Sweden, pp 3-11. Luleå: Centek Publishers.

Grossmann, N.F. 1988. About the Joint Set Intensity, in Romana, M., Ed. — Rock Mechanics and Power Plants, ISRM Symposium, Madrid, Spain, Vol. 1, pp 41-47. Rotterdam: A.A. Balkema.

Grossmann, N.F. and Muralha, J. 1987. About the Mean Area of a Joint Set, in Herget, G. and Vongpaisal, S., Eds. - Proceedings of the 6th International Congress on Rock Mechanics, Montréal, Canada, Vol. 1, pp 373-376. Rotterdam: Balkema.

LNEC (Grossmann, N.F.; Fialho Rodrigues, L.; Muralha, J. and Sousa, M.R.) 1989. Estudo das Fundações da Barragem do Fridão (Fridão Dam Foundations Study). Lisbon: LNEC.

Muralha, J. 1990. Evaluation of Mechanical Characteristics of Rock Joints under Shear Loads. Proceedings of the International Symposium on Rock Joints, Loen, Norway. Rotterdam: A.A. Balkema.

Pereira, E. 1989. Carta Geológica de Portugal. Notícia Explicativa da Folha 10-A - Celorico de Basto (Geological Map of Portugal). Lisbon: Serviços Geológicos de Portugal.

The terranes and the waters of the Roman aqueduct at Venosa (southern Italy)

Les terrains et les eaux de l'aqueduc romain à Venosa (Italie du Sud)

G. Baldassarre, B. Radina & F. Vurro
Bari University, Italy

ABSTRACT: Much of the roman aqueduct, built at the end of the first century BC to supply the latin colony of Venusia (now Venosa in the province of Potenza, southern Italy) is now in a state of disrepair. However, some stretches are still in a reasonable condition and others have been rebuilt to enable its waters to be used for irrigation. In the light of the geometric characteristics of the aqueduct, this paper deals with the engineering geology of the ground involved, the feed springs and water quality and the effect exerted by the waters which have always been conveyed by this public utility. Collapse of the works and of the formations involved is also treated.

RÉSUMÉ: L'aquéduc romain, construit à la fin du I siècle A.C. pour le ravitaillement d'eau de la colonie latine de Venusia (actuelment Venosa en province de Potenza–Italie du midi) est dérangé. Cependant, quelque partie est bien conservée et quelque autre a été réparée au fin d'utiliser les eaux pour l'arrosement. L'étude informe, en considération des caractéristiques géométriques des oeuvres, sur les conditions géologiques et tecniques des terres intéressées, sur les sources que l'alimentent, sur le chimisme des eaux et sur les effets des eaux que l'ont alimentées et que l'alimentent actuellement. Pour ce que concerne les dérangements, on a mise en évidence soit ceux des terres traversées soit ceux des oeuvres.

1 INTRODUCTION AND AIMS

Even in the case of towns built in olden times water supply was considered to be of great importance. For the Romans, especially, this was the primary feature of the infrastructure, as demonstrated by the numerous examples of aqueducts (Hauck 1988, Smith 1978) constructed between the fourth century BC and the sixth century AD.

The aqueduct in question was probably built at the end of the first century BC to replace the one that served the latin colony of Venusia in 291 BC (Cenna 1902), traces of which - drainage galleries - have been identified in the Historic Centre of present-day Venosa.

Some parts of the long surveyed stretch of the whole aqueduct have been rebuilt recently to carry irrigation waters.

Various surface signs of the aqueduct found along the roman Appian Way indicate that the water-supply system terminated on the site of the sixteenth century castle in the Historic Centre of Venosa. Several underground cisterns, which probably collected water from the aqueduct, have been identified here.

The studies performed included investigations on the alignments of the main and secondary conduits of the aqueduct, the engineering-geology of the various types of ground involved, the feed springs, the chemical quality of the waters and the effect of the latter on the structures concerned, as well as their role in on-going collapse.

The work was performed on the occasion of the celebrations marking the second millenium of the death of the roman poet Horace or, more properly, Quintus Horatius Flaccus. The aim was not only to identify the structure but also to ascertain the possibility and viability of restoring the aqueduct so that it can again perform the role for which it was originally built.

2 ALIGNMENT AND WORKS

The alignment of the main conduit of the aqueduct below ground was traced by means of eighty-four access shafts up to 17 m in depth. The line runs roughly SSW–NNE (Figure 1). The elevation of the aqueduct at the head end (Serra La Croce) is 744 m a.s.l., while that at the tail end (Vignali) is 437 m a.s.l.. The shafts (mainly of the inspection type) have also revealed the presence of secondary conduits running more or less straight, some draining from SE to NW and others in the opposite direction.

Virtually the whole length of the main conduit (4970 m) and that of the secondary conduits which reach the main one (970 m) have been inspected directly. It ensues that about 65% of the conduits consist of galleries, the remainder being channels of various depths. Where the conduits are lined they have a rectangular and/or circular cross-section but where there is no lining the cross-section is irregular. The cross-sectional area of the channels ranges from 1.5 to 7.0 m^2.

The linings of the conduits and some wells (Figure 2) consist mostly of dressed limestone blocks bonded by cement mortar, while limestone slabs are used to roof the channels. Deposits of travertine ranging in thickness from a few millimetres to several decimetres are found in several places on the bottom and sides of the conduits and the headworks.

3 FORMATIONS ENCOUNTERED BY THE AQUEDUCT

As has already been ascertained (Centamore and Lunari 1968, Centamore et al. 1970, Hiecke Merlin et al. 1971, Piccarreta and Ricchetti 1970, Servizio Geologico d'Italia 1970), the first 2.6 kilometres of the aqueduct run through the pre-Pliocene formations of the southern Apennine Range, while the remaining 1.8 kilometres traverse the Pleistocene series of the Bradano Trough. For a length of about 1.1 kilometres these formations are covered by volcanic Pleistocene deposits deriving from Mount Vulture.

The surveys made (Figure 1) indicate that the pre-Pliocene terrane consists of Faeto Flysch and Varicoloured Clays. Faeto Flysch is barely represented in the initial part of the aqueduct, being present for a little less than 100 m. It consists principally of hazel-white limestones in beds averaging 10 to 30 cm thick with rare variable 10-30 cm thick interbeds of greenish-white marly limestones having a conchoidal fracture, and subordinately of thin layers of greyish clays and greenish marls. The attitude is variable and quite irregular in places. The limestones are fractured and broken in some stretches, the fractures occasionally being filled with secondary calcite. Overall permeability is very variable, though mainly low. The local springs associated therewith have a total discharge of 0.2 to 0.5 l/s.

Overall strength of the ground can be quite reasonable but it depends on the presence or absence of clay interbeds and the fracture state. To a large extent the stability of the steepest outcrops is contingent upon these conditions.

There are some good exposures of Faeto Flysch in a number of abandoned quarries not far from the alignment, where it overlies the Varicoloured Clays. The thickness in the initial part of the outcrop ranges from a few metres to a dozen or more, while being considerably greater (several dozen metres) in the remainder.

The Varicoloured Clays, extending for about 1.7 km, are represented by successions of thin beds of indurated rock and coherent loose rock in no regular order. The formation generally consists of scaly greyish marly clays intercalated in a disorderly manner with limestones, sandstones, marly limestones and marls in beds of various thickness, generally not exceeding a few decimetres. The original attitude has been so strongly disturbed by tectonic events in several stretches that the indurated rocks occur in quite random lines or are reduced to small blocks or even minute fragments embedded in a plastic clay mass. The virtually chaotic conditions found in several places on the surface are very likely the result of old mass movements. The maximum thickness of Varicoloured Clays observed along the alignment is in the order of several dozen metres.

Taken as a whole these Clays are impermeable. However, where the rock element predominates in their composition small aquifers may be present, as evidenced by some of the shafts and wells. The mechanical properties of these Clays are very poor, rendering civil engineering work extremely difficult.

The Bradano Trough formations are represented by Mount Marano Sands, which are readily identifiable along the alignment, being directly traversed by the

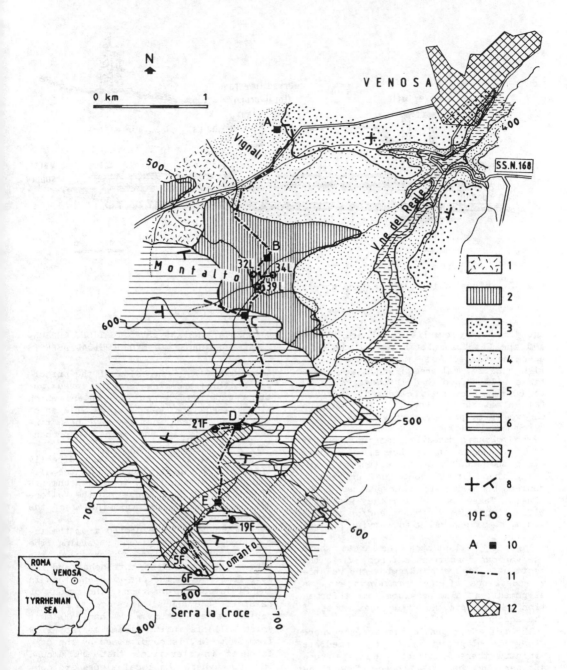

Fig. 1 Geological map
1 versant detritus; 2 volcanic deposits; 3 Irsina Conglomerate; 4 Mount Marano Sands;
5 Subapennine Clays; 6 Faeto Flysch; 7 Varicoloured Clays; 8 beds attitude; 9 spring;
10 sampling wells; 11 aqueduct alignment; 12 Venosa town

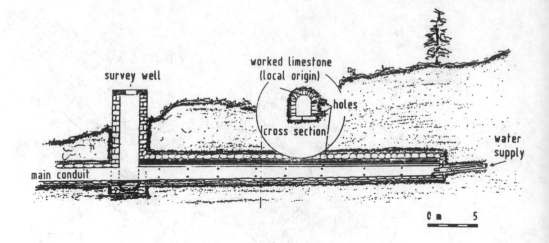

Fig. 2 Linings of the conduits and some wells

aqueduct. They overlie the Faeto Flysch and the Calabrian Formation of the Subapennine Clays. These medium-fine yellowish quartzy, calcareous sands are sometimes well-bedded (bed thichness from 20 to 60 cm) and partly cemented. They attain a maximum thickness of 20 m and are endowed with primary permeability of varying magnitude.

The volcanic deposits, ranging from one to ten metres in thickness, are found along about 600 m of the alignment overlying the Faeto Flysch. They are represented by medium-to fine-grained pyroclasts. Taken as a whole, they are stratified, moderately compact and/or indurated and are endowed with varying permeability.

Figure 1 illustrates the areas where the various formations outcrop, while the section in Figure 3 indicates the relative positions of the formations and the alignment; it also provides some information on the present structural state of the aqueduct.

The relief is gentle almost everywhere along the alignment and in the immediate vicinity thereof, though the slopes are steeper in the pre-Pliocene formations between El. 724 m and El. 550 m than in the Pleistocene terrane lying between El. 550 m and El. 445 m.

It should be noted that the area in question is seismically active, being classified as Category II according to the Italian Code.

4 OBSERVATIONS AND INVESTIGATIONS ON SURFACE DRAINAGE AND HYDROLOGICAL ASPECTS

An outline is given below of the principal findings stemming from the investigations made on the surface waters which flow towards the aqueduct, as well as on the ground waters that may have been tapped to supply it and which at present feed some stretches thereof.

Surface waters more or less directly affecting the area traversed by the alignment are attributable to the catchments of the Vallone Isca Lunga and the Vallone del Reale, which are tributaries of the Venosa River.

As might well be expected from the indications already given regarding the permeability of the ground concerned, the streams are much less entrenched and far more ramified in the pre-Pliocene than in the Pleistocene formations. The amount of rainfall (the annual mean over a 20-year period is a little more than 700 mm), which collects in the steams as they flow from SW to NE, is conditioned by the differences in elevations that are around 780-800 m where the local slopes are conconcerned and between 740 and 435 m along the line of the aqueduct.

The conformation of the two drainage networks is governed not only by the steepness of the slopes and the permeability of the ground but also by the plant cover, the lands being wooded especially in the central stretch of the alignment

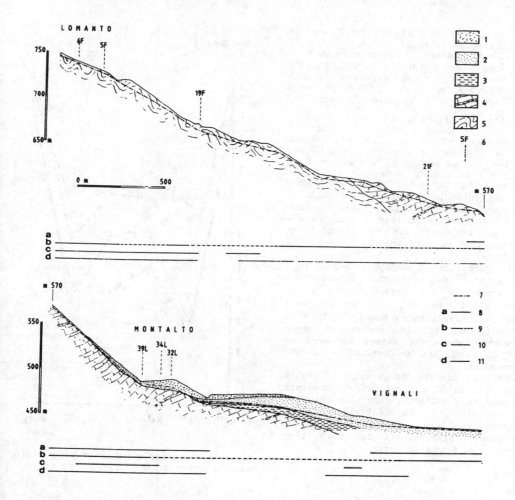

Fig. 3 Geological section
1 volcanic deposits; 2 Mount Marano Sands; 3 Subapennine Clays; 4 Faeto Flysch; 5 Varicoloured Clays; 6 aqueduct alignment; 7 still in use aqueduct; 8 trench and tunnel
cross-section; 9 obstructed stretches; 10 fractured and/or broken places

and cultivated in the terminal part.

The groundwaters that have been tapped to feed the aqueduct right from the time of its construction come from the perched aquifers found at various shallow depths mainly in the Faeto Flysch but also in the Mount Marano Sands and the volcanic deposits. The existence of the aquifers is attested to at the present time by lo-cal springs (no. 7 – Figure 1) whose di-scharges, however, are quite small (max 0.85 l/s). The presence of other aquifers is evidenced by wells which produce bet-ween 0.09 and 2 l/s with a drawdown of

about one metre.

The fact that the aquifers are of the shallow type is also borne out by the temperature of the water which does not appear to differ greatly from that of the air.

Chemical analyses have been conducted on samples of water from the springs and from the present-day feed to the aque-duct, the aim being to evaluate the pro-posal to rehabilitate the facility and also to check on the effects that the conveyed waters had in the past on the structures concerned. The results, which

are set forth in Tables 1 and 2, can be summarized as follows:
- The waters are of meteoric origin
- They come within the bicarbonate-alkaline-earth class (Figure 4) so they probably derive mainly from the Faeto Flysch that is principally carbonate in nature
- The variations in chemical composition (Samples A and B of Figure 5) encountered in the final stretch of the aqueduct are probably attributable to the mixing of the waters from the Flysch with those from the aquifers in the Mount Marano Sands and the volcanic deposits. It should be noted that the high nitrate content of the waters in this stretch is almost certainly due to pollution by fertilizers used locally in agriculture.

The chemistry of bicarbonate-alkaline-earth waters is, of course, dependent on variations in CO_2 partial pressure. In this regard, when CO_2 is given off, carbonate is precipitated and travertine is formed. This actually occurs in those stretches where the waters conveyed by the aqueduct are in contact with the atmosphere.

Analysis of the travertine indicates that between 87 and 96% (wt/wt) consists of carbonate (slightly magnesian calcite) and a residue that is insoluble in dilute hydrochloric acid; this is formed of amorphous silica, quartz, feldspars, clay minerals and organic matter.

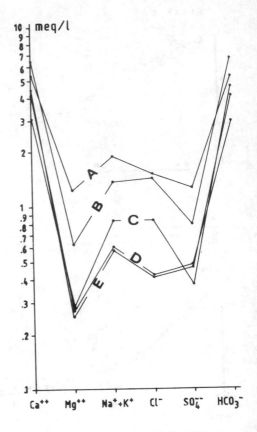

Fig. 5 Schoeller's diagram summarizing the chemistry of the aqueduct waters

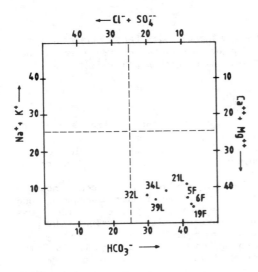

Fig. 4 Langelier–Ludwig diagram for the waters studied

5 CONCLUDING REMARKS ON PRESENT STATE OF THE WORKS

The survey of the aqueduct alignment and cross-sections thereof, together with an assessment of its structural state and a study of the construction materials, has revealed the presence of various kinds of collapse and numerous obstructions.

Regarding the first aspect – collapse- in many places the limestone blocks used to line various parts of the aqueduct are either very loose or badly fractured or have broken away completely. As regards obstructions, in addition to the presence of piles of debris on the bottom of the headworks and conveyance system, there are also many travertine encrustations of various thickness.

There are diverse reasons for this state of affairs depending, of course, on the type of terrane involved more or less directly. These can be summarized as

Table 1. Chemical data for the spring waters analysed.

No.	Date of sampling	Temperature (°C)	PH	Ca^{++} (mg/l)	Mg^{++} (mg/l)	Na$^+$ (mg/l)	K$^+$ (mg/l)	HCO$_3^-$ (mg/l)	Cl$^-$ (mg/l)	SO$_4^{--}$ (mg/l)	NO$_3^-$ (mg/l)	TDS (mg/l)
5F	Oct 1991	15.3	7.7	79.2	5.8	16.1	1.6	247.7	13.5	23.1	13.6	329
6F	Oct 1991	13.3	8.0	89.2	5.3	12.4	1.2	272.7	11.0	23.1	11.2	347
19F	Oct 1991	12.0	7.3	91.4	3.2	10.1	2.4	273.4	13.1	15.4	9.3	336
21F	Oct 1991	12.8	7.5	61.3	5.9	16.1	1.2	195.9	13.5	24.0	11.2	259
32L	Oct 1991	13.4	7.6	122.4	8.8	27.6	2.4	300.8	72.2	56.2	6.2	528
34L	Oct 1991	12.3	7.3	106.2	6.7	25.3	3.1	305.7	44.3	39.9	9.3	406
39L	Jun 1992	15.2	7.6	147.3	7.4	26.9	2.3	364.3	75.2	54.8	9.3	410

Table 2. Chemical analysis for the aqueduct waters.

No.	Date of sampling	Temperature (°C)	PH	Ca^{++} (mg/l)	Mg^{++} (mg/l)	Na$^+$ (mg/l)	K$^+$ (mg/l)	HCO$_3^-$ (mg/l)	Cl$^-$ (mg/l)	SO$_4^{--}$ (mg/l)	NO$_3^-$ (mg/l)	TDS (mg/l)
A	Oct 1991	13.2	7.8	107.8	18.6	37.9	9.0	318.5	53.2	60.0	42.2	543
B	Oct 1991	11.7	6.9	129.3	9.4	29.2	4.3	403.3	50.0	37.9	1.4	518
C	Oct 1991	12.6	7.7	89.0	5.8	17.9	2.4	281.3	29.8	17.8	6.2	371
D	Oct 1991	13.0	7.9	61.1	4.1	12.6	2.0	176.3	14.9	23.1	12.4	276
E	Oct 1991	12.3	7.7	86.4	3.8	12.2	2.0	249.0	14.5	22.6	13.6	329

follows:

- instability induced by the various ways in which surface waters have flowed over the years
- collapse of various types and intensity especially in the Varicoloured Clays and, particularly, those which may have affected the structure
- deterioration in the engineering properties of the ground on which the structure is founded
- seismic activity
- the high carbonate content of the waters feeding the aqueduct
- very probably the fact that for many years the aqueduct has not been used or has not been appropriately maintained.

REFERENCES

Cenna, G. 1902. Cronaca Venosina, a cura di G. Pinto, Vecchi, Trani.

Centamore, E. & G. Lunari 1968. Considerazioni sulla vergenza di alcune strutture nelle formazioni fliscioidi lucane. Boll. Soc. Geol. It. 87: 285-289.

Centamore, E., Chiocchini, U., Jacobacci, A., Lamari, G. and Santagati, G. 1970. Geologia della zona nord-occidentale del F° 187 "Melfi" (Lucania). Boll. Ser. Geol. d'It. 91: 113-148.

Hiecke Merlin, O., La Volpe, L., Nappi, G., Piccarreta, G., Redini, R. and Santagati, G. 1971. Note illustrative della Carta Geologica d'Italia, F° 186 "Sant'Angelo dei Lombardi" e F° 187 "Melfi". Serv. Geol. d'It., Roma.

Langelier, W. E. & H. F. Ludwig 1982. Graphical method for indicating mineral character of natural water. Amer. Water Works Assoc. Journ. 34: 335-352.

Piccarreta, G. & G. Ricchetti 1970. I depositi del bacino fluvio-lacustre della Fiumara di Venosa-Matinelle e del Torrente Basentello. Studio geologico-petrografico. Mem. Soc. Geol. It. 9 (1): 121-134.

Schoeller, H. 1935. Utilité de la notion des eschanges de bases pour la comparison des eaux souterraines. Soc. Geol. de France Bul, Serie 5: 651-657.

Servizio Geologico d'Italia 1970. Carta Geologica d'Italia alla scala 1:100.000, F° 187 "Melfi".

Smith, N. 1978. L'ingegneria idraulica romana. Le Scienze, 119: 98-107.

Cut slope recommendation

Recommendations pour le projet de talus d'excavation

I. Yilmazer
Sial Ltd, Ankara, Turkey

S. Selcuk & H. Tumer
Teksan A.S., Ankara, Turkey

ABSTRACT: Cut and embankment slopes are the main elements of a motorway, especially in rugged mountainous terrains which are typical of the Anatolia where the Turkish Motorway (TM) project within an approximate length of 3000 km was commenced in 1985 and half of it has already been completed. It is experienced that engineering characteristics of geological units and their areal distributions, discontinuity surveys, hydrogeology, back-analyses of persistent topographic slopes, climatic changes depending upon physiography, seismicity, and land-use form the essential basis for recommending slope and also to execute proper design of other elements of a motorway. Kinematic analysis, based upon a detailed discontinuity survey, is the most satisfactory method to predict failures, and accordingly, to recommend a stable slope and measures for cuts located in a tilted and stratified unit.

RÉSUMÉ: Les inclinaisons d'incision et de remblai, sont des elements essentiels de moteur-route, surtout en terrains montagneux qui sont caractéristiques en Anatolie où, le projet de moteur-route de la Turquie, dont la longueur est 3000 km, a commencé en 1985 et dont 50 pour cent a été terminé. On a expérimenté que les caractéristiques d'ingénierie des unités géologiques et leurs distributions selon les domains, les surveillances de discontinuité, l'hydrogéologie, reanalyser des inclinaisons topographiques et persistentes, les changements climatiques dépendants de la physiography, la sismicité et la forme, de l'utilisation de la terre, sont des bases essentielles à recommender l'inclinaison et même à exécuter des projets propres des autres éléments d'une moteur-route. Les analyses cinémathiques qui se basent sur des surveillances détaillées des discontinuités, sont des méthodes les plus satisfaisantes à prévoir des fautes et par conséquence, à recommender l'inclinaison stable pour des incisions qui se localisent en unités pliées et stratifiées.

1 INTRODUCTION

Turkey is characterised by highly dissected topography and rugged mountainous terrains trending nearly east-west direction due to the prevailing tectonic activities. As given in detail by Sengör and Yilmaz (1981), continental crust of the Eastern Anatolia is thickening due to compression exerted by the Arabian Plate moving northward whereas the Western Anatolia is subjected to a tensional regime directed Northwest (Fig. 1). Cut and embankment works constitute the essential part of the Turkish Motorway (TM) Project in which American, British, Italian, Japanese, and Austrian (mainly for tunnel design) companies are involved. No distinct consistency can be found in these projects with respect to ground classification, selection of motorway elements, slope

recommendation, and accompanying design works. The first author has worked as an investigator and geotechnical designer in the Ankara-Gerede and Ankara peripheral motorway (270 km), Izmir-Aydin and Izmir Peripheral motorway (200 km), and Tekir-Adana-Gaziantep

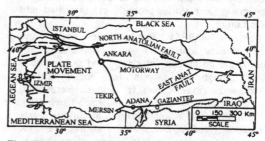

Fig. 1 Turkish motorway network and the major tectonic zones.

motorway (300 km) projects which are together representative of the whole country from the engineering geological viewpoint.

Various terrain evaluation systems have already been proposed. Some of which are mentioned in Khawlie & A'war (1992) who adopted the P.U.C.E (Province, Pattern, Unit, Component Evaluation) system over a 500 km² area in the Chouf (Lebanon) through which an 18 km highway was constructed. They found the P.U.C.E. system, which considers the geology, geomorphology, and the surficial cover essential for the purpose of dividing a terrain into four classes, as a practical and economical approach in the assessment of terrains of large scale engineering structures. However, slope recommendation and selection of alignment and motorway elements are dependent basically upon the site specific properties which could not be generalised. Route location is the most significant phase of a motorway project, which is the continuum of highway engineering between comprehensive transportation planning and highway design. The basic principles of route location study is detailed out by Baker (1975). 80% of instabilities encountered in TM is affected by the improper route location. Even reconnaissance work from the viewpoint of engineering geology could not be carried out because of the fast tract character of the very old fashioned tendering system (Design & Built) which is no more in use of tendering of TM after 7 years of practice. Priority has been given to geometry rather than geology and geotechnics. An extensive survey has been conducted concerning 750 km of the motorways in Turkey. 90% of cuts and embankments have been constructed in and over stratified and tilted geological units. 80% of cut slope failures and in places embankment slope failures where shear steps and proper drainage have not been implemented, have occurred due to outslope condition of discontinuities. Yilmazer et al. (1993a) and Zanbak (1983) suggested that kinematical analyses have a significant role in designing cut slopes properly in stratified and tilted units.

Romana (1985) distinguished how the discontinuity conditions and orientations are influential on stability of slopes and suggested a new adjustment ratings for application of Bianiawski classification to slopes. Excavation techniques including blasting methods (Swindells, 1985) have also adverse effects at various levels on slope stability. Discontinuity strength at first stage of dilation is lost during slope heave. Presplitting in stratified rocks is the best blasting method which does not create a thick disturbance zone over the blasted slope face.

Since over ninety percent of the rock outcrops are stratified and tilted, kinematic analyses based upon a detailed discontinuity survey are frequently used in designing cut slopes. Slope heights, especially over sloping grounds, has to be optimised (Yilmazer & Ertunç, 1993). Besides implementing effective engineering measures, uphillside carriageway could be designed as a Cut & Cover section whereas the downhillside embankment could be replaced via a short bridge to avoid excessive slope heights. A typical example can be given from the Gerede-Ankara Motorway where an embankment with a height of 114 m representing an earthfill dam and trunk type cuts with heights of >100 m were constructed in 1990 to cross the Yakakaya Saddle in spite of that the height of an embankment should be less than 50 m as specified in motorway standards.

2 DISCONTINUITY SURVEY

A detailed discontinuity survey could provide appreciable data about;

- subsurface geological condition,
- groundwater movement,
- distribution of geological units, seeps, and springs,
- classification of discontinuity systems to identify the ones which can adversely affect the stability,
- to select proper engineering structural elements,
- to prepare further comprehensive subsurface investigation programs to find out reliable geotechnical parameters,
- to produce an effective design, and
- to select practical engineering measures to alleviate ground instability problems.

The above items can be visualised through the following field examples. Table 1 is suggested to record main features effective in slope stability analyses and in other engineering works. The numbers in table 1 are described briefly in table 2. The discontinuity survey is based on mainly visual inspection and practical field tests (Tab.3). However, any engineering data obtained from field and laboratory tests could be used in the preparation of the table 1.

As a field example the Nurmountain Range crossed by the Adana-Gaziantep motorway is characterised by NW dipping homoclinal structures (beddings, schistosities, imbricated

Table 1 A sample data form used in discontinuity survey

LOCATION	DATA POINT	TYPE	DIP AMOUNT (°)	DIP DIRECTION (°)	SPACING (m)	APERTURE (mm)	ROUGHNESS	SHAPE	WATER	WEATHERING	SLAKING	DISCONTINUITY NUMBER	OUTCROP (m²)	INFILL TYPE	THICKNESS (mm) COMPACTNESS/ CONSISTENCY	LITHOLOGICAL DESCRIPTION	SOIL STRENGTH	ROCK STRENGTH	ROCK QUALITY	RESIDUAL VALUES OF COHESION (C_r, kPa) AND FRICTION	ANGLE (∅, °) OF DISCONTINUITY
Eastern portal	1	B	25	225	0.25	0	2	3	0	4	0	4	40	-		Slate	-	8	MQ	75	45
	3	U	10	130	150	25	1	1	1	1	9	5	90	CH		Caliche	C0	5	VPQ	20	8
	5	J	80	170	1.00	1	2	3	0	3	0	6	50	ML		Schist	C2	7	PQ	60	25

Table 2 Key for rock mass description based on visual inspection and practical field tests.

TYPE	SHAPE	SLAKING	OUTCROP AREA
B Bedding	1 Planar	0 No apparent absorbency and no detritus from cleaned surface when wetted and rubbed by thumb	S Small 0<A < 10
D Cleavage	1 Infill > Undulations		M Medium 10<A < 100
F Fault	2 Undulating	1 Some apparent absor. but no detritus	L Large 100<A < 200
G Fissure	3 Irregular	3 Absorbency and some detritus	V Very Large A > 200
H Flow layering	4 Discontinuous	6 Absorbency and much detritus	
I Foliation		9 Slow complete breakdown (2-5min.)	SOIL STRENGTH
J Joint	WATER	12 Rapid complete breakdown (<2min.)	C0 Very Soft: exudes between fingers when squeezed
L Lithol. contact	0 Dry		C1 Soft: easily moulded with fingers
P Shear plane	1 Damp		G1 Very loose: cohesionless
S Schistosity	2 Seep	DISCONT. SYSTEM NUMBER	C2 Firm: effort required to mould with fingers
T Tension crack	3 Flow	0 None or a few occasional sets	G2 Loose: pick can be pushed in fully with effort
U Unconformity	4 Artesian	1 One set 2 One set + random	C3 Stiff: can be intended with thumb
		3 Two sets 4 Two sets + random	G3 Dense: pick blow penetrates 50%
ROUGHNESS	WEATHERING	5 Three sets 6 Three sets + random	C4 Very stiff: intended by thumbnail with effort and some fracturing, easily carved with knife & excavated with hammer without fracturing
1 Slickensided/ Polished	1 Completely weath.	8 Four or more sets, random heavily dissected, "sugar cube" etc.	
2 Smooth	2 Highly weathered	10 Crushed rock, earthlike	
2 Infill > Roughness	3 Moderately weath.		G4 Very dense: pick blow penetrates 25% -50%
3 Rough	4 Slightly weathered	INFILL TYPE	
4 Very rough	5 Fresh	Unified Soil Classification System (ASTM D2487)	

ROCK STRENGTH

4 Extremely weak: intended by thumb-nail with effort and some fracturing, easily carved with knife, easily excavated with hammer without fracturing. If granular, pick blow penetrates 25%-50%

5 Very weak: indented by thumbnail

6 Weak: can be gouged with thumbnail, grooved deeply with knife, excavated with moderate difficulty with hammer

7 Mod. weak: can not be gouged with thumbnail, core sized pieces can be broken in hand with effort across fabric, >3mm indentations made with pick

8 Mod. strong: core sized pieces cannot be broken in hand across fabric, 1-3mm indentations made by pick

9 Strong: <1mm indentations made by pick, single heavy blow will break fist sized piece

10 V. strong: multiple blows needed

Table 3. Evaluation of the table 1 in terms of residual shear strength values, cohesion (C$_r$, kPa) and friction angle (∅$_r$,°) given in the columns 1 and 2 respectively.

ROUGH.		SHAPE		WEATH.		SLAKING		DISCON. SYS. NUM.		ROCK STRENGTH		SOIL STREN.	
10	0	10	< 5	20	20	>40	<35	>65	>40	30	10	5	0
<20	< 5	<20	<10	>25	>25	40	35	>45	40	10	35	<10	0
<40	<10	20	<15	>40	30	35	35	>45		25	<30	15	0
>45	>20	30	>15	>50	>30	30	35	>40		> 50	<30	0	20
>50	>30	40	>20	>100	>35	25	20	35		> 250	>35	20	< 5
						15	<10	35		> 500	35	0	25
								30		>1000	40	30	10
								20		>2000	>45	10	35
								>10	20				

thrusts, compressional joint systems, drag folds, recumbent folds, overturned folds, and so forth). Over the western flanks where the majority of the discontinuities are outslope or dipslope; perennial streams, springs, seepages, and unstable grounds are common whereas just opposite conditions prevail over the eastern side where discontinuities are mainly inslope. The motorway crosses 4 large landslides across the western side (6 km long portion) and does not cross any distinct unstable ground on the other side. 400 m long portion of the motorway with a 27 m high trunk-type cut was constructed into the first slide mass at immediate beginning of the western part, during the period of 1991-1993. The slide mass has initially been described as a rock unit in borehole logs and in related reports. The slide mass unusually consists of huge bedchunks, bed clasts, and boulders in fine matrix. Abrupt changes in any discontinuity system over the slide area are quite distinct although each system remains identical in large areas beyond the slide area. A half-a-day long discontinuity survey carried out in the vicinity area revealed that it is a slide mass in 1990. Instead of simply relocating over a nearby stable ground, very complicated treatment works including drainage shafts and galleries have been implemented. The slopes and other elements of the motorway across the slide mass are still prone to fail. Similar case studies can be found in Yilmazer et al. (1993b).

3 KINEMATICAL ANALYSES

About 90% of geological units are stratified and/or jointed systematically. Bedding is a common discontinuity type in sedimentary rocks, metadetritics, cumulates, lava flow rocks, and also in slopewash deposits. Joint systems could be observed in any kinds of rock units.

Compressional, conjugate, and tensional joint systems develop basically under the effect of tectonic stresses whereas cooling joints form in igneous rocks. Subvertical columnar joints are pertinent to thick lava flows and sills. Subhorizontal cooling joints are peculiar to dykes and fissure type eruption ridges. All of these cooling joints in volcanic rocks develop perpendicular to cooling surface. The cooling joints in dykes and fissure type eruption ridges looks like lava flows. However, subvertical lineation of prismatic minerals (phenocrysts) might help to prevent such confusion. Besides the above discontinuities, lithological contacts between igneous rocks and country rocks, tectonic contacts (faults), structural contacts (unconformities), and schistosities in metamorphic rocks constitute the other major discontinuity types.

Nomenclature of discontinuity systems with respect to cut slope face is depicted in Figure 2. It plays significant roles in engineering geological and geotechnical communication. The suggested nomenclature is as follows.

A. Acute angle between strikes of the discontinuity plane and the slope face <20°.
Outslope: dipping out of the slope face; unstable, plane failure is expected.
Inslope: dipping into the slope face, stable.
Dipslope: dips toward the same direction with the slope but steeper than the slope face; stable, toppling could be expected.

B. Acute angle between their strikes is >70°.
Rightslope: strike of the discontinuity system is subperpendicular to the slope face; stable.

C. Acute angle between their strikes lies in between 20° and 70°.
Obliqueslope: strike of the discontinuity system is oblique to the slope face; stable, it may cause

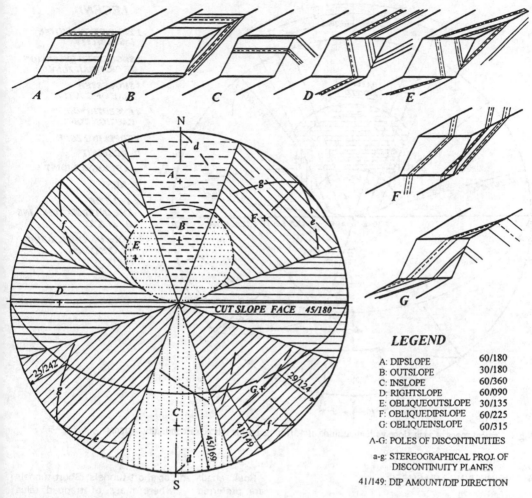

Fig.2 Suggested nomenclature for discontinuity systems in a cut slope of 45/180.

LEGEND

A: DIPSLOPE	60/180
B: OUTSLOPE	30/180
C: INSLOPE	60/360
D: RIGHTSLOPE	60/090
E: OBLIQUEOUTSLOPE	30/135
F: OBLIQUEDIPSLOPE	60/225
G: OBLIQUEINSLOPE	60/315

A-G: POLES OF DISCONTINUITIES

a-g: STEREOGRAPHICAL PROJ. OF
DISCONTINUITY PLANES

41/149: DIP AMOUNT/DIP DIRECTION

generally wedge failure together with other discontinuity systems.

In order to execute kinematical analyses, a stereographical projection of the given slope face has to be prepared. Figure 3 is presented as an example. All of the discontinuity system, recorded in engineering data sheets (see Tab.1) should be sorted and put into five groups.

I.   Poles of which daylight in critical zone which may cause plane failure.
II. Poles of which daylight in toppling zone.
III. Poles of intersection line of which daylight in critical zone which may cause wedge failure.
IV. Poles of which daylight in caution zones.

V. Poles of which and/or poles of their intersection lines do not daylight in critical, toppling, or caution zones.

Then, residual value of internal friction angle (∅) of discontinuities can be estimated by evaluating the engineering data sheets prepared. Back-analyses of similar field conditions may provide reliable shear strength parameters in terms of ∅ and cohesion (C). Table 3 may help to estimate ∅ and C values depending on the field conditions described in tables 1 and 2.

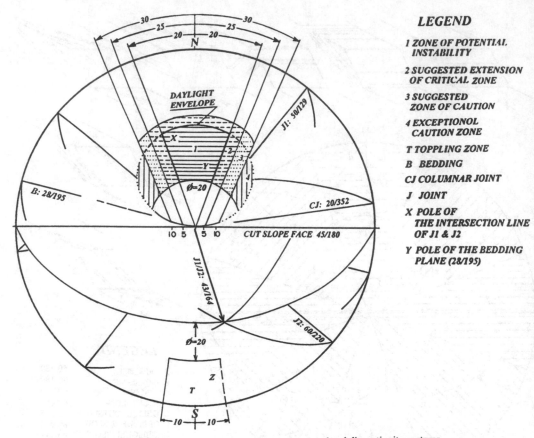

Fig.3 Critical zones of a cut slope with an attitude of 45/180 and some related discontinuity systems.

4 ENGINEERING MEASURES

Slope stability problems could be eliminated or their magnitudes could be reduced by locating the alignment properly and by selection of appropriate structural elements. Topographical constraints and limiting geometrical standards may force a motorway to cross some unstable grounds. In such cases, various slope stability measures could be effectively implemented and some of which are given below.

- Resloping: It is not practical on sloping grounds. Cut height increases with flattening.
- Wire mesh on slope to trap falling rock: It is expensive but practical for a short cut which does not have any overall stability problems other than rock fall.
- Fence to catch rolling rocks: Its economic life is short and could support light mass of rock debris.

- Rock sheds and/or short tunnels: Short tunnels are preferred where mass of trapped talus exceeds 1 m³/lin.m/year or overburden is thicker than 4 m. Otherwise rock sheds might be practical to convey rock avalanche, debris flow, rolling rocks, avalanche, and so forth downhill side.
- Rock bolts for dissected masses: It is advisable to fix wedges and blocks having tendency to fall or topple.
- Cable anchors: They are very useful to increase support length. Plane failure and huge wedges could be supported by using cable anchors.
- Shotcrete, wire mesh, rock bolts, and weep holes could be used jointly to prevent surface bleedings and slopewashes over intensely fractured rock slopes.
- Concrete pedestals for overhangs: If overhangs are very limited in areal extension, concrete pedestals could be used to support overhangs effectively.

- Dowels could be used to fix wedges on gentle slopes.
- Slope drainage and depressurization: This could not be implemented in impervious geological units. Permeability (k) should be greater than 10^{-3} m/s.
- Scaling and trimming: It is vital to execute prior to slope treatments.
- Benching and unloading: Benching might help to retain slope debris and to transmit runoff to nearby drainage system whereas unloading at top of slope reduces driving forces.
- Replacement with free draining material: It could be recommended to prevent circular failures in clayey units and to impede enlarging of linear trace of weak zones basically structural contacts, lithological contacts, and fault zones. It could be accompanied with inclined drain pipes.

The principles of engineering measures to treat a slope are

- to make slopes more stable by (a) increasing the resisting (supporting) forces, (b) reducing driving forces, and (c) draining the system properly,
- to minimise maintenance cost,
- to sustain aesthetic view, and
- last but not least to protect environment.

In order to clarify the above principles, an actual field case is depicted in figure 4. A simplified map and a section are presented to show alternative motorway elements. The geological units are represented by colluvial deposits, flysch, and melange which consists mainly of well preserved limestone bedchunks, deformed spilite, autobreccia, and some pyroclastics. In 1993, the slopes were constructed by cut & fill method without implementing any engineering measure. An area of 60 000 m² has been destroyed by cutting & filling, constructing access roads, and/or dumping over the downhill side. The northern slope has already started to fail in the form of rock falls, debris slides, small scale plane failures where outslope conditions prevail (KM 91+425 - 91+710), and bleeding along major discontinuities. If reinforcement, cut & cover, and a retaining wall were used as recommended by geotechnical engineers, 80 % of natural land would have been protected. Well preserved limestone bedchunks comprises calcilutite-calcarenite, argillaceous limestone, and fossiliferous limestone levels. All are dipping southward forming outslope condition in the northern slope. The berm edges disappeared during the first 3 months and an overall slope face has formed (Fig. 5). This kind of smooth slope obviously favours runoff, rock rolling, and avalanches. The northern slope, where outslope condition prevails, was kinematically found unstable at a slope greater than 24°. The practical solution was to design reinforced slope as presented in Figure 4. If a cut slope is in an outslope condition, "Berm Edge Fence" method (Yilmazer & Ertunç 1993) has to be implemented following the construction. It is a practical one to stabilise outsloping weak to strong rock alternation. Furthermore, it constitutes a major part of landscaping works. Besides the size and type of materials and steel bar intervals have to be defined as a site specific application. Steel bars could not be sheared under the given load condition (Fig. 5) and they act as rock bolts. However, they are susceptible to bending out above the slope face. The horizontal force due to wedge of backfill could be calculated by the equation

$$P = K_a * \gamma_w * (H * W/2) * L$$

where

K_a is the active pressure coefficient,

γ_w is the unit weight of the saturated soil, kN/m^3,

H is the length of the steel bar in contact with the backfill, m,

W is the width of the backfill at the berm level, m, and

L is the distance between the two nearest steel bars, m.

The probable displacement (δ) due to bending in a steel bar could be found by the equation

$$S = 11 * P * L^4/120 * E * I$$

where

E is the elastic modules of the steel bar,

I is the moment of inertia ($I = 11 * d^4/2$) of the steel bar,

d is the diameter of the steel bar.

δ is less than 0.01 m under the conditions given in Figure 5. Although it is a negligibly small value, it might be greater with time due to the corrosion effect of rusts on steel bars. For this reason, it is recommended to install steel bars inclined 5° toward the slope. Reinforced PVC pipes could substitute steel ones. Intertwining seedlings fix

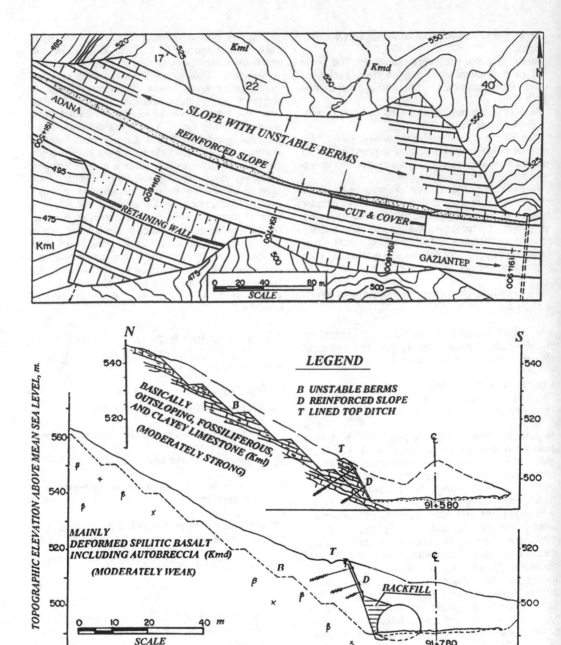

Fig.4 Maps and sections indicating the significance of engineering measures in a cut slope design to reduce earthworks.

the backfill into the slope and cover all slope faces in a few years depending upon basically climatic conditions and plant types.

As in the TM projects, rock bolts and anchors are frequently used in slope stability works. they are used to be installed with an inclination of 20° to 35° from the horizontal into the slope. This application is not effective in outslope conditions. Removal of overburden especially along the berm edges causes noticeable heaves due to stress relief at discontinuity surfaces. Heave and groundwater percolation rate favour each other

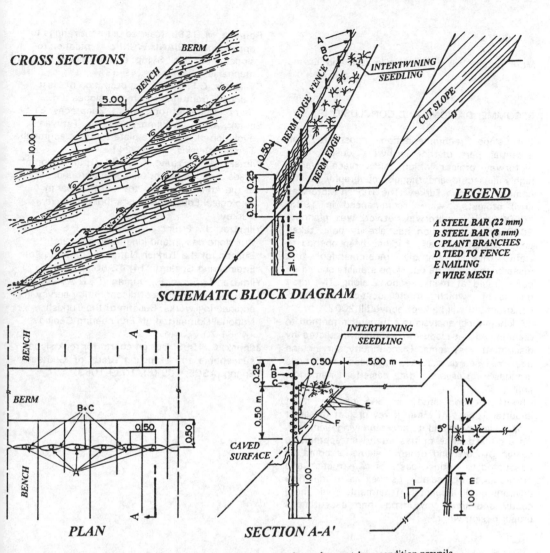

CROSS SECTIONS

BERM

BENCH

5.00

10.00

SCHEMATIC BLOCK DIAGRAM

BERM EDGE FENCE

BERM EDGE

INTERTWINING SEEDLING

CUT SLOPE

LEGEND

A STEEL BAR (22 mm)
B STEEL BAR (8 mm)
C PLANT BRANCHES
D TIED TO FENCE
E NAILING
F WIRE MESH

PLAN

BENCH

BERM

B+C

0.50

0.50

BENCH

SECTION A-A'

INTERTWINING SEEDLING

0.50 — 5.00 m

0.25

A
B
C

D

0.50

E

F

CAVED SURFACE

1.00

W

5°

84 K

E
1.00

Fig.5 "Berm Edge Fence" method and its implementation on a cut slope where outslope condition prevails.

and ultimately cause slope failures. Rock bolting for shallow (<10 m) fixation is one of the practical application. However, its inclination has to be predefined to increase bolting efficiency. Internal friction angle (∅) of a discontinuity plane controls resisting force produced by normal stresses. Tension (T) in a bolt has two components, namely T_n and T_t. F_r could be expressed in terms of T_n and T_t as follows.

$$T = T_n^2 + T_t^2$$
$$T_n = T \cos\alpha$$
$$T_t = T \sin\alpha$$
$$F_r = T \cos\alpha \tan\emptyset + T \sin\alpha$$

The extreme values of the F_r can be found by differentiating it with respect to α and then equating to zero as below.

$$\partial F_r/\partial \alpha = 0$$

From this equation "α" could be found and expressed as below.

$$\alpha = \tan^{-1} (1/\tan\emptyset)$$

The angle of inclination from the horizontal "γ" is given by the expression

$$\gamma = 90 - \alpha$$

More details are given in Yilmazer and Ertunç (1993).

RECOMMENDATIONS AND CONCLUSIONS

Cut slope recommendation constitutes an essential part of an earthwork design in a motorway project which takes place over a highly dissected and rugged topography as the common cases in Turkey. The Turkish Motorway (TM) projects were commenced in 1985. 10000 km long motorway network was planned. 3000 km long portion has already been taken under design and half of it has been opened to traffic. It is strongly recommended that researches about the cut slope stability should be commenced at many sections along TM three phases of which (route location, design, & construction) will be kept active till 2002.

A kinematical analysis is a thrustable method to recommend cut slope in rock units dissected by discontinuity systems. Discontinuity survey and back-analyses of naturally stabilised or unstable grounds form essential data basis for kinematical analyses.

Route location study is one of the most important phase of a motorway project to select proper alignment and its accompanying elements. As a consequence of this study, an appropriate design of cuts and other elements could be conducted to reduce costs of construction and maintenance appreciably as well as to minimise environmental impacts. Treatments are more costly and time consuming than executing a proper design work.

ACKNOWLEDGEMENTS

The authors are grateful to Mr. J. Gallerani, J. Tattersal, and M. Bethard who encouraged to carry out a research work on the slope stability.

REFERENCES

Baker, R.F. 1975. Handbook of highway engineering. Litton Educational Publishing, Inc.

Khawlie, M.R. & A'war, R. 1992. Terrain evaluation for assessment of highways in the mountainous Eastern Mediterranean of Lebanon. Bulletin of the Int. Ass. of Eng. Geology, 46: 71-78.

Romana, M. 1985. New adjustment ratings for application of BIANIAWSKI classification to slopes. Proc. Int. Symp. on he role of rock mechanics. Zacatecas, 49-53.

Swindells, C.F. 1985. The detection of blast induced fracturing to rock slopes. Int. Symp. on role of rock mech. Zacatecas, 81-86.

Sengor, A.M.C. & Yilmaz, Y. 1981. Tethyan evolution of Turkey: a plate tectonic approach. Tectonophysics, 75, 81-241.

Yilmazer, I. & Ertunç, A. 1993. Optimisation of made slope height on sloping grounds. Proc. of the Anniversary of the 60th Year in Geological Education in Istanbul University, Turkey.

Yilmazer, I., Ertunç, A. and Kaya, S. 1993a. Cut slope design and kinematic analysis. Bulletin of the Turkish National Committee of Engineering Geology, 14, 42-60, Turkey.

Yilmazer, I., Erhan, F., Duman, T., & Aytekin, E. 1993b. Significance of discontinuity survey in engineering works. Bulletin of the Turkish National Committee of Engineering Geology, 14, 61-70, Turkey.

Zanbak, C. 1993. Design charts for rock slopes susceptible to toppling. Jour. of Geotech. Engng. ASCE, 109 (8), 1039-1062.

Geotechnical research for the pumping storage station

Les recherches géotechniques pour une usine de pompage

O. Horský
Geotest, Ostrava, Czech Republic

ABSTRACT: The engineering geological research into the pumping storage power station in Centro Cuba in the Caracusey river represents the most exciting research for the construction in the history of Cuba. The relatively large complex of modern investigation methods enabled to prove that in given very complicated geological conditions the variant of the underground location of a power station and conduits is unsuitable. The superficial variant was accepted as optimal and without any risk. The geological situation of the locality is evaluated and reasons for the conception solution are mentioned in the article.

RESUME: La reconnaissance géologique du sol de fondation de l' usine d' accumulation par pompage Centro-Cuba prés de la riviére Caracusey réprésente la prospection la plus exigeante d' un ouvrage de construction dans l' histoire de Cuba, jusqu' aujourd'hui. Un complex de méthodes de prospection a montré l' incompatibilité de la variante souterraine de la disposition de l' usine d' accumulation et des arrivées d' eau et c' était la variante superficielle qui fut acceptée comme une variante optimale et sans risque. Dans cet article la situation géologique de la localité et des parties principales de l' ouvrage est introduite avec des raisons pour la conception acceptée.

1. INTRODUCTION

In 1981-89 Geotest Brno performed a complex engineering geological research into the pumping storage water power station in the central part of Cuba in the Escambray range. Following preliminary studies of four sites in the central part of Cuba carried out in 1981, the Caracusey river has been chosen as the most suitable site. Here in 1984-89 the complex research was realized in two phases: prefeasibility study (finished 1986) and feasibility study (1989).

The engineering geological research into feasibility study was carried out as a part of detailed design, i.e., investigation works are considered to be finished.

The aims of the research led to clarify geological and tectonic textures of rock massif in the area connected with a construction, to asses the degree of the development and the significance of karst phenomena, to obtain reliable and obligatory information about engineering geological conditions concerning the construction of main objects and to prove the safety of work with reference to the geological situation of the whole pumping power station.

The engineering geological research was realized by Czechoslovak and Cuban geological organizations. Czechoslovakia realized the methodic leading and supervision of investigation works, assistance at primary and final documents of research ensured the realization, evaluated tests of rock mechanics "in situ", checking laboratory tests of rock samples, the geological and geophysical documentation of pioneer gallery, well logging and special geophysics, the controlling and the evaluation of compaction tests (various types of rock materials for dams) and the final evaluation of all investigation works.

The main task for Cubans was to ensure all the boring and mining works, including a preparation of special tests in boreholes and in mining works, topographic works, a superficial geophysical research, regime hydrological measurements and realization of special hydrological tests, laboratory analysis of water samples of soils and rocks. To ensure the accessibility of the terrain for the engineering geological mapping and for the proper realizing of the investigation works.

2. THE CONCEPTION DESIGN OF THE PUMPING STORAGE POWER STATION

The complex of the station consists of the lower reservoir (i.e., the valley dam of expressway stream course of the Caracusey river) and of the upper reservoir (i.e., the natural water reservoir filling up the slope depression with removing other materials to enlarge the volume of the reservoir). The difference of height between water levels in both of the reservoirs reached 335 m. Bodies of both the dams are supposed to be of stones or boulders. There is a loamy sealing at the lower reservoir and the artificial sealing water jacket with asphaltic-concrete sealing of the whole bottom at the upper reservoir.

The dam of the lower reservoir was situated in two alternatives. It resulted from very complicated geological conditions of the construction. The height of the dam (considering the accepted alternative) reached 5O m (the elevation of backwater is E1 213.8 m). The height of the upper dam is 58 m (the elevation of backwater is E1 545.8 m). The power station was situated in two alternatives as well, i.e., underground location and surface location.

3. GEOLOGICAL CONDITIONS OF THE SPECIAL INTEREST AREA

In geological aspects the special interest area participates in the tectonic unit Guamahaya, which consists of two megatextures of domed type (the dome Trinidad and Sancti Spiritus). The geological development in early Jurassic with the terrigenic sedimentation lasting to late Jurassic. It is possible to put the development of volcanic-sedimentary formation on the interface of Jurassic and Cretaceous. During Cretaceous the huge folding with regional metamorphism of rocks occurs. This metamorphism finished the creation of the present geological state. From late Cretaceous the lifting and the creation of domed textures have appeared to the present time.

The whole primary petrological complex was (as a result of folding) affected with textural deformations. Napes and their overthrust were created. The last nape, overthrusted to the folding complex in Cretaceous, became external caloric supply which by the effect of the relation: heat pressure finished metamorphism of rocks. It is characteristic that the degree of metamorphism, metasomatism and folding is reducing in sections with the depth, because there is growing distance from external heating and pressure supplies. The result is the inversion arrangement of rock densities.

The complicated geological history of special interested area caused the following very complicated geological construction:

1. The texture of napes with regional subhorizontal contacts, which from the south to the north consists of the overthrusted slice of marble with silicates of the Mayari formation of the group San Juan (today eroded), the nape of serpentines and metabazits and the nape of the La Chispa formation from the group Grupo Naranjo (very reduced thickness in the area of the upper and the lower reservoirs).

The nape of the Corbito formation is substrate in the whole special interested area and it was verified (by our investigation works) to the depth of 450 m. The nape of the La Gloria Formation occurs in reduced thickness in the area of the lower reservoir.

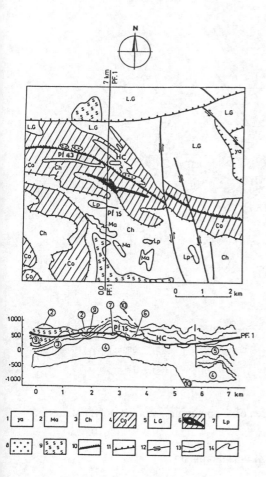

Fig. 1 Generalized geologic map of the area
1-Yayabo form (Jurrassic-Cretaceous), 2-Mayari form (J3), 3-Chispa form (J1-J3ox), 4-Cobrito form (J1-J3ox), 5-Loma la Gloria (J1-J3ox), 6-the hori of black brecciated marbles, 7 - metabasits metagabroidig, 8-metabasits eclogicital-galukophanic , 9-serpentines, 10-the edge between different of metamorphism (tectonic dislocation), 11-tectonic dislocation-decrease, 12-tectonic dislocation (shift), 13-tectonic deformated weathered edges, 14-contact normal lithological, Pf 15-the dam of the upper reservoir, Pf 43-the edge of the lower reservoir, HC- hydro-power station.

2. The postmetamorphic fault of the first order crosses the special interested area, separating zones with different degrees of metamorphism presenting themselves with the intensive detailed folding, with the rock fracturing and high degree of graphitization. In relation with proposed work, this tectonic disturbance appears as weakened zone of rock massif.

3. In the area connected with the construction of the work, tectonic lineations connecting three different geological phenomena occur. They emerge as tectonic dislocations of the primary premetamorphic substrate, covered with later geological processes. They also appear as plicative synmetamorphic deformation and finally as the lineation connected with manifestations of postmetamorphic copularegenesis which affected the concentric metamorphic zonation and which is connected with radial and concentric dislocations.

No floe-like shifting of blocks was found directly in our locality. Faults present themselves by small shifting in the direction of dislocations. The intensive folding and drags of foliation take place here. Fractures are filled in a secondary way and they do not present themselves mostly with an increasing permeability and neither with deteriorating technical qualities. With reference to repeated tectonic movements along these systems, it is necessary to consider them as weakened zones of a massif, which caused the differential erosion in geological history and today they are presented in morphology of the terrain as the depression. The radial dislocations present themselves very significantly in the slopes of the Cobrito formations, where they predispose the rock faces and systems of opened fractures in dorsal parts and the symptoms of development of a karst phenomena.

In the area connected with the work construction there were limited 30 tectonic dislocations and 12 concomitant fracture (Fig. 2).

It was very difficult to determine stratigraphic sequences of formations with reference to complicated lithofacies composition of units respecting the inner textural-tectonic structure and following regional metamorphism. The result of very complicated geotectonic development of these areas is lithology of individual rock formations.

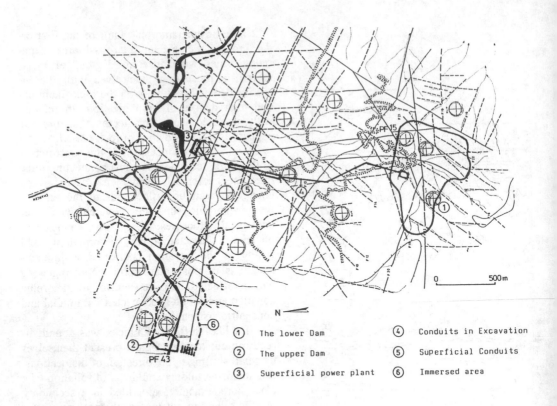

Fig. 2 The tectonic scheme of the special- interest area of the pumping - storage power station
.... axises of courses and depressions,
----------- tectonic dislocations,
= = = = = the zone between the different degree of metamorphism,
x_x_x_x_ course of contour lines.

The rocks of Chispa formation occur in the area of the upper reservoir. They bear on the weathered set of the Cobrito formation and their thickness mostly does not exceed 20 m. The Chispa consists of the complex of rocks of terrigenic genesis mainly of quartzitic, muscovitic in some places graphitic metaslates and of tuphitic slates. These rocks form the planar denudation relic of a continuous very weathered mantle with gradual transition from residual soils to very much weathered rocks. With reference to their small thickness, faultess rocks of Chispa were not found out. Rocks of the Cobrito formation are predominating in the whole area connected with the construction of the work. Conduits, a power station and the majority of the lower reservoir inclusively a profile of the dam will be founded. They are represented by a set ofcarbonate sequences of metacarbonates, including marbles, crystalline limestones and calcareous schists with variable share of graphite and graphitic slates with inserts of chloritic sericitic schists. Rocks of the Cobrito formation are weathered to the depth of 25-40 m under the surface of the terrain.

Rocks of the La Gloria formation occur especially on the right shore of the lower reservoir. Typical rocks of this formation are metaterrigenic quartzitic slates and quartzitic-muscovitic slates with relative abundance of graphite.

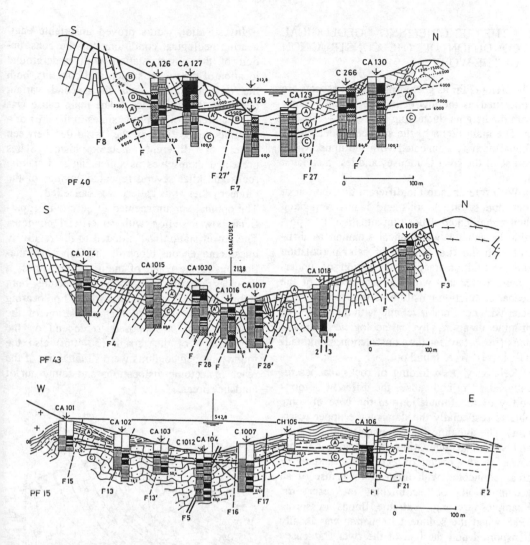

Fig.3 Geological and geotechnological conditions in the dam profiles.
Pf 40-the original variant for the lower reservoir
Pf 43-the definitive variant
Pf 15 - the upper reservoir
RQD-to the right from the hole
▦ lager 50% ▥ 50-25%   ▨ 25-10%   ■ smaller 10%
Water pressure test- 1/min/m/3at

▦	unpermeable	▥ little perm	▨ permeable	■ very well perm
	0.0-0.5	0.5-2.5	2.5-5.0	> 5.0

> 4000 . . . . the limited velocity from the superficial seismic
   4000----the izoline of the wave velocity from the radiography between holes.
A-the zone of loosened tensions, A°-decreased tensions B-concentrated tensions, C-normal tensions, D-the zone of understructive deformations.

3923

4. THE ENGINEERING GEOLOGICAL CONDITION OF THE CONSTRUCTION OF THE WORK

The area (as far as geomorphological aspects are concerned) is the low or middle rolling country with the heights fluctuating between E1 180-690 m. The main factor of the surface modelation is fluvial activity connected with the upper water course of the river Caracusey and less then karst activity.

With reference to the different distribution of rock and semirock soils and loams respecting their complicated depositing situation, it is possible to valuate hydrological situation as difficult with the first horizon of calm percolation and percolation-fracture circulation of underground water and with the second hori of the deeper circulation mostly with the fracture character with confined level and with higher degree of mineralization. Physical geological phenomenons (important for the construction of individual objects) were found out.

Relatively deep folding of rocks reaches locally 30-40 m and causes the different deformability of the foundations in the base of some objects (especially the dam of the upper reservoir). The dynamical development of the erosion and stream accumulation can mean a risk because of an erosion of earth constructions or areas connected with them and because of an accumulation of sediments in reservoirs. Examples are e.g., extreme floods in spring 1988 when the sedimentary mantle was locally transported and the bed of the river Caracusey as well as some of its affluents were changed.

The karst phenomena development with the water drainage by karst textures has local significance but it was one of the main reasons for forsaking the original profile of the lower reservoir. Also the slope deformations have local significance but respecting their occurrence it was possible to select the way of the location of superficial conduits. The strength decrease and stability as a result of migrated graphite to more tectonically exposed zones was proved during the research. Therefore the power station and boots of conduits were situated off the main tectonic lines. There is a chance of neotectonic movements with related phenomena (the degree of seismic activity VIMCS).

Investigation works proved unsuitable engineering geological conditions for the construction of the power station with underground location of the power station and conduits, both the original variant and the second variant deeply in the rock massif. The main cause was the occurrence of graphitic slates in the area of a tectonic dislocation of the 1st order between zones of different metamorphism. Slates presented themselves as squeezing and slaking rocks and after several repeated cavings of the pioneer gallery this gallery was cancelled.

The optimal variant seemed to be one of superficial power station with superficial conduits. The power station was situated to the relatively quasihomogeneous block of the Cobrito formation out of the reach of main tectonic lines. It provides suitable conditions respecting both stability of breakings and the foundation bearing capacity. The depth of the foundation of the power station is supposed to depend on the morphology of the terrain (30-40m) also the foundation bed conduits were situated out of the reach of tectonic dislocations and found out of landslide areas.

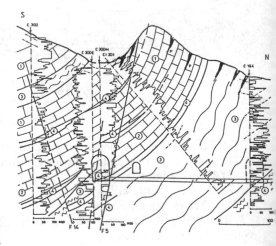

Fig. 4 Geological conditions in places of underground location of the power station
1-stratified and breccieted marbles
2-complex of marbles and calcareous schists
3-complex of calcareous graphitic schists
4-graphitic schists
F-tectonic dislocation
P-opened cracks with the karst appearance

Fig.5 View from the upper reservoir of the pumping storage power station to the line of the conduits.

Fig.6 The line of conduits is led on the left morphologic ledge to the valley where the lower reservoir is situated.

Fig.7 The occurrence of the karst texture on the right bank of the original profile of the dam of lower reservoir.

Considering the engineering geological conditions of the main dam at the upper reservoir, it is possible to say that the profile is suitable for the construction of earth and stony dam but it is necessary to respect the considerable depth of the rock folding connected with low values of deformation and fortification characteristics.

Limestone slates and marbles of the Cobrito formation are recommended as the building material.

Investigation works proved unsuitable geological and geotechnical conditions for the dam profile of the lower reservoir as a result of the relatively high permeability and the problematic stability of the rock block in the left bank and as a result of the occurrence of the karst texture in the right bank. The new bank profile (1 km downwards the stream course of the Caracusey river) appeared to be suitable because of the relatively lower permeability of rocks and respecting a foundation situation of the dam and connecting objects.

5. CONCLUSION

By the mining experiments of building materials in late 1989, all the geological and geotechnological ground control was finished by Geotest Brno. All these documents were necessary for finishing technical project of the pumping storage power plant in Centro-Cuba, the range Escambray, Czechoslovak participation in this project, i.e., in securing engineering geological control meant remarkable special contribution and a qualitative shift in the development of geology and geotechnics in Cuba. The considerably great extent of investigation works - especially the realization of 2 adits and several textural bore holes (the deeper one was 450 m) became a contribution for detailed investigation of geological structure of the special interest area and also to determine main engineering geological characteristics important for the selection of other constructions in that geological region.

Investigation works proved that it was very complicated to build that pumping storage power station (because of complicated engineering geological conditions). The result of these works was the recommendation to build the power station on the surface and with su-perficial conduits. This variant means the smallest risk respecting its construction and safe operation.

REFERENCES

Bláha,P.1994. Shallow refraction processing in geology. *In this Proc.*, Rotterdam: Balkema.

Horský,O.1987. *Informe sobre las Investigaciones Ingeniero Geologicas para la Hidroacumuladora Centro Cuba en el rio Caracusey para el Proyecto de Factibilidad.* Brno: Geotest.

Horský,O.1989. *Informe sobre las Investigaciones Ingeniero Geologicas para la Hidroacumuladora Centro Cuba para el Proyecto Technico.* Brno: Geotest.

Horský,O.1989. Methodics of Engineering Geological Investigation for a pumping storage power station Centro Cuba. *Yearbook Geotest,* 1-12.

Kelly,W.E.&S.Mareš(editors)1993:*Applied Geophysics in Hydrogeological and Engineering Practice.* Monography: Rotterdam: Balkema.

Éboulement rocheux dans le Paléozoïque du Rif: Présentation d'un cas sur la route Oued Lao-Jebha

Rock falls in the Rif Paleozoic: Case history on the road Oued Lao-Jebha

Bouchta El Fellah
Institut Scientifique, Rabat, Maroc

ABSTRACT: The falling, happened in mai 1993 (about 25 000 m3) was a result of the conjonction of multiple unfavourable natural parameters. The seism of 1933, mai 23rd, is supposed to be the direct cause of it unhook.
Consecutively to the man oversight, the equilibrium of this area, already precarious, was broken. The circulation on Oued Lao– Jebha road, unique line of regional communication, is confronted here and there to obstacles which induce hazard.

RESUME: L'éboulement, survenu fin mai 1993 (environ 25 000 m3), résulte de la conjonction d'une multitude de paramètres naturels défavorables. Le séisme du 23 mai 1993 est supposé être la cause directe de son déclenchement. Suite aux négligences de l'homme, l'équilibre – déjà précaire– de cette zone a été rompu. La circulation sur la route Oued Lao– Jebha, d'ailleurs unique voie de communication régionale, est confrontée, çà et là,à des obstacles induisant le risque.

INTRODUCTION

Le littoral au sud de la mer d'Alboran, objet de cette étude, présente un milieu fragile et vulnérable. Le risque d'instabilité menace la route côtière reliant Tetouan à Jebha(fig.1a). A une échelle plus vaste,l'analyse des instabilités de terrain dans le Rif a fait l'objet de nombreux travaux (El Gharbaoui 1981; Heusch 1968; Maurer 1968; Milles Lacroix 1965 et 1968). Leurs conséquences sur l'homme et son environnement méritent, de la part des spécialistes, une attention particulière, axée évidemment sur l'étude et la recherche dans la perspective d'engager des travaux de génie et d'aménagement et de prendre des mesures préventives au préalable (Benazzouz 1991).

1 MILIEU NATUREL ET ACTIVITES HUMAINES

1.1 *Traits particuliers du Rif méditerranéen*

Le littoral au sud d'Alboran, entre Oued Lao et Jebha, montre une suite de falaises vives et de baies étroites développées sur d'anciens golfes pliocènes (El Gharbaoui 1981). Les altitudes atteignent 200m près du littoral. Plus à l'intérieur, le relief est composé de croupes et de

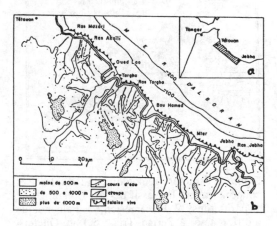

Figure 1 a et b. Situation et aspects morphologiques du Rif interne au SE de Tetouan.

sommets arrondis (500 à 1000m), séparés par des torrents encaissés et perpendiculaires à la côte (Fig. 1b).

La pluviomètrie de la région varie entre 300mm à Jebha, 400mm à Oued Lao et 600mm à Martil.

Cependant des pluies exceptionnelles s'y abattent sous forme d'averses brutales, de longues durées ou sériées. Milles Lacroix (1965) a bien souligné que l'année "1962/63 a été dans le Rif une remise en équilibre des plus brutales, en raison même de l'irrégularité temporelle de la répartition pluviomètrique". D'une façon générale, les stations de ce littoral s'encartent dans le bioclimat semi–aride tempéré, avec des (m) (moyennes des températures minimales du mois le plus froid) comprises entre 5 et 8°c.

1.2 *Géologie et géomorphologie*

Le Paléozoïque du Rif septentrional (zone–interne), bordant la mer d'Alboran à partir de Sebta jusqu'à Jebha (Pointe des Pêcheurs) est constitué dans sa partie méridionale par le massif des Bni Bouchra (péridotites) entouré par une auréole métamorphique à roches tendres (gneiss, micaschistes et schistes) en plus des argiles du Trias (Fig. 2).

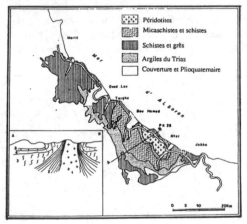

Figure 2. Structure géologique du Rif interne et coupe à travers le massif de Bni Bouchra.
1 péridotites; 2 Précambrien; 3 Paléozoique indifférencié; 4 Permo–Trias; 5 Plio–Quaternaire.

Les micaschistes présentent des plans de litage (schistosité) redressés et intercalés de bancs gréseux avec un pendage subvertical. Ils sont soumis aux dilatations et retraits dûs aux variations des températures et à l'infiltration des eaux, ce qui conditionne l'épaississement de la couche altérée. Des paquets de terrains sont fauchés (schistes et micaschistes) et évoluent en éboulement ou en glissement peu profonds mais dangereux.

Les entailles causées par la route et le sapement au pied de la falaise activent amplement ces mouvements. La raideur des pentes et la présence d'un manteau d'altérations tendre et imperméable en plus des manifestations climatiques agressives font de cette région un milieu propice à différentes formes d'érosion et de risques géologiques. Les processus d'instabilité dans le massif des Bni Bouchra émanent en effet de la conjonction de nombreux facteurs entreautres la lithologie, l'altitude, les pentes et l'exposition; on y reconnait:
– des éboulements par gravité;
– des solifluxions laminaires et des accumulations de débris colluvionnaires fins sur la chaussée;
– des loupes et des foirages;
– des glissements semi–profonds du type fauchage.

1.3 *Végétation*

Les versants surplombant la mer supportent des sols rouges colluvionnaires et acides relativement épais et riches en matière organique.

La végétation forestière ou préforestière (thuya de Berbérie, pin d'Alep..) souffre de l'extension récente des champs (mise en culture des versants à fortes pentes), ce qui se traduit actuellement par un taux de couverture qui ne cesse de reculer et un paysage d'érosion caractéristique (ruissellement diffus, griffures, ravinement..). Ces phénomènes s'accentuent davantage lors des séquences pluvieuses denses et concentrées. La dégradation de la forêt est, en effet, suivie rapidement par le départ des dépôts altérés en premier lieu (Maurer 1968) puis par l'instabilité des versants qui évoluent par régression vers l'amont tout en montrant une multitude de petites cicatrices d'arrachement.

1.4 *Infrastructure*

L'axe routier Tétouan–Jebha dessert toute la

frange littorale sur une distance d'environ 80km. Son tracé sinueux s'installe, en surplomb, sur le flanc des falaises dont la base est constamment balayée par le flot des vagues. Plusieurs localités sont reliées par cette route: Oued Lao, Targha, Bou Hamed, Mter et Jebha. La pêche et l'agriculture demeurent les principales activités, le tourisme y est encore timide mais en plein essor.

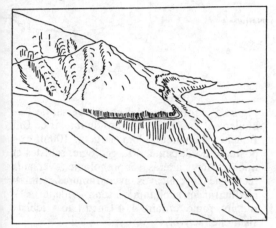

Figure 3. Vue à l'est de Bou Hamed, la route loge sur des versants en surplomb, très instables.

La route descend localement à moins de 10m d'altitude (plages d'embouchures), puis remonte aux environs de 200m sur les mêmes versants sans arriver à éviter les falaises subverticales (fig.3). Elle est sujette à de nombreuses déformations: gabions ruinés, drains colmatés, chaussée ensevelie sous les débris... Elle suit fidèlement l'ancienne piste, et parait n'avoir fait l'objet d'aucune étude géotechnique exhaustive. Durant ces dernières années, le trafic routier n'a cessé d'augmenter pour les raisons suivantes:
– cet axe routier est unique dans toute la région comprise entre Tétouan, Jebha et l'arrière pays de Chaouen;
– les densités de populations sont fortes, comprises entre 100 et 200 hab/km2;
– présence de plusieurs villages portuaires le long de la côte;
– développement remarquable du tourisme estival et balnéaire.
Le débit routier peut dépasser le millier de véhicules par jour pendant la haute saison. Poids lourds, autocars et voitures circulent

avec, parfois, des charges supérieures à 30 voire à 40 tonnes; ce qui constitue une contrainte supplémentaire qui s'ajoute à celles, déjà en jeu (poids de la masse, inertie, gravité..) à l'intérieur des secteurs vulnérables aux glissements.

2 EBOULEMENT DE MAI 1993

2.1 *Les faits*

L'éboulement du PK 26, situé entre Bou Hamed et Jebha a eu lieu vers fin mai 1993, après une saison à pluviométrie normale. Il a succédé à la secousse sismique du 23 mai 1993 (7h 40') qui a frappé toute la région (y compris l'Alboran et l'Espagne méridionale). le tremblement de terre a été ressenti localement par la majorité de la population. Selon le témoignage de personnes appartenant au service des Travaux Publics de Chaouèn, rencontrés sur place pour des opérations de restauration, l'éboulement qui s'est déclenché au PK 26 eut lieu à la suite d'un éclat accompagné de bruit assourdissant qui provenait du bas du remblai (10m).

Aussitôt, la route fut encombrée de débris en forme de cône de déjection, faisant obstacle au trafic routier (fig.3a). Une rupture se dessina à environ 40m en haut du versant, sa forme est semi-circulaire avec parfois plus d'un mètre de commandement.

Figure 3 a et b. Vue générale de l'éboulement au PK 26 et détails de la rupture au niveau de la chaussée.

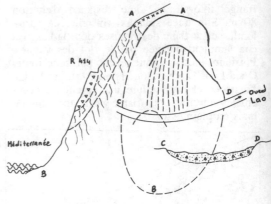

Figure 4. Les murs de soutènement destinés à bloquer la poussée des terres amont, le drainage pour évacuer les eaux. Opération peu réussie réalisée sur micaschistes à l'est de Targha.

En même temps une nuée de poussière se dégagea de la zone affectée. Une autre fissure, en arc de cercle indiquant la présence d'une rupture secondaire, s'installa sur la plate-forme routière (fig.3b) en annonçant le début d'une évolution critique.

2.**2** *Le risque*

Les versants du massif de Bni Bouchra, donnant sur la mer d'Alboran, montrent des symptômes d'aléas incontestables. Ils sont susceptibles d'engendrer des catastrophes de dimensions limitées mais importantes (Chardon 1990). Le risque dans certains de ces versants est omniprésent; celui du PK 26 en est un exemple qui présente une menace permanente. Il s'agit d'un risque géomorphologique littoral à caractère complexe; fauchage, éboulement, blocs qui chutent. La vitesse du déplacement de la masse est aléatoire, le volume a été évalué, par simple estimation, à 25.000 m3.

Ce milieu est marqué par une dynamique très active (Faugeres et Noyelle 1992). Les services des TP et les utilisateurs sont inquiets, surtout que cette route vient à peine (en 1985) d'amorcer une enclave jadis difficile à atteindre. Bien qu'elle n'ait pas encore affronté des années eccessivement pluvieuses, elle est soumise à des réfections fréquentes (fig.5).

Figure 5. Profils vertical et transversal de l'éboulement du PK 26, au nord du massif de Bni Bouchra. Longueur:300m, largeur: 100m, exposition: nord, pente: 40°, géologie: schistes et micaschistes altérés, morphologie: versants rectilignes à concaves avec ruptures, climat: méditerranéen subhumide à semi-aride, infrastructure: route tertiaire 414 Oued Lao – jebha sismicité: active.

D'ailleurs, un séisme de magnitude supérieure à 6° dans l'échelle MSK, renferme l'éventualité de sa suppression. Celui survenu fin mai 1993, avec 5°,2 était suffisamment fort pour ébranler l'équilibre au PK 26. Les pressions interstitielles, à la surface du glissement, ont probablement subi un accroissement qui s'est traduit par la destabilisation de la masse. Les glissements consécutifs aux séismes sont d'ailleurs connus à travers le monde. L'arrêt du trafic présenterait, dans le cas d'une réactivation fort probable de l'éboulement, des conséquences et des effets néfastes sur la structure socio-économique de tout le littoral; d'autant plus que les moyens à mettre en oeuvre pour contrecarrer ce phénomène sont énormes. Les restaurations effectuées, il y a deux années, à quelques kilomètres près de Targha (Fig.**4**), n'ont pas réussi à stabiliser le glissement.

CONCLUSION

Les glissements de terrain sont largement répandus dans la montagne rifaine. Il sont parfois accompagnés d'éboulements et de chutes de pierres. Ce qui, par conséquent, peut générer, lorsque les pentes dépassent 15 ou 20°, la modification de l'équilibre des versants, la réactivation de l'érosion, la déstabilisation des écosystèmes et le bouleversement des activités humaines.

Les remèdes apportés aux désordres que subit la route Oued Lao–Jebha sont, pour la plupart, ponctuels de type excavations de remblais, murs de soutènement et gabions. Ils se sont avérés, dans plusieurs cas, incertains et peu sûrs.

Toute intervention doit d'abord être précédée d'une prospection étroite et continue des zones d'aléas. Les secteurs les plus menacés par le risque, nécessitent des études exhaustives d'ingéneerie. Mais l'adoption d'une politique de reboisement pourrait permettre d'éviter le recours aux travaux de confortation et d'assainissement excessivement coûteux. D'où l'intérêt d'un plan d'aménagement qui soit en mesure de protéger cette zone côtière.

Pour les mesures d'urgence, il faut mettre l'accent sur le captage des eaux de surface en généralisant le drainage superficiel (dans le sens amont et aval). La réalisation de galeries drainantes, de tunnels ou l'ancrage de la masse glissée au substratum sous–jacent, sont autant de mesures possibles dans les secteurs les plus vulnérables.

REFERENCES

Benazzouz M.T. 1991. L'évaluation de l'aléa géomorphologique et des coûts des risques naturels à Canstantine. *Z. Für Géomorph.* 83:63–70

Chardon M. 1990. Quelques réflexions sur les catastrophes naturelles en montagnes. *R.G.A*, fas. 1/2/3, 78:193–213.

El Gharbaoui A. 1981. La terre et l'homme dans la Péninsule Tingitane, étude sur l'homme et le Milieu naturel dans le Rif occidental. *Trav. Inst. Scient, série géol. et géogr. phys.*, n°15, Rabat.

Faugeres L. et . Noyelle 1992. Risques natu rels, paysages et environnement en France. Inf. Géog., fasc.5, 56:194–209.

Heusch B. 1969. L'érosion dans le bassin du Sebou, une approche quantitative. *R. Géogr. Maroc*, 15:109–128.

Martin J. 1987. Les risques naturels au Maroc, réflexions préliminaires. in mélanges offerts à G.Maurer.

Maurer G. 1968. Les montagnes du Rif central; étude géomorphologique. *Trav. Inst. Scient. série géol. et Géog. phy.*, n°15, Rabat.

Milles Lacroix A. 1965. L'instabilité de ver sants dans le domaine rifain. *Rev. Géomor. Dyn.* n°7–8–9:97– 109.

Milles Lacroix A. 1968. Les glissements de terrain. Présentation d'une carte prévisionnelle des mouvements de masse dans le Rif. *Mines et Géologie*, Rabat, 27:45–54.

An engineering geomorphological appraisal of slope stability along a proposed by-pass in Natal

Une évaluation technique et géomorphologique de la stabilité des pentes au long d'un futur détour au Natal

G.J. Ivins & F.G. Bell

Department of Geology and Applied Geology, University of Natal, Durban, South Africa

ABSTRACT: Because of the steep gradients on Town Hill the national trunk road which at that point passes Pietermaritzburg is associated with traffic hazards. Consequently three by-passes have been suggested. Each has certain advantages and disadvantages, which include social and economic factors as well as geological.

One of the major problems associated with each of the proposed routes to by-pass this area is mass movements along the slopes of the terrain through which they would pass. The slopes in question normally are covered with colluvium, much of which is potentially unstable. Consequently in the past mass movements have taken the form of soil creep, rockfalls, rotational slides and translational slides.

An engineering geomorphological survey was undertaken to evaluate each routeway and to identify and map sensitive areas, especially those where problems of slope stability could arise, so that a preferred route could be chosen. The desk study made use of aerial photographs and published geomorphological and geological data. Geological mapping was largely confined to mapping the sub-soil conditions and geomorphological mapping recorded data on surface processes, surface form and types of surface materials. An assessment of slope stability involved determination of the magnitude and type of movement, as well as attempting to assess the risk of landslide activity.

RESUME: En raison dufort gradient du "Town Hill", la route principale qui passe par Pietermaritzburg a set endroit, présente des hasards au traffic. Conséquement trois détours possible ont été suggéré. Chacun d'entre eux ont certains avantages et désaventages, ceux-ci comprennent des facteurs sociaux, economique aussibien que geologiques.

Un des principaux problèmes associé a chacune de ces routes est le mouvement des masses de terre au long des rampes du terrain ou ces routes devront passer. Les rampes en question sont couvert par colluvium qui est géneralement potentiellement instable. Conséquement les mouvements de masse sont du type rampant, chute de rochers, eboulements rotationels et eboulements translationels.

Un arpentage technique et géomorphologique fut entrepris pour évolver chacun de ces détours et pour produire un plan des endroits problématiques spécialement ceux qui presentaient des problemes de rampes ou de stabilité. Dans le bût de determiner la meilleure de ces détours l'etude paite au bureau employe des photo aeriennes et public des données geomorphologiques et géologiques. L'arpentage géologique consiste largement a étudier les conditions du sous-sol. L'arpentage géomorphologique enregistra les données sur les phénomênes de surface, formations du terrain et les differents materiaux prévent sur ce terrain. Une évaluation de la magnitude et du type de mouvements est nécessaire pour déterminer la stabilité des rampes.

1 INTRODUCTION

Although the road between Durban and Johannesburg is a major national road, along some sections it does not meet the standards set by the Department of

Transport. Therefore upgrading to acceptable standards is a priority. One of the sections which needs to be dealt with is that which passes Pietermaritzburg, notably at Town Hill where the steep gradient is related to a poor safety record. Hence alternative routes have been investigated. This has involved mapping certain areas, especially with regard to slope stability, to identify a preferred corridor. Three corridors were chosen and an area of sensitive slope stability lay within each which required more detailed investigation. The three areas concerned were Game Valley, Shooters Hill and Claridge. The initial investigation involved a desk study which made use of orthophotos, aerial photographs and geological maps; geological mapping; geomorphological mapping which recorded details on surface form, surface processes and materials; slope stability assessment which attempted to determine the magnitude rate and type of movement, and the frequency with which landslides occur on a given slope (aerial photographs proved particularly helpful); and laboratory testing of representative soil samples. The maps and accompanying report were then presented to the engineers who were responsible for deciding which of the three routes was the most acceptable.

All three areas have an average annual rainfall in excess of 800 mm per annum, with most of this rainfall occurring during the summer months. In addition, in the Pietermaritzburg area maximum precipitation of 50 mm/h can be expected once every two years. This high intensity rainfall is enough to trigger large scale mass movements under the right conditions.

The three areas were targeted for investigation due to the presence of excessively thick and potentially unstable colluvial slopes which could pose engineering problems. Both rotational and translational slides have occurred in all three areas under investigation. Most slides occurred during periods of heavy rainfall. During such heavy rainfall there is an increase in shear stresses caused by an increase in unit weight and a reduction of effective stress. In other words, in the latter case, the frictional resistance is lowered by increased pore water pressures. In rotational slides shearing occurs relatively rapidly on well defined curved surfaces. Backward rotation of the sliding mass often causes fracturing of further blocks at the crown of the slide and sometimes is accompanied by ponding of water in the depressed, or backwards-tilted, area of the crown. This water

increases saturation which tends to promote continuing movement. Translational slides are mostly shallow features which have essentially planar failure surfaces. There may be some curvature towards the crown of a slide as most of these slides are in fact a combination of rotational and translational movements (Fig. 1). These landslides could be grouped into recent landslides, which are easily recognised by fresh erosional scars and their affect on vegetation, and old landslides which are usually overgrown or disturbed by erosion so that traces of movement are difficult to discern. Nonetheless slopes that have undergone movement are predisposed to reactivation, hence the identification of mass movement features, is an integral part of predicting potential instability.

2 GEOLOGY

The Ecca Group underlies much of the area and has been intruded by Karoo dolerites. Sediments of Quaternary age overlie the Ecca Group. The Pietermaritzburg Shale Formation is the lowermost unit of the Ecca Group and consists of light to dark grey silty shales which are up to 450 m thick. Within the shales harder, indurated horizons of up to 300 mm in thickness, occur fairly extensively over considerable distances. The shales become increasingly micaceous upwards, passing without break into the overlying Vryheid Formation. The Pietermaritzburg Formation also becomes sandier as it passes upwards into the Vryheid Formation. Clayey zones occur some 30 m above the base. The shales are

Fig 1. A translational slide in Game Valley.

generally well jointed, as well as being well bedded. The lack of common orientation of the joints suggest that they have formed as a result of drying out of sediment rather than tectonic processes. Deposition of iron minerals from circulating groundwater on joints in the unweathered rock, commonly in the form of pyrite, gives rise to "snuff box" weathering of the shale at some locations. On weathering the shale alters to a light brown or buff colour and becomes clayey. Soil developed on the shale generally comprises a shallow silty clay on upper slopes where very frequently, due to erosion, the weathered shale crops out at the ground surface. Soil thickness is appreciably greater on the lower slopes and in valley bottoms. The Vryheid Formation consists of cross-bedded, greyish-white feldspathic sandstones, gritty sandstones, blue-grey micaceous shales and mudstones. This formation varies up to 520 m thick. The sandstones generally consist mainly of quartz grains with subordinate feldspar and muscovite but some sandstones are arkosic. Minor coal seams are present. The Vryheid Formation is characterized by numerous lateral and vertical lithological variations. In other words, the sandstones in the Vryheid Formation, are generally closely intercalated with mudrocks. Problems have been encountered in alternating sandstones and siltstones of the Vryheid Formation where water flowing through the sandstones is impeded by relatively impermeable siltstones. Excess pore water pressures may develop on such interfaces and sliding may result. On weathering, the sandstones give rise to light grey sandy soil and the shales to darker clayey soils of varying thickness. With quartz often comprising more than 50% of the sandstone, disintegration appears to be the predominant mode of weathering. However, the feldspars decompose to produce a clayey material in the sandy residual soil. Feldspathic sandstones of the Vryheid Formation become collapsible when the feldspars weather and the resulting material consists of a quartz framework loosely held together by clay minerals derived from the weathered feldspar. An increase in the moisture content causes a rapid decrease in volume of this material.

The upper most unit of the Ecca Group is the Volksrust Formation which consists predominantly of light to dark grey, laminated, micaceous shales. It ranges up to 350 m in thickness. All three formations have a general easterly dip of between 5° and 15°.

The Karoo intrusions consist of dolerites which predominantly occur as sills. These range in thickness from 0.5 to 10 m or more, the thinner width predominating.

3 SENSITIVE AREAS

3.1 Game Valley

Game Valley is situated 4 km due east of Albert Falls reservoir (Fig. 2). The proposed route runs the length of the valley as it climbs over the Pietermaritzburg escarpment. The area is mainly underlain by fine grained sandstones of the Vryheid Formation. The mobilization zone, at the crest of the slope, is subject to soil creep and is predominantly underlain by Karoo dolerites, while the river valley to the north of the proposed route is underlain by shales of the Pietermaritzburg Formation. The soils consist predominant of silty and sandy loams of colluvial origin. The vegetation consists mainly of grassy slopes broken up by areas of dense vegetation which grows in drainage gullies.

Due to limitations on gradient and camber of a national road, the proposed route would have to negotiate several large scale mass movements which have occurred in Game Valley. Translational slides are particularly common on the thick colluvial slopes running the length of the valley.

3.2 Shooters Hill

Shooters Hill is situated about 3 km north

Fig 2. Location of sensitive areas in relation to the proposed corridors.

west of Albert Falls reservoir (Fig. 2).
The suggested route is predominantly
underlain by Karoo dolerite and
Pietermaritzburg Shale. Shooters Hill has a
hummocky topography caused by mass wasting.
Many *en echelon* tension cracks occur at the
margins of incipient debris slides on
Shooters Hill. They ranged in size from 5 m
to 50 m in length, some having a small
vertical displacement as well as a
horizontal displacement. Evidence of old
landslides can be clearly seen.

Hillwash derived from dolerite is
predominant on Shooters Hill. The hillwash
consists of sandy clays which are
potentially expansive. These sediments
contain ferricrete nodules and isolated
dolerite boulders. The soil profiles in
this area range from 2 m to 10 m thick.
Evidence of soil creep is present in the
form of terracettes.

3.3 Claridge

Claridge is situated about 3 km north of
Pietermaritzburg (Fig. 2). The proposed
route would run approximately east-west
through Claridge as it made its way around
the suburb of Northdale. This is
predominantly underlain by Pietermaritzburg
Shale and Karoo dolerites. The soils on the
slopes consist primarily of sandy loam
colluvium. The proposed route traverses the
transportational midslope to the south of
the Everthorpe Hill, a hill capped by Karoo
dolerite. A zone of instability lies to the
east of Claridge and to the north of
Northdale. Figure 3 shows the hummocky
topography indicating ongoing mass wasting
on thick colluvial slopes.

4 DISCUSSION AND CONCLUSIONS

The following types of map were produced
for each area under investigation, a
topographic map, a geological map, a map of
slope units and mass movements, and a
stability hazard assessment map (Figs. 4-
6). Basic map information is needed to
assess the feasibility of the proposed
roadways. The hazard zones of each area
were developed from an assessment of the
underlying geology, slope angles, and
evidence and magnitude of mass movements.
Slight hazard represents a zone apparently
or practically free from ground movements.
Moderate hazard represents a zone affected
by scattered occurrences of poorly defined
ground movements of moderate extent. The
zone of high hazard is affected by large
scale movements indicating well-defined

Fig 3. Hummocky topography to the east
of Claridge indicating ongoing mass
wasting.

aspects of instability.

The purpose of an engineering
geomorphological map is to portray surface
forms, the nature and properties of the
materials of which these surfaces are
composed and to indicate the type and
magnitude of the processes in operation
(Bell et al, 1987). As such they provide a
comprehensive, integrated statement of
landform and drainage. Geomorphological
maps also give a rapid appreciation of the
nature of the ground and thereby help the
design of more detailed investigations, as
well as focusing attention on problem
areas. Geomorphological mapping is well
suited to the investigation of road
alignments because of the speed at which
the work can be carried out; the extent to
which shape and origin of landforms can be
identified and the inclusion on the map of
features of importance to engineering
interpretation, such as ancient and recent
slope features, surface and groundwater
characteristics, and superficial deposits
(Doornkamp et al, 1979).

Slopes on which thick colluvial deposits
occur are present in all three areas and
mass movements occur in these soils.
Thicknesses of hillslope deposits vary from
5 m up to 24 m in places. The soils consist
of rock fragments, of quartz, feldspar,
kaolinite and illite, with smectitic clays
being found in the deeper soil profiles. On
the convex slope of hilltops movement of
soil material occurs as a result of creep.
Prominent cliff-like outcrops of dolerite
and sandstone provide a source of angular
blocks to talus slopes below. Transport of
material over the midslope areas is
dominated by mass movement and sheet wash.
Colluvial redeposition of weathered soil on

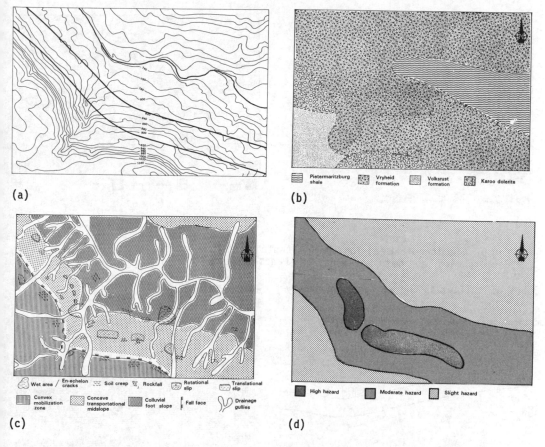

(a)

(b)

(c)

Wet area | En-echelon cracks | Soil creep | Rockfall | Rotational slip | Translational slip

Convex mobilization zone | Concave transportational midslope | Colluvial foot slope | Fall face | Drainage gullies

(d)

High hazard | Moderate hazard | Slight hazard

Pietermaritzburg shale | Vryheid formation | Volksrust formation | Karoo dolerite

Fig 4. Game Valley (a) Topography (contours in 20 m intervals) (b) Geology
(c) Geomorphology and mass movements (d) Landslide hazard.

the lower slope positions results in burial of soils in this area. Many colluvial hollows are present and possess a wide arcuate upper boundary along the debris slope and narrow downslope. The boundaries of the lower parts of colluvial depressions are less well defined due to shallow side-slopes and lateral merging of thin colluvial deposits. Because of the poorly sorted and unconsolidated nature of these

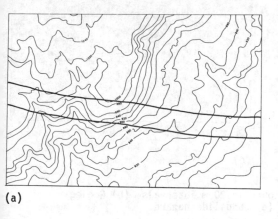

(a)

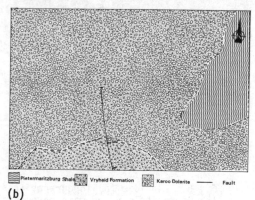

Pietermaritzburg Shale | Vryheid Formation | Karoo Dolerite | —— Fault

(b)

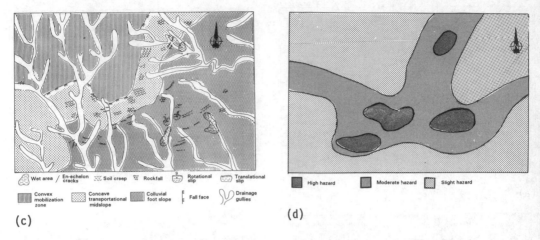

(c)

(d)

Fig 5. Shooters Hill (a) Topography (contours in 20 m intervals) (b) Geology (c) Geomorphology and mass movements (d) Landslide hazard.

lobes of colluvium, the surface topography   appears hummocky.

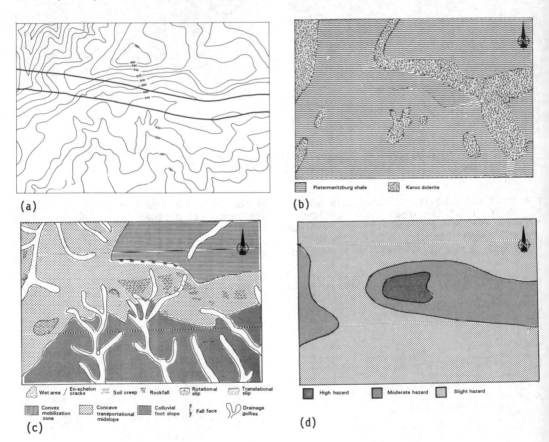

(a)

(b)

(c)

(d)

Fig 6. Claridge (a) Topography (contours in 20 m intervals) (b) Geology (c) Geomorphology and mass movements (d) Landslide hazard.

It has been suggested by Watson et al (1984) that landsliding in this area was at its most active when rainfall was greater than it is at present, that is, approximately 10 000 years ago when a period of warmer, wetter, sub-tropical conditions prevailed. This landslide material has consolidated over time and is reasonably stable while undisturbed. Nonetheless past landslides mean that talus in this region contains a number of shear surfaces. It is highly likely that future slope failure could occur along one or more of these shear surfaces.

The *in situ* moisture contents of soils are noticeably higher in summer than in winter (Table 1). This is in line with the summer rainfall experienced in this area. However, the moisture content of wet areas seemed unaffected by the seasons suggesting that these areas are constantly saturated and would need to be dealt with accordingly. It is also noticeable that the soils on Shooters Hill have a significantly greater *in situ* moisture content than the other two regions. This may be because it occurs in a mist zone which makes some contribution to soil moisture from heavy dew and also reduces evapotranspiration.

There are two main types of instability found on Karoo mudrocks. The first is the movement of completely weathered and colluvial material, and the second is the sliding of rock blocks on bedding planes. The ingress of water into the mudrock and the resulting high pore water pressure probably play a major part in these failures and therefore joints and fissures should be drained in road cuttings. In all areas where mudrocks occur the dip of the bedding planes and the orientation of other discontinuities have an important effect on the stability of the slope. Downslope dips or steep dips may contribute to instability, particularly when influenced by inadequate drainage and a change of slope morphology. Weathering weakens the intergranular bonding of the mudrock and leads to the removal of a large proportion of cementing material, leaving a residue of mainly quartz and illite of low shear strength. The shear strength of such a residual soil is further reduced by an increase in moisture content.

The median size, natural moisture content, consistency limits and linear shrinkage of some of the soils in the three areas investigated are given in Table 1. As remarked above, there is frequently a noticeable difference between the natural moisture content during the summer and winter seasons reflecting the differences in seasonal rainfall. Taking the liquid limit, most soils from Game Valley and Shooters Hill have a low plasticity, while those from Claridge have a medium plasticity (Fig. 7). The linear shrinkage rarely exceeds 10% and in some instances is less than 5%. The plasticity index or range of moisture contents through which a soil remains plastic is an important factor in road construction. Because soils forming the subgrade for highways are likely to be subject to moisture influxes, highway departments commonly require that the base course for roadways has a plasticity index less than 6 and in some cases an even lower figure is specified (Ahlvin and Smoots, 1988). With these specifications in mind, it is noticeable that the soils in the three areas would not be suitable for highway construction without some form of soil treatment.

Increased moisture contents can cause swelling in Karoo mudrocks as they have the ability to absorb water. Mudrock is very seldom used in the base course of roads, but has been used together with a stabilizer (usually lime) as a subbase.

Extensive use is made of dolerite, both fresh and weathered, in road construction. In bituminous surfacings, the adhesive properties of crushed fresh dolerite are usually satisfactory. However, dolerite from chill zones should be avoided due to insufficient adhesive properties. Crushed dolerite is also used in the bases or subbases of roads. Dolerite is available along all three routes.

Game Valley has a definite zone of instability which runs directly beneath the proposed route. This zone of instability, which occurs on the steepest slope, appears to be the most active and susceptible to landslide activity. Past records show that it was the particularly affected by the heavy rainfall of September, 1987, and has

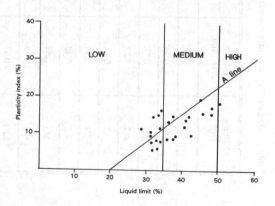

Fig 7. Plasticity chart of soils from the sensitive areas.

Table 1. Median Size, moisture content, consistency limits and linear shrinkage of colluvial and residual soils.

Sample No.	Median Size (mm)	Moisture Content (%)		Liquid Limit (%)	Plastic Limit (%)	Plasticity Index	Liquidity Index	Linear Shrinkage (%)
		January	July					
1G	0.29	22.3	11.1	31.6	21.5	10.1	0.08	5.8
2G	0.38	29.4	13.3	33.2	25.4	7.8	0.51	3.5
3G	0.52	29.9	14.6	36.6	23.1	13.5	0.50	7.1
4G	9.27	28.3	12.0	31.5	21.8	9.7	0.67	3.3
5G	0.28	99.4	94.4	38.4	29.4	9.0	7.78	5.4
6G	0.38	27.0	10.1	33.6	18.5	15.1	0.56	7.6
7G	0.40	20.2	12.4	28.8	17.6	11.2	0.23	6.1
8G	0.54	28.0	14.0	34.0	22.6	11.4	0.47	4.5
9G	0.59	20.1	13.8	32.1	17.6	14.5	0.17	6.2
10G	0.47	24.0	12.3	34.8	18.5	16.3	0.34	4.3
11S	0.27	34.3	16.2	31.2	23.4	7.8	1.40	7.6
12S	0.25	32.7	19.4	31.9	26.6	5.3	1.15	7.0
13S	0.28	99.3	98.6	37.1	28.2	8.9	7.99	10.5
14S	0.39	32.5	18.2	34.4	36.7	7.7	-0.54	14.4
15S	0.35	29.1	14.1	36.0	27.3	8.7	0.21	9.1
16S	0.37	29.4	20.6	40.8	28.8	12.0	0.05	5.9
17S	0.35	27.5	17.5	33.3	27.6	5.7	-0.02	4.6
18S	0.34	33.1	14.2	42.5	32.7	9.8	0.04	8.1
19C	0.53	14.5	11.2	42.9	28.3	14.6	-0.95	8.6

Table 1. continued

20C	0.56	18.0	13.1	41.2	27.6	13.6	-0.71	6.1
21C	0.51	23.0	12.2	48.4	33.3	15.1	-0.68	4.2
22C	0.62	22.2	10.5	50.2	32.0	18.2	-0.54	8.9
23C	0.55	26.1	12.6	46.8	31.4	15.4	-0.34	10.6
24C	0.24	98.0	99.5	47.8	32.6	15.2	4.30	7.1

G = Game Valley; S = Shooters Hill; C = Claridge

yet to fully recover. Shooters Hill has thick colluvial slopes with a gentler gradient. The hummocky topopgraphy suggests that the slopes are subject to continued mass wasting. The Claridge area seemed to be the least sensitive in terms of the proposed construction. Although the slopes showed a hummocky topography, the gradients were slightly shallower. In conclusion, it would appear that the Claridge route would be the most suitable for the proposed construction in terms of geomorphology and related slope stability.

REFERENCES

Ahlvin, R.G. & Smoots, V.A. 1988. Construction Guide for Soils and Foundations. Second Edition, Wiley, New York.

Bell, F.G., Cripps, J.C., Culshaw, M.G. & O'Hara, M. 1987. Aspects of geology in planning. In Planning and Engineering Geology, Engineering Geology Special Publication No 4, The Geological Society, London, 1-38.

Doornkamp, J.C., Brunsden, D, Jones, D.K.C., Cooke, R.U. & Bush, P.R. 1979. Rapid geomorphological assessments for engineering. Quarterly Journal Engineering Geology, 12: 189-204.

Watson, A., Price-Williams, D. & Goudie, A.S. 1984. The palaeoenvironmental interpretation of colluvial sediments and palaeosols of the Late Pleistocene hypothermal in southern Africa. Palaeogeography, Palaeoclimatology, Palaeoecology, 45: 225-249.

Geological and geotechnical models in the south-eastern plain of Cuneo, Italy

Modèles géologiques et géotechniques dans la partie sud-orientale de la plaine de Cuneo, Italie

G. Bottino, C. Cavalli & B. Vigna
Politecnico of Turin, Italy

A. Eusebio & P. Grasso
GEODATA SpA, Turin, Italy

ABSTRACT: The study reports some of the results of an extensive study carried out in the south-eastern area of the plain of Cuneo with the aim of defining the main geological, geotechnical and hydrogeological features of this zone of Piedmont of which there is scarce knowledge regarding large-scale geological structures and localised situations. Given that today this sector is an area of rapid growth, more detailed basic knowledge is now required in order to make appropriate plans for this development. In particular, this study reports the data and the characteristics of the main geotechnical units identified, comparing the principle features of the geological structure with the technical aspects which have the greatest influence on choices in planning and construction.

RESUME: Le modèlage géologico-géotechnique de la partie sud-orientale de la plaine de Cuneo a été réalisé pour obtenir les données nécessaires à l'élaboration du projet de l'autoroute entre les villes d'Asti et de Cuneo, en Piémont (Italie nord-ouest). Cette autoroute de 90km de long traverse la plaine de la rivière Stura et les collines des Langhe. Les caractéristiques très variées au point de vue géologico-géotechnique du sol nous placent dans une situation très complexe en ce qui concerne le projet de l'autoroute. Les formations de la plaine de Cuneo sont composées de silts et de marne (Bassin Tertiaire Ligure-Piémontais) et de gravier, silt et sable avec argile (continental et dépôts deltaïques de la plaine de Cuneo). Les tests ont indiqué des caractéristiques géotechniques qui permettent de quantifier les paramètres des données pour le projet d'ingénierie. Le modèle final a permis de mieux définir la resource hydrogéologique et son interaction avec la route. Les procédés décrits ici sont applicables pour développer les schémas des caractérisations géologico-géotechniques dans d'autres situations analogues.

INTRODUCTION

This study forms part of a wider study whose objective was to provide an organic framework of knowledge regarding the geological, stratigraphic, lithological, hydrogeological and geotechnical characteristics of the formations which form the structure of the plain of Cuneo.

In particular, this report centres on the presentation of the results of the geotechnical characterisation of both the overlying soils and the underlying sedimentary formations in the south-eastern sector of the plain and high terraces lying between the Stura and Ellero rivers. This characterisation was then compared with the stratigraphic units in the area with the aim of creating a model in which the various parameters integrated with one another, thus making the extrapolation of data locally acquired more reliable for the entire sector examined.

The study was mainly based on the interpretation and correlation of numerous stratigraphic data produced by surveys carried out recently in the area, a major contribution to which was given by the geognostic investigation performed for the construction of the Cuneo-Asti highway link road, data for which were kindly made available by SATAP (Società Autostrade Torino-Alessandria-Piacenza).

Geological setting

In morphological terms the area studied corresponds to a plain, gently sloping towards NE and connected to the surrounding mountainous areas by a series of "high terraces" (Fig. 1). There are also a series of deep cuts in the main plain whose genesis is related to the erosive action of the main rivers which is in turn conditioned by the deviation of the Tanaro river and by the general raising of the entire area.

Few geological and stratigraphic studies have been carried out in this zone and the majority concern specific local aspects (Sacco, 1912; Carraro et al., 1978). Recently, Bottino et al., 1992 and 1994 presented a substantially new overall situation of the stratigraphic and geological settlement of the area, identifying a series of Stratigraphic-Depositional Units dating back to between the Messinian and Quaternary periods.

In particular, the authors identified three stratigraphic units dating respectively from the Early Pliocenic (Unit ST1), Plio-Pleistocenic (Unit ST2) and Middle Pleistocenic-Olocenic (Unit ST3) ages. These units, which are separated by clear discontinuity surfaces, can be readily characterised by their geometry, orientation and the laterovertical distribution of the internal associations of facies.

The deposits in Unit ST1 are typical of a relatively deep marine environment, characterised by frequent episodes of resedimentation and gravitational translation, replaced towards the upper layers by a shallower environment, probably in the form of a shelf and outer beach. In lithological terms, this unit is predominantly composed of homogeneous grey-blue pelites, massive or weakly stratified and subordinated by sands, fine silty sand, sometimes with pebbles, in layer or more frequently in lenses.

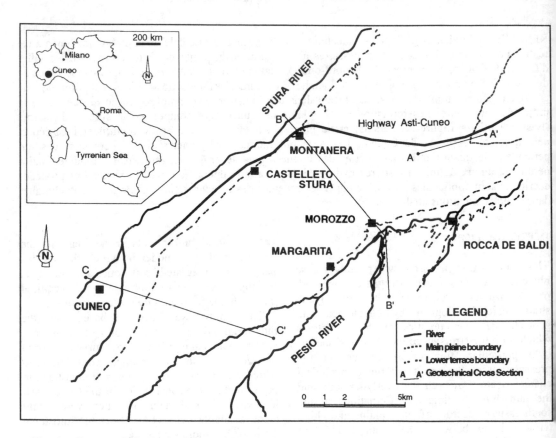

Fig. 1   Geomorphological scheme of Cuneo plaine

The deposits in Unit ST2 are typical of a series of different environments, associated to form a system of fans and alluvial plain, evolving into marginal-marine type environments towards the north. This unit is made up of a series of lithofacies with marked lateral variations, closely correlated to one another. The predominant rock types are coarse gravel in a silty-sandy matrix, often highly weathered, which are prevalent in the fans zones, gravels, sands and silts in the alluvial plain zone, and sand with silts predominantly in the transition facies.

The deposits in Unit ST3 can all be referred to a continental sedimentation environment; internally, it is possible to identify lithofacies which are differentiated in lithological and granulometric terms. The predominant rock types are medium and coarse gravels in a silty-sandy matrix which are highly weathered in correspondence to the high terraces of the plain and non-weathered and locally cemented in correspondence with the main level of the plain and on the lower terraces.

The orientation of sediments in Unit ST1 is virtually subhorizontal (3-4 degrees on average), usually with considerable thicknesses (at many points these exceed 100m). The deposits in Unit ST2 are likewise largely subhorizontal, and reach their maximum thickness (over 200m) in correspondence with an axis connecting Cuneo with Fossano, whereas, to the sides, within a 5-km radius, the thickness is reduced to approximately 10-15m. Lastly, Unit ST3 contains widely varying thicknesses in the different areas of the plain, ranging between 5-15m in correspondence with the high terraces, 10-60m in the zone of the main plain, and 2-5m in the secondary terraces and present day valley floors. This unit is normally covered by a level of agricultural soil with thicknesses varying between 0.5 and 1m. This level may, locally, be highly weathered in correspondence with the high terraces, and it even reaches the thickness of more than 3m.

Geotechnical tests performed

The geotechnical characterisation of the various soils was carried out using available data from the numerous geognostic surveys performed recently in the study area.

In this way data were utilised from over 100 boreholes made mostly for the construction of the Asti-Cuneo highway during the period 1990-91. These boreholes were drilled to depths of 20-50m by rotary-hydraulic method with continuous bore logging; SPT were also performed every 1.5m together with DACTEST-type tests in some cases, and significant samples were collected of both disturbed and undisturbed soils; in some instances the borehole was then fitted with a piezometer either of the open tube or Casagrande type.

In a few highly representative zones plate bearing tests were also carried out (using 300 mm diameter plates) and geophysical surveys were performed using vertical electrical profiling technique. The following parameters were assessed in laboratory: consistency limits, grain size, pocket penetrometer, volume weight, porosity, edometric module, uniaxial and triaxial compressive strength, and direct shear strength.

On the basis of these tests it was possible to characterise the various horizons crossed by boreholes; the study of the geological and geotechnical characteristics of the various geological units then allowed the geotechnical characterisation performed locally to be extrapolated to the entire sector of the plain covered by the study.

Geotechnical characterisation

Using the geological model reconstructed by geological-geotechnical survey and the correlation of stratigraphic data from boreholes as a starting point, and on the basis of the geotechnical characterisation carried out using in situ and laboratory tests, it was possible to identify the geotechnical units which are representative of different lithologies.

The geological model allowed standard sequences, representative of the vertical variations of geotechnical soil characteristics, to be identified in various sectors of the plain, on the basis of the geometry of the different horizons.

The following main standard geotechnical units were identified, the principal features of which are reported in Table 1:

- Unit 1 - Agrarian soil. This represents the

most superficial horizon, present virtually everywhere, with thicknesses only just over 1m. In granulometric terms, this unit is highly non-homogeneous, alternating between predominantly sandy zones and silty levels. In view of this heterogeneity, this formation can be classified using penetrometric tests to calculate the degree of shear resistance.

Electrical resistivity was also variable but on average was always relatively low, with mean values ranging between 40 - 60 ohm*m; owing to these low values, however, the formation can be readily distinguished from the underlying formation using electrical logging.

Unit 2 - Weathered soils. This forms the topmost level of the high terraces of the plain, on the summit of which there is extensive outcropping, with a thickness of 3-4m; the passage to the underlying gravels with varying degrees of weathering is normally gradual and progressive. Compared to the high percentage of fine fraction, these soils always demonstrated a low electrical resistivity varying between 20 and 40 ohm*m.

According to the USCS classification, these materials form part of the groups of inorganic silts and clays (subgroups CL and ML), whereas according to the AASHO classification (or CNR-UNI 10006 Italian Class.) they form part of groups A4 and A6.

In terms of shear strength they have rather poor characteristics with a peak angle of friction equivalent to 24° and residual ranging between 18° and 20°.

The plate bearing tests carried out to characterise this unit showed widely varying results; in particular, for loads of 2.4 kg/cm2 the corresponding average settlement is of 5 mm for the most superficial levels (0.5m below ground level) and 2 mm for the deeper part (2m below ground level).

This unit should be considered not suitable for shallow foundations.

• Unit 3 - alluvial complex. This unit outcrops extensively in correspondence with the main level and base level of the plain, with an average depth in the region of 40m in the zones near the outler of the main valleys, gradually lessening to 8-10m towards the plain. In granulometric terms, it is composed of gravels and sands in a sandy-silty matrix, with local horizons with a fine grain size and

discontinuous, cemented levels, above all in the zones close to the mountains. The component materials predominantly form part of the USCS class GW (A1a according to AASHO Class.), with local horizons which may be attributed to classes SM, SP, SC, SP and ML.

From a geotechnical point of view this is a poorly compacted soil, ranging between medium to strongly dense, whose behaviour can be broadly described, by the angle of internal friction deduced from the results of dynamic penetrometric tests SPT; it was found to range on average between 30 and 35 degrees.

Electric resistivity varied considerably depending on granulometry, density and saturation level with values ranging between 300 and 1,200 ohm*m.

• Unit 4 - complex of weathered gravels. This represents a deep unit, underlying Unit 3, in the most eastern sector of the plain, with thicknesses becoming increasingly shallower from west to east, until it disappears altogether. Moreover, the unit is present on the high terraces, with thicknesses of 8-10m, where it rests directly on the sands of Unit 5 and on the marls of Unit 6.

From a granulometric point of view this unit can be subdivided into two subunits (4a and 4b), characterised by predominant and strongly weathered sands and gravel (4a) and predominant sandy silts (4b) respectively, laid in lentiform intercalations, a few metres deep, within the gravels. In particular, according to the USCS classification, unit 4a forms part of the SM and SC subgroups (A2-4 and A1-b in the AASHO Class.); unit 4b, on the other hand, forms part of USCS classes ML and CL (A6 and A7 according to AASHO Class.).

Complex 4a may be defined in geotechnical terms solely by the angle of internal friction, which can be calculated from data obtained from dynamic penetrometric tests STP whose estimated values are around 30-34°. The clayey silts of the 4b complex, on the other hand, show a much more variable plasticity given that the values lying parallel to line A of Casagrande's plasticity diagram, dispersed in a very wide field, range between Wℓ=29 and Wℓ=65.

Depending on the extent of weathering and saturation, the level of electrical resistivity may range in a relatively wide field between 90 and 400 ohm*m.

The plate bearing tests performed showed settlement ranging between 0.5 and 1 mm for loads of 2.5 kg/cm2.

- Unit 5 - Sandy complex. This represents a level covered by the alluvial deposits of Unit 3 and Unit 4 in the high terraces. In the zone studied the thickness varied depending on the various areas but did not normally exceed 20m, although further north it reached over 100m.

In granulometric terms this unit can be subdivided into two subunits of which the most representative was that characterised by coarser grain sizes (5a) classifiable as USCS class SM (A2-4 according to AASHO); the second subunit, which was present in lentiform levels within the sand, consists of silty-clayey deposits belonging to USCS classes ML and MH (A7-5 according to AASHO).

On the basis of laboratory tests an angle of internal friction was measured varying between 29° and 32° for sands and between 22° and 26° for silts; moreover, the latter can be described as having medium-high plasticity, with a position close to line A, in Casagrande's diagram.

Electrical resistivity was relatively homogeneous and ranged between 80 and 160 ohm*m.

- Unit 6 - Marly complex. This represents the substratum of Units 4 and 5 and its depth increases moving towards the western sector. In the eastern part of the area examined there were buried of this complex covered by a thin level of alluvial deposits.

In lithological terms the unit can be subdivided into two different subunits: the first, most representative, is predominantly composed

Tab. 1: Main characters of geotechnics Units

Unit	Lithology	Wl	Δs	Wp	Δs	PI	Δs	UNI/CNR 10006 AASHO	USCS
1	clay-sand								
2	clay	38	5.7	23	3.9	15	3.5	A6/A4	CL-ML
3	gravel							A1a	GW
4a	weathered gravel	24	2.6	20	1.8	4	1.2	A1b/A2-4	SM-SC
4b	silt	44	12.5	26	4.2	18	8.4	A7/A6	ML-CL
5a	sand	26	2.0	24	1.8	2	1.0	A2-4	SM
5b	silt	48	10.0	29	2.9	19	7.7	A7-5	ML-MH
6a	siltstone/ marlstone	39	4.8	26	2.7	13	3.9	A6/A7-6	ML-CL
6b	sand	22	5.9	20	2.4	2	3.6	A2-4	SM

Unit	Lithology	Φ' (°)	c' (kPa)	cu (kPa)	<200 mesh (%)	E (MPa)	γ (kN/m³)	ρ (ohm*m)
1	clay-sand	24-28					19	40-60
2	clay	22-26	50	80/120	65	10	21	60-80
3	gravel	30-35			11	40	18	300-1.200
4a	weathered gravel	30-34			29	40-50	19	130-300
4b	silt	22-26	40	400	85	10	19	
5a	sand	28-32	20/40		28	20-40	19	80-160
5b	silt	22-26	30		96	10	19	
6a	siltstone/ marlstone	26-30	20/40	200/600	95	40-50	21	60
6b	sand	30-34			28	40-50	19	

of marlstone and siltstone (6a) belonging to USCS classes CL and ML (A6 and A7-6 according to AASHO), and the second is made up predominantly of silts and sands (6b) lying in lenses within the marls and belonging to USCS class SM (AASHO A2-4).

Laboratory tests showed a medium plasticity for marlstone, a degree of overconsolidation OCR greater than 10 and a shear strength (direct shear in consolidated and drained conditions) with peak values of internal friction angle between 26° and 32°, with a residual strength of approximately 20°. Peak values rise to 30-34° for the sands in subunit 6b. The characteristics of deformability and consolidation (evaluated using oedometric tests on undisturbed samples) revealed the state of overconsolidation of these marlstone with consolidation coefficients (Cv) ranging between $1.28*10^{-3}$ and $12.58*10^{-3}$, with mean values of $5.35*10^{-3}$.

Electrical resistivity was relatively homogeneous in general and was measured in the region of 60 ohm*m.

Correlation between the stratigraphic-depositional units and geotechnical units

If the subdivisions of the stratigraphic-depositional units identified within the area are compared to the geotechnical units it can be seen that there is a marked correlation between the two subdivisions. This enables the results of the geotechnical classification to be extrapolated to the entire area studied (Fig. 2-3).

Given that the stratigraphic-depositional units identify a set of sedimentary units which is well defined in lithological, geometric and stratigraphic terms, the former can be used to correlate the geotechnical characteristics of these sedimentary bodies throughout the area. Therefore, even if in an approximate and schematic manner, it is possible to extrapolate precise data, often obtained from a limited number of boreholes, to all the areas in which a geological model was defined based on the analysis of facies.

Generally speaking, it can be observed that the geotechnical subdivisions represent a simplification compared to the subdivisions obtained from facies analysis. The geotechnical

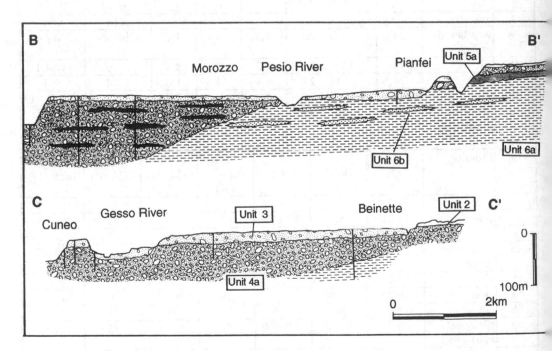

Fig. 2   Geotechnical Cross Section of Cuneo plaine

characteristics of sediments are generally conditioned by the rock type rather than by the deposition environment, even if in some cases it was seen that broadly similar rock types corresponded to different geotechnical characteristics in relation to the age and geological history. Due to these possible geotechnically "anomalous" patterns of behaviour, in order to extrapolate precise data it is important to start with an adequate number of tests, distributed evenly throughout the area, so as to assess the degree of homogeneity of the geotechnical behaviour of a particular unit.

The important aspects which emerged from the comparison between the stratigraphic and geotechnical units identified here, may be summed up as follows:

• Stratigraphic Unit ST1 corresponds to geotechnical Unit 6. The facies with a predominantly pelitic component is more recurrent and presents relatively homogeneous geotechnical characteristics (Unit 6a) and lies within easily identifiable lithological limits; on the other hand, the other facies with a prevalently sandy component have been grouped together in a single geotechnical unit referred to as 6b.

• Stratigraphic Unit ST2 broadly corresponds to geotechnical Units 4 and 5. In particular, in the zone studied, two main sedimentation environments were identified: the alluvial fan, with predominantly altered gravels, which corresponds to geotechnical unit 4a, and the deltaic environment made up of prevalently sandy facies, corresponding to geotechnical unit 5a. The border between geotechnical Units 4 and 5 is gradual and follows the transitional passage between continental and marginal marine facies. The results of the technical tests performed show the differences between geotechnical Units 5a and 6b; although similar in lithological terms, they belong to two distinct stratigraphic units and are characterised by different plasticity indexes and angles of internal friction (Table 1), namely a higher plasticity index in the former and a lower friction angle compared to the latter. The horizons made up of finer sediments, lying between the weathered gravels and sands (geotechnical Units 4b and 5b), also show considerable diversification in their technical behaviour due to different sedimentation environments.

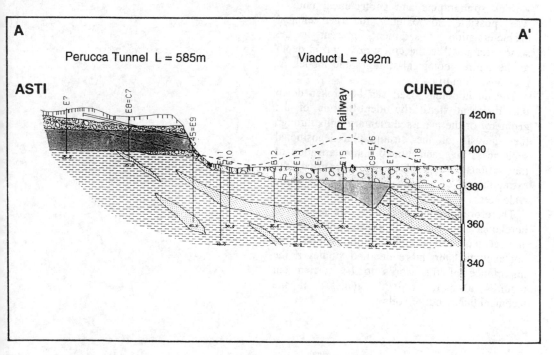

Fig. 3 Geotechnical Longitudinal Section of Highway Asti Cuneo

Stratigraphic Unit ST3 substantially corresponds to geotechnical Units 3 and part of 4. In particular, the profoundly weathered alluvial deposits lying on the high terraces reveal the same geotechnical parameters found in the weathered gravels in stratigraphic Unit ST2. For this reason, although differentiated in stratigraphic terms, these were grouped together in the same geotechnical unit (4a). The limits of geotechnical Unit 3 are always clearly defined both by boreholes and in geophysical surveys due to a distinct passage in relation to the underlying units.

The soils overlying the alluvial deposits were subdivided into separate geotechnical units (Units 1 and 2). These soils are clearly diversified in terms of both their composition and geotechnical behaviour depending on whether they cover the alluvial deposits on the main plain or secondary terraces (geotechnical Unit 1) or the high terraces where they are much deeper and argilliferous (geotechnical Unit 2).

Conclusions

Thanks to the good correlation between the various stratigraphic and geotechnical units it was possible to extend the geotechnical characterisation of sediments present in the sector examined to the entire area and in depth to the entire geological-stratigraphic model, as previously identified.

The methodology used can be broken down into different steps: the identification of the geometry of the main sedimentary units through the analysis of the stratigraphic-depositional sequences; the geotechnical characterisation of the various stratigraphic units identified; the extrapolation of precise data to a sufficently wide area.

Therefore, starting from a limited number of boreholes and tests, it is possible to obtain a general technical characterisation which does not exclude later, more detailed studies if the importance of the works to be carried out requires a more in-depth analysis of the technical behaviour of soils.

REFERENCES

Bottino G., Cavalli C., Vigna B. 1992. *L'analisi di facies nella prospezione idrogeologica.* Proceedings of the II Congress "Giovani Ricercatori di Geologia Applicata". Viterbo. pp 13.

Bottino G., Cavalli C., Eusebio A., Vigna B. 1994. *Stratigrafia ed evoluzione plio-quaternaria del settore sud-orientale della pianura cuneese.* Atti Ticinensi di Scienze della Terra.

Carraro F., Bortolami G.C., Campanino F., Clari P.A., Forno M.G., Ferrero E., Ghibaudo G., Maso V., Ricci B. 1978. *Dati preliminari sulla neotettonica dei Fogli 56 (Torino), 68 (Carmagnola) e 80 (Cuneo).* Contributi preliminari alla realizzazione della Carta Neotettonica d'Italia. C.N.R. Progetto Finalizzato Geodinamica.

Sacco F. 1912. *Geoidrologico dei pozzi profondi della Val Padana,* Ann.R. Acc. Agric. Torino.

Slope stability phenomena along the Egnatia highway: The part Ioannina-Metsovo, in Pindos mountain chain, Greece

Phénomènes de stabilité des pentes le long de l'autoroute Egnatia: La partie Ioannina-Metsovo dans la chaîne de Pindos, Grèce

B. Christaras, N. Zouros & Th. Makedon
Laboratory of Engineering Geology & Hydrogeology, School of Geology, Aristotle University of Thessaloniki, Greece

ABSTRACT: The Egnatia highway, will connect eastern with western Greece, by crossing Pindos mountain-chain. The presence of flysch in the area creates serious geotechnical problems related to the stability of the slopes. In the present investigation the geotectonic and geomechanical conditions of the area were studied by means of geological mapping and determination of probable unstable zones along the road axis. Furthermore an attempt has been made to interpret failure and sliding phenomena, in relation to the general geotectonic frame of the area.

RESUMÉ: L' autoroute Egnatia en construction, unit la Grèce de l' Est et de l' Ouest, en traversant la chaîne de Pindos. La présence du flysch dans la région provoque de problèmes géotecniques importants liés à la stabilité des pentes. Dans cet étude les conditions géotectoniques et géotecniques de la région ont été étudiées par cartographie géologique et caracterisation de zones de stabilité basse le long de l' axe de la route. L' interprétation des phénomènes de rupture et de glissement de pentes a été essayée dans le cadre géotectonique de la région.

1. INTRODUCTION

The Egnatia highway which is presently under construction, aside from being one of the most important traffic arteries in Greece. It is also the main artery linking trade and commerce from West Europe to the Middle East. The construction of the tunnels of Katara and Anilio (Metsovo area), in relation to the new tracing of the road signify the importance of the road axis especially after the division of Former Yugoslavia.

The area in study belongs to Pindos mountain range and consists mainly of flysch formations carbonate and ophiolitic rocks. The presence of flysch in the area create conditions that disturb the stability equilibrium of the hillsides along the road and triggers slope movements that vary from creep to landsliding. The problems arising from these phenomena, have been affecting the road since the early days of its construction.

The instability phenomena in the area started almost from the beginning of the road construction. Landslides occured on the cut slopes as well as on the the the road itself, along the tracing axis, even in sites where the intersection of discontinuities didn't daylight on the cut slopes.

The purpose of the present study was to investigate the cause of the creation of these landslides and to propose confronting measures for the problem.

2. THE EGNATIA HIGHWAY

The construction of the Egnatia highway was decided so as to connect the western Greek port of Igoumenitsa with Constantinople, via Thessaloniki (fig. 1). The Ioannina - Metsovo road,

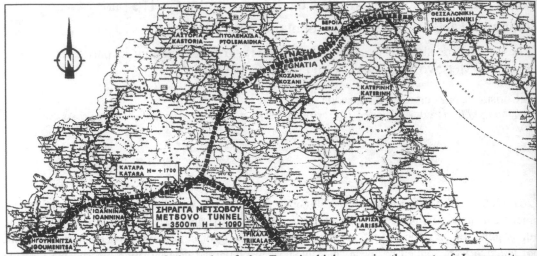

Fig. 1. Indicative tracing of the axis of the Egnatia highway, in the part of Igoumenitsa - Thessaloniki.

that crosses Pindos mountain-range, can be characterized as one of the most difficult parts for the construction, because the altitude is high, the mountain slopes are very steep and the flysch which is the dominating geological formation in the area causes important geotechnical problems. The existing mountainous national road of Trikala - Igoumenitsa, cannot serve the traffic demand of today any more, because it is narrow and passes through areas of high elevation. The bypass of Katara col, with two tunnels, in Metsovo (3.516 m long) and Anilio (about 700 m long), is a part of this future hifgway, which sortens significantly the existing route and lowers its elevation from +1700 m to +1090 m.

The benefits of the above project, as mentioned by the General Secretariat of Public Works, are: a) the fund-saving which results from the shortening of the above mentioned route, b) the impovement of communication (without the danger of being blocked during winter) between Epirus Thessaly and Macedonia, c) the essential improvement of traffic conditions along the international transportation axis connecting Western and North Europe with Middle East and d) the improvement of the explotation and transportation of the products of the region.

Fig. 2. Landslides are occuring a) on the cut slopes and b) the road itself, along the axis of the highway.

3952

Fig. 3. The landslide has broken the retaining wall in the existing national road, near Mikro Peristeri village.

The construction of the highway in the flysch formations causes important stability problems, related to the nature of these formations. The alternation of silts and clays with sandstones creates sliding conditions especially during rain periods, when the water content in the formation is high.

The landslides occured either on the cut slopes (fitg. 2a) or the roaad itself (fig. 2b) along the road axis almost from the beginning of the road construction. Some tension cracks, occured in the early stages of the phenomenon developing quickly in very important landslides. Similar slope stability problems were also determined along the old national road, in the area of Mikro Peristeri village, near Metsovo (fig. 3), on the northern bank of Metsovitikos river.

3. GEOLOGICAL SETTING

The study area is located in eastern Epirus, in northern Pindos mountain range, near the city of Metsovo (FIG. 4). The northern Pindos mountains expose a sequence of tertiary thrust sheets, including the Pindos nappe, which consist the Pindos geotectonic zone and overthrust westwards the flysch of the Gavrovo and Ionian zones. Further to the east the Pindos ophiolite sequence, a part of the Subpelagonian ophiolites, thrusts towards WSW over the Pindos zone

flysch.

Pindos zone is composed of mesozoic carbonate and silisiclastic rocks and the tertiary flysch. Spesifically Pindos zone consists of clastic and deep sea sediments (Middle Triassic), Micritic limestones with chert and siltstone (Late Karnian-Liass), Radiolarites of pure chert and mudstone (Early Doggerian-Tithonian), Calpionellid thin-beded limestone (Tithonian - Berriassian), flyschoid formation composed of clastic material, intercalated by mudstone and marlstone (Albian-Turonian), Platty limestone (Coniasian-Early Maestrichtian), Transition beds from limestone to flysch (Maestrichtian) and Tertiary flysch consisting of pelites, siltstones, marls and sandstones (Palaeocene-Late Eocene).

Three lithostratigraphic groups of flysh sendiments have been distinghuished in the study area. Politses group is very well exposed in the area, represents the nappe of the Pindos flysch and overthrusts the Zagori group sendiments. The last one represents the younger sendiments of the flysch of the Ionian zone. The Metsovon group appears as a tectonic window under the thrust-sheets of the Pindos Flysch nappe and the Subpelagonian ophiolite nappe.

Pindos flysch (Politses group) is divided into four formations, from base to top these are: the "red flysch", alternation of red shales, pelites and sandstones with maximum thickness 100 m, the second formation comprises thin grey micaceous sandstones alternating with grey shales and marls with an average thickness 70 m, the third formation comprises thick massive sandstones and interbedded grey shales and marls with an maximum thickness 350 m and the last one characterized as "wild flysch" composed of strongly tectonized grey siltstones and sandstones.

Beneath the Pindos nappe, along the thrust front appears a "tectonic formation" consisting a melange of strongly tectonized rocks, pelites, sandstones and blocks of limestones (Zouros and Mountrakis 1990). Different thrusting planes have been distinguished along the thrust front of

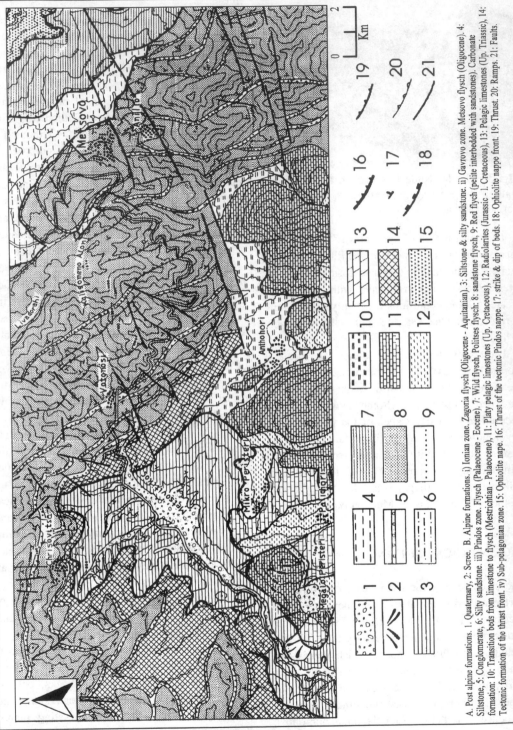

A. Post alpine formations. 1. Quaternary, 2: Scree. B. Alpine formations. i) Ionian zone. Zagoria flysch (Oligocene - Aquitanian). 3: Siltstone & silty sandstone. ii) Gavrovo zone. Metsovo flysch (Oligocene). 4: Siltstone, 5: Conglomerate, 6: Silty sandstone. iii) Pindos zone. Flysch (Palaeocene - Eocene). 7: Wild flysch, Polites flysch: 8: sandstone flysch, 9: Red flych (pelite interbedded with sandstones). Carbonate formation: 10: Transition beds from limestone to flysch (Mestrichtian - Palaeocene), 11: Platy pelagic limestones (Up. Cretaceous), 12: Pelagic limestones (Jurassic - l. Cretaceous), 13: Radiolarites (Jurassic - l. Cretaceous), 14: Tectonic formation of the thrust front. iv) Sub-pelagonian zone. 15: Ophiolite nape. 16: Thrust of the tectonic Pindos nape. 17: strike & dip of beds. 18: Ophiolite nappe front. 19: Thrust. 20: Ramps. 21: Faults.

Fig. 4. Geological map of the study area (Zouros, 1993).

the Pindos nappe, within the tectonic formation.

Tectonic windows of the Ionian and Gavrovo zones (Zagori and Metsovon groups respectively) has been established by regional mapping under the Pindos nappe. Zagori and Metsovon groups consist of silty-marls and interbedded fine-grained sandstones as well as conglomerates, mainly in the Metsovon group.

4. TECTONIC ANALYSIS

NW-SE to NNW-SSE trending inverse faults are the dominant tectonic features in the area and bound the tectonic slices with a movement direction towards SW. Strike slip faults with remarkable displacements of the deformational front of the Pindos nappe along them, are closely related with the above mentioned compressional features. These faults are either dextral or sinistral1 The largest exists along Metsovitikos river. It is a major transverse fracture zone, known as Kastaniotikos fault (Lyberis et al., 1982) that interrupts the continuation of the Pindos zone.

Successive tectonic events arise from the structural analysis in the area. The sense of movement was established by using shear criteria and kinematic indicators. Using the methods of quantitative analysis it was possible to provide a quantitative interpretation in terms of strain from the striations observed on the fault planes (Zouros, 1993).

Tertiary evolution started in Late Eocene times with a D_0 compressional event (maximum stress σ_1 axes ENE-WSW) which caused detachment, folding and thrusting of the Pindos flysch before the emplacement of the ophiolite over the flysch.

D_0 event was followed by an important D_1 extensional event (minimum σ_3 axes ENE-WSW) in Early Oligocene times, which caused a semi-ductile to brittle deformation in the area i.e. major extensional features in both ophiolites and flysch, the emplacement of the ophiolites over

the Pindos flysch and certainly the formation of the Meso-Hellenic Trough.

Two younger successive compressional events D_2 and D_3 are responsible for the refolding, imbrication and final shape of the Pindos nappe, with the maximum stress axes trending E-W and N-S respectively, took place during the Middle-Late Miocene (the second probably evolutionary to the first).

D_2 compressional event, caused inverse faults trending NNW-SSE, mainly dipping towards ENE as well as antithetic ones dipping towards WSW. The inverse faults usually accompanied by kink-banding with axes trending NW-SE. Additionally, some very important sinistral strike-slip faults trending NW-SE have also been caused by this compressional event.

Subsequently, D_3 compressional event took place in the area and caused inverse faults trending E-W, mainly dipping to the S. It also caused major dextral strike-slip faults trending WNW-ESE. These features are common particularly in the serpentinized peridotites.

The post-alpine evolution of the area during neotectonic times is rather complicated forming a transitional zone that shows both extensional and compressional patterns.

Large and mesoscale extensional features are observed in the area. NNW-SSE trending normal faults and low angle shear zones have been observed due to a NE-SW extentional event during Pleiocene. Large scale E-W trending normal faults are the younger tectonic structures coused by a N-S to NNW-SSE extention.

5. THE LANDSLIDES

The more important landslides occuring in the area are not only due to the geometry of the discontinuities of the flysch, in relation to the cut slopes directions, but mainly to the nature of the specific geological formation in which landslides are determined.

After our investigation, the more important

Fig. 5. The material of the tectonic melange. Rain water destroys the initial form (a) and the material behaves like a soil, well satured (b).

geological formation considered to be responsible for landslides creation is the "tectonic formation" that lies under the Pindos nappe overthrusting the Ionian flysch; it can be observed in many places along its front (Zouros & Mountrakis, 1990). It concerns a tectonic melange having a "chaos" structure and an appearance that reminds a "wild-flysch" formation. The matrix of the melange is mainly grey shales and

sandstones in most cases completely sheared (fig. 5a). Detached blocks of limestones and deep sea sediments such as thin bedded pelagic limestones, radiolarian cherts, and Late Cretaceous neritic limestones with dimensions from several centimetres up to several hundred meters, are observed within the matrix. The blocks are particularly tectonized and generally fault bounded.

Table 1. Particle size and Atterberg limits results, of the matrix of the tectonic formation and the pelites of Pindos flysch

Propery	Tectonic formation	Pelites
Liquid limit (LL)	36	41
Plastic limit (PL)	25	25
Plasticity index (PI)	11	16
Pass No 200 sieve	58 %	96 %
Group index (I_G)	5	17

Mechanically the material behaves differently way in dry and in wet conditions. In dry it behaves like a rock. In wet it loses rapidly its cohesion and its original structure and behaves like a saturated soil (fig. 5b). The matrix of this formation was characterized as silt to silty fine sand, with low to intermediate plasticity, according to the particle size distribution (fig. 6a) and the data of the plasticity chart (fig. 7) (Johnson & Graff, 1988). The small plastic range between plastic (PL) and liquid limits (LL), given in Table 1, determine the ability of the material to change rapidly from the semi-solid to the liquid state, improving the significant decrease of the cohesion, angle of internal friction and bearing capacity after raining (Lambe & Whitman, 1979). The Group Index (I_G), given in Table 1,

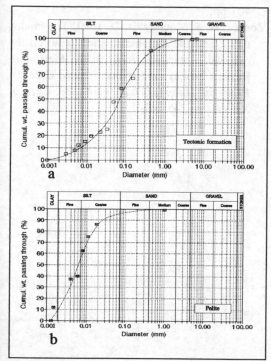

Fig. 6. Particle size distribution of a) the tectonic formation matrix and b) the pelites.

is not extremely high but is not also near to "0", however it determines rather poor or intermediate foundation conditions (Dunn et al., 1980).

The formation has a significant thickness of 20 to nearly 80 m, depending on the site, and extends under Pindos flysch, creating important foundation problems. The presence of important tectonic structures in relation to the bedding and the alternation of silts and clays with sandstones in the red Pindos flysch, strengthens the instability conditions in the area. It is mentioned that the most of the important landslides in the broader area are related to this formation, regardless of the slope face. In this framework the area corresponding with the tectonic formation should be excluded, for road construction. Furthermore the lower parts of Pindos flysch should also be excluded regardless of possibly stable local conditions because the sum of the

overlying strata could be sliding along the tectonic formation itself.

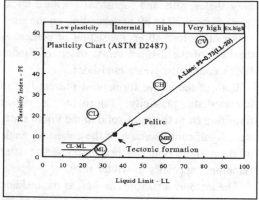

Fig. 7. Plasticity chart according to ASTM D 2487.

Another formation that could create important sliding phenomena is the pelites of the "red flysch" formation. Pelites are in alternation with sandstones at the base of the distinct thrust sheets of the Pindos flysch nappe. According to the particle size analysis (fig. 6b) and the data of the plasticity chart (fig. 7) the matrix of this material can be characterized as silt and clay with intermediate plasticity. The plasticity measured (Table 1) determines a rapid transition from solid to liquid state with significant decrease of the cohesion and angle of internal friction, during rain periods. The Group index (I_G), given in Table 1, is very high indicating a poor subgrade material with low bearing capacity. The thickness of the pelites is small, compared to the one of the tectonic formation; however the interbedding with sandstones creates serious unstable conditions.

6. CONCLUSIONS

Our investigation has shown the following:
The tectonic formation lying under the nappes of the red Pindos flysch, overthrusting the Ionian flysch, could be considered as mainly responsible for the creation of landslides along the

Egnatia highway, in the study area. This formation has a "chaotic" structure consisting of grey shales, silts and sandstones, with a thickness of about 20 to 80 m depending on the site.

An other formation, consisting of pelites, at the base of the distinct thrust sheets of Pindos flysch could also create landslides.

Both of the above formations present low to intermediate plasticity, improving a rapide transition from the semi-solid to the liquid state, with significant decrease of the cohesion, angle of internal friction and bearing capacity, after rainning, especially in winter time.

The tectonic structures as well as the bedding and the alternation of silts and sandstones in the above Pindos flysch, aggravate the instability phenomena.

According to the above observations we believe that the tectonic formation and the lower parts of the red flysch should be excluded from the tracing of the axis of the Egnatia highway.

We mention that the research of the geotechnical problems in the area is in evolution, with *in situ* investigation and laboratorial tests, so as more completed results will be presented in the future.

REFERENCES

Dunn, I.S., Anderson L.R. & Kiefer, F.W. 1980. Fudamentals of Geotechnical analysis. John Wiley & Sons: 414pp.

Johnson, R.B. 1988. Principles of Engineering Geology. John Wiley & Sons: 497 pp.

Lambe, T.W. & Whitman, R.V. 1979). Soil Mechanics, SI Version: 553 pp.

Lyberis, N., Chrowicz, J. & Papamarinopoulos, S. 1982. La paleofaille transformante du Kastaniotikos (Grece): telediction, donees de terrain et geophysiques., *Bull. Soc. Geol. France*, 7, XXIV, No 1: p. 73-85.

Zouros, N. 1993. Study of the tectonic phenomena of Pindos nappe overthrust, in Epirus area. Ph.D. Thesis, Univers. Thessaloniki: 407 pp.

Zouros, N. & Mountrakis, D. 1990. The Pindos thrust and the tectonic relation between the external geotectonic zones in the Metsovon-Eastern Zagori area (Northwestern Greece). *Proc. 5th congress, Bull. geol. Soc. Greece*, XXV/1: p. 245-262.

Engineering geological studies for the new Tagus crossing in Lisbon

Études de géologie de l'ingénieur pour la nouvelle traversée du Tage à Lisbonne

F.T.Jeremias & A.G.Coelho

Site Investigation Division, Laboratório Nacional de Engenharia Civil (LNEC), Lisbon, Portugal

ABSTRACT: The GATTEL-Gabinete para a Travessia do Tejo em Lisboa, is currently planning the development of a new crossing of the Tagus estuary in Lisbon with a total length of 18 km. The selection of the crossing site was made among four alternative sites after a decision-making process which took into account the local geological conditions and its implications for the feasibility of the undertaking. The general requirements of the proposed work and the geological conditions of the site were discussed in a preliminary desk study allowing to anticipate the major engineering geological issues. The paper describes the site investigation program carried out along the proposed site: seismic reflexion profiling, drilling, in situ tests (SPT, CPTU, SBP, Vane and Crosshole) and laboratory tests on disturbed and undisturbed samples (index properties, consolidated undrained triaxial tests and cyclic shear tests). Some conclusions are drawn from this case study concerning namely the performance of the techniques used and the cost of site investigation works.

RÉSUMÉ: GATTEL - Gabinete para a Travessia do Tejo em Lisboa développe actuellement une nouvelle traversée du Tage à Lisbonne avec une longueur total de 18 km. Le choix de la localisation de la traversée a été fait parmi quatre sites alternatifs après un processus de décision qui a pris en compte les conditions géologiques et leurs implications pour la faisabilité de l'entreprise. La confrontation des conditions de l'ouvrage proposé et des conditions géologiques du site ont permis d'anticiper les problèmes géotechniques plus importants. Cet article décrit le plan d'étude de la traversée: réflexion sismique, forages, essais in situ (SPT, CPTU, SBP, scissomètre, Crosshole) et essais en laboratoire sur des échantillons intacts et remaniés (propriétés indice, triaxial et cisaillement simple). On souligne quelques conclusions concernant la performance des techniques appliquées et le coût des travaux réalisés.

1 INTRODUCTION

The 25 Abril bridge, built in the sixties across the narrow outlet of the Tagus river mouth, is the only Tagus crossing in the Great Lisbon Area, nowadays completely saturated by heavy traffic volumes. The Ministery of Public Works represented by GATTEL is currently planning the development of a new crossing located in the wide Tagus estuary about 13 km upstream of the existing bridge, between Sacavém in the eastern border of Lisbon, and Alcochete in the left bank (fig. 1). The crossing will connect the northern road network of the Lisbon Area to Setúbal peninsula road network and the South motorway.

Conception of the crossing was constrained by existing urban areas and environmental factors, such as fluvial navigation through Cala do Norte and preservation of the tidal flats in the south bank. Accordingly, the crossing will have a total length of 18 km consisting of a main cable-stayed bridge 0.8 km long over Cala do Norte, with a main span of 400 m and a deck height of 40 m. This main bridge will be connected with Lisbon road network by an onshore North viaduct 1.2 km long, and to the South by an offshore viaduct 10.5 km long. The bridge will have two roadways each one with an operating width of 10.5 m.

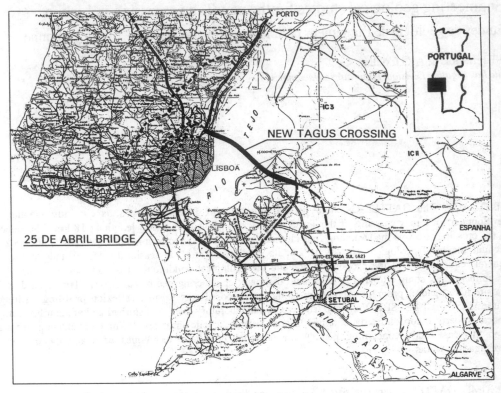

Fig. 1 - Geographical setting

The selection of the crossing site was made among four alternatives sites after a decision-making process which took into account the local geological conditions and its implications for the feasibility of the undertaking. Site investigation works for the basic project concept designs were planned taking into consideration both the general requirements of the crossing and the geological conditions of the site which allowed to antecipate the major engineering geological issues. The methods and techniques of site investigation were selected accordingly and aiming at giving a full account of the site conditions and to provide the parameters deemed pertinent in this phase for the engineering design.

The site investigation program carried out along the proposed site consisted of seismic reflexion profiling, drilling, sampling and in situ testing (SPT, CPTU, SBP, Vane and Crosshole) and laboratory testing on disturbed and undisturbed samples (index properties, consolidated undrained triaxial tests and cyclic shear tests).

Drilling, sampling, SPT, CPTU and Vane tests were performed by Seacore Continental under supervision of LNEC geologists who acted as client representatives during the field investigations. Seismic reflexion profiles, pressiometer tests, crosshole and laboratory tests were carried out by LNEC.

2 - GENERAL GEOLOGICAL SETTING

The lower Tagus valley is part of the Tagus Tertiary Basin, a subsident fault basin, elongated in the NW-SE direction, which started forming in the Paleogene. The channel was formed during the Wurm glaciation by the erosion that followed the retreat of the sea and the lowering of the base level about 100 m below the present sea level, and subsequently filled up by alluvium during the Flandrian transgression. The bedrock topography of the Wurmian channel underlying the alluvium was excavated in Miocene and Plio-pleistocene formations. Miocene formations (shelly calcarenites, sandstones, claystones and silty-clayey fine sands) outcrop in the right bank in the eastern area of Lisbon, forming a regular monoclinal structure dipping gently about 5° to ESE and SSE (towards the estuary). Within the estuary the Miocene formations are transgressively overlapped by Pliocene formations (poorly cemented silty-clayey sands with interbedded clay layers) characterized by a regular stratification gently dipping East, which outcrop in the left bank.

As the Flandrian transgression did not occur continuously but with minor forward and backward oscillations, the corresponding changes of base level caused sucessive erosions and depositions giving rise to bedrock irregularities mainly in stream mouths.

The alluvial deposits have shown a maximum thickness of 75.5 m and consist of an upper sequence of mud, muddy sand and carbonaceous clay, and of a lower sequence of sand layers and base gravel overlying the bedrock.

The new crossing is located in the larger zone of the Tagus estuary (more than 10 km wide) in the so-called "Mar da Palha" which shows several inner basins or breaks ("esteiros") and extends up to the parallel of Sacavém. A great part of this inner "sea" is limited by the 5 metres isobath. Upstream of "Mar da Palha" lies an area of islets ("mouchões") and shoals, furrowed by deeper streams with usually muddy bottoms forming a real inner delta with its base on the line Sacavém-Alcochete. Tidal flats crossed by tide and dredged channels are widely spread along the left bank reaching a width of about 4 km in front of Alcochete. These tidal flats are made of mud and muddy sands overlying Plio-pleistocene terraces (silty to clayey sand and interbedded sandy clay).

The lower Tagus valley has been the source of major intraplate earthquakes within continental Portugal (1344, 1531?, 1909). The Benavente earthquake of April 23, 1909, M = 6.5, is deemed to be the strongest historical intraplate earthquake that occured in all northwestern Europe to the North of the African-Eurasian plates boundary. This seismic activity apparently occurs along the lower Tagus valley probably related with faults or a fault zone corresponding to the western border of the Tertiary Tagus Basin. However, existing geological and seismological data are still insuficient to provide evidence on the location, direction, dip and seismotectonic parameters of this fault-source zone.

3 - SITE INVESTIGATIONS

3.1 - Planning

The site investigation program concerned the needs of engineering geological information for the inicial project conception. In this stage the final type of structure and its location was still unknown. As a first step, all the existing geological and geotechnical data were colected and analysed. The general requirements of the proposed work and the known geological conditions at the site were then discussed in a preliminary desk study. This allowed to anticipate the major engineering geological issues, among which the ccurrence of thick alluvial deposits, implying the necessity of deep foundations, and the evaluation of the seismic response of alluvial deposits, were by far the most relevant. Thus, within the estuary, the main points at issue were as follows:

i) the identification of the foundation conditions at the site of the main bridge and along the South viaduct;
ii) the performance of a site response analysis for the derivation of a design surface seismic motion;
iii) the evaluation of the liquefaction potential of the alluvial sand layers;
iv) the tentative identification of faults affecting the bedrock and eventually the alluvial deposits.

For the assessment of these points the following objectives were established:

1. the reconnaissance of the stratigraphic sequence and geological structure of both the alluvial deposits and the bedrock;
2. the continuous identification of the bedrock topography along the selected crossing alignment;
3. the determination of index properties of soil and rock units within the alluvial deposits and the bedrock;
4. the static and dynamic characterization of the alluvial and bedrock soil units;
5. the definition of representative geotechnical models - geometric modelling and laws of non-linear behaviour of its constituent soil units - to be incorporated in a seismic site response analysis.

The methods and techniques of site investigation were selected accordingly.

3.2 - Site investigation works

Site investigations were carried out onshore in the river banks and offshore within the estuary along a line of 17.5 km. Site investigation works consisted of drilling and sampling and in situ tests, namely, standard penetration test (SPT), selfboring pressumeter test (SBP), piezocone penetration test (CPTU), vane test and crosshole test, as well as laboratory tests on disturbed and undisturbed samples. Offshore, the locations of boreholes and in situ tests within the estuary were spaced at 0.4 to 1.1 km intervals along a line of 9.5 km (fig.2).

3.2.1 - Drilling

Boreholes were drilled in order to recognize the sequence, lithological composition and thickness of the alluvial deposits and the underlying bedrock. Within the estuary, boreholes were drilled from a jack-up platform and were put down by the rotary water flush method to a maximum depth of 85.5 m. Main drilling difficulties occurred to pass through a basal gravel bed of alluvium made of coarse sand, fine to coarse gravel and sparse cobbles, the thickness of which reached 17.5 m. The coarser elements are dispersed within the gravel bed or concentrated in layers about one to few metres thick in which the boreholes had to advance by chiselling.

Undisturbed sampling was carried out using piston samplers with Shelby tubes at every 5.0 m intervals in the mud. Besides SPT samples, disturbed samples were obtained in the alluvial sands and in the Plio-pleistocene and Miocene formations using split spoon and Shelby tube samplers.

3.2.2 - In situ Testing

In situ testing was performed in order to provide data on the engineering geological characterization of soil and rock units of the alluvial deposits and the underlying bedrock. In the estuary six boreholes locations were selected to carry out on each one, along the same profile, pressumeter tests, piezocone penetration tests and vane tests. Crosshole tests were also performed on two of these locations. The concentration of in situ testing in the same location allowed an easier interpretation of different site investigation methods and helped very much to establish correlations between the results obtained.

a) Standard penetration tests were carried out in all borings with a wire line SPT hammer, spaced at 1.0 to 1.5 m intervals, over the full depth of borings. The SPT's provided data on the in situ density and/or consistency of soils units as well as samples for the visual of classification lithology and water content determinations.

b) Selfboring pressumeter tests were performed onshore and offshore to a maximum depth of 21 m, allowing the computation of in situ horizontal stress, unload-reload modulus and undrained shear strength. Fig. 3 shows the record of a pressumeter test at a depth of 16.6 m.

c) Piezocone penetration tests were carried out to a maximum depth of 33.2 m, at a standard rate of penetration of 20 $mm.s^{-1}$ using an electrical cone. The electronic system used with the piezocone records continuously the cone resistance, the local sleeve friction and the pore pressure. Data were recorded on diskettes and later automatically processed to compute and to plot out the graphs of the net cone resistance, the friction ratio and the excess pore pressure ratio. The piezocone records allowed the identification of more or less sparse thin layers of sand interbedded in the upper mud complex. In this mud complex, the cone resistance increases linearly with depth from 0.2 to 1.25 MPa and reaches maximum values of about 8 MPa in the interbedded sandy levels.

d) Vane penetration tests were carried out only on the upper mud complex to a maximum depth of 28.0 m. The results obtained from vane tests have shown that peak and residual undrained shear strength increase linearly with depth from 20 to 45 $kN.m^{-2}$ and from 4 to 9 $kN.m^{-2}$, respectively.

e) Crosshole tests were performed in order to provide data on the dynamic parameters of the alluvial deposits and the underlying bedrock. The tests were carried out both onshore and offshore to a maximum depth of 80 m, between boreholes 5 m apart. Determination of longitudinal and shear wave velocities at several depths allowed to compute the dynamic shear modulus and the dynamic Young modulus.

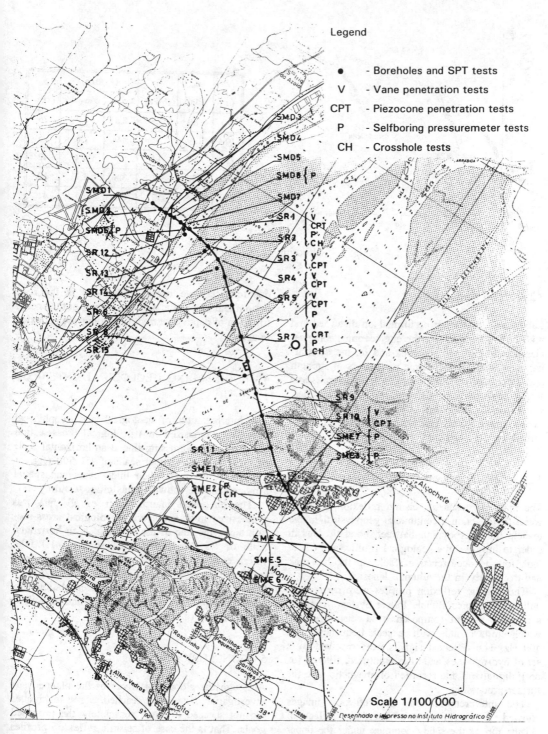

Legend

● - Boreholes and SPT tests
V - Vane penetration tests
CPT - Piezocone penetration tests
P - Selfboring pressuremeter tests
CH - Crosshole tests

Fig. 2 - Location map of site investigations works

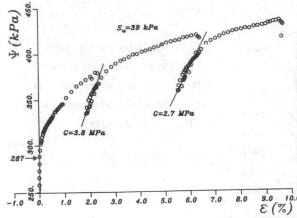

Z	σ_ho	G_i	G (MPa)		S_u
(m)	(kPa)	(MPa)	ε=2%	ε=6±1%	(kPa)
6.3	72	3.5	8.8	6.9	76
10.4	129	-	4.9	2.8	25
13.4	197	14	5.7	4.4	42
16.6	287	13	3.8	2.7	39
20.1	367	12	5.0	3.3	41

Fig. 3 - Selfboring pressumeter test

3.2.3 - Laboratory testing

Laboratory tests were performed on disturbed and undisturbed samples for the engineering geological classification of soil and rock and for the determination of static and dynamic parameters of strength and deformability. Grain size, Atterberg limits and water content were determined in all samples. Consolidated undrained triaxial tests (CU) and cyclic simple shear tests were performed on undisturbed samples of cohesive soil.

3.2.4 - Seismic reflexion profiling

The main purposes of seismic reflexion profiling were: (i) to obtain a continuous geophysical image to help the interpolation between borehole logs in order to improve the geological profile, and (ii) to look for the occurrence of faults in the bed-rock and eventually in the alluvial deposits.

The seismic reflexion profiles were carried out with a Datasonics Bubble Pulser System. These is a portable system composed by a seismic source which emits signals with a peak frequency of a 400 Hz. The reflected signals were received by a set of hydrophones and after amplified, filtered (or not), digitalized and recorded on magnetic tape for further processing.

The interpretation of the reflexion images allowed to recognize two interface reflectors: (1) the top of the sandy complex under the upper mud, and (2) the top of the base gravel. However, in some zones of the profiles, multiple reflections

due to successive energy reflections in the river bottom and absorption and attenuation of the energy till 30 m of depth occurred, degrading the reflexion images.

4 - ENGINEERING GEOLOGICAL INTERPRETATION

The interpretation of geological and geophysical data was synthesized in a geological cross-section along the Tagus crossing line at an horizontal scale of 1:15 000 and vertical scale of 1:1 500 (fig. 4). The geological units, their thicknesses and spatial arrangement are shown in this cross-section.

The integrated interpretation of all data was synthesized in an engineering geological cross-section along the line of the crossing at an horizontal scale of 1:5 000 and vertical scale of 1:1 000, where every individual borehole logs are represented and the results of in situ tests and laboratory tests are plotted (fig. 5). This representation displays a comprehensive picture of the data as well as of the density of investigation.

5 - CONCLUSIONS

Although the global objectives of the planned site investigation have been reached, some techniques did not succeed to reach completely the proposed goals. That is the case of seismic reflexion profiles which failed to provide data on the geological structure of the bedrock, the seismic energy being

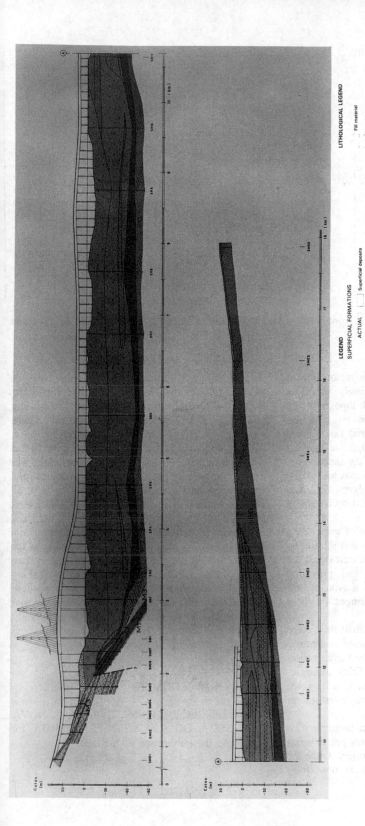

LEGEND

SUPERFICIAL FORMATIONS

ACTUAL
- Superficial deposits

HOLOCENE
- Very soft (mud) to stiff clay, clayey sand and gravelly sand

PLEISTOCENE
- Gravel with cobbles

BEDROCK FORMATIONS

PLIO-PLEISTOCENE | PQ
- Medium to coarse silty sand with subordinated clay and silty clay

MIOCENE
- M3 "Areolas de Cabo Ruivo e de Braço de Prata (M³ᵥₐ e M³ᵥₐ)" - Sand, silt, clay and limestone
- M2 "Calcários de Marvila e Arenitos de Grilos (M²ᵥₐ e M²ᵥₐ)" - Sandstone, limestone and calcarenite
- M1 "Argilas de Xabregas (M¹ᵥₐ)" - Clay, silt and limestone

LITHOLOGICAL LEGEND

- Fill material
- Superficial soil
- Mud
- Muddy sand
- Clay
- Sand
- Interbedded clay-sand
- Gravel with cobbles
- Limestone
- Silt

CONVENTIONAL SYMBOLS

- – – – Probable fault
- ——— Geological contact
- .... Inferred geological contact

Fig. 4 - Engineering geological profile along the crossing

3965

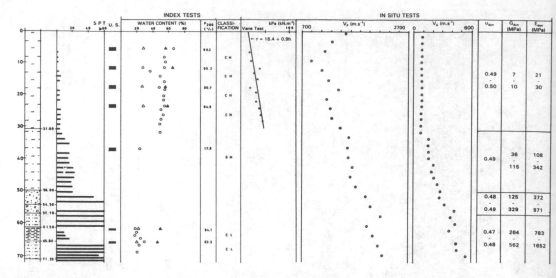

Fig. 5 - Correlation of data obtained from borelog description and in situ and laboratory tests for borehole location SR2

completely reflected by the basal layer of alluvial gravel. However, they were extremely useful in assisting the drawing of the bedrock topography by interpolation between borehole logs.

As undisturbed samples of alluvial sands could not be obtained and depth of investigation impeded the cone penetration tests to reach the sand layers, the characterization of the alluvial sands had to rely almost exclusively in grain size analyses, SPT data and shear wave velocities determined by crosshole tests.

Piezocone penetration tests were of great value in the characterization of the upper soft mud complex allowing to identify with significant detail the interbedded thin levels of sand and providing continuous logging of strength in a type of soil where SPT is not sensitive to changes in shear strength with depth.

At the interpretation level, main difficulties were related with problems of systematization and generalization of data, i. e., concerning the identification of units and the assumptions of validity of engineering geological parameters for each unit, given the considerable distance between site investigation locations (0.4 to 1.1 km).

After this stage of initial project conception, a complementary site investigation program is being planned to assist the final design project. Once the type of structure and its location is now fully defined, the site investigation program will be directed towards the investigation of specific points and will allow to confirm and complete the existing knowledge of site conditions along the new Tagus crossing.

The site investigation program carried out in this preliminary stage for such a long linear work, was a compromise between the needs of geological and engineering data for the initial project conception and the constraints in terms of costs of site investigation. Given the actual estimated cost of the work of about 120 billions PTE, the total cost of site investigations for the project and construction of the Tagus crossing may be now estimated in less than 0.3 % of the estimated cost of construction. Taking into account the benefits of site investigations, these figures show that the spending of a larger amount of money in site investigations should be completely recovered in terms of better projects and effecting a significant saving in costs of construction.

7 - ACKNOWLEDGMENT

The authors thank GABINETE PARA A NOVA TRAVESSIA DO RIO TEJO EM LISBOA for permitting the publication of the data.

REFERENCES

Coutinho, A.; J. Silva; A. Costa; J. Veiga, 1992. *Nova travessia sobre o rio Tejo em Lisboa - Ensaios com o pressiómetro autoperfurador*. LNEC, internal report, Lisboa.

Gomes Coelho, A., 1991 - *Travessia do Tejo em Lisboa síntese da informação geológica e geotécnica existente sobre o estuário do Tejo*. LNEC, internal report, Lisboa.1

Jeremias, F. T.; A. Gomes Coelho; M. Oliveira, 1993. *Nova travessia sobre o rio Tejo em Lisboa - Estudo geológico e geotécnico. Relatório de Síntese*. LNEC, internal report, Lisboa.

Pinto, S.; J. Portugal; J. Serra, 1993. *Nova travessia sobre o rio Tejo em Lisboa - Ensaios de laboratório*. LNEC, internal report, Lisboa.

Rodrigues, L.; M. Oliveira, F. T. Jeremias, 1993. *Nova travessia sobre o rio Tejo em Lisboa - Ensaios entre furos*. LNEC, internal report, Lisboa.

SEACORE CONTINENTAL, 1992. *Nova ponte do Tejo em Lisboa - Prospecção Geotécnica. Final Factual Report*. Seacore Continental, The Netherlands.

Drastic geotechnical variations of soils within a limited construction area

Variations géotechniques drastiques des sols dans un site de construction

M.A. El-Sohby & M.I. Aboushook
Faculty of Engineering, Al-Azhar University, Nasr City, Cairo, Egypt

S.O. Mazen
Ministry of Planning at Abu Dhabi, UAE

ABSTRACT: A site investigation was carried out in connection with the construction of car service and maintenance station at Nasr city – an arid region newly developed north east of Cairo. The general project area is 150 x 380 meters on which seven different types of structures are to be constructed. They range between one floor for the workshops to three floors for the administration building. Seventeen open pits were executed with depths varying from 10 to 25 meters below ground level. They indicated that stratigraphy and geotechnical properties of soils and rocks in this limited area are highly variable and characterized by rapid changes both vertically and horizontally. Furthermore, site survey of this site has indicated differences in ground surface levels of about 10 meters. In this arid area, it is a normal practice to execute open pits manually. The open pits penetrate the cemented soils to reach the clean sand. In some locations the clean sand was reached at only 7.0 meters. In some other locations it was only traced at 23 meters. In some borings highly cemented soils was met at 10 meters. It was so hard that it made the penetration to the sand strata not possible. This paper outlines the present state of knowledge of the geological environment of this region. The various types of soils and rocks and their distributions are identified from boreholes and related to this geological environment. Isoproperties maps showing different characteristics of the cemented clayey silty sand layers are presented. Finally, the properties of the various strata are identified and their engineering significance discussed.

RESUME : Une investigation d'un site a ete mise a execution dans connection avec la construction de sevice d'automobile et de station de maintien dans une bonlieue du Caire. La superficie de ce projet est 150 x 380 metres surlequel sept types de structures sont consttructees enter un et trois etages. Dix-sept fouilles ouvert sont executees avec de prefondeur de 10 a 25 metres. Elles indiquent que la stratigraphie et les propreties geotechniques des sols et des roches dans cette aire limitee sont haute variees et sont caracterisees avec des changements rapides dans les directions verticaux et horizontaux. La fouille ouverte ont ete penetree la couche cementee pour atteindre a sable proper. Dans quelques loclites, le sable proper a ete atteint a 7 metres seulement. Dans auters localites il a ete atteint a 23 metres. Dans auteres fouilles la couche tres haute cementee a ete rencontree a 10 meters. Elle est tellement dure que la penetration au sable proper ne sont pas possible. Cette papier contourne la condition de l'environnement geologique de cette region. Les types differentes des sols et des roches et leur distributions sont identifiees. Les cartes des isopropreties montrant les caracterstics differentes de la couch argileux cemente sont presentees.

1 SUBSURFACE INVESTIGATION & LABORATORY TESTING

The project site was selected at Nasr City- an arid region newly developed to the north east of Cairo. The location of the site is on an easterly sloping ground with a difference in levels as high as 10 meters (see figures 1 & 2).

Seventeen boring were drilled with depths ranging between 10 meters and 25 meters. Undisturbed samples were taken every one meter.

Initially, the subsurface investigation included drilling 10 boring, then increased to 17 boring due to the encountered significant variations in soil profile. These latter borings were performed in

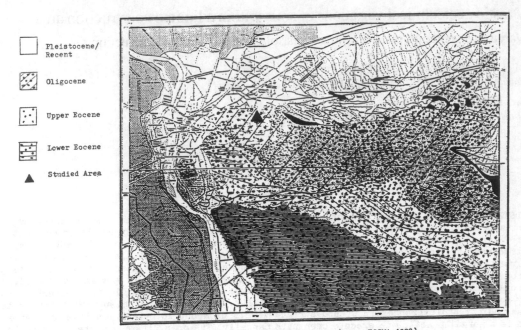

Fig. 1 Geological map of greater Cairo area (After EGSMA,1982)

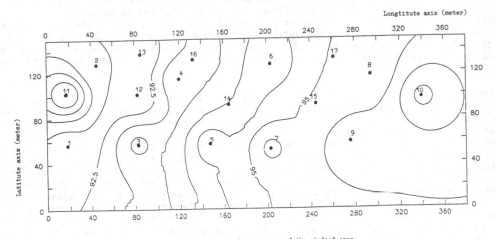

Fig. 2   Topographic and location map of the studied area

order to develop more accurate information relative to the configuration of strata surfaces. The soil samples generated were all examined and representative samples were tested for the following characteristics :

1. Description of each rock type using the geotechnical engineering terms(according to ASTM, 1985). Table 1 shows a summarized columnar section for the area of study.

2.   Determination   of   geotechnical properties of the encountered cemented strata [bed B] relevant of this study. Table 2 shows theses properties including:

-Percentage of clay content

-Liquid Limit (w_L)

-Plastic Limit (w_p)

3970

-Plasticity Index ($I_p = w_L - w_p$)

-Shrinkage Limit (w_S)

3. Drawing of the isolines maps, geological and geotechnical sections and the panel diagram showing distribution of the previous properties within the area of study as shown in figures from 3 to 15.

2 STRATIGRAPHY

Stratigraphy of the site is very erratic. However, it can be broken into three layers as shown in figure 7 and table 1. They are summarized as follows:

1. The Upper Layer [Bed A]
This is an upper surface layer of a thickness ranging between 0 and 3 meters and consisting mainly of slightly calcareous cemented sand with small pieces of stone and gravel.
2. The Middle Layer [Bed B]
This layer has a thickness ranging between 5 and 20 meters. It consists of calcareous cemented clayey silty sand with a degree of cementation ranging from slight to very high. This bed can be classified as shale although it does not exhibit the horizontally bedded structure characteristics of shale. For engineering purposes this bed is considered a problematic soil due to the changes of their behavior with the change of their moisture content (Gromko, 1974; Push, 1982 Popescu, 1983 and others).
3. The Lower Layer [Bed C]
This layer exists at a depth ranging between 7 meters to 22 meters and exists to the end of boring. It consists of fine to coarse sand.

3 GEOLOGICAL SETTING

The study area is essentially characterized by the Oligocene sediments Beds B & C] which are covered by a very thin bed of the Pliocene sediments [Bed A].

Two main lithotypes of Oligocene sediments are represented in the environs of Cairo. The first one, in which Bed C belongs, is of early Oligocene and is represented by coarse clastic of Gebel Ahmer formation. It consists generally of vividly colored gravel and sands with tubular structure having different sizes and tree trunks with common silicified wood fragments. In some parts it is highly ferruginous and / or siliceous. In other parts they are interbeded by thin shale beds consisting of indurated stratified silty clay (Attia, 1954; Said, 1962; Saleh et al., 1981 and EGSMA, 1982).

The second lithofacies type of the Oligocene sediments, in which Bed B belongs, is the basalt flows which is represented by about 20 meters thicknesses in average with different degrees of alterations. In the mostly fresh areas which had no any alteration or weathering processes, basalt quarries are famous around Cairo environs for building purposes (see figure 1). The basaltic flow during the late of Oligocene age filled up all the lower parts of the early Oligocene sand deposits [represented here by Bed C]. The alteration of this basaltic layer started from the Pliocene age to the Pleistocene age. The depositional environment during that ages was acidic condition which was convenient to the biochemical weathering of the basalt.

4 EVALUATION OF GEOLOGICAL AND GEOTECHNICAL CHARACTERISTICS

Stratigraphy of the sediments in this limited area is extremely variable and is characterized by rapid changes both vertically and horizontally. Although Nasr city area is known for variations in soil profile, this complexity of the stratigraphy has not been fully realized before.

4.1 Alteration

The cemented clayey sand layer in the present study area [bed B] is an example of the altered and weathered type of the basalt (Aboushook et al;, 1993). This altered bed can be recognized at different localities around Cairo environs. However, the degree of alteration differs from one locality to another and even for the same place (see figures 3,4,5&6).

According to the engineering classification of weathered rock (Fookes et al., 1971 and IAEG, 1981), this cemented clayey sand or altered basalt layer [bed B] is classified as moderate to highly altered grade. In this level the rock is discolored with alteration penetrating inward and the intact rock is noticeably weaker than the fresh rock and is considered as a problematic rock or soil.

The foregoing aspects of weathering concern processes which have occurred over long periods of geological time and have combined to produce weathered rocks in the

Table 1. Summarized columnar section through the area of study

Bed	Log.	Thickness (m)	Color	Description	Cementation	Age
A		0 - 3	Light to dark brown or yellowish brown	Fine to medium sand, lime, pieces of stones & gravel traces of silt	Non to slight	Pliocene
B		5- 20	Light to dark brown/grey or Yellowish/greenish brown/grey	Clayey silty sand, some stones [Altered basalt]	Slight to medium to high	Late Oligocene
C		1- 10	Light brown or yellowish brown	Fine to medium to coarse sand, traces of gravel	None to slight	Early Oligocene

Table 2. Geotechnical properties of the altered basalt [Bed B]

BH. No.	Thick m	w_e %	γ_d kN/m³	Clay %	w_L %	w_p %	I_p %	w_s %
1	17.5	17.6	17.3	11.7	63.3	33.7	29.6	20.3
2	17.0	18.7	18.0	8.0	71.0	32.0	39.0	18.7
3	10.1	16.8	18.3	11.0	63.0	39.0	24.0	19.0
4	12.3	12.4	17.1	9.5	67.0	35.0	32.0	21.0
5	4.0	11.2	19.8	6.5	73.5	34.0	39.5	12.5
6	8.0	13.0	17.8	10.0	73.5	29.5	44.0	18.0
7	12.0	7.9	21.0	6.0	52.5	27.5	25.0	11.0
8	19.0	6.2	23.0	4.0	42.0	20.0	22.0	10.0
9	16.5	3.7	24.0	4.0	42.0	20.0	22.0	10.0
10	18.0	10.2	17.1	8.3	54.0	20.0	34.0	20.3
11	17.0	16.2	16.7	9.5	57.0	28.0	29.0	22.3
12	12.0	16.9	16.7	9.0	57.5	32.5	25.0	30.0
13	18.0	14.9	16.4	12.0	77.5	35.0	42.5	27.0
14	10.0	12.2	16.8	11.3	80.3	34.0	46.3	25.0
15	19.5	5.4	20.5	7.0	62.0	29.0	33.0	22.0
16	14.0	12.9	17.0	11.7	67.3	29.3	38.0	21.0
17	20.0	5.7	20.6	3.0	35.5	20.0	15.5	10.0

* BH. No.= Borehole number, w_e= Natural moisture content,
γ_d= Natural dry density, w_L= Liquid limit, w_p= Plastic limit,
w_s= Shrinkage limit, PI= Plasticity index= $w_L - w_p$

state which is found by the engineer when he arrives on site.

4.2 Cementation

The degree of cementation varies considerably in both horizontal and vertical direction. This Could be due to the degree of alteration of the basaltic rock. As the degree of alteration is inversely proportional with the degree of cementation then the altered basalt has a medium to high degree of alteration in the upper part of the strata and becoming slight to medium with depth (see table 1 and figure 7).

4.3 Dry Unit Weight

Test results indicate that the dry unit

weight varies between 15.7 and 22.2 kN/m³. This variation can be attributed to the degree of alteration and the environmental conditions during the alteration process (see table 2 and figures 7&10).

4.4 Clay Content and Plasticity

Particle size distribution indicate that the clay content varies from 1% to 14%. Although this a low percentage of clay content the variation is significant (see figure 11). Also the tested soil showed a relatively high liquid limit values. This means that the clay in this soil consists of highly active mineral (Skempton, 1953; Seed et al., 1962 and El- Sohby and Rabba, 1981).

4.5 Distribution

The distribution of geotechnical characteristics within the area of study are shown in figures from 7 to 15 which indicate the considerable variations in all directions within such limited area of study.

5 CONCLUSIONS

Several significant points have emerged. They are either new finding or confirming previous observations:

1. Rock alteration is the product of several processes controlled by geology, geography and environmental conditions of the area under investigation.
2. The individual influence of each of the above factors on rock alteration can significantly affect geological and geotechnical characteristics and consequently variations even within a limited area.
3. For engineering purposes, each site should be thoroughly investigated if it is predetermined that alteration process is likely to be significant. This could be for better interpretation of results.
4. For altered rock formations, field and laboratory technique for engineering purposes should be developed from a geological, genesis and weathering based investigation.

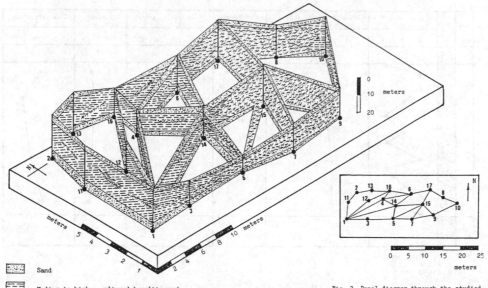

Sand

Medium to high   qltered basaltic rock

Medium to slight   altered basaltic rock

Fig. 3 Panel diagram through the studied
area.

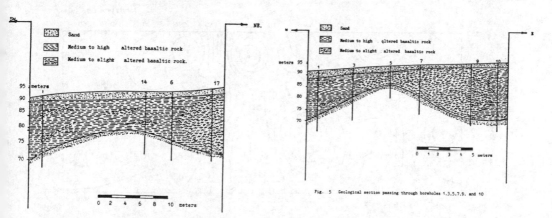

Fig. 4   Geological section passing through boreholes 1, 14, 6 & 17

Fig. 5   Geological section passing through boreholes 1,3,5,7,9, and 10

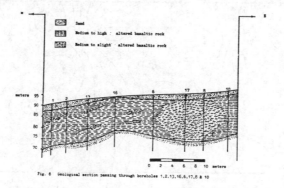

Fig. 6   Geological section passing through boreholes 1,2,13,16,6,17,8 & 10

Bed No.	Log	Degree of Cementation *			Dry unit weight (γ_d) kN/m3					Water content w_0 %		Plasticity index PI %				Shrinkage limit ws%		Clay percentage %		
		S	M	H	16	17	18	19	20	10	15	10	20	30	40	10	20	5	10	15
A																				
B																				
C																				

* S = Slightley cemented
M = Medium cemented
H = Highly cemented

Fig (7) Variation of geotechnical properties in function of depth for a typical borehole

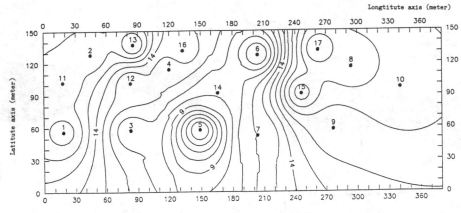

Fig. 8   Contour map of the thickness of the altered basalt bed "B"

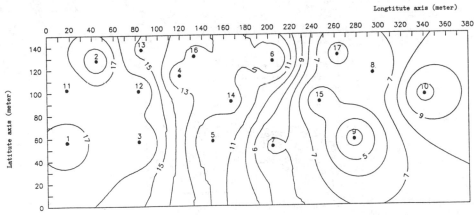

Fig. 9   Contour map of the water content of the altered basalt bed "B"

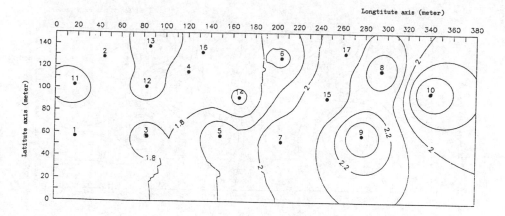

Fig. 10   Contour map of the natural dry density of the altered basalt bed "B"

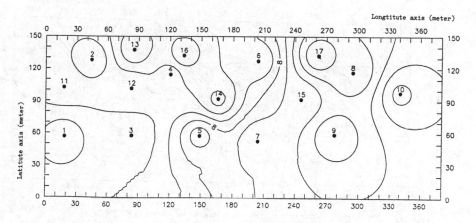

Fig. 11   Contour map of the clay percentage of the altred basalt bed "B"

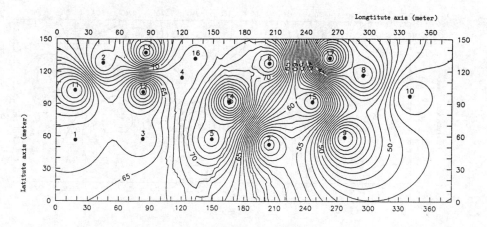

Fig. 12   Contour map of the liquid limit of the altered basalt bed "B"

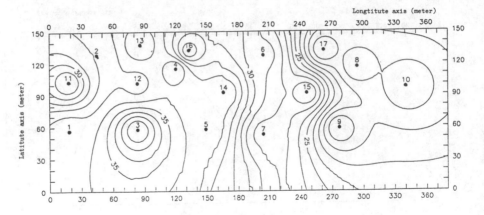

Fig. 13 Contour map of the plastic limit of the altered basalt bed "B"

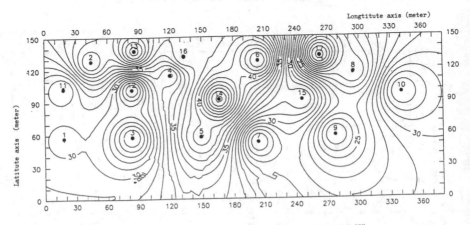

Fig. 14 Contour map of the plasticity index of the altered basalt bed "B"

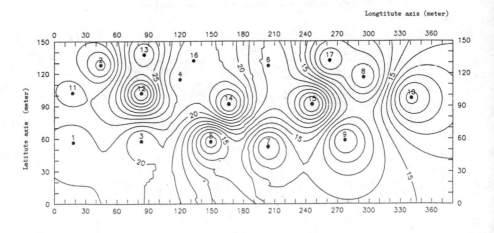

Fig. 15 Contour map of the shrinkage limit of the altered basalt bed "B"

REFERENCES

Aboushook, M. I.; Darwish, M. and Tisot, J. P. 1993. Fabrics and mineral compositions and their impacts on swelling/desiccation cycles of some expansive Egyptian rocks. Proc. of Int. Symp. on Geotech. Eng. on Hard Soils - Soft Rocks, Athens, 1:11-20.

ASTM. 1985. Standard practice for description and identification of soils (visual- manual) procedure. Annual Book of ASTM Standards, American Society for Testing and Materials, Philadelphia, Test Designation D 2488- 84, Vol.04.08, P.395-408.

Attia, M. I. 1954. Deposits in the Nile valley and the Delta. Mines and Quarries Dept., Egyptian Geological Survey, Cairo, 357 PP.

EGSMA. 1982. The geological map of greater Cairo. Egyptian Geological Survey and Mining Authority, Cairo.

El- Sohby, M. A. and Rabba, E. A. 1981. Some factors affecting swelling of clayey soils. Geotechnical Engineering, 12:19-39.

Fookes, P. G.; Dearman, W.R. and Franklin, J.A. 1971. Some engineering aspects of rock weathering with field examples from Dartmoor and elsewhere. Q. J. of Eng. Geol., London, 4:139-185.

Gromko, G. J. 1974. Review of expansive soils. Proc. of ASCE, J. of Geotech., Gt6:667- 687.

IAEG. 1981. Rock and soil description and classification for engineering geological mapping. Bull. of the Int. Ass. of Eng. Geol., 24:235-274.

Popescu, M. E. 1986. A comparison between the behaviour of swelling and collapsing soils. J. of Eng. Geol.. Amsterdam, 23: 145-163.

Pusch, R. 1982. Mineral water interactions and their influence on the physical beahviour of highly compacted Na bentonite. Canadian Geotech. J., 19:381-387.

Said, R. 1962. Geology of Egypt. El- Sevier, Amserdam, N.Y., 377 PP.

Saleh, S. A., El- Sohby, M. A. and Aboushook, M.I. 1981. X-ray diffraction analysis of some expansive soils from Nasr city area, Cairo, Egypt. Egyptian Desert Inst. Bull., Cairo, 1- 2: 1-15.

Seed, H. B.. Woodward, R. J. and Lundgren, R. 1962. Prediction of swelling potential for compacted clay. J. of the Soil Mech. and Found. Div., ASCE, 88,SM4:107-131.

Skempton, A. W. 1953. The colloidal activity of clay." Proc. of the 3rd ICSMFE, Swiss, 1: 57-61.

Isostatic foundation by Expanded Poly-Styrol and quick lime

Fondation de structures légères en utilisant 'Expanded Poly-Styrol, EPS' et chaux

Y. Iwao
Department of Civil Engineering, Saga University, Japan

K. Nishida
Matsuo Construction Co., Japan

Y. L. Liew
Graduate School of Science and Engineering, Saga University, Japan

ABSTRACT: Light weighted foundation and levee system which uses light weight material such as EPS (Expanded Poly-Styrol) is developed for poor subsoil area. It can reduce settlement and restrain sliding failure at low cost compare to the conventional method. EPS and lime were mixed with soft clay in the laboratory and at the test field. Samples from both laboratory and test field were tested. Semi actual size of foundations and levees were constructed and have been monitored for settlement and lateral displacement at the test field. Comparison of settlement for both the field measured results and the predicted values by simulation model is presented as well.

RÉSUMÉ: Un système d'une fondation léger et d'un dique qu'utilise un matériel léger comme EPS a été développé pour un subsol pauvre. Celui peut réduire l'affaissement et le glissement a bon prix confrontant avec le méthode conventionnel. L'EPS et le chaux vive ont été mélangés avec argile molle aux essais laboratoire et in situ (en place) et les échantillons de ces essais ont été essayées par ça fois. Les fondations et les diques ont été construits avec "semi-actual-size" et avaient été "monitored" en place pour les phénomènes d'affaissement et de glissement latéral. On présente aussi la comparaison des valeurs d'affaissement mesurés en place et des valeurs prévus par simulation.

1. INTRODUCTION

Ariake clay which distributes around the Ariake sea, is one of the most famous soft clay in Japan. It has been accumulated from Pleistocene till Recent age under the brackish or marine circumstance. The material was mainly supplied from Mt. Aso and Kujyu volcanos. The ground of Saga plain is mainly composed of Ariake clay and dilluvium which are composed of sand and clay.

The polder in Saga plain has spreaded by the reclamation day by day recently, for about two hundred years. The construction of levee or dyke was carried out by the following steps:

1) to put twigs or stones into the shallow marine clay, or to set wood piles in.

2) to wait for the accumulation of lucustrine mud around them.

3) to put sand or rock block on it.

4) to cast it into polder dyke.

After the construction, the polder dyke sunk gradually with time. Sometimes, the surge overflowed the polder dyke particularly during typhoon seasons. Though the crown of the ploder dyke has the height of 7.2m, it is still insufficient at all the points. This is the main problem of settlement due to the softness of Ariake clay layer and the heavy weighted levee.

2. LIGHT WEIGHTED FOUNDATION AND LEVEE SYSTEM

Discovered light weighted levee system was hinted by isostasy theory of Pratt and Airy. Usually the earth pressure exceeds the preconsolidation load of Ariake clay layer. The thickness of Ariake clay layer is about 20m and the settlement of the polder dyke is about 3 to 4 cm per year.

Consequently, levee raising and accompanying levee widening works are necessary. And it leads to the corpulent polder dyke and the endless raising process. This proposed new system is aimed to clear this vivious circle.

This system is based on the following points:

1) Using the light materials such as EPS and pumice etc.

2) These materials are mixed with soft clay.

3) The alternated unit weight of clay are 3 to 12 KN/m^3.

4) Sometimes the binder is used for mixing.

5) The unit weight is designed according to the existence of ground water and free water.

3. LIGHT WEIGHTED CLAY IN THE LABORATORY

3.1 Materials

Ariake clay for the laboratory test was sampled from the test field. The soil characteristics of the samples are shown in Table 1. The beeds of EPS were used to lighten the stabilized clay. Lime was used as the binder, because it is very effective for the Ariake clay according to the local experiences.

Unit weight(kN/m^3)	13.54 - 15.40
Water content(%)	140 - 150
Liquid limit(%)	117 - 120
Plastic limit(%)	43 - 48
Unconfined compression str.(kPa)	25 - 29
Coeffcient of permiability(cm/s)	7×10^{-7} - 2×10^{-6}
Cohesion(kPa)	20
Angle of internal friction( )	12.0

Table 1: The soil characteristics of the samples.

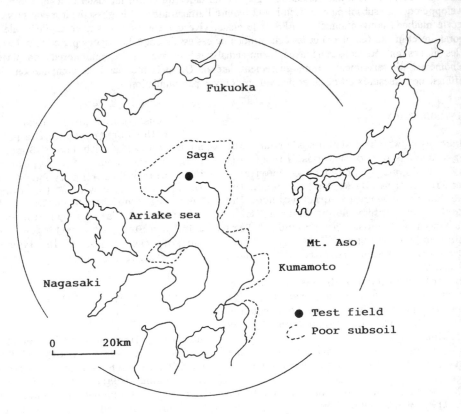

Figure 1: Map of Saga Plain and location of the test field.

3.2 Preparation of the test piece

Measured quantities of Ariake clay, EPS and the binder in correct proportion (or targeted proportion) are mixed in the mixing bowl. The mixed materials were packed in the sleeve and wrapped. Three days later, it was stripped out and slept in the air-controlled room.

3.3 Testing of the piece

The light weighted materials were tested for the following characteristics:
1) Unit weight
2) Poisson's ratio
3) Unconfined compression test
4) Bending test
5) Tension test
6) Repeated unconfined compression test
   In addition, erosion by artificial rain fall, leakage of alkaline and conductivity of heat were also tested.

3.4 Test results

The results of unconfined compression tests are shown in Figures 2 and 3.

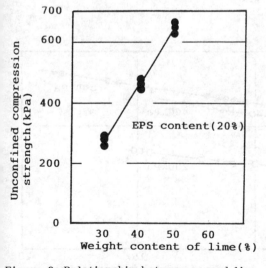

Figure 2: Relationship between q_u and lime content.

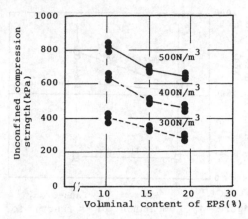

Figure 3: Relationship between qu and EPS content.

The compressive strength of the stabilized clay increases with the amount of lime added. The aimed strength is 200 to 300 KPa with considering the slip failure and the settlement of the levee or embankment. These compressive strength could be gained, even if the mixing condition is not very sufficient at the field work.

At the early stage of compression, the stress of the piece with EPS appears later than the one with only lime. Young's modulus was estimated by compression test and the Poisson's ratio was estimated by the P and S wave velocity. The relevant values are shown in Table 2.

| Young's modulus (MPa) | 64 – 80 |
| Poisson's ratio | 0.292 – 0.316 |

Table 2: Young's modulus and Poisson's ratio.

According to the age of the material, only the unconfined compression strength increases with time as shown in Figure 4. Figure 5 depicts a distinct strain appears at the early stage of the repeated load, but the rate of increment is constant for further numbers of repeated load.

When the stabilized clay was soaked in the bath, the pH of the water was a little alkaline. However the change of pH of the

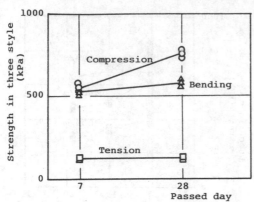

Figure 4: Comparison of strength for compression, tension and bending.

water after that was not significant and almost as usual condition.

The shrinkage of the stabilized clay is extremely small comparing with the usual dried clay. The samples were also being checked for their erosion level by the artificial rain fall. Four weeks later, the erosion effect was found not significant.

4. LIGHT WEIGHTED FOUNDATION AND LEVEE TRIAL

4.1 Design

The stabilized clay with EPS and lime are aimed to use for foundation and levee. The strength and the unit weight are designed according to the data which are obtained from the laboratory. In the case of the ploder dyke, the unit weight of levee must be more than the unit weight of the ground water and sea water. Usually 10.5 to 11 KN/m^3 will be adequate. The upper part of levee is possible to below 10 KN/m^3. (such as 6 to 9 KN/m^3)

4.2 Field test with semi actual scale foundation and levee

At the rice field in the polder, the following items were checked before carrying out the test:
1) The workability of mixing by the machine.
2) The movement of the earth pressure, the lateral displacement and settlement.

In this field test, 4 type of systems were designed for the comparison as shown in Figure 6. The levee or earth embankment is constructed using decomposed granite. It is 1m high, 3.5m wide at the top and 5.5m wide at the bottom of the embankment. Each system is having different type of foundation namely:
1) Levee on stabilised foundation with EPS and lime which is 2m deep and 7.5m wide.

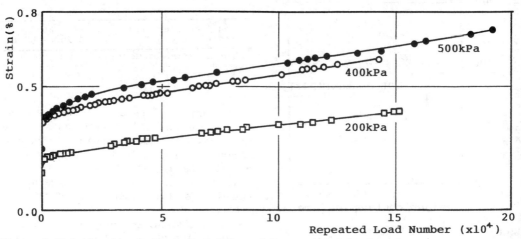

Figure 5: Relationship between the stain and the repeated load numbers.

2) Levee on stabilised foundation with EPS and lime which is 1m deep and 7.5m wide.

3) Levee on foundation of decomposed granite which is 1m deep and 7.5m wide.

4) Levee on original poor subsoil ground.

The workability of mixing was tested. Initially, lime was spreaded on the clay then the beeds of EPS were spreaded. After this, mixing was carried out. The capacity of mixing is about 70 cm deep by mixing machine. After mixing, it was compacted by backhoe, the swelling of the volume was recognized.

Measurement and monitoring of lateral earth pressure, lateral displacement and settlement had been carried out for one year. The relationship of settlement and time (days) for the field test is shown in Figure 7. It is obvious the light weighted system cases C and D settled much lesser

than cases A and B.

Figure 8 depicts the settlement pattern of the cross section of the embankments, it is cleared that the foundations which are stabilized with EPS and lime are more rigid than the foundations on decomposed granite and original ground. Hence the former gives more even settlement and less differential settlement.

The variation of lateral displacement with depth for the field test embankments are shown in Figure 9. It depicts cases C and D have lesser lateral displacement occurred than cases A and B. It is again the stabilized foundations (with EPS and lime) show its another merit i.e better lateral stiffness than the later. This will help to increase the stability of the embankment or levee against sliding or slip failure.

4.3 Quality of the mixing

The quality of mixing was checked by the trenching and boring. Figure 10 shows the variation of the unit weight of the stabilized clay by EPS and lime which were sampled at different positions i.e upper, middle and lower parts. The variance is very small. Hence this shows the good quality of mixing.

5. SIMULATION MODEL OF TOTAL SETTLEMENT

The total settlement consist of two parts namely the short and long term settlement which are occurred during and after construction of the levee or embankment. The short term or immediate settlement is calculated by using elastic solution and it can be simulated by a spring element. For the long term settlement, it is predicted by using the consolidation theory and it can be simulated by the dashboard element. Hence the simulation model is the combination of these elements as Ford Rheological Model for predicting the total settlement with time.

Based on the soil characteristics data from the laboratory and the test field, the coefficients and input parameters which are required for the model can be established. If the imposed load from the embankment is given or known, the total

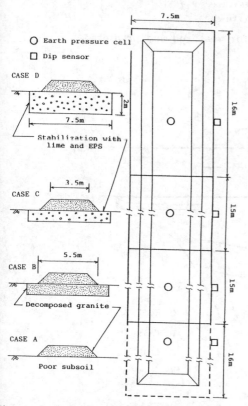

Figure 6: Layout of the test embankments or levees and the monitoring points.

settlement with time can be predicted reasonably well. Figure 7 depicts the predicted values (curves) lie closely with the measured results with time (days) for the four cases A, B, C and D. Hence with the simulation model, an adequate foundation and levee system can be evaluated.

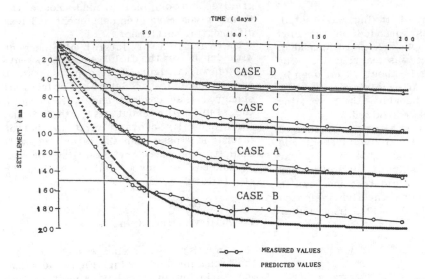

Figure 7: Relationship of settlement and time for measured and predicted values.

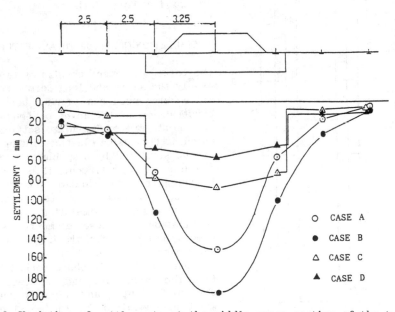

Figure 8: Variation of settlements at the middle cross section of the test embankment.

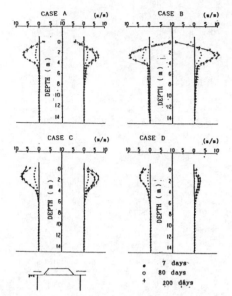

Figure 9: Variation of the lateral displacements with depth for the test embankments.

settlement and differential settlement.

b. It has better lateral stiffness than the conventional replacement method agianst sliding or slip failure.

c. It is not difficult to construct and can be constructed at low cost.

d. The light weighted system has better safety against earthquake.

e. The transportation of the waste soil is minimum.

f. Its application can be extended to foundation system for low rise building such as one to two-storey residential house.

Reference:

Iwao.Y, Nishida.K, Komokata.K and Saito.A 1993. The Characteristics of Stabilized Soft Clay with Expanded Polystyrol and Quicklime. Report of the Faculty of Science and Engineering, Saga University. Vol.21, No.2, pp.49-56.

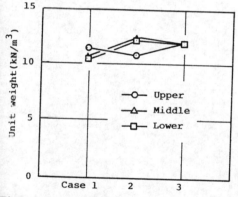

Figure 10: Quality of mixing.

6. CONCLUDING REMARKS

The light weighted foundation and levee system which reduces settlement and lateral displacement in the poor subsoil area has been tested and monitored in the field.

The advantages of the proposed system can be summarized as belows:

a. It is reasonably light and has good foundation stiffness to cause less

Renforcement des fondations et auscultation de l'église Sainte-Foy de Conques, France

Reinforcement of the foundations of the Sainte-Foy de Conques church, in France, and monitoring of the displacements of the monument

S. Fortier & P. Vandangeon
Simecsol, Le Plessis Robinson, France

RESUME : L'église abbatiale de Conques, chef-d'oeuvre de l'art roman du 12ème siècle, a été construite sur le versant d'un massif schisteux recouvert d'une épaisseur variable de terrains d'altération de médiocres caractéristiques. La partie centrale, la plus chargée de l'église, correspondant à la croisée des transepts, a subi au cours des âges des tassements importants qui ont engendré une forte fissuration. Après la consolidation des superstructures de la tour lanterne, les quatre points d'appui les plus chargés de l'église ont été repris en sous-oeuvre au moyen de micropieux ancrés dans le schiste sain.

Parallèlement, un dispositif de surveillance du monument comportant des mesures topographiques et extensométriques a été mis en place pour déterminer s'il y avait lieu d'étendre la reprise en sous-oeuvre à la totalité de l'église et s'il était nécessaire de compléter les consolidations des superstructures fortement déformées.

ABSTRACT : Conques abbey, a masterpiece of the twelfth century roman art, was built on a schistose hillside covered by variable thickness of highly weathered soils. The central part of the church, corresponding to the transept, is the most loaded and suffered in time important settlements which generated a large numbers of cracks. After the reinforcing of the "lantern tower" superstructures, the four bases of the church were underpinned using micropiles founded on sound schist.

Moreover, a permanent monitoring system was set up to measure movements of the monument to decide whether or not it would be necessary to underpin all the foundations, and reinforce the highly deformed structures. This system consists of the installation and monitoring of extensometers and regular topographical measurements.

1 INTRODUCTION

L'église abbatiale de Conques a subi au cours des âges un certain nombre de désordres liés au site, au sous-sol et au caractéristiques mêmes du monument. La chute d'un fragment important de chapiteau, à la croisée du transept, à la fin de 1979 a incité le Ministère de la Culture à entreprendre une étude des causes des désordres et à effectuer des travaux de consolidation de la tour lanterne, partie la plus endommagée de l'édifice. Ces travaux ont consisté, en premier lieu, à réaliser dans la superstructure des chaînages en béton armé afin de donner à cette tour lanterne une rigidité suffisante puis, en second lieu, à conforter les fondations afin d'arrêter les mouvements qui affectent l'édifice. Le présent exposé se rapportera exclusivement au renforcement des fondations de la tour lanterne et à l'auscultation de l'église.

2 CARACTERISTIQUES DE L'OUVRAGE ET DE SON ENVIRONNEMENT

L'église de Conques est un des chefs-d'oeuvre de l'art roman du 12ème siècle et constitue une des plus célèbres étapes sur le chemin de Saint-Jacques de Compostelle. En particulier, son tympan polychrome passe pour être un des plus beaux du monde. Caractéristique des églises dites de pélerinage, cette église comprend une nef avec un transept allongé, des collatéraux contournant les croisillons puis l'abside pour former un déambulatoire sur lequel s'ouvrent des chapelles rayonnantes. De dimension réduite par rapport aux autres églises de pélerinage (55 x 45 m environ), elle possède par contre un élancement exceptionnel, 22,10 m sous les voûtes de la nef, 10,50 m pour les collatéraux. (figure 1).

fig. 1 :   Perspective sur le chevet de l'église

Cette église, dont la maçonnerie est constituée de calcaire jaune de Lunel, de schiste et de grès rouge, a été construite sur le versant d'un massif schisteux recouvert d'une épaisseur variable de terrains d'altération constitués de blocs ou plaques schisteuses dans une matrice argileuse de médiocres caractéristiques. Les fondations de l'église sont constituées d'un assemblage hétérogène de blocs de schiste ; à l'amont elles reposent directement sur le schiste dur, à l'aval elles sont établies dans les terrains d'altération parcourus par des circulations d'eau. (figure 2).

Des réseaux enterrés (canalisations, drains, captages...) existent à proximité et sous l'église, certains d'entre eux ont pu être obstrués. Une fontaine célèbre à Conques appelée fontaine du Plô et captant des sources voisines se trouve juste devant l'église ; son débit est directement lié à la pluviosité.

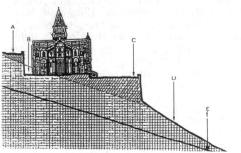

fig. 2 :

A – Remblais modernes (route tracée en 1874) avec mur de soutènement
B – Déblais récents (ancien emplacement du cimetière escavé à partir de 1836).
C – Remblais anciens pour assise de l'édifice et du cloître, retenus par l'ancien rempart
D – Talus naturel
E – Niveau probable du socle

Des désordres divers affectent l'église Sainte-Foy, en particulier, fissures en clés de voûtes de la nef et du transept, fissures dans le chevet, fissures de la façade sud du transept et fissures dans la zone des tours de façades réalisées au 19ème siècle par l'architecte Formigé.

Ces désordres sont visiblement dus à trois phénomènes :

1. Tassement des parties les plus chargées :

Il s'agit pour ce qui nous intéresse de la tour lanterne dont on peut évaluer le tassement à des valeurs comprises entre 14 et 27 cm d'après les relevés effectués (le tassement maximum correspondant au pilier nord-ouest). Ce tassement entraîne un enfoncement très net de la partie centrale de l'édifice avec des fissures à 45° dans le chevet, des faux-aplombs des colonnes de la galerie haute à la limite des collatéraux ou du chevet et du transept, et des lézardes dans les balustrades de tribune de transept. Les tours de façade au-dessus des collatéraux sont visiblement des zones de tassement, malgré les pieux en bois qui auraient été faits sous ces tours en 1878. Les symptômes (fissures des voûtes et des murs au-dessus des arcs) sont du même type que précédemment.

2. Contreventement insuffisant :

Ce phénomène se traduit par des fissures en clé de voûte de la nef, avec un aplatissement des arcs qui va en s'accentuant en allant du narthex au chœur. Les voûtes du transept sont affectées des mêmes désordres, la fissure du bras sud étant plus accentuée. Les façades du transept présentent des faux-aplombs et la façade sud est fortement ouverte.

3. Tassement différentiel :

Il apparaît que la partie sud de l'ouvrage, côté aval, est plus sinistrée que la partie nord à l'amont. Ces désordres proviennent de contraintes excessives sur le sol ; à la croisée du transept, les charges par pilier sont de 4000 kN environ pour une pression au sol de 0,74 MPa nettement trop forte pour les caractéristiques des terrains d'assise, mais aussi d'un pendage important du schiste sain induisant une épaisseur plus forte des terrains d'altération compressibles à l'aval. Les circulations d'eau ont pu contribuer à altérer le terrain d'assise des fondations et accentuer le phénomène de tassement différentiel.

4 TRAVAUX CONFORTATIFS

A la suite de la chute d'un élément de la maçonnerie, la tour lanterne a fait l'objet d'un étaiement complet. Ensuite, la tour a été consolidée au moyen d'anneaux en béton précontraint. Enfin, les fondations de la tour ont été reprises en sous-oeuvre. Cette reprise en sous-oeuvre des fondations des piliers devait permettre de reporter les charges dans le schiste sain incompressible. En raison de la présence d'eau, de la profondeur variable du schiste et des descentes de charges ponctuelles, il n'était pas raisonnable d'envisager cette reprise par des puits manuels. On a donc eu recours à des micropieux scellés en tête dans les massifs de fondations des 4 appuis et ancrés dans le schiste sain (figure 3). Les travaux de micropieux ont duré environ trois mois, de janvier à mars 1989.

Les charges par point d'appui étant de l'ordre de 4000 kN, il a été réalisé pour chaque pilier 8 micropieux subverticaux (5° maximum d'inclinaison) d'une charge portante unitaire de 500 kN. Pour le calcul de la charge portante, le frottement latéral a été négligé sur la hauteur des terrains d'altération. Seuls ont été pris en compte le frottement latéral sur la hauteur du schiste sain (200 kPa) et la résistance en pointe (4,3 MPa). Après vérification de la profondeur du toit du schiste au moyen de sondages carottés au droit de chaque pilier, les calculs ont donné une longueur de l'ordre de 20 m pour des micropieux ϕ 130 mm.

Les forages dans lesquels ont été mis en place les micropieux ont été réalisés en carottage sur la hauteur des maçonneries et des massifs de fondation afin d'éviter d'accentuer les désordres. Ils sont tubés sur la hauteur des terrains d'altération. Les micropieux sont armés d'une barre Gewi ϕ 50 mm en éléments de 10 m et munie de centreurs tous les 3 m. Cette barre est scellée par injection sans pression d'un coulis de ciment CLK 45 dosé à 1200 kg/m3 (C/E = 2). Le scellement des micropieux aux maçonneries (socle et fondation existante) est réalisé au moyen d'un mortier sans retrait de type Sikalatex.

Un essai d'arrachement d'un micropieu et deux essais d'adhérence des barres aux maçonneries (longueur 0,5 et 1 m) ont été effectués et ont donné des résultats satisfaisants. Il y a lieu de noter toutefois que l'essai d'arrachement prévu jusqu'à 1,4 fois la charge de service soit 700 kN a été arrêté à 500 kN. En effet, l'essai consistait à arracher l'armature du micropieu, non scellée dans le socle du pilier, en prenant appui sur ce socle. Or, la force appliquée a contribué à faire tasser un côté de ce socle et à faire basculer le pilier, ce qui a entrainé

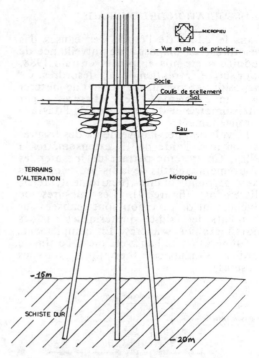

fig. 3 : Coupe de principe de la reprise en sous-oeuvre

une accentuation des désordres (fissures dans les arcatures et basculement d'une colonette de la tribune). Par ailleurs, au cours de l'exécution des micropieux, il a été noté une accentuation importante des désordres en tribune à proximité des piliers 2 et 4 malgré les précautions prises au niveau de l'exécution des forages. Les travaux de confortement ont donc contribué à accentuer le tassement des piliers, ce qui peut s'expliquer par la forte sensibilité des terrains d'altération sous l'effet des vibrations engendrées. Devant ces phénomènes, il a donc été décidé :
- d'étayer une travée du transept en tribune
- de suspendre l'essai d'arrachement de micropieu
- de poursuivre le chantier en alternant les micropieux d'un pilier à l'autre.

En fin de travaux, les carottes prélevées dans les maçonneries de socles ont été reposées au même emplacement en utilisant du ciment blanc mélangé avec les sédiments du forage des pierres afin de retrouver l'aspect originel. A la suite des travaux de reprise en sous-oeuvre, l'étaiement de la tour lanterne, qui avait été mis en place pour mettre l'édifice en sécurité, a pu être enlevé.

5 AUSCULTATION DE L'OUVRAGE

Dans le cadre de l'étude des causes des désordres, un dispositif de surveillance de l'édifice a été mis en place, en juin 1988, pour suivre l'évolution de ces désordres. Ce dispositif de surveillance permet de mesurer les mouvements des fissures par extensométrie et de tassement de l'ouvrage par nivellement.

L'évolution des mouvements des fissures est suivie à l'aide de 13 extensomètres à billes. Ce système permet de mesurer les mouvements relatifs de la fissure selon trois axes perpendiculaires (écartement, rejet, glissement). (figure 4). Les mesures de nivellement de précision sont réalisées sur 18 points de l'édifice, matérialisés par des plots métalliques ancrés. En complément, la température de l'air a été mesurée afin de vérifier l'influence thermique sur les mesures.

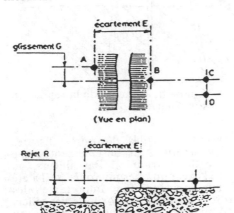

fig. 4 : Schéma de principe des fissuromètres à billes

Une première série de mesures a été réalisée entre juin 1988 et mai 1989, incluant la période de travaux de micropieux. Elle a montré une accentuation sensible des désordres au niveau du transept et du choeur pendant les travaux puis une relative stabilisation. Une deuxième série de mesures, entre août 1990 et juillet 1991 avec une périodicité de deux mois, a été effectuée afin de déterminer s'il y avait lieu d'étendre la reprise en sous-oeuvre à la totalité de l'église et s'il était nécessaire de compléter les consolidations des superstructures fortement déformées. Cette série de mesures a montré qu'il n'y a pas eu d'évolution sensible des désordres et par voie de conséquence qu'il ne paraissait pas nécessaire de prévoir, à court terme, un confortement complémentaire de l'église.

Toutefois, la déformation importante des arcs doubleaux de la nef, déformation ancienne mais qui semble évoluer, a conduit le Ministère de la Culture à procéder à l'auscultation de la fissure en clé de voûte de la nef au moyen de huit extensomètres électriques. Ces extensomètres sont reliés à un boîtier de lecture disposé à la base d'un des piliers de la tribune. Une troisième série de mesures concernant l'ensemble du dispositif de surveillance a démarré en septembre 1993 pour une période de 3 ans.

6 CONCLUSIONS

L'église Sainte-Foy à Conques est établie sur un terrain en pente avec des fondations, partie sur du schiste sain en place, partie sur du schiste très fortement altéré. Les tassements différentiels résultant de cette situation ont été de plusieurs dizaines de centimètres et les désordres ont affecté particulièrement la tour lanterne qui correspond aux charges les plus lourdes, au centre de l'édifice. Les quatre piliers supportant cette tour ont été repris en sous-oeuvre au moyen de micropieux forés ancrés dans le schiste sain. Malgré une accentuation des désordres pendant les travaux, cette reprise en sous-oeuvre a permis de stabiliser l'église au moins provisoirement. Un dispositif de surveillance comprenant des mesures de nivellement et des mesures extensométriques sur les fissures est régulièrement suivi pour déterminer s'il y a lieu d'étendre le confortement des fondations à l'ensemble de l'église. Bien que ce système de reprise en sous-oeuvre partielle ne soit pas conseillé, il semble ici donner satisfaction et autorise à différer les travaux de renforcement moins urgents. Parallèlement, la superstructure de la tour lanterne a été consolidée et l'étaiement provisoire mis en place suite aux désordres a pu être déposé pour restituer à la vue des visiteurs et des pélerins la pleine dimension de ce joyau de l'art roman du sud de la France.

Settlement and deformation of a nineteenth century building in Hong Kong

Tassement et déformation d'un édifice du 19ème siècle à Hong Kong

R.A. Forth
University of Newcastle upon Tyne, UK

C.B.B. Thorley
Gutteridge, Haskins & Davey Pty Ltd, Brisbane, Qld, Australia

ABSTRACT: Prior to construction of the Island Line of the Mass Transit Railway in Hong Kong, predictions were made of settlement and distortion of buildings close to the line of the underground railway and associated deep excavations. Careful monitoring of the performance of the buildings were made during construction. This paper describes the predictions and performance of a late nineteenth century brick building on shallow foundations.

RESUMÉ: Dans l'attention de constuire une ligne de métro sur l'île de Hong Kong, des prévisions out été faites sur le tassement et la déformation causés par les excavations profondes sur les édifices proches de cette ligne. Ces effets étaient en constant observation durant la construction. Cet exposé relate les prévisions et le comportement d'un bâtiment en briques sur des fondations peu profondes.

1 INTRODUCTION

A three-storey brick building about 50m by 25m in plan was situated close to the route of the Mass Transit Railway (MTR) construction in the western area of Hong Kong Island, which was reclaimed from the sea in the period 1873 to 1889. Careful monitoring of the settlement and tilt of the building was carried out during the construction of a 32 metre deep excavation immediately adjacent to the southern part of the building, and the construction of two large diameter tunnels close to the south eastern corner of the building. This paper reports the results of the monitoring exercise over a period of about two years.

2 THE BUILDING

The building is constructed on an area of reclaimed land and with an old sea wall passing beneath the building. Foundation details of the building could not be ascertained but it is likely that a system of spread footings on timber piles, probably China Fir, was adopted.

The building was designed to use load-bearing brick walls on all exterior faces and on the interior sides of the two end blocks (3 storeys in height) whilst the central block (2 storeys) is supported by 8 steel columns founded on the ground floor slab. The roof spans from column to wall using purlin trusses. Entrances to the building were brick crowned arches measuring 4 metres in width and 5 metres in height.

A location plan and a plan of the monitoring points is shown on Figure 1.

3 SITE GEOLOGY

Typical cross-sections of the site geology, determined from boreholes

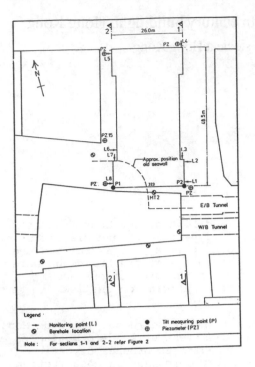

Fig. 1 Plan of monitoring points

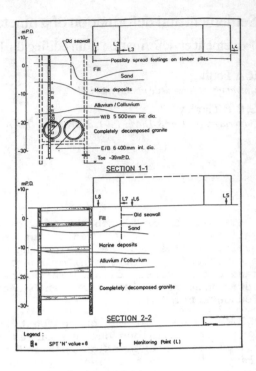

Fig. 2 Cross-sections

sunk in the area, are shown in
Figure 2. The stratigraphy is
typical of the reclamation areas
along the northern coastline of
Hong Kong Island. Bedrock is Hong
Kong granite, slightly weathered,
at depths of 25 to 40 metres below
current ground level. Weathering
of the bedrock has formed a
saprolite mantle composed of
highly to completely decomposed
granite. The upper surface of
this formation is some 15 to 20
metres below ground level.
Overlying the saprolites are
alluvial and marine deposits
ranging from 10 to 18 metres in
thickness, with fill forming the
upper 3 to 8 metres below ground
level.

The fill thickness varies
considerably and this may be due
to the presence of an old sea wall
(Figure 1). This wall however
represents only the break between
two stages of reclamation (the
original nineteenth century
coastline being about 100 metres
to the south).

Typical soil parameters
extracted from various Site
Investigations in the area are
listed in Table 1.

4 ADJACENT CONSTRUCTION WORK

The construction work undertaken
adjacent to the building from
September 1982 to July 1984 is
represented in Figure 2. A 32
metre deep 'box' was excavated
within strutted diaphragm walls,
followed by the driving of two 5.5
metre diameter tunnels from the
eastern box wall.

The excavation of the box was
undertaken in two main stages;
(a) diaphragm wall excavation
and installation and
(b) box excavation within the
diaphragm walls.

The box excavation was by the
"top-down" method whereby the
ground level slab was constructed
first and excavation and shoring
undertaken beneath it.

Table 1. Site investigation soil
parameters

Strata	Fill	Marine Deposits	Collu-vium/ Alluvium	CDG
Wet Density (kN/m³)	17–20	16–21	20–21	18–22
Cohesion c' (kN/m²)	0	0	0	14
Angle Friction ϕ' (0)	30–35	36	38	37
SPT 'N' values	6–13	4–8	8–66	32–200
Compression Index C_c	–	0.03–0.25	0.03–0.09	0.08–0.23
Permeability k (m/year)	8×10^{-4}	2×10^{-4}	2×10^{-4}	5×10^{-5}

Intermediate floor slabs were then case as excavation proceeded, with the base slab being the last to be installed. The ground water level within the excavation was maintained below the intermediate formation levels by pumping. The outside ground water levels were maintained by recharge wells where necessary.

The diaphragm walls varied from 36 to 42 metres below ground level, generally into a strata exhibiting an SPT 'N' value of 200. The main box excavation was to a depth of 32 metres below ground level.

The two tunnel drives were at depths of 26 metres below ground level and were constructed under compressed air. The initial drive from the diaphragm wall was excavated within an annuloid of grouted ground under free air, prior to installation of the air-locks. Final tunnel decompression occurred in July 1984.

5 ESTIMATED EFFECTS OF CONSTRUCTION

5.1 Box Construction

Estimation of the likely effects of construction of the box on the building was divided into three parts as follows:

(a) effects of diaphragm wall excavation and installation,
(b) effects of excavation of the box itself,
(c) effects of dewatering for the excavation of the box.

The excavation for, and installation of, a diaphragm wall by slurry trench methods has been shown to cause settlement of adjacent ground and structures (Davies and Henkel, 1980; Morton, Cater and Linney, 1980). Whilst the causes of these settlements are various, the main factors appear to be the breakdown of arching in the weathered granite around completed panels, overbreak during excavation, fluctuations in slurry level causing local instability and vibrations due to chiselling or other dynamic methods of boulder or rock removal.

Estimation of the likely settlements due to a diaphragm wall installation is empirical and very dependent on site conditions, building foundations and construction methods.

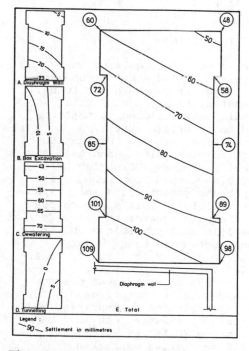

Fig. 3 Settlement contours – estimate

Dewatering of the main excavation causes settlements of adjacent structures, the magnitude depending on the extent of drawdown, the efficiency of recharge systems and the building foundation details. Estimation of drawdown induced settlement is usually carried out by conventional flow net and settlement analyses. Soil compressibility properties can be determined by either laboratory testing or in-situ field tests such as the SPT (Morton, Leonard and Cater, 1980).

The excavation of the main box usually causes settlement of adjacent ground and buildings as a result of lateral wall movements as excavation progresses. Several empirical methods are available to determine this settlement (Morton, Leonard & Carter, 1980; Lambe, 1970; Peck, 1969), most requiring a design determination of anticipated wall movements. Likely settlements are then usually expressed as a percentage of maximum horizontal wall deflection.

5.2 Tunnelling

The effect of tunnelling on overlying or adjacent ground and buildings is dependent on tunnel diameter, tunnel depth, soil conditions, building foundations and to a certain extent, tunnelling methods. Estimation of settlement magnitude and geometry usually assumes a Gaussian Error curve developed from tunnel size, depth and general site soil types (Peck, 1969). This approach can be expanded however to allow for specific soil conditions and construction methods (Howat & Cater, 1983; Mair, 1983; Fujita, 1983; Cater et al, 1984). Components of settlement due to ground water movement, on decompression of a compressed air tunnel for example, can also be estimated (Cater et al, 1984).

Figure 3 shows estimated settlements for the building, based on the methods described above. The diaphragm wall installation component has been adjusted to allow for decreased settlement adjacent the end of the wall, as has the settlement due to lateral wall movement during box excavation. The estimated settlement due to tunnelling is also adjusted to allow for the fact that tunnelling will commence after the diaphragm walls are completed and therefore have only a minor effect on the building.

6 SETTLEMENT MONITORING

6.1 Monitoring System

The monitoring of the building commenced in December 1982. A system of 8 settlement monitoring points and 2 tilt monitoring points was adopted as shown on Figure 1. The settlement points consisted of steel pins embedded in the brickwork of the exterior walls and monitored by conventional levelling techniques. The tilt points consisted of two pins embedded in the walls between 10 to 11 metres apart in height. Tilt was then calculated by measuring the horizontal distance apart of the pins (relative to the initial distance) in two directions.

6.2 Settlement Records

Figure 4 shows the settlement profile of the building at various dates during construction. Figure 5 shows time-settlement plots of the two points immediately adjacent the box. From these results and the settlement records of the other points, settlement contour plans can be constructed (Figure 6). Obviously the points represent settlements measured along the exterior walls and not necessarily the floor slabs. A comparison with internal settlement however, as measured on four steel columns, shows agreement to within 0 to 4 mm with the extrapolated contours in Figure 6. This close agreement

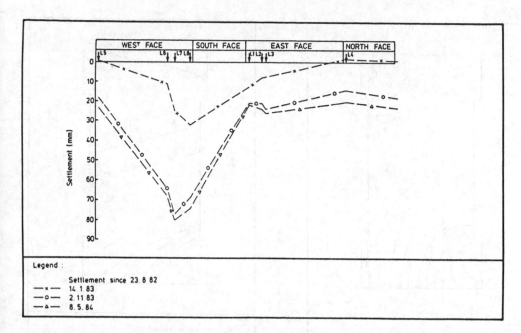

Fig. 4   Settlement profiles

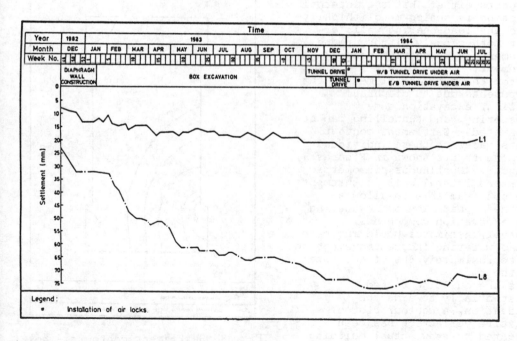

Fig. 5   Settlement-time points L1 and L8

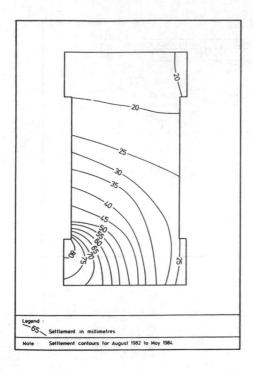

Fig. 6 Total settlement contours

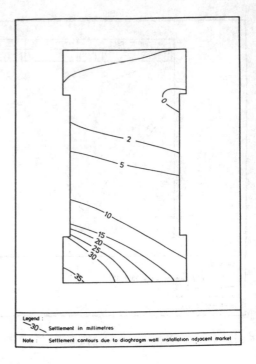

Fig. 7 Settlement contours - diaphragm wall installation

suggests either that the internal flooring is acting monolithically with the load-bearing walls or that both closely mirror ground movement or foundation base movement.

From the total settlements, the components due to diaphragm walling, excavation and dewatering, and tunnelling can be separated. Settlement contour plots based on these individual components are shown in Figures 7, 8 and 9. Continuous piezometric data and lateral wall measurements are not available to allow a further separation of dewatering and excavation components, however, estimates would indicate the dewatering induced movement to be approximately 80% of the total of the two.

As the building tilted substantially towards the excavation, a contour plot of relative deflection has been prepared to show actual building deformation (Figure 10). The plot

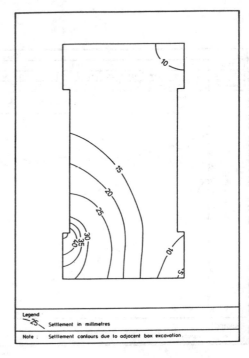

Fig. 8 Settlement contours - box excavation

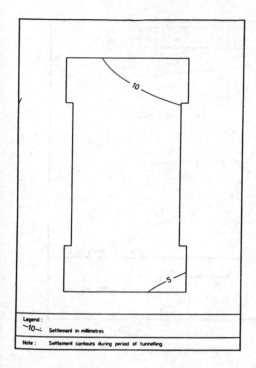

Fig. 9 Settlement contours – tunnelling

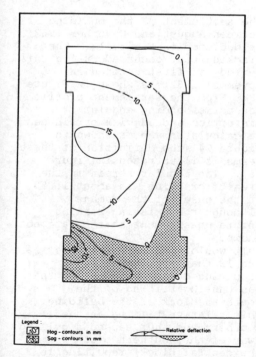

Fig. 10 Total relative deflection contours

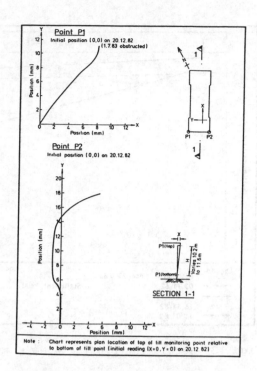

Fig. 11 Tilt measurements

was constructed by settlements relative to the plane formed by joining the four corners of the building with straight lines.

6.3 Tilt Records

Figure 11 represents the tilt measurements taken during the construction period. The tilt is represented as the movement, in two directions, of the top of the tilt monitoring point relative to the bottom of the tilt point. The top of point P2 for example (on the south-east corner of the building) is shown to have moved 19 mm (relative to the bottom of the point) in a north-easterly direction. This corresponds with the westerly tilt of that face as shown in the profile in Figure 5.

7 BUILDING BEHAVIOUR

The behaviour of the building during the adjacent excavation works can be considered in two stages;

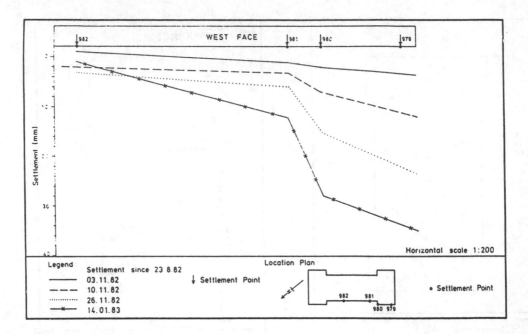

Fig. 12   Settlement profile - West Face - November 1982

(a) initial movement causing cracking, and
(b) continued movement with stabilisation of cracking.

Visible damage to the building first occurred in early November 1982 during excavation of the immediately adjacent diaphragm wall panels. This damage included cracking above the arches on the southern end of the west face, cracking of the first floor slab at its joint between the central section and the south block of the building, and "pulling out" of the roof purlins and tie rods from their connections with the wall at the south end.

Extensive and effective remedial works to the damaged sections were undertaken by the MTRC in November 1982, when maximum settlement was approximately 20 mm, with no further major damage occurring during the remainder of adjacent construction (with maximum settlement readings approximately 80 mm)

7.1 Initial Building Movement

The settlement of the building between August and November 1982 is depicted in Figure 12. The cracking that occurred during this period, as first noticed in November 1982, is shown in Figure 13. From the settlement profile the corresponding angular distortion and angular strain can be calculated and is shown in Figure 14 (the definition of these terms is illustrated in Figure 15). The angular strain in the area of cracking was about 1:850 at the onset of the damage although the existing strains before measurement started are not known.

It would appear from Figures 12 and 13 that settlement was occurring during this period that was causing tilting of the Southern block of the building. This relative tilting at the southern end thus caused a pulling away of the southern block from the central block, resulting in cracking at the first floor slab connection and the roof anchor

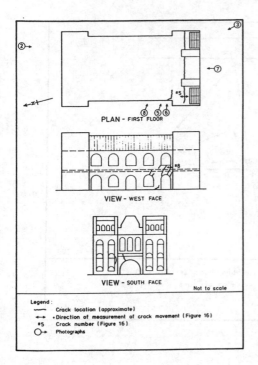

PLAN - FIRST FLOOR

VIEW - WEST FACE

VIEW - SOUTH FACE

Not to scale

Legend :
—— Crack location (approximate)
←→ •Direction of measurement of crack movement (Figure 16)
#5 Crack number (Figure 16)
O→ Photographs

Fig. 13 Crack locations

tie-rods. Further, the hogging of the southern end of the western wall has resulted in the development of shear cracking, these cracks forming at the minimum section, namely the first entrance arch. It is probable that the western arch cracking is aggravated by the tension due to the relative tilting of the end block, however, without horizontal measurements of ground and building movement the contributions of shear and tension cannot be determined.

The reason for the severe settlement gradient at the southern end (1:530 over a 13 metre length by January 1983) is not clear. A possible active wedge from a diaphragm wall panel would be expected to extend some 20 m along the buildings western face, whereas the severest gradient occurs in the first 10 metres of the western face. It is possible that the presence of the sea wall has an influence,

however, this is thought to be located under the southern block itself and would therefore not fully explain the settlement behaviour, particularly as the wall would need to extend to a depth of over 16 metres below ground level and to be at least 8 metres high, to be founded outside a theoretical active wedge and therefore act as a barrier to movement beyond it. This is not considered likely as the wall dates to the pre-1880 period during which sea-walls were generally poorly constructed and founded as evidenced by their frequent destruction in typhoons of the period (Tregear and Berry, 1959). A possible explanation of the settlement gradient is that the normal diaphragm wall induced settlement was complemented by:

(a) local vibration due to chiselling of the old wall in the adjacent panels

(b) extreme proximity of the timber piles to the trench walls and

(c) the acting of the southern block as a 'box' unit due to its four wall construction,

thus allowing it to behave as unit separate to the longer central sections. A brick building of this type on shallow foundations could be expected to have its deformation behaviour influenced as much by its above-surface structural design as by its sub-surface foundation detail.

7.2 Final Building Movement

Movement of the building continued until mid 1984 during which time the adjacent box was excavated and the two tunnels driven. The settlement trends during this period are shown in Figure 5. Figure 16, also shows a plot of measured crack widths, and typical settlement-time plot, for the period November 1982 - July 1984. Included are time plots of angular strain and distortion of various points along the western face.

Interestingly Figure 16 shows the crack #8 closing during the

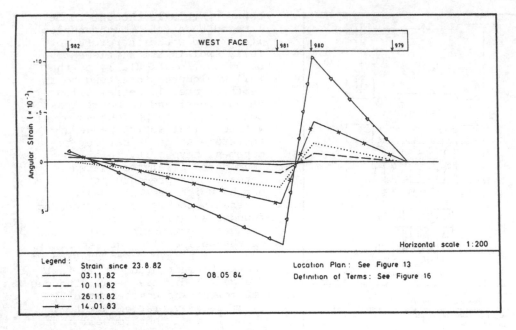

Fig. 14   Strain profiles - West Face

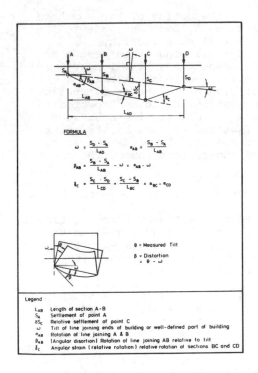

Fig. 15   Definition of terms

period April 1983 to December 1983. Settlement continued to increase during this period and angular strain at the crack location remained fairly constant at about its maximum value of 8.5 × 10⁻³ (or 1:120 approximately). It would appear therefore that the crack behaviour during this period is affected by relative tilt of the distinct sections of the building, namely the southern and central blocks (Figure 4). During this period the southern block tilted northwards whilst the central block continued to tilt southwards, therefore closing the cracks (see crack #5 - Figure 16) which had previously opened. This is regardless of the fact that the angular strain at the approximate crack location remained constant, although the effect of the remedial works will have contributed.

The total settlements of the building, shown as a contour plot, is included as Figure 6. Figures 7, 8 and 9 show the settlements recorded during the various periods of adjacent construction activity.

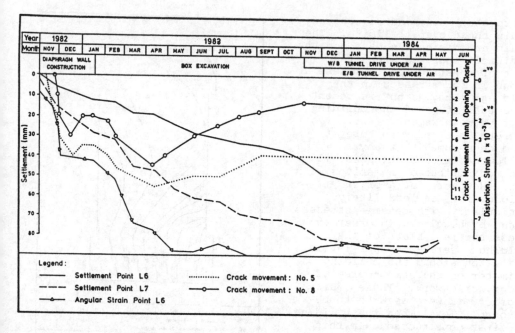

Fig. 16   Settlement and crack and strain plots: 1982-1984

The settlement occurring during diaphragm walling is illustrated in Figure 7. As can be seen it agrees reasonably with that estimated, particularly with regard to the distribution of settlement. The excavation and dewatering settlement shown in Figure 8 is however substantially less than estimated. Whilst a system of recharge wells were operated during this stage, therefore reducing total drawdown settlement, the distribution of movement is not as would be expected. Settlement appears to be concentrated at a point 10 metres along the west face (from the southern end). This could be explained by localised drawdown due to peculiarities in geology or alternatively uneven operation of the recharge/well systems. The action of the southern block as a unit may also have contributed to this behaviour as discussed earlier.

The settlement recorded during tunnelling is of the order of that estimated although appears to be the directionally opposite of that expected. Obviously, considering the total settlements involved (i.e. 5 to 10 mm) small errors will have a substantial effect. Also the grouting of the initial drive area may have substantially reduced actual tunnel induced movement resulting in the measurements over that period reflecting general fluctuations due to recharge operation, tunnelling and residual movement from ground water fluctuations. The use of compressed air can also be expected to cause some ground water movement, resulting in settlement, depending on the particular geology between the drive and building foundations.

The overall settlement was 82 mm compared to an estimated maximum of 109 mm, a difference of 25%. Similarly the overall tilt was 1:880 compared to an estimated 1:1000. More importantly, however, the maximum angular strain recorded was greater than 1:100 whereas the estimated strains were less than 1:6000. Obviously therefore the building, whilst settling and tilting much as expected, deformed severely within these overall limits. The strain and distortion based on the settlement measured since the start of construction represent a change in the stresses in the building. They do not represent the actual state of the building prior to measurement. For a concrete or steel structure it could be assumed that any settlement induced stresses will have redistributed themselves with time. However, for a brick building this is less likely, particularly with tensile stresses caused by hogging. In order to assess the real state of the building a series of spot levels were taken around the building perimeter on the same course of masonry/brickwork. These course of brickwork were assumed to be reasonably level when the building was first constructed. In this case it is a reasonable assumption as the building is three-storeys of brickwork high, the masonry end sections are well dressed, the brick and masonry arches are uniform and have not used separate keystones and the complexity of the facades would indicate a reasonable degree of control during construction.

To assess these relative levels the highest point on the brickwork (at the south-east corner) has been assigned a "settlement" value of 22 mm, equal to the settlement recorded at that point since August 1982. All other brick-work levels have been related to this value. A contour plan of these levels is shown as Figure 17. Figure 18 shows the relative shape of the brick course compared to a plane joining all four corners of the building. Figure 19 is a perspective view of the relative levels of the brickwork and Figure 20 shows the profiles of the four sides of the building.

From Figures 17 and 20 it is obvious that the building is much more severely deformed than the construction monitoring would indicate. Whereas adjacent construction works caused a maximum of 80 mm settlement at the

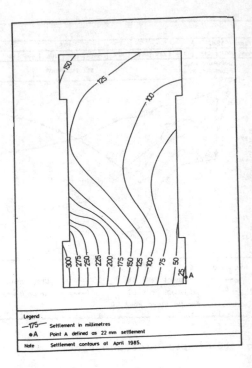

Fig. 17 Settlement contours – brickwork

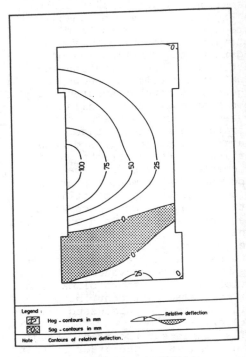

Fig. 18 Relative deflection contours – brickwork

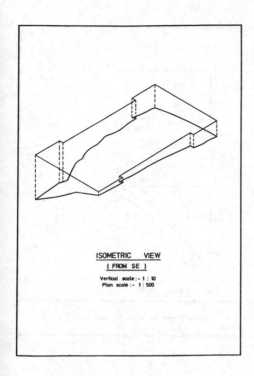

ISOMETRIC VIEW
(FROM SE)

Vertical scale :- 1 : 10
Plan scale :- 1 : 500

Fig. 19 Isometric view –
brickwork

south west corner, the brickwork
at that point is 315 mm below the
zero level. The profiles in
Figure 20 (the upper profiles
represent the inferred August 1982
levels) indicate that the
construction caused settlement
generally increased the
deformation of the building
without significantly changing its
existing state. The south face
for example appeared to have a
tilt of 1:109 prior to August
1982, increasing to 1:90 by 1985.
This compares to a tilt measured
during construction only of 1:520.
Similarly on the east face the
tilts in August 1982 and April
1985 appear to be 1:500 and 1:560
respectively whereas the
construction induced tilt is
measured as less than 1:10000.

The deformed shape of the
brickwork gives guide to the
building behaviour during
construction. The inferred pre-
construction profile shows the
building to be hogging severely on

the western side and dipping at
the south-west corner. During
construction the settlement
followed this same course, namely
dipping at the south-west corner
and hogging along the west side
(Figure 4). It is possible
therefore that either foundation
conditions, or possibly the sea
wall (although this cannot be
assumed without more detailed
information) caused the building
to deform in the manner it did.
As described the severe tilting of
the end block due to the adjacent
construction can be essentially
related to the works that
occurred. The pre-construction
deformed shape however, had it
been known beforehand, may have
highlighted the possible problems
and cracking with the southwestern area. The deflection
ratio of the west face prior to
construction appears to have been
about 1.1×10^{-3}, and also 1.3×10^{-3} for the southern face. These
ratios are extremely high and
would normally cause damage to
brickwork (Burland et al, 1977)
and therefore are probably
overestimated. This would infer
that:

(a) some settlement of the end
blocks may have occurred during
building construction and
therefore some hogging was built
into the side walls,

(b) the inferred pre-
construction profile (Figure 20)
is not as severe, and

(c) the tension forces have
distributed themselves with time.

In-built hogging is possible
during building construction as it
was constructed on newly reclaimed
land and the settlements could be
expected to be reasonably large.
The measured distortions at a
constructed tilt of say 1:1000 on
the end blocks would result in a
deflection ratio of say 0.3×10^{-3}
(15 mm relative deflection)
however the final deflection ratio
above this would still be 1.4×10^{-3} and well above what is
considered the damage limit.
Similarly the relative reflection
required to obtain a deflection
ratio of say 0.5×10^{-3} at

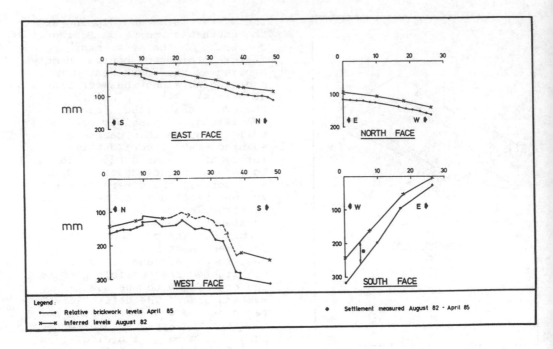

Fig. 20   Settlement profiles - Brickwork

cracking stage is approximately 60 mm (a ratio of 0.8×10^{-3}) corresponding to a tilt of about 1:400. It is unlikely this would have occurred for reasons previously described. Also the current tilt of the south block is 1:90 and was 1:109 prior to the adjacent construction works and this is well in excess of any construction errors likely to have been built in.

Only two points were monitored along the central western section during construction. There were no points at the region of the severest hogging suggested by the brickwork levels and therefore the pre-construction brickwork profile can only be estimated. It is possible that the profile shown in Figure 20 (dotted line) was not as severe as thought, although the available monitoring points suggest a relative deflection ratio of at least 1×10^{-3}.

It is possible that some load distribution has occurred along the western face since the building was constructed. (It is also possible that remedial works have been carried out during the last eighty years although no records of these exist and no obvious works are visible.) The complexity of the brickwork and the design of the internal first floor using anchored tie rods suggests that some internal reinforcement to the walls may have been used. In this case it is feasible that some load distribution may have occurred over the buildings lifetime provided the settlements developed slowly.

8 SUMMARY

Settlements and damage measurement during the period of construction of an adjacent excavation and tunnels have been presented. The behaviour of and damage to the building appears to have been substantially affected by its specific design and relative stiffness. Overall settlement and tilt recorded were within a reasonable margin of those estimated using current methods.

Strains and distortions within the building length however were orders of magnitude higher than expected and can be assumed to result from the state of the building prior to the adjacent construction, which may also be a function of the particular building/foundation interaction.

A survey of the true deformed shape of the building highlights the possible problem areas of the western and southern faces on which most of the visual brickwork damage occurred. It suggests that this could have been an indicator to imminent damage prior to the commencement of the adjacent construction works and that the structure may have already been at the limits of its serviceability.

The unfortunate lack of extensive borehole data, foundation detail and building design detail precludes the possibility of the determination, beyond the general trends indicated, of the detailed causes of movement damage described in this paper.

ACKNOWLEDGEMENTS

The authors wish to acknowledge the assistance of their former colleagues in the Geotechnical Control Office of the Hong Kong Government in data monitoring and drafting of figures.

REFERENCES

Burland, J.B., Broms, B.B. and De Mello, V.F.B. (1977). Behaviour of foundations and structures. Proceedings of the Ninth International Conference on Soil Mechanics and Foundation Engineering, Tokyo, Vol. 2, pp 495-546.

Cater, R.W., Shirlaw, J.N., Sullivan, C.A. and Chan, W.T. (1984). Tunnels constructed for the Hong Kong Mass Transit Railway. Hong Kong Engineer.

Cowland, J.W. and Thorley, C.B.B. (1984). Ground and building settlement associated with adjacent slurry trench excavation. Proceedings at the Third International Conference on Ground Movements and Structures, Cardiff.

Davies, R.V. and Henkel, D.J. (1980). Geotechnical problems associated with construction of Chater Station. Proceedings of the Conference on Mass Transportation in Asia, Hong Kong, Paper J3, 31 p.

Fujita, K. (1983). Tunnelling in soft soils. Proceedings of the Conference on Construction Problems in Soft Soils. N.T. 1, Singapore.

Howat, M.D. and Cater, R.W. (1983). Ground settlements due to tunnelling in weathered granite. Proceedings of the International Symposium on Engineering Geology and International Construction, Lisbon, Vol. 1.

Lambe, T.W. (1970). Braced excavations. Speciality Conference on Lateral Stresses in the Ground and Design of Earth-Retaining Structures, Ithaca, New York, State-of-the-Art Volume, pp. 149-218.

Mair, R.J. (1983). Geotechnical aspects of soft-ground tunnelling. Proceedings of the Conference on Construction Problems in Soft Soils, N.T. 1, Singapore.

Morton, K., Cater, R.W. and Linney, L. (1980). Observed settlements of buildings adjacent to stations constructed for the modified initial system of the mass transit railway, Hong Kong. Proceedings of the Sixth South-East Asian Conference on Soil Engineering, Taipei, pp. 415-429.

Morton, K., Leonard, M.S.M. and Cater, R.W. (1980). Building settlements and ground movements associated with the construction of two stations of the modified initial system of the mass transit railway, Hong Kong. Proceedings of the Second Conference on Ground Movements and Structures, Cardiff, pp. 788-802.

Peck, R.B. (1969). Deep excavations and tunnelling in soft-ground. Proceedings of the Seventh International Conference on Soil Mechanics and Foundation Engineering, Mexico City, State-of-the-Art volume, pp 266-290.

Tregear, T.R. and Berry, L. (1959). The Development of Hong Kong and Kowloon as told in Maps. Hong Kong University Press, 31 p.

Critical analysis of the geomechanical characterization results of an unusual basaltic lithology in Porto Primavera Power Plant

Analyse critique des résultats des études géomécaniques des formations basaltiques de la centrale hydroélectrique du barrage de Porto Primavera

M.C.A.B.Guimarães & G.Re
Themag Engenharia Ltda, São Paulo, Brazil

L.B.Monteiro
CESP Cia. Energética de São Paulo, Brazil

ABSTRACT: A low density basaltic lithology strongly affected some parts of the structure foundation of Porto Primavera Hydroelectric Plant, mainly the Intake and Navigation Lock Chamber. The results of all investigations carried out in order to determine the geomechanical parameters and the stress-strain behaviour of that lithology are related.
Design values for most important parameters are discussed in view of its geological characterization. The design alternatives for the structures are also presented emphasyzing adaptations implemented in order to preserve that lithology as foundation.

RESUMÉ: La découverte, pendant les travaux de construction du Barrage de Porto Primavera, sur le fleuve Paraná, d'un corp de roche basaltique (une bréche très altérée) avec des caracteristiques très faibles, a provoqué la réalisation de nombreuses recherches pour déterminer la valeur des paramétres a utiliser dans le projet de la fondation des structures.
Dans ce rapport sont présentées les résultats des essais réalisés en discutant les valeurs adoptées en vue de la caractérisation géologique du materiel. On décrit aussi la conception des solutions envisagées pour éviter l'excavation totale de la bréche.

1 INTRODUCTION

Porto Primavera Power Plant and its navigation facilities, owned by CESP-Companhia Energética de São Paulo - are among the most important hydroelectric schemes currently under construction in Brazil.

The damsite is located on the Paraná river on the borderline of the states of São Paulo and Mato Grosso do Sul, about 700km west from the City of São Paulo in Southeastern Brazil.

The description of the project layout as well as a brief synthesis of the site geology are presented by Re et al (1994a). During the excavation works of the foundation basaltic rock mass a peculiar lithology was identified characterized by its low natural density. It was called for this reason "light-weight" basalt, hereby abbreviated to LWB.

Due to the volume and the extension of such occurrence, that affected the foundation of some structural blocks of the Intake-Powerhouse, Spillway and Navigation Lock Chamber structures, it was necessary to carry out several "in-situ" and laboratory investigations in order to minimize excavations and foundation treatment costs.

Due to Federal Government economic policy, construction schedule of Porto Primavera Power Plant suffered several delays. It was consequently possible to investigate the LWB occurrences step by step beginning in the first area in which such lithology was exposed, i.e., in the riverbed.
The last investigations were concentrated in the Navigation Lock area in which LWB was overlain by almost 40m of Caiuá sandstone.

The results obtained in each investigation phase are presented and a critical analysis is done mainly in concern with deformability and strength characterization. Design solutions in each kind of structure are also presented.

2 HISTORICAL BACKGROUND

Relevant historical details of the Porto Primavera Hydroelectric Scheme design and construction may be summarized as follows.

During the basic design investigations, the foundation rock mass for the main concrete structures was characterized as a succession of partially recovered basaltic lava flows, in the extreme left side of the dam axis, by a few meters of Caiuá sandstone and, in the riverbed, by alluvial deposits (sands and conglomerates) up to 15m thick. Such basaltic lava flows belong to the Serra Geral geologic Formation, of juro-cretacic age. They usually constitute suitable support of concrete gravity dam structures as can be stated in the majority of the Southern Brazil hydroelectric schemes.

At Porto Primavera site however, the succession of lava flows was not uniform along the dam axis. An important structural discontinuity was found underneath the Intake-Powerhouse block U-16 throgh Spillway block BV-2. The basaltic rock mass on the South side presented large thickness flows, up to 25m. Its lateral continuity was very large too (several thousand meters) so they were called "macroflows".

On the northern side of the discontinuity, basalt lava flows were very thin (from a few decimeters up to some meters) and their lateral continuity was very small (limited to only some tens of meters). They was called "microflows" there. Between "macro" and "microflows" a singular basaltic lithology early named "cathaclastic basalt" occurred with clayey composition and peculiar physical and geomechanical characteristics.

Actual space distribution of such lithology was only recognized after the beginning of the excavation works with the complete removal of the alluvial layer that covered the basaltic rock mass. It was stated that the foundation of the upstream end of the Central Wall and some blocks of both Intake and Spillway structures was composed by LWB up to 15m in thickness.

As basaltic rock, this lithology presented unusually low dry density, high porosity, low deformability modulus and showed to be very sensitive to moisture changes, easily and rapidly weathering when exposed.

The removal of all LWB from the foundation was estimated to be very expensive, more than 4.5 US$ million at the time (1985). Some structural alternative were then studied in order to minimize costs. At the same time, several "in-situ" and laboratory tests and investigations were performed in order to better characterize the lithology under geomechanical standpoint. The results of such investigations could be used in the stress-strain analysis and computations to guide the choice of the most convenient solution for the structural arrangements.

From the structural standpoint, main problems consisted in differential deformability between upstream and downstream parts of the blocks, shear strength and durability of the foundations under high hydraulic gradients. As shown later on in this paper, the solution adopted in the main dam case was the partial removal of the LWB from the foundation of some blocks, protecting with an upstream concrete cut-off (which also had structural functions) the portion of such lithology left under the structures.

Fig.2 to fig.4 in Re et al. (1994a) show the structural solution adopted in each case.

The second important occurrence of LWB was found just underneath the upstream third part of the Navigation Lock Chamber structure. This occurrence was somehow different from that of the main dam because of the thick overburden of Caiuá sandstone (up to 40m) that exists in this case. On the other hand, LWB would be exposed only after the necessary excavations for the Navigation Lock Chamber structure construction. It would play a different and important role in the lateral rock wall stability and in the concrete structure performance.

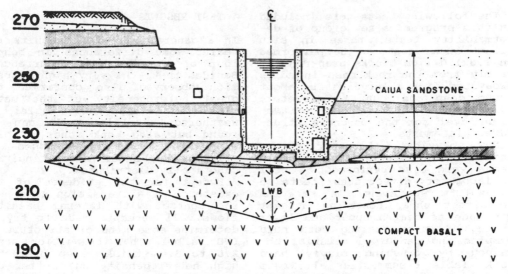

FIG. 1 - NAVIGATION LOCK CHAMBER - CROSS SECTION

Under the rock mass stability standpoint concerns were raised about the possibility of bottom failure due to low shear strength. As a consequence, the stability of the lateral sandstone walls (much more stiff than the foundation) would be compromised. Even in the case of no stability problems during construction, the different loadstransmitted to the foundations by the left and right structural concrete walls would bring up serious doubts about the adequate behaviour of the structure mainly in relation to the bottom slab.

Fig.1 shows a cross section of the Navigation Lock Chamber structure supported by LWB. It seems clear that the removal of the poor rock from the foundation would not be a reasonable solution in this case. So, a new investigation program was carried out to obtain further data to detail construction procedures.

3 THE INVESTIGATION PROGRAMS

As it was previously referred,first evidences of the existence of a geologic anomaly in the structure foundations occurred in the riverbed. It was necessary however to wait for the construction of the first stage second step cofferdam, dewatering of the work area and the removing of the thick alluvial layer before the first visual sur-

veys. Until that time some tests had been conducted on samples obtained from core borings of earlier geologic investigations, aiming at a preliminary physical characterization.

From these test results it was possible to program further geomechanical "in-situ" tests due to the need to characterize the deformability behaviour of the foundation. Three 0.80m diameter plate tests were performed up to 0.9MPa load. An intensive campaign of borehole dilatometric tests was undertaken to try to define any eventual structural pattern inside the body of the anomaly. New strength and characterization tests were also performed (see Table 1).

Geomechanical investigations on main structures occurrence of LWB were concluded with a creep tests program, in view of the results obtained from the "in-situ" plate tests. These first complete investigations were performed during the years of 1983 to 1985.

As regards the Navigation Lock Chamber, not only deformability parameters were considered important, but also the shear strength characteristics should be determined. On the other hand, despite the similarity of the LWB of this occurrence with that of the main structures, its different form of occurrence could confer different geomechanical parameters.

4009

The following tests were included in this program: a new group of deformability tests, three in situ direct shear tests to be performed on 1.0x1.0x1.0m blocks prepared directly from the rock mass in adits excavated from a large shaft, about 30.0m deep and 4.0m in diameter, just underneath the future excavation level for the concrete structures placements.

New physical characterization, uniaxial and triaxial strength tests were also performed as well as laboratory direct shear tests on 0.25x0.25m specimens. In relation to direct shear tests an attempt was made to search upper and lower strenght limits testing both hard samples and remoulded material obtained by crushing clayey hard rock. Table 1 summarizes all tests performed in all investigation surveys.

4 TEST RESULTS

In a general way, test results revealed some differences in behaviour of the two LWB occurrences. Besides that, the very high dispersion observed also confirmed the great heterogeneity of the material. Not only hard nuclei of weathered basaltic rock can be found but also weak zones of clay minerals frequently crossed by slickensided shear discontinuities cahotically distributed.

Except for the evidence of the existence of a transition zone near the contact with lateral basaltic flows, all attempts made to try to determine some kind of structure in LWB failed. This transition zone, 2.0 to 3.0m thick, also presented high hetrerogeneity but better geomechanical characteristics. For design purposes the most important results are described as follows.

Table 1. Summary of performed investigations

Test Performed	1983	1983/84	1985	1985/86	1990/91
Place	MS(1)	MS	MS	MS	NL(2)
Physical Charac.	X		X		X
Petrographic Charac.			X	X	
Sonic Velocity	X		X		X
Uniaxial Compression	X		X		X
Indirect Tension	X				
Permeability			X		X
In-situ Deformability		X			X
In-situ Direct Shear					X
Labor. Direct shear			X		X
Triaxial compression				X	
Bore hole dilatometric			X	X	
Creep				X	

Obs.: (1) Mains Structures occurrence
(2) Navigation Lock occurrence

Table 2. LWB characterization

PARAMETER	LWB (MS)	TZ. (MS)	LWB (NL)
Nat.unit weight	19.9	22.4	21.6
Dry unit weight	15.7	19.4	18.1
Specific gravity			29.0
Comp. Strenght	4.3	13.1	7.3
Sonic velocity	1900		2250
Deform. modulus	600	3600	3200

Values in kN/m3, MPa, m/s

4.1 Physical characterization

In regards to the physical characteristics of LWB a more comprehensive and detailed discussion is presented by Re et al. (1994b).

However, in order to better perceive further considerations related to deformability and strength characterization the average measured values for main parameters are presented in Table 2.

4.2 Deformability

More than 900 dilatometric tests were performed in boreholes following LNEC procedures. Fig.2 presents the histogram of the obtained results. Statistic mean for the dilatometric index was determined to be 1430MPa considering all tests but, if two different universes are considered (LWB and Transition Zone) characteristic mean values are

4010

980MPa and 2130MPa for LWB and Transition Zone respectively.

In main structures foundation, plate tests were performed in cycles, increasing loads up to 300, 600 and 900kPa respectively. They presented results in accordiance with those obtained from dilatometric tests. Fig.3 presents the "stress x vertical displacements" curves recorded in all tests.

In regards to Navigation Lock foundation , deformability tests performed in similar conditions to those of the main structures, indicated better characteristics for the rock mass. Fig.4 summarizes the computed values for the secant modulus at the end of each cycle of all tests.

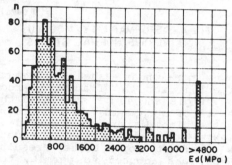

FIG. 2 - HISTOGRAM OF DILATOMETRIC INDEX

4.3 Strength characterization

Fig.5 to Fig.7 show the computed results from direct shear and triaxial compression tests performed in both occurrences of LWB in Porto Primavera. "In-situ" tests were performed on 1.0x1.0x 1.0m blocks directly cutted in the rock mass by means of diamond rotary cutters. In laboratory tests samples obtained from rotary drillings (cylinders of 75mm in diameter) were used or samples cut from rock blocks obtained from shaft and adit excavations (prismatic 25x25x25cm samples).

Normal stresses varied from 0.1 to 1.0MPa in direct shear tests meanwlile in triaxial tests confining,pressures varied from 0.2 to 4.0MPa.

It must be remarked that failure deformations, from 2 to 4% indicated nearly brittle behaviour and that a reasonable statistic,data

correlation in each kind of test group was observed (in opposition of that expected).

5 ANALYSIS OF THE RESULTS

Discussion about the results obtained from the tests performed must be carried out in view of the design requirements and in relation to the material characterization and its conduct during testing performance.

As it was already described, LWB appearence is of an extremely heterogeneous material, having been geologically classified as "hydrotermalized basaltic breccia" (Tressoldi 1986) Its mineralogical composition presents about 75 to 95% of clay minerals, but it must be considered as rock in character, not soil. So, despite of its heterogeneity in geological terms, it can be considered "tight" in geotechnical standpoint.

Jointing pattern is cahotic, but joints are commonly closed and welded by a white stiff clay mineral or by calcite. Any preferred direction of joint sets was not identified neither in strike or in dip. The most important structural feature in LWB is consists of fractures that present low to medium dip and only a few tens of meters in continuity. These do not have preferred defined attitudes, frequently changing both in dip and in strike. They always exhibit "slickensides" oriented according to the local dip even in sinuos stretchs and do not present a particular distribution pattern.

As it was stated before design requirements were of different nature because of the differnt needs of the structures in the two occurrences. In the Main Dam one was looking for support capacity and for uniformity in deformation distribution in order to obtain an harmonic interaction between the upstream and the downstream parts of each block. There were no concerns in regard to shear strength (assured by the downstream part of the structure). On the other hand, in the Navigation Lock Chamber foundation,one was just looking for shear strength to assure adequate stability conditions to the excavation slopes since the concrete

σ MPa

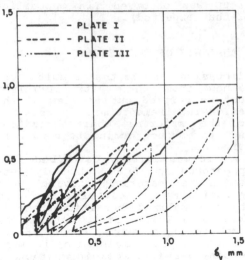

FIG. 3 - PLATE TEST RESULTS

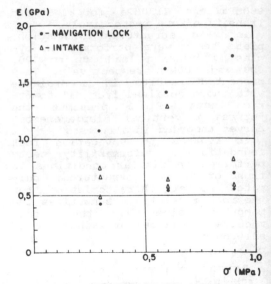

FIG. 4 - DEFORMABILITY MODULUS
VERSUS NORMAL STRESS

structure was considered able to withstand expected deformations.

Interpretation of results was done having in mind this picture.

As regards shear strength for that kind of rock, results obtained from small blocks were surprisingly high either in direct shear or in triaxial tests but they can be justified in view of the fact that the majority of samples had been obtained from excavated rock blocks or from core drillings. So they must represent the more resistent parts of LWB rock mass and such characteristics must be considered as upper limits of that parameter.

Detailed visual inspections on the sheared planes of "in-situ" tested blocks were done and they were considered representative of the LWB rock mass. Not any particular feature (either fracture, joint or clayey pocket) appears to have influenced the obtained results.

The tests performed on crushed material samples aimed to lower limits of the claymineral shear strength. Despite the register of a little cohesion value (0.05Mpa) friction results appear to be consistent with the residual strength obtained in post-failure tests on intact rock.

In order to verify the consistence of these results an attept was made to apply to them the Hoek-Brown non-linear shear strength criterion for a fractured rock mass. Fig.8 and fig.9 summarize the results of such analysis. Considering the high dispersion observed in uniaxial compression tests values, the highest of these results (26Mpa) was adopted to permit, together with the other parameters derived from the Authors' tables (see Hoek et al. 1992) the computation of the corresponding Mohr's envelope. The comparision of this envelope and that obtained from test results showed that there is not any similarity between them.

A parametric analysis performed adopting higher values for uniaxial compression strength (60 to 120MPa) showed that results are very sensitive to this value (see fig.8 and fig.9). Incorrect assumptions could consequently conduct to wrong evaluations of shear strength. It must be pointed out however that LWB rock mass is extremely heterogeneous and cannot be considered as a "jointed rock mass" as stated by the Authors' assumptions.

Regarding deformability characteristics, the better uniformity of results must be remarked, even for different kind of tests. In the Main Dam occurrence computed deformability modulus was slightly lower than those registered in Lock Chamber, the characteristic values

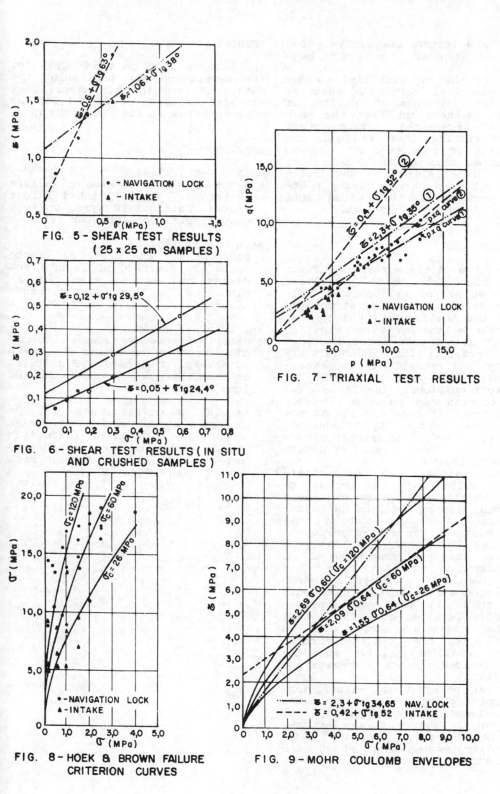

FIG. 5 – SHEAR TEST RESULTS
(25 x 25 cm SAMPLES)

FIG. 6 – SHEAR TEST RESULTS (IN SITU
AND CRUSHED SAMPLES)

FIG. 7 – TRIAXIAL TEST RESULTS

FIG. 8 – HOEK & BROWN FAILURE
CRITERION CURVES

FIG. 9 – MOHR COULOMB ENVELOPES

of 600 and 1000MPa respectively for design purposes having been adopted.

Despite the impossibility to define numerical values because of the great dispersion of results, it must be pointed out that the main dam LWB presented marked creep characteristics that obliged the exclusion of such kind of rock as structures support. Creep behaviour wasn't identified in Navigation Lock rock mass where, during plate test performing, final load were mantained for several weeks.

6 CONCLUSIONS

Despite the little extension of its occurrences at Porto Primavera site, LWB played an important role in the definition of the structural design and the foundation treatment measures. A long period of time and a great effort were spent in order to characterize that peculiar lithology, but not all of its parameters were well defined nor was its behaviour completely identified.

As regards its physical characterization, it seems clear that the statistical variability of the parameters is only the natural consequence of its geologic structure and composition.

In the same way, it was impossible to define only one characteristic envelope to represent shear strength, but one reached to establish the limit above which this value must be. For design purposes, in view of the FEM analysis results, this was considered to be sufficient.

Deformability characteristics are, among all parameters, the only ones that are not yet entirely cleared. Despite of plate test results, there are some lacks in knowledge of the complete stress-strain behaviour mainly regarding creep characteristics and swelling potential.

Further observations either during construction works, or by means of instrumentation readings, may fill out in future these lacunae. This may be very important under scientific standpoint, about the behaviour of so special rock mass.

ACKNOWLEDGEMENTS

The authors wish to thank CESP for its permission for publishing the data contained in this paper. They are also grateful to THEMAG for the collaboration in its preparation.

REFERENCES

Hoek,E. et al.1992.
A modified Hoek-Brown failure criterion for jointed rock masses. Intern.ISRM Symp. on rock characterization. Chester (UK).

Re,G. et al.1994a.
Performance analysis of Porto Primavera Power Plant foundations during the second stage diversion. VII IAEG Inter.Congr. Lisbon (P)

Re, G. et al.1994b.
Caracterização geomecânica do maciço rochoso da Usina de Porto Primavera. I Simpósio Brasileiro de Mecânica das Rochas. Foz do Iguaçu (BR)

Tressoldi, M. et al. 1986.
Ocorrências de basalto de baixa densidade na Usina de Porto Primavera e aspectos de interesse ao projeto. II Simp. Sulam. Mec. Roc. Porto Alegre (BR)

On utilization of rock foundation in high building engineering in Wuhan city of China

Étude d'utilisation de fondation rocheuse dans la construction de grand bâtiment à Wuhan de Chine

Su Jingzhong & Guo Lianju

Central-Southern Institute of Geotechnical Investigation and Design, Wuhan, People's Republic of China

ABSTRACT: Because of the heterogenous anisotropic physicomechanical properties and geological characteristics of both overburden soils and underlying rockbeds, the utilization of rock foundation in high building engineering is a complex problem. This paper comprehensively classifies rock foundation into six types, and elaborates the methods of overall evaluation and determining characteristic parameters of rock foundation. The tip bearing capacity of drilled pile is affected by the length-diameter ratio and thickness of rock debris at the bottom, accordingly a regulating factor method is proposed.

RESUME: Comme les propriétés physico-mecanigues et les caractéristigues geologiques de la fondation rocheuse sont anisotropes et hétérogenes, l'utilisation de la fondation rocheuse dans la construction de grand bâtiment est un problèm complexe. Cette exposé divise la fondation rocheuse en six types. Elle présente les méthodes de l'evaluation globale et détermination des paramètres caractéristiques de la fondation rocheuse. La capacité de résistance de pile drillé est affecté par le ratio de longueur-diamètre et le debris rocheux en bas, alors, la méthode de facteur modifiant est proposé.

1 INTRODUCTION

More than two hundred high buildings of twenty storegs or more have been or are being built in Wuhan city during recent years. Those buildings are mainly distributed in Hanko and Wuchang district. Most of them utilize rock mass as supporting layer. Because of the heterogeneous anisotropic physicomechanical properties and geological characteristics of both overburden soils and underlying engineering rockbeds, the utilization and evaluation of rock foundation in high building engineering have a great engineering significance.

2 SOIL FOUNDATION TYPES AND FEATURES

According to the formation, thickness and main properties of overburden soils, the soil foundation in Wuhan city can be divided into following two types.

2.1 Type I:

It is distributed on the first class accumulational terrace of the Changjiang River, composed of sediments of Holocene, Quaternary, as shown in Fig 1. Its thickness is about 35-50m. From the upper to the bottom there are mixed fill, slime peat, clay, silty soil, silty fine sand, sand with gravel or pebble.

2.2 Type II:

It is distributed on the third or the fouth class accumulational terrace of the Changjiang river, composed of sediments of the upper to middle Pieistocene, Quaternary, 8-40m thick, as shown in Fig.2. Except for mixed fill, local ooze mud and plastic clay in the upper, the majority of soil mass is brown yellow or brown red plastic to hard clay intercalated with rock debris in the lower. Its bearing capacity f_K =300-450 KPa and compressibility modulus Es=12-20MPa. Accordingly, it can be adopted as natural supporting layer for box or raft foundation of buildings which are about twenty storegs high.

3 ROCK FOUNDATION TYPES AND FEATURES

The underlying rockbeds in Wuhan region are principally composed of formations of Mesozoic group and Palaeozoic group. The

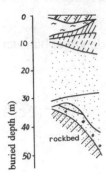

Fig.1 Type I soil foundation

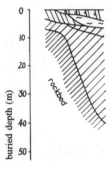

Fig.2 Type II soil foundation

formations strike SE110°-NW280° with dip angle of 65°-80°. Rock masses frequently encountered in engineering are shale and muddy sandstone of Silurian, quartz sandstone of Devonian, limestone and shale and mudstone of Carboniferous to Triassic, sandstone, mudstone and carbonaceous mudstone with coal of Jurassic, purple red mudstone and sandstone with conglomerate of Cretaceous to Tertiary. High buildings require rock mass as supporting layer, in terms of lithologic character, strength and orientation of the rock mass in situ, rock foundation is classified as follows:

3.1 Type I: Horizontal extremely soft to soft rock

Taking the project of Hanko New Centry Commecial Centre for example, see Fig.3, the formation is clay rock, silty sandstone with conglomerate of Tertiary The

dip angle is generally less than 30°. The axial compression strength is estimated 0.82-2.74MPa, at natural state(σ_{cn}) and 0.1-0.7MPa at saturated states(σ_{cs}). The point load test compression strength is approximately 4.0-16.74MPa for moderately and slightly weathered rocks. The rock mass is prone to soften while it is soaked in water. The variation of engineering rock property is not considerable in horizontal direction due to gentle dip angle.

This type of rock foundation can be adopted as the supporting layer of pile foundation, because the buried depth is generally larger than 40m in Hanko district. If moderately to slightly weathered rocks are selected as the supporting layer, the tip bearing capacity q_p =1500-2500KPa, and the lateral frictional force q_s =100-150KPa, when the thickness of rock debris at the bottom of drilled pile is less than 0.1m. As for relatively shallow rockbeds in Wuchang district, it can be used as the supporting layer for box or raft foundation, its f_k =700-1600KPa.

3.2 Type II: Horizontal soft to Slightly hard rock

A project in Hanko district can be taken as an illustrative example, as shown in Fig.4, where grey conglomerate intercalated with purple red silty sandstone and grey green silty sandy mudstone are main formations. The conglomerate is of calcareous and sandy cement, as a result the axial compression strength of moderate to slight weathered rock is about 16.2-43.4MPa at saturated state(σ_{cs}), and 36.2-55.9MPa at dry state(σ_{cd}). The softening coefficient(S_c) of rock mass is 0.43-0.81. However, the σ_{cn} of purple red silty sandstone with muddy cement is 1.4-3.9MPa. As for silty sandy mudstone, σ_{cn}, σ_{cd} and σ_{cs} are 3.8MPa, 5.1-10.7MPa and 1.8-31.MPa respectively, Sc=0.29-0.35.

In Hanko district the burried depth of rockbeds is larger than 35m. Depending on the practical test results, the q_p and q_s of this type of rock foundation are about 4000KPa and q_s =280KPa for drilled piles.

3.3 Type III: Highly inclined extremely soft to soft rock

This type of rock foundation is encountered in the project of Wuhan Torch building, as shown in Fig.5. Mudstone and shale were highly and extensively weathered, grey green to yellow green. Weathering

4016

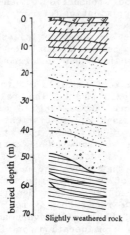

Fig.3 Type I rock foundation
(An example in Hanko)

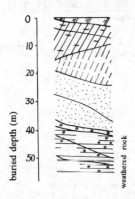

Fig.4 Type II rock foundation
(An example in Hanko)

fissures were considerably developed. The dip angle ranges from 55° to 85°, generally 70°, locally vertical or appearing compressive flexure fold zone. In Wuchang the influence of underground water is relatively weak, the σ_{cn} and σ_{cd} of moderately to slightly weathered rocks are 1.1-10.8MPa and 5.9-14.6MPa respectively. However, in Hanko due to the long-term action of underground water and the relax of geotress during sampling, its σ_{cn} =0.05-1.1MPa only, σ_{cd} =0.26-8.2MPa, and σ_{cs} =0.04-0.9MPa, S_c =0.02-0.45. This type of rock foundation is covered by the type I soil in Hanko, buried more than 40m. If the moderately weathered rock mass is adopted as the supporting layer of drilled pile, then q_P =

1000KPa and q_s =55KPa provided that the thickness of rock debris is less than 0.1m. It is covered by the type II soil in Wuchang buried less than 10m, hence the moderately to slightly weathered rock is taken as the supporting layer of excavated cylinder pile, its q_P =2400-3400KPa. If box or raft foundation is selected, the f_k ranges from 1800-2800KPa.

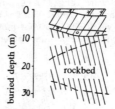

Fig.5 Type III rock foundation
(Wuhan Torch building)

3.4 Type IV: Highly inclined hard rock intercalated with extremely soft rock

The main formation is feldspathic sandstone intercalated with carbonaceous mudstone and coal. The dip angle of strata is about 60°-70°. The σ_{cs} of sandstone varies from 32 to 50MPa. Fig.6 shows a profile of Chuang-he building foundation. Because of the strength defference between sandstone and mudstone, the characteristic parameter of rock foundation greatly varies. The q_P and q_s of moderately to slightly weathered sandstone are 3500-4500KPa or more and 230-300MPa respectively. For carbonaceous mudstone q_P =800-1000KPa, q_s =50-60KPa.

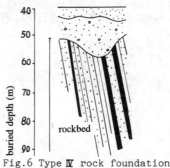

Fig.6 Type IV rock foundation

3.5 Type V : Highly inclined soft and hard rock

Fig.7 shows a geological profile of the rock foundation in Wuchang Luxiang, from which it can be seen that the shale and Quartz sandstone account for 1/3 and 2/3

4017

respectively. Because it is located at the axial part of an inverted anticline, the occurrence of strata greatly varies, and the X-shaped shear joints intensively developed. Highly weathered zone situated considerably deep.

It is appropriate to adopt excavated cylinder pile foundation for this type mainly because rock mass is exposed or less covered. Based on measured data for the pile of 22m long, the q_p and q_s of fractured shale are 700-1000KPa and 50-70KPa respectively, 2000-2500KPa and 100-150KPa for fractured quartz sandstone.

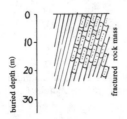

Fig.7 Type Ⅴ rock foundation
(Wuchang Luxiang)

3.6 Type Ⅵ: Hard karstified rock

This type of rock foundation is composed of grey muddy limestone intercalated with calcareous shale and thick limestone in which vertical joints and kartified caves were developed. Fig.8 provides a profile of a building foundation in Minglen street, Wuchang district. The strata highly inclined, locally vertical.

If the overburden soil is thin, then f_k=2500KPa for box foundation. If drilled pile is used, q_p=4500KPa and q_s =300KPa.

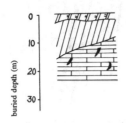

Fig.8 Type Ⅵ rock foundation
(Wuchang Minglen street)

4. DETERMINATION OF WEATHERED ZONE OF ROCK FOUNDATION

It is often required to determine the

thickness of weathered zone in the evaluation of rock foundation. Such measures as naked discrimination for rock core, analysis of core recovery and RQD, point load test in situ are used to investigate the mineral and cement composition of rock, fracture density and underground water condition.

Point load test is divided into longitudinal and axial test. The relationship between σ_c and sample depth Z can be obtained by means of statistical analysis for samples of different rock types. The test results of Hanko New Centry Commecial Centre are shown in Fig.9. From which it is found that above the depth of 57.5m all values of σ_c at various test points are less than 3MPa, from depth 57.5m to 61.2m, all σ_c range from 3 to 5MPa, blow the depth of 61.2m all σ_c vary among 5 to 10MPa. With consideration on lithologic character and core recovery, the classification of weathered zone is made that heavily weathered zone reaches the depth of 57.5m, moderately weathered zone ranges from 57.5m to 61.2m, then slightly weathered zone downwards.

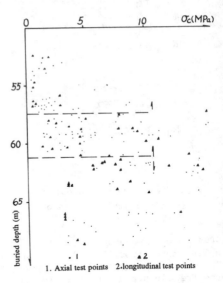

1. Axial test points   2.longitudinal test points

Fig.9 Relationship between σ_c
and sample depth Z

Fig.10 Shows the statistical results of σ_c and Z obtained from Wuhan Torch building project. The regressive line and envelop curve of test data reflect that σ_c increases with sample depth Z, that is to say, weathering degree of rocks reduces as increase in depth. Because the rocks belong to the type Ⅲ rock foundation, the

4018

thickness of weathered zone is relatively large. It causes difficult to strictly identify the interface between moderately and slightly weathered rock.

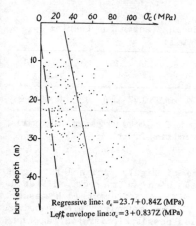

Regressive line: $\sigma_c = 23.7 + 0.84Z$ (MPa)
Left envelope line: $\sigma_c = 3 + 0.837Z$ (MPa)

Fig.10 Relationship between σ_c and sample depth Z

5. EVALUATION OF ROCKBEDS FOR PILE FOUNDATION

The selection of pile type is usually affected by the type of overburden soil mass. In Wuhan city drilled pile is generally adopted for the type I soil mass, and manually excavated cylider pile for the type II soil mass. The determination of rock mass parameter for pile foundation should consider such factors as pile type, pile length, technique of forming pile, rock characteristics around the tip and the boundary of pile.

Point load test in situ of the core taken from prospecting boreholes, or vertical compression test in Lab of the standard sample is required to measure the limit compression strength at various state. q_s and q_p are then comprehensively determined with consideration on the actual behaviour of rock mass, which are used as parameters of test pile design. The static load test in situ for experimental pile is performed to calculate the bearing capacity of a single pile. Modification for the q_s and q_p previously proposed by geological investigation is necessary to determine the parameters of practical pile.

It is proved that point load test is proper to statistically determine the compression stength of rock mass. There is a relation between σ_c and I_s:

$$\sigma_c = I_s \cdot C$$
$$I_s = p/1000D^2 \qquad (1)$$
$$C = 0.166(D-20) + 17.86$$

where σ_c (MPa) stands for the compression strength by point load test, I_s (MPa) point load strength index. p(KN) total failure load, D(mm) sample diameter and C coefficient.

In order to establish the relation between σ_c and q_s as well as q_p, the static load test in situ of pile is taken. There are

$$q_p = \sigma_c/8 \qquad (2)$$

$$q_s = \sigma_c/120 \qquad (3)$$

As far as extremely soft rock foundation is concerned, the parameters q_s and q_p can be estimated by the results of static load test (P_n). Following statistical expressions are usually used:

$$q = P_n/(0.42L + 1.57D) \cdot D \qquad (4)$$

$$q_s = q_p/15 \qquad (5)$$

where L(m) denotes the pile length inlaid into rock mass, D pile diameter, and P_n(KN) the limit load measured by the vertical static load test of experimental pile. Following two case studies are presented: case one is the project of Wuhan Torch building where manually excavated cylinder piles are adopted. The vertical static load tests for three pile groups were performed, as shown in Fig.11. The pile length ranges from 26m to 28m. Among three piles No.1 and NO.2 are inlaid into the folding zone of muddy shale, the NO.3 is inlaid into moderately weathered muddy shale. By using the expressions (4) and (5) it is calculated :

q_s=120KPa, q_p=1800KPa for the NO.1 pile

q_s=180KPa, q_p=2400KPa for the NO.2 pile

q_s=208KPa, q_p=3120KPa for the NO.3 pile

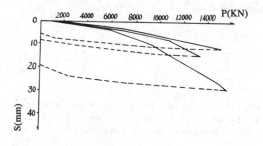

Fig.11 Relation between load P and settlement deformation S

4019

The second case is the project of Wuhan square where drilled piles were adopted and static load tests were performed for four pile groups, as shown in Fig.12. The pile G3 and G8 sat on a pebble bed of Quaternary, and G4 and G5 were inlaid into heavily to moderately weathered shale with the inlet length of 10m. The relavent parameters are listed in Table 1.

Table 1 Parameters of piles

pile numble	D (m)	L (m)	supporting layer	Hw (m)	Pn (KN)
G	1.0	51.18	pebble bed	2.0	8000
G	1.0	63.05	weathered shale	2.0	1600
G	1.0	63.32	weathered shale	2.0	13000
G	1.0	49.80	pebble bed	2.0	7000

note: D is the diameter of pile, L is the length of pile , Hw is underground water buried depth, Pn is pile limit load

It is clear that the increased limit load resulting from the pile inlying into rock mass is averaged about 7000KN. The parameters q_p and q_s are then calculated to be 1213KPa and 80.9KPa respectively.

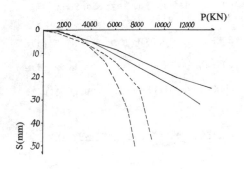

Fig.12 Relation between load p and settlement deformation S from the static load tests in Wuhan square project

It is also found that the thickness of rock debris deposited at the bottom of drilled piles affects on the tip bearing

capacity of pile foundation according to several cases in Hanko district. Following reducing factor is suggested, as listed in Table 2.

Table 2 Reducing factor for bearing capacity of pile

Th (cm)	≤5	8	10	13	15	20	25	30
Rf	1.00	0.94	0.86	0.77	0.70	0.55	0.40	0.25

note: Th stands for the thickness of rock debris, and R_f is the reducing factor.

In the case of highly inclined hard rock intercalated with soft rock, different piles may sit on different rock masses. Therefore, some measures are required to make piles posses similar bearing capacity. In general, it is realised by regulating the length-diameter ratio of pile or controlling the thickness of rock debris which should be controlled less than 0.1m, however, it can range from 0.25-0.3m for the piles on hard rock. The length of the pile on soft rock can be longer than that of the pile on hard rock by 1.5 times of pile diameter. The diameter of the pile on soft rock can also be larger than that of the pile on hard rock, generally by 10-20 percent.

6 CONCLUSIONS

On basis of practical experinces in Wuhan city, following conclusions can be achieved:

In light of the load, structural feature of high buildings, the strict requiments for the investigation of rock foundation are necessary. The types of rock mass should be classified according to fundamental geological conditions, and corresponding engineering measures should be taken into account.

In HanKo district drilled piles are adopted due to thick overburden soil mass and abundant underground water. Moderately to slightly weathered zone is selected as supporting layer. In Wuchang district (the third class terrace) manually excavated cylinder piles or manually excavated enlargered cylinder piles are used.

In the case of soft to highly soft rock mass, the division of weathered zone is especially importment. It can be quantitatively fulfilled by the analysis of core recovery, RQD and point load test in situ as well as measuring longitudinal wave velocity in boreholes.

The constuction quality of drilled pile relates to the normal working of pile. When the density of drilled mud is abnormal the lateral frictional force would be affected, so it is controlled no larger than 1.10. Moreover, the thickness of rock debris should be less than 0.1m.

As for the complex structure of rock mass foundation, the factors as length-diameter ratio of pile, the thickness of rock debris should be regulated to ensure piles from inhomogeneous settlement.

REFERENCES

Xiang, G.F. (1986) Proceedings of point load test of rock Chengdu Geology College Press(in chinese)

Chang, S.P. (1992) Handbook of engineering geology, Chinese architectural Press (in Chinese)

Structural damages caused by swelling soils: Experiences from case histories in Greece

Dommages structurals dues aux sols gonflants: Expériences des sites étudiés en Grèce

J. Christodoulias, C.V.Constantinidis, C.Giannaros & J.Vassiliou
Central Public Works Laboratory, Athens, Greece

ABSTRACT: In this paper an effort is made to determine geotechnical parameters and study the factors causing the structural damages in buildings from different areas of Greek territory. For this purpose a geotechnical programme consisting of site investigation and laboratory testing was carried out by Geotechnical Engineering department of Central Public Works Laboratory in Athens. After the completion of relevant works and evaluation of results it was found that swelling soils encountered in the foundation area in all three sites were mainly responsible for the structural damages in the buildings.

RESUMÉ: La présente étude s'efforce de déterminer les paramètres géotechniques et les facteurs qui sont al'origine aux dommages structurales des bâtiments, situés en differentes régions en Grèce. Dans ce but, un programme géotechnique constitué des essaies sur place et en laboratoire, a été réalisé par le Département Géotechnique du Laboratoire Central des Travaux Publics d'Athènes. Après l'accomplissement du projet et l'évaluation des résultats, il est montré que la présence des sols gonflants au niveau de fondation, dans les trois sites étudiés, engendre les dommages du structure des bâtiments.

1. INTRODUCTION

The engineering significance of swelling soils is now widely recognized and is well demonstrated by many case histories all over the world.

The existence of expansive soils in relation to the peculiar climatological conditions in Greece (certain dry months followed by periods of heavy rain) have resulted in activating the destructive action of clay minerals with all its unfavourable consequences.

Until recent years there was no investigation concerning the existence of such soils and consequently many stuctures(embankments, buildings and roads) are either resting or constructed by such soils. In this paper three case histories from different areas of Greece (Figure 1) are presented, concerning structural damages of buildings caused by swelling soils.

The first case (site 1) concerns the damages of some light built houses in the vicinity of Athens.The second (site 2) the structural damages of a regional medical

Fig. 1. Sites of investigation
1. Acharnes
2. Agia Anna (Evia)
3. Mytilini

center on the island of Evia. The third (site 3), the damages of a governmental building (Ministry) on the island of Lesvos.

At the beginning, a preliminary geotechnical investigation for all three areas was carried out, as it was at first believed that the damages were associated with earth movements. For this reason some bore holes were drilled and some inclinometers and stand-pipe piezometers were installed. The investigation revealed the existence of swelling clay material under the buldings of all three cases.

In every one of the above mentioned cases, the mechanism of swelling process in the soil was actually determined by evaluating field and laboratory data which further on allowed for some correlations to be made and remedial measures to be designed.

2. CASE HISTORIES

2.1. Site 1 (Acharnes)

In the town of Acharnes, 20 Km N.W. of Athens, extended distortions appeared, in many houses of an area along the crest of a railway cutting .The damaged houses were mainly single and two storey buildings with masonry walls and reinforced elements founded on isolated or continuous shallow footings. In that part of the town there was no sewage system, instead cesspools were provided for each building.

The distortions were mainly cracks on the basement slabs and diagonal cracks on masonry walls (Fig.2).

In the area of interest, the predominant formation is brownish-yellow, brownish-white, brownish-green to brown sandy to clayey marl (upper Neogene), with sparse calcareous concretions.

In order to identify the subsurface geology, to obtain samples for laboratory testing and to install two inclinometer tubes for determining the depth and shape of probable slides (as it was first considered that the damages were associated with earth movements), two borings were carried out.

2.2. Site 2 (Evia Island)

The building of regional medical center of Agia Anna village is located at the N.E. part of Evia Island, by the central regional road.

It was established during 1955 as a single storey building with masonry walls,

Fig. 2. Cracks on basement slabs (Acharnes).

reinforced concrete slabs and elements, founded at low depth (about 1.00 m).

During the Sixties, some hair cracks on the vertical masonry walls were observed.

Although necessary restorations were made, the cracks reappeared, along with new ones, on the floor (Fig.3) and the walls (Fig.4), which became wider with time. Some cracks were also observed on the ground surrounding the buildings, as well as the retaining walls downhill the medical center. Similar structure damages were also observed in the nearby old and recent buildings. The village of Agia Anna is located on the slope of a hilly area, with an average geomorphological inclination about 16% and is a part of a superficial drainage ravine.

The area of interest is covered, at its largest part (up to 10 m depth), by brown, brownish-red clay with many calcareous concretions and sparse gravel and cobbles. This layer represents the recent Quaternary scree, fans and landslide material, underlain by the Upper Miocene-Pliocene sequence of fluvial-lacustrine sediments (I.G.M.E., 1980). Groundwater table is generally found near the surface, in winter.

In order to identify the geology and to obtain samples for laboratory testing, five boreholes were drilled and two trial pits were excavated. Also three inclinometer tubes and four open stand pipe piezometers were installed, for monitoring probable slides and groundwater level respectively.

Fig. 3. Agia Anna. Floor having causes door's "shortening".

Fig. 4. Diagonal cracks on walls.

2.3. Site 3 (Lesvos Island)

The recently renewed buildings to house the Ministry of Aegean Region in the city of Mytilini (Lesvos island), have been severely affected by diagonal cracking on its masonry walls (Fig.5).

The complex of the Ministry consists of one and two storey buildings with masonry walls and reinforced concrete elements. Cesspools were provided instead of sewage system.

From a geomorphological point of view, the area of interest is located near the foot of the Castle Hill, with slopes dipping gently eastwards.

The geological structure of the site area, is consisted of a thin weathering mantle followed in depth by alterations of marly to clay horizons and locally marly limestones, which are all belonging to the Upper Pliocene (Andronopoulos, 1980). The predominant formation in the area, is a brown to brownish grey marly clay of high plasticity.

At the site investigation stage, a drilling programme consisted of four rotary borings was carried out to investigate the geotechnical conditions of the area where Ministry of Aegean is founded on. In addition, a series of laboratory tests on borehole representative and undisturbed samples were carried out to determine either index properties or engineering characteristics of clayey marly formations. Apart from the above investigations, inclinometers (slope indicators) were also installed in two boreholes, selected in a way

to give as much information as possible concerning the overall stability of the site.

3. LABORATORY TESTING RESULTS

3.1. Physical Properties

In order to identify the mineralogy and physical characteristics of the foundation soil at the three sites, sieve and hydrometer tests and x-ray analyses were

Fig. 5. Distortions on walls.

carried out. The Atterberg limits, linear shrinkage, shrinkage limits and carbonate content were also determined.

The results of sieve analyses, liquid limit and plasticity index tests on oven dried samples passing the US sieve No 40 (425 μm) are shown in Table 1.

In all the above sites the depth of interest concerning the soil characteristics is about 5-6 m, depending on the volume of soil influenced by the foundation of the structures.

Table 1. Physical properties of soils

Site	Grain Size Distribution %			Atterberg Limits %		
	grav.	sand	fines	L.L.	P.L.	P.I
1	0	1- 9	91-100	32- 75	18-24	13-54
2	0-26	6-39	51- 87	28- 83	11-27	15-56
3	0- 3	2-34	66- 98	51-105	17-34	34-75

Taking into account the grain size distribution as well as the Atterberg limits the material of Site 1 is classified according to Unified Soil Classification System (ASTM D2487-85) as clay (CL). The material of Site 2 as clay (CL) & (CH) and the material of Site 3 as clay (CH).

Twenty bar linear shrinkage tests (BS 1377, 1975) were carried out in all sites, using remoulded soil from liquid limit test in 15x15x140 mm semispherical moulds (Table 2). Twelve shrinkage limit tests were also carried out using the apparatus suggested by TRRL (1974). The specimens were remoulded in the form of discs 2 in. (5cm) diameter and 1 in. (2.5 cm) thick, in a constant volume mould (Table 2). Twenty one free swell tests were performed according to Holtz and Gibbs (1956), the obtained values are also presented in Table 2.

Correlations between linear shrinkage, free swell and liquid limit are presented in Figures 6, 7, 8.

The mineralogical composition of the clay fraction from Sites 2 and 3 was determined by x-ray diffraction analysis, by the semi-quantitative method described by Brindley & Brown (1980), using a Philips diffractometer with CuKa radiation and the quantitative analysis was obtained by the method described by Bayliss (1986). This indicated that for Site 2 montmorillonite (10-53%), chlorite (4-10%), illite (6-10%), quartz (24-40%), calcite (7-25%), were present. Also for Site 3 montmorillonite (20-25%), feldspar (5-10%),

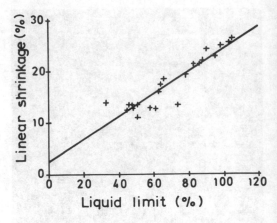

Fig. 6. Linear shrinkage vs. Liquid Limit.

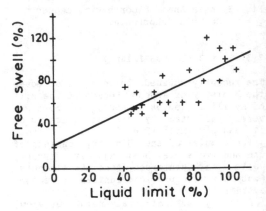

Fig. 7. Free swell vs. Liquid Limit.

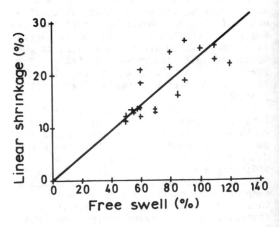

Fig. 8. Linear shrinkage vs. Free swell.

dolomite (5-10%), quartz (45-54%) and calcite (10-15%) were present.

Table 2. Swell characteristics

Site	Depth m	Linear Shrinkage %	Shrinkage Limit %	Free Swell %
1	2.5	13.6		60
	3.8	12.9		60
	4.4	11.0		50
2	0-1	12.5-13.6	10.7-10.9	50-70
	1-2	14.2-17.5	10.3	
	2-3	16.1	10.6	85
	4-5			75
	5-6	13.2	10.5	70
3	2-3	25.0		100
	3-4	21.4-24.3	13.0	80
	5-6	19.3	9.0	90
	6-7	22.9	8.0	110
	8-9	25.7	12.5	110
	9-10	22.1	10.5	120
	0-11	18.6-21.4		60
	14-15	26.4		90

3.2. Strength characteristics

In order to determine the shear strength characteristics of the soils, quick triaxial (UU) and unconfined compressive tests were carried out. The undrained shear strength ranges between 75-160 kPa for site 1, 50-160 kPa for site 2 and 90-210 kPa for site 3.
The swelling pressure of the soil was determined by tests carried out in the consolidometer. Half inch thick undisturbed soil samples were placed in the consolidometer ring. Initial dial reading was recorded after applying a seating load of 6.25 kPa. In order to identify the influence of moisture content changes on swelling pressure, samples from the same shelby have been tested :
(i) under natural moisture content.
(ii) after being dessicated for a few days in a silica gel laboratory dessicator.
The results of the above tests are illustrated in Table 3 and Fig.9.

4. DISCUSSION

For all three cases earth movements (creep or sliding) were first considered as the main cause of the distortions. However the evaluation of inclinometric data showed no

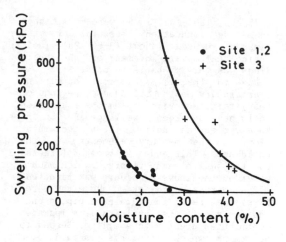

Fig. 9. Influence of moisture content on swelling pressure.

Table 3. Swelling pressure variations

Site	Depth (m)	Test m.c %	Swelling Pressure (kPa)	Remarks
1	1.10-1.40	19.3	85	n.m.c.
	2.10-2.50	19.0	75	n.m.c.
	3.80-4.10	17.0	125	n.m.c.
	3.80-4.10	15.7	163	dessic.
	4.10-4.40	18.7	110	n.m.c.
	4.10-4.40	15.5	185	dessic.
2	1.50-2.10	22.6	100	n.m.c.
	3.20-3.00	26.4	15	n.m.c.
	3.50-4.10	22.8	87	n.m.c.
	5.80-6.40	23.5	38	n.m.c.
3	4.60-5.20	36.9	325	n.m.c.
	4.60-5.20	27.8	510	dessic.
	4.60-5.20	25.4	625	dessic.
	5.00-5.60	41.6	100	n.m.c.
	5.00-5.60	38.2	180	dessic.
	5.00-5.60	30.0	340	dessic.
	7.10-7.70	40.2	120	n.m.c.

n.m.c. = natural moisture content
dessic.= dessicated

significant earth movements.
From the strength and deformation characteristics of the soil, it was concluded that no bearing capacity failure or settlement was the cause of the problem.
On the other hand the investigation indicated the presence of swelling soils at considerable depth in all sites. The swelling pressures exerted by the soils

(>75 kPa, Table 3, Fig.9) were much higher than the foundation pressures of the damaged houses (about 40-50 kPa) indicating that upward movement of the buildings was very probable to occur.

Due to the climatological conditions (certain dry months followed by periods of heavy rain and vice-versa) the foundation soils have undergone swelling and shrinkage cycles leading to stress accummulation in the masonry elements of the buildings. This ended to breaks (cracks) on the walls and floors, which became wider as the above procedure was repeated.

The corrective measures suggested were first to check the drainage piping and cesspools of the buildings, as leakage contributes greatly to swelling. Second to determine which footings showed the largest upward movement, then further attention could be directed there. In houses without reinforeced concrete resting on continuous footing, paving the ground surface and eliminating all plants within a few meters of the periphery of the building was suggested. Additionally in Site 2, peripheral drainage system was proposed, in order to minimize the natural moisture content fluctuation.

5. CONCLUSIONS

Many structural damages especially in light buildings were at first attributed to different causes as earth movements, settlements and bearing capacity failures. After a detailed geotechnical investigation it was found that the main causes of the problem was actually the existence of swelling soils under the foundations of the buildings. Identification of swelling soils, laboratory testing to predict their engineering behaviour as well as design of remedial measures are presented for three sites in different parts of Greece.

ACKNOWLEDGEMENTS

The authors would like to thank Mrs M. Koukoutsi for drawing the graphs as well as Mrs V. Florou and Mrs A. Lytra for their painstaking care in typing the text.

REFERENCES

Andronopoulos, P. 1980. Geological and geotechnical study in the Mytilini Museum foundation area. IGME, report No 13 (in Greek).

Bayliss, P. 1986. Quantitative analysis of sedimentary minerals by powder diffraction. Powder Diffraction, Vol.1, No 2, June, pp 37-39, USA.

Brindley, G.W. & Brown, G. 1980. Crystal structures of clay minerals and x-ray identification. Mineralogical Society, London, U.K.

Holtz, W.G. & Gibbs, H.J. 1956. Engineering properties of expansive clays. Transactions, ASCE, Vol.121.

HMSO, 1974. Soil Mechanics for road engineers. Determination of soil shrinkage, TRRL, pp 50-53, London U.K.

KEDE, 1989. Geotechnical investigation at the building of regional medical center of Agia Anna village (Evia Island). Central Public Works Laboratory, Athens, internal report (in Greek).

KEDE, 1992. Geotechnical investigation at Acharnes site. Central Public Works Laboratory, Athens, internal report (in Greek).

KEDE, 1993. Geotechnical investigation at the site of Aegean Ministry, Mytilini, Central Public Works Laboratory, Athens, internal report (in Greek).

IGME, 1980. Geological map of Greece; Limni sheet, scale 1:50.000.

Ministry of Culture and Science, 1987. Museum of Mytilini, Geotechnical study and Report, Geomechaniki Ltd (in Greek).

Schriener, H.D. 1987. State of the Art review on expansive soils. Report compiled for TRRL, Ministry of Transport, Crowthorne, Berkshire, U.K.

Evaluation of the abutment stability of a ninety-year-old bridge founded on serpentinite

Évaluation de la stabilité de l'appui d'un grand pont (age de 90 ans) fondé sur un serpentinite

Michael Kavvadas, Paul Marinos & Andreas Anagnostopoulos
National Technical University, Athens, Greece

ABSTRACT: A 210-meter, steel-truss valley bridge in Greece has suffered significant displacement of one abutment, founded on weathered saturated peridotite (serpentinite), which continue at a rate of several millimetres per year. The paper reviews the history of the bridge construction (in an attempt to present the practice of engineering geology and geotechnical engineering in Greece around the turn of the century), investigates the causes of the instability and evaluates the effectiveness of the rehabilitation measures taken over the last ninety years. It is shown that the instability is caused by an extensive creep-type slide of the slope of the valley due to the low residual shear strength of the weathered serpentinite in conjunction with the high porewater pressure regime in the slope.

RESUME: Un pont métallique de 210 m de longueur en Grèce a subit des deformations notables à l'un de ses appuis, fondé sur un serpentinite. L'article présente en revue l'histoire de la construction à partir du début du siècle et les mésures de protection prises le long de ces 90 ans. Les causes de l'instabilité sont discutées et les mésures de protection et d'assainissement utilisées sont evalués. Il est montré que l'instabilité est provoquée par une réptation de la pente de la vallée. Le phénomène est dûe à la faible résistance résiduelle au cisaillement de la roche alterée en combinaison avec un régime d'eau interstitielle élevée dans la pente.

1 INTRODUCTION

The bridge under consideration is part of a major link between Athens and northern Greece. This link, constructed around the turn of the century, crosses the mountainous terrain of Central Greece via tunnels and bridges. Most of these structures are founded on limestone and have not caused significant geotechnical problems in their century-long operation. However, a few tunnels and bridges are built on serpentinite and have shown significant convergence of the tunnel walls, creep-type movements of the piers, slides and instabilities in the abutments, etc. The present paper studies the behaviour of one such valley bridge in an attempt to: (i) review the state-of-the-art in engineering geology in Greece around the turn of the century and (ii) evaluate the effectiveness of various rehabilitation measures taken in the last ninety years. The long history of the bridge and the relatively good records of the observations and rehabilitation measures preserved up to the present, offer a good example of the long-term behaviour of a natural slope in serpentinite.

The bridge crosses a valley which is about 60 meters deep and 200 meters wide. At the location of the bridge (Fig. 1), the north slope of the valley consists of thick beds of Jurassic limestone, which is tectonically faulted with principal fault axes in the N-S direction. The north abutment of the bridge and three of the adjacent piers are founded on the limestone; there are no reports or other evidence of movements or instability in the foundation of these piers.

The south slope of the valley consists of peridotites, altered to serpentinites by weathering; the weathered material has completely lost its rock texture and has the appearance of a stiff dark-green clay. The serpentinites are occasionally interbedded with

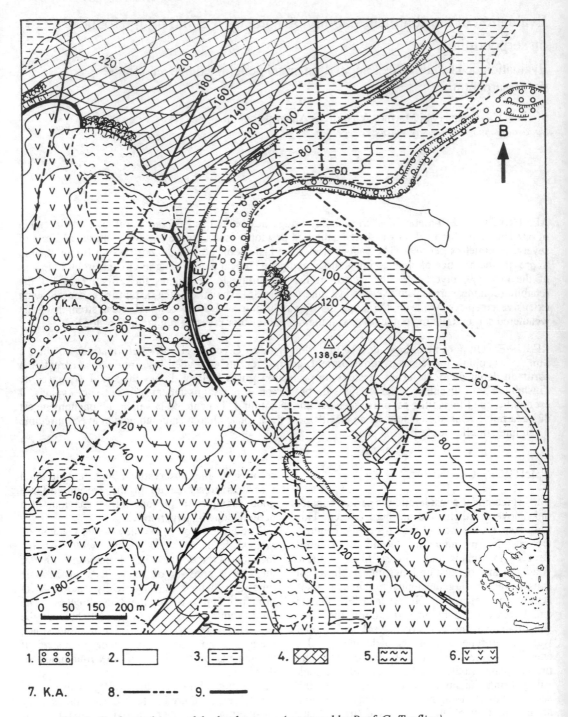

Fig. 1: *Geological map of the bridge area (prepared by Prof. G. Tsoflias).*
  1: Torrent deposits, 2: Plain alluvia, 3: Pleistocene, 4: Limestones (Jur.-Cret.)
  5: Cherts-Schists with ophiolithic bodies, 6: Weathered peridotite (serpentinite),
  7: K.A. Rock-fall, 8: Fault, 9: Tectonic contact

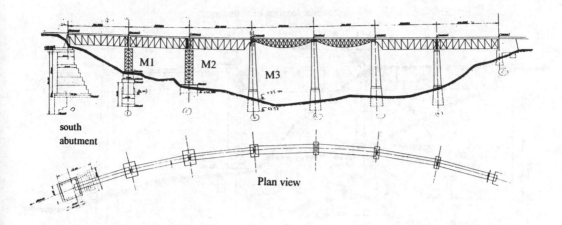

Fig. 2: Longitudinal section and plan view of the bridge. Each span is 30 metres. Piers M1 and M2 are steel trusses, while the rest are massive masonry structures.

brown-reddish schist-cherts also altered to a stiff clay-like material. According to recent data, the groundwater table in the area of the south abutment is very high (2-3 meters below the ground surface) and thus the material is practically saturated. The south abutment of the bridge and the three adjacent piers are founded on the serpentinites; they have suffered significant movements which have started short after the construction of the bridge and have continued until the present time, causing concern among the engineers and extensive re-positioning of the bridge trusses on the piers.

Fig. 2 presents a longitudinal section of the bridge which consists of seven spans, each about 30 meters long, resting on tall piers. The spans are steel truss structures, simply supported on the bridge piers via "fixed" Everall-type bearings which allow limited relative movement of the superstructure by frictional sliding. The end supports of the bridge are massive stone-masonry structures (typical engineering of the turn of the century). The two southern-most piers are steel trusses (15 and 21 meters high) founded on masonry footings while the remaining piers are stone-masonry tower-like structures (20-35 meters high) founded on masonry footings at a shallow depth.

2 HISTORICAL RECORDS ON THE BEHAVIOUR OF THE BRIDGE

This section presents a relatively brief account of the early history of the bridge with the objective to point out the practice of engineering geology and geotechnical engineering in Greece in the beginning of the twentieth century.

The original design of the bridge dates back to 1890 and construction started in 1902. The design called for a six-span bridge, i.e., it did not include the last southern-most span shown in Fig. 3. Furthermore, all piers were designed as stone-masonry structures with shallow foundations at an allowable bearing pressure equal to about 180 kPa. In the spring of 1903, the construction of the piers on the north side (those founded on the limestone) was completed, as well as the construction of pier M3 (Fig. 2) which is founded on the serpentinite via a masonry footing at a depth of about six meters below the ground surface. No significant geotechnical problems were encountered during the construction of the footing of M3, most probably due to the very mild slope of the serpentinite at that location. Works for the excavation of the foundation for pier M2 were underway when the first indications of a slowly progressing creep-type slide were observed. The foundation of pier M2 was located near the toe of the slope (see Fig. 3) and the excavation had evidently disrupted the equilibrium of the serpentinite at this location. Construction records indicate that when the excavation for the

4031

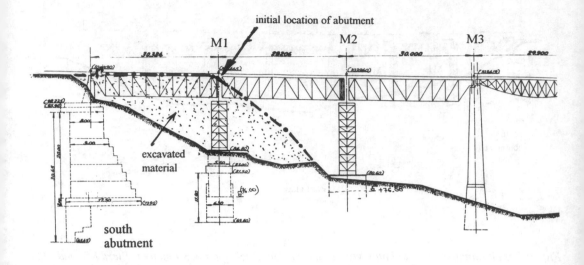

Fig. 3: Longitudinal section of the foundation of the bridge piers on the south slope of the valley. The figure shows the location of the original ground surface, before earth removal during the construction of the bridge.

foundation of pier M2 had reached the depth of four meters, a "moist red clay" was encountered and the uphill face of the excavation pit gradually moved inwards. A modern re-evaluation of this record indicates that the "red-clay" was actually an inclusion of schist-chert in the otherwise homogeneous mass of the serpentinite and that sliding probably occurred in the serpentinite along its interface (which was softened by groundwater) with the underlying schist. The sliding of the slope towards the excavation pit concerned the engineers who thought that the cause of the instability was the "moist red clay" layer. It was thus decided to take the following measures:
1. Excavate part of the slope in the area of the southern bridge abutment and make it less steep by removing earth material (see Fig. 3). This required to extend the bridge structure to the south by adding an extra span (the seventh), a new pier (M1) and to move the abutment by about 30 meters to the south.
2. Abandon excavation for the foundation of pier M2 and construct the footing at the current depth (about four meters) to avoid further disruption of the stability of the slope.
3. Substitute the massive masonry structures of piers M1 and M2 by much lighter steel trusses in order to decrease the pressure on the foundation soil.

4. Construct an extensive system of drainage galleries in the south abutment slope to locate the "red clay" layers and drain them.

The implementation of these measures resulted in:
1. The removal of about 10000 cubic meters of soil from the area of the south bridge abutment; a relatively mild slope (1:2) was thus formed.
2. The construction of a massive masonry structure for the foundation of the south abutment at a depth of about 30 meters below the ground surface and an equally massive structure for the foundation of pier M1 at a depth of 16 meters (see Fig. 3). Evidently, with these massive works the engineers attempted to transfer the bridge loads to depths "not affected" by the instability of the slope.
3. The construction of a long network of horizontal drainage galleries (about 800 meters) and twenty vertical shafts (each 10-20 meters deep from the surface to the gallery) in the greater area of the south abutment, in an attempt to drain the "red clay" (see Fig. 4). According to the construction records, the "red clay" layers were not continuous and, in searching for them, the galleries branched and meandered inside the mountain to an extraordinary length especially considering the technical facilities of the time.

Re-evaluation of these actions indicates that, with the exception of the attempted drainage of the slope, they were either irrelevant or that they resulted in a further deterioration of the stability of the south abutment of the bridge. In fact, the removal of earth material from the area of the abutment may have made the slope locally milder but, in the large scale, it undercut the toe of the serpentinite mass thus increasing the average slope of the mountain which was already in a state of limiting equilibrium. Furthermore, the products of the excavation were placed at a very short distance to the west of the excavated area and formed a fill about 14 meters high (see Fig. 4), thus aggravating the stability of the slope. This fill loaded the underlying serpentinite and also blocked the flow of a stream thus reducing the surface drainage of the slope. In addition to that, decreasing the bearing pressure on the foundation of the piers (by changing the masonry piers to the lighter truss structures) did not help since sliding was not caused by a shallow bearing capacity failure. Finally, increasing the foundation depth of the bridge abutment and of pier M1 did not help much, since the lateral earth pressure of the sliding mass on the foundation blocks was too high and eventually sheared through the masonry foundation. In fact, shortly after construction, the masonry lining of the inspection shaft inside the south abutment was cracked at a depth of about 24 meters, i.e., about six meters higher than the foundation level, indicating that the shear failure surface passed through that elevation. Since that time, the horizontal relative translation at that location has reached 40 centimetres and continues at an average rate of about five millimetres per year. The top of the abutment structure has also translated downhill but considerably less: an estimated 25-30 centimetres as concluded by the tilting of the inspection shaft in the abutment. In general, the engineers of the time did not realise that the instability was caused by an extensive deep-seated creep-type slide in the mass of the serpentinite and treated it as a local small-scale failure.

Since the end of the construction of the bridge (in 1905), the south abutment and the adjacent pier M1 have continued to move towards the NE and have required repeated repositioning of the three southernmost bridge spans, whenever the travel that could be accommodated by their bearings (: about 4-5 centimetres) was exceeded. While measurements of these movements only

started in 1975, the average rate of sliding before that time is believed to be about 3-4 millimetres per year; this estimate is based on the number of times it was required to reposition the bridge spans on the piers in the period 1905-1975. It should be pointed out that since all bearings were frictional (of the sliding type) and the bridge spans were weakly connected to each other, the required repositioning of three spans does not necessarily imply movement of all three piers; it is possible that fewer piers were directly affected by the sliding but the spans were pushed against each other. In 1966, it was decided that the main cause of the sliding was in the south abutment and it was decided to replace its frictional bearing by a roller-type bearing to isolate the movement of the abutment from that In the last twenty years the movements of the slope have been measured at relatively regular intervals and have indicated that:

1. The top of the south abutment (at ground level) and the base of pier M1 move gradually to the NE at an average rate of about 3 millimetres per year; the other piers seem not to have been affected by the slide. However, since the spans of the bridge are linked, the movement is transmitted to at least the first three spans.

2. The direction of the slide is to the NE, i.e., oblique with respect to the longitudinal axis of the bridge (N-S) by about 45 degrees. This is a result of the loading of the slope by the excavation products dumped to the west of the abutment at the time of the bridge construction (Fig. 4). Furthermore, since the sliding direction is not aligned with the axis of the bridge, the roller bearing at the south abutment is only effective in the longitudinal direction. As a result, the need to periodically reposition the three trusses of the bridge on their respective piers has not stopped since the time of the replacement of the bearing on the south abutment.

3 EVALUATION OF PRESENT STATE

The continued sliding of the south abutment slope has recently called for a geotechnical site investigation programme and a re-evaluation of its stability. The boreholes verified that the weathered zone of the peridotite (serpentinite) is quite deep (20-30 meters) and its lower limit approximately coincides with the location of the shear surface. The consistency of the

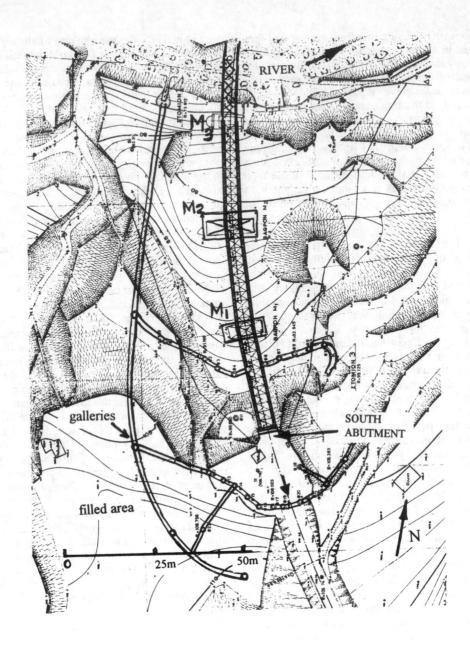

Fig. 4: Plan view of the area of the south abutment. The figure shows the area filled by the products of the excavation and the layout of the drainage galleries and vertical shafts (designated with circles).

serpentinite is very sensitive to its moisture content. More specifically, when the boreholes were advanced dry (i.e., without water circulation to reduce the wear of the cutting bit) the sampled material appeared to have a significant shear strength and gave practical refusal to the penetration of the Terzaghi sampler during the SPT test. On the contrary, when water was used for drilling, the material was softened to the consistency of a soft clay and the SPT sampler penetrated without any resistance with only the self-weight of the drilling rods. The low strength of the moist serpentinite was also manifested by the extremely low values of its shear strength parameters: a friction angle of 13-17 degrees and a cohesion intercept of about 10 kPa, which was reduced to practically zero when the residual strength was reached after large deformations.

The groundwater regime in the area of the south abutment is also peculiar. The serpentinite is practically saturated, an indication of the existence of a high groundwater table in the slope. While the serpentinite is generally impermeable, it occasionally contains relatively more permeable layers which were encountered during drilling. On one occasion, the borehole was dry up the depth of about ten meters, when a water bearing layer was drilled and the water level in the hole abruptly rose by eight meters and was stabilised at a depth of only two meters below the ground surface. Such high groundwater levels were observed in all boreholes while, in most of them, water table stabilisation required a few days due to the low permeability of the serpentinite. The high pore water pressures are indicative of incomplete drainage of the area despite the extensive network of the drainage galleries and shafts. This can be explained by the low permeability of the serpentinite which decreases the zone of influence of the drainage works to only a few meters, thus reducing the efficiency of the drainage works in the area of the abutment. Finally, the high pore water pressures are due to the blocking of the stream in the area to the west of the abutment by the excavation products dumped in that area during the construction of the bridge. In this way, surface waters (e.g. rainfall) are collected in the upper part of the slope and seep into the soil through the surface layers which have a relatively higher permeability.

The softening of the serpentinite upon wetting in conjunction with the highly elevated pore water pressure regime in the abutment area have resulted in the slope being in a condition of limiting equilibrium. This condition was triggered by the excavation at the toe of the slope for the construction of the bridge and, subsequently, the continued deformation reduced the shear strength of the material to its residual value, thus leading to a creep-type failure. The location of the shear failure surface was clearly defined by the cracking and the relative translation of the lining in the twenty vertical shafts in the area of the slope. According to the records, in most of the shafts, the relative translation at the level of the shear surface was 40-50 centimetres and the shafts had to be blocked to avoid accidents from falling pieces of the lining in the drainage gallery at their base. The location of wall cracking in all shafts is consistent with a roughly circular shear surface at a depth of about 25 meters. This shear surface extends below the whole area of the slope including the area filled by the products of the excavation for the foundation of the bridge piers in 1903. Back analysis of the stability of the slope using the above shear surface in conjunction with the residual shear strength parameters of the serpentinite and the measured high pore water pressures verified that the slope is in fact in a state of limiting equilibrium (factor of safety equal to unity).

4 PROPOSED REHABILITATION MEASURES

The previous discussion has concluded that the instability of the south abutment of the bridge is mainly due to the low residual shear strength of the saturated serpentinite, the high pore water pressures and finally the added load exerted on the slope by the fill placed during the construction of the bridge to the west of the abutment.

Drainage of the slope would be a very effective means to increase the factor of safety and provide an extra margin against a potential catastrophic failure. In fact, stability analyses of the slope have shown that a reduction of the piezometric head in the slope by two meters would increase the factor of safety to 1.10, from the present limiting value of unity. Unfortunately, drainage of the practically impermeable serpentinite is a difficult endeavour as has been proved by the inefficiency of the horizontal galleries and vertical shafts to drain the area. It was thus decided that a more efficient method of at least partial control of the pore pressure regime in the slope would be to prevent surface waters (e.g. rainfall) from seeping into the serpentinite mass in the upper part of the slope and gradually charge the groundwater table. To this objective, it was proposed (a) to construct a shallow drainage trench along the upper part of the slope to collect and discharge rainwater, and (b) to remove part of the fill that was placed to the west of the abutment at the time of the construction for the bridge and had blocked the flow of a stream. In fact, partial removal of the fill would also reduce the surcharge load on the abutment slope and thus further increase its margin of safety against shear failure.

5 CONCLUSIONS

The paper presents a relatively detailed account of the historical records on the behaviour of a ninety-year-old bridge which has suffered a slowly progressing creep-type instability of its abutment founded on weathered peridotite (serpentinite). The main causes of the instability, in addition to the geometry of the slope, are: the high pore water pressure regime in the slope which has proved difficult to drain despite the extensive drainage works, the extremely low shear strength of the saturated serpentinite and finally the surcharge that has been placed on the slope as a fill since the time of the construction of the bridge.

It was shown that while the instability of the abutment was identified even before the completion of the bridge construction (in 1903), the magnitude of the problem was not realised and the instability was treated as a small scale phenomenon by increasing the foundation depth and reducing the contact pressure of the bridge piers. On the contrary, the instability proved to be quite extensive with a shear failure surface at a depth of about 25 meters. In this way the gradual movement of the slope sheared through the foundation of the bridge piers and the vertical drainage shafts and has been advancing at an average rate of about 3 millimetres per year.

Due to the large scale of the phenomenon and the very low permeability of the serpentinite, it is very difficult to effectively drain the area and relieve the elevated pore water pressures in the slope which would certainly increase the factor of safety of the abutment and provide a margin of safety against a potential stability failure. It was thus decided to propose measures to control the seepage of surface waters in the upper part of the slope and limit their capacity to charge the groundwater table of the abutment slope. It was also proposed to monitor the pore water pressures in the slope by installing a set of piezometers and, as an ancillary precaution, install a few pumped deep wells (to depths below the level of the shear surface) to assist in the relief of the pore water pressures.

ACKNOWLEDGEMENTS

The authors would like to thank Prof. G. Tsoflias of the N. Tech. Univ. of Athens for the geological mapping of the bridge area.

The authors also evaluated previous work of IGME by G. Kallergis and E. Karageorgiou.

Choice of rational types of foundations in different permafrost conditions

Choix des types de fondations adéquates dans des conditions de permafrost

V. I. Grebenets & A. G. Kerimov

Research Institute of Bases and Underground Structures, Norilsk, Russia

ABSTRACT: Under technogenesis and global climatic heating aggravation of the frozen grounds properties occur, deformations of building take place, payment for foundations construction increase. Traditional for the Nothern regions methods of the foundations construction don't ensure the required safety. Rational types of foundations in dependence on the frozen ground properties are presented in the paper. Working methods in weak grounds, methods of artificial permafrost freezing, different types of cold ventilated foundations were worked out and explored.

RÉSUMÉ: Sous l'influence du progrès technique et du climat qui devient plus chaud s le globe terrestre a lieu l'aggravation des caractéristique solides de la congélation éternelle; la déformation des bâtiments augmente, les dépenses pour les fondations des bâtiments deviennent de plus en plus forts, les moyens traditionnels d'installation des fondations au Nord n'assurent pas la sécurité. Dans ce rapport on propose des types rationnels de fondations qui dépendent des caractéristique des terrains congelables. On a élaboré et on a donné son approbation aux moyens du passage des terrains faibles, à la congélation artificielle des terrains l'installation des fondations superficiels froids ventilés.

Conditions and properties of frozen rocks are being highly changed in permafrost regions under techogenic influence. So, working out of special foundations constructions and constructing methods which ensure long explotating period without damages is very actual task.

The foudations bearing capacity depends on the frozen grounds temperatures, salt concentrations, amount of ice involved. Aggravation on the main grounds engineering characteristics in the nearest 30-40 years under technogenesis is prognosticated. Meanwhile the safety of buildings with traditional and widely spread types of foundations is achived likely to freezing in the system: underground part of foundations - frozen ground. When the temperature of frozen ground and the depth of active layer rise, then the freezing forces decrease.

The traditional method of foundation is the pile foundations, the piles are installed in the holes and diametre of the hole is larger than that of the pile. The bearing capacity is ensured by the freezing forces between piles surface and the grounds. The less the permafrost temperature is - the more the piles bearing capacity.

Last 30-40 years in big nothern cities the aggradation permafrost tendencies are being observed. For example, on-location geothermal observations in Norilsk (nothern Siberia) showed, that at the depths 10...60 metres permafrost temperature rised on 0.5...1.0°C in last 30 years. The similar trend of the air temperarure is not observed. Degradation permafrost tendency is caused by technogenic factors: non-efficient exploitation of cold underground constructions, heating of permafrost by the underground collectors, artificial surface snow accumulation, sewage and technogenic waters filtration, chemical pollution, construction of technogenic artificial grounds, microclimatic changes. In different parts of the urban zone constructed in 50-60 years the bearing capasity decreased in 1.5-2 times in nowdays, many building are damaged.

Global permafrost degradation make stronger the degradation of local permafrost condition: rising of the active layer and permafrost temperatures, formation of chemically active underground waters, etc. The traditional constructing methods often don't ensure the projective bearing capacity and the costs for construction rise.

Invistigations were fulfilled and rational methods of foundations construction were worked out in three main directions: ground properties improving, foundations construction perfecting and working out of new technologies. Several methods of permafrost stabilising, lowering the ground temperatures and the bearing capacity rising were worked out. Artificial ground freezing methods whith the help of seasonally-cooling thermal piles, usage of non-filtrating cooling artificial grounds, snow-cleaning, new types of underground constructions are very effective. Usage this methods in Norilsk lead to the ground temperatures lowering on 0.8...1.2°C at the bottom of foundations. Looking throgh the new constructions of foundations there must be noted the following types: stuffed piles, composite piles and some others.

Bored-drived in piles, fig.1, are very effective: the hole is bored in the upper technogenically changed ground (the diametre of the hole is more than that of the pile), the pile is being put down in the hole and drived in to reach the persistent horisont or to the calculated depth. This methods is routher cheap. On-location investigation testified high results in achieving the projective values of the bearing capacity.

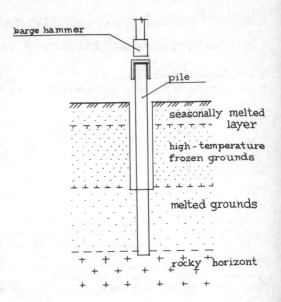

Fig.1 Bore-hitting technology
a) boring of holes with large diameter in high-temperature frozen grounds;
b) hitting of piles up to the rocky horizont.

Usage of autonomic seasonally cooling thermal piles are very perspective for the safety increasing, fig.2.

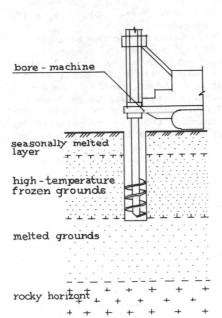

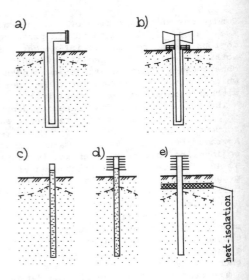

Fig.2 Seasonally cooling construction
a) artificial ventilation;
b) natural atmospheric air cooling;
c) fluid thermal pile;
d,e) steam-fluid thermal pile.

However, on-location observations testify that in the case of the ice high percentage in the grounds, thermal piles usage is unrational. Active usage of the thermal piles in Siberia and Alaska in clays, silts, loamy soils lead to bearing capasity decreasing. In sands and rocky grounds usage of thermal piles is effective.

The surface horizontally oriented cooling types of foundations are recomended to be constructed in the conditions of clays and loamy soil, fig.3.

a) surface foundation

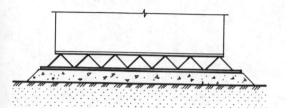

b) ventilated channel in the ground

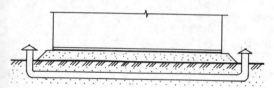

c) thermal piles

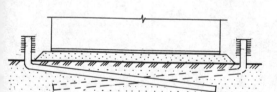

d) massife foundation with ventilated channels

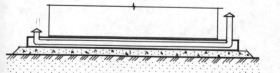

Fig.3 Surface cooling foundations.

Cooling is achived by two ways : artificial ventilation and natural air convection. Cooling (and, consequently, consolidation) of grounds occur along all the foundations surface and due to the constructions pressure vertical deformations of bases are practically absent. Newly formed ice layers are horizontally oriented, this rises the grounds stability due to excluding of shear-faulty deformations (with the vertical icy layers).

Construtions of structural regulated foundation is perspective for small buildings when the ground conditions are comprehensive, fig.4.

a)

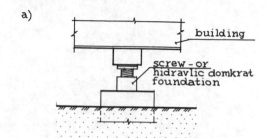

b)

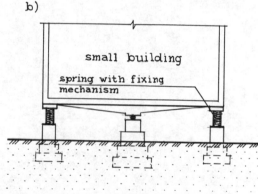

Fig.4 Regulated foundations for small buidings
a) domkrat-foundation;
b) foundations from blockes-boxes.

Light hollow foundations are recomended for the territories with mosy-lichen vegetation.

Changes of engineering-geocryological properties of frozen rocks under techogenesis influence and global climatic heating is a serious problem, working out of methods of ground bases preparation and new types of foudation is actual.

Engineering geology in the construction of the Chalkis cable bridge in Greece

Géologie de l'ingénieur et la construction du pont suspendu de Chalkis, Grèce

P. Marinos & N. Fytrolakis
National Technical University of Athens, Greece

M. Vainalis
Athens, Greece

ABSTRACT: The paper describes the results of the geological investigations which contributed to the knowledge of the geotechnical conditions for the foundation a cable bridge constructed in an area with complicated geological conditions and earthquake activity. The main problems were the understanding of the structural geometry of the area and the evaluation of the behaviour of a serpentinite, soil-like, mass at the foundation of one of the 90-meter-high piers which was finally founded inside this weak material on friction piles, due to the internal roughness of the mass. The recent activity of the faults in the area is also discussed.

RESUME: Les résultats de la reconnaisance géologique pour la construction d' un pont suspendu dans une région ayant une structure géologique compliquée, sont discutés. Les problèmes principaux étaient la définition de la géometrie structurale de la région et l' évaluation géotechnique d' un serpentinite, de comportement du type sol, au site de fondation de l' un des pilier, haut de 90 m. Le pillier étai finallement fondé dans ce matériel, très faible, par des pieux de friction, grace à la rugosité interne que pourrait se développer dans sa masse. L' activité des failles de la région est aussi descutée.

1. INTRODUCTION

The new cable-stayed high bridge over the Evripos Channel, opened for the traffic in 1993, is part of the bypass of the city of Chalkis and links the Boeotian coast of the mainland with the coast of the island of Euboean, NE of Athens. The bridge has a total length of 694.50 m, an operating width of 12.60 m, and a deck height of 34.20 m. The main part of the bridge, measuring 395 m, is cable-stayed, with a main span of 215 m and two side spans of 90 m each. Each of the two piers (M_5 and M_6) 90 m high, consists of two columns of variable section. For the foundation of the pier M_5, large concrete piles with a diameter of 1.20 m and bored to a depth of 28.0 m were used. This depth was selected after the geological evaluation, discussed in this paper, and based on the results from pile load tests (Stathopoulos, 1987 and Frank et al 1989).

2. THE GEOLOGICAL FORMATIONS

The geology of the region was studied by several authors (Guernet, 1971, Katsikatsos, 1979, and others). The region of the bridge was firstly examined by Papageorgakis, 1970. During our investigation a detailed geological research was performed. The expansion of the geological formations and the general structure of the region of the bridge can be seen in the geological map of fig. 1.

The southern bank and hill consists of older limestones and dolomites (Triassic-Jurassic). Their bedding ranges between moderate to thick, with an average dip of 20°-30°, to the NW. The northern bank and hill consists of younger limestones of Cretaceous age.

The alluvia are mainly sands and silts with some gravel. They cover the older formations, mainly at the zone of the strait. Their thickness does not exceed 3 m and they are loose.

The pleistocene sediments consists of red clays with sand and gravel in a clayey or sandy matrix.

Their thickness on land does not exceed 8 m, but in the bottom of the sea, on the northern side, it reaches 20 m. They have suffered a reasonable consolidation. However the cementation is weak and it becomes strong, only locally, where the calcite element prevails.

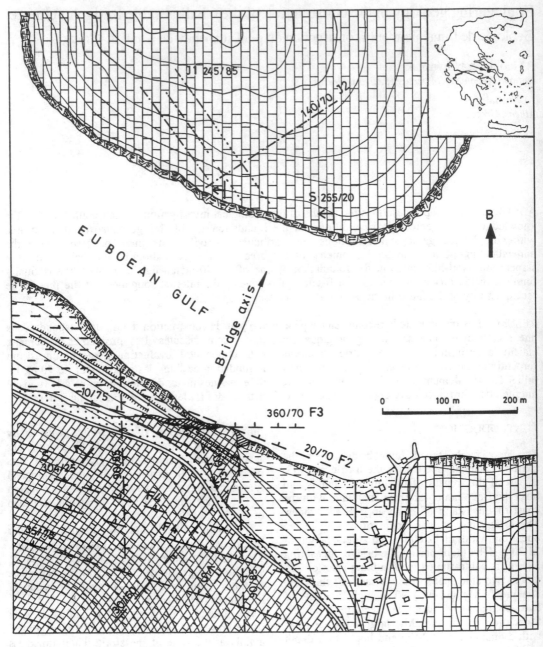

1. ⬚⬚⬚ Dumps 2. ≡≡≡ Neogene and Quaternary deposits

3. ⊞⊞⊞ Cretaceous limestone 4. ▨▨ Serpentine and schists

5. ▨▨ Triassic-Jurassic limestone 6. —⊥—F Faults 7. —·—·—J Joints

8. ⋏ S Bedding

Fig. 1. Geological map of the bridge area.

The upper cretaceous limestones are moderately to thickly bedded. The bed thickness ranges between 0.30 and 1.5 m. There is no intercalations of thinly layered marls, between the beds of the limestone and therefore there is good contact between them, reinforcing the friction resistance of the bedding planes. These limestones are crossed over by two systems of joints (J_1 and J_2), vertical to the bedding. The average density of the joints are usually between 2.5-3 m and the limestones are divided into rectangular blocks with dimensions of 1.5 to 5 m. The system J1 on the northern bank, is more obvious and corrosion due to the sea is more intense. The beds dip slightly to the west (265°/20°), the joints J_1 to the southwest (245°/85°) and J_2 to the southeast (140°/70°). These limestones present widened karstic opening, which were expected to become quickly narrow in depth. These cavities and the widened joints are usually filled with clay or gravels.

The ophiolites and cherts are found in limited outcrops with small thickness in the Boeotian (southern) shore. However, as it was presumed, in the bottom of the sea and beneath the quaternary deposit, they present considerable thickness of up to 30 m and are developed a great extent.

They are mostly made up of serpentinite and some layers of cherts, rich in clay elements. They compose a "melange" with an age ranging from upper Jurassic to lower Cretaceous. The outcrops are usually weathered and very rarely compact masses of serpentinite of small sized can be observed. Stratigraphically they are located between the upper Cretaceous and the Triassic limestones. Their contacts in this site is, although, a tectonic one.

The upper Triassic - lower Jurassic limestones are found in the southern side of the foundation area of the bridge and they also cover a great region. Close to the shore the limestones are massive with no bedding and an irregular pattern of high spaced joints. Further to the south, towards the hillside, they develop bedding but is that place they do not affect the foundation. The karstic erosion is present, but not to an extend to cause dangerous conditions. The karstic conduits and cavities are generally of small size in comparison to the extent covered by the base of the piers.

3. THE TECTONIC STRUCTURE

3.1 The pattern

The Euboean gulf is a neotectonic graben, which was formed by normal faults of a general direction NW-SE. Normal secondary ruptures also exist. The area of the foundation of the bridge forms a smaller segment of this vast

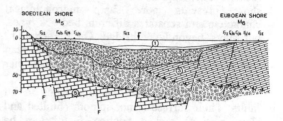

Fig. 2. Geological section of the strait
1. Recent marine deposits (sandy silt), 2. Clay, silt with limestone fragments, 3. Cretaceuous limestone, 4. Serpentine complexe, 5. Triassic-Jurassic limestone, F. faults, ⊔⊔ thrusts, Γ. bore holes, M. site of piers.

tectonic basin. The foundation of the bridge is related with two independent tectonic blocks of horst type and with the trench between them.

Although all details can not be seen within the trench and the northern horst, with the data gathered from the research on the southern horst and from the drilling within the trench, the structure were designed reliably enough along the axis of the bridge (fig. 2).

In the southern horst, two main fault systems are observed. The first has an ESE-WNW direction (120°-300°), with a dip to the NE (35°/75° to 10°/70°). The other system has a N-S direction, dipping to the east. The first system is parallel or almost parallel to the main faults of the Euboean graben.

The fault F_1 is a major one and has a length of many kilometres. Faults F_2 and F_3 are advancing NW, into a large rupture. The three F_4 faults are of smaller size.

The layers dip at a small angle (20°) to the W or WNW (265° on the north and 305 on the south horst).

3.2 The question of fault activity

Our remarks regarding the greatest possible movements along the faults due to an earthquake, are as follows:
- In the greek experience in the case of faults

with known recent activity, there is a preference for the large movements (several dm) in those sections that divide competent with less competent formations.

- The longitudinal faults were more active in the shaping of the Euboean Gulf area.

- There are no morphological indications of rupture activity in the area, where the bridge is located. The longitudinal fault of the Boeotian shore presents more sharp morphological characters and separates different bedrock types, but stays beyond the bridge. On the contrary these features are not present in the Euboean side, where a fault runs probably through the pier M_6 area.

- There are two main characteristics in all faults: Their surface is not smooth (limited and slow motion) and a friction factor can be developed. The other characteristic is that most rupture surfaces have been covered by calcite coatings and the remaining aperture have been filled with "fossil" earth material or, again, calcite, which means that in the recent past, these faults have not been activated.

Conclusively, the possibility of any relative movement, or rotation, of one shore towards the other, is very limited. A differential settlement equal to 0.20 m, has, however, been taken into consideration by the designers (Stathopoulos, 1987).

3.3 The older structural geometry of thrusts

In relevance to the main alpine tectonics, an emphasis has to be made on the fact that the ophiolites sequence with cherts is heterochthonous and its contact with the underlying Triassic limestones is tectonic (thrust). With the late, alpine, tangential movements, the upper Cretaceous limestones slipped on top of the ophiolitic complex too (fig. 2). Their contact is also a tectonic one. Both tectonic contacts present brecciated zones. Consequently the cherts-ophiolite sequence has been intensely stressed, during its thrusting and during its compression, between the two limestone masses and the result is the production of a very incompetent mass.

3.4 Seismic activity

The region experiences a relatively high seismic activity and a series of studies which were conducted by professor I. Drakopoulos led to the following conclusions (Stathopoulos, 1989):

Ground movement which has a probability of 90 to 95% not to be overcome within the next 100 to 200 years (which means a period of repetition in the order of 900-1000 years) are the following: a=0.209, v =0.15 cm/sec, d=5.0 cm. Using these values and following Seed's, Newmark's and ATC's methodologies, the appropriate spectrum was designed, where it is clearly shown that projects with a high fundamental period, like this bridge, are not highly stressed in earthquake loading.

4. ENGINEERING GEOLOGY OF THE SITE

Our evaluation on the geological conditions for the foundation of piers was based on the study of the drilling's cores, on the results of the laboratory and on the in situ tests, executed by "Geomechaniki Ltd".

4.1 Piers of the Boeotian (southern) coast (A_0, M_1, M_2, M_3, M_4)

The piers are set on the triassic limestones, where no structural disturbances are present.

The slight cover, with soil materials and the upper loose part of the bedrock, did not cause excavation problems in the foundation area. Part of the pier M_1 is an exception, because limestone screes are as deep as 5 m.

The limestones present good geotechnical behaviour. The high RQD values confirm such a good appearance. Locally, only, fractured sections are encountered and especially at the drilling site Γ_3 of pier M_2, but beyond 6 m the fractured sections interchange with massif ones.

Karstic cavities and internal erosion phenomena, with engineering significance, were not encountered. The results of permeability (packer) tests, if they can be considered as reliable, assigned values ranging between 10^{-3} - 10^{-4} cm/sec. These values do not correspond, at all, to a rock mass with wide discontinuities and most certainly with karstic voids. Wagon-Drill - Enpasol's investigation data confirmed this situation.

4.2 The pier M_5

The stratigraphy beneath this 90 m high pier is comprised of:

- the filling quaternary of the strait (with a thickness of about 13.5-19.5 m).

- the "chert" sequence with ophiolites (thickness of about 24-31 m, 17.5 to 50 m under the sea level).

- the triassic limestones

The quaternary deposits, from gravels and

pebbles in a clayey-silty environment, present a dense enough structure.

The sequence of cherts and ophiolites (peridotites mainly) presents a great petrographic heterogeneity and a structural disorder, with an intense sheared character. This condition, together with the weathering of the rock mass to serpentine, which also appears with a great disorder and loose character, provides, often, a soil-like than a rock-like mass. The extensive testing done on this weathered mass confirmed this condition.

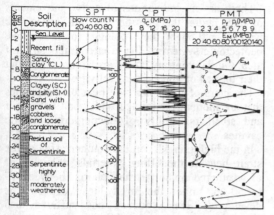

Fig. 3. Average results from Standard Penetration (SPT), Cone Penetration (CPT) and Ménard Pressuremeter (PMT), Tests (Frank et al, 1991).

The limestones beneath the ophiolites are massive with excellent geotechnical rating. However at the contact with the ophiolites they are slightly fissured, in a thickness of 1 to 3 m. At this contact the ophiolites are well crushed. This state is explained by our interpretation as a thrust (fig. 2).

Due to these unfavourable conditions at the foundation zone of pier M_5, a first solution was to use bearing end piles down to the limestones, at a level of almost -50 m (fig. 2, 4). The difficulty with such a solution with large piles, is obvious. However a second closer detailed examination showed that inside the weathered ophiolithic mass exist sectors with irregular blocks of less or even not weathered ophiolite, which "float" in the weak weathered environment. This situation increases the heterogeneity of the mass, but improves its overall behaviour, regarding both bearing capacity and settlement. After observation of the cores, the impression is that the mass, as an entity, will behave better than the type of behaviour implied by the results from the ponctual testing of the weak sections of the sequence. The in situ pressiometric tests give, indeed, the same impression in most cases. The values of the standard penetration tests confirms the heterogeneity and disorder, but they do not reach lower levels (fig. 3).

This heterogeneity of the weak ophiolites as described above, is assessing to the mass an

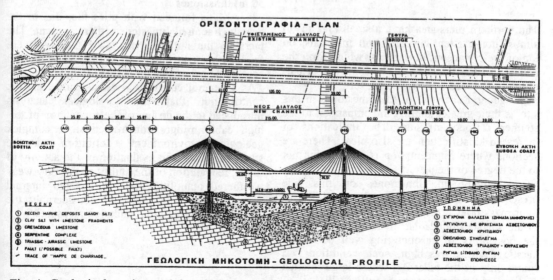

Fig. 4. Geological section and foundation features (Geology by the authors; design of the bridge from Stathopoulos, 1987).

internal roughness and this fact guided the final solution for the foundation.

The foundation solution of the pier, was finally friction piles of a diameter of 120 cm. They reside within the serpentinite, at a level of -27 to -28 m (fig.4).

The bearing capacity was estimated to be 800 t (operational load). The results of a reliable loading test which was conducted by the Greek Laboratory of Public Works, showed excellent results. A testing pile of 65 cm of diameter with a load of 800 t showed elastic behaviour with maximum settlement at the head less than 5 mm. The pile exhibited, again, exceptional behaviour under horizontal load up to 80 t; its behaviour remained elastic, with horizontal head displacement less that 30 mm (Frank et al, 1989).

4.4 The pier M_6

After the 3,5 m thick soil mantle, the entire 38 m depth drilled, at the Euboean shore, was with the cretaceous limestones. In this site the intense to complete fracturing, of the rock is the specific characteristic. The rock mass, so heavily fractured, was simulated to a rockfill, mainly near the surface. However, the inbrication between fragment gives a better overall behaviour against the foundation loads.

The condition improves, generally, in deeper sections. This was one of the reasons to adopt piles for the foundation of pier M_6, too.

4.5 Piers of the Euboean coast (M_7, M_8, M_9, M_{10}, A_{11})

The northern piers area have a small (1-2 m) to almost non-existent cover of soil materials, on top of the cretaceous limestones, which form the bedrock of the foundation. The loose upper part of the limestones is insignificant, too.

The general character of the limestones in this site, is their greater jointing in comparison to the southern (Boeotian) side. The indications of increased karstification are also clear. There are sections where fragmentation is intense, but not to the degree of the pier M_6.

Conclusively, the limestones on this side present moderate quality, but the condition of their mass does not cause any specific problems to the surface foundation of the piers. Although the possibility of encountering well developed karstic cavities was high, such condition was found only in the pillar A_1, where large internal sections of the rockmass were full of silty soil

5. CONCLUSIONS

The geological research allowed the distinction of the tectonical blocks of the area, created by the recent evolution of the Euboean Gulf area: The piers of the access bridge in the Boeotian shore, are founded within the same major tectonic block but the pier M_5, of the central bridge (cable bridge) is founded in an other, subsided, one. There is a strong possibility that one of the faults of the area runs through the site of pier M_6. The piers of the access bridge, in the Euboean side, belong to their own tectonic part.

There are no indications of active movements of the faults that define the area's tectonic pattern, although they belong to the active neotectonic system of the greater Euboean Gulf area. On the contrary, there are arguments about their limited or non existent activity.

The piers of the south side bear on limestones that are moderately jointed with good behaviour and small possibility of encountering karstic cavities.

In the high pier M_5, on the southern side of the strait, beneath the filling materials, a "melange" of weathered ophiolites and cherts (between 17 to 50 m, under the strait loose deposits) is developing. The sequence has suffered also sheared processes and exhibit a poor to very poor mechanical quality. However the differential behaviour of this sequence allowed the selection of a friction pile solution instead of using bearing end piles down to the underlying (-50 m) limestones.

Pier M_6 is founded with piles too, due to the heavily fractured limestones, there present. The piers on the northern side access bridge were founded also on limestone with moderately but acceptable behaviour for surface foundation.

As a general conclusion we can underline the contribution that the geological analysis introduced to the design of the foundation of this high cable bridge, constructed in a complex geological environment: definition of the structural geometry; evaluation of the activity of faults; recognition of the ability of the very weak weathered ophiolitic mass to develop internal roughness, fact that reduced considerably the size of the appropriate foundation measures.

ACKNOWLEDGEMENTS

The bridge was designed by the consulting office "Domi"; the geotechnical investigation was conducted by "Geomichaniki Ltd" and the structure was constructed by "Elliniki Technodomiki S.A." and "Techniki Etaeria Volou S.A.". The authors acknowledge the assistance provided during their collaboration.

REFERENCES

Guernet, C., 1971. Contribution à l' étude géologique de l' Eubée et des régions voisines. *Thèse Fac. Sciences,* Paris.

Frank, R., Kalteziotis, N., Bustamate, M., Christoulas, St., Zervogiannis, H., 1991. Evaluation of performance of two piles using pressuremeter method. *Journal of Geotechnical Engineering,* ASCE, 117, 5, 695-712.

Katsikatsos, G., 1979. La structure tectonique d' Attique et de l' île d' Eubée. *VI Colloq. Geology Aegean Region,* Athens.

Papageorgakis, J., 1970. Relations of the geological structure with the geotechnical conditions in the region of the foundation of the new bridge of Chalkis. *Scientific Annals of the National Technical University of Athens,* (in greek).

Stathopoulos, S., 1987. The high bridge of Chalkis. *Proc. 8th Greek Congress on Concrete.* (in greek)

Rock mass strength of basaltic breccia in underwater foundations

Résistance d'un massif de brèche basaltique entant que fondation d'un pont de grande dimension

Lineu A. Ayres da Silva
Rock Mechanics Laboratory, Mining Engineering Department, Polytechnic School, University of São Paulo, Brazil

Helmut Born
Mining Engineering Department, Polytechnic School, University of São Paulo, Brazil

ABSTRACT: In the studies of the geomechanic properties of a rock mass for foundations of a large bridge, a specific rock mass classification was adopted because of the difficult access conditions to the rock material. The behavior of the uniaxial compressive strength of a basaltic breccia, was observed. Determination of this property for the rock mass, based on laboratory tests, was made according to the methodology proposed by Ayres da Silva [8]. The Scale Effect of the compressive strength for the basalt and basaltic breccia was verified. For this last lithology a surprisingly high manifestation of that effect was noted. This paper presents all the adopted procedures, from sampling to the conclusions about mechanical behavior of the basaltic breccia.

RÉSUMÉ: Dans l'étude du massíf de fondation d'un pont de grand dimension, une classification spécifique a été adoptée du aux difficiles conditions d'accés au matériau rocheux. Le comportment de résistance à la compression d'une brèche basaltique est ici décrit. A partir d'essais de laboratoire, cette propriété a été déterminée à l'échelle du massíf, selon la méthodologie proposée par Ayres da Silva [8]. L'effet d'échelle a été vérifié pour les roches de ce massif; pour la brèche basaltique en particulier, une manifestation exceptionnellement élevée de cet effet a été observée. Cet article décrit les procedures adoptées depuis la collecte d'échantillons jusqu'à la description du comportement de la roche basaltique, dont les conclusions ont ultérieurement été confirmées lors des travaux sur le terrain.

1 INTRODUCTION

A 2.600 meter long road-railway bridge, spanning the Paraná River in the Ilha Solteira Reservoir area, is being constructed close to Santa Fé do Sul, at the border of São Paulo and Mato Grosso do Sul States, in Southern Brazil.

The Paraná River valley is situated in the domains of the Paraná Basin and is constituted by sub-horizontal flows of massive to vesicular/amygdaloidal basalt and basaltic breccia, from the upper portion of the Serra Geral Formation, overlain by sandstone.

To determine the geomechanical parameters of the foundation rocks, with emphasis on the compressive strength of the rock mass, a specific classification for underwater foundations was developed, to detect and demarcate areas with the same behavior, relatively to the intended use. The reservoir depth varies from 15 m, at the site of the now drowned Ilha Grande, to 50 m, at the bottom of the original river valley. This excludes the possibility of in-situ measurement of several parameters normally used by the mostly adopted geomechanical classifications [1]. In this particular classification, lithology, fracture intercept, weathering and RQD were utilized as rock parameters. Their association and ratings, according to the relative influence of each one in the stability and safety of the bridge foundations, were discussed and described by Ayres da Silva, Fujii and Jardim [2]. Therefore, to solve that specific problem, the proposed classification furnishes some important design data and leads to the best way to conduct the excavation of each rock mass type or class.

This paper discusses particularly the compressive strength characteristics of the basaltic breccia.

2 SAMPLING PROCEDURES

The basaltic breccia is constituted by irregular fragments of brown amygdaloidal basalt in a light colored fine to medium grained sandy matrix. The intact rock presents variable degrees of alteration in the studied area. The geological character of the rock mass is described in the above mentioned paper [2] and the basaltic breccia has an important role for the structure's foundations.

Test samples were obtained from drill cores from each foundation block, from several outcrops and from a basalt quarry at Porto Itamaraty, close to the bridge site. At the quarry, basalt blocks of up to one meter diameter were loosened by controlled blasting and randomly collected by local workers.

3 PREPARATION OF TEST SAMPLES

As the methodology used for rock mass strength determination requires cubic test samples, 5, 10 and 15 cm edge cubes were cut from the rock samples, with a special dimension stone saw, with laser alignment. In spite of some problems with the cutting machine, not usually used for such purposes, the final quality of the cubes was superior to that normally obtained for test specimens. Additionally the loading planes of the cubes were rectified to ensure proper regularity and parallelism of the cube faces.

The cylindrical test samples, used for intact rock strength determination, were obtained from drill cores and received the same treatment with respect their loading face parallelism.

As the position of the samples in the rock mass was marked and respected, all of them were tested according to their original spatial orientation.

4 TESTING METHODOLOGY

The uniaxial compressive strength properties of the tests samples, with 5, 10 and 15 cm edges, were measured by specialized and experienced technicians of the Rock Mechanics Laboratory.

Determination of the compressive strength of the different rock mass classes, recognized by the adopted classification, presupposes the utilization of the scale effect as proposed by Ayres da Silva and Hennies [3].

Thus, according to the above mentioned procedure, the scale effect was verified on cubic samples, while the compressive strength of the intact rock was measured on cylindrical samples. Results obtained for cylindrical specimens, previously published by Ayres da Silva and Born [4] together with data obtained by others authors, are reproduced in Table 1:

Table 1. Compressive strength values obtained for cylindrical specimens of intact basaltic breccia.

D (cm)	σ_D (MPa)
4.9	97
5.0	98
10.0	87
10.1	88

The results for cylindrical specimens of basaltic breccia, always refer to tests in dry conditions.

The basaltic breccia cubes were tested under three different conditions:

- Water saturated sample and loading perpendicular to the basalt flow bedding.
- Water saturated sample and loading parallel to the basalt flow bedding.
- Dry sample and loading perpendicular to the flow bedding.

These test conditions were established because of the following reasons:

a) The rock mass can be water saturated down to the loading depth of the bridge foundations.
b) It is not known whether the rock mass presents an anisotropic behavior in relation to the compression load orientation, i.e., if the rock mass is more resistant to horizontal or vertical compressive efforts with respect to the flow orientation.
c) Dry rock compressive strength behavior must be known for correlation with other data mentioned in the literature.

5 RESULTS

5.1 Shape effect

The shape effect reflects the influence of the width/height ratio of test specimens on the strength properties of the samples [5]. The results obtained for the test samples must be corrected by the equation that reproduces this dependency [6]. Therefore, the function of the basalt has been adopted for this purpose for the following reasons:

- The equation has been established for a rock mass of volcanic origin.
- The breccia and the basalt are lithological varieties within a rock body of volcanic origin.
- The equation obtained for basalt is very similar to that determined by Bieniawski [7] for coal and considered generically representative of that function.

Thus, the test results from cubic samples were corrected by the expression:

$$C_p = C_1(0.62 + 0.38 \, W/H) \quad \text{(Eq. 1)}$$

Were:

C_p = Uniaxial compressive strength of the prism with any W/H ratio.

C_1 = Uniaxial compressive strength of the prism with W/H ratio equal to one.

5.2 Strength variation due to the scale effect

The determinations were grouped according to the test conditions and the sample size in Table 2:

Table 2: Breccia tests

Side Dimension (cm)	Normal Charge (Saturated Specimen)		Parallel Charge (Saturated Specimen)		Normal Charge (Dry Specimen)	
	Specimen	Corrected Stress (MPa)	Specimen	Corrected Stress (MPa)	Specimen	Corrected Stress (MPa)
5	BR - 1	106	BR - 4	98	BR - 7	101
	BR - 3	102	BR - 6	98	BR - 8	97
	BR - 2	103	BR - 5	85	BR - 9	88
10	BR - 14	73			BR - 4	104
	BR - 15	76			BR - 4	101
15			CP - 3	37	CP - 1	29
			CP - 4	15	CP - 2	18

The corrected values for water saturated test samples were studied by numerical regression analyses. The representative function, for the variation of compressive strength as related to the size of the basaltic breccia test samples, corresponds to the equation:

$$C_D = 212.35 - 62.53 \ln D \text{ (MPa)} \quad \text{(Eq. 2)}$$

With a correlation index $R^2 = 0.90$

This equation is depicted in Figure 1.

5.3 Uniaxial compressive strength determination of rock mass portions with predominance of basaltic breccia (sub-classes B of the specific classification [2])

From Equation 2 and Protodyakonov's parametric equation:

$$\frac{C_D}{C_M} = \frac{\dfrac{D}{b} + m}{\dfrac{D}{b} + 1} \quad \text{(Eq. 3)}$$

Where:

C_D = Uniaxial compressive strength of cubic samples with size D.

C_M = Uniaxial compressive strength of the rock mass.

D = Cube edge in cm.

4052

b = Least spacing between discontinuities, in cm (10 cm in this case)

m = C_o/C_M Where C_o is the compressive strength of the unit cube.

The convergence of the two functions is studied, for maximum and least m values. It is shown that, starting from any value of m, the functions will converge when m → ∞, and, then, the compressive strength of the rock mass with predominant basaltic breccia will became negligible.

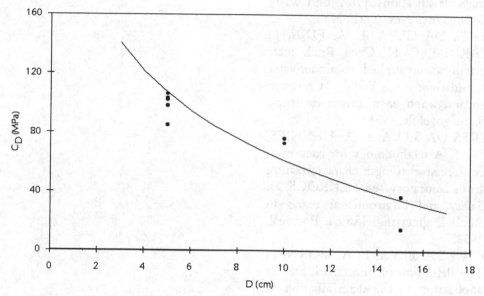

Figure 1: Scale effect for breccia specimens

6 CONCLUSIONS

- The uniaxial compressive strength values for dry test cubes of 5 and 10 cm edge are almost the same, as shown in Table 2. This indicates that the least distance between discontinuities in the dry basaltic breccia is in this range. This was also verified by the difficulty of cutting good intact cubic test specimens with more than 10 cm edge.
- The uniaxial compressive strength of the basaltic breccia is extremely influenced by the water saturation and, even more, by the alteration degree of the rock. With increasing sizes of test samples, the probability of finding very weathered material, which can not be considered rock anymore, also increases.
- The obtained function, expressed by Equation 2, indicates that, with increasing sizes of unconfined specimens of basaltic breccia, the uniaxial compressive strength became so low, that it behaves as regolith, rather than rock.

Thus, it is concluded that the portions of the rock mass constituted predominantly by basaltic breccia, when the confining pressure is released by excavation, certainly will present disintegrated material, which can not be considered rock. In the field it was verified that basaltic breccia, as predicted, could be manually excavated, while the use of explosives was required only in basaltic rock intervals.

REFERENCES

[1] BIENIAWSKI, Z. T. Engineering rock mass classifications. 1989, John Wiley & Sons Inc., New York, NY.

[2] AYRES DA SILVA, L. A; FUJII, J.; JARDIM, C. H. C. A Rock mass classification applied to underwater foundations. IV Congreso sudamericano de mecanica de rocas, Santiago, chile, 1994.

[3] AYRES DA SILVA, L. A; HENNIES, W. T. A methodology for rock mass compressive strength characterization from laboratory tests. EUROCK'93: Safety and Enviromental Issues in Rock Engineering, Lisboa, Portugal, 1993.

[4] AYRES DA SILVA, L. A.; BORN, H The definition of intact rock and its application to the determination of the smallest spacing between discontinuities in rock masses. EUROCK'93: Safety and Enviromental Issues in Rock Engineering, Lisboa, Portugal, 1993.

[5] OBERT, L; DUVAL, W. I. Rock Mechanics and the design of structures in rock. New York, Wiley, 1967.

[6] AYRES DA SILVA, L. A.; SANSONE, E. C.; FUJII, J.; Shape effects considerations on the uniaxial ccompressive strength of hard rocks. EUROCK'93: Safety and Enviromental Issues in Rock Engineering, Lisboa, Portugal, 1993.

[7] BIENIAWSKI, Z. T.; Rock mechanics design in mining and tunneling. Rotterdam, Balkema, 1984.

[8] AYRES DA SILVA, L. A. Proposta de metodologia para a determinação da resistência à compressão de maciços rochosos, a partir de ensaios de laboratório. São Paulo, 1992. 225p. thesis - Escola Politécnica, Universidade de São Paulo.

In-situ shear test for optimum slope design of a phosphorite mine

Essais de cisaillement in-situ pour le projet des talus de mines de phosphorite

V. K. Singh, B. D. Baliga & B. B. Dhar
Central Mining Research Station (CMRS), Dhanbad, Bihar, India

ABSTRACT: The shear strength is one of the important input parameter for optimum slope design. The in-situ jack shear test was conducted to determine the shear strength parameters of different lithological units, which was further used for optimum slope design in a block of phosphorite open cast mine.

RÉSUMÉ: La résistance au cisaillement est un important paramètre d'entrée pour l'étude optimal des pentes. L'essai de cisaillement en place permet calculer les paramètres de résistance au cisaillement des différentes unités lithologiques, lesquels ont été utilisés aussi pour le dessin optimal des pentes dans une exploitation à ciel ouvert de phosphorite.

1 INTRODUCTION

The present paper discusses how the data related to shear strength properties and groundwater condition can be precisely and quickly collected, evaluated and best utilized for optimum slope design of the D-block of Jhamarkotra phosphorite open pit mine. The mine is characterized by highly weathered impure limestone and dolomitic limestone towards footwall and hangingwall side respectively. The direct shear test and in-situ (Jack) shear test were conducted for the determination of shear strength parameters.

The water table is present at 490 MRL at most of the locations in the pit. The water table was considered to go down to 445 MRL after implementation of dewatering scheme and mining itself. Hence, stability analysis was done for two water tables, at 490 MRL without dewatering and also at 445 MRL with effective dewatering scheme.

Slope monitoring by Wild NA2 precise level and Wild DI4L electronic distance meter was done in and around the pit to know any movement.

2 EXPERIMENTAL WORK

Slope stability analysis requires the measurement of geomechanical properties of slope material by appropriate field and laboratory tests. Engineering properties of the litho units will influence the analysis for slope stability.

In most part of the pit, the main lithological unit towards footwall is impure limestone (I.L.), while the hanging wall mostly consists of dolomitic limestone (D.L.).

The samples of these litho-units were collected from different location and depth of the already developed part of the pit. Undisturbed specimens of these litho units were tested in the laboratory to determine density and shear strength parameters (Table 1). Consolidated undrained direct shear tests were conducted on direct shear machine.

A few in-situ flat jack shear tests were also conducted in the footwall and hangingwall side. In

Table 1. Geomechanical properties of the slope materials

Properties	Litho Units	
	I.L.	D.L.
Unit weight (kN/m^3)	22.16	24.32
Cohesion (kN/m^2)	13.7	17.6
Friction angle (deg.)	33	35

I.L. = Impure Limestone,
D.L. = Dolomitic Limestone.

this test, a block of weathered material of known dimension is made to fail by gradually pushing it from a fixed reaction face. A record was made of the loading at various stages with details of measurements of the failure surface. This test was conducted according to the Russian practice (Anand and Rao, 1967). A test block of 40cm height, 80cm width and of length 100 cm was cut in the pit at the site of investigation. The loading face of the block is kept at a distance equal to the total length of the equipment assembly. The sides of the test block were separated from the main soil mass by a narrow cut of width 15 to 20cm to the full depth and loosely backfilled by the excavated soil (Fig. 1). Both reaction face of the pit and the test block were truly vertical so that the load applied is horizontal.

The shear jack assembly was lowered into the pit and put into the testing position. The load was applied in increments of 0.5 tones was maintained for a period of 10 to 15 minutes after which the load was increased to the next stage. It was noticed that after the application of some load the block of soil started moving up along a sliding plane exhibiting cracks and heaving of the failed material. The application of load was continued till the test block moved by a distance approximately 10cm horizontally. The load at the start of movement (P_{max}) and at the time when block moved by 10cm (P_{min}) was recorded from the proving ring. After the test, the assembly was taken out from the pit. The true shape of the sliding surface was determined by removing the soil which sheared off along the sliding plane. Proper judgment is necessary in determining the plane of sliding. After the removal of failed soil the depth of the failure surface was measured at three locations along width of the block and at every 10 cm intervals along the length of the block. The average value (h) of the measurements at three locations was used to determine the shear strength parameters.

The shear strength parameters such as cohesion (C) and angle of internal friction(ϕ) are calculated as follows.

A cross section like Fig. 1 is drawn by making use of the average depth of the sliding surface. It was further subdivided into suitable number of slices. The weight of each slice (w) and length (l) along the sliding plane were determined. Further, the weight of the whole sliding mass (W) was determined by following eqation:

$$w = \tau * h_m * x \qquad (1)$$
$$W = \sum_{i}^{n} w_n \qquad (2)$$

where h_m = midheight of the slice (m); τ = unit weight (kN/m^3); n = no. of slices; and x = width of the slice (m).

Using the values of P_{max} and P_{min}, lengths, and weight of slices, the values of cohesion (C) and fiction angle (ϕ) were calculated by the following equation :

$$C = (P_{max} - P_{min})/(b*X) \qquad (3)$$
$$Tan\phi = ((m*A) - B - (C*X))/((m*B) + A) \qquad (4)$$

where

$$m = (P_{max}/(b*W)) \qquad (5)$$
$$A = W*Cos\ \Theta_n \qquad (6)$$
$$B = W*Sin\ \Theta_n \qquad (7)$$
$$X = \sum_{i}^{n} x_n \qquad (8)$$

where b = width of the test block; $\Theta_1, \Theta_2, \ldots$ = angles between the sliding surface and the horizontal; and ϕ = angle of internal friction.

The average values of the cohesion and friction angles for both the soil types, obtained by

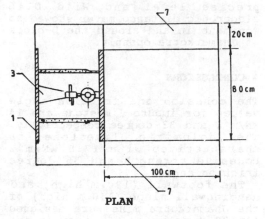

PLAN

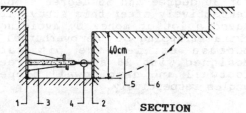

SECTION

1,2 Steel plates 80 cm×40 cm
3 Jack
4 Proving ring
5 Testing soil block sample
6 Failure plane
7- Loose earth filling

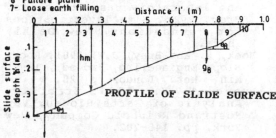

PROFILE OF SLIDE SURFACE

Fig.1 – In-situ shear (Jack) test

in-situ shear (Jack) test, are presented in Table 2.

Table 2. Shear strength parameters from in-situ jack shear test

Soil type	Average value	
	Cohesion (kN/m^2)	Friction angle (degree)
I.L.	14.03	33
D.L.	17.46	36

I.L. = Impure Limestone,
D.L. = Dolomitic Limestone.

3 SLOPE DESIGN

The interim pit slope (up to 500m RL depth) and ultimate pit slope design (up to 400m RL depth) was done considering the following conditions based on the findings of the geotechnical study:

(1) The mode of failure for the slopes of this mine is categorized as a circular failure due to the presence of highly weathered lithological units in the hangingwall and footwall.

(2) Stability analysis was done for two water table conditions. Initially the water table was considered to be at 490m RL without taking into consideration any dewatering. Later, the water table was considered to go lowered, to 445m RL, as a result of excavation and pumping.

(3) The ground water did not play any role during design of the pit up to 500m RL because the water table was at 490m RL in block-D.

(4) A 1.3 factor of safety was selected, for the pit slope to the 400 m RL depth, on the basis of the long term stability (Hoek and Bray, 1981) associated with the 190m deep pit on the footwall side and the 140m deep pit on the hangingwall side.

(5) A low value of factor of safety (1.15) was used for slope design upto 500 mRL depth, i.e. 90 and 40 m high footwall and hangingwall slopes, because short term stability is needed and strength properties of the slope material were known with certainty upto that depth (Hoek and Bray, 1981).

The circular failure analysis was done with the computer program "REAME" using the simplified Bishop method. REAME was developed by Huang (1983), which was suitably modified to meet the author's requirements and the computer (HP 9000) compatibility. The results of analyses performed by the program with different combinations of pit depth and slope angles of footwall and hangingwall are summarized in Tables 3 and 4 respectively. Piezometric surfaces I and II (Tables 3 and 4) mean the water table at 490m RL and 445m RL. Analysis without any piezometric surface condition means the analysis was done in dry condition.

Table 3. Factor of safety for footwall slope

Slope height (m)	Slope angle (deg.)	Piezometric surface no.	Factor of safety
190	30	1	1.26
190	30	2	1.32
190	31	1	1.224
190	31	2	1.279
140	32	1	1.198
140	32	2	1.301
140	33	1	1.15
140	33	2	1.26
90	35	–	1.17
90	36	–	1.15
90	37	–	1.11
50	39	–	1.15
50	40	–	1.11
40	42	–	1.159

Table 4. Factor of safety for hangingwall slope

Slope height (m)	Slope angle (deg.)	Piezometric surface no.	Factor of safety
140	34	1	1.24
140	34	2	1.34
140	35	1	1.22
140	35	2	1.33
140	36	1	1.201
140	36	2	1.303
90	40	–	1.149
50	43	–	1.154
40	45	–	1.158
40	46	–	1.143

The presence of groundwater in the slope is highly disadvantageous to slope stability, hence the influence of water pressure behind the slope is of prime importance. The drainage of the groundwater improves the stability of slopes.

Except the rainy season the pit remains almost dry. Suitable surface drains were provided to divert the rain water away from the pit. These surface drains are properly maintained, especially in rainy season, to keep them effective.

Slope monitoring by Wild NA2 precise level and Wild DI4L electronic distance meter showed no movement in and around the D-block of Jhamarkotra openpit.

4 CONCLUSIONS

The cohesion and friction angle values for impure limestone are 14 kN/m^2 and 33 degree respectively. The dolomitic limestone is characterized with 17.5 kN/m^2 cohesive strength and 35 degree friction angle.

The footwall (190 m high) and hangingwall slope (140 m high) of the Jhamarkotra mine were designed for 30 degree and 36 degree slope respectively after this study. It saved a lot of money by avoiding extra stripping of the overburden, because earlier the pit was designed with 28 and 32 degree footwall and hangingwall slope angles respectively.

REFERENCES

Anand, H.N. and Rao, Y.V.N., 1967. Comparative study of the field and laboratory tests on soils, Proc. Symp. Site Investigations for Foundations, New Delhi, India.

Hoek, E. and Bray, J.W. 1981. Rock Slope Engineering, 3rd ed., Inst. Min. Met., London, p. 28.

Huang, Y. H., 1983. Stability Analysis of Earth Slopes, Van Nostrand Reinhold Company, New York, pp. 145–282.

Rock excavation in Nyalau sandstone

Excavations du grès de Nyalau

J. Peuchen
Fugro Engineers BV, Netherlands

ABSTRACT: Excavatability predictions for Nyalau Sandstone, Sarawak, are compared with subsequent observations for a 50,000 m^3 excavation. The Nyalau Sandstone is folded and includes competent rock, classified as very strong and with very widely spaced jointing. Three conventional prediction procedures were adopted for the design geotechnical studies. The procedures are based on site-specific acquisition and interpretation of seismic refraction, rock coring and laboratory testing data. Application of the prediction procedures is discussed for D9G bulldozer ripping and G120 hydraulic impact breaking. Observations and actual production rates for excavation are presented. Comments on rock surface finish after excavation are also given.

RESUME: Les prédictions de facilité d'excavation du grès de Nyalau, Sarawak, sont comparées avec des observations faites pour une excavation de 50,000 m^3. Le grès de Nyalau est plissé et contient une roche competent classifiée comme très dure avec des joints largement espacés. Trois procedures de prédiction conventionelles ont été adoptées pour les études geotechniques. Ces procedures sont basées sur l'acquisition des données in-situ et l'interpretation de seismique réfraction, carrotage des roches et essais en laboratoire. L'application des procedures de prédiction est discuté pour le decapage par D9G bulldozer et rupture par impact hydraulique G120. Les observations et les taux de production actuels pour excavation sont presentés. Des commentaires sur la surface de la roche après excavation sont aussi donnés.

INTRODUCTION

Site constraints for civil engineering construction often preclude the use of drilling and blasting techniques for surface excavation of rock. Rock loosening by bulldozer ripping and hydraulic impact breaking is then applied, but there is a lack of success in prediction of feasibility and productivity. Cost overruns and claims result all too often.

The excavation documentation described here adds to, unfortunately sparse, case history data for calibration of prediction procedures. The case history concerns a 50,000 m^3 excavation in Nyalau Sandstone, Sarawak. The surface area of the excavation is approximately 10,000 m^2 and the maximum depth below existing ground surface level about 12 m.

Details are given on data acquisition, interpretation, prediction and performance.

GEOLOGICAL SETTING

A ground investigation strategy was developed on the basis of an engineering geological desk study. Results of this study are as follows:

- The Nyalau Formation is of Miocene age.
- Most of the Miocene sediments are considered to have been derived from previously compacted, folded, and uplifted rocks. The thickness of the Nyalau Formation is in excess of 3,000 metres.
- The excavation area is underlain by argillaceous sandstones, in alternation with siltstone and mudstone layers. The siltstone and mudstone layers are believed to be laid down in shallow marine conditions, possibly coastal swamp or deltaic conditions.
- Nyalau rock is slightly synorogenic or postorogenic. Later phases of folding and uplifting brought the upper rock above sea level. Subsequent weathering produced a layer of residual soil.
- Groundwater levels tend to follow the interface of residual soil and rock.

Parameter	Zone 1	Zone 2
Stratum Classification	Sandstone with alternating mudstone layers and seams	Sandstone with some mudstone layers and seams
Particles	50% fine grained quarz, angular; 10% microcrystalline quartz; 25% fine to medium grained feldspars, generally altered to clay minerals	50% fine grained quarz, angular; 10% microcrystalline quartz; 25% fine to medium grained feldspars, generally altered to clay minerals, some fresh particles; locally skeletal carbonate
Matrix	15% recrystallized clay minerals, some iron ore staining	15% recrystallized clay minerals, some iron ore staining; locally carbonate cement
Texture	Grain supported, predominant concave-convex and suture contacts, clay coating around particles	Grain supported, predominant concave-convex and suture contacts, clay coating around particles
Colour	Medium grey to reddish brown	Light to medium grey
Weathering	Highly and moderately weathered	Slightly weathered and fresh
Water Content	3% to 8%	2% to 4%
Bulk Density	2.3 t/m^3 to 2.5 t/m^3	2.4 t/m^3 to 2.6 t/m^3
Pulse Velocity	2 km/s to 3.5 km/s	3.5 to 4 km/s
Point Load Index (axial)	0.5 MPa to 2 MPa	Typically 3 MPa to 5 MPa, locally up to 10 MPa
Uni-axial Compressive Strength	Typically 15 MPa, locally up to 40 MPa	Typically 90 MPa, locally up to 120 MPa
Secant Elastic Modulus	--	Typically 20 GPa, locally up to 50 GPa
TCR[1]	60% to 95%	95% to 100%
SCR[2]	30% to 80%	50% to 100%
RQD[3]	30% to 80%, typically 50%	30% to 90%, typically 70%
Discontinuity Spacing in Borings	50 mm to 500 mm, typically 200 mm	50 mm to 1500 mm, typically 300 mm
Discontinuity Type	Bedding approximately planar and parallel to original topography, with dip angle of less than 10°; near-vertical orthogonal and conjugate joint set with average spacing of 1.5 m to 2 m	Bedding approximately planar and parallel to original topography, with dip angle of less than 10°; near-vertical orthogonal and conjugate joint set with average spacing of 1.5 m to 2 m; 1 m to 2 m thick zone with high angle joints
Discontinuity Features	Predominantly irregular, rough, tight and ironstained; locally crushed zones	Predominantly irregular, rough, tight and clean; high angle joint zone includes open discontinuities, locally with infill
Discontinuity Shear Strength	--	Sandstone surfaces: 35° to 55°, typically 45° (adjusted for dilation); mudstone surfaces: 35°
Downhole P-wave Velocity	--	1500 m/s to 2500 m/s, typically 2200 m/s
Seismic Velocity	1500 m/s	2000 m/s to 3500 m/s, typically 3000 m/s

Table 1: interpretive model

Notes for Table 1:
1. TCR = Total Core Recovery; total core length divided by core run length
2. SCR = Solid Core Recovery; all pieces of solid core that have a complete circumference added together, divided by core run length
3. RQD = Rock Quality Designation; all pieces of sound core over 100 mm long along centreline added together, divided by core run lengths per stratum or core run; drilling breaks are fitted together and treated as sound core

GROUND INVESTIGATION

Ground investigation procedures were based on ASTM standards, and ISRM recommendations. On-site ground investigation work was carried out in 1991 and included:
- engineering geological mapping of nearby rock outcrops
- logging, sampling and testing of 11 vertical borings including HMLC triple tube coring to below final excavation level, within and nearby the excavation area
- seismic refraction surveying along 8 lines
- seismic downhole testing at two locations
- monitoring of 6 standpipe piezometers.

Data acquisition was monitored progressively, by updating the initial geotechnical model developed from the desk study. In particular, attention was given to detection of potential faulting.

Laboratory testing of selected rock specimens included classification tests such as mineralogic description, porosity and pulse velocity. Uniaxial compressive strength with axial and lateral strain gauge measurements was determined for intact specimens, and shear strength of discontinuities was determined by direct shear testing including dilatancy correction.

It is noted that the programme of ground investigation also included data acquisition for slope and foundation design.

INTERPRETATION

The project team for the data integration and interpretation phase included an engineering geophysicist/geologist and a geotechnical engineer. The resulting interpretive model is presented in Table 1. Details refer to the predominant sandstone strata.

In summary, two zones can be identified. Zone 1 comprises highly and moderately weathered sandstone with alternating mudstone, typically moderately strong. Zone 2 is similar but slightly weathered and fresh, and typically strong to very strong.

Seismic refraction data integrated with boring data provided a reasonable delineation between Zone 1 and Zone 2.

PREDICTION

Three prediction procedures were applied to assess rippability of the two rock zones schematized by the interpretive model. These procedures are:
- Weaver (1975); essentially a fairly comprehensive rock mass rating system
- Modified Franklin (Franklin et al., 1971; Braybrooke, 1988); based on rock strength and discontinuity spacing
- Caterpillar (1988); primarily based on in-situ seismic wave velocity.

It is noted that none of the classification systems include factors such as competence of machine operator, maintenance condition of the machine, groundwater conditions, size of ripped product and production.

Rippability parameters selected for the design study are presented in Table 2. Conclusions drawn from the Class A prediction procedures (Lambe, 1973) were:
- Zone 1 is excavatable by using combinations of rippers and hydraulic excavators;
- slightly weathered sandstone of Zone 2 is generally rippable using powerful rippers; locally, some fracturing by hydraulic rock breaking will be necessary;
- fracturing by hydraulic rock breakers will be needed for fresh rock of Zone 2.

EXCAVATION

Excavation was carried out in May through July 1992. Ground conditions were found to be in general accordance with predictions, with no surprises. Water inflow into the excavation comprised mainly periodic rainfall and run-off, and minor groundwater.

Zone 1 material proved rippable by a Caterpillar D9G bulldozer equipped with a single shank. Supplementary hydraulic rock breaking was carried out by a Caterpillar 245B hydraulic excavator equipped with a Giant G120 hydraulic rock breaker and a 195 mm moil point. The G120 has a working mass of 5800 kg and an impact energy of 12000 J. Spoil removal was carried out by two hydraulic excavators. An average production rate of 130 m^3/h was achieved for a volume of about 35,000 m^3, essentially constant with time.

Two Cat-245B/G120 units provided the principal rock breaking equipment for Zone 2 material. Supplementary support was available from a smaller hydraulic rock breaker. Spoil removal was assisted by a D9G bulldozer and a hydraulic excavator. For this excavation set-up, an average production rate of 40 m^3/h, i.e. 20 m^3/h per Cat-245B/G120 unit was achieved. The excavation volume was about 15,000 m^3. The production rate decreased with time, due to stronger rock conditions and increased trimming activities.

The massive nature of Zone 2 led to a rough undulating surface finish for the base of the excavation, as rock fracturing took place across rock material rather than along discontinuities. Depressions were typically about 0.3 m deep.

REFERENCES

Braybrooke, J.C. (1988), "The State of the Art of Rock Cuttability and Rippability Prediction", Proc. 5th Australia-New Zealand Conf. on Geomechanics, Sydney, August, pp. 13-42.

Caterpillar Inc. (1988), "Handbook of Ripping:, 8th Edition, Peoria, Illinois, December.

Franklin, J.A., Broch, E. and Walton, G. (1971), "Logging the Mechanical character of Rock", Trans. Inst. Min. & Met. 80A, pp. 1-9.

Lambe, T.W. (1973), "Predictions in Soil Engineering", Geotechnique, Vol. 23, No. 2, pp. 149-202.

Weaver, J. (1975), "Geological Factors Significant in the Assessment of Rippability", Civ. Engr. in South Africa 17 (12), pp. 313-316.

WEAVER METHOD			
CLASSIFICATION	Fresh	SW[1]	MW-HW[2]
Seismic Velocity	26[3]	20	5
Rock Hardness	10	10	5
Rock Weathering	9	7	4
Defect Spacing	20	10	5
Defect Continuity	5	5	3
Defect Gouge	5	5	5
Strike and Dip	15	15	15
Total Rating	90	72	42
Rock Classification	I	II	IV
Assessment	Blasting	Extremely hard ripping and blasting	Hard ripping
MODIFIED FRANKLIN METHOD			
Rock Strength	Very Strong UCS 50 - 120 MPa	Very Strong UCS 50 - 90 MPa	Strong UCS 20 to 50 MPa
Fracture Spacing	100 - 1000 mm	100 - 300 mm	30 - 300 mm
Assessment	Blasting	Marginal ripping and blasting	Digging and ripping
CATERPILLAR METHOD			
Seismic Velocity	2200 - 3700 m/s	1500 - 2200 m/s	1500 m/s
Assessment	Blasting	D10N ripper	D9N/D8 ripper

Notes for Table 2:
1. SW = slightly weathered rock
2. MW-HW = moderately weathered to highly weathered rock
3. Values apply to Weaver rating system

Table 2: rippability parameters

The dip slope problem on the mining districts

Problèmes de glissement potentiel de formations argileuses de Ilan

Hongey Chen
Department of Geology, National Taiwan University, Taiwan

ABSTRACT: This paper focuses on the potential unstable dip slopes on the china clay mining districts at Ilan area. This problem was found by stereonet analysis of the discontinuities measuring. The results of experimental tests showed that the dip angle on the discontinuities is larger than the basic friction angle of the geomaterials. In terms of the slope stability, the stereonet analysis of discontinuities demonstrated that half of the four areas are in a potentially unstable condition. Only one area, the Tsailien mining district, has a lower than stable safety factor. The analytical results also indicate that some of the necessary conditions for the instability on the dip slope in that the dip angle of the slope surface should be larger than the dip angle of the slip surface and that the dip angle of the slip surface should be larger than the friction angle of the geomaterials.

RÉSUMÉ: Cette étude présente le problème de glissement potentiel en utilisant sur les pentes instables, dans les carrières d'argiles de la région d'Ilan, l'analyse de la fracturation (projection Schmidt). Les résultats expérimentaux des expériences mécaniques de roche montrent que l'angle de "glissement sur fracture" est supérieur à l'angle de friction. En terms de stabilité de pentes, l'analyse de la fracturation montre que deux des quatre régions étudiées sont potentiellement instables. Seule la carrière Tsailien présente des valeurs de sécurité. Les études analytiques indiquent que passoire les conditions nécessaires pour créer des pentes instables, l'angle de pendage de fracture est plus grand que l'angle de pente de glissement qui est lui même plus grand que l'angle de friction des matériaux.

1 INTRODUCTION

The stability analysis follows the typical failure models on the different soil and rock formations. The stability analysis on a soil formation is normally ran on various types of failure estimates--such as circular, non-circular, plain failure and wedge failure perspectives. For rock formations, plane and wedge failures are mainly used to analyze slope stability by the quantitative method (Bromhead 1986).

These types of analyses should be accompanied by the relevant geological data, the mechanical behavior of the geomaterials and the geomorphological characteristics gained from suitable investigation or measurements (Chen 1992). The actual slip surface will be precisely applied together with the analysis and planning work. If these factors are not carefully considered, the preliminary results could be unavailable to support proper planning for the slope engineering project.

The dip slope failure is a comparatively unusual sight in rock slopes. It is only occasionally that all the discontinuity conditions required, occur together sufficiently to produce an actual failure on such a slope (Hoek and Bray 1981).

This study focuses on the dip slope stability of the mining district of the china clay which are

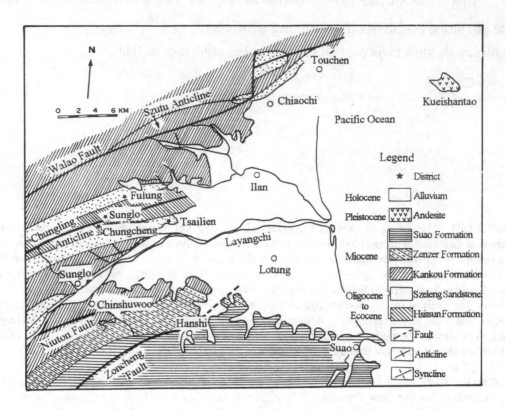

Figure 1   Geological map of the study area (Modified from Tang and Yang 1976)

included the Tsailien, Sunglo, Chungcheng and Fulung areas. These four mining districts are located on the west hillside of Ilan city. Most of the mining districts are situated in mountainous regions where little industrialization has occurred.

The site investigation and stability analysis are described below. This site investigation involved a major exploration and understanding of the geological distribution and the mechanical behaviour of the geomaterial.

2 GEOLOGY CONDITIONS

The study areas are situated in northeastern Taiwan. The four china clay mining districts of the study area, Tsailien, Sunglo, Chungcheng and Fulung, are mainly distributed on the Hsitsun formation and the Szeleng Sandstone formations from the Neogene Miocene age (Ho

1986). The Szeleng sandstone formation is mainly composed of white, coarse quartz sandstone and dark-gray slate or hard argillite. The Hsitsun formation is mainly composed of dark-gray slate. The slate sometimes has interbedded quartz sandstone. The china clay derived from the surrounding rocks in the Ilan area are mostly distributed on the Szeleng sandstone formation. Figure 1 shows the geological map of the study areas. Ho (1986) pointed out that most of the china clay in the mining districts are distributed on the sandstone formation.

2.1 Fulung and Tsailien mining districts

The Fulung mining district is located on the west side of the Ilan area. The Tsailien mining districts on the southwest side of the Ilan area and is nearest to the Ilan city center. The stratum

4064

of the two mining districts areas are distributed on the Szeleng Sandstone formation of the Neogene Miocene (Tsan 1976). The rock type is mainly derived from the characteristics of the white with coarse quartz sandstone. This Szeleng quartz sandstone is intervened with dark-gray slate or hard shale. The Szeleng quartz sandstone becomes large and rough sand after weathering, due to an uncemetation peculiarity. The effloresced rough sand easily comes off and is loosely arranged. However, the stratum has developed exfoliation and is often disintegrated after efflorescence.

The strike of the stratum is north to east from 23° to 70°, the dip is toward the south from 35° to 58°. The discontinuities are well developed. The area receives abundant rain and contains evidence of significant ground water, seepage and weathering. These water forces leave a rust coloration which becomes yellowish brown in some parts.

2.2 Sunglo and Chungcheng mining districts

The Sunglo and Chungcheng mining districts are situated in the western portion of the Ilan area were the mountainous elevation is at about 500 meters. The Sunglo mining district is at around 200 meters elevation and this area covers about 20 hectares. These mining districts are distributed on the Hsitsun formation from the early Oligocene (Tang and Yang 1976). The components of the rocks are dark-gray slate and phyllite slate; mixed with hard orthoquartzite sandstone. This sandstone is dark-gray and sized from medium to coarse.

The gray sandstone of the Hsitsun formation is

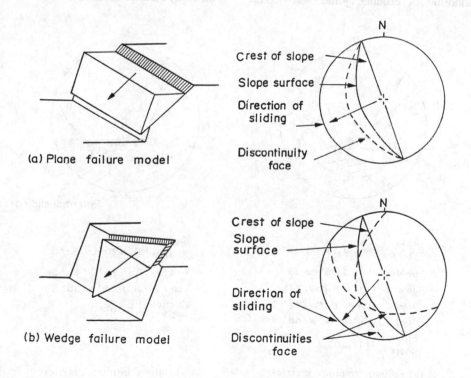

(a) Plane failure model

(b) Wedge failure model

Crest of slope
Slope surface
Direction of sliding
Discontinuity face

Crest of slope
Slope surface
Direction of sliding
Discontinuities face

Figure 2   The main failure models of rock slopes with its corresponding stereonet diagrams

well cemented and the quality of the rock is hard and solid. The dark-gray slate and the phyllite slate are also very hard and solid. Their grain arrangement is tightly compacted. Where these formations contain well developed foliation and large portions of clay, the clay is easily softened from hydration (Chung et al. 1975). The water seepage leaves the discontinuities of the stratum with a rust colored discoloration.

The slate has a well developed foliation. The stratum set is from about 66° to 72° northeast and dips toward the south 42° or dips toward the north 32° as caused by the Chung-Ling Anticline.

3 DISCONTINUITIES ANALYSIS

The geological attitude was measured in onsite investigations and applied in the stereonet method. The geological attitudes include the discontinuities of bedding, joints and diverse slope surfaces. These discontinuities comprise various geometric shapes and set up the possibility of various potential failure models at the points of intersections of these discontinuities (Figure 2). The cross section of the stratum was selected and drawn by tracing the position of the potential failure (Leonards 1982). Therefore, the number of safety factors evident on the slope stability analysis could be totaled from the cross section of the stratigraphy distribution. The stability problems and their necessary solutions could then be considered and discussed.

Geometric analyses of potential collapse modes were examined by their respective configurations based on the field data. These geometric types of discontinuities revealed themselves as potential failure models and clearly showed a capacity for instability or even collapse of the slope (Hencher 1987). The involved stereonet analysis on this structural data revealed the possible directions of displacement of the study areas as depicted in Figure 3.

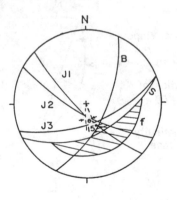

Slope
Surface (S): N 64°E / 54° S

Bedding (B): N 25°E / 58° S

Joint 1 (J1): N 48°W/ 74° S

Joint 2 (J2): N 62°W/ 72° S

Joint 3 (J3): N 66°W/ 58° S

Friction
angle (f) : 30°

(a) Fulung mining district

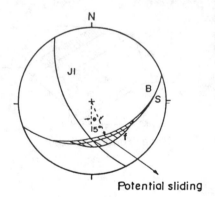

Potential sliding

Slope
Surface (S): N 70°E / 32° S

Bedding (B): N 70°E / 35° S

Joint 1 (J1): N 38°W/ 65° S

Friction
angle (f): 30°

(b) Tsailien mining district

Figure 3 The diagram of stereonet analysis of study areas

4 ENGINEERING PROPERTIES

In this project, around 20 samples were taken from each slope at different mining district above. 5 samples were taken from each geomaterial formation. Each 5 samples was taken variously from the top, middle and toe of each study slope. The nature and properties of the samples taken from the different parts of each slope could than be compared. The engineering properties test include physical properties, slake durability index and shear strength test. The results of physical properties show that the water content of china clay is quite low which around 5.51%. The unit weight is around 2.62 kg/cm^3. The specific gravity is around 3.03 and the porosity is around 17.78% (Table 1). These physical properties were tested by following the procedure of the BS 1377:1975.

Table 1 The mean results of the physical properties at different mining districts

Study area	Wn (%)	γt (g/cm^3)	G	n (%)
F.	4.78	2.63	3.02	16.83
T.	5.82	2.57	3.05	18.07
S.	5.93	2.64	3.03	18.32
C.	5.72	2.62	3.02	17.90
Average	5.5	2.62	3.03	17.78

* F. : Fulung mining district
  T. : Tsailien mining district
  S. : Sunglo mining district
  C. : Chungcheng mining district
  Wn: Water content
  γt: Unit weight
  G: Specific gravity
  n: Porosity

The slake durability index normally offer a quantitative method of discriminating between the rock and soil geomaterial. The second cycle yielded a result of 95.1% durability. It means that the intact rock involved the china clay has a very high slake durability property on the Gamble (1971) classification. In general, rocks getting lower slake durability results should be subjected to the soils classification. Franklin and Chandra (1972) pointed out that the indexes taken after three or more cycles of slaking process may be useful when evaluating rock samples of higher durability.

The china clay was evidently produced from the surrounding rocks deposited in the slope of the mining district. Given the hazard potential of elevated rocks on a slope apt to slip after excavation, it is crucial to accurately understand the shear strength of the materials involved. The direct shear equipment was used to perform the standard shear test on the rock. The highest applicable weight of the direct shear equipment is 5 tons in this study. The equipment was calibrated right before the experiment. The diameter of the drilling core rock samples was 5.50 cm. The standard filler of gypsum was poured and allowed to dry to fill out the space in the shear box. The driving speed used in this test was 0.215 mm/min (Head 1984). All the standard test procedures suggested in the ISRM (1981) were followed in the execution of this experiment. The normal stress was applied at 83 kN/m^2, 168 kN/m^2, and at 253 kN/m^2. These changes of normal stress were applied on the sample to simulate different in-situ normal depths.

The experimental tests of shear box involved on both of the basic frictional angles and frictional angles of the various rock types. The basic frictional angle was obtained by resistance of the cutting surface on the two cut samples of the intact rock. The frictional angle was obtained by the resistance of the slip surface on the sample's discontinuities of the different rocks. The discontinuity surface along which shearing occurs in exactly parallel to the direction of the shear stress. An extremely important aspect of shearing on discontinuities which are inclined to the direction of the applied shear stress z is that any shear displacement u must be accompanied by a normal displacement v. Patton (1966) and Barton (1976) found that the inclination of the bedding plane trace was approximately equal to the sum of the average angle i and the basic friction angle found from laboratory test on planner surfaces.

4.1 The Fulung mining district

In Fulung mining district, the results of experimental tests with the direct shear test equipmental reveals that the basic friction angle of the slate is ranging between 30º to 36º and none cohesion. The shale is ranging between 30º to 34º and the cohesion is ranging between 2.6 to 3.4 kN/m². The basic friction angle of the quartz sandstone is ranging between 48º to 54º, with none cohesion. The friction angle of the bedding plane on the slate and shale interbedded, one of the discontinuity, ranges from 32º to 36º and cohesion ranges from 2.2 to 3.0 kN/m² (Table 2). The water content of slate is around 4.83% and the specific gravity is around 3.04. The water content of shale is around 4.95% and the specific gravity is around 2.99. The water content of quartz sandstone is around 4.54% and the specific gravity is around 3.03. The water content of slate and shale interbedded is around 4.79% and the specific gravity is around 3.01.

Table 2 The results of engineering properties at Fulung mining district

Rock	G	Wn	C	φ
Slate	3.04	4.83	0	30-36
Shale	2.99	4.95	2.6-3.4	30-34
Q. sandstone	3.03	4.54	0	48-54
Sl/Sh	3.01	4.79	2.2-3.0	32-36

* G: Specific gravity
  Wn: Water content, %
  C: Cohesion, kN/m²
  φ: Friction angle, degree
  Q. sandstone: Quartz sandstone
  Sl/Sh: Bedding plane on both the slate and shale interbedded

4.2 The Tsailien mining district

In Tsailien mining district, the results of experimental tests with the direct shear test equipmental reveals that the basic friction angle of the slate is ranging between 22º to 25º and that the argillite is ranging between 23º to 26º.

There are none cohesion. The basic friction angle of the quartz sandstone is ranging between 25º to 29º, with none cohesion. The friction angle of the bedding plane on the slate and argillite interbedded, one of the discontinuity, ranges from 30º to 36º and cohesion ranges from 2.5 to 3.0 kN/m² (Table 3). The water content of slate is around 5.60% and the specific gravity is around 3.05. The water content of shale is around 5.93% and the specific gravity is around 3.03. The water content of quartz sandstone is around 5.72% and the specific gravity is around 3.02. The water content of slate and argillite interbedded is around 6.20% and the specific gravity is around 3.07.

Table 3 The results of engineering properties at Tsailien mining district

Rock	G	Wn	C	φ
Slate	3.05	5.60	0	22-25
Shale	3.03	5.93	0	24-28
Q. sandstone	3.02	5.72	0	34-38
Sl/Argi.	3.07	6.02	2.5-3.0	30-36

* G: Specific gravity
  Wn: Water content, %
  C: Cohesion, kN/m²
  φ: Friction angle, degree
  Q. sandstone: Quartz sandstone
  Sl/Argi.: Bedding plane on both the slate and argillite interbedded

In general, the shear strength of the geomaterial is deeply related with its physical properties (Horn and Deere 1962). The low water content of the china clay could have higher shear strength than high water content . The influence of the cohesive and frictional properties of the rock discontinuities depends upon the nature of the filling or cementing material.

5 QUANTITATIVE STABILITY ANALYSIS

The dip slope failure type to analysis is where sliding occurs along a single plane (Figure 4).

The safety factor is calculated by resolving all forces along the failure plane and dividing the forces resisting sliding by the forces inducing sliding (Hoek and Bray 1981). Figure 3 shows that the geometry of the slop slip surface is well defined then stability can be assessed graphically by constructing a force diagram.

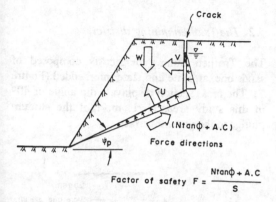

$$F = \frac{Ntan\phi + A.C}{S}$$

Factor of safety

Figure 4 Force diagram for slope stability analysis

$$F = \frac{C \times A + (Wcos\phi_\rho - U - Vsin\phi_\rho) \times Tan\phi}{Wsin\phi_\rho + Vcos\phi_\rho}$$

Where

C: Cohesion, kN/m^2
ϕ_o: Failure angle, degree
A: Length of slip surface, m
W: Weight of failure slope, t/m
U: Uplift force, t/m
V: Lateral water pressure, t/m

The choice of appropriate shear strength values for use in a rock slope design depends upon a sound understanding of the basic mechanics of shear failure and of the influence of various factors which can alter the shear strength characteristics of a rock mass (Hencher 1987). Fulung and Tsailien mining districts were chosen for a slope stability analysis after reviewing the above four stereonet analysis. These two areas showed significant hazard potential. This slope stability analysis used the plane failure mode to count the data related to the safety factor.

The safety factor counting of the stability analysis on the plane failure model was considered using a four way analysis of various. The four conditions were: saturated/earthquake, saturated / non-earthquake, unsaturated/earthquake and unsaturated/non-earthquake conditions.

5.1 The Fulung mining district

The Fulung mining district is mainly composed of slate, shale, quartz sandstone, slate and shale interbedded (Figure 5). The quartz sandstone has better permeability than the slate and shale interbedded. The slate and shale interbedded easily becomes soft and loose after submergence in water. The slate and shale interbedded formation could create a new failure surface with relatively normal increases in water saturation. The evidence strongly suggests a high possibility that a landslide of this slate and shale interbedded formation could occur in the reasonably near nature.

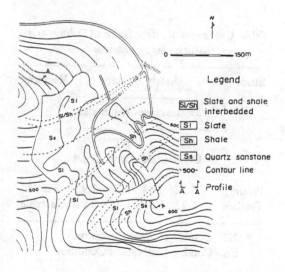

Figure 5 The geological map of Fulung mining district

4069

Fulung area has a 64° dip angle toward the south. Most of the stratum plunge to the ground surface (Figure 6). The slate and shale interbedded formation only displays on the outcrop of the ground surface and presents the potential failure formation.

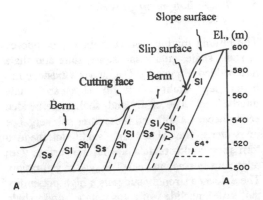

Figure 6   The geological profile of Fulung mining district

Table 4   Results of safety factor at Fulung and Tsailien mining districts

Study area	Safety Factor				Remarks
	Non-water state		Water state		
	N1. Cond.	Eq. Cond.	N1. Cond.	Eq. Cond.	
Fulung	3.52	3.04	2.28	1.93	Stable
Tsailien	1.22	1.08	1.12	0.97	Unstable

* N1. Cond. = Normal Condition
  Eq. Cond. = Earthquake Condition

The results of the safety factor are listed in the Table 4. In general, the safety factor of this mining district scored higher than the standard data previously available. The safety factor was 1.93 for a saturated earthquake condition. This safety factor result is higher than the security requirement on the excavated portions of the mining district. Therefore, it is possible to say that this area is a reasonably safe area under the conditions studied. Excavations in this study area should not result in instability when carried out with normal procedures and care.

5.2 The Tsailien mining district

The Tsailien mining district is composed of sandstone, argillite and slate interbedded (Figure 7). The cross section displays a dip angle of 40° in this study area. Here, most of the stratum plunge to the ground surface.

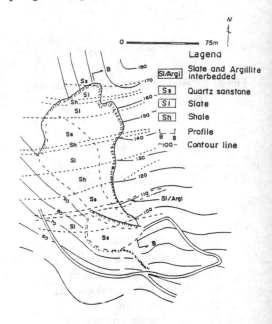

Firure 7 The geological map of Tsailien mining district

The mining district of the study region has excavated areas exposing 20 meters of sandstone, argillite and slate interbedded--overhanging the excavations done at ground level. As can be seen in Figure 8, such cut slopes at the ground level of the stratum will easily create a plane failure by slipping surfaces. That is, the sandstone bedding plane will easily slip down the slope over the argillite and slate bedding planes--particularly in saturated states.

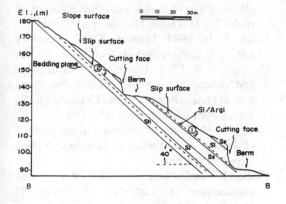

Figure 8   The geological profile of Tsailien mining district

The results shown in Table 4 reveal that the safety factor was under the critical standard value of 1.0. This indicates that the slope is in a stable condition under normal unsaturated, non-earthquake condition. However, when a saturated and/or earthquake condition exists, then special attention must be paid to these hazardous areas.

6 DISCUSSIONS

Determination of reliable shear strength values is a critical part of a slope design because relatively small changes in shear strength can result in significant changes in the safe height or angle of a slope. The determination of the orientation and inclination of the discontinuities for subsequent inclusion in stability analysis is important as is the sampling for shear strength testing. The slake durability index test is intended to assess the resistance offered by a rock sample to weakening and disintegration when subjected to two standard cycles of drying and wetting. The strength characteristics of the individual discontinuity surface as well as that of the overall rock mass. If the presence of the discontinuities is important to suspected, a special effort should be made during the site investigation programme to check whether they do exist.

The stratum of the mining district and their discontinuities appear to readily result in dip slope and wedge type failures--particularly in the presence of saturation and earthquakes. For the sake of a stable environment, the mining district should carry out its excavations stage by stage in the following manner.

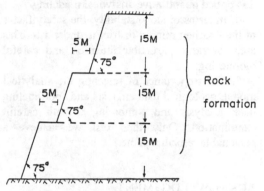

Figure 9   A schematic diagram for the rock cutting slope

The cut procedure should follow the angle of the stratum formation. That is, cut and excavate the top layer of the strata, in line with the angle of the strata. This should be done without cutting below the strata until all of that strata has been removed. The removal of the top strata should be to a degree beyond any possibility of creating a hazard when the strata below it is cut (Figure 9). This is the most economical method for cutting the slope. This is particularly true when one considers the high costs in resources and lives a hazard can cause in a few seconds of disaster. Part of this safety procedure should include a proper drainage system to help provide a necessary decrease in the pore water pressure within the slope during periods of precipitation. Such a drainage system may not always prevent a hazard in all storm conditions but should lesson the cases where moisture alone would cause a hazard.

7 CONCLUSIONS

1. The China clay of the Ilan area contains 5.61% water. The unit weight is 2.60 kg/cm³. Porosity is 17.90% and the slaking durability factor is 94.3%.

2. From the investigation of the geological attitude and stereonet analysis, we can comprehend the possible unstable area created by natural discontinuities.

3. The importance of understanding in detail the morphology of the failure surface in order to do applied quantitative analytical modeling.

4. In terms of slope stability, the safety factor of the Tsailien mining district is under 1.0. This area warrants particular attention and careful monitoring.

5. The importance of selecting the analytical model, geological data analysis and coordinating back analysis and combining it with careful examination. Only then can we achieve a reasonable remedial plane.

ACKNOWLEDGEMENT

The author would like to express his thanks to Ms. Hsu for the drawings and Ms. Yu for the correcting in this manuscript. This research project is partilly funded by the Taiwan Province Mining Bureau and National Science Council, Republic of China.

REFERENCES

Barton, N. 1976. The shear strength of rock and rock joints. Intl. J. Rock Mech. Min. Sci. Geomech., V.13. 9: 225-279.

Bromhead, E. N. 1986. The stability of slope. Surrey University Press.

BS 1377, 1975. Methods of test for soils for civil engineering purposes: British Standards Institution, London.

Chen, H. 1992. The stability and mechanical behaviour of colluvium slopes. Jl. Geol. Soc. of China, V.35. 1: 95-114.

Chung, C. T., Yang, Y. T. & Wu, Y. C. 1975. Geological investigation at Wulai-Ilan New cross road. Mining Tech. V.13. 9: 377-383.

Franklin, J. A. & Chandra, R. 1972. The slake-durability test. Intl. J. Rock Mech. Min. Sci. 9: 325-341.

Gamble, J. C. 1971. Durability-plasticity classification of shales and other argillaceous rocks. Ph. D. Thesis, University of Illinois.

Head, K. H. 1984. Manual of soil laboratory testing. V.2. Pentech Press, London.

Hencher, S. R. 1987. The implications of joints and structures for slope stability, In: Slope stability (Anderson M. G., and Richards K. S., Eds.,), John Wiely & Sons, 145-186.

Ho, C. S. 1986. An introduction to the geology of Taiwan. The Ministry of Economic Affairs, Republic of China.

Hoek, E. and Bray, J. 1981. Rock slope engineering: The Institution of Mining and Metallurgy, London.

Horn, H. M. and Deere, D. U. 1962. Frictional characteristics of minerals. Geotechnique. 12: 318-319.

ISRM 1981. Rock characterization testing and Monitoring, (Brown, E. ed.) Pergaman Press, Oxford.

Leonards, G. A., 1982. Investigation of failures: J. Geotech. Engng. Div., ASCE. V.108. GT2: 187-246.

Patton, F. D. 1966. Multiple modes of shear failure in rock and related materials. Ph. D. Thesis University of Illinois.

Tang, C. H. and Yang, C. Y. 1976. Mid-Tertiary stratigraphic break in the northeast Hsuehshan range of Taiwan: Petrolum Geology of Taiwan, 13: 139-147.

Tsan, S. F. 1976. Neogene geology at Ilan area. Mining Tech. V.14. 7: 252-257.

Excavated slope stability in lateritic soils: Contributions of natural factors

Stabilité de talus d'excavation dans des sols latéritiques: Contribution des facteurs naturels

A. B. Paraguassú
University of São Paulo, São Carlos, Brazil

S. A. Röhm
Federal University of Viçosa, Brazil

ABSTRACT: A study of the mechanism of hardened crust formation, that occurs in exposed surfaces of man-made slopes in laterised cenozoic sediments, at the central region of the State of São Paulo, Brazil, shows the crust and its hardening evolution. The quick begining and developement of the slope surface cementation are a natural erosion protection of this perculiar material. This surface strenght reaches values several times higher than those for the sediments in their original unexposed conditions.

This paper presents conclusions about data from several road cuts and from an experimental cut excavated in those sediments, and their physical conditions during many years of weathering exposition.

RESUMÉ: La formation d'une croûte cimentante et sa evolution en talus et en superficies exposées de sols lateritiques de la region centrale de l'état de São Paulo, Brasil, a été etudiée par des mesures efectueés sur quelques talus de routes et sur un talus experimental. La rapide formation de cette croûte cimentante est importante car elle est capable d'améliorer la estabilité des talus, principalement sa résistance à l'érosion. Il est montré que la résistance de cette croûte surpasse beaucoup la résistance du sol d'origine et il est montré aussi quelques conclusions sur les conditions physiques de cette croûte depuis plusiers années.

1 INTRODUCTION

Hardened crusts formed at the exposed surfaces of laterised sandy sediments slopes in the Paraná Basin are a common and systematic phenomenon (Paraguassú, 1972; Rodrigues 1982; Vilar et al., 1986; Paraguassú & Röhm, 1992; and Gaioto & Queiroz, 1993).

It plays a significant role in preventing the erosion of slope surfaces produced during road construction, which are directly responsible for the stability of the soil mass.

The present investigation was conducted in several excavations in cenozoic sediments (TQ) in the central region of the State of São Paulo (Fig. 1). This region is of great economic importance and, as such, possesses a dense road network.

These sediments occupy an extensive area of the State of São Paulo and complete the stratigraphic sequence; they always occur in erosional unconformity with respect to other regional geologic units.

The average geotechnical characteristics of cenozoic sediments of the studied slopes are as follows: specific mass of solids: 2.8 g/cm^3; void ratio: 0.9; liquid limit: 38%; plasticity limit: 24%; and permeability coefficient: 5 x 10^{-4} cm/s. Grain-size distribution, according to ASTM, shows 8% of medium sand, 45% of fine sand, 19% of silt and 28% of clay. Mineral composition of these materials is as follows: quartz: 54%; iron and aluminum oxides and hydroxides: 17%; kaolinite: 15%; gibbsite: 12%; and others: 2%. The average chemical characteristics are shown in Tabel I.

These data are typical of materials having a non expansive behavior due to mineralogy as well as due to low cationic capacity. From pedologic point of view, the materials correspond to well evolved ferralitic soils of low fertility with predominance of negative charges on particle surfaces.

2 ORIGIN OF THE CRUST

The process of origin of hard crusts starts soon after the execution of a cut in the sediments, which remain exposed to the agents of weathering resulting from interaction between divers factors: lithological, geomorphological, climatic, hydrological and pedological (Paraguassú & Röhm, 1991; Paraguassú & Röhm, 1992).

The regional climate is of Cwb type according to Koeppen (Oliveira & Prado, 1984; Mattos, 1982). Pluviosity is limited by isohyets of 1200 mm and 1500 mm, with dry periods between May and September, corresponding to values of about 18.5% of the total annual precipitation. The average temperatures of the hot and cold month are, respectively, 18.1° and 23.1°C (DAEE, 1974).

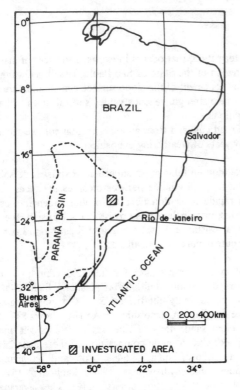

Fig. 1 - Location map of investigated area

The manner in which water percolates through the soil mass is the main factor in the superficial cementation of the studied slopes (Paraguassú, 1972; Vilar et al., 1986). The cementation develops rather quickly and is associated to the alternate rainy and dry periods with insolation (Paraguassú, 1989; Paraguassú & Röhm, 1991).

In this investigation, field observations on experimental slopes were conducted in order to understand how the hardening process initiates and how the factors responsable for it interact among themselves during crust evolution.

3 EXPERIMENTAL WORKS

Field observations were conducted in various slopes located at road construction sites from which information on geological and geotechnical characteristics of excavated soil mass and the execution schedules were known.

In the majority of slopes, inclinations were greater than those that Classical Soil Mechanics would recommend for such construction. Futhermore, no erosive processes were observed on the slopes, as expected in view of the pluviometric regime of the region.

In some of the "old" slopes, the crusts are quite resistant, similar to sandstones with thickness of up to 80 mm although thicknesses of few millimeters are more common. The accelerated rate at which the crusts develop makes it possible to maintain the original geometry of the slope, even in the most inclined ones. This fact was confirmed by the observation of many stable slopes in the region, which were constructed more than twenty years ago and which still exhibit scars left by the earthmoving machinery during construction, establishing that surface cementation occured soon after excavation (Paraguassú, 1989).

The construction of experimental slopes followed the same geometric and geologic characteristics as those found in various excavations in the region.

Table I - Average chemical parameters of sediments in experimental slopes

chemical parameter	meq/100g soil	chemical parameter	meq/100g soil
Na	0.043	Al^{+3}	0.128
K	0.038	$(H^+ + Al^{+3})$	1.76
Ca	0.095	S^*	0.199
Mg	0.023	CEC	1.9059
m^{**}	39.1 %	V^{***}	10.2 %

* Na + K + Ca + Mg
** percentage base saturation
*** exchangeable bases

The construction of experimental slopes followed the same geometric and geologic characteristics as those found in various excavations in the region. This permitted the direct and representative measurements of the phenomenon and observation of its evolution (Paraguassú & Röhm, 1992). The measurements of temperature of the sediment were made with thermocouples installed in a continous manner down to a depth of 0,8 m. The variation of matric potential, a function of moisture content, was measured by means of tensiometers with the porous bulbs installed at a depth of 20 cm in a vertical section of the slope. The variation of moisture content from surface to freatic surface was measured by a neutron probe (501 DR - Hydroprobe moisture gauge). The evolution of the mechanic resistence to penetration of the exposed surface of the slopes was measured with the field mini-CBR device (Röhm & Correa, 1980).

4 CONCLUDING REMARKS

The soil water transference processes associated to the thermal gradients and the evaporation and suction phenomena result in the alternate moviments of water near the surface (Philip & De Vries, 1957; Raudkivi & Van U'u, 1976; Geraminegad & Saxena, 1986; Nielsen et al., 1986), where the soluble material is deposited and the hardening takes place.

The soil water, subjected to diurnal and nocturnal thermal gradients, moves repetitively from the surface into the soil mass or in the opposite direction, tending to the condition of energy equilibrium. This phenomenon was quantitatively established on the basis of data from equipment installed in the soil mass. When the surface is heated, the concentration of the vapor phase tends to increase and consequently an efflux to the atmosphere occurs. A movement also occurs toward the colder interior of the soil mass, where part of the vapor condenses. During hotter periods of the day the matric suction increases while it diminishes during falling temperatures. During rain, the matric suction diminishes depending upon the amount of rainfall. One of the basic factors that determines the natural cementation of the slopes in sandy soil mass is the volume of water (solution) evaporated from the surface. The position of the slope with respect to the sun determines the intensity of evaporation and thus the rate at which the cementation process advances.

As a quantitative example of the phenomenon of the natural hardening of the crust developed in an experimental slope constructed in the field with the geotechnical characteristics previously referred (Paraguassú & Röhm, 1992), the evolution of its mechanical resistance to penetration is as follows: soon after the excavation of the soil mass (before the formation of crust), 0.25 MPa to 4 MPa, after 18 months. This shows a 16 fold gain in the original strength.

This mechanism of cementation progresses, as follows: the diurnal and seasonal variations of temperature and rainfall cause the movements of water in the soil mass as a whole, especially at exposed surface where the evaporation takes place and leaves there the dissolved materials. At the same time, also due to solar heating, there occur the hardening of the fine materials and oxides which are present together with the sand fraction occurs.

REFERENCES

Departamento de Águas e Energia Elétrica e Faculdade de Filosofia Ciências e Letras - USP. 1974. Estudo de águas subterrâneas - região administrativa 6 - Ribeirão Preto. São Paulo, 4 v.

Gaioto, N. & Queiroz, R.C. 1993. Taludes naturais em solos. In: Solos do Interior de São Paulo. Associação Brasileira de Mecânica dos Solos, São Paulo: 207-242.

Geraminegad, M. 7 Saxena, S.K.1986. A coupled thermoelastic model for saturated-unsaturadet porous media. Géotechnique, 36:539-550.

Mattos, A. Método de previsão de estiagens em rios perenes usando poucos dados de vazão e longas séries de precipitação. Tese de Doutoramento, Escola de Engenharia de São Carlos - USP, 182 p.

Nielsen, D.R.; Van Genuchten, M.T. & Biggar, J.W. 1986. Water flow and solution transport porocesses in the unsaturated zones. Water Resourses Research, 22:157-160.

Oliveira, J.B. & Prado, H. 1984. Levantamento semidetalhado do Estado de São Paulo: quadrícula de São Carlos - II memorial descritivo. Bol. Tec. Inst. Agronom., Campinas - SP, 98, 188 p.

Paraguassú, A.B. 1972. Experimental silicification of sandstone. Geological Society of America Bull., 83: 2853-2858.

Paraguassú, A.B. 1989. A cimentação natural como processo de estabilização de taludes arenosos: I Simpósio de Geologia do Sudeste da Sociedade Brasileira de Geologia:198.

Paraguassú, A.B. & Röhm, S.A. 1991. O movimento da água no solo sob efeito da temperatura e a sua influência na cimentação de superfícies de sedimentos arenosos: II Simpósio de Geologia do Sudeste da Sociedade Brasileira de Geologia:309-313.

Paraguassú, A.B. & Röhm, S.A. 1992. Evolução da resistência da superfície de taludes em sedimentos cenozóicos provocada pela cimentação natural. Geociências, São Paulo, 11(2):181-190.

Philip, J.R. & De Vries, D.A. 1957. Moisture movement in porous materials under temperature gradientes. Trans. Amer. Geophys. Union, 38:222-232.

Raudkivi, A.J. & Van U'u, N. 1976. Soil moisture movement by temperature gradient. Jour. Geotechnical Engineering Division of ASCE, 102:1225-1244

Rodrigues, J.E. 1982. Estudo de fenômenos erosivos acelerados (Boçorocas). Tese de Doutoramento, Escola de Engenharia de São Carlos - USP.

Röhm, S.A. & Correa, F.C. 1980. Ensaio de mini-CBR de campo: XV Reunião Anual de Pavimentação da Associação Brasileira de Pavimentação, 13p.

Vilar, O.M.; Rodrigues, J.E.; Bjornberg, A.J. & Paraguassú, A.B. 1986. The phenomenon of silicification in slopes: V International Congress of International Association of Engineering Geology, Vol. 3.4: 931-932.

Slopes stability at Alpetto quarry (Northern Italy)
Stabilité des pentes de la carrière Alpetto (nordest d'Italie)

A.Clerici
Earth Sciences Department, Milan University, Italy

L.Griffini
Milan, Italy

E.Rey
Cementeria di Merone SpA, Milan, Italy

ABSTRACT: The need for an optimized exploitation of a marly limestone deposit in the district sites of Cesana Brianza and Suello (Como, Northern Italy) in safe conditions for both operators and residential centres located nearby, led the licencee Company to plan and carry out an extensive, non conventional study on the stability of the solpes. The study, concerning both the whole crop area and single rock parts, is constantly updated according to present conditions and allows to forecast future conditions on the basis of the excavation plan adopted.

RÉSUMÉ : Afin d'optimiser l'exploitation d'un dépôt de calcaire marneux dans les sites de Cesana Brianza et de Suello (Como, Italie du nord) en des conditions de sécurité tant pour les opérateurs que pour les centres résidentiels situés dans la zone, la Société titulaire de la licence d'exploitation décida de planifier et d'exécuter une étude extensive non conventionelle sur la stabilité des versants. Concernant aussi bien la globalité de la zone de culture que des parties de roche individuelles, cette étude est constamment mise à jour suivant les conditions actuelles et permet la prévision des conditions futures sur la base du plan d'excavation adopté.

1 INTRODUCTION

The need for optimizing the exploitation of the deposit of marly-limestone for concrete located in the Cesana Brianza and Suello (Como) district in safe conditions for operators and for the uninhabited centres of that area led the contractor company - i.e. Cementeria di Merone S.p.A. - to plan and execute a non conventional and continuous in-depth study on the stability of exploitation fronts (fig. 1).

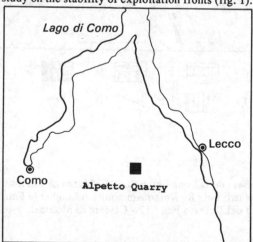

Fig. 1 - Location of the studied area

The study was begun in 1987 and is still being carried out both on the whole exploitation area (global stability) by using a numerical model and on single rock parts considered in detail (local stability) with a limit equilibrium criteria.

Regular site inspections, measurements taken by on-site equipment and constant updating of the calculation model enable a safe conduction of mining activity and, besides that, they also enable researchers to foresee the conditions that will be found, from time to time, as the adopted exploitation plan progresses and, where needed, to make partial modifications of the plan.

2 STUDY METHODS

As regards global stability analysis, the study approach foresees several subsequent stages, according to the following operation plan:
a) geomechanical surveys on excavation fronts in order to acquire basic parameters on the rock mass;
b) executing low-depth seismic profiles for measuring dynamic parameters of elasticity; c) setting an initial finite difference (EFD) model for foreseeing as far as possible the stress-deformation conditions within the rock mass and for establishing an optimal location for direct measurement bases; d) executing dilatometer tests and door-stopper tests in-hole; e)

laying multiple extensimetric bases and pressure cells inside holes; f) executing triaxial, monoaxial and ultrasonic laboratory tests on rock material and shear tests on discontinuity; g) implementing the finite difference model by inputting all directly mea-sured data on the resistance features of the rock material and of discontinuities, and on the deforma-tion parameters, and stress and deformation conditions of the rock mass.

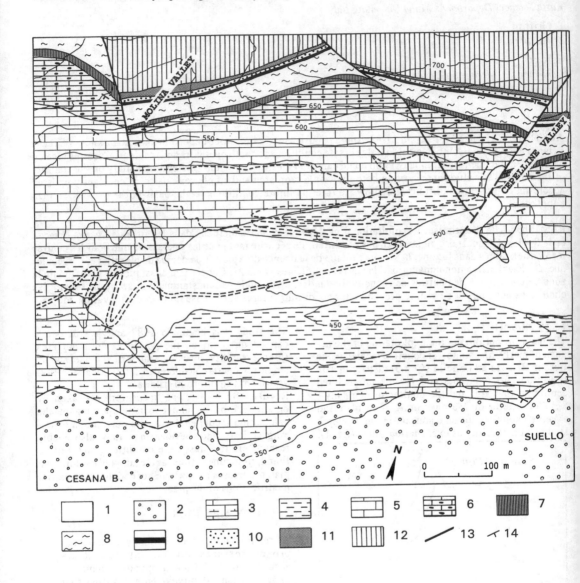

Fig. 2 - Geological pattern

1 - Debris of anthropic origin; 2 - Moraine deposits; 3 - Sass de la Luna Formation; 4 - Marna di Bruntino Fm.; 5 - Maiolica Fm.; 6 - Rosso ad aptici Fm.; 7 - Radiolariti Fm.; 8 - Rosso Ammonitico Lombardo Fm.; 9 - Sogno Fm.; 10 - Calcare di Morbio Fm. ; 11 - Calcare del Domaro Fm. ; 12 - Calcare di Moltrasio Fm. ; 13 - Main faults; 14 -Bedding dip.

As regards local stability, the study approach was the following:

a) shootings with a high resolution thermographic technique (HRT) for preliminary detection of released bodies on the exploitation front; b) detecting and characterizing rock portions showing evidence of instability by executing very detailed surveys, even on the wall by using mountain climbing techniques; c) calculating stability conditions by using standard procedures (limit equilibrium analysis); d) designing consolidation techniques.

3 GEOLOGICAL PATTERN

The lithostratigraphic units outcropping in the exploitation area are part of the sedimentary succession of the Lombard Basin of Southern Alps in a time interval between upper Jurassic (Hettangian) and lower Cretaceous (Cenomanian) ages.

In particular, the borelog series includes the following formations - from most ancient to recent times: Calcare di Moltrasio (Upper Hettangian - Upper Sinemurian), Calcare del Domaro (Carixian - Lower Domerian), Calcare di Morbio (Domerian - Lower Toarcian), Sogno (Lower and Middle Toarcian), Rosso Ammonitico Lombardo (Upper Toarcian - lower Bajocian), Maiolica (Upper Titonian - Lower Aptian), Marna di Bruntino (Lower Aptian - Middle Albian) and Sass de la Luna (Upper Albian - Lower Cenomanian).

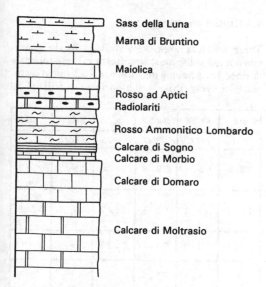

Fig. 3 - Typical stratigraphic sequence

The cover consists of moraine deposits and debris deposits due to anthropic activity.

From a structural point of view, the mine is located on the southern side of the anticlinal of Mount Cornizzolo/Mount Rai, with an East-West axis, a slight westward inclination and an immersion in the North.

The presence of faults almost perpendicular to the fold implies the presence of wide rock bodies and interrupts the regularity of the plicative motive.

The geological pattern of the area is summarized in fig. 2, while the stratigraphic series is shown in fig. 3.

4 SITE INVESTIGATION AND LABORATORY TESTS

4.1 Geomechanical surveys

The first stage of surveys consisted in executing a series of detailed geomechanical surveys (Clerici et al., 1988, 1990). These surveys enabled an initial characterization of rock masses, a preliminary detection of geomechanical units (fig. 4) and a detailed plan to be drawn for surveys, on-site tests and laboratory tests.

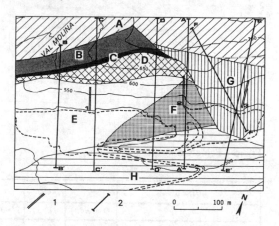

Fig. 4 - Lithological-technical pattern
1 - Measure axes; 2 - Numerical model sections; A-H - Geomechanical units

4.2 Laboratory tests on rock material

Tests were carried out on the five lithotypes more directly concerned by the exploitation, namely: marly calcilutite (Maiolica), marly cherty limestone (Rosso ad Aptici), calcareous chert (Radiolariti), nodular marly limestone (Rosso Ammonitico Lombardo) and marly limestone (Marna di Bruntino).

The following tests were carried out: determination of volume weight, measuring the dynamic modulus of elasticity and dynamic Poisson ratio, monoaxial and triaxial compressive tests with determination of elastic-static parameters of deformability. The results of the tests are summarized in table 1.

On the basis of monoaxial and triaxial compressive test results, failure envelops were calculated according to the Hoek-Brown (1980) criterion:

$$\sigma'_1 = \sigma'_3 + (m_i \, Co \, \sigma'_3 + Co^2)^{1/2}$$

As is known, besides defining the m_i parameter on the $\sigma1/\sigma3$ plan (see fig. 5), this criterion enables the failure envelop to be calculated on the σ/τ plan (see fig. 6) and, therefore, also enables calculating instantaneous cohesion and instantaneous friction angle values to be used in stability analyses in function of the stress condition on site.

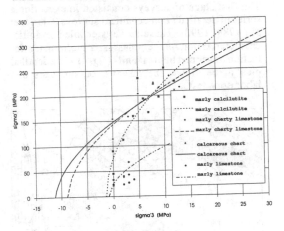

Fig. 5 - Principal stresses at failure

4.3 Laboratory tests on natural discontinuity

Direct shear resistance tests were carried out on the discontinuity plans of the joint sets neatly prevailing (layer plans) and more directly influential on the stability conditions of the mining wall both for their orientation with respect to the front and for their high persistence. Layer joints of Radiolariti and Maiolica were tested with and without the presence of clayey filling material.

Table 2 summarizes the shear resistance features of these discontinuity sets.

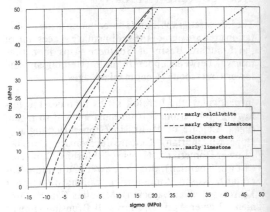

Fig. 6 - Failure envelops

4.4 On-site measures

Three measure axes were defined in the points considered as the most significant in three formerly defined areas having uniform geomechanical features. On every axis, a sub-horizontal drill of 50

Unit	γ	Co	σ	mi	Et50	Es50	μ	Edin	din
	[kN/m^3]	[MPa]	[MPa]	[-]	[GPa]	[GPa]	[-]	[GPa]	[-]
marly calcilutite (MAIOLICA)	26.32	76.1	1.4	53.6	58.6	59.9	0.26	16.5	0.43
marly cherty limestone (ROSSO AD APTICI)	26.60	142.2	8.9	15.9	58.3	62.3	0.27	17.8	0.48
calcareous chert (RADIOLARITI)	25.36	147.3	11.2	13.0	70.9	78.1	0.33	-	-
nodular limestone (ROSSO AMMONIT. L.)	26.58	56.6	-	-	29.1	26.6	-	-	-
marly limestone (MARNA DI BRUNTINO)	25.53	26.1	0.96	27.2	20.8	33.8	0.25	-	-

Tab. 1 - Synthesis of the results of laboratory tests on rock material

Unit	φ	C	φ	C
	[°]	[kPa]	[°]	[kPa]
MAIOLICA clay-free joints	47	400	26.5	270
MAIOLICA joints with clay filling	25	350	21	210
RADIOLARITI	41	100	35	80

Tab. 2 - Direct shear test results

meters in length was made perpendicular to the exploitation front.

During drillings, dilatometric tests were executed for measuring the elastic modulus of the rock mass and door-stopper tests were executed for measuring the stress condition.

The dilatometric tests were executed at a distance of 19, 20, 29 and 30 m from the wall. They enabled the elastic modulus - as calculated on the linear portion of the last loading cycle executed - and the deformation modulus to be measured both in vertical and in horizontal directions. In brief, we have: $Eh=5000$ MPa; $Ev=3200$ MPa; $Eh/Ev=1.6$; $D/Eh=0.4$; $D/Ev=0.6$.

The door-stopper tests executed 20 to 30 m far from the front measured major principal stresses - ranging from 3.18 and 7.72 MPa (and, therefore, considering site geometry, these stresses were consistent with the theoretical values of lithostatic pressure) - and minor principal stresses ranging from 0.52 and 0.77 MPa. The rotation axis of the stress ellipse results to be rotated about 25° from the vertical axis.

4.5 In-hole equipment

In every borehole, 4 extensimetric bases were positioned 1.5, 15 and 20 m, and a triaxial pressure cell was positioned 5 m far from the wall. Measures are taken weekly.

5 A STUDY ON GLOBAL STABILITY CONDITIONS

The analysis of global stability conditions was executed by using a finite difference computing programme (fig. 7). The constitutive model used is an anisotropic model of elastic-plastic deformation of an 'ubiquitous joint' type. This model considers both rock mass characteristics and the characteristics of the main joint system - a system that, in the

case being examined, coincides with bedding. In two of the six calculation sections considered, a Mohr-Coulomb constitutive model was also used because in that area an intensely fractured rock mass exists, the behavior of which can be schematized as an 'equivalent soil'.

The model was processed by defining first of all the geometries and the characteristics of the geomechanical units detected on the basis of on-site surveys and of laboratory tests. The model was calibrated by using regularly executed deformation measures and stress variation measures.

The fundamental parameters ascribed to the main units are shown in table 3.

In every case, modeling was executed by simulating first of all natural slope conditions, that is, the conditions before the beginning of mining activity, and by proceeding in accordance with the exploitation plan adopted. For every calculation section, present conditions were checked (with maximum wall height equal to about 150 m) as well as the conditions in view of exploitation development, by foreseeing a progressive lowering of the yard and, as a result, an increase in wall height. In particular, the entity and the direction of deformations, as well as the stress variations induced by excavations were calculated.

UNIT	E	K	G	γ	φ	c
	MPa	MPa	MPa	KN/m^3	°	MPa
H	4500	3000	1200	2.4	60	2
F	5000	3300	2000	2.5	65	8
D	4000	2650	1600	2.4	60	4
B	5000	3300	2000	2.5	58	5

Tab. 3 - Typical parameters of main geomechanical units

Therefore, through a series of successive simulations, optimal conditions were detected which would assure safety conditions and an optimized exploitation of the deposit until the completion of exploitation, in compliance with the limits imposed by the environment recovery project to be implemented at the end of exploitation. That was obtained by profiling the excavation front by inserting benches of an adequate geometry into the different mine portions in function of the different geomechanical characteristics of rock masses.

4081

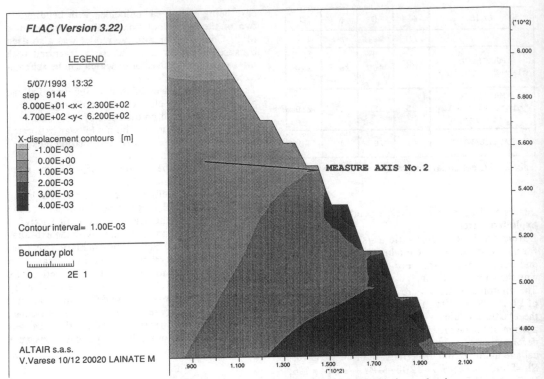

Fig. 7 - Example of model output with foreseen deformation values on the horizontal axis

6 LOCAL STABILITY PROBLEMS

By 'local stability problems', we mean all actual or potential instability phenomena that involve rock volumes lower or equal to 10,000 cu.m and that can by no means concern areas external to the mining license area.

Local stability problems of most significant sizes concentrate close to major faults and joints and to contacts between formations featuring different mechanical characteristics. Especially as regards smaller sized problems, these phenomena are difficult to foresee much in advance and, therefore, they are verified with regular specialized site inspections, even by using mountain climbing techniques. Besides that, they are also daily controlled by licensee company personnel.

The most frequent instability models are ascribable to blocks slipping along layer plans, to plates toppling and to en-echelon failure surface models. In every case, the problem is considered analytically in terms of limit equilibrium.

Where needed, consolidation and drainage are conceived according to a priority criterion apt to assure workers safety.

7 CONCLUSIONS

The analyses executed up to present, along with on-site measurements and regular site inspections, enabled and enable a good knowledge of both general and specific conditions of stability within the mining area.

They also enable realistic estimates to be made on the evolution of conditions as the mining activity progresses.

As the mine bottom progressively lowers -as is foreseen by the exploitation plan - the work schedule that was adopted up to date will be maintained by regularly updating calculation sections and by integrating existing equipment with other measu rement bases located on lower levels compared to present bases.

In conclusion, the survey and control procedure described before is an example of an articulated non conventional and continuous geomechanical study that, with an investment fully compatible with the mining activity being carried out, enables an optimized exploitation of the deposit in safe conditions both for workers and for the areas located close to the exploitation area, within an area that is compelling from a geological, morphological and geomechanical viewpoint.

SYMBOLS

As referred to the rock material:

γ_n natural volume weight
Co   uniaxial compressive strength
σ_t tensile resistance
m_i Hoek-Brown failure criterion constant
Et'_{50} tangent modulus of elasticity at 50% failure stress
Es_{50} secant modulus of elasticity at 50% failure stress
μ Poisson ratio in static conditions
E_{din} dynamic modulus of elasticity
μ_{din} dynamic Poisson ratio
σ_1,σ_3'major and minor principal stresses

As referred to discontinuity:

φ_p,φ_r peak friction angles and residual friction angles
c_p,c_r peak cohesion and residual cohesion

As referred to rock mass:

Eh,Ev horizontal and vertical modulus of elasticity
D   modulus of deformation
K   volumetric modulus of deformation
G   shear modulus
γ_n natural unit weight
φ friction angle
c   cohesion

ACKNOWLEDGMENTS

Many thanks to Prof. Renato Pozzi for his critical reading of the text and to Mr Riccardo Bersezio for his support in reconstructing site geology.
This study was conducted within the research foreseen by MURST contribution by 60% (Pozzi, 1992).

REFERENCES

Bernoulli, D. 1964. *Zur Geologie des Monte Generoso (Lombardische Alpen). Ein Beitrag zur Kenntnis der Sudalpine Sedimente.* Mat. Carta Geol. Svizzera, pp. 134

Bersezio, R. 1992. *La successione aptiano-albiana del Bacino Lombardo (Alpi Meridionali).* Giornale di Geologia, vol. 54 n. 1, pp. 125-146; Bologna.

Cantaluppi, G. & L. Montanari 1969. *La serie domeriana della Val Cepelline (alta Brianza).* Atti Soc. Ital. Sc. Nat., 109-223

Clerici, A., L. Griffini & R. Pozzi 1988. *Procedura per l'esecuzione di rilievi strutturali geomeccanici di dettaglio in ammassi rocciosi a comportamento rigido.*Geologia Tecnica 3/88, pp.21-31

Clerici, A., L. Griffini & R. Pozzi 1990. *Procedure for the execution of detailed structural geomechanical surveys on rock masses with a rigid behaviour.* Mechanics of jointed and faulted rock, Rossmanith (ed.), pp. 87-94

Gaetani, M. & E. Erba 1990. *Il Bacino Lombardo: un sistema paleoalto/fossa in un margine continentale passivo durante il Giurassico.* 75° Congr. Naz. S.G.I.

Hoek,E. & E.T. Brown 1980. *Underground excavation in rock.* The Institution of mining and Metallurgy. London, 527 pp.

Licthensteiger,T.C. 1986. *Zur Stratigraphie und Tektonik der südöstlidhen Alta Brianza (Como, Lombardei).* Ph.D. Thesis, Univ. Basel (not published).

Engineering geological assessment of an abandoned andesite quarry in the city centre of Izmir, Western Turkey

Évaluation géotechnique d'une carrière abandonnée dans le centre ville de Izmir (Turquie)

N. Türk, M. Y. Koca & F. Çalapkulu
Dokuz Eylül University, Geological Engineering Department, Bornova, Izmir, Turkey

ABSTRACT: Asansör quarry is one of the largest abandoned andesite quarries worked in the 19th century and now located within the overbuilt city center of Izmir, in western Turkey. As this quarry has been left by itself without taking any precautions against slope instabilities, some of the buildings built in and around it have been destroyed by the slope failures took place in the quarry over the years. A detailed engineering geological investigation involving large scale engineering geological mapping, discontinuity and weathering grade mapping, tilt testing of andesite blocks and back analysis of the failed parts and stability analysis of the standing parts of the quarry was carried out to characterize and make a stability assessment of the andesite quarry face.

RÉSUMÉ: La carrière "Asansör (IZMIR) a été exploitée en 19. siècle, laquelle se trouve actuellement dans un endroit où il ya une très dense agglomeration cette carrière est abandonnèe maintenant. Sous les negligeances très differentes, il ya actuellement des glissements de terrains causés par les eaux de fosse-septiques et cette carrière ne possède aucune protection prise pour empêcher ces glissements. Aux alentour, il ya eu aussi la destruction de certains bâtiments soumis aux glissements. La cartographie des discontinuities des roches volcaniques (andésites) et la relevée de grande echelle en cartes de gcnie civil et de desagrégation des roches ont été mis en question et on a precisé l'angle de frottement entre deux blocs d'andesite. On a effectué des analyses de stabilité des talus qui sont boulversés ou non. Avec ce travail on a caracterisé les talus de l'ancienne carriére d'andesite et on a mis au jour les nouveaux dangers potentiels.

INTRODUCTION

Izmir is the third largest city located in Western Turkey. There are numerous abandoned andesite quarries in the city center which were active in the 19th century supplying construction raw materials to the city. Andesite quarry at Asansör was one of such quarries which was abandoned at the beginning of the 19th century (Fig.1). Although, most of these quarries including the one at Asansör were than located in the outskirts of the city, they now remain within the city center due to expansion of the city over the years. Buildings were constructed in and around these quarries without taking any precautions against slope failures, even before they were abandoned.

Asansör quarry is one of the largest abandoned andesite quarry having 40-42m. high slope face, 350m. length, and slope face striking E-W and

having face angle of 70°-87°. Buildings were constructed in and around this quarry, since its abandonment towards the end of 19th century. One of these buildings was constructed as an elevator in 1904 to carry people and materials from bottom of the quarry to its crest and is still in operation. As this quarry face has been left by itself without taking any precautions against rock falls and sliding, some of the buildings built in and around it have been destroyed by the slope failures in the past. Recently, some of the buildings around this quarry have been restored and opened to public as an art center and restaurant and the quarry ground was made a recreation center because of its panaromic view and historic value.

A detailed engineering geological study of the quarry was carried out to characterize and make an assessment of the andesite rock face stability. Large

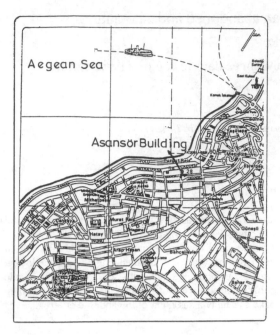

Figure 1. Location map of study area.

scale engineering geological mapping of the quarry and its surroundings, discontinuity and weathering grade mapping of the quarry face, testing of andesite with different weathering grades, back analysis of the failed parts and the stability analysis of the standing parts of the quarry face were carried out, thus, critical parts of the quarry slope from slope stability point view were identified.

I. GENERAL GEOLOGY

Upper Cretaceous aged Bornova Melange is found in the basement in and around Izmir. Bornova melange is composed of flysch matrix which is developed intercalating sandstone and shale and allocthoneous limestone or/and silicified limestone blocks floating in the matrix. Additionally, the melange is extremely folded, laminated and quartz veined. Neogene sedimentary sequences consisting of conglomerate, siltstone, claystone/mudstone, marl, clayey limestone layers and having varying thicknesses and weakly cemented and slightly consolidated lake sediments discordantly overlie the Bornova melange. Sedimentary rocks dip at low angles and have well developed joints vertical to the bedding. Volcanic rocks were laid down on top of the Neogene sedimentary rocks formed as a result of the several volcanic activities in and around Izmir city center.

1. Alluvium   2. Alluvial fan   3. Volcanic rocks (Andesite, Agglomerate, Tuff) 4. Limestone 5. Clay, Marl, Conglomerate 6. Massive Limestone 7. Flysch 8. Faults.

Figure 2. Geology map of Izmir region.

Q.		Alluvium/Slopewash
NEOGENE	MIDDLE-UPPER MIOCENE 30-115m.	Andesite, including agglomerate (autobreccia) levels.
	10-55m.	Agglomerates, containing andesite gravels and blocks in glassy and tuff matrix. Lithic tuff having thickness of 7-20 m.
	100-290m.	Conglomerate, sandstone, siltstone, claystone/mudstone, marl.
PALEOGENE PALEOCENE	1000 m.	Flysch composed of interclating sandstone, shale and allocthoneous limestone blocks floating in the matrix.

EXPLANATION

Figure 3. Stratigraphic columnar section is taken from south of the Izmir gulf.

Volcanic rocks found in north of the Izmir gulf and south of the Izmir gulf belong to different volcanic sequences (Fig.2). While the presence of a dasitic succession is observed at the bottom of volcanic complex in north of the gulf, the same dasitic level is not observed in south of the gulf. The volcanic rocks in ascending order consist of lithic tuffs having thickness of 7-20m., agglomerates having thickness of 30-115m. in south of the gulf (Fig.3). The order in the succession of these volcanic rocks may change locally and andesite may lie on top of the Neogene units directly (Fig.3). Similar situation have also been observed for the other volcanic rock types in and around Izmir. Agglomerates in south of the gulf include andesite blocks within a glassy and tuff matrix and have a interfingering contact with andesite vertically and laterally. Additionally, the unit includes basaltic lava levels having thickness of 4-15m. (Fig.3). Andesite concordantly overlie the agglomerates, including agglomerate (otobreccia) levels composed of andesite blocks laying within a volcanic glassy matrix.

I.1. Local geology

The Upper-Miocene andesites which is located in the Asansör quarry, have three different discontinuities as shown on the map (Fig.6). The main structural features seen in the andesitic lava within the study area are flow bands,cooling joints and tectonic joints. Two different flow structures were noticed within the Asansör quarry. The first flow structures is observed at the bottom of the slope face (E-W) and these discontinuities (190/23 and 200/17), daylight in the quarry face. The second flow structures overlay the firs ones. On the contrary, dip direction of the second ones is into the slope. Cooling joints are well developed vertical to the first and second flow bands. 245/70, 260/62 and 345/70, daylight in the quarry face and are important from the slope stability point of view (Fig.4). Additionally, in some cases, tectonic joints cut the cooling joints running vertically and in others run the parallel to them (245/70).

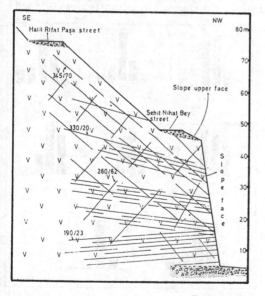

Scale : 1/1100

Figure 4. A cross-section taken across the Asansör Quarry in SE-NW.

I.2. Petrography

Microscopic studies show that andesites have a vitrophyric-porphyritic texture having large phenocrysts laying within a microcrystalline and glassy matrix. The modal analysis showing the volumetric ratio of minerals of eleven samples from andesite lava units is presented in Table 1.

Table 1: Modal Analysis of Andesites at Asansör

Phenocrysts : 42%	
Matrix : 58%	
Minerals	Volume
Plagioclases..........	58
Pyroxenes.............	16
Amphibole............	9
Biotite..................	8
Opaque................	5
Free Quartz...........	3
Kaolen+Clorite.....	1

I.3. Water situation at Asansör quarry

Water seepages are observed at various points in the quarry. The source of this water is waste and domestic water. Additionally, ground water level was noticed to be standing at 2m. below the surface in the waterwell, located 8m. east of the Asansör building in the bottom of the quarry. Apparently, there was an other well in the west of Asansör building, but it is filled now.

There are a number of waste and infiltrated water seepage points in the quarry. The waste water flows along the topographic slope and reaches high values in the quarry. The quantity of the waste water reaches up to 5-6lt/sn along the slope at the east of 305th street in N-S direction. The locations of the waste water seepage points are also shown on the fig.6.

II. ROCK MASS CHARACTERIZATION

The characterization of the andesite rock mass in the quarry face involved detailed discontinuity property measurements along a determined horizontal line, discontinuity and weathering grades mapping of the quarry face and tilt testing of the andesite blocks.

II.1. Detailed discontinuity measurements on the slope

A detailed discontinuity survey was made in the study area to carry out stereonet analysis and identify the possible rock failure types and their locations in the study area.

II.1.1. Orientation

The most readily apparent influence of the orientation of discontinuities on rock mass strength is evident in the failure of rock slopes along one or more discontinuities. There are two flow bands approximately striking in the same direction and while one of them dips north and the other one dips south, the cooling joints and tectonic joints cut these flow structures vertically in the Asansör area. Tectonic joints are long running joints and they are connected to one another by rock bridges and form structures prone to sliding and toppling in the area.

II.1.2. Discontinuity apertures, types and thicknesses of the fill materials

This survey was carried out along a horizontal line, on the rock slope, at about 1.5m. above the quarry floor. Surveying line was divided into 2m. long spacings and kept parallel to the rock slope surface. The points where the discontinuity cross the surveying line were recorded. Discontinuity apertures, types and thicknesses of the fill materials were also recorded. The second surveying line was established by crossing the first line vertically, using a tape measure. An area of 2m² was mapped using this technique at several locations. Thus, the discontinuity data was collected in this fashion along the quarry slope. Discontinuity frequency was estimated by producing 1m² grids using the tape measures and counting the number of joints falling into the each grid. Classification of the rock mass was made based on discontinuity frequency. Andesites were classified in terms of discontinuity aperture and the number of discontinuities found in 1m² using the joint spacing and classification proposed by John (1962). The cooling joints are "closely spaced", the tectonic joints are "widely spaced" and the flow bands are "very closely spaced" based on the discontinuity aparture. According to the number of discontinuity falling in 1m² the andesites are classified as "poor quality rock". Additionally, while the cooling joints are smooth and filled with thin fills (1-4mm. thick), tectonic joints are moderately rough, open and filled with thick clay materials formed as a result of shearing and weathering of andesites. Roughness classification proposed by Barton and Choubey (1977) was used in this work. It was also noted that the rock mass was prone to sliding along the cooling joints even at very low angles. However in the case of the tectonic joints, the movement of the rock blocks were controlled by the dip angle of the block surfaces and the cohesion and internal friction angle of the clay fills.

II.1.3. Joints spacing measurements

Measurements of the discontinuity spacings in three different weathering grade zones which exposed on the andesite quarry slope is presented in Figure.8. Dominant modes of the discontinuities in andesite are cooling joints, flow bands, and tectonic joints. Spacing of above mentioned discontinuities were obtained from 1/100 scale weathering map and analysized using a personal computer. Spacing versus frequency relation were plotted for each discontinuity type for three different weathering grades (Fig.5) Since weathering process were developed after the formation of the discontinuities, as expected there is no relation established between the weathering grades and the discontinuity spacings. While, spacing of the cooling joints vary between 25 and 50cm, the flow bands vary between 1 and 5cm, and the spacing of the tectonic joints vary between 0.25 and

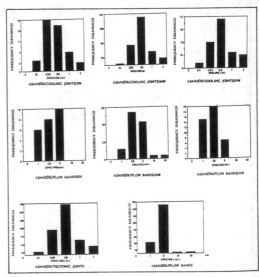

Figure 5. Discontinuity spacing versus frequency relations for three different weathering grades.

3.0m but their dominant values vary between 0.25-0.5m. as shown in fig.5. The tectonic joints, which is trending to NE-SW, are clearly seen developed from the cooling joints. As a result of the detailed joint survey and the computer analysis, andesites were classified in terms of joint spacing as "frequently jointed" (5-30cm). Using the joint joint classification proposed by John (1962).

III. DISCONTINUITY MAP

The discontinuity map covering most part of the abandoned andesite quarry was prepared at 1/100 scale and presented in Figure 6. Figure 7 shows that the area 70m away from the beginning of mapped quarry slope to the east and this area is marked on the 1:3000 scale map of quarry. Initially, 1/100 scale discontinuity map of the abandoned andesite quarry were produced by means of grids formed using 40m. long ropes kept at 3m. apart and tied to one another at 2 m. spacing with colored strings (Fig 6). Using this grid system, the position of the

discontinuities were recorded on a millimetric paper from distance. Then, 1/100 scale topographic section and plan of the Asansör area was prepared using theodolite. Thus, the location of the critical points on the slope were established with good accuracy. This increased the reliability of the discontinuity measurements taken during the grid survey as some of the critical points used were common in both surveys.

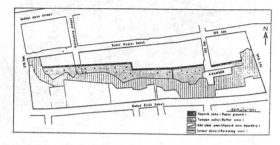

Figure 7. 1/3000 scale hazard map of the Asansör area-Izmir.

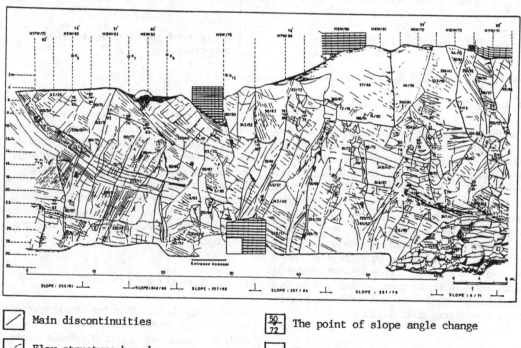

▱ Main discontinuities	50/72 The point of slope angle change
▱ Flow structure boundary	N10E/81 The strike and dip of the grids
71/7 Dip direction and dip angle of discontinuities	⊙ Groundwater leakage points

Figure 6. Discontinuity map of the Asansör quarry slope face.

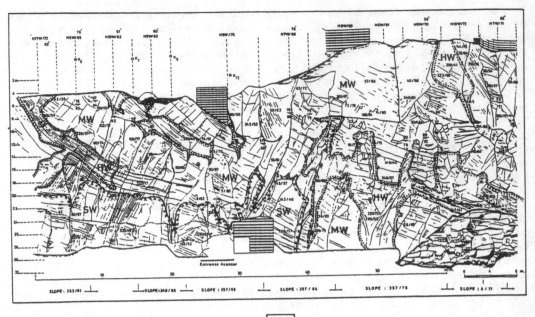

| SW | Slightly weathered zones. | MW | Moderately weathered zones. |
| HW | Highly weathered zones. | | Contact of two different flow structures. |

Figure 8. Weathering grade map of the Asansör quarry slope face.

IV. WEATHERING GRADE MAPS

Andesites outcropping in the abandoned quarry at Asansör are found in different colors because of their varying weathering grades. Andesites are dark grey when they are fresh and changes to brick red, pink and brown, due to weathering. Andesite were graded into different zones based on weathering grades as slightly, moderately, highly and completely weathered, and boundaries of the different weathering grades were mapped with accuracy. The weathering classification by Dearman et al. (1974) was used in this work. Three different weathering zones were differentiated on the quarry slope face and different symbols were used in this map to show the each weathering grade (Fig.8).

V. FIELD TEST

Tilt tests were carried out in the field to determine surface friction angle of the discontinuities by sliding an andesite block over another one. Andesite blocks were taken from the slope face used for the tilt test. In carrying out this test, two andesite blocks were placed on top of each other and the lower block was tilted until it reached to an appropriate angle at which the upper block started sliding. Then, the dip angle of the lower block surface was measured by a compass in the field. Block sliding tests carried out in the field has shown that while the friction angle of the rough surface were found to be 38°±5°, the friction angle of the smooth surfaces were found to be 20°±2° and these values were used in the stability analysis of the slopes using the stereographic projections.

VI. BACK ANALYSIS

Three wedge failures have been back analyzed in order to determine their failure conditions. The back analyzed wedges are located at different points on the slope face. Location of one of the wedges analyzed shown on western of the upper parts of the discontinuity map (Fig.6). The principles of the back analysis used here is based on the comparison of the dip and direction of the intersection-line of the wedge forming planes with that of the slope angle and the surface friction angle of the discontinuities.

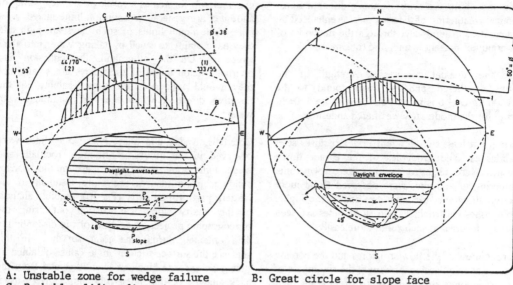

A: Unstable zone for wedge failure   B: Great circle for slope face
C: Probable sliding direction   0: Poles for discontinuities

Figure 9. Dip and dip direction of the intersection-line obtained from back analysis of the discontinuities.

Case. 1.   Wedge forming discontinuities are 342/52 and 50/62

        Rock slope face: 357/67
        Intersection of the discontinuity :
N2E/50NW
        Q_{i1} = 50° (Fig.9)

Case. 2 Wedge forming discontinuities: 333/55 and 44/70

        Rock slope face: 356/73
        Intersection of the discontinuity:
N13W/53NE
        Q_{i2} = 53° (Fig.9)

Case. 3 Wedge forming discontinuities: 340/61 and 57/68

        Rock slope face: 357/78
        Intersection of the discontinuity:
N11E/57NW
        Q_{i3} = 57°

The following friction angle values were obtained from the back analysis of the above mentioned wedges Q_{i1}=50°, Q_{i2}=53 and Q_{i3}=57°, Q_{imean}=53°. The difference between above found friction angles and the surface friction angle obtained from the tilt tests is due to roughness and wedging effect of the wedge forming surfaces.

VII. GENERAL ASSESSMENT

In this study, the rock mass features such as discontinuities, weathering state and groundwater seepage points were identified and mapped (Fig.6). Thus, the probable failure types, their locations and the factors influencing them were identified with accuracy. Initially, mapping of the discontinuities were attempted directly by ascending and/or descending a person along equally spaced vertical profiles at the outset of the study. However, this method was found to be highly dangerous, and thus, it was not used. The position of the discontinuities and the weathering states of the quarry face was mapped by forming 2x3m. grids using 45m long 4 pieces of ropes and recording the geological features from the distance. The field measurements made by this technique were checked by the direct measurements taken at the bottom and the upper part of the slope face and they were found to be in good agreement with one another. The discontinuities and weathering profiles of the rock slope face were mapped to a great extend using this grid system. These maps were presented in Fig.6 and 8. Thus, the dimensions of the blocks and boundaries of the different weathering zones were determined accurately.

Andesites are completely weathered at the alteration

zones and along the water running discontinuities. Chemical weathering of the rock mass is enhanced by movement of groundwater through the networks of discontinuities present in andesitic rock mass.

While the uniaxial compressive strength of the slightly weathered andesites was found to be 650 ± 200 kg/cm^2 this value was found to be 420 ± 80 kg/cm^2 for the moderately weathered andesites.

There are at least two different flow structures seen in andesites. The properties of the earlier flown andesites were changed by the later flown one. Weathering of andesites were observed to be more effective along the discontinuities and in the highly altered zones and rockfalls and slope slides are seen to have developed in such zones more easily.

The roughness of the flow structures and the cooling joints are at small scale. Although, the roughness of the tectonic joints are at relatively a larger scale, their influences are decreased along such surfaces because of weathering and the presences of fill materials. The

clay minerals formed as a result of the weathering of andesites are generally, montmorillinite-mixed layer type. The liquid limits of such clays are high and they have ability to swell by taking water into their structures. On the otherhand, the cohesion and internal friction angle of such clays are small, thus they would influence the slope stability in the negative direction when they are saturated with water.

Rockfalls, wedge and plane type slope slides were observed in the quarry. Although, rockfalls and wedge failures are more frequent around the Asansör building, plane failures are more frequent in the western part of the quarry (Fig.10). Stability analysis of the quarry slope was carried out using the stereographic projection analysis of the discontinuity data obtained during the joint survey. In this analysis, the surface friction angle values obtained by sliding an andesite block over another and from the back analysis of the slided rock slopes, were used. The influence of clay fills and groundwater were not taken into consideration. The expected slope failure

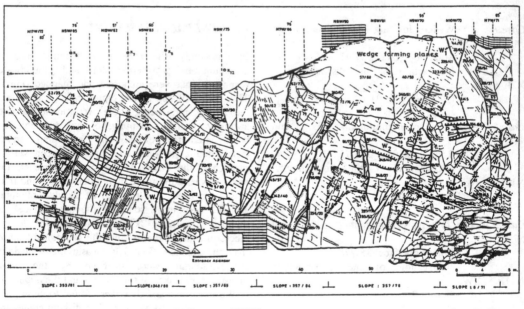

	Rock blocks having possible plane failure		Rock blacks having possible wedge failure
P		W_{12}	
$T_{1,2}$	Rock blocks having possible toppling failure		Direction of the plane failure

Figure 10. The expected rock mass failure modes and their degrees of importance in the andesite quarry

4092

types determined from the field observation and stereographical analysis of the discontinuity data of the quarry are given in fig.10. The slope failure types were identified as plane and wedge sliding, rockfalls and toppling. The rock slope failure types were also graded as 1, 2 and 3 in terms of their importance from the stability point of view (Fig.10). Grade 1 slopes are at critical state of stability. The slopes graded as 2 and 3 indicate that they may fail as a result of the action of weathering, earthquakes, blasting and any change in the slope geometry in the long term. Thus, this map will be useful in deciding about the preventive measures to be taken against the rock slope failures.

It is observed that while the rockfalls and wedge sliding type of slope failures effect the limited areas in the quarry the discontinuities forming them have high dip angles, plane failures effect relatively wider areas. Rockfalls and wedge failures generally developed in the upper parts and in the undercut areas of the slope, the plane failures developed in the lower part of the quarry slopes. Additionally, in the areas where the multiple wedge failures have developed between two near vertical discontinuities, the upper wedges have been observed to have failed first. If the quarry slope is left as it is without taking any preventive measures against sliding, then the falling andesite blocks are expected to damage the buildings nearby the quarry. Additionally, fallen and slided rock blocks would unload the slope and form new faces. Then, discontinuities may daylight in these newly formed slope faces and initiate new rock slope failures southwards. Thus, the houses built between the Sehit Nihat Bey and Mithatpasa streets will be damaged (Fig.1) and as a result, the boundary of the geologically hazardous area will expand. If the stability of the quarry slope is improved by reinforcement, then there will be little damage given to the buildings around the quarry. Although, the second solution is more expensive in the short term, it will be more economical and safe in the long term. Additionally, if the protection of the Asansör and other historic buildings are taken into consideration, then the second alternative is more practical solution.

VIII. CONCLUSION

Engineering geological assessment of the abandoned andesite quarry at Asansör-Izmir, have shown that the rock mass properties degrade with time and this may result in causing instabilities in the rock slopes. In andesite the older flown bands are altered by the younger ones, thus increasing the rate of weathering.

While the cooling and tectonic joints having high dip angles cause rockfalls and wedge type failures, daylighting flow bands laying at relatively low angles cause plain failures. Weathering grades of andesites are found to increase with increasing joint frequency. Additionally, discontinuity and weathering grading maps have given good correlation in characterizing and identifying the critical areas in the quarry slopes. Any improvement to be undertaken should only be made after carrying out detailed assessment of the slope as presented in this paper.

IX. REFERENCES

John, K.W., 1962, An approach to rock mechanics: Proc.Am.Soc.Civ.Eng., J. Soil Mech. Found. Eng. Div., Vol.88, No. SM4, pp.1-30.

Barton, N., and Choubey,V., 1977, The shear strength of rock joints in theory and practice: Rock Mech., Vol.10, NOS.1-2, pp.1-54.

Dearman, W.R., Baynes, F.J., and Irfan, T.Y., 1978, Engineering grading of weathered granite: Eng.Geol., Vol.12, pp.345-374.

Rock mass movement and deformation under the conditions of open excavation

Déplacement et déformation des massifs rocheux soumis à des excavations à ciel ouvert

Xu Jiamo

Institute of Geology, Chinese Academy of Sciences, Beijing, People's Republic of China

ABSTRACT: So for the research into rock mass movement and deformation under the conditions of open excavation, as a new subject in the field of engineering geomachanics, has had only a beginning. This kind of literature concerned with the respect of content could be hardly found. Based on the research results of "Influences of open excavation and underground extraction on deformation of the north slope rock mass of the Fushun west open coal mine (in China)", one of the national scientific research projects of focal point during the seventh Five-Year, using some experiences of other mines for reference, some regularities and problems concerned with the rock mass movement and deformation have been given and briefly discussed according to the following twelve subheadings in the paper.

RÉSUMÉ: Jusqu'ici la recherche du déplacement et de la déformation de la masse de la roche sous les conditions du creusage ouvert, comme un objet nouveau dans l'aspect de la géomécanique ingénieuse, eut un commencement. Nous pouvons obtenir cette référence difficilement. D'après le résultat de la recherche de "L'influence du creusage ouvert et souterrain à la déformation de la masse de la roche à la pente norde de la mine ouverte ouest à Fushun (Chine)",un des projets de la recherche scientifique focale nationale dans les Cinq-Années septièmes, l'auteur se réfère à l'expérience des autres mines, donne plusieurs régularités et problèmes relatives au déplacement et à la déformation de la masse de la roche, et discute brièvement d'après 12 topiques dans l'article.

1 EXTENT OF GROUND MOVEMENT FROM OPEN EXCAVATION

so long as long-term monitoring is kept to, we can find that some measuring points far from an open pit being mined have moved. Other conditions being equal, the extent of the rock mass surrounding an open pit is mainly related to compliance of the rock mass. Besides rock media forming the rock mass and discontinuities in it, its compliance is mainly related to depth of an open pit (or slope height) and slope angle.

An extent of ground movement calculated according to a constitutive model is different from results according to other constitutive models. According to the linear elastic relation the author derived the following displacement equation under the condition of self-weight stresses from the principle of similitude and the theory of elasticity

$$\Delta_H = \frac{L_H^2 \rho_H E_M}{L_M^2 \rho_M E_H} \Delta_M \tag{1}$$

where L, P and E are respectively characteristic size, unit weight and elasticity modules. Their signs H and M represent "prototype" and "model" respectively. This equation represents a relationship between displacement of prototype Δ_H and that of model Δ_M. If $E_M = E_H$ and $\rho_M = \rho_H$, from the equation we obtain the relationship between Δ_H and Δ_M:

$$\frac{\Delta_H}{\Delta_M} = \frac{L_H^2}{L_M^2} \tag{2}$$

That is to say, on this condition a displacement ratio of a slope body to another similar one is equal to a ratio between the two squares corresponding slope heights. In other words, if the height of a slope body increases to n time, displacements of corresponding points will go up to n^2 times. This conclusion is correspond approximately to a result calculated in accordance with the finite element method.

According to other constitutive relationships (or constitutive models) calculated results would be different from this one, and one of them would also be different from the others. So the extent of ground movement from open excavation can not be expressed a unified formula or can not be formalized. In brief, it is related to engineering-geological conditions of the rock mass surrounding an open pit.

According to a oral data furnished by professor L. V. Shaumian (1991), there is a open pit in the Soviet Union, where the movement of a upper surface of a slope rock mass of toppling failure had exceed a extent of 3 times h (the then depth of the open pit) from the slope crest. In all the information on rock mass movement from open excavation, as far as I know, this is the greatest extent relative to the depth of a open pit.

2 MODEL CURVES OF DISPLACEMENT DISTRIBUTION OF SLOPE ROCK MASS

Based on the result of an experimental study on deformation of the model slopes of a soft photoelastic material (Xu Jiamo, 1983) and monitoring data on displacement of slope rock mass of toppling failure-deformation, the model curves are given in figure 1. They represent characteristics of movement and deformation of a section of two slopes through an open pit excavated once. It should be noted that for the slope rock masses of non-toppling failure and of low slope angle, monitoring data on slope displacement show the slope crest point being in the slope body of dashed line. From figure 1 it can be seen that there are both two points of the maximum of vertical displacement and two points of the maximum of horizontal displacement, their positions coincide with the slope crest.

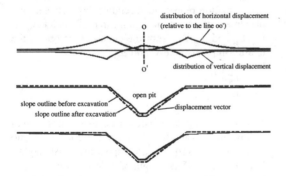

Figure 1   Model curves and diagram respectively illustrating displacement distribtion and deformation of slope rock mass surrounding an open pit after excavation

Any of the following factors can cause geometrical characteristics of the model curves to change, For example, effects of faulting on movement and deformation of rock mass from the excavation, various asymmetries of the rock mass behind the two slope walls, differences in behavior of deformation due to differential properties of rocks forming a slope rock mass, changes in ratio of slope (or curvature of slope) along a slope, and so on. Comparing the model curves with a (practical) measured curve of displacement distribution, we can find that there may be some differences between the two.

As for a longitudinal profile parallel to the crest line of slope, if all vertical sections perpendicular to the line satisfy plane strain conditions, the model curve of displacement distribution for both upper slope surface and slope face would be a straight line. But actually, because of difference of engineering geological conditions along a longitudinal profile, a measured curve of vertical (or horizontal) displacement distribution would be different from a straight line.

3 CHARACTERISTIC OF DISPLACEMENT DISTRIBUTION FROM THE OUTSIDE TO THE INSIDE

A lot of data of borehole displacements show that under the conditions of open excavation a distribution regularity of displacement along a borehole is that magnitude of displacement from any surface of a slope body vertically to the inside decreases gradually. But usually a rapid change occurs at the point that a observation hole intersects a fault (of other great discontinuity) at. This phenomenon is right an indication of the effect of faulting on movement and deformation of rock mass from open excavation. That is to say, along the fault (or other large discontinuity) its two wells have moved relatively to each other under the conditions of the excavation. But if the movement, to be more exact, a slipping occurs along a circular bedding surface, or when the circular failure occurs in a slope mass, the distribution regularity of displacement along a borehole would be contrary to the above mentioned.

4 MOVEMENT DIRECTION OF MEASURING POINT

When a geologic body is in the engineering state of slope, the part of slope face is in a state of compression and the part of upper slope surface is in a state of tension. If the slope face is excavated, the rock outside excavation area will move towards the area. The shorter the distance between a marker point and an excavation area, the more obvious the characteristic of rock movement towards the excavation area. Range of rock mass movement of the characteristic influenced by open excavation is much wider above the excavation area than below that. Movement direction of a marker point on the upper slope surface is related to attitude of discontinuities and soft-and-weak surfaces, along which its two walls have moved relatively to each other. For example, the movement direction of a marker point on the upper surface of rock slope body with well developed steeply dipping discontinuities of antidip is towards the open pit. If bedding surfaces in a slope rock mass are flat-lying or gently dipping, the movement direction of a marker point on the upper slope surface will be situated between the vertical direction and that towards the open pit. If there is no fault in a slope rock mass, generally speaking, the farther apart from an open pit an observation point is, the higher the ratio of the vertical component of the displacement vector to its horizontal is.

In plan movement of a observation point is related to the following directions:

a. towards the nearest excavated face;

b. towards the geometrical center of the excavated area of maximum volume (or weight);

c. in the direction of the maximum compliance of a slope rock mass (the direction is related to attitude of discontinuities and weakness planes in the rock mass);

d. in the direction of the maximum weight component parallel to the slope face;

e. towards a free face, or in a direction of displacement being been allowable.

The actual movement direction of a measuring point depend on the many factors that control the above mentioned directions.

In the course of excavation movement direction of a measuring point is changeable in detail, in varying degrees, or to a great extent. But in plan the overall movement route of the point trends invariably towards a certain direction. It is obvious from monitoring data of the Fushun west open coal mine in China over the years that the movement directions of the measuring points in the slope have been being between the dip of the slope face and the contrary dip of the strata (the angle between the two is about 20°).

5 POSITION OF THE TENSILE STRAIN POINT OF MAXIMUM

Supposing constitutive behavior of materiels forming a slope rock mass is line-elastic, the result calculated in the accordance with the finite element method shows that the point of maximum tension strain in slope rock mass is located at the slope crest. It is observed that a tension crack in the upper surface of an isotropic slope soil mass of a high slope angle occurs at a distance from the slope crest.

As for a slope rock mass, if the distribution of discontinuities and weakness planes in it is inhomogeneous, it is difficult for us to determine the position of the crack. An experimental study (Xu jiamo, 1983) with the model slope of a soft material has shown that after a model slope (in which there is a lot of discontinuities artificial, short, parallel to each and all, of a constant spacing, of a constant length, of a dip angle equal to 60°, and acrossing the two free surfaces of it) deformed under the action of self-gravity, the widest tension crack utilizing one of the discontinuities in the upper surface also occurs at a distance from the crest of the model slope. Based on the above - mentioned facts, the author feels that the result calculated may be out of accord with the objective reality; and the position of the point of the maximum tension strain may be located at a distance of L from the slope crest, where L is mainly related to the slope height and the slope angle. But it must be pointed out that position of actual crack in the upper surface of a slope rock mass are also related to the excavation procedure of the slope rock mass.

6 INFIUENCE OF EXCAVATION IN DIFFERENT CONDITIONS ON MOVEMENT AND DEFORMATION

1. The movement of the upper slope surface is obviously correlative with the amount of material dug out only after the size of an open pit (related to the depth of the open pit and the slope angle) increase to a certain value. The higher the value, the stronger the susceptibility of the movement and deformation of the slope rock mass to the open excavation.

2. If same amount of rock material is dug out separately in both a higher area of a slope face and in a lower area of it, then the corresponding movement and deformation have each different characteristics: in the former case the extent of movement and deformation of the upper slope surface is smaller, but the tilt and curvature of the surface is larger; in the latter case all of them are separately contrary to the above mentioned (shown in figure 2).

3. When an excavation along strike of a part slope face between two elevations is kept to, influence of the excavation on movement and deformation of a point on slope face or upper slope surface above the excavation line (or zone) is either to a great extent or to a certain

extent related to a length (or distance) in the direction of the excavation. The greater (or longer) the length (or distance) of the excavation, the stronger the influence of the excavation on the point; conversely, the smaller (or shorter) the length (or distance), the weaker the influence of the excavation on the point. But influence of the excavation exceed a certain extent on the point will get to be light and disappearing. That is to say, the higher the position of a measuring point, the greater (or larger) the extent of the excavation along strike of the slope face monitored with the point; the lower the position, the smaller the extent.

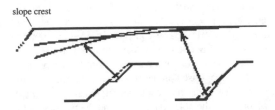

Figure 2 Vertical displacement distribution of slope upper surface in the two different conditions of excavation

7 REGULORETIES OF MOVEMENT AND DE-FORMATION IN THE CAURSE OF THEIR DE-VELOPMENT

1. A response of a point on a slope face to a lower excavation is that the point will go through a state of tension, and then turn into a state of compression. So the curve of the vertical component of movement velocity of the point versus time will have a upwards - convex - change, and then a downwards - convex - change. Thus we can image that the deformation of the slope face along a section will have a dynamic characteristics just like the back of a worm wriggling towards the excavation area, or just like a very gently advance wave.

2. The synchronism deviation and amplitude difference between displacement - time curves of two measuring points depend on to a great extent the ratio of the distance between the two points to the average distance from them to an excavation area: the smaller the ratio, the smaller the synchronism deviation and amplitude difference too; the greater the ratio, the greater the synchronism deviation and amplitude difference too. Conversely, according to the ratio we can judge or estimate that great or small is the average distance from the two point to the geometric center of the excavation area where the excavation can influence movements of

the two points.

8 ACCOMPANYING PHENOMENA IN THE COURSE OF MOVEMENT AND DEFORMATION

1. The change in volume force acting on a slope rock mass surrounding an open pit is mainly caused by uninterrupted excavation, or by uninterrupted change of geometrical conditions of the open pit. Even if geometrical conditions of a slope (or an open pit) gradually and continuously change, displacement - time curve of a point on the slope rock mass also has a characteristic of some steps. The essence of the phenomenon of the steps is that some of sudden dislocations along some of weakness planes in the slope rock mass result in some sudden increase in its compliance. In some cases due to a change in geometrical conditions of a slope rock mass, some of new fractures would produce and also result in a increase in compliance of the slope rock mass. In either case we can say that sometimes movement and deformation of a slope rock mass from the open excavation are accompanied with dislocating along weakness planes and fracturing, and then the slope rock mass remain stable.

2. When movement and defomation of a slope rock mass develop, emptying-phenomenon also occurs in the rock mass (Xu Jiamo, 1981). Emptying-phenomenon of rock mass in engineering state is bacause in the course of movement and deformation along some weakness planes (or discontinuities) in a rock mass a lot of couples of walls have moved relatively to each other and the rock mass further lose its corrdinate conditions of deformation. The phenomenon of emptying can occur in the surfaces of a slope rock mass and shallow rock mass, and also in the depths of it. This is shown by model experiments. Because there are a lot of discontinouites in a rock mass, its average deformation module in tension is much less than in compression. And in consideration of the emptying effect (or action), the author inferred that after a slope rock mass deforms from the excavation the volume surrounded by enveloping surface of it would increase.

9 LAG TIME OF MOVEMENT AND DEFORMATION RELATIVE TO EXCAVATION

When a slope rock mass deforms from the open excavation it will show obvious viscosity sooner or later. The viscosity is related to the following factors: mechanical effects (especially, the shearing) of discontinuities and weakness surfaces in the rock mass, self - rheological behaviors of the rocks, effects (actions, and influences) of the water in the rock mass, the volume force which depend on the geometry and unit weight of the rock mass, and so on . That the movement and deformation of a slope rock mass caused by open pit mining propagate from the excavation area to far surrounding rock mass is a progressive, developing and enlarging progress. So lag time of a movement and deformation relative to an excavation is related to the distance from a measuring point to the excavation area and is related to the position of the point relative to the area besides the viscosity of the rock mass. A movement of a measuring point during a term does not depend on the excavation during the term to a certain (or great) extent or not at all but depend on a history of open excavation before the excavation. It is not difficult to measure and roughly estimate the lag time from the beginning of a excavation to that of movement of a measuring point which the excavation has influenced. By the way, it is preferred to measure the lag time that before the measuring no excavation in a very wide area around the area to be excavated has kept for a long time. From the monitoring data of the Fushun west open coal mine in China, a lag time has been calculated by the author. Both the measuring point 9 on the upper slope surface and point 3 on the slope face are on the top edge of the W200 - section. The point 9 is at the distance of 270m from the point 3; and the point 9 is 32m higher than the point 3. A rock mass movement of a peak rate propagates from the point 9 to 3 in about 47 days. When a slope rock mass disturbed by a excavation is excavated, movement and deformation of the rock mass will propagate more quickly and farther than one not disturbed.

10 THE PROBLEM ABOUT THE MOVEMENT AND DEFORMATION WHETHER PROGRESSIVELY TO A STOP AFTER THE END OF EXCAVATION

Soon after the end of excavation in an open pit the volume force acting on the rock mass surrounding the open pit will not change, but movement and deformation of the rock mass still continues and then, generally speaking, will come to a stop in the course of time. The hysteresis (or hysteresis effect) lasts for as long as about several (or many) months. It is difficult to define the duration. The complicacy of the problem is due to complex and different engineering conditions for a rock mass surrounding an open pit. But the author assumed that the duration would be approximate to (or related to) the interval of time between the beginning of the excavation of the lowest level (assumed to be the latest excavation) and the beginning of the movement

of the farthest point (awy from the excavation face) which is influenced by the excavation. This may be a preliminary and rough supposition. So far we have not obtained any satisfying monitoring data on the rock movement after the ending of the open excavation to prove the author's view out, but according to the facts that length and width of cracks in the upper surface of slope will not further increase in many weeks after the ending of the open excavation, the author's view about above mentioned, as a supposition for prediction of movement and deformation of a slope rock mass, should be hold on by himself.

Considering that time-dependent factors such as weathering, creep play a part in a process of movement and deformation of a slope rock mass, some have had another way of saying the problem, that is, movement and deformation of a slope rock mass will never stop from a long-term point of view. It seems that the problem is philosophized. The author would rather hold on to his own view than speak highly of the other.

The author does not oppose that in the long term, a new and more secular movement and deformation will occur and even slope failure may happen due to secular variation of the slope engineering geological conditions, but those phenomena occur long after the movement and deformation of a slope rock mass from the excavation come to a stop and should be discussed without the paper.

11 INFIUENCE OF BACKFILLING IN AN OPEN PIT ON ROCK MASS MOVEMENT

It is worth discussing the problem that due to a backfilling in an open pit whether the slope rock mass will move back or not after ending of the open excavation. Considering some present buildings on the upper slope surface to be whether in danger or in safety in the days to come, the discussion of the problem would not be of no little significance. According to the fact that after water storage in a impounding reservoir for the first time the rock mass surrounding the reservoir would move back obviously from some monitoring data on rock movements, a backward movement would also occur under the conditions of backfilling in an open pit. But, however, it is not a reversible movement process. Because the backfilling is much more slower than the water storage, it is not easy to find such a phenomenon from a short - tern monitoring.

12 INFIUENCE OF ATTITUTE OF STRATA ON MOVEMENT AND DEFORMATION OF SLOPE ROCK MASS

Whether or not discordant deformation (or discontinuous displacement) occurs along a weakness plane (or surface) in a rock mass which is acted on by a force (or forces) is a important criterion or a ample condition to determine that whether or not the weakness plane (or surface) influences compliance of the rock mass. The author has obtained a concept from observation of and research into deformation of slope rock mass: attitude of a weakness plane (or surface) or discontinuity and geometry relation between the plane and a slope face are two very important factors in terms of influence of them on deformation of the slope rock mass. Orientations of a weakness plane (or surface) relative to principal compression stress trajectories intersecting the plane in stress field of a slope rock mass is defined by the above two factors, thus deformation type (or mode) and mechanism of a slope rock mass, even its compliance in different orientations are defined also by the two factors. For example, toppling-deformation occurs in a slope rock mass of steeply dipping weakness planes of the antidip and shows that compliance of the slope rock mass is very high under the condition of the deformation; buckling-deformation occurs in a slope rock mass of weakness planes parallel to the slope face; the slope rock mass of antidip strata of dip angle equal to about 20°-40° is of a strong stability; and so on.

CONOLUDING REMARKS

The monitoring of slope movements is the only practical tool currently available for dealing with the prediction of slope failure. The interpretation of the results obtained from slope movement monitoring is based on experience and theory. There is little doubt, however, that this interpretation would benefit greatly from a sounder understanding of regularities (or laws) of movement and deformation of slope rock mass. As you know, only be comparing can one distinguish. Only by knowing well ordinary regularities of movement and deformation of a slope rock mass surrounding an open pit being excavated can you recognize abnormal phenomena well. Thus and thus when critical slope problems are anticipated one would not go so far as to misunderstand a ordinary movement or deformation of a slope rock mass as a premonitory phenomenon of the slope failure. Especially, when there are some buildings or constructions on the upper slope surface near an open pit the research into the problem on the movement and deformation would be of great significance. So we can say that it would be of theoretical significance dealing with the subject for making a deep research into some environmental engineering geological problems of a mining area and going further into their essences of rock mechanics.

REFERENCES

Xu Jiamo, 1983. An experimental study on deformation of rock slope with the use of a soft photoelastic material model. Edited by the Institute of Geology, Academia Sinica, PP. 191-195, Sciences and technology Publishing house, Beijing, China. (in Chinese)

Xu Jiamo, 1988. A method and examples for the determination of cause of ground movement according to characteristics of displacement vectors. The selection of the articles in the third Chinese National Engineering Geological Congress, PP. 1205-1211, Chengdu Univ. Publishing house. (in Chinese).

Xu Jiamo, 1990. Some of major distinctions between the two kind of rock movements caused by open pit mining and underground mining. Proc. IAEG 6th Int. Cong. Amsterdam, NetherLands, Vol. 4, 2561-2566.

Leaky cut-off walls for excavations under groundwater: A case history

Parois moullée pour confiner une excavation sous la nappe

A. Focardi & G. Focardi
Studio Focardi, Firenze, Italy

P. Focardi & A. Gargini
Dipartimento di Scienze della Terra, Università di Firenze, Italy

ABSTRACT: A real case hystory concerning hydraulic conductivity testing of a cement-bentonite cut-off wall, devoted to confine an excavation in ground water, is presented. Hydraulic conductivity values obtained directly on the wall by laboratory and in situ infiltration tests were compared with indirect estimates derived from pumping tests perfomed, inside the confined area, either at the beginning and at the end of the excavation works. Results obtained showed that the bulk hydraulic conductivity value of the wall was more than one order of magnitude higher than that one obtained by lab tests.

RESUME': On présente un cas d' étude qui regarde la conductivité hydraulique d' un cloison en ciment-béntonite réalizé pour l' excavation sous falde. Les valeures obtenues directement à travers les tests de laboratoire et les tests d' infiltration in situ ont été comparées avec les valeures calculées par test de pompage dans l' aire d' excavation au début et à la fin des travaux. Les resultats obtenus montrent que la valeur moyenne de conductivité hydraulique du cloison était plus que d' un ordre de grandeur élévée que ça donnée à travers des ' tests de laboratoire.

1 INTRODUCTION

The hydraulic confinement of large portions of terrain is a fundamental operation when one wants to operate in dry conditions for the construction of foundation or underground works and also in the case of containment of masses of buried refuse or contaminated terrains.

In many cases plastic cut-off walls composed of mixtures of cement and bentonite are utilized. When the hydraulic conductivity value of these works is measured in laboratory tests the resulting value lies in the range of $10^{-8,-9}$ m/s and guarantees the efficacy of the confinement.

In reality the seepage capacity of the water through the wall is often much greater than the projected parameter and this results in all the difficulties which such a situation implies. This problem arises as a result of the fact that the laboratory samples do not sufficiently represent the hydrodynamic features of the medium from a volumetric point of view and also because of the difficulty in keeping the material homogeneous during construction .

This paper illustrates a real case history concerning the construction of a large excavation in alluvial terrains of the Florence plain, on the banks of the Arno River. The studies carried out allowed us to evaluate the real hydraulic conductivity of the wall and consequently to assess how the projected intervention should be carried out .

2 OUTLINE OF THE SITE

The projected intervention concerns the hydraulic confinement of a large excavation under groundwater for the laying of a mat for the sewage lifting plant of the purifier of the city of Florence.

The location, shown in Fig. 1, lies about 10 km west of the centre of Florence and is situated between the Arno River and a drainage trench in a level area where there are ancient gravel quarries with outcropping groundwater.

From a geological point of view the plain is formed by recent alluvial deposits which were deposited by the Arno. The local stratigraphic succession was studied by Capecchi et al. (1976). In particular we have a surface horizon , 6m thick , composed of sandy silts, which can be attributed to the fine overbank sediments (Horizon FIRENZE 1). Below, to a depth of 13.5 m from the local ground level, we have a horizon composed of pebbles, gravel and sand which is totally lacking in fine matrix; sandy horizons alternate with horizons of pebbles and gravel in an abundant sandy matrix (fig. 2). The pebbles are typically round and imbricate; forms of channel erosion and cross bedding can be seen along the excavation fronts. The macroclastites can be attributed to channel, bar and bank deposits which were deposited by the water course in the past (Horizon FIRENZE 2). Below we have lacustrine-palustrine clayey silts (Horizon FIRENZE 4).

The macroclastic horizon forms the main aquifer of the area and is the site of unconfined groundwater with phreatic level variyng from - 7 m (dry season), to - 6 m (wet season); the local piezometric gradient is directed towards the river which therefore drains the groundwater.

3 THE PROJECTED INTERVENTION

An excavation, with an approximately square section (average side 50 m) and a maximum depth of 12m, was to be made 60 m from the left bank of the Arno.

A rigid reinforced concrete cut-off wall (thickness 1 m, height 15 m) was built on the east side. The rigid cut-off wall has a dual purpose. Firstly it has to assure total hydraulic

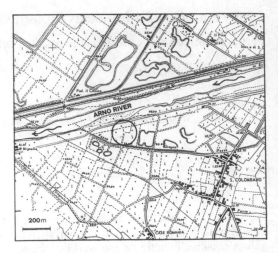

Fig 1. Geographical outline of the studied area.

Fig. 2 Arno River alluvial sediments.

confinement with respect to a former quarry of inert materials adjacent to the intervention area which has now become a pond where water table outcrops. When the excavation is completed it has to constitute a retaining wall for the terrain outside the site. In the following simulations this cut-off wall is considered as being practically impermeable, and in fact it was possible to verify this directly once the excavation was completed.

A plastic cut-off wall made of cement and bentonite was built on the other sides of the site. The main dimensions were as follows: thickness: 0.8 m; height: 15 m; total length : 235 m. The height of the two cut-off walls allows 1.5 m of the walls to penetrate the clayey horizon which lies below the macroclastites. In this way risks of water siphoning at the base of the work are avoided. The planimetry of the cut-off walls is shown in Fig. 3.

On the basis of the results of laboratory permeability tests carried out with a falling head permeameter , the plastic cut-off wall was to be considered as having quite a low degree of permeability, with a mean saturated hydraulic conductivity value of 1.1×10^{-8} m/s. In these conditions it would have been easy to empty the water out of the confined area and carry out the foundation works in dry conditions.

It must be said, however, that the firm who built the plastic cut-off wall pointed out some difficulties during its construction. These were attributed to the high flow speed of the groundwater which therefore tended to wash away the fluid cement-bentonite mixture. This washing away process was said to have led to an incomplete homogenization of the cut-off wall with resulting leaks in places.

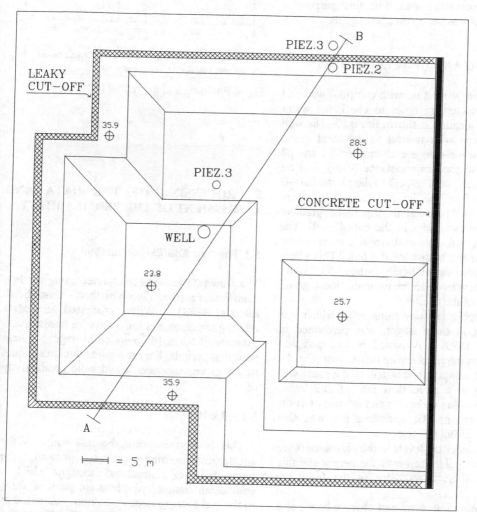

Fig. 3 Excavation area map.

The probable poor hydraulic confinement capacity of the work was highlighted by the saturated hydraulic conductivity value obtained by a Nasberg-Terletskata infiltration test (Celico, 1986).

$Ks = 9 \times 10^{-7} m/s.$

Given the limited size of tested volume with respect to the work size as a whole, we felt that it was necessary to assess the permeability of the cut-off wall on an even greater volume, in order to plan the water emptying system for the confined excavation area. For this purpose a pumping test was set up on a test station.

4 PUMPING AND UPWELLING TEST

A test station was set up with one pilot well and three piezometers in order to check the water levels. The location is shown in Fig. 3. The well is 18 m deep with respect to the local grund level; the piezometers are 15 m deep. P1 and p2 are placed at the same distance (4 m) but on opposite sides, with respect to the plastic cut-off wall. Therefore the level measured in them allows us to determine the hydraulic gradient between the two sodes of the cut-off wall. The well is equipped with a submersible pump which is placed at the bottom of the hole. This pump switches off automatically when the water reaches a level of -15 m from the local grund level (base of the aquifer).

The pumping and upwelling test, which was carried out in three stages, was performed in September 1993. It consisted in two pumping tests with an intermediary upwelling test (Fig. 4).

The first pumping test lasted 15 days and was carried out at a mean flow rate of 282 l/min. There were some variations and interruptions for technical reasons. An upwelling test was also performed during the test.

If we consider the levels in the piezometers we can note a regular drop over the period and this can be used as a reference in the following evaluations.

If we examine Fig. 4 we can observe how, at a rate of 282 l/min, the area delimited by the cut-off wall tends to empty, albeit quite slowly.

The processing of the data obtained from the flow rate test allowed us to assess the hydraulic conductivity value of the plastic cut-off wall in two ways, i.e. processing of the test data with the *non equilibrium* method and a water mass balance within the confined area.

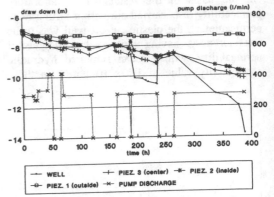

Fig. 4 Pumping and upwelling test.

5 PROCESSING OF THE DATA AND ASSESSMENT OF THE PERMEABILITY

5.1 The *Non Equilibrium* method

The flow rate test was interpreted using the *non equilibrium* method (Jacob method - Custodio & Llamas, 1983). When processed, the data relating to piezometer no. 3 gave a transmission capacity of $6.7 \times 10^{-4} m^2/s$ on average, for long pumping periods. For an aquifer thickness equal to 7.5 m this produces a hydraulic conductivity of

$Ks = 9 \times 10^{-5} m/s.$

This is obviously not the real value of the aquiferous medium but is, instead, the permeability of a simulated medium which is equivalent, from a hydrodynamic point of view, to the real heterogeneous medium composed of the aquifer and the plastic cut-off wall. In this hypothesis the simulated radius of action, for a draining flow rate of 282 l/m and a drop in the pilot well of 7.5 m, is 199m.

In this situation, a hypothetical water particle, placed at a distance from the well equal to the radius of action, should cover first a linear stretch of 158.2 m in the aquifer outside the cut-off wall, then 0.8 m inside the cut-off wall and finally a stretch of 40 m in the aquifer inside the cut-off wall in order to reach the well. In this case we have 3 hydraulic conductors in succession. In this situation we can write that :

$$199/Kspm = (158.2/Ka)+(0.8/Kw)+(40/Ka) \quad (1)$$

where Kspm, Ka and Kw are the hydraulic conductivity of the simulated porous medium, the aquifer and the cut-off wall respectively.

If, on the basis of the grain size characteristics, we give the aquifer a range of possible hydraulic conductivity values which goes from $Ka = 1 \times 10^{-4}$ to 1×10^{-3} m/s , we obtain a value for Kw which lies between 4×10^{-7} and 3×10^{-6} m/s.

The approximations are many as we were not able to take into account the contrasting effects due to the supply of the Arno River and the drainage trench and to the presence of the impermeable barrier of the reinforced concrete wall.

5.2 Water mass balance within the confined area

Within the context of the pumping test the time intervals considered were as follows:

	START (date - time)	END (date - time)
A	16/9 - 14.30	28/9 - 11.00
B	20/9 - 18.00	21/9 - 09.00
C	21/9 - 18.00	23/9 - 10.40
D	27/9 - 17.00	29/9 - 08.00

Fig. 5 gives the permeability values of the cut-off wall as a function of the effective porosity of the aquifer for the 4 time intervals.

The analytic expression of the lines shown in the figure was obtained from the following expression.

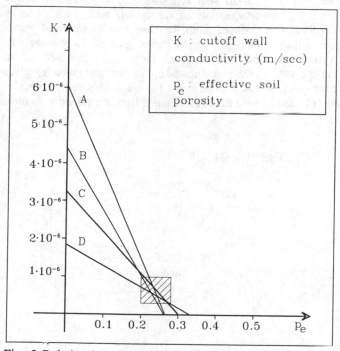

Fig. 5 Relationships between cut-off wall hydraulic conductivity and macroclastites effective porosity (water mass balance).

ne x dVa = Vp - Vf

with:
Vf = Kw x i x S x dt

and where :
Vf = volume filtered through the cut-off wall,
Kw = hydraulic conductivity of the cut-off wall,
i = hydraulic gradient across the cut-off wall,
s = seepage front within the cut-off wall,
dt = time,
ne = effective porosity,
dVa = apparent volume variation of the aquifer,
Vp = volume of water extracted

Fig. 7 Wiew of excavation area and well-points system.

If, for the aquiferous medium in question, we consider the range of effective porosity as between 0.20 and 0.28, we obtain a range of possible values for Kw which go from 3 x 10 $^{-7}$ to 1 x 10 $^{-6}$ m/s.

The results conform with the ones obtained using the previous method and the range is better specified.

6 EXPERIMENTAL VALIDATIONS

The excavation was carried out in February 1994. On the basis of the above evaluations they were advised to set up a pumping system. A system of well points was set up along the the entire perimeter of the deepest excavation (Figs. 3,6 and 7). These keep the level of the groundwater 12.5 m from the local grund level below the deepest part of the excavation (base is -12 m from the local grund level). The pumping rate of the well points required in order to stabilize the water table at this depth was 158 l/m.

This flow rate obviously corresponds to the one filtering through the plastic cut-off wall. The profile of the water table in the aquifer and the cut-off wall is shown in Fig. 6. The hydraulic conductivity of the cut-off wall, which can be deduced from the flow rate of the well points and from the hydraulic gradient across the cut-off wall (the latter is obtained from measurements taken in the piezometers) gives Kw a value = 4 x 10 $^{-7}$ m/s. This value coincides ‑ith the ones obtained from the previous studies.

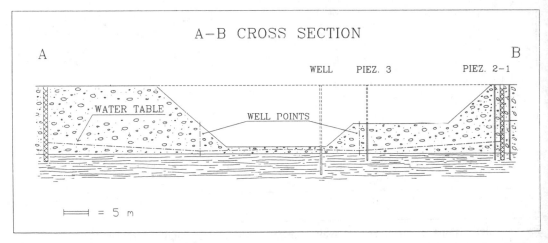

Fig. 6 Excavation area cross-section.

The determination of the real value of Kw allows us, among other things, to determine the hydraulic conductivity and the effective porosity value of the aquifer. The values obtained from (1) and from the diagrams in Fig. 5 respectively are:

$Ka = 1 \times 10^{-3}$ m/s
$ne = 0.24$.

7 CONCLUSIONS

We have illustrated a real example of the assessment of the hydraulic conductivity of a plastic cement-bentonite cut-off wall by means of an in situ pumping test. The evaluation method used, albeit with a certain degree of uncertainty, allowed us to assess the order of magnitude of the hydraulic conductivity of the cut-off wall and thus dimension the draining system of the confined area appropriately in the excavation phase.

The following considerations can be made:

- the real Kw value is about 1.5 order greater than the one determined in laboratory tests. This fully confirms what was experimentally proven on fine grain porous mediums , both artificial ones such as clay liners (Daniel, 1989) and natural ones (Gargini, 1990). The small size of the laboratory samples leads to an underevaluation of the real datum, also in the case of cement-bentonite mixtures.

- in situ tests, such as the Nasbeg-Terietskata ones or, even better, the ones carried out with the "high sensivity ring infiltrometers" (Focardi et al. , 1993) , certainly arrive nearer the real value for the permeation of a larger volume of porous medium. However, this work confirm that the infiltration tests in borehole, as Nasbeg-Terietska, tend to overestimate hydraulic conductivity as they do not take into account the capillary tension of the medium, and they have flow conditions (vertical infiltration surface) which are more difficult to represent with a single equation.

- the methodology used allowed us to determine the hydraulic conductivity and the effective porosity of the aquifer macroclastites deposited by the Arno River (1×10^{-3} m/s and 0.24 respectively).

REFERENCES

Capecchi, F., G. Guazzone & G. Pranzini 1976. Ricerche geologiche e idrogeologiche nel sottosuolo della pianura di Firenze. *Bollettino Società Geologica Italiana,* 94:661-662.

Celico, P. 1986. *Prospezioni Idrogeologiche.* Liguori Editore (Italy).

Custodio, E & M.R. Llamas 1983. *Hydrologia subterranea.* Ediciones Omega S.A., Barcelona (Spain).

Daniel, D.E. 1989. In situ hydraulic conductivity tests for compacted clay. *Journal of Geotecnical Engineering,* 115, 9:1205-1226.

E.P.A. 1984. Slurry trench construction for pollution migration control. *EPA 540/2-84-01,* Cincinnati.

Focardi, P., A. Gargini & G. Gabbani1993. An experimental study of the flow of organic pollutants throgh compacted clays. *SARDINIA 93 Fourth International Landfill Symphosium,* S Marghcrita di Pula, Cagliari (Italy), 1:313-324.

Gargini, A. 1990. Permeabilità del terreno non saturo e vulnarabilita' all' inquinamento di acquiferi a porosita' primaria; proposta di un nuovo metodo di valutazione. University of Florence (Italy), *Doctoral thesis (unpubl).*

Modification of a quarry face: Stabilisation criteria and environmental reclamation

Modification de la surface d'une carrière: Critères de stabilité et récupération environnementale

O. Del Greco, M. Fornaro & C. Oggeri
Dipartimento di Georisorse e Territorio, Politecnico di Torino, Italy

ABSTRACT: The study here presented refers to a large sliding movement in the rock mass of a quarry face located in a mountainous area near the town of Varese (Italy). A high quality quartziferous porphyrite is quarried. The quarry face is 180 m high, 220 m wide and is divided into 15 m high benches.
Some instability phenomena has developed on the mountain slope at the back of the top of the quarry with the occurrence of tension cracks, spreading over a length of some tens of meters and with an aperture which reaches 120 cm. The study of the sliding phenomenon was carried out by means of investigations and monitoring over a two year period, in order to select and develop an effective stabilization method.

RESUME: L'étude présentée concerne un glissement dans le rocher d'une carrière près de Varese (Italie), dans un zone montagneuse. Une porphyrite quartzeuse de haute qualité est extraitedela carrière. Le talus de carrière mesure 180 mètres de hauteur, 220 de largeur et est divisé par des banquettes de 15 mètres de hauteur.
Le glissement c'est produit dans le versant montagneux au-dessus du bord de la carrière, avec des fissures de traction, étendues sur plusieurs dizaines de mètres, et ayant une largeur de 120 centimètres. Le problème a été étudié avec des mesurations et des relevés en situ de la durée de deux ans, pendant les quels l'évolution du phénomène de glissement a été suivie. Les interventions de stabilizations du rocher ont été prédisposées et en partie réalisées.

1 INTRODUCTION

The here studied quarry is located in Northern Italy, in the Pre-Alpine range, not far from the town of Varese. A porphyritic andesite has been exploited in the quarry for many years. This type of material is well suited for use in railroad embankments and to make high quality bituminous road concrete and runway concrete. A year's production amounts to 150.000 m3 (bulk volume).

The exploitation has always been carried out by blasting. After the choice of a first classical scheme of cylindrical blastholes in the 15 meter high benches (subvertical slices exploitation, advancing upwards by simultaneously retrating all the excavation benches), a new method was adopted, consisting of a slight fragmentation using preliminary blasting and the subsequent use of a hydraulic power shovel (horizontal slices exploitation advancing downwards, by creating berms and benches at progressively lower elevations) (Fornaro et al. 1992). The blasted material is then gravity conveyed to the quarry yard at an elevation of 704 meters, where a preliminary comminution allows the hauling of the material, by a conveyor belt, to the sieving plant, located at a lower yard at an elevation of 670 meters. The material is finally stockpiled there.

The quarry face was extended, in January 1992, to a length of 250 meters at the bottom and 150 m at the top, at an elevation of 860 meters, and was 180 meters high. The inclination of the overall quarry face was of about 50°, the topographic inclination of the natural slope was of about 28° (figure 1 and 2).

A large instability phenomenon, consisting of the slump of the overburden formation, was observed in the mountainous area at the back of the quarry (Walker 1987).

At that moment the excavation activity was interrupted, for safety measures, by the Environmental and Quarry Department of Varese. A study concerning the behaviour and the characteristics of the sliding phenomenon was meanwhile started by the authors. The purpose of the study was to obtain information on the extension of the unstable area and the evolution of the sliding. The possibility of coordinating the geotechnical

Figure 1. Overall view of the quarry.

investigations with the procedures that were to follow in order to ensure the stability of the mountainous slope and the quarry face was also defined.

Two aspects of the stability were considered: the stability of the exploitation area, linked to the quarrying activities, and the possible impact caused by the acceleration of the sliding phenomenon on the surrounding land. In addition, a rehabilitation technique had to be considered, as the quarry is located in an area of particular naturalistic importance.

The geotechnical investigations, started in February 1992, furnished knowledge on the situation of the overall slope and, in particular, on the quarry face. These results permitted the choice of useful interventions for a final land reclamation (Del Greco et al. 1994).

2 GEOLOGICAL FEATURES

The rock mass of the quarry belongs to a Permian volcanic formation, embedded between a Paleozoic crystalline bedrock (gneiss, schist and amphibolites) of the "Serie dei Laghi", and a Mesozoic carbonatic formation of Eastern Lombardy. The stratigrafy of the quarry is characterized by three units. The sequence is formed by fine gneiss and micaschist which are overlapped by typical volcanic rocks, such as pyroclastites and lavas; there is a granitoide

quartziferous porphyry at the top of the series.

The exploited rocks belong to the volcanic series, and they are represented by green-violet coloured andesites-dacites of a fine texture, and by pyroclastic rocks cemented in an agglomerated structure. These rocks are highly weathered for a great depth, as the glacial erosion activity did not involve this area, and they were subjected to an induced fracturation over a large area, due to strong regional tectonical activities. The fracturation is evident both in the rock mass and in the presence of tectonical faults; the main faults follow the WSW-ENE, N-S and NE-SW orientations These geological actions determined the outcrop of the effusive series of the quarry.

3 SLOPE INSTABILITY ANALYSIS

The sliding phenomenon was observed at the back of the quarry face edge. The falling of a few tens of cubic meters of the detritic overdurden down the quarry face occurred at the beginning of the phenomenon.

The unstable area spread over the cohesionless overburden and also involved the volume of the underneath layer of weathered and heavy fractured rock. The total thickness of the unstable volume was evaluated as being about 20 meters, with the length of the front border of the slide being about 180 meters, the length in the normal direction about 90 meters and, finally, the difference in elevation about

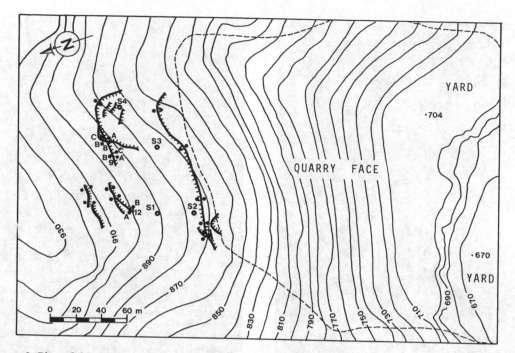

Figure 2. Plan of the quarry area: the dotted line borders the excavation zone, above which the fractures are indicated. Borehole points (S1/S4) and jointmeter points are also indicated.

50 meters. The estimated unstable volume was about 120.000 m3 (in situ volume) (figure 2).

The discontinuities of the overburden rock, which are more or less persistent, are situated along the main WSW-ENE directions, and they are grouped into two principal sets. As two typical crown failures with circular shapes and with the concavity oriented towards the quarry were observed, the landslide was preliminarly classified as a slump or rotational slide. The features of the landslide have been studied since the beginning following the evolution of the absolute and relative displacements of the soil elements contoured by the cracks. The structure of the area is very irregular and complex, also due to the presence of large cracks inside the rock formation and the occurrence of very altered material. This situation determined the fall of a relatively limited volume of debris material down the quarry face. Furthermore, the amount of the displacement rates has always been very high over the whole unstable area, for the same reason.

In order to understand the mechanical behaviour of the landslide, two hypothesis could and can still be considered as, at the moment of occurrence, the hypothesis were not the univocal and exclusive interpretations of the real situation. The phenomenon, in fact, showed features of both hypothesized types of movement, that is the landslide seems to be a combination of two principal types of movement (figure 3).

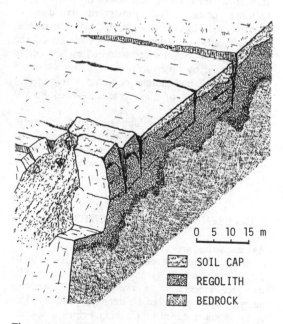

Figure 3. Sketch of instability features at the top of the quarry face.

Figure 4. Open fractures in the overburden above the quarry face: a) normal throw; b) thrust throw.

The first hypothesis involves a rotational and traslatory movement of rigid units surrounded by the tectonic cracks. This movement is due to gravitational forces that cause an overturning moment and it determines, together with the quarrying activities, the formation of surface cracks characterized by noteworthy vertical throws (more than 1 meter), both normal or thrust throws (figure 4). The aperture of these cracks reached 80-120 centimeters.

The second hypothesis supposes a deep settlement of the thick rock formations, irregularly jointed and weathered, accompanied by a differential compaction of the overburden and with a total volume reduction of the rock mass.

4 INVESTIGATIONS AND MONITORING

As landslides are evolutive phenomena, the field investigations cannot be considered as isolated or defined activities. This study, which is constrained by time reqirements for a preventive or corrective design, should therefore be iterative (table 1). This study requires adequate planning before a specific landslide is instrumented. This process consists of:
- the determination of the types of measurements;

- the selection of the best suited types of instruments to make the required measurements;
- the planning of the location, number and depth of the instrumentation;
- data acquisition.
The described study deals with the following features (Dunnicliff 1988):
- assessment of the extension of the unstable area;
- definition of the lithostratigraphy of the rock mass;
- geomechanical and structural rock mass characterization;
- dominant landslide type classification;
- assessment of the unstable rock mass volume and of the potential zone of accumulation above the quarry yard;
- forecasting the evolution of the landslide.
A series of on site investigations were carried out for these purposes:
- topographic survey of the surface cracks;
- geophysical surveys, by means of a seismic refraction method, to determine the thickness of various rock mass layers and the depth of the surface of rupture;
- four geotechnical drillings with core recovery;
- large scale structural and geomechanical survey.
Some monitoring techniques were also performed:

Table 1. Scheme for planning investigations and remedials in quarry landslides.

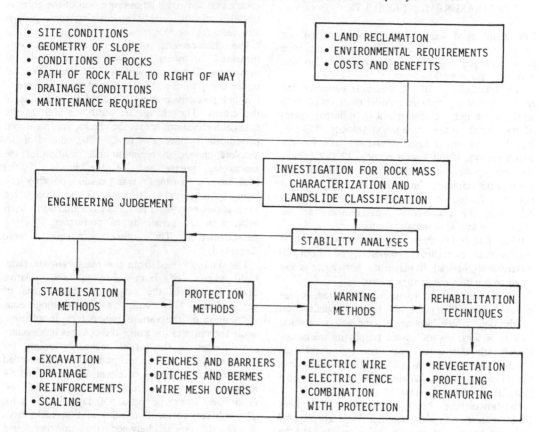

- monitoring of relative displacements across the top cracks and transverse cracks, by means of the setting up of 21 large surface jointmeters;

- inclinometric measurements in the four drilled boreholes, using a probe inclinometer;

- periodic topographic surveys of the absolute planimetric and altimetric positions of targets located at the main crack sides and the heads of the four inclinometric casings;

- recording of meteoric precipitations.

A careful inspection of the water drainage conditions of the rock mass was carried out; the results of this inspection showed a good drainage of the whole mass and the localisation of an intermittent local water flow at the top of the quarry face.

Finally, some rock fall tests were carried out at the top berm of the quarry by ISMES of Bergamo. These tests were carried out in order:

- to verify the kinetic effects at the base of the quarry;

- to recognize the experimental trajectories of the rock blocks of 0.2-2.6 m3 and the standstill boundaries of the rock block fragments;

- to define an interdiction area at the base of the quarry face.

The investigations usually permitted the identification of the areas subjected to sliding and the unstable volumes. They also permitted the scheduling and carrying out of the removal of the unstable mass, with the contemporaneous setting up of passive protections at the base of the quarry face. These planned stabilization methods were subjected to environmental restrictions - the quarry is located in the "Campo dei Fiori" natural park - and safety requirements, both for short and long term. These methods had the purpose of developing corrective measures to minimize the evolution of the movements and to obtain a final land reclamation, without excluding, while proposing, a future quarrying activity.

5 DISCUSSION AND INTERPRETATION OF THE MEASUREMENT RESULTS

The geophysical seismic surveys were carried out using eight refraction measurements, which have been arranged parally and ortogonally to the maximum slope surface direction.

The interpretation of the measures permitted the recognition of a geophysical model made up of three stratigraphic units, characterized by different values of the seismic wave propagation velocity: 350 m/s for the overburden layer; 600-1500 m/s for the weathered and highly fractured rock; 1200-2000 m/s for the unweathered rock mass, that is, the porphyritic refracting bedrock. This modelling has been confirmed by the core logging, that is calibrating the geophysical measurements on the basis of the available statigraphic data.

It was possible to show a surface formation, that is cohesionless or scarcelly cohesive, of about 20 meters deep; beneath this layer the porphyrite is less weathered with a closer rock matrix.

In the above mentioned surface formation, it was possible to distinguish a loose organic soil overburden, which becomes a sandy-clayed natural shot rock with several clastic porphyritic elements. Underneath this quarry-run rock the porphyrite, which could be locally named regolith, is nevertheless weathered and highly fractured (figure 4), showing a granular soil-like behaviour.

The unweathered bedrock, located at a depth of 16-20 meters, was also identified by a geophysical measurement carried out on a similar rock base of the quarry yard.

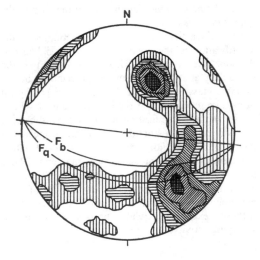

Figure 5. Stereonet of the discontinuities in the rock mass (Fq: overall quarry face; Fb: bench face).

The geostructural survey, carried out on the entire rock mass, showed a large dispersion of the poles of the discontinuities, while maintaining some general structures (figure 5).

The displacements of the unstable mass were measured in order to determine the rate of movement. Twentyone surface jointmeters were set up for this purpose; they were made simply made of two hub sets driven into the ground on both sides of the cracks. The change in width, as well as the change in elevation, across the cracks, was measured periodically by means of tape. The control of the absolute change in elevation and position of the monitoring points and of the heads of the inclinometric casings was made possible by topographic surveys. Because of the progressive excavations, only eight hub sets have afterwards been kept, with the possibility of performing 7 crack measurements. The other jointmeters were destroyed.

The data obtained from the measurements show high values of both absolute and relative displacements, in the range of several tens of centimeters, but with a non-homogeneous distribution on the various jointmeters. In addition, while the range is the same, the changes in elevation are different for the various measuring points: for example, the S1 and S2 inclinometer heads recorded a 25 centimeter change of elevation, the S3 and S4 inclinometer heads recorded values of 60 centimeters, over a period of 550 days (figure 6 a-b). The hub sets at the two sides of the cracks have shown differences between the upstream and downstream anchor pin displacements, as the downstream anchor pins generally recorded higher movements. The ratio of horizontal to vertical components of the surface displacements of the various set ups were not homogeneous, confirming the complexity of the type of movement. This behaviour could be explained by the crown failure shape of the instrumented cracks.

The available data, if related to the planimetric location of the monitoring points, showed higher movement rates towards the East side of the quarry. A good correlation between the sliding movements and the precipitation has been observed. The time based graph of both the landslide displacement and the accumulated rainfall and snowmelting indicates a visual relationship, as is shown in fig. 7. The area was subjected to heavy precipitation during the spring and autumn of both 1992 and 1993.

The displacement rates increased during the mentioned periods, excluding a spontaneous stabilisation. Because of the drainage and permeability conditions of the surface rock formation, the water content changes did not cause

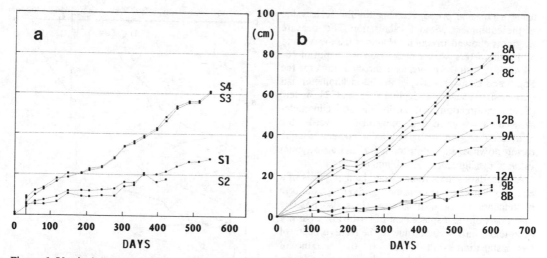

Figure 6. Vertical displacements vs. time, since 29.07.92: a) inclinometric casing heads; b) pins of jointmeters.

dependent movements, such as earth or debris flows.

The planned investigations did not foresee the use of piezometers, as no groundwater table was found during the drilling investigations and the excavation phases. On the other hand, it was observed, both in the past and during this last study period, that as a consequence of heavy rainfalls and water infiltration, the intersection of the aquifer with the quarry face produced a flow situated at the second upper bench, in the central part of the quarry.

Four inclinometer casings were installed in the exploration drill holes in order both to control the displacement evolution of the overburden formation and obtain an indication of the position of the surface of rupture. These were extended through the formation suspected of movement, and well into the materials that, in the best judgement at the time, were assumed to be stable. The casings were in fact installed two meters inside the porphyritic bedrock. The measurements were performed periodically

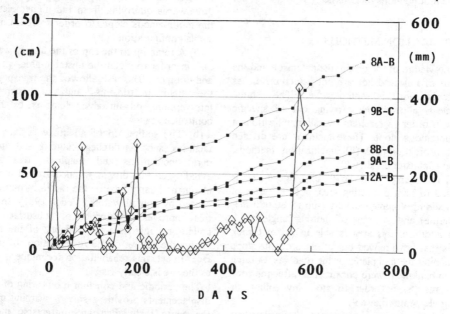

Figure 7. Relative displacements of pins of some monitored fractures vs. time. Cumulative precipitations (◇) vs. time are also plotted in the same period. Days are counted since 31.03.92.

4115

using a portable probe, reaching the full scale value of the equipment without stabilisation. The measure diagrams showed irregular behaviour for the S1 and S2 inclinometers, with a peak value at a depth of 10-14 meters, with the movements directed towards the South and East respectively; the S3 inclinometer had two peak values at a depth of 15 and 21 meters (Southern direction) and, finally, the S4 inclinometer showed a very regular movement towards the Southern direction, with a gradual alignment of guide casing down to the reference point, and a maximum value of disalignement of 110 mm (figure 8).

All this data seems to confirme the hypothesis of a complex slide phenomenon without uniform movements on the unstable area. It is also possible that a part of the displacement of the bedrock has not been taken into account, due to the relative depth of the installation with respect to the amount of movement rates and volume involved in the landslide.

As it had been planned to shape using benches the area where the inclinometers were installed, a new installation of at least two very deep guide casings was foreseen, for after the completion of the excavation activities in the surrounding areas, starting from the intermediate yard at the elevation 850 meters. This solution seemed to be the most suitable for the development and monitoring of the long term quarrying and stability features. The future inclinometric measures, coupled with topographic surveys, should indicate any possible deep movement inside the mountainous rock mass.

6 STABILIZATION METHODS

The knowledge of the origin, the extension and the behaviour of this landslide are not yet complete, as the phenomenon is still active. The study, nevertheless, allowed the carrying out of specific methods in order to reach an overall stabilization of the mountainous slope. These methods are divided into (a) protection and (b) stabilization methods, including (Schuster et. al 1978):

(a) The quarry yard (670 m., figure 2) is considered to be a protecting area, both in terms of the high sliding volume capacity (up to 80.000 m3, bulk volume) and in terms of shutting single block falls. In addition, this area is able to prevent rock materials that have moved out of place on the slope from reaching the mill plants or the roadway. A ditch and an embankment were constructed, after the rock fall test results, in order to stop any rolling or bouncing elements (figure 9).

(a-b) An intermediate quarry yard, at an elevation of 850 meters, corresponding to the original top of the quarry, was excavated, with the purpose of

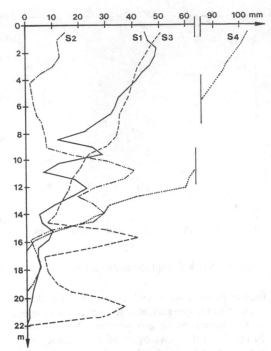

Figure 8. Inclinometric measured values.

acting as a first passive protection. The quarry yard at 704 meters of elevation was also considered as a potential zone of accumulation. This arrangement foresees the subsequent profiling using benches, with downwards quarrying from the higher elevation of the mountainous slope, in order to reach an overall stable configuration.

(b) A ramp up to the top of the quarry was carried out, in order to reach the unstable area with shovels and dumpers. This also allowed the transportation of the field instruments and to operate local interventions on unstable elements by means of controlled scaling.

(b) The setting up of adequate surface drainage facilities, such as ditches, culverts and appropriate berm inclinations and shapings, was aimed at avoiding erosion due to an uncontrolled surface flow along the quarry face or the ramp, in particular that caused by rainfall waters (Veder 1981). In addition, this improves the surface or subsurface drainage conditions, increases the stability of the slope and allows the setting of the revegetation of the quarry face. Infact, this rehabilitation technique is necessary as the site is clearly visible.

The periodic and constant monitoring of measured displacements provide a sort of warning method, in the sense that when an appreciable increase of movements is noticed, due for example to heavy rain, the stabilization excavation activities are stopped.

Figure 9. The ditch and the embankment, in the protection area, constructed in the quarry yard.

7 CONCLUSIONS

The slope stabilisation and land reclamation, planned for the period from 1992 until 1995, foresee, at their completion, the removal of the overburden soil and weathered porphyrite for a thickness variable from 15 to 25 meters, between elevations 940 and 850. The total volume excavated will amount to 480.000 m3. A final slope profile will be shaped with a series of benches (maximum 8 meters high and maximum 45° inclined) and 8 meter wide berms. The overall slope inclination will result to be 35° maximum. A 6 meters wide ramp will connect the various upper berms, with a maximum slope of 15% in order to have the possibility of reaching the top area for inspection and maintenance.

The outcrops, pointed out by the already profiled benches at the top of the slope, have shown a rock mass characterized by irregular contacts between the overburden and the bedrock. Local asperities determine, on a metric scale, the alternance of porphyrite relict rock elements with inlets of soil filling. These soils, as a result of the processes by which they were formed, influence the slope stability. The modifications of slope profiles, the bench inclinations and the related stability analyses were carried out assuming that some mechanical properties of the rock formation inside the mass, such as rock mass structure features, are better than those of the surface materials. The situation obtained by an effective slope profiling excavation, instead of a superficial excavation, is expected to improve the stability, even though the slope shows less inclined benches and overall profile inclination. The modification of the slope determines a top unloading and a consequent improvement of the stability conditions (Giani 1992, Hoek et al. 1981).

The cost of stabilisation and protection measures applied during the excavation phase can be absorbed in the initial construction cost, especially if a contingency item is included in the construction budget. In the case of a quarry, law requirements foresee a previsional cost for land reclamation and rehabilitation techniques, while the consequences of a landslide become an additional cost. In the described case, the comprehensive reclamation cost has been charged to the quarrying company, and amounts to about some hundreds of thousands of dollars, a part of which has been used for the track construction, as the original quarry ramp was eliminated for "aesthetic" reasons. In order to avoid any charge for the collectivity, the availability of these financial resources has been achieved by the selling of a large amount of excavated granular material from the described operations. This confirmed the idea, which was not completely accepted by the public, that environmental protection and mineral resource exploitation, if correctly applied, can further pursue convergent interests (Mancini et al. 1986).

Acknowledgements. The Authors are grateful to the Environmental and Quarry Department of Varese for their kind cooperation.

REFERENCES

Del Greco, O., Fornaro, M., Oggeri, C., Traversi, G.
...& R. Carimati 1994. Indagini e controlli per la
...bonifica di un fronte instabile di cava. Proc. IV
...Geoengineering Int. Congr. on Soil and
...Groundwater Protection: 109-116. Associazione
...Mineraria.Subalpina, Turin.
Dunnicliff, J. 1988. Geotechnical instrumentation for
monitoring field performance. Wiley, New York.
Fornaro M. & R. Mancini 1992. The treatment of
...rock faces of the abandoned quarries and open cast
...mines. Criteria and examples from Italian
...experience. Proc. ISRM Regional Symp. on Rock
...Slopes: 295-302... Oxford. &.. IBH .Publ. Co.,
...New Delhi.
Giani, G.P. 1992. Rock slope stability analysis.
...Balkema, Rotterdam.
Hoek, E. & J.W.Bray 1981. Rock slope engineering.
...The Institution of Mining and Metallurgy, London.
Mancini, R. & M. Fornaro 1986. Impostazione e
...conduzione dei lavori di coltivazioni minerarie per
...un corretto sfruttamento delle risorse nel rispetto
...dell'ambiente. Bollettino Associazione Mineraria
...Subalpina, Turin, 1, 51-69.
Schuster, R.L. & R.J. Krizek ed. 1978. Landslides:
...analysis and control. Transportation Research
...Board, National Research Council, Special Report
...176, Washington.
Veder, C. 1981. Landslides and their stabilization.
...Springer- Verlag, Wien.
Walker, B.F. & R. Fell ed: 1987 Soil slope instability
...and stabilisation. Balkema, Rotterdam.

Engineering geology applied to mine planning at N4E Mine, Serra dos Carajas, Brazil

Géologie de l'ingénieur appliquée à la planification de la mine N4E, Serra dos Carajas, Brésil

J.A.Hilario, J.R.C.Cordeiro & R.G.Ferreira Jr
CVRD, Companhia Vale do Rio Doce, Brazil

D.J.L.Rocha & L.M.Ojima
Figueiredo Ferraz Consultoria e Engenharia de Projeto, Brazil

ABSTRACT: N4E Mine is an iron ore open pit mine. The final pit is about 400m deep, 4000m extent and 2000m wide. The total iron ore production is around 1.200 million tons, with 600 million m^3 of waste material. One of the five waste dumps predicted will have a volume of 100 million m^3 and a maximum height of 330m. A tunnel 2.400m extent and with a 10m^2 cross-section, connected with four drilled wells, is being excavated to provide the pit drainage. Engineering Geology has been widely applied to mine planning, specially at the studies dealing with pit slopes stability, waste dumps and drainage gallery designs. This paper comments the used methodology, as well as the benefits derived from its appliance to mining activities.

RÉSUMÉ: La Mine N4E est une mine de minerai de fer à cavité ouverte. La cavité finale a presque 400m de profondeur, 4.000m d'extension et 200m de largeur. La production totale de minerai de fer est de presque 1.200 million de tonnes, avec un rejet de 600 million de m^3. Une des cinq piles de rejets prédictes présentera un volume de 100 million de m^3 et une hauter maxime de 300m. Un tunnel avec 2400m d'extension connecté à quatre fosses excavés est en train d'être construit pour proportionner le drainage. La Géologie de l'ingénieur a été largement utilisé dans la planification de la mine spécialement dans les études sur la stabilité du déclive, les dessins de la galerie de drainage et de le piles de rejet. Ce travail montre la méthodologie utilisée et les bienfaits de son application dans les projects de mineralisation.

1. INTRODUCTION

N4E open pit mine is located in Serra dos Carajas, state of Pará, Brazil, with an iron ore production of 35 million tons a year and an annual volume of waste material of about 20 million m^3 (fig. 1).

The final pit will present very significant dimensions: 4.000m extension, 2.000m width and 450m depth.

The mining firm, COMPANHIA VALE DO RIO DOCE - CVRD, decided to develop Engineering Geology studies since the initial stages of mine planning, due to different and complex geological and geotechnical aspects connected to such a huge excavation. N4E Mine, as an exception to usual geomechanical models found in iron ore mines, presents the first 100 to 180m of the cut slopes in "saprolitic soil".

The paper presents the methodology used for performing the studies related to the following itens, as well as its main results: stability of final pit slopes, waste dumps and drainage gallery designs.

The final pit layout, the approximate location of the waste dumps and of drainage gallery axis are shown in fig. 2.

2. PIT SLOPES STABILITY ANALYSES

2.1 *Preliminary Evaluations and Investigation Programs*

Stability studies have been managed in three distinct stages: Preliminary Evaluations and Investigation Programs, Basic Studies and Risk Analysis.

The main purpose of the initial phase was to develop geomechanical model for mine rock mass, so that more realistic information on its geological and geotechnical features should be used for general slope angles evaluation. As a final result, a preliminary mapping of different slope angle zones was made for the pit, allowing CVRD to use these data to reappraise mine planning and waste/iron ore relation. Besides, it was also possible to evaluate the operational and economical consequences of introducing the slope angles designed to assure the desirable safety factors.

The geological and geotechnical features that came out and could not be elucidated in this stage, due to the short period of time and undetailed degree of information available, were used as a guideline for programming field and laboratory investigations, to be executed next phase.

The preliminary studies were developed in 45 days, envolving the following tasks: geomechanical description and classification of rock and soil samples from drillholes, previously executed to provide data for iron ore reserve evaluation; structural mapping of exposed pit benches; data analysis to allow geomechanical modelling, based on Bieniawski rock mass classification (1981); programming field and laboratory investigation including 19 geotechnical drillholes (1517m), 102 triaxial tests, 14 consolidation tests; installing 16 piezometers in the drillholes planned; pit mine zoning according to different geotechnical domains and pit mapping, with slope general angle indication for each zone previously designed.

2.2 Basic Studies

The main purpose was to achieve a more detailed geomechanical zoning of the pit mine. The applied methodology comprised: geotechnical description of the drillholes programmed in the previous stage; interpretation of piezometric data and laboratory tests results to reevaluate water level conditions and strength parameters; redefinition of geomechanical pit zoning and selection of geotechnical cross-sections for more detailed stability studies (fig. 3); slopes computational analyses by using the software "CLARA", available by FIGUEIREDO FERRAZ; redefinition of slope angles map (fig. 2).

Highly altered layers as thick as those existing in N4E Mine rock mass (more than 100 meters), demanded a specific field and laboratory investigation program to enable their geotechnical evaluation. This subject is presented with details in Cruz., Ojima & Rocha (1992). The criterium suggested by Hoek (1988) was used to determine the strength parameters of less altered rock mass layers.

2.3 Risk Analysis

After defining the geomechanical sectors and the general slope angles for each zone, CVRD designed the definite pit mine.

Studies related to risk evaluations in specific pit sectors are now being developed by FIGUEIREDO FERRAZ and CVRD. Complementary stability analyses are going to be made, specially in west face sectors, trying to achieve stronger slope angles by foreseeing eventual rock mass drainage. More detailed studies on the effect of natural water level lowering due to pit excavation are essential to assure the real need of using artificial drainage systems to achieve stronger slope angles.

3. WASTE DUMPS DESIGN

3.1 Preliminary Design

The total volume of waste material to be originated from pit operations is around 200 million m³. Its disposal will be made in 5 waste dumps shown in fig.2, which are actually being designed or built.

Each dump design is being developed in two distinct phases: preliminary and executive.

The main purpose of preliminary design was to appreciate the influence of foundation and waste materials geotechnical characteristics on future embankments safety, so that ways of stabilizing the dumps could be preliminarly designed and their costs estimated.

The comparison between waste disposal capacity of each site previously selected and costs, periods of time and eventual specific operational practices needed to guarantee safety conditions, sometimes leaded to searching alternative places for the embankments in N4E Mine. Some of the designed dumps will be built in stages to enable piezometric monitoring. The interpretation of measurements will guide future decisions on changing building practices, reinforcing the embankments or even limiting disposal volumes.

For instance, West Dump is placed near the top of pit slopes at N4E Mine west face (fig. 2). It was previously designed to enable the disposal of 125 million m³. Stability analyses were developed, revealing low safety factors related to potential circular sliding surfaces envolving both dump body and foundation rock mass. The total volume and embankment height were consequently limited to 20 million m³ and 50m, respectively.

Northwest Dump presents an estimated volume of 100 million m³, 330m height and about 2.300m extension (fig. 4). It is being built in three different stages to achieve the desirable safety degrees.

The site initially selected for the dump building presented a 20m thick and widely spread stratum of a very soft sedimentary layer, originated from a previous non controlled waste disposal at the natural slopes of the drainage area. The site had not been protected against erosive processes acting on waste slopes and consequent deposition on natural drainage system. Removing this layer was impossible due to the necessary period of time involved and the access difficulties. Building the embankments in three stages with different limiting volumes and dump heights, enabled CVRD to initiate the immediate disposal on that area without compulsory removal of soft material until the second phase comes to an end (fig. 4).

The methodology adopted for preliminary waste dumps studies was based on FIGUEIREDO FERRAZ and CVRD designers previous experiences. Field observations, existing data gathering and analyses were also considered to manage embankments design in a short period of time, about 30 to 60 days. Aspects related to internal and external layouts, drainage systems and building techniques were preliminary defined.

The following tasks were developed: field inspections; design criteria definition, together with CVRD technical crew (desirable safety factors, benches dimensions and slope angles currently adopted by CVRD, construction materials available, access dimensions and dips); preliminary geomechanical classification of waste and foundation materials based on field observations and existing investigation results; stability analyses; different layout alternatives design, intending to associate CVRD practice and experience in building waste dumps with environmental impact reduction, desirable safety levels achievement and disposal capacity needed; selection of the best solution to waste dump building; preliminary design of drainage systems; costs evaluation; indication of monitoring plan, field and laboratory investigation programs.

3.2 *Executive Design*

The main purpose was to detail the waste disposal scheme designed in the previous phase. The following aspects were carefully approached:

- Confirming the technical and economical viability of using the materials previously selected for filter drainage. Granular materials are rarely available in N4E Mine, specially in the present stage of mine operations. Filter drainage design had to associate this operational difficulty with the volumes needed for building the embankments, so that dumps safety should not be damaged and construction costs very much increased. Samples of waste materials and ore produced in mine operations were collected and submitted to laboratory tests to confirm their applicability in drainage systems.

- Reevaluating foundation and waste materials strengh parameters through tests performed. The same process was valid to foundation water levels previously admitted, which were compared to available piezometric records, originated from monitoring programmed at preliminary design stage. All those information were considered and updated computational analyses were made to verify the suitability of original geometric definitions for the embankments.

- Waste dumps are being built applying controlled disposal techniques. Embankment benches are constructed from the lower levels to the upper ones, so that all layers can be compacted at least by mechanical equipment traffic when building the dumps.

- The idea of a detailed monitoring plan was reinforced and equipment installation and measuring practices were specified.

4. DRAINAGE GALLERY DESIGN

4.1 *Reasons for Building a Drainage Gallery*

Significant contributions from runoff and waste ground seepage would tend to be retained in N4E Mine bottom pit. The outflow is now going to occur through a gravitational system made up by four wells connected to a drainage tunnel 2.400m extent, which is placed bellow the final bottom pit.

Developed economic studies showed that building a gallery to manage gravitational drainage of mine would be significantly less costing than continuous pumping. Installing and operating this system would demand more than 300m high pumping in the final stages of mine operations.

The total water volume that tends to be retained in the bottom pit is extremely relevant, due to a significative drainage area, with medium annual precipitation of about 2.100mm, focused on the period from October to April. Adicted to it, a continuous outflow of 2.000m³/h from ground water seepage is pumped through 16 wells installed to lower water level and manage mine operations.

4.2 *Drainage System: Wells Connected to the Gallery*

Four wells with 45cm diameter each will be drilled, provided with metalic covering of 34cm, 300m deep.

Drilling the wells will come to an end before the gallery reaches them. The connection will be achieved more easily by topographic fixing of wells depth.

The tunnel presents a 10m² cross-section, 2.400m long, with a 0,65% slope. The schematic gallery profile connection is shown in figure 5.

4.3 *Drainage Gallery Design*

Three rotary drillholes were made, two at the tunnel exit portal area and one at the wells connection region, providing information for geomechanical classification of rock mass.

Drillholes were not programmed at intermediate gallery zones, due to very thick existing overburden (more than 300m) and planned horizontal drillholes to be made from excavation head, during tunnelling operations.

Rock mass was classified according to Bieniawski (1981). The results correspondent to tunnel profile are shown in fig. 5.

Two building methods were considered, even in executive design stage: NATM (New Austrian Tunnelling Method) and TBM (Tunnel Boring Machine). Five typical cross-sections were predicted for both, according to different classes of rock mass: NATM with 10m² and TBM with 5,6m². Figure 5 also presents typical sections for each of the tunnelling techniques to be applied in soil and fractured rock conditions.

Commercial proposals presented by international companies to build drainage gallery, proved that NATM method is cheaper than TBM, considering N4E Mine aspects. The minimum price for TBM was twice the value presented for NATM (4 million dollars).

5. CONCLUSIONS

Engineering Geology applications to mining enterprises are very recent. In spite of that, the quality of the informations achieved indicates that its correct use contributes to a well based mine planning, considering technical, economical and operational aspects.

This paper intended to show some of the technical areas in which Engineering Geology has already been successfully applied. For this purpose N4E Mine, owned by CVRD - Companhia Vale do Rio Doce, was selected due to its huge dimensions, varied and complex geomechanical aspects.

The results of slope stability analyses developed for the final pit will arise a relevant economic reduction, due to less waste material that will be produced in mine excavation.

Tunnelling was revealed as a technical and economical attractive alternative to solve drainage problems in mine.

The identification of geological and geotechnical aspects connected to waste dump building, mostly in the preliminary stages of design, allowed the prediction and adoption of suitable technical and operational behaviours to guarantee safety levels required. Besides, eventual changing
places to build the embankments or even limiting disposal volumes can futurely turn into appropriate ways of preventing accidents, abruptly altering mine operations and spending significant amount of money.

Nowadays the environmental bureaus do not allow anymore to begin or continue waste disposal without submitting a detailed design for approval. ABNT - Associação Brasileira de Normas Técnicas also recently prepared rules connected to the design and building of waste dumps.

CVRD - Companhia Vale do Rio Doce appears as an example of an international distinguished mining company conscious of the benefits derived from the correct application of Enginnering Geology in mine planning.

6. ACKNOWLEDGEMENTS

The authors wish to express their gratitude for permission to publish this paper to Companhia Vale do Rio Doce (CVRD).

Figueiredo Ferraz Consultoria e Engenharia de Projeto, engaged as consultants with regard to the design of the slopes, waste dumps and drainage gallery, provided all 'technical information and the necessary resources for preparing the paper.

Dr. Paulo Teixeira da Cruz developed geotechnical consultive activities in waste dumps design and slopes stability analyses.

The authors wish to specially thank Eng. Mosze Gitelman, Eng. Marcello Pucci, Eng. Carlos Augusto Campanha and Eng. Luiz Augusto Boisson Santos, from Figueiredo Ferraz, who have been responsible for tunnel design coordination, development and building, for kindly contributing with valuable technical comments.

7. REFERENCES

Bieniawski, Z.T. 1983 - The Geomechanical Classification (RMR System) in Design Application to Underground Excavations. In: Proc. International Symposium of Engineering Geology and Underground Construction, A.A. Balkema, Rotterdam, pp. I.33-I.47.

Cruz, P.T., Ojima, L.M. & Rocha, D.J.L. 1992 - Stability of
Cut Natural Slopes N4E Mine - Carajas. In: US/Brazil Geotechnical Workshop on Applicability of Classical Soil Mechanics Principles to Structured Soils, Belo Horizonte.

Hoek, E. & Brown, E.T. 1988 - The Hoek-Brown Failure Criterium - A 1988 Update. In: Proc. 15th Canadian Rock Mechanics Symposium, Toronto.

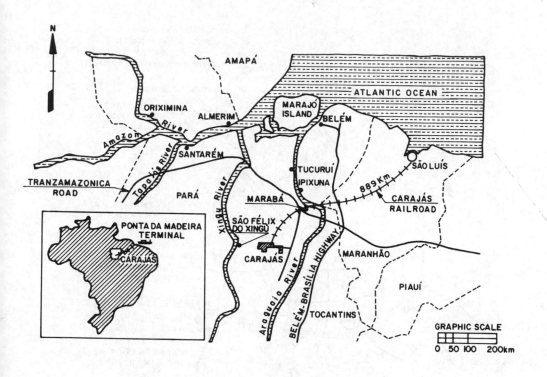

FIGURE 1 - LOCATION OF N4E MINE - CARAJÁS

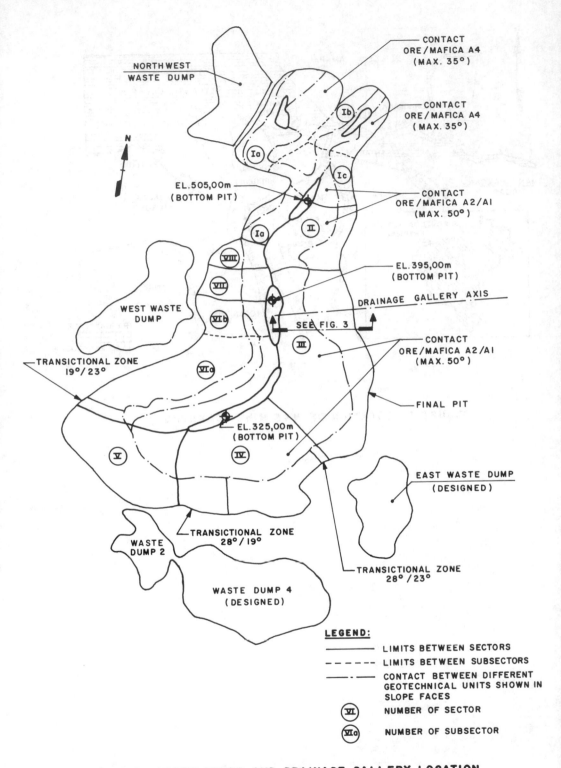

FIGURE 2 - WASTE DUMPS AND DRAINAGE GALLERY LOCATION

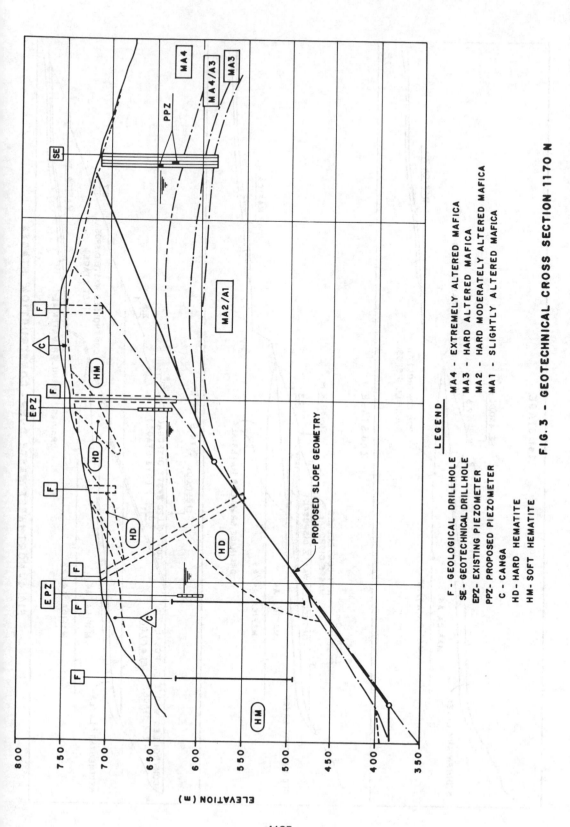

LEGEND

F - GEOLOGICAL DRILLHOLE   MA4 - EXTREMELY ALTERED MAFICA
SE - GEOTECHNICAL DRILLHOLE   MA3 - HARD ALTERED MAFICA
EPZ - EXISTING PIEZOMETER   MA2 - HARD MODERATELY ALTERED MAFICA
PPZ - PROPOSED PIEZOMETER   MA1 - SLIGHTLY ALTERED MAFICA
C - CANGA
HD - HARD HEMATITE
HM - SOFT HEMATITE

FIG. 3 - GEOTECHNICAL CROSS SECTION 1170 N

ELEVATION (m)

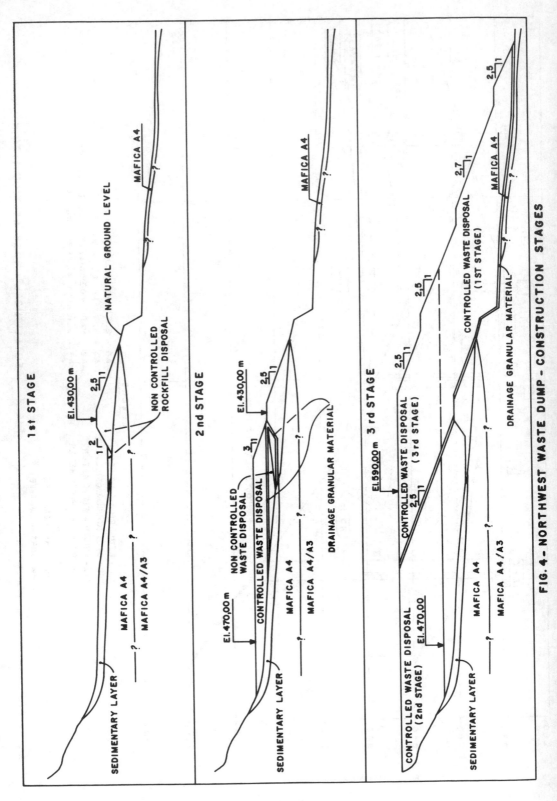

FIG. 4 – NORTHWEST WASTE DUMP – CONSTRUCTION STAGES

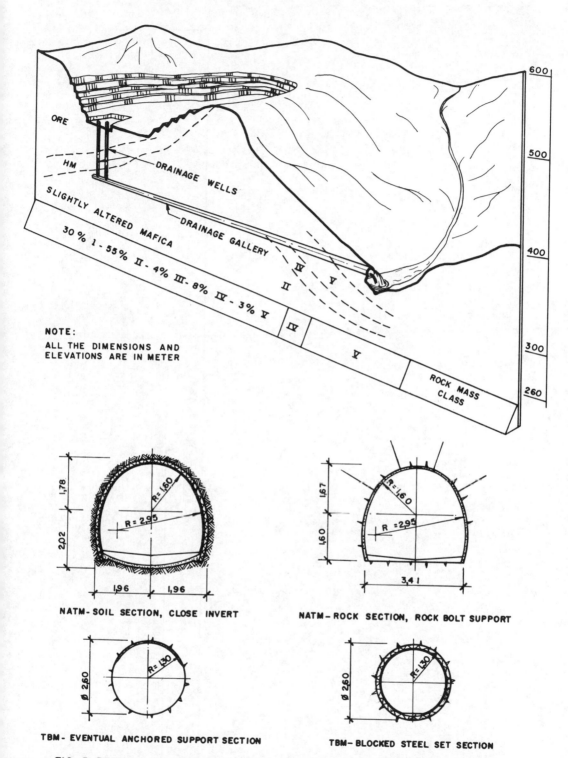

ORE

HM

DRAINAGE WELLS

DRAINAGE GALLERY

SLIGHTLY ALTERED MAFICA

30 % I - 55% II - 4% III - 8% IV - 3% V

IV

II

V

IV

V

NOTE:
ALL THE DIMENSIONS AND
ELEVATIONS ARE IN METER

ROCK MASS
CLASS

600

500

400

300

260

1,78

2,02

R= 1,60

R = 2,95

1,96   1,96

NATM - SOIL SECTION, CLOSE INVERT

1,57

1,60

R= 1,6 0

R = 2,95

3,41

NATM - ROCK SECTION, ROCK BOLT SUPPORT

R = 1,30

Ø 2,60

TBM - EVENTUAL ANCHORED SUPPORT SECTION

R = 1,30

Ø 2,60

TBM - BLOCKED STEEL SET SECTION

FIG. 5 PROFILE AND DRAINAGE GALLERY - PROFILE AND CROSS SECTION

The effects of engineering geology on open-cut coal mining and deep quarry flange stability

Conditions géologiques et géotechniques d'une excavation à ciel ouvert en région minière

V.E.Olkhovatenko & G.M.Rogov
Tomsk State Academy of Architecture and Civil Engineering, Russia

ABSTRACT: The effects of engineering geology on open-cut coal mining and deep quarry flange stability are assessed. As an example, the Kuznetsk coal basin is considered. Structural and tectonic peculiarities, lythologic composition and coal deposit flooding are surveyed. Systematic occurrences in the formation of phisical and mechanical rock properties while under lythogenesis are described. Engineering and geological coal deposit type designs and assessment of deep flange stability are given.

RESUME: On apprecie l'influence de la geologie du genie sur le dehouillement par un mode ouvert des gisements et la fermete des bords des carrieres profondes. On donne des gisements houillers du Bassin Kouznezk a titre d'exemple. On eclaire des particularites techtoniques, la composition lithologique et l'irrigation des couches des gisements houillers. On descrit des regularites de la formation des proprietes physiques et mecaniques des couches houilleres d'une lithogenese. On donne la typisation des gisements houillers et on apprecie la fermete des bords des carrieres profondes.

1 ENGINEERING AND GEOLOGICAL CONDITIONS OF OPEN-CUT COAL MINING IN THE KUZNETSK BASIN

Open-cut coal mining is accompanied by a great deal of excavation that involves geological medium disturbance and gives rise to various gcodynamic processes and phenomena in deep quarry flanges. To avoid negative consequences of this intervention it is important to predict the engineering and geological conditions and stability of flanges. The impact of engineering geology on open-cut coal mining is assessed using the example of the Kuznetsk coal basin.

1.1 *Structural and tectonic features, lythological composition and flooding of coal deposit rocks*

The Kuznetsk coal basin is a major synclinorium, with a structural plan shaped like an irregular quadrangle. From the point of view of geomorphology it is a basin surrounded by mountain structures. These are the Kuznetskiy Alatau from the east, the Gornaya Shoria from the south, the Salair Ridge from the south-west and west, and the Kolyvan-Tomsk fold zone from the north and north-west.

The present structural and tectonic conditions of the Kuznetsk coal basin are the result of its long formation accompanied by several tectogenesis phase manifestations. Two structural stages have developed. They are the lower one, including intensivly dislocated rocks of the Higher Paleozoic, and the higher one, including Mesozoic rocks. The latter has less changed in the process of tectogenesis. Tangential movements of the Salair Ridge and the Kolyvan-Tomsk folded zone have effected the basins structural plan and tectonics. Arched lineal folded structured complicated by disjunctive tectonics have developed on thes region borders. Tectonics become simpler towards the basin centre. Intense lineal folding gives way to brachy-synclines. The latter give way to monoclinal roch folding in the south-east and east of the basin. According to the tectonic zonation diagram (Koudelniy 1976), five geotectonic zones have been determined within the Kuznetsk basin. Thay are the following: the Near-Salair zone of lineal folding and breaks, Near-Kolyvan-Tomsk zone of folding and breaks, Central brachy-fold zone, the

Near-Gornaya Shoria and Near-Kuznetskiy Alatau monocline and gentle fold zone.

Tectonic structure formation in the Salair zone is connected with repeated tectonic movement from the Salair side. In first stages of formation thay were of a tangential and clumpy character, but later on the latter prevailed over the former. The distinguishing features of the zone are steep limbs of major folds with dip angles 60° to 90°, and fold and break structures of different sizes. Folded thrusts, compresion faults, and near-fault zones are abundant.

Lineal folding and rupture formation in the Near-Kolyvan-Tomsk zone owes its origin to tangential movements from the side of the Tomsk-Kolyvan folding arch. The main structures of this zone are major echelon-like brachy-folds with gentle limb sloping 5° to 30° westward and steep limb sloping 25° to 80° eastward. Dislocations with a break in continuity have appeared due to concordant and disconcordant compression faults, and also due to thrusts of different shapes and amplitudes.

The central brachy fold zone is the zone of transition between the lineal fold zone and the monocline zone. Major syncline folds of oval form and asymetric structure are wide spread within this zone. Dip limb angles of syncline folds range from 10° to 30°, dip limb angles of anticline folds range to 80°.

The Near-Gornaya Shoria zone of monoclines and gently sloping folds has developed under the influence of vertical and tangential movements from the side of the Kuznetskiy Alatau and the Gornaya Shoria. This zone is characterized by gentle monoclinal roch folding with dip angles from 10° to 30°.

The characteristic feature of the Near-Kuznetskiy Alatau zone of monoclines and domings is its structural development under the influence of released tangential movements from the side of the Kolyvan-Tomsk folding arch and block movements. Gently pitching rocks with angles from 5° to 10°, sometimes from 30° to 40°, and tectonic dislocations are characteristic of this zone. Among tectonic structure elements, structural forms, rock bedding condition, rock dislocations and tectonic failure are thought to effect open-cut mining and quarry flange stability. The most favourable conditions are in the Near-Gornaya Shoria monocline zone and less favourable conditions are in the Near-Salair and Near-Kolyvan-Tomsk zones of intensive folding and ruptures.

The geological section of the Kuznetsk basin is presented as lagoon and continental deposits of the Balakhon series (C_{2-3}-P_1 bl), continental deposits of the Kolchugin series (P_2) and continental Mesozoic deposits (Mz). Lythologically, the upper Balakhon subseries (P_1 bl) consists of sandstones, aleurites, hard coals, and rarely as conglomerates, gritstones, and argillites. Sandstones and aleurites are characterized by their complex polimictic composition. Fragmental parts of those rocks contain from 10 to 25 per cent quartz, from 10 to 30 per cent feldspar, from 10 to 20 per cent effusive rocks, and from 5 to 15 per cent carbonates. Cement is kaolinite-hydromicaceous, sericitic, carbonaceous, or mixed. Kolchugin series continental deposits consist of sandstones, aleurites and hard coals. Studies of sandstone mineral composition have shown that quartz amounting to from 14 to 40 per cent, feldspar amounting to from 12 to 18 per cent, effusive rocks amounting to from 10 to 25 per cent, sedimentary rocks amounting to from 2.5 to 23.5 per cent prevail. Fragmental parts include carbonates amounting to from 1.7 to 16.6 per cent. Sandstone cement is classified into clayey (Kaolinite-hydromicaceous), mixed (clayey-carbonaceous), and carbonaceous; more often it is dolomitic and more rarely calcareous. The aleurolites of this series contain from 16 to 20 per cent quartz, from 14 to 20 per cent feldspar, from 3 to 21 per cent effusives, from 3 to 20 per cent sedimentary rocks, and from 1.3 to 29 per cent carbonates. Tarbagan series continental deposits contain conglomerates, sandstones, aleurites, argillites, and brown coals. The hydrogeological conditions of the Kuznetsk basin are characterized by the propagation of four underground water complexes typical for lagoon-continental deposits of the Balakhon, Kolchugin and Tarbagan series, and also by current alluvial Quaternary deposits.

The Near-Gornaya Shoria zone rocks are more floded; water inflow amounts to from 600 to 900 m³/hour.

The Near-Salair zone rocks are less flooded; water inflow into quarries is not more than 315 m³/hour. In the brachy-fold Central zone water inflow into mines ranges from 200 to 800 m³/hour.

1.2 *Formation of physical and mechanical rock features of the Kuzbass coal deposits while lythogenesis*

The development of physical and mechanical properties of the Kuzbass rocks is connected with its geotectonical development, sedimentation facies conditions, sediment diagenetic transformations and rock katagenetic changes. Among the Kuzbass coal-bearing deposits the Tarbagan series rocks have undergone fewer transformations in the post-diagenetic stage. This is demonstrated by the presence of coaly inclusions with pronounced plant tissue cellular structure. The Kolchugin series rocks are in the middle Katagenesis stage characterised by replacement of feldspar and effusives by carbonates. Extensive corrosion of plagioclases and quartz grains results in the development of stronger structural linkage as compared with initial cementing linkage.

The Balakhon series rocks are the final katagenesis stage that has resulted in extensive replacement of feldspar and effusives by carbonates and pseudomorphs. Katagenetic dolomite is widespread and presented in effusive fragments. Feldspar is replaced by carbonates with the simultaneous formation of sericites and muscovites. The clayey cement substance is replaced by a carbonaceous one. There are various carbonaceous fragment replacements in a wide range of rock katagenetic transformations. Early in the development they are at an insignificant level of corrosion, later on total replacement occurs with the formation of pseudomorphs. Secondary changes in rocks exert influence on fundamental physical and mechanical properties as given in Tables 1 and 2.

Table 1. Sandstone property changes at various lythogenesis stages

Lithogenesis stage	Density, g/cm³	Porosity, %	Compressive strength, MePa
Initial diagenesis	2.01	36.0	1.6
Katagenesis			
initial	2.36	14.0	31.3
medium	2.47	8.2	49.1
final	2.53	6.2	75.7
Metagenesis	2.56	4.8	52.6

Table 2. Aleurite property changes at various lythogenesis stages

Lithogenesis stage	Density, g/cm³	Porosity, %	Compressive strength, MePa
Initial diagenesis	2.01	36.0	1.6
Katagenesis			
initial	2.36	14.0	31.3
medium	2.47	8.2	49.1
final	2.53	6.2	75.7
Metagenesis	2.56	4.8	52.6

Rock density rises steeply while going from initial diagenesis to initial katagenesis. The value increase within the given range amounts to 0.35 g/cm³ for sandstones and to 0.28 g/cm³ for aleurites. Increments between initial and final katagenesis are significantly lower. Rock consolidation in the final katagenesis stages becomes difficult and can't provide essential rock stability. Porosity changes are shown in Tables 1 and 2. In contrast to physical properties rock strength has close increments in the various lythogenesis stages. The value increment between the final diagenesis and initial katagenesis stages amounts to 27.7 MePa for sanstones and 30.3 MePa for aleurites. In the final katagenesis stage the increment is higher than in the medium stage by 26.6 MePa for sandstones and 26.4 MePa for aleurites. Along with katagenesis changes, tectonic processes of the postinversion period greatly effected rock consolidation. That is why rock consolidation in the Near-Salair and Near-Kolyvan-Tomsk zones of intensive lineal folding does not exceed 60 MePa and reaches 120 MePa in the Near-Gornaya Shoria monocline zone. Rock stability studies show that the stability is much lower in solid than in samples. Owing to extensive effects and surface weakness in solid rocks, structural slackening ratio is 0.01 to 0.06. These abovementioned regularities of formation and changes of rock strengh were used for quantitative assessment of quarry flange stability.

1.3 *Geodynamic processes and phenomena in Kuzbass operating quarry flanges*

Open-cut mining in the Kuzbass causes geological medium deteriotion and instigates various geodynamic processes and phenomena. These are widespread in coal quarries of the Near-Salair and Near-Kolyvan-Tomsk zones of intensive lineal folding. Withing the Bachat coal quarry, block landslides have occured on the inherited slickensides. Landslide body mass reaches 13000 m³. Rapid deformation results in simultaneous diplacement of several blocks. These deformations take place when quarry depth is 200 m and flange slope angle is 26° to 36°. The main reasons for these occurences are steeply-dipping bed shearing inside cutting, tectonic dislocation zones, mountain mass strained state changes and quarry low-strength flange rocks. A landslide took place at the South-West flange in Vakhrushev quarry whose mass amounted to about 1 mln m³. The main reasons for the landslide were the presence of the Tyrgan thrust, surfase water penetration, and poor near-flange zone drainage. Natural cyclical landslides have been observed in the Krasnobrod coal quarry. They originate at the Afonino-Kiselevsk compression fault zone and result from intensive rock dislocation within this zone and also from high flooding and increasing flange dip angles.

Flange landslides, talus and mud flows are found in Near-Kolyvan-Tomsk zone coal quarries. The largest landslides have been observed in Kedroviy coal quarry where total landslide body mass is 2.7 mln m³. The landslide resulted from the appearance of a weakness zone when making contact with the coal bed, from intensive rock moistening by atmospheric precipitation, and increasing flange dip angles. Quarry flange deformations were observed in some other coal quarries of the zone considered.

Coal quarries of the Near-Gornaya Shoria monocline zone are free of deformations, and their flanges are stable. Minor landslides occur in spoil bank slopes and are caused by intensive moisture penetration into dislocated rocks.

Thus, the main reasons for operating quarry flange dislocation are weakness zones when making contact with coal beds, rock moistening due to intensive atmospheric precipitation and high flange dip angles. Studies show that as a rule quarry flange formation is of an inherited nature and occurs on surfaces weakened during geological formation. To prevent geodynamic processes and flange stability dislocation we have carried out engineering and geological type designs of coal deposits, and predicted design quarry flange stability assessment.

1.4 Engineering and geological type designs of the Kuzbass coal deposits and quarry flange stability assessment

Engineering and geological type designs consist of establishing the general and most essential features of coal deposit engineering and geological conditions. This makes it possible to present their variety as engineering and geological models having common requirements to be studied and industrially promoted (Olkhovatenko, 1992). Qualitative and quantitative criteria justification for engineering and geological type designs has been carried out on the basis of factor analysis. As a result we have determined impacts on quarry flange depth stability, rock composition and strength, mode of occurrance, lythogenetic transformation level and coal deposit dislocation and water content. Rocks are divided into 3 categories, based on uniaxle compression strength: low strength category with σ_{com} < 30.0 MePa, medium strength category with σ_{com} = 30.0 to 60.0 MePa and high strength category with σ_{com} > 60 MePa. Based on bedding conditions, rocks are divided into masses with steep dipping amounting to more than 60°, medium dipping amounting to from 60° to 30° and gentle dipping amounting to less than 30°. Rock composition was determined by the predominance of sandstone, aleurite and argillite in rock sections. Deposits containing sandstones amounting to less than 40 per cent, amounting to from 40 to 60 per cent and amounting to more than 60 per cent were separated out. Massif dislocation degree was assessed by dislocation ratio. For values higner than 0.75 the dislocation was considered to be low; within the range of 0.5 to 0.75 to be medium and less than 0.5 to be high. Jointing was assessed by specific jointing value, and water conductivity by the water conductivity coefficient. Also taken into account during engineering and geological type designing were: special stratigraphic and genetic features of the coal deposit, character and intensity of manifestation of geodynamic processes in operating quarry flanges, and physical and geographic zone conditions. For this purpose maps of engineering and geological conditions and engineering and geological zonation of Kuznetsk basin made by Olkhovatenko (1991 and 1976 respectively) were used. In the Kuznetsk basin area five engineering and geological coal deposit types of the first order and thirteen types of the second order were determined using quantitative and qualitative criteria. The Near-Salair western type is the most complicated among them. It is characterized by high dislocation and degree of rock tectonic failure. Rocks of this zone have dipping more than 60° and compression strengths from 30 to 60 MePa. The Near-Salair eastern type is characterized by more regular bedding and a lower degree of lythogenesis and higher flooding. Rock strength from uniaxial compression is less than 40 MePa. The Near-Kolyvan-Tomsk type Balakhon and Kolchugin series rocks having different bedding (from gentle ammounting to 10° to step amounting to more that 60°), medium lythogenesis and degree of rock dislocation. Near-Gornaya Shoria type deposits are of the most simple tectonic structure. Carbonaceous rock have gentle bedding from 10° to 30°, and a high degree of lythogenesis which determines high rock strength from uniaxial compression amounting to more than 80 MePa. For the deposit types discussed, typical structural section models were designed, and physical and mechanical property values and quarry flange stability were calculated.

To make the calculations, we used diagrams taking into account engineering and geological conditions of coal deposits including structural massif weaknesses and anysotropy of its physical and mechanical properties. The calculation procedure is presented in detail in professor V.E. Olkhovatenko's paper (Olkhovatenko 1976) and the results are given below. The most stable are coal quarry flanges of the Near-Gornaya Shoria monocline zone with stable slope angles from 36° to 42° having 150 to 200 m depth. The lowest flange slope angle values were calculated for the Near-Salair zone of intensive lineal folding. Flange slope angles amount to from 30° to 32° having from 150 to 200 m depth. In the Near-Kolyvan-Tomsk zone flange slope angles do not exceed 35° with 150 to 500 m depth. And in brachy-fold Central zone they range from 31° to 33° with 150 m depth. One can avoid negative consequences resulting from intensive impact on geological environment using calculated flange slope angles when designing and reconstructing quarries.

2 CONCLUSIONS

The results of the studies carried out allow us to make conclusion about the greatest effects on structural and tectonic peculiarities of deposits, lythologic composition, degree of rock lythogenetic transformation, their dislocation and strength in open-cut coal mining in the Kuzbass. Coal quarry flange stability highly depends on the above-mentioned factors.

REFERENCES

Koudelniy, V.Ya. 1976. Tectonic zonation of the Kuzbass basin. *Tectonics of coal basins and deposits of the USSR*. Moscow: Nedra.

Olkhovatenko, V.E. 1976. *Engineering and geological conditions of major quarry construction in the Kuznetsk coal basin*. Tomsk: Tomsk University.

Olkhovatenko, V.E. 1991. *Engineering geology of Siberia and Far East coal deposits*. Vol.1. *Regularities of engineering and geological condition formation of coal deposits*. Tomsk: Tomsk University.

Olkhovatenko, V.E. 1992. *Engineering geology of Siberia and Far East coal deposits*. Vol.2. *Engineering and geological type designs of coal deposits and assessment of quarry flange stability*. Tomsk: Tomsk University.